JAGUAR XJ-S

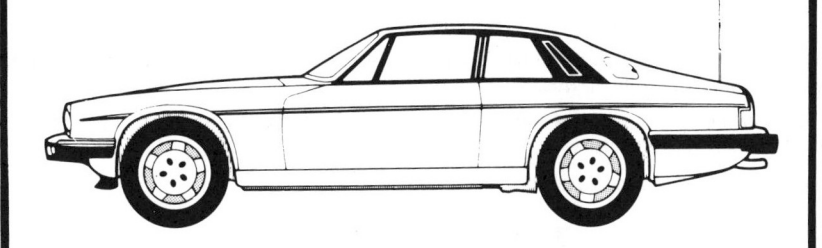

This manual covers all V12 5.3 and 6.0 litre powered XJS models fitted with coupé, cabriolet and convertible body styles.

Repair Operation Manual
Incorporating XJ-S HE, 5.3 & 6.0 Supplements

Published by
Jaguar Cars Ltd

Publication Part No. AKM 3455 Ed 4/3

INTRODUCTION

The purpose of this Manual is to assist skilled mechanics in the efficient repair and maintenance of Jaguar vehicles. Using the appropriate service tools and carrying out the procedures as detailed will enable the operations to be completed in the time stated in the 'Repair Operations Times'.

SUPPLEMENTS

Three supplements have been added to the original manual to extend its coverage.

Supplement A 1979-1984. This supplement contains information on the HE engine models with the 'P' Type fuel injection system, plus other changes to specification incorporated between 1979 and 1984.

Supplement B 1984-1988^1/2. This supplement contains information on the 5.3 Litre V12 engined cars fitted with the Lucas 'P' Type fuel injection system and also the 'cabriolet' hood (targa top) models from VIN Number 112586.

Supplement C 1988^1/2 -. This supplement contains information on all later vehicles mainly dealing with those modifications brought about with the introduction of the 6.0 Litre engine and 4-speed electronically-controlled transmission, but also covering ABS brake systems, body modifications, and more.

When dealing with cars of these years first turn to the supplements. Failure to locate information will indicated that the information given in the main body of the book is still relevant.

The supplements are self contained with an index at the front. The fault diagnosis section of Supplement A can be used on all cars where the circuit is the same as the later models.

Operation Numbering

A master index of numbered operations has been compiled for universal application to all vehicles manufactured by Jaguar Cars Ltd. and, therefore, because of the different specifications of various models, continuity of the numbering sequence is not maintained throughout this manual.

Each operation described in this manual is allocated a number from the master index and cross-refers with an identical number in the 'Repair Operation Times'. The number consists of six digits arranged in three pairs.

Each operation is laid out in the sequence required to complete the operation in the minimum time, as specified in the 'Repair Operation Times'.

Service Tools

Where performance of an operation requires the use of a service tool the tool number is quoted under the operation heading and is repeated in, or following, the instruction involving its use. A list of all necessary tools is included in Section 99.

References

References to the left- or right-hand side in the manual are made when viewing from the rear. With the engine and gearbox assembly removed the timing cover end of the engine is referred to as the front. A key to abbreviations and symbols is given on page 04—1.

REPAIRS AND REPLACEMENTS

When service parts are required it is essential that only genuine Jaguar or Unipart replacements are used.

Attention is particularly drawn to the following points concerning repairs and the fitting of replacement parts and accessories.

1. Safety features embodied in the vehicle may be impaired if other than genuine parts are fitted. In certain territories, legislation prohibits the fitting of parts not to the vehicle manufacturer's specification.

2. Torque wrench setting figures given in this Service Manual must be strictly adhered to.

3. Locking devices, where specified, must be fitted. If the efficiency of a locking device is impaired during removal it must be replaced.

4. Owners purchasing accessories while travelling abroad should ensure that the accessory and its fitted location on the vehicle conform to mandatory requirements existing in their country of origin.

5. The vehicle warranty may be invalidated by the fitting of other than genuine Jaguar or Unipart parts. All Jaguar and Unipart replacements have the full backing of the factory warranty.

6. Jaguar Distributors and Dealers are obliged to supply only genuine service parts.

JAGUAR

SPECIFICATION

Purchasers are advised that the specification details set out in this Manual apply to a range of vehicles and not to any one. For the specification of a particular vehicle, purchasers should consult their Distributor Dealer.

The Manufacturers reserve the right to vary their specifications with or without notice, and at such times and in such manner as they think fit. Major as well as minor changes may be involved in accordance with the Manufacturer's policy of constant product improvement.

Whilst every effort is made to ensure the accuracy of the particulars contained in this Manual, neither the Manufacturer nor the Distributor or Dealer, by whom this Manual is supplied, shall in any circumstances be held liable for any inaccuracy or the consequences thereof.

COPYRIGHT

© Jaguar Cars Ltd. 1984, 1989 and 1995

All rights reserved. No part of this publication may be reproduced, stored in a retrieval system or transmitted in any form, electronic, mechanical, photocopying, recording or other means without prior written permission of Jaguar Cars Ltd.

CONTENTS

	Page No.
General Specification Data	04
Engine Tuning Data	05
Torque Wrench Settings	06
General Fitting Instructions	07
Lifting and Towing	08
Recommended Lubricants, Fuel and Fluids—Capacities	09
Anti-freeze	09–1
Automatic Transmission Fluid	09–1
Brake Fluid	09–2
Capacities	09–1
Coolant Additives	09–2
Dimensions and Weights	09–1
Fuel Requirements	09–1
Power Steering Fluid	09–1
Recommended Lubricants	09–1

MAINTENANCE   10

	Page
Lubrication Chart	10–1
Routine Maintenance Operations	
3,000 miles (5 000 km)	10–1
6,000 miles (10 000 km)	10–3
12,000 miles (20 000 km)	10–4
Summary Chart	10–5
Additional maintenance operations	10–6

ENGINE

Camshaft	Operation No.	Page No.
– overhaul	12.13.26	12–1
– remove and refit	12.13.01	12–1
Camshaft cover		
– remove and refit – left-hand	12.29.43	12–8
– right-hand	12.29.44	12–8
Camshaft oil feed pipes – remove and refit	12.60.83	12–17
Connecting rod bearings – remove and refit (set)	12.17.16	12–3
Crankshaft – remove and refit	12.21.33	12–4
Crankshaft damper and pulley – remove and refit (engine in situ)	12.21.01	12–3
Crankshaft front oil seal – remove and refit	12.21.14	12–4
Crankshaft – general	12.41.05	12–14
Crankshaft main bearings – remove and refit (set)	12.21.39	12–4
Cylinder head		
– gasket – remove and refit	12.29.02	12–4
– overhaul	12.29.18	12–7
– remove and refit – left-hand	12.29.11	12–4
– right-hand	12.29.12	12–6
Cylinder liners – checking	12.25.27	12–4
Cylinder pressures – check	12.25.01	12–4
Drive plate – remove and refit	12.53.13	12–15
Engine – dismantle and reassemble	12.41.05	12–11
Engine and gearbox assembly – remove and refit	12.37.01	12–9
Engine mounting – front set – remove and refit	12.45.04	12–14

continued

	Operation No.	Page No.
Engine mounting – rear centre – remove and refit	12.45.08	12–14
Flywheel – remove and refit	12.53.07	12–14
Oil cooler – remove and refit	12.60.68	12–16
Oil filter assembly – remove and refit	12.60.01	12–15
Oil pick-up strainer – remove and refit	12.60.20	12–15
Oil pressure relief valve – remove and refit	12.60.56	12–16
Oil pump – overhaul	12.60.32	12–15
– remove and refit	12.60.26	12–15
Oil sump pan – remove and refit	12.60.44	12–16
Piston and connecting rod – overhaul	12.17.10	12–3
– remove and refit	12.17.01	12–2
Sandwich plate assembly – remove and refit	12.60.45	12–16
Spark plug inserts – fitting	12.29.78	12–8
Tappet block – remove and refit – left-hand	12.13.29	12–1
– right-hand	12.13.30	12–2
Tappets – adjust	12.29.48	12–8
Timing chain – remove and refit	12.65.12	12–17
Timing chain dampers – remove and refit	12.65.50	12–17
Timing chain tensioner – remove and refit	12.65.28	12–17
Timing cover – remove and refit	12.65.01	12–17

EMISSION CONTROL

	Operation No.	Page No.
Adsorption canister – remove and refit	17.15.13	17–7
Air delivery pump – remove and refit	17.25.07	17–7
Air delivery pump/compressor drive belt – remove and refit	17.25.15	17–8
– tensioning	17.25.13	17–7
Air rail – remove and refit – left-hand	17.25.17	17–8
– right-hand	17.25.18	17–8
Catalytic converter – remove and refit – left-hand	17.50.01	17–9
– right-hand	17.50.03	17–9
Check valve – remove and refit	17.25.21	17–8
Diverter valve – remove and refit	17.25.25	17–8
Emission control system – description	17.00.00	17–1
– evaporative loss control system	17.00.00	17–3
– fault finding	17.00.00	17–4
– testing	17.00.00	17–3
Engine breather filter – remove and refit	17.10.02	17–7
Exhaust gas recirculation control unit – remove and refit	17.45.07	17–9
Exhaust gas recirculation valve – remove and refit	17.45.01	17–8
Exhaust gas recirculation valve, transfer pipe – remove and refit	17.45.11	17–9

FUEL SYSTEM

	Operation No.	Page No.
Air cleaners		
– remove and refit – left-hand	19.10.01	19–11
– right-hand	19.10.02	19–11
– renew element	19.10.08	19–11
Air temperature sensor		
– remove and refit	19.22.22	19–15
– test	19.22.23	19–15
Auxiliary air valve		
– remove and refit	19.20.16	19–13
– test	19.20.17	19–13
Cold start injector		
– remove and refit	19.60.06	19–22
– test	19.60.07	19–22
Cold start relay – remove and refit	19.22.31	19–16
Cold start system – test	19.22.32	19–16
Coolant temperature sensor		
– remove and refit	19.22.18	19–14
– test	19.22.19	19–14
Electronic control unit (E.C.U.) – remove and refit	19.22.34	19–17
Electronic fuel injection system		
– component location and description	19.00.00	19–1
– fault finding – diagnostic test (Lucas 'EPITEST')	19.00.00	19–5
– initial diagnosis	19.00.00	19–3
– good practice	19.00.00	19–2
– maintenance	19.00.00	19–3
Fuel cut-off inertia switch – remove and refit	19.22.09	19–14
Fuel main filter – remove and refit	19.25.02	19–19
Fuel pressure regulators		
– adjust	19.45.12	19–19
– remove and refit	19.45.11	19–19
Fuel pump – remove and refit	19.45.08	19–19
Fuel rails – remove and refit – left-hand	19.60.05	19–22
– right-hand	19.60.04	19–21
Fuel system		
– depressurize	19.50.02	19–20
– pressure test	19.50.13	19–20
Fuel tank		
– drain	19.55.02	19–21
– remove and refit	19.55.01	19–20
Idle speed – adjust	19.20.18	19–3
Injectors		
– remove and refit – left-hand	19.60.03	19–21
– right-hand	19.60.01	19–21
– test	19.60.08	19–22
Main relay – remove and refit	19.22.38	19–18
Manifold pressure sensor		
– remove and refit	19.22.29	19–16
– test	19.22.28	19–16
Over-run valve		
– remove and refit	19.20.22	19–14
– test	19.20.21	19–14
Power amplifier		
– remove and refit	19.22.33	19–17
Pump relay		
– remove and refit	19.22.39	19–18
– test	19.22.40	19–18
Thermotime switch		
– remove and refit	19.22.20	19–14
– test	19.22.21	19–15

continued

	Operation No.	Page No.
Throttle butterfly valve – check and adjust	19.20.11	19–13
Throttle cable – remove and refit	19.20.06	19–12
Throttle linkage – check and adjust	19.20.05	19–12
Throttle pedal – remove and refit	19.20.01	19–11
Throttle pedestal – overhaul	19.20.03	19–12
Throttle switch		
– adjust	19.22.35	19–17
– remove and refit	19.22.36	19–17
– test	19.22.37	19–18
Trigger unit		
– remove and refit	19.22.26	19–15
– test	19.22.27	19–16

COOLING SYSTEM

	Operation No.	Page No.
Coolant		
– description	26.00.00	26–1
– drain and refill	26.10.01	26–1
Fan belt		
– adjust	26.20.01	26–2
– remove and refit	26.20.07	26–2
Fan blades – remove and refit	26.25.06	26–2
Fan and motor unit – remove and refit	26.25.23	26–3
Fan motor and fan motor relay – test	26.25.25 / 26.25.29	26–3
Fan motor relay – remove and refit	26.25.31	26–3
Fan and Torquatrol unit – remove and refit	26.25.21	26–3
Idler pulley housing – remove and refit	26.25.15	26–2
Jockey pulley – remove and refit	26.25.16	26–2
Radiator block – remove and refit	26.40.04	26–4
Radiator drain tap – remove and refit	26.40.10	26–4
Remote header tank – remove and refit	26.15.01	26–1
Thermostat – remove and refit – left-hand	26.45.01	26–5
– right-hand	26.45.04	26–5
– test	26.45.09	26–5
Thermostat housing – remove and refit – left-hand	26.45.10	26–5
– right-hand	26.45.11	26–5
Thermostat switch – remove and refit	26.25.35	26–4
Water pump		
– overhaul	26.50.06	26–6
– remove and refit	26.50.01	26–6
Water pump pulley – remove and refit	26.50.05	26–6

MANIFOLD AND EXHAUST SYSTEM

	Operation No.	Page No.
Exhaust manifold – remove and refit – left-hand	30.15.10	30–3
– right-hand	30.15.11	30–3
Exhaust trim – remove and refit	30.10.23	30–1
Front pipe – remove and refit – left-hand	30.10.09	30–1
– right-hand	30.10.10	30–1
Induction manifold – remove and refit – front left- and right-hand	30.15.02	30–2
– rear left- and right-hand	30.15.03	30–2

continued

	Operation No.	Page No.
Intermediate pipe – remove and refit – left-hand	30.10.11	30–1
– right-hand	30.10.12	30–1
Mounting rubber – front – remove and refit – left- or right-hand	30.20.02	30–3
Mounting rubber – rear – remove and refit – left- or right-hand	30.20.04	30–4
Rear intermediate pipe – remove and refit – left-hand	30.10.24	30–1
– right-hand	30.10.25	30–1
Silencer assembly – remove and refit – left-hand	30.10.15	30–1
– right-hand	30.10.16	30–1
Tail pipe and silencer – remove and refit – left- or right-hand	30.10.22	30–1

CLUTCH

	Operation No.	Page No.
Clutch assembly – remove and refit	33.10.01	33–1
Clutch hydraulic system – bleed	33.15.01	33–1
Clutch master cylinder		
– overhaul	33.20.07	33–2
– remove and refit	33.20.01	33–2
Clutch pedal		
– overhaul	33.30.07	33–3
– remove and refit	33.30.02	33–2
Clutch slave cylinder		
– overhaul	33.35.07	33–3
– remove and refit	33.35.01	33–3
Clutch slave cylinder push-rod – check and adjust	33.10.03	33–1
Release assembly – overhaul	33.25.17	33–2
Withdrawal assembly – remove and refit	33.25.12	33–2

MANUAL GEARBOX

	Operation No.	Page No.
Bell housing – remove and refit	37.12.07	37–1
Gearbox assembly		
– overhaul	37.20.04	37–3
– remove and refit	37.20.01	37–3
Gear change selectors – remove and refit	37.16.31	37–3
Oil seal		
– front – remove and refit	37.23.06	37–6
– rear – remove and refit	37.23.01	37–6
Rear extension – remove and refit	37.12.01	37–1
Speedometer drive gear		
– pinion – remove and refit	37.25.05	37–6
– remove and refit	37.25.01	37–6
Top cover		
– overhaul	37.12.19	37–2
– remove and refit	37.12.16	37–1

AUTOMATIC TRANSMISSION (Borg Warner)

	Operation No.	Page No.
Converter – remove and refit	44.17.07	44–14
Converter housing – remove and refit	44.17.01	44–13
Description	44.00.00	44–1
Front brake band – adjust	44.30.07	44–26
Gear selector cable – remove and refit	44.15.08	44–12

continued

	Operation No.	Page No.
Hydraulic flow charts	44.00.00	44–4
Kick-down solenoid		
– remove and refit	44.15.24	44–13
– test	44.30.11	44–26
Kick-down switch		
– check and adjust	44.30.12	44–26
– remove and refit	44.15.23	44–13
Lubrication system – drain and refill	44.24.02	44–25
Oil pan – remove and refit	44.24.04	44–25
Oil strainer – remove and refit	44.24.07	44–25
Rear brake band – adjust	44.30.10	44–26
Reverse light switch		
– check and adjust	44.15.21	44–13
– remove and refit	44.15.22	44–13
Stall speed – test	44.30.13	44–26
Starter inhibitor switch		
– check and adjust	44.15.18	44–13
– remove and refit	44.15.19	44–13
Transmission assembly		
– overhaul	44.20.06	44–14
– remove and refit	44.20.01	44–14
Vacuum control unit		
– diaphragm test	44.30.09	44–26
– line pressure check and adjustment	44.30.05	44–25
– remove and refit	44.15.27	44–13
Valve block – remove and refit	44.40.01	44–27

GM 400 Automatic Gearbox

	Operation No.	Page No.
Band apply pin – selection check	44.30.21	44–30
Data and description	44.00.00	44–1
Fault finding and diagnosis	44.00.00	44–4
Front unit end-float – check and adjust	44.30.22	44–31
Gearbox assembly		
– overhaul	44.20.06	44–18
– remove and refit	44.20.01	44–17
Hand selector assembly – overhaul	44.15.05	44–17
Kickdown switch – check and adjust	44.30.12	44–30
Oil filter – remove and refit	44.27.07	44–30
Oil pan – remove and refit	44.24.04	44–29
Rear extension housing – remove and refit	44.20.15	44–29
Rear unit end-float – check and adjust	44.20.23	44–31
Reverse light switch		
– check and adjust		
– remove and refit	refer to 86.65.20	86–25
Selector cable – adjust	44.30.04	44–30
Speedometer drive pinion – remove and refit	44.38.04	44–32
Starter inhibitor switch		
– check and adjust	44.15.18	44–17
– remove and refit	refer to 86.65.28	86–25
Transmission fluid level – check	44.00.00	44–3
Valve block assembly – remove and refit	44.40.01	44–32

PROPELLER AND DRIVE SHAFTS

	Operation No.	Page No.
Drive shaft		
— overhaul	47.10.08	47–1
— remove and refit	47.10.01	47–1
Propeller shaft assembly		
— overhaul	47.15.10	47–2
— remove and refit	47.15.01	47–2

FINAL DRIVE

Description	51.00.00	51–1
Differential assembly (Powr Lok) — overhaul	51.15.30	51–2
Drive flange — remove and refit	51.15.36	51–2
Final drive unit		
— overhaul	51.25.19	51–4
— remove and refit	51.25.13	51–3
Hypoid casing rear cover gasket — remove and refit	51.20.08	51–3
Output shaft assembly — remove and refit	51.10.20	51–1
Output shaft bearings — remove and refit	51.10.22	51–1
Output shaft oil seal — remove and refit	51.20.04	51–2
Pinion bearings — remove and refit	51.15.19	51–2
Pinion oil seal — remove and refit	51.20.01	51–2

STEERING

Camber angle — check and adjust	57.65.05	57–10
Castor angle — check and adjust	57.65.04	57–10
Control valve and pinion		
— remove and refit	57.10.19	57–3
— test	57.10.20	57–4
Control valve and pinion port inserts — remove and refit	57.10.24	57–4
Control valve pinion seal — remove and refit	57.10.23	57–3
Diagnosis chart	57.00.01	57–1
Front wheel alignment — check and adjust	57.65.01	57–9
Power steering rack		
— adjust	57.10.13	57–3
— overhaul	57.10.07	57–2
— remove and refit	57.10.01	57–1
Power steering rack gaiters — remove and refit	57.10.27	57–4
Power steering oil system		
— test	57.15.01	57–4
— bleed	57.15.02	57–5
Steering column adjusting clamp — remove and refit	57.40.07	57–8
Steering column lock and ignition/starter switch assembly — remove and refit	57.40.31	57–8
Steering column lower — remove and refit	57.40.05	57–7
Steering column universal joint — upper — remove and refit	57.40.25	57–8
Steering column — upper — remove and refit	57.40.02	57–7
Steering lever — remove and refit	57.55.29	57–9
Steering oil cooler — remove and refit	57.15.15	57–5
Steering oil cooler return hose — remove and refit	57.15.16	57–5

continued

	Operation No.	Page No.
Steering pump		
— overhaul	57.20.20	57–6
— remove and refit	57.20.14	57–6
Steering pump drive belt		
— adjust	57.20.01	57–5
— remove and refit	57.20.02	57–6
Steering rack feed hose — remove and refit	57.15.21	57–5
Steering rack return hose — remove and refit	57.15.22	57–5
Steering wheel — remove and refit	57.60.01	57–9
Steering wheel pad — remove and refit	57.60.03	57–9
Tie-rod ball joint inner — remove and refit	57.55.03	57–8
Tie-rod ball joint outer — remove and refit	57.55.02	57–8

FRONT SUSPENSION

	Operation No.	Page No.
Accidental damage		60–1
Anti-roll bar — remove and refit	60.10.01	60–3
Anti-roll bar link — remove and refit	60.10.02	60–3
Anti-roll bar link bushes — remove and refit	60.10.03	60–3
Anti-roll bar rubbers — remove and refit	60.10.04	60–4
Ball joint — lower		
— adjust	60.15.04	60–4
— overhaul	60.15.13	60–4
Ball joint — upper — remove and refit	60.15.02	60–4
Bump stop — remove and refit	60.30.10	60–8
Front damper — remove and refit	60.30.02	60–8
Front hub assembly — remove and refit	60.25.01	60–5
Front hub bearings — remove and refit	60.25.14	60–6
Front hub bearing end-float — check and adjust	60.25.13	60–6
Front hub grease seal — remove and refit	60.25.15	60–6
Front hub stub axle — remove and refit	60.25.22	60–6
Front hub stub axle carrier — remove and refit	60.25.23	60–7
Front hub studs — remove and refit	60.25.29	60–8
Front spring — remove and refit	60.20.01	60–5
Front suspension riding height — check and adjust	60.10.18	60–3
Front suspension unit — remove and refit	60.35.05	60–9
Rebound stops — remove and refit	60.13.14	60–8
Suspension unit mounting bush — remove and refit	60.35.06	60–11
Suspension unit mounting — rear — remove and refit	60.35.07	60–11
Wishbone — lower		
— overhaul	60.35.09	60–12
— remove and refit	60.35.02	60–9
Wishbone — upper		
— overhaul	60.35.08	60–11
— remove and refit	60.35.01	60–8

REAR SUSPENSION

	Operation No.	Page No.
Accidental damage		64–1
Anti-roll bar – remove and refit	64.35.08	64–5
Anti-roll bar rubbers – remove and refit	64.35.18	64–7
Bump stop – remove and refit	64.30.15	64–5
Camber angle – check and adjust	64.25.18	64–4
Inner fulcrum mounting bracket – remove and refit	64.35.21	64–7
Mounting bracket – remove and refit	64.35.20	64–7
Radius arm – remove and refit	64.35.28	64–8
Radius arm bushes – remove and refit	64.35.29	64–8
Rear anti-roll bar link		
– overhaul	64.35.27	64–8
– remove and refit	64.35.26	64–7
Rear hub and carrier assembly		
– overhaul	64.15.07	64–1
– remove and refit	64.15.01	64–1
Rear hub bearing end-float – check and adjust	64.15.13	64–2
Rear hub oil seals – remove and refit	64.15.15	64–3
Rear hub wheel studs – remove and refit	64.15.26	64–3
Rear hydraulic damper units – remove and refit	64.30.01	64–5
Rear road springs (one side) – remove and refit	64.20.01	64–3
Rear suspension assembly – remove and refit	64.25.01	64–3
Rear suspension mounting (pair) – remove and refit	64.25.03	64–4
Rear suspension mounting (single) – remove and refit	64.25.02	64–4
Rear suspension unit – overhaul	64.25.06	64–4
Riding height – check	64.25.12	64–4
Wishbone – remove and refit	64.35.15	64–6
Wishbone bearings – remove and refit	64.35.16	64–7
Wishbone oil seals – remove and refit	64.35.17	64–7

BRAKES

	Operation No.	Page No.
Brake caliper		
– front – remove and refit	70.55.02	70–9
– front – overhaul	70.55.13	70–10
– rear – remove and refit	70.55.03	70–10
– rear – overhaul	70.55.14	70–11
Brake disc – front – remove and refit	70.10.10	70–3
Brake disc – rear – remove and refit	70.10.11	70–3
Brake disc shields – front – remove and refit	70.10.18	70–3
Brake pads		
– front – remove and refit	70.40.02	70–8
– handbrake – remove and refit	70.40.04	70–9
– rear – remove and refit	70.40.03	70–9
Brake system		
– bleed	70.25.01	70–4
– description	70.00.00	70–1
– drain, flush and bleed	70.25.17	70–5
Check valve – remove and refit	70.50.15	70–9
Fluid reservoir – master cylinder – remove and refit	70.30.16	70–6

continued

	Operation No.	Page No.
Handbrake		
— assembly — remove and refit	70.35.08	70–7
— cable — adjust	70.35.10	70–8
— cable assembly — remove and refit	70.35.14	70–8
— mechanism — remove and refit	70.55.04	70–10
Hoses — remove and refit	70.15.00	70–4
Pedal box		
— overhaul	70.35.04	70–7
— remove and refit	70.35.03	70–7
Pipes — remove and refit	70.20.00	70–4
Pressure differential warning actuator		
— check and reset	70.25.08	70–4
— operational check	70.25.14	70–5
— remove and refit	70.25.13	70–5
Reservac tank — remove and refit	70.50.04	70–9
Servo assembly		
— check and test	70.50.03	70–9
— overhaul	70.50.06	70–9
— remove and refit	70.50.01	70–9
Tandem master cylinder		
— overhaul	70.30.09	70–6
— remove and refit	70.30.08	70–5
Unions — rear three-way — remove and refit	70.15.34	70–4

BODY

	Operation No.	Page No.
Ashtray		
— front — remove and refit	76.67.13	76–12
— rear — remove and refit	76.67.14	76–13
Automatic transmission selector quadrant — remove and refit	76.25.08	76–7
Bonnet		
— hinges — remove and refit	76.16.12	76–2
— lock — adjust	76.16.20	76–2
— lock — remove and refit — left-hand	76.16.21	76–3
— right-hand	76.16.26	76–3
— lock control cable — adjust	76.16.28	76–3
— release handle — remove and refit	76.16.30	76–3
— remove and refit	76.16.01	76–2
— stay — remove and refit	76.16.14	76–2
Boot lid		
— hinge — remove and refit	76.19.07	76–4
— lock — remove and refit	76.19.11	76–4
— lock handle — remove and refit	76.19.17	76–4
— lock striker — remove and refit	76.19.12	76–4
— remove and refit	76.19.01	76–3
— seal — remove and refit	76.19.06	76–3
Bumper		
— centre section — front — remove and refit	76.22.11	76–4
— centre section — rear — remove and refit	76.22.12	76–5
— cover — front — remove and refit	76.22.28	76–5
— cover — rear — remove and refit	76.22.29	76–5
— energy absorbing beam — front — remove and refit	76.22.26	76–5
— energy absorbing beam — rear — remove and refit	76.22.27	76–5
— energy absorbing strut — rear — remove and refit	76.22.32	76–6
— quarter section — front — remove and refit	76.22.16	76–5
— side section — rear — remove and refit	76.22.13	76–5
Console assembly — remove and refit	76.25.01	76–6
Door		
— arm-rest — remove and refit	76.34.22	76–9
— glass — remove and refit	76.31.01	76–8
— glass weatherstrip — remove and refit	76.31.50	76–8
— hinge — remove and refit	76.28.42	76–7
— inside handle — remove and refit	76.58.18	76–12

continued

	Operation No.	Page No.
– lock – adjust	76.37.01	76–9
– lock – remove and refit	76.37.12	76–9
– lock striker – adjust	76.37.27	76–10
– lock striker plate – remove and refit	76.37.23	76–10
– outside handle – remove and refit	76.58.01	76–12
– quarter-light – remove and refit	76.31.29	76–8
– remove and refit	76.28.01	76–9

Facia

– glove box lid – remove and refit	76.52.02	76–11
– glove box – remove and refit	76.52.03	76–11
– panel – remove and refit	76.46.01	76–10
– switch panel – remove and refit	76.46.16	76–11
– underscuttle casing – driver – remove and refit	76.46.11	76–10
– underscuttle casing – passenger – remove and refit	76.46.15	76–11

Fuel filler flap – remove and refit	76.10.25	76–1

Grille

– radiator – remove and refit	76.55.03	76–11
– radiator lower – remove and refit	76.55.06	76–11

Headlining – remove and refit	76.64.01	76–12
Parcel shelf – rear – remove and refit	76.67.06	76–12

Rear quarter

– trim pad – lower – remove and refit	76.13.12	76–1
– trim pad – upper – remove and refit	76.13.13	76–2

Rear quarter-light – remove and refit	76.31.31	76–8

Seat

– assembly – remove and refit	76.70.01	76–13
– belt – front – remove and refit	76.73.10	76–14
– belt – rear – remove and refit	76.73.18	76–14
– cushion – front – remove and refit	76.70.02	76–13
– cushion – rear – remove and refit	76.70.37	76–13
– runner and adjuster – remove and refit	76.70.24	76–13
– squab – rear – remove and refit	76.70.38	76–14

Spoiler – front – remove and refit	76.10.46	76–1

Trim casing

'A' post trim casing – remove and refit	76.13.07	76–1
'B' post trim casing – remove and refit	76.13.08	76–1
Door trim casing – remove and refit	76.34.01	76–9

AIR CONDITIONING

Air conditioning

– charge	82.30.08	82–23
– demist and defrost test	82.30.15	82–25
– depressurize	82.30.05	82–22
– description	82.00.00	82–1
– evacuate	82.30.06	82–23
– leak test	82.30.09	82–24
– preliminary test	82.30.16	82–25
– sweep (purge)	82.30.07	82–23
– test operation	82.30.11	82–24
– valve core – remove and refit	82.30.12	82–25

Blower motor

– assembly – remove and refit – left-hand	82.25.14	82–19
– right-hand	82.25.13	82–18
– overhaul	82.25.30	82–22
– power relay – remove and refit	82.20.06	82–15
– relay – remove and refit	82.20.27	82–18
– resistor unit – remove and refit	82.20.26	82–17

Charging/testing equipment – remove and refit	82.30.01	82–22

Compressor

– drive belt – adjustment	82.10.01	82–13
– drive belt – remove and refit	82.10.02	82–13
– oil level – check	82.10.14	82–13
– remove and refit	82.10.20	82–14

continued

	Operation No.	Page No.
Condenser – remove and refit	82.15.07	82–14
Control system		
– ambient sensor – remove and refit	82.20.02	82–15
– amplifier unit – remove and refit	82.25.29	82–22
– description	82.00.00	82–1
– in-car sensor – remove and refit	82.20.03	82–15
– mode control – remove and refit	82.20.11	82–16
– servo motor assembly – remove and refit	82.25.24	82–21
– temperature selector – remove and refit	82.20.10	82–16
– thermostat – adjust	82.20.19	82–17
– thermostat – remove and refit	82.20.18	82–17
– vacuum solenoid – remove and refit	82.25.23	82–21
– water valve – remove and refit	82.20.33	82–18
– water valve temperature switch – remove and refit	82.20.29	82–18
Evaporator – remove and refit	82.25.20	82–20
Expansion valve – remove and refit	82.25.01	82–18
Fault finding charts	82.00.00	82–4
Heater/cooler unit – remove and refit	82.25.21	82–20
Procedure for flap link adjustment	82.20.17	82–16
Receiver drier unit – remove and refit	82.17.01	82–15
Ventilator outlet, facia, centre – remove and refit	82.20.38	82–18
Ventilator outlet, facia, right – remove and refit	82.20.39	82–18
Ventilator outlet, facia, left – remove and refit	82.30.40	82–18
Wiring diagram		86–39

WINDSCREEN WASHERS AND WIPERS

	Operation No.	Page No.
Washer jets – remove and refit	84.10.09	84–1
Washer pump – remove and refit	84.10.21	84–2
Washer reservoir – remove and refit	84.10.01	84–1
Washer reservoir bracket – remove and refit	84.10.02	84–1
Windscreen wiper arms – remove and refit	84.15.01	84–2
Windscreen wiper blades – remove and refit	84.15.05	84–2
Windscreen wipers/washer – description		84–1
Windscreen wipers/washer – operation		84–1
Windscreen wiper/washer control switch – remove and refit	84.15.34	84–3
Windscreen wiper motor gear assembly – remove and refit	84.15.14	84–2
Windscreen wiper motor harness – bulkhead connector – remove and refit	84.15.35	84–3
Windscreen wiper motor – remove and refit	84.15.12	84–2
Windscreen wiper rack drive – remove and refit	84.15.24	84–2
Windscreen wiper wheel boxes – remove and refit – left-hand	84.15.28	84–3
– right-hand	84.15.29	84–3

ELECTRICAL

	Operation No.	Page No.
Alternator		
– bench test	86.10.14	86–4
– control unit – remove and refit	86.10.26	86–6
– description	86.10.00	86–1
– drive belt – adjust	86.10.05	86–2
– remove and refit	86.10.03	86–2
– overhaul	86.10.08	86–3
– pulley – remove and refit	86.10.04	86–2
– remove and refit	86.10.02	86–2
– test (in situ)	86.10.01	86–1
Battery		
– battery tray – remove and refit	86.15.11	86–7
– lead – negative – remove and refit	86.15.19	86–7
– remove and refit	86.15.01	86–6
– terminals – disconnect and reconnect	86.15.20	86–6
– test	86.15.02	86–6
Cigar-lighter assembly – remove and refit	86.65.60	86–27
Electrically operated windows and door locks		
– circuit breakers – remove and refit	86.25.31	86–8
– description	86.25.00	86–7
– door lock selector switch – remove and refit	86.25.14	86–8
– door lock solenoids – remove and refit	86.25.32	86–8
– door lock solenoid relays – remove and refit	86.25.33	86–9
– front panel switch – remove and refit	86.25.13	86–8
– motors – remove and refit	86.25.05	86–7
– relay – remove and refit	86.25.28	86–8
Horns		
– adjust	86.30.08	86–9
– circuit check	86.30.17	86–9
– description	86.30.00	86–9
– push – remove and refit	86.30.01	86–9
– relay – remove and refit	86.30.18	86–9
– remove and refit	86.30.09	86–9
Ignition		
– amplifier unit – remove and refit	86.35.30	86–15
– ballast resistor unit – remove and refit	86.35.33	86–15
– coil – remove and refit	86.35.32	86–15
– description	86.35.00	86–9
– distributor – remove and refit	86.35.20	86–12
– distributor cap – remove and refit	86.35.10	86–10
– distributor leads – remove and refit	86.35.11	86–10
– distributor – overhaul	86.35.26	86–11
– electronic timing rotor – remove and refit	86.35.17	86–11
– ignition protection relay – remove and refit	86.35.36	86–14
– ignition system – check	86.35.29	86–12
– pick-up module – remove and refit	86.35.18	86–11
– rotor arm – remove and refit	86.35.16	86–11
Lighting – external		
– flasher repeater assembly – remove and refit	86.40.53	86–17
– bulb – remove and refit	86.40.52	86–17
– lens – remove and refit	86.40.51	86–17
– front flasher assembly – remove and refit	86.40.42	86–17
– bulb – remove and refit	86.40.41	86–16
– lens – remove and refit	86.40.40	86–16
– headlamp		
– pilot bulb – remove and refit	86.40.11	86–16
– sealed beam unit/bulb – remove and refit	86.40.09	86–16
– headlamp alignment	86.40.18	86–16
– assembly – outer – remove and refit	86.40.02	86–16
– rim finisher – remove and refit	86.40.01	86–15
– number-plate and lamp assembly – remove and refit	86.40.87	86–18
– number-plate lamp bulb – remove and refit	86.40.85	86–18
– lens – remove and refit	86.40.84	86–18
– reverse lamp bulb – remove and refit	86.40.90	86–18
– lens – remove and refit	86.40.89	86–18
– side front marker assembly – remove and refit	86.40.59	86–17
– bulb – remove and refit	86.40.58	86–17
– lens – remove and refit	86.40.57	86–17
– side rear marker assembly – remove and refit	86.40.64	86–17
– bulb – remove and refit	86.40.63	86–17
– lens – remove and refit	86.40.62	86–17
– tail/stop/flasher assembly – left-hand – remove and refit	86.40.70	86–18
– bulb – remove and refit	86.40.69	86–18
– lens – remove and refit	86.40.68	86–17

continued

	Operation No.	Page No.
Lighting – internal illumination		
– cigar lighter illumination – bulb – remove and refit	86.45.55	86–6
– luggage compartment assembly – remove and refit	86.45.16	86–19
– bulb – remove and refit	86.45.15	86–19
– map/courtesy lamp assembly – remove and refit	86.45.04	86–19
– bulb – remove and refit	86.45.03	86–19
– opticel – remove and refit	86.45.27	86–19
– bulb – remove and refit	86.45.28	86–19
– panel switch illumination – bulb – remove and refit	86.45.31	86–19
– roof lamp bulb – remove and refit	86.45.01	86–18
– assembly – remove and refit	86.45.02	86–19
Lighting indicator		
– automatic transmission – bulb – remove and refit	86.45.40	86–19
– clock illumination – bulb – remove and refit	86.45.54	86–20
– warning lamp bulbs – remove and refit	86.45.61	86–20
Lighting instrument		
– instrument illumination bulb – remove and refit	86.45.48	86–20
– speedometer – bulb – remove and refit	86.45.49	86–20
– revolution counter (tachometer) – bulb – remove and refit	86.45.53	86–20
Relays, flasher and alarm units		
– hazard/turn signal flasher unit – remove and refit	86.55.12	86–20
– headlamp relay – remove and refit	86.55.17	86–21
– ignition/starter controlled relay – remove and refit	86.55.28	86–21
– low coolant control unit – remove and refit	86.55.33	86–21
– park lamp failure warning sensor – remove and refit	86.55.22	86–21
– starter relay – remove and refit	86.55.05	86–20
– steering lock/safety belt audio unit – remove and refit	86.55.13	86–20
– stop light failure sensor – remove and refit	86.55.34	86–21
Starter		
– bench test	86.60.14	86–24
– overhaul	86.60.13	86–23
– remove and refit	86.60.01	86–22
– solenoid – remove and refit	86.60.08	86–23
– solenoid – test	86.60.09	86–23
– starter motor roller clutch drive unit – remove and refit	86.60.05	86–22
Switches and rheostats		
– combined headlight/direction/flasher/dip switch – remove and refit	86.65.55	86–26
– door pillar switch – remove and refit	86.65.15	86–25
– door pillar switch – key alarm – remove and refit	86.65.27	86–26
– fuel cut-off inertia switch – remove and refit	86.65.58	86–27
– fuel cut-off inertia switch – reset	86.65.59	86–27
– handbrake switch		
– adjust	86.65.46	86–26
– remove and refit	86.65.45	86–26
– ignition switch – remove and refit	86.65.03	86–25
– luggage compartment light switch – remove and refit	86.65.22	86–26
– master lighting switch – remove and refit	86.65.09	86–25
– oil pressure switch – remove and refit	86.65.30	86–26
– panel light rheostat – remove and refit	86.65.07	86–25
– panel switches – remove and refit	86.65.06	86–25
– reverse lights switch – remove and refit	86.65.20	86–25
– stop light switch		
– adjust	86.65.56	86–27
– remove and refit	86.65.51	86–26
– windscreen wiper/washer switch – remove and refit	86.45.41	86–26
Seat belt warning – belt switches – remove and refit	86.57.25	86–22
Wiring and fuses – wiring diagram		End of section

INSTRUMENTS

Battery condition indicator – remove and refit	88.10.07	88–1
Clock – remove and refit	88.15.07	88–1
Coolant temperature gauge – remove and refit	88.25.14	88–1

continued

	Operation No.	Page No.
Coolant temperature transmitter – remove and refit	88.25.20	88–2
Fuel gauge – remove and refit	88.25.26	88–1
Fuel tank unit – remove and refit	88.25.32	88–2
Instrument panel (module) – remove and refit	88.20.01	88–1
Instrument panel lens assembly – remove and refit	88.20.17	88–1
Instrument panel printed circuit – remove and refit	88.20.19	88–1
Oil gauge – remove and refit	88.25.01	88–1
Oil pressure transmitter – remove and refit	88.25.07	88–1
Oil pressure warning switch – remove and refit	88.25.08	88–2
Revolution counter (tachometer) – remove and refit	88.30.21	88–3
Speedometer – remove and refit	88.30.01	88–2
Speedometer cable assembly – remove and refit	88.30.06	88–2
Speedometer cable – inner – remove and refit	88.30.07	88–2
Speedometer, right angle drive, gearbox – remove and refit	88.30.16	88–3
Speedometer, right angle drive, instrument – remove and refit	88.30.15	88–2
Speedometer trip reset – remove and refit	88.30.12	88–2

SERVICE TOOLS

Engine	99.00.01	99–1
Automatic Transmission	99.00.01	99–1
Final Drive	99.00.02	99–2
Steering	99.00.03	99–3
Front Suspension	99.00.03	99–3
Rear Suspension	99.00.04	99–4
Brakes	99.00.04	99–4
Body	99.00.04	99–4

STANDARDISED ABBREVIATIONS AND SYMBOLS

Term	Abbreviation or Symbol	Term	Abbreviation or Symbol	Term	Abbreviation or Symbol
Across flats (bolt size)	A.F.	Gallons (Imperial)	gal.	Miles per gallon	m.p.g.
After bottom dead centre	A.B.D.C.	Gallons (U.S.)	U.S. gal.	Miles per hour	m.p.h.
After top dead centre	A.T.D.C.	Grammes	g	Millimetres	mm
Alternating current	a.c.			Millimetres of mercury	mm Hg
Amperes	A	High compression	h.c.	Minimum	min.
Ampere-hour	Ah	High tension (electrical)	h.t.	Minus (of tolerance)	—
		Horse power	hp	Minute (of angle)	′
		Hundredweight	cwt		
Before bottom dead centre	B.B.D.C.			Negative (electrical)	—
Before top dead centre	B.T.D.C.	Inches	in	Number	No.
Bottom dead centre	B.D.C.	Inches of mercury	in Hg		
Brake horse power	b.h.p.	Independent front suspension	i.f.s.	Ounces (force)	ozf
Brake mean effective pressure	b.m.e.p.	Infinity	∞	Ounces (mass)	oz
British Standards	B.S.	Internal diameter	i.dia.	Ounce inch (torque)	ozf in
				Outside diameter	o.dia.
Carbon monoxide	CO	Kilogrammes (force)	kgf	Overdrive	O/D
Centigrade (Celsius)	C	Kilogrammes (mass)	kg		
Centimetres	cm	Kilogramme centimetre	kgf cm	Paragraphs	para.
Cubic centimetres	cm³	Kilogramme metres	kgf m	Part Numbers	Part No.
Cubic inches	in³	Kilogrammes per square	kgf/cm² or	Percentage	%
Cycles per minute	c/min	centimetre	kg/cm²	Pints (Imperial)	pt.
		Kilometres	km	Pints (U.S.)	U.S. pt.
Degree (angle)	deg. or °	Kilometres per hour	km/h	Plus or minus	±
Degree (Temperature)	deg. or °	Kilovolts	kV	Plus (tolerance)	+
Diameter	dia.	King pin inclination	k.p.i.	Positive (Electrical)	+
Direct current	d.c.			Pounds (force)	lbf
		Left hand	L.H.	Pounds (mass)	lb
Fahrenheit	F	Left hand steering	L.H. Stg.	Pounds feet (torque)	lbf ft
Feet	ft	Left hand thread	L.H. Thd.	Pounds inches (torque)	lbf in
Feet per minute	ft/min	Low compression	l.c.	Pounds per square inch	lbf/in²
Fifth	5th	Low tension	l.t.		or lb/in²
Figure (illustration)	Fig.				
First	1st	Maximum	max.	Radius	r
Fourth	4th	Metres	m	Ratio	:
		Miniature Edison Screw	MES	Reference	ref.

Term	Abbreviation or Symbol
Revolutions per minute	rev/min
Right-hand	R.H.
Right-hand steering	R.H. Stg.
Second (angle)	″
Second (numerical order)	2nd
Single carburetter	SC
Society of Automobile Eng.	S.A.E.
Specific gravity	sp. gr.
Square centimetres	cm²
Square inches	in²
Standard	std.
Standard wire gauge	s.w.g.
Synchroniser/synchromesh	synchro.
Third	3rd
Top dead centre	T.D.C.
Twin carburetters	TC
United Kingdom	UK
Volts	V
Watts	W
Screw threads	
American Standard Taper Pipe	N.P.T.F.
British Association	B.A.
British Standard Fine	B.S.F.
British Standard Pipe	B.S.P.
British Standard Whitworth	B.S.W.
Unified Coarse	U.N.C.
Unified Fine	U.N.F.
Metric	M (e.g. M20)

GENERAL SPECIFICATION DATA

Final Drive Unit — Standard
Type. . . . . Hypoid, with 'Powr-Lok' differential
Ratios. . . . . 3.31 : 1 43/13 – USA – Australia
3.07 : 1 All others

Automatic Gearbox
Make and Type. . . . . **Borg-Warner Model 12**   **GM 400**
Ratios. . . . . First gear   2.40 : 1   2.48 : 1
Second gear   1.46 : 1   1.48 : 1
Third gear   1.00 : 1   1.00 : 1
Reverse   2.00 : 1   2.07 : 1
Torque converter   2.00 : 1 max.

Clutch (On Manual Transmission Cars Only)
Make and Type. . . . . Borg and Beck single dry plate
Clutch pedal diameter. . . . . 10.50 in (266,7 mm)
Facing material. . . . . Raybestos WR7
No. of damper springs. 6
Damper spring colour. . . . . Dark green/cream
Clutch release bearing. . . . . Carbon thrust type
Clutch fluid. . . . . Castrol-Girling Universal

Manual Gearbox
Type. . . . . 4-speed with baulk-ring synchromesh on forward gears
Ratios. . . . . First gear   3.238 : 1
Second gear   1.905 : 1
Third gear   1.389 : 1
Fourth gear   1.00 : 1
Reverse   3.428 : 1

Cooling System
Water pump – Type. . . . . Centrifugal (2 outlet ports)
Drive. . . . . Belt
No. of cooling fans. . . . . 2 (1–12 blade, belt driven through 'Holset' coupling plus 1–4 blade, electrically driven, thermostatically controlled)
Cooling system control. . . . . 2 Thermostats
Thermostat opening temperature. . . . . . 79°C to 83°C (174°F to 181°F)
88°C (190°F) – Later cars
Thermostat fully open at. . . . . 93°C to 96°C (200°F to 205°F)
Filler cap pressure rating. . . . . 15 lb/in² (1,05 kgf/cm²)
Filler cap – Make. . . . . AC Delco

Fuel Injection Equipment
Type. . . . . Bosch–Lucas electronically controlled port injection
Fuel pressure. . . . . 28.5 to 31.3 lb/in² (2,0 to 2,2 kg/cm²)

Braking System
Front Brakes – Make and Type. . . . . . . Girling; ventilated discs, bridge type calipers
Rear Brakes – Make and Type. . . . . With normal final drive: Girling, damped discs, bridge type calipers incorporating handbrake friction pads
Handbrake – Type. . . . . Mechanical, operating on rear disc pads
Disc diameter – Front. . . . . 11.18 in (284 mm)
– Rear. . . . . 10.375 in (263,5 mm)
Disc thickness – Front. . . . . 0.95 in (24,13 mm)
– Rear, standard final drive
– Rear, with optional overdrive on final drive   0.50 in (12,7 mm)
Disc pad total area – Front. . . . . 0.95 in (24,13 mm)
– Rear main. . . . . . 40.8 in² (263 cm²)
– Rear handbrake. . 25.1 in² (162 cm²)
6.9 in² (44.5 cm²)
Pad material – Front and Rear main. . . Ferodo 2430 slotted
handbrake. . . . . Mintex M68/1
Swept area of discs – Front. . . . . 256 in² (1652 cm²)
– Rear. . . . . 200 in² (1290 cm²)
Master cylinder bore diameter. . . . . 0.938 in (23,83 mm)
Brake fluid. . . . . Castrol-Girling Universal Brake and Clutch Fluid – exceeding specifications S.A.E. 1703D
Servo unit ref. R.H. Stg cars. . . . . Girling ref. 64049748
L.H. Stg cars. . . . . Girling ref. 64049747

Front Suspension
Type. . . . . Independent – coil spring
Castor angle. . . . . 3½° ± ¼° positive
Camber angle. . . . . ½° ± ¼° positive
Front wheel alignment. . . . . 0 to 1/16 in (0 to 1,6 mm) toe-out
Dampers. . . . . Telescopic, gas filled

Rear Suspension
Type. . . . . Independent – coil springs, coaxial with dampers
Camber angle. . . . . ¾° ± ¼° negative
Rear wheel alignment. . . . . Parallel ± 1/32 in (± 0,08 mm)
Dampers. . . . . Telescopic, gas filled

Power Assisted Steering
Type . . . . Rack and pinion
Number of turns, lock to lock 2.9
Turning circle, kerb to kerb . . . . 36 ft 3 in (11,05 m)

Electrical Equipment
Battery – Make and Type . . . . Lucas CP13/11-8, GBY 218
Voltage 12V
No. of plates per cell 13
Capacity at ten hour rate . . . . 60 Ah
Capacity at twenty hour rate . . . . . . 68 Ah

Alternator
Make and Type . . . . Lucas 20ACR; 25ACR or Motorola
Nominal voltage 12V
Cut-in voltage . . . . 13.5V at 1500 r.p.m.
Earth polarity . . . . Negative
Maximum output 66A
Maximum operating speed . . . . 15,000 r.p.m.
Rotor winding resistance . . . . 3.6 ohms @ 20°C
Brush spring pressure . . . . 9–13 oz (255–368 g)

Starter Motor
Make and Type . . . . Lucas M45 pre-engaged
Lock torque . . . . 29 lbf ft (4,01 kgf m) at 940 A
Torque at 1000 r.p.m. . . . . 11 lbf ft (1,52 kgf m) at 540 A
Light running current . . . . 100 A at 5000 to 7500 r.p.m.

Distributor
Make and Type . . . . Lucas Opus 36 DE 12

LAMP BULBS

LAMP	BULB No.	WATTS	NOTES
Headlamp – halogen bulbs	H1	55	R.H.Stg. except Japan (Cibie Lamps)
Headlamp – halogen bulbs	H1	55	L.H.Stg. except North America, France (Cibie Lamps)
Headlamp – halogen bulbs		55	France only (Cibie Lamps)
Headlamp – pilot bulb	15602	4	Osram miniature bayonet
Headlamp – light unit inner		37.5	North American only (sealed beam)
Headlamp – light unit outer		37.5/50	
Headlamp – light unit inner		37.5	Japan only (sealed beam)
Headlamp – light unit outer		37.5/50	
Front flashers and side lamps	GLB 380	21/5	Double filament – Italy only
Front flasher	GLB 382	21	All other countries
Side lamps	GLB 207	5	
Reverse lamp (festoon)	GLB 273	21	
Number plate lamp	GLB 254	6	
Tail flasher	GLB 382	21	
Stop/tail	GLB 382	21	
Tail	GLB 207	5	
Interior lamps (festoon bulbs)	GLB 272	10	
Luggage compartment lamp	GLB 989	6	
Clock illumination (MES)	GLB 987	2.2	
Warning lamps (capless)		1.2	
Instrument lamps (capless)		2.2	
Cigar lighter (mini bayonet)	G70112	2.2	
Fibre optic light source	6253	6	
Auto transmission selector indicator	GLB 281	2	Smith 40-621-109-86 Smith 40-621-109-74 Vitality bulb
Warning light, catalyst (when fitted)	GLB 281	2	AAU 9276 bulb (Wotan)

04–2

TYRE DATA

TYRE PRESSURES

N.B. Check pressures with tyres cold, and not when they have attained their normal running temperature.

Tyre Type SP Sport Super 205/70 VR 15 (SP Radial Tubeless)

	PRESSURE	
	FRONT	**REAR**
For normal use up to 120 m.p.h. (193 km/h) with driver and one passenger and 60 lb (27 kg) luggage.	26 lbf/in² 1,83 kgf/cm² 1,79 bars	24 lbf/in² 1,69 kgf/cm² 1,65 bars
For normal use up to 120 m.p.h. (193 km/h) with full load (inclusive of luggage) of 720 lb (327 kg).	26 lbf/in² 1,83 kgf/cm² 1,79 bars	26 lbf/in² 1,83 kgf/cm² 1,79 bars

For sustained speeds above 120 m.p.h. (193 km/h) add 6 lbf/in² (0,42 kgf/cm² or 0,41 bars) to the above pressures.

Tyre Replacement and Wheel Interchanging

When replacement of tyres is necessary, it is preferable to fit a complete car set. Should either front or rear tyres only show a necessity for replacement, new tyres must be fitted to replace the worn ones. No attempt must be made to interchange tyres from front to rear or vice-versa as tyre wear produces characteristic patterns depending upon their position and if such position is changed after wear has occurred, the performance of the tyre will be adversely affected. It should be remembered that new tyres require to be balanced.

The DUNLOP radial ply tyres specified are designed to meet the high speed performance of which this car is capable. Only tyres of identical specification as shown above must be fitted as replacements and, if of different tread pattern, should not be fitted in mixed form.

UNDER NO CIRCUMSTANCES SHOULD CROSS-PLY TYRES BE FITTED.

Repair of small injuries to tubeless tyres

a) Small penetrations of up to 1.5 mm diameter by nails or other slim objects which occur in the centre tread area i.e. middle five bars of tread can be repaired permanently so long as any repair carried out complies with British Standard BS.AU.159A. This standard requires the tyre to be removed from the rim and to be carefully inspected both externally and internally for any additional damage. If the tyre penetration is the only damage it should then be permanently repaired by vulcanisation of a rubber patch applied internally. Three such damages in any one tyre may be so repaired provided that no two of them occur within one quarter of the tyre circumference.

b) Damage of up to 5 mm should be repaired by vulcanisation of a rubber patch internally and the hole filled with rubber dough which should also be vulcanised. Two repairs of this size may be carried out in any one tyre provided that they occur in different halves of the tyre.

c) Steel braced tyres should only be repaired by patching internally. For these tyres – as fitted to XJ-S models – plug/patches should never be used.

d) On tubed tyres where the tyre casing is not seriously damaged the fitment of a new inner tube is recommended. Provided the above recommendations are followed the speed capability of the car is unaffected.

Where a tyre is more excessively damaged than is noted above no repairs of any sort should be carried out. A new tyre must be fitted.

In such cases owners must be advised of our policy and accept responsibility for any deviation from the above recommendations.

ENGINE DATA

General Data

Number of cylinders	12
Bore	3.543 in (90 mm)
Stroke	2.756 in (70 mm)
Cubic capacity	5343 cc (326,0 cu in)
Compression ratio	'S' (Standard) or 'L' (Low)
Ignition timing, static	10° B.T.D.C. (0° B.T.D.C. for Sweden only)
Firing order 'A' Bank – Right Hand 'B' Bank – Left Hand	1A, 6B, 5A, 2B, 3A, 4B, 6A, 1B, 2A, 5B, 4A, 3B (Cylinders numbered from front of engine)

Cylinder Block

Material (Cylinder Block)	Aluminium alloy
Angle of cylinders	60° Vee
Type of cylinder liner	Slip fit, wet liner
Material (cylinder liners)	Cast iron
Nominal size of bore after honing	
GRADE 'A' – RED	3.543 in (89,992 mm)
GRADE 'B' – GREEN	3.544 in (90,018 mm)
Outside diameter of liner – both grades	3.858 in + 0.001 in – 0.00 in (97,99 mm + 0,02 mm – 0,00 mm)
Main line bore for main bearings	3.1665 in to 3.1667 in (80,429 mm to 80,434 mm)

Cylinder Heads

Material	Aluminium alloy
Valve seat angle – Inlet	44½°
Valve seat angle – Exhaust	44½°

Crankshaft

Material	Manganese molybdenum steel
Number of main bearings	7
Main bearing type	Vandervell V.P.3
Journal diameter	3.0007 in to 3.0012 in (76,218 mm to 76,231 mm)
Journal length – Front	1.170 in to 1.180 in (29,72 mm to 29,97 mm)
– Centre	1.425 in to 1.426 in (36,20 mm to 36,22 mm)
– Intermediate	1.198 in to 1.202 in (30,43 mm to 30,53 mm)
– Rear	1.425 in to 1.426 in (36,20 mm to 36,22 mm)
Thrust taken	Centre bearing thrust washers
Thrust washer thickness	0.101 in to 0.103 in (2,57 mm to 2,62 mm)
Permissible end float	0.004 in to 0.006 in (0,10 mm to 0,15 mm)
Width of main bearing – Front	0.963 in to 0.973 in (24,40 mm to 24,65 mm)
– Centre	1.190 in to 1.200 in (30,2 mm to 30,5 mm)
Side clearance of oil control rings in groove	Nil (self expanding rings)
Top compression ring gap in bore	0.014 in to 0.020 in (0,36 mm to 0,51 mm)
Second compression ring gap in bore	0.010 in to 0.015 in (0,25 mm to 0,38 mm)
Gap of oil control ring rails in bore	0.015 in to 0.045 in (0,38 mm to 1,14 mm)

Gudgeon Pins

Type	Fully floating
Length	3.120 in to 3.125 in (79,25 mm to 79,38 mm)
Outside diameter – Grade 'A' Red	0.9375 in (23,81 mm)
– Grade 'B' Green	0.9373 in (23,76 mm)

Camshafts

Number of journals	7
Number of bearings	7 per shaft (14 half bearings)
Type of bearings	Aluminium alloy – Camshafts run direct in caps
Journal diameter – All journals	1.0615 in + 0.0005 in – 0.000 in (26,93 mm + 0,013 mm – 0,000 mm)
Diametrical clearance	0.001 in to 0.003 in (0,03 mm to 0,07 mm)
Thrust taken	Front end of shaft

Jackshaft

Number of bearings	3
Diametrical clearance in block	0.0005 in to 0.003 in (0,013 mm to 0,07 mm)
Thrust taken	Front end of shaft
Permissible end float	0.005 in (0,13 mm)
Line bore of front bearing	1.251 in to 1.252 in (31,78 mm to 31,80 mm)
Line bore of centre and rear bearing	1.190 in to 1.191 in (30,23 mm to 30,25 mm)

Valve and Valve Springs

Inlet valve material	Silico chrome steel
Exhaust valve material	Austenitic steel
Inlet valve head diameter	1.620 in to 1.630 in (41,15 mm to 41,40 mm)
Exhaust valve head diameter	1.355 in to 1.365 in (34,42 mm to 34,67 mm)
Valve stem diameter – Inlet and Exhaust	0.3092 in to 0.3093 in (7,854 mm to 7,856 mm)
Valve lift	0.375 in (9,5 mm)
Inlet valve clearance	0.012 in to 0.014 in (0,305 mm to 0,356 mm)
Exhaust valve clearance	0.012 in to 0.014 in (0,305 mm to 0,356 mm)
Outer valve spring free length	2.103 in (53,4 mm)
Inner valve spring free length	1.734 in (44,0 mm)

Valve and Valve Springs — (contd.)
    — Intermediate . . . . 0.963 in to 0.973 in (24,40 mm to 24,65 mm)
    — Rear . . . . 1.190 in to 1.200 in (30,2 mm to 30,5 mm)
Diametrical clearance — All bearings . . . 0.0015 in to 0.003 in (0,04 mm to 0,07 mm)
Crankpin diameter . . . . 2.2994 in to 2.3000 in (58,40 mm to 58,42 mm)
Crankpin length . . . . 1.699 in to 1.701 in (43,15 mm to 43,20 mm)

Connecting Rods
Length between centres . . . . 5.96 in ± .002 in (151,38 mm ± 0,05 mm)
Big end bearing material VP2C
Bore for big end bearing . . . . 2.441 in + 0.0005 in − 0.0001 in (62,00 mm + 0,013 mm − 0,0025 mm)
Width of big end bearing . . . . 0.720 in to 0.730 in (18,3 mm to 18,5 mm)
Big end diametrical clearance . . . . 0.0015 in to 0.0034 in (0,04 mm to 0,09 mm)
Big end side clearance . . . . 0.007 in to 0.013 in (0,17 mm to 0,33 mm)
Small end bush material VP.10
Bore for small end bush . . . . 1.062 in + 0.001 in − 0.000 in (26,98 mm + 0,025 mm − 0,00 mm)
Width of small end bush . . . . 1.03 in to 1.05 in (26,2 mm to 26,7 mm)
Bore diameter of small end bush . . . . . . 0.9375 in to 0.9377 in (23,813 mm to 23,818 mm)

Pistons
Type . . . . Solid skirt
Skirt clearance (measured midway down bore across bottom of skirt) . . . . 0.0012 in to 0.0017 in (0,03 mm to 0,04 mm)

Piston Rings
Number of compression rings 2
Number of oil control rings 1
Top compression ring width . . . . 0.150 in to 0.160 in (3,81 mm to 4,06 mm)
Second compression ring width . . . . 0.150 in to 0.160 in (3,81 mm to 4,06 mm)
Oil control ring width . . . . Self expanding
Width of oil control ring rails . . . . 0.103 in ± 0.003 in (2,62 mm ± 0,07 mm)
Top compression ring thickness . . . . 0.062 in to 0.063 in (1,58 mm to 1,60 mm)
Second compression ring thickness . . . . 0.077 in to 0.078 in (1,96 mm to 1,98 mm)
Side clearance of top compression ring in groove . . . . 0.0029 in (0,07 mm)
Side clearance of second compression ring in groove . . . . 0.0034 in (0,09 mm)

Valve Guides and Seats
Valve guide material . . . . Cast iron
Inlet valve guide length . . . . 1.910 in (48,5 mm)
Exhaust valve guide length . . . . 2.125 in (54,0 mm)
Inlet valve guide outside diameter . . . . . As exhaust valve guide
Exhaust valve guide outside diameter
    Standard . . . . 0.502 in to 0.501 in (12,75 mm to 12,72 mm)
    First oversize (2 grooves) . . . . 0.507 in to 0.506 in (12,88 mm to 12,85 mm)
    Second oversize (3 grooves) . . . . 0.512 in to 0.511 in (13,01 mm to 12,98 mm)
Inlet valve guide finished bore . . . . 0.311 in to 0.312 in (7,90 mm to 7,92 mm)
Exhaust valve guide finished bore . . . . . 0.311 in to 0.312 in (7,90 mm to 7,92 mm)
Maximum clearance between valve stem and guide . . . . 0.0020 in to 0.0023 in (0,05 mm to 0,06 mm)
Interference fit in cylinder head . . . . 0.002 in to 0.006 in (0,05 mm to 0,15 mm)
Valve seat insert material . . . . Sintered iron
Inlet valve seat insert outside diameter . 1.7282 in $^{+0.0005}_{-0.0000}$ in (43,9 mm $^{+0.01}_{-0.00}$ mm.)
Exhaust valve seat insert outside diameter . . . . 1.4882 in $^{+0.0005}_{-0.0000}$ in (37,8 mm $^{+0.01}_{-0.00}$ mm.)
Inlet valve seat insert inside diameter (before cutting seat) . . . . 1.350 in $^{+0.010}_{-0.000}$ in (34,3 mm $^{+0.25}_{-0.00}$ mm.) to 1.490 in $^{+0.010}_{-0.000}$ in (37,84 mm $^{+0.25}_{-0.00}$ mm.)
Exhaust valve seat insert inside diameter (before cutting seat) . . . . 1.160 in $^{+0.008}_{-0.000}$ in (29,5 mm $^{+0.25}_{-0.00}$ mm.) to 1.325 in $^{+0.010}_{-0.000}$ in (33,7 mm $^{+0.25}_{-0.00}$ mm.)

Service Replacements
Inlet valve seat insert diameter . . . . 1.744 in $^{+0.0005}_{-0.0000}$ in (44,30 mm $^{+0.013}_{-0.000}$ mm.)
Exhaust valve seat insert diameter . . . . 1.503 in $^{+0.0005}_{-0.0000}$ in (38,17 mm $^{+0.01}_{-0.00}$ mm.)
Inlet valve seat inside diameter . . . . 1.400 in $^{+0.003}_{-0.000}$ in (35,56 mm $^{+0.07}_{-0.00}$ mm.) to 1.565 in $^{+0.010}_{-0.000}$ in (39,74 mm $^{+0.25}_{-0.00}$ mm.)
Exhaust valve seat inside diameter . . . . 1.185 in $^{+0.003}_{-0.000}$ in (30,1 mm $^{+0.07}_{-0.00}$ mm.) to 1.315 in $^{+0.005}_{-0.000}$ in (33,4 mm $^{+0.12}_{-0.00}$ mm.)

Tappets and Tappet Guides
Tappet material . . . . Cast iron (chilled)
Outside diameter of tappet . . . . 1.373 in to 1.374 in (34,87 mm to 34,90 mm)
Diametrical clearance . . . . 0.001 in to 0.002 in (0,025 mm to 0,05 mm)

Lubrication System

Oil pump	Epicyclic gear type	
Oil pump gears	Diametrical Clearance	Radial Clearance
Driving gear O/D	0.005 to 0.012 in (0,127 to 0,305 mm)	0.0025 to 0.006 in (0,065 to 0,152 mm)
Driven gear O/D	0.007 to 0.010 in (0,178 to 0,254 mm)	0.0035 to 0.005 in (0,09 to 0,13 mm)
Driven gear I/D	0.011 to 0.018 in (0,28 to 0,46 mm)	0.0055 to 0.009 in (0,14 to 0,23 mm)
Side clearance — Driving and Driven Gear	0.0045 in to 0.0065 in (0,115 mm to 0,165 mm)	
Oil filter type	Full flow, renewable element or disposable canister	

Timing Chains and Sprockets

Type of chain	Duplex endless
Pitch	0.375 in (9,5 mm)
Number of pitches	180
Camshaft sprockets — Number of teeth (each)	42
Crankshaft sprocket — Number of teeth	21
Jackshaft sprocket — Number of teeth	21

Sparking Plugs

Make	Champion
Type	N10Y
Gap	0.035 in (0,89 mm)

ENGINE TUNING DATA

Model: XJ-S   **Year** 1975 on

ENGINE
Type	V-12
Capacity	326.10 m³ 5344 cm³
Compression ratio: S	9.0:1
L	7.8:1
Firing order from front of engine	1A-6B-5A-2B-3A-4B-6A-1B-2A-5B-4A-3B 'A' is R.H. back, 'B' is L.H. back
Cranking pressure: S	165 lbf/in² 11,6 kgf/cm²
L	135 lbf/in² 9,5 kgf/cm²
Idling speed	750 rev/min
Ignition timing, static (crank degrees and rev/min)	10° B.T.D.C. up to 900 rev/min, vacuum disconnected 4° A.T.D.C. — 750 rev/min, vacuum connected — 1977-78 cars to Australia and California
Timing marks	T.D.C. mark on crank damper rim and degree scale on timing cover
Valve clearance (Inlet and Exhaust)	0.012 to 0.014 in (0,305 to 0,355 mm)

SPARK PLUGS
Make/type	Champion N10Y	
Gap (except North America)	0.025 in	0,64 mm (early cars only)
(North America) All 1978 cars	0.035 in	0,89 mm

IGNITION COIL
Make/type	Lucas 22 C.12
Primary resistance at 20° C (68° F)	0.9 to 1.1 ohms
Consumption: stationary	5.0 to 6.5 amps
running	2.5 to 3.0 amps
Ballast resistance	0.9 to 1.0 ohms

DISTRIBUTOR
Make/type	Lucas Opus 36 DE 12
Rotation of rotor (locking down at rotor)	anti-clockwise
Pick-up module to timing rotor gap	0.020 to 0.022 in (0,51 to 0,55 mm)
Centrifugal advance	Crank degrees and rev/min
Decelerating check, vacuum pipe disconnected	34° to 38° at 6200 29° to 33° at 4000 22° to 26° at 2000 No advance below 900
Vacuum advance (crank degrees)	
Maximum	4° at 10 in Hg 254 mm Hg
Starts	6 in Hg 152 mm Hg

FUEL INJECTION EQUIPMENT
Make/type	Bosch-Lucas electronically controlled

EXHAUST EMISSION
Exhaust gas analyser reading at engine idle speed	1–2% max. CO at 750 rev/min without air injection

TORQUE WRENCH SETTINGS

ENGINE

ITEM	DESCRIPTION	lbf.ft.	kgf.m.
Cylinder head	7/16 in UNF nut	52 max.	7,19 max.
Main bearing	3/8 in UNF nut	27–28	3,73–3,87
	1/2 in UNF nut	62,5 max.	8,64 max.
	3/8 in UNF nut	27–28	3,73–3,87
Con rod big end	3/8 in UNF nut	40–41	5,53–5,67
Flywheel	7/16 in UNF bolt	66,7 max.	9,22 max.
Crankshaft bolt	3/4 in UNF bolt	125–150	17,28–20,74
Camshaft cap	5/16 in UNF nut	9 max.	1,24 max.
Camshaft cover	1/4 in UNC screws	8.3 max.	1,15 max.
Oil filter canister (early cars)	7/16 in UNC special bolt	15–20	2,07–2,76
Disposable canister oil filter	1 in × 12 UNF thread	6–8	0,83–1,10
P.A.S. pump to mtg. bracket	3/8 in UNC nut	37 max.	5,12 max.
Oil sump setscrews	5/16 in UNC screws	12–15	1,66–2,07
Pulleys to crank damper	5/16 in UNF bolts	12–15	1,66–2,07
Air pump pivot	3/8 in UNF bolts	15–20	2,07–2,76
Union block to compressor	3/8 in UNF bolt	10–25	1,38–3,46
Gemi hose clips up to No. 16	4 mm thread	0.25–0.50	0,035–0,069

ENGINE MOUNTINGS

ITEM	DESCRIPTION	lbf.ft.	kgf.m.
Rear mounting bracket to body fixing	5/16 in UNF bolt	14–18	1,94–2,49
	3/8 in UNF bolt	27–32	3,73–4,42
Front bracket to beam	5/16 in UNF nut	14–18	1,94–2,49
Rear mounting peg	1/2 in UNF nut	25–30	3,46–4,15
Rear rubbers	3/8 in UNF nut	27–32	3,73–4,42
Tie bolts	1/2 in UNF nut	25–30	3,46–4,15

EXHAUST AND HEAT SHIELDS

ITEM	DESCRIPTION	lbf.ft.	kgf.m.
Front down pipes to manifold	3/8 in UNF nut	22–26	3,04–3,59
'U' bolt clips	5/16 in UNF nut	11–13	1,52–1,80
Coupling flange	5/16 in UNF nut	11–13	1,52–1,80
Clip, outer exhaust pipe	5/16 in UNF nut	11–13	1,52–1,80
Exhaust mounting ring assy. to cross beam	5/16 in UNF nut	14–18	1,94–2,49
Tail pipe mounting assy. to body	5/16 in UNF nut	14–18	1,94–2,49
Tail pipe grub screw	1/4 in UNF screw	6–7	0,83–0,97
Exhaust mtg. ring assy. to cross beam	5/16 in UNF bolt	14–18	1,94–2,49
Catalyst stays to converter	1/4 in UNC bolt	6–7	0,83–0,97
Catalyst stays to exhaust pipe tie plate	1/4 in UNF nut	6–7	0,83–0,97

ENGINE COOLING SYSTEM

ITEM	DESCRIPTION	lbf.ft.	kgf.m.
Expansion tank attachment to valance	5/16 in UNF nut	14–18	1,94–2,49
Fan cowl to body	1/4 in UNF nut	6–7	0,83–0,97
Fan cowl fixing bracket to lower cross member	1/4 in UNF bolt	6–7	0,83–0,97
Mounting bracket to fan motor	6 mm nut	6–7	0,83–0,97
Fan mounting to cowl	1/4 in UNF nut	6–7	0,83–0,97
Engine oil cooler pipes	1.1/16 in UNS nut	40–45	5,53–6,22

FUEL SYSTEM

ITEM	DESCRIPTION	lbf.ft.	kgf.m.
Fuel pipe unions	5/16 in dia. pipe	6.3–7	0,87–0,97
Fuel pipe unions	3/8 in dia. pipe	8–10	1,11–1,38
Fuel pipe unions	1/2 in dia. pipe	20–25	2,77–3,46
Petrol gauge locking ring	Special	4–10	0,55–1,38
Banjo bolts for fuel pipes	—	22–26	3,04–3,59
Petrol tank mtg. bkt. lower to body	3/8 in UNF bolt	22–26	3,04–3,59
Petrol tank mtg. bkt. upper to body	3/8 in UNF nut	22–26	3,04–3,59
Sump tank to body	1/4 in UNF bolt	6–7	0,83–0,97
Sump tank drain plug	5/8 in UNF bolt	27–32	3,73–4,42
Petrol pump mtg. bkt. to battery support panel	1/4 in UNF bolt	6–7	0,83–0,97
Petrol pump clip	1/4 in UNF bolt	6–7	0,83–0,97
Carbon canister to body	5/16 in UNF nut	14–18	1,94–2,49
Clamp bracket	1/4 in UNF nut	6–7	0,83–0,97
Vapour separator to body	1/4 in UNF bolt	6–7	0,83–0,97

TRANSMISSION

ITEM	DESCRIPTION	lbf.ft.	kgf.m.
Propeller shaft flange bolts	3/8 in UNF nut	27–32	3,73–4,42

AUTOMATIC GEAR CONTROLS

ITEM	DESCRIPTION	lbf.ft.	kgf.m.
Mounting plate to body	1/4 in UNF bolt	6–7	0,83–0,97
Selector lever return spring locknut	No. 10 UNF nut	4–4.5	0,55–0,62
Reverse switch	16 mm special	14–18	1,94–2,49
Adjusting ball end to gearbox lever	1/4 in UNF nut	6–7	0,83–0,97
Cable to adjustable ball end	7/16 in UNF nut	11–13	1,52–1,80
Cable abutment bkt. to selector gate	1/4 in UNF nut	6–7	0,83–0,97
Bracket to trans. unit	3/8 in UNF nut	22–26	3,04–3,59
Cable to upper abutment bkt.	5/8 in UNF nut	14–18	1,94–2,49
Cable to location block lower	5/16 in UNF bolt	14–18	1,94–2,49

AUTOMATIC TRANSMISSION (Borg-Warner Model 12)

ITEM	DESCRIPTION	lbf.ft.	kgf.m.
Pump to gear case	bolt	17–22	2,35–3,04
Gear case to bell housing	bolt	30–35	4,15–4,84
Front servo to gear case	bolt	40–50	5,53–6,91
Rear servo to gear case	bolt	20–25	2,77–3,46
Centre support to gear case	1/4 in bolt	5–8	0,69–1,11
Valve body to gear case	5/16 in bolt	17–20	2,35–2,77
Extension housing to gear case	bolt	28–33	3,87–4,56
Oil pan to gear case	bolt	10–13	1,38–1,80
Pressure check point	plug	13–17	1,80–2,35
Filler tube	nut	20–25	2,77–3,46
Rear band adjusting screw	locknut	40–44	5,53–6,08
Gear case to bell housing	bolt	58–65	8,02–8,99
Manual lever attachment	nut	35–40	4,84–5,53
Front pump cover attachment	screw	2–3	0,28–0,42
Governor inspection cover attachment	screw	5–6	0,69–0,83
Governor valve body to counterweight	bolt	4–5	0,55–0,69
Governor valve body cover	screw	2–2.5	0,28–0,35
Valve body and strainer	screw	2–2.5	0,28–0,35

ITEM	DESCRIPTION	lbf. ft.	kgf. m.
AUTOMATIC TRANSMISSION—(contd.)			
Vacuum control unit	bolt	13–17	1,80–2,35
Rear seal cover to housing extension	3/8 in UNF bolt	30–35	4,15–4,84
Banjo bolt (conn. to M20 o'drive control)	1.1/16 in – 16 UNS	8–10	1,11–1,38
Dipstick tube to oil pan	nut	34–38	4,70–5,25
(GM 400 Automatic Transmission)			
Solenoid to case	1/4 in dia. × 20	12	1,66
Control valve unit to case	1/4 in dia. × 20	8	1,1
Line pressure plug	1/8 in dia. pipe	10	1,38
Pump body to cover	5/16 in dia. × 18	18	2,49
Pump to case	5/16 in dia. × 18	18	2,49
Rear servo cover	5/16 in dia. × 18	18	2,49
Governor cover to case	5/16 in dia. × 18	18	2,49
Parking pawl bracket	1/4 in dia. × 28	10	1,38
Vacuum modulation retainer to case	5/16 in dia. × 18	12	1,66
Speedometer drive shaft nut	3/8 in dia. × 16	23	3,18
Sump			
Rear extension			
Manual shaft to detent lever	3/8 in dia. × 24	18	2,49

ITEM	DESCRIPTION	lbf. ft.	kgf. m.
MANUAL GEARBOX			
Top cover	5/16 in UNF bolt & setscrew	12–14	1,66–1,94
Rear cover	5/16 in UNF nut	12–14	1,66–1,94
Clutch housing	7/16 in UNF bolt & setscrew	40–44	5,53–6,08
Slave cyl. to clutch housing	3/8 in UNF nut	22–25	3,04–3,46
Filler plug	3/8 in B.S.P.	25–27	3,46–3,73
Drain plug	3/8 in B.S.P.	25–27	3,46–3,73
Coupling flange	3/4 in B.S.F. nut	100–	13,83–
		120	16,59
Reverse lever	3/8 in UNF bolt	22–25	3,04–3,46
Gear lever securing	5/16 in UNF nut	12–14	1,66–1,94
Constant pinions	1.1/2 in 20 UNS nut	140–	19,36–
		160	22,12
Mainshaft – rear	1.3/8 in 20 UNS nut	140–	19,36–
		160	22,12
Mainshaft – front	1.1/2 in 20 UNS nut	140–	19,36–
		160	22,12
FRONT SUSPENSION			
Stub axle to vertical link	5/8 in UNF nut	80–90	11,06–12,44
Tie rod lever to vertical link	M12 bolt	50–55	6,91–7,60
Front disc to hub	7/16 in UNF bolt	30–36	4,15–4,98
Caliper to vertical link	M12 bolt	50–60	6,91–8,30
Ball socket cap to vertical link	5/16 in UNF bolt	15–20	2,07–2,77
Lower ball pin to lower wishbone	9/16 in UNF bolt	45–55	6,22–7,60
Upper ball pin to vertical link	1/2 in UNF nut	35–50	4,84–6,91
Fulcrum shaft upper wishbone	1/2 in UNF nut	45–55	6,22–7,60
Fulcrum shaft lower wishbone (slotted nut)	9/16 in UNF nut	32–50	4,42–6,91

ITEM	DESCRIPTION	lbf. ft.	kgf. m.
FRONT SUSPENSION—(contd.)			
Upper ball joint to wishbone	3/8 in UNF nut	26–32	3,59–4,42
Upper fulcrum to crossmember	7/16 in UNF nut	49–55	6,77–7,60
Clamp and shield to vertical link	1/4 in UNF nut	6–7	0,83–0,97
Spring pan bolts	3/8 in UNF bolt	27–32	3,73–4,42
Damper mounting bracket	3/8 in UNF nut	27–32	3,73–4,42
Damper – front upper	3/8 in UNF nut	27–32	3,73–4,42
Damper – front lower	7/16 in UNF nut	32–36	4,42–4,98
Rubbers to spring pan	5/16 in UNF nut	8–10	1,11–1,38
Rebound rubbers to upper wishbone	5/16 in UNF bolt	8–10	1,11–1,38
Anti-roll bar bracket to body	3/8 in UNF bolt	27–32	3,73–4,42
Anti-roll bar to link	3/8 in UNF nut	14–18	1,94–2,49
Link to lower wishbone	3/8 in UNF nut	14–18	1,94–2,49
Clamp bolt, front mounting	1/2 in UNF nut	25–30	3,46–4,15
Front mounting bolt	3/4 in UNF nut	95–115	13,13–15,90
Vee mounting to body	3/8 in UNF nut	22–26	3,04–3,60
Vee mounting to beam	3/8 in UNF nut	14–18	1,94–2,49
Wheel nuts (set spanner to 45 lbf. ft.)	1/2 in UNF nut	40–60	5,53–8,30
REAR SUSPENSION AND MOUNTINGS			
Bottom plate to cross beam and inner fulcrum mounting	5/16 in UNF bolt & nut	14–18	1,94–2,49
Inner fulcrum mounting to drive unit	7/16 in UNC bolt	60–65	8,30–8,99
Drive unit to crossbeam	1/2 in UNC bolt	70–77	9,68–10,64
Fulcrum pin (inner)	1/2 in UNF nut	45–50	6,22–6,91
Outer pivot pin	5/8 in UNF nut	97–107	13,41– 14,79
Drive unit to drive shaft	7/16 in UNF nut	49–55	6,78–7,60
Drive shaft to hub carrier	3/4 in UNF nut	100–	13,83–
		120	16,59
Radius rods to wishbones	1/2 in UNF bolt	60–70	8,30–9,68
Safety straps and radius rods to body	7/16 in UNF bolt	40–45	5,53–6,22
Safety strap to floor panel	3/8 in UNF bolt	27–32	3,73–4,42
Rear damper upper	7/16 in UNF nut	32–36	4,42–4,98
Rear damper lower	3/8 in UNF nut	32–36	4,42–4,98
Vee mounting to body	3/8 in UNF nut	27–32	3,73–4,42
Vee mounting to beam	5/16 in UNF nut	14–18	1,94–2,49
Bump stop rubber to body	M12 bolt	8–10	1,11–1,38
Calipers to drive unit flange (opt. axle)	7/16 in UNF bolt	50–60	6,91–8,30
Calipers to drive unit flanges (std. axle)	7/16 in UNF nut	49–55	6,78–7,60
Anti-roll bar bracket to body	5/16 in UNF nut	14–18	1,94–2,49
FINAL DRIVE UNITS			
STANDARD 4 HU UNIT			
Bearing housings to unit	7/16 in UNC bolt	60–69	8,30–9,54
Differential bearing caps	1/2 in UNC bolt	63–70	8,71–9,68
Differential through-bolts	3/8 in UNC bolt	43–50	5,94–6,91
Drive gear to differential housing	7/16 in UNF bolt	70–88	9,68–12,17
Pinion shaft nut (with new spacer)	3/4 in UNF nut	120–	16,59–
		130	17,97
Rear cover attachment (Loctited)	5/16 in UNC bolt	18–20	2,49–2,77

ITEM	DESCRIPTION	lbf. ft.	kgf. m.
STEERING COLUMN, RACK AND MOUNTINGS			
Pinion housing cover plate	5/16 in UNF nut	14–18	1,94–2,49
Rack to ball end of track rod	13/16 in UNF nut	45–55	6,22–7,60
	1.1/8 in UNF locknut		
Track adjustment	5/8 in UNF locknut	140–	19,36–
		155	21,43
Track rod end ball joint	1/2 in UNF locknut	45–50	6,22–6,91
Bolt in tie rod end assy.	1/4 in UNF bolt	6–7	0,83–0,97
Mounting bolts	5/16 in UNF nut	14–18	1,94–2,49
Universal joints	5/16 in UNF bolt	14–18	1,94–2,49
Column adaptor bolt	1/4 in UNF bolt	10–12	1,38–1,66
Steering wheel to shaft	5/8 in UNF nut	25–32	3,46–4,42
P.A.S. Hose assy. valve to pump	5/8 in UNF pipe nut	18–20	2,49–2,77
	1/2 in UNF pipe nut	14–16	1,94–2,21
Locknut — collet adaptor retaining screw	1/4 in UNF nut	6–7	0,83–0,97
P.A.S. oil cooler to L.H. front engine mtg.	1/4 in UNF bolt	6–7	0,83–0,97
Bracket, vertical strut to body	5/16 in UNF bolt	14–18	1,94–2,49
Vertical strut to bracket	5/16 in UNF nut	14–18	1,94–2,49
Upper column to vertical strut	5/16 in UNF nut	14–18	1,94–2,49
Longitudinal strut to vertical strut	5/16 in UNF nut	14–18	1,94–2,49
Lower column & longitudinal strut to body	5/16 in UNF bolt	14–18	1,94–2,49
Transverse strut to vertical strut	5/16 in to UNF bolt	14–18	1,94–2,49
Rack hydraulic connections	nuts	8–9	1,11–1,24
Rack locking bolt	3/8 in UNF bolt	27–32	3,73–4,42
Adaptor (rack)	7/16 in UNF	49–55	6,78–7,60
BRAKE AND CLUTCH SYSTEMS			
Brake pedal box to body	5/16 in UNF bolt	11–13	1,52–1,80
Brake pedal box to body	5/16 in UNF bolt	11–13	1,52–1,80
Clutch master cyl. to pedal box	5/16 in UNF nut	11–13	1,52–1,80
Brake reservoir to bracket	1/4 in UNF nut	2.0–2.5	2,7–3,4
Hydraulic connections	3/16 in & 1/4 in pipe nuts		
Rear 3-way connection	1/4 in UNF nut	6.3–7	0,87–0,97
Handbrake to body	5/16 in UNF bolt	6–7	0,83–0,97
Front and rear hoses to bracket	M10 nut	14–18	1,94–2,49
Master cylinder to booster	M10 nut	10–12	1,38–1,66
		15.5–	2,14–
		19.5	2,70

ITEM	DESCRIPTION	lbf. ft.	kgf. m.
BRAKE AND CLUTCH SYSTEMS—(contd.)			
Booster to pedal box	M8 nut	8–10	1,11–1,38
P.D.W.A. to body	1/4 in UNF nut	6–7	0,83–0,97
Hydraulic connections	Metric	6.3–7	0,87–0,97
Pedal boot retainer to body	No. 10 UNF bolt	4–4.5	0,55–0,62
Brake light switch to bracket	1/4 in UNF bolt	3.5–4.5	0,47–0,61
Rear flexible hose bkt. to crossbeam	5/16 in UNF nut	14–18	1,94–2,49
Vacuum tank to nut	1/4 in UNF nut	6–7	0,83–0,97
Vacuum tank bkt. to body	5/16 in UNF nut	14–18	1,94–2,49
Vacuum pipe clips to valance tie tubes	No. 10 UNF nut	4–4.5	0,55–0,62
Top bolt — handbrake	5/16 in UNF bolt	14–18	1,94–2,49
Handbrake to body	5/16 in UNF bolt	14–18	1,94–2,49
ACCELERATOR CONTROL			
Accelerator cable to body	5/16 in UNF nut	8–10	1,11–1,38
Accelerator mtg. bkt. to toe box	1/4 in UNF bolt	6–7	0,83–0,97
AIR CONDITIONING			
Mounting bracket to drier bottle	No. 10 UNF nut	2–2.5	0,28–0,35
Pipe connections 0,38 in dia. pipe	nut	11–13	1,52–1,80
Pipe connections 0.50 in dia. pipe	nut	21–27	2,90–3,73
Strut, air con. unit to tunnel	5/16 in UNF nut	14–18	1,94–2,49
Strut, air con. unit	1/4 in UNF nut	6–7	0,83–0,97
Air con./heating assy. and heat shields to body	5/16 in UNF nut	14–18	1,94–2,49
Stays to heater/air convolute	1/4 in UNF bolt	6–7	0,83–0,97
Stays to body	1/4 in UNF nut	6–7	0,83–0,97
Heater valve to mtg. bkt.	No. 10 UNF nut	4–4.5	0,55–0,62
Heater valve bkt. to dash upper panel	No. 10 UNF nut	4–4.5	0,55–0,62
Upper mtg. bkt. to condenser	1/4 in UNF nut	6–7	0,83–0,97
Spacing bkt. on oil cooler to radiator bracket	5/16 in UNF bolt	14–18	1,94–2,49
Condenser bkt. to upper radiator	1/4 in UNF nut	6–7	0,83–0,97
X-member	1/4 in UNF bolt	6–7	0,83–0,97
Fuel cooler to compressor	No. 10 UNF nut	4–4.5	0,55–0,62
Hose (evap. to comp.) to L.H. stay	1/4 in UNF bolt	6–7	0,83–0,97
Hose (comp. to cond.) to compressor	1/4 in UNF nut	6–7	0,83–0,97
Vacuum heater control to cover plate			

GENERAL FITTING INSTRUCTIONS

Precautions against damage

1 Always fit covers to protect wings before commencing work in engine compartment.
2 Cover seats and carpets, wear clean overalls and wash hands or wear gloves before working inside car.
3 Avoid spilling hydraulic fluid or battery acid on paint work. Wash off with water immediately if this occurs. Use polythene sheets in boot to protect carpets.
4 Always use a recommended Service Tool, or a satisfactory equivalent, where specified.
5 Protect temporarily exposed screw threads by replacing nuts or fitting plastic caps.

Safety Precautions

1 Whenever possible use a ramp or pit when working beneath car, in preference to jacking. Chock wheels as well as applying handbrake.
2 Never rely on a jack alone to support car. Use axle stands or blocks carefully placed at jacking points to provide rigid location.
3 Ensure that a suitable form of fire extinguisher is conveniently located.
4 Check that any lifting equipment used has adequate capacity and is fully serviceable.
5 Inspect power leads of any mains electrical equipment for damage and check that it is properly earthed.
6 Disconnect earth (grounded) terminal of car battery.
7 Do not disconnect any pipes in air conditioning refrigeration system, if fitted, unless trained and instructed to do so. A refrigerant is used which can cause blindness if allowed to contact eyes.
8 Ensure that adequate ventilation is provided when volatile de-greasing agents are being used.
CAUTION: Fume extraction equipment must be in operation when trichlorethylene, carbon tetrachloride, methylene chloride, chloroform, or perchlorethylene are used for cleaning purposes.
9 Do not apply heat in an attempt to free stiff nuts or fittings; as well as causing damage to protective coatings, there is a risk of damage to electronic equipment and brake lines from stray heat.
10 Do not leave tools, equipment, spilt oil etc., around or on work area.
11 Wear protective overalls and use barrier creams when necessary.

Preparation

1 Before removing a component, clean it and its surrounding area as thoroughly as possible.
2 Blank off any openings exposed by component removal, using greaseproof paper and masking tape.
3 Immediately seal fuel, oil or hydraulic lines when separated, using plastic caps or plugs, to prevent loss of fluid and entry of dirt.
4 Close open ends of oilways, exposed by component removal, with tapered hardwood plugs or readily visible plastic plugs.
5 Immediately a component is removed, place it in a suitable container; use a separate container for each component and its associated parts.
6 Before dismantling a component, clean it thoroughly with a recommended cleaning agent; check that agent is suitable for all materials of component.
7 Clean bench and provide marking materials, labels, containers and locking wire before dismantling a component.

Dismantling

1 Observe scrupulous cleanliness when dismantling components, particularly when brake, fuel or hydraulic system parts are being worked on. A particle of dirt or a cloth fragment could cause a dangerous malfunction if trapped in these systems.
2 Blow out all tapped holes, crevices, oilways and fluid passages with an air line. Ensure that any O-rings used for sealing are correctly replaced or renewed if disturbed.
3 Mark mating parts to ensure that they are replaced as dismantled. Whenever possible use marking ink, which avoids possibilities of distortion or initiation of cracks, liable if centre punch or scriber are used.
4 Wire together mating parts where necessary to prevent accidental interchange (e.g. roller bearing components).
5 Wire labels on to all parts which are to be renewed, and to parts requiring further inspection before being passed for reassembly; place these parts in separate containers from those containing parts for rebuild.
6 Do not discard a part due for renewal until after comparing it with a new part, to ensure that its correct replacement has been obtained.

Inspection – General

1 Never inspect a component for wear or dimensional check unless it is absolutely clean; a slight smear of grease can conceal an incipient failure.
2 When a component is to be checked dimensionally against figures quoted for it, use correct equipment (surface plates, micrometers, dial gauges, etc.) in serviceable condition. Makeshift checking equipment can be dangerous.
3 Reject a component if its dimensions are outside limits quoted, or if damage is apparent. A part may, however, be refitted if its critical dimension is exactly limit size, and is otherwise satisfactory.
4 Use 'Plastigauge' 12 TypePG-1 for checking bearing surface clearances; directions for its use, and a scale giving bearing clearances in 0.0001 in (0,0025 mm) steps are provided with it.

Ball and Roller Bearings

NEVER REPLACE A BALL OR ROLLER BEARING WITHOUT FIRST ENSURING THAT IT IS IN AS-NEW CONDITION.

1 Remove all traces of lubricant from bearing under inspection by washing in petrol or a suitable de-greaser; maintain absolute cleanliness throughout operations.
2 Inspect visually for markings of any form on rolling elements, raceways, outer surface of outer rings or inner surface of inner rings. Reject any bearings found to be marked, since any marking in these areas indicates onset of wear.
3 Holding inner race between finger and thumb of one hand, spin outer race and check that it revolves absolutely smoothly. Repeat, holding outer race and spinning inner race.
4 Rotate outer ring gently with a reciprocating motion, while holding inner ring; feel for any check or obstruction to rotation, and reject bearing if action is not perfectly smooth.
5 Lubricate bearing generously with lubricant appropriate to installation.
6 Inspect shaft and bearing housing for discolouration or other marking suggesting that movement has taken place between bearing and seatings. (This is particularly to be expected if related markings were found in operation 2.) If markings are found, use 'Loctite' in installation of replacement bearing.
7 Ensure that shaft and housing are clean and free from burrs before fitting bearing.

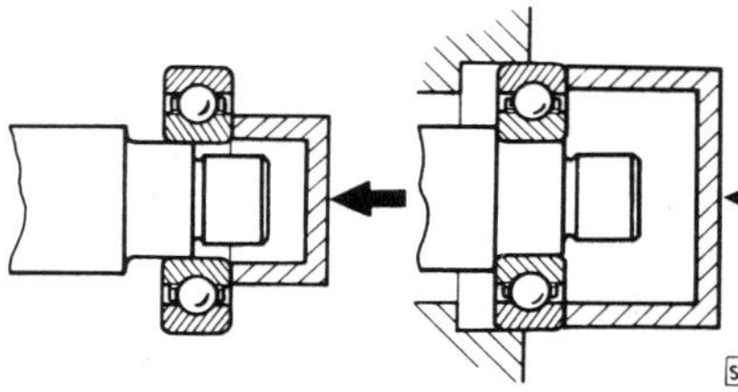

8 If one bearing of a pair shows an imperfection it is generally advisable to renew both bearings; an exception could be made only if the faulty bearing had covered a low mileage, and it could be established that damage was confined to it.

9 When fitting bearing to shaft, apply force only to inner ring of bearing, and only to outer ring when fitting into housing.

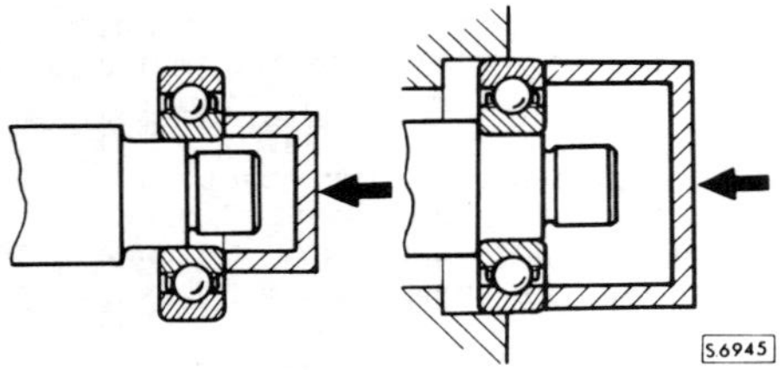

10 In the case of grease-lubricated bearings (e.g. hub bearings) fill space between bearing and outer seal with recommended grade of grease before fitting seal.

11 Always mark components of separable bearings (e.g. taper roller bearings) in dismantling, to ensure correct reassembly. Never fit new rollers in a used cup.

Oil Seals

1 Always fit new oil seals when rebuilding an assembly. It is not physically possible to replace a seal exactly as it had bedded down.

2 Carefully examine seal before fitting to ensure that it is clean and undamaged.

3 Smear sealing lips with clean grease; pack dust excluder seals with grease, and heavily grease duplex seals in cavity between sealing lips.

4 Ensure that seal spring, if provided, is correctly fitted.

5 Place lip of seal towards fluid to be sealed and slide into position on shaft, using fitting sleeve when possible to protect sealing lip from damage by sharp corners, threads or splines. If fitting sleeve is not available, use plastic tube or adhesive tape to prevent damage to sealing lip.

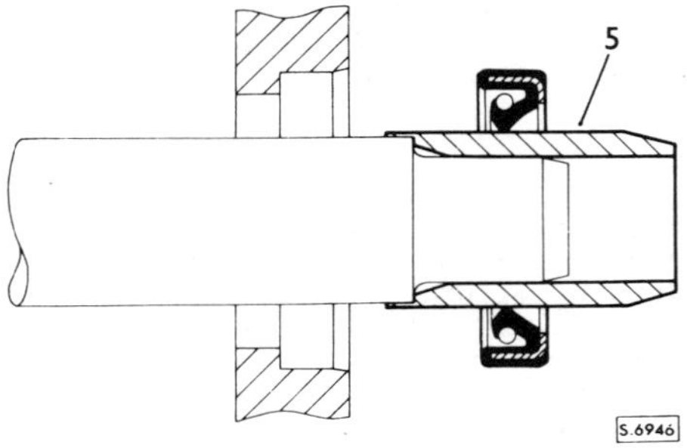

6 Grease outside diameter of seal, place square to housing recess and press into position, using great care and if possible a 'bell piece' to ensure that seal is not tilted. (In some cases it may be preferable to fit seal to housing before fitting to shaft.) Never let weight of unsupported shaft rest in a seal.

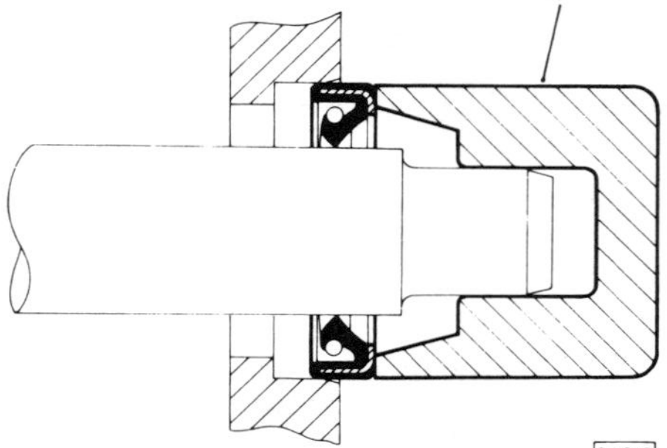

7 If correct service tool is not available, use a suitable drift approximately 0.015 in (0,4 mm) smaller than outside diameter of seal. Use a hammer VERY GENTLY on drift if a press is not suitable.

8 Press or drift seal in to depth of housing if housing is shouldered, or flush with face of housing where no shoulder is provided.

NOTE: Most cases of failure or leakage of oil seals are due to careless fitting, and resulting damage to both seals and sealing surfaces. Care in fitting is essential if good results are to be obtained.

Joints and Joint Faces

1 Always use correct gaskets where they are specified.

2 Use jointing compound only when recommended. Otherwise fit joints dry.

3 When jointing compound is used, apply in a thin uniform film to metal surfaces; take great care to prevent it from entering oilways, pipes or blind tapped holes.

4 Remove all traces of old jointing materials prior to reassembly. Do not use a tool which could damage joint faces.

5 Inspect joint faces for scratches or burrs and remove with a fine file or oil stone; do not allow swarf or dirt to enter tapped holes or enclosed parts.

6 Blow out any pipes, channels or crevices with compressed air, renewing any O-rings or seals displaced by air blast.

Flexible Hydraulic Pipes, Hoses

1 Before removing any brake or power steering hose, clean end fittings and area surrounding them as thoroughly as possible.

2 Obtain appropriate blanking caps before detaching hose end fittings, so that ports can be immediately covered to exclude dirt.

3 Clean hose externally and blow through with airline. Examine carefully for cracks, separation of plies, security of end fittings and external damage. Reject any hose found faulty.

4 When refitting hose, ensure that no unnecessary bends are introduced, and that hose is not twisted before or during tightening of union nuts.

5 Containers for hydraulic fluid must be kept absolutely clean.

6 Do not store hydraulic fluid in an unsealed container. It will absorb water, and fluid in this condition would be dangerous to use due to a lowering of its boiling point.

7 Do not allow hydraulic fluid to be contaminated with mineral oil, or use a container which has previously contained mineral oil.

8 Do not re-use fluid bled from system.

9 Always use clean brake fluid, or a recommended alternative, to clean hydraulic components.

10 Fit a blanking cap to a hydraulic union and a plug to its socket after removal to prevent ingress of dirt.

11 Absolute cleanliness must be observed with hydraulic components at all times.

12 After any work on hydraulic systems, inspect carefully for leaks underneath the car while a second operator applies maximum pressure to the brakes (engine running) and operates the steering.

Metric Bolt Identification

1 An ISO metric bolt or screw, made of steel and larger than 6 mm in diameter can be identified by either of the symbols ISO M or M embossed or indented on-top of the head.

2 In addition to marks to identify the manufacture, the head is also marked with symbols to indicate the strength grade e.g. 8.8, 10.9, 12.9, or 14.9, where the first figure gives the minimum tensile strength of the bolt material in tens of kg/sq mm.

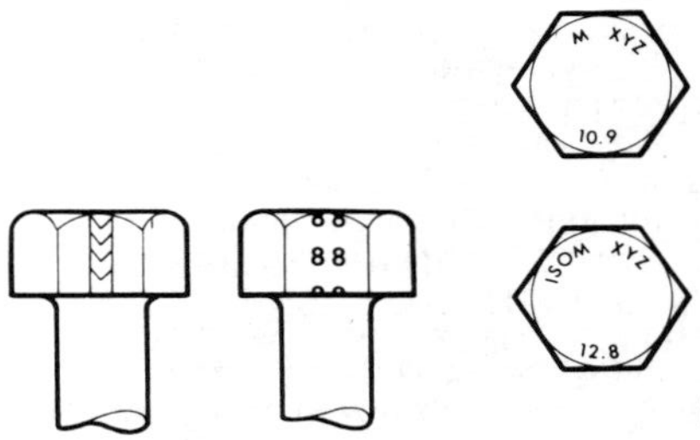

3. Zinc plated ISO metric bolts and nuts are chromate passivated, a greenish-khaki to gold-bronze colour.

Metric Nut Identification

1. A nut with an ISO metric thread is marked on one face or on one of the flats of the hexagon with the strength grade symbol 8, 12 or 14. Some nuts with a strength 4, 5 or 6 are also marked and some have the metric symbol M on the flat opposite the strength grade marking.

2. A clock face system is used as an alternative method of indicating the strength grade. The external chamfers or a face of the nut is marked in a position relative to the appropriate hour mark on a clock face to indicate the strength grade.

3. A dot is used to locate the 12 o'clock position and a dash to indicate the strength grade. If the grade is above 12, two dots identify the 12 o'clock position.

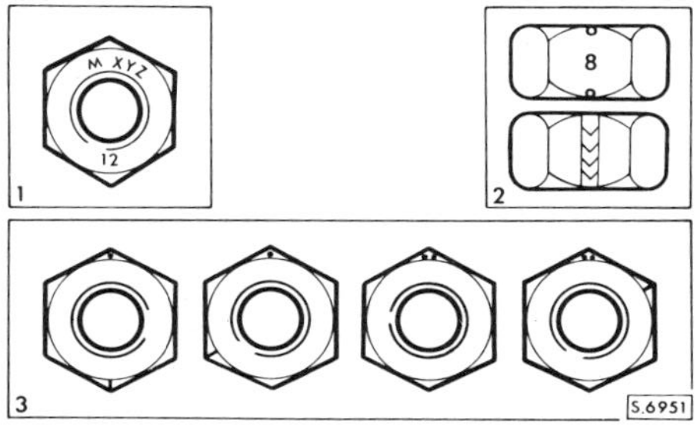

Hydraulic Fittings – Metrication

WARNING: Metric and Unified threaded hydraulic parts. Although pipe connections to brake system units incorporate threads of metric form, those for power assisted steering are of UNF type. It is vitally important that these two thread forms are not confused, and careful study should be made of the following notes.

Metric threads and metric sizes are being introduced into motor vehicle manufacture and some duplication of thread sizes must be expected. Although standardisation must in the long run be good, it would be wrong not to give warning of the dangers that exist while UNF and metric threaded hydraulic parts continue together in service.

Fitting UNF pipe nuts into metric ports and vice-versa should not happen, but experience of the change from BSF to UNF indicated that there is no certainty in relying upon the difference in thread size when safety is involved.

To provide permanent identification of metric parts is not easy but recognition has been assisted by the following means.

1. All metric pipe nuts, hose ends, unions and bleed screws are coloured black.
2. The hexagon area of pipe nuts is indented with the letter 'M'.
3. Metric and UNF pipe nuts are slightly different in shape.

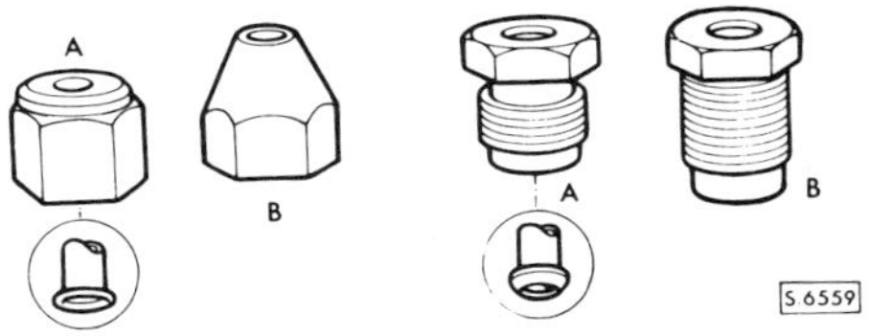

The metric female nut is **always** used with a trumpet flared pipe and the metric male nut is **always** used with a convex flared pipe.

4. All metric ports in cylinders and calipers have no counterbores, but unfortunately a few cylinders with UNF threads also have no counterbore. The situation is, all ports with counterbores are UNF, but ports not counterbored are most likely to be metric.

5. The colour of the protective plugs in hydraulic ports indicates the size and the type of the threads, but the function of the plugs is protective and not designed as positive identification. In production it is difficult to use the wrong plug but human error must be taken into account.
The Plug colours and thread sizes are:–

	UNF
RED	3/8" x 24 UNF
GREEN	7/16" x 20 UNF
YELLOW	1/2" x 20 UNF
PINK	5/8" x 18 UNF

BLACK	METRIC
GREY	10 x 1 mm
BROWN	12 x 1 mm
	14 x 1.5 mm

6. Hose ends differ slightly between metric and UNF.
Gaskets are not used with metric hoses. The UNF hose is sealed on the cylinder or caliper face by a copper gasket but the metric hose seals against the bottom of the port and there is a gap between faces of the hose end and cylinder.
Pipe sizes for UNF are 3/16 in, 1/4 in, and 5/16 in outside diameter.
Metric pipe sizes are 4.75 mm, 6 mm and 8 mm.
4.75 mm pipe is exactly the same as 3/16 in pipe. 6 mm pipe is .014 in smaller than 1/4 in pipe. 8 mm pipe is .002 in larger than 5/16 in pipe.
Convex pipe flares are shaped differently for metric sizes and when making pipes for metric equipment, metric pipe flaring tools must be used.

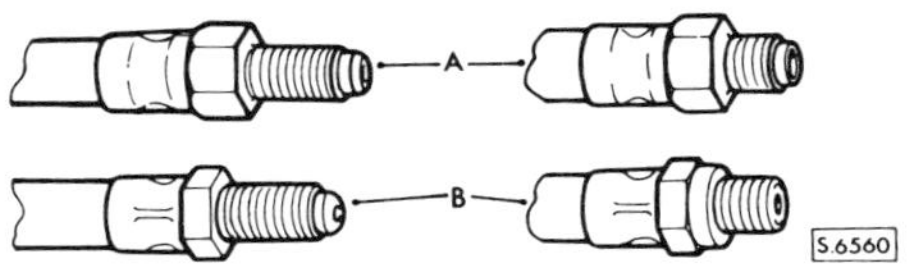

The greatest danger lies with the confusion of 10 mm and 3/8 in UNF Pipe nuts used for 3/16 in (or 4.75 mm) pipe. The 3/8 in UNF pipe nut or hose can be screwed into a 10 mm port but is very slack and easily stripped. The thread engagement is very weak and cannot provide an adequate seal. The opposite condition, a 10 mm nut in a 3/8 in port, is difficult and unlikely to cause trouble. The 10 mm nut will screw in 1½ or two turns and seize. It has a crossed thread 'feel' and it is impossible to force the nut far enough to seal the pipe. With female pipe nuts the position is of course reversed.

The other combinations are so different that there is no danger of confusion.

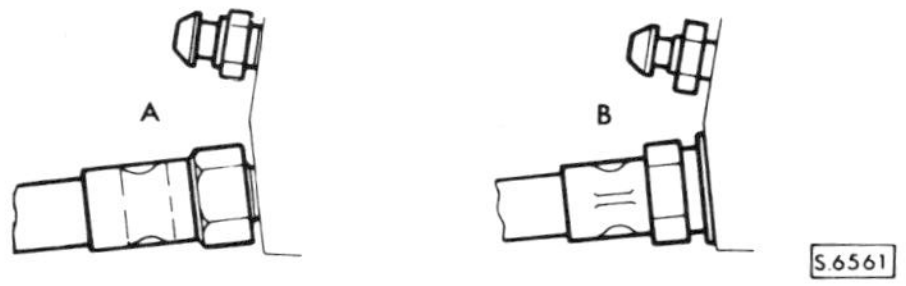

Keys and Keyways

1. Remove burrs from edges of keyways with a fine file and clean thoroughly before attempting to refit key.
2. Clean and inspect key closely; keys are suitable for refitting only if indistinguishable from new, as any indentation may indicate the onset of wear.

Split Pins

1. Fit new split pins throughout when replacing any unit.
2. Always fit split pins where split pins were originally used. Do not substitute spring washers: there is always a good reason for the use of a split pin.
3. All split pins should be fitted as shown unless otherwise stated.

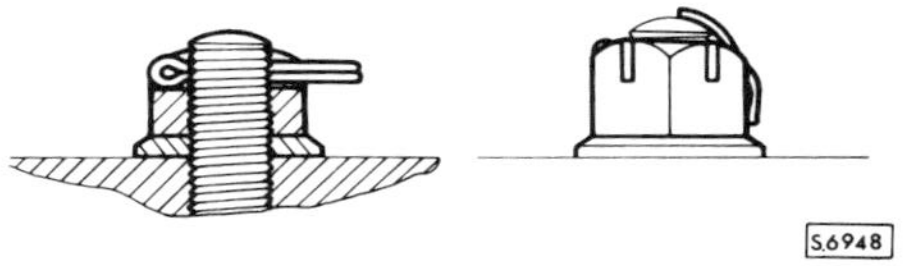

Tab Washers

1. Fit new tab washers in all places where they are used. Never replace a new tab washer.
2. Ensure that the new tab washer is of the same design as that replaced.

Nuts

1. When tightening up a slotted or castellated nut **never slacken it back** to insert split pin or locking wire except in those recommended cases where this forms part of an adjustment. If difficulty is experienced, alternative washers or nuts should be selected, or washer thickness reduced.

2 Where self-locking nuts have been removed it is advisable to replace them with new ones of the same type.
   NOTE: Where bearing pre-load is involved nuts should be tightened in accordance with special instructions.

Locking Wire

1 Fit new locking wire of the correct type for all assemblies incorporating it.

2 Arrange wire so that its tension tends to tighten the bolt heads, or nuts, to which it is fitted.

2 **Nuts**
   A continuous line of circles is indented on one of the flats of the hexagon, parallel to the axis of the nut.

3 **Studs, Brake Rods, etc.**
   The component is reduced to the core diameter for a short length at its extremity.

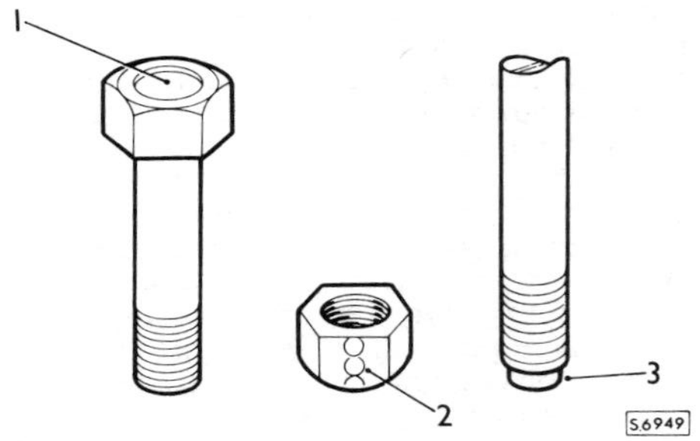

Screw Threads

1 Both UNF and Metric threads to ISO standards are used. See below for thread identification.

2 Damaged threads must always be discarded. Cleaning up threads with a die or tap impairs the strength and closeness of fit of the threads and is not recommended.

3 Always ensure that replacement bolts are at least equal in strength to those replaced.

4 Do not allow oil, grease or jointing compound to enter blind threaded holes. The hydraulic action on screwing in the bolt or stud could split the housing.

5 Always tighten a nut or bolt to the recommended torque figure. Damaged or corroded threads can affect the torque reading.

6 To check or re-tighten a bolt or screw to a specified torque figure, first slacken a quarter of a turn, then re-tighten to the correct figure.

7 Always oil thread lightly before tightening to ensure a free running thread, except in the case of self-locking nuts.

Unified Thread Identification

Bolts

1 A circular recess is stamped in the upper surface of the bolt head.

07—4

LIFTING AND TOWING

STANDS

When carrying out any work on the car which requires a wheel to be raised (apart from a simple wheel change) it is essential that the jacking spigot is replaced by a stand, located by the jacking spigot, to provide a secure support for the car.

JACKING POINTS

The jack provided in the car's tool kit engages with spigots situated below the body side members, in front of the rear wheels and behind the front wheels. Always chock wheels as well as applying handbrake when using the jack.

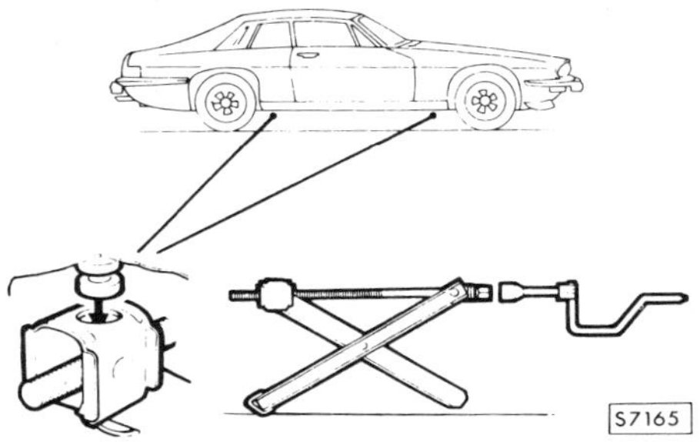

WORKSHOP JACK

Front – one wheel

Jack under the lower spring support pan, using a suitable wooden block on the jack head. Place a stand in position at the adjacent spigot when the wheel is raised.

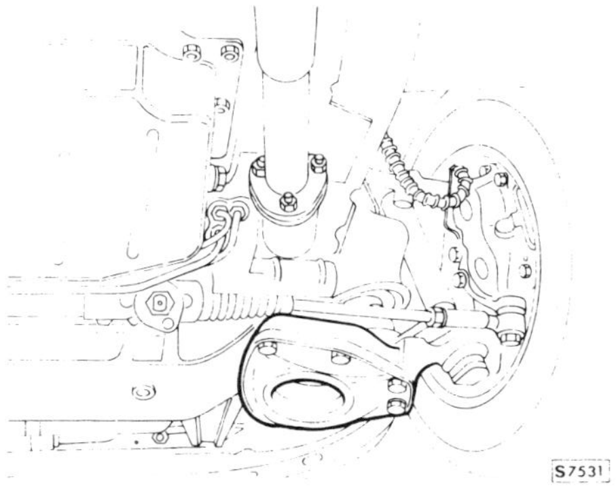

Rear – one wheel

Locate the jack with a wooden block on its head, under the outer fork of the wishbone at the wheel to be raised. Take care to avoid damage to the aluminium alloy hub carrier or to the grease nipple fitted to it. Place a stand under the adjacent jacking spigot when the wheel is raised.

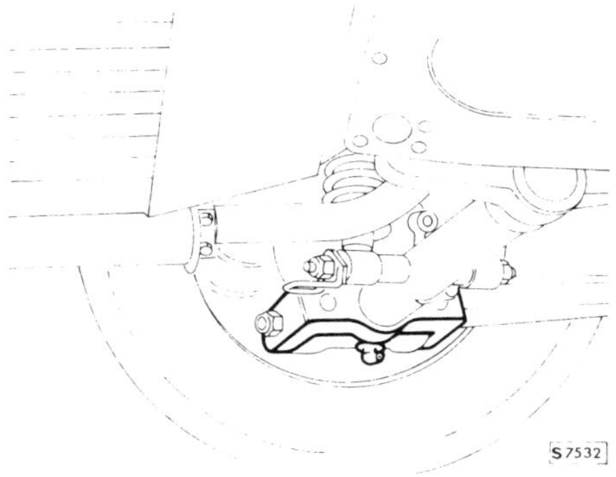

Front – both wheels

Place the jack, with a wooden block on its head, centrally under the front suspension. Place stands under both front jacking spigots when the car is raised.

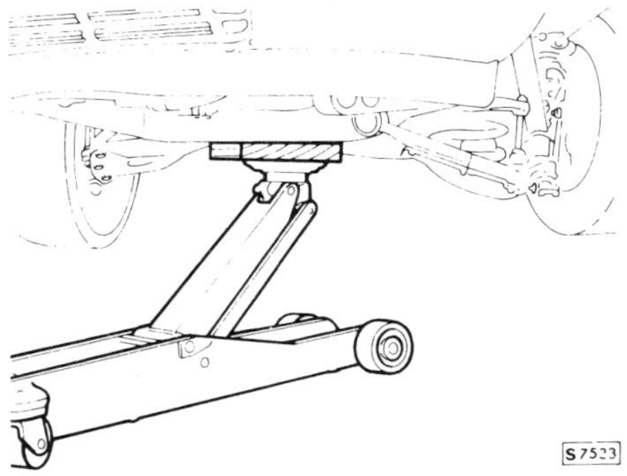

Rear – both wheels

Place a suitable shaped wooden block between the jack head and the plate in the centre of the rear cross-member, ensuring that the jacking load is not applied to the flanges of the plate. Place stands under both rear jacking spigots when the car is raised.

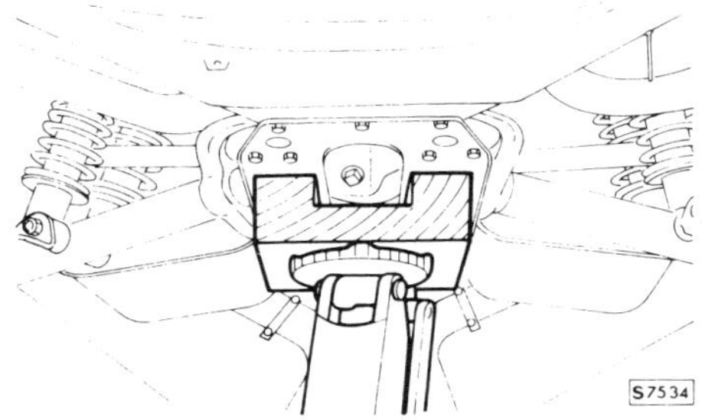

TOWING

Two towing eyes are provided, below the front bumper, to enable the car to be towed from the front.
The tie-down lugs at the rear damper lower attachments are not suitable for towing.
When towing an automatic transmission car, observe the following procedure:–

1. Add an additional 3.6 pints of fluid of correct specification to gearbox.
2. Position selector lever at 'N'.
3. Car may be towed for a distance no greater than 30 miles (48 km.) at a speed not exceeding 30 m.p.h. (48 k.p.h.).
4. Ensure fluid added to gearbox in Operation 1 is drained and oil brought to correct level prior to car being driven.
5. The ignition key MUST be in the 'ACC' position.

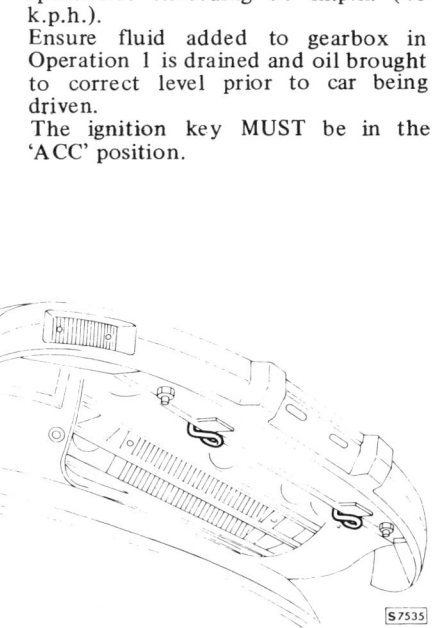

LIFTING

Locate lifting pads at the four jacking spigots.

Towing – Automatic Gearbox Defective

The car should be towed with rear wheels clear of the ground or propellor shaft disconnected at the final drive input flange.
If propellor shaft is disconnected, firmly secure shaft to one side of final drive input flange.

CAUTION

It must be remembered that steering is no longer power-assisted when engine is not running, and that brake servo will become ineffective after a few applications of the brakes. Be prepared therefore, for relatively heavy steering and the need for increased pressure on the brake pedal. This applies to manual transmission cars as well as to those with automatic transmission.

B. With automatic transmission defective, either tow car with rear wheels clear of ground, or disconnect propeller shaft at final drive input flange and firmly secure rear end of shaft to one side of flange. Restrictions on towing distance do not apply when output shaft of gearbox is not being turned, but it is still essential that ignition key is turned to 'ACC' and the cautionary note above still applies.

Manual transmission cars require no special precautions in towing, but it is essential that the ignition key is turned to 'ACC' and the cautionary note above must be remembered.

TRANSPORTING

Automatic transmission cars only

CAUTION

WHEN VEHICLE IS BEING TRANSPORTED SELECTOR LEVER MUST BE IN 'N' OR 'D'. NEVER IN 'P', TO OBVIATE POSSIBILITY OF DAMAGE TO PAWL MECHANISM. HAND BRAKE SHOULD BE APPLIED.

SERVICE LUBRICANTS, FUEL AND FLUIDS – CAPACITIES

RECOMMENDED LUBRICANTS, FLUIDS, FUEL CAPACITIES AND DIMENSIONS

ENGINE Distributor and Oil Can	MANUAL GEARBOX	POWR-LOK DIFFERENTIAL	AUTO TRANSMISSION POWER ASSISTED STEERING	GREASE POINTS
	E.P. 90 (MIL-L-2105B)	NOTE: For re-fill, use only approved brands of fluid specially formulated for Powr-Lok. For toppings up ONLY EP 90 can be used	GM 400 Dexron 2D Borg Warner Specification type 'F' (M2C 35F)	Multipurpose Lithium Grease (N.L.G.I. Consistency No. 2.)

RECOMMENDED SAE VISCOSITY RANGE/AMBIENT TEMPERATURE SCALE.

Use a well known brand of oil to BLSO.OL02, MIL-L-2104B or APLSE specification with a viscosity band spanning the temperature range of locality in which the car is used.

COMPONENT	ADDITIVE
Cooling system	1. Radiator leak inhibitor (Jaguar Part No. 12953) – 14-oz container per car. 2. BP Type H21 or Union Carbide UT 184, Prestone 2 or Texaco Anti-freeze – 55% concentration (5 parts Anti-freeze to 4 parts water).
Windscreen Washer	Windscreen washer antifreeze fluid (proprietary brands).

FUEL REQUIREMENTS

Only cars with 'S' compression ratio engines require 97 octane fuel. Cars with 'L' compression ratio engines should use 94 octane fuel. In U.S.A., catalyst equipped cars require unleaded fuel with minimum octane rating of 91 R.O.N.

In the United Kingdom use '4 STAR' FUEL.

If, of necessity, the car has to be operated on lower octane fuel do not use full throttle otherwise detonation may occur with resultant piston trouble.

CAPACITIES

	Pint	U.S. Pint	Litres
Engine (refill including filter)	20	24	11,4
Model 12 Automatic transmission	16	19.2	9,1
GM 400 Automatic transmission	16	19.2	9,1
Manual gearbox	3	3.6	1,7
Final drive unit	2¾	3.3	1,6
Cooling system	37	44.4	21
	gal.	U.S. gal.	Litres
Fuel tank	20	24	90
Boot volume	15 cu. ft.		425 litres

DIMENSIONS AND WEIGHTS

Wheelbase	102.0 in (2,591 mm)
Track – Front	58.0 in (1,473 mm)
– Rear	58.6 in (1,488 mm)
Overall length	190.2 in (4,831 m)
Overall width	70.6 in (1,793 m)
Overall height (kerb wt.)	49.6 in (1,260 m)
Kerb weight (approx.)	
(European specn.)	3795 lb (1721 kg)
(N. American specn.)	4085 lb (1853 kg)
Gross vehicle weight	
(European specn.)	4545 lb (2062 kg)
(N. American specn.)	4656 lb (2112 kg)
Turning circle	36 ft 3 in (11,05 m)
Ground clearance (mid-laden)	5 in (127 mm)

NOTE: Weights reduce by 106 lb (48 kg) if auto transmission is replaced by manual.

09–1

IMPORTANT NOTES

1 CHANGE COOLANT ANNUALLY. In places where UNIPART Universal, Prestone 2 or Texaco Anti-freeze is not available, drain the cooling system, flush and refill with a solution of an anti-freeze which complies with Specification BS.3150.

2 ALWAYS top up the cooling system with recommended strength of anti-freeze, NEVER WITH WATER only.

3 Where difficulty is experienced in obtaining recommended antifreezes or satisfactory 'soft water', a complete coolant fill known as CARBUROL RADMASTER is recommended. It is stressed however that the use of Carburol Radmaster should be confined to those countries NOT subject to temperatures below −10°C (14°F). Before Carburol Radmaster is used, the original antifreeze should be drained and the system flushed. Once in use, the cooling system should be topped-up with Carburol Radmaster ONLY.

RECOMMENDED HYDRAULIC FLUID

Braking System

Castrol-Girling Universal Brake and Clutch fluid. This fluid exceeds S.A.E. J1703/D specification. NOTE: Check all pipes in the brake system at the start and finish of each winter period for possible corrosion due to salt and grit used on the roads.

CAUTION: Additional maintenance operations which must be carried out on cars built to Federal Specifications are detailed on pages 10–6 and 10–7.

LUBRICATION CHART

Daily

1. Engine – check oil level and top up with recommended oil if necessary.
2. Cooling system – check level and top up if necessary with correct coolant.

Weekly

3. Battery – check electrolyte level and top up with distilled water if necessary.
4. Windscreen washer – check fluid level in reservoir and top up with suitable fluid if necessary.
5. Windscreen wiper blades – inspect and clean.
6. Tyres – check pressures and inspect for damage; adjust to correct pressures (including spare) if necessary.

Every 3,000 miles (5 000 km)

7. Brake fluid reservoir and clutch fluid reservoir, if fitted – check fluid level and top up with Castrol-Girling brake fluid.
8. Power assisted steering – check fluid level in reservoir, and top up with recommend fluid if necessary.

Every 6,000 miles (10 000 km)

9. Carry out operations 7 and 8.
10. Engine – drain and refill sump.
11. Engine – renew oil filter element or change disposable canister.
12. Distributor – lubricate with two or three drops of clean engine oil on rotor carrier shaft oil pad.
13. Gearbox – check oil level and top up with recommended lubricant if necessary.
14. Overdrive (if fitted) – check oil level and top up with recommended lubricant if necessary.
15. Final drive unit – check oil level and top up with recommended lubricant if necessary.

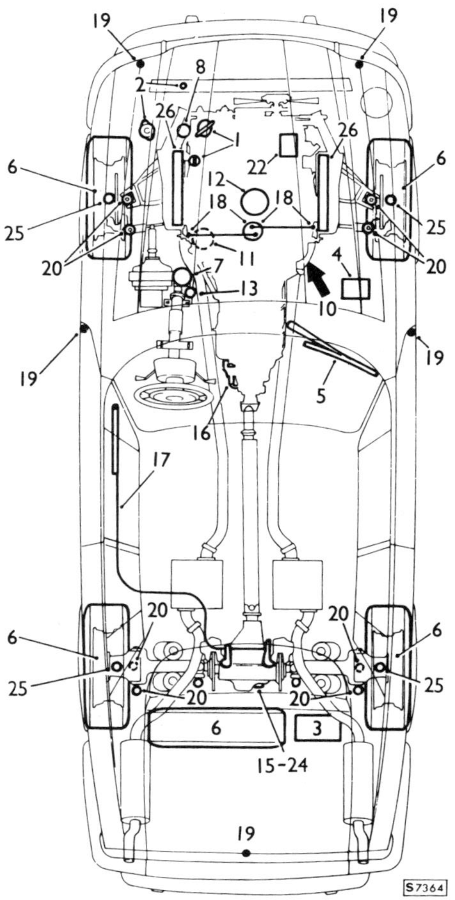

16. Gearbox selector linkage – lubricate exposed parts with engine oil.
17. Handbrake – lubricate mechanical linkage and cable with engine oil.
18. Accelerator linkage – lubricate with engine oil.
19. Sparingly lubricate bonnet, boot, door locks, hinges and petrol filler cap flap.
20. Grease all points excluding wheel hubs.

Every 12,000 miles (20 000 km)

21. Carry out operations 9 to 13 and 16 to 20.
22. Engine – renew fuel filter.
23. Overdrive (if fitted) – drain and refill with recommended oil.
24. Final drive unit – drain and refill with recommended oil.
25. Grease all points including wheel hubs.
26. Renew air cleaner elements and seals.

3000 mile (5000 km) Service   10.10.06

NOTE: Before undertaking any operation which involves access to the interior of passenger or luggage compartments, operators must wear gloves, or ensure that their hands are clean. Fit seat cover and place protective covers on carpets and steering wheel; these covers are to remain in place until completion of service. Drive car on lift (ramp) or over pit.

PASSENGER COMPARTMENT

1. Check function of original equipment i.e. interior and exterior lamps, indicators, horns and warning lights. Report any faults.
2. Check operation of window controls; check electrically operated windows for smoothness of operation and complete closure.
3. Check operation of handbrake; if a normal hand load causes the brake handle to move more than 8 inches (200 mm) adjustment is necessary; proceed as follows:
   a Move driver's seat fully forward.
   b Chock wheels and set handbrake fully off.
   c Raise carpet at rear of driver's seat to give access to adjusting nuts.
   d Slacken locknut.
   e Screw up adjuster nut until only a slight amount of slack is apparent in cable.
   NOTE: Do not eliminate all slack from cable.
   f Tighten locknut.
   g Lightly smear undersurface of carpet with adhesive and replace in position.

Check footbrake operation and action of servo; check that pedal does not move under maintained foot pressure, indicating a fluid leak. Refer to item 30.

4. Check that clock is running and set to time; reset if necessary, by means of a small knob protruding through the glass.
5. Check operation of windscreen washers and wipers; ensure that jets are clear, and reposition if necessary for optimum efficiency.

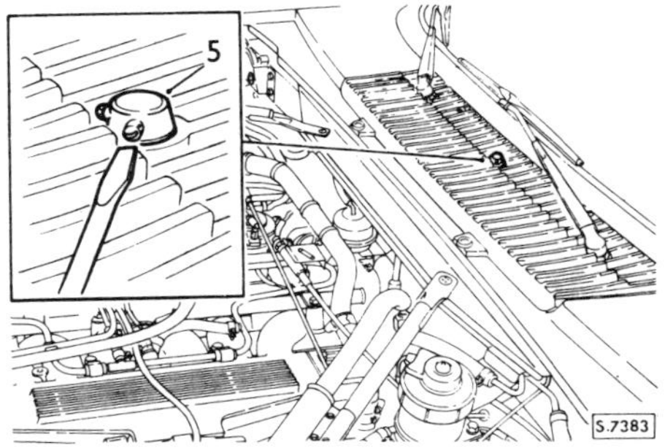

6. Check condition and security of seats and seat belts; check seat adjustment mechanisms for smooth operation, and inertia reels and buzzer alarm (if fitted) for function.
7. Check rear view mirrors for cracks and crazing; check operation of dipping lever and of door mirror remote control (if fitted).

EXTERIOR AND LUGGAGE COMPARTMENT

8. Check door and luggage compartment locks for correct operation; if door closing requires noticeable effort, or is in any way unsatisfactory, refer to 76.37.01.

continued

10–1

9 Switch on sidelights, open luggage compartment lid and check operation of inside light. If failure to operate is found, not corrected by bulb replacement, and red lead to light unit is 'live', then automatic switch, which is an integral part of light unit, must be replaced. Refer to 86.45.16.

10 Remove cover from battery (in luggage compartment) and check level of electrolyte, visible through translucent cell walls, and if one or more cells is below tops of separators, remove manifold, ensure that all six sleeves are raised, and add distilled water slowly to trough until all sleeves are filled. Replace manifold, wipe dry, lightly grease both battery terminals but do not refit cover.

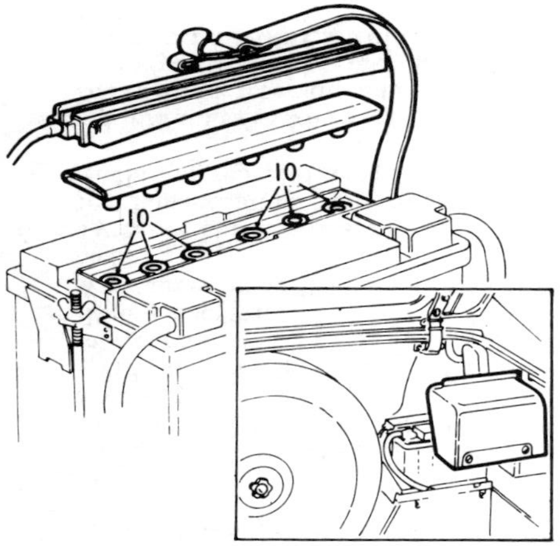

11 After confirming that tyres are cool, test pressures of all tyres (including spare) and adjust if necessary to specified figures. Remove spare wheel.

12 Check that tyres are of types and size quoted in 'General Specification Data' and report if other tyres are fitted.

13 Measure minimum tread depth of all tyres, and examine carefully for cuts in fabric, exposure of ply or cord, and lumps or bulges in sidewalls. Report any defects.

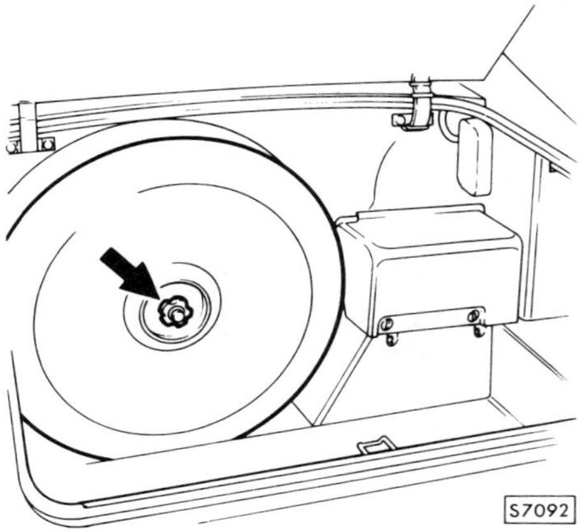

14 Before replacing spare wheel or battery cover, remove two screws securing metal cover below battery. Switch on ignition, raise carpet on front wall of luggage compartment and check for fuel leaks at pipes and joints to tank, pump and sump. Ensure that all connections are tight, switch off ignition and replace covers and spare wheel.

15 Set torque wrench to 45 lbf ft (6,2 kgf m) and check that all wheel nuts are tightened to this figure. Verify that spare wheel is correctly stowed.

16 Clean windscreen wiper blades and examine carefully; renew if worn, split or perished.

17 Using approved equipment, check beam patterns of headlights on Full Beam and Dipped settings. Adjust if necessary; refer to 86.40.17 or 18, for two or four headlamp systems.

ENGINE COMPARTMENT

18 Operate bonnet release from within car, open bonnet and fit protective covers to front wings.

19 Check engine oil level and if necessary add oil to bring level up to cross-hatched area of dipstick; record quantity of oil used.

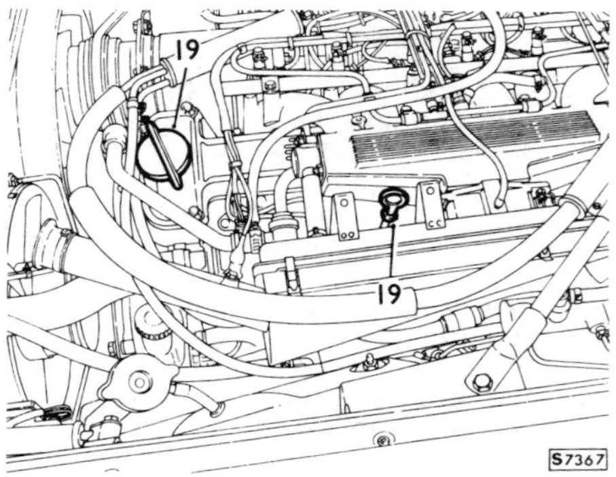

20 Ensure that engine is cold, and that car is standing on a level surface. Remove coolant filler from expansion tank filler tube and remove plug or open radiator bleed tap at left-hand side of radiator top rail. Air will enter expansion tank and coolant level will rise in expansion tank. Measure approximate vertical distance between final coolant level in filler tube and cap seating ring; if this distance is less than 2 in (50 mm) cooling system does not require topping up. Close bleed tap and replace filler cap. If coolant level is more than 2 in (50 mm) below seating ring, add correct coolant SLOWLY to filler until either system is full or coolant escapes from bleed. Close bleed and replace filler cap.
CAUTION: If more than 4 pints (4,8 U.S. pints or 2,3 litres) of coolant is required to replenish system, start engine and run at 1500 r.p.m. for 3 minutes with heater control on 'defrost' to clear trapped air. Then allow to cool and repeat 'top up' if necessary.
NOTE: Ensure that specific gravity of coolant is maintained by only adding correct mix of anti-freeze and water.

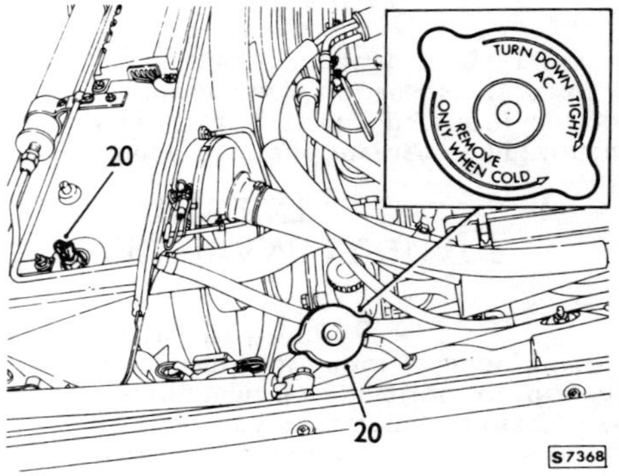

21 Inspect level of fluid in windscreen washer reservoir, and top up if necessary.

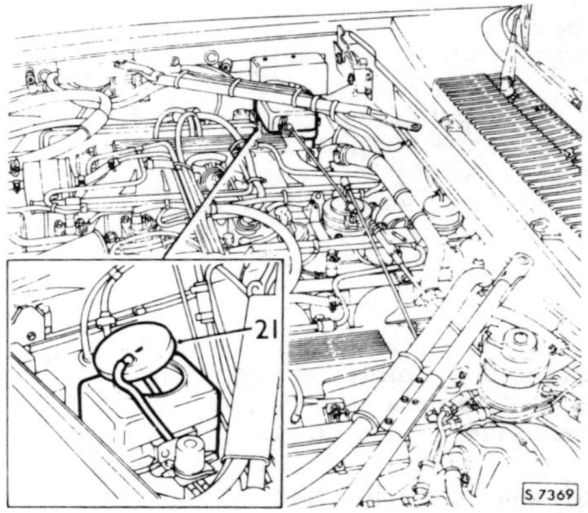

10—2

22 Inspect level of fluid in brake reservoir, and report if low; add specified fluid as necessary to bring up to 'Max. level' line, but do not overfill.

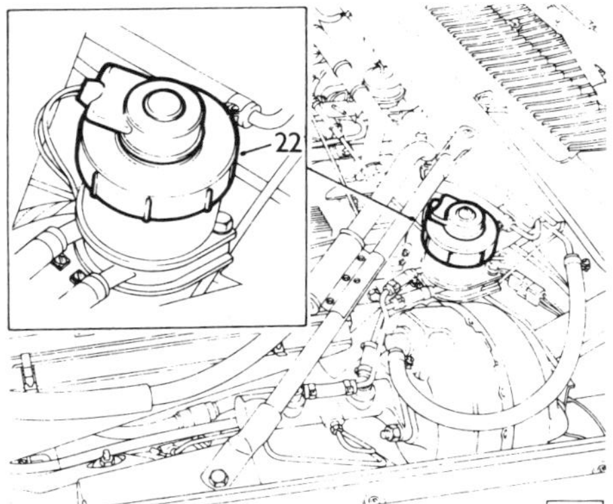

23 On manual transmission cars only, inspect level of fluid in clutch reservoir, and top up if necessary.

24 Remove filler cap from power steering oil reservoir (situated below front end of left-hand air cleaner) taking great care to prevent any foreign matter from entering it. Inspect oil level on dipstick and top up if necessary with recommended fluid. Level must be on 'Full' when oil is warm.

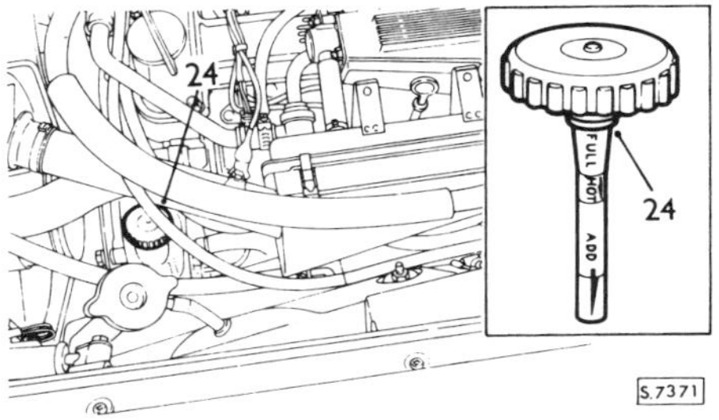

25 Check tensions of all drive belts to data below, which quote deflection caused in longest accessible run of each belt when a specified load is applied at right angles to belt. Adjust as necessary to obtain correct deflection.

Driving belt for:	Deflecting force lb.	kg.	Deflection in.	mm.	Location of adjustment data
P.A.S. pump and water pump	6.5	2,95	0.16	4,06	57.20.01
Alternator	3.2	1,5	0.16	4,06	86.10.05
Air pump and compressor	6.4	2,9	0.22	5,6	17.25.13
Fan drive	6.4	2,9	0.13	3,3	26.20.01

NOTE: It is not necessary to slacken air pump pivot bolt before adjusting belt tension.

26 Inspect all pipes and joints in engine cooling and car heating systems which are visible from above for leaks and security of connections and attachments.

27 Inspect all visible hydraulic pipes and unions for signs of chafing, leaks or corrosion.

28 Inspect all visible joints for evidence of petrol, oil or air leaks.

29 Check security of exhaust manifold mountings and inspect for signs of gasket leakage.

UNDERBODY

30 Raise ramp. Inspect brake pads and discs for wear and condition; minimum permitted thickness for brake pads is ⅛ in (3,2 mm). If less than this, pads must be changed; see 70.40.02 and 03 (footbrake pads) and 70.40.04 (handbrake pads). Discs should be free from deep scoring and signs of distortion or uneven wear.

31 Check mountings and joints of undercar exhaust system for security, and inspect for signs of damage, leaking or corrosion.

32 Inspect carefully for signs of oil leaks from engine, power assisted steering, gearbox or final drive. Report any leakage.

33 Check condition and security of steering joints; examine gaiters carefully for splits, cracking or damage, and ensure that they are firmly secured.

34 Carefully inspect all visible brake pipes and unions for leaks, chafing or corrosion, and for possible damage to flexible pipes.

35 Check all joints in visible external pipes for evidence of damage and of petrol, oil or air leaks. Lower ramp.

36 Remove protective covers from front wings, close bonnet and check all three bonnet catches for correct operation.

ROAD OR DYNAMOMETER TEST

Clean hands or wear gloves before carrying out the following items. Ensure that protective covers are still in place on seats and carpets. Remove steering wheel cover. Drive car off lift (ramp) or pit.

37 Carry out road or roller test and check that all instruments and controls are functioning correctly. Check that footbrake performance exceeds all requirements of current regulations, and will provide a retardation of 10 ft/sec² (3 m/sec²) with a pedal load not exceeding 30 lbf (13,5 kgf). Check that handbrake holds car securely on a hill, facing up or down, and that retardation it provides as an emergency brake meets legal requirements with a hand load not exceeding 88 lbf (40 kgf). Check that brake servo performance is consistent, and that there is no evidence of air leak into vacuum system.

38 Remove protective covers from seats and carpets, ensure that steering wheel, controls and car interior have not been marked, and complete Passport to Service.

6 month or 6000 mile (10000 km) Service, 10.10.12.

1 Carry out items 1 to 15 of 3000 mile (5000 km) Service.
2 Check front wheel alignment, using approved equipment. Correct figures are 1/16 to 1/8 in (1,6 to 3,2 mm) toe-in, measured at rims. Refer to 57.65.01.
3 Lightly lubricate locks and hinges of doors, bonnet and boot using engine oil. Do not lubricate steering lock, and ensure that all surplus oil is removed.
4 Carry out items 16 to 18 and 20 to 25 of 3000 mile (5000 km) service.
5 On automatic transmission cars only. Carefully clean area of top of transmission dipstick, set handbrake firmly and select 'P'. Start engine and run at under 750 rev/min for several minutes, passing selector through full range until normal operating temperature has been reached. With engine still idling, remove dipstick, wipe with clean paper or non-fluffy rag, replace fully and immediately withdraw. Fluid level should be at

continued

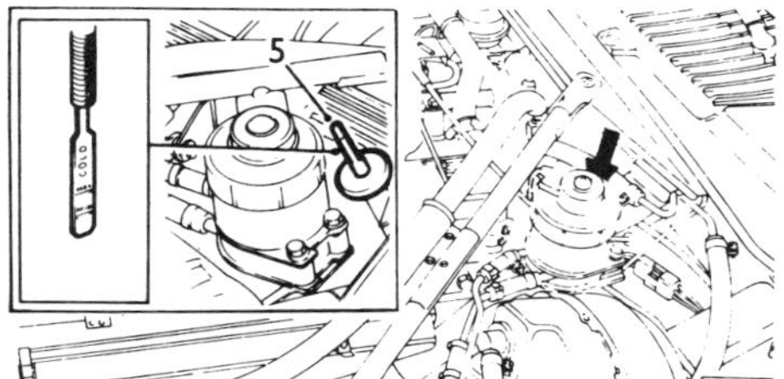

'Full' or 'High' mark. Top up if necessary using a specified fluid with engine still idling. Do not overfill. Report quantity of fluid required.

6 Remove all spark plugs. Check that they are to specification, clean on an approved plug cleaning machine and set points to a gap of 0.025 in (0,64 mm) or 0.035 in (0,89 mm) on cars fitted with emission control to N. American requirements. Refit plugs.

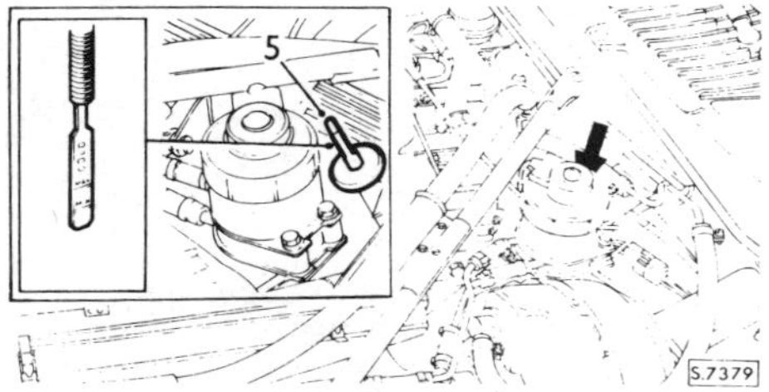

7 Remove distributor cap and rotor arm. Apply a few drops of light machine oil to felt pad in recess below rotor keyway. Replace rotor arm and distributor cap.

8 Lubricate all pivots and joints in throttle control linkage with engine oil. Check that operation is smooth, that throttles close correctly, and that full opening is obtained (before 'kickdown' on automatic transmission cars).

9 Check ignition timing characteristics, using electronic equipment. Static setting is 10° crankshaft BTDC. Advance characteristics are checked at ACCELERATING speeds with vacuum pipe disconnected, and are:

Engine rev/min	Advance,° crankshaft
600	0
800	0
1200	1–5
2000	12–14
2600	16–20
5200	22–26
7000	24–28

10 Check engine idling speed, which should be 650 to 750 rev/min with engine warm. If necessary adjust settings of fuel injection equipment to obtain this speed. Refer to 19.20.18.

11 Carry out items 26 to 29 of 3000 mile (5000 km) Service.

12 Raise ramp. Clear area around engine sump drain plug, remove plug, discard sealing washer, and drain oil into a container of at least 2 gallons (9 litres) capacity. Unscrew bolt securing oil filter canister; withdraw canister with bolt, sealing ring, filter element and by-pass valve. Discard filter and sealing ring and wash canister, bolt, and valve in clean petrol; when quite dry, reassemble with new element and sealing ring; replace sump plug, with new sealing washer, and refill sump with a specified oil to knurled area of dipstick. Run engine at a fast tickover for 2 to 3 minutes, stop, wait one minute, withdraw dipstick, wipe clean, re-insert and withdraw again to read oil level. Add oil as necessary to bring level to middle of knurled area.
If a replacement canister type of filter is to be renewed ensure that mounting is clean and sealing ring correctly seated. Screw replacement canister fully home BY HAND ONLY. On no account use any form of wrench or tourniquet in fitting canister or its subsequent removal will be very difficult.

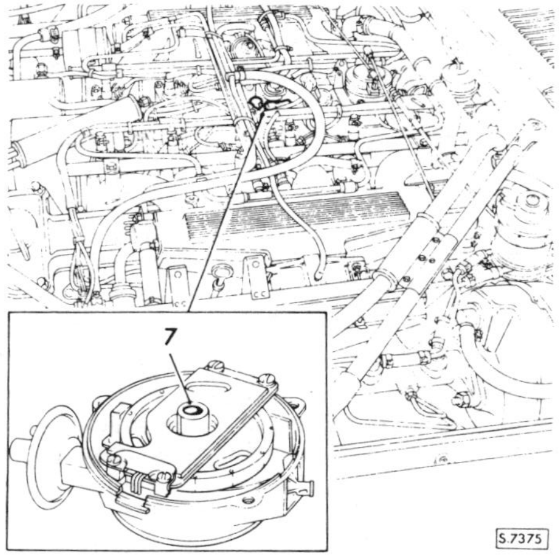

13 On manual transmission cars only. Clean right hand side of gearbox in area of filler/level plug, and remove plug. Oil level should be up to plug hole. Top up slowly, if necessary, with a recommended S.A.E. 90 oil, taking care to ensure that only clean oil enters the gearbox.
Replace plug and tighten to 25–27 lbf ft (3,46–3,73 kgf m).

14 Clean face of final drive rear cover around filler/level plug, and remove plug. Oil level should be up to plug hole. Top up if necessary, using oil of the same type as that already present. If brand of oil in unit is not known, it is preferable to drain and refill rather than to introduce another oil which may not mix satisfactorily. Use only recommended oils, which are specially formulated for 'Powr-Lok' differentials.

15 On manual transmission cars only. Check free movement of clutch operating rod, which should be ⅛ in (1.5 mm), on rod which couples slave cylinder to clutch withdrawal lever. Adjust if necessary by slackening locknut, turning operating rod until required free movement is obtained, and re-tightening locknut. Free travel is increased by screwing rod into knuckle joint. Replace return spring, if disturbed, after adjustment.

16 On manual transmission cars, lightly lubricate end joint of clutch operating rod with engine oil.

17 On automatic transmission cars, lubricate joints of exposed selector linkage with engine oil.

18 Using an approved grease, lubricate front wheel swivel joints (4 nipples) steering tie rods (2 nipples) steering pinion housing, rear suspension outer bearings (2 nipples) rear suspension inner bearings (4 nipples) and rear axle universal joints (4 nipples). Access to these joint nipples is obtained by removing rubber plugs from joint covers. Ensure that plugs are correctly refitted after greasing. Before greasing outer suspension bearings ensure that bleed holes, diametrically opposite to grease nipples, are clear; when sufficient grease has been applied to the joints any excess will appear at the bleed holes. Do not lubricate hubs.

19 Carry out items 31 to 34 of 3000 mile (5000 km) Service.

20 Inspect all pipes and joints in cooling and heating systems for evidence of leaks.

21 Carry out items 35 to 38 of 3000 mile (5000 km) Service.

12 month or 12000 mile (20000 km) Service, 10.10.24.

1 Carry out items 1 to 5 of 6000 mile (10000 km) Service.

2 Remove all spark plugs and replace with new Champion N10Y plugs with gaps set to 0.025 in (0,64 mm) or 0.035 in (0,89 mm) on cars fitted with emission control to N. American requirements.

3 Carry out items 7 and 8 of 6000 mile (10000 km) Service.

4 Clean engine breather filter.
 a Remove 'O' clip securing rubber elbow to filter housing at front of left-hand cylinder head; remove elbow.
 b Lift out gauze, wash in petrol, inspect and if satisfactory refit.
 c Replace rubber elbow, using new 'O' clip.

5 Release clips securing air cleaner covers, remove and discard elements and seals.
Fit new elements and seal assemblies, ensuring that blank sections of inner faces of elements are opposite throttle ports. Replace covers and secure clips.

SUMMARY CHART

Item No.	OPERATION	Operation Number Interval in Miles × 1000 Interval in Kilometres × 1000	10.10.06 3 5	10.10.12 6 10	10.10.24 12 20
	PASSENGER COMPARTMENT (Clean hands or fit gloves when carrying out items 1 to 8) **Fit seat cover, place protective cover on carpets.** **Drive car on lift (ramp).**				
1	Check function of original equipment, i.e. interior and exterior lamps, indicators, horns and warning lights		x	x	x
2	Check operation of window controls		x	x	x
3	Check handbrake operation		x	x	x
4	Check footbrake operation		x	x	x
5	Check clock is running and set to time		x	x	x
6	Check windscreen washers and wipers for correct operation and that jets are clear and correctly positioned		x	x	x
7	Check condition and security of seat belts		x	x	x
8	Check rear view mirrors for cracks and crazing		x	x	x
	EXTERIOR AND LUGGAGE COMPARTMENT				
9	Check door locks for correct operation		x	x	x
10	Check luggage compartment light for correct operation.		x	x	x
11	Check for fuel leaks at tank, sump tank, pump, pipes and connections		x	x	x
12	Check/adjust tyre pressures including spare		x	x	x
13	Check that tyres comply with manufacturer's specification		x	x	x
14	Check tyres for tread depth, visually for cuts in fabric, exposure of ply or cord structure, lumps or bulges.		x	x	x
15	Check tightness of road wheel fastenings and that spare is correctly stowed			x	x
16	Check front wheel alignment		x	x	x
17	Lubricate all locks and hinges (not steering lock)		x	x	x
18	Check, if necessary renew windscreen wiper blades		x	x	x
19	Check/adjust headlight alignment		x	x	x
	ENGINE COMPARTMENT				
20	Open bonnet, fit wing covers		x	x	x
21	Check/top up engine oil		x	x	x
22	Renew petrol filter				x
23	Check/top up cooling system		x	x	x
24	Check/top up windscreen washer reservoir		x	x	x
25	Check/top up brake fluid reservoir		x	x	x
26	Check security of H.T. leads		x	x	x
27	Check/top up fluid in power steering reservoir		x	x	x
28	Check/top up automatic gearbox fluid		x	x	x
29	Clean/adjust spark plugs		x	x	
30	Renew spark plugs				x
31	Lubricate distributor		x	x	x
32	Lubricate accelerator control linkage and check operation		x	x	x

continued

6 Depressurise fuel system and change main filter.

   a Remove right-hand luggage compartment trim.

   b Pull cable from terminal 85 of fuel pump relay.

   c Disconnect H.T. lead from ignition coil.

   d Switch ignition on and crank engine for a few seconds.

   e Clamp inlet and outlet hoses at filter.

   f Release hose clips and pull hoses from fuel filter.

   g Remove bolt and washer securing filter assembly to inlet manifold mounting bracket, remove filter assembly.

   h Remove screw and spire nut from fuel filter clamp, draw fuel filter clear of clamp. Discard filter.

   i Fit replacement filter by reversing operations a to g.

   j Reconnect H.T. lead to ignition coil.

   k Reconnect cable to terminal 85 of fuel pump relay.

   l Fit right-hand luggage compartment trim.

7 Carry out items 9 to 13 of 6000 mile (10000 km) Service.

8 Remove oil filler/level plug and drain plug from 4 HU final drive unit and collect oil in a suitable receptacle. This is best done after a run, while the oil in the unit is still warm. Replace drain plug and slowly fill with a specified oil up to filler hole. Replace filler/level plug. To drain 20 HU final drive units, fitted to overdrive equipped cars, loosen rear cover. Detach rear cover and fit new gasket before refilling with oil.

9 Carry out items 15 to 18 of 6000 mile (10000 km) Service.

10 Remove front wheels, exposing grease nipples. Inject specified lubricant until it appears at bleed holes at hub centres. Remove rear wheels, exposing holes in hub carriers closed by dust caps. Clean caps and surrounding area, prise out dust caps and inject grease until no more will enter; do not build up pressure in hubs. Clear vent holes in dust caps and replace. Replace rear wheels.

11 Clean areas of engine mountings and of front and rear crossbeam and suspension attachments. Inspect rubber mountings and bushes carefully for signs of deterioration or distortion and check all fixings for security; tightening torques are quoted in section 06.

10–5

SUMMARY CHART

Item No.	OPERATION	10.10.06 Miles×1000: 3 / km×1000: 5	10.10.12 6 / 10	10.10.24 12 / 20
33	Clean engine breather filter			X
34	Renew air cleaner elements and seals			X
35	Check/adjust ignition timing and distributor characteristics using electronic equipment		X	X
36	Check/adjust idle speed	X	X	X
37	Check/adjust driving belts		X	X
38	Check/top up battery electrolyte; clean and grease terminals	X	X	X
39	Check cooling and heating systems for leaks	X	X	X
40	Check visually hydraulic pipes and unions for chafing, leaks and corrosion	X	X	X
41	Check visually all joints for petrol, oil or air leaks	X	X	X
42	Check exhaust system for leaking and security	X	X	X

UNDERBODY
Raise ramp

Item No.	OPERATION	10.10.06	10.10.12	10.10.24
43	Renew engine oil and filter		X	X
44	Check/top up manual gearbox oil		X	X
45	Check/top up final drive oil		X	X
46	Renew final drive oil			X
47	Lubricate automatic gearbox exposed selector linkage		X	X
48	Lubricate handbrake mechanical linkage and cable		X	X
49	Lubricate all grease points excluding hubs		X	X
50	Lubricate all grease points including hubs			X
51	Inspect brake pads for wear and discs for condition	X	X	X
52	Check security of engine and suspension fixings		X	X
53	Check exhaust system for leakage and security	X	X	X
54	Check engine, power assisted steering, gearbox and final drive for oil leaks	X	X	X
55	Check condition and security of steering unit joints and gaiters	X	X	X
56	Check cooling and heating system for leaks	X	X	X
57	Check visually hydraulic pipes and unions for chafing, leaks and corrosion	X	X	X
58	Check visually all joints for petrol, oil or air leaks	X	X	X

Lower ramp
59. Remove wing covers, close bonnet and check bonnet catches for correct operation

ROAD OR DYNAMOMETER TEST
(Clean hands before carrying out following items)

Ensure seat cover and protective cover on carpets are in place

Drive car off lift (ramp)
60. Carry out road/roller test, check function of all instrumentation and inertia reel safety harness — X, X
61. Remove seat cover and protective cover from carpets — X, X

ADDITIONAL MAINTENANCE OPERATIONS – ALL VEHICLES

Drain braking system, retract pistons, renew fluid, bleed brakes and check operations of P.D.W.A. unit — Every 18,000 miles
Overhaul complete braking system — Every 36,000 miles

ADDITIONAL MAINTENANCE OPERATIONS – FEDERAL VEHICLES

MAINTENANCE SUMMARY

Detailed maintenance instructions will be found in the operation indicated in brackets after each item.

Weekly
Check/top up engine oil (10.10.06/19).
Check/top up brake (22), clutch (23) and automatic transmission fluid reservoirs.
Check/top up battery electrolyte (10).
Check/top up cooling system (20).
Check/top up washer reservoir(s) (21).
Check function of original equipment, i.e. exterior lights (1), wipers (5) and warning indicators (1).

Check tyres for tread depth, visually for external cuts in fabric, exposure of ply or cord, structure, lumps or bulges.
Check/adjust tyre pressures including spare.
Check tightness of road wheel fastenings.

MINIMUM MAINTENANCE SCHEDULES

NOTE: The maintenance schedules are based on an annual mileage of 12,000 miles. Should the vehicle complete substantially less miles than this, then it is recommended that a lubrication service is completed at six month intervals and a major service annually.

● SPECIFIED OTHERWISE

North American specification only

SERVICE	1	3	6	9	12.5	16	19	22	25	28	31	34	37.5	41	44	47	50
									MILEAGE × 1000								
A	X																
B		X		X		X		X		X		X		X		X	
C			X				X			X					X		
D					X								37.5 X				
E								24 X									

SERVICE	1	3	9	15	18	21	27	30	33	36	39	42	45	48
					MILEAGE × 1000									
A														
B		3	9	15		21	27		33		39		45	
C			6		18			30				42		
D				12			24			36				48

* These items are emission related.

10–6

KILOMETRES × 1000

A	1.6	5		15		25		35		45						
B		5	10								50			65		
C													55			75	
D					20			30		40				60	70		80

LUBRICATION

	A	B	C	D	E
Lubricate all grease points.	x	x	x	x	x
*Renew engine oil (10.10.12/12)	x	x	x	x	x
*Renew engine oil filter (12)	x			x	x
*Lubricate accelerator control linkage (and pedal pivot) – check operation					x
Check level of fuel in: Brake reservoir (10.10.06/22), battery (10), engine (19), final drive (10.10.12/14), transmission (10.10.12/5), cooling system (10.10.06/20), power steering (24) and windshield washer (21).		x	x	x	x
Lubricate all locks and hinges (not steering locks).		x	x	x	x

ENGINE

	A	B	C	D	E
*Check all driving belts – adjust or renew as necessary	x			x	x
Check brake servo and cooling system hoses for condition and tightness	x	x	x	x	x
*Check/rectify crankcase breathing and evaporative system hoses, pipes and restrictors for blockage, security and deterioration.					x
Clean crankcase breather filter (10.10.24/4).				x	x
*Renew air cleaner elements (10.10/24/5).				x	x
*Check/rectify operation of diverter and check valves.				x	x
*Check security of all vacuum pipes.				x	x
*Clean EGR valves, pipes and ports.					x
*Renew catalytic converters.					x
*Renew charcoal canister.	x				x
*Check/clean throttle housing bore.				x	x

FUEL SYSTEM

	A	B	C	D	E
*Renew fuel filter (10.10.24/6).	x	x		x	x
*Check fuel system for leaks.	x	x	x	x	x
*Check condition of fuel filler cap seal.				x	x

OSCILLOSCOPE AND COMBUSTION CHECK

	A	B	C	D	E
*Check ignition wiring for fraying, chafing and deterioration	x			x	x
*Check distributor cap, check for cracks and tracking				x	x
*Lubricate distributor (10.10.12/7).				x	x
*Renew spark plugs.				x	x
*Check coil performance on oscilloscope.	x			x	x
*Check operation of distributor vacuum unit.				x	x
*Check/adjust ignition timing using electronic equipment	x			x	x
*Check/adjust idle speed.	x			x	x
*Check/adjust exhaust CO at idle.	x			x	x
*Check exhaust system for security and leaks.	x	x	x	x	x
Check charging system output.			x	x	x

SAFETY

	A	B	C	D	E
Check condition of steering unit joints and gaiters			x	x	x
Check security of suspension fixings			x	x	x
Adjust front hub bearing end float				x	x
Check visually hydraulic pipes and unions for leaks chafing and corrosion			x	x	x
Inspect brake pads for wear and discs for condition (10.10.06/30).		x		x	x
Inspect brake linings/pads for wear, drums/discs for condition (10.10.06/30)			x	x	x
Check/adjust headlamp alignment ,¡0.10.06/17).			x	x	x
Check/adjust tyre pressures (including spare).			x	x	x
Check/adjust front wheel alignment.				x	x
Check tyres visually for tread depth, cuts in fabric, exposure of ply or construction, lumps or bulges.	x	x	x	x	x
Check/adjust foot and handbrakes (10.10.06/3).				x	x
Check operation of all door locks and window controls				x	x
Check condition, security and operation of seats, seat belts/interlock.				x	x
Check, if necessary renew wiper blades.				x	x
Check for fuel leaks at filter, pump and pipes, ensure all connections tight.	x	x	x	x	x

ROAD TEST

	A	B	C	D	E
Ensure that operation of vehicle is satisfactory and report all items requiring attention.	x	x	x	x	x

BRAKES

At 19,000 and 37,500 miles brake fluid in the hydraulic brake system should be renewed. All hydraulic seals and hoses in the brake system should be renewed at 37,500 miles.

● At 50,000 mile intervals – Renew EGR valves.

10–7

CAMSHAFT

Remove and refit – left hand or right hand 12.13.01

Service tools: Sprocket retaining tool JD.40; valve timing gauge C.3993; Timing chain tensioner retractor tool and release tool JD.50.

Removing

1. Remove both camshaft covers – left hand camshaft removal only, see 12.29.43.
2. Remove right hand camshaft cover, see 12.29.44.
3. Remove right hand air cleaner, see 19.10.01.
4. Remove rubber grommet from timing cover.
5. Insert blade of release tool through hole to release locking catch on timing chain tensioner.
6. Using special tool JD.50 retract timing chain tensioner. Locking catch will engage on step. Remove tools. 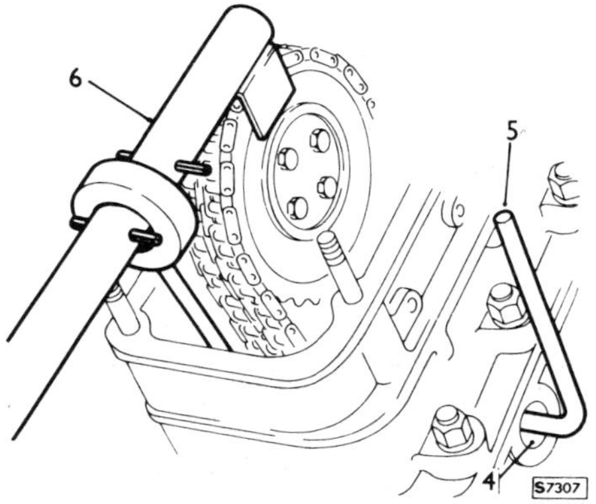
7. Bend back locking tabs and remove two camshaft sprocket retaining bolts.
8. Rotate engine until timing gauge C.3993 can be inserted in slot in camshaft to be removed.
9. Remove two bolts and fit sprocket retaining tool JD.40.
   CAUTION: DO NOT ROTATE ENGINE WHILE CAMSHAFT DISCONNECTED.
10. Progressively slacken camshaft bearing cap nuts starting with centre cap and working outwards; lift off bearing caps.
11. Lift camshaft out of tappet block.

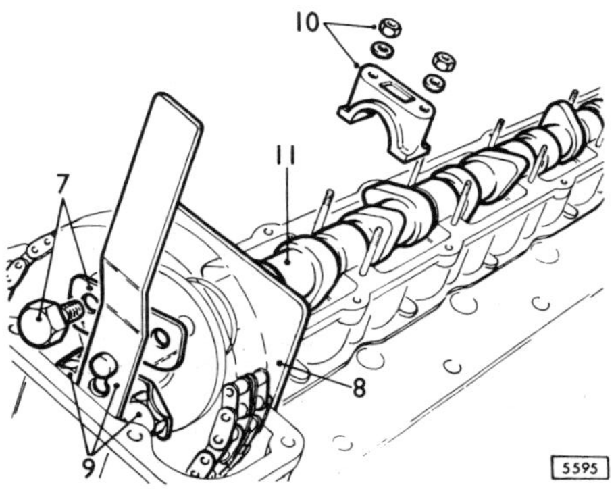

Refitting

12. Smear camshaft journals and tappets with clean engine oil.
    NOTE: Use gauge C.3993 to position camshaft correctly before fitting bearing caps.
13. Position camshaft in tappet block, refit bearing caps, washers and nuts.
14. Progressively tighten bearing cap nuts, working from the centre outwards, to a torque of 9.0 lb. ft (1,2 kg. m).
15. Check timing gauge is still correctly positioned.
16. Engage camshaft sprocket with shaft and fit one retaining bolt through 'fit' hole on tab washer. Turn up tabs. 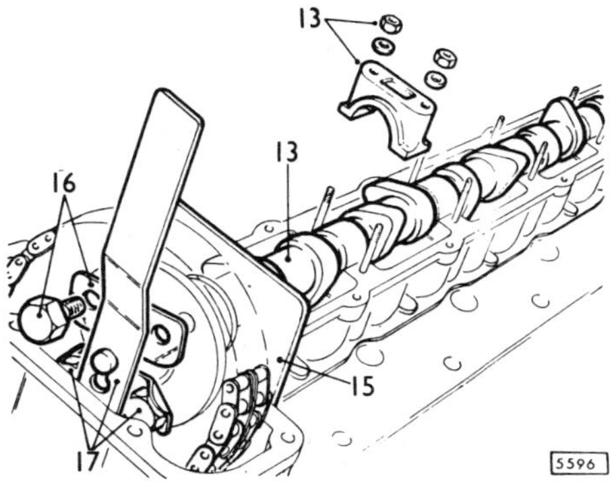
17. Rotate engine, fit remaining bolts and tabwasher. Turn up tabs.
18. Using tool JD.50 raise timing chain tensioner slightly, insert releasing tool in hole in timing chain cover. Release locking catch and allow tensioner to expand. Refit rubber grommet.
19. Refit camshaft covers.
20. Refit right hand air cleaner.

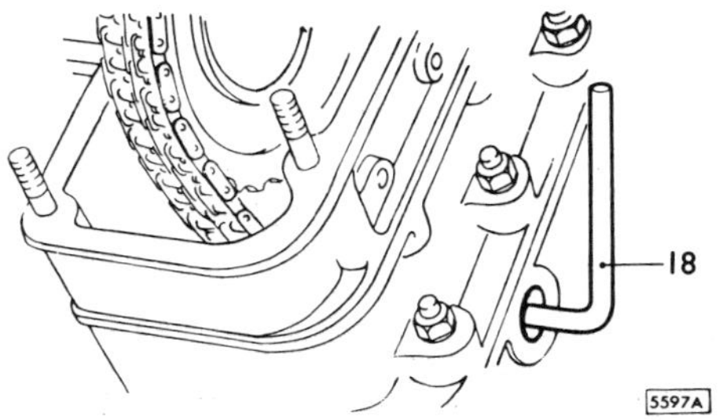

TAPPET BLOCK – Left hand 12.13.29

Remove and refit

Removing

1. Remove camshaft, see 12.13.01.
2. Remove banjo bolt securing oil feed pipe to tappet block.
3. Disconnect breather pipe.
4. Progressively slacken retaining nuts and cap screws, working from centre outwards.
5. Lift off tappet block carefully, retrieve tappets and valve adjusting pads.
   NOTE: Record from which valve each tappet and pads are removed. Failure to do this will result in incorrect valve adjustment upon reassembly.

continued

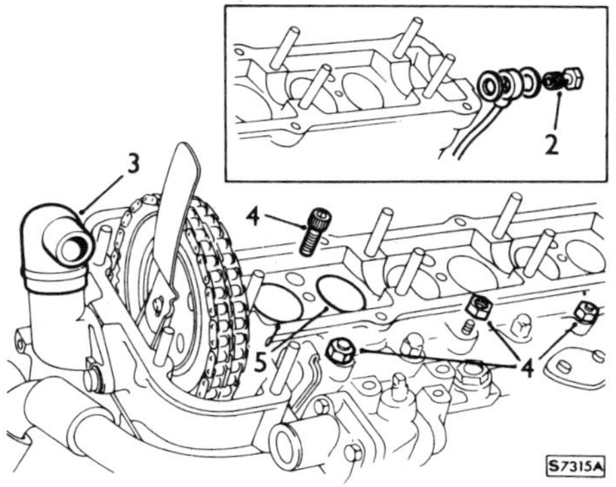

CAMSHAFT

Overhaul 12.13.26

1. Check that journal diameters are within limits – see Engine data 05.
2. Ensure all oil passages are unobstructed, blow through with dry, clean compressed air.

12–1

PISTON AND CONNECTING ROD 12.17.01

Remove and refit

Service tool: Piston ring clamp 38U.3

Removing

1. Remove engine and gearbox assembly, see 12.37.01.
2. Remove cylinder heads – right hand, see 12.29.12 – left hand, see 12.29.11.
3. Carry out items 2 to 13, see 12.60.45.
4. Rotate crankshaft until bearing cap to be removed is accessible.
5. Remove nuts, bearing cap and shell.
6. Remove any carbon deposit from top of cylinder bore. Push connecting rod up cylinder bore, withdraw piston together with connecting rod.

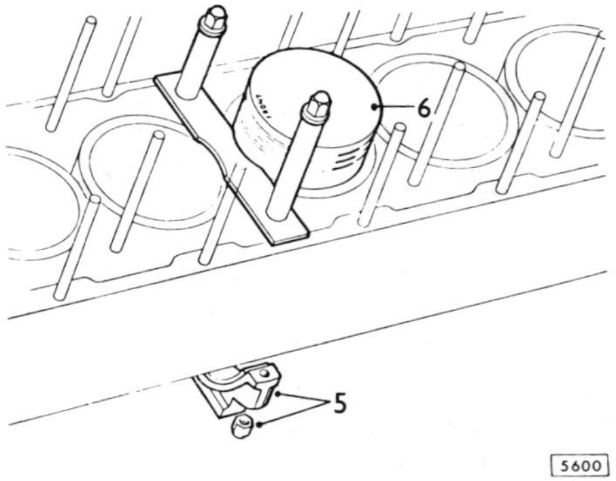

Refitting

7. Retrieve remaining bearing shell.
8. Ensure that cylinder bore, piston and all bearing surfaces are scrupulously clean.
9. Coat piston rings, gudgeon pin, big end bearing shell and cylinder bore liberally with clean engine oil.
10. Ensure that piston ring gaps are spaced evenly around circumference of piston.
11. Compress piston rings with Service Tool 38U.3.
12. Enter piston and connecting rod into bore ensuring that word stamped 'Front' on piston faces front of engine.
13. Push piston and connecting rod down bore, do not use undue force.
14. Check that big end shell bearing tab is correctly located in connecting rod.
15. Fit other half of big end bearing shell to cap; oil shell and crankshaft journal.
16. Refit bearing cap and nuts: tighten nuts to torque of 40.5 lb. ft (5,5 kg. m).
17. Refit sandwich plate.
18. Refit cylinder head – 'A' right hand, 'B' left hand.
19. Refit engine and gearbox assembly.

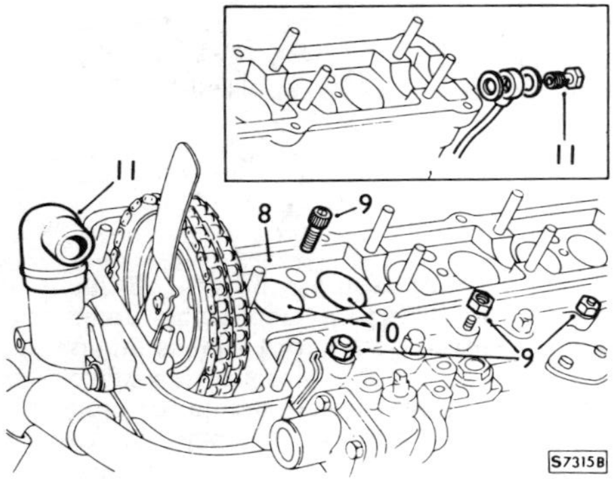

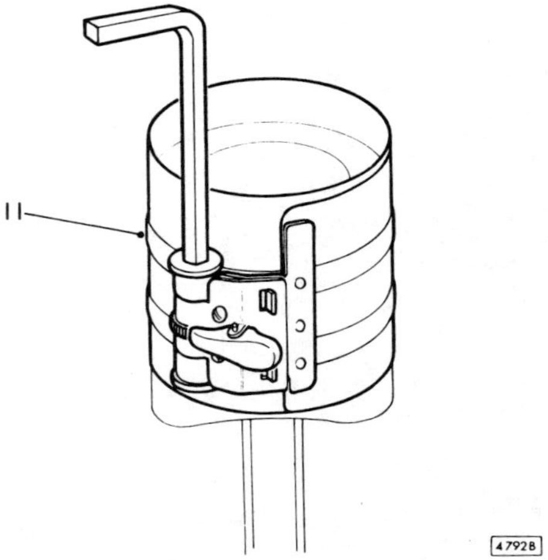

TAPPET BLOCK – Right hand 12.13.30

Remove and refit

Removing

1. Remove camshaft, see 12.13.01.
2. Remove banjo bolt securing oil feed pipe to tappet block.
3. Progressively slacken retaining nuts and capscrews, working from centre outwards.
4. Lift off tappet blocks carefully, retrieve tappets and valve adjusting pads.

NOTE: Record from which valve each tappet and pads are removed. Failure to do so will result in incorrect valve adjustment upon reassembly.

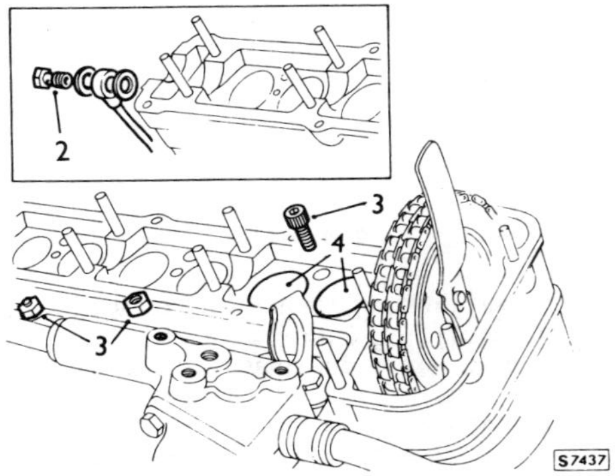

Refitting

5. Ensure that mating surfaces of tappet block and cylinder head are clean.
6. Smear mating surfaces of tappet block and cylinder head with Hylomar.
7. Fit tappet block, ensuring that dowels are correctly located.
8. Tighten retaining nuts and capscrews by diagonal selection working from centre outwards.
9. Lubricate tappets and adjusting pads with clean engine oil, fit to their respective valves.
10. Reverse operations 1 and 2.

NOTE: If tappet block has been renewed, it will be necessary to check valve clearance, see 12.29.48.

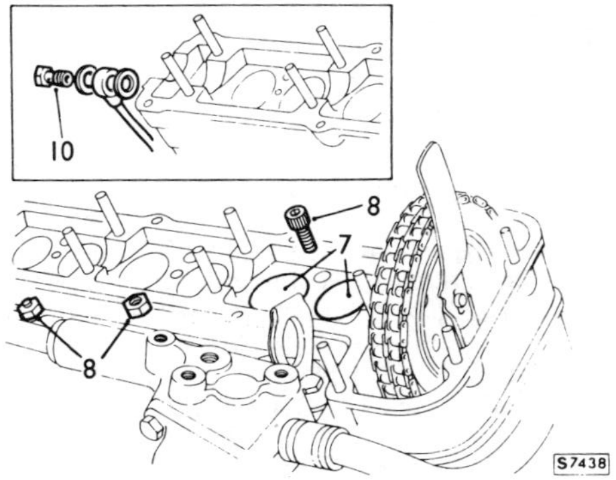

Refitting

6. Ensure that mating surfaces of tappet block and cylinder head are clean.
7. Smear mating surfaces of tappet block and cylinder head with Hylomar.
8. Fit tappet block, ensuring that dowels are correctly located.
9. Tighten retaining nuts and capscrews by diagonal selection working from centre outwards.
10. Lubricate tappets and adjusting pads with clean engine oil, fit to their respective valves.
11. Reverse operations 1 to 3 inclusive.

NOTE: If tappet block has been renewed, it will be necessary to check valve clearances, see 12.29.48.

12–2

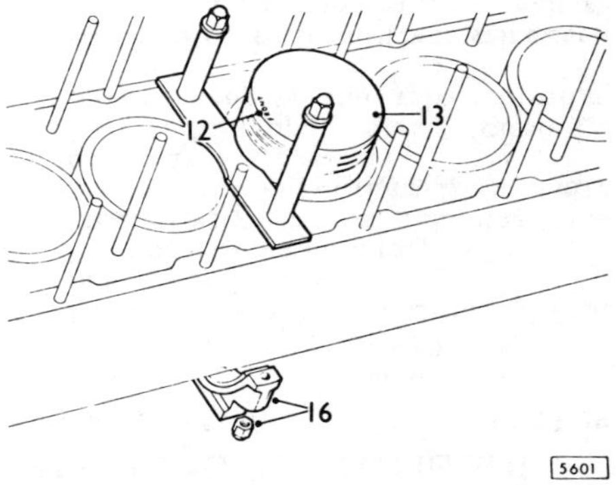

CRANKSHAFT DAMPER AND PULLEY

Remove and refit (Engine in situ) 12.21.01

Removing
1. Remove radiator, see 26.40.04.
2. Remove fan and torquatrol unit, see 26.25.21.
3. Remove power assisted steering pump drive belt, see 57.20.02.
4. Remove alternator drive belt, see 86.10.03.

CARS FITTED WITH AIR CONDITIONING/EMISSION CONTROL
5. Remove compressor/air pump drive belt, see 82.10.02.

ALL CARS
6. Remove bolts securing pulley to damper, withdraw pulley.
7. Remove crankshaft damper bolt.
8. Strike damper sharply with hide mallet to loosen.
9. Withdraw damper and cone, taking care to recover a Woodruff key from cone and from crankshaft.

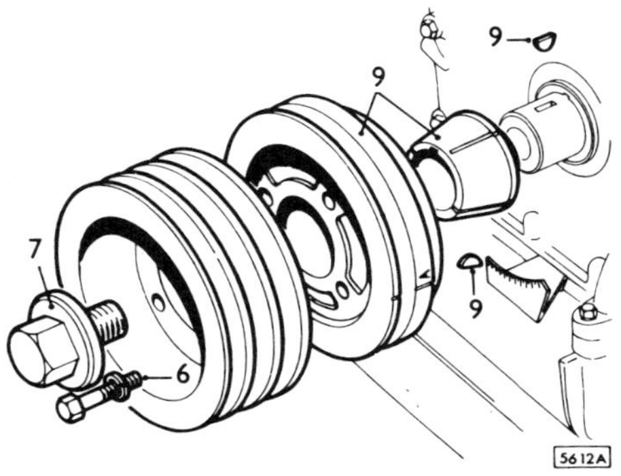

Refitting
Reverse operations 1 to 9; tighten bolt to 125 to 150 lbf ft (17,3 to 20,7 kgf m).

CONNECTING ROD BEARINGS

Remove and refit (set) 12.17.16

It is stressed that this operation should only be carried out if crankshaft damage is not suspected and the failure is due to bearing shell wear.

If, upon inspection, journals are found to be damaged, the engine must be removed and the crankshaft renewed.

Removing
1. Remove sandwich plate assembly, see 12.60.45.
2. Remove four setscrews and washers securing suction pipe clips and bracket. Draw suction pipe from 'O' ring at elbow.
3. Remove two setscrews and washers securing crankshaft undershield and delivery pipe clips. Draw undershield clear.
4. Rotate crankshaft until two journals are at bottom dead centre.
5. Remove big end bearing cap nuts from connecting rods on one journal and renews shells. DO NOT ROTATE CRANKSHAFT WHILE CAPS ARE DETACHED.
6. Repeat operation on other journal.
7. Rotate crankshaft until two more journals are at bottom dead centre and renew shells.
8. Repeat for last pair of journals.

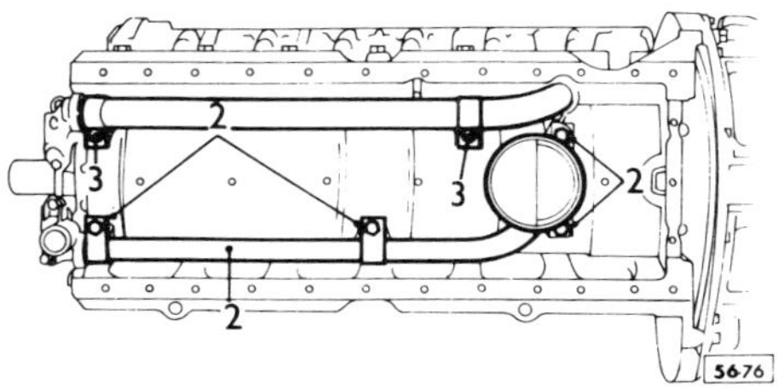

Refitting
Reverse operations 1 to 3 renewing suction pipe 'O' ring seal and cleaning strainer before replacement. Tighten bearing cap nuts to 40.5 lb. ft (5,5 kg. m).

PISTON AND CONNECTING ROD

Overhaul 12.17.10

NOTE: Pistons are supplied complete with gudgeon pin. As pins and pistons are matched assemblies, it is not permissible to interchange component parts.

1. Remove circlips.
2. Push gudgeon pin out of piston.
3. Withdraw connecting rod.

Refitting
4. Fit gudgeon pin in piston.
CAUTION: Connecting rods must be refitted to pistons in such a way that when installed in engine, word 'FRONT' on piston crown faces front of engine and chamfer on big end eye faces crank pin radius.
5. Align small end with end of gudgeon pin and push pin home.
6. Use new circlips to retain gudgeon pin.
NOTE: Gudgeon pin is a push fit in piston at 20 deg. C (60 deg. F). Fit will vary with ambient temperature. Three piston rings are fitted, they are as follows:
   a. Top ring — Compression.
   b. Second ring — Compression.
   c. Bottom ring — Oil control.

Both top and second rings have tapered peripheries and second rings are marked 'TOP' to ensure correct fitting. In addition, the top ring has a chrome plated periphery and is also cargraph coated. This coating is coloured RED and must not be removed.

The bottom ring consists of an expander sandwiched between two rails, the assembly being held together by an adhesive.

7. Check piston ring gap in bore. Push ring to a point midway down bore, check that ring is square and measure gap — see Engine data 05.
8. Fit bottom ring ensuring that expander ends are not overlapping.
9. Fit second and top rings ensuring that they are fitted the correct way up.
10. Position rings so that gaps are in positions shown.
11. Check side clearance of rings in piston groove — see Engine data 05.
12. Check connecting rods for alignment on a suitable jig.
13. Check bore of small end bush — see Engine data 05.
CAUTION: If small end bush is worn beyond acceptable limits; a service exchange connecting rod must be fitted. It is NOT possible to renew bushes as specialised equipment is needed to hone bushes to finished size.

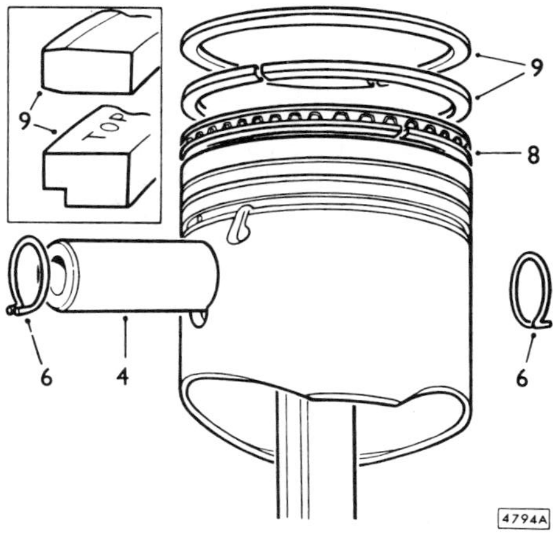

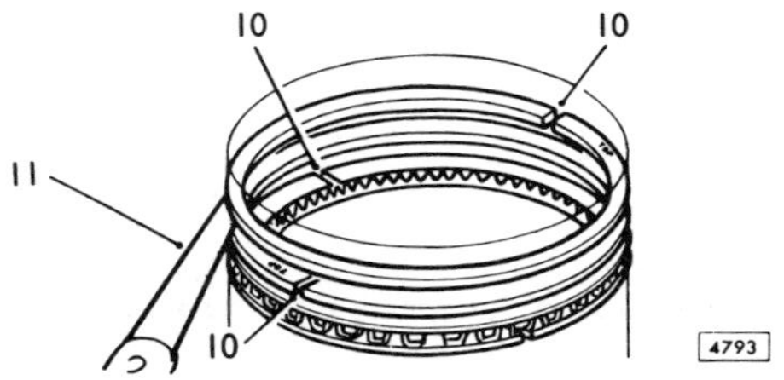

12—3

CRANKSHAFT FRONT OIL SEAL 12.21.14

Remove and refit

Removing
1. Remove crankshaft damper and pulley, see 12.21.01.
2. Prise seal out of timing cover and discard.
3. Withdraw distance piece.

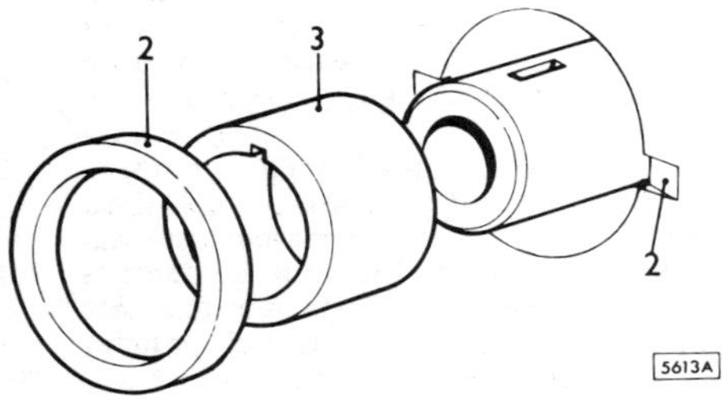

Refitting
4. Ensure that seal recess in timing cover is absolutely clean.
5. Smear new oil seal with clean engine oil.
6. Position oil seal squarely in recess and tap gently home using a hide mallet.
7. Refit distance piece.
8. Reverse operations 1 and 2.

CRANKSHAFT 12.21.33

Remove and refit

Follow procedures detailed under Engine – Dismantle and reassemble, see 12.41.05 as necessary.

CRANKSHAFT MAIN BEARINGS (set) 12.21.39

Remove and refit

It is stressed that this operation should only be carried out if crankshaft damage is suspected and the failure is due to bearing shell wear.

If upon inspection, journals are found to be damaged, the engine must be removed and the crankshaft renewed.

10. Reverse operations 1 to 6, renewing suction and delivery pipe 'O' ring seals, and cleaning suction pipe strainer before replacement.

CYLINDER PRESSURES 12.25.01

Check

1. Set transmission selector at 'P' or gear lever in neutral.
2. Run engine until normal operating temperature is reached. Switch off engine.
3. Remove H.T. cable from ignition coil.
4. Remove all spark plugs.
5. Fit approved pressure gauge at one plug hole and, with throttle held fully open, crank engine. Note highest steady pressure reading achieved and repeat at each plug hole in turn; the pressure noted must be between 120 and 140 lbf/in^2 (8,44 and 9,84 kgf/cm^2). The reading taken at each cylinder must not differ from the reading taken at any other cylinder by more than 5 lbf/in^2 (0,35 kgf/cm^2).
6. Refit spark plugs.
7. Refit H.T. cable to ignition coil.

CYLINDER LINERS 12.25.27

Checking

1. Check bore of liner and compare dimension obtained with dimensions quoted in group 04.
2. Bore grade of liner e.g. 'A' or 'B' is stamped on top of liner. When liners are to be renewed: the new liner must be of the same grade as the old one.

Cylinder Block – General

1. Following engine dismantling operation crankcase must be thoroughly cleaned.
2. Check all Welch washers and renew any showing signs of corrosion.
3. Ensure that all galleries are unobstructed. Blow through with dry, clean compressed air.
4. Check condition of studs, renew any showing signs of corrosion.

Removing
1. Remove sandwich plate assembly, see 12.60.45.
2. Remove four setscrews and washers securing suction pipe clips and bracket. Draw suction pipe from 'O' ring elbow at elbow.
3. Remove two setscrews and washers securing crankshaft undershield and delivery pipe clips. Draw undershield clear.
4. Remove bolt and washer and nut and washer securing oil delivery pipe elbow to oil pump casting.
   NOTE: Leave bolt at outboard fastening in position in oil pump casting. If bolt is removed for any reason, it must be replaced in a downward direction.
5. Draw oil delivery pipe and elbow downwards from bore in crankcase and from oil pump casting. Remove and discard gasket.
6. Draw undershield clear.
   NOTE: Record position of pillar nuts.
7. Remove four nuts and washers securing front main bearing cap. Remove bearing cap and renew shells.

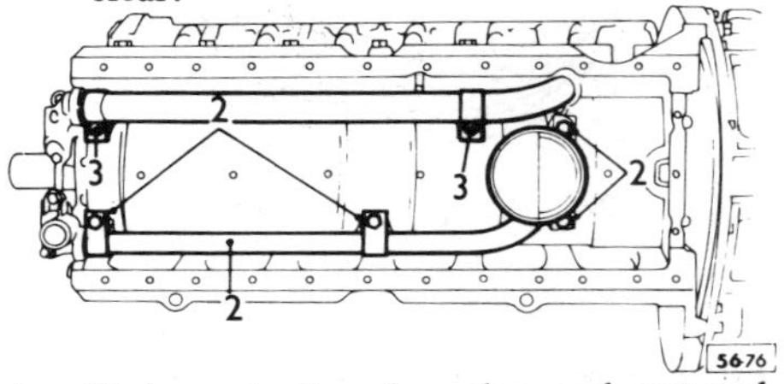

Refitting
8. Fit bearing cap and tighten securing nuts. Torque to: 27.5 lb. ft (3,7 kg. m) for 3/8 in (9,5 mm) studs, 62.5 lb. ft (8,6 kg. m) for 1/2 in (12,7 mm) studs.
9. Repeat operations 7 and 8 for all main bearings.
   NOTE: Rear main bearing is secured with four small nuts and washers and two large nuts and washers.
   CAUTION: Centre and rear main bearing shells must not be confused with each other; rear main bearing shell has an oil groove whilst centre main bearing shell is plain.

CYLINDER HEAD GASKET 12.29.02

Remove and refit

Removing
Follow procedure given for removing relevant cylinder head, see 12.29.11 or 12.29.12. Check cylinder head and faces of block and liners for damage that caused, or was the result of, gasket failure; rectify as necessary.

CYLINDER HEAD 12.29.11

Remove and refit – Left hand – 'B' bank

Service tools: Timing chain tensioner retractor tool and release tool JD.50; Camshaft sprocket retaining tool JD.40; Cylinder liner retaining tool JD.41; Valve timing gauge C.3993.

Removing
1. Disconnect battery, see 86.15.20.
2. Remove left hand camshaft cover, see 12.29.43.
3. Remove right hand camshaft cover, see 12.29.44.
4. Slacken clips securing hose to thermostat housing and coolant cross over pipe.
5. Remove bolts securing thermostat housing, auxiliary air valve and lifting eyes to cylinder head.
6. Move housing clear of cylinder head, remove and discard gaskets.
7. Remove bolts securing fan and torquatrol unit to spindle.
8. Rotate engine until valve timing gauge C.3993 can be fitted in slot 'A' bank camshaft front flange.

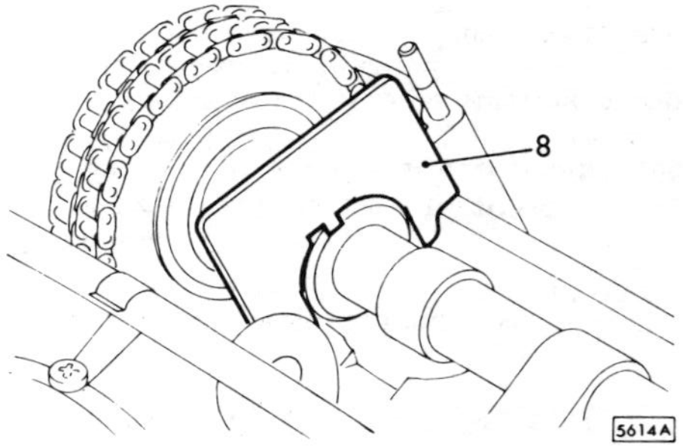

9 Remove rubber grommet from timing cover.
10 Insert blade of release tool through hole to release locking catch on timing chain tensioner.
11 Using special tool JD.50, retract timing chain tensioner. Locking catch will engage when tensioner is correctly retracted. 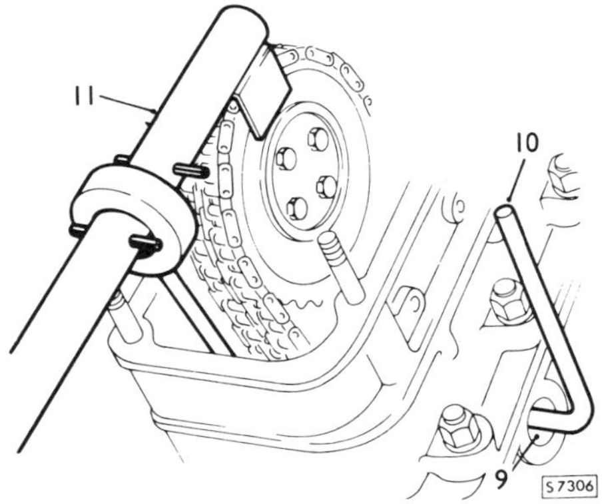
12 Disconnect camshaft sprocket from camshaft, fit sprocket retaining tool JD.40.
13 Remove steering rack gaiter heat shield.
14 Remove screws securing front heat shield.
15 Remove nuts securing down pipe heat shield.
16 Remove nuts securing down pipe to exhaust manifold; discard sealing rings.
17 Remove nuts securing heat shield to exhaust manifolds, lift off heat shield.
18 Remove nuts securing rear exhaust manifold to cylinder head. Lift off manifold, remove and discard gasket.

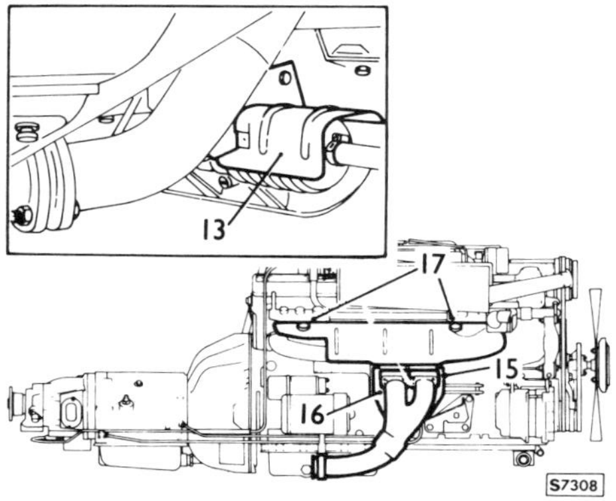

NOTE: On cars fitted with emission control, it will be necessary to disconnect E.G.R. pipe.

CARS FITTED WITH AUTOMATIC TRANSMISSION ONLY
19 Remove nut and bolt clamping dipstick tube; swing tube away from engine.

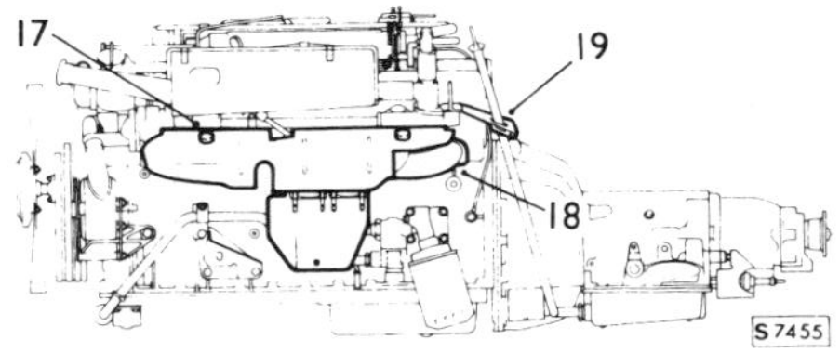

ALL CARS
20 Disconnect fuel hoses from cooler, plug broken connections to prevent ingress of dirt.
21 Remove nuts securing amplifier unit; move amplifier away from cylinder head.
22 Remove camshaft oil feed banjo bolt; discard washers. 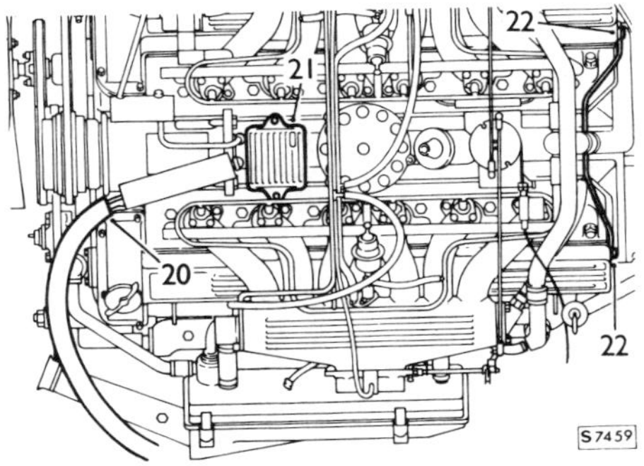
23 Remove three nuts and washers securing cylinder head to timing cover.
24 Progressively slacken cylinder head nuts working from centre outwards.

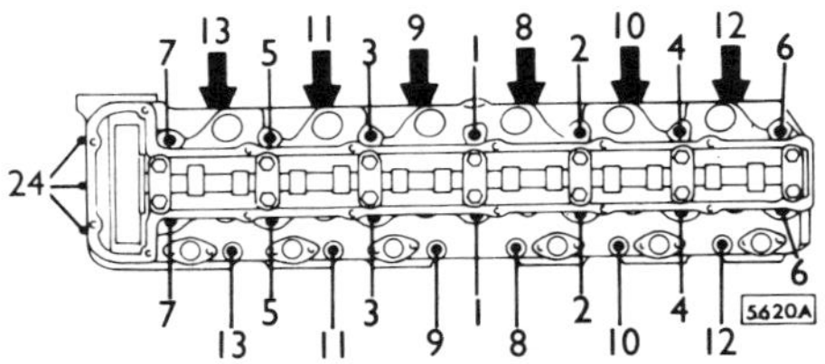

25 Lift off cylinder head and place on blocks of wood. This prevents damage to valves which, when open, protrude below cylinder head face.
CAUTION: Do not remove valve timing gauge from 'A' bank camshaft.

Refitting
26 Rotate 'B' bank camshaft until timing gauge C.3993 can be fitted in slot in camshaft. 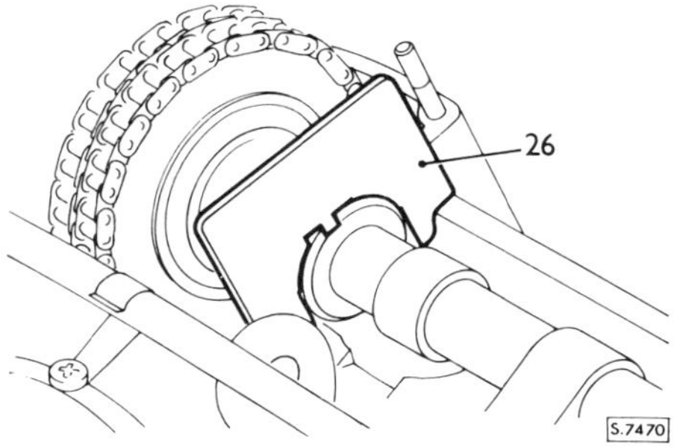
27 Ensure mating surface of cylinder head and block are clean.
28 Fit gasket, ensuring side marked 'TOP' is uppermost. DO NOT use jointing compound or grease.
29 Fit cylinder head and retaining nuts.
30 Tighten retaining nuts in order shown to a torque of:
27.5 lb. ft (3,7 kg. m) for $\frac{3}{8}$ in nut.
52 lb. ft (7,2 kg. m) for $\frac{7}{16}$ in nuts.
31 Tighten cylinder head to timing cover nuts to a torque of 9 lb. ft (1,2 kg. m).

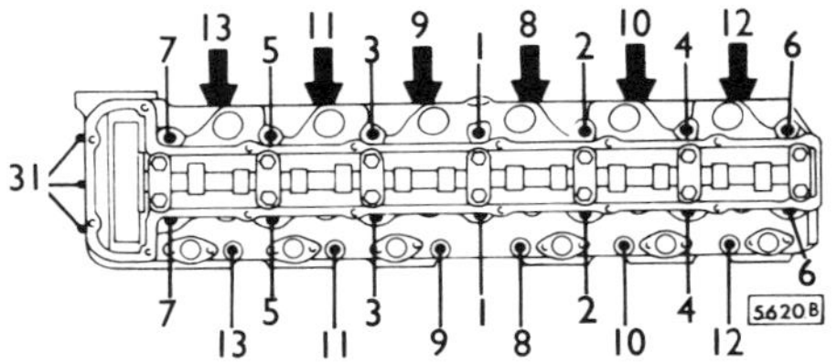

32 Remove sprocket retaining tool, JD.40, and check alignment of retaining bolt holes. If camshaft and sprocket holes are not in alignment, remove circlip retaining camshaft coupling to sprocket and disengage coupling from splines. Press sprocket on to camshaft shoulder.
33 Rotate coupling until access to retaining bolt holes is obtained.
34 Bolt coupling to camshaft using new tab washers.
35 Refit circlip and remove gauges C.3993.

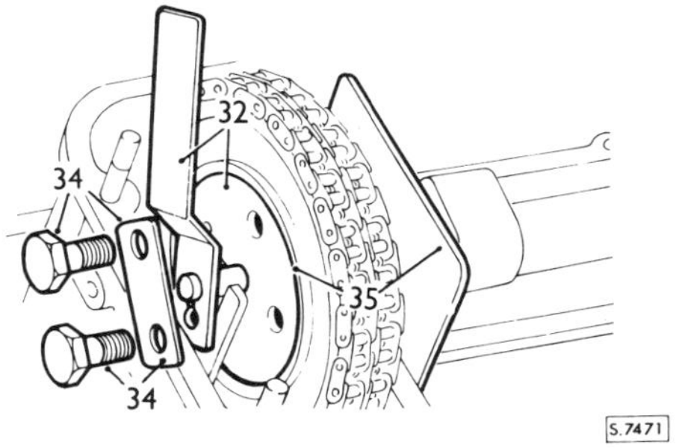

36 Rotate engine until remaining camshaft sprocket retaining bolts can be fitted; secure bolts with tab washers.
37 Reverse operations 1 to 7, 9 to 11 and 13 to 22; use new gaskets, sealing rings and washers.
38 Adjust fan belt tension, see 26.20.01.
39 Check ignition timing, see 86.35.29.

CARS FITTED WITH EMISSION CONTROL ONLY
40 If components mentioned in Group 17 have been disturbed, the appropriate emission test MUST be carried out.

12—5

CYLINDER HEAD

Remove and refit – Right hand 'A' bank — 12.29.12

Service tools: Timing chain tensioner retractor tool and release tool JD.50; Camshaft sprocket retaining tool JD.40; Cylinder liner retaining tool JD.41; Valve timing gauge C.3993.

Removing

1. Disconnect battery, see 86.15.20.
2. Remove right hand camshaft cover, see 12.29.44.
3. Remove fan and torquatrol unit, see 26.25.21.
4. Slacken clips securing hose to thermostat housing and coolant cross over pipe.
5. Remove bolts securing thermostat housing, lifting eyes and auxiliary air valve to cylinder head.
6. Move housing clear of cylinder head, remove and discard gaskets.
7. Slacken nuts locking air pump/auxiliary pulley adjusting bolt.
8. Remove screws securing adjusting bolt bracket to cylinder head.

CARS FITTED WITH AIR CONDITIONING ONLY

9. Remove compressor securing bolts, swing disengage drive belt from pulley, swing compressor away from right hand cylinder head.

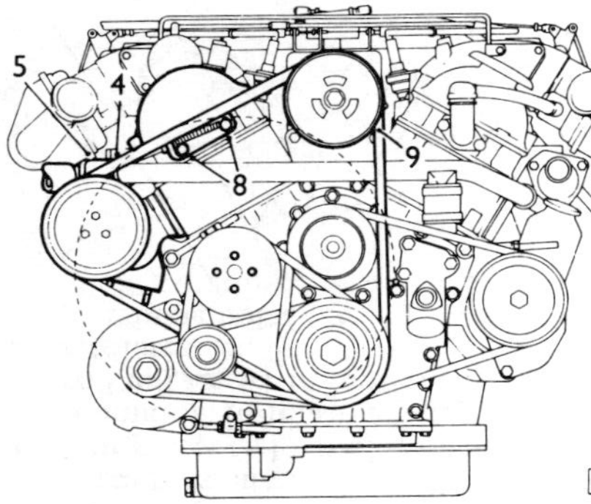

10. Rotate engine, using crankshaft damper nut, until timing gauge C.3993 can be fitted in slot in camshaft front flange.

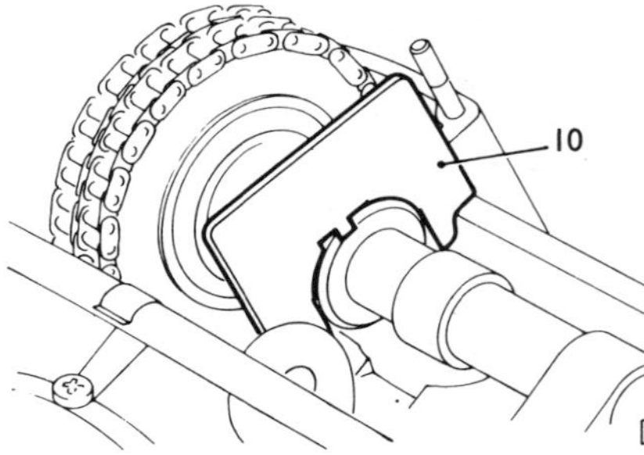

NOTE: On cars fitted with emission control, it will be necessary to disconnect E.G.R. pipe.

11. Remove rubber grommet from timing cover.
12. Insert blade of release tool through hole to release locking catch on timing chain tensioner.
13. Using special tool JD.50 retract timing chain tensioner. Locking catch will engage when tensioner is correctly retracted.
14. Disconnect camshaft sprocket from camshaft, fit sprocket retaining tool JD.40.
15. Remove setscrews securing heat shield to exhaust manifold.
16. Disconnect down pipes from exhaust manifold. Remove and discard sealing rings.
17. Remove screws securing right hand heat shield.
18. Remove starter motor securing bolts.
19. Remove starter motor heat shield securing bolts.
20. Manoeuvre starter motor and heat shield away from engine.
21. Remove rear manifold securing nuts; withdraw manifold remove and discard gasket.
22. Remove camshaft oil feed banjo bolt, discard washers.
23. Remove three nuts and washers securing front of cylinder head to timing cover. Move cable clips from studs.
24. Progressively slacken cylinder head nuts working from centre outwards.
25. Lift off cylinder head and place on blocks of wood. This prevents irreparable damage to valves which, when open, protrude below cylinder head face.
   CAUTION: Do not rotate engine until cylinder liner retaining tools JD.41 have been fitted to cylinder head studs.

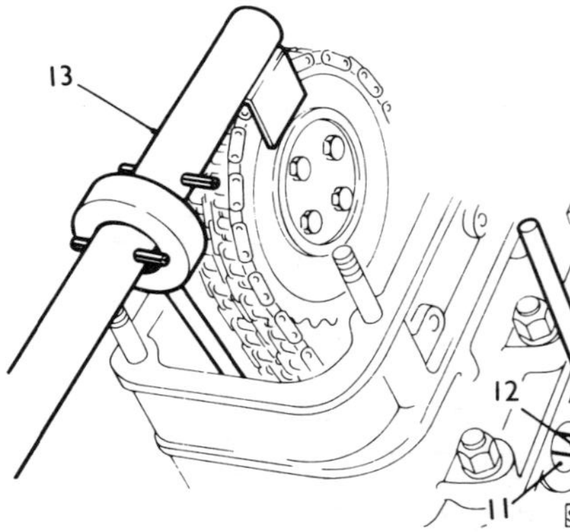

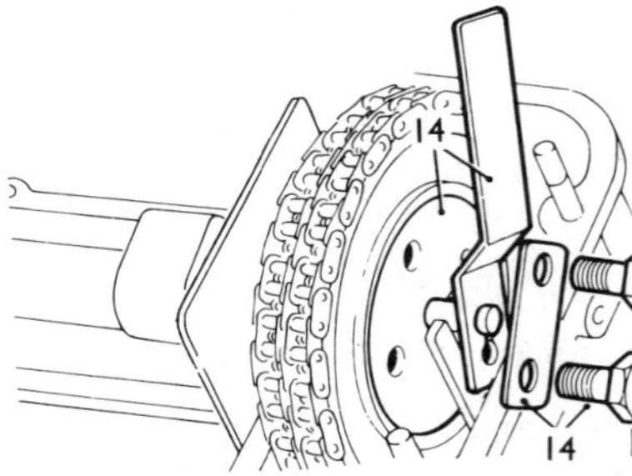

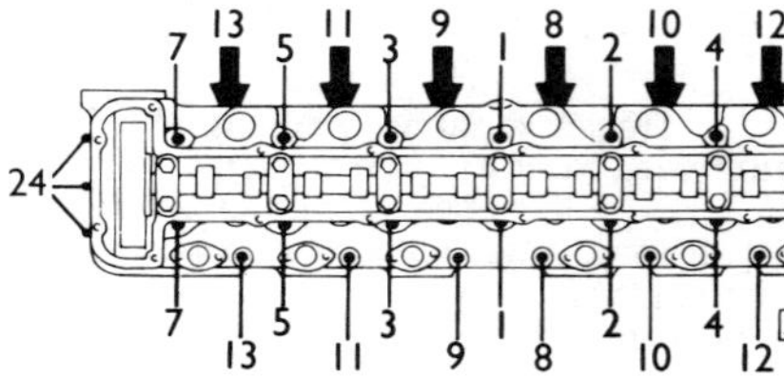

NOTE: If the cylinder head has been removed for the purpose of changing the gasket, and neither crankshaft nor camshaft are moved, continue with item 31. If either shaft is moved accidentally, or if cylinder head overhaul and/or piston de-carbonisation has been carried out, continue with item 26.

Refitting

26. Remove distributor cover.
27. Attach a suitable clock gauge to cylinder head stud.
28. Rotate engine and by means of clock gauge, set number one piston 'A' bank at T.D.C. firing stroke. Remove gauge.
    NOTE: Timing mark 'No. 1 cyl.' on electronic timing rotor will not be pointing directly at pick up module but should be 5° from coincidence.
29. Turn camshaft until valve timing gauge C.3993 can be fitted into slot in camshaft front flange.
30. Remove cylinder liner retaining tools JD.41 from cylinder block.
    CAUTION: Do not rotate engine until cylinder head is fitted.
31. Ensure mating surfaces of cylinder head and block are clean.
32. Fit gasket, ensuring side marked 'TOP' is uppermost. DO NOT use jointing compound or grease.
33. Fit cylinder head and retaining nuts.
34. Tighten retaining nuts in order shown to torque of
    27.5 lb. ft (3,7 kg. m) for ⅜ in nut.
    52 lb. ft (7,2 kg. m) for ⁷⁄₁₆ in nuts.
35. Tighten cylinder head to timing cover nuts to a torque of 9 lb. ft (1,2 kg. m).

12–6

36 Remove sprocket retaining tool, JD.40, and check alignment of retaining bolt holes. If camshaft and sprocket holes are not in alignment, remove circlip retaining camshaft coupling to sprocket and disengage coupling from splines. Press sprocket on to camshaft shoulder.

37 Rotate coupling until access to retaining bolt holes is obtained.

38 Bolt coupling to camshaft using new tab washers.

39 Refit circlip and remove gauge C.3993.

40 Rotate engine until remaining camshaft sprocket retaining bolts can be fitted; secure bolts with tab washers.

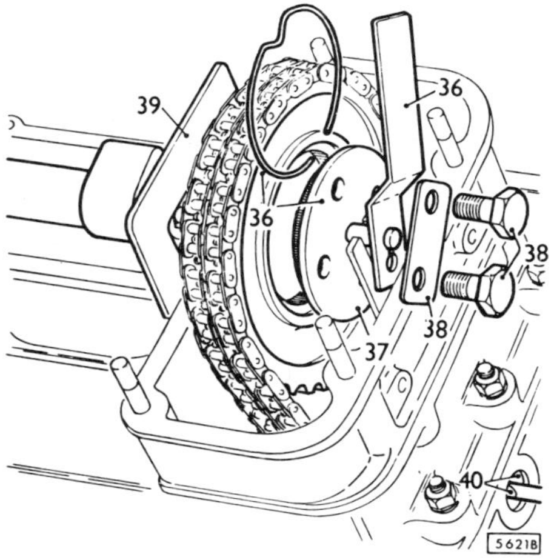

41 Refit camshaft oil supply banjo bolt using new sealing washers.

42 Reverse operations 1 to 9, 11 to 13 and 15 to 21. Use new gaskets and sealing rings.

43 Adjust fan belt tension, see 26.20.01.

44 Check compressor/air pump drive belt tension —CARS FITTED WITH AIR CONDITIONING/EMISSION CONTROL ONLY, see 82.10.01.

45 Check ignition timing, see 86.35.29.

CARS FITTED WITH EMISSION CONTROL ONLY

46 If components mentioned in Group 17 have been disturbed, the appropriate emission test MUST be carried out.

CYLINDER HEAD

Overhaul   12.29.18

Service tool: Valve spring compressor J.6118B and adaptor J.6118C-2.

1 Support valves by means of wooden block.

2 Compress valve spring using Service Tool J.6118B and adaptor J.6118C-2. Retrieve collars, cotters and spring retaining plates.

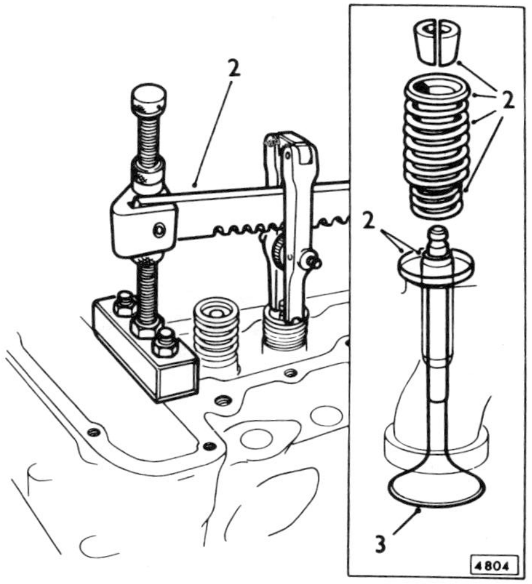

3 Remove valves ensuring that they can be fitted in their original guides.

4 Remove all traces of carbon from cylinder head and deposits from induction and exhaust ports. Great care must be taken to avoid damaging head, use worn emery cloth and paraffin only.

5 Check clearance between valve guide and stem; this should be 0.001 in to 0.004 in (0,025 mm to 0,10 mm). Should guides be worn, proceed as follows:—

6 Immerse head in boiling water for 30 minutes.

7 Using a piloted drift, drive the guide out of head from combustion chamber end.

8 Coat new valve guide with graphite grease and refit circlip.

9 Heat cylinder head.

10 Using a piloted drift, drive in guide from top until circlip is seated in groove.

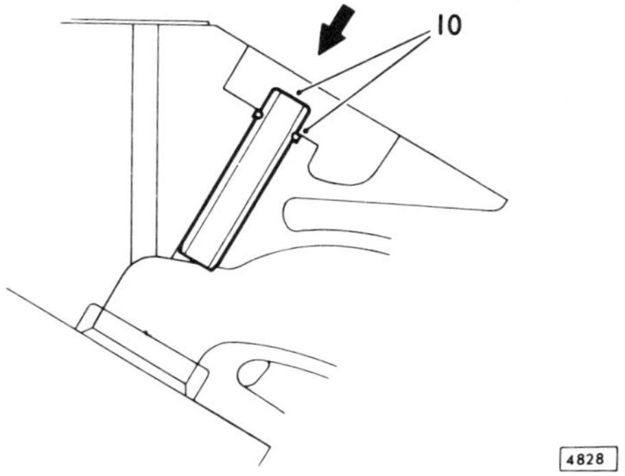

11 Examine valve seat inserts for pitting or excess wear. If renewal is necessary, proceed as described at items 12 to 16.

12 Remove inserts by machining, leaving approximately 0.010 in (0,025 mm) of metal which can easily be removed by hand without damaging cylinder head.

13 Measure diameter of insert recess in cylinder head.

14 Grind down outside diameter of new insert to a dimension 0.003 in (0,08 mm) larger than insert recess.

15 Heat cylinder head for half an hour from cold at a temperature of 150°C (300°F).

16 Fit insert ensuring that it beds evenly in the recess.

17 Renew or reface valves as necessary. Correct valve seat angles are:

Inlet   Exhaust
44½ deg.   44½ deg.

18 Check valve stems for distortion or wear, renew valves with stems worn in excess of 0.003 in (0,08 mm) see group 05.

19 Using a suitable suction tool, grind the valves into their respective seats.

20 If new valve inserts have been fitted, the clearance between valve stem and cam must be checked; this should be 0.320 in (8,13 mm) plus the valve clearance. The dimension must be taken between valve stem and back of cam. Should this dimension not be obtained, metal must be ground from valve seat of insert.
NOTE: Only suitable grinding equipment should be used.

21 Fit valves and place cylinder head on wooden blocks.

22 Fit valve spring seats, inlet valve guide oil seals, springs and collars.

23 Compress springs using Service Tool No. J.6118B and adaptor J.6118 C-2; insert split cotters.

DATA

Replacement guides are available in two sizes and have identification grooves machined in the shank, sizes are as follows:

First oversize (2 grooves) 0.507 in to 0.506 in (12,88 mm to 12,85 mm) diameter.
Second oversize (3 grooves) 0.511 in to 0.512 in (13,00 mm to 12,98 mm) diameter.

When new guides are to be fitted, they should always be one size larger than the old guide. Cylinder head bores will require reaming as follows:

1st oversize – two grooves
0.505 in + 0.0005 in – 0.0002 in.
(12,83 mm + 0,012 mm – 0,005 mm).

2nd oversize – three grooves
0.510 in + 0.0005 in – 0.0002 in.
(12,95 mm + 0,012 mm – 0,005 mm).

12–7

CAMSHAFT COVER

Remove and refit – Left hand 12.29.43

Removing
1. Remove left hand induction manifold, see 30.15.02.
2. Disconnect cable from coolant temperature sensor.
3. Move harness clear of cover.
4. Slacken clip securing cross pipe hose to auxiliary air valve; disconnect hose.
5. Remove nuts, bolts and washers securing cover. Lift off cover; remove and discard gasket and neoprene sealing plug.

Refitting
6. Reverse operations 1 to 5, use new gasket and neoprene sealing plug.

CAMSHAFT COVER

Remove and refit – Right hand 12.29.44

Removing
1. Depressurise air conditioning system (if fitted), see 82.30.05.
2. Remove right hand induction manifold, see 30.15.03.
3. Remove bolt and spring washer securing compressor silencer to compressor.
4. Slacken but do not remove bolt clamping silencer union to compressor.
5. Withdraw union, discard sealing ring and plug open connections in union and compressor.
6. Remove nuts, bolts and washers securing cover. Lift off cover; remove and discard gasket and neoprene sealing plug.

Refitting
7. Reverse operations 1 to 6. Use new gasket, neoprene sealing plug and silencer union to compressor sealing ring.

TAPPETS

Adjust 12.29.48

1. Ensure valve adjusting pads are fitted; fit tappets to their respective valves.
2. Fit camshaft, bearing caps, washers and nuts.
3. Tighten bearing cap nuts evenly to a torque of 9 lb. ft (1,2 kg. m).
4. Check and record clearance between each tappet and heel of each cam. For correct clearance see group 5.
5. Subtract appropriate valve clearance from dimension obtained and select suitable adjusting pads which equal this new dimension. Adjusting pads are available rising in 0.001 in (0,03 mm) sizes from 0.085 in to 0.110 in (2,16 mm to 2,79 mm) and are etched on the surface with letter 'A' to 'Z' each letter indicating an increase in size of 0.001 in (0,03 mm).
6. Remove camshaft and tappets.
7. Fit adjusting pads.

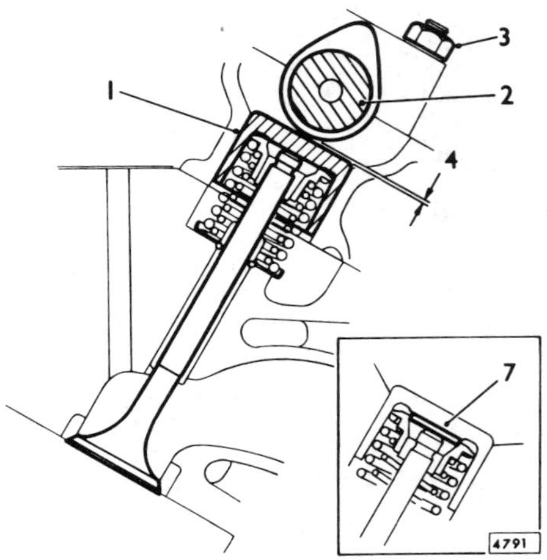

SPARK PLUG INSERTS 12.29.78

Fitting

1. Remove cylinder head, see 12.29.11 or 12.
2. Remove inlet and exhaust valves, see 12.29.62.
3. Bore out stripped thread to 0.750 in (19,05 mm) diameter and tap out to 16 UNF – 2B – Dimension 'A'.
4. Counterbore to 0.95 in (24,13 mm) – Dimension 'B' and Dimension 'C' = 0.875 in (22,23 mm).
   Dimension 'D' = 0.560 in to 0.570 in (14,22 mm to 14,48 mm).
   Dimension 'E' = 0.425 in (10,78 mm).
   Dimension 'F' = 0.575 in (15,75 mm).
   Dimension 'G' = 0.465 in to 0.470 in (11,81 mm to 11,94 mm).
5. Fit screwed insert ensuring that it sits firmly at bottom of thread.
6. Drill and ream a 0.125 in (3,17 mm) diameter hole 0.19 in (2,83 mm) deep between side of insert and head.
7. Drive in locking pin and secure by peening edge of insert and locking pin. Dimension 'A' = 0.40 in (10,16 mm).
8. Reverse operations 1 to 3.

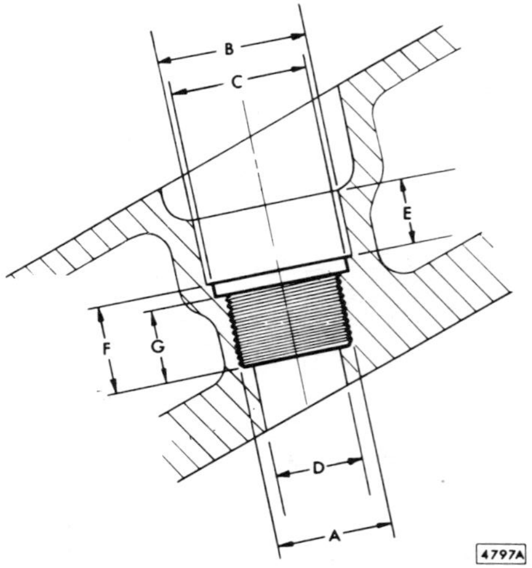

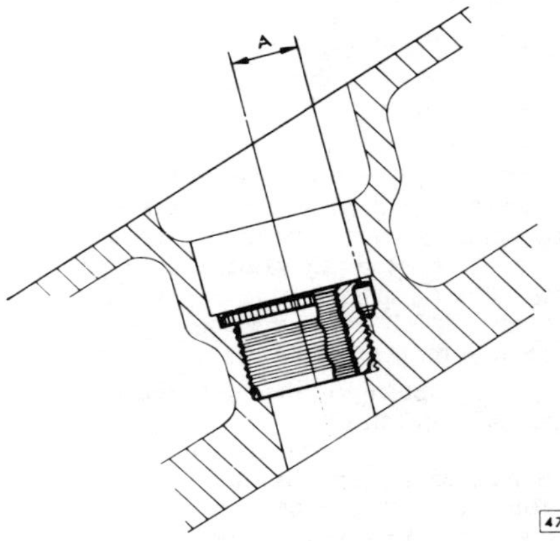

ENGINE AND GEARBOX ASSEMBLY

Remove and refit   12.37.01

Service tool: Engine support tool MS.53A.

Removing

NOTE: Car must be on ramp or over pit.

1. Remove bonnet and lower grille, see 76.16.01 and 76.55.06.
2. Disconnect battery.
3. Drain coolant, see 26.10.01.
4. Depressurise fuel system, see 19.50.02.

CARS FITTED WITH AIR CONDITIONING ONLY

5. Depressurise air conditioning system, see 82.30.05.
6. Disconnect hoses from fuel cooler and air conditioning unit. IMMEDIATELY seal all broken connections using clean, dry plugs.

ALL CARS

7. Remove Pozidriv screws securing rear of left-hand wing valance stay.
8. Remove bolts, spring and plain washers securing front of stay; withdraw stay together with air conditioning hose.
9. Release clips securing harness to right-hand wing valance stay.
10. Remove Pozidriv screw securing rear of stay.
11. Remove bolt, spring and plain washers securing front of stay; withdraw stay.
12. Disconnect cables from ambient and coolant temperature sensors.
13. Release clips securing cold start relay harness to left hand fuel rail and cross over pipe.
14. Disconnect cable from throttle switch and trigger unit.
15. Remove cables from kickdown switch.
16. Remove bolt securing earth leads to right-hand induction manifold.
17. Release clips securing main harness, swing harness clear of engine.
18. Remove screws and washers securing clips retaining harness to right hand wing valance; withdraw harness.
19. Disconnect harness at block connector.
20. Remove two self tapping screws securing relay cover on right hand wing valance; lift off cover.
21. Note cable connections at cold start injection relay, disconnect cables; manoeuvre harness through valance bracket.
22. Disconnect right hand water rail feed pipe from heater valve.
23. Disconnect brake vacuum pipes from left and right hand induction manifolds.
24. Disconnect heater vacuum pipe from right-hand induction manifold.
25. Disconnect throttle cable from throttle pedestal.
26. Disconnect starter solenoid and starter motor feed cables from bulkhead connector.
27. Release clips; withdraw left and right hand air cleaner covers and elements.
28. Release clips securing pressure sensor feed pipe.
29. Disconnect pipe from tee piece; plug broken connections to prevent ingress of dirt.
30. Release clips securing fuel pipes to filter; withdraw pipes; plug broken connections to prevent ingress of dirt.

CARS FITTED WITH AIR CONDITIONING ONLY

31. Release clip securing pipe to silencer; withdraw pipe. IMMEDIATELY seal all broken connections using dry, clean plugs.

ALL CARS

32. Release clip securing fuel pipe to cooler, withdraw pipe; plug broken connections to prevent ingress of dirt.
33. Release clips securing left and right hand radiator top hoses to thermostat housings; disconnect hoses.
34. Release clip securing header tank hose; disconnect hose.
35. Release clip securing heater return hose to tee-piece; disconnect hose.

CARS FITTED WITH AUTOMATIC TRANSMISSION ONLY

36. Release clips securing transmission oil cooler hoses to pipes, withdraw hoses; plug broken connections to prevent ingress of dirt.

ALL CARS

37. Disconnect snap connectors securing alternator harness.
38. Disconnect two pin connector from thermostatic switch.
39. Release clips securing harness.
40. Remove nut and washer securing fan motor line fuse retaining clip. Withdraw clip; disconnect fuse at snap connector.
41. Disconnect smaller of two block connectors at front of right hand wing valance.
42. Remove banjo bolt and washer securing expansion pipe to radiator.
43. Remove bolts, spring and plain washers securing radiator top rail noting fitted positions of earth lead terminals and harness clip.

CARS FITTED WITH AIR CONDITIONING ONLY

44. Disconnect outlet pipe from receiver drier unit. IMMEDIATELY plug broken connections with clean, dry plugs.

ALL CARS

45. Remove self locking nuts and plain washers securing fan cowl to top rail.
46. Remove nut, bolt, washer and spacers securing front stays to right hand bracket.
47. Remove nut, bolt and washers securing front stay to left hand bracket.
48. Lift off radiator top rail together with evaporator and receiver drier units.
49. Release condenser pipe from clip.
    Recover two copper washers and spacer from top of radiator.

NOTE: Washers and spacer are located over tapping for expansion pipe banjo bolt.

continued

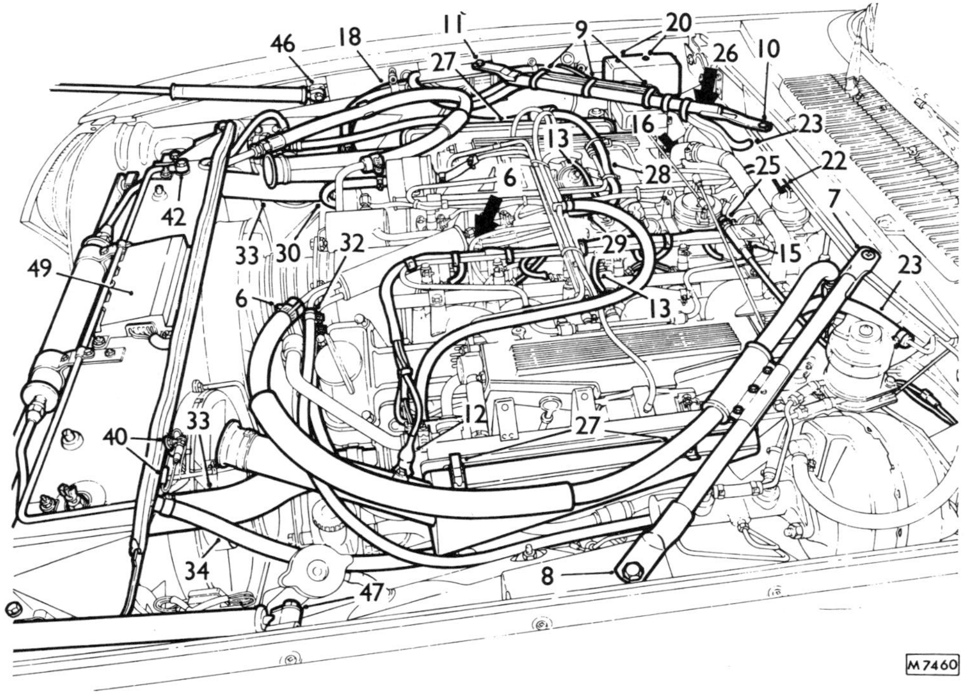

12–9

50 Fit engine support tool MS.53A.
51 Disconnect cable from low coolant sensor unit located in right hand side of radiator.
52 Remove nuts securing bottom of fan cowl to radiator.
53 Slacken clip, disconnect bottom hose.
54 Lift radiator slightly to facilitate access to oil cooler connections.
55 Disconnect oil cooler pipes; plug broken connections to prevent ingress of dirt.
CAUTION: When carrying out this operation, two spanners must be used i.e. one on hose union and one on oil cooler union.
56 Carefully lift radiator, fan cowl and electric fan out of car.
57 Disconnect exhaust intermediate pipes from down pipes.
58 Remove fixings securing left and right hand front heat shields.
59 Remove fixings securing rear heat shields.
60 Remove fixings securing intermediate heat shields.
61 Remove nuts securing collision plate to transmission unit.
62 Position jack and suitably formed block of wood beneath rear engine mounting plate.
63 Remove self locking nut and plain washer securing collision plate to rear engine mounting stud; withdraw plate.
64 Remove fixings securing rear engine mounting plate noting fitted positions of spacers (if fitted) also long and short bolts.
65 Lower jack, recover mounting plate, spacers, washers and spring, seating washer and rubber.
66 Remove stud and seating plate.
67 Remove nuts, bolts and spring washers securing propeller shaft to output flange.

CARS FITTED WITH MANUAL TRANSMISSION ONLY
68 Remove gear knob.
69 Remove screws securing rubber gaiter to console, withdraw gasket and gaiter.
70 Place gear lever in first gear position.

71 Disconnect pipe from clutch slave cylinder or, if fitted, at bracket on bulkhead. Plug broken connections to prevent ingress of dirt.

CARS FITTED WITH AUTOMATIC TRANSMISSION ONLY
72 Remove bolt, spring and plain washers securing selector cable to trunnion block.
73 Remove self locking nut securing selector lever to shaft; slide lever off shaft.

ALL CARS
74 Disconnect speedometer drive cable; position cable to side of transmission tunnel.
75 Disconnect earth strap from torque converter/bell housing.

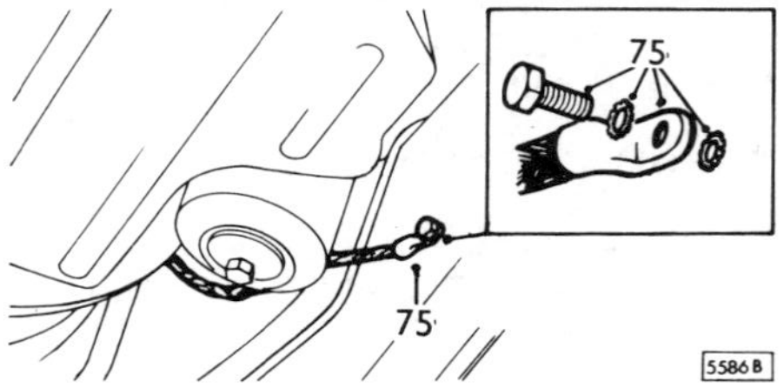

76 Remove bolts and spring washers securing steering gaiter heat shields to mounting brackets; withdraw shields taking care not to damage steering rack gaiters.
77 Remove nuts and plain washers securing heat shield to left hand exhaust manifold.
78 Remove nuts and plain washers securing left hand down pipes to exhaust manifolds, withdraw down pipes, remove and discard sealing rings.
NOTE: On cars fitted with exhaust gas recirculation system, it will be necessary to disconnect E.G.R. pipe from down pipe.
79 Remove nuts and plain washers securing right-hand down pipes to exhaust manifold. Rotate down pipes through 180 degrees to facilitate withdrawal. Remove and discard sealing rings.
NOTE: On cars fitted with exhaust gas recirculation system, it will be necessary to disconnect E.G.R. pipe from down pipe.
80 Remove bolts and spring washers securing power assisted steering oil cooler to mounting bracket; swing cooler away from engine.
81 Remove nuts, bolts and washers securing power assisted steering pump to mounting bracket; move pump away from bracket and release drive belt. Lay pump in an upright position away from engine.
82 Remove car from ramp or from over pit.
83 Support transmission with a suitable trolley jack.
CAUTION: On cars fitted with automatic transmission, jack MUST be located under rear extension housing not under oil pan.
84 Attach suitable lifting chains to front and rear lifting eyes.
NOTE: The removal operation will be facilitated if rear lifting chains are approximately 6.0 in (154 mm) longer than front chains.

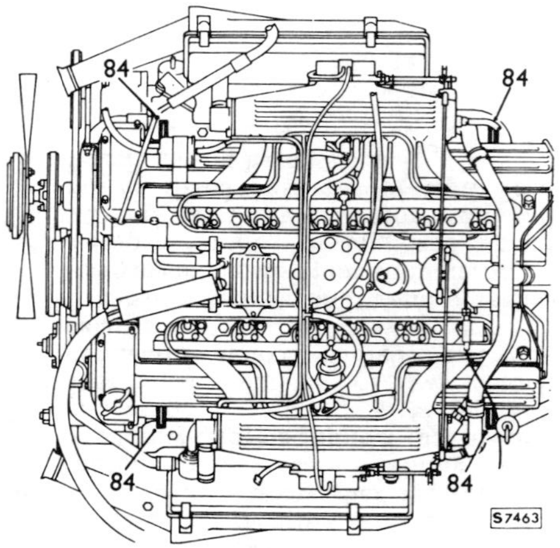

MANUAL TRANSMISSION CARS ONLY
Ensure that gear lever is below console and clear of transmission tunnel before engine is lifted out.
WARNING: MAINTAIN CONTINUAL WATCH ON POWER STEERING RACK, PIPES AND ANY COMPONENTS LIABLE TO FOUL ENGINE.

Refitting
86 Attach suitable lifting chains to front and rear engine lifting eyes, see operation 84.
87 Fit new sealing rings, chamfered side upwards, to exhaust down pipes.
88 Position right hand down pipes in engine compartments.
89 Locate foam pad on top of transmission casing.
90 Lower engine into frame until trolley jack can be positioned beneath transmission.
CAUTION: On cars fitted with automatic transmission, jack MUST be located under rear extension housing not under oil pan.
91 Continue lowering engine and raising jack until engine can be located on mounting studs. Refit nuts and washers.
WARNING: MAINTAIN CONTINUAL WATCH ON POWER STEERING RACK, PIPES AND ANY COMPONENTS LIABLE TO FOUL ENGINE.
92 Fit engine support beam and remove trolley jack.
93 Loosely secure right hand down pipe to exhaust manifold. DO NOT attempt to fully tighten nuts at this stage.
94 Place car on ramp or over pit.
95 Tighten right hand down pipe securing nuts by diagonal selection. Refit E.G.R. pipe – CARS FITTED WITH EMISSION CONTROL ONLY.
96 Position fan and cowl assembly in engine compartment.
97 Refit power assisted steering pump.
98 Tension steering pump drive belt, see 57.20.01.

85 Withdraw engine taking care to ensure that engine does not foul or damage components.

12–10

99 Refit radiator and reconnect oil cooler pipes.
CAUTION: When carrying out this operation, two spanners must be used i.e. one on hose union and one on oil cooler union.
100 Loosely fit fan cowl to radiator.
101 Reconnect propeller shaft.
102 Reconnect speedometer drive cable.

CARS FITTED WITH AUTOMATIC TRANSMISSION ONLY
103 Place selector lever in 'D'.
104 Move gearbox selector lever until selector cable can be refitted. Refit trunnion block securing bolt.

ALL CARS
105 Remove any traces of Loctite from threaded portion of rear engine mounting stud.
106 Smear threads of stud with Loctite Grade AV. Screw stud and seating washer into transmission.
107 Reassemble rear engine mounting. DO NOT fit collision plate at this stage.
108 Position jack and suitably formed block of wood beneath rear engine mounting.
109 Raise jack until rear mounting can be bolted to body. Tighten securing bolts by diagonal selection.
110 Remove jack, remove self locking nut from centre stud and refit collision bracket.
111 Refit left hand exhaust down pipe; tighten nuts by diagonal selection. Refit E.G.R. pipe – CARS FITTED WITH EMISSION CONTROL ONLY.
112 Refit front, intermediate and rear heat shields.
113 Refit earth strap.
114 Reconnect left and right hand intermediate pipes to down pipes. Smear joints with Holts Firegum.
CAUTION: Tighten securing clamp nuts by diagonal selection; DO NOT overtighten.
115 Refit power steering bellows heat shields.
116 Refit power assisted steering oil cooler.
117 Reconnect bottom radiator hose.
118 Secure fan cowl to mounting brackets.

119 Remove engine support beam.
120 Secure lower end of angled front strut.
121 Refit top rail assembly ensuring that copper washers and spacers are not displaced.
NOTE: A thin rod passed through above components will assist in this operation.
122 Fit banjo bolt, pipe and sealing washer.
123 Refit earth lead.
124 Fit and tighten top rail securing bolts.
125 Refit upper cross strut and upper end of angled cross strut.

CARS FITTED WITH AIR CONDITIONING ONLY
126 Reconnect receiver drier unit following procedures in Group 82.

ALL CARS
127 Secure fan cowl line fuse and harness clip, reconnect harness at snap connector.
128 Reconnect otter switch at water pump.
129 Reconnect heater and expansion hoses.

CARS FITTED WITH AIR CONDITIONING ONLY
130 Reconnect air conditioning hoses following procedures in Group 82.

ALL CARS
131 Reconnect radiator top hoses.
132 Reconnect all fuel pipes.
133 Reconnect all vacuum hoses.
134 Refit air cleaner elements and covers.
135 Reconnect alternator.
136 Refit throttle cable.
137 Refit valance stays.
138 Relocate and connect all harnesses.
139 Refill cooling system, see 26.10.01.
140 Check/top up engine and transmission oil.

CARS FITTED WITH AIR CONDITIONING ONLY
141 Recharge air conditioning system, see 82.30.08.

CARS FITTED WITH AUTOMATIC TRANSMISSION ONLY
142 Adjust kickdown switch, see 44.30.09.

CARS FITTED WITH MANUAL TRANSMISSION ONLY
143 Refit rubber boot, gasket and gear knob.
144 Bleed clutch, see 35.15.01.

ALL CARS
145 Refit bonnet and lower grille.
146 Reconnect battery.
147 Run engine and check for leaks.
148 Bleed power assisted steering by turning steering from lock to lock with engine running. Top up reservoir.
149 Recheck coolant level also engine and transmission oil levels.

CARS FITTED WITH EMISSION CONTROL ONLY
150 If components mentioned in Group 17 have been disturbed, the appropriate emission test MUST be carried out.

ENGINE

Dismantle and reassemble 12.41.05

NOTE: All instructions unless otherwise stated, apply to both 'A' bank right hand and 'B' bank left hand cylinder head assemblies.

Service tools: Camshaft sprocket retaining tools JD.40; Cylinder liner retaining tools JD.41; Jack shaft retaining tool JD.39 and adaptor JD.17B; Piston ring clamp 38U.3; Valve seal presizing gauge C.3993; Special screwdriver JD.42.2; Timing chain tensioner retractor tool JD.44.

Dismantling

1 Drain lubricating oil sump.
2 Remove engine and gearbox, see 12.37.01.
3 Remove induction manifolds – left hand 30.15.02; right hand 30.15.03.
4 Remove exhaust manifolds, see 30.15.10 – 30.15.11.

5 Remove gearbox, see 44.20.01 or 37.20.01.
6 Remove hose clip securing oil pressure relief valve bleed pipe to top of sandwich plate.
7 Remove setscrews securing clips on oil cooler feed pipe.
8 Disconnect oil cooler feed pipe at pipe union.
9 Release through bolt at base of oil filter bowl and remove filter element (early cars) or unscrew and discard filter element (later cars).
10 Remove three setscrews securing by-pass valve union to sandwich plate.
11 Remove four setscrews securing filter head to cylinder block.
12 Remove torque converter, see 44.17.07 or clutch, see 33.10.01.
13 Release oil cooler suction pipe union nut.
14 Slacken fan belt and remove setscrews and nuts securing fan belt idler pulley assembly. Remove assembly.
15 Remove power steering pump.
16 Remove power steering pump mounting bracket and adjustment link.
17 Remove four setscrews securing compressor – CARS FITTED WITH AIR CONDITIONING ONLY.
18 Remove compressor mounting bracket – CARS FITTED WITH AIR CONDITIONING ONLY.
19 Remove air pump if fitted.
20 Remove air pump mounting bracket and adjustment link.
21 Release transmission oil cooler pipe clips and remove pipes from engine.

continued

12–11

22 Remove starter motor.
23 Remove four bolts securing crankshaft pulley to damper, withdraw pulley.
24 Remove damper bolt, strike damper sharply with hide mallet; withdraw damper. Recover Woodruff key. 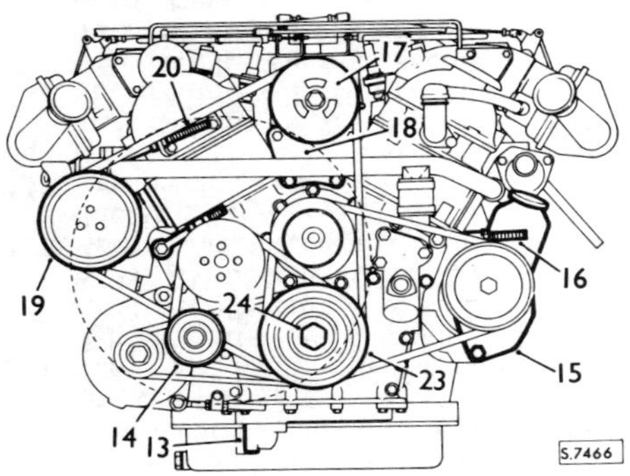
25 Remove damper cone. Recover Woodruff key.
26 Remove setscrews and washers retaining water inlet spout to water pump. Carefully break seal.
27 Remove setscrews, studs and bolt retaining water pump. Carefully break seal and remove water pump and backplate assembly complete with engine cross pipe. 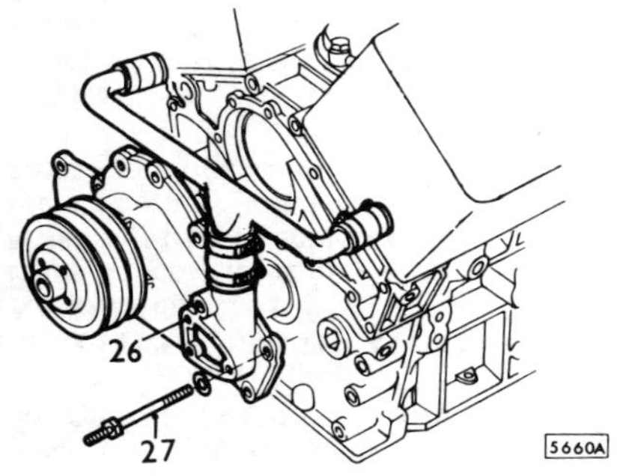
28 Remove distributor and amplifier unit.
29 Remove banjo bolts securing camshaft oil feed pipes to rear of tappet blocks and oil gallery. Remove throttle pedestal, amplifier bracket, heater return pipe and engine cable.
30 Remove alternator and bracket.
31 Remove domed head nuts, copper washers and setscrews securing camshaft covers to cylinder head; lift off covers.
32 Remove and discard gaskets and neoprene plugs. 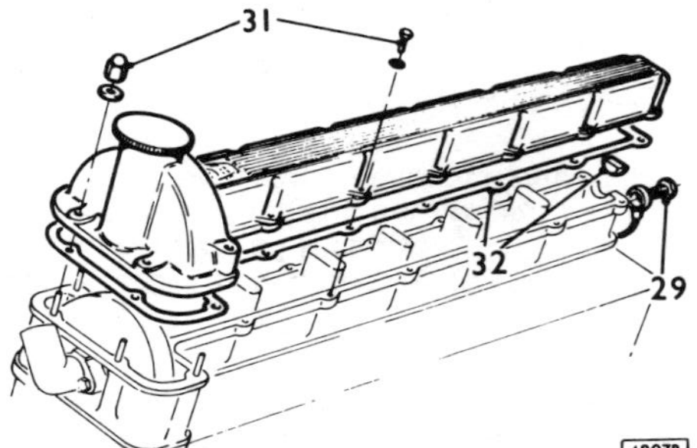
33 Bend back locking tabs securing camshaft sprocket retaining bolts.
34 Remove two bolts from each sprocket. Use timing chain tensioner retractor tool JD.44 to extend tensioner to full extent.
35 Rotate engine until remaining bolts are accessible and remove.
36 Fit retaining tool JD.40 to each sprocket and remove JD.44.
37 Slacken off camshaft bearing cap nuts working from centre outwards.
38 Remove camshafts.
39 Remove nuts and capscrews securing tappet block to cylinder head; remove tappet block together with tappets. 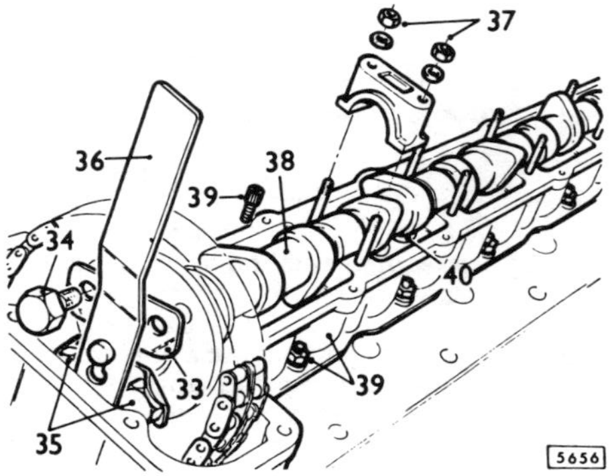
40 Retrieve valve adjusting pads. Note location.
41 Progressively slacken cylinder head nuts, working from centre outwards.
42 Lift off cylinder heads and place on blocks of wood.
43 Fit cylinder liner retainers JD.41.
44 Remove bolts securing timing cover to cylinder block noting relative positions of long, short and dowelled bolts.
45 Remove timing cover, lift off gaskets and discard.
46 Remove oil seal from cover and discard.
47 Withdraw spacer from crankshaft.
48 Move chain tensioner clear of locating bracket and slide off dowel pin.
49 Disengage timing chain from sprockets and remove.
50 Remove bolts securing jackshaft cover to cylinder block, lift off cover. NOTE: Note position of spacing washers and long bolts to assist replacement – non air conditioned cars only. 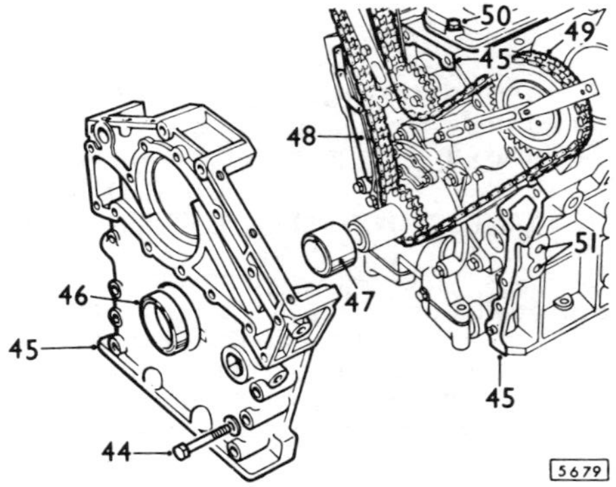
51 Remove engine mounting brackets, invert engine and remove bolts securing oil sump pan. NOTE: Note position of oil cooler pipe clip.
52 Remove bolts securing sandwich plate baffle and sandwich plate; recover crankshaft angle indicator scale.
53 Carefully break joint and remove all traces of gasket. Remove suction union elbow from sandwich plate. 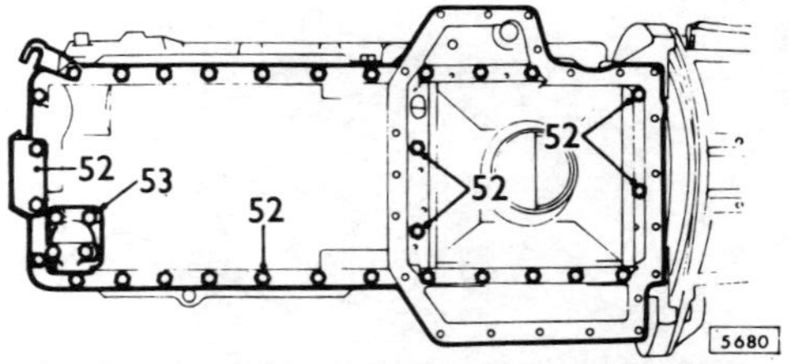
54 Remove four setscrews and washers securing suction pipe clips and bracket. Draw suction pipe from 'O' ring at elbow.
55 Remove two setscrews and washers securing crankshaft undershield and delivery pipe clips. Draw undershield clear. 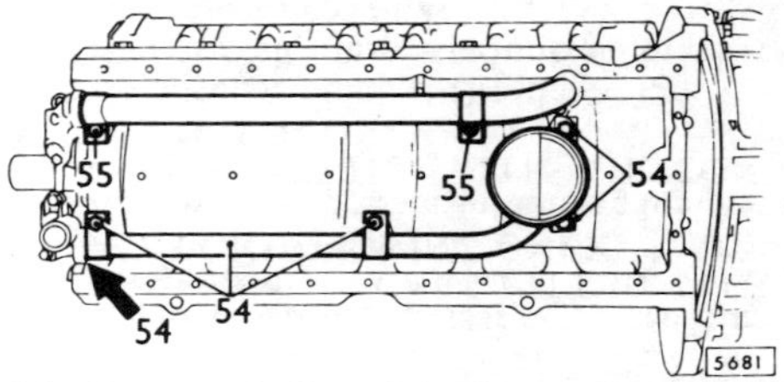
56 Remove bolt and washer and nut, bolt and washer securing oil delivery pipe elbow to oil pump casting. Lift oil delivery pipe from crankcase. Remove and discard gasket.
57 Withdraw crankshaft sprocket and Woodruff key.
58 Remove four bolts securing oil pump to cylinder block, withdraw pump, drive gear and Woodruff key.
59 Remove setscrews and tabwasher securing sprocket to jackshaft; withdraw sprocket, discard tabwasher.
60 Remove bolts securing jackshaft locking plate to cylinder block; lift plate out of groove in jackshaft flange.

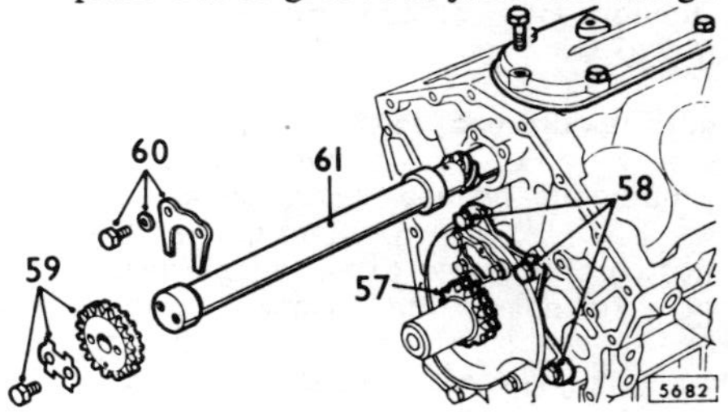

12–12

61 Withdraw jackshaft.
NOTE: Care must be taken to identify pistons with their respective bores, big end caps should be marked 'front' and fitted to connecting rods immediately after removal.
62 Remove nuts securing connecting rod bearing cap; lift off cap together with shell bearing.
63 Remove carbon deposit from top of bore; push connecting rod and piston up cylinder bore and withdraw.
64 Repeat operations 62 and 63 on remaining pistons.
65 Remove bolts on locking plate securing drive plate to crankshaft.
66 Lift off drive plate.
67 Remove small nuts securing main bearing caps, starting from centre bearing.
68 Remove pillar nuts and large nuts securing main bearing caps starting from centre bore.
NOTE: Note location of pillar nuts to assist replacement.
69 Lift off bearing caps and shell, slide rear main bearing casting out of cylinder block, remove and discard seals.
70 Lift crankshaft out of cylinder block, retrieve upper half of main bearing shells.
NOTE: If for any reason cylinder liners are to be removed and re-used, they should be marked 'front' and refitted in their original bore.
71 Remove cylinder liner retaining tools JD.41. 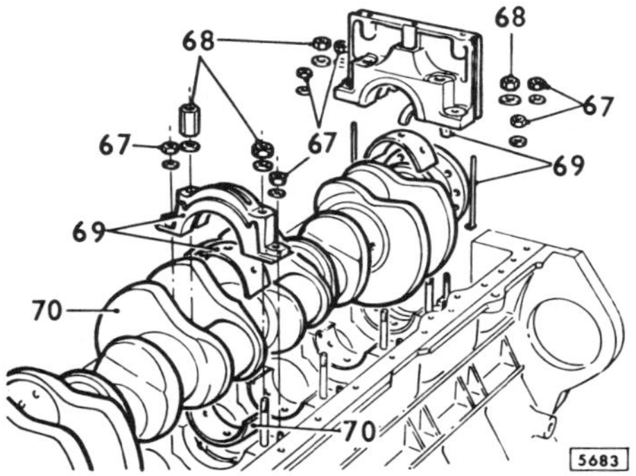
72 Position a suitable mandrel between cylinder liner and press arbor.
73 Press out cylinder liners from below.

Reassembling

CAUTION: Ensure that all components are scrupulously clean, blow out all oil galleries in crankshaft, camshafts etc. with dry clean compressed air.
74 Smear shoulders of cylinder liners with Hylomar and slip them into cylinder block. Remove excess sealant.
NOTE: Cylinder liners must be fitted dry.
75 Ensure liners are correctly seated and fit retaining tools JD.41.
76 Fit new sealing strips to grooves of rear main bearing casting.
77 Fit new crankshaft rear oil seal, applying one drop of red Hermetite into both sealing grooves, top and bottom, before fitting seal halves into grooves.
78 Fit main bearing casting to cylinder block and tighten retaining nuts.
79 Pre-size rear oil seal using Service Tool JD.17B together with adaptor JD.17B-1. 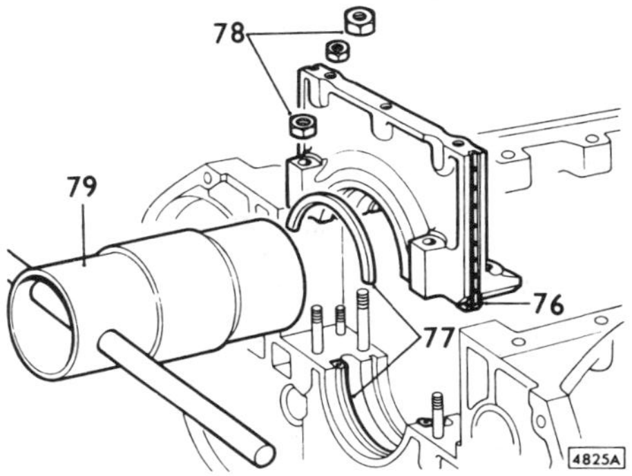
80 Remove rear main bearing casting.
81 Liberally oil upper main bearing shells and fit in cylinder block. Smear rear oil seal with Dag Colloidal Graphite.
CAUTION: Centre and rear main bearing shells must not be confused with each other; rear main bearing shell has an oil groove whilst centre main bearing shell is plain.
82 Liberally oil upper main bearing shells and fit in cylinder block.
83 Position crankshaft in cylinder block.
84 Fit bearing shells to caps; fit caps, using pillar nuts as noted.
85 Tighten securing nuts to a torque of: 27.5 lb. ft (3,7 kg. m) for ⅜ in (9,5 mm) studs. 62.5 lb. ft (8,6 kg. m) for ½ in (12,7 mm) studs.
86 Check crankshaft end float.
87 Select thrust washers which will reduce end float to .004 in to .006 in (,10 mm to ,15 mm). For sizes of thrust washers available see Engine data 05.
88 Remove bearing caps and fit thrust washers selected to groove in block.
NOTE: Grooved side of washers must face outwards.
89 Fit bearing shells to caps, oil shells and crankshaft journals.
90 Fit main bearing caps and nuts; smear oil shells in rear main bearing casting with oil before assembly.
NOTE: Ensure that reference marks on bearing caps face marks on cylinder block.
91 Tighten bearing caps one at a time, working from centre outwards to a torque of: 27.5 lb. ft (3,7 kg. m) for ⅜ in (9,5 mm) studs. 62.5 lb. ft (8,6 kg. m) for ½ in (12,7 mm) studs. 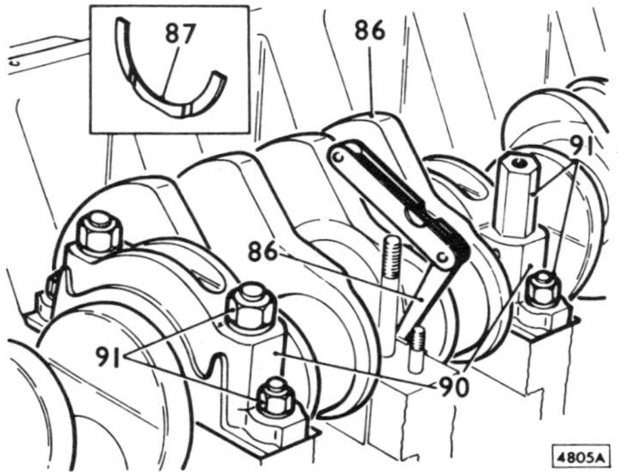
92 Liberally smear bore of number one cylinder with clean engine oil.
93 Ensure that piston ring gaps of number one piston are evenly spaced around circumference of piston. 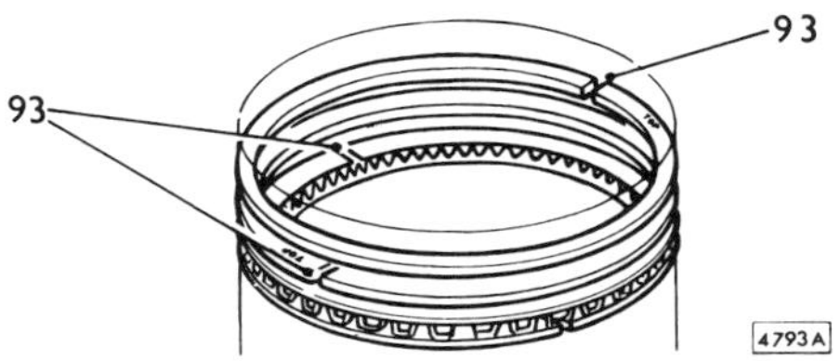
94 Smear piston rings with oil and compress using Service Tool 38U.3. 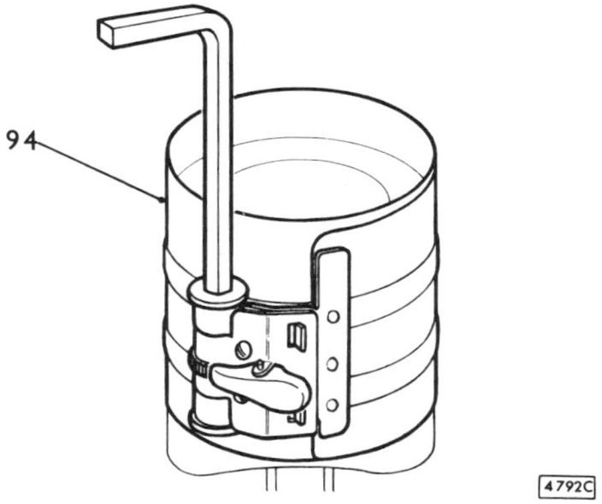
95 Enter piston and connecting rod into top of bore ensuring that 'FRONT' stamped on piston faces forward. DO NOT use undue force when fitting piston.
96 Fit big end bearing shell to connecting rod and bearing cap, ensure that locking tabs on shells are correctly located.
97 Oil shells and crankshaft journal, fit bearing cap ensuring that it is correct way round. Tighten connecting rod nuts to 37.5 lb. ft (5,1 kg. m).
98 Repeat operations 92 to 97 on remaining pistons.
99 Check that engine rotates freely.
100 Refit oil pump to drive gear and secure.
101 Use new 'O' ring seals at both ends of oil delivery pipe.

continued

12—13

ENGINE MOUNTING – FRONT SET   12.45.04

Remove and refit

Removing
1. Disconnect battery, see 86.15.20.
2. Remove air cleaners, see 19.10.01/19.10.02.
3. Remove nuts and spring washers securing top and bottom of both engine mountings.
4. Using chains and a spreader, lift front of engine sufficiently to free mountings.
5. Recover fibre washers.

Refitting
6. Locate replacement mountings with fibre washer on top surface on front suspension cross member brackets.
7. Fit plain nut and washer to secure each mounting to brackets on cross member.
8. Lower engine to locate on mountings. Fit plain nuts and spring washers to secure engine to mountings.
9. Remove chains.
10. Refit air cleaners.
11. Reconnect battery.

ENGINE MOUNTING – REAR CENTRE   12.45.08

Remove and refit

Follow relevant procedures detailed under Engine – Remove and refit – 12.37.01 as necessary.

FLYWHEEL   12.53.07

Remove and refit

Removing
1. Remove clutch, see 33.10.01.
2. Lift locking tabs and remove bolts securing flywheel to crankshaft; discard lock plate.
3. Tap flywheel sharply with a hide mallet; withdraw flywheel.

Refitting
4. Reverse operations 1 to 3 use a new lock plate.
5. Turn up locking tabs.

---

102. Locate crankshaft undershield on pillar nuts, and place oil delivery pipe into position.
103. Loosely secure undershield and delivery pipe using two setscrews and serrated washers.
104. Use new 'O' ring seal at suction elbow and locate suction pipe.
105. Secure suction pipe clips and bracket using four setscrews and serrated washers. Fully tighten all six setscrews securing undershield.
106. Refit drive plate to crankshaft; use new locking plate, see 12.53.13; tighten bolts to 66.5 lb. ft (9,1 kg. m).
107. Refit crankshaft sprocket.
108. If a timing chain guide has been renewed, reset all guides, see 12.65.50.
109. Refit timing chain tensioner ensuring that it is fully retracted.
110. Smear journals of jackshaft with clean engine oil and fit jackshaft.
111. Refit jackshaft locking plate.
112. Refit camshaft sprockets, use retaining tools JD.40.
113. Attach clock gauge to number one 'A' bank cylinder head stud.
114. Turn engine over and by means of clock gauge set number one 'A' piston at T.D.C.
115. Refit jackshaft sprocket and timing chain, ensuring equal amount either side of crankshaft sprocket and jackshaft. Ensure that centre punch marks on sprocket and jackshaft are at 180° and that mark on jackshaft is at top. Fit jackshaft retaining tool JD.39. Tighten bolts and turn up tabs.
116. Remove jackshaft retaining tool JD.39.

CAUTION: Engine must on no account be rotated until camshaft sprockets are coupled to camshafts.

117. Refit timing cover, use new crankshaft oil seal and gaskets, tighten bolts by diagonal selection.
118. Refit crankshaft spacer.
119. If necessary, renew 'O' ring seal in oil pump suction elbow.
120. Fit sandwich plate to crankcase, using new gasket, and secure using setscrews and washers. Secure crankshaft angle indicator scale at front.
121. Fit baffle plate to sandwich plate.
122. Fit oil sump pan, securing oil cooler pipe bracket as noted.
123. Fit suction union elbow to sandwich plate and secure. Use new seal.
124. Remove cylinder liner retaining tools JD.41.
125. Smear mating faces of tappet block and cylinder head with Hylomar.
126. Refit tappet block, tighten nuts and capscrews by diagonal selection working from centre outwards.
127. Refit camshafts.
128. Refit bearing caps ensuring that reference marks correspond, tighten nuts by diagonal selection working from centre outwards to a torque of 9.0 lb. ft (1,2 kg. m).
129. Adjust tappets, see 12.29.48.
130. Fit cylinder head gasket with 'TOP' uppermost. Do not use jointing compound or grease.
131. Turn each camshaft until valve timing gauge C.3993 can be fitted to slot in front flange.
132. Refit cylinder heads, tighten nuts in order shown to a torque of: 27.5 lb. ft (3,7 kg. m) for ⅜ in nuts. 52 lb. ft (7,2 kg. m) for ⅞ in nuts. 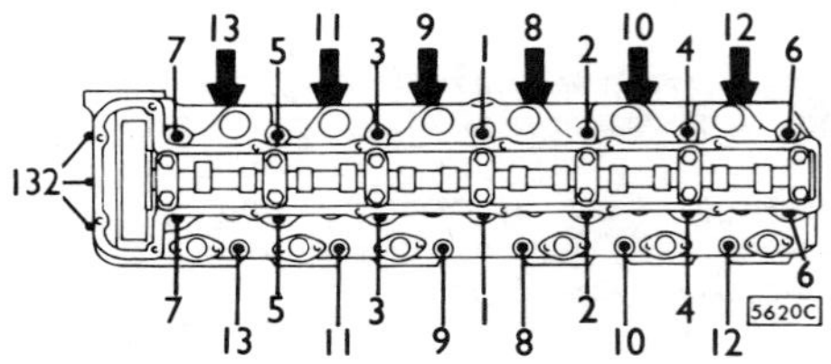
133. Reconnect camshaft oil feed pipe.
134. Remove retaining tools JD.40.
135. Remove circlip retaining camshaft sprocket couplings, press sprocket on to camshaft shoulder. Rotate coupling until two bolt holes align with holes in camshaft.
136. Refit couplings to camshaft sprockets, refit circlip, remove gauge C.3993.
137. Bolt couplings to camshafts on tab washers.
138. Insert screwdriver JD.42-2 through hole in timing cover and release chain tensioner locking catch; refit rubber grommet. 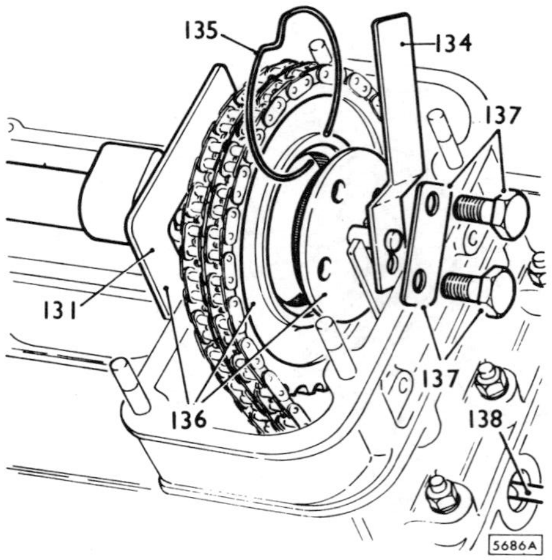
139. Rotate engine until remaining bolt holes in coupling are visible.
140. Fit remaining bolts, secure all bolts with tabwashers.
141. Use new camshaft cover gaskets and neoprene sealing plugs.
142. Refit camshaft cover, tighten bolts to a torque of 8.0 lb. ft. (1,1 kg. m).
143. Refit jackshaft cover.
144. Reverse operations 1 to 30.

NOTE: Tighten crankshaft bolt to 125 lb. ft to 150 lb. ft (17,3 kg. m to 20,7 kg. m.).

145. Check engine timing by means of a stroboscope.

CRANKSHAFT – GENERAL

Following crankshaft removal, journals surface of crankshaft journals, should be checked in accordance with dimensions given in group 04.

NOTE: Due to the extremely hard surface of crankshaft journals, it is not possible to grind crankshafts satisfactorily.

DRIVE PLATE

Remove and refit   12.53.13

Removing

1. Remove torque converter, see 44.17.07.
2. Lift locking tabs and remove bolts securing drive plate to crankshaft. Remove stiffener plate.
3. Draw two locating dowels from drive plate.

CAUTION: If a draw bolt is to be used, turn crankshaft to react bolt against rear main bearing cap, NOT cylinder block.

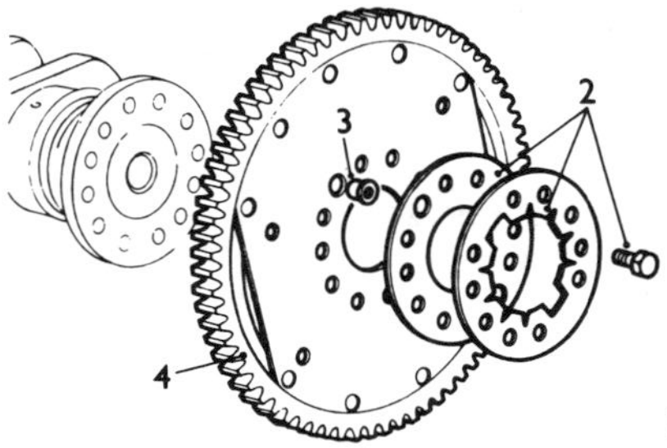

4. Remove drive plate.

Refitting

5. Tap dowels partially through drive plate to locate in crankshaft flange.
6. Refit stiffener plate and lock plate and secure using ten setscrews; tighten to 66.5 lb. ft (9,1 kg. m). Turn up lock tabs.
7. Tap dowels fully home.
8. Refit torque converter.

OIL FILTER ASSEMBLY

Remove and refit   12.60.01

Removing

1. Remove left hand air cleaner, see 19.10.01.
2. Remove eight special nuts and washers securing left hand down pipe to manifold.

NOTE: On cars fitted with emission control, disconnect E.G.R. Pipe from down pipe.

3. Remove down pipe to intermediate pipe flange and remove down pipe. Discard sealing rings.

NOTE: On left hand drive car, full left steering lock must be applied and handbrake must be fully on. Manoeuvre pipe clear.

4. Disconnect union on oil cooler supply pipe.
5. On pressure relief valve vent hose, cut clip nearest valve.
6. Remove filter bowl (early cars) and element.
7. Remove five self tapping screws retaining left hand front heat shield.
8. Remove setscrews retaining filter head.

NOTE: It is not possible to fully withdraw lower right hand setscrew.

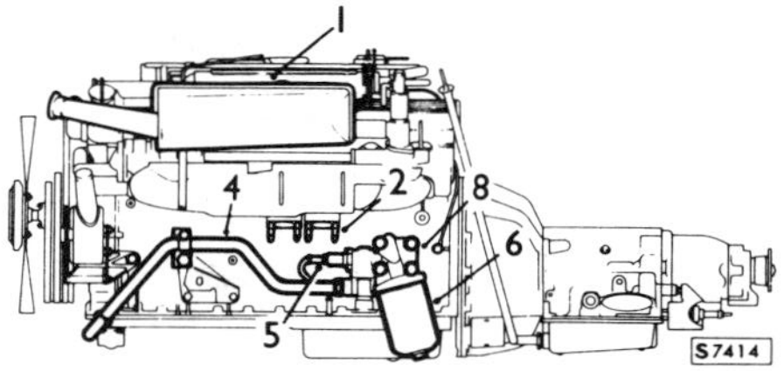

9. Carefully break filter head to cylinder block seal and press filter head upwards until by-pass valve stub pipe clears housing.
10. Separate vent pipe from relief valve.
11. Manoeuvre filter head clear.

Refitting

NOTE: If new filter assembly complete is being fitted, the bowl/element must first be removed from the head.

12. Check condition of 'O' ring seal on by-pass valve stub pipe, and, if necessary renew.
13. Use new filter head to cylinder block gasket.
14. Position setscrew and serrated washer in lower right hand fixing hole, and offer filter head into position.
15. Press stub pipe down into by-pass valve housing and locate fixing setscrew.
16. Fit remaining setscrews and serrated washers to secure filter head.
17. Fit new clip to relief valve vent hose and secure.
18. Secure union nut and oil cooler supply pipe.
19. Fit filter bowl (early cars) and element ensuring gasket ring in position and undamaged.
20. Fit heat shield.
21. Fit exhaust down pipe, using new sealing rings. 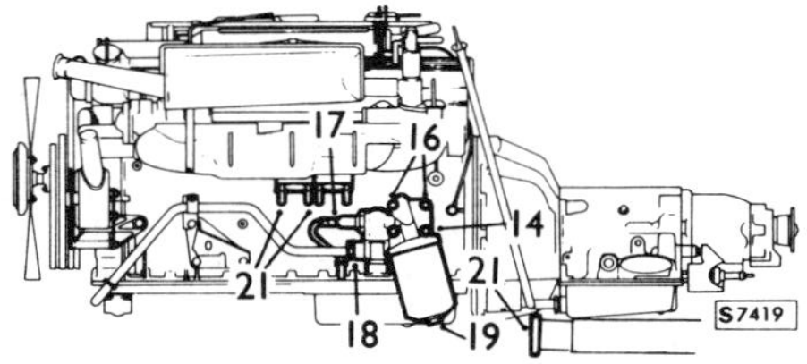
22. Fit air cleaner.
23. Run engine, check oil level and top up as necessary.
24. Carry out appropriate emission test as necessary.

OIL PICK-UP STRAINER

Remove and refit   12.60.20

Removing

1. Remove sandwich plate assembly, see 12.60.45.
2. Remove four setscrews and serrated washers securing suction pipe clips and bracket.
3. Draw suction pipe from elbow.

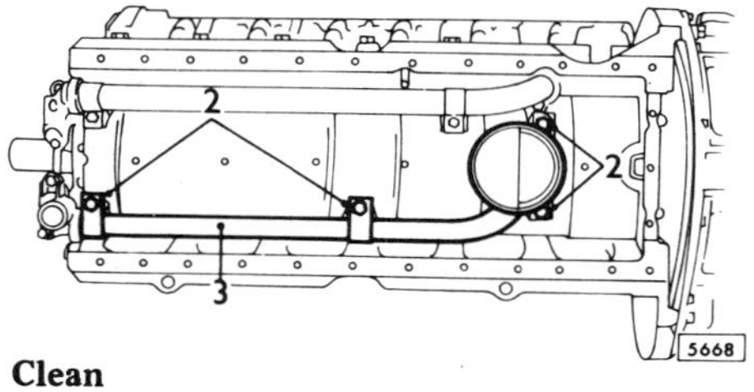

Clean

4. Wash suction strainer in clean paraffin or petrol, and dry thoroughly.

Refitting

5. Renew 'O' ring seals at elbow.
6. Offer suction pipe into elbow and secure using four setscrews and serrated washers at pipe clips and bracket.
7. Refit sandwich plate assembly.

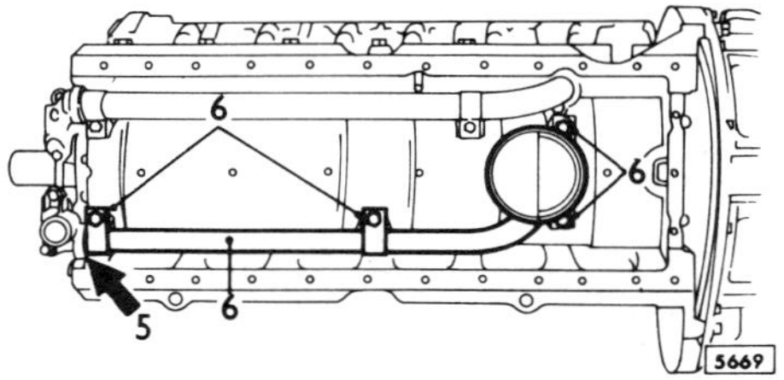

OIL PUMP

Remove and refit   12.60.26

If the oil pump is to be removed, it is necessary to use procedures from Engine – dismantle and reassemble – 12.41.05 – that will provide access. As the work involved is so extensive, it is recommended that the oil pump is inspected for wear or damage whenever it is accessible for other reasons.

The crankshaft MUST NOT be rotated while oil pump is removed.

OIL PUMP

Overhaul   12.60.32

1. Remove eight bolts and lockwashers and detach pump cover from gear housing.
2. Mark drive and driven gear faces to ensure that when reassembled the gears are replaced in the same position as prior to removal.

continued

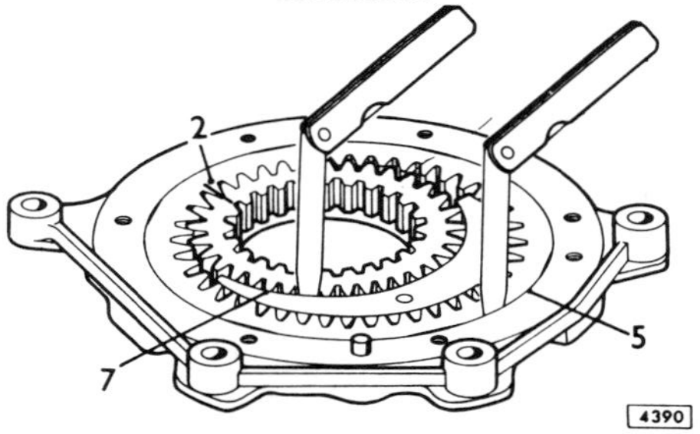

OIL PRESSURE RELIEF VALVE

Remove and refit 12.60.56

Removing

1. Cut clip on oil pressure relief valve bleed hose and prise clear.
2. Slacken relief valve and unscrew from oil filter head.
   NOTE: On cars fitted with catalytic converters, it will be necessary to remove exhaust down pipe.

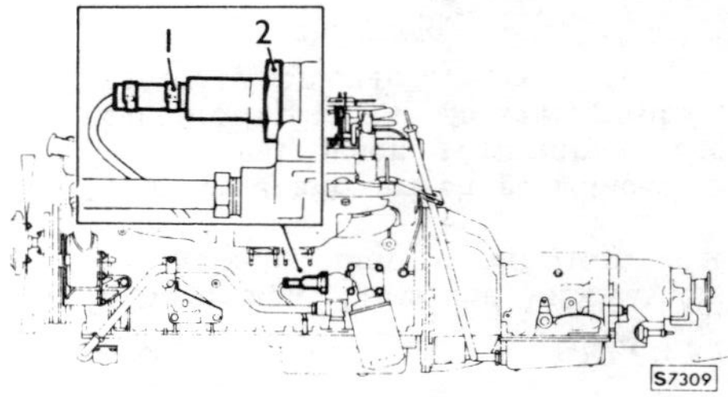

Refitting

3. Screw relief valve into oil filter head, using new sealing washer.
4. Fit vent hose to relief valve and secure using a new clip.
5. Carry out exhaust emission test if catalytic converter fitted.

OIL COOLER

Remove and refit 12.60.68

Removing

1. Remove screws securing lower splash panel.
2. Undo feed and return pipe union nuts.
   CAUTION: Two spanners, one on oil cooler union, the other on hose union MUST be used when carrying out this operation.

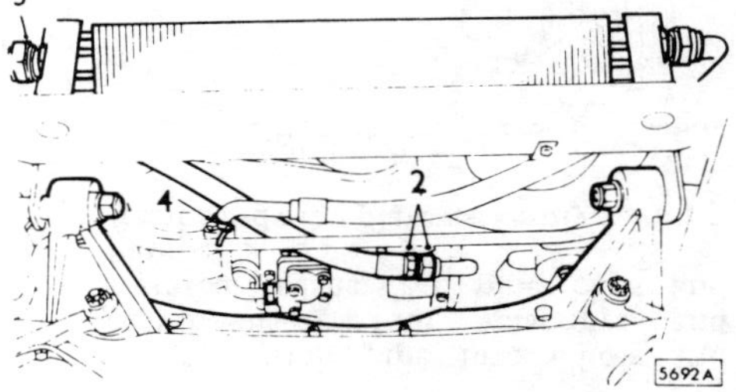

OIL SUMP PAN

Remove and refit 12.60.44

Removing

1. Remove sump plug and drain oil into a suitable container.
2. Remove setscrews and serrated washers securing sump pan.

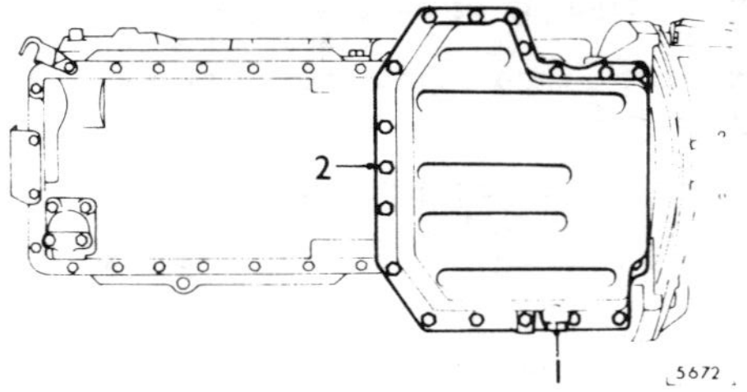

Refitting

3. Lower sump.

4. Use new gasket lightly coated with Hylomar and secure oil sump pan using setscrews and serrated washers.
   NOTE: Secure transmission oil cooler pipe clip.

5. Remove four setscrews and serrated washers securing suction elbow to sandwich plate. Carefully break seal and draw elbow downwards until coupling tube is clear.
6. Release clips and stay securing delivery pipe.
7. Cut clip securing vent pipe hose to sandwich plate spigot.
8. Remove setscrews and washers securing sandwich plate. Recover crankshaft angle indicator from beneath front two setscrews.
   NOTE: Take careful note of location of setscrews, as several different lengths are used.
9. Remove four setscrews securing baffle plate.
10. Remove three setscrews and washers securing by-pass valve housing to sandwich plate.
11. Remove by-pass valve assembly from housing; clean, inspect, replace or renew as necessary.
12. Thoroughly clean all mating faces.
13. Renew 'O' ring seals at by-pass valve housing and spigot.

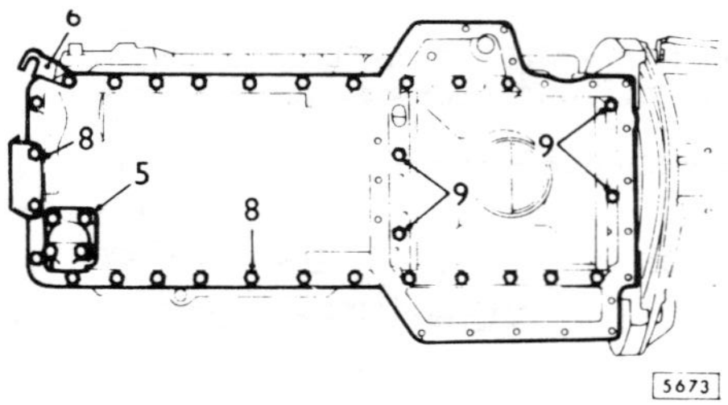

Refitting

14. Reverse operations 1 to 10 using clean engine oil to lubricate by-pass valve spigot 'O' ring seal.

SANDWICH PLATE ASSEMBLY

Remove and refit 12.60.45

Removing

1. Carry out operations 1 to 26 – operation 60.35.05. Slacken front mounting bolts and allow suspension to pivot forward.
2. Remove oil sump pan, see 12.60.44.
3. Remove strapping on oil cooler pipes.
4. Release oil cooler suction and delivery pipe unions from elbow and delivery pipe.

3. Remove both gears, wash all parts in clean petrol and dry with compressed air.
4. Check the condition of all gear teeth and remove any burrs with a fine file.
5. Refit driven gear and check radial clearance between gear and housing. Checks should not be taken at the six radial flats on the gear.
6. Clearance should not exceed 0.005 in (0,127 mm).
7. Refit drive gear and check radial clearance between gear and crescent. Clearance should not exceed 0.006 in (0,152 mm).
8. Check gear end float by placing a straight edge across joint face of housing and measuring clearance between straight edge and gears.

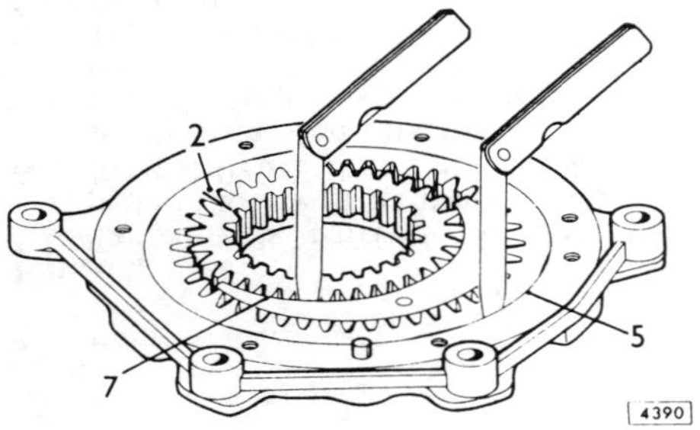

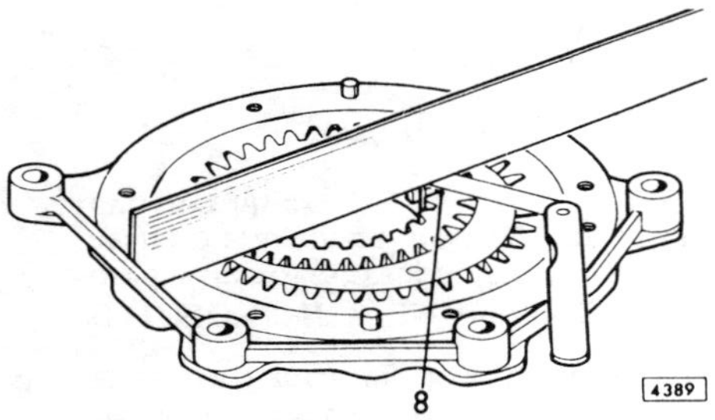

9. Figure obtained should not exceed 0.005 in (0,127 mm).
10. Reassembly is the reverse of items 1 to 3.
11. Lubricate gears with clean engine oil before refitting pump assembly and check that all surfaces are clean.

12–16

3   Remove bolts securing oil cooler to radiator.
4   Lift oil cooler until clear of packing pieces; withdraw cooler. Note number and fitted position of packing pieces.
5   Manoeuvre oil pipes clear.
6   Remove oil pressure switch from four-way banjo.

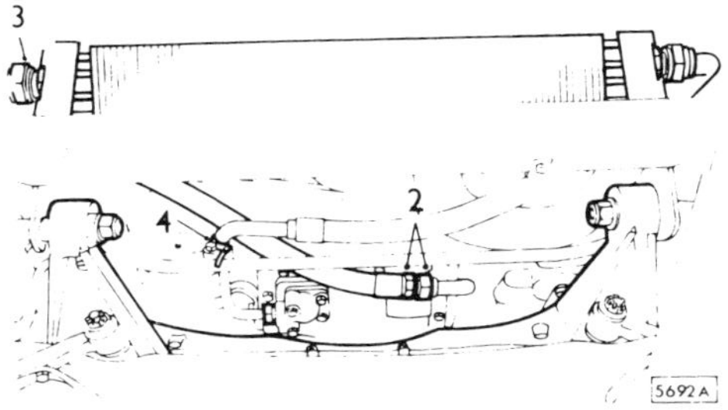

Refitting
7   Reverse operations 1 to 6; use new sealing washers; check oil level and top up as necessary.

TIMING COVER
Remove and refit   12.65.01

Removing
1   Remove engine and gearbox assembly, see 12.37.01.
2   Remove cylinder heads, see 12.29.12 – 12.29.11.
3   Remove sandwich plate, see 12.60.45.
4   Remove alternator, see 86.60.45.
5   Remove power assisted steering pump, see 57.20.14.
6   Remove emission control air pump – cars fitted with emission control only, see 17.25.07.
7   Remove air conditioning compressor and compressor bracket – cars fitted with air conditioning only.
8   Remove water pump, see 26.50.01.
9   Carefully spread cone and draw from crankshaft.
10 Remove bolts, washers and spacers securing alternator and air pump mounting bracket.
11 Remove bolts and serrated washers securing timing cover to cylinder block noting relative positions of different length bolts also dowel bolts.
12 Remove timing cover together with oil seal.
13 Remove gaskets and oil seal and discard.

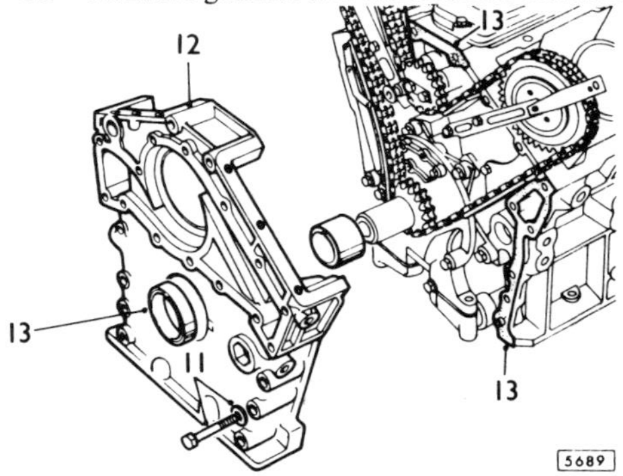

CAMSHAFT OIL FEED PIPES
Remove and refit   12.60.83

Removing
1   Remove banjo connector bolts at rear of each camshaft.
2   Remove four-way banjo connector bolt at throttle pedestal.
NOTE: On cars fitted with catalytic converters it will be necessary to remove exhaust down pipe.
3   Remove oil filter bowl/element.
4   Remove banjo connector bolt at oil gallery on crankcase.

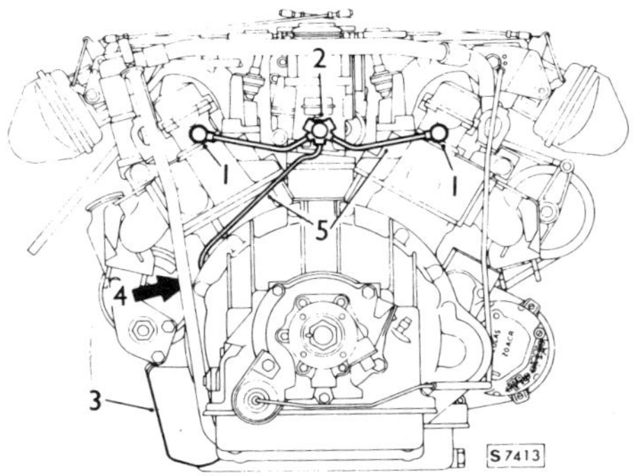

Refitting
14 Ensure mating surfaces of timing cover and cylinder block are scrupulously clean.
15 Immerse new oil seal in clean engine oil and press into timing cover.
16 Smear both sides of each new gasket with suitable jointing compound and position on timing cover.
17 Fit timing cover ensuring that lip on seal is not distorted or damaged.
18 Fit bolts and serrated washers, tighten bolts by diagonal selection.
19 Reverse operations 1 to 10.

TIMING CHAIN
Remove and refit   12.65.12
Service tool: Jackshaft retaining tool JD.39.

Removing
1   Remove timing cover, see 12.65.01.
2   Fit jackshaft retaining tool JD.39.
3   Disconnect timing chain from camshaft and jackshaft sprockets; withdraw crankshaft sprocket and chain.
DO NOT ROTATE ENGINE.

Refitting
4   Reverse operations 1 to 3; check engine timing both statically and by means of a stroboscope.

TIMING CHAIN TENSIONER
Remove and refit   12.65.28

Removing
1   Remove timing cover, see 12.65.01.
2   Move chain tensioner clear of locating bracket and slide off dowel pin.

Refitting
3   Reverse operations 1 and 2.

TIMING CHAIN DAMPERS
Remove and refit   12.65.50
Service tool: Timing chain damper setting jig JD.38.

Removing
1   Remove engine and gearbox assembly, see 12.37.01.
2   Remove timing chain, see 12.65.12.
3   Remove oil pump, see 12.60.26.
4   Remove bolts securing camshaft sprocket hangers and timing chain dampers to cylinder block.

Refitting
5   Fit camshaft sprocket hangers and timing chain dampers to cylinder block; do not fully tighten bolts at this stage.
6   Note relative position of jackshaft sprocket to jackshaft sprocket retaining tool (fitted to engine under operation 12.65.12).
7   Remove jackshaft sprocket retaining tool JD.39.
8   Position damper setting jig JD.38 (shown in skeleton form opposite) on front of cylinder block; do not overtighten retaining bolts.

continued

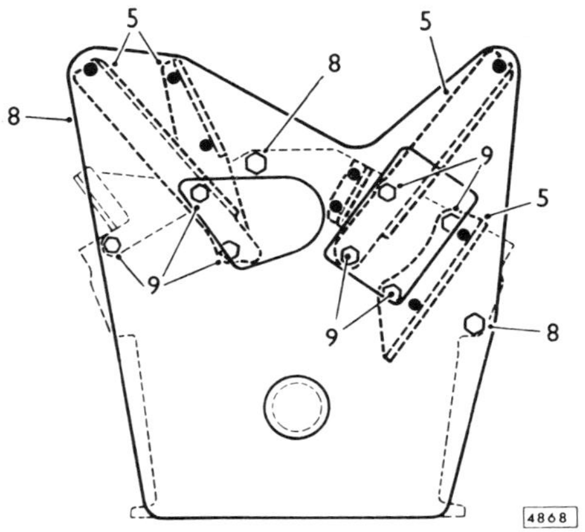

12–17

9   Position camshaft sprocket hangers and timing chain dampers so that they are in even contact with locating dowels; tighten securing bolts.
10 Remove damper setting jig JD.38.
11 Refit jackshaft sprocket retaining tool JD.39.
12 Reverse operations 1 to 3.

EMISSION CONTROL SYSTEM

Description

The emission control system fitted is designed to comply with local legislative requirements. Some or all of the following components may be fitted depending on those requirements. The description that follows refers to cars with an emission control system that complies with North American Federal specification.
The system is of the air injection type and comprises the following major components. The components act upon the engine and interact with each other as detailed.

Air injection system — An air delivery pump 'A', supplies air under pressure, air being passed through a diverter valve 'B', and a check valve 'D' through air rails 'E' to the exhaust ports just above the exhaust valve heads.

This air combines with the exhaust gas to continue the oxidisation process in the exhaust system. The check valve prevents flow in the air rails when exhaust gas pressure exceeds air supply pressure. The diverter valve operates in response to an abrupt fall in manifold pressure, i.e. sudden closure of 'throttle, and diverts secondary air to atmosphere for a period of 2-3 seconds. This reduces the fuel/air ratio which would otherwise be too rich to burn and would pass through the engine to mix with secondary air and become combustible. The next firing cycle would then ignite the mixture causing a backfire in the exhaust system. The diverter valve is actuated by manifold pressure via a rubber tube connected to a tapping in the inlet manifold.

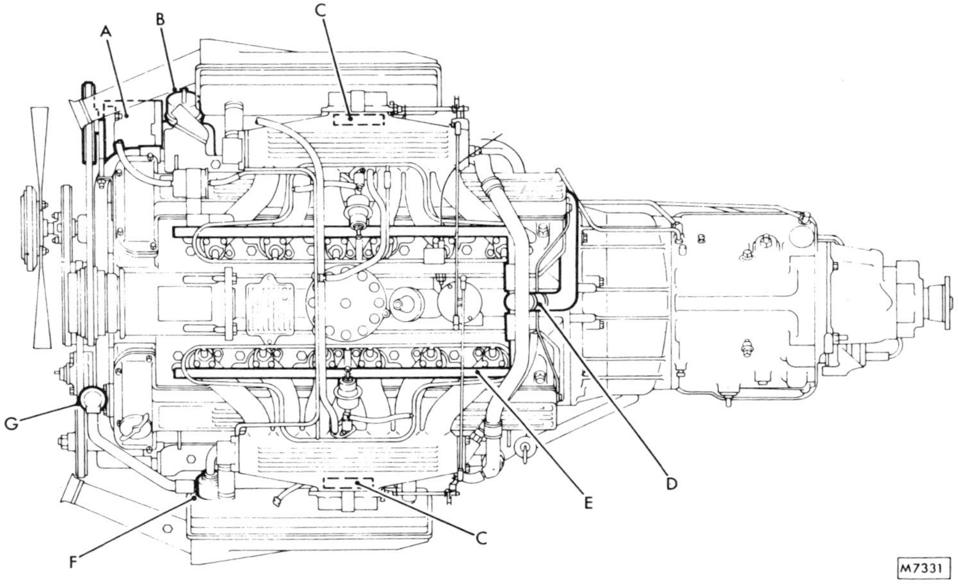

Thermostatic Vacuum System

In conjunction with the Emission Control four conditions of the thermostatic vacuum system are fitted:

A For cars to Canada and USA (Federal) except California. Fig. 1.
B For cars other than A.C.D. Fig. 1.
C For cars to California (1977 and 1978). Fig. 2-3.
D For cars to Australia (1977 and 1978). Fig. 4-5.

For ignition timing see 86.35.20.
The various components used in these systems are described below:
1 A distributor is fitted with:
   A vacuum retard capsule (1B) for 1977 Australia and California cars.
   A vacuum advance/retard capsule (1C) for 1978 California and Australia cars.
   A vacuum advance capsule (1A) for Federal and all others as B above.
2 Throttle edge ports.
3 Vacuum supply.

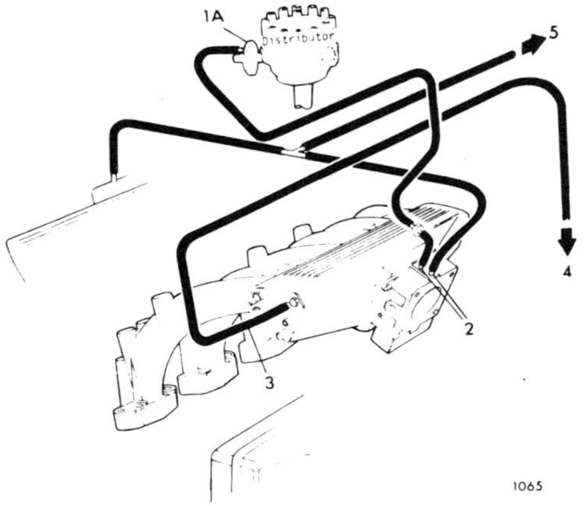

Fig. 1 A and B Emission

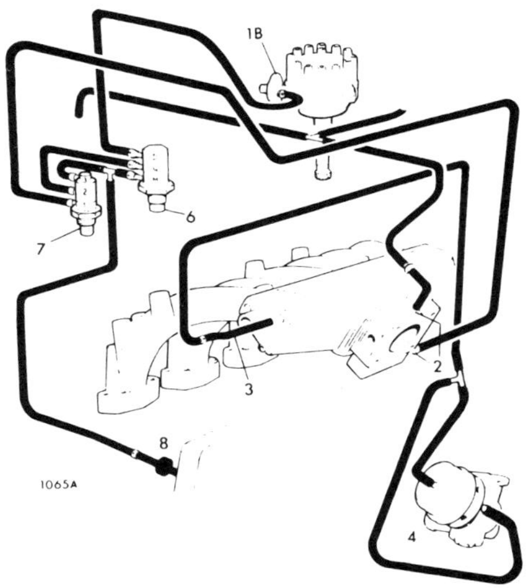

Fig. 2 C Emission California 1977

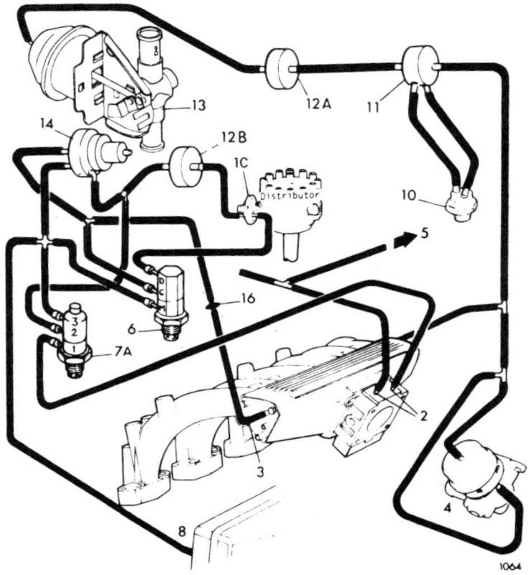

Fig. 3 C Emission California 1978

A Air delivery pump
B Diverter valve
C Exhaust gas recirculation (E.G.R.) valves
D Check valve
E Air rails
F Positive crankcase ventilation valve
G Engine breather filter and housing

17–1

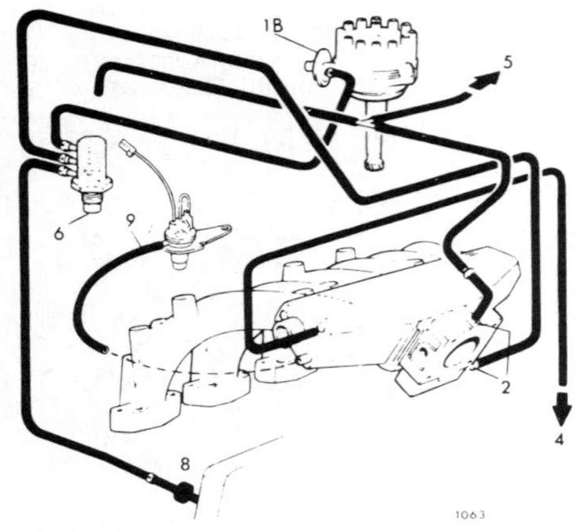

The function of the valve is to reduce emissions of unburnt hydrocarbons during engine overrun modes.

Fast Idle System (California and Australia 1978)

A vacuum operated supplementary air valve (13) supplies the extra air required to support a fast idle of up to 1200 rev/min at coolant temperatures up to 122°F (50°C). The control system operates as follows. The operating vacuum source is the inlet manifold, therefore in order to prevent the supplementary air valve (which is normally open) closing immediately the inlet vacuum builds up following engine starting, a vacuum delay valve (11) is fitted to the signal pipe.

In order to inhibit the operation of the supplementary air valve, after a hot start, it is necessary to by-pass the vacuum delay valve (11). This is accomplished by means of a thermal vacuum valve (10), sensing engine coolant temperature. The thermal vacuum valve (10) connects the inlet and outlet ports of a by-pass incorporated in the delay valve (11).

A second vacuum delay (12A) is placed in the signal pipe between the supplementary air valve and delay valve (11) with the resistance to flow in the opposite direction to that of valve (11).

The function of this second valve is to delay the loss of vacuum signal to the supplementary air valve during short periods of engine operation at or near wide open throttle.

Part Throttle Vacuum Ignition Retard System (Australia 1978)

At idle the ignition timing is controlled by a vacuum operated capsule attached to the ignition distributor (1C). The vacuum source is a port located downstream of or in close proximity to the throttle disc. In the event of the engine overheating the vacuum signal is dumped by means of a thermostatic vacuum switch (6). The switch is an on/off valve controlled by a wax capsule sensing engine coolant temperature. The valve advances idle ignition timing, hence increasing idle speed, should the engine coolant temperature exceed 104°C ± 1.5°C (220°F).

---

be inhibited under fully warm conditions.

The delay valve is controlled by inlet manifold vacuum and engine coolant temperature, without this device the cold fast idle will not function.

12A Vacuum delay valve (full throttle).

The valve delays the fall of the vacuum signal to the supplementary air valve when the inlet vacuum drops towards zero, e.g. when the throttles are fully opened; flow in the opposite direction is made possible by means of a one way valve.

The function of the valve is due to delay operation of supplementary air valve during short periods of engine operation at or near wide open throttle.

12B Vacuum delay valve (ignition).

The valve delays rise of vacuum signal at ignition advance vacuum capsule signal port; flow in the opposite direction is made possible by means of a one way valve. The valve is controlled by throttle edge vacuum signal.

The function of the valve is to reduce emissions of oxides of nitrogen during vehicle acceleration modes.

13 Supplementary air valve.

A vacuum operated valve delivers extra air to support cold engine fast idle; the operating vacuum is controlled by a vacuum delay valve.

The valve is controlled by engine coolant temperature, inlet manifold vacuum and delay valve calibrations.

The supplementary air valve increases the rate of engine and catalytic converter warm-up so reducing the emission of unburnt hydrocarbons.

14 Ignition retard dump valve.

The valve senses rising throttle edge vacuum signal and dumps the vacuum retard signal.

The valve is controlled by throttle edge vacuum signal which enables improvements to be made to part throttle fuel economy.

15 Overrun valve.

A spring throttle valve bleeds air into the inlet manifold during engine overrun modes.

The valve is controlled by spring setting, atmospheric pressure and inlet manifold pressure.

The overrun valve opens at depression 19.2 ± 0.4" Hg flow 1.60 cu ft/min at 21.2" Hg depression.

---

6 Thermostatic vacuum switch — (overtemperature).

The switch incorporates a capsule controlled valve sensing engine coolant temperature.

The valve advances ignition timing should the coolant temperature exceed 220°F (105°C). At this temperature port D switches from port C to port M.

7 Thermostatic vacuum switch (retard inhibit. California 1977 only.

The switch incorporates a wax capsule controlled valve sensing engine coolant temperature. The valve inhibits part throttle ignition retard until the coolant exceeds the specified temperature.

The switch is controlled by coolant temperature to switch port 2 from port 3 to port 1 at (fully open) 168.8 ± 3.6°F (76°C ± 2°). This was fitted to 1977 California cars only.

7a Thermostatic vacuum switch (advance inhibit). The switch incorporates a wax capsule controlled valve sensing engine coolant temperature which inhibits part throttle ignition advance until engine coolant exceeds the specified temperature.

The switch is controlled by engine coolant temperature and switches port 2 from port 3 to port 1 at (fully open) 169 ± 4°F (76°C ± 2°). The use of this valve improves the efficiency of the exhaust emission control system during the engine warm up phase by inhibiting the part throttle ignition advance.

8 Air cleaner.

9 Full throttle switch (Australia 1977)

The vacuum operated switch initiates a signal to the ECU at full throttle and modifies the injection pulse.

10 Thermal vacuum valve.

The valve, controlled by a bimetal element sensing engine coolant temperature provides a signal to the vacuum delay valve bypass during fully warm engine operation so inhibiting the fast idle system.

11 Vacuum delay valve (by-pass).

The valve delays the rise of the vacuum signal received by the supplementary air valve; flow in the opposite direction is made possible by means of a one way valve; a by-pass of the restriction is present in order to allow the fast idle function to

---

Fig. 4 D Emission Australia 1976-7

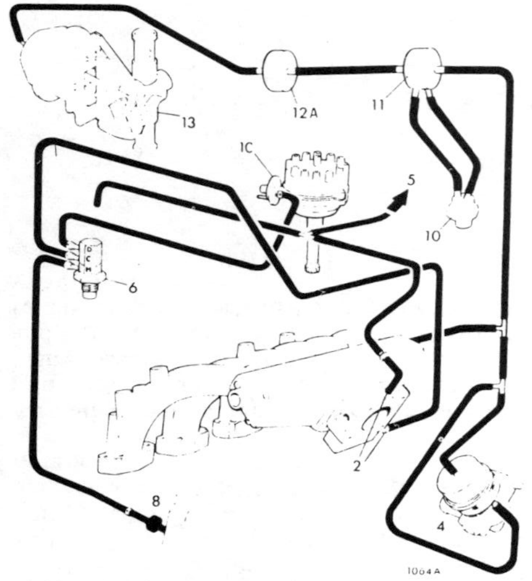

Fig. 5 D Emission Australia 1978

4 Divertor valve with pressure relief valve (PRV).

The PRV incorporated in the divertor valve is controlled by secondary air supply back pressure.

8.2 – 10.5 lbs/in²

The PRV provides the safety for the secondary air pump protection.

5 Carbon canister (A.C.D.) only.

17–2

EMISSION CONTROL SYSTEM

Description

The emission control system fitted is designed to comply with local legislative requirements. Some or all of the following components may be fitted depending on those requirements. The description that follows refers to cars with an emission control system that complies with North American Federal specification.

The system is of the air injection type and comprises the following major components. The components act upon the engine and interact with each other as detailed.

Air injection system — An air delivery pump 'A', supplies air under pressure, air being passed through a diverter valve 'B', and a check valve 'D' through air rails 'E' to the exhaust ports just above the exhaust valve heads.

This air combines with the exhaust gas to continue the oxidisation process in the exhaust system. The check valve prevents flow in the air rails when exhaust gas pressure exceeds air supply pressure. The diverter valve operates in response to an abrupt fall in manifold pressure, i.e. sudden closure of throttle, and diverts secondary air to atmosphere for a period of 2-3 seconds. This reduces the fuel/air ratio which would otherwise be too rich to burn and would pass through the engine to mix with secondary air and become combustible. The next firing cycle would then ignite the mixture causing a backfire in the exhaust system. The diverter valve is actuated by manifold pressure via a rubber tube connected to a tapping in the inlet manifold.

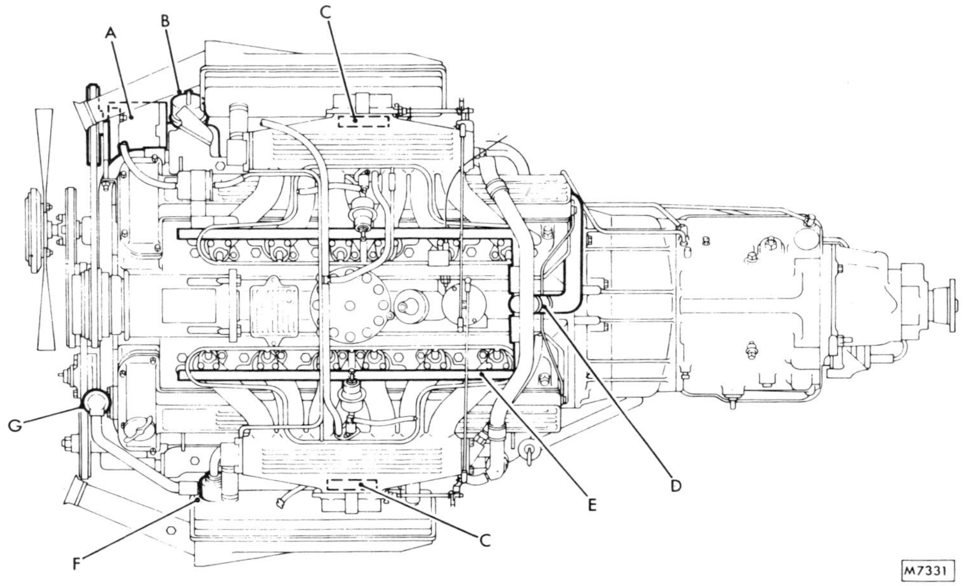

Thermostatic Vacuum System

In conjunction with the Emission Control four conditions of the thermostatic vacuum system are fitted:

A   For cars to Canada and USA (Federal) except California. Fig. 1.
B   For cars other than A.C.D. Fig. 1.
C   For cars to California (1977 and 1978). Fig. 2-3.
D   For cars to Australia (1977 and 1978). Fig. 4-5.

For ignition timing see 86.35.20.
The various components used in these systems are described below:
1   A distributor is fitted with:
    A vacuum retard capsule (1B) for 1977 Australia and California cars.
    A vacuum advance/retard capsule (1C) for 1978 California and Australia cars.
    A vacuum advance capsule (1A) for Federal and all others as B above.
2   Throttle edge ports.
3   Vacuum supply.

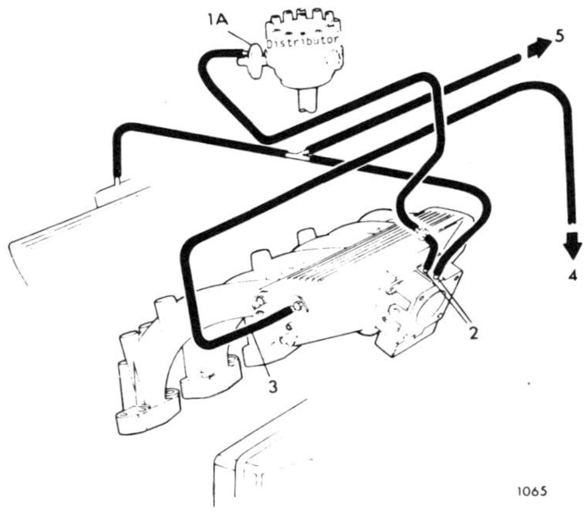

Fig. 1   A and B Emission

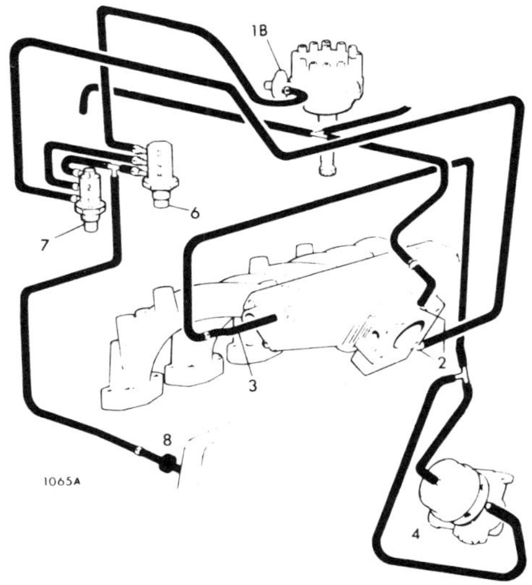

Fig. 2   C Emission California 1977

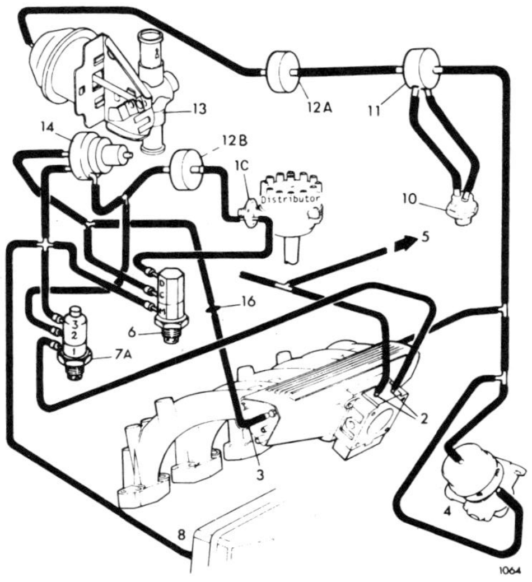

Fig. 3   C Emission California 1978

A   Air delivery pump
B   Diverter valve
C   Exhaust gas recirculation (E.G.R.) valves
D   Check valve
E   Air rails
F   Positive crankcase ventilation valve
G   Engine breather filter and housing

17–1

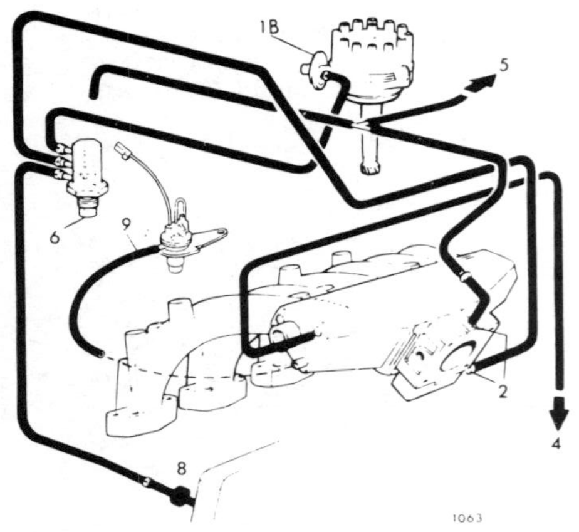

be inhibited under fully warm conditions.

The delay valve is controlled by inlet manifold vacuum and engine coolant temperature, without this device the cold fast idle will not function.

12A Vacuum delay valve (full throttle).

The valve delays the fall of the vacuum signal to the supplementary air valve when the inlet vacuum drops towards zero, e.g. when the throttles are fully opened; flow in the opposite direction is made possible by means of a one way valve.

The function of the valve is due to delay operation of supplementary air valve during short periods of engine operation at or near wide open throttle.

12B Vacuum delay valve (ignition).

The valve delays rise of vacuum signal at ignition advance vacuum capsule signal port; flow in the opposite direction is made possible by means of a one way valve. The valve is controlled by throttle edge vacuum signal.

The function of the valve is to reduce emissions of oxides of nitrogen during vehicle acceleration modes.

13 Supplementary air valve.

A vacuum operated valve delivers extra air to support cold engine fast idle; the operating vacuum is controlled by a vacuum delay valve.

The valve is controlled by engine coolant temperature, inlet manifold vacuum and delay valve calibrations.

The supplementary air valve increases the rate of engine and catalytic converter warm-up so reducing the emission of unburnt hydrocarbons.

14 Ignition retard dump valve.

The valve senses rising throttle edge vacuum signal and dumps the vacuum retard signal.

The valve is controlled by throttle edge vacuum signal which enables improvements to be made to part throttle fuel economy.

15 Overrun valve.

A spring throttle valve bleeds air into the inlet manifold during engine overrun modes.

The valve is controlled by spring setting, atmospheric pressure and inlet manifold pressure.

The overrun valve opens at depression 19.2 ± 0.4″ Hg flow 1.60 cu ft/min at 21.2″ Hg depression.

The function of the valve is to reduce emissions of unburnt hydrocarbons during engine overrun modes.

16 Restrictor.

Fast Idle System (California and Australia 1978)

A vacuum operated supplementary air valve (13) supplies the extra air required to support a fast idle of up to 1200 rev/min at coolant temperatures up to 122°F (50°C).

The control system operates as follows. The operating vacuum source is the inlet manifold, therefore in order to prevent the supplementary air valve (which is normally open) closing immediately the inlet vacuum builds up following engine starting, a vacuum delay valve (11) is fitted to the signal pipe.

In order to inhibit the operation of the supplementary air valve, after a hot start, it is necessary to by-pass the vacuum delay valve (11). This is accomplished by means of a thermal vacuum valve (10), sensing engine coolant temperature. The thermal vacuum valve (10) connects the inlet and outlet ports of a by-pass incorporated in the delay valve (11).

A second vacuum delay (12A) is placed in the signal pipe between the supplementary air valve and delay valve (11) with the resistance to flow in the opposite direction to that of valve (11).

The function of this second valve is to delay the loss of vacuum signal to the supplementary air valve during short periods of engine operation at or near wide open throttle.

Part Throttle Vacuum Ignition Retard System (Australia 1978)

At idle the ignition timing is controlled by a vacuum operated capsule attached to the ignition distributor (1C). The vacuum source is a port located downstream of or in close proximity to the throttle disc. In the event of the engine overheating the vacuum signaf is dumped by means of a thermostatic vacuum switch (6). The switch is an on/off valve controlled by a wax capsule sensing engine coolant temperature. The valve advances idle ignition timing, hence increasing idle speed, should the engine coolant temperature exceed 104°C ± 1.5°C (220°F).

---

6 Thermostatic vacuum switch – (overtemperature).

The switch incorporates a capsule controlled valve sensing engine coolant temperature.

The valve advances ignition timing should the coolant temperature exceed 220°F (105°C). At this temperature port D switches from port C to port M.

7 Thermostatic vacuum switch (retard inhibit. California 1977 only.

The switch incorporates a wax capsule controlled valve sensing engine coolant temperature. The valve inhibits part throttle ignition retard until the coolant exceeds the specified temperature.

The switch is controlled by coolant temperature to switch port 2 from port 3 to port 1 at (fully open) 168.8 ± 3.6°F (76°C ± 2°). This was fitted to 1977 California cars only.

7a Thermostatic vacuum switch (advance inhibit). The switch incorporates a wax capsule controlled valve sensing engine coolant temperature which inhibits part throttle ignition advance until engine coolant exceeds the specified temperature.

The switch is controlled by engine coolant temperature and switches port 2 from port 3 to port 1 at (fully open) 169 ± 4°F (76°C ± 2°). The use of this valve improves the efficiency of the exhaust emission control system during the engine warm up phase by inhibiting the part throttle vacuum ignition advance.

8 Air cleaner.

9 Full throttle switch (Australia 1977)

The vacuum operated switch initiates a signal to the ECU at full throttle and modifies the injection pulse.

10 Thermal vacuum valve.

The valve, controlled by a bimetal element sensing engine coolant temperature provides a signal to the vacuum delay valve bypass during fully warm engine operation so inhibiting the fast idle system.

11 Vacuum delay valve (by-pass).

The valve delays the rise of the vacuum signal received by the supplementary air valve; flow in the opposite direction is made possible by means of a one way valve; a by-pass of the restriction is present in order to allow the fast idle function to

---

Fig. 4 D Emission Australia 1976-7

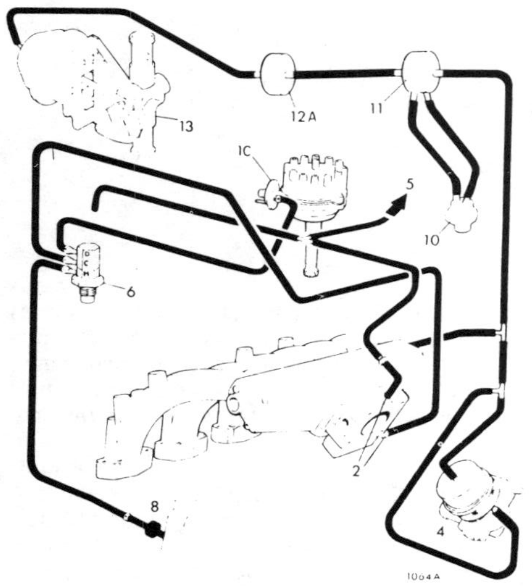

Fig. 5 D Emission Australia 1978

4 Divertor valve with pressure relief valve (PRV).

The PRV incorporated in the divertor valve is controlled by secondary air supply back pressure. 8.2 – 10.5 lbs/in²

The PRV provides the safety for the secondary air pump protection.

5 Carbon canister (A.C.D.) only.

Part Throttle Vacuum Ignition Advance/Retard System (California 1978)

At idle the ignition timing is controlled by a vacuum operated capsule attached to the ignition distributor (1C). The vacuum source is the inlet manifold (3), via a 0.040" restrictor (16).

In the event of the engine overheating the vacuum signal is dumped by means of a thermostatic vacuum switch (6) sensing engine coolant temperature. This results in the ignition timing being advanced and the idle speed increased.

The ignition vacuum advance system is inhibited by a second thermostatic vacuum switch (7A) until the engine coolant temperature exceeds 169.8°F, at which point the vacuum signal sourced at the throttle edge (2) is connected to the ignition advance capsule (1C) via a delay valve (12B).

In order to inhibit the vacuum ignition retard the throttle edge vacuum signal is also fed to a vacuum switch (14) which bleeds air at atmospheric pressure into the inlet manifold vacuum pipe, thus destroying the inlet manifold vacuum retard signal.

Crankcase breather system – To ensure that piston blow-by gas does not escape from the crankcase to atmosphere, a depression is maintained in the crankcase under all operating conditions. This is achieved by connecting the crankcase breather housing 'G' to a chamber 'F' in the left-hand air cleaner backplate; this chamber has two outlets, one of which is connected to the inlet manifold balance pipe and the other to the inlet side of the air cleaner. In the former there is fitted a variable orifice valve that controls the part throttle crankcase ventilation. A depression is maintained in the crankcase at full throttle by the depression on the inlet side of the air cleaner.

Exhaust gas recirculation (E.G.R.) system – Solenoid operated valves 'C' meter a proportion of exhaust gas into the induction system. The gas is diverted from the exhaust downpipe at a tapping upstream of the catalytic converters and fed via valves and fixed orifices in the induction system upstream of the throttle butterfly valves.

The signal that operates the E.G.R. valves is determined by the position of the throttle switch, such that there is no recirculation at idle and full throttle. The E.G.R. valves are further inhibited until engine coolant temperature exceeds 35°C (95°F) and at road speeds in excess of 60-65 mph. These functions are controlled by the E.G.R. controller mounted inside the luggage compartment behind the right-hand rear light cluster.

Catalytic converters – Fitted into the exhaust system to further reduce carbon monoxide and hydrocarbon emissions. A catalyst/E.G.R. system maintenance indicator is built into the centre switch panel. The indicator illuminates at 25,000 miles to indicate the necessity for the renewal of catalytic converters and maintenance to the E.G.R. system. The switch mechanism is built into the speedometer flexible drive line. The unit is a gear driven mechanical reduction device and is fitted with a magnetically operated reed switch. The device is re-set to zero after maintenance has been performed, by means of a special key. Unleaded fuel MUST be used on catalyst equipped cars and labels to indicate this are displayed on the instrument panel and below the filler flap. The filler cap is designed to accommodate unleaded fuel pump nozzles only. Also the anti-surge flap 'A' prevents leaded fuel from being added to the fuel tank in that it does not open when a leaded fuel pump nozzle is entered into the filler neck, up to the position of the restrictor 'B' and the pump is switched on.

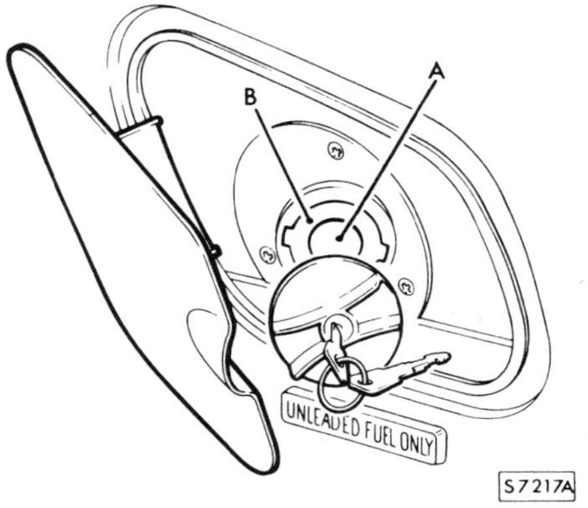

The emission control system fitted to this engine is designed to keep emissions within legislated limits providing ignition timing is correctly maintained and that the engine is in sound mechanical condition. It is essential that routine maintenance operations detailed in the handbook and manual are carried out at the specified mileage intervals.

Testing

In order that engine emissions are kept within legislated limits an emission test MUST be carried out after completing certain operations. The table below lists examples of the operations together with the type of emission test required.

CAUTION: CO content MUST NOT exceed 2% or be less than 1% with air injection system inoperative, i.e. by removing blanking plug from the diverter valve diaphragm housing. It is essential that the equipment used for testing purposes is of the following type.

1. An infra-red CO exhaust gas analyser.
2. Engine and ignition diagnostic equipment.
3. Lucas 'EPITEST' fuel injection diagnostic equipment.

OPERATION	EMISSION TEST REQUIRED
Air delivery pump – remove and refit.	1. Check that pump delivers air.
Crankcase breather, valve, pipe – remove and refit.	1. Check filter for obstruction. 2. Exhaust gas CO content analysed.
Catalytic converters – remove and refit.	1. Exhaust gas CO content analysed.

Fuel tank evaporative loss control system

An evaporative emission control system is fitted to cars to U.S. Federal and Australian design rule specifications. The system utilizes canister 'A', containing activated charcoal to retain fumes given off the engine crankcase assembly 'C' and the fuel tank 'H' while the engine is at rest. When the engine is running, a stream of air is drawn through the air pipe 'B', by means of connections to the throttle edge tappings on the inlet manifold throttle bodies 'D', carrying the adsorbed fumes into the engine and purging the canister.

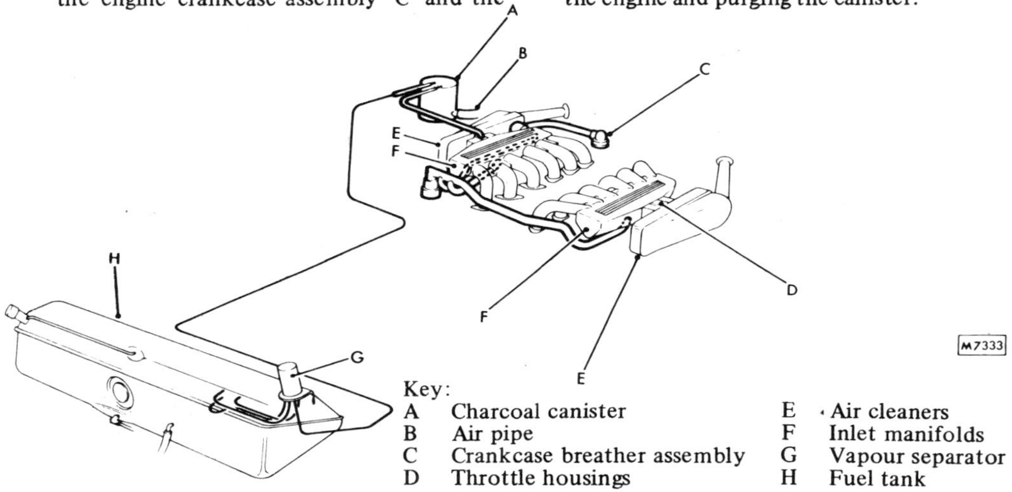

Key:
A Charcoal canister
B Air pipe
C Crankcase breather assembly
D Throttle housings
E Air cleaners
F Inlet manifolds
G Vapour separator
H Fuel tank

Fault finding

SYMPTOM	CAUSE		CURE	
Engine will not start	Low battery or poor connections	1	Check battery, recharge. Clean and secure terminals.	1a
			Check for short circuit or low charge from alternator	1b
	Start system malfunction	2	Clean and check main starter circuit and connections	2
	Incorrect or dirty fuel	3	Check grade of fuel. If contamination suspected drain and flush fuel tank, flush through system, renew fuel filter.	3
	Fuel starvation	4	Check fuel pressure, see 19.50.13. If not satisfactory check feed pipes for leaks or blockage. Renew connectors if damaged or deteriorated.	4
	Fuel injection equipment electrical connections	5	Ensure all connector plugs are securely attached. Pull back rubber boot and ensure plug is fully home. While replacing boot, press cable towards socket. Ensure ECU multi-pin connector is fully made. Check that all ground connections are clean and tight.	5
	Auxiliary air valve inoperative	6	Remove valve and test, see 19.20.17.	6
	Cold start system inoperative	7	Check function of cold start system, see 19.22.32.	7
	Pressure sensor	8	Ensure manifold pressure pipe is attached to sensor, and is not twisted, kinked or disconnected elsewhere.	8
	Trigger unit	9	Check function of trigger unit, see 19.22.27.	9
	Temperature sensors	10	Check sensors for open and short circuit.	10
	H.T. circuit faults	11	Check for sparking.	11
	Power faults	12	Carry out ignition checks.	12
	LT. switching faults	13	Check pick-up module.	13
	Ignition timing incorrect	14	Check and adjust as necessary.	14
	E.G.R. valve malfunction	15	Check function of E.G.R. valve on vehicle. If not satisfactory remove manifold and clean ports. Renew valve if spring is broken, solenoid failed or other fault is obvious.	15

Fault finding

SYMPTOM	CAUSE		CURE	
Engine will not start	ECU/amplifier	16	As a last resort check by substitution.	16
Poor or erratic idle	Check items 3, 4, 5 and 9 above	17	If trouble still persists proceed with item 18.	17
	Throttle switch	18	Check function of idle and full load switches, see 19.22.37.	18
	Incorrect idle speed	19	Adjust auxiliary air valve by-pass bleed screw, see 19.20.18.	19
	Check items 8 and 13 above	20	If trouble still persists proceed with item 21.	20
	Ignition system deterioration	21	Check ignition wiring for fraying, chafing and security. Inspect distributor cap for cracks and tracking and rotor condition. Renew as necessary.	21
	Spark plug faults	22	Clean, reset and test plugs, renew as necessary.	22
	Check item 14	23	If trouble still persists proceed with item 24.	23
	Vacuum system faults	24	Check operation of vacuum unit and condition of vacuum pipes. Renew as necessary.	24
	Advance or retard mechanism faults	25	Check operation of advance/retard mechanism. Lubricate or renew as necessary.	25
	Throttle by-pass valves	26	Check and adjust as necessary.	26
	Exhaust system leaking or blocked	27	Check and rectify as necessary.	27
	Incorrect idle mixture	28	Check CO level, see 17.35.01 and adjust to specified levels using knurled screw on side of E.C.U. Air injection system should be disconnected for this operation.	28
	Poor compressions	29	Check compressions and rectify as necessary.	29
	Air leaks at inlet manifold	30	Check inlet manifold to cylinder head joint. Remake with new gasket if necessary. Check manifold tappings for leaks.	30
	Check item 6 above	31	If trouble still persists proceed with item 32.	31
	Engine oil filler cap loose or leaking	32	Check cap for security. Renew seal if damaged.	32

Fault finding

SYMPTOM	CAUSE		CURE	
Poor or erratic idle	33	Engine breather pipe restrictors missing or blocked	33	Check and clear or renew as necessary.
	34	Engine breather hoses blocked or leaking	34	Check and clear or renew as necessary.
	35	Charcoal canister restricted or blocked	35	Inspect and renew as necessary.
	36	Check items 15, 10 and 9	36	Check in order shown.
Hesitation or flat spot	37	Check items 3, 4 and 5	37	If trouble still persists proceed with item 38.
	38	Check item 7 with engine cold	38	If trouble still persists proceed with item 39.
	39	Throttle butterfly	39	Adjust as necessary, see 19.20.04.
	40	Check item 8	40	If trouble still persists proceed with item 41.
	41	Brakes and clutch	41	Check for binding brakes and slipping clutch.
	42	Check items 13, 21, 22, 14, 24 and 25	42	If trouble still persists proceed with item 43.
	43	Air cleaner blocked	43	Inspect element and renew as necessary.
	44	Check items 27, 29, 30, 32, 33, 34, 35, 15, 10, 9 and 16	44	Check in the order shown.
Excessive fuel consumption	45	Leaking fuel	45	Check fuel system for leaks, rectify and renew connectors as necessary.
	46	Check items 18, 7, 8, 41, 27, 29 and 30	46	If trouble still persists proceed with item 47.
	47	Cylinder head gasket leaking	47	Check cylinder head to block joint for signs of leakage. Renew gasket as necessary.
	48	Cooling system blocked or leaking	48	Flush system, check for blockage. Check hoses and connections for security and leaks, renew as necessary. Check function of thermostats, renew if necessary.

Fault finding

SYMPTOM	CAUSE		CURE	
Excessive fuel consumption	49	Check items 15, 30, 32, 33, 34, 35, 10 and 16	49	Check in the order shown.
Lack of engine braking or high idle speed	50	Air leaks	50	Any air leak into the manifold will appear as an equivalent throttle opening; correct fuel will then be supplied for that apparent degree of throttle and the engine will run faster. Ensure all hose and pipe connections are secure. Check all joints for leakage and remake as necessary.
	51	Throttle sticking	51	Lubricate, check for wear and reset, see 19.20.05.
	52	Check items 6, 24, 26, 14, 28, and 41	52	If trouble still persists proceed with item 53.
	53	Throttle spindle leaks	53	Check seals, bearings and spindles for wear, renew as necessary.
	54	Check item 30		
Lack of engine power	55	Check items 3, 4, 5 and 7	55	If trouble still persists proceed with item 56.
	56	Throttle inhibited	56	Check throttle operation, free off and reset as necessary.
	57	Check items 41, 43, 8, 13, 21, 22, 14, 25, 27, 29, 15, 30, 24, 32, 33, 34, 35, 39 and 16	57	Check in order shown.
Engine overheating	58	Check items 48, 47, 14, 24 and 15	58	Check in order shown.
Engine cuts out or stalls	59	Check items 3, 4, 5, 10, 19, 7, 43, 8, 27, 13, 21, 22, 14, 25, 28, 18, 30, 24, 32, 33, 34, 35, 15, 29 and 16	59	Check in order shown.
Engine misfires	60	Check items 3, 4, 5, 7, 9, 8, 13, 21, 22, 14, 25, 43, 27, 28, 29, 30, 24, 32, 33, 34, 35, 15 and 16	60	Check in order shown.

Fault finding

SYMPTOM	CAUSE		CURE	
Fuel smells	61	Check items 45, 7 and 34	61	If trouble still persists proceed with item 62.
	62	Fuel filler cap defective	62	Check seal and cap for deterioration, renew as necessary.
	63	Check items 33, 35, 28, 43 and 16	63	Check in the order shown.
Engine runs on	64	Check items 3, 19, 51, 26, 34, 33, 48, 47, 14, 24, 25 and 15	64	Check in the order shown.
Engine knocking or pinking	65	Check items 3, 14, 25, 24, 48, 47 and 15	65	Check in the order shown.
Arcing at plugs	66	Check items 21 and 22		
Lean running (Low CO)	67	Check items 5, 53, 18, 10, 3, 4, 30, 24, 32, 33, 34 and 35	67	Check in the order shown.
Rich running (Excess CO)	68	Check items 7, 28, 35 and 16	68	Check in the order shown.
E.G.R./catalyst warning light illuminates	69	E.G.R./catalyst warning indicator malfunction	69	Ensure 25,000 mile service is not due. Check electrical connections. If satisfactory check speedometer drive cable and gears. Renew unit if necessary ensuring that it is set to mileage indicated on unit being replaced.
Back-firing in exhaust	70	Check items 3, 4, 5, 43, 27, 30, 14, 48 and 32	70	If trouble still persists proceed with item 71.
	71	Diverter valve malfunction	71	Check valve line for condition and security, rectify as necessary. Check that air is dumped on deceleration by disconnecting air outlet pipe at diverter valve and feeling the operation of the valve when the throttle is opened and closed quickly.
	72	Check item 16 above		

Fault finding

SYMPTOM	CAUSE		CURE	
Noisy air injection	73	Incorrectly tensioned air pump drive belt.	73	Check and adjust drive belt tension, renew belt if necessary.
	74	Relief valve faulty or low pump pressure	74	Check that valve operates at 8.2 to 10.5 lb/in^2. If pump fails to produce enough pressure to lift the valve check item 73. If satisfactory renew the pump.
	75	Check item 71 above	75	If trouble still persists proceed with item 76.
	76	Check valve sticking	76	Check valve operation and hoses for security or blockage. Rectify or renew as necessary.

ENGINE BREATHER FILTER

Remove and refit   17.10.02

Removing
1. Remove hose clip securing rubber cover to breather housing.
2. Disconnect breather pipe from rubber cover.
3. Remove rubber cover. 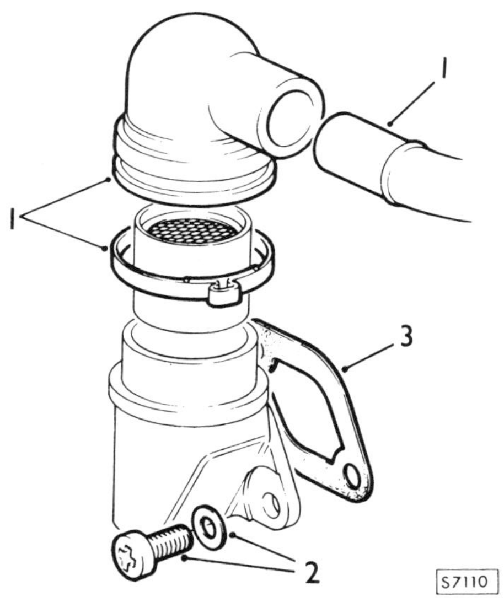
4. Lift out filter.

Refitting
Reverse operations 1 to 4.

ADSORPTION CANISTER

Remove and refit   17.15.13

Removing
1. Remove front left-hand road wheel.
2. Remove three screws and washers securing access cover to spoiler, release three studs, remove cover.
3. Note position of and remove hoses from canister stub pipes.
4. Remove screw, nut and washers securing canister clamp. 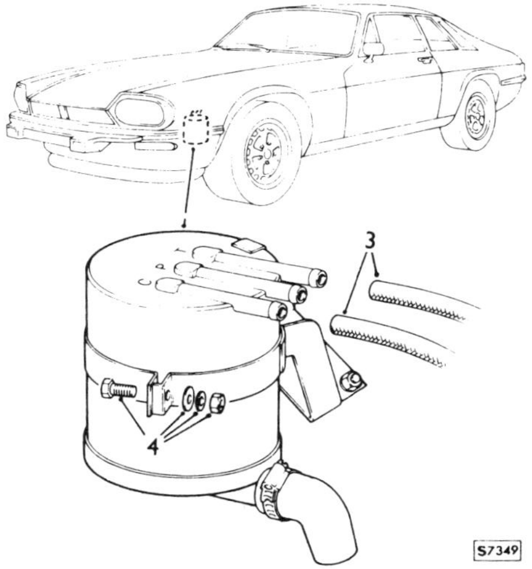
5. Prise open clamp and remove canister.

Refitting
Reverse operations 1 to 5.

AIR DELIVERY PUMP

Remove and refit   17.25.07

NOTE: No servicing or overhaul of the air delivery pump is possible. In event of failure a service exchange unit must be fitted.

Removing
1. Remove right-hand air cleaner cover.
2. Remove three bolts securing air pump pulley to drive shaft.
   CAUTION: A screwdriver or wedge MUST NOT be used to prise pulley off drive shaft as extensive damage to filter element will result.
3. Remove pump pulley.
4. Slacken bolt securing adjuster rod trunnion to pump.
5. Remove locknut from adjuster rod.
6. Slacken nut securing air pump mounting bolt.
7. Pivot pump away from engine.
8. Support pump, remove mounting nut, flat and spacer washer and bolt.
9. Disconnect vacuum hose from diverter valve. 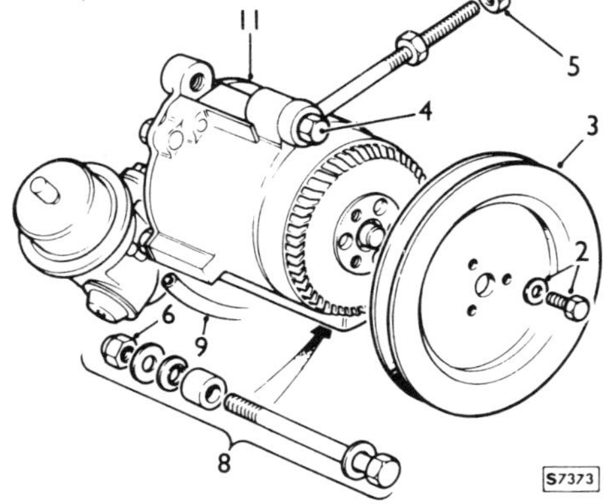
10. Release secondary air pipe from diverter valve. Discard sealing ring.
11. Remove air delivery pump.
12. Remove two bolts and washers securing diverter valve elbow to air pump, remove elbow and diverter valve.
13. Remove and discard gasket.

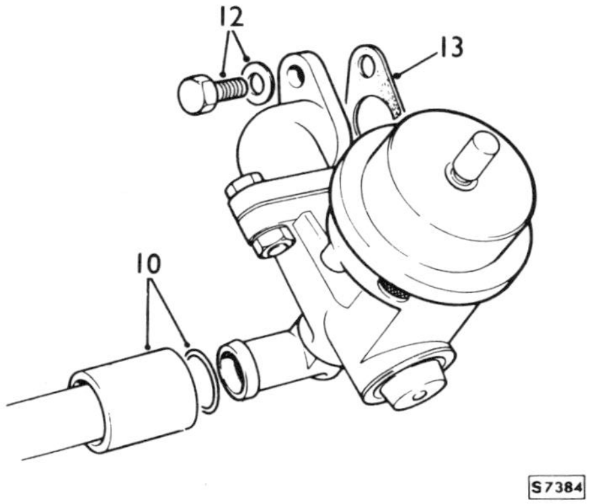

Refitting
14. Reverse operations 1 to 13 using new gasket and sealing ring.
15. Check tension of air pump/compressor drive belt, see 17.25.13.

AIR DELIVERY PUMP/ COMPRESSOR DRIVE BELT

Tensioning   17.25.13

1. Remove right-hand air cleaner cover.
2. Slacken nut securing air pump mounting bolt.
3. Slacken adjusting link securing bolt.
   CAUTION: Ensure head of bolt does not foul pulley.
4. Slacken adjusting link locknut.
5. Adjusting belt tension by means of adjusting link nut; correct tension is as follows:
   A load of 6.4 lb (2,9 kg) must give a total belt deflection of 0.22 in (5,6 mm) when applied at point A. 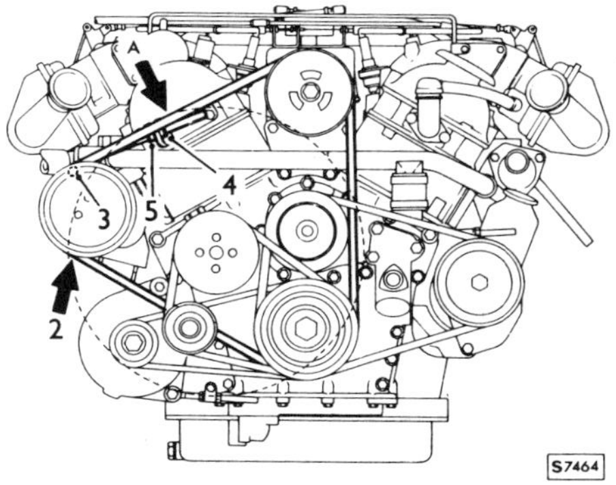
6. Reverse operations 1 to 5.

17—7

AIR DELIVERY PUMP/COMPRESSOR DRIVE BELT 17.25.15

Remove and refit

Removing

1. Remove right-hand air cleaner cover.
2. Remove fan belt, see 26.20.07.
3. Remove power assisted steering/water pump drive belt, see 57.20.02.
4. Slacken nut securing air delivery pump mounting bolt.
5. Slacken adjusting link securing bolt.
   CAUTION: Ensure head of bolt does not foul pulley.
6. Slacken adjusting link locknut.
7. Slacken adjusting link nut and wind along thread until belt can be manoeuvred clear.

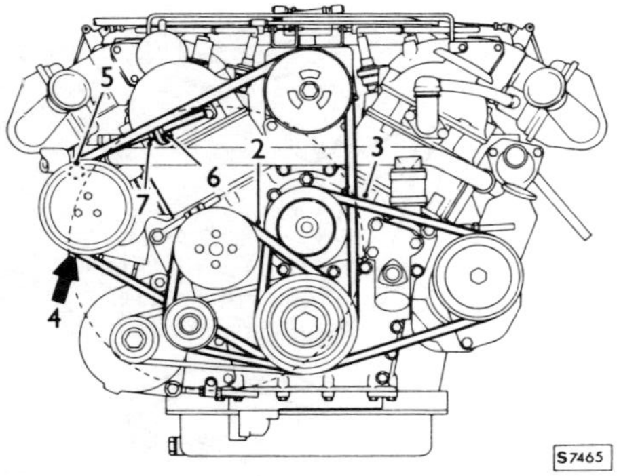

Refitting

8. Manoeuvre replacement belt into position.
9. Set drive belt tension, operation 5 – 17.25.13.
10. Secure adjusting link locknut.
11. Tighten adjusting link securing bolt.
12. Tighten nut securing air delivery pump mounting bolt.
13. Refit power assisted steering/water pump drive belt.
14. Refit fan belt.

AIR RAIL   left-hand 17.25.17
                right-hand 17.25.18

Remove and refit

Removing

1. Depressurise fuel system, see 19.50.02.
2. Disconnect battery, see 86.15.20.
3. Remove fuel rail, see 19.60.04/5.
4. Remove fuel pressure regulator, see 19.45.11.
5. Remove two screws securing air rail bracket to inlet manifold ram tubes.
6. Remove earth strap from rear inlet manifold ram tube – right-hand side only.
7. Remove six nuts and serrated washers securing air rail and manifold stud spacers to cylinder head.
8. Release hose clip securing air rail to check valve connecting hose.
9. Lift out air rail and disconnect from check valve connecting hose.
10. Remove and discard rubber sealing rings.

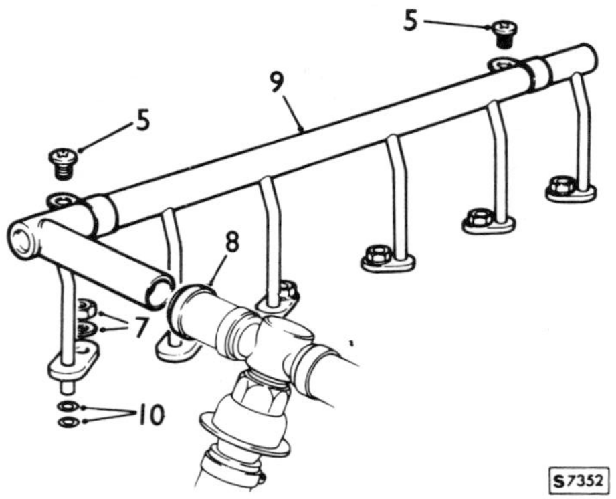

Refitting

Reverse operations 1 to 10 using new rubber sealing rings.

CHECK VALVE 17.25.21

Remove and refit

Removing

1. Release hose clips securing air delivery rail hoses to non-return valve outlet; slide hoses clear of valve outlets.
2. Remove lower clip securing hose to check valve inlet pipe.
3. Remove check valve.

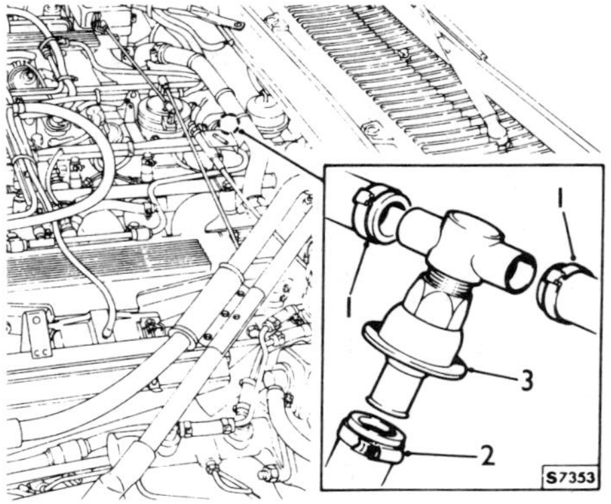

Refitting

Reverse operations 1 to 3 using new hose clips.

DIVERTER VALVE 17.25.25

Remove and refit

Removing

1. Remove air cleaner and element.
2. Disconnect vacuum hose from diverter valve stub pipe.
3. Remove two bolts, nuts and spring washers securing diverter valve to elbow.
4. Release air rail pipe from diverter valve.
5. Remove diverter valve.
6. Remove and discard diverter valve to elbow gasket.
7. Remove and discard air rail pipe sealing ring.
8. Remove air rail pipe to diverter valve union from diverter valve, remove and discard sealing ring.

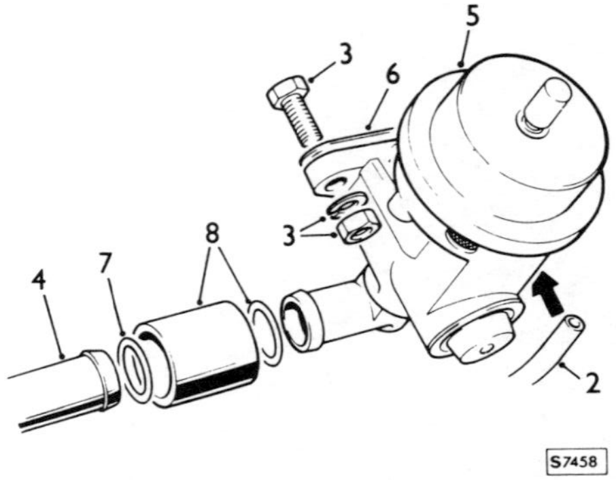

Refitting

Reverse operations 1 to 8 using new gasket and sealing rings.

EXHAUST GAS RECIRCULATION (EGR) VALVE 17.45.01

Remove and refit

Removing

1. Remove air cleaner cover.
2. Disconnect EGR valve transfer pipe from down pipe take-off stub.
3. Remove electrical connectors from rear of EGR valve.
4. Remove two bolts and spring washers securing EGR valve to elbow.
5. Remove and discard EGR valve to elbow gasket.
6. Remove transfer pipe from EGR valve.

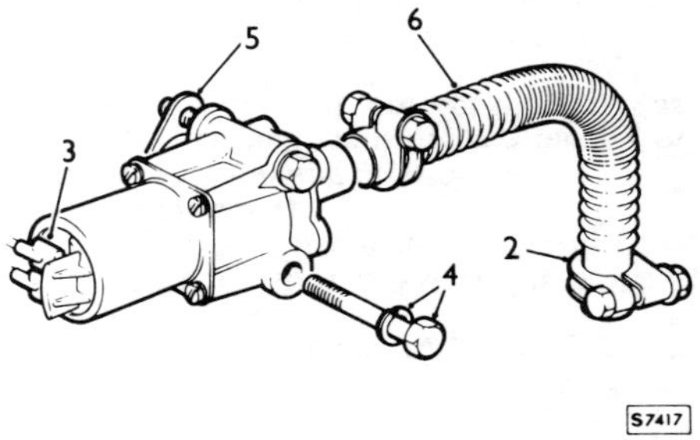

Refitting

Reverse operations 1 to 6 using new gasket and EGR valve transfer pipe.

17–8

EGR CONTROL UNIT

Remove and refit   17.45.07

Removing
1. Disconnect battery, see 86.15.20.
2. Remove two nuts and washers securing right-hand rear light cluster.
3. Disconnect earth lead from light cluster stud.
4. Remove light cluster from housing.
5. Remove two nuts and bolts securing EGR control unit to body.
6. Disconnect EGR control unit block connector, remove EGR control unit.

Refitting
Reverse operations 1 to 6.

EGR VALVE TRANSFER PIPE

Remove and refit   17.45.11

Removing
1. Remove air cleaner cover.
2. Slacken nuts securing EGR valve transfer pipe clamps.
3. Disconnect EGR valve transfer pipe from take-off pipe.
4. Beneath car remove bolts and spring washers securing power steering bellows heatshield, remove heatshield.
5. Remove nuts, plain washers and bolts securing flanges, separate intermediate pipe from down pipe. Ensure intermediate pipe is adequately supported.
6. Remove nuts and plain washers securing heatshield and down pipe to exhaust manifolds; withdraw heatshield.
7. Withdraw down pipe/catalyst assembly.

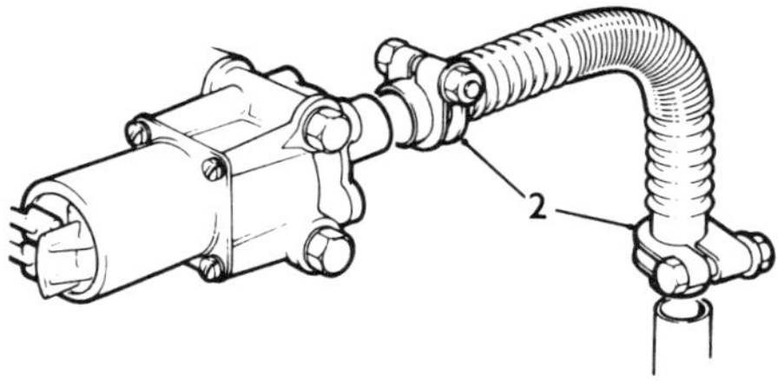

NOTE: This operation will be greatly facilitated if steering is turned when manoeuvring assembly clear.

Refitting
Reverse operations 1 to 7; coat all joints with Holts Firegum. Tighten down pipe and clamping flange fixing. by diagonal selection to avoid distortion.

3. Release transfer pipe from EGR valve and down pipe stub connections.

Refitting
4. Clean EGR valve and down pipe take-off stub connections.
5. Reverse operations 1 to 3 using new EGR valve transfer pipe.

CATALYTIC CONVERTER

Remove and refit   left-hand   17.50.01
                    right-hand   17.50.03

Removing
1. Remove air cleaner element.
2. Slacken nut securing EGR valve transfer pipe clamp at take-off pipe.

ELECTRONIC FUEL INJECTION

Description

The electronic fuel injection system can be divided into two separate systems interconnected only at the injectors.

The systems are:

1. A fuel system delivering to the injectors a constant supply of fuel at the correct pressure.
2. An electronic sensing and control system which monitors engine operating conditions of load, speed, temperature (coolant and induction air) and throttle movement. The control system then produces electrical current pulses of appropriate duration to hold open the injector solenoid valves and allow the correct quantity of fuel to flow through the nozzle for each engine cycle.

As the fuel pressure is held constant, varying the electrical pulse duration increases or decreases the amount of fuel passed through the injector to comply precisely with engine requirements. Pulse duration and therefore fuel quantity, is also modified to provide enrichment during starting and warming up and at closed throttle, full throttle and while the throttle is actually opening.

The injectors are operated by the Electronic Control Unit (ECU) in two groups of six following the engine firing order. Each group is further broken down into two sub-groups of three by a Power Amplifier Unit although each pair of sub-groups is operated simultaneously to make up the two groups of six.

KEY TO LOCATION DIAGRAM

A   Manifold pressure sensor.
B   Thermotime switch
C   Fuel pressure regulator
D   Cold start injector
E   Cold start relay
F   Electronic control unit (ECU)
G   Fuel pump relay
H   Main relay
J   Induction manifolds
K   Throttle switch
L   Trigger unit
M   Auxiliary air valve
N   Idle speed regulating screw
P   Overrun valve
Q   Air temperature sensor
R   Coolant temperature sensor
S   Fuel cooler
T   Power amplifier

The induction system is basically the same as that on a carburetted engine, tuned ram pipes, air cleaners, plenum chambers and induction ports. The air is drawn through paper element cleaners to a single throttle butterfly valve for each bank and to individual ports for each cylinder leading off the plenum chamber. The injectors are positioned at the cylinder head end of each port so that fuel is directed at the back of each inlet valve.

Fuel System

Fuel is drawn from tank at the rear of the car by a fuel pump via expansion tank and passed through a filter to the fuel rails. Fuel is maintained at a constant pressure of 30 lbf/in^2 (2,1 kgf/cm^2) by pressure regulators 'C'. Any fuel in excess of this pressure is returned to the fuel tank via a cooler 'S'.

The twelve injectors are connected to the fuel rails. They are solenoid operated and respond to electrical current pulses from the ECU, via a power amplifier, to open and inject fuel into each inlet port. Fuel is also supplied to two cold start injectors 'D' that are operated only during the initial starting of a cold engine.

Electronic System

The main criteria governing the injection of fuel into the engine are manifold depression (engine load) and engine speed.

continued

Firing order
A – Right-hand bank   B – Left-hand bank
1A 6B 5A 2B 3A 4B   6A 1B 2A 5B 4A 3B
Cylinders numbered from front of engine.

Injection in two groups of six
1st group   2nd group
1A 3A 5A 2B 4B 6B   1B 3B 5B 2A 4A 6A

19–1

Engine load sensing – The driver controls engine power output by varying the throttle opening and therefore the flow of air into the engine. The air flow determines the pressure that exists within the plenum chamber, the pressure being a measure of the demand upon the engine. This pressure is used to provide the principle control of fuel quantity being converted by the pressure sensor 'A' into an electrical signal to be passed to the ECU 'F'. The signal varies the duration of the injector operating pulse as appropriate.

The pressure sensor is fitted with a separate diaphragm system that compensates for ambient barometric variations.

Engine speed sensing – The trigger unit 'L' fitted within the distributor has two reed switches mounted 180° apart so that they are closed alternatively, one each revolution of the crankshaft. Each switch 'triggers' the ECU to produce the timed electrical current pulse to a group of six injectors, although the trigger switch itself has no part in determining the pulse length. In addition to this primary function of initiation of injection, the trigger unit switching is monitored by the ECU for frequency of operation. From this signal an engine speed function is determined that modifies the pulse width, already established by manifold pressure, to take account of engine speed – dependant resonances in induction and exhaust.

Temperature sensors – The temperature of the air being taken into the engine through the inlet manifold and the temperature of the coolant in the cylinder block is constantly monitored. The information is fed directly to the ECU.

The air temperature sensor 'Q' has a small effect on the injector pulse width and should be looked upon as a trimming rather than a control device. It ensures that the fuel supplied is directly related to the weight of air drawn in by the engine. As with falling temperature, so the amount of fuel supplied is also increased to maintain the optimum fuel/air ratio.

The coolant temperature sensor 'R' has a much greater degree of control although its main effect is concentrated while the engine is initially warming up. The coolant temperature sensor operates in conjunction with the cold start system and the auxiliary air valve 'M' to form a completely automatic equivalent to a carburetter automatic choke.

Cold start system – For cold starting, additional fuel is injected into the inlet manifolds by two cold start injectors 'D'. These are controlled by the cold start relay 'E' and thermotime switch 'B'. The thermotime switch senses coolant temperature and depending on that temperature, interrupts or completes the earth (ground) connection for the relay. When the starter is operated, the cold start relay is energised with its circuit completed via the thermotime switch. The thermotime switch also limits the length of time for which the relay is energised, to a maximum of eight seconds under conditions of extreme cold. This enrichment is in addition to that provided by the coolant temperature sensor. If the coolant temperature is above the rated value of the thermotime switch, the thermotime switch does not operate; no starting enrichment being required.

Cranking enrichment – The ECU provides an increased pulse duration during engine cranking in addition to any enrichment due to the coolant temperature sensor or the cold start injectors. The additional signal reduces slightly when cranking stops but does not fall to normal level for a few seconds. This temporary enrichment sustains the engine during initial running.

Throttle switch – The throttle switch 'K' is a rotary switch directly coupled to the throttle pulley. It contains sets of contacts that provide information for the ECU regarding the position and movement of the throttle butterfly valves. The operation of these contacts is as follows:

1   **Throttle closed (idle) contacts**
    These contacts establish a specific, slightly richer, level of fuelling while the throttle is completely closed and the engine is running at idling revolutions. While the throttle is in this position, the exhaust CO level can be varied using the idle mixture control knob on the ECU. This knob **MUST NOT** be moved unless correct test equipment and skilled personnel are in attendance to monitor changes made.

2   **Throttle movement contacts**
    Immediately the throttle is opened a series of 20 make-and-break contacts are put into circuit. If the throttle butterfly is opened quickly a slight delay occurs before the pressure sensor reacts to the change in manifold pressure. This period of delay, is overcome by the throttle switch contacts which transmit a series of voltage spikes to the ECU. These signals produce an increased pulse duration while the throttle is moving. On certain cars full load enrichment is provided by revising the response curve of the manifold pressure sensor. In these instances the full throttle enrichment contacts necessary for cars to USA Federal specifications are not used.

Flooding Protection System

When the ignition is on but the engine not cranking, the fuel pump will run for one to two seconds to raise the pressure in the fuel rail; it is then automatically switched off by the ECU. Only after cranking has started is the pump switched on again. Switching control is built into the ECU circuitry. This system prevents flooding should any injectors become faulty (remain in the open position) and the ignition is left switched on.

Auxiliary Air Valve

The auxiliary air valve 'M' is controlled by coolant temperature. To prevent stalling at cold start and cold idle conditions due to the increased drag of the engine, the valve opens to allow air to by-pass the throttles and so increase engine speed. In addition to the main coolant temperature regulated air passage, the auxiliary air valve has a by-pass controlled by an adjusting screw 'N'. This screw controls the idle speed by regulating the air flow.

GOOD PRACTICE

The following instructions must be strictly observed.

1   Always disconnect the battery before removing any components.
2   Always depressurise the fuel system before disconnecting any fuel pipes.
3   When removing fuelling components always clamp fuel pipes approximately 1.5 in (38 mm) from the unit being removed. Do not overtighten clamps.
4   Ensure rags are available to absorb any spillage that may occur.
5   When re-connecting electrical components always ensure that good contact is made by the connector before fitting the rubber cover. Always ensure that earth (ground) connections are made on to clean bare metal, and are tightly fastened using correct screws and washers.

WARNING

1   **Do not let the engine run without the battery connected.**
2   **Do not use a high-speed battery charger as a starting aid.**
3   **When using a high-speed battery charger to charge the battery, the battery MUST be disconnected from the rest of the vehicle's electrical system.**
4   **When installing, ensure that battery is connected with correct polarity.**
5   **No battery larger than 12V may be used.**

S7207

MAINTENANCE

There is no routine maintenance procedure laid down for the Electronic Fuel Injection System other than that at 12,000 mile (20,000 km) intervals the fuel filter must be discarded and a replacement component fitted. At all service intervals, the electrical connectors must be checked for security.

Fault finding

The fault finding procedures are divided into two sections. The first section considers initial roadside diagnosis and rectification, while the second section gives a complete test layout and usage procedure for the Lucas 'EPITEST' equipment.

Initial diagnosis

Fault conditions in this section can be further divided into three types as follows:

1. Faults that prevent the engine from starting.
2. Faults that allow the engine to start, but stop it either immediately or after a short delay.
3. Faults that allow starting and continued running, but cause incorrect fuelling at some stage of a driving cycle.

Examples of all three classes of fault are given in the 'Symptoms' column of the Initial Diagnosis and Rectification chart, together with a list of possible causes in the order in which they should be checked. This is followed by Procedures for Rectification which details the effect that each possible failure will have upon the engine and its remedy. It is assumed that the vehicle has sufficient fuel in the tanks, and that purely engine functions, e.g. ignition timing, valve timing, and the ignition system as a whole are operating satisfactorily. If necessary, these functions must be checked by following the relevant procedures in the Repair Operations Manual before the fuel injection system is suspected.

INITIAL DIAGNOSIS AND RECTIFICATION

POSSIBLE CAUSES IN ORDER OF CHECKING

	Will not start*	Difficult cold start	Difficult hot start	Starts but will not run	Misfires and cuts out	Runs rough	Idle speed too fast	Hunting at idle	Low power and top speed	High fuel consumption
Battery	A	A	A							
Connections	B	B	B	A	A	A			A	A
Ignition system	C	C	C	B	B	B			C	B
Fuel system	D	E	E	D	C	C			D	D
Trigger unit	E	F								
Pressure sensor	F	G	F	E	D	D	B	B	B	C
Cold start system	G	D	D	C	E	F			F	F
ECU/amplifier	H	J	H	G	G	H	D		D	H
Air leaks		G				A				
Temperature sensors		H	G	F	F	G			G	G
Auxiliary air valve			F	E	D				F	E
Throttle switch					E					
Throttle butterfly						C				
Overrun valve						D				
Compression							A	E		
Idle fuel control setting							C			
Air filters									B	
Throttle linkage									F	

SYMPTOMS

* Before proceeding with checks, hold throttle fully open and attempt a start. If the engine then starts and continues to run, no further action is necessary.

19—3

PROCEDURES FOR RECTIFICATION OF CAUSES SHOWN IN TABLE

Battery:	Battery depleted, giving insufficient cranking speed or inadequate spark. Check battery condition with hydrometer; Recharge, clean and secure terminals, or renew as necessary.	Auxiliary Air Valve:	Check opening throttle. If engine immediately starts, unscrew idle speed adjustment, and re-check start with closed throttle. Re-set idle speed when engine hot. Check cold start. Check throttle return springs and linkage for sticking or maladjustment as a sticking throttle may have enforced incorrect idle speed adjustment on a previous occasion.
Connections:	Ensure all connector plugs are securely attached. Pull back rubber boot and ensure plug is fully home. While replacing boot press cable towards socket. Ensure Electronic Control Unit (ECU) multi-pin connector is fully made. Ensure all ground connections are clean and tight.	Throttle Switch:	Check operation of throttle switch. Incorrect function or sequence of switching will give this fault.
Ignition System:	Check ignition system as detailed in electrical section.	Throttle Butterfly:	Check adjustment of both throttle butterfly valves, ensure return springs correctly fitted, and throttle not sticking open.
Fuel System:	Check for fuel pipe failure (strong smell of fuel). Check inertia switch closed. If necessary, clear fuel tank vents or supply pipe.	Overrun valve:	Check operation of overrun valve—19.20.21.
Trigger Unit:	Check operation of reed switches; engine will not run unless both reed switches are satisfactory.	Compression:	Low compressions; a general lack of engine tune could cause this fault. Check engine timing, ignition timing, and function of ignition system complete. If necessary, check valve condition.
Pressure Sensor:	Ensure manifold pressure reference pipe is attached to sensor and is not twisted, kinked or disconnected elsewhere. Engine may start but will run badly.	Idle Fuel Control Setting:	Set butterfly valve (see 19.20.11). Set idle speed adjustment screw — 19.20.16/17/18 to obtain 750 rev/min with engine fully warm. Remove cap from air injection diverter valve to interrupt air injection. Run engine at 2,000 rev/min for 15 seconds and return to idle speed. Adjust idle potentiometer (see CAUTION) on electronic control unit to obtain required CO reading (1–2%) at sampling point in the end of each air injection rail. Refit cap to diverter valve and reset idle speed if necessary. CAUTION: This knob MUST NOT be moved unless correct test equipment and skilled personnel are in attendance to monitor changes made.
Cold Start System:	Fault conditions could cause cold start system to be inoperative on a cold engine, or operative on a hot engine. If engine is either very hot, or cold, these particular faults will cause the engine to run very rich. Check cold start system, see 19.22.32.		
ECU/Amplifier:	If either of these components is faulty it is possible that various groups of injectors will be inoperative. This will range from barely detectable, one group, to very rough or no start, two, three or four groups. The ECU may also be responsible for any degree of incorrect fuelling problems, however, all other likely components should be proved good.	Air Filters:	Remove air filters and check for choked filter element.
		Throttle Linkage:	Check throttle linkage adjustment and ensure that throttle butterfly valves can be fully operated.
Air Leaks:	Ensure all hose and pipe connections are secure. Engine is, however, likely to start more easily with air leaks if cold, as air leaking augments that through the auxiliary air valve. A leak, or failed air valve is shown up, however, by a very high idle speed when engine is warm and air valve main passage should be closed.		
Temperature Sensors:	If either sensor is short-circuited, starting improves with higher engine temperature. Engine will run very weak, improving as temperature rises, but still significantly weak when fully hot. If a sensor is open circuit, or disconnected, engine will run very rich, becoming worse as temperature rises. Engine may not run when fully hot, and will almost certainly not re-start if stalled. Effect of air temperature sensor will be less marked than coolant temperature sensor.		

19–4

DIAGNOSTIC TEST (LUCAS 'EPITEST')

PREPARATION

Ensure all connections are clean and tight, particularly battery connections and engine to chassis bonding strap. Check battery condition and ensure ignition system satisfactory.

If engine still refuses to run, carry out the following test procedure commencing with Part A.

Operating Instructions for 'EPITEST' Test Box

CAUTION: Ensure ignition is switched off at all times when making or breaking main wiring connections.

Part A
1. Disconnect harness multi-plug from ECU.
2. Connect test box 25-way multi-plug adaptor to vehicle harness multi-plug.
3. Select 'Switch II' on switch I.
4. Select 'Volts 1' on switch II.
5. Switch on vehicle ignition and leave on for all tests in Part A. Warning lamp on test box should illuminate, otherwise check ignition supply, multi-plug, etc.
6. Continue testing until end of Part A, moving to new position on switch II, and using push switches when indicated on test chart.
7. Rectify any faults found before continuing with further tests.
8. Switch off ignition on completion of tests in Part A.

TEST CHART – PART A

TEST 1 – CHECK VOLTAGE SUPPLIES TO ECU – SWITCH II AT 'VOLTS 1'

METER READING		POSSIBLE FAULTS AND REMEDIES
CORRECT	INCORRECT	
11–12.5V	No reading	(a) Open circuit or poor connections between pin 16 of ECU and terminal 87 of main relay. (b) Main relay not energised; check voltage at terminal 86 of main relay. If 0V, check feed from ignition switch. If satisfactory, check relay (terminal 85) and its ground connections. (c) Check voltage at terminal 30 of relay. If 0V, check battery supply.
	Low–below 11V	Battery flat, or high resistance in cable to pins 11 and 16 of ECU, or across relay contacts.

TEST 2 – CHECK VOLTAGE SUPPLIES TO ECU – SWITCH II AT 'VOLTS 2'

METER READING		POSSIBLE FAULTS AND REMEDIES
CORRECT	INCORRECT	
11–12.5V	No reading	As test 1, but check for open circuit or poor connections between pin 24 of ECU and terminal 87 of main relay.

TEST 3 – CHECK VOLTAGE AT START FEED TO ECU (TERMINAL 18) – CRANK ENGINE WITH STARTER MOTOR – SWITCH II AT 'CRANK VOLTS'

METER READING		POSSIBLE FAULTS AND REMEDIES
CORRECT	INCORRECT	
9.0–12V while cranking	No reading, starter operates	Open circuit between starter relay terminal C4 and ECU (check cable to terminal 18 on ECU). Also check operation of starter relay.
	No reading, starter does not operate	Check ignition/start switch, starter relay, solenoid, all connections and associated wiring.
	Below 9.0V	Battery flat, or excessive voltage drop. Check all cables and connections including ignition/start switch, and starter relay circuit (check cable with voltmeter).

19–5

TEST CHART – PART A

CALIBRATE METER CIRCUITS TO BATTERY VOLTAGE/CHECK BATTERY VOLTAGE – SET METER TO '00' WITH 'ADJUST 00' CONTROL – SWITCH II AT 'PRESSURE SENSOR ADJUST 00'

METER READING		
CORRECT	INCORRECT	POSSIBLE FAULTS AND REMEDIES
'∞'	Other than '00'	Battery voltage too low, replace with charged unit.

TEST 4 – CHECK PRESSURE SENSOR PRIMARY WINDING RESISTANCE – PUSH 'PRIMARY SWITCH' – SWITCH II AT 'PRESSURE SENSOR ADJUST'

METER READING		
CORRECT	INCORRECT	POSSIBLE FAULTS AND REMEDIES
00.8–1.2 ohms	Below nominal value	Damage to insulation; pull plug from pressure sensor. If meter shows '∞' replace pressure sensor.
	'0'	Short circuit to ground, or short circuit in primary winding. Pull plug from pressure sensor. If meter shows '∞', replace pressure sensor.
	Above nominal value	High resistance connections; check plugs and cables for poor connections or open circuits.
	'∞'	Open circuit; disconnect plug and bridge terminals 15 and 17. If meter shows '0', replace pressure sensor. If '∞' is indicated, check cables.

TEST 5 – CHECK PRESSURE SENSOR SECONDARY WINDING RESISTANCE – PUSH 'SECONDARY SWITCH' – SWITCH II AT 'PRESSURE SENSOR ADJUST'

METER READING		
CORRECT	INCORRECT	POSSIBLE FAULTS AND REMEDIES
3–4 ohms	As test 4 except '∞'	Open circuit, disconnect plug and bridge terminals 10 and 8. If meter shows '0', replace pressure sensor. If '∞' is indicated, check cables.

TEST 6 – CHECK PRESSURE SENSOR WINDINGS FOR SHORT CIRCUIT TO GROUND – PUSH 'GROUND' SWITCH – SWITCH II AT 'PRESSURE SENSOR ADJUST'

METER READING		
CORRECT	INCORRECT	POSSIBLE FAULTS AND REMEDIES
'∞'	'0'	Short circuit to ground in cables or at pressure sensor. Pull plug from pressure sensor, if meter shows '00' replace pressure sensor. If meter remains at '0', check cables between plug and terminals 7, 8, 10 and 15 of ECU for short circuit.
	Below '∞' but not '0'	Damage to insulation either in sensor or cables as above.

TEST 7 – CHECK TRIGGER CONTACTS IN DISTRIBUTOR – ROTATE DISTRIBUTOR BY CRANKING ENGINE – SWITCH II AT 'DISTRIBUTOR 1'

METER READING		
CORRECT	INCORRECT	POSSIBLE FAULTS AND REMEDIES
Alternating between '∞' and '1'	Constant at '∞' or '1'	Check terminals 12, 21 and 22 at trigger. Check for faulty in-line cable connector. If terminals and cable satisfactory, replace trigger unit.

TEST 8 – CHECK TRIGGER CONTACTS IN DISTRIBUTOR – ROTATE DISTRIBUTOR BY CRANKING ENGINE – SWITCH II AT 'DISTRIBUTOR 2'

METER READING		
CORRECT	INCORRECT	POSSIBLE FAULTS AND REMEDIES
Alternating between '∞' and '1'	Constant at '∞' or '1'	Check terminals 12, 21 and 22 at trigger. Check for faulty in-line cable connector. If terminals and cable satisfactory, replace trigger unit.

TEST 9 – CHECK TEMPORARY ENRICHMENT DEVICE – OPEN THROTTLE SLOWLY – SWITCH II AT 'THROTTLE 1'

METER READING		
CORRECT	INCORRECT	POSSIBLE FAULTS AND REMEDIES
Needle should swing between '1' and '∞' approx. 20 times	See next column	As the fully open throttle is released, the meter needle must remain at '∞'. If '1' is shown the throttle switch is faulty and must be replaced. If some of the swings are missed as the throttle is opened, re-check. If still not satisfactory, replace throttle switch.

19–6

TEST CHART – PART A

TEST 10 – CHECK IDLE ENRICHMENT CONTACTS IN THROTTLE SWITCH – THROTTLE CLOSED – SWITCH II AT 'THROTTLE 2'

METER READING		POSSIBLE FAULTS AND REMEDIES
CORRECT	INCORRECT	
'0'	'∞'	Throttle switch incorrectly adjusted or open circuit in cable. Check adjustment. Remove plug and bridge terminals 17 and 12/47. If still at '∞', check for short circuit to ground in cable harness, otherwise replace throttle switch.

TEST 11 – CHECK IDLE ENRICHMENT CONTACTS IN THROTTLE SWITCH – THROTTLE OPEN (5° APPROX.) – SWITCH II AT 'THROTTLE 2'

METER READING		POSSIBLE FAULTS AND REMEDIES
CORRECT	INCORRECT	
Should fall to '0' when throttle released	'0'	Throttle switch incorrectly adjusted or open circuit in cable. Remove plug if meter still shows '0', check cable harness, otherwise adjust or replace throttle switch.

TEST 12 – CHECK FULL LOAD ENRICHMENT DEVICE – FEDERAL CARS ONLY – THROTTLE FULLY OPEN – SWITCH II AT 'THROTTLE 3'

METER READING		POSSIBLE FAULTS AND REMEDIES
CORRECT	INCORRECT	
'0'	'∞'	Throttle switch incorrectly adjusted or open circuit in cable. Remove plug and bridge terminals 12/47 and 2/14. If still at '∞', check for short circuit to ground in cable harness, otherwise replace throttle switch.

TEST 13 – CHECK FULL LOAD ENRICHMENT DEVICE – FEDERAL CARS ONLY – THROTTLE 5° BEFORE FULLY OPEN – SWITCH II AT 'THROTTLE 3'

METER READING		POSSIBLE FAULTS AND REMEDIES
CORRECT	INCORRECT	
'∞'	'0'	Throttle switch incorrectly adjusted or short circuit in cable. Remove plug; if meter still shows '0', check cable harness, otherwise adjust or replace throttle switch.

TEST 14 CHECK RESISTANCE OF AIR TEMPERATURE SENSOR – SWITCH II AT 'TEMPERATURE 1'

METER READING		POSSIBLE FAULTS AND REMEDIES
CORRECT	INCORRECT	
Affected by temp. See table below	'∞'	Open circuit; remove plug and bridge terminals. If meter shows '0', replace sensor, otherwise check and rectify fault in cables.

In case of doubt, remove temperature sensor and measure resistance with ohmmeter. The nominal values, which are dependent on temperature, are listed in the table below (tolerance ±10%).

AIR TEMPERATURE SENSOR	METER READING
– 10° C corresponds to 960 ohms	9.6
0° C corresponds to 640 ohms	6.4
+ 10° C corresponds to 435 ohms	4.4
+ 20° C corresponds to 300 ohms	3.0
+ 30° C corresponds to 210 ohms	2.1
+ 40° C corresponds to 150 ohms	1.5
+ 50° C corresponds to 108 ohms	1.1
+ 60° C corresponds to 80 ohms	0.8

TEST 15 – CHECK RESISTANCE OF COOLANT TEMPERATURE SENSOR – SWITCH II AT 'TEMPERATURE 2'

METER READING		POSSIBLE FAULTS AND REMEDIES
CORRECT	INCORRECT	
Affected by temp. See table below	'∞'	Open circuit; remove plug and bridge terminals. If meter shows '0', replace sensor, otherwise check and rectify fault in cables.
	'0'	Short circuit; remove plug. If meter then shows '∞', cables are faulty. If meter still shows '0', replace sensor.

19–7

TEST CHART – PART A
TEST 15 (cont'd.)

In case of doubt, remove temperature sensor and measure resistance with ohmmeter. The nominal values, which are dependent on temperature, are listed in the table below (tolerance ±10%).

COOLANT TEMPERATURE SENSOR	METER READING
−10° C corresponds to 9.2k ohms	9.2
0° C corresponds to 5.9k ohms	5.9
+10° C corresponds to 3.7k ohms	3.7
+20° C corresponds to 2.5k ohms	2.5
+30° C corresponds to 1.7k ohms	1.7
+40° C corresponds to 1.18k ohms	1.2
+50° C corresponds to 840 ohms	0.85
+60° C corresponds to 600 ohms	0.6
+70° C corresponds to 435 ohms	0.4
+80° C corresponds to 325 ohms	0.3
+90° C corresponds to 250 ohms	0.25
+100° C corresponds to 190 ohms	0.2

TEST 16 – CHECK RESISTANCE OF EACH INPUT TO POWER AMPLIFIER – PRESS 'SWITCH 2' – SWITCH II AT 'AMPLIFIER'

METER READING		POSSIBLE FAULTS AND REMEDIES
CORRECT	INCORRECT	
6–12 (7.5 ohms approx)	'0'	Short circuit in cable or amplifier. Separate in-line plug and socket adjacent to amplifier, and if meter then shows '∞', replace amplifier, otherwise replace cable harness.
	'∞'	Open circuit in cable harness or amplifier. Separate in-line plug and socket adjacent to amplifier and bridge harness socket: (i) Pins 11 to 9 – press 'Switch 2' (ii) Pins 12 to 9 – press 'Switch 3' If meter then shows '∞' cable harness is faulty, or if '0', amplifier is faulty.
	Below '6' and over '12'	Separate in-line plug and socket adjacent to amplifier. If meter then shows '∞', replace amplifier, otherwise check cables.

TEST 17 – CHECK RESISTANCE OF EACH INPUT TO POWER AMPLIFIER – PRESS 'SWITCH 3' – SWITCH II AT 'AMPLIFIER'

METER READING		POSSIBLE FAULTS AND REMEDIES
CORRECT	INCORRECT	
As test 16	As test 16	As test 16

---

7 Switch on vehicle ignition and leave on for all tests in Part B.
8 Continue testing to end of Part B, moving to next position on switch I and using push switches as indicated in test chart.
9 Rectify any faults found before continuing further tests.
10 Switch I, position 'Fuel Check 2' should be ignored.
11 Switch off vehicle ignition on completion of Part B.

WARNING:
1 Do not operate switches 'pump' and 1, 2, 3 and 4 when switch is at 'Fuel Check' unless indicated on chart.
2 Start engine ONLY when switch I is in 'Run Engine' position.

TEST CHART
PART B – SWITCH OFF IGNITION

The position of 'Switch II' is immaterial during these tests.
1 Select 'Volts 3' on switch I.
2 Crank engine to depressurise fuel system.
3 Contact pressure gauge to fuel rail.

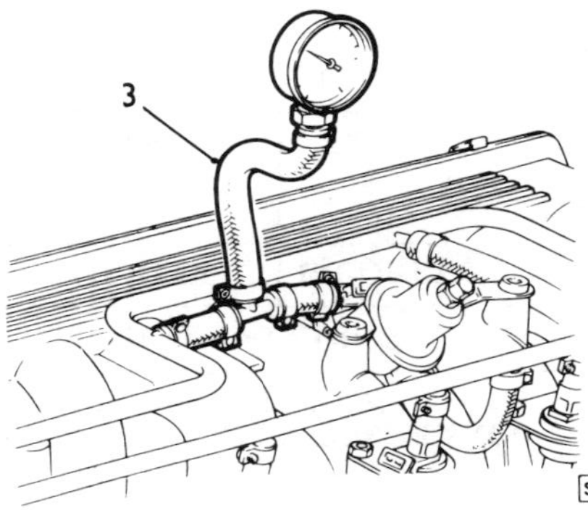

4 Locate in-line connectors adjacent to power amplifier at right-hand valance in engine compartment.
5 Connect multi-plug of 'EPITITEST' to in-line socket connector using adaptor supplied.

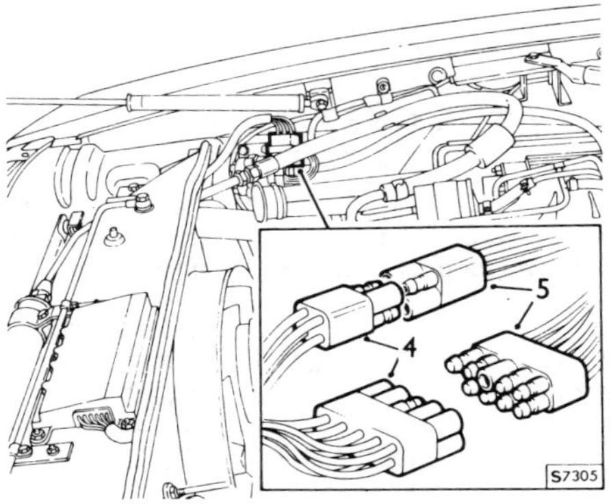

6 Connect 'EPITEST' 25-way multi-plug adaptor (still connected to vehicle harness multi-plug) to ECU.

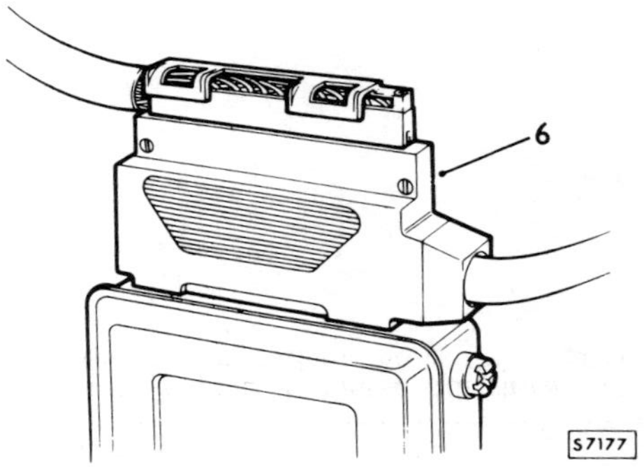

19–8

TEST CHART – PART B

TEST 1 – CHECK VOLTAGE SUPPLY TO POWER AMPLIFIER – SWITCH 1 AT 'VOLTS 3'

METER READING		POSSIBLE FAULTS AND REMEDIES
CORRECT	INCORRECT	
11–12.5V	No reading	(a) Open circuit or poor connections between pin 10 of power amplifier and terminal 87 of main relay.
		(b) Main relay not energised; check voltage at terminal 86 of main relay. If 0V, check feed from ignition switch, or if voltage is satisfactory, check relay (terminal 85) and its ground connections.
	Below 11V	Battery flat, or high resistance in either cables to power amplifier terminals 9 and 10, or across main relay terminals.

TEST 2 – CHECK RESISTANCE OF INJECTOR WINDINGS AND CONNECTIONS – SET METER TO '00' – SWITCH 1 AT 'INJECTOR'

1 Injector switch 1 energises injectors, A1, A3 and A5. Check each injector by disconnecting the plugs from the other two and pressing 'Injector Switch 1'.
2 Injector switch 2 energises injectors, B2, B4 and B6. Check each injector by disconnecting the plugs from the other two and pressing 'Injector Switch 2'.
3 Injector switch 3 energises injectors, A2, A4 and A6. Check each injector by disconnecting the plugs from the other two and pressing 'Injector Switch 3'.
4 Injector switch 4 energises injectors, B1, B3 and B5. Check each injector by disconnecting the plugs from the other two, and pressing 'Injector Switch 4'.

METER READING		POSSIBLE FAULTS AND REMEDIES
CORRECT	INCORRECT	
2–3 (2.4 ohms at 20°C) for each injector	'0'	Short circuit in cable or injector, disconnect injector, if meter shows '∞' (with appropriate switch pressed), replace injector, otherwise check and rectify cable faults.
	'∞'	Open circuit injector or cables: Bridge contacts of injector harness plug, if meter shows '0' (with appropriate switch pressed), cable is faulty. If meter shows '0', replace injector.
	'3'	High resistance connections between terminal 9 of power amplifier and engine (ground).

TEST 3 – CHECK FUEL LINE PRESSURE – PRESS 'PUMP' SWITCH – SWITCH 1 AT 'FUEL CHECK 1'. DISCONNECT L.T. LEAD AT IGNITION COIL

PRESSURE GAUGE/METER READING		POSSIBLE FAULTS AND REMEDIES
CORRECT	INCORRECT	
28.5–30.8 lbf/in² 2,0–2,2 kgf/cm²	No pressure (pump does not run) 12V	Disconnect cables from pump, press 'Pump' switch and measure voltage at cable ends. Ensure changeover valve selecting correct pump. Pump defective – replace.
	'0'	Disconnect feed to changeover switch. Check voltage; if satisfactory, check for open circuit switch. Check, by listening, that pump relay energises. If yes: Open circuit in cable between plug relay terminal 87 to changeover switch, or the positive pump connection, or from the negative pump connection to ground. Ensure fuel switch connections and contacts satisfactory. If connecting cable and plug connections are satisfactory, pump relay is defective – replace. If no: Break in cable from main relay terminal 85 to pump relay terminal 86, or from pump relay terminal 85 to pin 19 of ECU. If satisfactory, replace pump relay.
	Over 30.8 lbf/in² Under 28.5 lbf/in²	Pressure regulator(s) incorrectly adjusted; re-adjust. If adjustment not possible, pressure regulator(s) defective – replace.

TEST 4 – CHECK FOR LEAKS IN FUEL LINE – BRIEFLY PRESS 'PUMP' SWITCH (AS REQUIRED) – SWITCH 1 AT 'FUEL CHECK 1'

PRESSURE GAUGE READING		POSSIBLE FAULTS AND REMEDIES
CORRECT	INCORRECT	
17 lbf/in² min. (1.2 kgf/cm²)	See next column	If pressure drops quickly below 17 lbf/in² (1.2 kgf/cm²) as soon as 'Pump' switch is released, there is a leak in the fuel line (from pump to pressure regulator). Clamp fuel hose from pump, after fuel filter. If no pressure drop occurs, the leak is at the pump (faulty non-return valve) or in the line between the pump, and temporary clamp. If pressure continues to drop, check all pipe connections to fuel rails, injectors, cold start injectors, pressure regulators, and pressure gauge. Pinch off the fuel line before each pressure regulator. If pressure continues to drop, check injectors and cold start injectors for leaks. If pressure remains constant, release clamps at each regulator in turn to ascertain which is faulty. In order to ascertain in which injector group a known leak is to be found, the injectors must be removed (see Test 5). Operate fuel changeover switch and repeat test.

19–9

TEST CHART – PART B

TEST 5 – VISUAL CHECK OF INJECTOR SPRAY – REMOVE INJECTORS – SWITCH 1 AT 'FUEL CHECK 1'

NOTE: Carry out this test only if injectors are suspect. To remove injectors see operation 19.60.01/03.

WARNING: THIS TEST RESULTS IN FUEL VAPOUR BEING PRESENT IN THE ENGINE COMPARTMENT. IT IS THEREFORE IMPERATIVE THAT ALL DUE PRECAUTIONS ARE TAKEN AGAINST FIRE AND EXPLOSION.

Disconnect plugs from two injectors in each bank. Press 'Pump' switch, then switches 1, 2, 3 and 4 in sequence, visually checking one injector in each bank. Repeat for other injectors, collect sprayed fuel.

The injector orifice may become wet, but not more than two drops should form per minute on the nozzle. If no leaks have been determined, change the pressure regulator.

Press 'Pump' switch and check visually for leaks.

TEST 6 – CHECK COLD START INJECTOR AND THERMOTIME SWITCH – COOLANT TEMPERATURE ABOVE RATED VALUE OF SENSOR

PROCEDURE	CORRECT FUNCTION	POSSIBLE FAULTS AND REMEDIES
Press and hold 'Pump' switch, operate starter for one second with switch 1 at 'Fuel Check 1'	Very small or no pressure drop.	Pressure continues to fall: Thermotime switch faulty – replace.
Join connection 'W' of thermotime switch cable to ground. Operate starter for one second with switch 1 at 'Fuel Check 1'	Injector valve injects, pressure drops after engine has stopped.	Pressure does not drop: Check cables from starter relay (terminal C4) to terminal 85 (cold start relay), from thermotime switch (G or W) to terminal 85 (cold start relay), from terminal 30 of cold start relay. Also check cables from terminal 87 (pump relay) to cold start injectors and from injectors to ground. If cables and connections satisfactory, check cold start injectors. Resistance value of winding-in cold start coils is 4.2 ohms at 20° C.

TEST 7 – CHECK COLD START INJECTOR AND THERMOTIME SWITCH – COOLANT TEMPERATURE BELOW RATED VALUE OF SENSOR

PROCEDURE	CORRECT FUNCTION	POSSIBLE FAULTS AND REMEDIES
Press 'Pump' switch, operate starter for one second with Switch 1 at 'Fuel Check 1'	Pressure drops slowly	Pressure does not drop: Replace thermotime switch or test cold start injector as described in test 6.

TEST 8 – CREATE VACUUM IN PRESSURE SENSOR – SWITCH 1 AT 'RUN ENGINE' – RECONNECT L.T. LEAD AT IGNITION COIL

1. Reconnect vehicle in-line connector to power amplifier.
2. Switch off ignition and connect 25-way multi-plug from 'EPITEST' (harness plug attached) to ECU.
3. Connect vacuum gauge (no longer required to check fuel pressure).
*4. Start and run engine at 3,000 to 4,000 rev/min.
*5. Release throttle, and very quickly, before engine revs. drop, clamp pipe from pressure sensor between sensor and 'T' piece of balance pipes.

VACUUM GAUGE READING		POSSIBLE FAULTS AND REMEDIES
CORRECT	INCORRECT	
16 in Hg (400 mm Hg) with pipes clamped	Below 16 in Hg (400 mm Hg)	Repeat operations 1 to 5: If still unsatisfactory, check that gauge will read high vacuum without clamping when throttle quickly released from high revs. If still not possible to obtain high reading, check engine valve clearances, air leaks, etc.
	High, but cannot be maintained when pipe clamped.	Pressure sensor suspect or clamping operation not satisfactory (too late or leaking). If in doubt, test sensor with pump* to pull sufficient vacuum.

*If vacuum pump available, disconnect sensor feed pipe at sensor. Connect pump to sensor and pull approximately 16 in Hg (400 mm Hg).

TEST 9 – CHECK MECHANICAL OPERATION OF PRESSURE SENSOR – SWITCH 1 AT 'PRESSURE SIGNAL' – ENGINE WILL STOP

METER READING		POSSIBLE FAULTS AND REMEDIES
CORRECT	INCORRECT	
Swings between 2–4 and 12–14 on volts scale	Irregular swing	Pressure sensor faulty – replace.
	Final reading below 14	Check battery voltage (see 'EPITEST' diagnostic test Part A, Switch II position 'Volts 1' and 'Volts 2').

19–10

AIR CLEANER

Remove and refit—Left-hand 19.10.01

Removing
1. Disconnect battery, see 86.15.20.
2. Pull electrical connector from air temperature sensor.
3. Release two toggle clips securing air cleaner cover and pull cover from backplate.
4. Remove air cleaner element.
5. Release hose clip and remove auxiliary air valve hose from backplate.
6. Disconnect P.C.V. to auxiliary air valve feed pipe hose at P.C.V. valve.
7. Release hose clip and disconnect crankcase breather hose at P.C.V. valve. 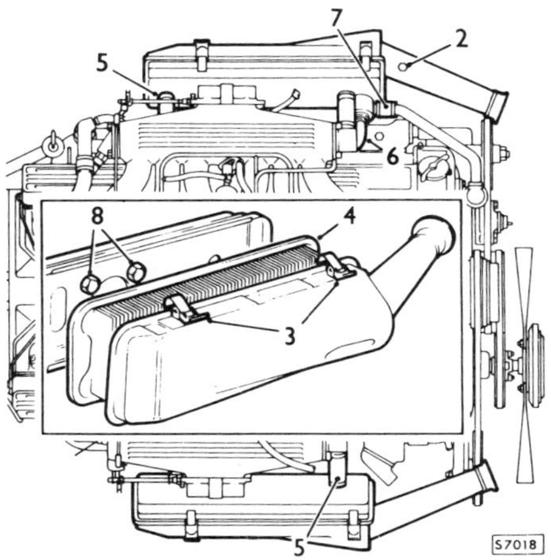
8. Remove four setscrews and washers securing air cleaner backplate to throttle housing.
9. Remove air cleaner backplate complete with overrun valve hose from throttle housing.
10. Remove and discard gasket and overrun valve hose from rear of backplate.

Refitting
NOTE: If necessary transfer components to replacement air cleaner cover and backplate.

11. Fit new overrun valve hose to rear of backplate and secure using new hose clip.
12. Locate new gasket to throttle housing using suitable sealing compound.
13. Fit air cleaner backplate to throttle housing locating overrun valve hose to valve and auxiliary air valve hose to backplate. Secure using new hose clip.
14. Reconnect P.C.V. to auxiliary air valve feed pipe hose at P.C.V. valve.
15. Secure air cleaner backplate to throttle housing using four setscrews, and spring washers.
16. Reconnect crankcase breather hose at P.C.V. valve. Secure using new hose clip.
17. Ensure air cleaner element seal in good condition and locate air cleaner element.
    NOTE: Ensure element correctly orientated with metal plate opposite throttle housing.
18. Locate air cleaner cover and secure two toggle clips.
19. Fit electrical connector to air temperature sensor.
20. Reconnect battery.

AIR CLEANER

Remove and refit – Right-hand 19.10.02

Removing
1. Release two toggle clips securing air cleaner cover and pull cover from backplate.
2. Remove air cleaner element.
3. Remove four setscrews and washers securing air cleaner backplate to throttle housing.
4. Remove air cleaner backplate complete with overrun valve hose from throttle housing.
5. Remove and discard gasket and overrun valve hose from rear of backplate.

Refitting
6. Fit new overrun valve hose to rear of backplate and secure using new hose clip.
7. Locate new gasket to throttle housing using suitable sealing compound.
8. Fit air cleaner backplate to throttle housing locating overrun valve hose to valve.
9. Secure air cleaner backplate to throttle housing using four setscrews and spring washers.
10. Ensure air cleaner element seal in good condition and locate air cleaner element.
    NOTE: Ensure element correctly orientated with metal plate opposite throttle housing.
11. Locate air cleaner cover and secure two toggle clips.

AIR CLEANERS

Renew element 19.10.08

1. Disconnect battery, see 86.15.20.
2. Pull electrical connector from air temperature sensor – left-hand air cleaner only.
3. Release two toggle clips securing air cleaner cover and pull cover from backplate. 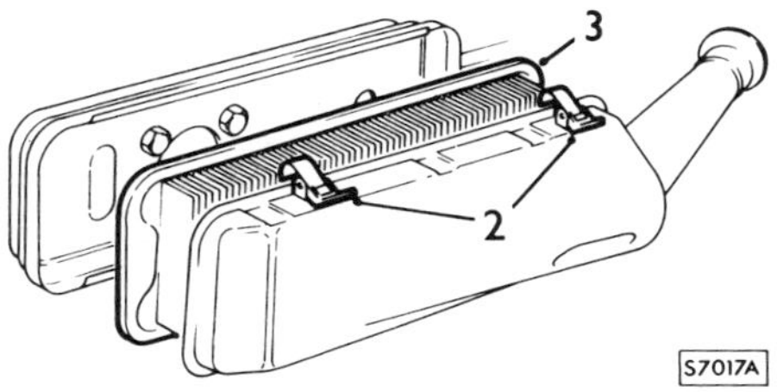
4. Remove and discard air cleaner element.
5. Ensure seal on new air cleaner element in good condition and locate air cleaner element.
   NOTE: Ensure element correctly orientated with metal plate opposite throttle housing.
6. Locate air cleaner cover and secure two toggle clips.
7. Fit electrical connector to air temperature sensor – left-hand air cleaner only.
8. Reconnect battery.

THROTTLE PEDAL

Remove and refit – Left-hand drive 19.20.01

Removing
1. Remove and discard split pin securing throttle cable to throttle pedal.
2. Pull off sleeve and disconnect throttle cable.
3. Remove two screws securing throttle pedal assembly to bulkhead.
4. Remove throttle pedal assembly.
5. Remove two circlips securing nylon bushes to throttle pedal bracket.
6. Turn nylon bushes through 45° and remove from throttle pedal bracket.
7. Remove pedal from bracket.

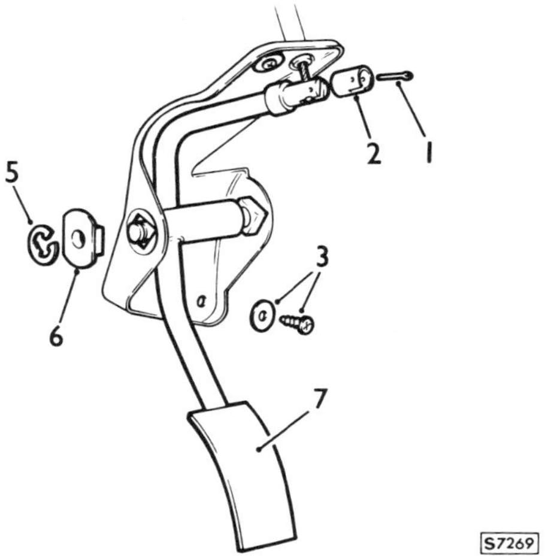

Refitting
Reverse operations 1 to 7.

THROTTLE PEDAL

Remove and refit – Right-hand drive   19.20.01

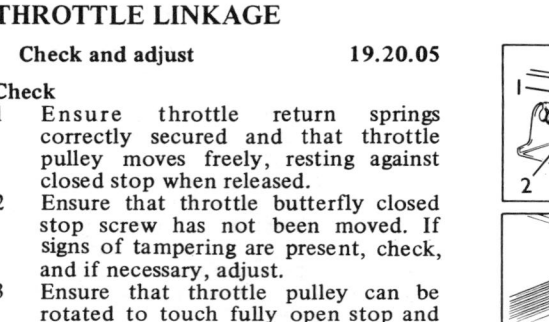

Removing
1. Remove and discard split pin securing throttle cable to throttle pedal.
2. Disconnect throttle cable from throttle pedal.
3. Remove three bolts and spring washers securing throttle pedal assembly to bulkhead.
4. Remove throttle pedal assembly.
5. Remove spring clip and plain washer securing nylon bush to throttle pedal bracket.
6. Turn nylon bushes through 45° and remove from throttle pedal bracket.
7. Remove pedal from bracket.

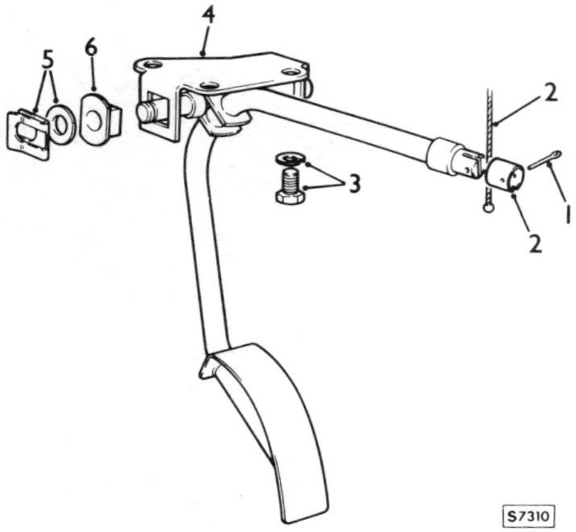

Refitting
Reverse operations 1 to 7.

THROTTLE PEDESTAL

Overhaul   19.20.03

Dismantling
1. Disconnect battery, see 86.15.20.
2. Disconnect throttle push rods from throttle pulley.
3. Remove nut securing throttle cable to pedestal bracket.
4. Rotate throttle pulley and release throttle cable nipple.
5. Disconnect multi-pin connector from throttle switch.
6. Remove four nuts and spring washers securing platform to pedestal, remove platform from pedestal.
7. Remove two taptite screws securing switch operating rod to throttle pulley, remove rod.
8. Remove circlip and spring washer securing throttle pulley to platform, remove pulley.
9. Remove two bushes from throttle pulley bore.
10. Remove and discard return spring.
11. Remove ball pins.
12. Clean all parts in clean petrol and inspect for wear.

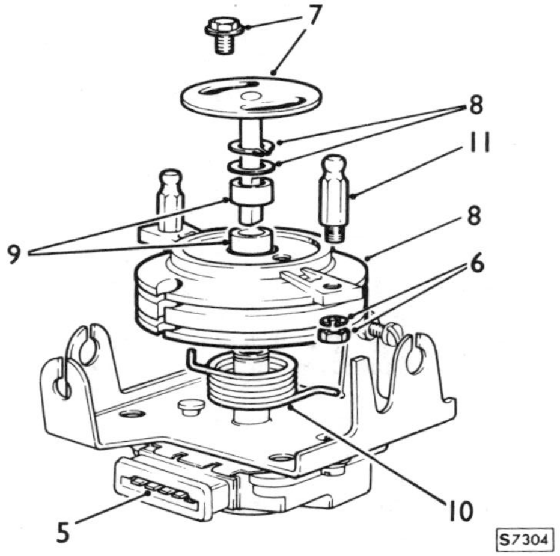

THROTTLE LINKAGE

Check and adjust   19.20.05

Check
1. Ensure throttle return springs correctly secured and that throttle pulley moves freely, resting against closed stop when released.
2. Ensure that throttle butterfly closed stop screw has not been moved. If signs of tampering are present, check, and if necessary, adjust.
3. Ensure that throttle pulley can be rotated to touch fully open stop and that throttle butterfly valve stop arm is touching throttle housing.
4. If conditions of operations 1, 2 and 3 are not satisfied, proceed with operation 5 – Adjust.

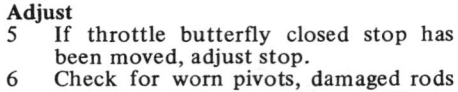

Adjust
5. If throttle butterfly closed stop has been moved, adjust stop.
6. Check for worn pivots, damaged rods or linkage and trace of any stiffness. Renew items as necessary.
7. Release throttle cross-rods from throttle pulley.
8. Slacken clamps securing levers to rear of throttle shafts.
9. With butterfly valve against closed stop, bellcrank against stop, and play in coupling taken up in opening direction, tighten clamp to lock lever to throttle shaft.
10. Repeat for other side of engine.
11. Offer cross-rods to ball connectors on pulley; rods must locate without moving pulley or linkage. If adjustment is necessary, continue with operations 11 and 12. If adjustment is not necessary, continue with operation 13.
12. Slacken locknuts on cross-rods and adjust length of rods to locate pulley ball connector while bellcrank against closed stop.
13. Tighten locknuts and ensure ball joints remain free.
14. Slacken locknut on throttle pulley, fully open stop and wind back adjustment screw.
15. Hold throttle pulley fully open and ensure that both throttle butterfly stop arms are against throttle housing.
16. Set fully open stop to just touch throttle pulley and tighten locknut.
17. Check operation of throttle switch, see 19.22.37.
18. Check kickdown switch adjustment, see 44.30.12.

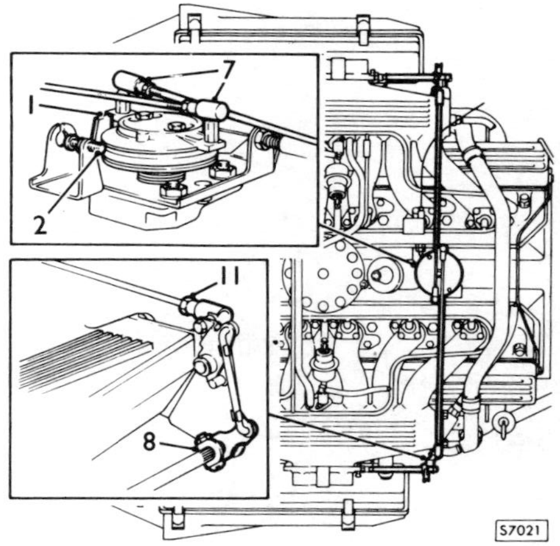

THROTTLE CABLE

Remove and refit   19.20.06

Removing
1. Disconnect battery, see 86.15.20.
2. Remove and discard split pin securing throttle cable to throttle pedal, pull off sleeve and release cable from pedal.
3. Slacken locknut and disconnect throttle cable from pulley.
4. Disconnect cable assembly from bulkhead tube. Pull cable through to engine compartment.
NOTE: On right-hand drive cars the cable is secured to the bulkhead by a nut above the driver's footwell.
5. Disconnect electrical connectors from kickdown switch.
6. Release cable from pedestal.
7. Slacken throttle cable locknut at pedestal bracket, remove throttle cable assembly.

19–12

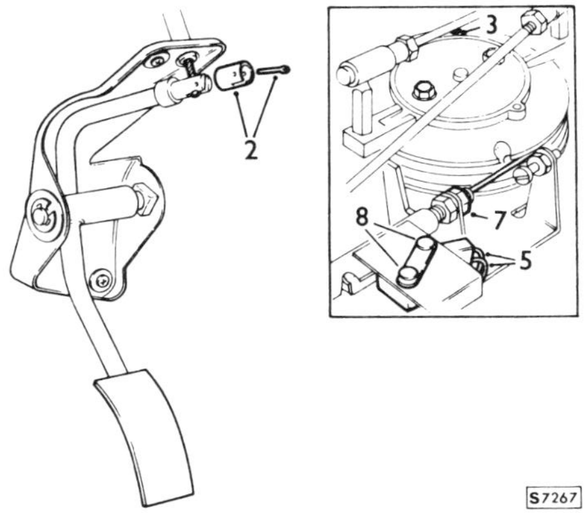

5 Adjust stop screw to just touch stop arm and tighten locknut with feeler in position.
6 Press stop arm against stop screw and withdraw feeler.

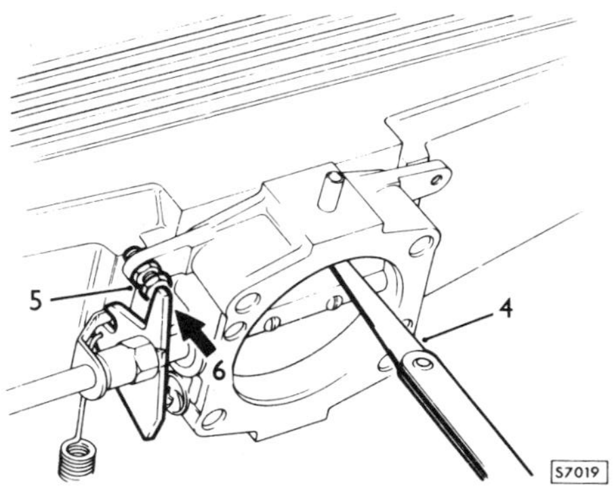

Refitting

7 Repeat on other side of engine.
8 Seal threads of adjusting screws and locknuts using a blob of paint.
9 Refit air-cleaners.
10 Check throttle linkage adjustment, see 19.20.25.
11 Check operation of throttle switch, see 19.22.37.
12 Check kickdown switch adjustment, see 44.30.12.

AUXILIARY AIR VALVE 19.20.16

Remove and refit

Removing
CAUTION: This procedure MUST ONLY be carried out on a cold or cool engine.

1 Carefully remove pressure cap from remote header tank to release any cooling system residual pressure. Replace cap tightly.
2 Slacken hose clip securing air balance pipe to auxiliary air valve and crankcase breather pipe.
3 Slacken hose clip securing manifold bleed pipe to auxiliary air valve.
4 Release air balance pipe from auxiliary air valve, manifold bleed pipe and crankcase breather pipe.

5 Slacken hose clip securing air cleaner pipe to auxiliary air valve, disconnect pipe.
6 Remove two screws and washers securing auxiliary air valve to coolant pipe, lift valve clear.
7 Clean all traces of gasket from coolant pipe, taking care not to damage seating area.
8 Scribe a line on idle speed adjusting screws and note number of turns required to screw fully in.

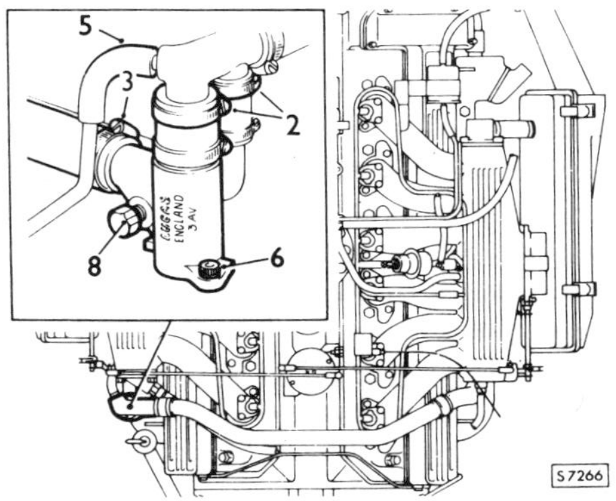

Refitting

9 Examine bulkhead grommet for damage, renew as necessary. Fit grommet to cable.
10 Reverse operations 2 to 8, apply suitable sealing compound around cable and grommet at bulkhead (right-hand drive cars).
11 Check throttle linkage adjustment, see 19.20.05.
12 Check kickdown switch adjustment, see 44.30.12.
13 Re-connect battery.

THROTTLE BUTTERFLY VALVE

Adjust – Both 19.20.11

CAUTION: Any adjustment must be carried out on both butterfly valves. It is NOT permitted to adjust one valve only.

1 Remove both air cleaners, see 19.10.01/02.
2 Slacken locknut on throttle butterfly stop screw. Wind back screw.
3 Ensure throttle butterfly valve closes fully.
4 Insert a 0.002 in (0.05 mm) feeler gauge between top of valve and housing to hold valve open.

4 Reset idle speed, replace air cleaner element.
5 Remove auxiliary air valve, see 19.20.16.
6 Fully close idle speed adjustment screw.
7 Immerse auxiliary air valve bulb in a container of boiling water and observe valve head through side port. Valve should move smoothly to closed position.
8 Quickly blow through side port; no air should pass.
9 Allow valve bulb to cool. Valve head should move smoothly back to open main air passage.
10 If valve performance is satisfactory reset idle speed adjustment screw and refit valve.
11 If valve performance is not satisfactory, fit replacement component.

IDLE SPEED

Adjust 19.20.18

1 Ensure engine is at normal operating temperature.
2 Check throttle linkage for correct operation, check that return springs are secure and effective.
3 Start engine, run for two to three minutes.
4 Set idle speed adjustment screw to achieve 750 rev/min.
NOTE: If it proves impossible to reduce idle speed to specified level proceed as detailed in operations 5 to 9.
5 Check ALL pipes and hoses to inlet manifolds for security and condition.
6 Check security of injectors and cold start injectors.
7 Ensure that all joints are tight and that inlet manifold to cylinder head fastenings are tight.
8 Ensure that both throttle butterfly closed stops show no signs of tampering; if they do, adjust throttle butterfly valves, see 19.20.11.
9 Check operation of overrun valves, see 19.20.21.
10 If operations 5 to 9 do not reduce idle speed, check operation of auxiliary air valve, see 19.20.17.

Refitting

9 Set idle speed adjustment screw of replacement valve to the number of turns open as noted in operation 8.
10 Locate valve, orientated correctly, secure using two screws and washers.
11 Reverse operations 2 to 5.
12 Check coolant level at remote header tank, if necessary top up.
13 Check, if necessary adjust idle speed, see 19.20.18.

AUXILIARY AIR VALVE

Test 19.20.17

1 Remove left-hand air cleaner element, see 19.10.01.
2 Fully close idle speed adjustment screw.
3 With engine at normal running temperature, blocking auxiliary air valve inlet should NOT affect idling speed. If idle speed affected, continue with operations 5 to 11. If idle speed is not affected continue with operation 4.

19–13

OVERRUN VALVE

Test 19.20.21

1 Remove both air cleaners, see 19.10.01/02.
2 Block overrun valve inlet pipes, start engine.
3 If idle speed now correct, stop engine, unblock one valve only.
4 Start engine. If idle speed still correct, stop engine and unblock second valve.
5 Start engine. If idle speed too fast renew relevant valve.

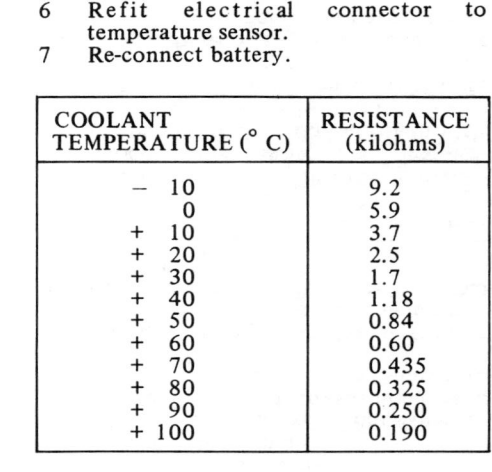

OVERRUN VALVE

Remove and refit 19.20.22

Removing
1 Remove three screws securing overrun valve to inlet manifold, retrieve spacer (left-hand side).
2 Remove valve and body by releasing air cleaner hose.
3 Remove and discard gasket, remove all traces of gasket, ensuring mating surfaces are not damaged. 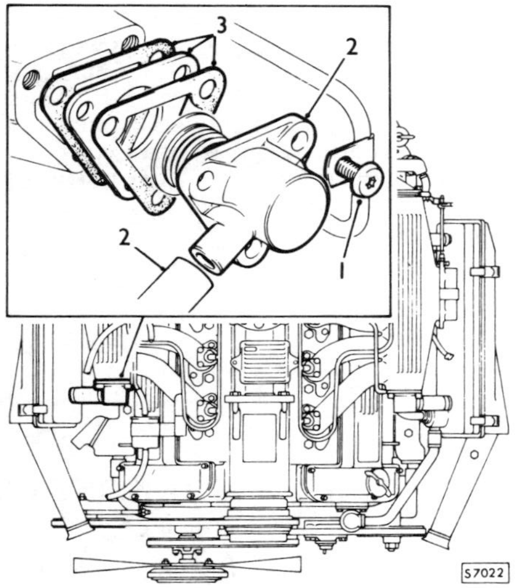
4 Remove valve from body.

Refitting
Reverse operations 1 to 4; use a new gasket.

FUEL CUT-OFF INERTIA SWITCH

Remove and refit 19.22.09

Removing
1 Disconnect battery, see 86.15.20.
2 Remove switch cover.
3 Pull electrical connectors from switch.
4 Pull switch from spring clips.

Refitting
5 Press switch into spring clips, ribs towards rear of car and terminals at bottom.
NOTE: Ensure switch is raised in clips to abut on top lip of bracket.
6 Fit electrical connectors to switch; polarity is unimportant.
7 Press in plunger at top of switch. 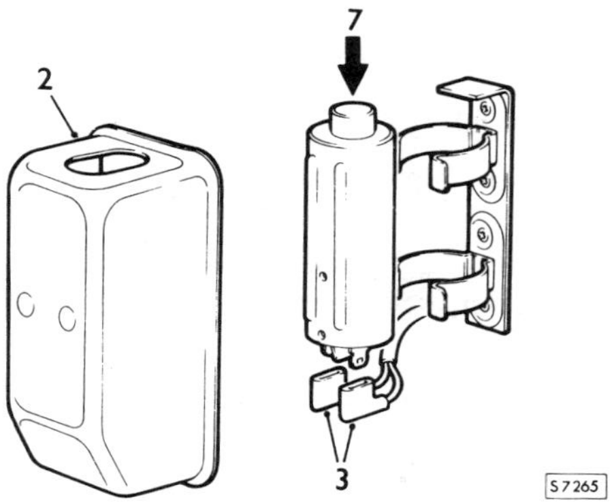
8 Fit switch cover and secure.
9 Re-connect battery.
NOTE: For fuel cut-off inertia switch test see operation 19.22.32.

COOLANT TEMPERATURE SENSOR

Remove and refit 19.22.18

Removing
NOTE: This procedure **MUST ONLY** be carried out on a cold or cool engine.
1 Disconnect battery, see 86.15.20.
2 Pull electrical connector from coolant temperature sensor. 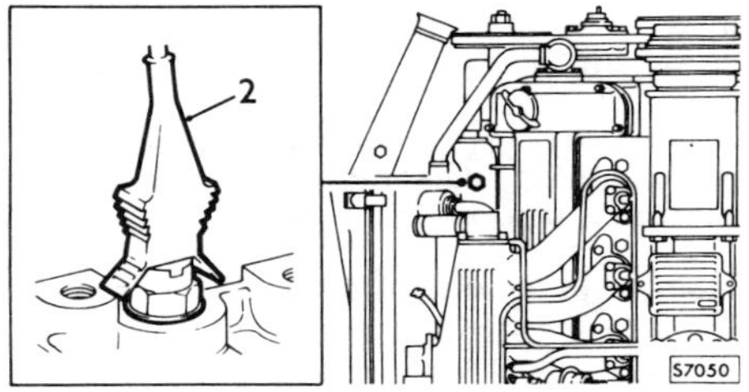
3 Carefully remove pressure cap from remote header tank to release any cooling system residual pressure. Replace cap tightly.
NOTE: The replacement component is prepared at this point and the transfer made as quickly as possible.
4 Ensure sealing washer located on replacement temperature sensor and coat threads with suitable sealing compound.
5 Remove existing temperature sensor from thermostat housing.

Refitting
6 Screw replacement temperature sensor into thermostat housing.
7 Refit electrical connector to temperature sensor.
8 Re-connect battery.
9 Check coolant level at remote header tank. If necessary, top up.

COOLANT TEMPERATURE SENSOR

Test 19.22.19

1 Disconnect battery, see 86.15.20.
2 Pull electrical connector from coolant temperature sensor.
3 Connect suitable ohmmeter between terminals, note resistance reading. The reading is subject to change according to temperature and should closely approximate to the relevant resistance value given in the table.
4 Disconnect ohmmeter.
5 Check resistance between each terminal in turn and body of sensor. A very high resistance reading (open circuit) must be obtained.
6 Refit electrical connector to temperature sensor.
7 Re-connect battery.

COOLANT TEMPERATURE (° C)	RESISTANCE (kilohms)
− 10	9.2
0	5.9
+ 10	3.7
+ 20	2.5
+ 30	1.7
+ 40	1.18
+ 50	0.84
+ 60	0.60
+ 70	0.435
+ 80	0.325
+ 90	0.250
+ 100	0.190

THERMOTIME SWITCH

Remove and refit 19.22.20

Removing
NOTE: This procedure **MUST ONLY** be carried out on a cold or cool engine.
1 Disconnect battery, see 86.15.20.
2 Pull electrical connector from thermotime switch.
3 Carefully remove pressure cap from remote header tank to release any cooling system residual pressure.
4 Remove thermotime switch.

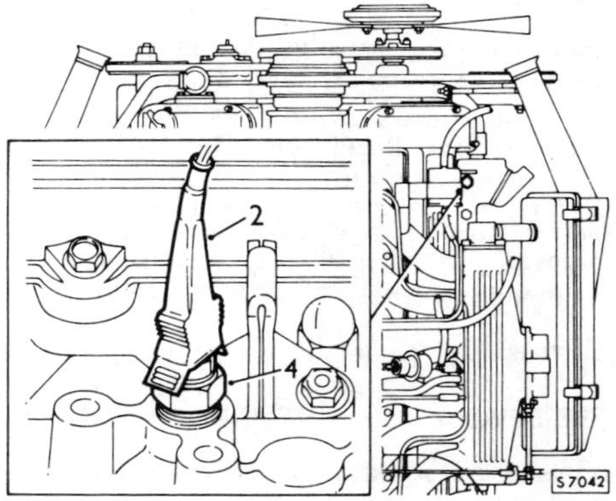

19−14

Refitting

5   Fit replacement thermotime switch using new copper washer.
6   Check coolant level at remote header tank, top up if necessary.

THERMOTIME SWITCH

Test 19.22.21

Equipment required: Stop watch, ohmmeter, single-pole switch, jump lead for connecting switch to battery and thermotime switch and a thermometer.
NOTE: Check coolant temperature with thermometer and note reading before carrying out procedures detailed below. Check rated value of thermotime switch (stamped on body flat).

1   Disconnect battery earth lead.
2   Pull electrical connector from thermotime switch.

'A' coolant temperature higher than switch rated value

3   Connect ohmmeter between terminal 'W' and earth. A very high resistance reading (open circuit) should be obtained.
4   Renew switch if a very low resistance reading (short circuit) is obtained.

'B' coolant temperature lower than switch rated value

5   Connect ohmmeter between terminal 'W' and earth. A very low resistance reading (short circuit) should be obtained.
6   Connect 12V supply via isolating switch to terminal 'G' of thermotime switch.
7   Using stop watch, check time delay between making isolating switch and indication on ohmmeter changing from **low** to **high** resistance. Delay period must closely approximate to time indicated in table for specific coolant temperature (noted above).
8   Renew thermotime switch if necessary.
9   Reconnect thermotime switch.
10 Reconnect battery earth lead.

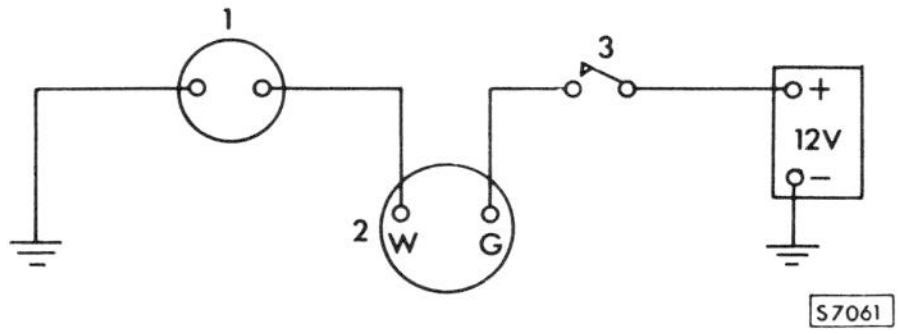

1   Ohmmeter.
2   Thermotime switch.
3   Single-pole switch.

COOLANT TEMPARATURE	DELAY – 15°C SWITCH	DELAY – 35°C SWITCH
–20°C	8 secs	8 secs
–10°C	5.7 secs	6.5 secs
0°C	3.5 secs	5 secs
+10°C	1.2 secs	3.5 secs
+15°C	0 secs	2.7 secs
+20°C	–	2.0 secs
+30°C	–	0.5 secs
+35°C	–	0 secs

AIR TEMPERATURE SENSOR

Remove and refit 19.22.22

Removing
1   Disconnect battery, see 86.15.20.
2   Pull electrical connector from air temperature sensor. 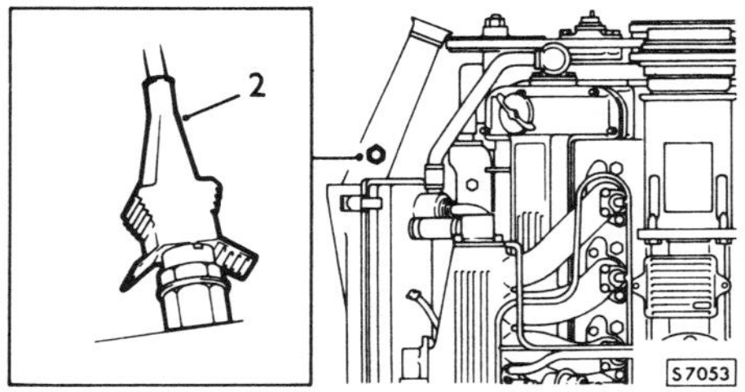
3   Remove sensor from air cleaner ram pipe.

Refitting
Reverse operations 1 to 3.

AIR TEMPERATURE SENSOR

Test 19.22.23

1   Disconnect battery, see 86.15.20.
2   Pull electrical connector from air temperature sensor.
3   Connect suitable ohmmeter between terminals, note resistance reading. The reading is subject to change according to temperature and should closely approximate to the relevant resistance value given in the table.
4   Disconnect ohmmeter.
5   Check resistance between each terminal in turn and body of sensor. A very high resistance reading (open circuit) must be obtained.
6   Refit electrical connector to temperature sensor.
7   Reconnect battery.

AMBIENT AIR TEMPERATURE (°C)	RESISTANCE (ohms)
–10	960
0	640
+10	435
+20	300
+30	210
+40	150
+50	108
+60	80

TRIGGER UNIT

Remove and refit 19.22.26

Removing
1   Disconnect battery, see 86.15.20.
2   Ensure spark plug leads adequately identified and remove leads from spark plugs on both banks of cylinders.
3   Remove H.T. lead from distributor.
4   Remove three securing screws and manoeuvre distributor cover clear.
5   Remove H.T. rotor arm.
6   Disconnect trigger unit cable from main harness at in-line connector.
7   Remove four nylon screws securing trigger unit to platform.
8   Remove nylon clamp from right-hand side of unit.
9   Remove trigger unit by releasing rubber seal from distributor body.

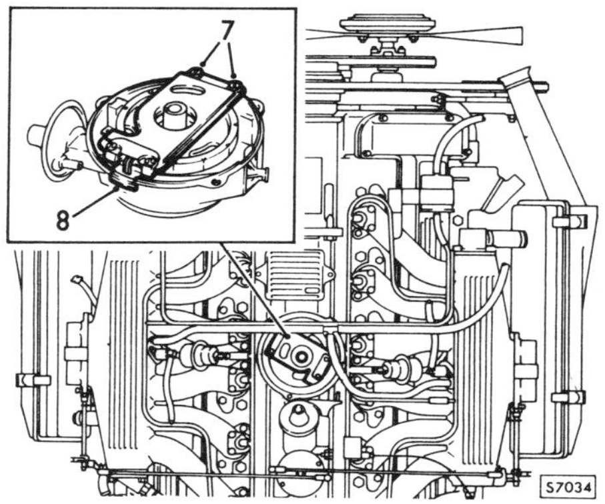

Refitting
Reverse operations 1 to 9.

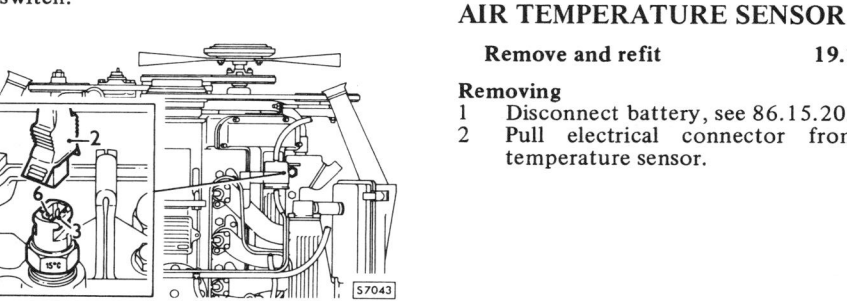

19–15

TRIGGER UNIT

Test   19.22.27

1 Disconnect cable from pump relay terminal 85.
2 Disconnect '–ve' cable at ignition coil.
3 Separate in-line connector to trigger unit.
4 Connect suitable ohmmeter between terminals 21 and 12 of distributor side in-line connector.
5 Switch 'on' ignition and crank engine. Ohmmeter should show regular even swing between low resistance (a current limiting resistor is fitted within the trigger unit) and a very high resistance (open circuit).
6 Transfer ohmmeter to terminals 22 and 12. An identical result to that shown in operation 5 must be obtained.
7 If ohmmeter displays a steady reading during cranking or the swing is uneven or intermittent, the trigger unit must be replaced.
8 Switch off ignition.
9 Re-connect '–ve' cable at ignition coil.
10 Re-connect cable to pump relay terminal 85.

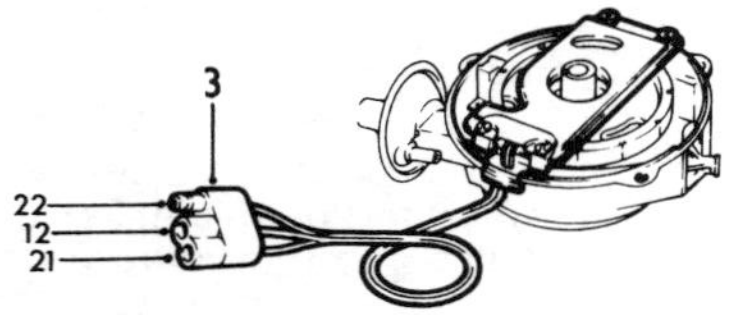

MANIFOLD PRESSURE SENSOR

Test   19.22.28

1 Disconnect battery, see 86.15.20.
2 Pull electrical connector from pressure sensor.
3 Connect suitable ohmmeter between pressure sensor terminals and note resistance readings.
If readings are outside the limits indicated in the table, pressure sensor is faulty and must be replaced.

4 Refit electrical connector to pressure sensor.
5 Re-connect battery.

SENSOR TERMINALS	RESISTANCE (Ohms)
Between 7 and 15	85.5 to 94.5
Between 8 and 10	346.5 to 353.5
Between 7 or 15 and ground	Open circuit (00)
Between 8 or 10 and ground	Open circuit (00)

MANIFOLD PRESSURE SENSOR

Remove and refit   19.22.29

Removing

1 Disconnect battery, see 86.15.20.
2 Pull electrical connector from pressure sensor.
3 Release hose clip and draw pressure pipe from pressure sensor.
4 Remove three bolts and washers securing pressure sensor bracket to wing valance, remove pressure sensor complete with bracket.
5 Remove two setscrews and washers securing pressure sensor to bracket.

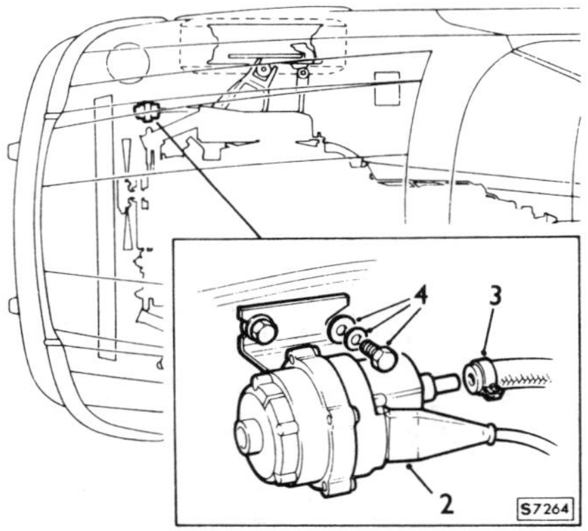

COLD START RELAY

Remove and refit   19.22.31

Removing

1 Disconnect battery, see 86.15.20.
2 Remove two screws securing cover to relay mounting bracket. Remove cover.
3 Remove two screws, one plain and one fibre washer securing mounting bracket to wing valance. Lift bracket and relays complete to gain access — right-hand drive cars only.
4 Remove nut, spring washer and plain washer securing relay to mounting bracket.

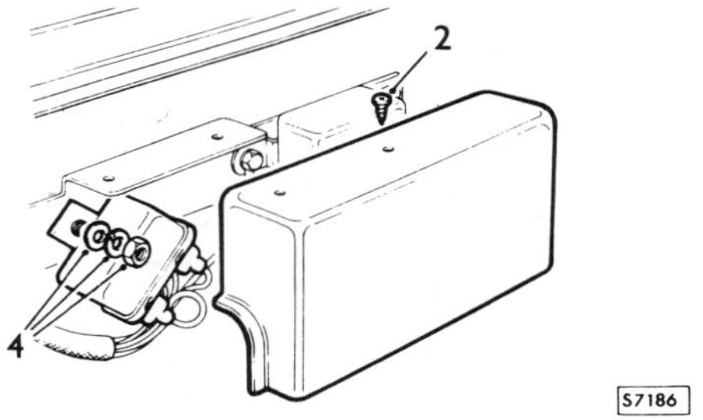

3 Disconnect '–ve' terminal from ignition coil.
   NOTE: If engine is cold (below 15°C), proceed with operations 4 and 5. If engine is hot (above 15°C) proceed with operations 7 to 11.
4 Switch on ignition and observe cold start injectors for fuel leakage. Crank engine one or two revolutions. Injectors should spray while engine cranks.
   NOTE: Do not operate starter motor longer than necessary to make this, and the following observations.

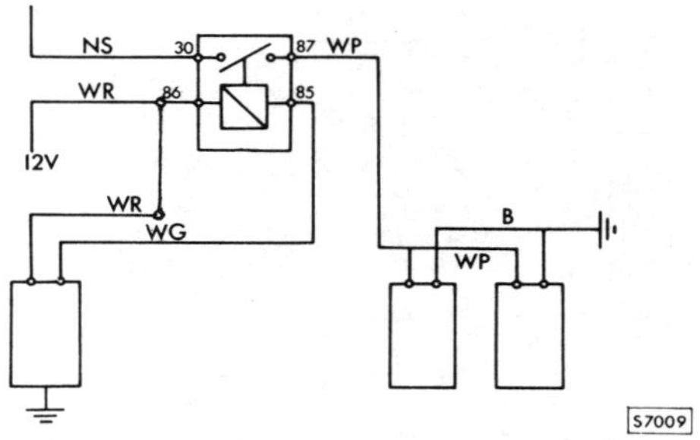

5 If injectors do not spray, carry out the following tests.
  a Crank engine. Check for battery voltage at cold start injector supply (White/Pink cable).
    If yes: check all electrical plug and earth connections to cold start injectors. If satisfactory, cold start injectors are suspect.
    If no: Proceed to test (b).
  b Crank engine. Check for battery voltage at terminal 87.
    If yes: Check cables between cold start relay and cold start injectors.
    If no: Proceed to test (c).
  c Crank engine. Check for battery voltage at terminal 30.
    If yes: Cold start relay not energised or contacts faulty, proceed to test (d).
    If no: Check supply from pump relay.

5 Note position of and remove cable connectors from relay.
6 Remove relay.

Refitting
Reverse operations 1 to 6.

COLD START SYSTEM

Test   19.22.32

WARNING: THIS TEST RESULTS IN FUEL VAPOUR BEING PRESENT IN THE ENGINE COMPARTMENT. IT IS THEREFORE IMPERATIVE THAT ALL DUE PRECAUTIONS ARE TAKEN AGAINST FIRE AND EXPLOSION.

1 Remove two setscrews and washers securing cold start injectors to inlet manifolds. Remove cold start injectors.
2 Arrange containers to collect sprayed fuel.

19–16

d Crank engine. Check for battery voltage at terminal 86.
   If yes: Cold start relay not energised or contacts faulty. Proceed to test (e).
   Disconnect lead at terminal 85.
e Check for battery voltage at terminal 85.
   If yes: Bridge terminal 85 to earth. Cold start relay should energise – check voltage at terminal 87. If 0V, replace relay. If satisfactory, check cables and connections to thermotime switch. If satisfactory, thermotime switch is suspect.
   If no: Replace cold start relay.
6 Replace cold start injectors and all cables and connectors removed.
7 Crank engine. Voltage at terminal 87 of cold start relay must be 0V.
8 If battery voltage present, disconnect terminal 85. Crank engine and re-check voltage at terminal 87.
9 If voltage at terminal 87 is 0V, thermotime switch is faulty and should be replaced.
10 If battery voltage is present, replace cold start relay.
11 If cold start injectors pass fuel while voltage at terminal 87 is 0V, injector(s) must be renewed.

POWER AMPLIFIER

Remove and refit 19.22.33

Removing
1 Disconnect battery, see 86.15.20.
2 Remove screw securing harness clip to radiator top-rail.
3 Remove bolt and washer securing harness clip to manifold pressure sensor mounting bracket.
4 Remove four nyloc nuts securing power amplifier to radiator top-rail.
5 Disconnect multi-plug connector from main harness.
6 Remove grommet from elongated hole in radiator top-rail.

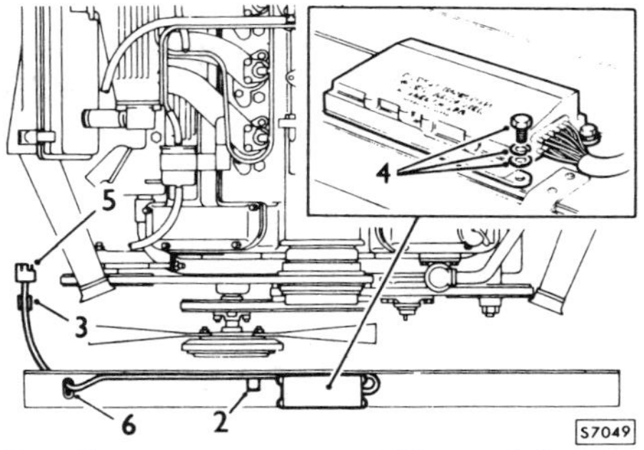

7 Remove power amplifier and harness passing multi-plug connector through elongated hole in radiator top-rail.

Refitting
Reverse operations 1 to 7.

ELECTRONIC CONTROL UNIT (ECU)

Remove and refit 19.22.34

Removing
1 Disconnect battery, see 86.15.20.
2 Remove right-hand luggage compartment trim panel.
3 Remove screw securing ECU harness clamp, move clamp aside.
4 Unclip ECU and cover.
5 Locate handle on harness plug and withdraw plug.
6 Pull ECU clear of mounting bracket.

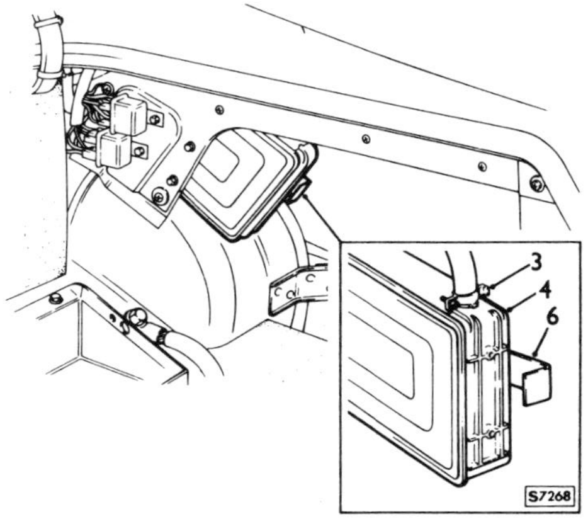

13 Refit throttle cross-rods to throttle pulley. Ball connectors must fit without moving throttle bell-cranks. Adjust length of rods if necessary.
   NOTE: When tightening locknuts ensure ball joints remain free.
14 Re-connect electrical connector to throttle switch.
15 Re-connect battery.

Refitting
CAUTION: THE IDLE FUELLING POTENTIOMETER IS PRE-SET AND MUST NOT BE MOVED.
7 Reverse operations 1 to 6.
8 Check idle CO level, see 17.35.01.

THROTTLE SWITCH

Adjust 19.22.35

1 Disconnect battery, see 86.15.20.
2 Remove throttle cross-rods from throttle pulley.
3 Pull electrical connector from throttle switch.
4 Remove four nuts and spring washers securing throttle platform to pedestal and lift clear.
5 Slacken two screws securing throttle switch.

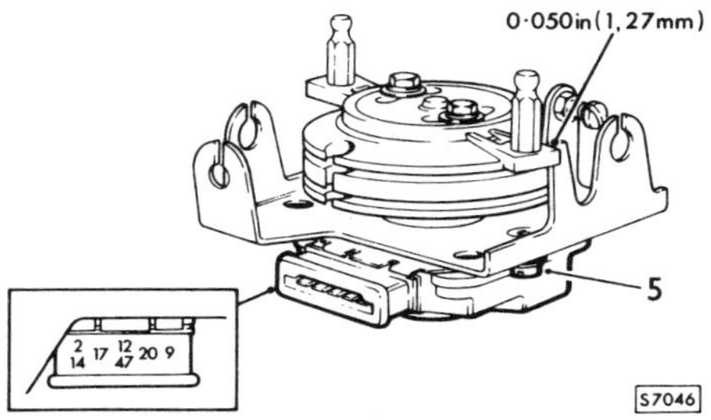

6 Insert a 0.050 in. (1,27 mm) feeler gauge between pulley and closed throttle stop.
7 Connect a suitable ohmmeter between terminals 12 and 17 of throttle switch.
8 Turn throttle switch slowly until ohmmeter flicks to very high resistance.
9 Tighten two fixing screws.
10 Remove feeler gauge; ohmmeter should read very low resistance (short circuit) when pulley is against closed stop.
11 Disconnect ohmmeter.
12 Fit throttle pulley platform on pedestal and secure using four nuts and spring washers.

THROTTLE SWITCH

Remove and refit 19.22.36

Removing
1 Disconnect battery, see 86.15.20.
2 Disconnect throttle push rods from throttle pulley.
3 Slacken nut securing throttle cable to pedestal bracket.
4 Pull electrical connector from throttle switch.
5 Remove four nuts and spring washers securing platform to pedestal, lift platform from pedestal.

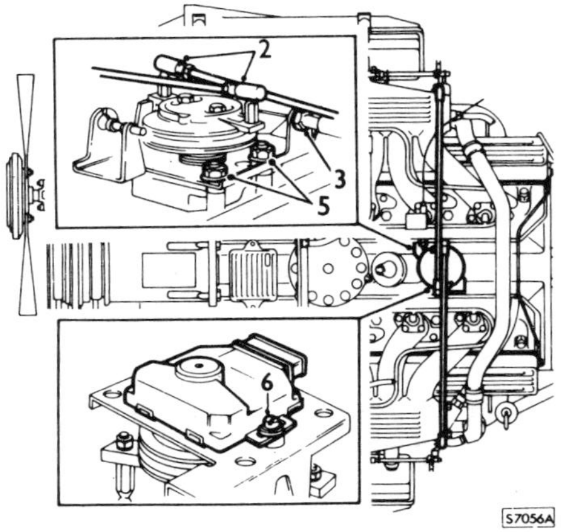

6 Remove two screws, plain and shakeproof washers securing throttle switch and lift switch from the spindle.

Refitting
7 Locate switch on spindle with connector socket facing to rear.

continued

19–17

8 Secure switch using two screws, plain and shakeproof washers, do not tighten.
9 Set up switch following operations 6 to 14 of Throttle Switch – Adjust – 19.22.35.
10 Re-connect battery.

THROTTLE SWITCH

Test 19.22.37

NOTE: Before commencing the following tests ensure that the throttle butterfly valve and throttle linkage are correctly set.

1 Disconnect battery, see 86.15.20.
2 Pull electrical connector from throttle switch.

Idle positions

3 Connect ohmmeter between terminals 12/47 and 17 of switch. Meter reading must show very low resistance (short circuit). If not, adjust before proceeding.

Temporary enrichment

4 Connect ohmmeter between terminals 9 and 12/47 of switch.
5 Observe ohmmeter and turn throttle pulley by hand. Meter needle should swing regularly between very low resistance (short circuit) and very high resistance (open circuit) ten times.
NOTE: If meter needle does not swing consistently or remain steady, replace throttle switch.
6 Refit electrical connector to throttle switch.
7 Re-connect battery.

MAIN RELAY

Remove and refit 19.22.38

Removing

1 Disconnect battery, see 86.15.20.
2 Remove right-hand luggage compartment trim panel.
3 Note position of and remove electrical connections from main relay.
4 Remove screw securing main relay to luggage compartment side panel, remove relay.

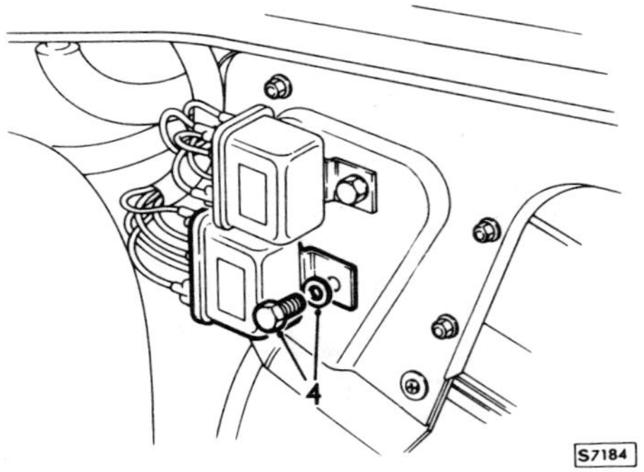

Refitting
Reverse operations 1 to 4.

PUMP RELAY

Remove and refit 19.22.39

Removing

1 Disconnect battery, see 86.15.20.
2 Remove right-hand luggage compartment trim panel.
3 Note position of and remove electrical connections from main relay.
4 Remove screw securing main relay to luggage compartment side panel, remove relay.

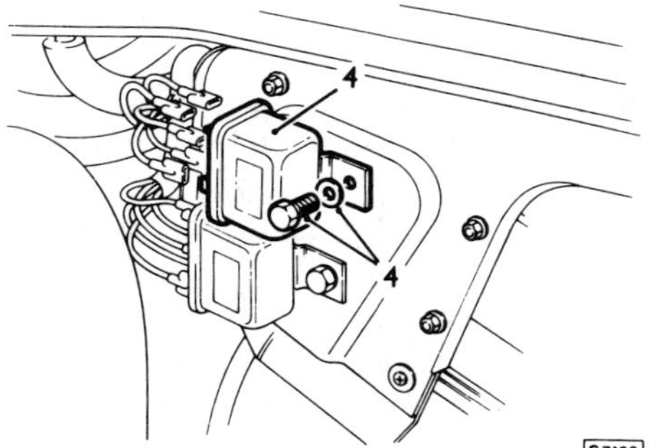

Refitting
Reverse operations 1 to 4.

PUMP RELAY

Test 19.22.40

1 Switch on ignition. Pump should run for one to two seconds, then stop.
NOTE: If pump does not run, or does not stop, check systematically as follows:
2 Check inertia switch cut-out button pressed in.
3 Detach inertia switch cover and ensure both cables secure.
4 Pull electrical connectors from switch and check continuity across terminals.

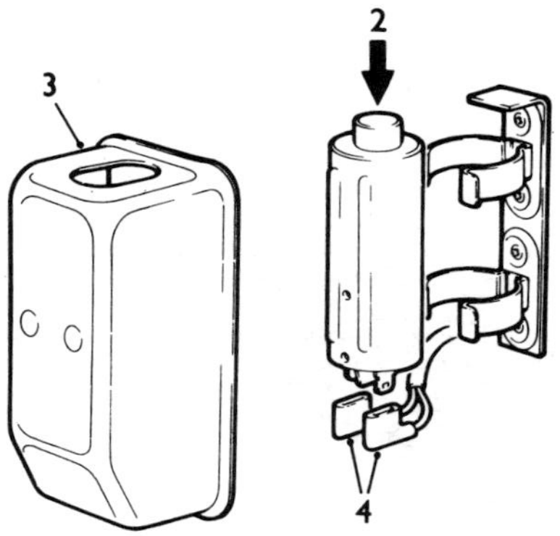

5 Pull button out and check open circuit.
6 Remove ohmmeter, replace electrical connectors, re-set button and refit cover.
7 If inertia switch satisfactory earth (ground) pump relay terminal 85, switch on ignition and check circuit systematically as detailed below.

a Check for battery voltage at terminal 86 of main relay.
  If yes: Proceed to test b.
  If no: Check battery supply from ignition switch via inertia switch.
b Check for battery voltage at terminal 87 of main relay.
  If yes: Proceed to test c.
  If no: Check for battery voltage at earth (ground) lead and connection from terminal 85 of main relay. If satisfactory renew main relay.
c Check for battery voltage at terminal 86 of pump relay.
  If yes: Proceed to test d.
  If no: Open circuit between terminals 87 of main relay and 86 of pump relay. If satisfactory proceed to test d.
d Check for battery voltage at terminal 87 of pump relay.
  If yes: Proceed to test e.
  If no: Check for battery voltage to earth (ground) lead and connections from terminal 85 of pump relay. If satisfactory, renew pump relay.
e Check for battery voltage at supply lead (NS) and connections to fuel pump.
  If yes: Faulty pump or earth (ground) connections.
  If no: Open circuit between terminal 87 of pump relay and supply lead connection to fuel pump.

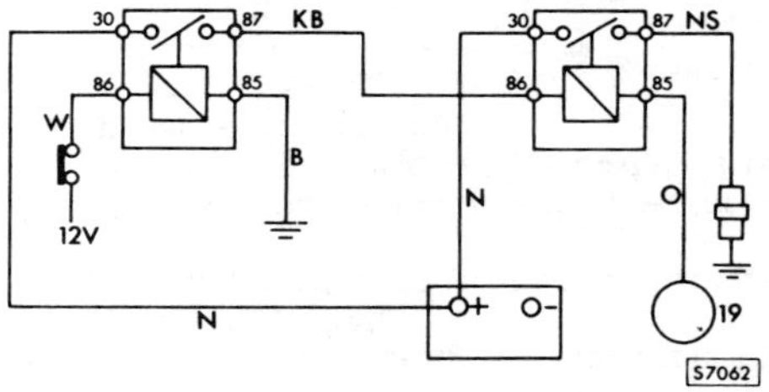

FUEL MAIN FILTER

Remove and refit   19.25.02

Removing
1. Depressurise fuel system, see 19.50.02.
2. Clamp inlet and outlet hoses.
3. Release hose clips and pull hoses from fuel filter.
4. Remove bolt and washer securing filter assembly to inlet manifold mounting bracket, remove filter assembly.
5. Remove screw and spire nut from filter clamp, draw fuel filter clear of clamp.

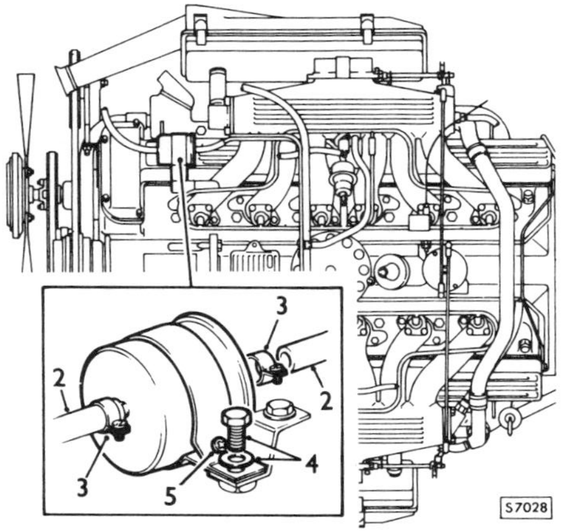

Refitting
Reverse operations 1 to 5.

FUEL PRESSURE REGULATORS

Remove and refit   19.45.11

Removing
1. Depressurise fuel system, see 19.50.02.
2. Disconnect battery, see 86.15.20.
3. Disconnect electrical connector from cold start injector.
4. Remove two setscrews and washers securing pressure regulator mounting bracket and carefully pull regulator upwards. Note orientation of regulator in bracket.
5. Clamp inlet and outlet pipes of regulator.
6. Release hose clips and pull hoses from regulator unions.
7. Remove nut and washer and release regulator from bracket.

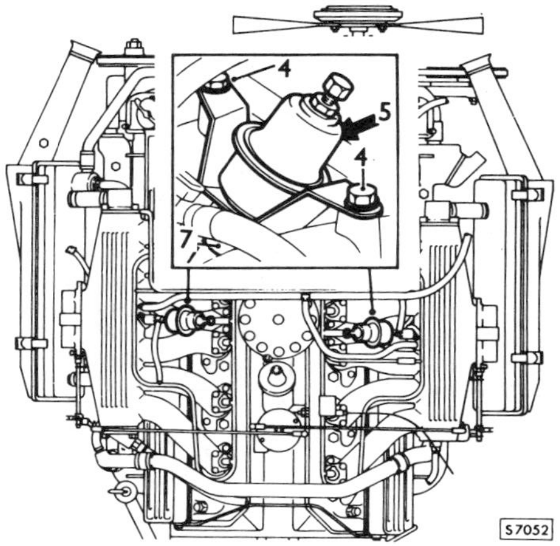

FUEL PUMP

Remove and refit   19.45.08

Removing
1. Depressurise fuel system, see 19.50.02.
2. Disconnect battery, see 86.15.20.
3. Remove spare wheel.
4. Remove two screws securing fuel pump cover to battery tray.
5. Peel back floor carpet, release fuel pump cover from floor clips.
6. Fit clamps to inlet and outlet hoses.
7. Remove two nyloc nuts and plain washers securing fuel pump retaining band to mounting, retrieve spacer washers and remove fuel pump assembly.
8. Remove electrical connector from fuel pump socket.
   NOTE: Place suitable receptacle beneath car to collect spilled fuel.
9. Release two hose clips securing inlet and outlet hoses to fuel pump, disconnect hoses.
10. Separate and remove two halves of fuel pump retainer band.
11. Remove two foam rubber insulation bands.

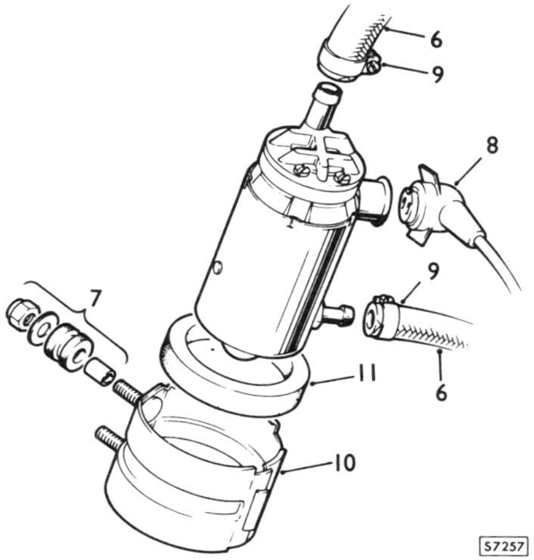

Refitting
Reverse operations 1 to 11.

FUEL PRESSURE REGULATORS

Check and adjust   19.45.12

Check
1. Depressurise fuel system, see 19.50.02.
2. Slacken pipe clip securing left-hand cold start injector supply pipe to fuel rail and pull pipe from rail.
3. Connect pressure gauge pipe to fuel rail and tighten pipe clip.
   CAUTION: Pressure gauge must be checked against an approved standard at regular intervals.
4. Pull '–ve' L.T. lead from ignition coil and switch ignition on. Connect terminal 85 of pump relay to ground. Check reading on pressure gauge; reading must be between 28.5 and 30.8 lbf/in^2 (2,0 – 2,2 kgf/cm^2).
   NOTE: The pressure reading may slowly drop through either the regulator valve seating or the pump non-return valve. A slow steady drop is permissible; a rapid fall MUST be investigated.
5. Remove ground connection to terminal 85 of pump relay.
   NOTE: If satisfactory results have been obtained, depressurise fuel system and continue with operations 6 to 9. If satisfactory results have not been obtained, continue with 'Adjust' procedure below.
6. Slacken pipe clip and remove pressure gauge from fuel rail.
7. Re-connect cold start injector supply pipe and secure pipe clip.
8. Switch ignition on and check for leaks.
9. Switch ignition off.

Adjust
NOTE: Fuel pressure should only be adjusted after complete system has been thoroughly checked.

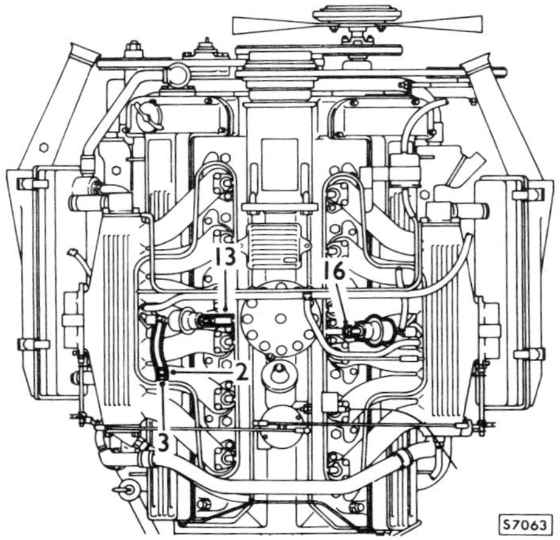

19–19

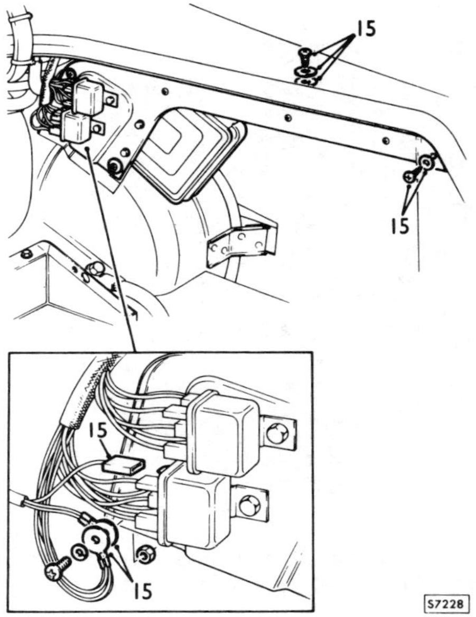

17 Remove expansion tank supply pipe.
18 Remove fuel gauge tank unit.

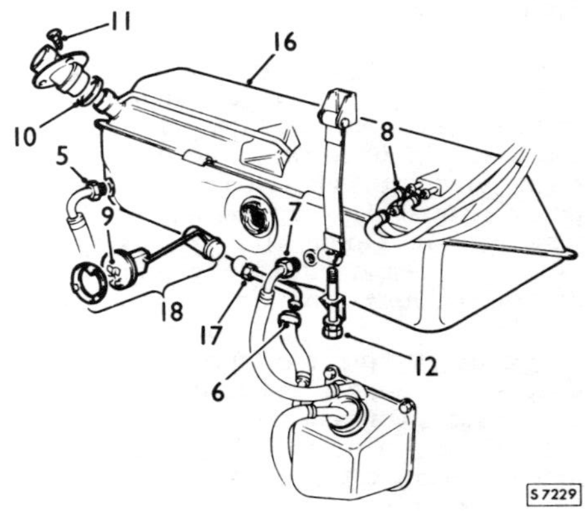

Refitting
19 Fit fuel gauge tank unit.
20 Fit expansion tank supply pipe.
21 Locate fuel tank, do not secure.

10 Switch off ignition, depressurise fuel system and remove pressure gauge.
11 Switch on ignition and check for leaks.
12 Remove earth (ground) connection from terminal 85 of pump relay.
13 Re-connect L.T. lead at ignition coil.

FUEL TANK

Remove and refit   19.55.01

Removing
1 Remove spare wheel.
2 Remove battery, see 86.15.01.
3 Drain fuel tank, see 19.55.02.
4 Fold back fuel tank carpet.
5 Clamp fuel return hose and disconnect union nut at fuel tank.
6 Release hose clip securing expansion tank supply hose to pipe, disconnect hose.
7 Release union nut securing expansion tank return pipe to fuel tank.
8 Release three hose clips securing vent hoses to fuel tank vent pipes, note position of and disconnect hoses.
9 Note position of and disconnect electrical connectors from fuel gauge tank unit.
10 Release hose clip securing fuel filler assembly to fuel filler neck.
11 Remove three screws securing fuel filler assembly to body, remove fuel filler assembly.
12 Remove two bolts and release fuel tank securing straps.
13 Remove left- and right-hand luggage compartment trim panels.
14 Remove seven screws, plain washers and four fibre washers securing left-hand luggage compartment side panel to body, lower panel.
15 Remove seven screws, plain washers and four fibre washers securing right-hand luggage compartment side panel to body, disconnect earth (ground) leads from relay bracket and white cable from pump relay, lower panel.
16 Remove fuel tank.

FUEL SYSTEM

Depressurise   19.50.02

CAUTION: The fuel system MUST always be depressurised before disconnecting any fuelling component.

1 Remove right-hand luggage compartment trim.
2 Pull cable from terminal 85 of fuel pump relay.
3 Disconnect H.T. lead from ignition coil.
4 Switch ignition on and crank engine for a few seconds.

NOTE: The fuel system will now be depressurised. On completion of operations to fuelling components and prior to starting the engine, proceed as follows:

5 Re-connect H.T. lead to ignition coil.
6 Re-connect cable to terminal 85 of fuel pump relay.
7 Fit right-hand luggage compartment trim.

FUEL SYSTEM

Pressure test   19.50.13

1 Depressurise fuel system, see 19.50.02.
2 Slacken clip securing left-hand cold start injector hose to fuel rail, disconnect hose.
3 Connect pressure gauge pipe to fuel rail and secure using hose clip.
4 Pull '–ve' L.T. lead from ignition coil.
5 Connect terminal 85 of pump relay to earth (ground).
6 Switch on ignition and check pressure gauge reading. Reading must be between 28,5 and 30.8 lbf/in^2 (2,0 – 2,2 kgf/cm^2).
7 If reading low, check for blockage or choked filter in supply line or pump suction pipes.
8 If reading high, check for blockage in return line.
9 If satisfactory reading cannot be obtained carry out operation 19.45.12. (Pressure regulator – adjust) and if necessary renew pressure regulator or fuel pump.

10 Remove setscrew securing both pressure regulators to inlet manifolds.
11 Clamp fuel pipe between 'B' bank pressure regulator inlet and fuel rail.
12 Re-connect '–ve' L.T. lead to ignition coil and start engine. If 'EPITEST' box fitted, depress 'PUMP' button.
13 Slacken locknuts at both pressure regulators.
14 Set adjuster bolt on 'A' bank regulator to set reading on pressure gauge of 30 lbf/in^2 (2,1 kgf/cm^2).
15 Release clamp at 'B' bank regulator and transfer to 'A' bank pressure regulator inlet.
16 Set adjuster bolt on 'B' bank regulator to set reading on pressure gauge to 30 lbf/in^2 (2,1 kgf/cm^2).
17 Release clamp and ensure pressure gauge reading is between 28,5 and 30.0 lbf/in^2 (2,0 – 2,2 kgf/cm^2).
18 Restrain adjuster bolts on each regulator in turn and tighten locknuts.
19 Switch off ignition.
20 Depressurise fuel system, see 19.50.02.
21 Slacken pipe clip and remove pressure gauge hose from fuel rail.
22 Re-connect cold start injector supply pipe and secure pipe clip. Secure both regulators using two setscrews.
23 Switch ignition on and check for leaks.
24 Switch ignition off.

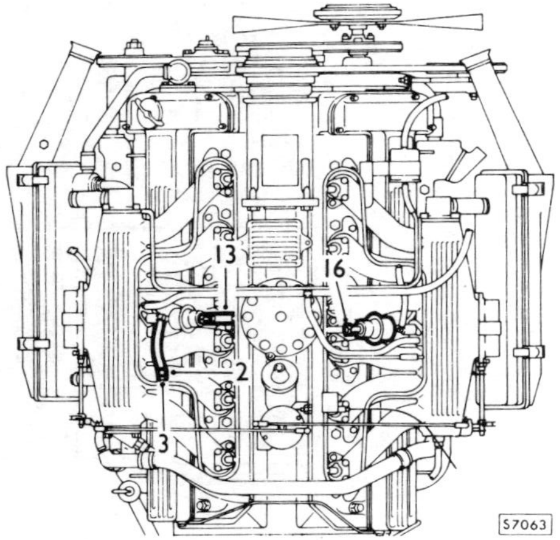

19–20

16 Lift injector clamps clear of mounting studs.
17 Lift fuel cross-over pipe assembly clear and remove fuel rail complete with injectors.

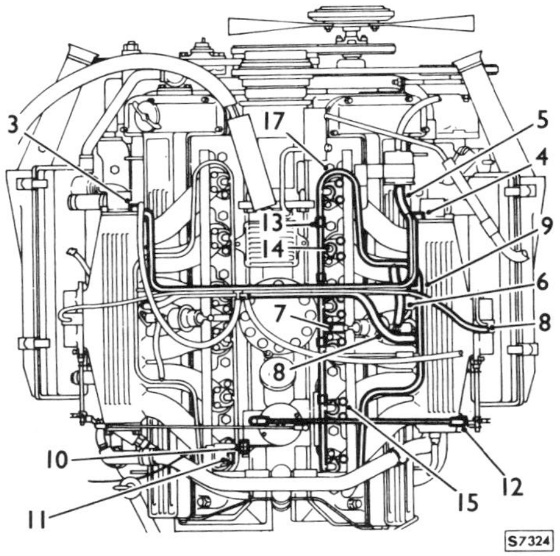

Refitting
7 Locate new sealing rings at **ALL** injectors.
8 Reverse operations 2 to 5.

FUEL RAIL – RIGHT-HAND 19.60.04

Remove and refit

Removing
1 Depressurise fuel system, see 19.50.02.
2 Disconnect battery, see 86.15.20.
3 Remove screw and spacer securing pressure regulator cross-over pipe to left-hand overrun valve.
4 Remove screw securing pressure regulator cross-over pipe to right-hand overrun valve.
5 Release hose clip securing fuel supply cross-over pipe to fuel filter hose, disconnect pipe from hose.
6 Release hose clip securing cold start injector hose to fuel rail, disconnect hose.
7 Release hose clip securing pressure regulator hose to fuel rail, disconnect hose.
8 Disconnect manifold pressure sensor hoses at T-piece and right-hand inlet manifold, also vacuum hoses from throttle housing.
9 Release hose clip securing fuel supply cross-over pipe to fuel rail, disconnect hose.
   NOTE: On right-hand drive cars proceed with operations 10 to 11, on left-hand drive cars proceed to operation 12.
10 Disconnect electrical connectors from kick-down switch.
11 Remove throttle cable locknut and throttle cable from pedestal, lift switch and cable clear.
12 Disconnect right-hand throttle cross-rod.
13 Remove six cable clips securing harness to fuel rail.
14 Remove electrical connectors from right-hand injectors and cold start injector.
15 Remove twelve nuts and spring washers securing injector clamps to inlet manifold.

18 Remove and discard injector sealing rings.
19 Remove injectors, see 19.60.01.
NOTE: If necessary transfer clips and insulation rubbers to replacement fuel rail.

Refitting
20 Remove sealing plugs from replacement fuel rail.
21 Fit injectors to fuel rail.
22 Fit new injector sealing rings.
23 Locate fuel rail and injector assembly to inlet manifold and cylinder head ports.
24 Fit injector clamps to mounting studs and secure using twelve nuts and spring washers.
25 Fit electrical connectors to injectors and cold start injectors.
26 Refit harness to fuel rail using six cable clips.
27 Reconnect right-hand throttle cross-rod.
   NOTE: On right-hand drive cars proceed with operations 28 and 29, on left-hand drive cars proceed to operation 30.

continued

9 Insert a suitable pipe into pump inlet hose, secure with hose clip and push pipe through vent hole in luggage compartment floor.
10 Place suitable receptacle beneath car and protruding pipe to collect fuel.
11 Remove fuel filler cap, release clamp and drain tank.

Refilling
12 Reverse operations 3 to 10.
13 Refill fuel tank and replace fuel filler cap.
14 Re-connect battery.

INJECTORS – EACH

Remove and refit – Left-hand **19.60.01**
                   – Right-hand **19.60.03**

Removing
1 Depressurise fuel system, see 19.50.02.
2 Disconnect battery, see 86.15.20.
3 Remove fuel rail, see 19.60.04/05.
4 Release hose clip(s) of injector(s) to be removed.

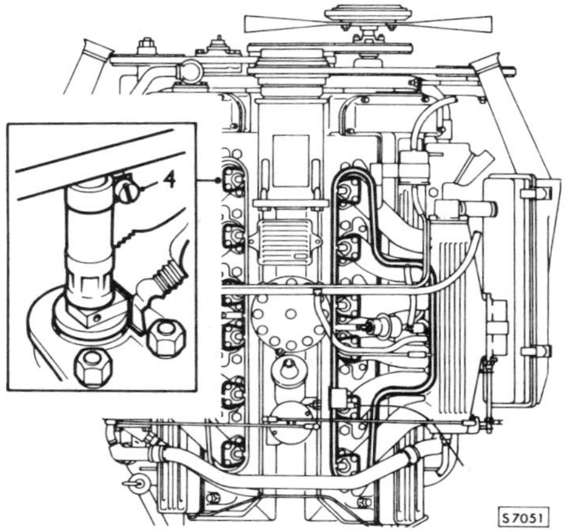

5 Note orientation of electrical sockets and pull injector(s) from fuel rail.
6 Remove rubber sealing rings from **ALL** injectors and discard.
CAUTION: Sealing rings MUST be renewed each time the injectors are removed.

22 Locate right-hand luggage compartment side panel, connect earth leads to relay bracket and white cable to pump relay. Secure panel using seven screws, plain washers and four fibre washers.
23 Locate left-hand luggage compartment side panel and secure using seven screws, plain washers and four fibre washers.
24 Fit two bolts to fuel tank securing straps and tighten.
25 Connect three vent hoses to fuel tank vent pipes and secure using new hose clips.
26 Fit expansion tank return pipe to fuel tank and secure with union nut.
27 Connect expansion tank supply hose to pipe and secure using new hose clip.
28 Connect fuel return pipe to fuel tank and secure with union nut. Remove clamp.
29 Fit electrical connectors to fuel gauge tank unit.
30 Locate fuel filler neck, secure using three screws and new hose clip.
31 Position carpet over fuel tank.
32 Refit left- and right-hand luggage compartment trim panels.
33 Refit battery.
34 Refit spare wheel.
35 Refill fuel tank.

FUEL TANK

Drain and refill 19.55.02

Draining
1 Depressurise fuel system, see 19.50.02.
2 Disconnect battery, see 86.15.20.
3 Remove spare wheel.
4 Remove two screws securing fuel pump cover to battery tray.
5 Peel back floor carpet, release fuel pump cover from floor clips.
6 Clamp inlet hose to fuel pump.
7 Release hose clip securing inlet hose to fuel pump, disconnect hose from pump.
8 Remove grommet from vent hole in luggage compartment floor.

19–21

22 Locate fuel rail and injector assembly to inlet manifold and cylinder head ports.
23 Fit injector clamps to mounting studs and secure using twelve nuts and spring washers.
24 Fit electrical connectors to injectors and cold start injectors.
25 Refit harness to fuel rail and secure using seven cable clips.
26 Re-connect left-hand throttle cross-rod.
NOTE: On right-hand drive cars proceed to operation 29, on left-hand drive cars proceed with operations 27 and 28.
27 Refit throttle cable to pedestal, tighten locknut.
28 Refit electrical connectors to kick-down switch.
29 Connect fuel supply cross-over pipe hose to fuel rail, secure with new hose clip.
30 Connect manifold pressure sensor hose to left-hand inlet manifold.
31 Connect pressure regulator hose to fuel rail, secure with new hose clip.
32 Connect cold start injector hose to fuel rail, secure with new hose clip.
33 Fit pressure regulator cross-over pipe to overrun valves using two screws and spacer on left-hand valve.
34 Re-connect battery.

COLD START INJECTORS – EACH

Remove and refit   19.60.06

Removing
1 Depressurise fuel system, see 19.50.02.
2 Disconnect battery, see 86.15.20.
3 Remove electrical connector from cold start injector.
4 Release clip securing cold start injector hose at injector. Disconnect hose.
5 Remove two screws securing fuel pressure regulator bracket to inlet manifold, move regulator clear of cold start injector.
6 Remove two Allen screws securing cold start injector to inlet manifold.

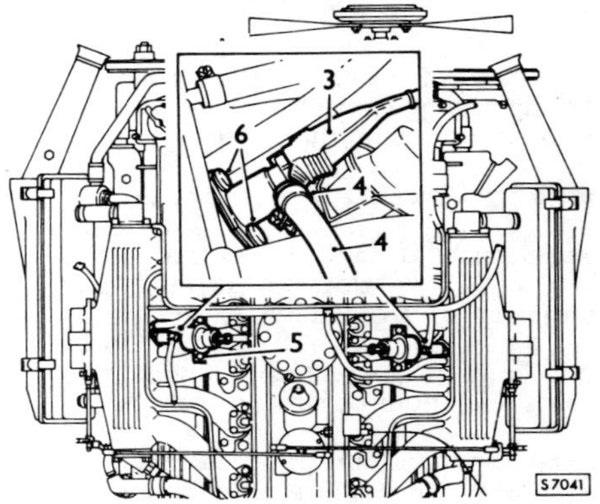

7 Remove cold start injector.
8 Remove and discard gasket.

Refitting
Reverse operations 2 to 8 using new gasket.

COLD START INJECTORS
Test   19.60.07

See operation 19.22.32.

INJECTORS
Test   19.60.08

Injector winding
1 Use ohmmeter to measure resistance value of each injector winding which should be 2.4 ohms at 20° C (68° F).
2 Check for short circuit to earth (ground) on winding by connecting ohmmeter prods between injector terminal and injector body. Meter should read '∞'. If any injector winding is open or short circuited, replace injector.

---

28 Refit throttle cable to pedestal, tighten locknut.
29 Refit electrical connectors to kick-down switch.
30 Connect fuel supply cross-over pipe hose to fuel rail, secure with new hose clip.
31 Connect manifold pressure sensor hoses to T-piece and right-hand inlet manifold.
32 Connect pressure regulator hose to fuel rail, secure with new hose clip.
33 Connect cold start injector hose to fuel rail, secure with new hose clip.
34 Connect pressure regulator cross-over pipe hose to fuel filter, secure with new hose clip.
35 Fit pressure regulator cross-over pipe to overrun valves using two screws and spacer on left-hand overrun valve.
36 Re-connect battery.

FUEL RAIL – LEFT-HAND

Remove and refit   19.60.05

Removing
1 Depressurise fuel system, see 19.50.02.
2 Disconnect battery, see 86.15.20.
3 Remove screw and spacer securing pressure regulator cross-over pipe to left-hand overrun valve.
4 Remove screw securing pressure regulator cross-over pipe to right-hand overrun valve.
5 Release hose clip securing fuel supply cross-over pipe hose at left-hand fuel rail, disconnect hose.
6 Release hose clip securing cold start injector hose to fuel rail, disconnect hose.
7 Release hose clip securing pressure regulator hose to fuel rail, disconnect hose.
8 Disconnect manifold pressure sensor hose at left-hand inlet manifold, also vacuum hoses from throttle housing.
NOTE: On right-hand drive cars proceed to operation 11, on left-hand drive cars proceed with operations 9 and 10.

9 Disconnect electrical connectors from kick-down switch.
10 Remove throttle cable locknut and throttle cable from pedestal, lift switch and cable clear.
11 Disconnect left-hand throttle cross-rod.
12 Remove seven cable clips securing harness to fuel rail.
13 Remove electrical connectors from left-hand injectors and cold start injector.
14 Remove twelve nuts and spring washers securing injector clamps to inlet manifold.
15 Lift injector clamps clear of mounting studs.
16 Lift fuel cross-over pipe assembly clear and remove fuel rail complete with injectors.

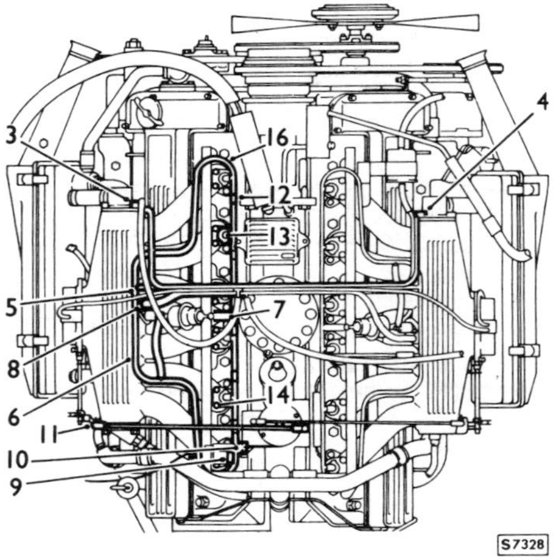

17 Remove and discard injector sealing rings.
18 Remove injectors, see 19.60.03.
NOTE: If necessary transfer clips and insulation rubbers to replacement fuel rail.

Refitting
19 Remove sealing plugs from replacement fuel rail.
20 Fit injectors to fuel rail.
21 Fit new injector sealing rings.

19–22

Supply pulses and cables
NOTE: Injectors are connected in two groups from the power amplifier, each group divided into parallel sets of three.

Group 1	Group 2
1A 6B	6A 1B
5A 2B	2A 5B
3A 4B	4A 3B

1 Disconnect coolant temperature sensor and terminal 85 of pump relay.
2 Disconnect electrical connector from each injector.
3 Connect voltmeter or 2.5V 0.03A bulb across terminals of plug of each injector lead. Start at 1A.
4 Crank engine. Bulb should flash or voltmeter register 2-3 volts as amplifier supplies each pulse.
5 Repeat for all injectors in first set of group 1 before connecting with second set.
If only one result per set is correct, suspect faulty cables or connections.
If all three results in same set are incorrect, suspect faulty cables or connections or power amplifier.
If all six results in same group are incorrect, suspect faulty cables or connections, power amplifier or ECU.

NOTE: On completion of tests, re-make all electrical connections.

COOLING SYSTEM

Description

The cooling system consists of a radiator matrix 'A', a water pump 'B' – belt driven by the engine crankshaft – and a remote coolant header tank 'C'. Two thermostatic valves 'D', are fitted – one to each cylinder bank – to ensure rapid warm up from cold. Under cold start conditions (see insets) coolant is forced by the water pump equally through each cylinder block and cylinder head ('E' and 'F') to the thermostatic valve housings. The valves are closed and coolant is therefore returned via the engine cross-pipe 'G', to the water pump inlet.

During this period the radiator is under pump suction and air is bled by jiggle pins 'H', in each thermostatic valve.

NOTE: When fitting a replacement thermostat the thermostat MUST be fitted with the jiggle pin at the top of the housing.

The engine contains air pockets which have to be purged before effective cooling is possible. The air entrained by the coolant rises to the highest point on each side of the engine, the thermostatic housings, then through the jiggle pins to the top of the radiator.

During this phase the thermotime switch 'J', the coolant temperature sensor 'K' and the auxiliary air valve 'L' function as an automatic choke and warm up the system. Full pump suction draws coolant from the base of the radiator and starts the full cooling circuit.

At this time pump suction also appears at the heater matrix 'M' and the remote header tank, purging both the matrix and the radiator via pipes 'N' and 'P'. The remote header tank carries out an air separation function in addition to providing a reservoir of coolant.

When coolant temperature rises to a predetermined level the thermostatic valves open and allow coolant to flow into the top of the radiator.

A thermostatic switch 'Q' is fitted in the water pump suction inlet elbow. The switch starts the radiator electric cooling fan 'R' should the temperature of the coolant leaving the radiator rise above a predetermined level.

A cooling tube coil 'S' is included in the fabrication of the right-hand end tank of the radiator, and is connected in series with the automatic transmission hydraulic fluid circulation.

The radiator is fitted with a bleed tap 'T' through which, during initial cold fill, the radiator is vented.

A drain tap 'U' is located in the base of the right-hand end tank.

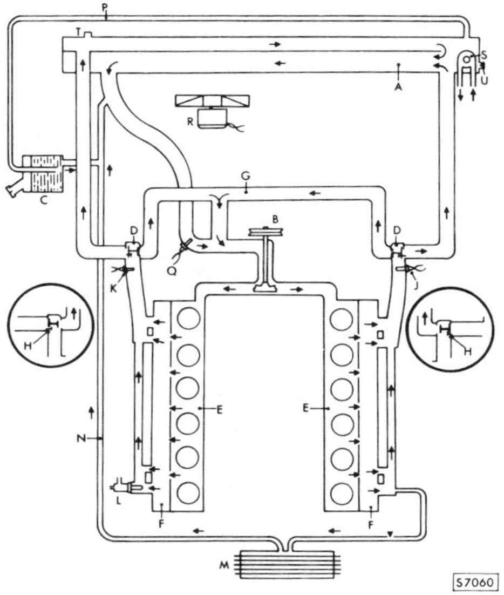

ANTIFREEZE

BP Type H21 or Union Carbide UT 184 (Prestone 2 or Texaco in USA/Canada) must be used at all times. This is a specially formulated anti-freeze which is designed to afford the maximum corrosion protection to all metals normally found in engine cooling systems as well as having the normal frost protection properties necessary during winter months. It should not therefore be mixed with other anti-freezes. In places where Unipart Universal is not available for top-up or replenishment, drain the system, flush and fill with anti-freeze which complies with specification BS.3150.

A solution of recommended antifreeze (40% Great Britain; 55% all other countries) must at all times be used either when topping up or replenishing the cooling system. For maximum corrosion protection, the concentration should never be allowed to fall below 25%. Always top-up with specified strength of antifreeze, **NEVER WITH WATER ONLY. CHANGE COOLANT ANNUALLY.**

COOLANT

Drain and refill   26.10.01

NOTE: A topping up procedure is given at the end of this operation.

CAUTION: Do not remove the pressure cap at the remote header tank unless engine is cold.

Drain

1. Carefully remove pressure cap at remote header tank.
2. Open bleed tap at left-hand side of radiator top rail.
3. Place suitable receptacle beneath radiator and open drain tap at right-hand side.
4. Close radiator drain tap at right-hand side.
5. Ensure pressure cap at remote header tank is removed and bleed tap is open.
6. **SLOWLY** add specified coolant at remote header tank filler neck until no more can be added. Wait 1 – 2 minutes, and if necessary add more coolant. Repeat until no more coolant can be added.
7. Close bleed tap, fit pressure cap at remote header tank.
8. Set air conditioner controls to '80' and 'DEF'.
9. Run engine at approximately 1,000 rev/min for three minutes. Switch off engine.
10. Open bleed tap. When air has purged from system, close bleed tap, remove pressure cap at remote header tank and slowly add coolant until no more can be added. Refit pressure cap at remote header tank.

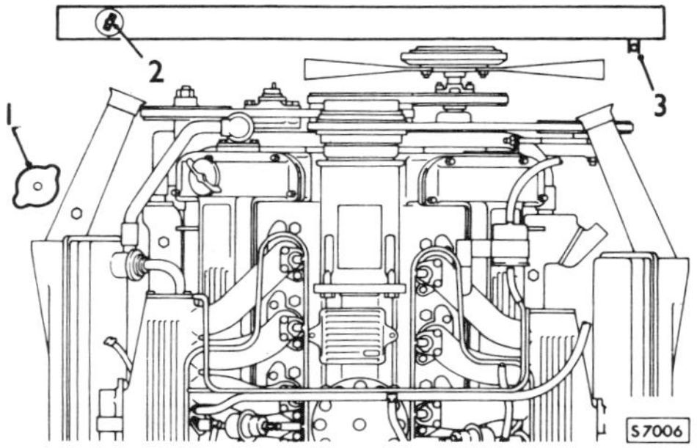

Topping up and generally checking coolant level

CAUTION: This procedure must only be carried out when engine is cold.

1. Carefully remove pressure cap at remote header tank. If coolant level sufficient to prevent the addition of more coolant, the system does not require topping up. Replace cap.
2. If coolant is below this level, **SLOWLY** add specified coolant to remote header tank until no more can be added.
3. Refit pressure cap at remote header tank.

REMOTE HEADER TANK

Remove and refit   26.15.01

Removing

1. Remove left hand air cleaner, see 19.10.01.
2. Slacken clip on pipe at top of remote header tank, remove pipe and secure, pointing upwards, as high as possible.
3. Repeat for pipe at base of tank.
4. Pull vent pipe from filler neck.
5. Apply full left hand steering lock.

continued

26–1

6 Beneath wheel arch, remove two nyloc nuts and plain washers securing front of remote header tank.
7 Remove setscrew, washers and nyloc nut securing rear of tank. Lift tank clear.

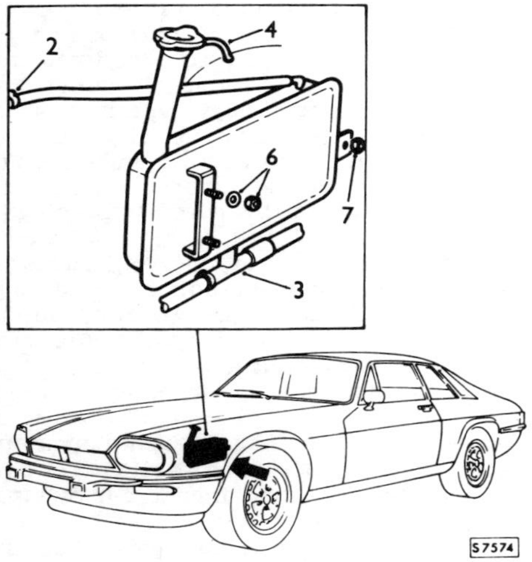

Refitting
8 Fit tank and secure using nuts, washers and setscrews.
9 Fit pipes at base and top of tank.
10 Fit vent pipe to filler neck.
11 Refit left hand air cleaner.
12 Carry out filling procedure given under 26.10.01, cold fill.

FAN BELT

Adjust   26.20.01

1 Slacken bolt securing jockey pulley carrier.
2 Slacken nut securing trunnion block.
3 Slacken nut securing adjusting link bolt.
4 Slacken adjusting link lock nut.
5 Adjust belt tension by means of adjusting link nut; correct tension is as follows:
   A load of 6.4 lb (2,9 kg) must give a total belt deflection of .17 in (4,3 mm) when applied at mid-point of belt.
6 Reverse operations 1 to 4.

FAN BLADES

Remove and refit   26.25.06

Removing
1 Remove fan and torquatrol unit, see 26.25.21.

Refitting
Reverse operation 1.

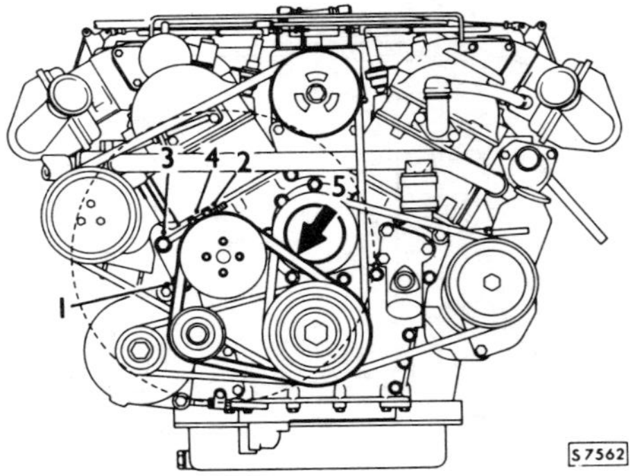

FAN BELT

Remove and refit   26.20.07

Removing
1 Slacken bolt securing jockey pulley carrier.
2 Slacken nut securing trunnion block.
3 Slacken nut securing adjusting link bolt.
4 Slacken adjusting link lock nut.
5 Wind back adjusting link nut until belt can be removed. Manoeuvre clear.

Refitting
6 Fit new belt to pulley.
7 Set belt tension, operation 5 of 26.20.01.

IDLER PULLEY HOUSING

Remove and refit   26.25.15

Removing
1 Remove fan belt, see 26.20.07.
2 Remove fan and torquatrol unit, see 26.25.21.
3 Remove locknut, washers and bolt securing adjustment bolt eye to pulley arm.
4 Remove two setscrews and washers, and two nuts and washers securing idler pulley housing. Remove housing complete with jockey pulley and arm.

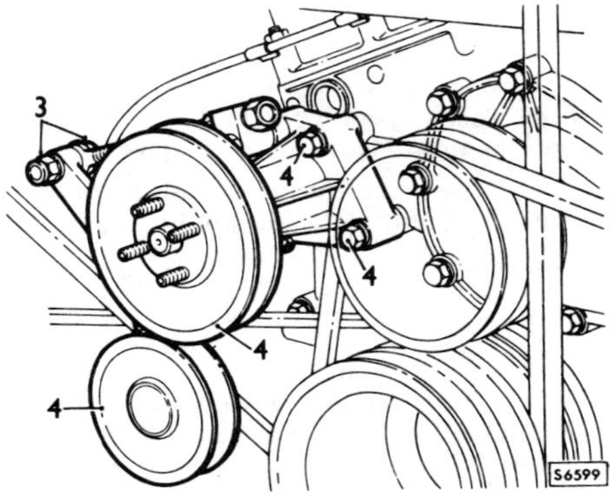

Refitting
Reverse operations 1 to 4.

JOCKEY PULLEY

Remove and refit   26.25.16

Removing
1 Remove right-hand air cleaner, see 19.10.02.
2 Wind back locknut at fan belt adjustment bolt.
3 Slacken locknut at adjustment bolt trunnion.
4 Slacken jockey pulley arm pivot bolt.
5 Remove locknut, washers and bolt securing adjustment bolt eye to pulley arm.
6 Release bolt.
7 Remove nut and special washer securing jockey pulley spindle in arm.

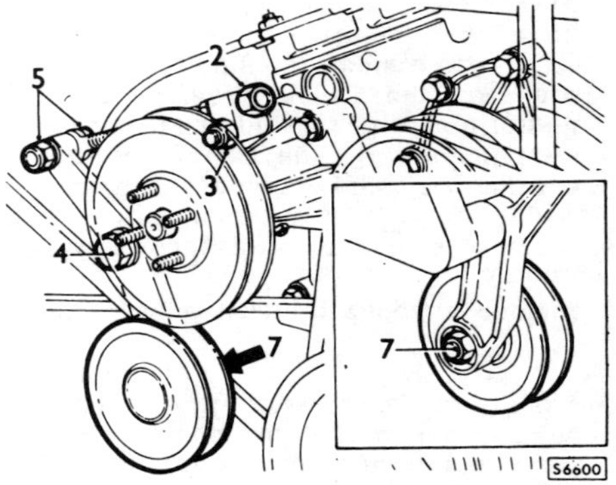

Refitting
Reverse operations 1 to 7.

26—2

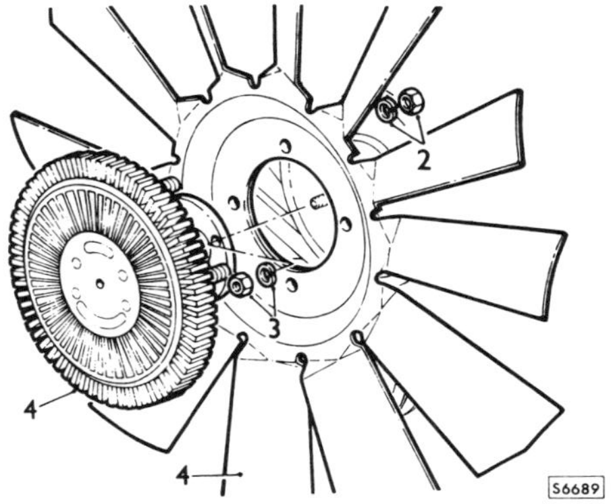

FAN AND MOTOR UNIT

Remove and refit   26.25.23

Removing
1. Release clips and remove left hand air cleaner cover and element.
2. Disconnect motor feed cable at snap connector.
3. Remove nut securing earth terminal to top rail.
4. Remove nuts securing motor mounting plate.
5. Withdraw fan and motor.

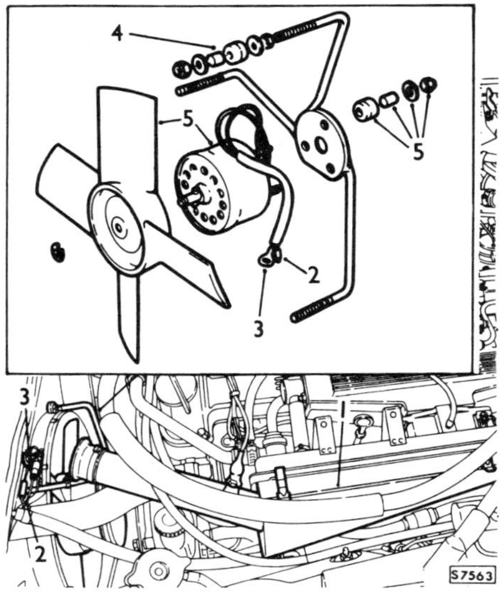

FAN MOTOR AND FAN MOTOR RELAY

26.25.25
26.25.29

Test

Test also checks function of air conditioning system and cooling fan override relay – air conditioned cars only. Reference should be made to location diagram – Group 86.

Cars not fitted with air conditioning
1. Pull connector plug from thermostatic switch in front of water pump.
2. Switch on ignition. **DO NOT ROTATE ENGINE.**
3. Short together sockets in plug; relay should be heard to operate and fan should start. Remove short.
4. Switch off ignition.
5. Reconnect thermostatic switch connector.

Cars fitted with air conditioning
6. Ensure right hand air conditioning switch set to 'OFF'.
7. Carry out operations 1 to 5 inclusive.
8. Switch on ignition. **DO NOT ROTATE ENGINE.**
9. Set left-hand air conditioning control to '65', and right-hand control to 'AUTO'; air conditioning relay should be heard to operate, fan should start and compressor clutch should engage.
10. Switch off ignition and set right hand air conditioning system control to 'OFF'.

FAN MOTOR RELAY

Remove and refit   26.25.31

Removing
1. Remove screws securing relay cover, move cover to one side.
2. Note electrical connections and pull cable terminals from relay.
3. Remove two nuts; bolts and washers securing relay; withdraw relay.

Refitting
Reverse operations 1 to 5.

FAN AND TORQUATROL UNIT
(Later Cars)

Remove and refit   26.25.21

Removing
1. Remove setscrews securing top section of fan cowl.
2. Remove four nuts and washers securing fan to Torquatrol unit.
3. Restrain pulley and remove centre bolt from Torquatrol unit, recover special washer.
4. Gently tap Torquatrol unit forward from pulley spigot.
5. Lift fan from fan cowl.

Refitting
6. Locate four bolts in Torquatrol unit.
7. Locate fan over pulley nose.
8. Ensure locating pin in position in pulley spigot and lightly grease spigot.
9. Offer Torquatrol on to pulley.
10. Secure Torquatrol unit to pulley using centre bolt and special washer.
    NOTE: Ensure special washer locates on pin in pulley spigot before tightening bolt.
11. Secure fan to Torquatrol unit using spring washers and plain nuts.
12. Refit top section of fan cowl.

FAN AND TORQUATROL UNIT
(Early Cars)

Remove and refit   26.25.21

NOTE: Torquatrol flange secured to water pump pulley with four plain nuts.

Removing
1. Slacken fan belt. See adjust, 26.20.01.
2. Remove four plain nuts and washers securing fan to Torquatrol unit.
3. Remove four plain nuts and washers securing Torquatrol unit to water pump pulley.
4. Manoeuvre fan and Torquatrol unit clear.

Refitting
Reverse operations 1 to 4 inclusive.

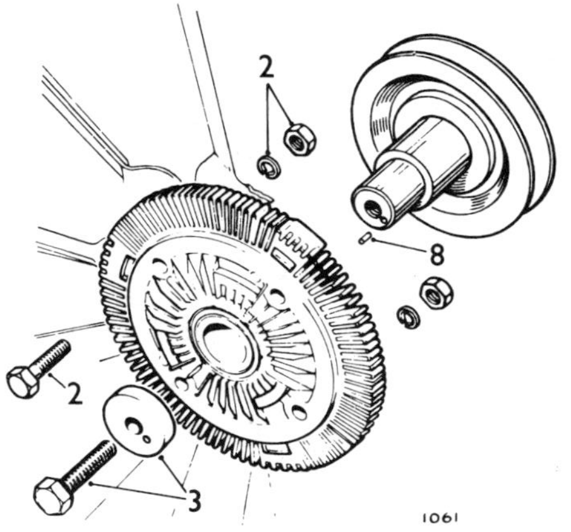

continued

26–3

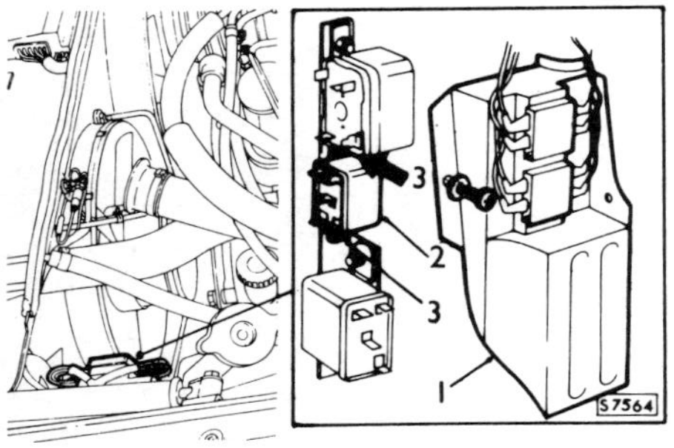

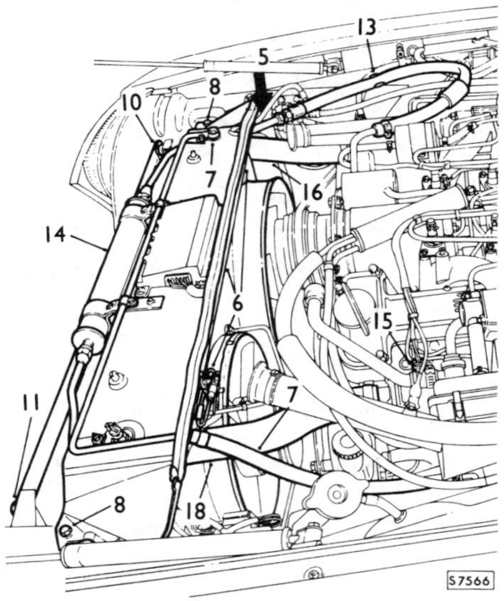

18 Disconnect left hand top hose from radiator.
19 Disconnect hose from expansion tank 'Tee' piece.
20 Disconnect hoses from oil cooler.
 CAUTION: When carrying out this operation, two spanners MUST be used, one on hose union, the other on oil cooler union.
21 Disconnect bottom hose.
22 Remove nuts securing fan cowl to mounting bracket.
23 Release cowl from brackets.
24 Lift out radiator.

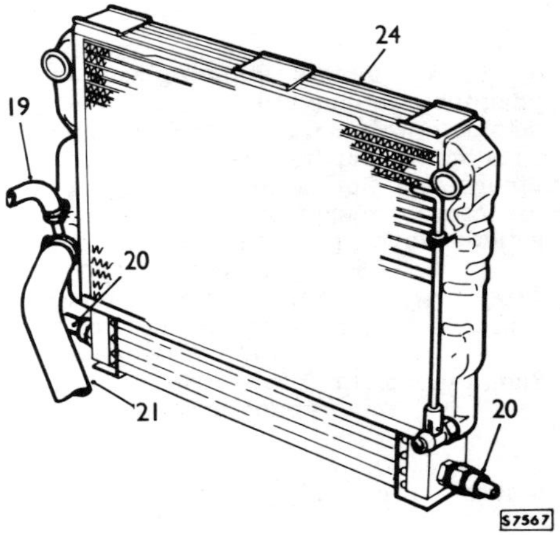

Refitting
Reverse operations 1 to 24.

RADIATOR DRAIN TAP 26.40.10
Remove and refit

Removing
1 Drain coolant, see 26.10.01.
2 Pull split pin from fork end of control rod.
3 Unscrew tap from radiator block. Discard seals.

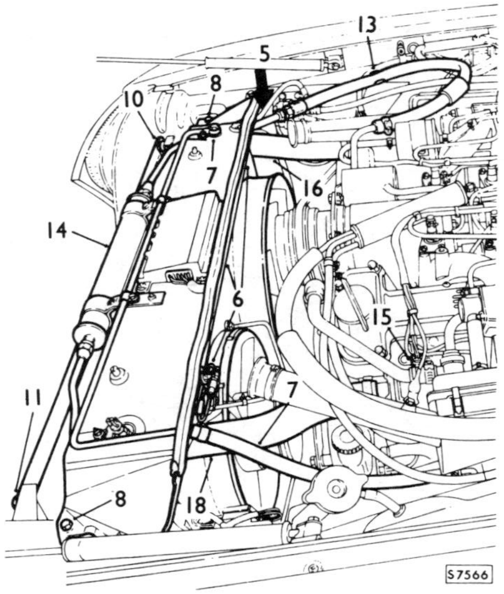

7 Disconnect hose from expansion pipe. Remove banjo bolt securing expansion pipe to radiator; note fitted position of spacer and sealing washers. Discard sealing washers.
8 Remove bolts securing radiator top rail.
9 Disconnect earth leads from left and right hand harnesses.
10 Remove bolt and spacer securing right hand front stay.
11 Remove bolt securing left hand wing valance stay.
12 Disconnect hoses from receiver drier unit — Cars fitted with air conditioning only.
 CAUTION: IMMEDIATELY PLUG BROKEN CONNECTIONS USING DRY, CLEAN PLUGS.
13 Unclip condenser hose from right hand wing valance.
14 Lift off top rail together with receiver drier unit.
15 Disconnect cable from low coolant level sensor.
16 Disconnect right hand top hose from radiator.
17 Disconnect transmission oil cooler hoses — Cars fitted with automatic transmission only. Plug broken connections to prevent ingress of dirt.

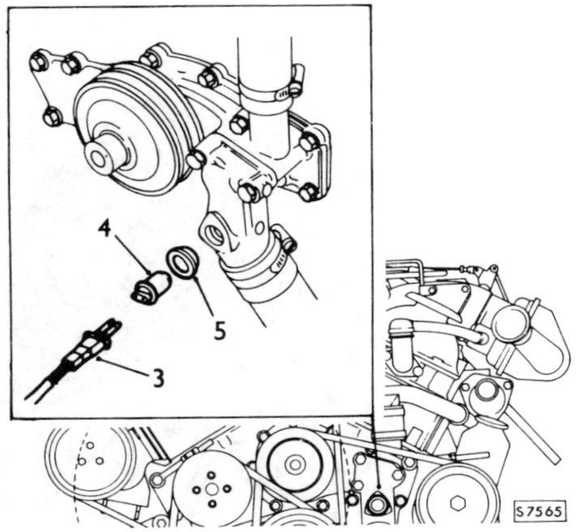

Refitting
6 Fit new rubber seal into housing.
 WARNING: NO LUBRICANT OF ANY DESCRIPTION MAY BE USED WHEN FITTING SWITCH TO SEAL. USE OF LUBRICANT WILL RESULT IN EJECTION OF SWITCH AND TOTAL LOSS OF COOLANT.
7 Press new switch into seal; ensure correctly seated.
8 Fit electrical connector.
9 Reconnect battery.
10 Refill with coolant — 'cold fill', see 26.10.01.

RADIATOR BLOCK 26.40.04
Remove and refit

Removing
1 Remove lower grille, see 76.55.06.
2 Remove bonnet, see 76.16.01.
3 Drain coolant, see 26.10.01.
4 Depressurise air conditioning system — Cars fitted with air conditioning only, see 82.30.05.
5 Disconnect amplifier block connector.
6 Remove nuts securing fan cowl to radiator top rail, noting fitted positions of earth lead and line fuse.

THERMOSTATIC SWITCH 26.25.35
Remove and refit

Removing
1 Disconnect battery, see 86.15.20.
2 Drain coolant, see 26.10.01.
3 Pull connector plug from thermostatic switch.
4 Use pair of pliers to withdraw switch from rubber seal.
5 Remove rubber seal from housing.
 CAUTION: A new rubber seal MUST be used. Under no circumstances shall a new switch be fitted in an old seal.

Refitting
Reverse operations 1 to 3.

26–4

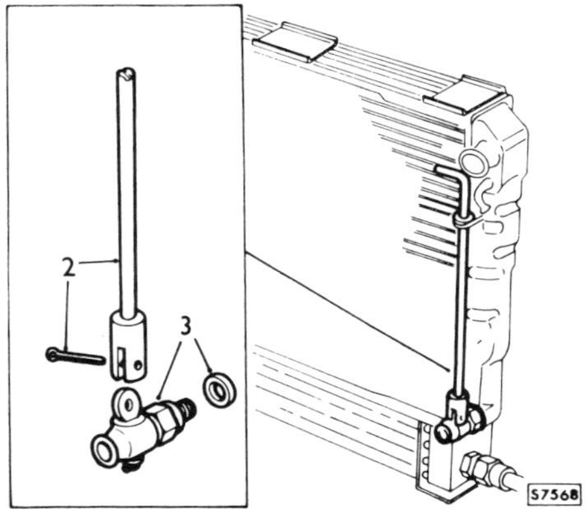

Refitting

4 Lightly coat threads of tap with non hardening sealing compound, and screw home on new seals to locate beneath control rod.
5 Couple control rod to tap, orientated correctly, and secure using new split pin.
6 Check operate tap.
7 Refill with coolant – 'cold fill', see 26.10.01.

THERMOSTAT

Remove and refit   Left hand 26.45.01   Right hand 26.45.04

Removing

1 Partially drain cooling system, see 26.10.01.
2 Disconnect hose from thermostat cover.
3 Remove thermostat cover retaining set screws, lift off cover.
4 Lift out thermostat.

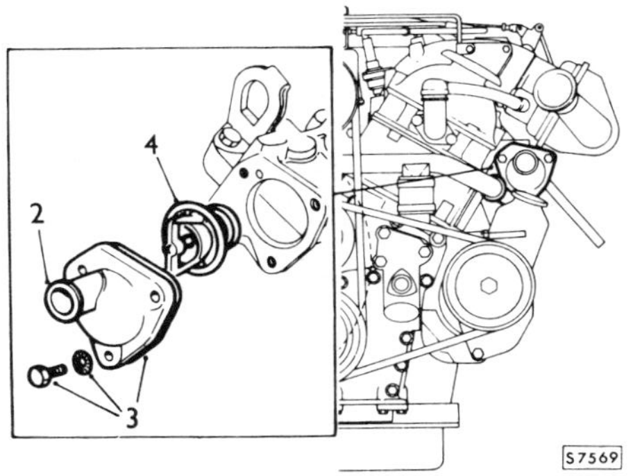

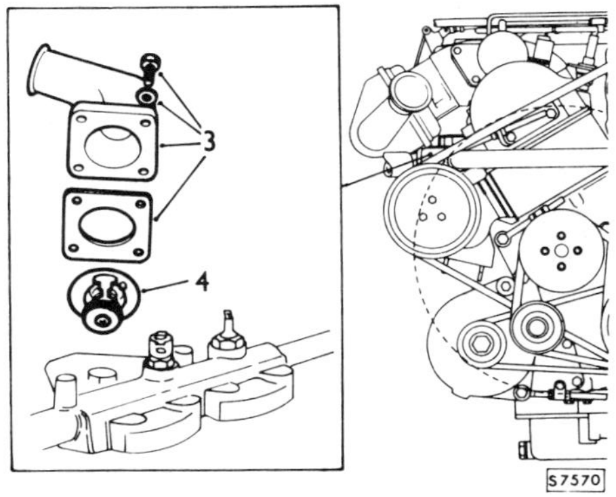

Refitting

CAUTION: It is imperative that the correct type of thermostat is used. Thermostats with no jiggle pin prevent correct purging and will inevitably damage the engine.
Reverse operations 1 to 4; use a new gasket. Tighten screws progressively to avoid distorting cover.

THERMOSTAT

Test   26.45.09

1 Remove thermostat – left hand, see 26.45.01; right-hand, see 26.45.04.
2 Thoroughly clean thermostat.
3 Place thermostat in a container of water together with a thermometer.

4 Heat water and observe if thermostat operates in accordance with data given in Group 4.

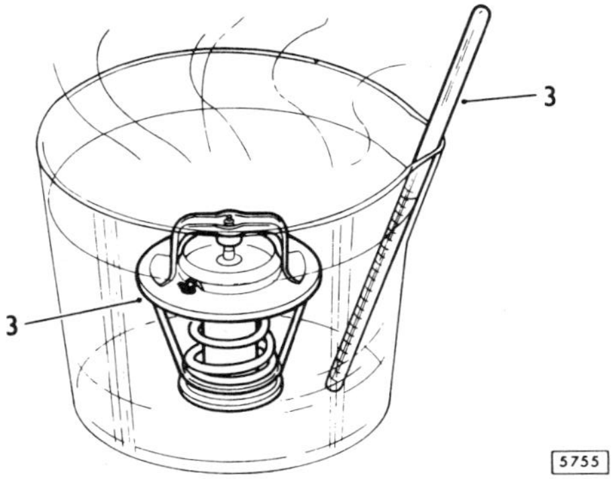

THERMOSTAT HOUSING

Remove and refit   Left hand 26.45.10

Removing

1 Remove air cleaner, see 19.10.01.
2 Open radiator tap and partially drain coolant.
3 Disconnect cables from sensors.
4 Disconnect crankcase breather tube from elbow.
5 Disconnect crankcase breather valve pipe from water pipe.
6 Disconnect top hose from thermostat cover.
7 Disconnect cross pipe hose from thermostat housing.
8 Remove bolts securing thermostat housing; disconnect housing from water rail.
9 Remove and discard gasket and water rail sealing ring.

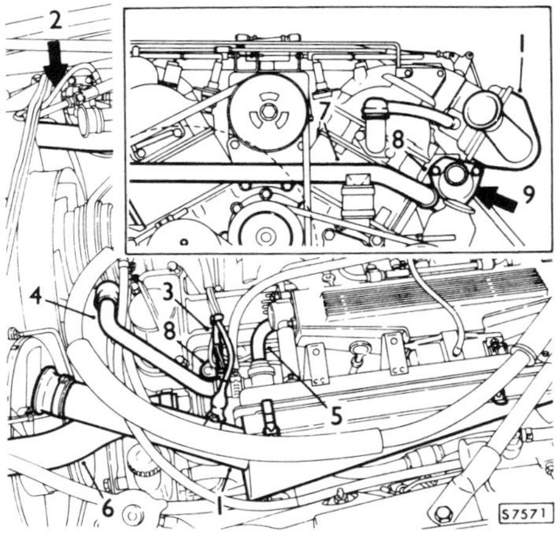

Refitting

10 Reverse operations 1 to 9; use new gasket and sealing ring.
11 Top up coolant.

THERMOSTAT HOUSING

Remove and refit   Right hand 26.45.11

Removing

1 Open radiator tap and partially drain coolant.
2 Disconnect cables from sensors.
3 Disconnect hose from thermostat cover.
4 Disconnect cross pipe hose.

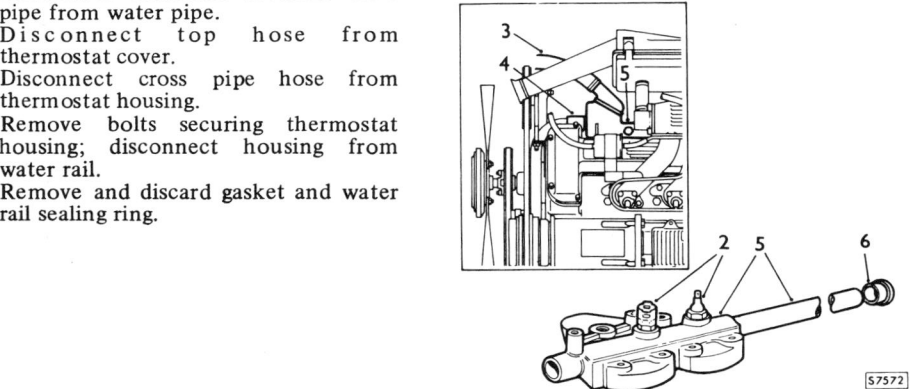

continued

26–5

5 Remove bolts securing thermostat housing; disconnect housing from water rail.
6 Remove and discard gasket and water rail sealing ring.

Refitting
7 Reverse operations 1 to 6; use new gasket and sealing ring.
8 Top up coolant.

WATER PUMP

Remove and refit 26.50.01

Removing
1 Remove fan and Torquatrol unit, see 26.25.21.
2 Remove fan belt.
3 Remove adjuster trunnion bolt.
4 Remove two setscrews and washers and two nuts and washers securing idler pulley housing.
5 Extract two studs.
6 Remove steering pump belt.
7 Remove air pump belt – emission control cars only.
8 Remove compressor pump belt – air conditioned cars only.
9 Slacken steering pump pivot bolts sufficient to draw adjustment bolt from special stud.
10 Remove special stud.
11 Remove three setscrews and washers securing thermostatic switch housing.
12 Remove housing and bottom hose complete.
13 Slacken upper clip on hose to engine cross pipe.
14 Remove special setscrews and washers securing water pump; draw pump out and downwards clear of cross pipe hose.

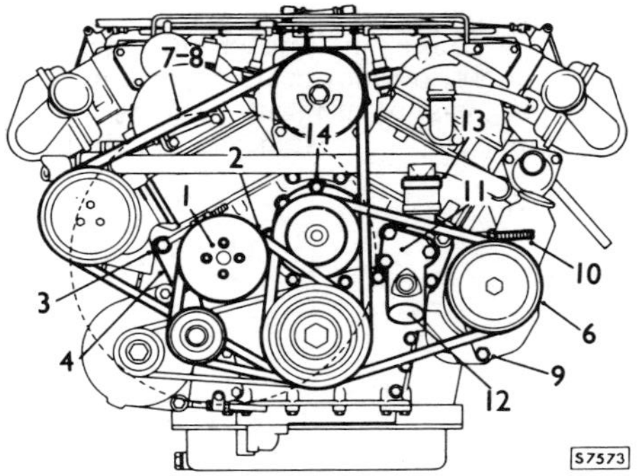

8 Carefully separate water pump body and sandwich by supporting front face of sandwich and gently tapping pump body through impeller orifice and on suction boss.
   CAUTION: Under no circumstances must the water pump body and sandwich be prised apart.
9 Remove all traces of old gasket from joint faces and thoroughly clean out water pump body.

Reassembling
10 Position water pump body front face upwards on press.
11 Offer shaft and bearing assembly, smaller shaft foremost, into water pump body. Ensure location holes in water pump body and bearing are aligned.
12 Using a mandrel acting on case of bearing, press bearing into body until location holes line up.
13 Fit Allen retaining screw and tighten firmly using Allen key only. Fit locknut and secure.
14 Fit seal over impeller end of shaft and drift squarely to seat on shoulder in body.
15 Fit new gasket, dry, over ring dowels, assemble sandwich and press firmly together.
16 Place impeller on bed of press, blades upwards. Locate shaft to impeller and press on shaft to fit.
17 Press pulley on to shaft until boss of pulley is flush with end of shaft.
   NOTE: Check that dimension of 0.025 in (0,635 mm) exists between impeller blades and sandwich face. Use puller and press to adjust as necessary.

Refitting
Reverse operations 1 to 14 inclusive; use new gaskets. Tighten setscrews progressively to avoid distorting water pump.

WATER PUMP PULLEY

Remove and refit 26.50.05

Removing
CAUTION: Irreparable damage to the impeller and water pump body will result if an attempt is made to carry out this operation without removing water pump from engine.
1 Remove water pump, see 26.50.01.
2 Fit puller to pulley using two 5/16 in U.N.F. × 2 in long bolts into tapped holes.
3 Draw off pulley.

Refitting
4 Position pump on bed of hand press.
5 Locate pulley on impeller shaft.
6 Position block of wood between press mandrel and pulley.
7 Press pulley on shaft until pulley boss is flush with end of shaft.
   NOTE: Check that dimension of 0.025 in (0,635 mm) exists between impeller blades and sandwich face. Use puller and press to adjust as necessary.
8 Refit water pump to engine.

WATER PUMP

Overhaul 26.50.06

1 Remove water pump from engine, see 26.50.01.
2 Remove water pump pulley, see 26.50.05.

Dismantling
3 Fit puller to impeller using two 5/16 in U.N.F. × 2 in long bolts into tapped holes.
4 Draw off impeller. 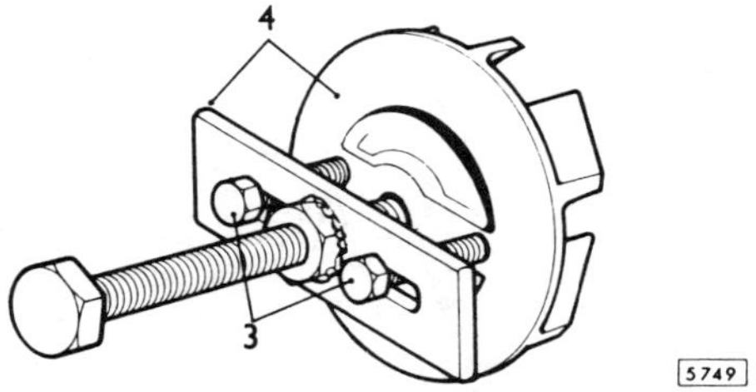
5 Slacken locknut and remove Allen retaining screw from water pump body.
6 Support water pump body and, using a suitable mandrel, carefully press out bearing and shaft assembly from impeller end to pulley end.
7 Carefully tap seal assembly from body.

26–6

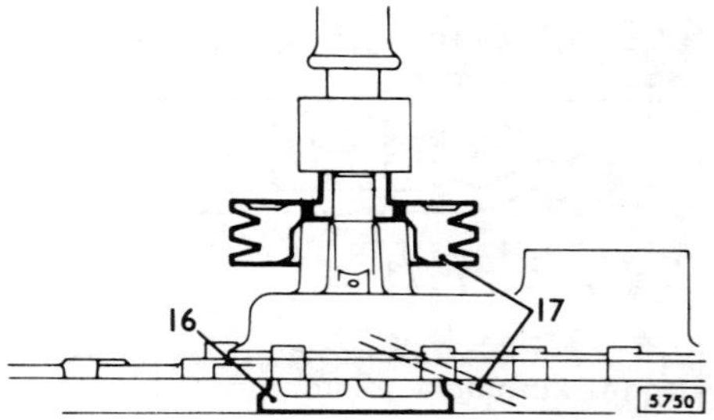

FRONT PIPE – LEFT-HAND 30.10.09

Remove and refit

NOTE: On cars to USA Federal specification, see Catalytic Converter – Remove and refit – 17.50.01.

Removing
1. Remove left-hand air cleaner element, see 19.10.11.
2. Remove bolts and spring washers securing power steering bellows heat shield.
3. Remove nuts, plain washers and bolts securing flanges; separate intermediate pipe from down pipe. Ensure intermediate pipe is adequately supported.
4. Remove nuts and plain washers securing heat shield and front pipe to exhaust manifolds; withdraw heat shield.
5. Withdraw front pipe.
   NOTE: On L.H. drive cars, this operation will be greatly facilitated if steering is turned when manoeuvring pipe clear.

Refitting
6. Reverse operations 1 to 5; coat all joints with Holts Firegum.
7. Tighten front pipe and clamping flange fixings by diagonal selection to avoid distortion.

FRONT PIPE – RIGHT-HAND 30.10.10

Remove and refit

NOTE: On cars to USA Federal specification, see Catalytic Converter – Remove and refit – 17.50.01.

Removing
1. Remove right-hand air cleaner element, see 19.10.08.
2. Remove bolts and spring washers securing power steering bellows heat shield; withdraw shield.
3. Remove nuts, plain washers and bolts securing flanges; separate intermediate pipe from front pipe. Ensure intermediate pipe is adequately supported.
4. Remove nuts and plain washers securing front pipe to exhaust manifolds; withdraw pipe.
   NOTE: On R.H. drive cars, this operation will be greatly facilitated if steering is turned to the left when manoeuvring pipe clear.

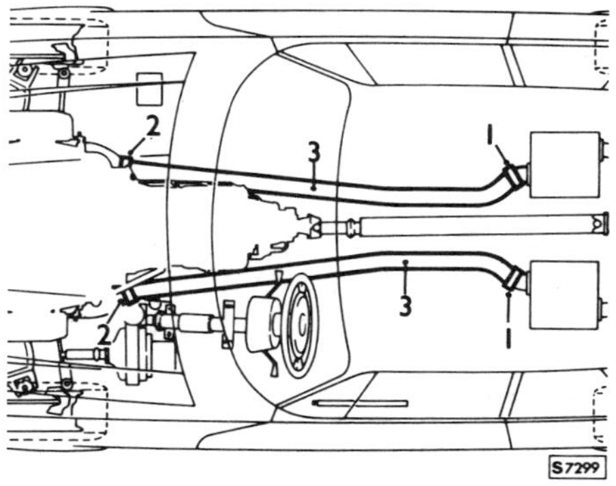

Refitting
4. Reverse operations 1 to 3; smear joints with Holts Firegum.
   CAUTION: Tighten flange fixings progressively to avoid distortion.

FRONT SILENCER

Remove and refit Left-hand 30.10.15
**   Right-hand 30.10.16**

Removing
1. Slacken nuts securing 'U' bolt clamps.
2. Separate silencer from rear intermediate pipe.
3. Separate silencer from front intermediate pipe.
   CAUTION: Front intermediate pipe must be adequately supported.

Refitting
Reverse operations 1 to 3; smear joints with Holts Firegum.

INTERMEDIATE PIPE

Remove and refit Left-hand 30.10.11
**   Right-hand 30.10.12**

Removing
1. Slacken nuts securing 'U' bolt clamp.
2. Remove nuts, washer and bolts securing front flanges.
3. Separate intermediate pipe from down pipe and silencer.

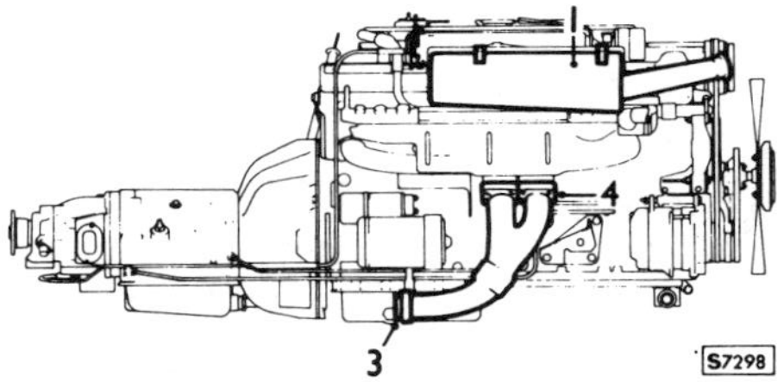

Refitting
5. Reverse operations 1 to 4; coat all joints with Holts Firegum.
6. Tighten front pipe and clamping flange fixings by diagonal selection to avoid distortion.

TAIL PIPE AND SILENCER – LEFT- OR RIGHT-HAND 30.10.22

Remove and refit

Removing
1. Slacken nut securing 'U' bolt clamp.
2. Using a soft faced mallet, tap silencer rearwards until silencer is separated from intermediate pipe.
3. Manoeuvre tail pipe until spigot can be withdrawn from mounting. 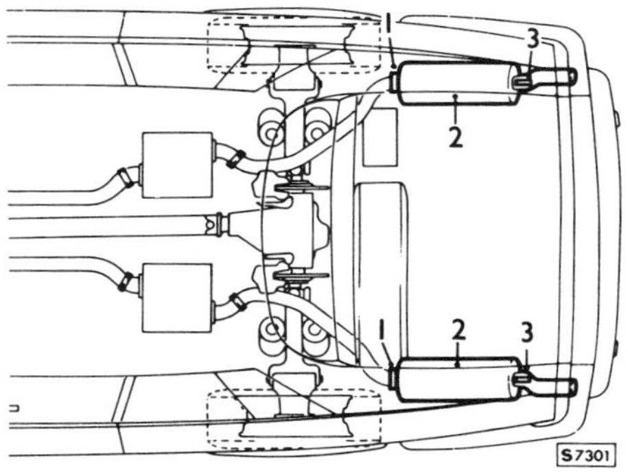
4. Check condition of mounting rubber and renew if necessary.

Refitting
Reverse operations 1 to 3; smear joint with Holts Firegum.

EXHAUST TRIM – LEFT- OR RIGHT-HAND

Remove and refit 30.10.23

Removing
1. Remove grub screw using Allen key; withdraw trim.

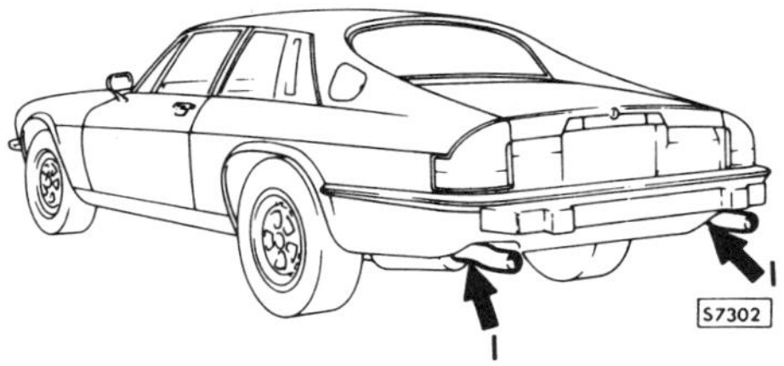

Refitting
Use Holts Firegum to seal joint; ensure trim is in horizontal plane before tightening screw.

REAR INTERMEDIATE PIPE

Remove and refit Left-hand 30.10.24
**   Right-hand 30.10.25**

Removing
1. Slacken nuts securing 'U' bolt clamps.
2. Using a soft faced mallet, tap rear silencer rearwards until intermediate pipe can be withdrawn.
3. Separate intermediate pipe from front silencer.
   CAUTION: Front silencer and pipe must be adequately supported.
4. Pull intermediate pipe forwards to disengage spigot from mounting rubber.
5. Manoeuvre intermediate pipe rearwards through suspension.
6. Check condition of mounting rubber and renew if necessary.

Refitting
Reverse operations 1 to 5; smear joints with Holts Firegum.

30–1

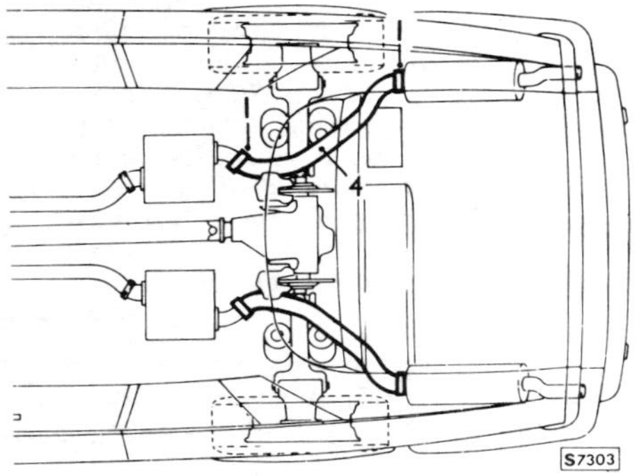

INDUCTION MANIFOLD

Remove and refit Right-hand 30.15.03

Removing

1. Disconnect battery, see 86.15.20.
2. Remove air cleaner, see 19.10.02.
3. Depressurise fuel system, see 19.50.02.
4. Remove screw securing fuel filter bracket to manifold ram tube.
5. Remove screws securing fuel pipe to overrun valves, retrieve spacer from left-hand valve.
6. Disconnect manifold pressure hose from manifold and 'T' piece.
7. Release hose clip securing fuel cross-over pipe to fuel rail, disconnect hose.
8. Disconnect electrical connectors to kick-down switch – right-hand drive cars only.
9. Remove locknut and disconnect throttle cable assembly from throttle pedestal – right-hand drive cars only.
10. Release throttle cross-rod from bell-crank.
11. Disconnect vacuum hoses from throttle housing.
12. Release hose clip securing air balance pipe to rubber elbow, disconnect pipe from elbow.

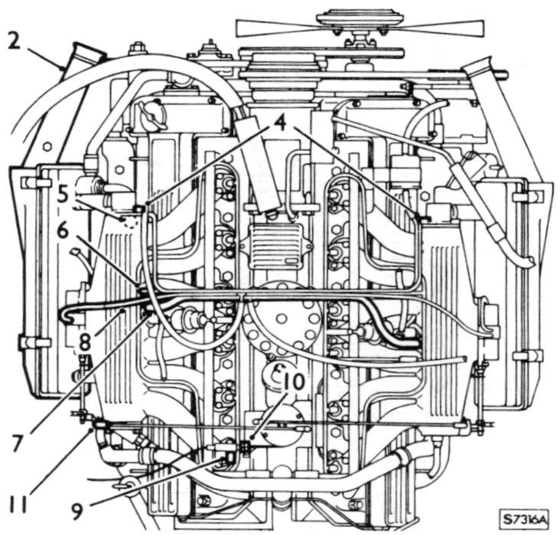

12. Remove plastic clips securing harness to fuel rail.
13. Disconnect electrical connectors at injectors and cold start injector.
14. Release hose clip securing brake vacuum hose to non-return valve, disconnect hose.
15. Remove throttle return spring.
16. Release hose clip securing manifold bleed pipe at rubber elbow, disconnect pipe from elbow.
17. Remove twelve nuts and serrated washers securing manifold assembly to cylinder head.
18. Remove two screws securing air rail clips to manifold ram tubes – Emission control cars to USA Federal specification only.
19. Release hose clips securing air rail to check valve connecting hose, release air rail from hoses. Remove air rail and discard rubber 'O' rings – Emission control cars to USA Federal specification only.
20. Remove two bolts and washers securing EGR valve to throttle housing flange, release EGR valve, remove and discard gasket – Emission control cars to USA Federal specification only.
21. Remove six manifold stud spacers.

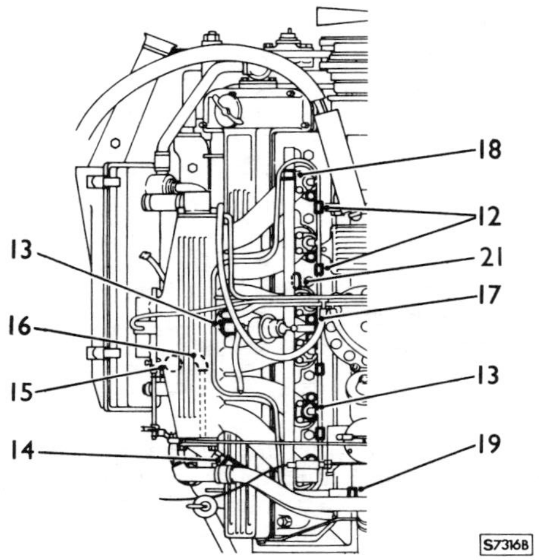

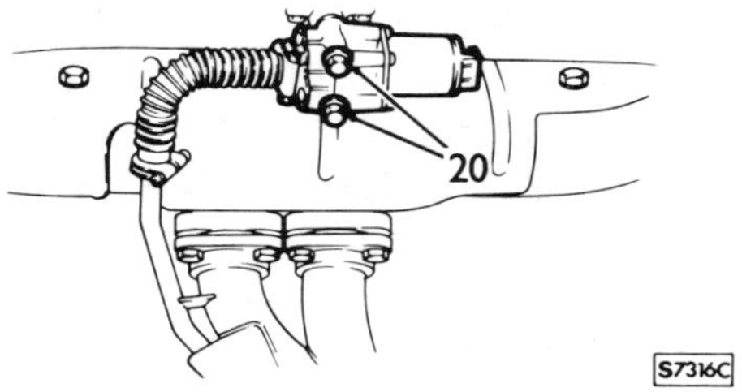

22. Remove induction manifold assembly from cylinder head easing aside air balance pipe and fuel pipes.
23. Remove and discard manifold gaskets.
24. Plug inlet ports.

Refitting

25. Transfer components to replacement induction manifold as necessary.
26. Reverse operations 1 to 22 using new gaskets, hose clips, cable clips and sealing rings.
    CAUTION: Sealing rings MUST be renewed each time the injectors or air rail are removed from the manifold.
27. Check throttle linkage adjustment, see 19.20.05.

INDUCTION MANIFOLD

Remove and refit Left-hand 30.15.02

Removing

1. Disconnect battery, see 86.15.20.
2. Remove air cleaner, see 19.10.01.
3. Depressurise fuel system, see 19.50.02.
4. Remove screws securing fuel pipe to overrun valves, retrieve spacer from left-hand valve.
5. Release hose clip securing pressure regulator return hose to fuel rail, disconnect hose.
6. Release hose clip securing fuel cross-over pipe to fuel rail, disconnect hose.
7. Disconnect manifold pressure hose from manifold.
8. Disconnect vacuum hose from throttle housing – Emission control cars to USA Federal specification only.
9. Disconnect electrical connectors from kickdown switch – left-hand drive cars only.
10. Remove locknut and disconnect throttle cable assembly from throttle pedestal – left-hand drive cars only.
11. Release throttle cross-rod from bell-crank.

30–2

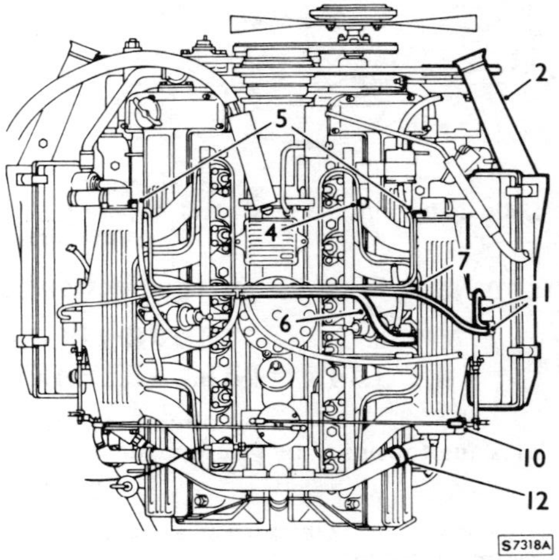

7 Remove six nuts and shakeproof washers securing upper half of manifold flanges to cylinder head.
8 Remove front manifold.
9 Remove rear manifold.
10 Remove and discard manifold gaskets.
11 Remove all traces of gasket material ensuring no damage to mating surfaces.

Refitting
Reverse operations 1 to 10 using new gaskets.

EXHAUST MOUNTING – FRONT
Remove and refit   Either   30.20.02

Removing
1 Remove two bolts securing mounting to suspension unit.
2 Turn mounting through 180° and remove from rear intermediate pipe spigot.

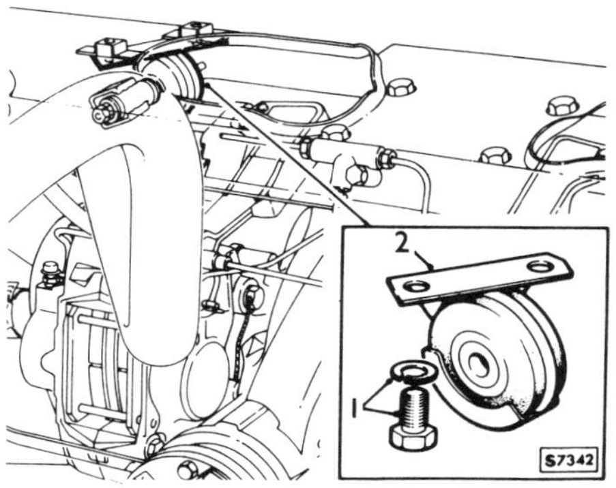

Refitting
NOTE: Mounting brackets are handed.

3 Fit mounting to rear intermediate pipe spigot and turn through 180° to align with fixing holes in suspension unit.
4 Fit but do not tighten front securing bolt.
5 Fit and tighten rear securing bolt.
6 Tighten front securing bolt.

8 Remove six nuts and shakeproof washers securing upper half of manifold flanges to cylinder head.
9 Remove front manifold.
10 Remove rear manifold.
11 Remove and discard manifold gaskets.
12 Remove all traces of gasket material ensuring no damage to mating surfaces.

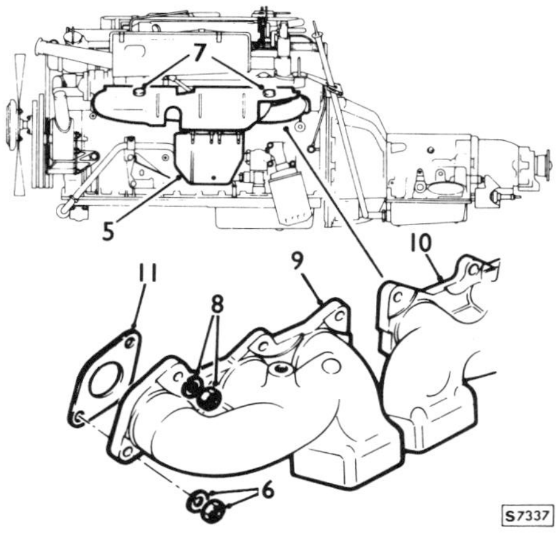

Refitting
Reverse operations 1 to 11 using new gaskets.

EXHAUST MANIFOLDS
Remove and refit   Right-hand   30.15.11

Removing
1 Remove air cleaner, see 19.10.02.
2 Remove front pipe, see 30.10.10.
3 Beneath car remove three screws securing underbody heatshield, remove heatshield.
4 Remove two nuts and washers securing starter motor heatshield, remove heatshield.
5 Remove six nuts and shakeproof washers securing lower half of manifold flanges to cylinder head.
6 Above car remove two bolts and washers securing upper manifold heatshield to manifolds, remove heatshield.

EXHAUST MANIFOLDS
Remove and refit   Left-hand   30.15.10

Removing
1 Disconnect battery, see 86.15.20.
2 Remove air cleaner, see 19.10.01.
3 Remove front pipe, see 30.10.09.
4 Underneath car remove three screws securing underbody heatshield, remove heatshield.
5 Remove oil pipe heatshield.
6 Remove six nuts and shakeproof washers securing lower half of manifold flanges to cylinder head.
7 Above car remove two bolts and washers securing upper manifold heatshield to manifolds, remove heatshield.

13 Remove plastic clips securing harness to fuel rail.
14 Disconnect electrical connectors at injectors and cold start injector.
15 Remove bolt and washer securing harness earth strap to manifold ram tube.
16 Release hose clip securing brake vacuum hose to non-return valve, disconnect hose from valve.
17 Disconnect heater vacuum hose from manifold stub pipe.
18 Disconnect diverter valve vacuum hose from manifold stub pipe – Emission control cars to USA Federal specification only.
19 Release hose clip securing automatic transmission vacuum pipe to connecting hose, disconnect pipe from hose.
20 Remove throttle return spring.
21 Remove twelve nuts and serrated washers securing manifold assembly to cylinder head.
22 Release hose clip securing air rail to check valve connecting hose, release air rail from hose. Remove air rail and discard rubber 'O' rings – Emission control cars to USA Federal specification only.
23 Release hose clip securing fuel cross-over pipe to fuel rail, disconnect hose at fuel rail.
24 Remove two bolts and washers securing EGR valve to throttle housing flange, release EGR valve, remove and discard gasket – Emission control cars to USA Federal specification only.
25 Remove six manifold stud spacers.
26 Remove induction manifold assembly from cylinder head easing aside air balance pipe and fuel pipes.
27 Remove and discard manifold gaskets.
28 Plug inlet ports.

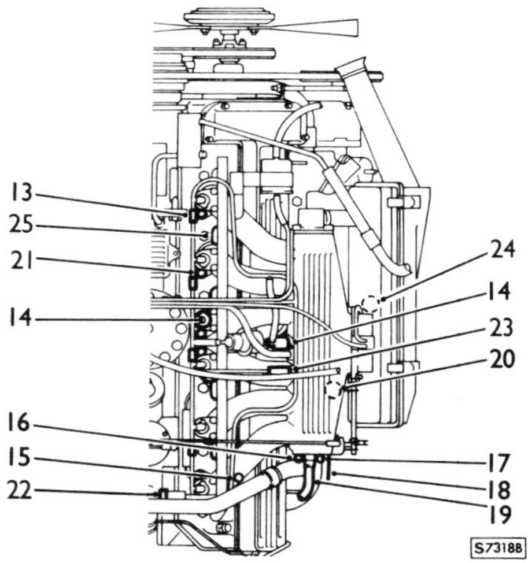

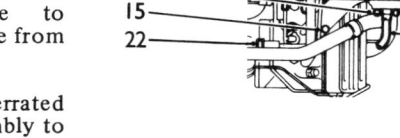

Refitting
29 Transfer components to replacement induction manifold as necessary.
30 Reverse operations 1 to 26 using new gaskets, hose clips, cable clips and sealing rings.
CAUTION: Sealing rings MUST be renewed each time the injectors or air rail are removed from the manifold.
31 Check throttle linkage adjustment, see 19.20.05.

30–3

EXHAUST MOUNTING – REAR

Remove and refit 30.20.04

Removing
1. Peel back carpet from side of luggage compartment.
2. Support exhaust system and remove nyloc nuts securing mounting. 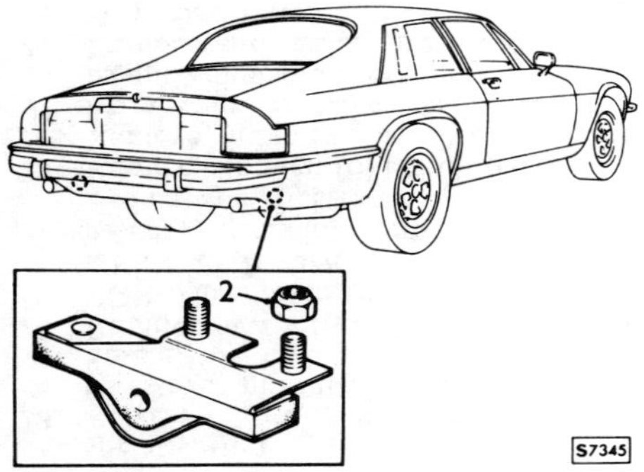
3. Carefully lower exhaust system until mounting is clear of body and can be withdrawn from spigot. Ensure system is adequately supported.

Refitting
Reverse operations 1 to 3; secure carpet with suitable petroleum based adhesive.

CAUTION: The hydraulic fluid used in the clutch hydraulic system is injurious to car paintwork. Utmost precautions MUST at all times be taken to prevent spillage of fluid. Should fluid be accidentally spilled on paintwork, wipe fluid off immediately with a cloth moistened with denatured alcohol (methylated spirits).

CLUTCH ASSEMBLY

Remove and refit   33.10.01

Removing

NOTE: At the time of going to press, procedures for removal of clutch with engine in situ are not available. These will be published at a later date.

1. Remove engine and gearbox assembly from car, see 12.37.01.
2. Remove bolts and spring washers securing flywheel cover to bell housing.
3. Remove bolts and spring washers securing starter motor, withdraw motor from bell housing.
4. Remove remaining bolts securing bell housing to cylinder block, noting positions of long bolts; disconnect gearbox breather pipe. Withdraw bell housing and gearbox.
5. Mark relative positions of clutch cover to flywheel and balance weights to clutch cover. 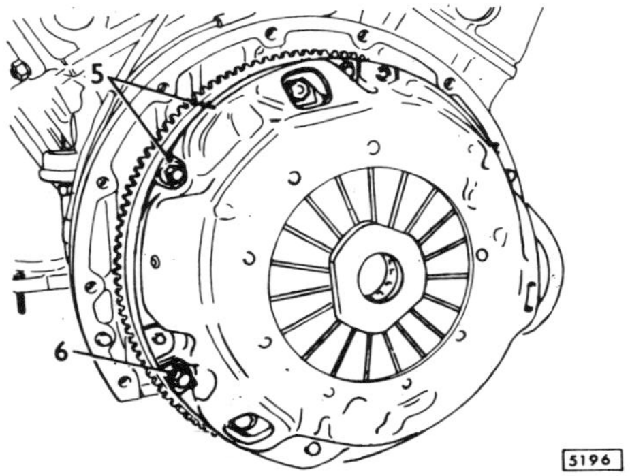
6. Remove bolts and spring washers securing clutch cover to flywheel; withdraw cover together with clutch plates.
7. Examine flywheel face for scoring. If scoring is excessive, flywheel must be renewed.
8. Examine clutch plates for oil contamination or evidence of slipping. If oil contamination is evident, crankshaft rear oil seal should be examined and, if necessary, replaced.

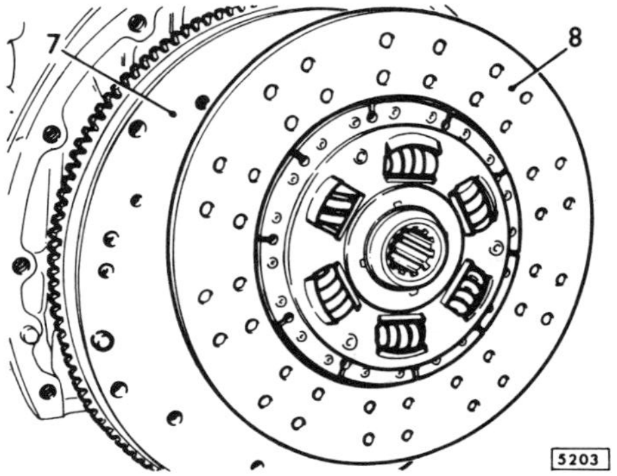

CLUTCH HYDRAULIC SYSTEM

Bleed   33.15.01

CAUTION: Only Castrol-Girling brake fluid (GREEN) may be used in the hydraulic system.

Bleeding

1. Remove reservoir filler cap.
2. Top up reservoir to correct level with hydraulic fluid.
3. Attach one end of a bleed tube to slave cylinder bleed nipple. 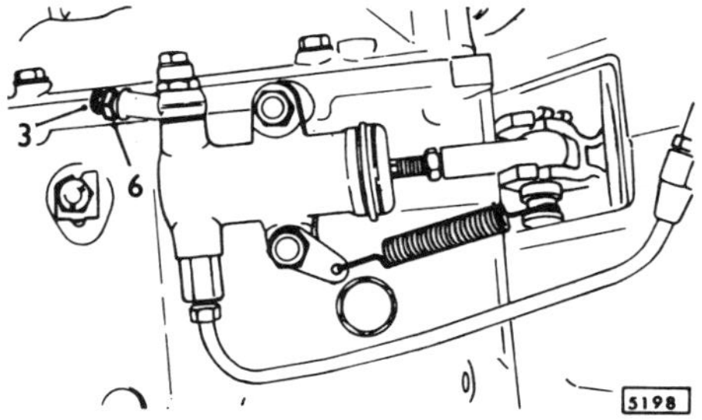
4. Partially fill a clean container with hydraulic fluid.
5. Immerse other end of bleed tube in fluid.
6. Slacken slave cylinder bleed nipple.
7. Pump clutch pedal slowly up and down, pausing between each stroke.
8. Top up reservoir with fresh hydraulic fluid after every three pedal strokes.
   CAUTION: Do not use fluid bled from the system for topping up purposes as this will contain air. If fluid has been in use for some time it should be discarded. Fresh fluid bled from system may be used after allowing it to stand for a few hours to allow air bubbles to disperse.
9. Pump clutch pedal until pedal becomes firm, tighten bleed nipple.
10. Top up reservoir.
11. Refit filler cap.
12. Apply working pressure to clutch pedal for two to three minutes and examine system for leaks.

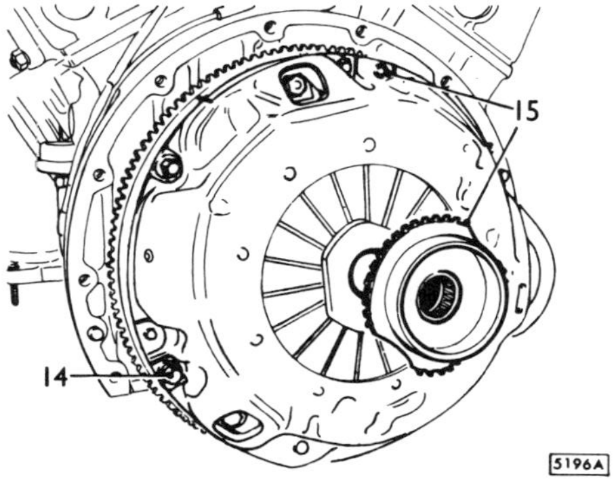

CLUTCH SLAVE CYLINDER PUSH ROD

Check and adjust   33.10.03

Checking

1. Move push rod backwards and forwards, measure total free movement of rod which, when correctly adjusted, is .125 in (3,2 mm).

Adjusting

2. Slacken locknut.
3. Screw push rod in or out of trunnion until correct free amount of movement is obtained.
   NOTE: Flats are machined on shank of push rod to enable spanner to be used.
4. Tighten locknut.
5. Operate clutch pedal several times and recheck amount of free travel.

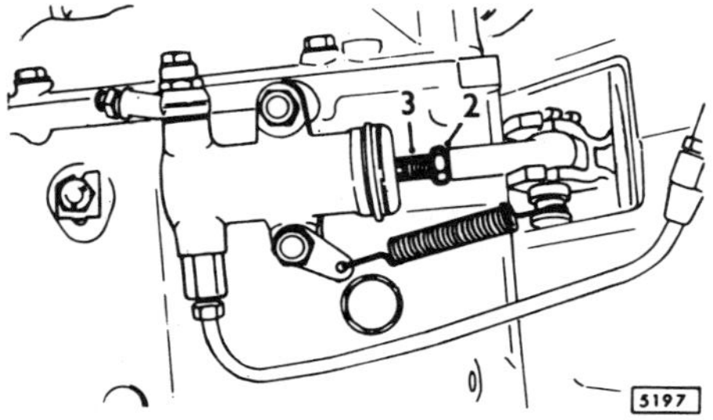

9. Release spring clips securing release bearing to withdrawal lever.
10. Disengage lugs from withdrawal lever.

CAUTION: It is always advisable when removing clutch to fit a new release bearing. To do this, proceed as follows:

Refitting

11. Position lugs of release bearing in withdrawal lever.
12. Fit spring clips; ensure that lips are correctly seated in recesses.
13. Position clutch plates and cover on flywheel, ensure reference marks made during dismantling are in alignment.
14. Fit balance weights, bolts and washers; do not tighten bolts at this stage.
15. Using dummy shaft, align clutch plates, ensure clutch cover is correctly located on dowels.
16. Tighten bolts by diagonal selection.
17. Reverse operations 1 to 4.
18. Check slave cylinder push rod adjustment, see 33.10.03.

33–1

MASTER CYLINDER

Remove and refit 33.20.01

Removing
1. Remove pedal box, see 70.35.03.
2. Note fitted position of return spring; disconnect spring from pedal.
3. Remove and discard split pin.
4. Withdraw clevis pin, recover plain washer.
5. Remove nuts and spring washers securing master cylinder to pedal box.

Right-hand drive cars only
6. Remove setscrew, nut and washers securing brake fluid reservoir bracket to pillar.
7. Withdraw brake fluid reservoir bracket; lift off master cylinder and shims (if fitted).

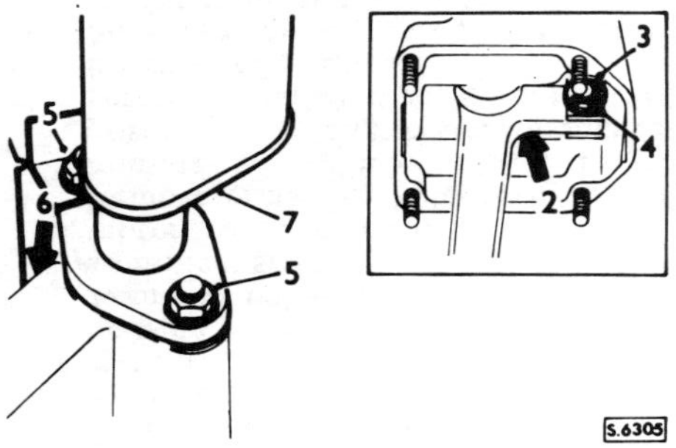

Refitting
Reverse operations 1 to 5 or 7.

MASTER CYLINDER

Overhaul 33.20.07

WARNING: USE ONLY CLEAN BRAKE FLUID OR DENATURED ALCOHOL (METHYLATED SPIRITS) FOR CLEANING. ALL TRACES OF CLEANING FLUID MUST BE REMOVED BEFORE REASSEMBLY. ALL COMPONENTS SHOULD BE LUBRICATED WITH CLEAN BRAKE FLUID AND ASSEMBLED USING THE FINGERS ONLY.

Dismantling
1. Remove master cylinder, see 33.20.01.
2. Detach rubber boot from end of barrel and move boot along push rod.
3. Depress push rod and remove circlip.
4. Withdraw push rod, piston, piston washer, main cup, spring retainer and spring.
5. Remove secondary cup from piston.

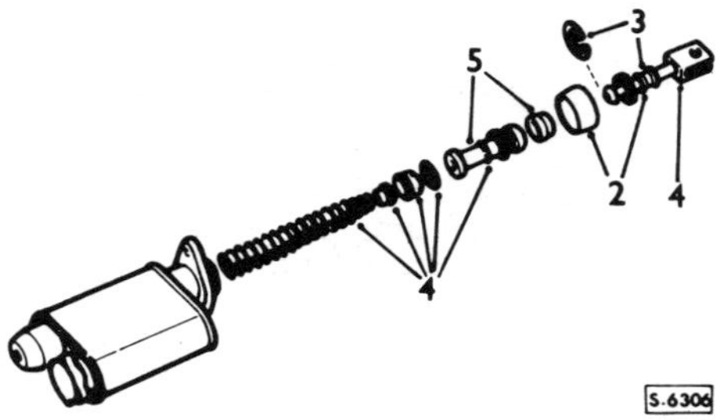

Inspection
6. Examine cylinder bore for scores.
7. Thoroughly wash out reservoir and ensure by-pass hole in cylinder bore is clear. Dry using compressed air or lint-free cloth.
8. Lubricate replacement seals with clean brake fluid.

Reassembling
9. If necessary, fit end plug on new gasket.
10. Fit spring retainer to small end of spring. If necessary, bend over retainer ears to secure.
11. Insert spring, large end leading, into cylinder bore; follow with main cup, lip foremost. Ensure lip is not damaged on circlip groove.
12. Using fingers only, stretch secondary cup on to piston with small end towards drilled end and groove engaging ridge. Gently work round cup with fingers to ensure correct bedding.
13. Insert piston washer into bore, curved edge towards main cup.
14. Insert piston in bore, drilled end foremost.
15. Fit rubber boot to push rod.
16. Offer push rod to piston and press into bore until circlip can be fitted behind push rod stop ring.
   CAUTION: It is important to ensure circlip is correctly fitted in groove.
17. Locate rubber boot in groove.

WITHDRAWAL ASSEMBLY

Remove and refit 33.25.12

Removing
1. Remove engine and gearbox assembly from car, see 12.37.01.
2. Remove clutch assembly, see 33.10.01.
3. Release spring clips securing release bearing to withdrawal fork.
4. Disengage lugs from withdrawal fork.
5. Remove clip and pivot pin at clutch slave cylinder push rod.
6. Drift roll pin from withdrawal fork fulcrum shaft.
7. Drill out upper seal plug from fulcrum boss.
8. Drift shaft downwards through withdrawal fork.
9. Remove withdrawal fork from bell housing.
   NOTE: Examine withdrawal fork shaft bushes for wear and renew if necessary. Lightly ream new bushes to size using shaft as guide.

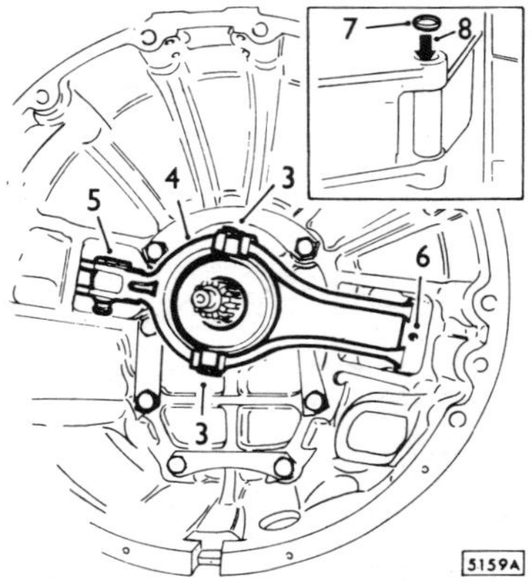

Refitting
10. Position withdrawal fork between bosses on bell housing.
11. Tap pivot shaft into position, locate roll pin holes and fit new roll pin.
12. Fit new seal plugs top and bottom of shaft.
13. Refit slave cylinder push rod pivot pin and retain with spring clip.
14. Position lugs of release bearing in withdrawal fork.
15. Fit spring clips; ensure that lips are correctly seated in recesses.
16. Refit clutch assembly, see 33.10.01.
17. Refit engine and gearbox to car, see 12.37.02.

RELEASE ASSEMBLY

Overhaul 33.25.17

1. Remove release assembly, see 33.25.12.
2. Check bearing bushes and pivot shaft for wear.
3. If replacement bearing bushes are necessary, use suitable pieces of tube for mandrels and press new bushes into position to displace the old.
4. Lightly ream to size, using pivot shaft for gauge.

CLUTCH PEDAL

Remove and refit 33.30.02

Removing
1. Remove pedal box, see 70.35.03.
2. Use large screwdriver to disengage return spring from pedal.

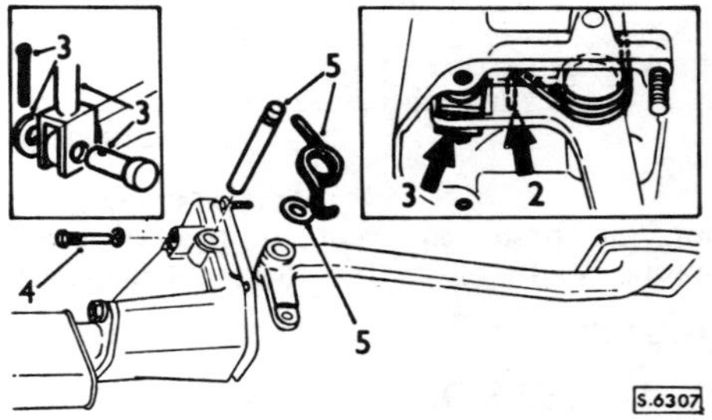

33–2

3 Remove split pin, washer and clevis pin securing pedal arm to master cylinder push rod.
4 Remove locating pin.
5 Tap clutch shaft from pedal box casting and recover pedal return spring and washers.

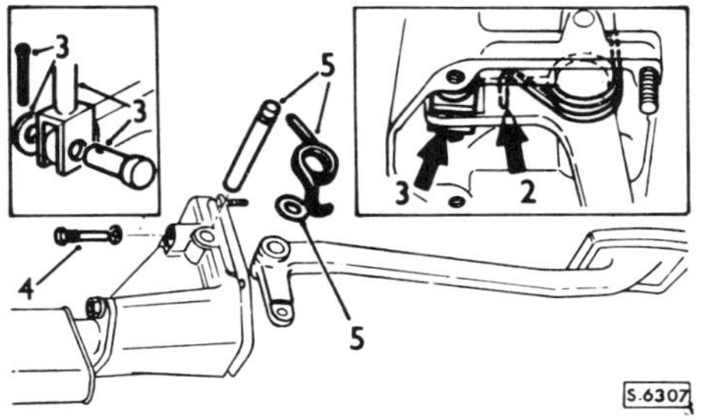

Refitting

6 Locate return spring on pedal boss.
7 Smear clutch shaft with suitable grease.
8 Tap clutch shaft through casting – locating groove to outside edge – positioning a washer at either side of pedal boss.
9 Fit pedal arm to master cylinder push rod and secure using clevis pin, plain washer and new split pin.
10 Fit pedal box.
11 Bleed hydraulic system, see 33.15.01.

CLUTCH SLAVE CYLINDER

Remove and refit   33.35.01

Removing

1 Disconnect pipe from slave cylinder; plug or tape pipe to prevent ingress of dirt.
2 Note relative position of spring anchor plate to gearbox.
3 Remove nuts and spring washers securing slave cylinder to gearbox.
4 Slide spring anchor plate off mounting stud.
5 Slide slave cylinder off mounting studs; slide rubber boot along push rod, withdraw cylinder from push rod.
6 Examine piston and slave cylinder bore for signs of scoring. Should scoring be evident, components must be renewed.
7 Examine spring for signs of distortion, renew if necessary.
8 Check condition of rubber boot on push rod. If distorted or perished in any way, boot must be renewed.

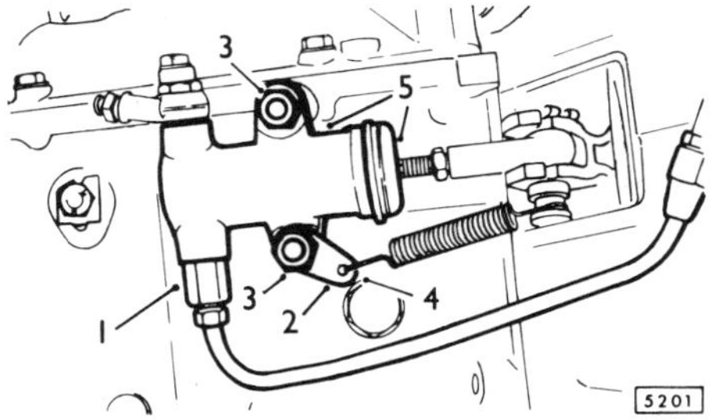

Reassembling

CAUTION: All components must be liberally coated with Girling rubber grease. Always assemble cup and cup filler using fingers only.

9 Reverse operations 1 to 4; use new cup.

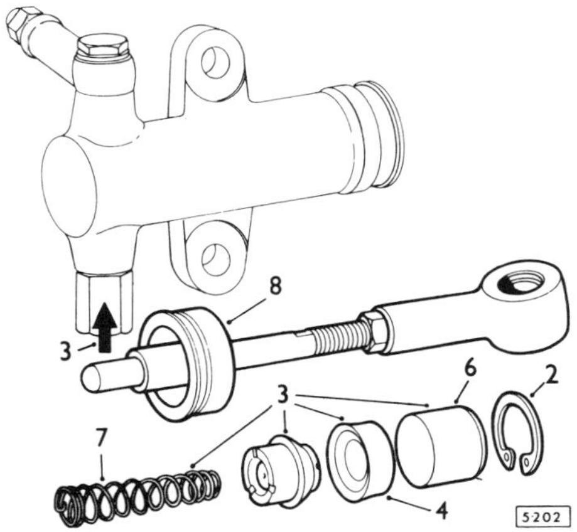

Refitting

6 Reverse operations 1 to 5.
7 Bleed clutch hydraulic system, see 33.15.01.
8 Check slave cylinder push rod adjustment, see 33.10.03.

CLUTCH SLAVE CYLINDER

Overhaul   33.35.07

Dismantling

1 Remove slave cylinder, see 33.35.01.
2 Remove circlip.
3 Apply low air pressure to inlet port and expel piston, cup, cup filler and spring.
4 Discard cup.

Inspecting components

5 Wash all components in denatured alcohol (methylated spirits) and dry using clean, lint free cloth.

CLUTCH PEDAL

Overhaul   33.30.07

Dismantling

1 Remove clutch pedal, see 33.30.02.
2 Using suitable mandrel, press bearing bushes from pedal boss.

Reassembling

3 Using suitable mandrel press new bearing bushes in from each side. Press until bushes are flush with sides of pedal.
4 Lightly ream bearing bushes to size using pedal shaft to check fit.
5 Smear bushes with suitable grease.
6 Fit clutch pedal, see 33.30.02.

33–3

REAR EXTENSION

Remove and refit 37.12.01

Removing
NOTE: At the time of going to press, procedures for this operation with engine in situ are not available. These will be published at a later date.

1. Remove engine and gearbox, see 12.37.01.
2. Remove drive flange. Remove setscrew and spring washer and withdraw speedometer drive assembly.
3. Remove seven setscrews and spring washers retaining rear extension.
4. Draw rear extension from third motion shaft. Remove all traces of old gasket from rear face of gearbox.
5. On front face of rear extension break staking and remove three countersunk head screws securing oil pump gear housing.
6. Enter two screws in tapped holes and evenly tighten down to withdraw gear housing.
7. Mark gears with ink to ensure correct refitting.
8. Recover oil pump drive sleeve and distance tube.
9. Drift rear oil seal from rear extension.
10. Recover speedometer drive gear.
11. Remove ball bearing race, examine, and replace if necessary.

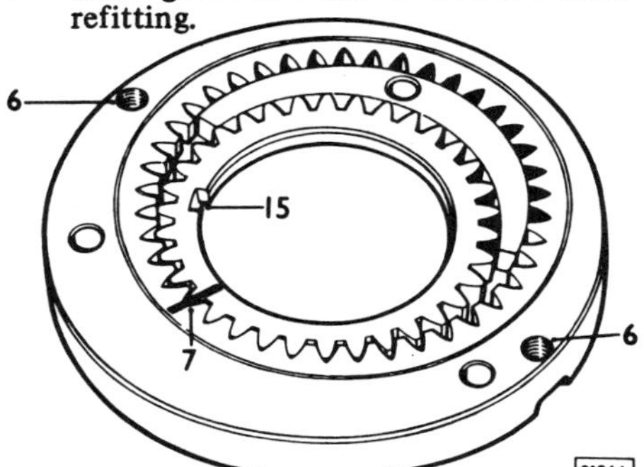

Refitting
12. Assemble gears to oil pump housing, as marked in operation number 7, and secure housing to gearbox extension using three countersunk head screws.
13. Stake screw heads.
14. Fit oil pump drive sleeve to oil pump gear, long shoulder first.
15. Using new paper gasket, fit rear extension to gearbox.
    NOTE: Ensure oil pump drive sleeve pin engages with gear.
16. Retain rear extension with seven setscrews and spring washers.
17. Place distance tube into rear extension on shaft.
18. Place ball bearing race on third motion shaft and gently drift home into rear extension.
19. Fit speedometer drive to third motion shaft.
20. Place rear oil seal in position, and tap into position. Smear seal with clean engine oil.
    NOTE: Ensure oil seal seats square.
21. Fit speedometer drive assembly and secure using one setscrew and spring washer.
22. Refit drive flange; use a new split pin to retain nut.
23. Reverse operation 1.

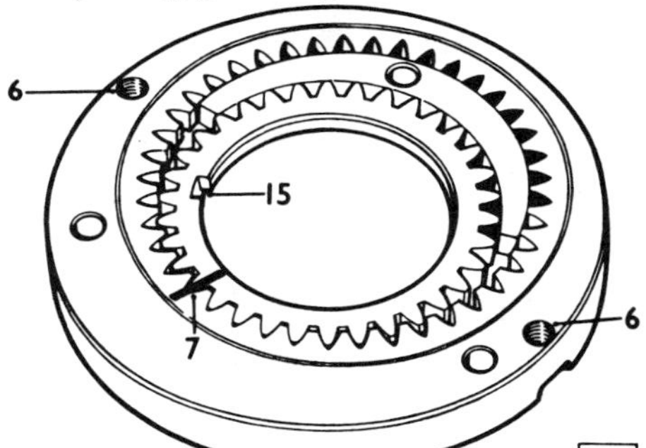

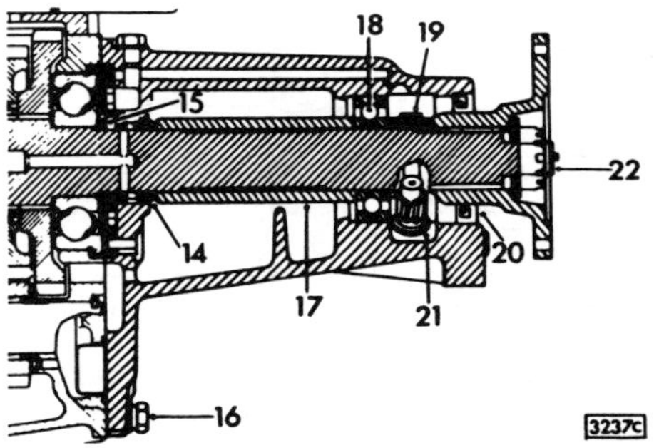

BELL HOUSING

Remove and refit 37.12.07

Removing
NOTE: At the time of going to press, procedures for removal of gearbox with engine in situ are not available. These will be published at a later date.

1. Remove engine and gearbox, see 12.37.01.
2. Remove clutch assembly, see 33.10.01.
3. Remove withdrawal assembly, see 33.25.12.
4. Remove and discard locking wire.
5. Knock back locking tabs and remove setscrews securing bell housing to gearbox; discard locking tabs.
6. Remove bell housing, remove and discard gasket.
7. Remove and discard oil seal.

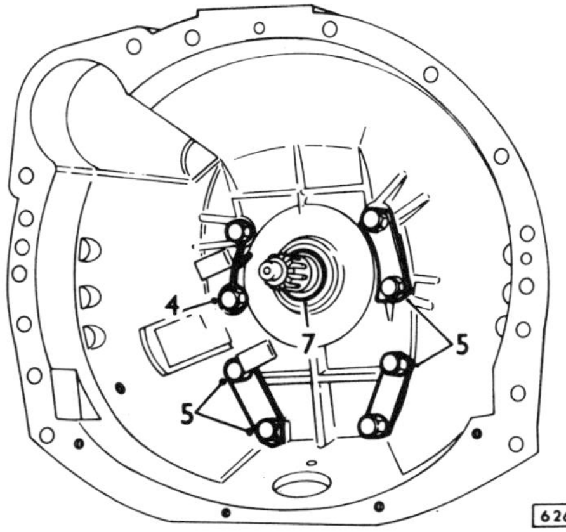

Refitting
8. Smear new oil seal with clean engine oil.
9. Push oil seal into bell housing with lip of seal towards gearbox.
10. Smear new gasket with grease; position gasket on front of gearcase.
11. Cover splines of first motion shaft with adhesive tape to prevent damage to oil seal.
12. Slide bell housing over first motion shaft; use new lock plates when refitting setscrews.
13. Tighten setscrews by diagonal selection.
14. Turn up tabs of lock plates and wire lock setscrews.
15. Reverse operations 1 to 3.

TOP COVER

Remove and refit 37.12.16

Removing
1. Remove engine and gearbox, see 12.37.01.
2. Slacken hose clip securing breather pipe to elbow; disconnect pipe.
3. Plug elbow to prevent ingress of dirt.
4. Place gear lever in neutral.
5. Note fitted position of top cover securing bolts; remove bolts and detach cable clips.
   NOTE: Top cover is dowelled to gearbox case.
6. Lift top cover off gear case; remove and discard gasket.

Refitting
7. Ensure reverse idler gear is out of mesh by pushing lever towards rear of gearbox.

continued

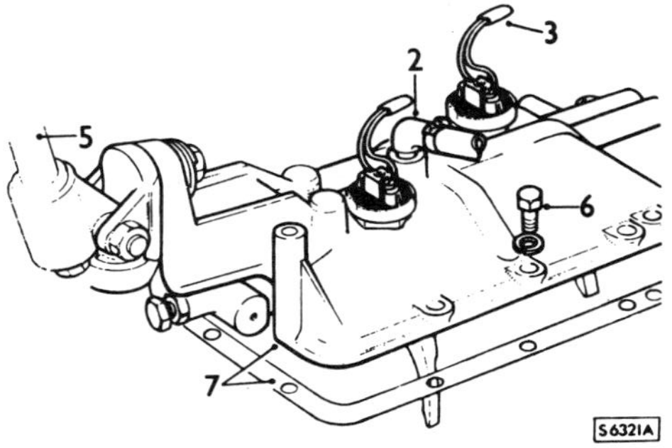

37–1

8 Reverse operations 1 to 6 ensuring selector levers are correctly engaged. Use a new gasket when fitting top cover.

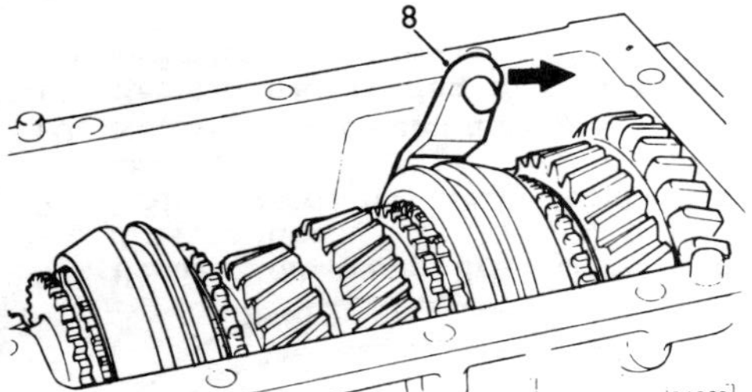

9 Withdraw reverse selector rod, fork and locating arm.
NOTE: Collect reverse rod detent plunger, ball and spring. Collect interlock ball.

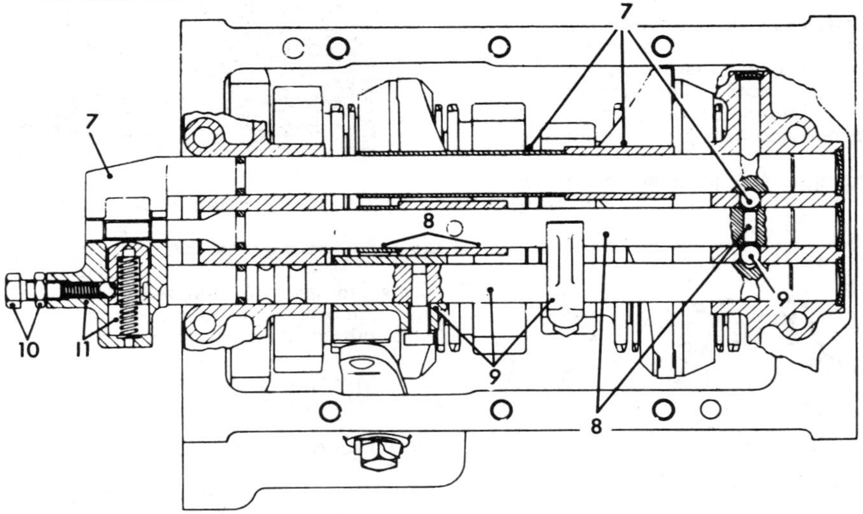

10 Release locknut on reverse selector, and slacken setscrew.

11 Recover detent plunger, stop spring, detent ball and spring.

Inspection

12 Examine all components for scores and undue wear. Examine ball bearings for pitting. Check springs for length against new items.

13 Remove 'O' ring seals from selector rods, and fit new. Examine gear change lever spigot bush for wear and renew if necessary.

Reassembling

14 Fit detent plunger and spring to reverse selector rod.

15 Fit detent ball, spring, setscrew and locknut.

16 Press detent plunger fully home, and tighten setscrew to lock plunger.

17 Slowly slacken setscrew until plunger is released and detent ball engages in groove.

18 Restrain setscrew and tighten locknut.

19 Place reverse selector rod detent spring, plunger and ball in its housing. Depress plunger and fit selector rod.

TOP COVER

Overhaul   37.12.19

Dismantling

1 Remove top cover, see 37.12.16.

2 Note fitted positions of cable; disconnect cables from reverse switch.

3 Remove self locking nut, remove coil spring, plain and fibre washer securing gear lever to top cover.

4 Withdraw gear lever together with harness; recover second fibre washer.

5 Press bush from pivot jaw housing.

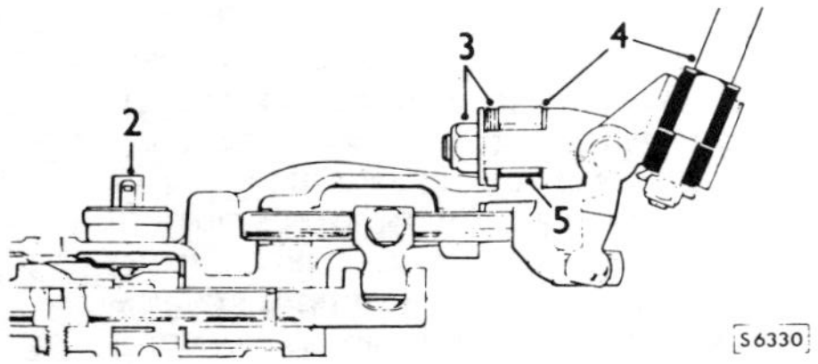

6 Remove locking wire and remove retaining screws from selector forks and locating arm.

7 Withdraw 3rd/top selector rod and collect selector fork, spacing tube and interlock ball.

8 Withdraw 1st/2nd selector rod and collect selector fork, spacing tube and loose interlock pin in rod.

37—2

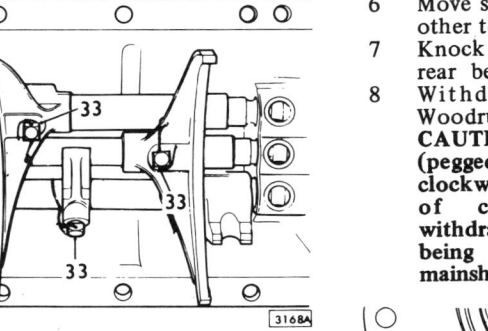

6 Move synchro sleeves away from each other to engage gears.
7 Knock back tab washer and remove rear bearing nut; discard tab washer.
8 Withdraw countershaft, collect Woodruff key.
   CAUTION: Ensure rear thrust washer (pegged to casing), drops down in a clockwise direction, viewed from rear of casing, as countershaft is withdrawn. This will prevent washer being trapped by reverse gear when mainshaft is driven forward.
9 Remove and discard fibre plug from front of countershaft.
10 Rotate first motion shaft until cutaway portions of driving gear are facing top and bottom of casing.
11 Using two levers, ease first motion shaft and front bearing forward until assembly can be withdrawn.
   continued

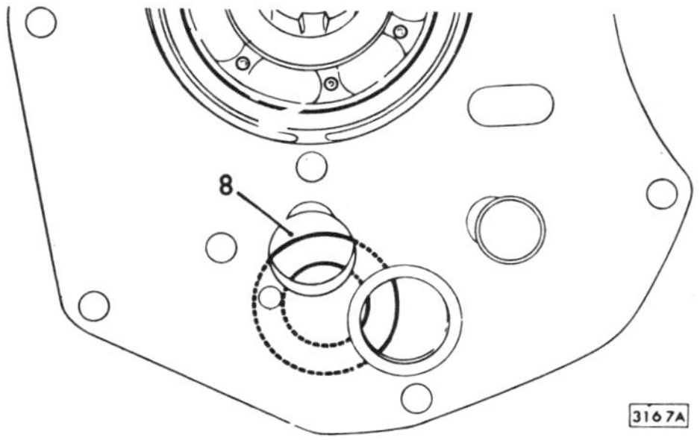

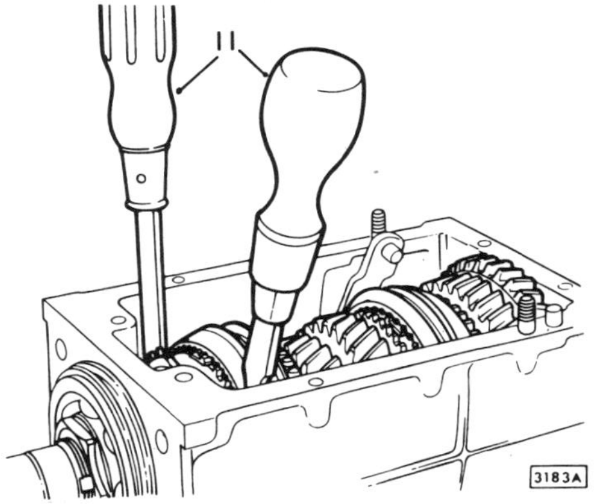

20 Fit selector fork and locating arm to reverse selector rod.
21 Fit retaining screws.
22 Fit 1st/2nd selector rod into cover far enough to collect selector fork and spacing tube.
23 Place top cover on its side, reverse selector rod downwards, and carefully push an interlock ball through 1st/2nd selector rod hole at front of top cover. Locate ball in hole in casting, against groove in reverse selector rod.
24 Push 1st/2nd selector rod into hole to retain ball in groove.
   NOTE: Take care not to push ball ahead of rod.
25 Fit retaining screw to 1st/2nd selector fork.
26 Check interlock by ensuring selector rods will not move together, and that moving one locks the other.
27 Carefully place interlock plunger through 3rd/top selector rod hole at front of top cover. Gently manipulate 1st/2nd rod until plunger drops through.
28 Ensure reverse 1st/2nd selector rods level and carefully push second interlock ball through 3rd/top hole to rest in hole in casting.
29 Pass 3rd/top selector rod through hole in top cover and fit selector fork and spacing tube. 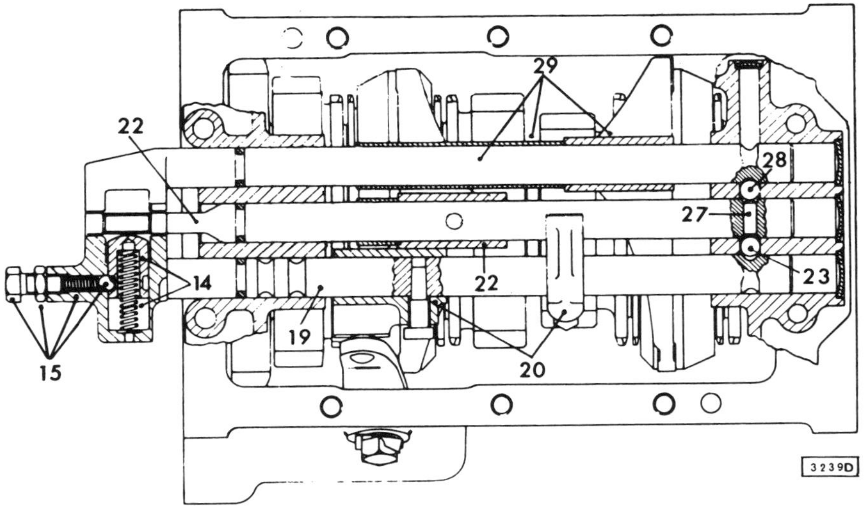
30 Push 3rd/top selector rod into hole to retain ball in groove.
   NOTE: Take care not to push ball ahead of rod.
31 Fit locating screw.
32 Check interlock by ensuring selector rods will not move together, and that moving one locks both the others.
33 Check tighten locating screws and wirelock each to its fork.
34 Fit new bush to pivot jaw housing. Lightly ream to size, using jaw pivot pin as a gauge.
35 Fit fibre washer to gear selector pivot jaw pin.
36 Push pin through bush in top cover and secure with fibre washer, flat washer, coil spring and self-locking nut.
37 Refit top cover; use a new gasket.

GEAR CHANGE SELECTORS

Remove and refit 37.16.31

Follow procedure given in operation 37.12.29 – Top Cover Overhaul.

GEARBOX

Remove and refit 37.20.01

Removing
1 Remove bell housing, see 37.12.07.

Refitting
Reverse operation 1.

GEARBOX

Overhaul 37.20.04

Dismantling
1 Remove engine and gearbox, see 12.37.01.
2 Carry out items 2 to 11, operation 37.12.01.
3 Remove bell housing, see 37.12.07.
4 Remove top cover, see 37.12.16.
5 Remove bolts securing adaptor plate, withdraw plate; remove and discard gasket.

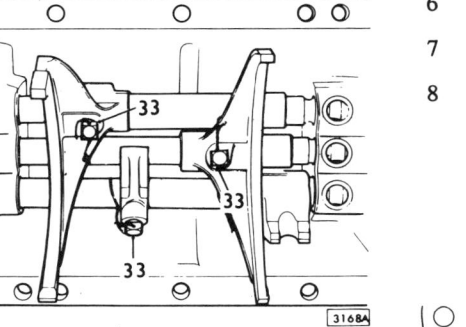

12 Recover spacer ring and top gear synchro ring.
13 Remove bearing from inside first motion shaft.
14 Knock back tab washer and remove nut, tab washer, and oil thrower; discard tab washer.
15 Tap shaft sharply against a metal plate to dislodge bearing.
16 Tap reverse idler shaft out of rear of gearbox, collect Woodruff key.
17 Knock back tab washer and remove bolt securing reverse lever; discard tab washer.
18 Withdraw reverse lever.
19 Rotate mainshaft until one of the large cutaway portions of third/top synchro hub is in line with countershaft gear. 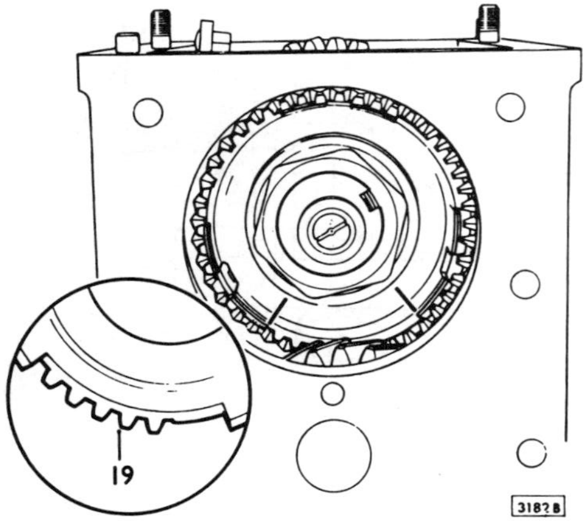
20 Tap mainshaft forward through rear bearing ensuring that reverse gear is kept pressed against first gear. 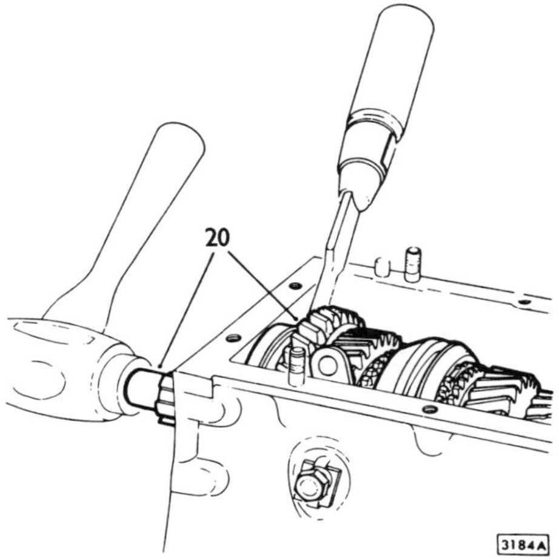
21 Withdraw rear bearing from casing and fit hose clip to mainshaft to prevent reverse gear from sliding off as mainshaft is withdrawn.
22 Lift out mainshaft and gears. 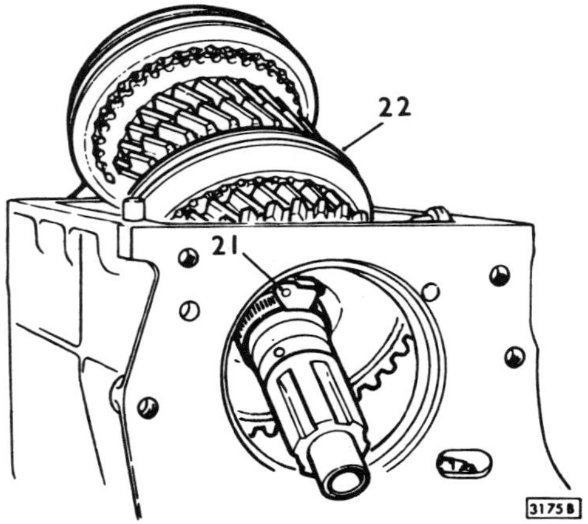
23 Lift out reverse idler gear.
24 Lift out countershaft gears and collect needle rollers, inner and outer thrust washers and retaining rings. Discard thrust washers.
CAUTION: Needle rollers are graded by size. If it is necessary to replace any rollers, complete sets MUST be obtained and old rollers discarded.
25 Remove hose clip retaining reverse gear; withdraw gear.
NOTE: To assist in identifying components, mainshaft is shown in position.
26 Withdraw first speed gear; collect needle rollers, spacer and sleeve.
27 Note fitted position of first/second synchro assembly to rings, remove synchro assembly and collect loose rings.
28 Withdraw second speed gear, collect needle rollers, remove narrow spacer.
29 Knock back tab washer and remove nut retaining third/top synchro assembly.
30 Note fitted position of third/top synchro assembly to rings, remove synchro assembly and collect loose rings; remove and discard thrust washer.
31 Withdraw third speed gear, collect needle rollers; remove wide spacer.
32 Push synchro hub from operating sleeve, collect synchro detent balls and springs, thrust members, plungers and springs.

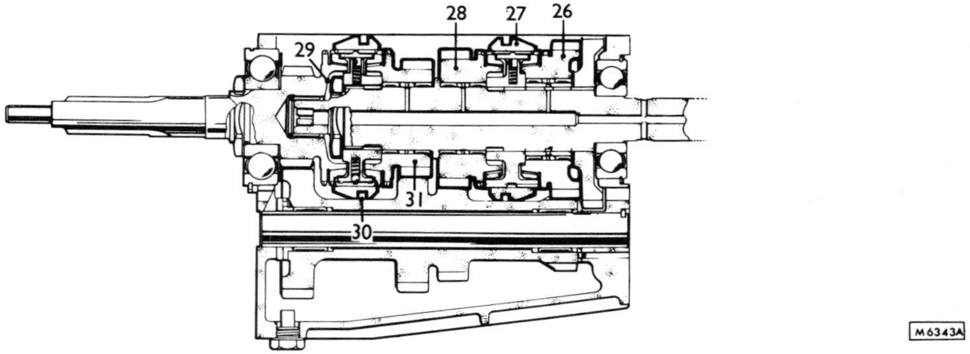

Inspection
33 Check all components for signs of damage, undue wear, overheating, scoring or excessive clearances.
34 Check that oil-ways in mainshaft are unobstructed.
CAUTION: Needle rollers are graded by size. If any rollers are displaced or damaged, all rollers MUST be renewed.
It is advisable to renew synchro assemblies especially if gearbox has seen considerable service. It is, however, permissible to re-use hubs and sleeves providing no damage is evident but springs, balls, plungers and thrust members MUST always be renewed.

37—4

Re-assembling

35 Assemble synchro hub to operating sleeve with wide boss of hub on opposite side to wide chamfer end of sleeve.
NOTE: Hub should slide smoothly over sleeve with no trace of binding.

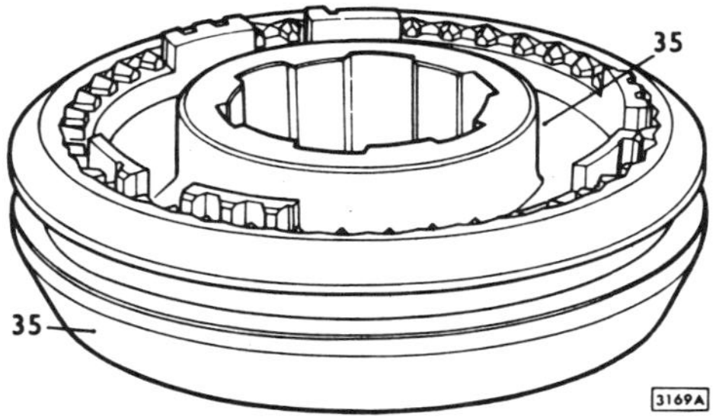

36 Line up hub so that holes for balls and springs are exactly level with top of operating sleeve.

37 Fit three springs, plungers and thrust members in synchro hub.
NOTE: These springs are colour coded red.

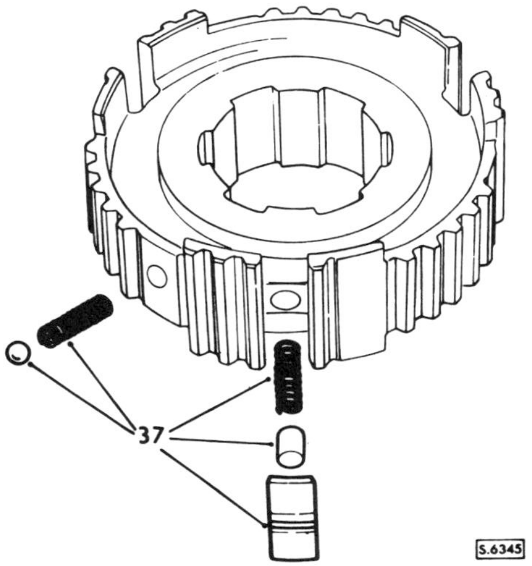

38 Ensure lip on each thrust member is facing outwards; press thrust members down as far as possible.

39 Assemble three balls and springs in line with operating sleeve teeth having three detent grooves.
NOTE: These springs have no colour coding.

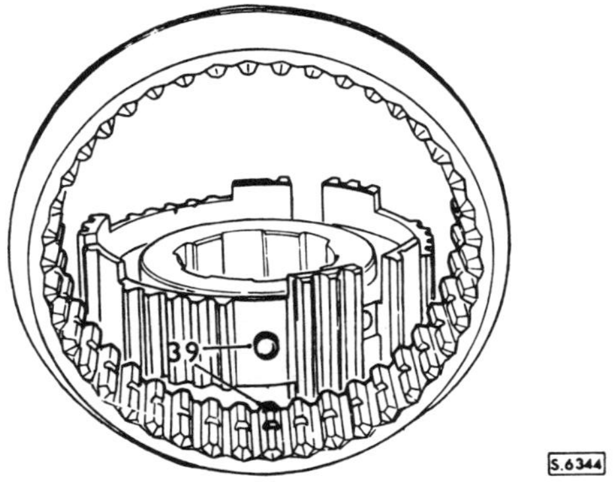

40 Compress springs with large hose clip or piston ring clamp.

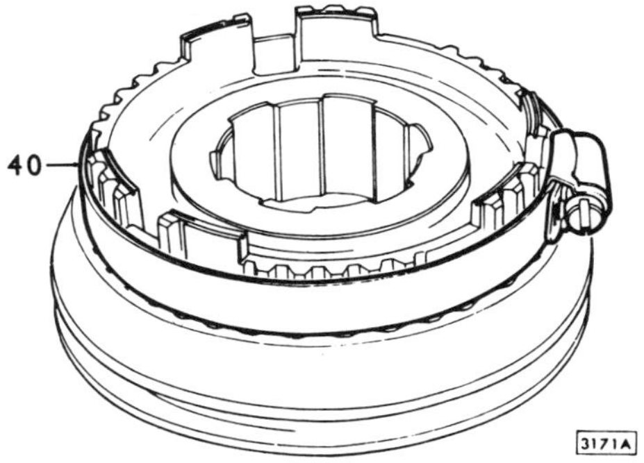

41 Depress hub slightly and push thrust members down until they engage neutral groove in operating sleeve.

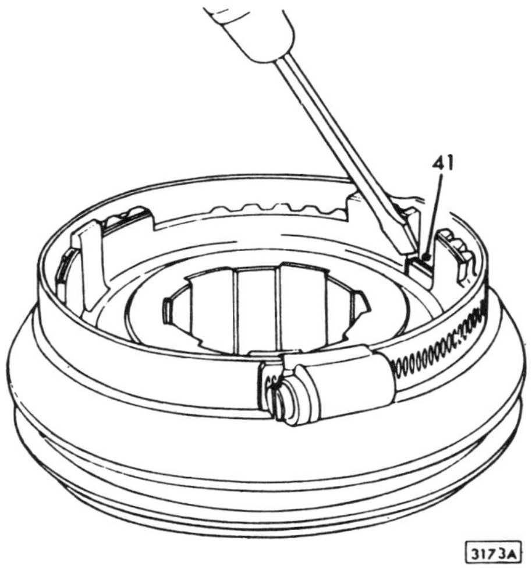

42 Tap hub down using a hide or lead hammer until balls can be heard and felt to engage with neutral groove (second click).

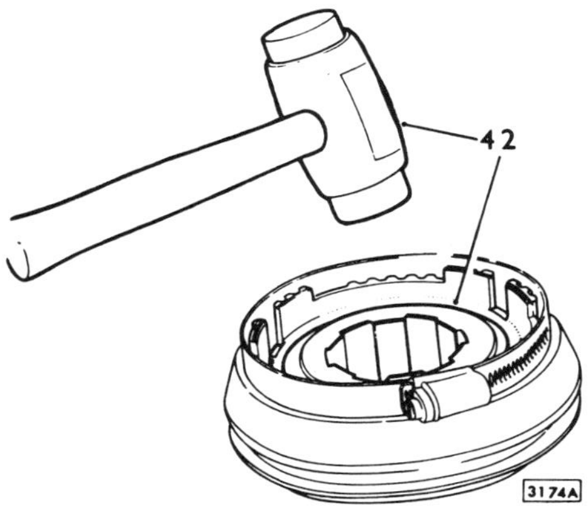

43 Repeat operations 35 to 42 on remaining synchro assembly.

44 Fit one retaining ring in front (large gear) end of cluster gear.

45 Fit needle rollers and inner thrust washer; retain rollers with grease.
CAUTION: Ensure peg on washer locates in groove in cluster gear.

46 Fit a retaining ring, needle rollers and second retaining ring in rear of cluster gear; retain rollers with grease.

47 Position new pegged rear thrust washer on boss; retain washer with grease.

48 Position new outer thrust washer on front of cluster gear; retain washer with grease.

49 Thoroughly clean interior and exterior of gearbox casing.

50 Carefully lower cluster gear into position, large gear to front of casing and insert dummy countershaft.

51 Tap countershaft backwards and forwards to ensure that it is seated correctly.

52 Check clearance between rear thrust washer and cluster gear. Dimension 'A' must be 0,004 in to 0.006 in (0,10 mm to 0,15 mm). If this dimension is not obtained, remove cluster gear and select thrust washer which will give correct clearance.

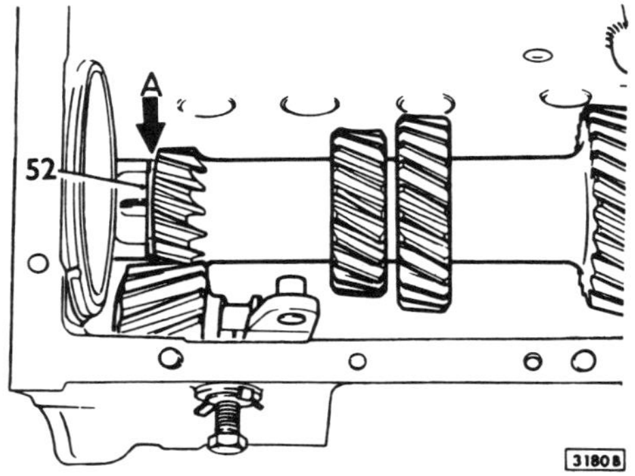

53 Withdraw dummy countershaft and simultaneously substitute a thin rod. Keep rod in contact with countershaft until it is clear of casing.

54 Place Woodruff key in reverse idler shaft.

55 Fit reverse idler gear, lever and idler shaft. DO NOT fit retaining bolt at this stage.

continued

37–5

SPEEDOMETER DRIVE GEAR
Remove and refit   37.25.01

Removing
1. Remove engine and gearbox, see 12.37.01.
2. Remove drive flange, see 37.10.01.
3. Remove rear oil seal.
   NOTE: While removing oil seal take care not to damage oil seal seat in rear extension.
4. Screw speedometer drive gear from pinion and withdraw from rear extension.

Refitting
Reverse operations 1 to 4.
Use new oil seal.

SPEEDOMETER DRIVE PINION
Remove and refit   37.25.05

Removing
1. Beneath car, release sleeve securing right angle drive to pinion.
2. Remove setscrew and plain washer and draw pinion and housing from gearbox.

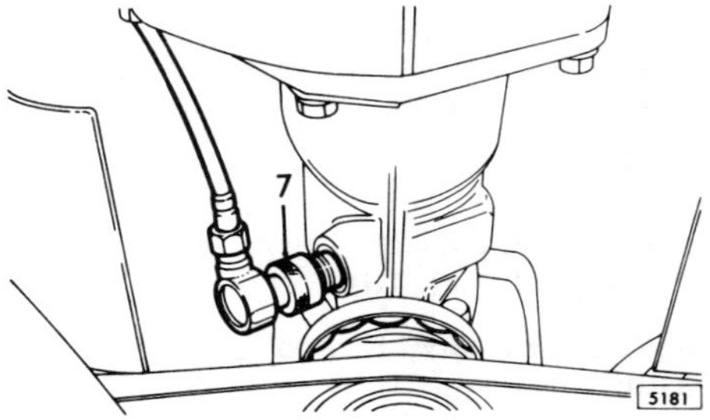

Refitting
Reverse operations 1 and 2.

FRONT OIL SEAL
Remove and refit   37.23.06

Removing
1. Remove engine and gearbox, see 12.37.01.
2. Remove bell housing, see 37.12.07.
3. Remove and discard oil seal.
4. Ensure oil seal recess is perfectly clean.

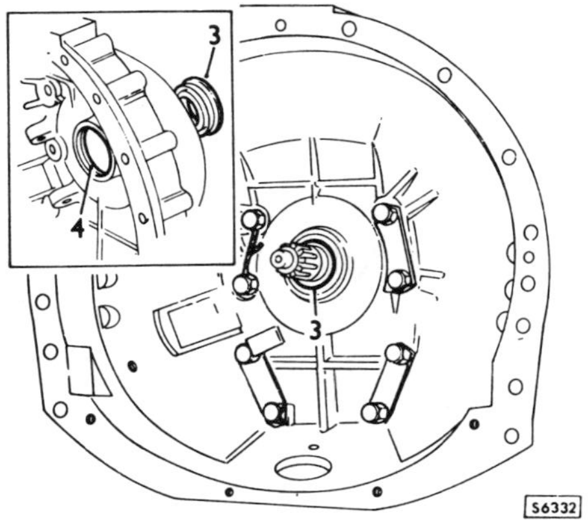

Refitting
5. Smear new oil seal with clean engine oil.
6. Push oil seal into bell housing with lip of seal towards gearbox.
7. Reverse operations 1 and 2; cover splines of first motion shaft with adhesive tape to prevent damage to oil seal.

REAR OIL SEAL
Remove and refit   37.23.01

Removing
1. Remove drive flange.
2. Remove oil seal.
   NOTE: While removing oil seal, take care not to damage oil seal seat in rear extension.

Refitting
3. Smear new seal with clean engine oil. Fit seal lip inwards.
4. Fit drive flange; use a new split pin to retain nut.

---

56. Fit bearing race in rear of first motion shaft.
57. Position oil thrower on first motion shaft with raised centre portion of thrower facing upwards towards bearing.
58. Fit snap ring in groove in front bearing.
59. Press bearing on first motion shaft ensuring that it is square to shoulder of gear.
60. Fit new tab washer and nut; turn up tab.
61. Reverse operations 25 to 32, retain needle rollers with grease; use new tab washer.
62. Enter mainshaft assembly through top of casing and pass rear of shaft through bearing aperture; remove hose clip.
63. Smear a new gasket with grease and position it on front of gearcase.
64. Position synchro ring to locate on front face of third/top synchro hub.
65. Enter first motion shaft through front of casing with cutaway portions of driving gear at top and bottom. 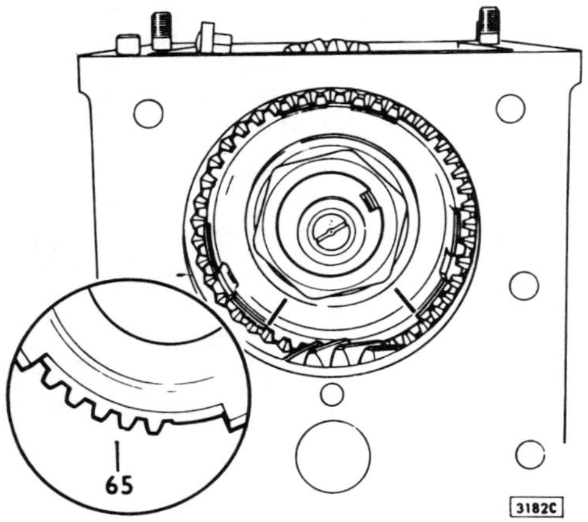
66. Tap first motion shaft into position ensuring that spigot on mainshaft enters bearing.
67. Fit snap ring in groove in rear bearing.
68. Hold first motion shaft in position and with a hollow drift, tap rear bearing into position. Ensure bearing is seated correctly.
69. Withdraw thin rod and simultaneously substitute dummy countershaft.
    CAUTION: Ensure that gasket is not damaged and needle rollers are not displaced.
70. Place Woodruff key in countershaft.
71. Enter countershaft through hole in rear of gearcase. Tap countershaft home at the same time withdrawing dummy shaft.
    CAUTION: Ensure that gasket is not damaged and needle rollers are not displaced.
72. Fit new fibre plug at front of countershaft.
73. Fit reverse lever bolt; use a new tab washer.
74. Fit rear bearing nut use a new tab washer.
75. Reverse operations 1 to 5.

37–6

SERVICE TOOLS – GM400 GEARBOX

18G 677-2	Adaptor Pressure Take Off	
18G 1295	Piston Accumulator Control Valve Compressor	
18G 1296	Front Pump Remover Screws	
18G 1297	Front Pump and Tailshaft Oil Seal Replacer	
18G 1298	Forward and Direct Clutch Piston Replacer Inner and Outer Protection Sleeve	
18G 1309	Intermediate Clutch Inner Seal Protection Sleeve	
18G 1310	Band Application Pin Selection Gauge	
18G 677 ZC	Pressure Test Equipment	
18G 1016	Clutch Spring Compressor	
18G 1004	Circlip Pliers	
18G 1004 J	Circlip Pliers Points	

ADDITIONAL TORQUE WRENCH SETTINGS FOR GM400 GEARBOX

GM 400 Automatic Transmission

	lbf ft	kgf m
Solenoid to case ¼ in dia. × 20	12	1,66
Control valve unit to case ¼ in dia. × 20	8	1,1
Line pressure plug ⅛ in dia. pipe	10	1,38
Pump body to cover 5/16 in dia. × 18	18	2,49
Pump to case 5/16 in dia. × 18	18	2,49
Rear servo cover 5/16 in dia. × 18	18	2,49
Governor cover to case 5/16 in dia. × 18	18	2,49
Parking pawl bracket 5/16 in dia. × 18	18	2,49
Vacuum modulation retainer to case ¼ in dia. × 28	10	1,38
Speedometer drive shaft nut 5/16 in dia. × 18	12	1,66
Sump ⅜ in dia. × 16	23	3,18
Rear extension ⅜ in dia. × 24	18	2,49
Manual shaft to detent lever		

DATA AND DESCRIPTION 44.00.00

The Hydramatic transmission is fully automatic and consists of a three-element-type torque convertor and a compound epicyclic planetary gear set.
Three multiple disc clutches, two roller clutches and two brake bands provide the friction elements required to obtain the necessary gear ratios.
The torque convertor couples the engine power to the transmission and hydraulically provides additional torque multiplication when the engine and transmission are subjected to high loads.
The compound planetary gear set provides three forward ratios and one reverse. Gear-changing is fully automatic relative to vehicle and engine speed and engine torque input. A vacuum modulator is used to automatically sense engine torque input to the transmission.

Engine torque sensed by the modulator is transmitted to the pressure regulator, thus ensuring that the correct gear-shifts are obtained at the relevant throttle positions.
Gear or torque ratios of the transmission are as follows:

First 2.48 : 1
Second 1.48 : 1
Third 1.1
Reverse 2.07 : 1

The gear selection quadrant has six positions 'P', 'R', 'N', 'D', '2', '1'.
An easily recognizable feature on cars fitted with this transmission is the increased length of selector lever travel between 'P' and 'R'.

- **'P'** Park enables the transmission output shaft to be locked, thereby preventing movement of the vehicle, 'P' **must not** be engaged whilst the vehicle is in motion. The engine can be started in this position.

- **'R'** Enables the vehicle to be driven in the reverse direction.

- **'N'** Neutral position, enables the engine to be started and run without driving the transmission.

- **'D'** Drive for all normal driving conditions and maximum economy. It has three gear ratios. Forced down-shifts are available for safe and rapid acceleration by quickly depressing the accelerator pedal to the full throttle position.

- **'2'** '2' has the same starting ratio as 'D' but prevents the transmission changing up from second gear, thereby retaining this gear for acceleration and engine braking. '2' can be selected at any road speed, as there is no safety override.

- **'1'** '1' first gear ratio can be selected at any speed from 'D' or '2' but the transmission will shift to second gear and will remain in this gear until the vehicle speed is reduced sufficiently to allow first gear to be engaged.

44–1 GM

DATA AND DESCRIPTION

APPLICATION CHART

Selector Position	Forward Clutch	Direct Clutch	Front Band	Intermediate Clutch	Intermediate Roller Clutch	Rear Roller Clutch	Rear Band
Park – Neutral							
Drive 1	●				●	●	
D 2	●			●	●		
3	●	●		●			
Intermediate 1	●				●	●	
2 2nd	●		●	●			
Lock-up 1st	●				●	●	●
Reverse		●					●

CLUTCH PLATE IDENTIFICATION

	Forward Clutch		Direct Clutch		Intermediate Clutch	
	Pressure Plates	Friction Plates	Pressure Plates	Friction Plates	Pressure Plates	Friction Plates
Flat	5	5	5	5	3	3
Dished	1	–	1	–	–	–

NOTE: The direct clutch has a plate thickness of 0.915 in (2.2 mm) the other four being of 0.077 in (1.9 mm) thickness.

OLD CONDITION UP TO GEARBOX No. 77ZA 4609

	Light Throttle			Kickdown			Part Throttle		Zero Throttle 1 Selected
	1–2	2–3	1–2	2–3	3–2	3–1	3–2		1st gear engages at
Home Market	10–15 m.p.h.	17–18 m.p.h.	57–60 m.p.h.	87–93 m.p.h.	77–86 m.p.h.	31–32 m.p.h.	52 m.p.h.		12 m.p.h.
Federal	8 m.p.h.	15–18 m.p.h.	60 m.p.h.	93 m.p.h.	82–84 m.p.h.	30–35 m.p.h.	52 m.p.h.		13–15 m.p.h.

GLOSSARY OF TERMS

1 ACCUMULATOR

Controls shift quality by delaying the full drive pressure applied to a clutch or band.

2 MANUAL VALVE

The main line fluid pressure distributing valve, directing fluid to all main components.

3 GOVERNOR ASSEMBLY

Responsible for timing the gear-changes in accordance with output shaft speed.

4 VACUUM MODULATOR VALVE

The vacuum modulator valve, activated by manifold vacuum, senses engine torque. The modulator ensures that the correct gear-shifts are obtained at relevant throttle positions.

Pressure from the modulator is applied to the 1-2 shift valve, compensating governor pressure, and to the pressure regulator valve in order to vary line pressure.

Governor pressure directed to the modulator reduces line pressure at high road speeds, when engine torque is minimal, thereby making it unnecessary for high pump output and resulting in a greater fuel economy for the unit.

5 PRESSURE REGULATOR

Controls main line pressure.

6 1-2 SHIFT VALVE

Controls the 1-2 and 2-3 shift patterns.

7 1-2 REGULATOR VALVE

Regulates modulator pressure to a proportional pressure and tends to hold the 1-2 shift valve in the down-shift position.

8 1-2 DETENT VALVE

This valve serves regulated modulator pressure and tends to hold the 1-2 shift valve in the down position shift and provides an area for detent pressure for 2-1 detent changes.

9 2-3 MODULATOR VALVE

This valve is sensitive to modulator pressure and applies a variable force on the 2-3 shift valve; tending to hold the valve in the down position.

10 3-2 VALVE

This prevents modulator pressure from acting on the shift valves after the direct clutch has been applied, thereby preventing a down-shift from third gear should wide throttle openings be used. If detent or modulator pressure rises above 92 lbf/in² (6.5 kgf/cm²) however, this pressure will then be directed to the shift valves to effect a down-shift.

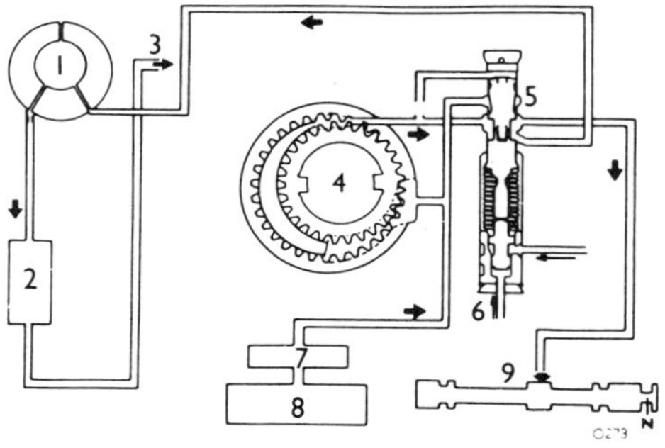

1 Torque convertor
2 Oil cooler
3 Oil to transmission
4 Oil pump
5 Pressure regulator valve
6 From modulator or throttle valve
7 Oil filter
8 Oil sump
9 Manual valve

NEW CONDITION FROM GEARBOX No. 77ZA 4610

	Light Throttle		Full Throttle		Full Throttle Kickdown		Kickdown			Downshift			Roll Out	
	1–2	2–3	1–2	2–3	1–2	2–3	3–2	2–3	3–1	Manual 2–1	PTKD* 3–2		3–2	2–1
	5–10 m.p.h.	10–20 m.p.h.	45±5 m.p.h.	60±5 m.p.h.	55±5 m.p.h.	85±5 m.p.h.	80±5 m.p.h.	136±8 m.p.h.	22–30 m.p.h.	13–18 m.p.h.	40–50 m.p.h.		8–12 m.p.h.	3–8 m.p.h.
	8–16 k.p.h.	16–32 k.p.h.	72±8 k.p.h.	96±8 k.p.h.	88±8 k.p.h.	136±8 k.p.h.	128±8 k.p.h.		35–48 k.p.h.	21–29 k.p.h.	64–80 k.p.h.		13–19 k.p.h.	5–13 k.p.h.

* PTKD = Part Throttle Kickdown

DATA AND DESCRIPTION   44.00.00

SPRING IDENTIFICATION CHART

Function	Colour	Free Length	No. of Coils	O.Dia.
1–2 accumulator valve	Dark Green	1.648 in	12.5	0.480 in
Pressure regulator	Light Blue	3.343 in	13	0.845 in
Front servo piston	Natural	1.129 in	4	1.257 in
Rear accumulator	Yellow	2.230 in	8.5	1.130 in
Governor	Dark Green	0.933 in	9.5	0.316 in
	Red	0.987 in	8.5	0.306 in
1–2 regulator	Pink	0.936 in	13.5	0.241 in
2–3 valve	Red	1.491 in	17.5	0.328 in
2–3 valve	Gold	1.555 in	18.5	0.326 in
3–2 valve	Green	2.017 in	16.5	0.400 in
Front accumulator piston	Natural	2.927 in	8.5	1.260 in
Detent regulator	Green	2.735 in	26.5	0.340 in

TRANSMISSION FLUID LEVEL

Check   44.00.00

1 Ensure that the transmission is at normal operating temperature by either:
   a Running the vehicle on a rolling road utilizing all the gear positions, or
   b Conducting a road test.
2 Check that the vehicle is on level ground.
3 Firmly apply the hand and footbrakes, and run the engine at a maximum speed of 750 rev/min for several minutes. To ensure that the valve block is primed, slowly move the selector lever through all the gear positions.
4 With the engine still running, engage 'P' (park) and withdraw the dipstick.
5 Wipe it clean with a lint-free cloth and replace it.
6 Immediately remove the dipstick again, and note the level indicated on the 'HOT' scale. It should be between the 'MAX' and 'MIN' marks.
7 Carefully top up the fluid to the correct level, using only a Dexron 2D type fluid. Take care not to overfill.

RECOMMENDED TRANSMISSION FLUID

Dexron 2D type fluid only should be used, which must not be mixed with other transmission fluids.
NOTE: Dexron 2D Fluid is reddish brown in colour and has a carbonized odour.

Fluid quantity
Transmission completely dry, but a quantity of fluid still remains in the torque convertor; fill with approximately 16 pints (9.12 litres).

Fluid condition
Any coolant or moisture in the transmission fluid will cause the transmission seals to swell and will also soften friction material. If this fault is found early, the leak repaired and the fluid changed, no overhaul is needed unless there are obvious defects in the operation of the transmission.

Varnished fluid
This gives the fluid a light to dark colour. With Dexron fluid it will intensify the already brown colour, and usually attaches itself to the dipstick.
Once varnish starts forming it builds up on all the valves, servos, clutches, etc., and causes sticking and hardening of the seals. Eventually it will clog the filter, and pump pressure will drop. When this happens the torque converter will not fill and there will be insufficient pressure for the clutch or band to hold torque, hence the transmission will not operate.
An evaluation of the degree of varnish will decide whether an overhaul is required or just a fluid change.

Low fluid level
This can result from the pump drawing air along with the fluid, thereby making fluid spongy and compressible due to air bubbles. This can result in a low build-up of pressure and can cause delayed engagement of drive and slipping on up-shifts.
Another possible fault is pump wear or a governor malfunction indicated by a buzzing noise emanating from the output shaft.

High fluid level
This can cause foaming and overheating of the transmission fluid resulting in the same faults occurring as in low fluid level. Overheating causes rapid oxidation of the fluid, leading to varnish formation.

44–3 GM

FAULT-FINDING AND DIAGNOSIS

PRELIMINARY FAULT-FINDING PROCEDURE

CHECK TRANSMISSION OIL LEVEL
↓
CHECK SELECTOR CABLE ADJUSTMENT
↓
CHECK ENGINE TUNE
↓
FIT TACHOMETER
↓
FIT OIL PRESSURE GAUGE
↓
STALL TEST

CAUTION: Total running time for this combination must not exceed 2 minutes

Check oil pressures in the following manner

	Range	Oil Pressure lbf/in²	Oil Pressure kgf/cm²
1	Neutral — brakes applied engine at 1000 rev/min	55 to 70	3,8 to 4,9
2	Drive — idle set engine idle to specifications	60 to 85	4,2 to 5,9
3	Drive — brakes applied engine at 1000 rev/min	*60 to 90	*4,2 to 6,3
4	2 or 1 — brakes applied engine at 1000 rev/min	135 to 160	9,5 to 11,2
5	Reverse — brakes applied engine at 1000 rev/min	95 to 150	6,7 to 10,5
6	Drive — brakes applied engine at 1000 rev/min down-shift switch activated	90 to 100	6,3 to 7,7
7	Governor check see below	Drop of 10 lbf/in² or more	Drop of 0,7 kgf/cm² or more
8	Drive — 30 M.P.H. — closed throttle on road, or on hoist	**55 to 70	**3,8 to 4,9

* If high line pressures are experienced, check vacuum and, if necessary, the modulator.
** Vehicle on hoist, driving wheels off ground, selector in drive, brakes released; raise engine speed to 3000 rev/min, close throttle and read pressure between 2000 and 1200 rev/min.

GOVERNOR CHECK 44.00.00

1. Disconnect vacuum line to modulator.
2. With the vehicle on hoist (driving wheels off ground), foot off brake, in drive, check line pressure at 1000 rev/min.
3. Slowly increase engine rev/min to 3000 and determine if a line pressure drop occurs (10 lbf/in² or more).
4. If pressure drop of 10 lbf/in² or more occurs, dismantle, clean and inspect valve block assembly.
5. If pressure drop is less than 10 lbf/in² inspect governor and governor feed system.

The results you obtain should then be related to the following PRELIMINARY DIAGNOSTIC CHART.

PRELIMINARY DIAGNOSTIC CHART
TRANSMISSION MALFUNCTION RELATED TO OIL PRESSURE

MALFUNCTION	1 NEUTRAL – BRAKES APPLIED 1000 REV/MIN — OIL PRESSURE	2 DRIVE – IDLE — OIL PRESSURE	3 DRIVE – BRAKES APPLIED 1000 REV/MIN — OIL PRESSURE	4 '1' – BRAKES APPLIED 1000 REV/MIN — OIL PRESSURE	5 REVERSE – BRAKES APPLIED 1000 REV/MIN — OIL PRESSURE	6 DRIVE – BRAKES APPLIED 1000 REV/MIN DOWN-SHIFT SWITCH ACTIVATED — OIL PRESSURE	7 PRESSURE DROP OCCURS WHILE ENGINE REV/MIN INCREASES FROM 1000 to 3000 REV/MIN WHEELS FREE TO MOVE*	8 DRIVE – 30 M.P.H. (48 KM/H) CLOSED THROTTLE — OIL PRESSURE	POSSIBLE CAUSE OF MALFUNCTION
NO 1-2 UP-SHIFT AND/OR DELAYED UP-SHIFT	Normal	Normal	Normal	Normal	Normal	Normal	10 lbf/in² (0,7 kg/cm²) drop or more	Normal	Malfunction in control valve assembly
	Normal	Normal	Normal	Normal	Normal	Normal	Less than 10 lbf/in² (0,7 kg/cm²) drop	Normal	Malfunction in governor or governor feed system
	Normal	High	High	Normal	Normal	Normal	Drop	High	Malfunction in detent system
	High	High	High	Normal	High	–	–	–	Malfunction in modulator or vacuum feed system to modulator
SLIPPING – REVERSE	Normal	Normal	Normal	Normal	Low	Normal	Drop	Normal	Oil leak in feed system to the direct clutch
SLIPPING – 1st GEAR	Normal	Low to Normal	Low to Normal	Low to Normal	Normal	Low to Normal	–	Low to Normal	Oil leak in feed system to the forward clutch
DOWN-SHIFT WITH ZERO THROTTLE AND NO ENGINE BRAKING IN DRIVE	Normal	High	Normal	Normal	Normal	–	–	High	Detent wires switched
NO DETENT DOWN-SHIFTS	Normal	Normal	Normal	Normal	Normal	Low	Normal	Normal	Malfunction in detent system

* Drive range, vacuum line disconnected from modulator.
NOTE: A dash (–) in the above chart means that the oil pressure reading has no meaning under the test condition.

LOW LINE PRESSURE

1 **LOW TRANSMISSION OIL LEVEL**

2 **MODULATOR ASSEMBLY** – Carry out 'bellows comparison check'

3 **FILTER**
   a Blocked or restricted.
   b 'O' ring on intake pipe and/or grommet missing or damaged.
   c Split or leaking intake pipe.
   d Wrong filter assembly.

4 **PUMP**
   a Pressure regulator or boost valve stuck.
   b Gear clearance, damaged, worn. (Pump will become damaged if drive gear is installed backwards, or if converter pilot does not enter crankshaft freely.)
   c Pump to case gasket wrongly positioned.
   d Pump body and/or cover machining error or scoring of pump gear pocket.

5 **INTERNAL CIRCUIT LEAKS**
   a Forward clutch leak. (Pressure normal in neutral and reverse – Pressure low in drive.)
      1 Check pump rings for damage.
      2 Check forward clutch seals for damage.
      3 Check turbine shaft journals for damage.
      4 Check stator shaft bushings for damage.
   b Direct clutch leak. (Pressure normal in neutral, low, intermediate and drive – Pressure low in reverse.)
      1 Check centre support oil seal rings for damage.
      2 Check direct clutch outer seal for damage.
      3 Check rear servo and front accumulator pistons and rings for damage.

6 **CASE ASSEMBLY**
   a Porosity in intake bore area.
   b Check case for intermediate clutch plug leak.
   c Low line pressure in reverse or '1'. If '1' – reverse check ball missing. This will cause no reverse and no over-run braking in '1'.

HIGH LINE PRESSURE

1 **VACUUM LEAK**
   a Full leak, vacuum line disconnected.
   b Partial leak in line from engine to modulator.
   c Improper engine vacuum.
   d Vacuum operated accessory leak (hoses, vacuum advance, etc.).

2 **DAMAGED MODULATOR**
   a Stuck valve.
   b Water in modulator.
   c Not operating properly.

3 **DETENT SYSTEM**
   a Detent switch actuated (plunger stuck) or shorted.
   b Detent wiring shorted.
   c Detent solenoid stuck open.
   d Detent feed orifice in spacer plate blocked.
   e Detent solenoid loose.
   f Detent valve bore plug damaged.
   g Detent regulator valve pin short.

4 **PUMP**
   a Pressure regulator and/or boost valve stuck.
   b Pump casting porous.
   c Pressure boost valve installed backwards.
   d Pressure boost bushing broken.
   e Wrong type of pressure regulator valve.

5 **CONTROL VALVE ASSEMBLY**
   a Control valve to spacer gasket wrongly fitted.
   b Gaskets installed in reverse order.

44–6 GM

BURNED CLUTCH PLATES

1 **FORWARD CLUTCH**
   a Check ball in clutch housing damaged, stuck or missing.
   b Clutch piston cracked, seals damaged or missing.
   c Low line pressure.
   d Manual valve wrongly fitted.
   e Restricted oil feed to forward clutch. (Examples: Clutch housing to inner and outer areas not drilled, restricted or porosity in pump.)
   f Pump cover oil seal rings missing, broken or undersize; ring groove oversize.
   g Case valve body face not flat or porosity between channels.
   h Manual valve bent and centre land not properly ground.

2 **INTERMEDIATE CLUTCH**
   a Constant bleed orifice in centre support blocked.
   b Rear accumulator piston oil ring, damaged or missing.
   c 1-2 accumulator valve stuck in control valve assembly.
   d Intermediate clutch piston seals damaged or missing.
   e Centre support bolt loose.
   f Low line pressure.
   g Intermediate clutch plug in case missing.
   h Case valve body face not flat or porosity between channels.
   i Manual valve bent and centre land not properly ground.

3 **DIRECT CLUTCH**
   a Restricted orifice in vacuum line to modulator (poor vacuum response).
   b Check ball in direct clutch piston damaged, stuck or missing.
   c Leaking modulator bellows.
   d Centre support bolt loose. (Bolt may be tight in support but not holding support tightly to case.)
   e Centre support oil rings or grooves damaged or missing.
   f Clutch piston seals damaged or missing.
   g Front and rear servo pistons and seals damaged.
   h Manual valve bent and centre land damaged.
   i Case valve body face not flat or porosity between channels.
   j Intermediate sprag clutch or roller clutch installed backwards.
   k 3-2 valve, 3-2 spring or 3-2 spacer pin installed in the wrong sequence in 3-2 valve bore.
   l Incorrect combination of front servo and accumulator parts.
   m Replace intermediate clutch piston seals.
   NOTE: If direct clutch plates and front band are burned, check selector cable adjustment, see 44.30.04.

OIL LEAKS

NOTE: Make sure underside of transmission is clean in order to isolate oil leaks and diagnose them correctly.

1 **TRANSMISSION OIL PAN LEAKS**
   a Attaching bolts not correctly torqued.
   b Improperly installed or damaged oil pan gasket.

2. **CASE EXTENSION LEAK**
   a Attaching bolts not correctly torqued.
   b If the rear seal is suspected:
      1 Check seal for damage or wrong installation.
      2 Check slip yoke for damage.
      3 If oil is coming out the vent hole in end of the slip yoke, inspect output shaft 'O' ring for damage.
   c Extension to case gasket or seal damaged.
   d Porous casting.

3 **CASE LEAK**
   a Filler pipe 'O' ring seal damaged or missing.
   b Modulator assembly 'O' ring seal damaged.
   c Electrical connector 'O' ring seal damaged.
   d Governor cover, gasket and bolts damaged or loose; case face damaged or porous.
   e Leak at speedometer driven gear housing or seal.
   f Manual shaft seal damaged.
   g Line pressure tap plug stripped.
   h Vent pipe (refer to item 5).
   i Porous case, or cracked at pressure plug boss.

4 **FRONT END LEAK**
   a Front seal damaged (check convertor neck for nicks, etc., also for pump bushing moved forward), garter spring missing.
   b Pump attaching bolts loose. Sealing washers damaged.
   c Converter leakage.
   d Large 'O' ring pump seal damaged. Also check case bore.
   e Porous casting (pump or case).
   f Pump drainback hole blocked.

44—7 GM

5 **OIL LEAKS FROM VENT PIPE**

   a Transmission overfilled.
   b Water in oil.
   c Filter 'O' ring damaged or improperly assembled causing oil to foam.
   d Foreign material between pump and case or between pump cover and body, or variable stator solenoid screws too long — holding pump halves apart.
   e Case porous, pump face improperly machined.
   f Pump wrongly fitted.
   g Pump to case gasket faulty.
   h Pump breather hole blocked or missing.
   i Hole in intake pipe.
   j Check ball in forward clutch housing stuck open or missing.
   k Drainback hole in case blocked or restricted.
   l Inspect turbine shaft bushing journals and stator bushings for scoring or other faults.

6 **OIL COOLER LINES**

   a Connections at cooler loose or stripped.
   b Connections at case loose or stripped.

7 **MODULATOR ASSEMBLY**

   a Vacuum diaphragm leaking.

IMPROPER VACUUM AT MODULATOR

1 **ENGINE**

   a Engine tune.
   b Loose vacuum fittings.
   c Vacuum operated accessory leak (hoses, vacuum advance, etc).
   d Engine exhaust system restricted.
   e Faulty exhaust gas recirculation (E.G.R.) valve (if fitted).

2 **VACUUM LINE TO MODULATOR**

   a Leak.
   b Loose fitting.
   c Restricted orifice, or incorrect orifice size.
   d Carbon build up at modulator vacuum fitting.
   e Vacuum pipe trapped or collapsed.
   f Grease in pipe (none or delayed upshift-cold).

MODULATOR ASSEMBLY DIAGNOSTIC PROCEDURE

1 **VACUUM DIAPHRAGM LEAK CHECK**

Insert a pipe cleaner into the vacuum connector pipe as far as possible and check for the presence of transmission oil. If oil is found, replace the modulator. Transmission oil may be lost through diaphragm and burned in engine.

NOTE: Fuel or water condensation may settle in the vacuum side of the modulator. If this is found without the presence of oil the modulator should **not** be changed.

2 **ATMOSPHERIC LEAK CHECK**

Apply a liberal coating of soap solution to the vacuum connector pipe seam, the crimped upper to lower housing seam, and the threaded screw seal. Using a short piece of rubber tubing, apply air pressure to the vacuum pipe by blowing into the tube and observing for bubbles. If bubbles appear, replace the modulator.

NOTE: Do not use any method other than human lung power for applying air pressure, as pressures over 6 lbf/in^2 (0.4 kgf/cm^2) may damage the modulator.

3 **BELLOWS COMPARISON CHECK**

Where modulator bellows are suspect, the unit should be checked by substituting a new modulator assembly.

4 **SLEEVE ALIGNMENT CHECK**

Roll the main body of the modulator on a flat surface and observe the sleeve for concentricity to the cam. If the sleeve is concentric and the plunger is free, the modulator is acceptable.

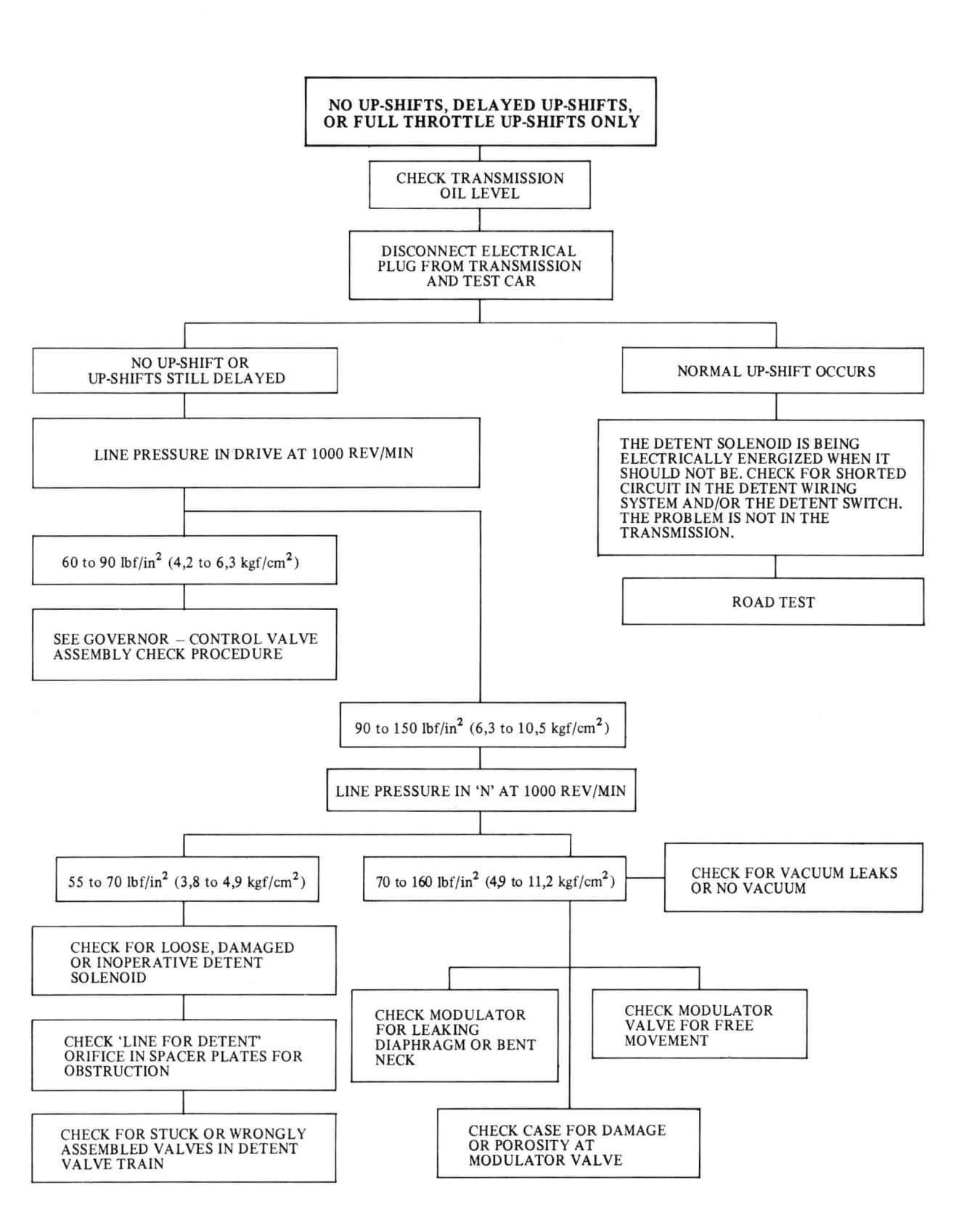

```
┌─────────────────────────────────────┐
│ NO UP-SHIFTS, DELAYED UP-SHIFTS,    │
│ OR FULL THROTTLE UP-SHIFTS ONLY     │
└─────────────────────────────────────┘
                  │
       ┌──────────────────────┐
       │ CHECK TRANSMISSION   │
       │ OIL LEVEL            │
       └──────────────────────┘
                  │
       ┌──────────────────────┐
       │ DISCONNECT ELECTRICAL│
       │ PLUG FROM TRANSMISSION│
       │ AND TEST CAR         │
       └──────────────────────┘
            │                │
┌───────────────────────┐  ┌──────────────────────┐
│ NO UP-SHIFT OR        │  │ NORMAL UP-SHIFT OCCURS│
│ UP-SHIFTS STILL DELAYED│  └──────────────────────┘
└───────────────────────┘             │
            │              ┌──────────────────────────────┐
┌───────────────────────────────┐ │ THE DETENT SOLENOID IS BEING │
│ LINE PRESSURE IN DRIVE AT     │ │ ELECTRICALLY ENERGIZED WHEN IT│
│ 1000 REV/MIN                  │ │ SHOULD NOT BE. CHECK FOR SHORTED│
└───────────────────────────────┘ │ CIRCUIT IN THE DETENT WIRING │
            │                     │ SYSTEM AND/OR THE DETENT SWITCH.│
                                  │ THE PROBLEM IS NOT IN THE    │
                                  │ TRANSMISSION.                │
                                  └──────────────────────────────┘
                                              │
                                    ┌──────────────┐
                                    │ ROAD TEST    │
                                    └──────────────┘

┌─────────────────────────────────────┐
│ 60 to 90 lbf/in² (4,2 to 6,3 kgf/cm²)│
└─────────────────────────────────────┘
                  │
       ┌──────────────────────────┐
       │ SEE GOVERNOR – CONTROL VALVE│
       │ ASSEMBLY CHECK PROCEDURE │
       └──────────────────────────┘

┌──────────────────────────────────────┐
│ 90 to 150 lbf/in² (6,3 to 10,5 kgf/cm²)│
└──────────────────────────────────────┘
                  │
       ┌──────────────────────────────┐
       │ LINE PRESSURE IN 'N' AT 1000 REV/MIN │
       └──────────────────────────────┘
            │                              │
┌──────────────────────────────┐  ┌──────────────────────────────┐  ┌─────────────────────┐
│ 55 to 70 lbf/in²             │  │ 70 to 160 lbf/in²            │  │ CHECK FOR VACUUM LEAKS│
│ (3,8 to 4,9 kgf/cm²)         │  │ (4,9 to 11,2 kgf/cm²)        │  │ OR NO VACUUM        │
└──────────────────────────────┘  └──────────────────────────────┘  └─────────────────────┘
            │                              │
┌──────────────────────────────┐  ┌──────────────────────┐  ┌──────────────────────┐
│ CHECK FOR LOOSE, DAMAGED     │  │ CHECK MODULATOR      │  │ CHECK MODULATOR      │
│ OR INOPERATIVE DETENT        │  │ FOR LEAKING          │  │ VALVE FOR FREE       │
│ SOLENOID                     │  │ DIAPHRAGM OR BENT    │  │ MOVEMENT             │
└──────────────────────────────┘  │ NECK                 │  └──────────────────────┘
            │                     └──────────────────────┘
┌──────────────────────────────┐              │
│ CHECK 'LINE FOR DETENT'      │  ┌──────────────────────────┐
│ ORIFICE IN SPACER PLATES FOR │  │ CHECK CASE FOR DAMAGE    │
│ OBSTRUCTION                  │  │ OR POROSITY AT           │
└──────────────────────────────┘  │ MODULATOR VALVE          │
            │                     └──────────────────────────┘
┌──────────────────────────────┐
│ CHECK FOR STUCK OR WRONGLY   │
│ ASSEMBLED VALVES IN DETENT   │
│ VALVE TRAIN                  │
└──────────────────────────────┘
```

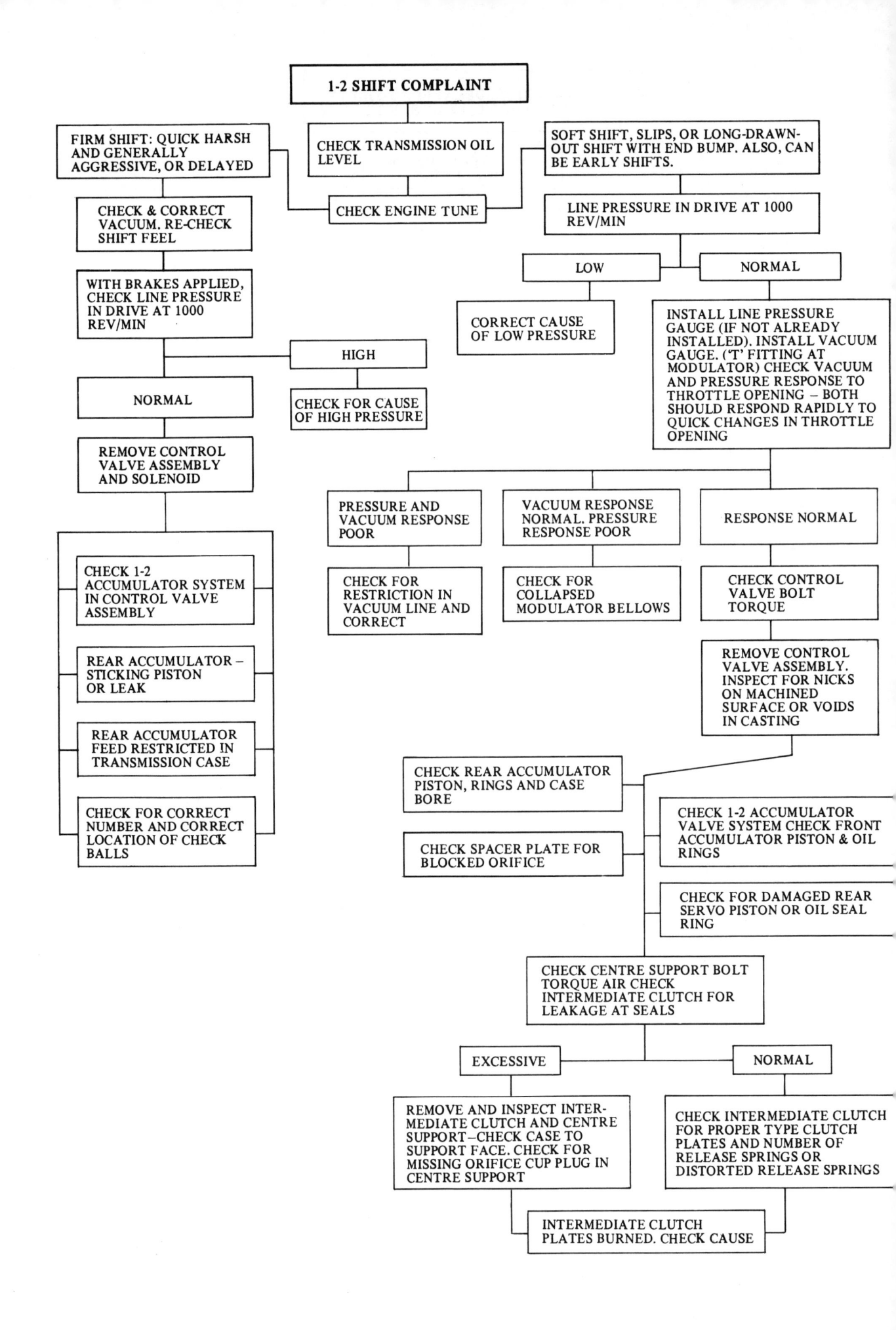

1-2 SHIFT COMPLAINT

CHECK TRANSMISSION OIL LEVEL → **CHECK ENGINE TUNE**

FIRM SHIFT: QUICK HARSH AND GENERALLY AGGRESSIVE, OR DELAYED

- CHECK & CORRECT VACUUM. RE-CHECK SHIFT FEEL
- WITH BRAKES APPLIED, CHECK LINE PRESSURE IN DRIVE AT 1000 REV/MIN
  - **HIGH**: CHECK FOR CAUSE OF HIGH PRESSURE
  - **NORMAL**: REMOVE CONTROL VALVE ASSEMBLY AND SOLENOID
    - CHECK 1-2 ACCUMULATOR SYSTEM IN CONTROL VALVE ASSEMBLY
    - REAR ACCUMULATOR – STICKING PISTON OR LEAK
    - REAR ACCUMULATOR FEED RESTRICTED IN TRANSMISSION CASE
    - CHECK FOR CORRECT NUMBER AND CORRECT LOCATION OF CHECK BALLS

SOFT SHIFT, SLIPS, OR LONG-DRAWN-OUT SHIFT WITH END BUMP. ALSO, CAN BE EARLY SHIFTS.

- LINE PRESSURE IN DRIVE AT 1000 REV/MIN
  - **LOW**: CORRECT CAUSE OF LOW PRESSURE
  - **NORMAL**: INSTALL LINE PRESSURE GAUGE (IF NOT ALREADY INSTALLED). INSTALL VACUUM GAUGE. ('T' FITTING AT MODULATOR) CHECK VACUUM AND PRESSURE RESPONSE TO THROTTLE OPENING – BOTH SHOULD RESPOND RAPIDLY TO QUICK CHANGES IN THROTTLE OPENING
    - **PRESSURE AND VACUUM RESPONSE POOR**: CHECK FOR RESTRICTION IN VACUUM LINE AND CORRECT
    - **VACUUM RESPONSE NORMAL. PRESSURE RESPONSE POOR**: CHECK FOR COLLAPSED MODULATOR BELLOWS
    - **RESPONSE NORMAL**: CHECK CONTROL VALVE BOLT TORQUE
      - REMOVE CONTROL VALVE ASSEMBLY. INSPECT FOR NICKS ON MACHINED SURFACE OR VOIDS IN CASTING
        - CHECK REAR ACCUMULATOR PISTON, RINGS AND CASE BORE
        - CHECK SPACER PLATE FOR BLOCKED ORIFICE
        - CHECK 1-2 ACCUMULATOR VALVE SYSTEM CHECK FRONT ACCUMULATOR PISTON & OIL RINGS
        - CHECK FOR DAMAGED REAR SERVO PISTON OR OIL SEAL RING
      - CHECK CENTRE SUPPORT BOLT TORQUE AIR CHECK INTERMEDIATE CLUTCH FOR LEAKAGE AT SEALS
        - **EXCESSIVE**: REMOVE AND INSPECT INTERMEDIATE CLUTCH AND CENTRE SUPPORT – CHECK CASE TO SUPPORT FACE. CHECK FOR MISSING ORIFICE CUP PLUG IN CENTRE SUPPORT
        - **NORMAL**: CHECK INTERMEDIATE CLUTCH FOR PROPER TYPE CLUTCH PLATES AND NUMBER OF RELEASE SPRINGS OR DISTORTED RELEASE SPRINGS
        - INTERMEDIATE CLUTCH PLATES BURNED. CHECK CAUSE

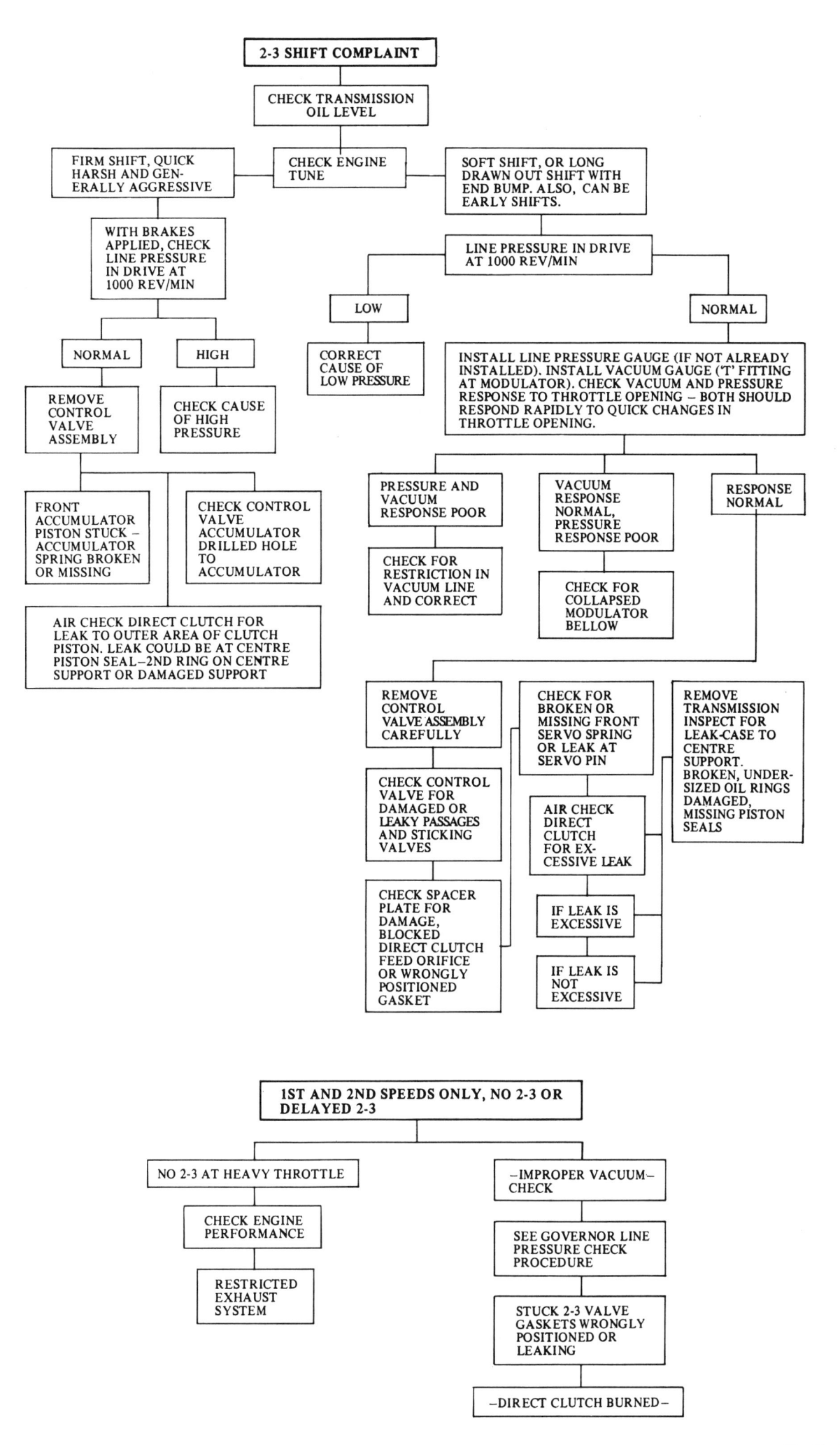

2-3 SHIFT COMPLAINT

```
                          2-3 SHIFT COMPLAINT
                                  │
                      CHECK TRANSMISSION OIL LEVEL
                                  │
                          CHECK ENGINE TUNE
                                  │
        ┌─────────────────────────┴──────────────────────────┐
FIRM SHIFT, QUICK HARSH                        SOFT SHIFT, OR LONG DRAWN
AND GENERALLY AGGRESSIVE                       OUT SHIFT WITH END BUMP.
                                               ALSO, CAN BE EARLY SHIFTS.
        │                                                   │
WITH BRAKES APPLIED,                           LINE PRESSURE IN DRIVE
CHECK LINE PRESSURE IN                         AT 1000 REV/MIN
DRIVE AT 1000 REV/MIN                                       │
        │                                      ┌────────────┴────────────┐
   ┌────┴────┐                                LOW                     NORMAL
 NORMAL    HIGH                                │                         │
   │         │                           CORRECT CAUSE         INSTALL LINE PRESSURE
 REMOVE   CHECK CAUSE                    OF LOW PRESSURE       GAUGE (IF NOT ALREADY
 CONTROL  OF HIGH                                              INSTALLED). INSTALL VACUUM
 VALVE    PRESSURE                                             GAUGE ('T' FITTING AT
 ASSEMBLY                                                      MODULATOR). CHECK VACUUM
                                                               AND PRESSURE RESPONSE
                                                               TO THROTTLE OPENING – BOTH
                                                               SHOULD RESPOND RAPIDLY TO
                                                               QUICK CHANGES IN THROTTLE
                                                               OPENING.
```

FRONT ACCUMULATOR PISTON STUCK – ACCUMULATOR SPRING BROKEN OR MISSING

CHECK CONTROL VALVE ACCUMULATOR DRILLED HOLE TO ACCUMULATOR

AIR CHECK DIRECT CLUTCH FOR LEAK TO OUTER AREA OF CLUTCH PISTON. LEAK COULD BE AT CENTRE PISTON SEAL–2ND RING ON CENTRE SUPPORT OR DAMAGED SUPPORT

PRESSURE AND VACUUM RESPONSE POOR
→ CHECK FOR RESTRICTION IN VACUUM LINE AND CORRECT

VACUUM RESPONSE NORMAL, PRESSURE RESPONSE POOR
→ CHECK FOR COLLAPSED MODULATOR BELLOW

RESPONSE NORMAL

REMOVE CONTROL VALVE ASSEMBLY CAREFULLY

CHECK CONTROL VALVE FOR DAMAGED OR LEAKY PASSAGES AND STICKING VALVES

CHECK SPACER PLATE FOR DAMAGE, BLOCKED DIRECT CLUTCH FEED ORIFICE OR WRONGLY POSITIONED GASKET

CHECK FOR BROKEN OR MISSING FRONT SERVO SPRING OR LEAK AT SERVO PIN

AIR CHECK DIRECT CLUTCH FOR EXCESSIVE LEAK
- IF LEAK IS EXCESSIVE
- IF LEAK IS NOT EXCESSIVE

REMOVE TRANSMISSION INSPECT FOR LEAK-CASE TO CENTRE SUPPORT. BROKEN, UNDERSIZED OIL RINGS DAMAGED, MISSING PISTON SEALS

1ST AND 2ND SPEEDS ONLY, NO 2-3 OR DELAYED 2-3

```
              ┌────────────────────┴────────────────────┐
    NO 2-3 AT HEAVY THROTTLE              –IMPROPER VACUUM– CHECK
              │                                         │
    CHECK ENGINE PERFORMANCE               SEE GOVERNOR LINE PRESSURE
              │                            CHECK PROCEDURE
    RESTRICTED EXHAUST SYSTEM                           │
                                           STUCK 2-3 VALVE GASKETS WRONGLY
                                           POSITIONED OR LEAKING
                                                        │
                                              –DIRECT CLUTCH BURNED–
```

44–11 GM

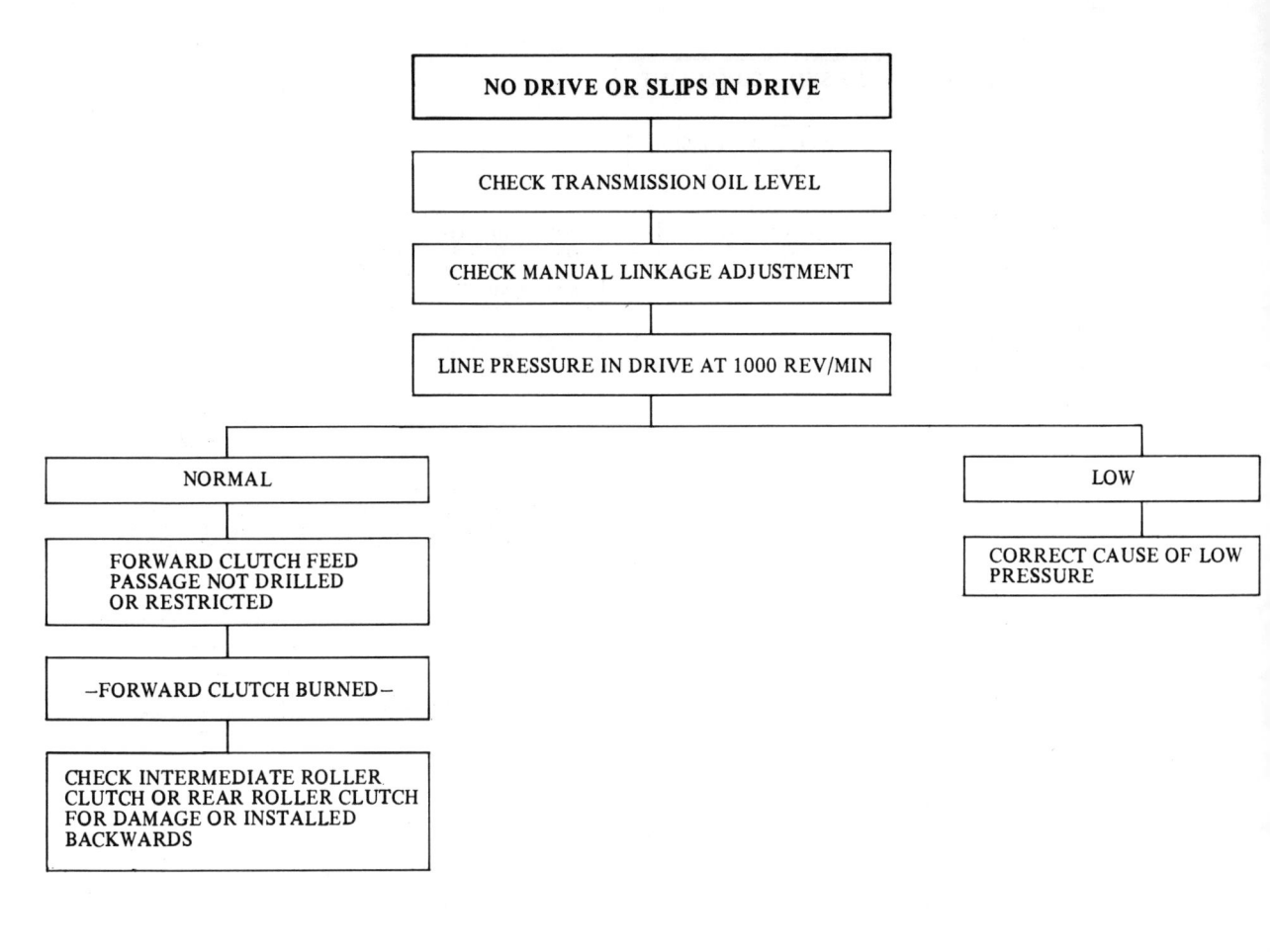

```
                    NO DRIVE OR SLIPS IN DRIVE
                              |
                    CHECK TRANSMISSION OIL LEVEL
                              |
                    CHECK MANUAL LINKAGE ADJUSTMENT
                              |
                    LINE PRESSURE IN DRIVE AT 1000 REV/MIN
                        /                    \
                    NORMAL                    LOW
                      |                         |
        FORWARD CLUTCH FEED          CORRECT CAUSE OF LOW
        PASSAGE NOT DRILLED          PRESSURE
        OR RESTRICTED
              |
        —FORWARD CLUTCH BURNED—
              |
        CHECK INTERMEDIATE ROLLER
        CLUTCH OR REAR ROLLER CLUTCH
        FOR DAMAGE OR INSTALLED
        BACKWARDS
```

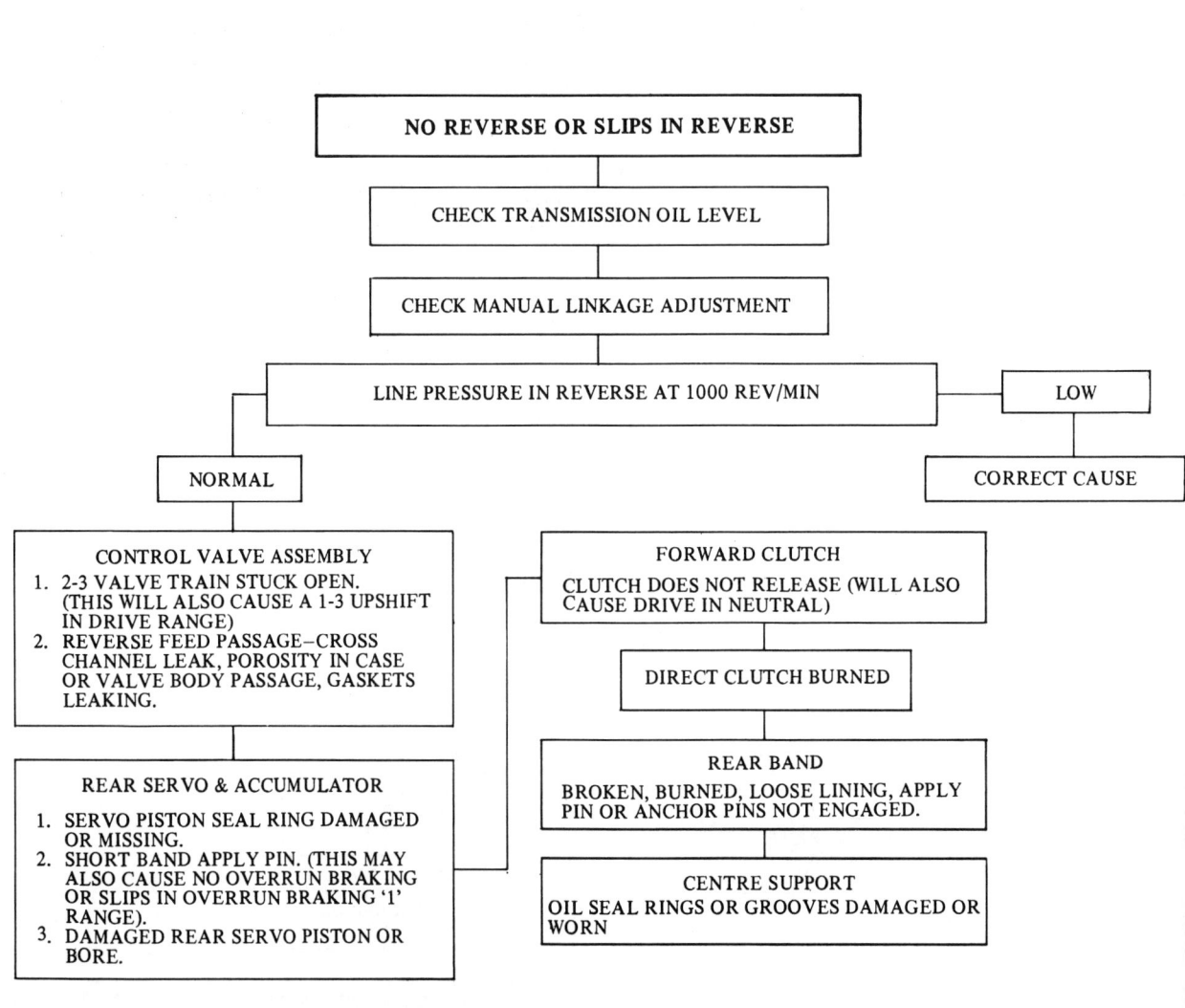

```
                    NO REVERSE OR SLIPS IN REVERSE
                              |
                    CHECK TRANSMISSION OIL LEVEL
                              |
                    CHECK MANUAL LINKAGE ADJUSTMENT
                              |
                    LINE PRESSURE IN REVERSE AT 1000 REV/MIN ——— LOW
                              |                                    |
                          NORMAL                            CORRECT CAUSE
```

CONTROL VALVE ASSEMBLY	FORWARD CLUTCH
1. 2-3 VALVE TRAIN STUCK OPEN. (THIS WILL ALSO CAUSE A 1-3 UPSHIFT IN DRIVE RANGE)	CLUTCH DOES NOT RELEASE (WILL ALSO CAUSE DRIVE IN NEUTRAL)
2. REVERSE FEED PASSAGE–CROSS CHANNEL LEAK, POROSITY IN CASE OR VALVE BODY PASSAGE, GASKETS LEAKING.	

DIRECT CLUTCH BURNED

REAR SERVO & ACCUMULATOR
1. SERVO PISTON SEAL RING DAMAGED OR MISSING.
2. SHORT BAND APPLY PIN. (THIS MAY ALSO CAUSE NO OVERRUN BRAKING OR SLIPS IN OVERRUN BRAKING '1' RANGE).
3. DAMAGED REAR SERVO PISTON OR BORE.

REAR BAND
BROKEN, BURNED, LOOSE LINING, APPLY PIN OR ANCHOR PINS NOT ENGAGED.

CENTRE SUPPORT
OIL SEAL RINGS OR GROOVES DAMAGED OR WORN

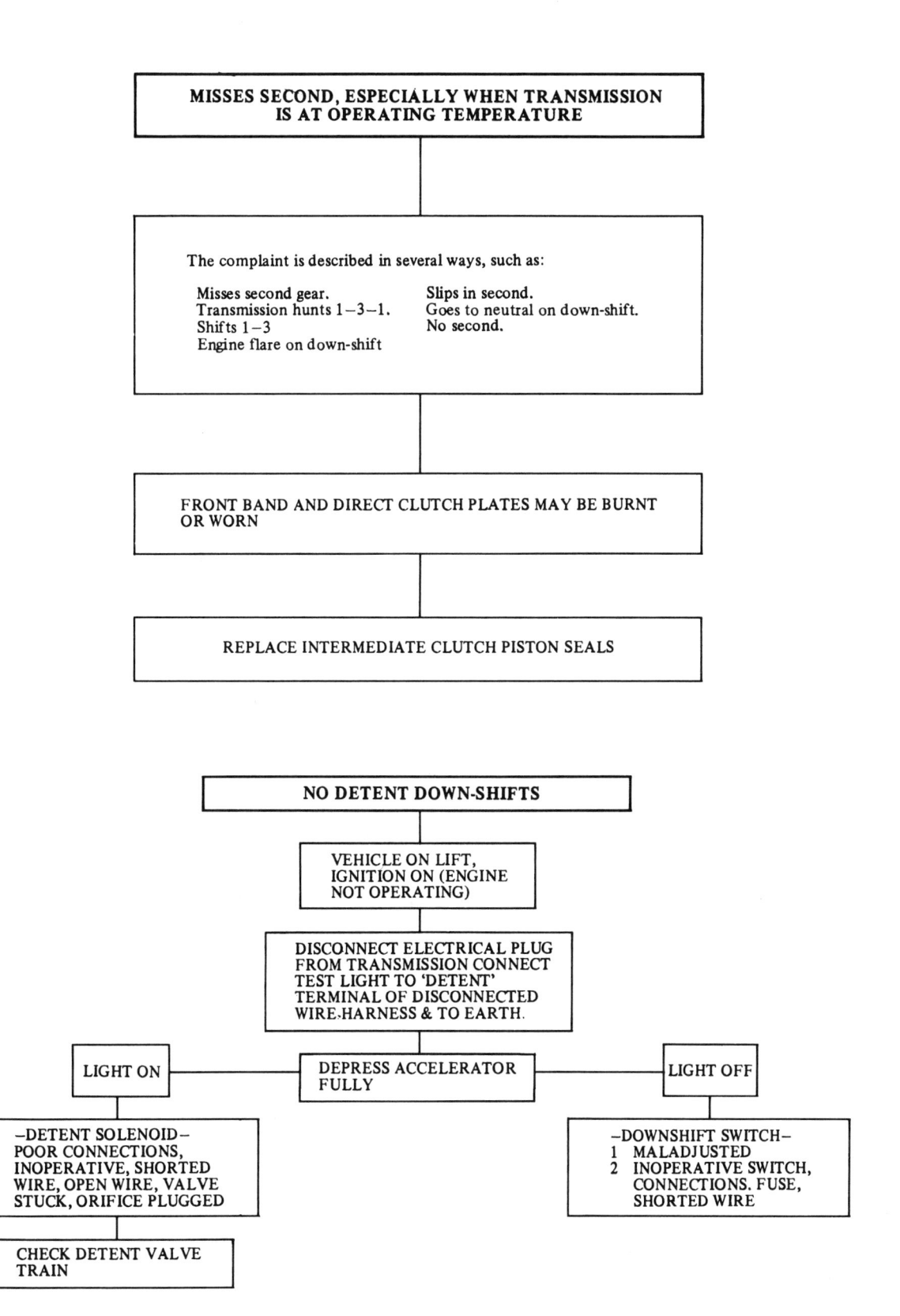

```
┌─────────────────────────────────────────────────────┐
│   MISSES SECOND, ESPECIALLY WHEN TRANSMISSION       │
│           IS AT OPERATING TEMPERATURE               │
└─────────────────────────────────────────────────────┘
                          │
                          ▼
┌─────────────────────────────────────────────────────┐
│  The complaint is described in several ways, such as: │
│                                                     │
│   Misses second gear.         Slips in second.      │
│   Transmission hunts 1–3–1.   Goes to neutral on down-shift. │
│   Shifts 1–3                  No second.            │
│   Engine flare on down-shift                        │
└─────────────────────────────────────────────────────┘
                          │
                          ▼
┌─────────────────────────────────────────────────────┐
│  FRONT BAND AND DIRECT CLUTCH PLATES MAY BE BURNT   │
│  OR WORN                                            │
└─────────────────────────────────────────────────────┘
                          │
                          ▼
┌─────────────────────────────────────────────────────┐
│         REPLACE INTERMEDIATE CLUTCH PISTON SEALS    │
└─────────────────────────────────────────────────────┘
```

```
                ┌───────────────────────────┐
                │    NO DETENT DOWN-SHIFTS  │
                └───────────────────────────┘
                              │
                              ▼
                ┌───────────────────────────┐
                │  VEHICLE ON LIFT,         │
                │  IGNITION ON (ENGINE      │
                │  NOT OPERATING)           │
                └───────────────────────────┘
                              │
                              ▼
                ┌───────────────────────────┐
                │ DISCONNECT ELECTRICAL PLUG│
                │ FROM TRANSMISSION CONNECT │
                │ TEST LIGHT TO 'DETENT'    │
                │ TERMINAL OF DISCONNECTED  │
                │ WIRE-HARNESS & TO EARTH.  │
                └───────────────────────────┘
                              │
          ┌──────────┬────────┴────────┬──────────┐
          │          │                 │          │
     ┌─────────┐   ┌───────────────────────┐   ┌──────────┐
     │LIGHT ON │───│ DEPRESS ACCELERATOR   │───│LIGHT OFF │
     └─────────┘   │       FULLY           │   └──────────┘
          │       └───────────────────────┘        │
          ▼                                        ▼
┌──────────────────────────┐          ┌──────────────────────────┐
│ –DETENT SOLENOID–        │          │ –DOWNSHIFT SWITCH–       │
│ POOR CONNECTIONS,        │          │ 1 MALADJUSTED            │
│ INOPERATIVE, SHORTED     │          │ 2 INOPERATIVE SWITCH,    │
│ WIRE, OPEN WIRE, VALVE   │          │   CONNECTIONS. FUSE,     │
│ STUCK, ORIFICE PLUGGED   │          │   SHORTED WIRE           │
└──────────────────────────┘          └──────────────────────────┘
          │
          ▼
┌──────────────────────────┐
│ CHECK DETENT VALVE       │
│ TRAIN                    │
└──────────────────────────┘
```

44–13 GM

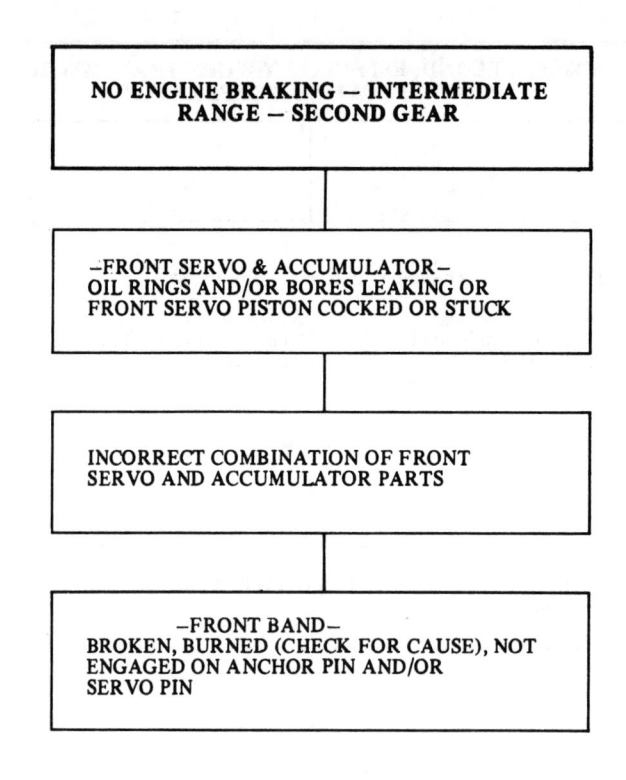

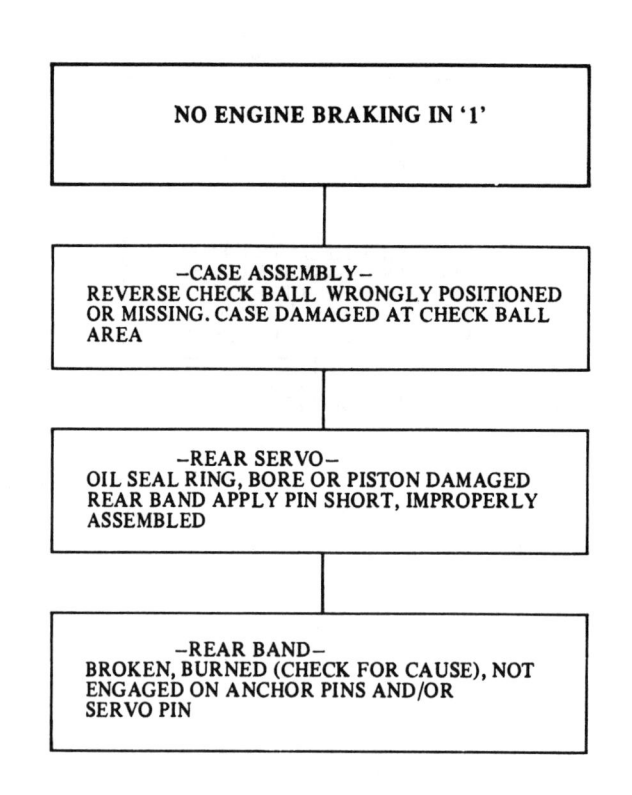

```
┌─────────────────────────────┐
│ NO ENGINE BRAKING – INTERMEDIATE │
│      RANGE – SECOND GEAR        │
└─────────────────────────────┘
              │
┌─────────────────────────────┐
│  –FRONT SERVO & ACCUMULATOR–    │
│  OIL RINGS AND/OR BORES LEAKING OR│
│  FRONT SERVO PISTON COCKED OR STUCK│
└─────────────────────────────┘
              │
┌─────────────────────────────┐
│  INCORRECT COMBINATION OF FRONT │
│  SERVO AND ACCUMULATOR PARTS    │
└─────────────────────────────┘
              │
┌─────────────────────────────┐
│           –FRONT BAND–          │
│  BROKEN, BURNED (CHECK FOR CAUSE), NOT│
│  ENGAGED ON ANCHOR PIN AND/OR   │
│  SERVO PIN                      │
└─────────────────────────────┘


┌─────────────────────────────┐
│     NO ENGINE BRAKING IN '1'    │
└─────────────────────────────┘
              │
┌─────────────────────────────┐
│          –CASE ASSEMBLY–        │
│  REVERSE CHECK BALL WRONGLY POSITIONED│
│  OR MISSING. CASE DAMAGED AT CHECK BALL│
│  AREA                           │
└─────────────────────────────┘
              │
┌─────────────────────────────┐
│           –REAR SERVO–          │
│  OIL SEAL RING, BORE OR PISTON DAMAGED│
│  REAR BAND APPLY PIN SHORT, IMPROPERLY│
│  ASSEMBLED                      │
└─────────────────────────────┘
              │
┌─────────────────────────────┐
│           –REAR BAND–           │
│  BROKEN, BURNED (CHECK FOR CAUSE), NOT│
│  ENGAGED ON ANCHOR PINS AND/OR  │
│  SERVO PIN                      │
└─────────────────────────────┘
```

NO HOLD IN PARK OR NO RELEASE FROM PARK

↓

CHECK SELECTOR CABLE ADJUSTMENT

↓

—INTERNAL LINKAGE—
1. PARKING BRAKE ROD ASSEMBLY. (CHECK ACTUATOR FOR CHAMFER.)
2. PARKING PAWL BROKEN, CHAMFER OMITTED.
3. PARKING BRAKE BRACKET LOOSE, BURR OR ROUGH EDGES, OR INCORRECTLY INSTALLED.
4. PARKING PAWL RETURN SPRING MISSING, BROKEN OR INCORRECTLY FITTED.

DRIVE IN NEUTRAL

↓

CHECK SELECTOR CABLE ADJUSTMENT

↓

—INTERNAL LINKAGE—
MANUAL VALVE DISCONNECTED OR END BROKEN, INSIDE DETENT LEVER PIN BROKEN

↓

— PUMP ASSEMBLY —
FLUID PRESSURE LEAKING INTO FORWARD CLUTCH APPLY PASSAGE

```
                    ┌─────────────────────┐
                    │ TRANSMISSION NOISY  │
                    └──────────┬──────────┘
                               │
┌──────────────────────────────┴──────────────────────────────────┐
│ CAUTION: BEFORE CHECKING TRANSMISSION FOR WHAT IS BELIEVED TO   │
│ BE 'TRANSMISSION NOISE', MAKE CERTAIN THE NOISE IS NOT FROM THE │
│ WATER PUMP, ALTERNATOR, AIR CONDITIONER, POWER STEERING, ETC.   │
│ THESE COMPONENTS CAN BE ISOLATED BY REMOVING THE APPROPRIATE    │
│ DRIVE BELT AND RUNNING THE ENGINE FOR NOT MORE THAN TWO         │
│ MINUTES AT ONE TIME.                                            │
└─────────────────────────────────────────────────────────────────┘
```

PARK, NEUTRAL AND ALL DRIVE RANGES

—PUMP CAVITATION—
OIL LEVEL LOW.
BLOCKED OR RESTRICTED FILTER.
WRONG FILTER.
INTAKE PIPE 'O' RING DAMAGED.
INTAKE PIPE SPLIT, POROSITY IN CASE INTAKE PIPE BORE.
WATER IN OIL.
POROSITY OR VOIDS IN TRANSMISSION CASE (PUMP FACE) INTAKE PORT.
PUMP TO CASE GASKET WRONGLY POSITIONED.

—PUMP ASSEMBLY—
GEARS DAMAGED.
DRIVING GEAR ASSEMBLED BACKWARDS.
CRESCENT INTERFERENCE.
BUZZING NOISE – ORIFICE CUP PLUG IN PRESSURE REGULATOR DAMAGED OR MISSING.
SEAL RINGS DAMAGED OR WORN.

—CONVERTER—
LOOSE BOLTS (CONVERTER TO DRIVE PLATE).
CONVERTER DAMAGE.
CRACKED OR BROKEN DRIVE PLATE.

DURING ACCELERATION— ANY GEAR

SQUEAL AT LOW VEHICLE SPEEDS, ESPECIALLY AT NORMAL OPERATING TEMPERATURE

SPEEDOMETER DRIVEN GEAR SHAFT SEAL– SEAL REQUIRES LUBRICATION OR REPLACEMENT

IF SPEEDOMETER DRIVEN GEAR SHAFT APPEARS TWISTED, CHECK FOR PRESENCE OF ENGINE COOLANT IN TRANSMISSION. CHECK TRANSMISSION COOLER FOR LEAKS.

FIRST, SECOND AND/OR REVERSE

—PLANETARY GEAR SET—
1 THOROUGHLY CLEAN, DRY & INSPECT CLOSELY THE ROLLER THRUST BEARINGS AND THRUST RACES FOR PITTING OR ROUGH CONDITION.
2 INSPECT GEARS FOR DAMAGE, WEAR, PITTING AND PINIONS FOR TILT.
3 INSPECT FRONT INTERNAL GEAR FOR RING DAMAGE.

HAND SELECTOR ASSEMBLY

Overhaul   44.15.05

Removing

1. Remove the selector quadrant, see 76.25.08.
2. Remove the nuts securing the indicator bulb mounting and remove the bulb mounting bracket.
3. Remove the circlip and circlip washer from the lever pivot shaft. 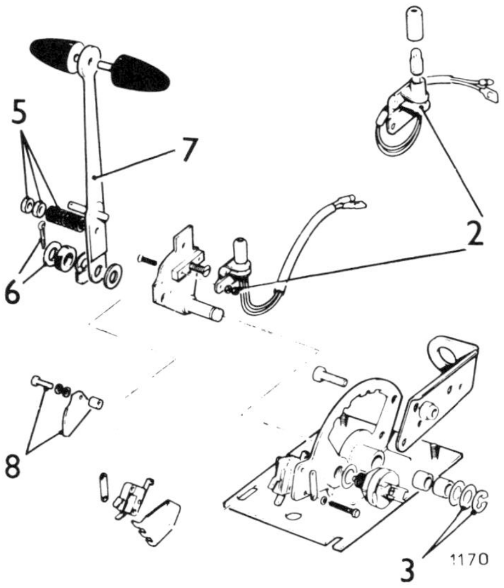
4. Remove the selector lever assembly from the car.
5. Remove the nuts securing the tension spring and remove the spring.
6. Remove the split pin and washers securing the lever to the cam plate pivot.
7. Remove the selector lever.
8. Holding the lever mounting plate in a vice, remove the screws securing the tapped block and illumination bulb bracket to the cam plate.
9. Remove the tension spring screw from the tapped block and remove the block from the cam plate.
10. Clean all parts.

Refitting

11. Holding the mounting plate in a vice, refit the tapped block to the cam plate by loosely fitting the tensioning spring screw.
12. Align the holes of the block with those in the cam plate and fit the illumination bulb bracket without fully tightening the retaining screws.
13. Tighten the tensioning spring screw.
14. Tighten the bulb bracket screws.
15. Secure the tensioning spring screw with two centre dots on the mating surface of the tapped block.
16. Remove the mounting plate assembly from the vice.
17. Refit the lever to the mounting plate assembly and secure to the lever pivot with the pivot washer, washer, rubber washer, washer and split pin.
18. Refit the tensioning spring and spring securing nut. Reset the spring to the correct tension and refit the locknut.
19. Lubricate the selector lever pivot shaft.
20. Refit the assembly to the car.
21. Refit the circlip washer, shim and circlip to the lever pivot.
22. Refit selector quadrant, see 76.25.08.

STARTER INHIBITOR SWITCH

Check and adjust   44.15.18

Adjusting

1. Disconnect the battery, see 86.15.20.
2. Unscrew the gear selector knob.
3. Carefully prise the electric window switch panel away from the centre console; do not disconnect the window switches.
4. Remove the screws securing the control escutcheon; withdraw the escutcheon slightly to obtain access to the cigar lighter and door lock switch terminals.
5. Note the fitted position of the cigar lighter and door lock switch terminals; detach the terminals and withdraw the escutcheon.
6. Detach the feed cable from the inhibitor switch.
7. Connect a test lamp and battery in series with the switch.

   NOTE: Switch is in earthed position. Place the selector lever in 'N' position.
8. Detach the feed cable from the inhibitor switch. Connect a test lamp and battery in series with the switch. Switch is in earthed position. 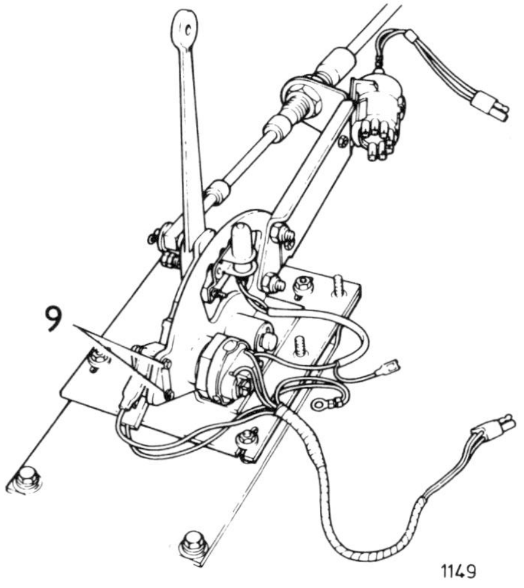
9. Slacken the locknuts securing the switch and adjust position of switch until the lamp lights.
10. Tighten the locknuts, check that lamp remains on with lever in 'P' position and is off with lever in drive position.
11. Remove the battery and test lamp, reconnect feed cable to switch.
12. Reverse operations 1 to 5.
13. Check operation of window switches, door lock switch and cigar lighter.

GEARBOX ASSEMBLY

Remove and refit   44.20.01

Service tools: MS 53 A Engine support bracket, Epco V1000 Unit lift

Removing

1. Drive the vehicle onto a ramp.
2. Remove the transmission dipstick.
3. Unscrew and remove the bolt securing the dipstick upper tube to the lifting eye bracket.
4. Remove the dipstick upper tube.
5. Slacken the wing stay to bulkhead securing bolt.
6. Remove the wing stay to wing securing bolts.
7. Remove the pipe to wing stay clamps.
8. Swing the wing stays away from the wings.
9. Unscrew and remove the handles from the engine lifting hooks – Tool No. MS 53 A.
10. Fit the hooks to the rear lifting eyes.
11. Fit the engine support tool.
12. Fit and tighten the handles to take the weight of the engine.
13. Raise the ramp.
14. Unscrew and remove the nuts/bolts securing the intermediate exhaust pipes, rotating the flanges for access.
15. Disconnect the exhaust pipes and remove the sealing olives.
16. Remove the intermediate heat shields.
17. Remove the rear heat shield.
18. Pull aside the exhaust pipes and secure.
19. Remove the front heat shields.
20. Unscrew and remove the rear mounting centre nut.
21. Remove the spacer.
22. Using a suitable block of wood interposed between the jack head and the gearbox rear mounting, support the mounting plate.
23. Remove the bolts securing the rear mounting.
24. Remove the rear spacers.
25. Lower the jack.
26. Remove the mounting assembly.
27. Remove the wooden block and jack.
28. Unscrew the bolts securing the cross-member.

continued

44–17 GM

29 Remove the cross-member.
30 Disconnect the propeller shaft from the transmission and move the shaft clear.
31 Working from above the engine compartment, slacken the hooks – 10 turns only.
32 From beneath the vehicle, disconnect the speedometer cable from the transmission.
33 Unscrew the nut securing the selector pin to the lever and disconnect the cable.
34 Unscrew the bolt securing the selector cable to the support bracket and move the cable away from the transmission.
35 Disconnect the kick-down solenoid feed-wire and remove the clamp bolt securing the feed wire to the transmission.
36 Disconnect the modulator capsule vacuum tube.
37 Remove the bolt and clamp plate securing the modulator.
38 Place a suitable receptacle under the modulator, withdraw the modulator and partially drain the transmission fluid.
39 Remove and discard the modulator 'O' ring.
40 Unscrew the cooler pipe union nuts from the unions.
41 Unscrew the bolt securing the cooler pipes bracket to the engine sump, remove the spacer.
42 Disconnect and plug the cooler pipes.
43 Unscrew the bolts/nuts securing the convertor access cover (and catalysts – where fitted) and remove the cover.
44 Unscrew the bolts securing the convertor to the drive plate, turning the drive plate for access.
45 Remove the bolts securing the right/hand rack gaiter heat shield and remove the heat shield (USA vehicles only).
46 Unscrew the nuts securing the right-hand catalyst (where fitted) and displace the catalyst from the manifold.
47 Remove the engine/transmission securing bolts with the exception of two lower left-hand bolts and lower starter motor securing bolt.
48 Remove the dipstick tube and reposition the tube/vacuum pipe mounting bracket along the vacuum pipe.
49 Utilizing an Epco V1000 unit lift:
 a Remove the front and rear clamps.
 b Traverse the lift under the transmission unit.
 c Take the weight of the transmission on the lift.
 d Adjust the tilt angle and side clamps.
 e Tighten the clamps.
 f Fit the chain assembly to the right-hand arm, fit the chain over the peg and pass the chain over the transmission into the front arm.
 g Tighten the chain adjuster.
50 Remove the remaining bolts securing the engine/transmission and starter motor.
51 Disconnect the transmission unit from the engine, lower the unit (easing the catalyst aside – where fitted) and traverse the transmission/unit lift from beneath the vehicle.

Extra operations for replacing the transmission assembly

52 Unscrew and remove the rear mounting spigot securing bolts, and remove the mounting.
53 Remove the selector cable mounting collar.
54 Unscrew and remove the cooler pipe unions.
55 Slacken the chain adjuster and release the chain from the front arm.
56 Slacken the clamp wing nuts and release the clamps.
57 Place the transmission unit aside to drain.
58 Fit the replacement transmission unit to the lift.
59 Reposition the clamps and tighten the wing nuts.
60 Refit the chain to the front arm and tighten the chain adjuster.
61 Remove all the blanking plugs from the new transmission unit.
62 Remove the convertor strap from the replacement unit and fit to the displaced unit.
63 Fit and tighten the cooler pipe unions to the replacement unit.

64 Fit the selector cable collar.
65 Fit the rear mounting; fit and tighten the securing bolts.
66 Clean the relevant mounting and attachment faces.

Refitting

67 Traverse the transmission/unit lift beneath the vehicle, raise the unit into position (easing aside the catalyst – where fitted) and place the speedometer cable, selector cable, kick-down solenoid feed wire and vacuum pipe into suitable positions.
68 Align the transmission mating flange over the locating dowels.
69 Fit and tighten three lower left-hand transmission/engine securing bolts.
70 Locate the starter motor in position and fit and tighten the securing bolts.
71 Release the unit lift clamps, slacken the chain tensioner and remove the pin from the left-hand arm, release the chain from the front arm and remove the chain assembly.
72 Lower the unit lift and remove from the working area; refit the clamps to the lift.
73 Fit and tighten the remaining engine/transmission securing bolts.
74 Position the dipstick pipe clamp on the torque converter housing and fit the lower dipstick tube.
75 Connect the dipstick tube to the transmission.
76 Pull the vacuum pipe through the bracket.
77 Fit two accessible torque convertor/drive plate bolts. Do not tighten.
78 Turn the drive plate, fit two further torque convertor/drive plate bolts. Do not tighten.
79 Turn the drive plate, fit and tighten final two torque convertor/drive plate bolts.
80 Turn the drive plate and tighten first four bolts.
81 Fit the torque convertor cover-plate (and the strap to the catalyst – where fitted).
82 Slacken the left-hand nut securing the strap to the cover and swing the strap aside.

83 Remove the blanking plugs from the cooler pipes and connect the pipes to the transmission.
84 Position the cooler pipe mounting bracket, fit the spacer and bolt and secure the bracket to the engine sump.
85 Fit a new 'O' ring to the modulator capsule and fit the modulator to the transmission unit with the clamp plate and bolt.
86 Connect the vacuum pipe to the modulator.
87 Connect the kick-down solenoid feed wire and secure to the transmission with the clamp and bolt.
88 Fit and secure the selector cable bracket to the mounting and connect the cable to the lever. Fit and tighten the selector pin securing nut.
89 Working from above the engine compartment, tighten the hook handles to raise the engine.
90 Working from beneath the vehicle, connect the propeller shaft to the transmission flange.
91 Position and align the cross-member, fit and tighten the upper securing bolts.
92 Fit and tighten the lower securing bolts.
93 Place the ramp jack under the rear mounting, and locate the wooden block and mounting assembly in position, raise the jack and align the attachment holes.
94 Fit the rear spacers, fit and tighten the securing bolts.
95 Remove the jack and wooden block.
96 Fit the rear mounting spacer and centre nut.
97 Position the right-hand catalyst (where fitted) into the manifold and secure with the nuts.
98 Secure the convertor cover strap to the catalyst (where fitted).
99 Refit the rack gaiter heat shield (where fitted).
100 Refit the front heat shields.
101 Untie the exhaust pipes and fit the rear heat shield.
102 Refit the intermediate heat shields.

continued

Dismantling

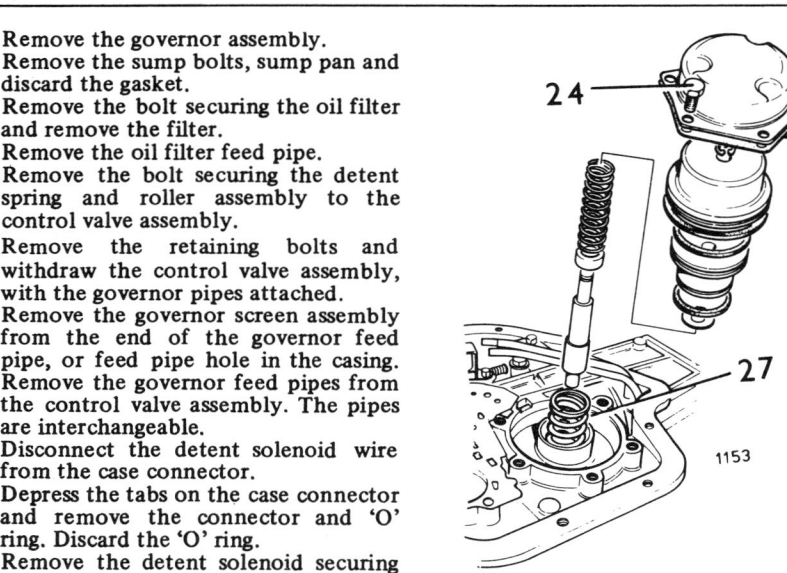

1. Thoroughly clean the gearbox casing.
2. Remove the gearbox assembly from the vehicle, see 44.20.01.
3. Remove the torque convertor.
4. Invert gearbox onto a suitable bench cradle.
5. Using a suitable flange retaining tool, undo and remove the drive flange securing bolt and remove the drive flange.
6. Remove the bolt securing the speedometer drive pinion clamp plate.
7. Remove the clamp plate and withdraw the speedometer pinion assembly.
8. Remove the four governor cover-plate securing bolts and remove the cover-plate and gasket.
9. Discard the gasket. 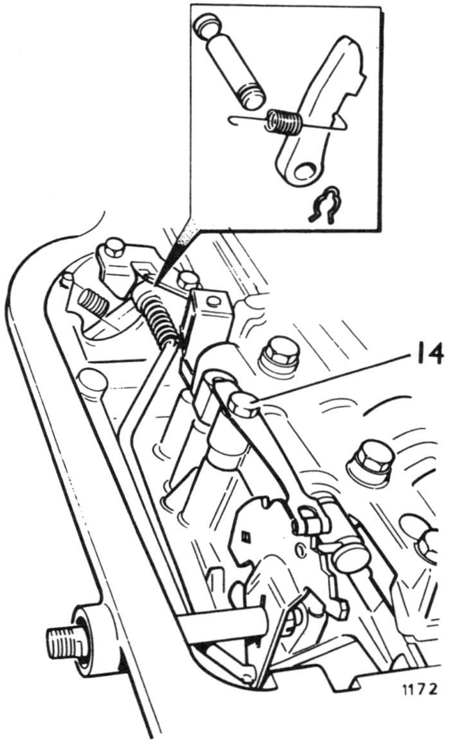
10. Remove the governor assembly.
11. Remove the sump bolts, sump pan and discard the gasket.
12. Remove the bolt securing the oil filter and remove the filter.
13. Remove the oil filter feed pipe.
14. Remove the bolt securing the detent spring and roller assembly to the control valve assembly.
15. Remove the retaining bolts and withdraw the control valve assembly, with the governor pipes attached.
16. Remove the governor screen assembly from the end of the governor feed pipe, or feed pipe hole in the casing.
17. Remove the governor feed pipes from the control valve assembly. The pipes are interchangeable.
18. Disconnect the detent solenoid wire from the case connector.
19. Depress the tabs on the case connector and remove the connector and 'O' ring. Discard the 'O' ring.
20. Remove the detent solenoid securing bolts and remove the solenoid and gasket.
21. Remove the control valve spacer plate from the casing.
22. Remove the six check balls from the transmission casing.
23. Lift the front servo piston assembly from the transmission case. 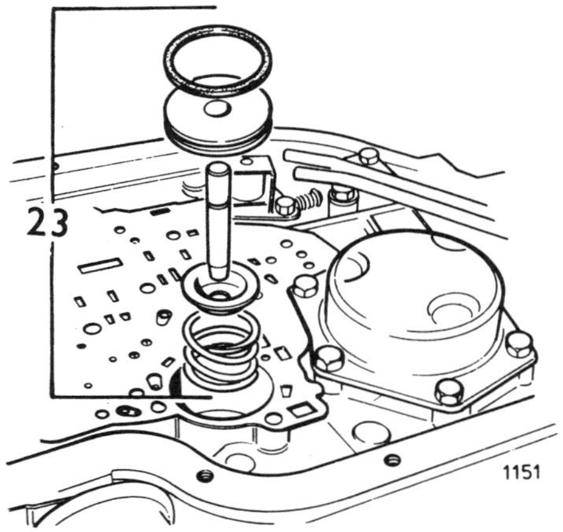
24. Remove the rear servo cover retaining bolts.
25. Remove the cover and gasket. Discard the gasket.
26. Remove the rear servo assembly from the transmission case.
27. Remove the rear servo accumulator spring.
28. Remove the modulator valve from the case.
29. Undo and remove the six rear extension securing bolts.
30. Remove the rear extension and gasket.
31. Discard the gasket.
32. Turn box over.
33. Insert service tools 18G 1296 into the two threaded holes in the pump body.
34. Using service tools 18G 1296, extract the pump.
35. Remove the pump assembly and discard the gasket.
36. Remove the service tools 18G 1296 from the pump body.
37. Remove the input shaft and forward clutch assembly.
38. Remove the direct clutch and intermediate roller assembly.
39. Remove the front band assembly.

103. Smear the exhaust sealing olives with 'Firegum', fit the olives, connect and secure the exhaust system.
104. Lower the ramp.
105. Unscrew and remove the support tool hook handles.
106. Remove the support tool, remove the hooks, fit and tighten the handles to the hooks.
107. Refit the dipstick upper tube and secure to the lifting eye bracket.
108. Reposition the wing stays to secure to the wings.
109. Tighten the wing stay/bulkhead attachment and refit the pipe clamps.
110. Fill the transmission unit with fluid and refit the dipstick.

GEARBOX ASSEMBLY

Overhaul 44.20.06

NOTE: Before commencing this operation, it is strongly recommended that the following checks are carried out and all readings noted:

Front Unit End-Float Check, see 44.30.22.
Rear Unit End-Float Check, see 44.30.23.
Rear Servo Band Apply Pin Selection Check, see 44.30.21.

Cleaning

Gamlen 265 is recommended for all cleaning purposes.

Service tools: 18G 1295 Compressor piston accumulator control valve; 18G 1296 Front pump remover screws; 18G 1297 Front pump and tailshaft oil seal replacer; 18G 1298 Forward and direct clutch piston replacer, inner and outer protection sleeve; 18G 1309 Intermediate clutch inner seal protection sleeve; 18G 1310 Band application pin selection gauge; 18G 677 ZC Pressure test equipment; 18G 1016 Clutch spring compressor; 18G 1004 Circlip pliers; 18G 1004 J Circlip plier points.

44—19 GM

40 Remove the intermediate clutch snap-ring.
41 Remove the intermediate clutch backing plate and clutch assembly.
42 Remove the chamfered snap-ring.
43 Undo and remove the centre support retaining peg and remove the centre support/roller clutch assembly.
44 Remove the sun gear shaft.
45 Remove the snap-ring from the bottom groove of the centre support.
46 Remove the rear brake band.
47 Remove the planet carrier assembly.
48 Remove the front internal gear ring.
49 Remove the rear thrust washer.
50 Remove the manual shaft retaining pin.
51 Release the manual shaft from the manual detent lever, remove the lever and shaft.
52 Remove the actuator rod assembly. 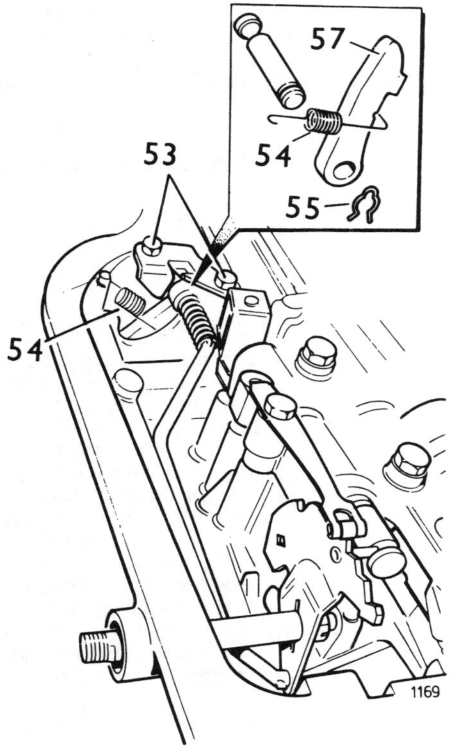
53 Remove the parking pawl bracket securing bolts and remove the bracket.
54 Remove the parking pawl spring.
55 Remove the spring clip securing the parking pawl to the pivot shaft.
56 Press the pivot shaft to displace the plug and remove the plug.
57 Remove the parking pawl.
58 Remove the pivot shaft.
59 Remove the pressure take-off plug.
60 Remove the filter pick-up seal.
61 Remove the selector shaft seal.
62 Clean and inspect the casing.

44—20 GM

CONTROL VALVE ASSEMBLY

During the dismantling procedure, carefully identify all valves, bushes and springs, noting their relative positions.

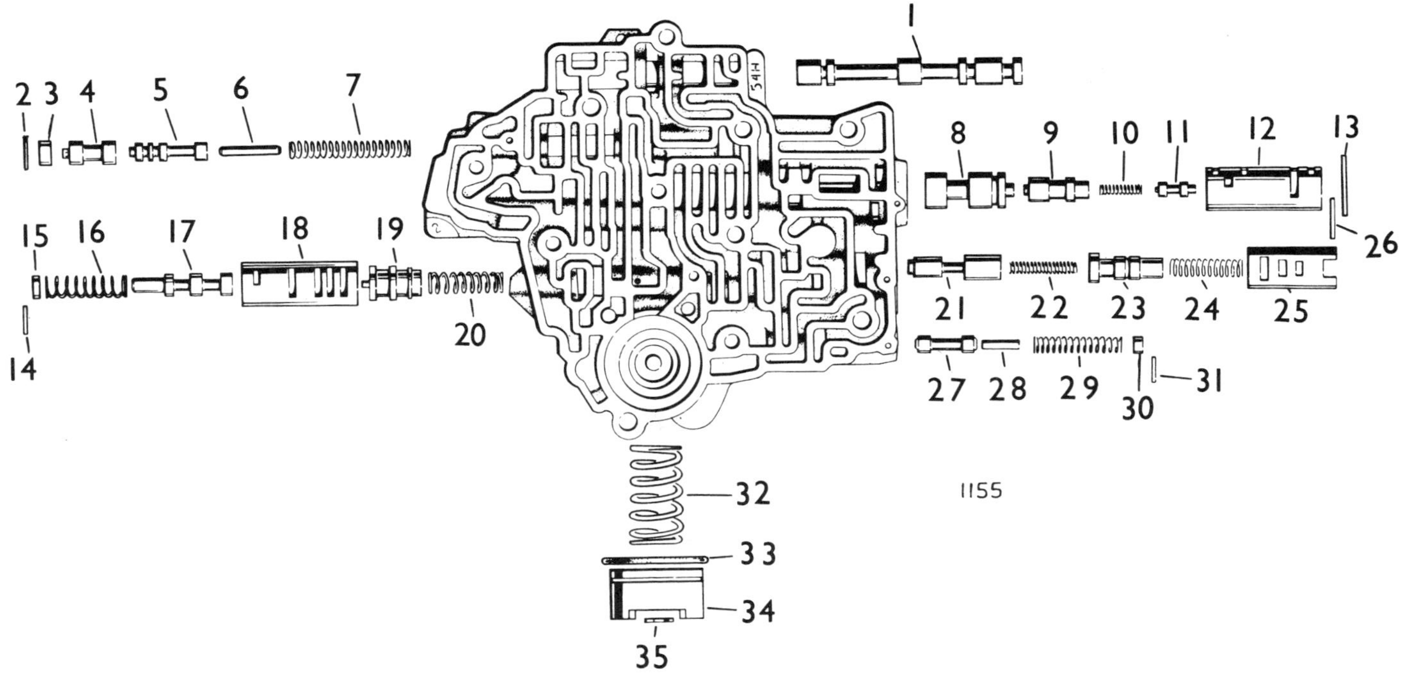

1. Manual Valve
2. Retaining Pin
3. Bore Plug
4. Detent Valve
5. Detent Regulator Valve
6. Spacer Pin
7. Detent Regulator Spring
8. 1–2 Shift Valve
9. 1–2 Detent Valve
10. 1–2 Regulator Spring
11. 1–2 Regulator Valve
12. 1–2 Modulator Bushing
13. Retaining Pin
14. Grooved Retaining Pin
15. Bore Plug
16. 1–2 Accumulator Secondary Spring
17. 1–2 Accumulator Secondary Valve
18. 1–2 Accumulator Bushing
19. 1–2 Primary Accumulator Valve
20. 1–2 Accumulator Primary Spring
21. 2–3 Shift Valve
22. 2–3 Intermediate Spring
23. 2–3 Modulator Valve
24. 2–3 Valve Spring
25. 2–3 Modulator Bushing
26. Retaining Pin
27. 3–2 Valve
28. Spacer Pin
29. 3–2 Valve Spring
30. Bore Plug
31. Retaining Pin
32. Accumulator Spring
33. Accumulator Piston Oil Ring
34. Accumulator Piston
35. 'E'-ring Retainer

Dismantling

63 Position the control valve assembly with the gasket face uppermost and the accumulator at the bottom. This position will be used to identify the components.

64 Remove the manual valve.

65 Using service tool 18G 1295, compress the accumulator piston and spring, remove the 'E' ring retainer.

66 Remove the service tool, accumulator piston and spring.

67 Using a pin punch remove the 1-2 modulator bushing retaining pin, upper right-hand bore.

68 Remove the 1-2 modulator bushing, 1-2 regulator valve and spring, 1-2 detent valve and the 1-2 shift valve.
NOTE: The 1-2 regulator valve and spring may be inside the 1-2 modulator bushing.

69 Using a pin punch remove the 2-3 modulator bushing retaining pin, centre right-hand bore.

70 Remove the 2-3 modulator bushing, 2-3 shift valve spring, 2-3 modulator valve, 3-2 intermediate spring and the 2-3 shift valve.

71 Using a pin punch remove the 3-2 valve retaining pin, lower right-hand bore.

72 Remove the bore plug, 3-2 valve spring, spacer and the 3-2 valve.

73 Using a pin punch remove the detent valve retaining pin, upper left-hand bore.

74 Remove the bore plug, detent valve, detent regulator valve, spacer and detent regulator valve spring.

75 Using a pair of long-nosed pliers remove the 1-2 accumulator valve retaining pin, lower left-hand bore.

76 Remove the bore plug, 1-2 accumulator valve and primary valve.

Inspection

77 Wash all components in a clean solvent. Do not allow valves to bump together, as this might cause nicks and burrs.

continued

78 Carefully check all valves and bushings for burrs and damage. Burrs should be removed with a fine stone, taking care not to round off the shoulders of the valves.
79 Check all valves and bushings for free movement in their respective bores.
80 Check the valves housing for cracks and the bores for damage and scoring. NOTE: If any valves or bores are found to be damaged beyond repair, then a new control valve assembly must be fitted.
81 Check all the springs for distortion.
82 Check the front accumulator piston and oil ring for damage; renew as necessary.

Reassembling
83 Fit the accumulator spring and piston into the valve body.
84 Using service tool 18G 1295, squarely compress the spring and piston. NOTE: Ensure that the piston pin is correctly aligned with the hole in the piston and the oil seal ring does not foul the lip of the bore when fitting the piston.
85 Fit the 'E' ring retainer and remove the service tool.
86 Fit the 1-2 accumulator primary spring, lower left-hand bore.
87 Fit the 1-2 accumulator secondary valve, stem end out.
88 Fit the 1-2 accumulator bore plug to the 1-2 accumulator bore.
89 Turn over control valve assembly and remove the grooved retaining pin from the cast surface side of the body, with grooved end of pin entering the hole last.
90 Tap retaining pin into control valve housing until pin is flush with cast surface. Return control valve assembly to its original position.
91 Fit spacer to detent regulator valve spring and fit spring and spacer into upper left-hand bore; ensure that spring seats correctly.
92 Compress the detent regulator valve spring, fit the detent regulator valve, stem end last, and detent valve, band first.
93 Fit the bore plug, hole outermost, and secure with the retaining pin, from the cored side of the body.
94 Fit the 3-2 valve, bottom right-hand bore.
95 Fit spacer to the 3-2 valve spring and fit the spring and spacer, bottom right-hand bore.
96 Compress the 2-3 valve spring, and fit the bore plug, hole end outermost; secure with retaining pin, from the cored side of the body.
97 Fit the 3-2 intermediate spring in the open end of the 2-3 shift valve, fit valve and spring to the centre right-hand bore. Ensure that the valve seats correctly.
98 Fit the 2-3 modulator valve, hole end first, to the 2-3 modulator bushing and fit both parts to the centre right-hand bore.
99 Fit the 2-3 shift valve spring into the 2-3 modulator valve, compress the spring and fit the retaining pin, from the cored side of the control valve.
100 Fit the 1-2 shift valve, stem end outermost ensuring that the valve seats correctly, to the upper right-hand bore.
101 Fit the 1-2 regulator valve, large stem first, spring and the 1-2 detent valve, hole end first, into the 1-2 bushing and fit all the components to the upper right-hand bore.
102 Compress the bushing against the spring and fit the retaining pin from the cored side of the control valve body.
103 Fit the manual valve, with the detent pin groove to the right.

REAR SERVO ASSEMBLY

Dismantling
104 Remove the rear accumulator piston from the rear servo piston.
105 Remove the 'E' ring retaining the rear servo piston to the rear band apply pin.
106 Remove the rear servo piston and seal from the band apply pin.
107 Remove the washer, spring and retainer.

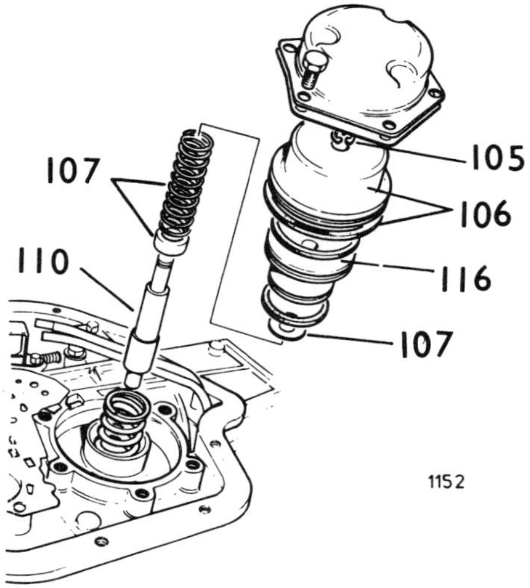

Inspection
108 Check the freeness of the oil seal rings in the piston grooves. Renew as necessary.
109 Check the fit of the band apply pin in the servo piston.
110 Check the band apply pin for cracks and scoring.
111 Check that band apply pin is the correct size as determined by the pin selection check.

Reassembling
112 Fit the spring retainer, cup side towards the band apply servo pin, spring and washer to the servo pin.
113 Fit the servo piston to the pin and secure with the 'E' ring retainer.
114 Renew piston oil seals as necessary.
115 Renew accumulator piston oil seals as necessary.
116 Fit the accumulator piston into the bore of the servo piston.

FRONT SERVO ASSEMBLY

Inspection
117 Check the servo pin for damage.
118 Check the piston and oil seal ring for damaged oil ring groove, check that the oil ring is free to move.
119 Check the piston for cracks and porosity.
120 Check the fit of the servo pin to the piston.

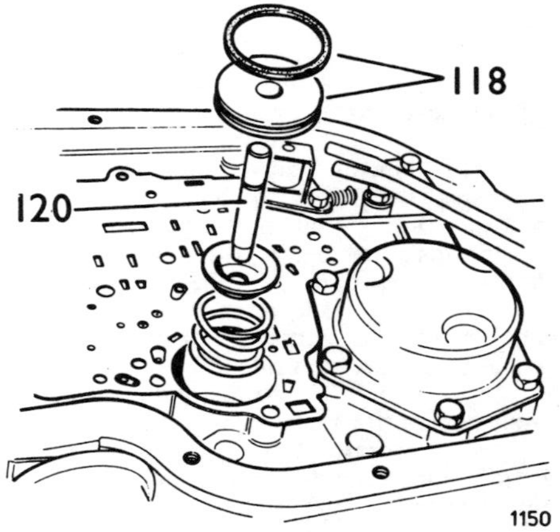

Reassembling
121 Refit the parts of the front servo; ensure that the tapered end of the servo pin points through the spring and retainer; Ensure that the retainer ring is in the servo pin groove.

44—22 GM

OIL PUMP

Dismantling

122 Remove the outer oil seal.
123 Compress the regulator boost valve bushing against the regulator spring and remove the snap-ring.
124 Remove the regulator boost valve bushing, boost valve, pressure regulator spring, spring retainer regulator valve and spacer(s).
125 Remove the pump body securing bolts and remove the pump cover from the body.
126 Mark the relevant positions of the drive and driven gears to ensure correct positioning when assembling.
127 Remove the retaining pin and bore plug from the end of the regulator bore.
128 Remove the two oil rings from the pump cover.
129 Remove the pump to forward clutch housing thrust washer.
130 Remove the front oil seal from the pump body.

Inspection — pump body

131 Check the gears for scoring, chafing and other damage.
132 Position the pump gears in the pump body, lay a straight-edge over the gears and casing and check the clearance between the gears and the underside of the straight-edge. Clearance should be 0.0008 to 0.0035 in.
133 Check the face of the pump body for scores and damage.
134 Ensure that all the oil passages are clean and free from any obstructions.
135 Check the threads in the pump body for damage.
136 Check that the pump body face is flat and free from warps

Inspection — pump cover

137 Check that the pump cover face is of uniform flatness and free from warps.
138 Check the pressure regulator bore for scoring, wear and dirt.
139 Ensure that all the oil passages are clean and free from any obstructions.
140 Check the pump gear face for scoring and damage.

141 Check the stator shaft for damaged splines or scored bushings.
142 Check the oil ring grooves for damage and wear.
143 Check the thrust washer face for wear and damage.
144 Fit the pump cover oil rings into the counterbore of the forward clutch housing and check for correct fit.
145 Ensure that the pressure regulator and boost valve operate freely.
146 Ensure that the air breather hole is free of any obstruction.

Reassembling

147 Fit the pump drive and driven gears into the pump body, aligning the marks previously made.
148 Fit the pressure regulator spacer(s), spring retainer and spring into the pressure regulator bore.
149 Fit the boost valve into the bushing, stem end out, and fit both parts into the pump cover by compressing the bushing against the spring.
150 Fit the retaining snap-ring.
151 Fit the pressure regulator valve from the opposite end of the bore, stem end first.
152 Fit the pressure regulator valve bore plug and retaining pin into the end of the bore.
153 Fit the front unit selective thrust washer over the pump cover delivery sleeve.
NOTE: The correct thickness was determined at the time the Front Unit End-Float Check (see 44.30.22) was carried out.
154 Fit the two oil seal rings to the pump cover.
155 Lubricate the pump gears with transmission fluid and fit the pump cover to the pump body.
156 Fit the pump securing bolts; do not tighten at this stage.
157 Using a suitable Jubilee clip around the pump assembly, tighten to align the pump cover with the pump body.
158 Fully tighten the securing bolts to 18 lbf ft (2,49 kgf m).
159 Fit a new square-cut 'O' ring to the pump.
160 Fit a new pump oil seal, using service tool 18G 1297.

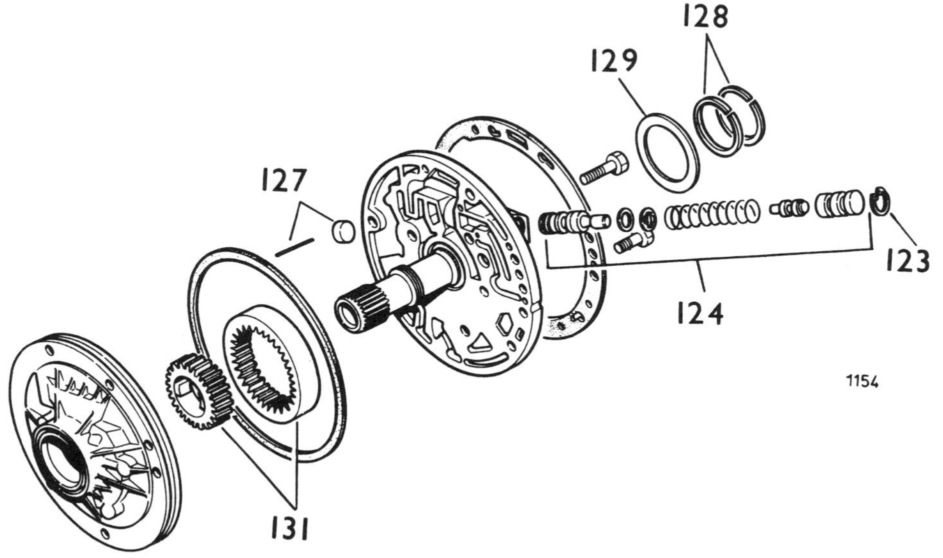

44-23 GM

FORWARD CLUTCH ASSEMBLY

Dismantling

161 Carefully secure the turbine shaft in a soft-jawed vice and remove the snap-ring securing the forward clutch housing to the direct clutch hub.
162 Remove the direct clutch hub.
163 Remove the outer thrust washer, forward clutch hub and inner thrust washer.
164 Remove the five composition and five steel clutch plates.
165 Using service tool 18G 1016, compress the spring retainer and remove the snap-ring securing the forward clutch piston assembly to the housing.
166 Remove the service tool 18G 1016 and withdraw the spring retainer and 16 clutch release springs.
167 Remove the forward clutch piston from the forward clutch housing.
168 Remove the seals from the piston.
169 Remove the centre piston seal from the forward clutch housing and withdraw the clutch housing and turbine shaft from the vice.

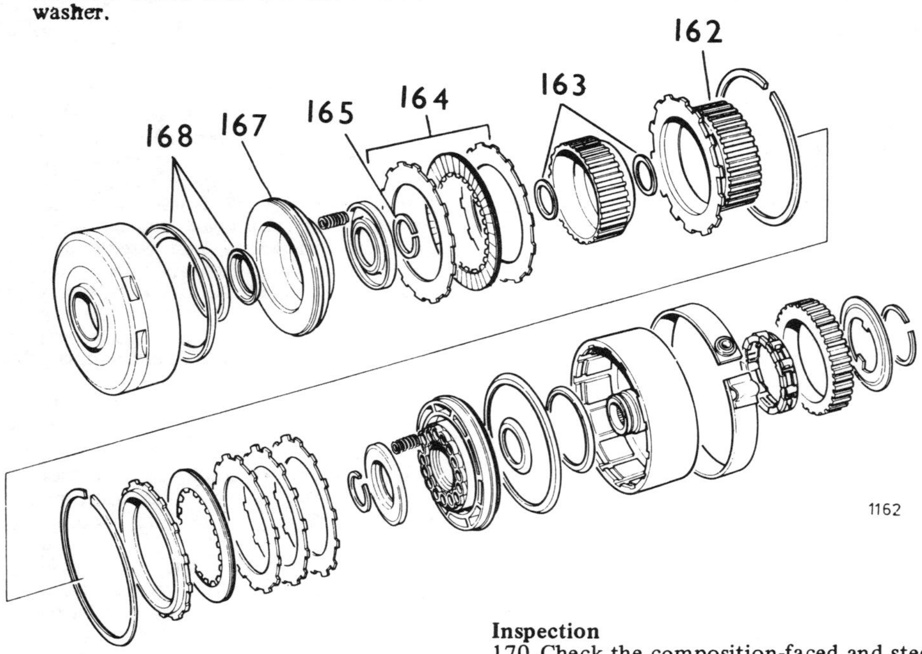

Inspection

170 Check the composition-faced and steel clutch plates for signs of burning, scoring and wear.
171 Check the forward clutch hub and direct clutch hub for wear on the splines and thrust faces; ensure that the lubrication holes are not blocked.
172 Check the piston for cracks.
173 Check the clutch housing for wear, scoring and cracks. Ensure that the oil passages are free from any obstructions.
174 Check the turbine shaft for cracks and distortion and the splines for damage.
175 Check the clutch release springs for signs of distortion.

Reassembling

176 Carefully secure the turbine shaft in a soft-jawed vice.
177 Lubricate new inner and outer clutch piston seals with new transmission fluid and fit the seals to the forward clutch piston, lips of seals facing away from spring pockets.
178 Fit a new centre piston seal to the forward clutch housing, lip facing upwards; lubricate with new transmission fluid.
179 Fit part of service tool 18G 1298 Inner seal protector, to the forward clutch hub.
180 Fit other part of service tool 18G 1298 Outer seal protector, to the clutch piston, and insert assembly in forward clutch housing.
181 Fit the clutch piston by rotating it in a clockwise direction until seated.
182 Remove service tools.
183 Fit the 16 clutch release springs to the spring pockets in the clutch piston.
184 Using bench press and service tool 18G 1016 fit the spring retainer, ensuring that retainer does not foul the snap-ring groove.
185 Fit the snap-ring and remove the service tools.
186 Ensure that the clutch release springs are correctly seated and are not leaning.
187 Fit the thrust washer to the outside face of the forward clutch hub. The bronze washer is fitted to the side of the hub which faces the forward clutch housing.
188 Fit the forward clutch hub to the forward clutch housing.
189 Fit the waved steel plate to the forward clutch housing and alternate composition and steel plates.
190 Fit the direct clutch hub in the forward clutch housing and secure with the snap-ring.
191 Fit the forward clutch housing to the pump delivery sleeve, and applying air to the forward clutch passage in the pump, check operation of forward clutch.

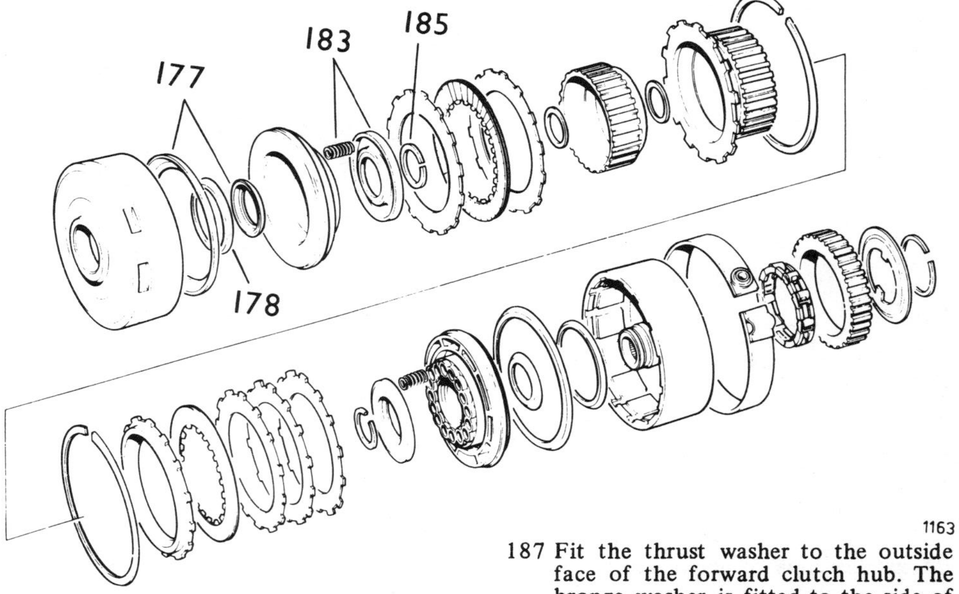

44—24 GM

DIRECT CLUTCH AND INTERMEDIATE ROLLER

Dismantling

192 Remove the roller retainer snap-ring and remove the clutch retainer.
193 Remove the roller outer race and remove the roller assembly.
194 Remove the snap-ring securing the direct clutch backing plate to the clutch housing.
195 Remove the direct clutch backing plate and the six composition and six steel clutch plates.
196 Using service tool 18G 1016, compress the spring retainer and remove the snap-ring.
197 Remove the tool, spring retainer and 14 clutch release springs.
198 Remove the direct clutch piston from the direct clutch housing.
199 Remove the seals from the piston.
200 Remove the centre piston seal from the direct clutch housing.

Inspection

201 Check the roller clutch assembly for damage to the rollers, cage and springs.
202 Check the direct clutch housing outer race for wear and scoring.
203 Check the direct clutch housing for cracks, wear and blocked oil passages; also check the clutch plate drive lugs for wear.
204 Check the composition-faced and steel clutch plates for signs of wear and burning.
205 Check the back plate for scratches, scoring and other damage.
206 Check the piston for cracks, ensure that the check ball operates freely.
207 Check the springs for wear and distortion.

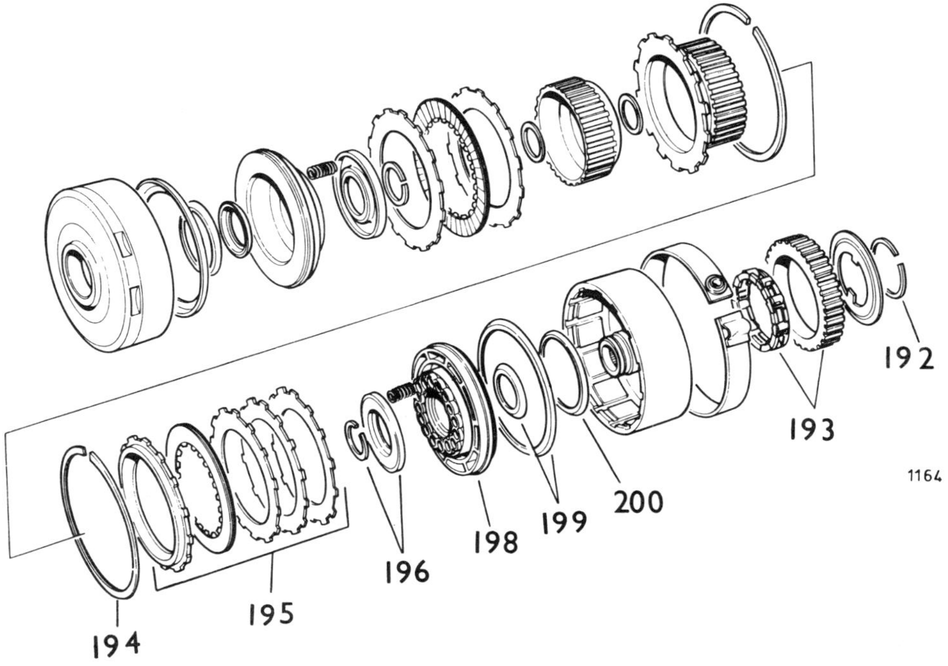

Reassembling

208 Lubricate new inner and outer clutch piston seals with new transmission fluid, fit the seals to the piston, seal lips facing away from spring pockets.
209 Fit a new centre piston seal to the direct clutch housing, lip facing upwards, and lubricate with new transmission fluid.
210 Fit part of service tool 18G 1298 forward and direct clutch inner seal protector, over the direct clutch hub.
211 Fit other part of service tool 18G 1298 forward and direct clutch piston outer seal protector to the clutch piston and inset assembly in the direct clutch housing.
212 Fit the clutch piston by rotating it in a clockwise direction.
213 Remove service tools.
214 Fit the 14 clutch release springs to the spring pockets in the clutch piston.
215 Using bench press and service tool 18G 1016, fit the spring retainer. Ensure that the retainer does not foul the snap-ring groove.
216 Fit the snap-ring and remove the service tools.
217 Ensure that the clutch springs are correctly seated and are not leaning.
218 Lubricate the five flat and one waved steel clutch plates, and the six composition-faced clutch plates with clean transmission fluid.
219 Fit the waved washer to the direct clutch housing and alternate composition and steel plates.
220 Fit the direct clutch backing plate and secure with the snap-ring.
221 Fit the roller clutch assembly to the intermediate clutch inner race, on the direct clutch housing.
222 Fit the intermediate clutch outer race. Outer race should not turn in an anti-clockwise direction.
223 Fit the roller clutch retainer and snap-ring.
224 Fit the direct clutch assembly to the centre support and check operation using compressed air.

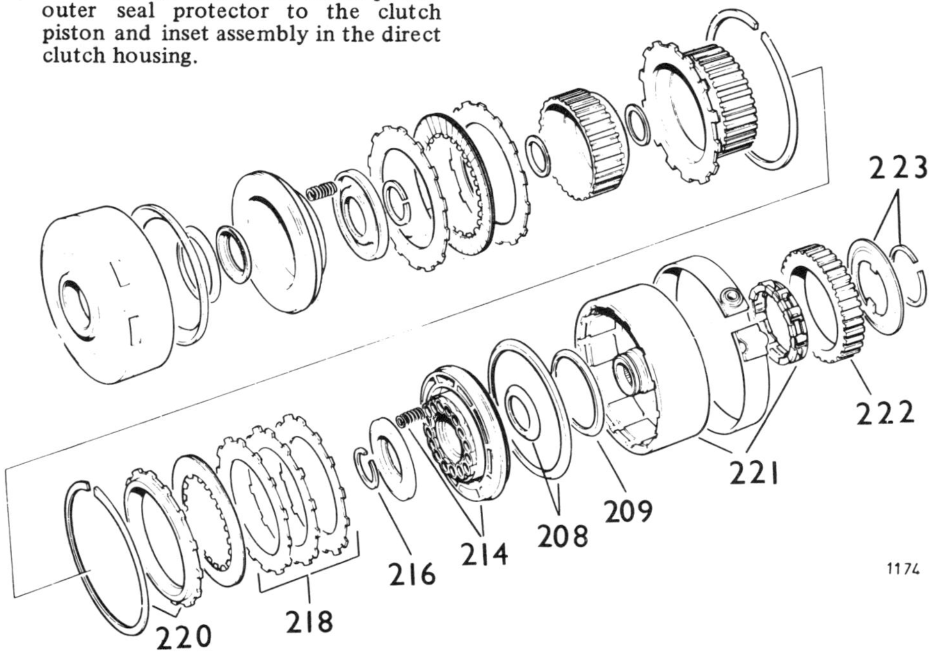

44—25 GM

PLANET GEAR CARRIER/ OUTPUT SHAFT ASSEMBLY

Dismantling
225 Remove the sun gear from the output carrier assembly.
226 Remove the reaction carrier to output carrier thrust washer, and remove the front internal gear ring.
227 Remove the snap-ring securing the output shaft to the output carrier and remove the output shaft.
228 Remove and discard the 'O' ring from the output shaft.
229 Remove the thrust bearing and races from the rear internal gear.
230 Withdraw the rear internal gear and mainshaft from the output carrier.
231 Remove the thrust bearing and races from the inner face of the rear internal gear.
232 Remove the snap-ring from the end of the mainshaft and remove the rear internal gear.
233 Remove the speedometer drive gear.

Inspection
234 Check the splines, 'O' ring grooves, bushes and gear teeth for burrs or signs of damage. Minor burrs can be removed with a very fine abrasive.
235 Check all oil drillings for obstructions and clear only with compressed air.
236 Examine the needle-roller assemblies, and renew if there are any signs of wear or damage.

Reassembling
237 Fit the rear internal gear to the end of the mainshaft that has the snap-ring groove and fit the snap-ring.
238 Fit the large diameter race, with flanged outer edge facing outwards, to the inner face of the rear internal gear.
239 Fit the thrust bearing to the race.
240 Fit the small diameter race, with flanged inner edge facing inwards, to the bearing.
241 Lubricate the pinion gears in the output carrier with new transmission fluid and fit the output carrier to the mainshaft, meshing the pinion gears with the rear internal gear.
242 Insert the assembly and hold the mainshaft in a soft-jawed vice. Be careful not to damage the shaft.
243 Fit the small diameter race, with flanged inner edge facing outwards, to the outer face of the rear internal gear.
244 Fit the thrust bearing to the race.
245 Fit the large diameter race, with flanged outer edge facing inwards, to the bearing.
246 Fit the output shaft into the outer carrier and fit the snap-ring.
247 Fit a new 'O' ring to the output shaft.
248 Fit the thrust washer to the output carrier, engaging the tabs of the washer with the slots in the carrier.
249 Fit the sun gear, chamfered internal diameter first.
250 Fit the sun gear shaft, long splined end first.
251 Fit the front internal gear ring to the output carrier.

44—26 GM

CENTRE SUPPORT AND INTERMEDIATE CLUTCH

Dismantling
252 Remove the four Teflon oil rings from the centre support.
253 Compress the spring retainer and remove the snap-ring.
254 Remove the spring retainer and the three intermediate clutch release springs.
255 Remove the spring guide.
256 Remove the intermediate clutch piston from the centre support.
257 Remove the seals from the clutch piston.

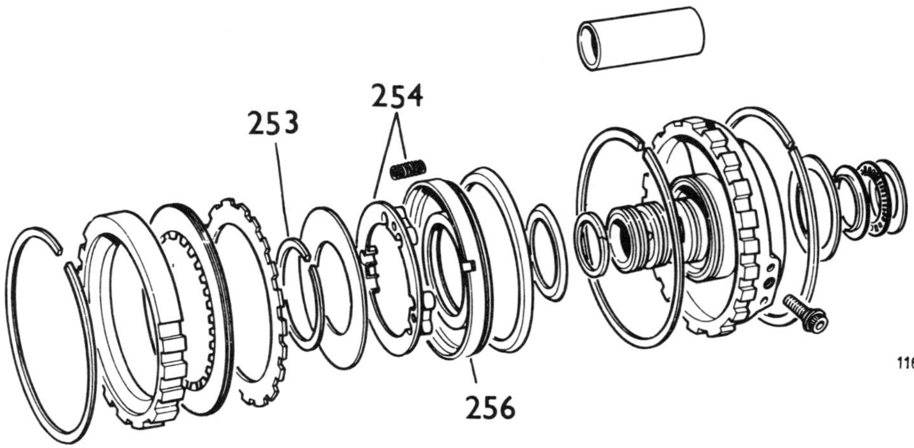

Inspection
258 Check the roller clutch inner race for wear or damage. Ensure that the lurication hole is clear.
259 Check bushes for wear, scoring and chafing.
260 Check the oil ring grooves for wear or damage.
261 Using compressed air, check oil passages and clear any obstructions.
262 Check the piston sealing surfaces for scratching and piston seal grooves for damage.
263 Check piston for cracks and seals for wear or damage.
264 Check the springs for distortion.

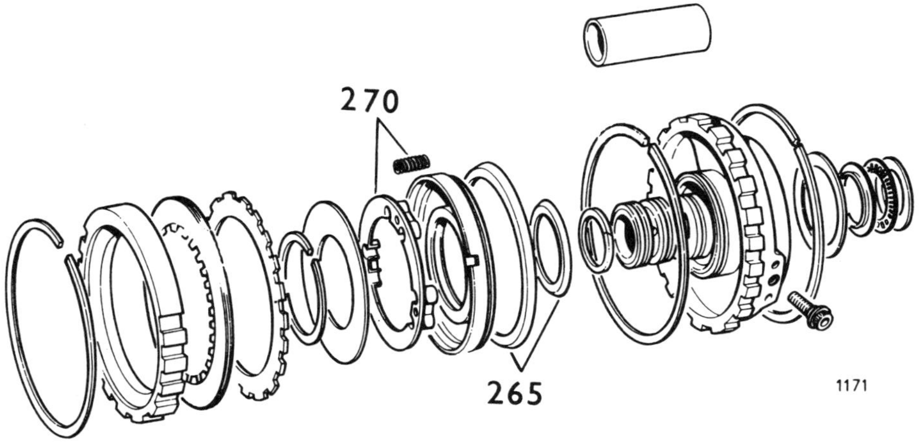

Reassembling
265 Lubricate the new inner and outer clutch piston seals with clean transmission fluid.
266 Lubricate the seal grooves in the intermediate clutch piston and fit the seals to the piston, with the lips facing away from the spring guide.
267 Fit 18G 1309 intermediate clutch oil seal protector sleeve over the centre support hub, fit the intermediate clutch piston to the centre support. Ensure that it seats fully.
268 Remove service tool 18G 1309.
269 Fit the spring guide.
270 Fit the three clutch release springs, equally spaced in the holes in the spring guide.
271 Fit the spring retainer and snap-ring.
272 Compress the spring retainer, ensuring that the retainer does not foul in the snap-ring groove; fit snap-ring.
273 Fit the four Teflon oil seal rings to the centre support.
274 Using compressed air, check the operation of the intermediate clutch. Apply air to the centre oil feed hole to activate the piston.

GEARBOX ASSEMBLY

Reassembling

275 Fit the parking pawl, tooth towards the centre of the transmission case, and fit the parking pawl shaft.
276 Fit the parking pawl shaft retaining clip.
277 Tap the parking pawl shaft plug into position, using a ⅜ in diameter rod, until the pawl shaft contacts the case rib.
278 Fit the parking pawl return spring, square end to pawl.
279 Fit the parking pawl bracket and secure with the two bolts.
280 Check the rear brake band for distortion, cracks, damage to the ends of the anchor lugs and apply lugs. Also check the lining for cracks, flaking, burning and looseness.
281 Fit the rear band assembly to the transmission case, locating the band lugs with the anchor pins.
282 Fit the rear unit thrust washer, the correct size having been determined in the Rear Unit End-Float Check, see 44.30.23. Engage the lugs of the washer with the slots in the transmission case.
283 Lubricate the pinion gears in the reaction carrier with clean transmission fluid and fit the reaction carrier to the output carrier; engage the pinion gears with the front internal gear.
284 Fit the large diameter race, flanged outer edge facing outwards, to the sun gear.
285 Fit the thrust bearing to the race.
286 Fit the small diameter race, flanged inner edge facing inwards, to the thrust bearing.
287 Lubricate the reaction carrier to centre support thrust washer with petroleum jelly and fit the washer to the recess in the centre support.
288 Fit the roller clutch to the reaction carrier.
289 Fit the centre support assembly to the roller clutch in the reaction carrier.
 NOTE: Ensure that the centre support to reaction carrier thrust washer is correctly positioned before fitting the centre support to the roller clutch in the reaction carrier. With the reaction carrier held, the centre support should only rotate in an anti-clockwise direction.
290 Fit the gear unit with centre support and reaction carrier to the transmission case. Align the centre support bolt hole with the hole in the casing.
291 Lubricate and fit the centre support to case snap-ring, flat face against centre support. Locate the gap adjacent to the front band anchor pin.
292 Fit the centre support to case bolt.
293 Check the intermediate clutch plates for scoring, wear and signs of burning.
294 Lubricate the three steel and three composition clutch plates with clean transmission fluid.
295 Fit the clutch plates commencing with the waved steel plate and alternate composition and steel plates.
296 Fit the intermediate clutch backing plate, flat machined face against clutch plates.
297 Fit the backing plate to case snap-ring, locate the ring gap adjacent to the front band anchor pin.
298 Re-check the Rear Unit End-Float, see 44.30.23.
299 Check the front band for cracks and distortion damage to the ends of the anchor lugs and apply lugs. Also check the lining for cracks, flaking, burning and looseness.
300 Fit the front band, aligning the band anchor hole and the band anchor pin, with the apply lug facing the servo hole.
301 Fit the direct clutch housing and intermediate roller assembly. Ensure that the clutch housing hub locates on the bottom of the sun gear shaft and that the splines on the forward end of the sun gear shaft are flush with those in the direct clutch housing.
302 Fit the forward clutch hub to the direct clutch housing thrust washer, to the forward clutch hub.
303 Fit the forward clutch assembly and turbine shaft. Ensure that the end of the mainshaft locates fully in the forward clutch hub. The distance between the forward clutch and pump mounting face should be 1 to 1¼ inches (25,4 to 31,8 mm).
304 Lubricate the turbine shaft journals and Teflon oil rings on the pump delivery sleeve.
305 Fit a new outer seal.
306 Fit a new gasket to the pump.
307 Fit the pump to the gearbox casing and secure with the bolts.
308 Re-check the Front Unit End-Float, see 44.30.22.
309 Fit a new manual shaft lip seal to the transmission case; use a 0.75 in (19 mm) diameter rod to seat the seal.
310 Fit the actuator rod to the manual detent lever from the side opposite the pin.
311 Fit the actuator rod plunger under the parking bracket and over the parking pawl.
312 Fit the manual shaft to the case, and insert through the detent lever.
313 Fit and tighten the locknut to the manual shaft.
314 Fit the retaining pin.
315 Fit a new extension housing gasket.
316 Check the 'O' ring on the output shaft for nicks and flattening, and renew as required.
317 Fit the extension housing to the case and secure with the six bolts.
318 If required, fit a new extension housing oil seal.
319 Fit the six check balls into their seat pockets in the casing.
320 Using two guide pins in the smaller diameter holes in the valve block casing, fit the control valve housing spacer plate-to-case gasket, 'C' towards case.
321 Fit the control valve spacer plate.
322 Fit the detent solenoid gasket.
323 Fit the detent solenoid assembly, with the connector facing the outer edge of the casing. Do not tighten the bolts.
324 Fit the front servo spring and spring retainer to the casing.
325 Fit the retaining ring to the front servo pin and fit the pin to the case, tapered end to contact band.
326 Fit the servo piston to the pin.
327 Fit a new 'O' ring to the solenoid connector.
328 Fit the connector, locate lock tabs to the connector terminal.
329 Connect the detent solenoid wire to the connector.
330 Lubricate the rear servo inner and outer bores. Fit the rear accumulator spring.
 NOTE: Ensure that the rear band apply lug aligns with the servo pin.
331 Fit the rear servo assembly, ensure proper sealing in the bore, and fit the rear servo cover and gasket. Secure with the six bolts; 'VB' to valve block.
332 Fit the control valve housing assembly-to-spacer gasket.
333 Fit the governor pipes to the control valve assembly.
334 Fit the governor screen assembly, open end first to the feed pipe hole in the casing.
335 Fit the control valve assembly and governor pipes to the transmission, carefully align the governor feed pipe with the screen. Ensure that all gaskets and spacers are correctly positioned.
 NOTE: Ensure that the manual valve properly locates with the pin on the detent lever. Check that the governor pipes are located correctly.
336 Fit the securing bolts.
337 Remove the two guide pins and fit the detent roller spring assembly and remaining bolts.

continued

338 Tighten the detent solenoid attachment screws.
339 Fit the modulator valve, stem end outermost, into the case.
340 Fit a new 'O' ring to the vacuum modulator.
341 Fit the vacuum modulator to the case.
342 Fit the modulator retainer, curved face inboard, fit and tighten the attachment bolt.
343 Fit the governor to the case.
344 Fit a new gasket, and secure the governor cover to the case with the four bolts.
345 Fit the speedometer driven gear assembly and secure with the clamp bolt.
346 Fit a new 'O' ring to the intake pipe and fit the pipe to a new filter assembly.
347 Fit the filter and pipe assembly to the casing.
348 Fit and tighten the filter retaining bolt.
349 Fit a new gasket to the oil pan and fit the pan to the casing, secure with the attaching bolts.
350 Fit the torque converter to the turbine shaft fully engage the converter drive hub slots with the pump olive gear lugs.

REAR EXTENSION HOUSING 44.20.15

Remove and refit

Service tool: MS 53 A – engine support bracket

Removing
1 Drive the vehicle onto a ramp.
2 Remove the transmission dipstick.
3 Unscrew and remove the bolt securing the dipstick upper tube to the lifting eye bracket.
4 Remove the dipstick upper tube.
5 Slacken the wing stay to bulkhead securing bolt.
6 Remove the wing stay to wing securing bolts.
7 Remove the pipe to wing stay clamps.
8 Swing the wing stays away from the wings.
9 Unscrew and remove the handles from the engine lifting hooks – Tool No. MS 53 A.
10 Fit the hooks to the rear lifting eyes.
11 Fit the engine support tool.
12 Fit and tighten the handles to take the weight of the engine.
13 Raise the ramp.
14 Unscrew and remove the nuts/bolts securing the intermediate exhaust pipes, rotating the flanges for access.
15 Disconnect the exhaust pipes and remove the sealing olives.
16 Remove the intermediate heat shields.
17 Remove the rear heat shields.
18 Pull aside the exhaust pipes and secure.
19 Remove the front heat shields.
20 Unscrew and remove the rear mounting centre nut.
21 Remove the spacer.
22 Using a suitable block of wood interposed between the jack head and the gearbox rear mounting, support the mounting plate.
23 Remove the bolts securing the rear mounting.
24 Remove the rear spacers.
25 Lower the jack.
26 Remove the mounting assembly.
27 Remove the wooden block and jack.
28 Unscrew the bolts securing the cross-member.

29 Remove the cross-member.
30 Disconnect the propeller shaft from the transmission and move the shaft clear.
31 Remove the drive flange retaining bolt, and remove the flange.
32 Remove the extension housing bolts, remove the housing and discard the gasket.

Refitting
33 Fit a new gasket to the extension housing.
34 Refit the extension housing to the transmission case, secure with the bolts.
35 Refit the drive flange and tighten retaining bolt, ensure that the propeller shaft bolts are fitted to the flange.
36 Connect the propeller shaft to the transmission drive flange.
37 Position and align the cross-member, fit and tighten the upper securing bolts.
38 Fit and tighten the lower securing bolts.
39 Place the ramp jack under the wooden mountings, and locate the wooden block and mounting assembly in position, raise the jack and align the attachment holes.
40 Fit the rear spacers, fit and tighten the securing bolts.
41 Remove the jack and wooden block.
42 Fit the rear mounting spacer and centre nut.
43 Position the right-hand catalyst (where fitted) into the manifold and secure with the nuts.
44 Secure the converter cover strap to the catalyst (where fitted).
45 Refit the rack gaiter heat shield (where fitted).
46 Refit the front heat shields.
47 Untie the exhaust pipes and fit the rear heat shield.
48 Refit the intermediate heat shields.
49 Smear the exhaust sealing olives with 'Firegum', fit the olives, connect and secure the exhaust system.
50 Lower the ramp.
51 Unscrew and remove the support tool hook handles.

52 Remove the support tool, remove the hooks, fit and tighten the handles to the hooks.
53 Refit the dipstick upper tube and secure to the lifting eye bracket.
54 Reposition the wing stays and secure to the wings.
55 Tighten the wing stay/bulkhead attachment and refit the pipe clamps.
56 Fill the transmission unit with fluid and refit the dipstick.

OIL/FLUID PAN 44.24.04

Remove and refit

Removing
1 Raise the vehicle on a ramp.
2 Remove the vacuum capsule clamp bolt and clamp.
3 Disconnect the capsule and drain the oil into a suitable container.
4 Reconnect the capsule.
5 Refit the clamp and tighten the clamp bolt.
6 Remove ten of the sump bolts, leaving two at the front and one at the rear to secure the sump while a suitable container is placed below the sump.
7 Slowly slacken the remaining sump bolts, allowing the oil to drain into the container.
8 Remove the remaining sump bolts and carefully remove the sump.
9 Drain the oil from the sump.
10 Remove and discard the sump gasket.
11 Clean the gasket faces and the sump.

Refitting
12 Fit a new sump gasket.
13 Fit the sump and replace the securing bolts.
14 Lower the vehicle on the ramp.
15 Refill the gearbox.

44–29 GM

OIL FILTER

Remove and refit 44.24.07

Removing
1. Remove the oil/fluid pan, see 44.24.04.
2. Remove the oil filter securing bolt.
3. Remove the oil filter.

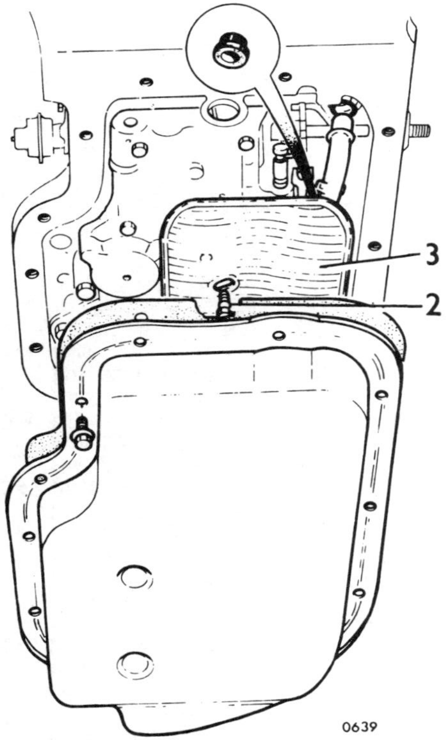

Refitting
4. Replace the oil filter.
5. Refit and tighten the filter securing bolt.
6. Replace the oil/fluid pan, see 44.24.04.

SELECTOR CABLE

Adjustment 44.30.04

1. Remove the centre console.
2. Slacken the locknuts on the outer cable at the abutment bracket.
3. Position the gearbox selector lever in neutral.
4. Position the centre console selector lever in 'N'.
5. Tighten the locknuts, ensure that the selector levers do not move.
   NOTE: A certain amount of selector lever free play should be evident.

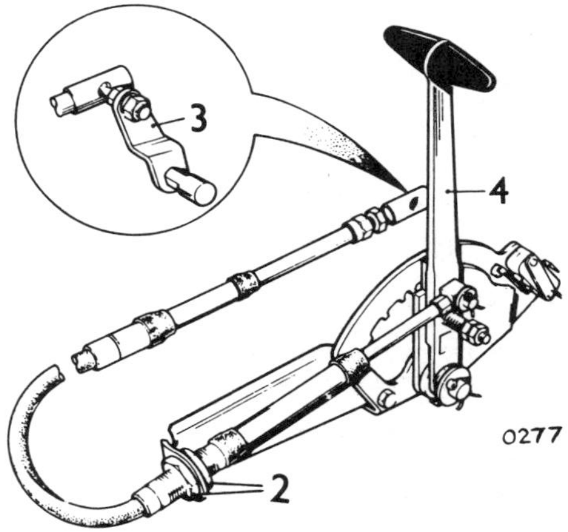

KICK-DOWN SWITCH

Check and adjust 44.30.12

1. Switch on the ignition and check that there is current at the input terminal of the switch (cable colour – green).
2. Connect an earthed test lamp to the output terminal (cable colour – green/white).
3. Fully depress the throttle pedal.
4. If the test lamp fails to light, release the throttle pedal and gently depress the switch arm.
5. If the test lamp still does not light, renew the switch.
6. If, however, the lamp lights when the switch arm is depressed, slacken the securing bolts and move the switch towards the cable until at full throttle opening the lamp lights.
7. Tighten the securing bolts and re-check.

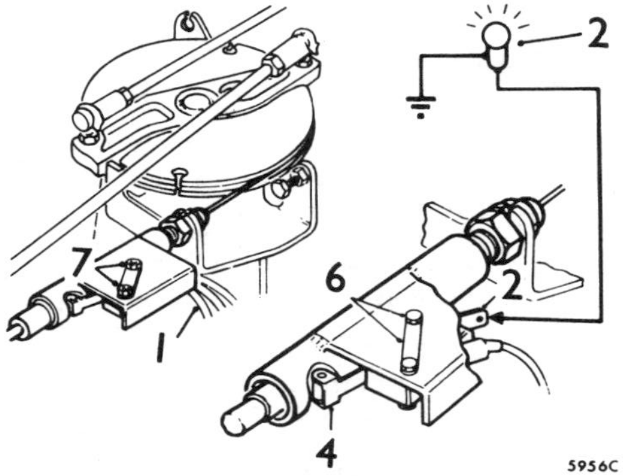

BAND APPLY PIN

Selection check 44.30.21

Service tools: 18G 1310, band apply selection gauge and gauge pin; Torque wrench

1. Remove the fluid pan and allow the fluid to drain.
2. Remove the control valve assembly and governor pipes.
3. Remove the six rear servo cover fixing bolts, remove the cover and gasket.
4. Remove the rear servo assembly from the transmission case.
5. Remove the servo accumulator spring.
6. Fit service tool 18G 1310, band apply selection gauge and gauge pin, secure with two bolts.
7. Tighten the two bolts, ensure that the gauge pin is free to move up and down in both the tool and the servo pin bore.
8. Fit a ⅝ in A.F. socket to the torque wrench.
9. Apply a torque of 25 lbf ft (3.46 kgf m) to the hexagon nut on the gauge.
10. Identify the land and letter on the gauge pin and select the appropriate size pin.
11. Remove the service tool 18G 1310.
12. Fit the servo pin as selected in the above check.
13. Refit the servo accumulator spring and servo assembly.
14. Refit the servo cover and gasket, secure with the six bolts.
15. Refit the governor pipes and control valve assembly.
16. Refit the fluid pan.
17. Refill the box with the appropriate transmission fluid.

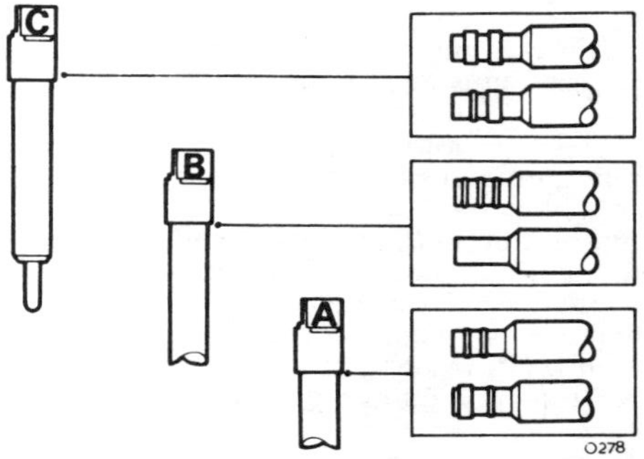

44–30 GM

FRONT UNIT END-FLOAT

Check and adjust 44.30.22

Service tools: CBW 33 (or CBW 87), 18G 1269

Thickness	Colour
0,060 to 0,064 in (1,52 to 1,63 mm)	Yellow
0,071 to 0,075 in (1,803 to 1,905 mm)	Blue
0,082 to 0,086 in (2,08 to 2,18 mm)	Red
0,093 to 0,097 in (2,36 to 2,46 mm)	Brown
0,104 to 0,198 in (2,64 to 2,74 mm)	Green
0,015 to 0,119 in (2,92 to 3,02 mm)	Black
0,126 to 0,130 in (3,20 to 3,30 mm)	Purple

Checking

1. Remove the gearbox, see 44.20.01.
2. Remove the torque convertor.
3. Remove the front pump attaching bolt and seal.
4. Attach clock gauge CBW 33; the end-float can now be checked. Alternatively, CBW 87 can be used, the movement of the turbine shaft being measured with feeler gauges.
5. Hold the output shaft forward, whilst pushing the turbine shaft rearward to its stop.
6. Set dial gauge to zero.
7. Pull the turbine shaft forward and note the reading obtained. Correct end-float is 0.003 to 0.025 in (0,076 to 0,610 mm).

The selective washer which controls the end-float is a phenolic resin washer, located between the pump cover and the forward clutch housing.
If the end-float is not within the above limits (preferably work to a mean tolerance reading between the above), select a new washer, referring to the chart.

NOTE: An oil-soaked washer may lead to discolouration. If necessary, measure the washer to ascertain the thickness.

8. To remove the pump, remove all the locating screws, removing them diagonally opposite each other.
9. Insert 18G 1269 into the two tapped holes in the pump body.
10. Apply a gradual equal force on each bolt until the pins force the pump out.
11. Fit the correct selective washer.
12. Reverse operations 10, 6 and 3.

REAR UNIT END-FLOAT

Check and adjust 44.30.23

Service tool: CBW 33

Thickness	Identification Notch
0,008 to 0,082 in (1,981 to 2,083 mm)	None
0,086 to 0,090 in (2,184 to 2,286 mm)	On side of 1 tab
0,094 to 0,098 in (2,358 to 2,489 mm)	On side of 2 tabs
0,102 to 0,106 in (2,591 to 2,692 mm)	On end of 1 tab
0,110 to 0,114 in (2,794 to 2,896 mm)	On end of 2 tabs
0,118 to 0,122 in (2,997 to 3,099 mm)	On end of 3 tabs

Checking

1. Remove the rear extension, see 44.20.19.
2. Fit a suitable extractor bolt to the end casing to enable a clock gauge to be fitted, or use CBW 33 with the slide bar inserted into the side of the governor holes (governor bolt removed).
3. Mount a dial gauge test indicator onto the bolt, or slide the connector on CBW 33 as near the block as possible and extend the dial gauge rod out as far as possible.
4. Ensure the indicator stem registers with the end of the output shaft.
5. Set gauge to zero.
6. Move the output shaft to and fro, noting the indicator reading to enable the correct end-float adjusting washer to be used when the transmission is assembled.

The end-float should be between 0.003 and 0.019 in (0,076 and 0,483 mm).
The adjusting washer, which controls this end-float, is the steel washer with three tabs, located between the thrust washer and the rear face of the transmission case. The notches on the tabs serve to identify washer thickness.
Select the correct washer from the table.

44—31 GM

SPEEDOMETER DRIVE PINION
Remove and refit   44.38.04

Removing
1. Raise the vehicle on a ramp.
2. Slacken the union connecting the speedometer cable angle drive to the pinion.
3. Disconnect the speedometer cable and place to one side.
4. Remove the pinion clamp bolt and clamp plate.
5. Remove the pinion assembly.
6. Remove the pinion from the housing.
7. Remove and discard the housing seals.
8. Clean the pinion and the housing.

Refitting
9. Refit new seals to the housing.
10. Lubricate the pinion.
11. Lubricate the 'O' ring seal.
12. Refit the pinion to the housing.
13. Refit the pinion assembly to the gearbox.
14. Refit the clamp plate and tighten the clamp bolt.
15. Reconnect the speedometer cable.
16. Re-tighten the union connecting the speedometer cable angle drive to the pinion.
17. Lower the vehicle on the ramp.

VALVE BODY ASSEMBLY
Remove and refit   44.40.01

Removing
1. Remove the oil/fluid pan, see 44.24.04.
2. Remove the oil filter, see 44.24.07.
3. Disconnect the pressure switch.
4. Remove the bolts securing the valve body.
5. Remove the detent spring.
6. Remove the valve body assembly.
7. Remove and discard the gasket.
8. Remove the conical filter.
9. Remove the oil feed pipes.
10. Remove the front servo piston assembly.
11. Slacken and remove the pressure switch.
12. Place the valve body to one side.
13. Clean all the relevant parts and faces.

Refitting
14. Refit and tighten the pressure switch.
15. Lubricate the front servo piston.
16. Refit the servo piston assembly.
17. Refit the oil feed pipes.
18. Refit the conical oil filter.
19. Fit a new valve block gasket and refit the valve body assembly.
20. Align the oil feed pipes.
21. Align the front servo.
22. Refit the detent spring.
23. Refit and tighten valve body securing bolts.
24. Re-connect the pressure switch.
25. Refit the oil filter, see 44.24.07.
26. Refit the oil/fluid pan, see 44.24.04.

44—32 GM

DESCRIPTION

TORQUE CONVERTER

The torque converter is of the three element, single phase type. The three elements are: Impeller, connected to the engine crankshaft; Turbine, connected to the gearbox input shaft, and Stator, mounted on a one-way clutch on the stator support projecting from the gearbox case. The converter provides torque multiplication of from 1:1 to 2:1 and the speed range during which this multiplication is obtained varies with the accelerator position.

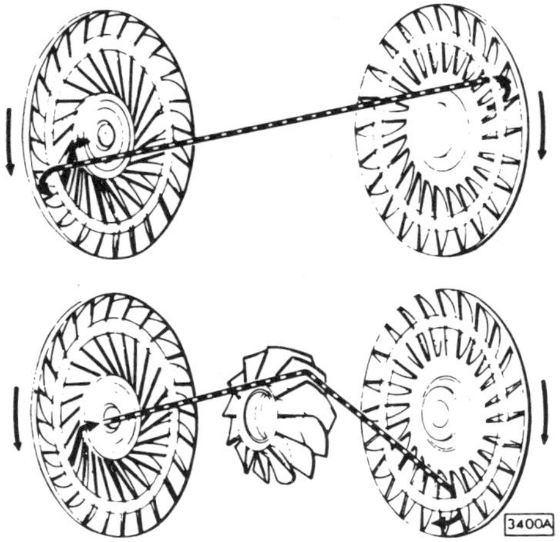

CLUTCHES

The gearbox input shaft is connected to the torque converter turbine at the front end and is therefore known as the turbine shaft. The rear end of the shaft is connected to the front and rear clutches (the clutches are of the multi-disc type operated by hydraulic pressure). Engagement of the front clutch connects the turbine shaft to the forward sun gear. Engagement of the rear clutch connects the turbine shaft to the reverse sun gear.

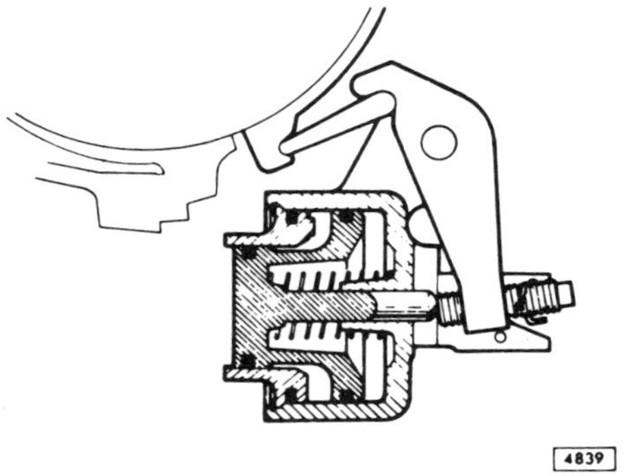

BRAKE BANDS

The brake bands operated by hydraulic servos, are used to hold drive train components stationary in order to obtain low, intermediate and reverse gears. The front band is clamped around the reverse sun clutch outer drum to hold the reverse sun gear stationary. The rear band is clamped around the planet carrier to hold the planet carrier stationary.

ONE-WAY CLUTCH

The 'one-way' clutch is situated between the planet carrier and the gearbox case. Rotation of the planet carrier against engine direction is prevented so providing the reaction member for low gear (drive). Rotation of the planet carrier in engine direction is allowed (free-wheeling) providing smooth changes from low to intermediate and intermediate to low gears.

MECHANICAL POWER FLOWS

Neutral and Park

In neutral the front and rear clutches are off, and no power is transmitted from the converter to the gear set. The front and rear bands are also released. In 'P' the Rear Servo circuit is pressurised while the engine is running, so that the rear band is applied, and a mechanically operated pawl is engaged with the ring gear.

First Gear ('D' selected)

The front clutch is applied, connecting the converter to the forward sun gear. The one-way clutch is in operation, preventing the planet carrier from rotating anti-clockwise. When the vehicle is coasting the one-way clutch over-runs and the gear set free-wheels.

First Gear ('1' selected)

The front clutch is applied, connecting the converter to the forward sun gear. The rear band is applied, holding the planet carrier stationary. The reverse sun gear rotates freely in the opposite direction to the forward sun gear.

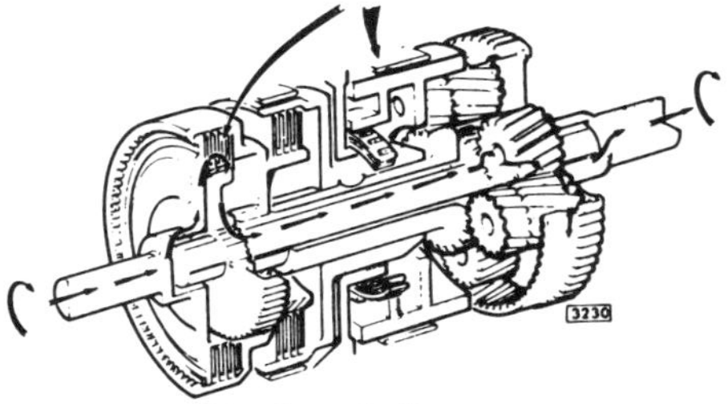

'1' selected; first gear

Second Gear ('2' or 'D' selected)

Again the front clutch is applied, connecting the converter to the forward sun gear. The front band is applied, holding the reverse sun gear stationary.

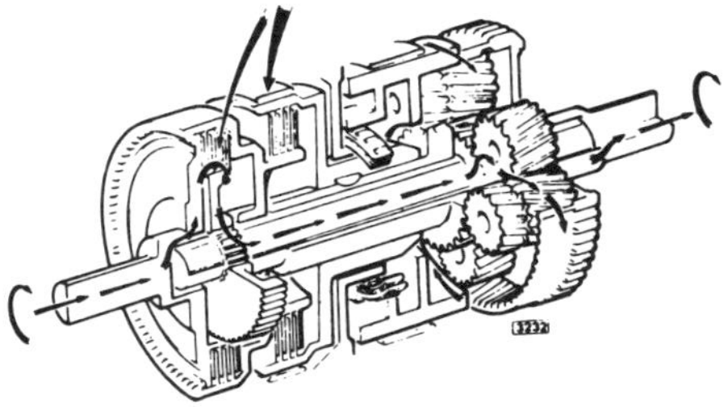

'2' or 'D' selected; second gear

'D' selected; first gear

Front servo operation

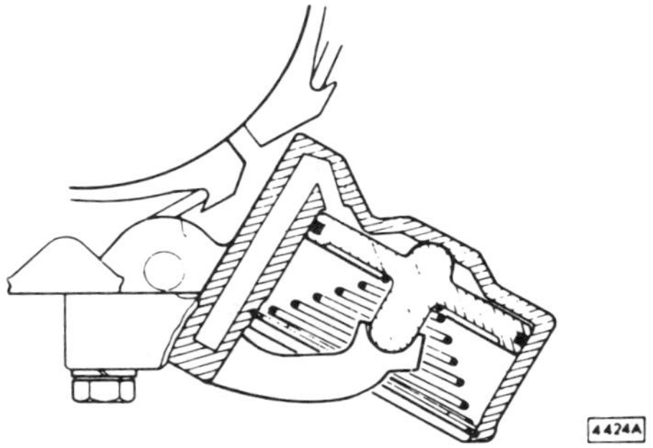

Rear servo operation

Torque converter – principle of operation

GEAR SET

The planetary gear set consists of two sun gears, two sets of pinions, a pinion carrier and a ring gear.
Power enters the gear set via the two sun gears, the forward sun gear driving in forward gears, the reverse sun gear driving in reverse gear. The ring gear, attached to the output shaft, is the driven gear. The planet wheels connect driving and driven gears, two sets of planet wheels being used in forward gears and one set in reverse.
The planet carrier locates the planet wheels relative to sun and ring gears, also serving as a reaction member.

Automatic Transmission–Model 12   44–1

GENERAL DATA

Gear Train End Float	0.008 to 0.044 in (0,20 to 1,12 mm)
Pinion End Float	0.010 to 0.020 in (0,25 to 0,51 mm)
Minimum Rear Clutch Plate Coning	0.010 in (0,25 mm)
One Way Clutch Spring to Lever Clearance	0.125 to 0.188 in (3,2 to 4,8 mm)
Thrust Washer Sizes	0.061 to 0.063 in (1,55 to 1,60 mm)
	0.067 to 0.069 in (1,70 to 1,75 mm)
	0.074 to 0.076 in (1,88 to 1,93 mm)
	0.081 to 0.083 in (2,08 to 2,11 mm)
	0.092 to 0.094 in (2,34 to 2,39 mm)
	0.105 to 0.107 in (2,67 to 2,72 mm)
Control Pressure at 9 to 10 in (23 to 25 cm) Hg	75^{+20}_{-5} lb in^2 ($5,27^{+1,40}_{-0,35}$ kg cm^2)
Stall speed (Normal)	$2,000 \pm 50$ r.p.m.

GEAR CHANGE SPEEDS
3.07:1 axle ratio

Throttle position	M.P.H.	K.P.H.
Light Throttle Upshifts		
1 – 2	11 – 12	18 – 20
2 – 3	28 – 31	45 – 50
Kickdown	nominal 55	89
1 – 2	97 – 100	157 – 161
2 – 3	82 – 90	132 – 145
3 – 2	23 – 30	37 – 48
3 – 1		
Full Throttle Upshifts		
1 – 2	21 – 27	34 – 43
2 – 3	62 – 70	100 – 113
Manual Shut Throttle Downshifts		
2 – 1	18 – 23	29 – 37
3 – 1	21 – 32	35 – 53
Part Throttle Downshifts		
3 – 2	19 – 23	31 – 37

Note: The figures given in this table are theoretical and actual figures may vary slightly from those quoted due to such factors as tyre wear, pressures etc.

Clutch and band application chart

A Front clutch
B Rear clutch
C Front band
D Rear band
E One-way clutch

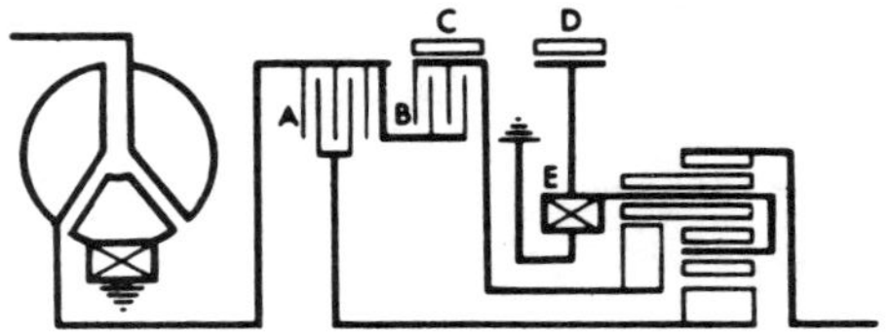

	A	B	C	D	E
1 (first gear)	●				●
D (first gear)	●				●
2 & D (sec. gr)	●		●		
D (third gear)	●	●			
R (rev. gear)		●		●	

Third Gear ('D' selected)
Again the front clutch is applied, connecting the converter to the forward sun gear. The rear clutch is applied, connecting the converter also to the reverse sun gear; thus both sun gears are locked together and the gear set rotates as a unit, providing a ratio of 1:1.

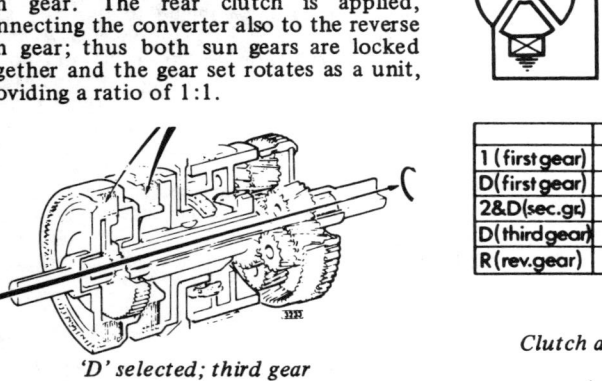

'D' selected; third gear

Reverse Gear ('R' selected)
The rear clutch is applied, connecting the converter to the reverse sun gear. The rear band is applied, holding the planet carrier stationary.

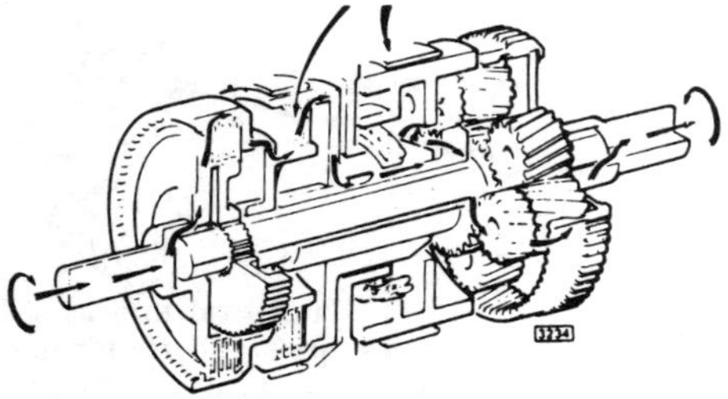

'R' selected; reverse gear

KEY TO COMPONENTS SHOWN ON HYDRAULIC CHARTS

A Torque converter
B Pump
C Front clutch
D Rear clutch
E One-way clutch
F Front servo
G Rear servo
H Primary regulator
J Secondary regulator
K Downshift valve
L Throttle valve
M Manual valve
N Throttle modulator valve
P Orifice control valve
Q 2-3 shift valve
R 1-2 shift valve
S Servo regulator timer
T Throttle modulator cut-back valve
U Governor modulator valve
V Servo regulator
W Governor
X Rear servo regulator valve

KEY TO HYDRAULIC CHART COLOUR CODE

Red line — Pump pressure
Cross hatch — To torque converter
Blue line — Governor line pressure
Yellow line — Governor modulator valve
Broken line — Kickdown – additional modulator throttle pressure
Green line — Rear servo regulated line pressure

The following spring identification table is given to assist in identifying valve springs when overhaul work is being carried out. When valve block is dismantled, springs should be compared with dimensions given. Any spring which is distorted or coil bound MUST be replaced.

DESCRIPTION	LENGTH	DIAMETER	NO. OF COILS	COL.
Primary regulator valve	2.032 in (51,6 mm)	.740 – .750 in (18,8 – 19,0 mm)	11	White
Primary regulator valve (inner)	1.557 in (39,6 mm)	.640 – .650 in (16,3 – 16,5 mm)	8	Plain
Throttle modulator valve	1.338 in (34,0 mm)	.463 – .473 in (11,7 – 12,1 mm)	11	Red
Secondary regulator valve	1.00 in (25,4 mm)	.228 – .238 in (5,7 – 6,0 mm)	17	Orange
Modulator valve (governor)	1.310 in (33,3 mm)	.443 in – .458 in (11,2 – 11,6 mm)	13	Plain
Orifice control and throttle valve	.686 in (17,4 mm)	.175 – .185 in (4,4 – 4,6 mm)	14	Pink
Downshift valve	.812 in (20,6 mm)	.178 – .188 in (4,5 – 4,7 mm)	19	Plain
Governor safety valve	.638 in (16,2 mm)	.220 – .240 in (5,6 – 6,1 mm)	12	Yellow
1 – 2 Shift valve	1.406 in (35,7 mm)	.620 – .635 in (15,7 – 16,1 mm)	8	Plain
2 – 3 Shift valve (inner)	.905 in (23,0 mm)	.140 – .150 in (3,6 – 3,8 mm)	13	Plain
2 – 3 Shift valve (outer)	1.81 in (45,9 mm)	.440 – .450 in (11,2 – 11,4 mm)	11.5	Plain
Modulator valve	1.370 in (34,8 mm)	.428 – .442 in (10,8 – 11,3 mm)	14	Plain
Servo regulator valve	1.784 in (45,2 mm)	.461 – .471 in (11,7 – 11,9 mm)	13.5	Plain
Rear servo regulator valve	1.380 in (35,0 mm)	.377 – .387 in (9,5 – 9,8 mm)	14	Plain
Servo regulator timer valve	.638 in (16,2 mm)	.220 – .240 in (5,6 – 6,1 mm)	10	Plain

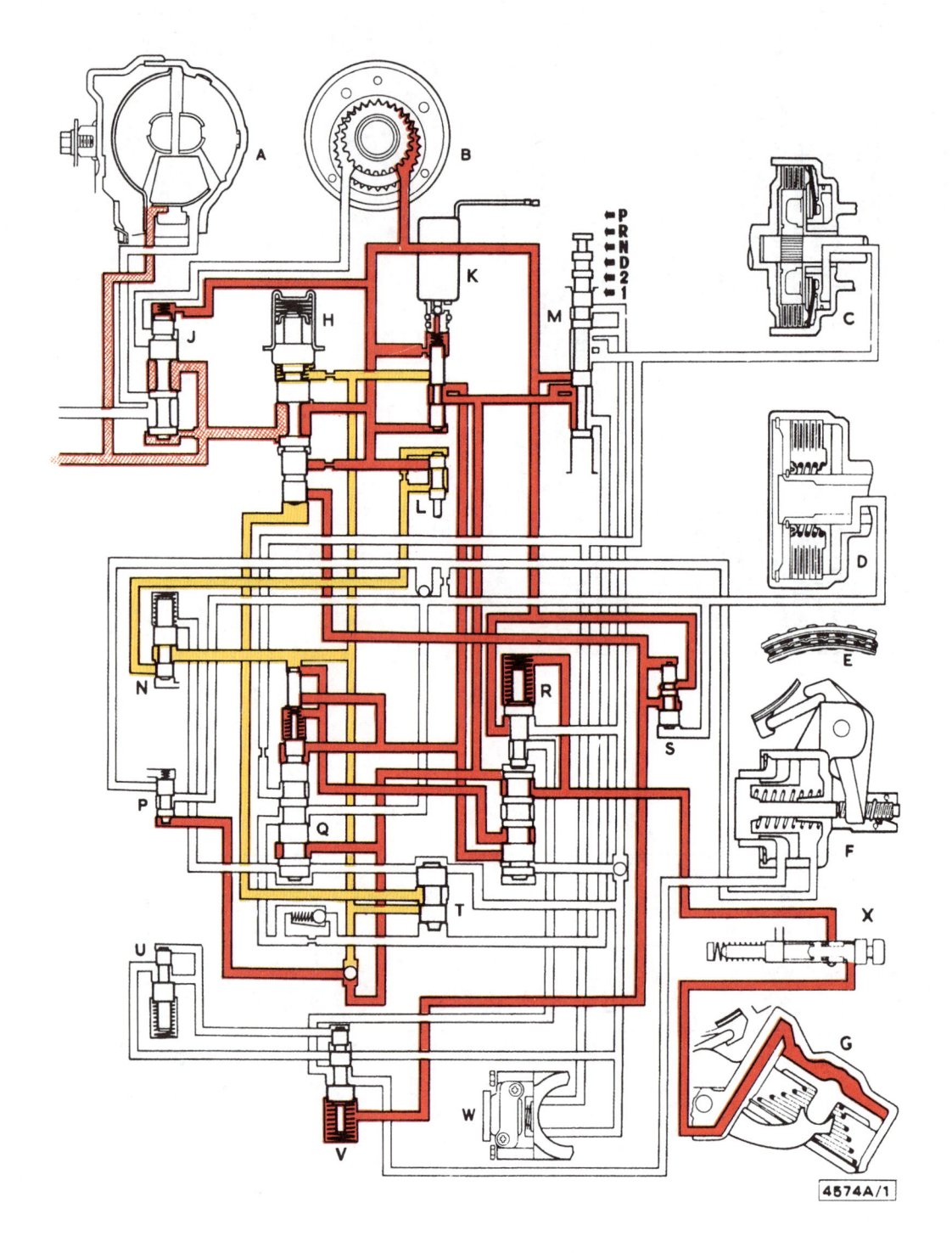

HYDRAULIC OPERATION IN 'N' (NEUTRAL)

With the engine running, the pump supplies fluid to the primary regulator which regulates line pressure.

Spill from the primary regulator supplies the torque converter and lubrication requirements. This supply is regulated by the secondary regulator.

The line pressure supplied to the manual valve is blocked by a land on the valve so that neither governor, clutches or servos are energised.

Line pressure at the throttle valve is converted to throttle pressure, dependent on manifold depression, i.e. throttle pedal position.

HYDRAULIC OPERATION IN 'P' (PARK)

Coupled to the manual valve operating lever is a linkage incorporating a pawl; movement of this lever to the 'Park' position engages the pawl with the toothed outer surface of the ring gear, so locking the output shaft to the transmission case. The rear servo is energised in 'P' selection but, as both the front and rear clutches are not energised, drive is impossible and the transmission remains inoperative.

44—4 Automatic Transmission—Model 12

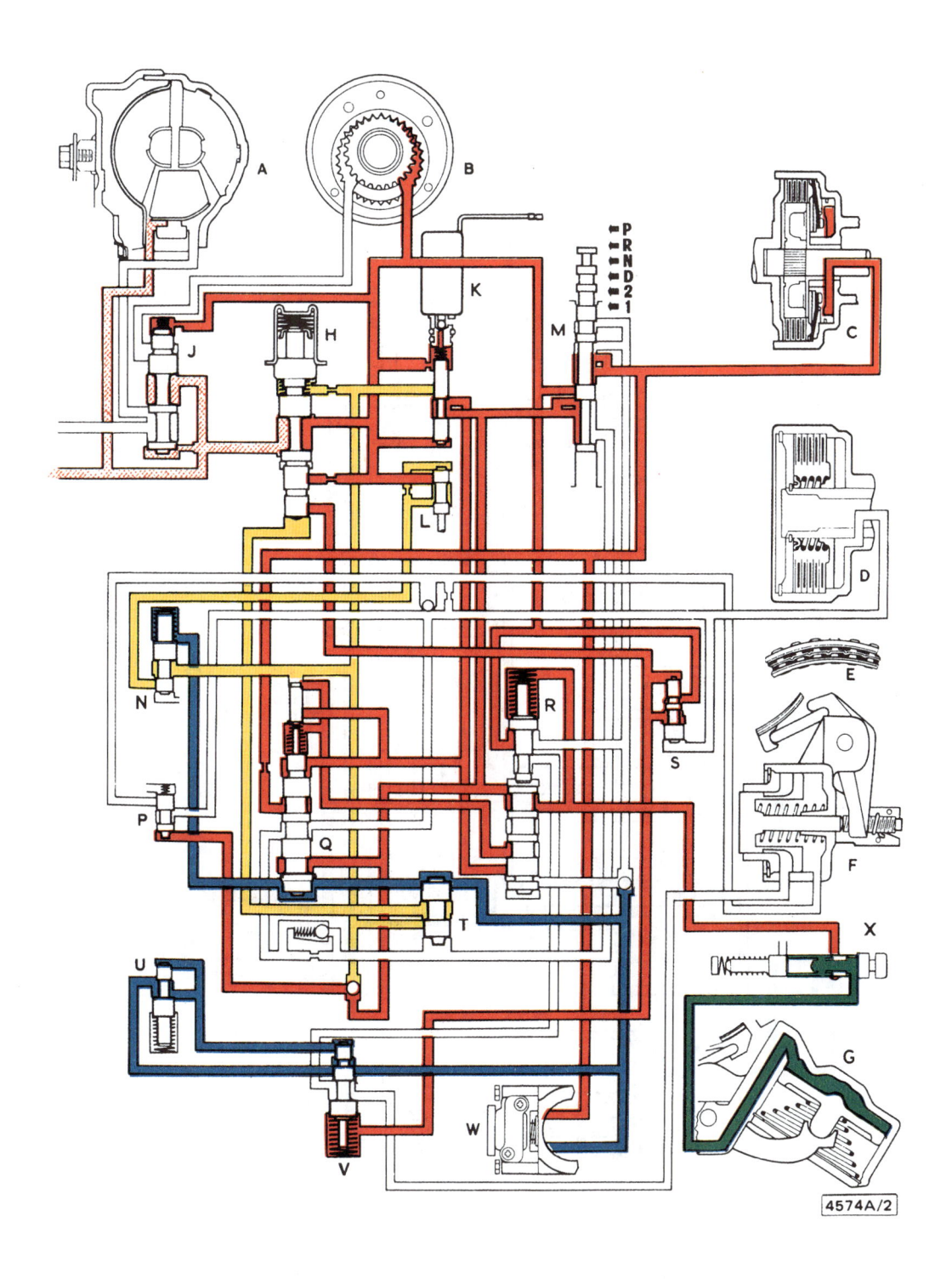

HYDRAULIC OPERATION IN 'D' (FIRST GEAR)

Movement of the manual valve to the 'D' position opens the front clutch and governor to line pressure. No other component is required to engage first gear. Line pressure is supplied to the top of the secondary regulator to control converter pressure. Throttle pressure is applied to the top and bottom of the primary regulator to modulate line pressure in the interests of shift quality.

HYDRAULIC OPERATION IN '1' (FIRST GEAR)

Application of the front clutch and rear servo are required in '1' (Manual) selection. The rear band is applied to provide engine braking. Line pressure applied to the lands of the 1-2 shift valve opposes governor pressure. There is, therefore, no upshift and the transmission remains in '1' (Low) ratio.

Automatic Transmission—Model 12   44–5

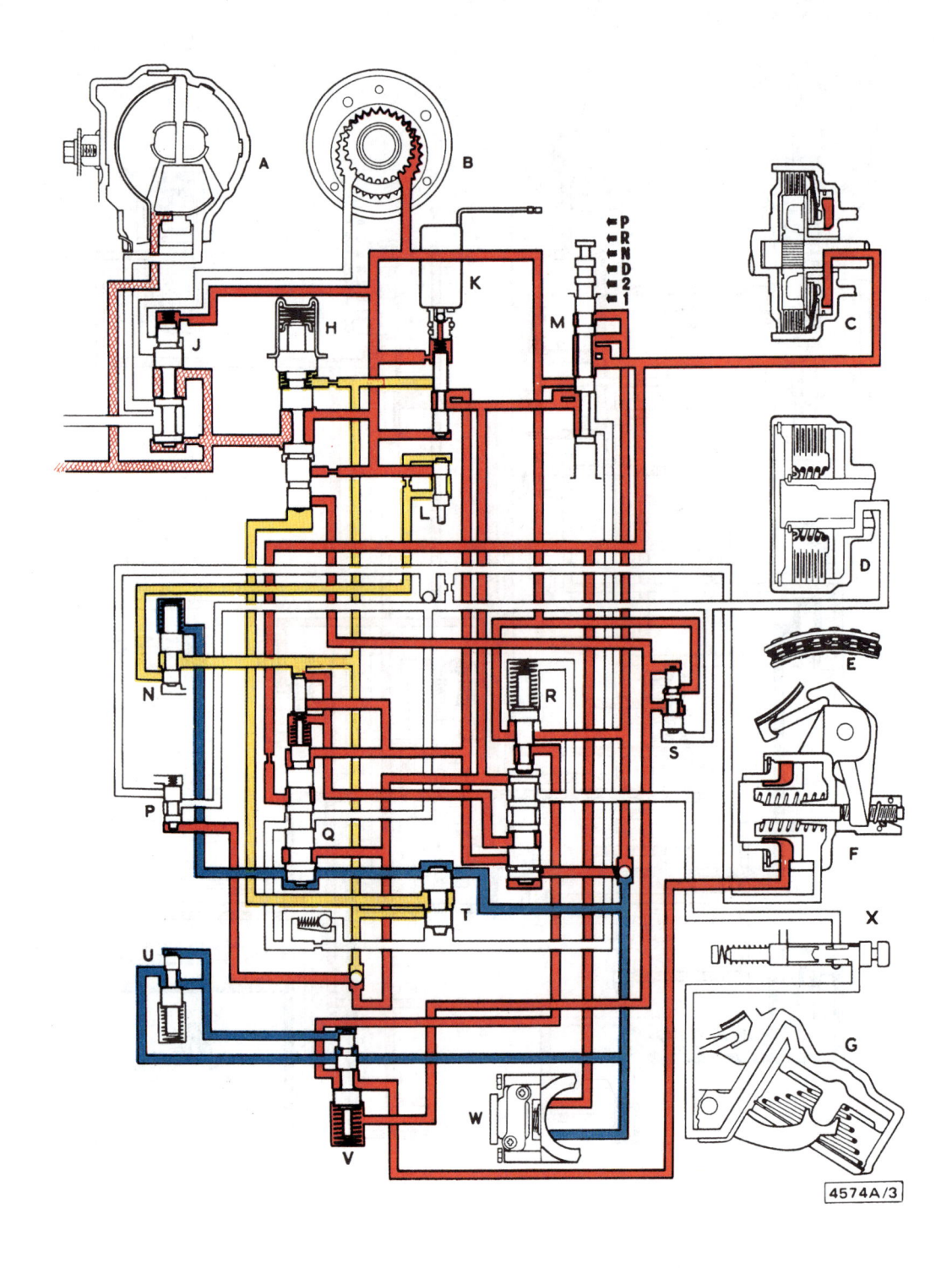

HYDRAULIC OPERATION IN 'D' (SECOND GEAR)

Increasing road speed results in a corresponding increase in governor pressure which will move the 1-2 shift valve to the 2nd gear position. The exact speed at which this change takes place is dependent upon the throttle pressure opposing governor pressure at the 1-2 shift valve.

With the 1-2 shift valve in the 2nd gear position, the line to the front servo apply side, through the servo regulator, is open to line pressure and the front band is applied.

HYDRAULIC OPERATION IN '2' (SECOND GEAR)

Once '2' (Manual) is engaged, there are no upshifts or downshifts. If either are required, a change to '1' (Manual) or 'D' must be made. Movement of this manual valve to the '2' (Manual) position allows line pressure to flow to a ball valve where it closes the governor line and introduces line pressure to the base of the 1-2 shift valve to retain it in the '2' (Intermediate) ratio.

44—6 Automatic Transmission—Model 12

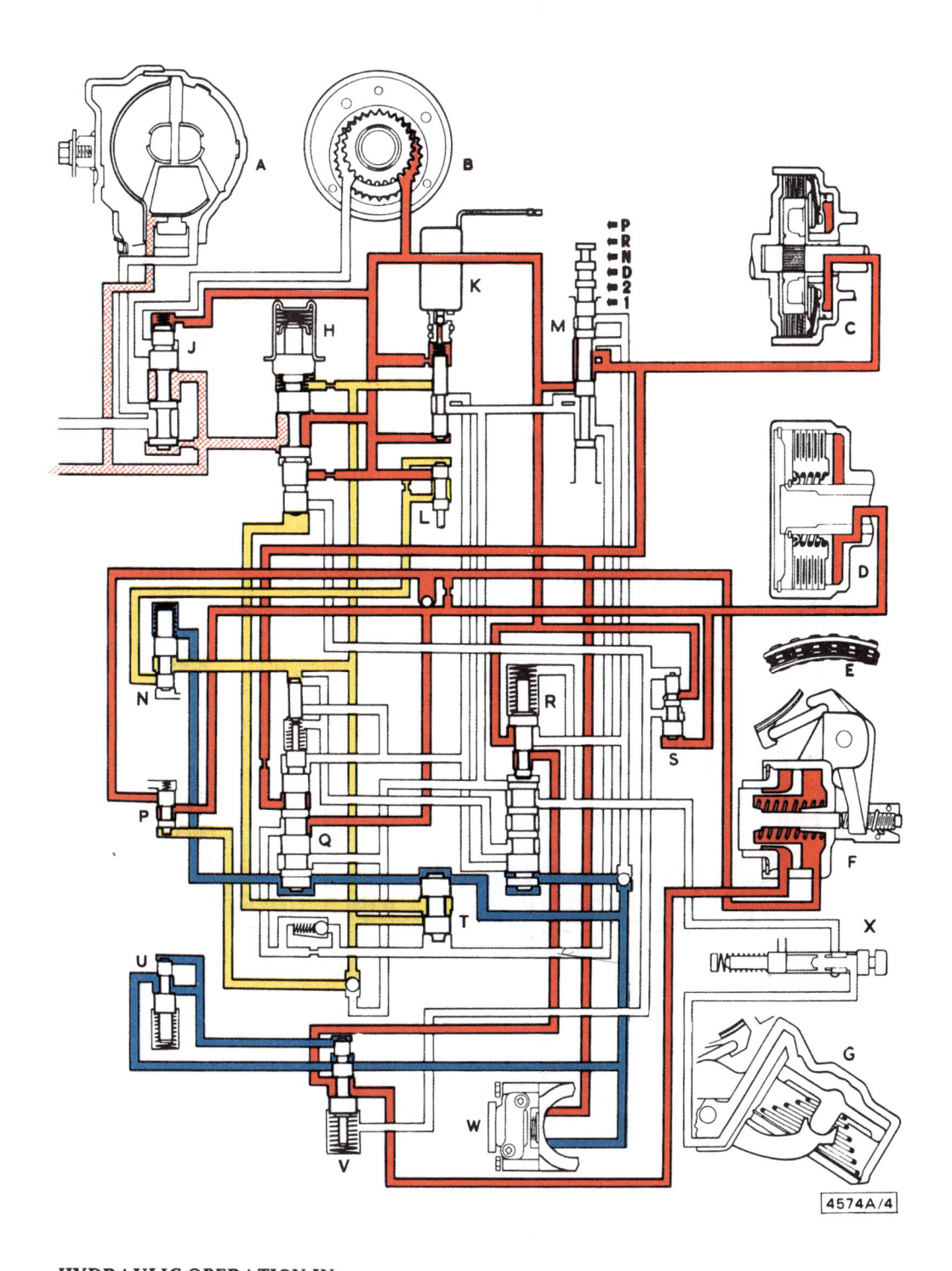

HYDRAULIC OPERATION IN 'D' (THIRD GEAR)

In order to change to third gear, the front band must be released and the rear clutch applied. With the movement of the 2-3 shift valve to the third gear position the lines to the front servo release and the rear clutch are open to line pressure.

The lines now supplied with line pressure are the front clutch, rear clutch, front servo apply and front servo release. As the front servo release has a greater area than the apply side, the front band is released, therefore the front and rear clutches remain applied.

Automatic Transmission—Model 12   44–7

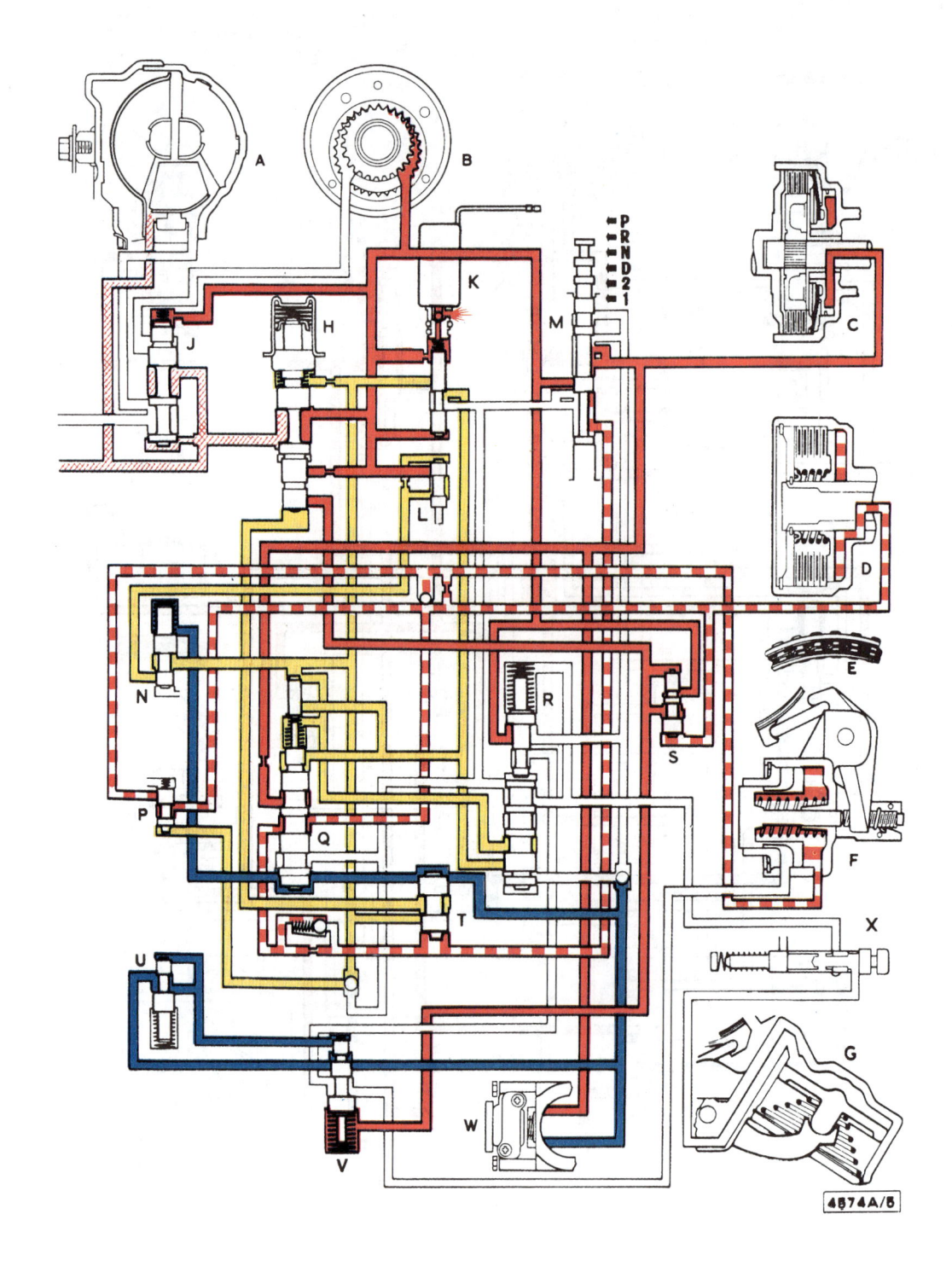

HYDRAULIC OPERATION IN 'D' (THIRD GEAR TO SECOND GEAR) KICKDOWN

Depression of the accelerator to the kickdown position actuates the kickdown switch to energise the kickdown solenoid at the valve block. With the solenoid energised, the plunger is lifted, allowing line pressure to escape past the ball valve. This sudden pressure drop allows the line pressure at the base of the downshift valve to move the valve upwards so introducing extra modulated throttle pressure to the 2-3 shift valve. This extra pressure opposing governor pressure at the 2-3 shift valve assists in overcoming governor pressure so returning the shift valve to the 2nd gear position.

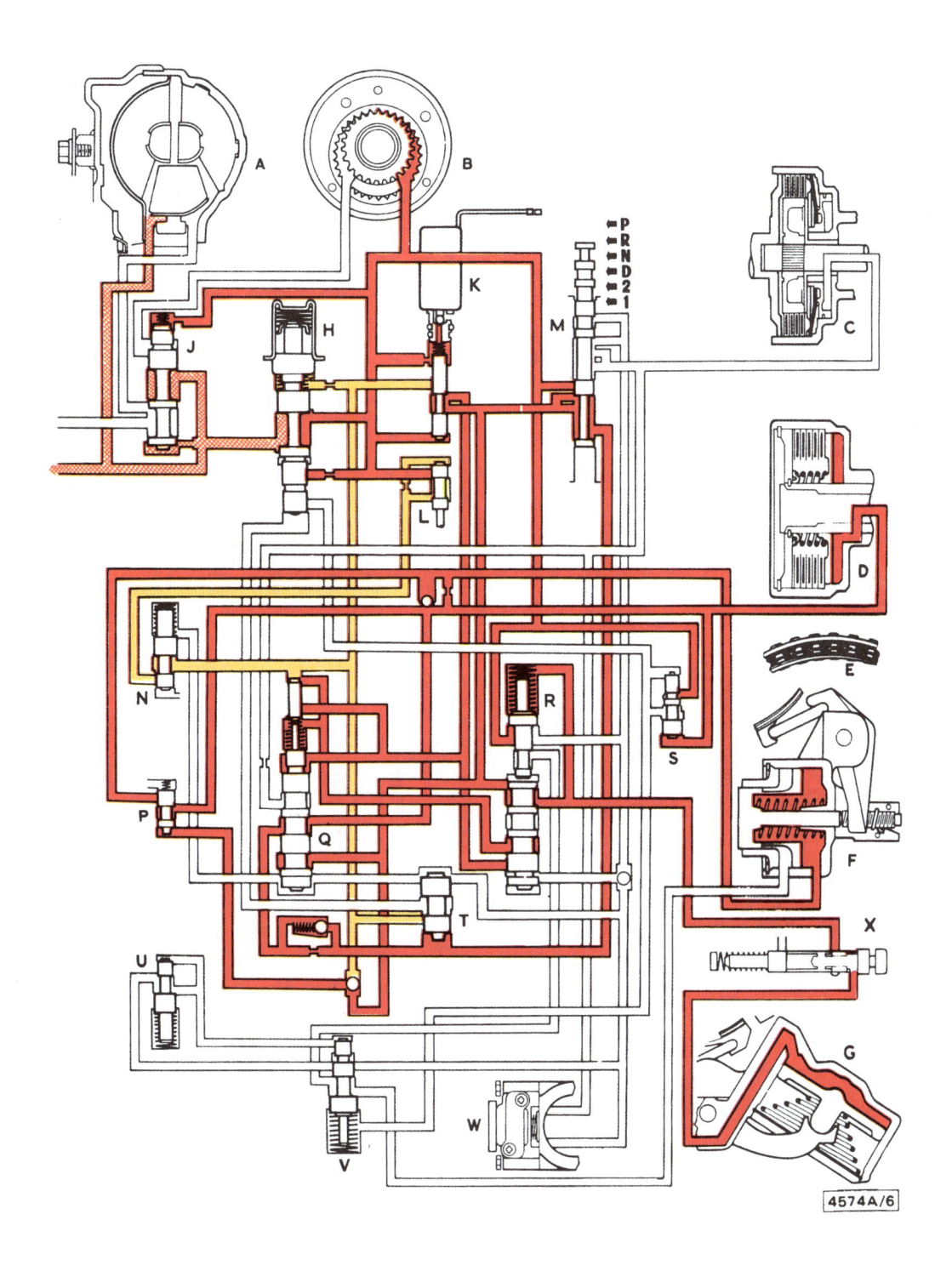

HYDRAULIC OPERATION IN 'R'

Movement of the manual valve to the reverse position closes the lines to the front clutch and governor. The rear clutch and rear servo are energised so reversing the direction of rotation of the output shaft.

As the front servo release and rear clutch are interconnected, the front servo release will also be energised. This has no effect on the operation of reverse gear.

Automatic Transmission—Model 12   44–9

ROAD TEST AND FAULT DIAGNOSIS 44.00.00

The following points should be checked before proceeding with the road test.
1. Fluid level.
2. Engine idle speed.
3. Manual lever adjustment.
4. Manifold vacuum of 9 to 10 in (23 to 25 cm) Hg.

ROAD TEST

The road speed figures for the tests listed below are to be found under 'GENERAL DATA – GEAR CHANGE SPEEDS'. Road testing should follow the complete sequence detailed below. Transmission should be at normal working temperature, i.e. after being driven on road or rollers.

1. With brakes applied and engine idling, move selector from:
   - 'N' to 'R'
   - 'N' to 'D'
   - 'N' to '2'
   - 'N' to '1'

Engagement should be felt with each selection.

2. Check stall speed.
3. Select 'D', accelerate with minimum throttle opening and check speed of first gear to second gear shift.
4. Continue with minimum throttle and check speed of second gear to third gear shift.
5. Select 'D', accelerate with maximum throttle opening (kickdown) and check speed of first gear to second gear shift.
6. Continue with maximum throttle and check speed of second gear to third gear shift.
7. Check for kickdown shift third gear to second gear.
8. Check for kickdown shift second gear to first gear.
9. Check for kickdown shift third gear to first gear.
10. Check for 'roll-out' downshift with minimum throttle, second gear to first gear.
11. Check for part throttle downshift, third gear to second gear.

FAULT DIAGNOSIS

Should a fault be apparent during road test, first identify the problem from the list printed in the Fault Diagnosis Chart. The reference numbers shown opposite each fault may be translated by reference to the page headed 'Transmission Fault Key'.

Problem	Checks in Vehicle	Checks on Bench
Engagements		
Harsh	A1, A3, A4, A5, M2, V1, V2, V3, V4	T4 (in reverse only)
Soft or delayed	M1, A2, A3, A4, A5, V1, V2, V3, V9	T14
None in all positions	V16, M1, A2, V2, C3	T9, T10, C2
No forward		
in 1 position	V16, M1, A2, V1, V2	T1, A4, T7, T14
in 2 position	V16, M1, A2, T16, T13, V1, V2, V10	T1, A4, T14
in D position	V16, M1, A2, V1, V2, A5	T1, A4, T7, T14
in all positions	V16, M1, A2, V1, V2	C2, T9, T10, T14
No reverse	V16, M1, A2, A5, T15, T6, V1, V5, V6	T2, T3, T14
Jumps in forward	A2, A3, A4, A5	T4, T7
Jumps in reverse	A2, A3, A4, A5	T2, T3
No neutral	A2, V1, V16	T2
Upshifts		
No 1-2	M1, A2, A4, G1, T5, T13, T16, V1, V2, V4, V5	T14
No 2-3	M1, A2, G1, T13, V1, V2, V4, V5, V6	T3, T14
Shift points too high	A1, A2, M2, G1, V1, V2, V4, V5, V8, V12, V14	T14
Shift points too low	A1, M3, G1, M4, V1, V5, V6, V12, V14	T14
Upshift quality		
1-2 slips or runs-up	M1, A1, A2, M3, G1, T13, V1, V2, V4, V9, V10, V5	T10, T5, T14
2-3 slips or runs-up	M1, A1, A2, A4, M3, G1, T13, V1, V2, V4, V5, V6, V9, V10, V12	T10, T5, T14
1-2 harsh	A1, A2, A4, A5, M2, V1, V2, V3, V4, V5, V9	T1, T7, T8
2-3 harsh	A1, A2, A4, M2, V1, V2, V4, V9	T4
1-2 ties-up or grabs	A4, A5, V1, V5, T16	T4, T7, T8
2-3 ties-up or grabs	A2, A4, T13, V17, T15	

continued

44-10 Automatic Transmission—Model 12

FAULT DIAGNOSIS — *continued*

Problem	Checks in Vehicle	Checks on Bench
Downshifts		
No 2-1	A1, A2, A6, M3, M4, G1, V1, V5, V14	T7
No 3-2	A1, A2, A6, M4, G1, V1, V6, V14, T13	T4
Shift points too high	A1, A2, M2, G1, V1, V4, V5, V6, V12	T14
Shift points too low	A1, A2, M3, G1, V1, V4, V5, V6, V12	T14
Downshift quality		
2-1 slides	T5, A1, A2, A4, T13, G1, V1, V2, V4, V9, V11, V18	T7
3-2 slides	A1, A2, A4, T13, G1, V1, V2, V4, V9, V11, T5	
2-1 harsh	A1, A2, A4, A5, M2, V1, V2, V3, V4, V9, V18	T1, T7, T14
3-2 harsh	A1, A2, A4, G1, V1, V2, V3, V4, V9	T3, T4
Reverse		
Slips or chatters	M1, A1, A2, A5, T6, V1, V2, V4, V10	T14, T2, T3
Line pressure		
Low idle	M1, A1, A2, A3, T13, V1, V2, G1	T10, T14
High idle	A1, M2, V1, V2, V3, V4	
Low stall	M1, A1, M3, M4, T13, G1, V1, V2, V4, V14	T10
High stall	A1, V1, V2, V3	
Stall speed		
Too low (200 rpm or more below)		C1
Too high (200 rpm or more above)	M1, A1, A2, A4, T13, V1, V4	T14, T1, T3, T6, T7, T9, T10
Others		
Transmission overheats	M1, A4, A5, V2, V3, M5, T13, V1	T1, T2, T3, T4, T5, T6, T7, T10, T14
Drag in neutral	A2, A3,	T2, T4
Poor Acceleration	M1, V2, V3	C1
Noisy in neutral	V13	T2, T4, V15

Problem	Checks in Vehicle	Checks on Bench
Noisy in park	V13, V15	T10
Noisy in all gears	V13, C3, V15	T10, C1, C2
Noisy during coast (30-20 mph)		T12
Park brake does not hold	A2, T11	
Ties up in 1 or low	A4, T15, T13, V1	T4, T14
Ties up in 2 or intermediate ratio D or 2 selected	A5, T16, A2, V1	T4, T14, T8
Ties up in direct drive	A5, T16, A2, V1, A4, M1, T15, T12	T17, T14, T8
Poor acceleration	M6, A1, A2, A4, M1, T15, T16, V1, V2	C1, T10, T14
Oil out breather	T18, M1, G1, T13, V1	T14
Oil out fill tube	T18, M1, G1, T13, M7	T14

Automatic Transmission—Model 12   44—11

GEAR SELECTOR CABLE

Remove and refit   44.15.08

Removing
1. Remove console, see 76.25.01.
2. Place quadrant selector lever in '1'.
3. Unscrew gear selector knob.
4. Remove screws securing selector indicator; withdraw indicator over selector lever.
5. Remove split pin and washer securing cable to selector lever; detach cable.
6. Unscrew front locknut securing cable to abutment bracket.
7. Lift carpet from left-hand side of transmission tunnel.
8. Remove screws securing cable shroud to transmission tunnel; withdraw shroud.
9. Withdraw cable from abutment bracket.
10. Remove screws securing access panel to transmission tunnel.
11. Withdraw panel, clean off old sealing compound.
12. Ensure gearbox selector lever is in '1'.
13. Remove nut securing selector cable to gearbox selector lever; detach cable.
14. Remove bolt and spring washer securing trunnion block.
15. Withdraw cable.

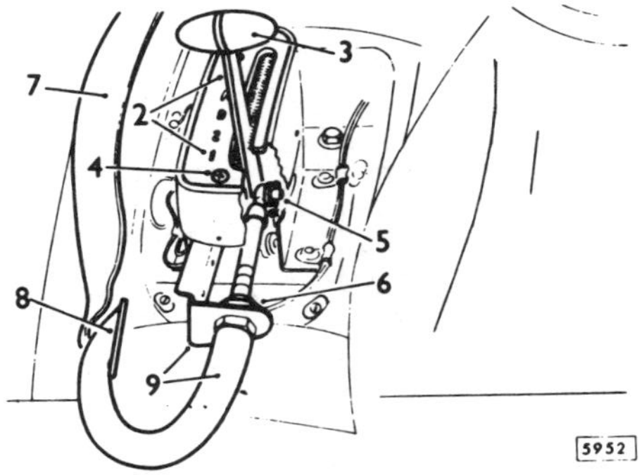

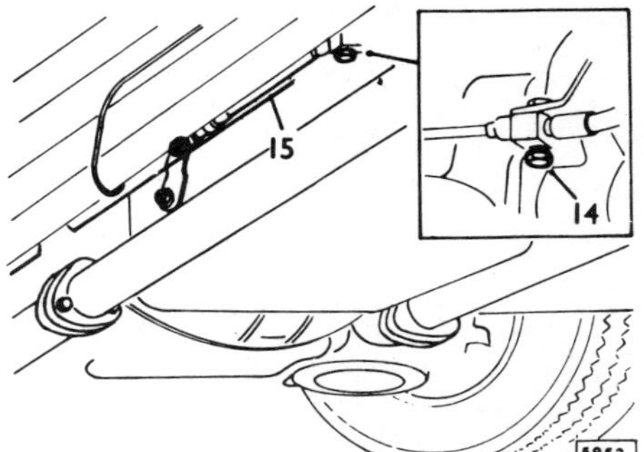

Refitting
16. Reverse operations 7 to 15, seal access panel and hole in the cable shroud with suitable non-hardening sealing compound.
17. Fit front locknut to cable but do not tighten at this stage.
18. Ensure gearbox selector and quadrant selector levers are in '1'.
19. Adjust front and rear locknuts until cable can be connected to quadrant lever without either quadrant or gearbox lever being disturbed.
20. Tighten locknuts, secure cable with new split pin.
21. Reverse operations 1 to 4.

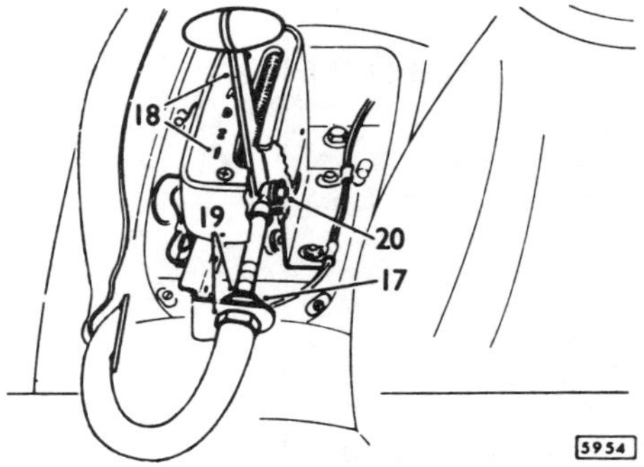

TRANSMISSION FAULT KEY

Adjustments
A1 Vacuum control adjustment.
A2 Manual control adjustment.
A3 Engine idle speed.
A4 Front band adjustment.
A5 Rear band adjustment.
A6 Kickdown switch adjustment.

Miscellaneous
M1 Fluid level.
M2 Vacuum leak.
M3 Vacuum line restricted.
M4 Broken kickdown wire or blown fuse.
M5 Oil cooler, lines and connections.
M6 Engine tune-up.
M7 Breather plugged.

Converter
C1 Converter blading or one-way clutch failed.
C2 Pump drive tangs on converter hub broken.
C3 Broken converter drive plate.

Governor
G1 Governor, sticking, leaking or incorrectly assembled.

Transmission
T1 Front clutch slipping due to worn or faulty parts.
T2 Front clutch seized or plates distorted.
T3 Rear clutch slipping due to worn plates or faulty parts.
T4 Rear clutch seized or plates distorted.
T5 Front clutch slipping due to a faulty servo, broken or worn band.
T6 Rear band slipping due to a faulty servo, broken or worn band.
T7 One-way clutch slipping or incorrectly installed.
T8 One-way clutch seized.
T9 Broken input shaft.
T10 Front pump worn or defective.
T11 Parking linkage.
T12 Planetary assembly.
T13 Oil tubes missing or broken.
T14 Sealing rings missing or broken and other oil leaks.
T15 Front band locked in the applied condition.
T16 Rear band locked in the applied condition.
T17 Rear clutch piston ball check valve leaking.
T18 Dipstick length.

Valve body
V1 Valve body improperly assembled or screws missing.
V2 Primary valve sticking.
V3 Secondary valve sticking.
V4 Throttle valve sticking.
V5 1-2 shift valve sticking.
V6 2-3 shift valve sticking.
V7 Governor modulator valve sticking.
V8 Throttle modulator valve sticking.
V9 Cutback valve sticking.
V10 Servo regulator valve sticking.
V11 Orifice control valve sticking.
V12 2-3 shift valve plug sticking.
V13 Regulator valve buzz.
V14 Defective solenoid.
V15 Dirty oil screen.
V16 Manual valve not connected to shift control.
V17 Ball check valve stuck.
V18 Rear servo regulator valve sticking.

44–12 Automatic Transmission—Model 12

STARTER INHIBITOR SWITCH

Check and adjust 44.15.18

Refer to operation 86.15.20.

Adjusting
1. Disconnect battery, see 86.15.20.
2. Unscrew gear selector knob.
3. Remove screws securing selector indicator; withdraw indicator over selector lever.
4. Carefully prise electric window switch panel away from centre console; do not disconnect window switches.
5. Remove screws securing control escutcheon; withdraw escutcheon slightly to obtain access to cigar lighter and door lock switch terminals.
6. Note fitted position of cigar lighter and door lock switch terminals; detach terminals and withdraw escutcheon.
7. Detach feed cable from inhibitor switch.
8. Connect a test lamp and battery in series with switch.
   NOTE: Switch is in earthed position.
9. Place selector lever in 'N' position.
10. Slacken locknuts securing switch and adjust position of switch until lamp lights. 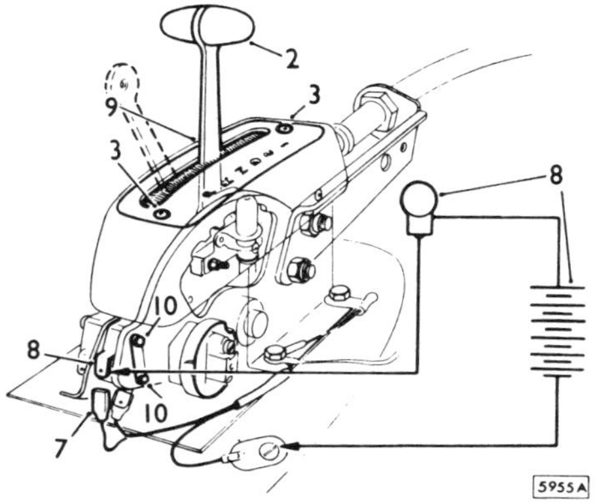
11. Tighten locknuts, check that lamp remains on with lever in 'P' position and is off with lever in drive position.
12. Remove battery and test lamp, reconnect feed cable to switch.
13. Reverse operations 1 to 6.
14. Check operation of window switches, door lock switch and cigar lighter.

STARTER INHIBITOR SWITCH

Remove and refit 44.15.19

Refer to operation 86.65.28.

REVERSE LIGHT SWITCH

Check and adjust 44.15.21

Refer to operation 86.65.20.

REVERSE LIGHT SWITCH

Remove and refit 44.15.22

Refer to operation 86.65.20.

KICKDOWN SWITCH

Remove and refit 44.15.23

Removing
1. Note location of input and output wires; disconnect wires from switch.
2. Remove bolts securing switch; withdraw switch.

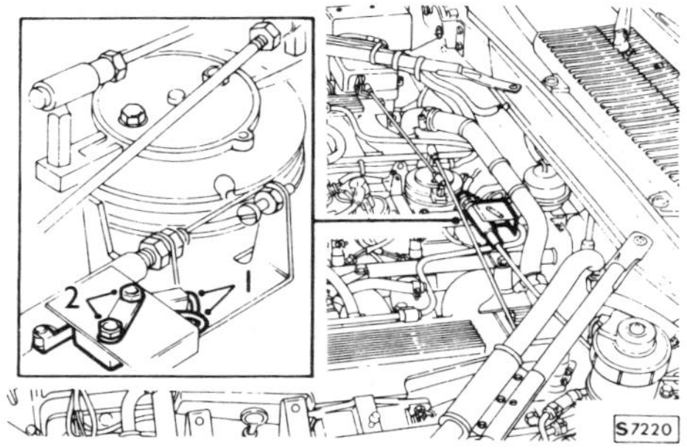

Refitting
3. Reverse operations 1 and 2.
4. Check switch adjustment, see 44.30.12.

KICKDOWN SOLENOID

Remove and refit 44.15.24

Removing
1. Remove oil pan, see 44.24.04.
2. Disconnect cable at connector.
3. Rotate solenoid through 180° and withdraw.

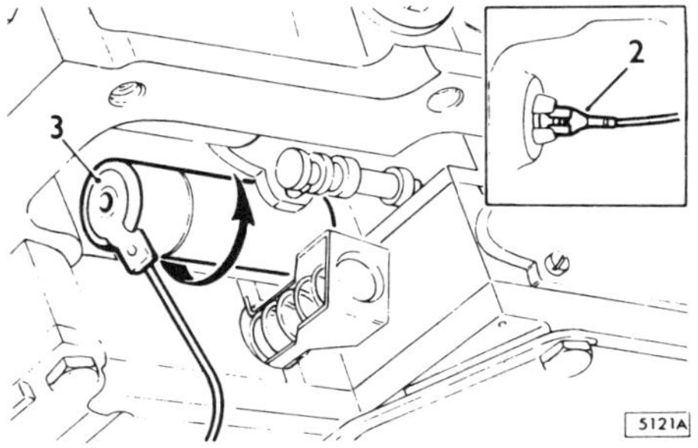

Refitting
Reverse operations 1 to 3, lubricate 'O' ring on solenoid prior to refitting.

VACUUM UNIT

Remove and refit 44.15.27

Removing
1. Disconnect hose from vacuum unit.
2. Unscrew vacuum unit.
3. Withdraw push rod.

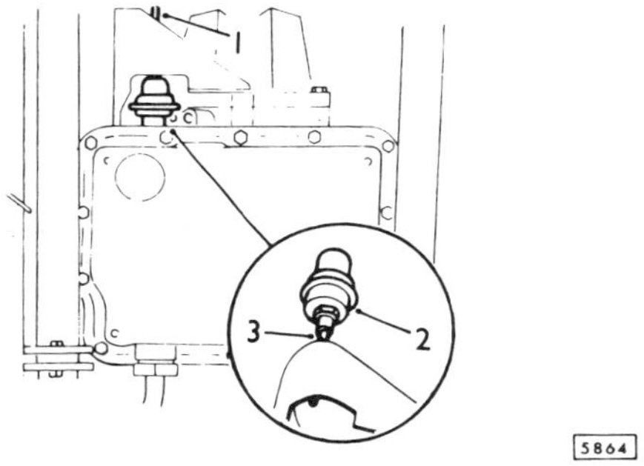

Refitting
4. Reverse operations 1 to 3.
5. If new vacuum unit has been fitted carry out operation 44.30.05.

CONVERTER HOUSING

Remove and refit 44.17.01

Removing
1. Remove engine and transmission assembly, see 12.37.01.
2. Remove transmission unit, see 44.20.01.
3. Remove bolts and spring washers securing starter motor; support or withdraw starter motor.
4. Remove bolts and washers securing front cover to converter housing; withdraw cover.
5. Remove bolts and lockwashers securing converter housing to cylinder block; withdraw housing.

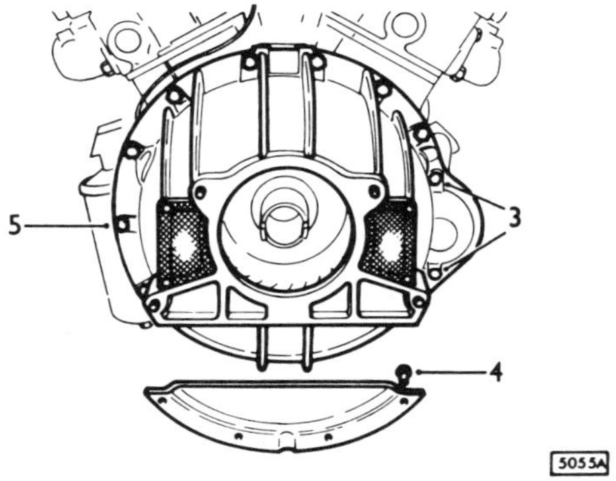

Refitting
Reverse operations 1 to 5.

Automatic Transmission—Model 12 44—13

CONVERTER

Remove and refit 44.17.07

Removing
1. Remove engine and transmission assembly, see 12.37.01.
2. Remove transmission unit, see 44.20.01.
3. Remove converter housing, see 44.17.01.
4. Remove oil filter assembly, see 12.60.01.
5. Remove plug from rear left-hand of cylinder block.
6. Rotate engine until one of four setscrews securing converter to drive plate is visible.
   CAUTION: On no account may engine be rotated if camshafts are disconnected.
7. Knock back tab washer and remove setscrew. 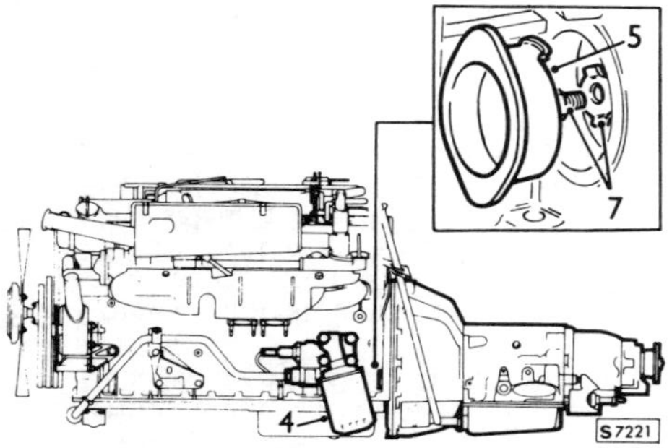
8. Repeat operations 6 and 7 until all setscrews are removed.
9. Withdraw converter.

Refitting
Reverse operations 1 to 9, use new tab washers; tighten bolts to a torque of 35 lb ft (4,8 kg m).
CAUTION: The torque converter is a sealed unit and no overhaul is possible. In the event of a defect arising, the assembly must be replaced.
WARNING: ON NO ACCOUNT SHOULD TORQUE CONVERTERS BE CLEANED EITHER EXTERNALLY OR INTERNALLY WITH FLAMMABLE LIQUIDS.

TRANSMISSION UNIT

Remove and refit 44.20.01

Removing
NOTE: At the time of going to press, procedures for removal of the transmission with engine in situ are not available. These will be published at a later date.

1. Remove engine and transmission assembly, see 12.37.01.
2. Disconnect oil cooler pipes, plug pipes and unions to prevent ingress of dirt.
3. Pull breather pipe off stub pipe.
4. Withdraw dipstick; disconnect oil filler tube at oil pan.
5. Drain and discard fluid.
6. Pull vacuum pipe off vacuum unit.
7. Remove four nuts and lockwashers securing gearbox to torque converter housing; withdraw gearbox.
CAUTION: Do not allow gearbox to hang on input shaft.

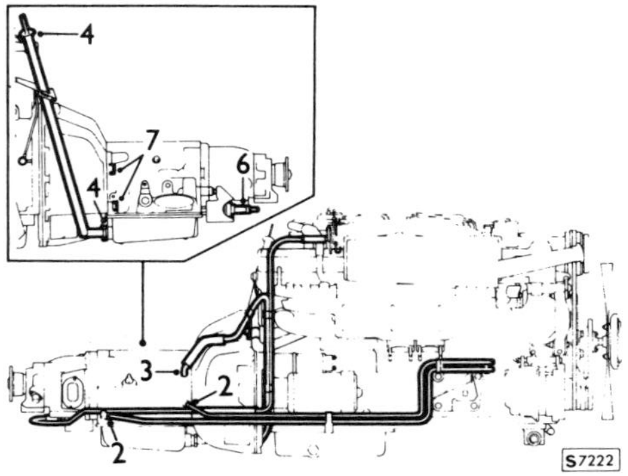

Refitting
8. Fit converter to converter housing and transmission unit assembly: use new tab washers and tighten bolts to a torque of 35 lbf ft (4,8 kgf m).
9. Fit assembly to engine.
10. Attach converter to drive plate.

TRANSMISSION ASSEMBLY

Overhaul 44.20.06

Service Tools: Bench cradle CWG.35; Mainshaft end float gauge CBW.33; Circlip pliers 7066; Clutch spring compressor CWG.37; Clutch spring compressor adaptor CBW.37A; Torque screwdriver CBW.548; Screwdriver bit adaptors CBW.547A-50-5 and CBW.548-1; Torque wrench CBW.547A-50; Rear clutch piston replacer CWG.41; Front clutch piston replacer CWG.42; Front servo adjuster adaptor CBW.548-2A; Gauge block CBW.34.

Dismantling
1. Thoroughly clean gearbox casing.
2. Fit bench cradle CWG.35.
3. Remove bolts and washers securing oil pan, lift off oil pan; remove and discard gasket.
4. Remove magnet from rear servo mounting bolt.

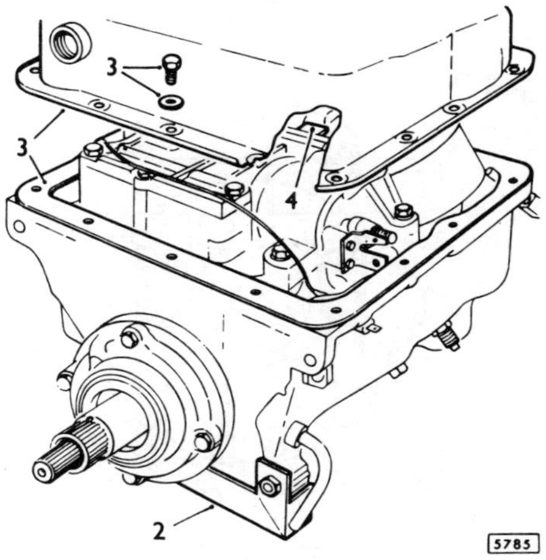

44—14 Automatic Transmission—Model 12

5 Unscrew vacuum control unit, remove and discard 'O' ring.
6 Withdraw push rod.
7 Disconnect cable from connector.
8 Compress retaining lugs and withdraw connector; remove and discard 'O' ring.
9 Slacken off front servo adjusting screw.
   CAUTION: This screw has a left-hand thread.
10 Slacken off rear servo adjusting screw locknut; remove screw.
11 Slacken off, but do not remove front servo retaining bolts.
12 Remove six valve body retaining bolts; note relative sizes and fitted positions of bolts.
13 Lift valve body clear of transmission case and withdraw servo tubes.
14 Remove front servo retaining bolts, carefully lift off servo at the same time noting fitted positions of brake band struts.
15 Remove rear servo retaining bolts, carefully lift off servo at the same time noting fitted positions of brake band struts.
16 Recover front and rear brake band struts from transmission case.
17 Remove speedometer driven gear housing together with driven gear; remove and discard 'O' ring.
18 Remove and discard split pin locking output flange securing nut.
19 Move manual lever until parking pawl engages ring gear.
20 Remove output flange securing nut; withdraw flange. 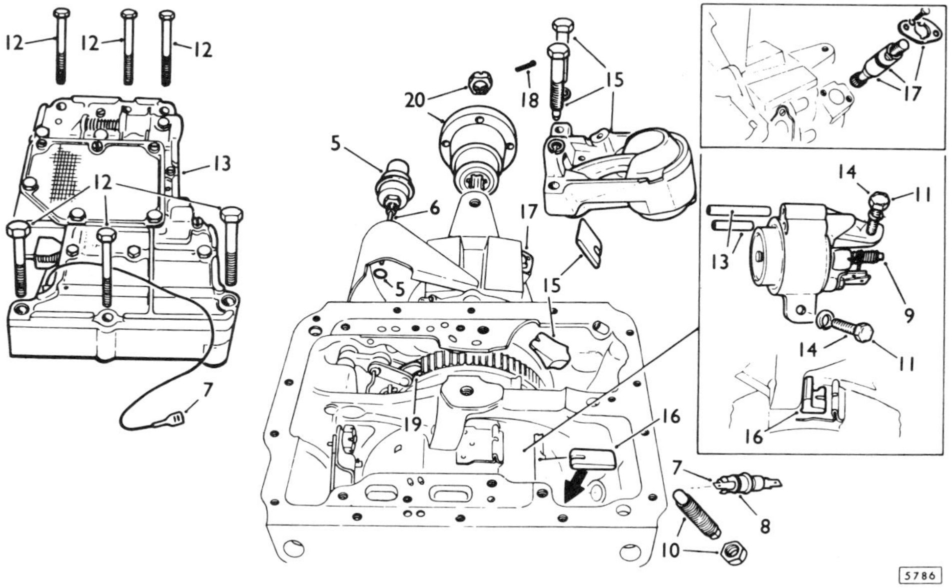
21 Remove bolts and spring washers securing speedometer drive gear housing to rear extension, withdraw housing; remove and discard gasket.
22 Prise oil seal out of housing and discard.
23 Slide speedometer drive gear off output shaft.
24 Remove bolts and spring washers securing rear extension housing and rear adaptor to gearbox case.
25 Remove stud bolts and lockwashers, lift off vacuum unit guard; withdraw rear extension housing.
26 Remove gasket and discard.
27 Remove snap ring retaining bearing, push bearing out of housing. 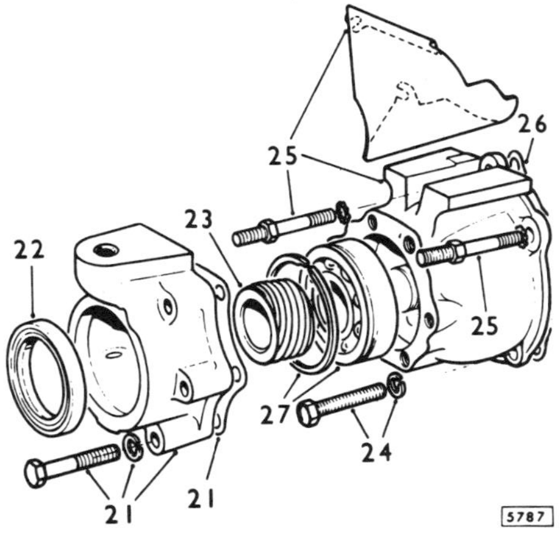
28 Remove snap ring retaining governor, slide governor off output shaft; recover drive ball.
29 Withdraw rear adaptor, remove and discard gasket.
30 Hold rear clutch drum steady and withdraw output shaft and ring gear assembly.
31 Remove selective thrust washer and discard.
32 Remove snap ring and withdraw output shaft from ring gear.
33 Hold forward sun gear shaft steady and withdraw planet carrier assembly.
34 Remove needle thrust bearing and steel backing washer.
35 Note fitted position of rear brake band; compress and withdraw brake band.

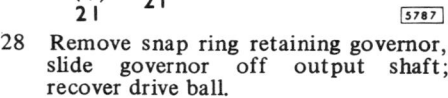

continued

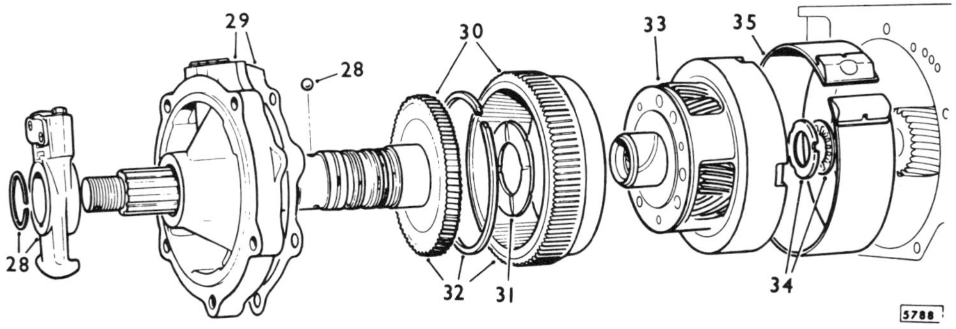

Automatic Transmission—Model 12   44—15

VALVE BLOCK

Overhaul

CAUTION: Ensure that all working surfaces are clean. Use only lint free cloth and clean transmission fluid for cleaning and lubricating.

Dismantling

42 Remove three pan head screws and three ¼ UNC x 2.25 in (57 mm) bolts: lift off filter screen.

43 Remove one ¼ UNC x 2.00 in (50,8 mm), one ¼ UNC x 2.75 in (60 mm) and four ¼ UNC x 2.25 in (57 mm) bolts securing upper valve body to valve block.
CAUTION: Do not allow components to separate at this stage.

44 Holding upper valve body and valve block together, invert assembly and carefully lift off upper valve body and separator plate.
CAUTION: Governor safety valve ball is spring loaded and is retained by the separator plate; great care should, therefore, be exercised when removing separator plate to ensure that ball is not misplaced.

45 Recover governor safety valve ball and withdraw spring.

36 Remove pump mounting bolts; withdraw pump.
NOTE: It may be necessary to free pump by tapping pump body with hide mallet.

37 Remove and discard sealing ring.

38 Remove centre support bolts and lockwashers.

39 Support forward sun gear shaft, push turbine shaft rearwards and withdraw centre support, rear clutch and front clutch.

40 Recover steel thrust washer and discard.

41 Note fitted position of front brake band; compress and withdraw brake band.

44—16 Automatic Transmission—Model 12

46 Invert valve body and recover three nylon ball valves.
47 Remove retaining pin and extract plug, secondary regulator valve and spring.
48 Remove spring retainer, seat and spring. Extract plug sleeve, primary regulator valve and spring.
49 Remove retaining pin and extract plug, governor modulator valve and spring.
50 Remove retaining pin, extract plug, spring and servo regulator valve.
51 Remove retaining pin, extract plug, spring and 1-2 shift valve.
52 Remove retaining pin, extract plug, 2-3 shift valve and spring.
53 Remove four 10 UNF x .375 in (9,5 mm) cheese head screws and detach valve body end plate.
54 Extract modulator spring and valve.
55 Extract throttle modulator cut back valve.
56 Extract 1-2 shift valve.
57 Extract 2-3 shift valve and spring.
58 Rotate kickdown solenoid through 180° and withdraw, remove and discard 'O' ring.
59 Withdraw retaining pin, extract spring and downshift valve.
60 Extract manual valve.
61 Remove three 10 UNF x .375 in (9,5 mm) cheese head screws and detach upper valve body end plate.
62 Extract orifice control valve and spring.
63 Extract servo regulator timer and spring.
64 Extract throttle valve.
65 Remove three 10 UNF x .56 in (14,3 mm) cheese head and two 10 UNF x .56 in (14,3 mm) pan head screws securing lower valve body to valve block.
66 Lift off lower valve body.
67 Remove separator plate.
68 Remove retaining pin and extract rear servo regulator valve and spring.
69 Renew any valves showing signs of scoring or burrs.
70 Check springs with data shown in spring identification tables, see page 44–3; renew springs which are distorted or shorter than specified length.

71 Thoroughly clean all components and lubricate with clean transmission fluid of the correct specification.

Re-assembling
Reverse operations 42 to 68, use new 'O' ring for kickdown solenoid.
CAUTION: Tightening torque figures must be adhered to, see group 06.

REAR CLUTCH

Overhaul

Dismantling
CAUTION: **The clutch units as removed from the gearcase are assembled together with the forward sun gear shaft. To facilitate separation of front and rear clutch assemblies a suitable stand is required and the planet carrier should be used for this purpose.**

72 Insert rear end of forward sun gear shaft in planet carrier.

continued

73 Carefully lift front clutch assembly off forward sun gear shaft.
74 Remove steel thrust washer and backing plate; discard thrust washer.
75 Carefully lift rear clutch assembly off forward sun gear shaft; recover needle roller bearings and bronze thrust washer.
76 Remove forward sun gear shaft from planet carrier.
77 Remove and discard sealing rings. 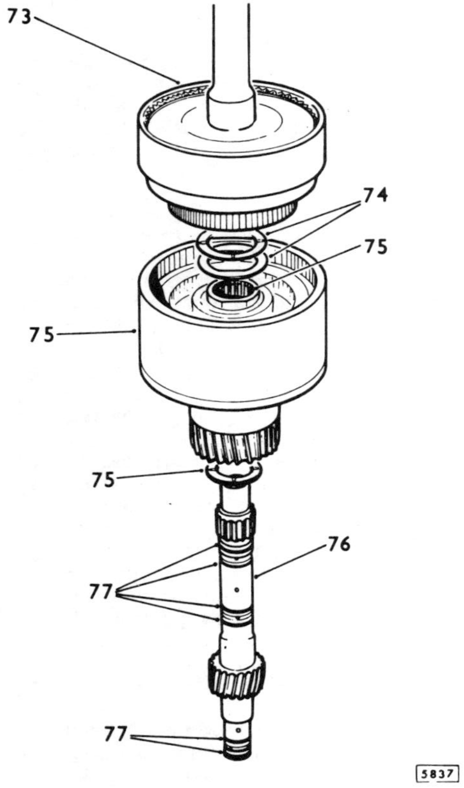
78 Check fluid passages in forward sun gear shaft for obstructions, clear passages with compressed air only.
79 Place rear clutch assembly over central spindle of clutch spring compressor CWG.37, reverse sun gear down.
80 Fit spring compressor CBW 37A over spindle. 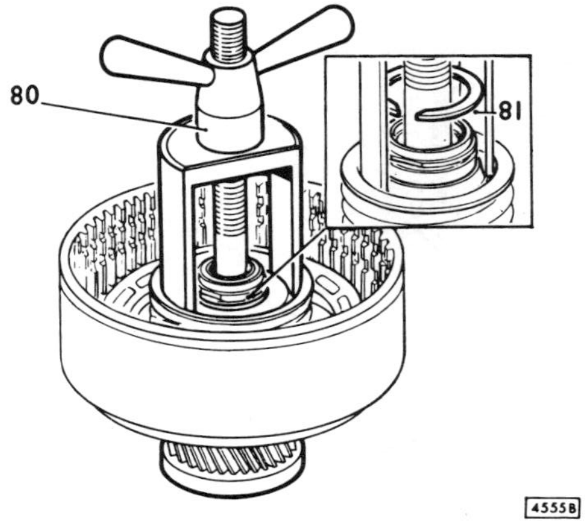
81 Compress spring and remove spring ring.
82 Slowly release pressure and remove compressor.
83 Remove retainer and spring.
84 Remove spring retaining pressure plate.
85 Remove pressure plate.
86 Remove inner and outer clutch plates.
87 Remove piston by applying air pressure to supply hole in centre bore.
88 Remove and discard piston ring and 'O' ring.

Inspection
89 Check clutch drum and bearing surfaces for scores or burrs; replace drum if damaged.
90 Check fluid passages for obstructions, clear passages with compressed air only.
91 Inspect piston check valve for free operation.
92 Check clutch release spring for distortion, renew if distorted.
93 Check inner clutch plates for flatness and that facings are undamaged. 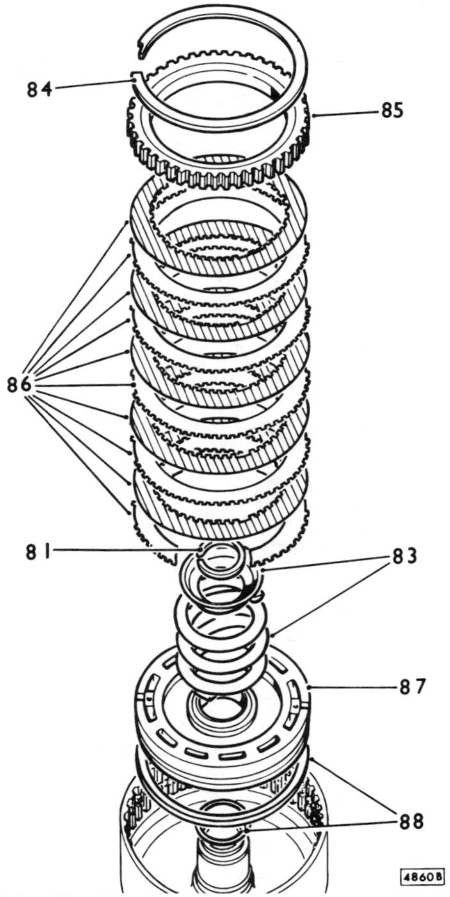
94 Check that coning on outer clutch plates is not less than .010 in (.25 mm). 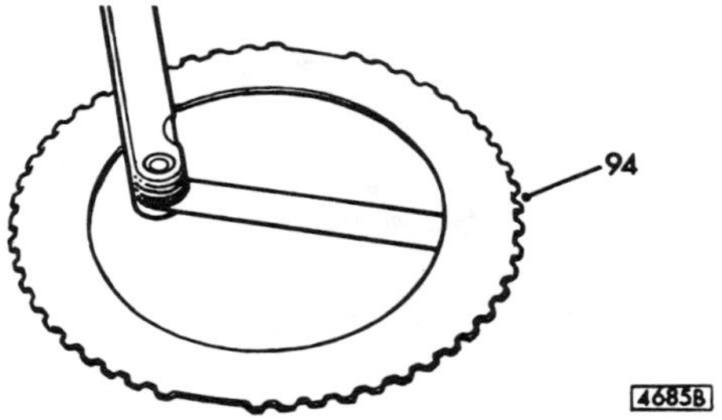
95 Check outer clutch plates for scores or burrs; renew if damaged. Minor scores or burrs may however be removed with a very fine abrasive.

Re-assembling
96 Fit new sealing rings on forward sun gear shaft.
97 Insert rear end of forward sun gear shaft in planet carrier.
98 Fit new bronze thrust washer.
99 Fit needle rollers; retain rollers with light grease.
100 Lubricate piston ring and 'O' ring with petroleum jelly.
101 Fit piston ring and 'O' ring.
102 Position rear clutch piston replacer tool CWG.41 in clutch drum.
103 Lubricate piston and install; remove replacer tool.

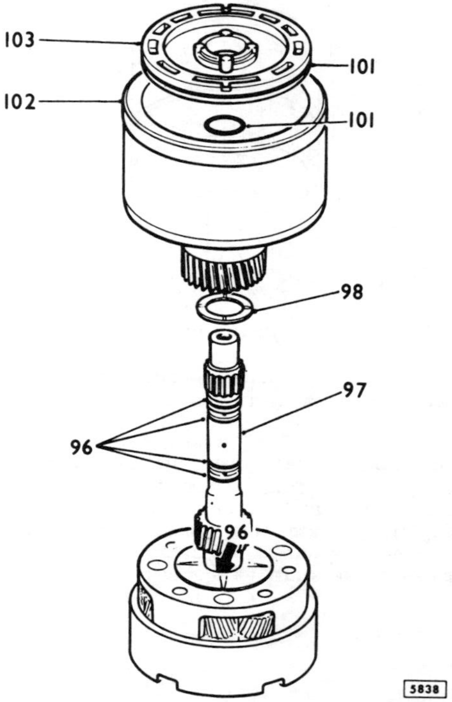

44–18 Automatic Transmission—Model 12

104 Reverse operations 79 to 86.
    CAUTION: Outer clutch plates must be assembled with cones facing same direction.
105 Position rear clutch assembly on forward sun gear shaft; refit steel backing plates followed by new steel thrust washer.
    CAUTION: Do not remove rear clutch assembly and forward sun gear shaft from planet carrier.

FRONT CLUTCH

Overhaul

Dismantling
106 Remove spring and withdraw turbine shaft.
107 Remove fibre washer and discard.
108 Remove clutch hub.
109 Remove inner and outer clutch plates and ring gear.
110 Remove spring ring.
111 Remove clutch release diaphragm and spring.
112 Remove piston by applying air pressure to supply hole.
113 Remove piston ring and discard.
114 Remove 'O' ring from clutch housing pedestal and discard.

Inspection
115 Check clutch drum and bearing surfaces for scores or burrs; replace drum if damaged.
116 Check fluid passages for obstruction; clear passages with compressed air only.
117 Inspect piston check valve for free operation.
118 Check clutch release diaphragm for cracks or distortion; renew if damaged.
119 Check inner clutch plates for flatness and that facings are undamaged.
120 Check outer clutch plates for flatness, scores or burrs, renew if damaged. Minor scores or burrs may, however, be removed with a very fine abrasive.

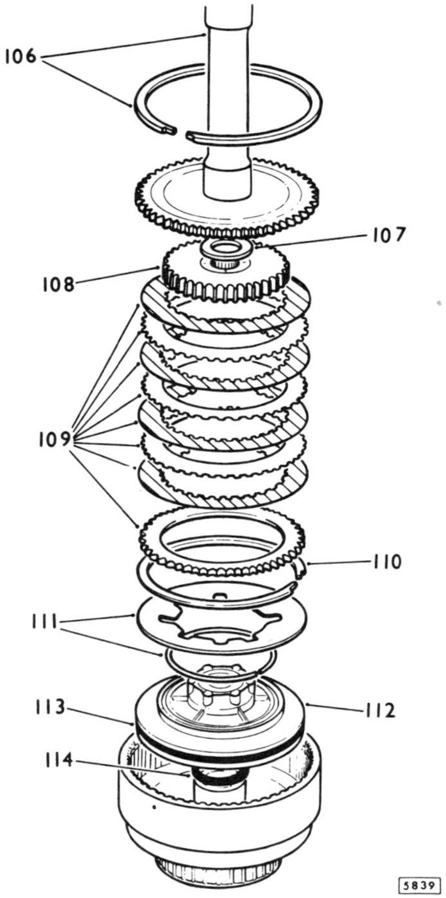

Re-assembling
121 Lubricate piston ring and 'O' ring with petroleum jelly.
122 Fit 'O' ring to clutch housing pedestal.
123 Fit piston ring.
124 Position front clutch piston replacer tool CWG.42 in clutch drum.
125 Lubricate piston and install; remove replacer tool.
126 Fit release diaphragm and spring. 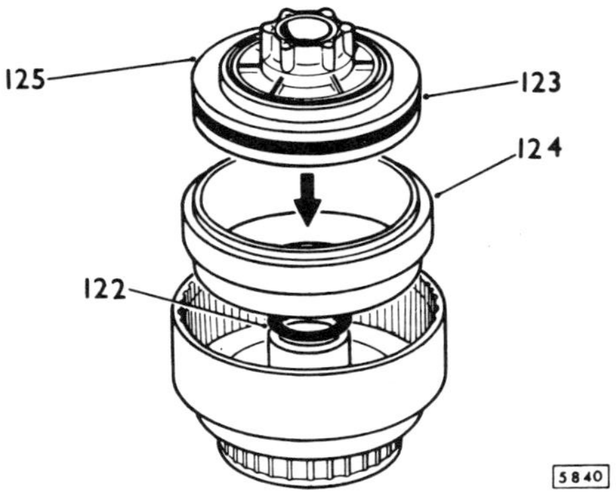
127 Fit spring ring; ensure ring is correctly seated in groove.
128 Fit backing plate and new steel thrust washer.
129 Carefully lower front clutch over forward sun gear shaft and into rear clutch.
130 Ensure that front clutch is fully seated in rear clutch.
131 Fit ring gear.
132 Fit inner and outer clutch plates.
133 Check to ensure that teeth of inner clutch plates are in alignment.
134 Fit clutch hub.
135 Place new fibre washer over forward sun gear shaft.
136 Fit turbine shaft and spring ring; ensure spring ring is correctly seated in groove.
    CAUTION: On no account should front and rear clutch assemblies be separated as damage to sealing rings on forward sun gear shaft will result.
137 Fit new steel thrust washer.

continued

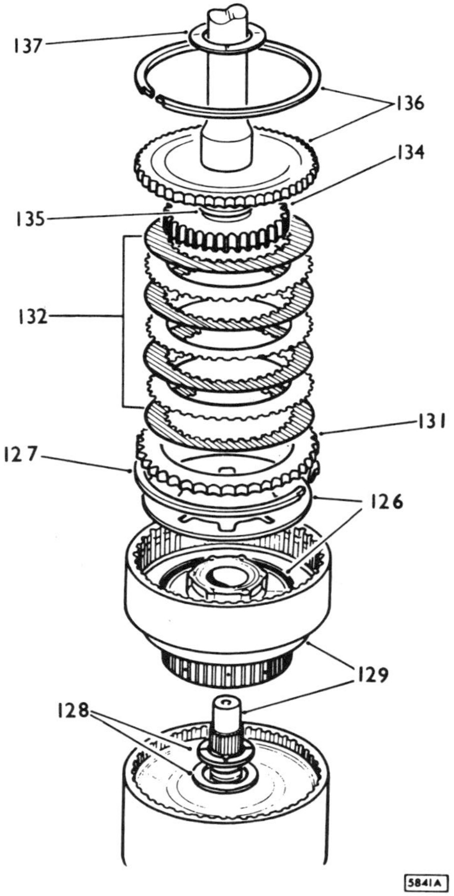

Automatic Transmission—Model 12   44—19

PUMP

Overhaul

Dismantling

138 Remove locking screw, separate pump halves.
139 Mark mating surfaces of gears with die marker.
140 Remove gears from pump body.
141 Prise oil seal out of pump body; discard seal.

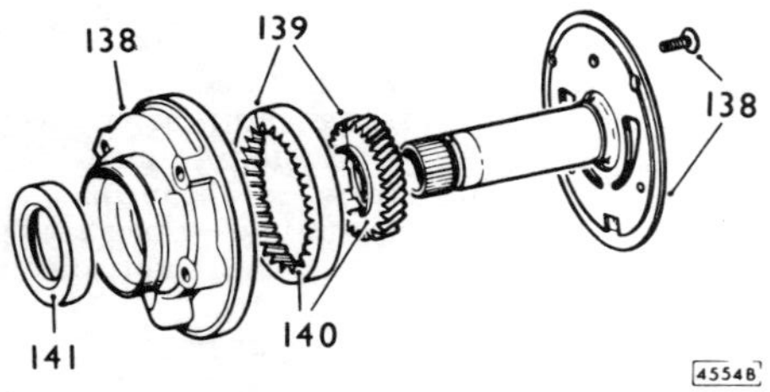

Inspection

142 Check bearing surfaces, gear teeth and splines for wear or damage; renew if necessary.
CAUTION: If gears are damaged or worn, they must be renewed as a set, individual gears must not be fitted.

Re-assembling

143 Soak new oil seal in clean transmission fluid and press carefully into pump body; ensure seal is squarely seated.
144 Reverse operations 138 to 140.

FRONT SERVO

Overhaul

Dismantling

145 Depress piston and sleeve, remove spring ring.
146 Withdraw piston and stop plate, collect spring.
147 Remove setscrew, lockwasher and plain washer.
148 Drift piston out of stop plate and piston sleeve.
149 Withdraw stop plate from piston sleeve.
150 Remove and discard sealing rings. 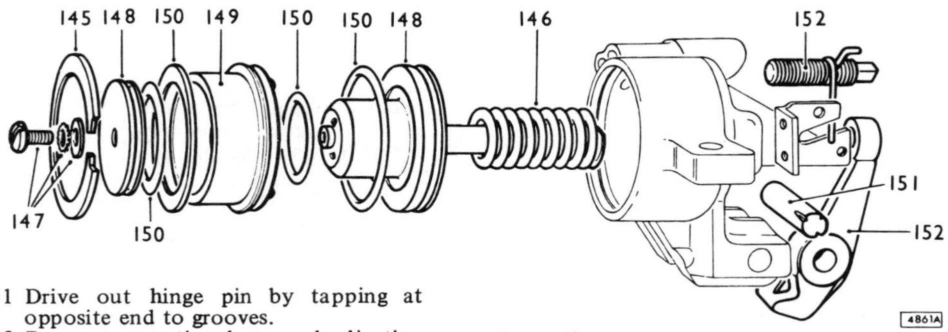
151 Drive out hinge pin by tapping at opposite end to grooves.
152 Remove operating lever and adjusting screw.
NOTE: Adjusting screw has left-hand thread.

Inspection

153 Check that hinge pin is a tight fit in servo body.
154 Check servo body for cracks or scoring; renew if necessary.
155 Check spring for distortion; renew if necessary.
156 Check fluid passages for obstruction; clear passages with compressed air only.

Re-assembling

Reverse operations 145 to 152, coat new sealing rings with petroleum jelly prior to fitting.

REAR SERVO

Overhaul

Dismantling

157 Drive roll pin out of hinge pin with a .125 in (3 mm) punch.
158 Remove hinge pin and operating lever, collect thrust washers and needle rollers.
159 Compress piston return spring by pressing down on cover plate.
160 Note fitted positions of spring ring and cover plate; remove spring ring.
161 Slowly release pressure on cover plate; remove plate, return spring and piston.
162 Remove and discard piston sealing ring.

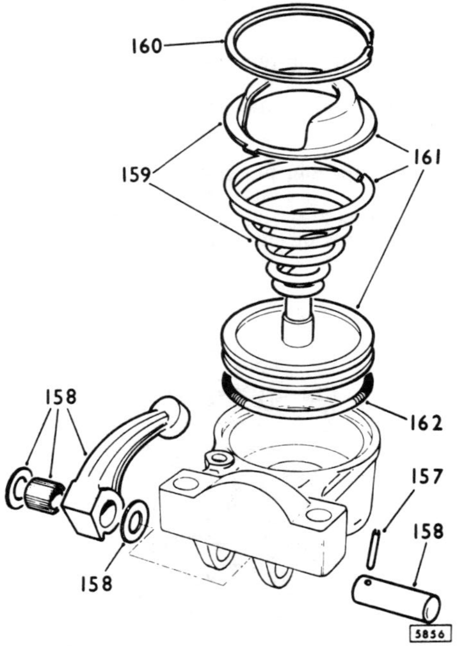

167 Check fluid passages for obstruction; clear passages with compressed air only.

Re-assembling

168 Reverse operations 159-162; coat piston sealing ring with petroleum jelly prior to fitting.
169 Position hinge pin in one side of servo body and fit thrust washer.
170 Lightly smear needle rollers with grease and position in operating lever.
171 Position operating lever in servo body and carefully push hinge pin half way through, ensuring that needle rollers are not displaced.
172 Fit remaining thrust washer.
173 Push hinge pin through operating lever, thrust washer and servo body.
174 Fit roll pin.

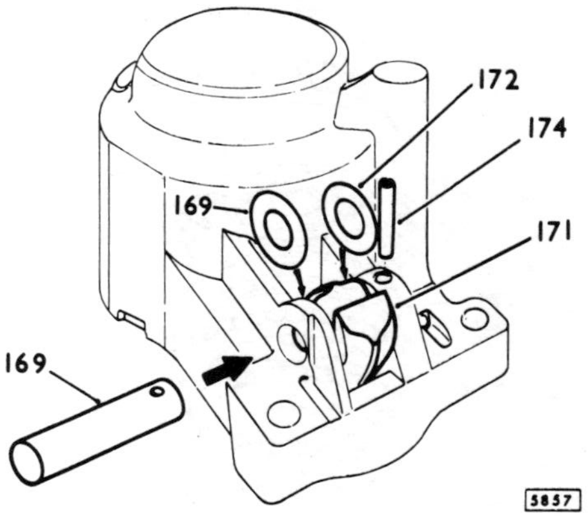

Inspection

163 Check servo body for cracks or scoring; renew if damaged.
164 Check piston return spring for distortion; renew if distorted.
165 Check hinge pin for scoring, burrs or ovality. Light scoring or burrs may be removed with a fine abrasive; pins showing signs of ovality must be renewed.
166 Check fluid passage plug for tightness.

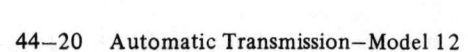

GOVERNOR

Overhaul

Dismantling
175 Remove screws securing cover plate; detach cover plate.
176 Remove bolts and spring washers securing body assembly to counterweight.
177 Slide retaining plate off weight; remove spring, valve and weight from body.

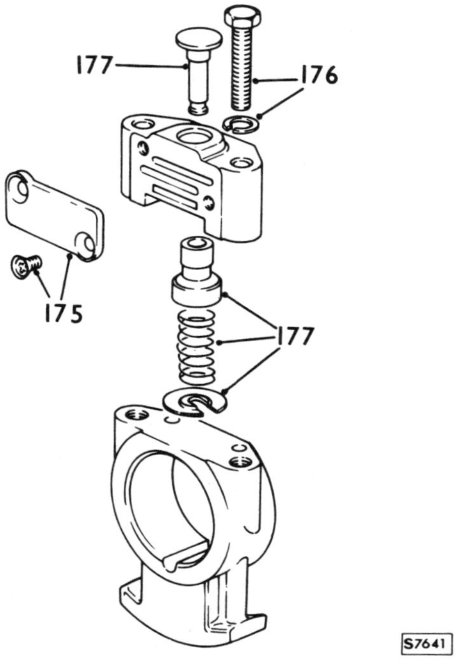

Inspection
178 Check weight and valve body for scoring or burrs. Minor scoring or burrs may be removed with a fine abrasive.
179 Check spring for distortion; renew if necessary.
180 Check fluid passages for obstruction; clear passages with compressed air.

Re-assembling
181 Reverse operations 175 to 177.
182 Check weight and valve for free movement.
CAUTION: If weight or valve show signs of sticking governor body assembly must be renewed.

PARKING BRAKE PAWL ASSEMBLY

Overhaul

Dismantling
183 Remove clips retaining link rod; withdraw rod.
184 Remove nut securing manual shaft lever; withdraw lever.
185 Remove lever detent, ball and spring.
186 Withdraw and discard seal.
187 Remove retaining clip, washer and torsion lever.
188 Note fitted position of pawl return spring.
189 Withdraw pawl return spring and retaining clip.
190 Remove toggle link and pins.
191 Move pawl back and forth until pin protrudes, then withdraw pin and pawl.
192 Drive toggle lever towards gearbox casing; withdraw toggle lever pivot pin and toggle lever.

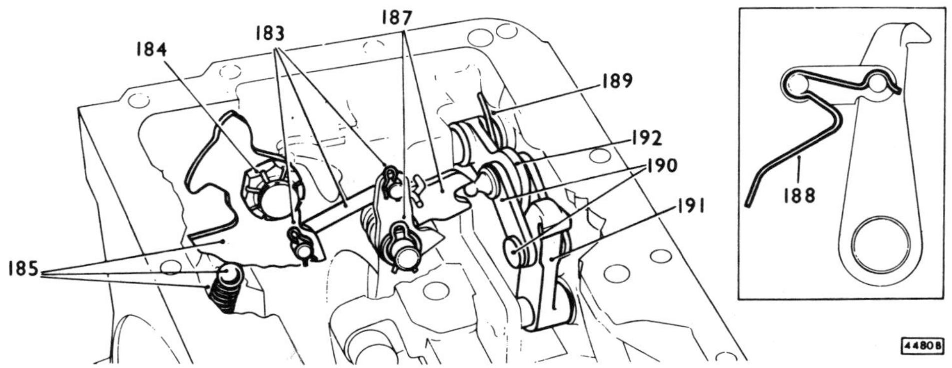

Inspection
193 Check components for excess wear, renew if necessary.
194 Check springs for distortion; renew if necessary.
195 Check pawl tooth for signs of chipping or uneven wear, renew if damaged.

Re-assembling
196 Lightly grease all pivots and pins.
197 Reverse operations 183 to 192.

BRAKE BANDS

Inspection
198 Check front and rear brake bands for damage or distortion.
199 Check linings for uneven, excess wear or damage.
CAUTION: Should doubt exist as to condition of brake bands or bands show any of the conditions detailed above, they must be renewed.

continued

Automatic Transmission—Model 12   44—21

correct end float to be obtained after re-assembly. Available selective thrust washer sizes are as follows:

Inches	Millimetres
1. .061 – .063	1,55 – 1,60
2. .067 – .069	1,70 – 1,75
3. .074 – .076	1,88 – 1,93
4. .081 – .083	2,06 – 2,11
5. .092 – .094	2,34 – 2,39
6. .105 – .107	2,67 – 2,72

Experience has shown that if thrust washer number 4 – .081-.083 in (2,06-2,11 mm) is selected, correct end float of .008 in to .040 in (,20 to 1,01 mm) is usually obtained and it is recommended that this washer be used.

210 Fit thrust washer.
211 Position output shaft in ring gear; fit snap ring ensuring that it is correctly located in groove.
212 Fit new sealing rings on output shaft.
213 Fit output shaft and ring gear assembly.
214 Position sealing rings so that gaps are staggered.
215 Lightly grease new rear adaptor gasket and position on rear adaptor.
216 Fit rear adaptor ensuring that holes in gearcase and adaptor are in alignment.
217 Position governor drive ball in blind hole in output shaft; retain drive ball with a smear of grease.
218 Slide governor on to output shaft with oil holes in governor towards gearcase.
219 Fit snap ring; ensure that it is correctly located in groove.

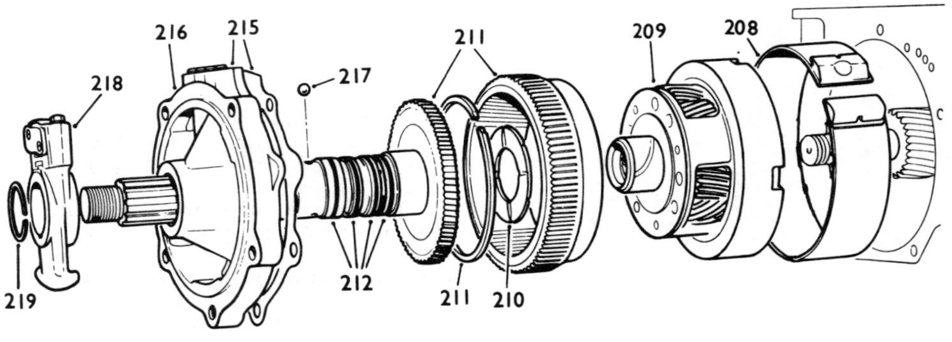

205 Smear new oil pump oil seal with clean transmission fluid; press oil seal into pump body using hand pressure only.
206 Lightly grease new oil pump sealing ring with petroleum jelly; position ring in recess in gearcase.
207 Position oil pump assembly on gearcase; tighten bolts evenly to a torque of 17 to 22 lb ft (2,35 to 3,0 kg m).

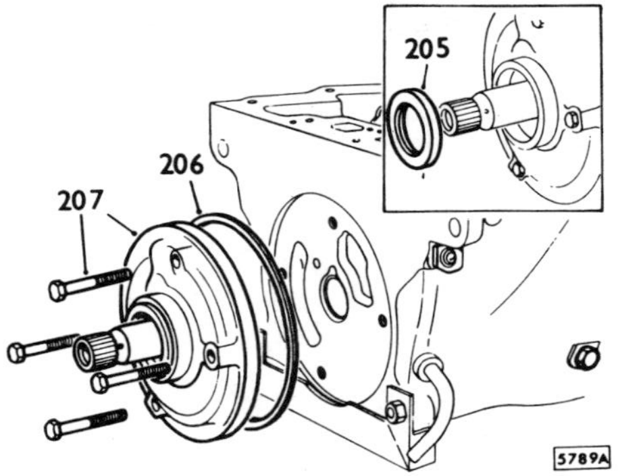

208 Position rear brake band in gearcase.
209 Fit planet gear carrier.
NOTE: If one way clutch has been removed, it must be refitted with lip on periphery of clutch facing towards gears.
CAUTION: The thrust washer located between the planet gear carrier and output shaft is selective to enable

TRANSMISSION ASSEMBLY

Re-assembling

200 Thoroughly clean gearcase; blow all oil holes through with dry, clean compressed air.
201 Position front brake band in transmission case.
202 Ensure that thrust washer, fitted in operation 137 is correctly positioned on turbine shaft.
203 Hold front and rear clutch assemblies firmly together and enter them into gearcase through rear aperture; fit needle thrust bearing and steel backing washer.
CAUTION: On no account must clutch assemblies be allowed to separate as damage to sealing rings on forward sun gear shaft will result.
204 Position centre support in gearcase, fit two side retaining bolts and lockwashers; DO NOT tighten bolts at this stage.

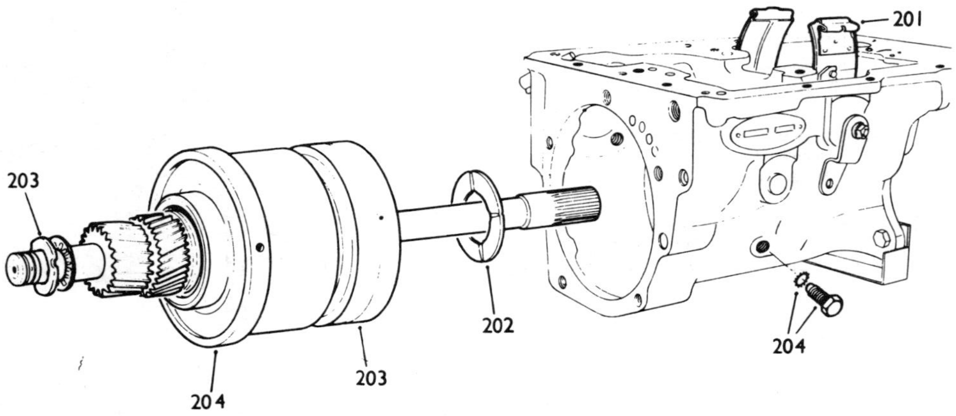

44–22 Automatic Transmission—Model 12

220 Fit snap ring on rear extension housing bearing, ensuring snap ring is correctly located in groove.
221 Lightly grease bearing.
222 Lightly grease new rear extension housing gasket; position gasket on housing.
223 Fit rear extension housing ensuring that holes in housing and rear adaptor are in alignment.
224 Fit securing bolts and lockwasher; do not tighten bolts at this stage.
225 Fit vacuum unit guard, stud bolts and lockwashers; do not tighten stud bolts at this stage.
226 Push bearing into rear extension housing until snap ring abuts against housing.
227 Tighten securing bolts and stud bolts by diagonal selection to a torque of 28-33 lb ft (3,9-4,5 kg m).
228 Slide speedometer drive gear on to output shaft.
229 Lightly oil new seal and press into speedometer drive gear housing.
230 Lightly grease new speedometer drive gear housing gasket; position gasket on housing.
231 Fit speedometer drive gear housing to rear extension housing ensuring that bolt holes are in alignment.
232 Fit securing bolts and spring washers; tighten bolts by diagonal selection to a torque of 28-33 lb ft (3,9-4,5 kg m).

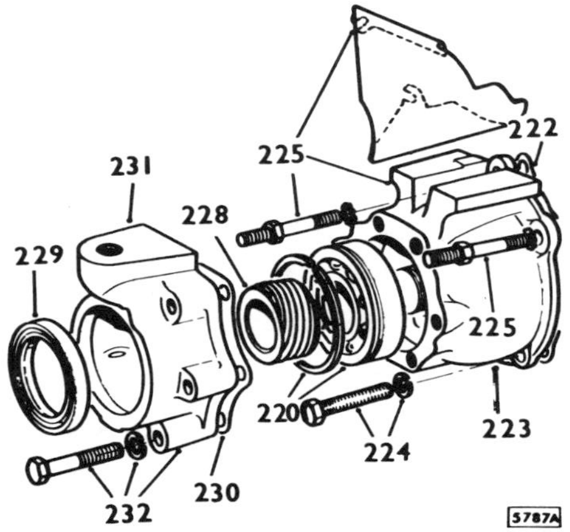

233 Slide output flange on to output shaft; fit castellated nut.
234 Move manual lever until parking pawl engages ring gear.
235 Tighten output flange securing nut; ensure that split pin hole in output shaft is in alignment with one of the castellations in the nut. DO NOT fit split pin at this stage.
236 Assemble end float gauge CBW.33 to gearcase with stylus contacting end of turbine shaft; insert a suitable lever between front clutch and front of gearcase. Ease gear train to rear of gearcase and zero end float gauge.
237 Insert lever between ring gear and rear clutch and ease gear train to front of gearcase.
238 Note reading on gauge which should be between .008-.040 in (,20 mm-1,01 mm).
CAUTION: If end float is not within above limits, reverse operations 213 to 235 and select thrust washer which will bring end float within limits. For example: If end float exceeds .040 in (1,01 mm) select a thicker thrust washer, number 5 or 6. If, however, end float is less than .008 in (,20 mm) select a thinner washer, number 1, 2 or 3. Repeat operations 213 to 238.

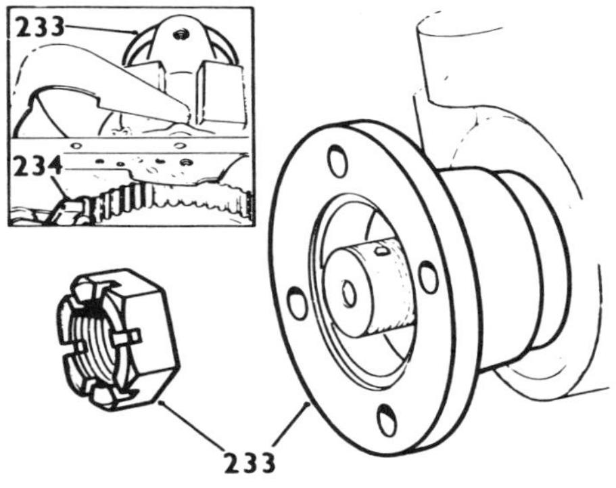

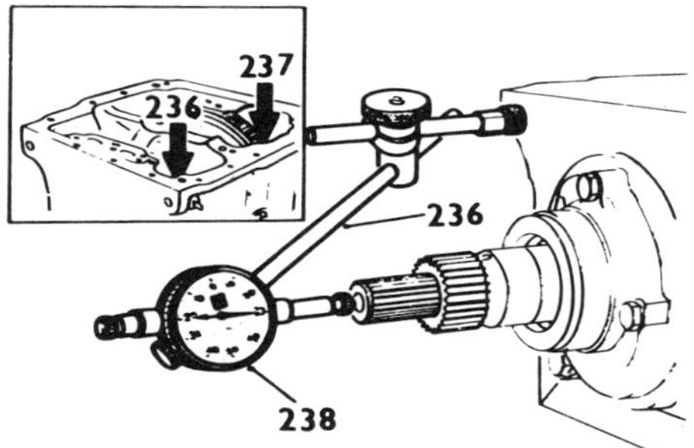

239 Remove end float gauge.
240 Fit new split pin to lock output flange nut.
241 Lubricate new 'O' ring with petroleum jelly and fit to speedometer driven gear shaft.
242 Fit speedometer driven gear ensuring that gear meshes correctly with drive gear.
243 Fit cover and securing bolts and spring washers.

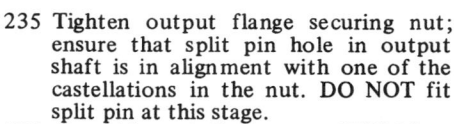

continued

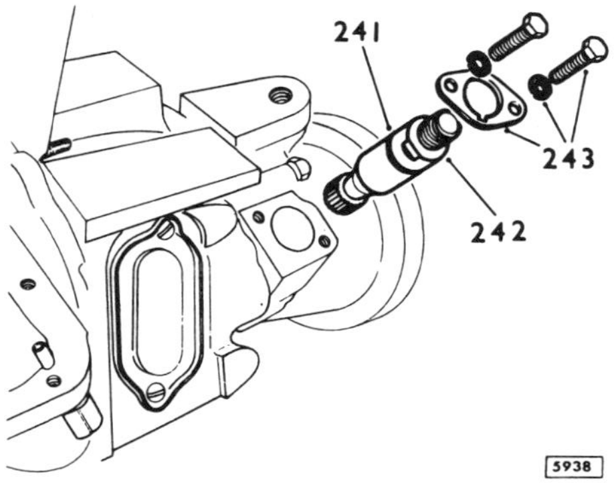

Automatic Transmission—Model 12 44—23

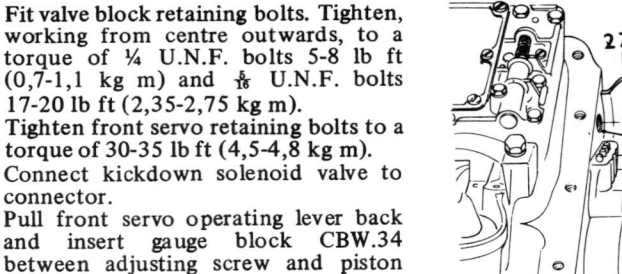

244 Locate rear servo adjusting screw in tapped hole in gearcase; turn screw until ball end protrudes into gearcase. Fit adjusting screw locknut but do not tighten at this stage.
245 Fit adjusting screw locknut but do not tighten at this stage.
246 Position cast brake band strut on ball end of rear servo adjusting screw and rotate rear servo adjusting screw until it is in contact with strut; ensure that staking on brake band locates in slot in strut.
247 Position steel brake band strut on rear servo operating lever.
NOTE: Strut should be retained in position with a smear of grease.
248 Position rear servo on gearcase ensuring that staking on rear brake band locates in slot in strut.
249 Fit servo retaining bolts and spring washers but do not tighten at this stage.
250 Screw in rear servo adjusting screw until finger tight.
251 Ensure that steel brake band strut has not been displaced and tighten securing bolts to a torque of 40-50 lb ft (5,5-6,9 kg m).
252 Tighten two centre support side bolts to a torque of 20-25 lb ft (2,8-3,45 kg m).
253 Position front brake band upper strut in gearcase ensuring that spigot on rear of strut is located in hole in gearcase.
254 Fit remaining brake band strut to front servo ensuring that staking on servo operating lever locates in slot in strut.
NOTE: Strut should be retained in position with a smear of grease.
255 Position front servo on gearcase and compress front brake band slightly until strut can be fitted under lip of brake band.
256 Fit servo retaining bolts and spring washers but do not tighten at this stage.
257 Ensure that brake band strut has not been displaced.
258 Lubricate new 'O' ring with petroleum jelly and fit to cable connector.
259 Insert cable connector into gearcase.
260 Slide servo tubes into holes in front servo.
261 Move manual lever to 'R' position.
262 Position valve block on gearcase and insert servo tubes.
263 Ensure that spigot on detent lever is located in groove in manual valve. 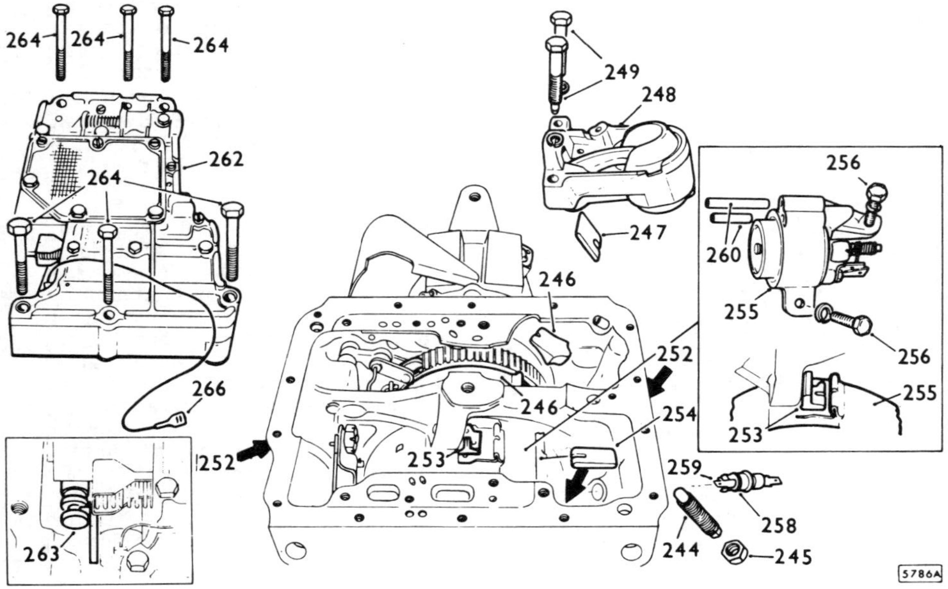
264 Fit valve block retaining bolts. Tighten, working from centre outwards, to a torque of ¼ U.N.F. bolts 5-8 lb ft (0,7-1,1 kg m) and 5/16 U.N.F. bolts 17-20 lb ft (2,35-2,75 kg m).
265 Tighten front servo retaining bolts to a torque of 30-35 lb ft (4,5-4,8 kg m).
266 Connect kickdown solenoid valve to connector.
267 Pull front servo operating lever back and insert gauge block CBW.34 between adjusting screw and piston pin.
268 Using torque screwdriver CBW.548 and adaptors CBW.547A-50-5 and CBW.548-1, tighten adjusting screw to 10 lb in (,12 kg m).
CAUTION: Adjusting screw has a left-hand thread.
269 Ensure that locking spring is located in retaining slot.
270 Remove gauge block.
271 Place magnet on rear servo mounting bolt. 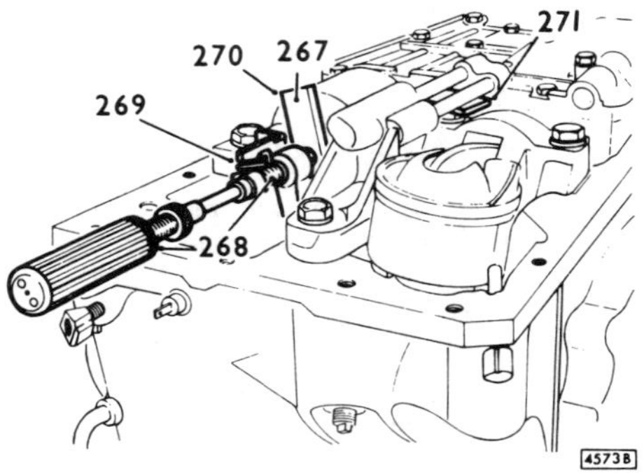
272 Insert vacuum unit push rod through gearcase and into valve block.
273 Lubricate new 'O' ring with petroleum jelly and fit on to vacuum unit.
274 Insert end of push rod in vacuum unit; screw unit into gearcase. DO NOT overtighten. 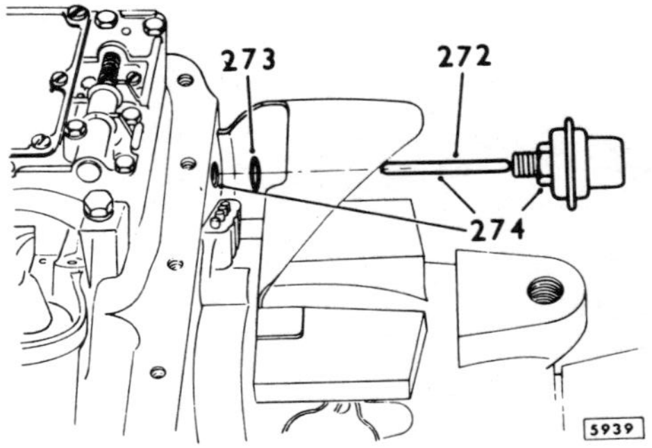
275 Tighten rear servo adjusting screw to a torque of 10 lb ft (1,4 kg m).
276 Loosen adjusting screw 1.25 turns and tighten locknut.
CAUTION: Ensure screw does not turn when tightening locknut as severe damage to transmission unit will result if screw is not backed off exactly 1.25 turns. 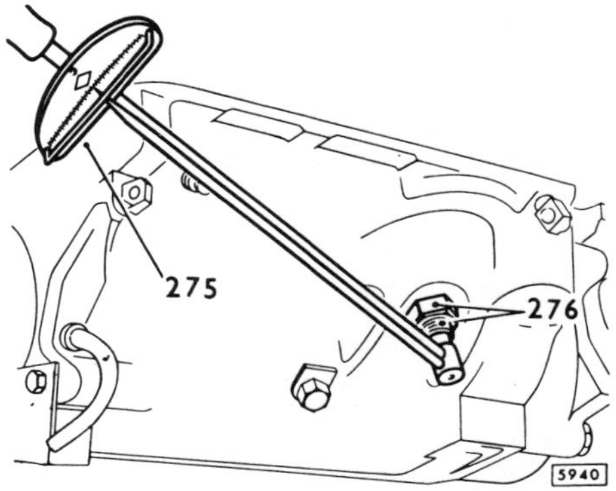
277 Lightly grease new oil pan gasket and position on gearcase.
278 Fit oil pan retaining bolts and washers. Tighten bolts progressively to a torque of 10-13 lb ft (1,4-1,8 kg m).
279 Remove bench cradle.

44—24 Automatic Transmission—Model 12

LUBRICATION SYSTEM

Drain and refill　44.24.02

CAUTION: Due to method of construction, it is not possible to completely drain transmission fluid, and this should be taken into account when transmission is being refilled. As it should only be necessary to carry out the following operations preparatory to carrying out work on the transmission which will involve removal of oil pan, the following procedure should be followed. It is possible however, to drain a smaller amount of fluid by removing dipstick/filler tube.

Draining
1. Remove oil pan, see 44.24.04.
2. Discard fluid.

Refilling
3. Refit oil pan, fill transmission to 'MAX' mark on dipstick.
4. Apply handbrake and select 'P' position.
5. Run engine until it reaches normal operating temperature.
6. With engine still running, withdraw dipstick, wipe clean and replace.
7. Immediately withdraw dipstick and note reading on 'HOT' side of dipstick.
8. If necessary, add fluid to bring level on dipstick to 'MAX'.
NOTE: Difference between 'MAX' and 'MIN' marks on dipstick represents approximately 1½ pints (2 U.S. Pints, 0,75 litre).

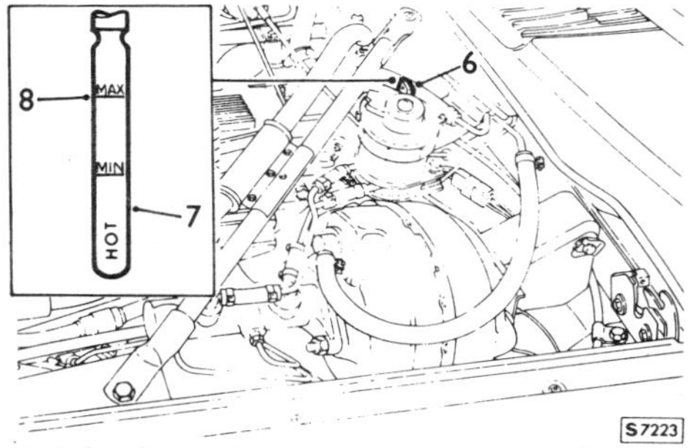

OIL PAN

Remove and refit　44.24.04

Removing
1. Unscrew union nut securing dipstick/filler tube to oil pan; withdraw tube. Drain and discard fluid.
2. Remove nut securing collision plate to rear engine mounting.
3. Remove nuts securing collision plate to locating studs, lower plate noting fitted position of washers.
4. Remove stud nuts, bolts and washers securing oil pan to transmission case; manoeuvre pan clear of exhaust system.
5. Remove and discard gasket.

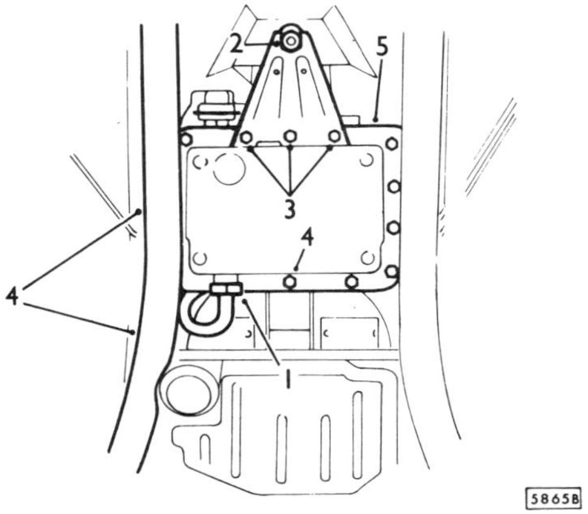

Refitting
6. Reverse operations 1 to 5; use a new gasket.
7. Refill transmission, see operation 44.24.02.

OIL STRAINER

Remove and refit　44.24.07

Removing
1. Remove oil pan, see 44.24.04.
2. Remove pan head screws and bolts securing strainer to valve block; detach strainer. 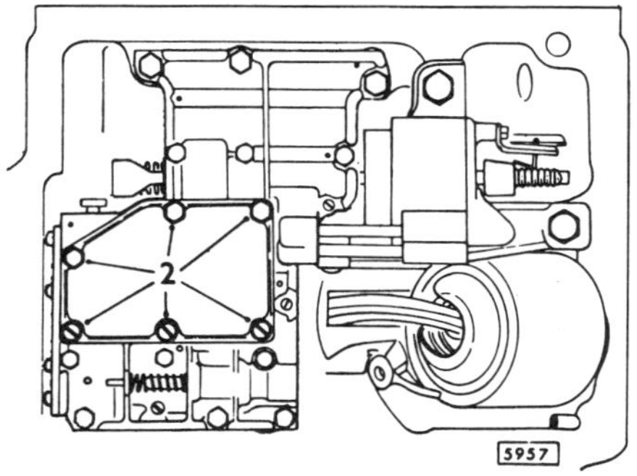
3. Thoroughly clean strainer.

Refitting
Reverse operations 1 and 2.

VACUUM CONTROL UNIT

Line pressure check and adjustment　44.30.05

Special tools: Pressure gauge CBW.1A-642; Adaptor STN.6752

1. Check engine tune, i.e. cylinder compressions, spark plugs, ignition timing.
2. Carry out stall speed test, see 44.30.13.
3. Lift carpet from left-hand side of gearbox housing.
4. Remove access plate.
NOTE: On early cars it will be necessary to cut around perforated portion of sound deadening material. This should be lightly glued to the access panel after completing the operation.
5. Disconnect pipe from vacuum control unit. Insert 'T' piece union and re-connect pipe. Connect a vacuum gauge to centre junction.
6. Remove the 0.125 in (3,2 mm) plug at gear case front left-hand side and connect the pressure gauge CBW.1A-642 using adaptor STN.6752.
7. With engine and transmission at normal running temperature select 'D', apply hand and foot brakes. 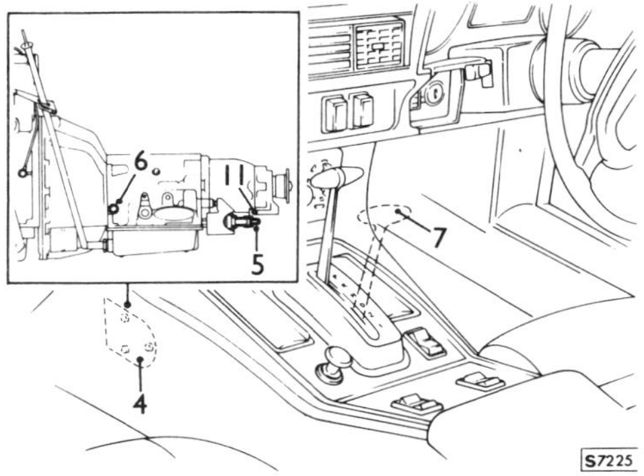
8. Accelerate engine until vacuum gauge reads 9-10 in (23-25 cm) Hg at 1,200 rev/min.
9. Check reading of line pressure gauge which should be:
$75 ^{+20}_{-5}$ lbf/in^2 ($5,27 ^{+1,40}_{-0,35}$ kgf/cm^2)
NOTE: Pressure reading below 70 lbf/in^2 (4,9 kgf/cm^2) at 1,200 rev/min will result in possible clutch slip and damage to transmission.
10. To gain access to vacuum control unit control screw, remove vacuum hose.
11. Insert screwdriver and turn clockwise to increase line pressure and anti-clockwise to decrease line pressure.
CAUTION: Approximately two full turns on screw will vary pressure about 10 lbf/in^2 (0,7 kgf/m^2). THERE IS NO LOCKNUT ON THE SCREW. Therefore, when all feel of loading on screw has been removed, it must be turned clockwise half-a-turn to ensure contact with the servo actuating rod. If contact is not maintained a rapid knocking noise will be evident between 600 to 800 rev/min.

continued

Automatic Transmission–Model 12　44–25

12 After each adjustment replace vacuum hose and re-check line pressure as before.
CAUTION: To avoid overheating of transmission, do not stall for more than 10 seconds at a time or for a total of one minute in any half-hour period.
13 Remove vacuum gauge, refit hose.
14 Remove pressure gauge, refit plug.
15 Refit access plate.
16 Refit carpet.

FRONT BRAKE BAND

Adjustment 44.30.07

Service tools: Torque screwdriver CBW.548; Screwdriver bit adaptors; Gauge block CBW.34.

1 Remove oil pan, see 44.24.04.
2 Slacken adjusting screw until it no longer contacts piston pin.
CAUTION: Adjusting screw has a left-hand thread.
3 Pull front servo operating lever back and insert gauge block CBW.34 between adjusting screw and piston pin.
4 Using torque screwdriver CBW.548 and adaptors, tighten adjusting screw to 10 lb in (,12 kg m).
CAUTION: Adjusting screw has a left-hand thread.
5 Ensure that locking spring is located in retaining slot.
6 Remove gauge block.
7 Refit oil pan.

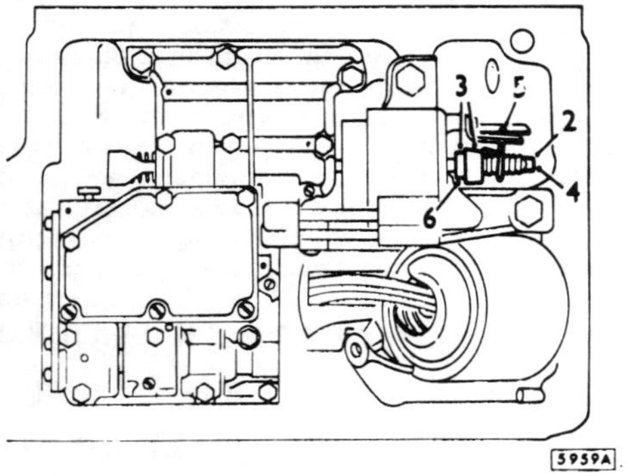

VACUUM CONTROL UNIT DIAPHRAGM

The presence of a rupture in the diaphragm will be indicated by one or all of the following:
– Engine exhaust will show increased smoking due to oil being drawn from gearbox, through diaphragm up to inlet manifold.
– Low oil level in transmission.
– Rough gear changes.

Test 44.30.09

1 Remove vacuum unit, see 44.15.27.
2 Couple control unit to a suitable vacuum source.
3 If control unit can hold 18 in (46 cm) Hg diaphragm is satisfactory, if however, vacuum falls below this figure, diaphragm is damaged and unit MUST be replaced.
4 Refit vacuum unit.

REAR BRAKE BAND

Adjustment 44.30.10

Service tool: Torque wrench CBW.547A-50.

1 Slacken locknut on adjusting screw.
2 Loosen adjusting screw approximately two full turns.
3 Tighten adjusting screw to a torque of 10 lb ft (1,4 kg m).
4 Loosen adjusting screw 1.25 turns and tighten locknut.
CAUTION: Ensure screw does not turn when tightening locknut as severe damage to transmission unit will result if screw is not backed off exactly 1.25 turns.

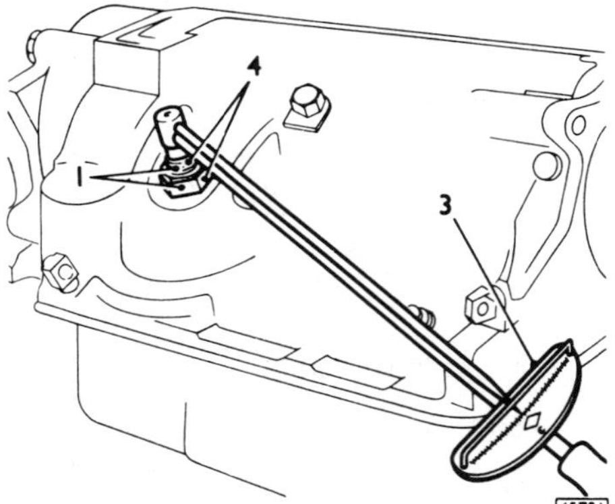

KICKDOWN SOLENOID

Test 44.30.11

1 Disconnect solenoid wire at connector.
2 With jumper lead, momentarily connect battery positive to connector.
3 Solenoid should operate with an audible click if functioning correctly.
4 Refit wire to connector.

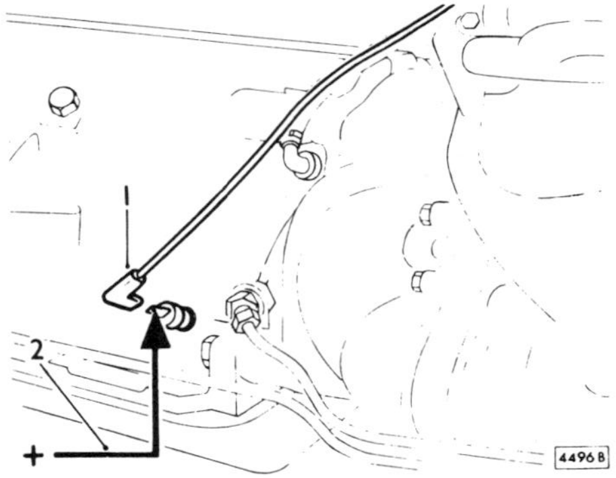

KICKDOWN SWITCH

Check and adjust 44.30.12

1 Switch on ignition and check that current is available at input terminal (cables coloured green).
2 Connect earthed test lamp to output terminal (cable colour green/white).
3 Fully depress throttle pedal.
4 If lamp does not light, release throttle pedal and gently depress switch arm.
5 If lamp still does not light, renew switch; see operation 44.15.23. If, however, lamp lights, proceed as follows:
6 Slacken securing bolts and move switch towards cable until at full throttle opening, lamp lights.
7 Tighten securing bolts and re-check.

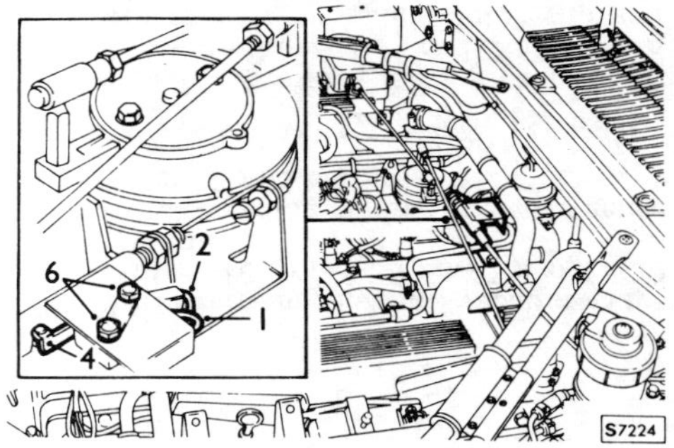

44–26 Automatic Transmission–Model 12

STALL SPEED

Test　　　　　　　　　　44.30.13

The results of this test indicate condition of gearbox and converter.
Stall speed is maximum engine revolutions recorded whilst driving impeller against stationary turbine. Stall speed will vary with both engine and transmission conditions so before attempting a stall speed check, engine condition must be determined. Engine and transmission must be at normal operating temperature before commencing check.

1　Apply handbrake.
2　Apply footbrake.
3　Start engine.
4　Select 'D'.
5　Fully depress accelerator.
6　Note tachometer reading.

CAUTION: To avoid overheating of transmission do not stall for more than 10 seconds at a time or for a total of one minute in any half-hour period.

RPM	CONDITION INDICATED
Under 1,300	Stator free wheel slip
1,950 to 2,050	Normal
Over 2,400	Clutch slip

VALVE BLOCK

Remove and refit　　　　44.40.01

Removing

1　Remove oil pan, see 44.24.04.
2　Remove kickdown solenoid, 44.15.24.
3　Remove vacuum unit and push rod, see 44.15.27.
4　Select 'R' on transmission selector.
5　Slacken, BUT DO NOT REMOVE front servo retaining bolts.
6　Remove bolt securing valve block to transmission.
7　Slide valve block sideways, withdraw servo tubes; lower valve block.
CAUTION: Ensure manual valve is not displaced.

Refitting

Reverse operations 1 to 7, ensure that spigot on detent lever is located in groove in manual valve.
Tighten fixing bolts from centre outwards to a torque of:
¼ U.N.F. bolts 5-8 lb ft (0,7-1,1 kg m).
⅟₁₆ U.N.F. bolts 17-20 lb ft (2,3-2,7 kg m).

DRIVE SHAFT

Remove and refit – Left or right hand 47.10.01

Service tool: Hub remover JD.1D

Removing

1. Jack up rear of vehicle and support on stands.
2. Remove wheel.
3. Slacken inner shroud clip and slide shroud away from flange. Remove nuts.
4. Withdraw split pin and remove castellated nut and plain washer from drive shaft in hub. Discard split pin.
5. Remove grease nipple from hub carrier.
6. Fit hub puller JD.1D to hub and secure. Withdraw hub assembly from splined shaft and remove puller.
7. Pivot hub assembly on fulcrum shaft.
8. Remove nut at damper mounting, detach lash down bracket, drift pin forward and release lower end of damper.
9. Remove drive shaft assembly; collect shims fitted between flange and brake disc.

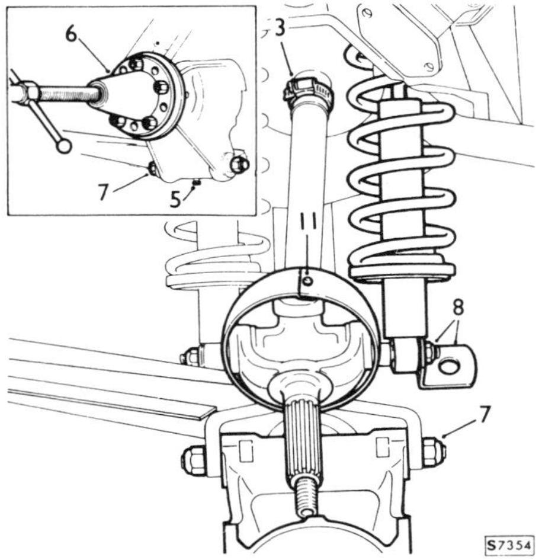

NOTE: If new drive shaft assembly is to be fitted it is now necessary to detach shrouds and remove oil seal track and spacer for fitting to replacement shaft; carry out operations 10 to 14. If the same shaft is to be refitted these operations may be omitted, and the shrouds, spacer and oil seal track left in position.

10. Remove phosphor bronze spacer and inner oil seal track from splined shaft.
11. Drill out pop rivets from shrouds, remove clips and collect halves of shrouds.
12. Place shroud halves on replacement shaft, line up holes and pop rivet halves together.
13. Seal shroud joints with underseal, fit clips, place outer shroud in position and tighten clip.
14. Fit inner oil seal track (note that chamfer clears radius on drive shaft) and spacer to splined shaft.

Refitting

15. Clean and refit camber shims, place flanged end of shaft in position and refit nuts but do not tighten at this stage.
16. Refit damper to lower mounting pin, replace lash down bracket and tighten nut to 32 to 36 lbf. ft (4,43 to 4,98 kgf. m).
17. Apply Loctite to splines, check that spacer and oil seal track are installed, raise hub carrier and enter splined shaft into hub.
18. Tap hub home on splines, replace washer, tighten nut to 100 to 120 lbf. ft (13,83 to 16,60 kgf. m) and fit split pin.
19. Check hub bearing end float, see 64.15.13.
20. Refit fulcrum grease nipple in hub carrier.
21. Tighten drive shaft nuts to 49 to 55 lbf. ft (6,7 to 7,60 kgf. m).
22. Slide inner shroud into position and tighten clip. 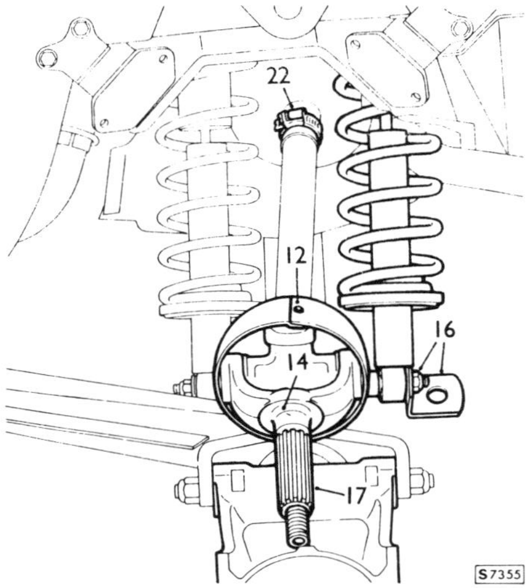
23. Replace wheel.
24. Remove car from stands.
25. Check and if necessary adjust camber angle of rear wheel, see 64.25.18.

DRIVE SHAFT

Overhaul 47.10.08

1. Remove drive shaft, see 47.10.01, operations 1 to 9.

Dismantling

2. Remove grease nipples.
3. Place shaft in vice and remove two opposed circlips.
   NOTE: Tap bearings slightly inwards to assist removal of circlips.
4. Tap one bearing inwards to displace opposite bearing.
5. Trap displaced bearing in vice and remove shaft and joint from bearing.
6. Replace shaft in vice, displace second bearing by tapping joint spider across and extract second bearing.
7. Remove two grease seals from spider.
8. Detach spider, with end section of shaft, from centre section of shaft.

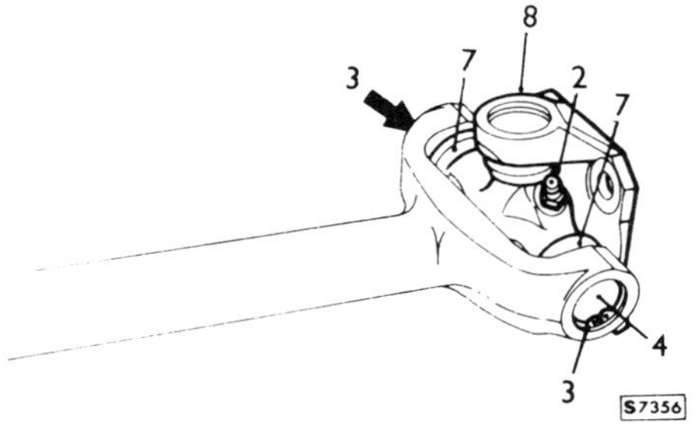

9. Place end section of shaft in vice and repeat operations 3 to 7.
10. Remove spider from end section of shaft.
11. Repeat operations 3 to 10 on opposite end of shaft.

Inspection

12. Wash all parts in petrol.
13. Check splined yoke for wear of splines.
14. Examine bearing races and spider journals for signs of looseness, load markings, scoring or distortion.
    NOTE: Spider or bearings should not be renewed separately, as this will cause premature failure of the replacement.
    It is essential that bearing races are a light drive fit in yoke trunnion.

Re-assembling

15. Remove bearing assemblies from one replacement spider; if necessary, retain rollers in housings with petroleum jelly. Leave grease shields in position.
16. Fit spider to one end section of shaft.

continued

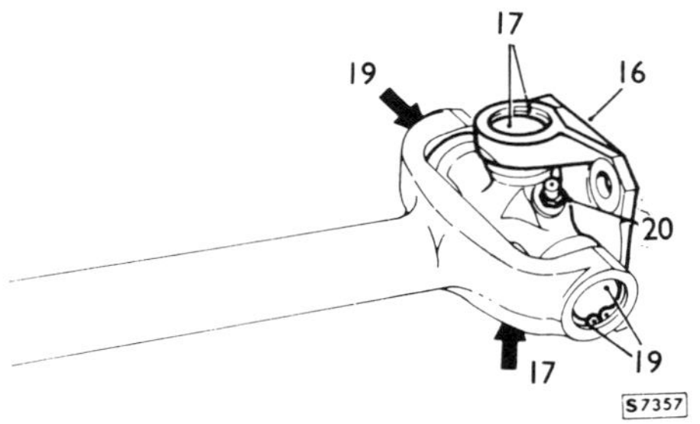

47–1

17 Fit two bearings and circlips in end section trunnions. Use a soft round drift against bearing housings.
18 Insert spider in trunnions of centre section of shaft.
19 Fit two bearings and circlips in centre section trunnions.
20 Fit grease nipple to spider.
21 Repeat operations 15 to 20 on opposite end of drive shaft.
22 Grease joints with hand grease gun.
23 Refit drive shaft, see 47.10.01, operations 15 to 25.

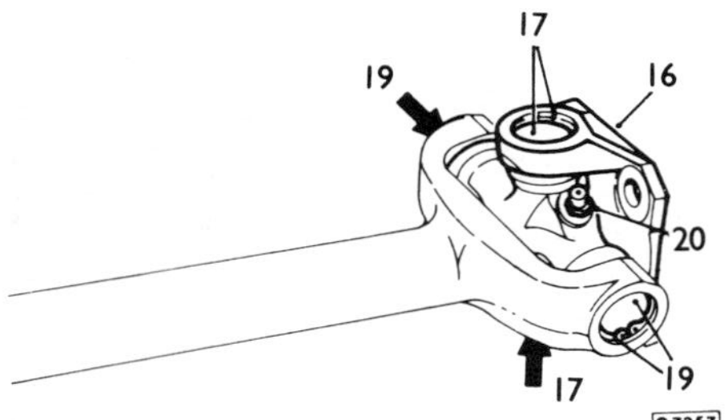

PROPELLER SHAFT

NOTE: Two different types of propeller shaft are fitted, dependent on which form of final drive is specified. Cars with standard final drive units are fitted with propeller shafts of 3 inches (76,2 mm) diameter, but if optional 2-speed final drive units are specified rubber damped shafts of 3.5 in (88,9 mm) diameter are fitted. These shafts are not interchangeable but operation instructions which follow apply to either type of shaft.

PROPELLER SHAFT

Remove and refit   47.15.01

Service tool: Engine support bracket MS.53A

Removing

1 Place car on ramp (hoist) or over pit and apply handbrake.
2 Fit engine support bracket hooks through rear lifting eyes, assemble crossbar and nuts and tighten nuts until rear end of engine is very slightly raised.
3 Raise ramp.
4 Remove three nuts, bolts and washers securing each intermediate exhaust pipe to downpipe flanges and separate pipes from flanges.
5 Remove four screws securing rear heat shield to underfloor and detach shield.
6 Remove bolt securing intermediate heat shields to rear engine mounting.
7 Remove four screws securing each intermediate heat shield to underfloor, and detach shields.
8 Remove three nuts and washers securing strengthening plate to rear of gearbox.
9 Remove nut securing strengthening plate to rear mounting spigot and detach plate.
10 Place jack under rear mounting spring plate with wooden block between jack head and plate.
11 Operate jack to support rear mounting.
12 Remove four bolts securing rear mounting to underfloor; collect washers, and two spacers from rear bolts.
13 Lower jack and remove rear mounting assembly. Remove jack.
14 Remove rear mounting spigot to improve access to propeller shaft flange bolts.
15 Remove four nuts securing forward universal joint flange to gearbox output flange.
16 Remove four bolts securing rear universal joint flange to final drive input flange.
17 Remove propeller shaft.

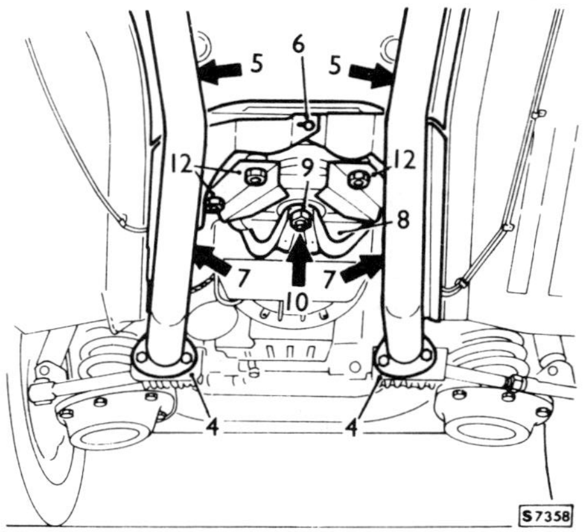

NOTE: To protect rubber gaiter over splines from damage, retain shaft in fully compressed position by wiring universal joint yokes together with 18 s.w.g. (1,2 mm) wire.

Refitting

NOTE: When fitting the revised propeller shaft for the Model 12 Transmission, part No. CAC 2190/3, the sliding coupling must be to the **FRONT** of the car. The revised propeller shaft for the GM 400 Transmission, part No. CAC 2191/1, must be fitted with **sliding coupling to the REAR of the car**.

Check that the battery cable has been repositioned and for the GM 400 gearbox that the mounting bracket is modified.

18 Lift propeller shaft into place, with splined end as above, and enter bolts into forward universal joint flange. Remove wire from joint yokes.
19 Fit nuts and tighten to 27 to 32 lbf. ft (3,73 to 4,42 kgf. m).
20 Raise rear end of shaft, insert bolts and tighten nuts to 27 to 32 lbf. ft (3,73 to 4,42 kgf. m).
21 Replace rear mounting spigot and tighten.
22 Position jack under rear engine mounting location, with wooden block on jack head.
23 Replace mounting spring and seating rubber on spigot.
24 Place mounting assembly under spigot and raise jack to compress spring.
25 Fit lower distance tube into mounting bush and fit spigot nut. Do not tighten.
26 Line up mounting bolt holes and fit but do not tighten front two bolts.
27 Lower jack and remove with wooden block.
28 Fit rear bolts and distance pieces and tighten to 27 to 32 lbf. ft (3,73 to 4,42 kgf. m).
29 Tighten front bolts to 14 to 18 lbf. ft (1,94 to 2,49 kgf. m).
30 Remove spigot nut and place strengthening plate in position on spigot and transmission studs.
31 Replace spigot nut and tighten to 25 to 30 lbf. ft (3,46 to 4,15 kgf. m).
32 Fit washer and nuts to transmission studs and tighten nuts.
33 Replace intermediate heat shields and refit securing screws and bolt.
34 Replace rear heat shield and refit securing screws.
35 Refit exhaust intermediate pipes to down pipes. Make joints with Firegum, and tighten nuts to 11 to 13 lbf. ft (1,52 to 1,80 kgf. m).
36 Lower ramp.
37 Remove engine support bracket and hooks.

47–2

PROPELLER SHAFT

Overhaul 47.15.10

1. Remove propeller shaft assembly, see 47.15.01.
2. Thoroughly clean shaft and secure in vice.

Dismantling
3. Remove circlip from both sides of fork at one end of shaft.
   NOTE: Circlip removal will be assisted if bearings are tapped inwards.
4. Tap spider across fork to displace one bearing. 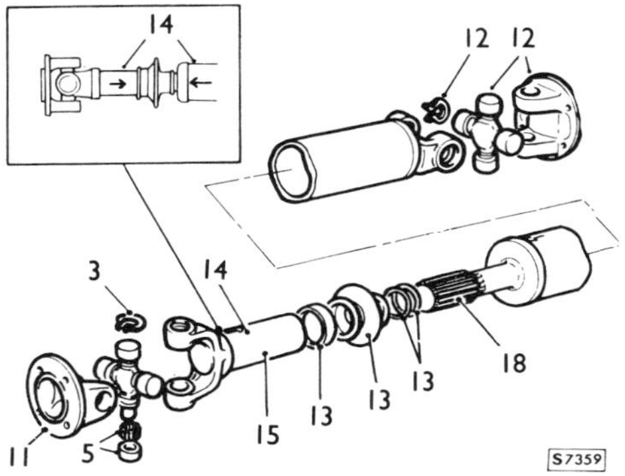
5. Trap cup of displaced bearing in vice and extract cup from fork. Collect rollers and grease seal; discard seal.
6. Tap spider across fork to displace second bearing.
7. Trap cup of second bearing in vice and extract cup from fork. Collect rollers and grease seal; discard seal.
8. Separate spider and flange from shaft.
9. Secure flange in vice and remove both circlips.
10. Repeat operations 4 to 7 on flange.
11. Separate spider from flange.
12. Repeat operations 3 to 11 on opposite end of shaft.
13. Remove metal and plastic clips from gaiter and slide gaiter along shaft.
14. Mark sleeve and shaft for reassembly (if not already marked).
15. Withdraw sleeve yoke from spline.

Inspection
16. Wash all parts in petrol, taking care not to transpose bearings and cups from mating journals.
17. Examine all bearing races and spider journals for signs of looseness, load markings, scoring or distortion.
    NOTE: Spiders or bearings may only be renewed as a complete set. Bearing cups must be a light drive fit in trunnions.
18. Examine spline for wear. Check circumferential play. If this exceeds 0.004 in (0,1 mm) measured on spline outside diameter, complete propeller shaft must be renewed.

Reassembling
19. Place two new grease seals over opposed journals of spider and insert spider through trunnions of one flange.
    NOTE: If new units are to be fitted, carefully remove cups and rollers before fitting spider to trunnions.
20. Pack cup and roller assemblies with recommended grease and press into position in trunnions.
21. Fit seals over inner ends of cups and retain cups with circlips.
22. Place new grease seals on second pair of spider journals and insert into shaft trunnions.
23. Grease second pair of cup and roller assemblies and press into position in shaft trunnions. Fit seals and circlips.
24. Repeat operations 19 to 23 on second universal joint.
25. Check stiffness of each joint which should be between 3 and 8 lbf. in (0,035 to 0,092 kgf. m) about either axis. Exercise joints if necessary to reduce stiffness.
26. Fit new gaiter and plastic and metal clips to sleeve yoke.
27. Grease splines.
28. Fit sleeve yoke to splines, ensuring that alignment marks coincide.
29. Secure gaiter clips.
    NOTE: If shaft is not to be fitted immediately, retain in fully compressed position with 18 s.w.g. (1,2 mm) wire between flange yokes.

47–3

DESCRIPTION

The standard transmission unit is a Salisbury 4HU final drive incorporating a 'Powr Lok' differential.

The unit is rigidly attached to a fabricated sheet steel crossbeam which is flexibly mounted to the body structure by four rubber and metal sandwich mountings.

Noises coming from the vicinity of the final drive unit usually originate from incorrect meshing of drive gear and pinion, or from bearings on differential or pinion shafts developing play. Operation procedures for the correction of these noise sources are fully covered in operation 51.25.19, but a noise occurring at low speeds only, under braking, could be caused by loss of pre-load in the output shaft bearings. Bearing inspection involves the removal and renewal of an oil seal before resetting pre-load, and is covered in 51.20.04, while if inspection indicates that bearing renewal is advisable this is detailed in 51.10.22.

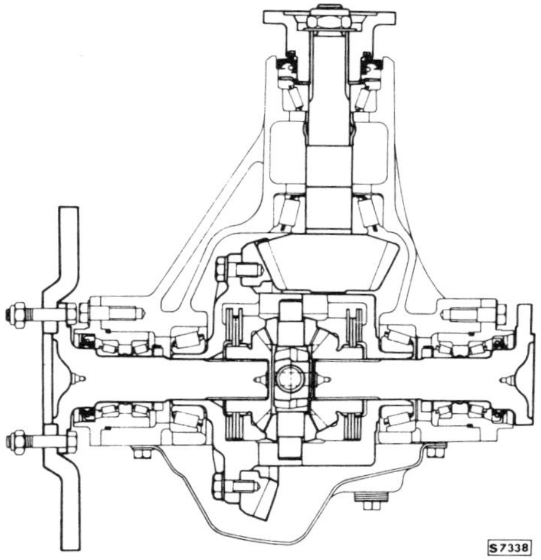

4HU Final Drive

OUTPUT SHAFT ASSEMBLY

Remove and refit (one side)   51.10.20

Removing

1. Remove brake disc attached to flange of shaft, see 70.10.11. Collect and retain separately any shims fitted at inner and outer faces of disc.
2. Cut locking wire and remove five set bolts securing caliper mounting bracket to final drive casing.
3. Withdraw output shaft assembly. Discard 'O' ring.

Refitting

NOTE: Before fitting replacement output shaft assembly check length of shaft on replacement unit against that removed, and ensure that four disc attachment bolts are in position in flange.

4. Fit new 'O' ring seal into groove in bearing housing.
5. Lightly oil splines and outside of bearing housing and replace assembly in final drive casing.
6. Fit five setbolts with spring washers to secure caliper mounting bracket to final drive and tighten to 60 to 69 lbf. ft (8,3 to 9,5 kgf. m).
7. Secure setbolts by wire-locking bolt heads together; arrange wire tension to pull bolt heads in a clockwise direction.
8. Replace brake disc, see 70.10.11, ensuring that shims are correctly replaced on both sides of disc.
9. Check camber angle of rear wheels and adjust if necessary, see 64.25.18.

OUTPUT SHAFT BEARINGS

Remove and refit   51.10.22

Service tools: Torque screwdriver CBW.548, Adaptor, Spanner SL.15 or 15A, Output shaft bearing remover/replacer SL.47

Removing

1. Remove output shaft assembly incorporating bearing to be removed, see 51.10.20 operations 1 to 3.
2. Clean assembly and clamp caliper mounting bracket between suitably protected jaws of vice.
3. Turn down tabs of lock washer and remove nut from shaft, using spanner SL.15. Remove and discard lock washer.
4. Withdraw output shaft from caliper mounting bracket. Collect inner bearing and cone. Discard collapsed spacer.
   NOTE: If outer bearing remains on shaft and pushes oil seal out of caliper mounting bracket on withdrawal, remove it from shaft using tool SL.47.
5. Prise oil seal from caliper mounting bracket. Collect outer bearing and cone. Discard oil seal.
6. Using a brass drift, gently tap bearing cups out of housing. 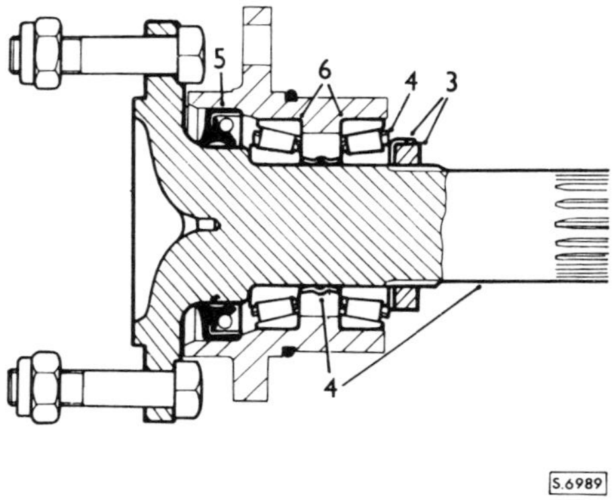
7. Remove caliper mounting bracket from vice and carefully clean internally.

NOTE: When bearings are to be renewed, always replace complete bearings. Never fit new cone and roller assemblies into used cups. Pack bearings with No. 2 grease (preferably with E.P. additive) before fitting.

Refitting

8. Press cups of replacement bearings into housing, using suitable press and adaptors to ensure that cups are pressed fully home in housing.
9. Place roller and cone assembly of outer bearing (already greased) in position.
10. Press replacement oil seal into position, ensuring that spring-loaded sealing edge is adjacent to bearing. Load seal with grease between sealing edges.
11. Clamp caliper mounting bracket between protected jaws of vice.
12. Check that four special bolts for brake disc are in position in output shaft flange and enter shaft through seal and outer bearing.
13. Fit new collapsible spacer, roller and cone assembly of inner bearing and new lock washer on to shaft.
14. Place nut on shaft, grease face next to washer and tighten finger-tight only.
15. Using torque screwdriver CBW.548 and suitable adaptor, check torque required to turn shaft in caliper mounting bracket against resistance of oil seal. Record torque.

NOTE: Set screwdriver initially to 4 lbf. in (0,05 kgf. m). Setting should then be progressively increased until torque figure is established at the point when shaft commences to turn.

16. Using spanner SL.15 and a tommy-bar at disc attachment bolts to oppose torque, tighten nut on shaft just sufficiently to almost eliminate play from bearings. Repeat operation 15. Torque required to turn shaft should be unchanged; if it has increased, slacken nut very slightly and re-check. Tighten nut very slightly further (not more than a thirty-second of a turn — about $\frac{3}{16}$ in (5 mm) at perimeter of nut) and re-check torque required to turn shaft. If this torque exceeds by between 4½ and 5½ lbf. in (0,052 and 0,063 kgf. m) the torque recorded in operation 15, correct bearing pre-load has been achieved. Otherwise, continue to tighten nut in very small increments, measuring torque after each increment, until correct figure is reached.

CAUTION: If torque required to turn shaft exceeds by more than 6 lbf. in (0,007 kgf. m) torque recorded in operation 15, it is necessary to dismantle assembly, discard collapsed spacer and rebuild with new collapsible spacer. It is not permissible

51–1

3. Turn down tabs of lock washer and remove nut from shaft, using spanner SL.15. Remove and discard lock washer.
4. Withdraw output shaft from caliper mounting bracket. Collect inner bearing and cone and mark for correct reassembly. Discard collapsed spacer.
5. Prise oil seal from caliper mounting bracket and discard. Collect outer bearing and cone.
6. Remove caliper mounting bracket from vice and thoroughly clean internally.

NOTE: Carefully inspect taper roller bearing components before refitting. If any fault is found in either bearing, replace both complete bearings. Refer to 50.10.22, operations 6 and 8 to 19. Never fit new cone and roller assemblies into used cups. If bearings are satisfactory, grease with Silkolene grade 904 petroleum grease or if this is not available, final drive oil.

Refitting

7. Place roller and cone assembly of outer bearing (already greased) in position.
8. Press replacement oil seal into position, ensuring that spring-loaded sealing edge is adjacent to bearing. Load seal with grease between sealing edges.
9. Clamp caliper mounting bracket between protected jaws of vice.
10. Check that four special bolts for brake disc are in position in output shaft flange and enter shaft through seal and outer bearing.
11. Fit new collapsible spacer, followed by marked roller and cone assembly for inner bearing and new lock washer.
12. Place nut on shaft, grease face next to washer and tighten finger-tight only.
13. Using torque screwdriver CBW.548 and adaptor check torque required to turn shaft in caliper mounting bracket against resistance of oil seal. Record torque.

NOTE: Set screwdriver initially to 4 lbf. in (0,05 kgf. m). Setting should then be progressively increased until torque figure is established at the point when shaft commences to turn.

13. Repeat operation 3. Torque required to turn pinion shaft through backlash should exceed by 5 to 10 lbf. in (0,06 to 0,12 kgf. m) the torque recorded in operation 3. If torque is below 25 lbf. in (0,29 kgf. m), tighten flange nut further until torque figure is between 25 and 30 lbf. in (0,29 and 0,35 kgf. m). If, however, torque required to turn pinion shaft exceeds 45 lbf. in (0,52 kgf. m), final drive overhaul, operation 51.25.19 MUST be carried out.
14. Lift propeller shaft into position, replace bolts, fit and tighten nuts to 27 to 32 lbf. ft (3,73 to 4,42 kgf. m).
15. Check oil level in final drive unit and top up if necessary.
16. Remove car from ramp and road test. If final drive is noisy, overhaul procedure 51.25.19 MUST be carried out.

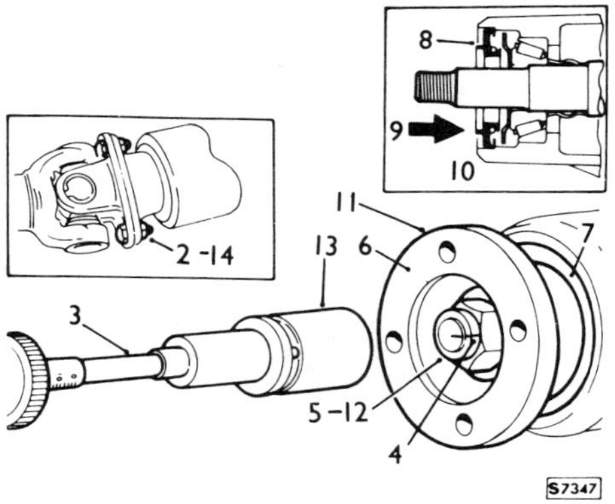

OUTPUT SHAFT OIL SEAL

Remove and refit　　51.20.04

Service tools: Torque screwdriver CBW.548, Adaptor, Spanner SL.15 or SL.15A.

Removing

1. Remove output shaft assembly, see 51.10.20, operations 1 to 3.
2. Clean assembly and clamp caliper mounting bracket between suitably protected jaws of vice.

to slacken back nut after collapsing spacer as bearing cones are then no longer rigidly clamped.

18. Turn down tab washers in two places to lock nut and remove assembly from vice.
19. Refit output shaft assembly to final drive unit, see 51.10.20, operations 4 to 10.

PINION BEARINGS

Remove and refit　　51.15.19

Pinion bearing replacement entails stripping of final drive assembly and resetting of mesh of pinion and drive gear. Refer, therefore, to Final Drive – Overhaul, 51.25.19.

DIFFERENTIAL ASSEMBLY (POWR LOK)

Overhaul　　51.15.30

Refer to 51.25.19, operations 1 to 7, 14 to 20 and 24 to 32. Refit after overhaul as detailed in later operations.

DRIVE FLANGE

Remove and refit　　51.15.36

CAUTION: In order that correct pinion bearing pre-load is obtained, carry out operation 51.20.01.

DRIVE PINION SHAFT OIL SEAL

Remove and refit　　51.20.01

Service tool: Torque screwdriver CBW.548, Hand press SL.14 with adaptor.

Removing

1. Place car on ramp, set handbrake, chock wheels and raise ramp.
2. Remove nuts securing propeller shaft flange to final drive flange; remove bolts and disconnect propeller shaft.
3. Using torque screwdriver CBW.548, suitable adaptor and socket, check and record torque required to turn final drive flange clockwise through backlash movement.

NOTE: Set screwdriver initially to 15 lbf. in (0,17 kgf. m) and increase setting progressively until torque figure is reached at which flange commences to move. Flange MUST be turned fully anti-clockwise through backlash between each check.

4. Mark nut and pinion shaft so that in refitting, nut may be returned to its original position on shaft.
5. Unscrew nut and remove washer and place both washer and nut aside for refitting.
6. Draw flange off pinion shaft using extractor.
7. Prise oil seal out of final drive casing.

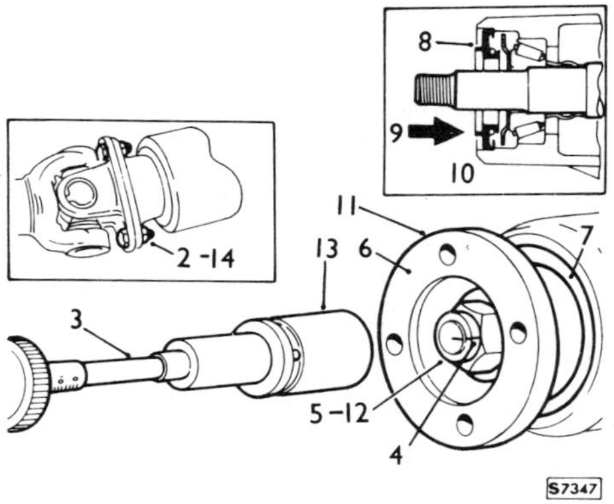

Refitting

8. Thoroughly clean splines on pinion shaft and flange, removing all old Loctite. Clean oil seal recess and coat internally with Welseal liquid sealant.
9. Using suitable 'bell-piece' tap new oil seal squarely into position with sealing lip facing to rear. Smear sealing lip with grease.
10. Apply Loctite grade A.V.V. to outer two thirds of pinion shaft splines.
11. Lightly tap flange back on pinion shaft, using wooden mallet.
12. Refit washer and nut not removed in operation 5 and tighten nut until it exactly reaches position marked in operation 4.

3 Remove 14 bolts and setscrews securing bottom tie plate to crossbeam and inner fulcrum brackets.

HYPOID CASING REAR COVER GASKET

Remove and refit   51.20.08

Removing
1. Place car on ramp, set handbrake, chock wheels and raise ramp.
2. Remove 14 bolts and setscrews securing bottom tie plate to crossbeam and inner fulcrum brackets.
3. Remove drain plug and collect oil in suitable container. Discard oil.
4. Remove 10 setscrews securing rear cover. Retain identification tabs carried by three screws.
5. Detach rear cover and remove gasket.
6. Clean faces of final drive casing and cover.

Refitting
7. Smear both faces of new gasket with grease and place on final drive casing.
8. Place cover over gasket and insert two screws and spring washers to retain, coating threads with Loctite.
9. Replace remaining 8 setscrews, coating threads with Loctite and replacing spring washers and tabs.
10. Tighten screws by diagonal selection to 18 to 20 lbf. ft (2,49 to 2,77 kgf. m).
11. Refit drain plug.
12. Remove filler plug from cover and fill with recommended lubricant until oil escapes from filler hole.
    NOTE: Car must be level for lubricant replenishment.
13. Refit filler plug.
14. Replace bottom tie plate and tighten bolts and setscrews to 14 to 18 lbf. ft (1,94 to 2,49 kgf. m).
15. Remove car from ramp.

FINAL DRIVE UNIT

Remove and refit   51.25.13

Service tool: Dummy shaft JD.14.

Removing
1. Remove rear suspension unit, see 64.25.01.
2. Invert unit and place on workbench.
3. Remove locking wires from setscrews securing fulcrum brackets to final drive unit. Remove setscrews and remove brackets, noting position and number of shims at each attachment point.
4. Remove nuts and washers securing dampers to wishbone.
5. Drift out damper mounting pins; recover spacers and tie-down brackets.
6. Slacken clips securing inner universal joint shrouds and displace shrouds outwards.
7. Remove four self locking nuts securing drive shaft inner universal joint to brake disc and output flange on one side.
8. Remove nut from inner wishbone fulcrum shaft.
9. Drift out shaft, collecting spacers, seals and bearings for wishbone pivots.
10. Remove drive shaft, hub and wishbone assembly from rear suspension.
11. Remove camber shims from drive shaft flange studs at brake disc.
12. Repeat operations 7 and 11 on opposite side assembly.
13. Tap spacer tubes from between lugs of fulcrum brackets.
14. Turn suspension assembly over on bench.
15. Disconnect brake feed pipes at calipers. Seal ports and ends of pipes.
16. Release brake return springs at handbrake levers.
17. Cut locking wire and remove bolts attaching final drive to crossbeam.
18. Lift suspension crossbeam off final drive unit. 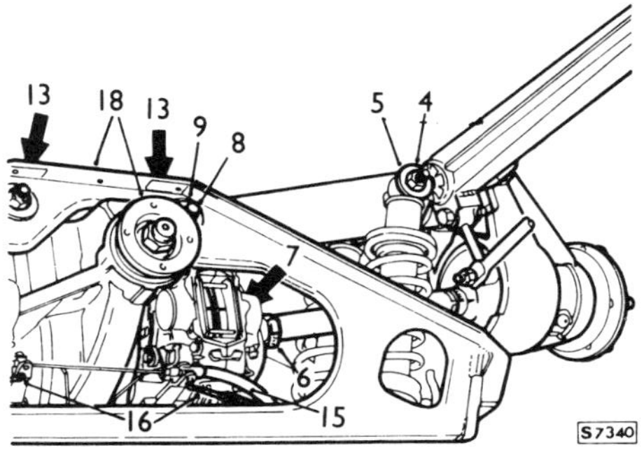
19. Invert final drive unit.
20. Remove locking wires from setscrews securing caliper mounting bolts; remove bolts and calipers.
21. Cut locking wires from caliper mounting bolts; remove bolts and calipers.
22. Remove brake discs, noting number of shims between discs and flanges.

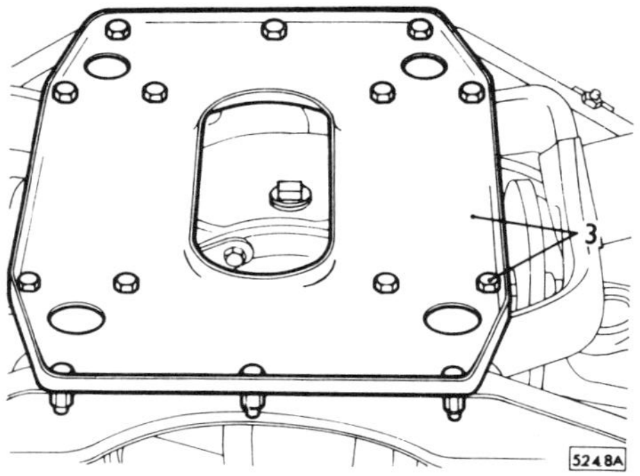

Refitting
23. Replace shims and disc on one drive shaft flange; secure with two nuts.
24. Replace caliper, tighten mounting bolts and check centering and run-out of disc.
25. Correct centering of disc, if necessary, to within ± 0.010 in (0,25 mm) by transferring shims from one side of disc to the other.
26. Tighten caliper bolts to 49 to 55 lbf. ft (6,77 to 7,60 kgf. m) and wire lock.
27. Repeat operations 24 to 27 on opposite side.
28. Remove nuts from both discs.
29. Place crossbeam over final drive, align, replace bolts, tighten to 70 to 77 lbf.ft (9,68 to 10,64 kgf. m) and wire lock.
30. Slacken brake feed pipes at centre union.
31. Unseal brake pipes and ports, align pipes and tighten union nuts to 6.3 to 7 lbf. ft (0,88 to 0,96 kgf. m).

continued

14. Using spanner SL.15 and a tommy-bar at disc attachment bolts to oppose torque, tighten nut on shaft just sufficiently to almost eliminate play from bearings. Repeat operation 13. Torque required to turn shaft should be unchanged; if it has increased, slacken nut very slightly and recheck.
15. Tighten nut very slightly further (not more than $\frac{3}{16}$ in (5 mm) at perimeter of nut and recheck torque required to turn shaft. If this torque exceeds 2½ to 3 lbf. in (0,03 and 0,035 kgf. m) the torque recorded in operation 13, correct bearing pre-load has been achieved; otherwise continue to tighten nut in very small increments, measuring torque after each increment, until correct figure is reached.
    CAUTION: If torque required to turn shaft exceeds more than 4 lbf. in (0,045 kgf. m) torque recorded in operation 13, it is necessary to dismantle assembly, discard collapsed spacer and rebuild with new collapsible spacer. It is not permissible to slacken back nut after collapsing spacer as bearing cones are then no longer rigidly clamped.
16. Turn down tab washers in two places to lock nut and remove assembly from vice.
17. Refit output shaft assembly to final drive unit, see 51.10.20, operations 4 to 9.

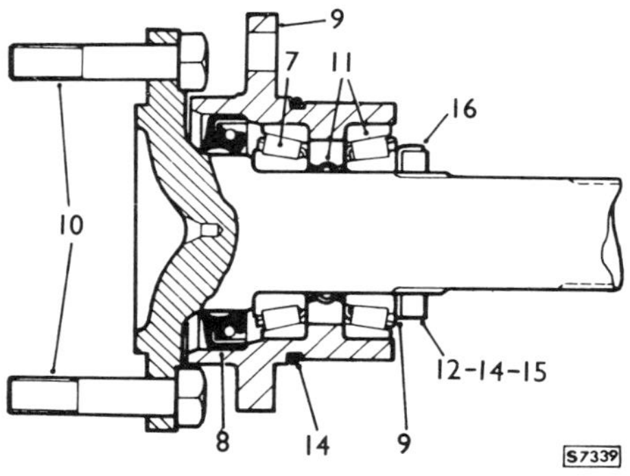

51–3

23 Press pinion inner bearing (or its replacement) into position against pinion, using tool No. SL.47-1.
24 Press differential bearings, without shims, against sides of differential halves, using replacer tool No. 550-1
25 Assemble drive gear to flange, fit set bolts and tab washers and tighten to 70 to 88 lbf. ft (9,68 to 12,17 kgf. m). Secure tab washers.
26 Replace clutch plates and discs, as shown, in differential halves. Belleville-type plates fit with their convex sides against differential casing.
27 Assemble rings and side gears to differential halves.
28 Assemble cross shafts together, fit pinions, and position on flanged half of differential casing, ensuring that ramps on shafts coincide with mating ramps on case.
29 Offer up differential halves together in marked position and insert eight bolts but do not tighten.
30 Check alignment of splines in differential casing and side gears by inserting drive shafts.

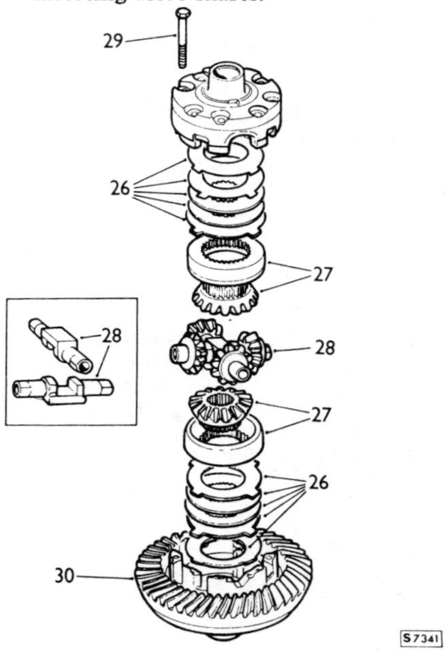

13 Tap outer bearing from housing; collect complete bearing, oil thrower and oil seal.
14 Mount differential unit in vice and remove eight through-bolts.
15 Mark halves of differential casing to ensure correct reassembly, and separate halves.
16 Remove clutch plates, side ring and gear from one side of differential casing.
17 Remove cross shafts and gears, mark for reassembly and separate.
18 Remove clutch plates, side ring and gear from opposite half of differential casing.
19 Release locking tabs and remove bolts securing drive gear to flange. Discard tab washers and remove gear.
20 Remove side bearings using tool No. SL.47 (formerly SL.14) with cone remover SL.47-2.
21 Remove collapsible spacer from pinion shaft and remove inner bearing using tool No. SL.47-1.
22 Clean all parts and inspect.

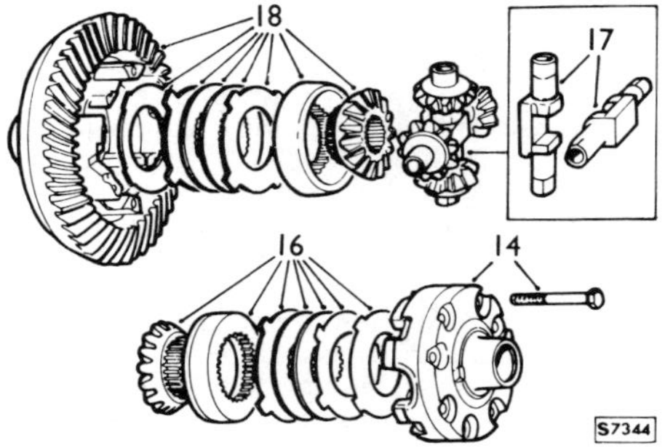

Reassembling
NOTE: Before commencing assembly, check from reference numbers and letters that pinion and drive gear are a matched pair, mounted in the correct case. The same serial number must be marked on pinion end and outer periphery of drive gear, and a letter and number (e.g. 1H or 2L) stamped on the casing rear face, must be duplicated on the pinion end. If these requirements are not met the unit must be exchanged.

FINAL DRIVE UNIT

Overhaul 51.25.19

Service tools: Puller SL.47 (formerly SL.14) with adaptor SL.47-2 and SL.47-1. Gauge mount SL.3. Gauge block 4.HA, Handle 550, Adaptor SL.550-1, SL.550-8 and SL.550-9.

Dismantling
1 Cut locking wire and remove five bolts securing caliper mounting bracket on one side.
2 Withdraw output shaft assembly.
3 Carry out operations 1 and 2 on opposite side output shaft assembly.
4 Remove ten bolts securing rear cover and detach cover. Discard gasket but retain identification tabs.
5 Mount unit in vice and unscrew two bolts securing each differential bearing cap. Mark caps for refitting.
6 Remove bearing caps complete with bolts.
7 Using two levers suitably padded to prevent damage to differential or carrier, prise out differential assembly. Collect bearing cups and mark to identify with bearings.
8 Re-position unit in vice with jaws gripping pinion flange.
9 Remove pinion flange nut and washer.
10 Using suitable press, extract pinion from housing. Collect flange.
11 Using drift, tap inner bearing cup out of housing. Collect shims.

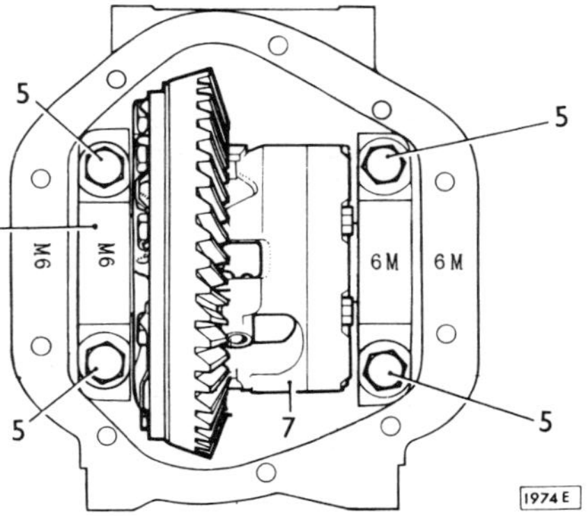

32 Replace handbrake lever return springs.
33 Invert assembly on bench.
34 Position fulcrum brackets against final drive unit and locate each bracket loosely with two setscrews.
35 Replace shims removed in operation 20 between brackets and final drive unit.
36 Tighten setscrews to 60 to 65 lbf. ft (8,30 to 8,98 kgf. m) and wire lock.
37 Refit camber shims, removed in operation 11, to drive shaft studs on one side.
38 Replace drive shaft on studs, and fit nuts loosely, then tighten to 49 to 55 lbf. ft (6,78 to 7,60 kgf. m).
39 Replace spacer tube between lugs of fulcrum bracket.
40 Clean, inspect and grease wishbone bearings, thrust washers etc. Refit with new oil seals.
41 Offer up wishbone fulcrum bracket lugs and locate with dummy shafts, tool No. JD.14. Take great care not to displace any component during this operation.
42 Drift dummy shafts from fulcrum bracket with fulcrum shaft. Restrain dummy shafts to prevent spacers or thrust washers dropping out of position.
43 Tighten fulcrum shaft nut to 45 to 50 lbf. ft (6,23 to 6,91 kgf. m).
44 Reposition and secure drive shaft shroud.
45 Line up damper lugs with wishbone bosses and replace damper shaft, including spacer and tie-down bracket. Tighten nuts to 32 to 36 lbf. ft (4,43 to 4,97 kgf. m).
46 Repeat operations 31 to 39 on opposite side.
47 Replace bottom tie plate and tighten bolts and setscrews to 14 to 18 lbf. ft (1,94 to 2,48 kgf. m).
48 Replace rear suspension unit, see 64.25.01.
49 Check rear wheel camber, see 64.25.18.
50 Bleed brakes, see 70.25.01.
51 Check final drive oil level and top up as necessary.

51—4

31 Tighten bolts to 43 to 50 lbf. ft (5,94 to 6,91 kgf. m) with drive shafts in position.
32 Check backlash in differential; with one drive shaft locked, the other shafts must not turn more than 0.75 in (19 mm) measured on a 6 in (152 mm) radius. Remove drive shafts.
33 Place cups in position on differential bearings and fit assembly into housing, without pinion installed.
34 Install dial gauge (tool No. SL.3) as shown, with stylus against back face of drive gear. 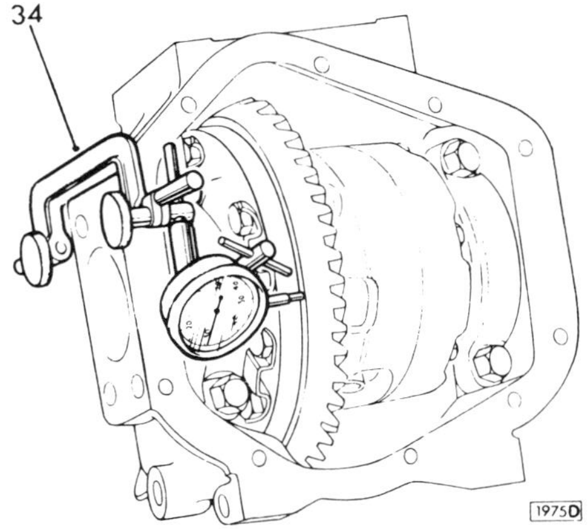
35 Using two levers between differential casing and bearing housing, move differential assembly to one side of carrier.
36 Set dial gauge to zero.
37 Lever differential assembly to opposite side of carrier and record dial gauge reading, which represents total clearance between bearing cones and abutment faces of differential casing. NOTE: This dimension plus 0.009 in (0,23 mm) for pre-load, represents the total shim thickness required for installation of differential bearings. Shims are divided between left and right hand bearings to give correct pinion mesh and backlash.
To obtain a silent-running final drive it is essential that pinion and drive gear are correctly meshed. Procedure to achieve this is as follows:

a Mount pinion at correct 'cone setting distance' by inserting correct shims behind cup of inner pinion bearing to bring pinion face to specified position.
b Install differential and drive gear with shims divided between left and right to achieve correct backlash as etched on drive gear.
c Check run-out of drive gear and correct if more than 0.005 in (0.013 mm).
d Re-check backlash of gears if run-out has been corrected and adjust again if necessary to obtain figure etched on gear.
e Check mesh of gears and rectify if necessary.
f Re-check backlash and adjust if necessary to maintain a minimum of 0.004 in (0,10 mm).
Above procedure is detailed in operating instructions which follow.

38 Remove dial gauge, unbolt bearing caps and remove differential assembly with bearings and drive gear.
39 Fit pinion outer bearing cup using tool No. SL.550-8 with 550 handle. 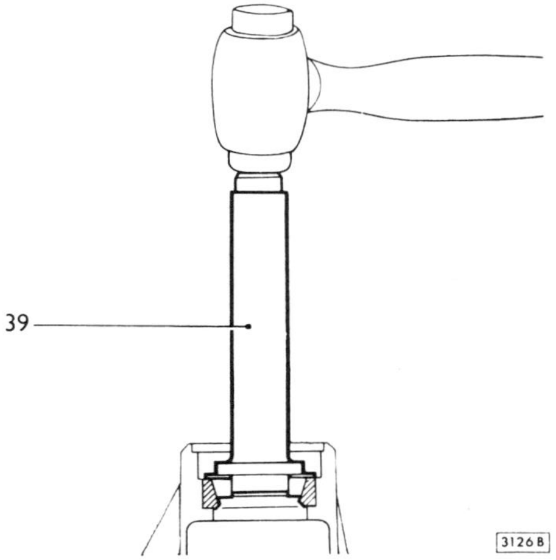
40 Replace shims removed from under inner bearing cup and fit inner bearing cup using tool No. SL.550-9
41 Replace pinion, with inner bearing, in casing.

42 Fit pinion outer bearing, flange, washer and nut, omitting collapsible spacer, oil thrower and seal. Tighten nut just sufficiently to eliminate all end play.
43 Set dial gauge, tool No. SL.3 to zero using setting block 4HA.
44 Set up zeroed gauge as shown, with assembly firmly seated on ground face of pinion. Move assembly slightly to obtain minimum reading on dial. This should coincide in thousandths of an inch, with a figure etched on pinion end face, as shown. Note magnitude and sense of any variation. 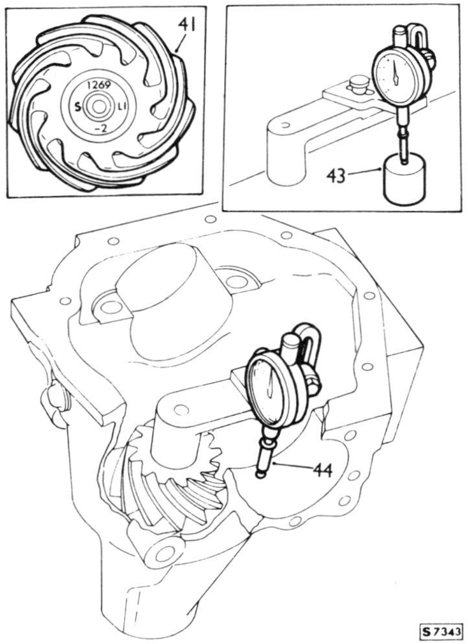
45 If pinion setting is incorrect it is necessary to dismantle pinion assembly and remove pinion inner bearing cup. Add or remove shims as required from pack locating bearing cup and re-install shim pack and bearing cup. Adjusting shims are available in thicknesses of 0.003 in, 0.005 in and 0.010 in (0,076 mm, 0,127 mm and 0,254 mm). Repeat operations 44 and 45 until satisfactory result is obtained.

46 Extract pinion shaft from gear carrier sufficiently far to enable outer bearing cone to be removed from pinion.
47 Fit collapsible spacer to pinion ensuring that it seats firmly on machined shoulder on pinion shaft.
48 Insert pinion into gear carrier.
49 Refit outer-bearing cone, oil thrower and oil seal using suitable 'bell-piece' to fit seal.
50 Lightly grease splines of pinion shaft and fit flange.
51 Fit new washer, convex side facing towards end of shaft, and nut, but DO NOT tighten.
52 Refit drive gear and differential assembly, with drive gears but without shims, as removed in operation 38, carefully leading drive gear teeth into mesh with pinion.
53 Refit dial gauge with stylus against back face of drive gear. 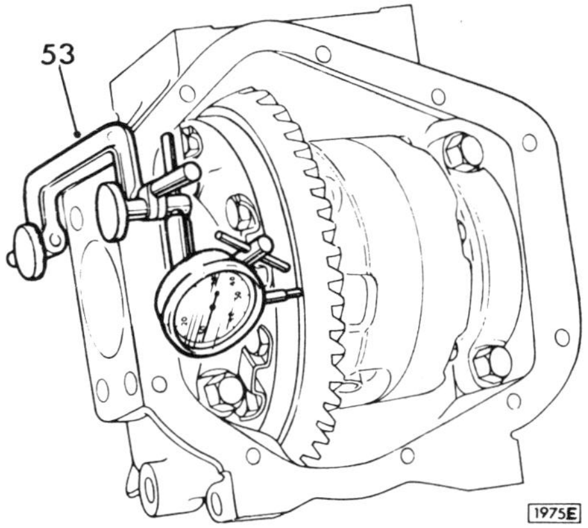
54 Using two levers between differential casing and bearing housing, move assembly away from pinion until drive gear side bearing is against housing. Zero dial gauge.
55 Lever assembly towards pinion until drive gear will mesh no more deeply with pinion. Record dial gauge reading.

continued

67 In **High Tooth Contact** it will be observed that the tooth contact is heavy on the drive gear face or addendum. To rectify this condition, move the pinion deeper into mesh, that is, reduce the pinion cone setting distance, by adding shims between the pinion inner bearing cup and the housing. This condition has a tendency to move the tooth bearing towards the toe on drive and heel on coast, and it may therefore be necessary after making this change to adjust the drive gear as described in operations 69 and 70.

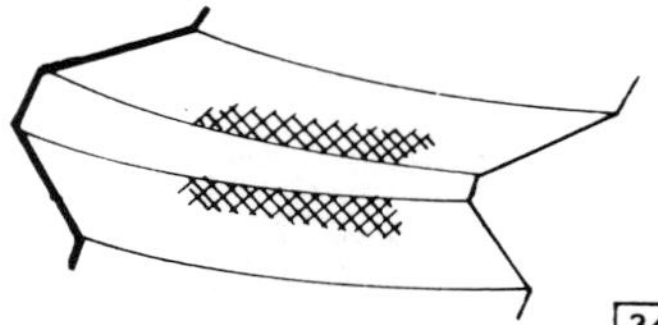

68 In **Low Tooth Contact** it will be observed that the tooth contact is heavy on the drive gear flank or dedendum. This is the opposite condition from that shown in 67 and is therefore corrected by moving the pinion out of mesh, that is, increase the pinion cone setting distance by removing shims from between the pinion inner bearing cup and housing. This correction has a tendency to move the tooth bearing towards the heel on drive and toe on coast, and it may therefore be necessary after making this change to adjust the drive gear as described in operations 69 and 70.

NOTE: To increase backlash remove shims from drive gear side of differential and install on opposite side. To decrease backlash transfer shims to drive gear side from the opposite side of differential case.

65 After setting backlash to required figure use a small brush to paint eight or ten gear teeth sparingly with a stiff mixture of marking raddle or with engineers blue. Move painted gear teeth in mesh with pinion until a good impression of tooth contact is obtained. The result should conform with ideal impression given. Correction procedure of poor meshing is also given.

66 The ideal tooth bearing impression on the drive and coast sides of the gear teeth is evenly distributed over the working depth of the tooth profile and is located nearer to the toe (small end) than the heel (large end). This type of contact permits the tooth bearing to spread towards the heel under operating conditions when allowance must be made for deflection.
NOTE: If 'ideal' impression obtained, proceed with operation 71, otherwise continue with operations 67, 68, 69 or 70 as applicable.
Nomenclature referring to gear teeth is as follows:
The HEEL is the large or outer end of the tooth (see 'A').
The TOE is the small or inner end of the tooth (see 'B').
The DRIVE side of the drive gear tooth is CONVEX (see 'C').
The COAST side of the drive gear tooth is CONCAVE (see 'D').

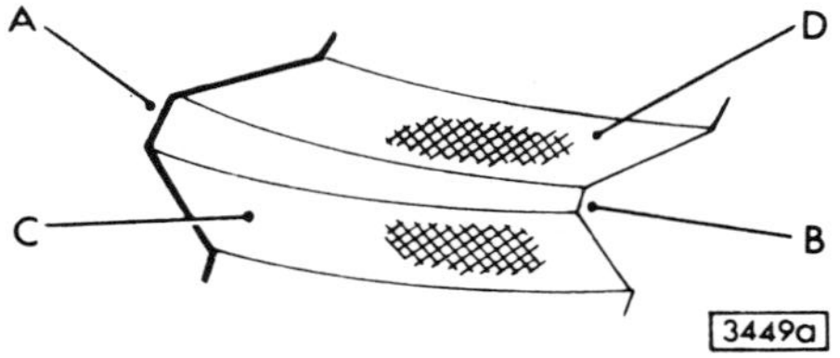

61 Mount a dial indicator in gear carrier housing with stylus against back face of gear as for operation 53.

62 Turn pinion by hand and check run out on back face of gear. Run out should not exceed .005 in (,13 mm). If run out excessive, strip assembly and rectify by cleaning surfaces locating drive gear. Any burrs on these surfaces must be removed.

63 Remount dial indicator on gear carrier housing with stylus tangentially against one of drive gear teeth.

64 Move drive gear by hand to check backlash which should be as etched on gear. If backlash is not to specification, transfer shims from one side of differential case to the other to obtain desired setting.

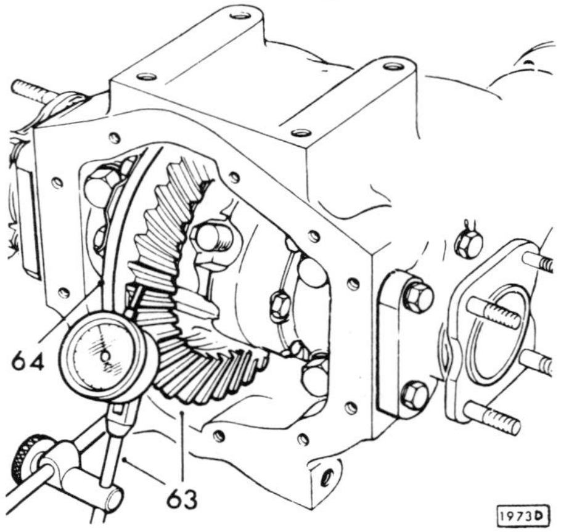

NOTE: This reading, less the backlash allowance etched on the drive gear (e.g. B/L 007) gives the thickness, in thousandths of an inch, of shims to be placed between the flange side of the differential and the bearing cone. This thickness of shims is to be removed from the total established after operation 37, the remaining shims being installed between the other bearing cone and the differential.
Thus, supposing operation 37 established a total sideways movement of 0,080 in (2,03 mm), the total thickness of shims required is then 0.080 in + 0.009 in (2,03 mm + 0,23 mm) = 0.089 in (2,26 mm), the 0.009 in (0,23 mm) being required for pre-load. If the reading from operation 55 was 0.042 in (1,067 mm) and the backlash allowance 0.007 in (0,178 mm) the thickness of shims to be fitted at the flange bearing would be 0.042 in − 0.007 in (1,067 mm − 0,178 mm) = 0.035 in (0,89 mm). At the other side, the total thickness of shims required is then 0.089 in − 0.035 in (2,26 mm − 0,89 mm) = 0.054 in (1,37 mm).

56 Remove differential unit, detach bearing cones and fit shims of total thickness now determined at appropriate sides of unit.

57 Refit bearing cones.

58 Carefully refit differential and drive gear assembly, leading teeth into mesh with pinion and lightly tapping bearings home with a hide hammer.

59 When refitting side bearing caps, ensure that position of numerals marked on gear carrier housing face and side bearing cap coincide.

60 Tighten caps to a torque of 63 to 70 lbf. ft (8,72 to 9,68 kgf. m).

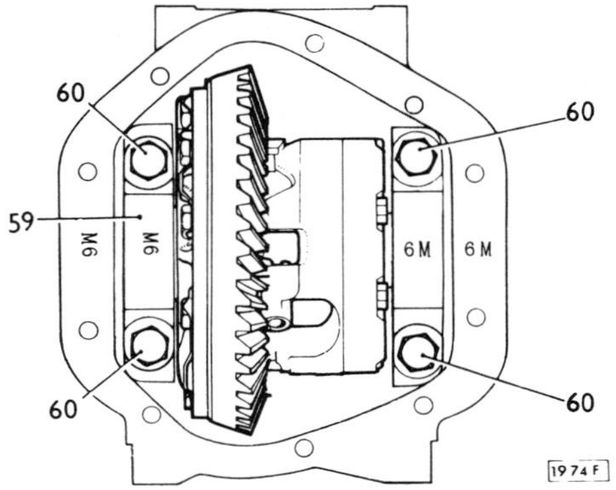

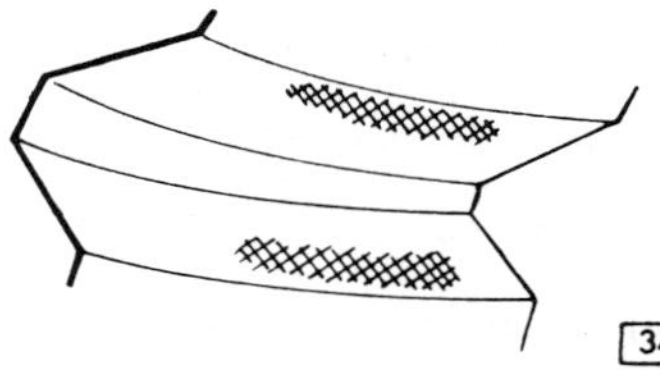

51-6

69 Toe Contact occurs when the bearing is concentrated at the small end of the tooth. To rectify this condition, move the drive gear out of mesh, that is increase backlash, by transferring the shims from the drive gear side of the differential to the opposite side.

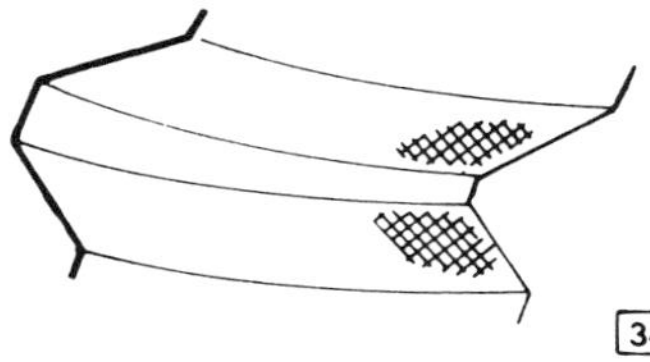

70 Heel Contact is indicated by the concentration of the bearing at the large end of the tooth. To rectify this condition move the drive gear closer into mesh, that is, reduce backlash, by adding shims to the drive gear side of the differential and removing an equal thickness of shims from the opposite side.
NOTE: It is most important to remember when making this adjustment to correct a heel contact that sufficient backlash for satisfactory operation must be maintained. If there is insufficient backlash the gears will at least be noisy and have a greatly reduced life, whilst scoring of the tooth profile and breakage may result. Therefore always maintain a minimum backlash requirement of 0.004 in (0,10 mm).
CAUTION: Exercise great care when tightening drive flange nut on pinion shaft, as overtightening will necessitate almost complete dismantling of final drive to replace collapsed spacer.

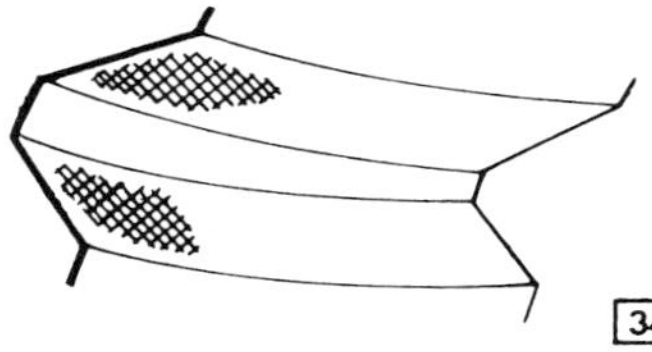

71 Tighten drive flange nut to 120 to 130 lbf. ft (16,6 to 18,0 kgf. m). During tightening, rotate drive flange to seat taper roller bearings. Should nut be overtightened a new collapsible spacer must be fitted. On no account may nut be slackened off and re-torqued, as collapsed spacer will then permit excessive pre-load of pinion bearings.

72 Fit end cover, using new gasket coated with Hylomar. Replace axle ratio tab and tighten bolts to 14 to 18 lbf. ft (1,94 to 2,48 kgf. m).

73 Replace output shaft assemblies, see 51.10.20, operations 4 to 9.

SYMPTOM AND DIAGNOSTIC CHART

SYMPTOM	CAUSE	CURE
External oil leaks from steering rack unit.	Damaged or worn seals. Loose unions. Damaged union sealing washers.	Replace seals. Tighten unions. Replace sealing washers.
Oil leak at pump shaft.	Damaged shaft seal.	Replace shaft seal.
Oil leak at high pressure outlet union.	Loose or damaged union. Damaged pipe end.	Tighten union. Replace pipe.
Oil leak at low pressure inlet connection.	Loose or damaged hose connection.	Remove and refit or renew hose and clip.
Oil overflowing reservoir cap.	Reservoir overfull. Sticking flow control valve (closed).	Reduce level in reservoir. Remove valve, renew and refit.
Oil leak at reservoir edge.	Damaged 'O' ring.	Replace 'O' ring.
Noise from hydraulic system.	Air in system.	Bleed system, 57.15.02.
Noise from pump.	Slack drive belt (squealing). Internal wear and damage.	Adjust drive belt tension, 57.20.01. Overhaul pump, 57.20.20.
Noise from rack (rattling).	Worn rack and pinion gears. Worn inner ball joints. Lower column universal joint loose	Adjust rack damper, 57.10.13. Replace inner ball joints, 57.55.03. Tighten clamping bolts.
Steering veering to left or right.	Unbalanced tyre pressures Incorrect tyres fitted. Incorrect geometry. Steering unit out of trim.	Inflate to correct pressure. Fit tyres of correct specification. Reset geometry to correct specification. Replace valve and pinion assembly, 57.10.19.
Heavy steering when driving.	Low tyre pressures. Tightness in steering column. Tightness in steering joints.	Inflate to correct specification. Grease or replace. Grease or renew joints.

SYMPTOM	CAUSE	CURE
Heavy steering when parking.	Low tyre pressures. Tightness in steering column. Tightness in steering joints. Slack driving belt (squealing). Restricted hose. Sticking flow control valve (open). Internal leaks in steering unit.	Inflate to correct specification. Grease or replace. Grease or renew joints. Adjust drive belt tension, 57.20.01. Replace hose. Remove and renew valve. Replace seals.
Steering effort too light.	Valve torsion bar dowel pins worn. Valve torsion bar broken.	Replace valve assembly. Replace valve assembly.

POWER ASSISTED STEERING RACK ASSEMBLY

Remove and refit   57.10.01

Service tools: Ball joint separator JD.24, Rack alignment tool JD.26A.

Removing
1. Remove lower steering column, see 57.40.05.
2. Using suitable syringe remove all fluid from steering pump.
3. Remove clip securing feed and return pipes to rack assembly, recover nylon bridge piece.
4. Release nuts securing feed and return pipes to pinion housing; fit blanking plugs to open ends of pipe.
5. Remove nut securing right-hand tie-rod to steering arm.
6. Using ball joint separator (JD.24) separate tie-rod from steering arm.
7. Repeat operations 5 and 6 on left-hand tie-rod.
8. Remove upper bolt, securing pinion side of rack to suspension unit, recover heatshield mounting bracket and spacers.
9. Remove remaining rack mounting bolts, recover packing washers.
10. Withdraw rack assembly from suspension unit.

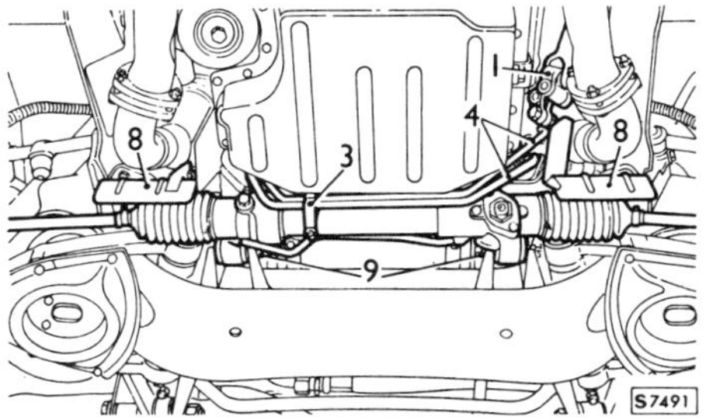

Refitting
11. Offer rack mounting brackets on to suspension beam, locate rack on brackets with suitable drifts.
12. Adjust rack assembly so that mounting lugs are central of brackets.
13. Ensuring sufficient shims are at hand, support rack and remove locating drift from mounting bracket at opposite side to pinion.
14. Select shims and position between rack lug and mounting bracket, sufficient shims should be fitted to obtain a 0.05 in gap (1,25 mm) on either side of rack lug and mounting bracket. Locate shims with rack mounting bolts and nut, do not fully tighten nut.

continued

57–1

POWER STEERING RACK

Overhaul   57.10.07

Dismantling

1. Remove rack assembly, see 57.10.01.
2. Clean exterior of rack and pinion housing.
3. Remove clips securing tie rod gaiters to rack tube, fold back gaiters to expose inner ball joints.
   NOTE: Do not disturb outer ball joints unless replacement is necessary. If joints are to be renewed, measure accurately and record length between tie rod locknuts before removing ball joint.
4. Knock back inner tab washer securing tie rod inner locknuts.
   CAUTION: Do not disturb tab washer between inner locknut and ball pin housing.
5. Release inner locknuts and unscrew tie rod assemblies from rack, recover thrust springs and packing washers. 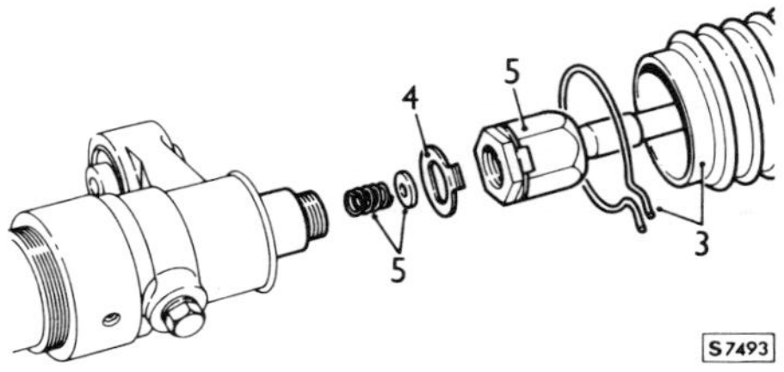
6. Release nuts securing feed pipes to pinion housing and rack body, remove pipes from rack assembly.
7. Remove nuts securing pinion housing to rack tube.
8. Remove pinion housing.
9. Remove and discard pinion seals in housing recess, recover metal washer.
10. Remove bolts locating rack tube to end cap and rack tube to inner sleeve. 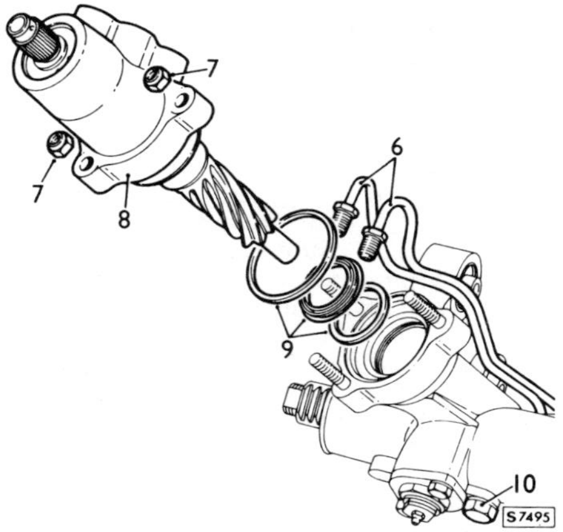
11. Remove Allen screw from end cap.
12. Using suitable drift unscrew locking ring from end cap.
13. Withdraw end cap from rack tube.
14. Remove air transfer pipe from rack.
15. Remove end cap 'O' ring seal, washer, nylon spacer, seal and seal seat. 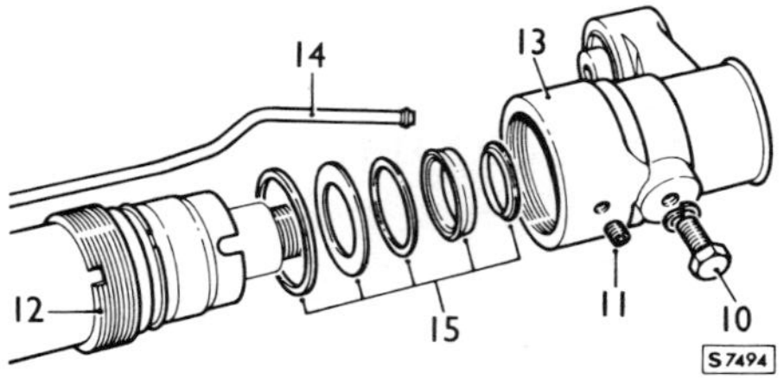
16. Remove bolts securing rack damper plate.
17. Remove damper plate, seal and spring.
18. Remove damper from rack tube.
19. Carefully remove rack from tube. 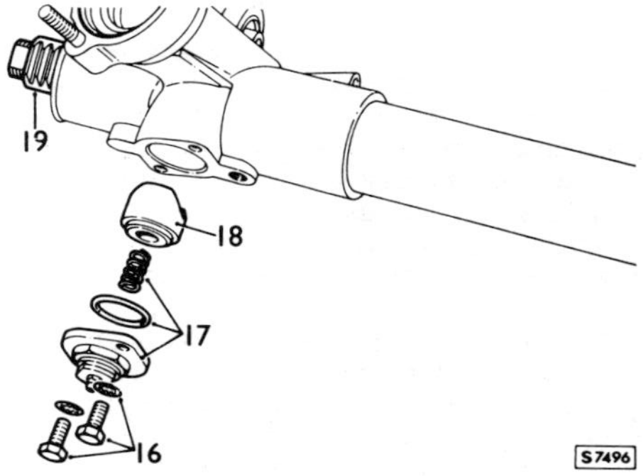
20. Using suitable bent drift remove rack inner sleeve from rack tube.
21. Remove circlip from end of inner sleeve and withdraw nylon washer, seal and seal seat.
22. Remove 'O' ring seal from periphery of inner sleeve. 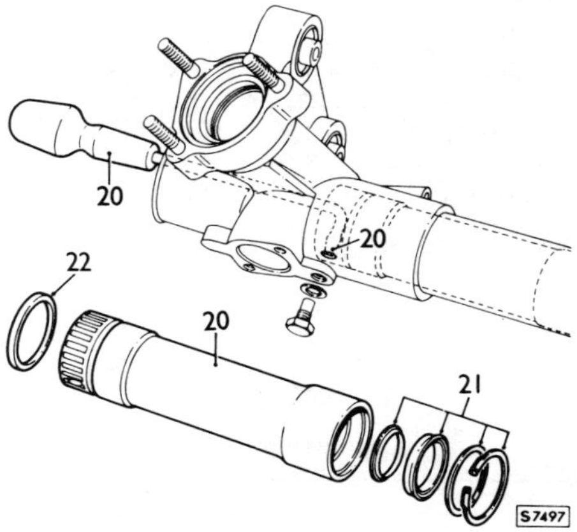
23. Remove seal from seal carrier on rack bar.

Inspection

24. All components of rack should be cleaned and examined for wear, fatigue or fracture. Should doubt on condition of any component exist a new item must be fitted.
25. Discard all old items, seals etc. that are replaced by new components contained in the repair kits.

15. Carry out operations 13 and 14 or upper and lower mountings on pinion side of rack, ensure heatshield mounting bracket is located on upper mounting bolt, ensure locking nuts are not fully tightened.
16. Remove locking wire securing gaiters to rack housing, fold back gaiters to expose inner ball joints.
17. Locate attachment brackets of alignment tool JD.36A over large hexagon heads of lower wishbone fulcrum shafts.
    NOTE: It may be necessary to bend heat shields slightly in order to locate tool correctly.
18. Slacken locking screw and slide collar along tool to front of suspension unit, locate groove in collar leg over flange at front of suspension cross-beam, lock collar in position.
19. Rotate alignment tool legs until legs rest on tie-rods.
20. If one leg fails to contact tie-rod, rack is out of alignment and must be adjusted.
21. To adjust rack slacken locknut of single bolt mounting and lower or raise same side of rack assembly.
22. Remove alignment tool. Restore heat shields to original condition.
23. Fully tighten rack mounting locknuts.
24. Re-position gaiters and secure with locking wire.
25. Reverse operations 3 to 6.
26. Refill and bleed steering system, see 57.40.05.
27. Refit and bleed steering system, see 57.15.02.
28. Check wheel alignment, see 57.65.01.

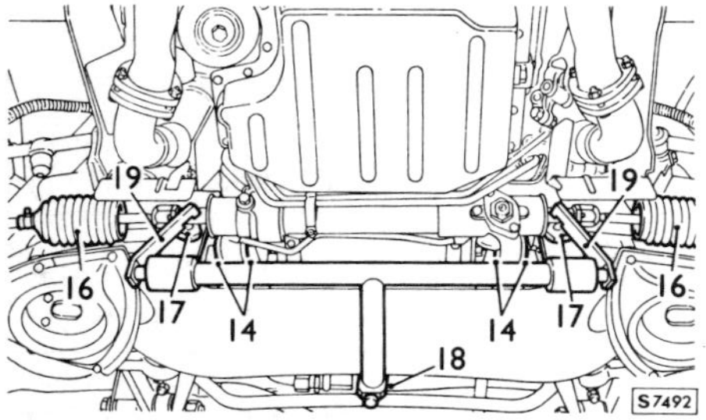

57–2

6 Firmly grip track rod and pull horizontally against adjusting screw until spring pressure is felt. Total play of rack at this point should not exceed 0.1 in (,25 mm). A correct minimum clearance should permit smooth travel of rack without binding.
7 Should rack play exceed 0.1 in (,25 mm) adjust screw as necessary, repeat operation 6.
8 Remove dial indicator, replace grease nipple and tighten adjusting screw locknut.
9 Secure track rods to steering arms.
10 Check wheel alignment, see 57.65.01.

CONTROL VALVE AND PINION 57.10.19

Remove and refit

Removing
1 Remove rack assembly, see 57.10.01.
2 Clean rack and pinion housing.
3 Release nuts securing oil feed and return pipes.
4 Remove nuts securing pinion housing to rack tube.
   NOTE: For reference when refitting, note position of recess on pinion shaft in relation to pinion housing.
5 Remove pinion assembly from rack, discard joint gasket.

continued

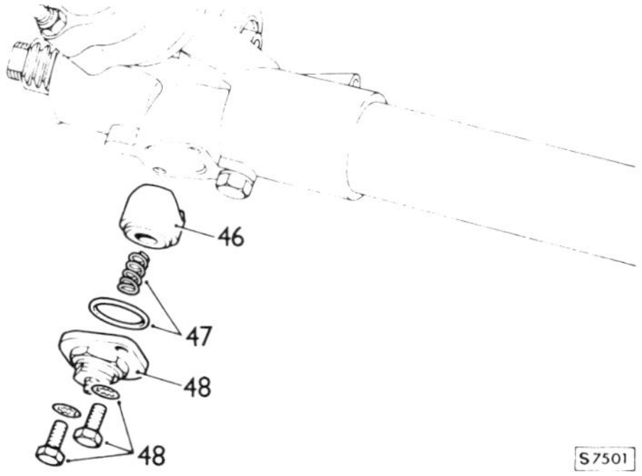

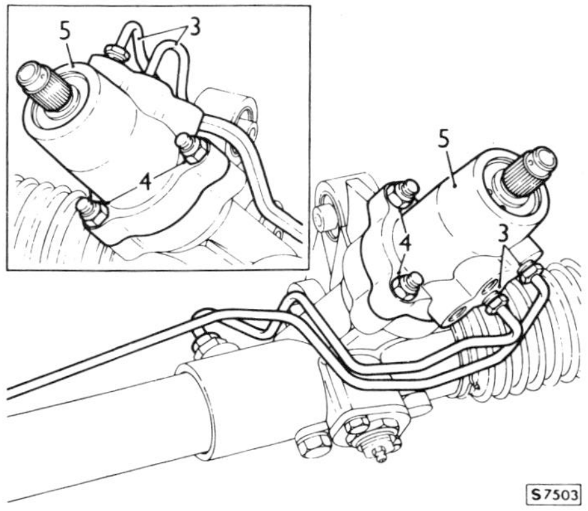

STEERING RACK

Adjust 57.10.13

Service tool: Ball joint separator JD.24.

Adjusting
1 Release locknut securing rack damper adjusting screws.
2 Screw in adjusting screw until firm resistance is felt then turn back one sixth of a turn.
3 Disconnect track rods from steering arm.
4 Remove grease nipple from adjuster screw.
5 Insert stem of dial gauge into grease feed hole of adjuster screw, ensuring stem passes through spring and pad to contact back of rack.

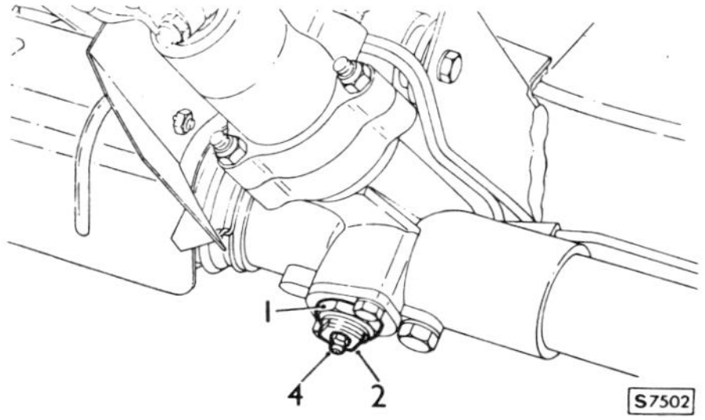

41 Fit Allen screw which secures threaded sleeve.
42 Remove tape from rack bar.
43 Lubricating new seal fit metal washer, seal and 'O' ring seal to pinion housing.
44 Place pinion assembly into pinion housing, slightly rotate rack bar to align rack teeth with pinion shaft.
45 Secure pinion to housing.
46 Fit rack damper to housing.
47 Fit new 'O' ring seal and spring to damper.
48 Locate damper cover plate in position, and secure with bolts.
49 Refit rack feed and return pipes.
50 Reverse operations 1 to 5, ensuring to coat inner ball joints with 1½ to 2 oz (44 to 55 g) of recommended grease.
51 Adjust rack damper, see 57.10.13.
52 Inject 1 oz (28 g) of grease into rack damper.

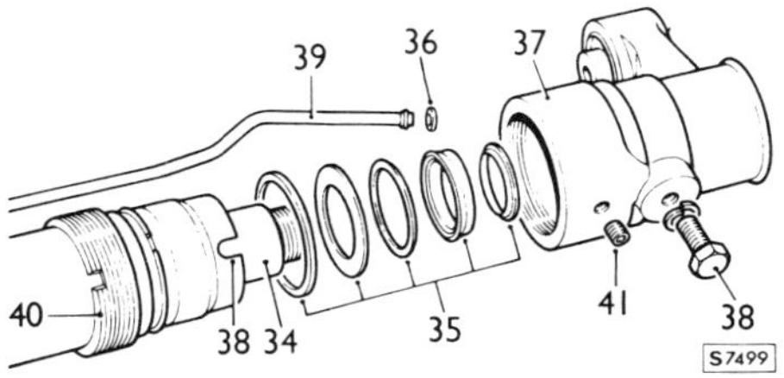

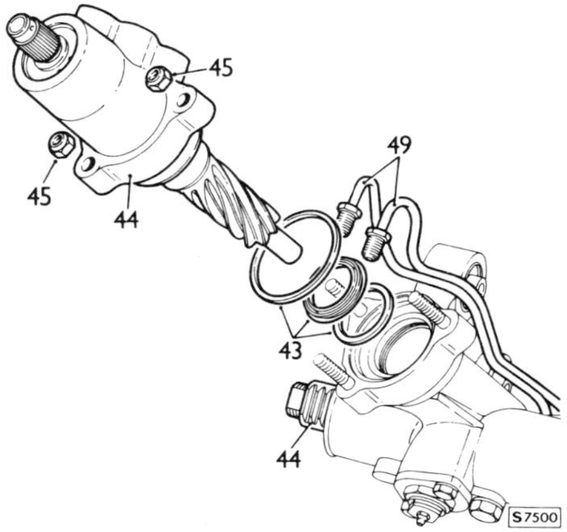

Re-assembly
26 Ensure work area is clean.
27 Lubricate new seal for inner sleeve.
28 Fit seal seat, seal and nylon washer to inner sleeve, secure with circlip.
29 Lubricate and fit new 'O' ring seal to periphery of inner sleeve.
30 Ensure castellated end of inner sleeve enters rack tube first and locating peg hole in inner sleeve is aligned with hole in rack end casing, push inner sleeve into rack tube until hole in end casing and inner sleeve are aligned.
31 Fit locating bolt to end casing.
32 To protect inner sleeve seal from damage cover teeth of rack with suitable adhesive tape.
33 Carefully fit new nylon seal to seal carrier on rack bar.
34 Carefully enter rack bar into rack tube and inner sleeve, taking care not to damage inner sleeve seal. Position rack bar central of rack tube.
35 Ensuring new seals are lubricated fit seal seat, seal, nylon washer, metal washer and 'O' ring seal to rack tube end cap.
36 Fit new 'O' rings to air transfer pipe.
37 Position end cap over rack tube.
38 Ensuring to align cut out in rack tube with locating bolt hole, fit locating bolt to end cap.
39 Fit air transfer pipe in position.
40 Rotate threaded sleeve and secure end cap to rack tube.

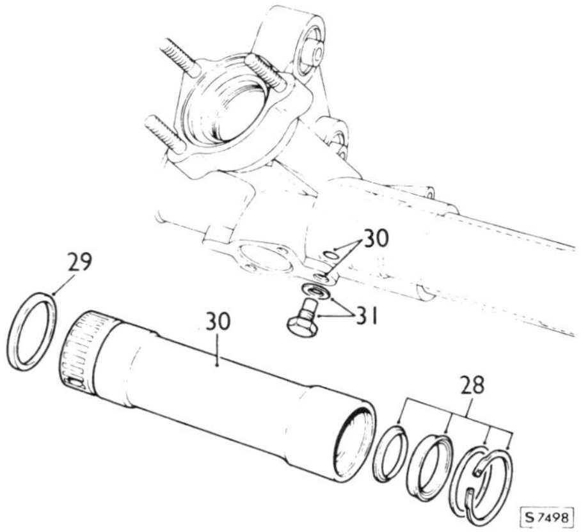

57—3

Refitting

6 Reverse operations 1 to 5, fit new joint gasket and ensure pinch bolt recess is in same position prior to removal.

CONTROL VALVE AND PINION

Test 57.10.20

Service tools: Pressure gauge JD.10; adaptors JD.10-3.

NOTE: Faults developing in the control valve and pinion assembly as indicated in the following test or symptoms or diagnostic chart will necessitate renewal of control valve and pinion. No adjustment or repair is permissible.

Check tyres, tyre pressures and steering geometry are all satisfactory before carrying out the following test.

1 Fit a 100 lbf/in² (7 kgf/cm²) pressure gauge in pump to pinion feed line.
2 Start engine and allow to idle; gauge should read approximately 40 lbf/in² (2,8 kgf/cm²).
3 Turn steering wheel on slight left or right lock.
CAUTION: Do not turn steering wheel excessively as high pressure will result and possibly cause damage to gauge.
Pressure should increase on either lock by similar amounts. Where a slight fall in pressure occurs before rising, a defective valve is indicated.
4 Stop and re-start engine, check steering does not kick to one side.

PINION SEAL

Remove and refit 57.10.23

Removing
1 Remove control valve and pinion, see 57.10.19.
2 Using suitable mallet gently tap pinion from housing.
3 Remove circlip securing nylon washer and seal to pinion housing.
4 Remove nylon washer from housing.
5 Prise seal free from housing, discard seal.

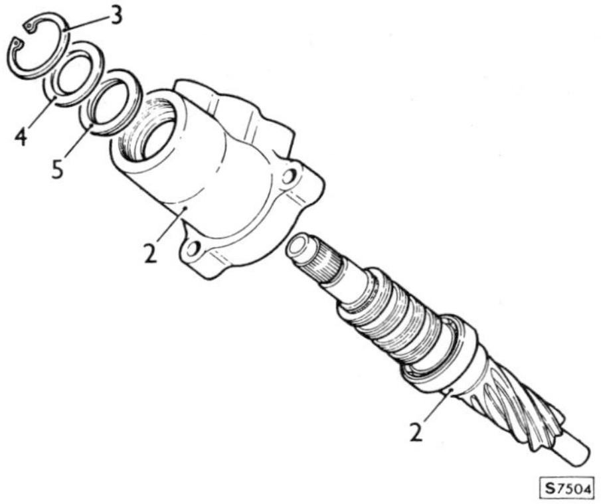

4 Remove clip securing gaiter to inner tie rod.
5 Ease gaiter from rack tube and slide from tie rod.

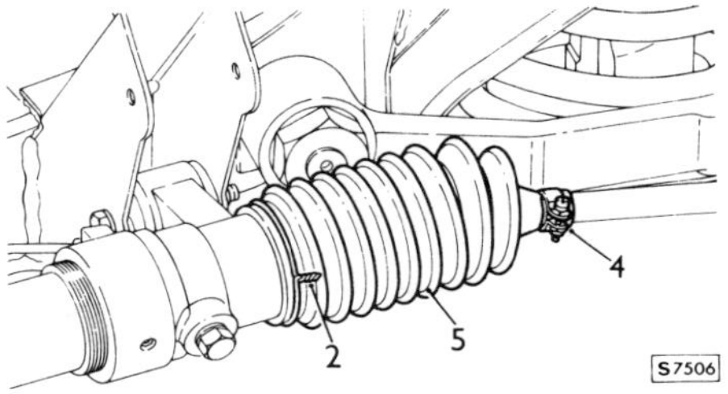

6 Clean grease from inner tie rod joints.

Refitting
7 Reverse operations 1 to 6. Coat inner ball joint with 2 oz (50 g) of recommended grease.
8 Check wheel alignment, see 57.65.01.

PORT INSERTS

Remove and refit 57.10.24

Removing
1 Tap suitable thread in bore of seat.
2 Insert setbolt with nut, washer and distance piece.
3 Tighten nut and withdraw seat.

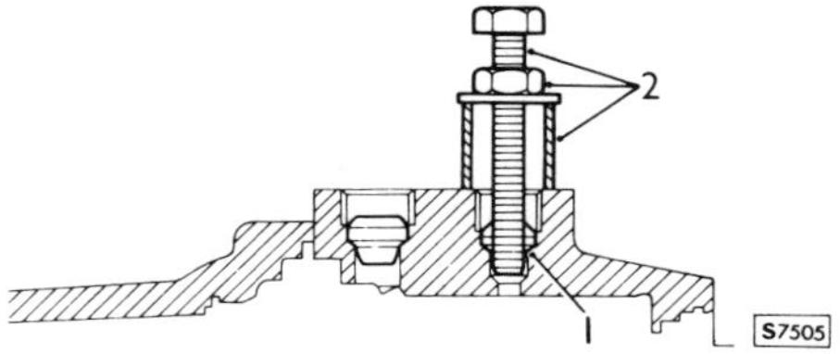

Refitting
4 Insert new seat and tap home squarely with suitable mandrel.

STEERING RACK GAITER

Remove and refit 57.10.27

Removing
1 Remove outer ball joint, see 57.55.02.
2 Remove clip securing gaiter to rack tube.
3 Noting position on inner tie rod remove outer ball joint locknut.

4 Check that recorded pressure lies between 1100 and 1200 lbf/in² (77,5 kgf/cm² and 84,4 kgf/cm²) at tickover, but rises to correct figure with increased engine speed, the reason is, a defective internal leakage in rack and pinion unit. Carry out following test to establish location.
5 Fit tap JD.10-2 between pump and pressure gauge, arranging connections as shown, so that pressure gauge is at all times connected to pump, but rack unit can be isolated from it.

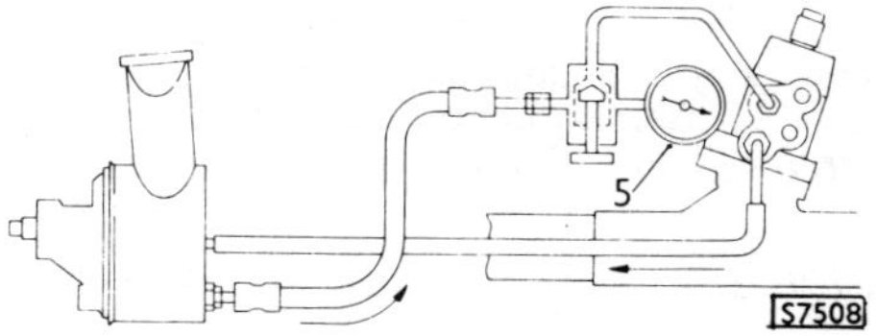

NOTE: If pressure is below 1100 lbf/in² (77,5 kgf/cm²) at tickover, but internal leakage in rack and pinion control valve in pump, or excessive unit. Carry out following test to establish location.

6 With tap OPEN, start engine and allow to run at idling speed.
7 Turn steering to full lock.
8 Check that gauge reading exceeds 1100 lbf/in² (77,5 kgf/cm²).
9 If pressure does not reach this figure, CLOSE TAP AT ONCE, noting gauge reading as tap reaches 'OFF' position.
CAUTION: Tap must not remain closed for more than 5 seconds when engine is running.
If reading of pressure gauge increases to at least 1100 lbf/in² (77,5 kgf/cm²) when tap is turned off, leaks are confined to steering unit, which must be overhauled, see 57.10.07.
If pressure reading exceeds 1200 lbf/in² (84,4 kgf/cm²) remove pump discharge port, withdraw valve

SYSTEM TESTING

57.15.01

Service tool: Tap JD.10-2; Pressure Gauge JD.10; Adaptors JD.10-3.

Faults in system can be caused by inefficiencies in the hydraulic system, see 'Diagnostic Chart'. The following test may be carried out without removing any components from the car. Before commencing work, fluid should be checked for correct level and freedom from froth.

Pump blow off pressure
1 Fit pressure gauge reading to 1500 lbf/in² (100 kgf/cm²) in pressure line from pump.
2 Start engine and allow to run at idling speed.
3 Turn steering to full lock and continue to increase steering effort until pressure recorded on gauge ceases to rise.

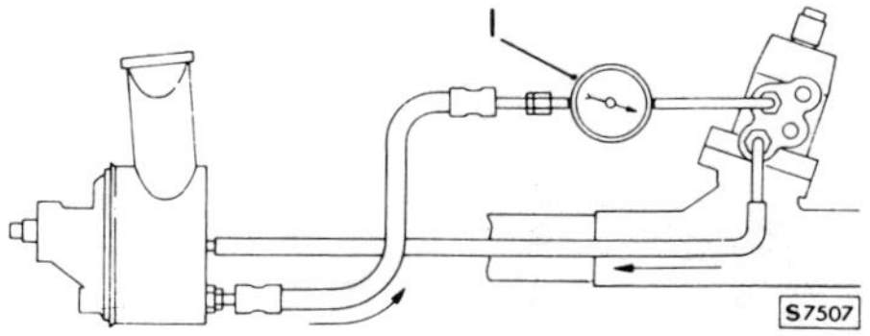

57—4

assembly located behind it, and inspect a small hemispherical gauze filter carried at its inner extremity, which may be found to be blocked. Clean filter by airline or other means, and replace valve and discharge port.

POWER STEERING SYSTEM 57.15.02

Bleed

Bleeding
1. Turn wheels to full left lock, add fluid to reservoir to bring level to 'COLD' mark on dipstick.
2. Start engine and run at fast idle, re-check fluid level. If necessary add fluid to bring level to cold mark on dipstick.
3. Turn steering wheel from lock to lock avoiding steering lock stops, constantly check fluid level and top-up as necessary.
4. Carry out operation 3 until all air is removed from fluid. Presence of air within system can be established by 'lumpiness' being apparent during turning of wheel from lock to lock.
5. Position wheels in central position, continue to run engine for two or three minutes then switch off engine. After system has settled and is at normal operating temperature, re-check fluid level. Level of fluid should be up to 'HOT' mark on dipstick; add fluid if necessary.

STEERING OIL COOLER

Remove and refit 57.15.15

Removing
1. Remove filler cap from pump assembly.
2. Using suitable syringe remove all fluid from steering pump.
3. Slacken clips securing return and feed hoses to cooler.
4. Ensuring waste rag is positioned beneath cooler, slide hose from cooler.
5. Remove two bolts securing oil cooler to suspension unit.
6. Lift cooler from car.

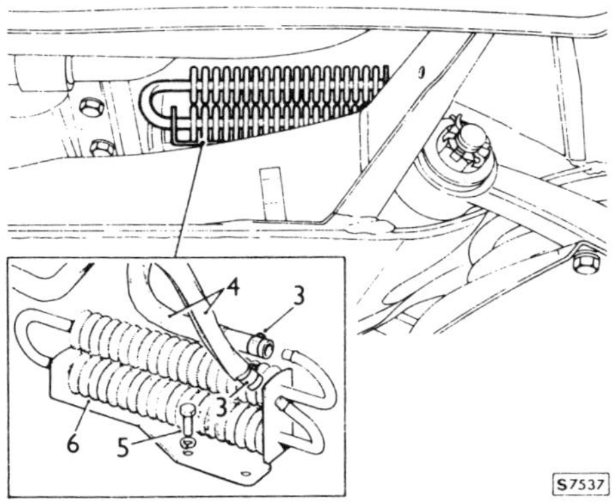

2. Release union securing feed hose to rear of steering pump, collect oil spillage with waste rag.

Right-hand drive cars only
3. Remove clip securing feed and return pipes to rack assembly, recover nylon bridge piece.

All cars
4. Release nut securing feed pipe to pinion housing.
5. Lift feed pipe from car and slide insulation sleeve from pipe.

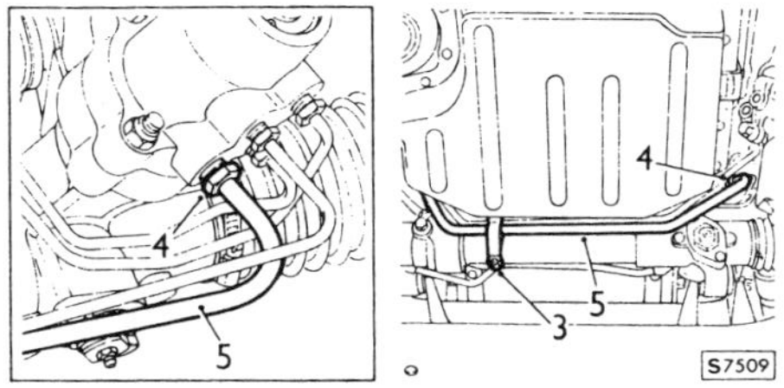

Refitting
7. Reverse operations 1 to 6.
8. Bleed system, see 57.15.02.

OIL COOLER RETURN HOSE

Remove and refit 57.15.16

Removing
1. Remove air cleaner element, see 19.10.08.
2. Using suitable syringe remove all fluid from steering pump.
3. Slacken clip securing return hose to oil cooler. NOTE: To catch oil spillage ensure sufficient waste rag lays beneath oil cooler and pump.
4. Slide return hose from oil cooler.
5. Slacken clip securing return hose to pump, slide hose from pump.
6. Lift hose from car.
7. Remove clips securing insulation sleeve from hose.

Refitting
8. Reverse operations 1 to 7.
9. Bleed system, see 57.15.02.

STEERING RACK FEED HOSE/PIPE

Remove and refit 57.15.21

Removing
1. Using suitable syringe remove all fluid from steering pump.

Refitting
7. Reverse operations 1 to 7.
8. Bleed system, see 57.15.02.

STEERING PUMP DRIVE BELT

Adjust 57.20.01

Adjusting
1. Slacken nut securing pump lower pivot mounting pin.
2. Slacken outer locknut on adjusting rod.
3. Tighten inner nut to increase belt tension, correct tension is such that a load of 6.5 lb (3 kg) applied to mid-point of upper belt run will give a total deflection of 0.16 in (4,0 mm).
4. Tighten adjuster rod outer locknut and pivot mounting locknut.

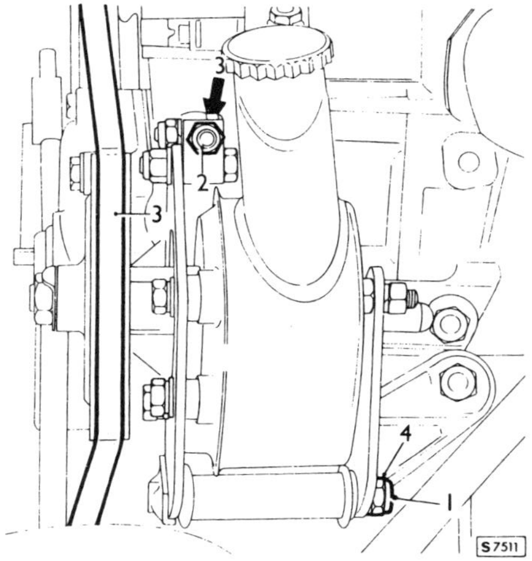

STEERING RACK RETURN HOSE/PIPE

Remove and refit 57.15.22

Removing
1. Using suitable syringe remove all oil from steering pump.
2. Slacken clip securing return hose to oil cooler.
3. Slide hose from cooler, collect oil spillage in waste rag.

Right-hand drive cars only
4. Remove clip securing feed and return hose to rack assembly, recover nylon bridge piece.

All cars
5. Release union securing return hose to pinion housing.
6. Lift hose from car and slide insulation cover from hose.

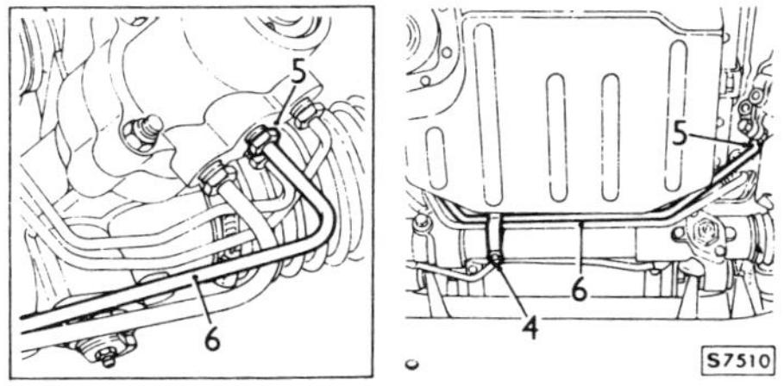

57—5

STEERING PUMP DRIVE BELT

Remove and refit 57.20.02

Removing
1. Remove fan belt, see 19.10.01.
2. Turn back inner locknut at pump adjuster trunnion.
3. Slacken lower pivot bolt securing nut.
4. Slacken nut securing trunnion block.
5. Disconnect wires from thermo switch in front of coolant pump.
6. Swing pump towards engine.
7. Disengage belt from pulley and lift from car.

Refitting
Reverse operations 1 to 7.

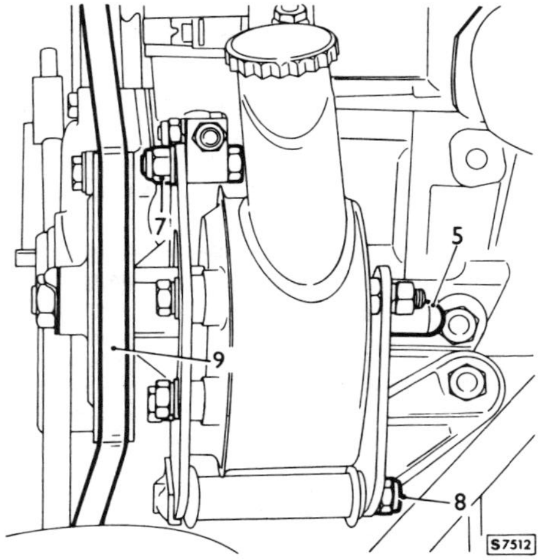

STEERING PUMP

Remove and refit 57.20.14

Removing
1. Remove right-hand air cleaner element, see 19.10.08.
2. Partially drain coolant.
3. Release clips securing top coolant hose, remove hose.
4. Using suitable syringe remove all oil from steering pump.
5. Disconnect return and feed hoses from rear of pump. Collect oil spillage in waste rag.
6. Slacken nut securing adjuster rod to timing cover.
7. Remove bolt securing adjuster rod trunnion to pump, swing adjuster rod clear of pump.
8. Remove nut securing pump lower pivot bolt.
9. Swing pump to engine and disengage drive belt.
10. Remove lower pivot mounting bolt.
11. Lift pump from car.

Refitting
12. Reverse operations 1 to 11.
13. Adjust drive belt, see 57.20.01.
14. Bleed system, see 57.15.02.

STEERING PUMP OVERHAUL

Dismantling and reassembling 57.20.20

Dismantling
1. Remove steering pump, see 57.20.14.
2. Remove rear mounting plate from pump.
3. Remove pump pulley.
4. Noting position and relative size of spacers remove front mounting plate from pump.
5. Thoroughly clean pump exterior. 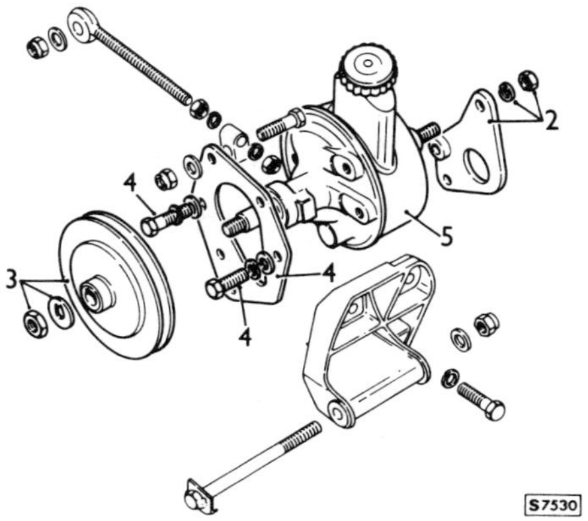
6. Remove high pressure outlet union and mounting plate studs from rear of pump.
7. Tip pump to one side to remove flow control valve and spring.
8. Position spindle body in vice and carefully tap pump casing from body. 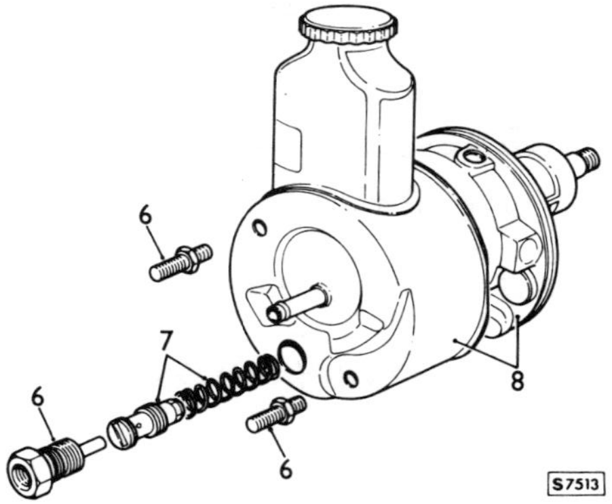
9. Remove three 'O' rings from ports in pump body and magnet on pump body flange.
   NOTE: If large amounts of fibrous metal material are deposited on magnet, damage or considerable wear within pump is indicated.
10. Insert suitable pin punch in pump body, using pin, push retaining ring free from groove and lever from body.
11. Remove spring retaining plate, if plate sticks give slight tap with mallet.
12. Remove spring.
13. Remove top 'O' ring from recess in pump body. 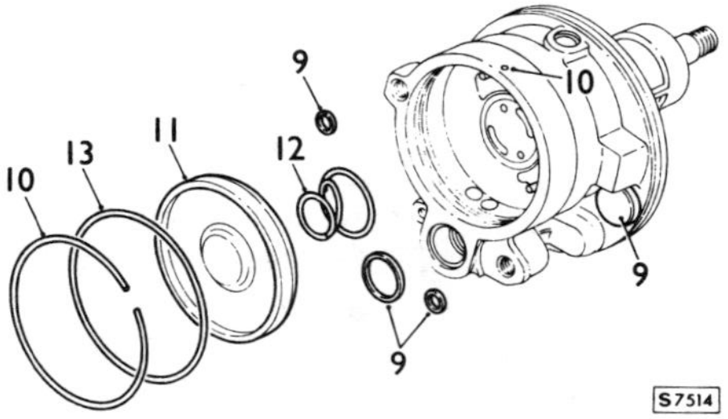
14. Remove Woodruff key from rotor drive shaft.
15. Gently tap roller spindle towards pump body and carefully remove pump assembly from body, lay assembly to one side.
16. Remove lower 'O' ring from recess in pump body.
17. If during operation 15, pump dowel pins remain in body, remove pins.
18. Remove rotor housing top plate from pump assembly.
19. Remove rotor housing.
20. Remove dowel pins if not removed in operation 17.
21. Remove rotor vanes.
22. Remove circlip securing rotor to drive shaft.
23. Remove drive shaft oil seal from pump body.

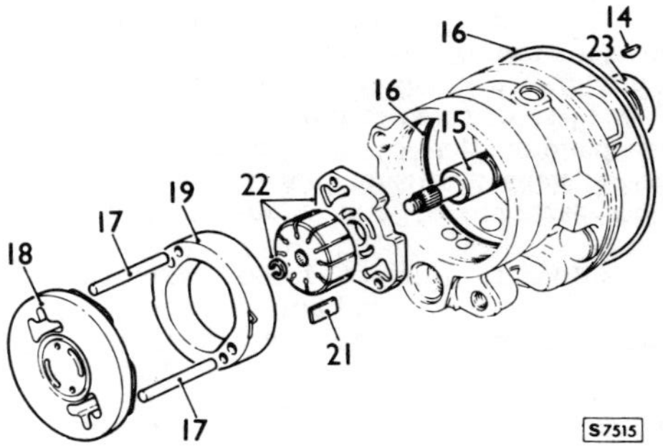

Inspection
24. Clean all parts with lint free cloth.
25. Discard old seals, 'O' rings, rotor housing and rotor vanes. All these items are contained within the repair kits.
26. Should any item other than those contained within the repair kit be defective or doubt as to their condition exist, a new pump must be fitted.

Reassembling
27. Lubricate new drive shaft seal and fit to pump shaft housing.
28. Fit new 'O' ring to lower recess in pump body.

Refitting

19 Reverse operations 14 to 18.
20 Refit column lower mounting bolts, do not fully tighten bolts.
21 Refit upper mounting bolts, spacers and washers, do not fully tighten bolts.
22 Adjust column so groove on inner splined shaft aligns with bolt locating hole in universal joint.
23 Align upper section of column to lay central of housing in fascia.
24 Tighten upper and lower mounting bolts.
25 Reverse operations 1 to 10.

LOWER STEERING COLUMN

Remove and refit   57.40.05

Removing

1 Adjust wheels to straight ahead position.
2 Remove screws securing right hand heatshield adjacent to exhaust downpipe, remove heatshield.
3 Remove bolts securing right hand gaiter heatshield to steering rack, remove heatshield.
4 Remove nut and bolt securing lower universal joint to pinion shaft.
5 From inside driver's footwell remove pinch bolts securing universal joint to lower and upper column.
6 Slide universal joint free from upper column.
7 Remove universal joint from lower shaft.
8 Slacken clip securing draught excluder to lower column.
9 From underneath car push lower column through bulkhead until lower universal joint disengages from pinion shaft.
10 Manoeuvre universal joint clear of pinion and withdraw lower column assembly from draught excluder housing.

continued

---

2 Remove screw securing steering column lower shroud, lift shroud from column.
3 Remove ignition key and lock steering wheel in straight ahead position.
4 Remove bolt securing column adjusting mechanism to upper column.
5 Slacken nut and release grub screw securing collet adaptor to upper column, withdraw adjuster assembly from column.
6 Remove screw securing upper section of ignition switch shroud.
7 Slacken screws securing lower portion of shroud, remove shroud from switch.
8 Disconnect block connector from ignition switch to main harness.
9 Disconnect two block connectors between main harness and auxiliary controls.
10 Remove bolt securing universal joint to upper column.
11 Remove bolts securing column at lower mounting.
12 Disconnect horn electrical feed from column.
13 Remove bolts securing column upper mounting, recover distance pieces and washers.
14 Carefully manoeuvre column free from universal joint, lift column from car.

NOTE: If column has been damaged and a new column is to be fitted carry out following operations. If column was removed for access only proceed to refitting operations.

15 Remove remaining half of shroud from column.
16 Slacken screw securing auxiliary switches to column, noting relative position of switches, slide assembly from column.
17 Using suitable punch remove two shear bolts securing ignition switch and lock assembly to column.
18 Remove bolts securing horn feed and contact assembly to column mounting bracket, remove feed and contact units from column.

NOTE: No further overhaul of column is possible as items still remaining on column are not available on a separate basis.

---

29 Fit dowel pins into locating holes in pump body.
30 With cutaway face uppermost fit bottom plate to drive shaft.
31 With counterbored face of rotor facing bottom plate, fit rotor over splines of drive shaft. Secure rotor and bottom plate with circlip.
32 Insert vanes in rotor slots, ensuring radiused edges are outwards.
33 Carefully fit drive shaft and rotor assembly to pump body, ensure dowel pins locate through smallest holes of bottom plate.
34 With arrowed section uppermost fit pump ring chamber over rotor assembly and dowel pins.
35 With spring recess uppermost fit chamber top plate over dowel pins.
36 Push complete pump assembly home.
37 Lubricate new 'O' ring and fit to upper recess in pump body.
38 Fit spring to recess in top plate.
39 Position retaining plate over spring in pump body.
40 Push plate into pump body and fit retaining ring, ensure ring is fully home in locating recess.
41 Fit new 'O' rings to port recesses in pump body.
42 Fit large 'O' ring to periphery of pump body and magnet to pump body flange.
43 Position pump outer casing over body.
44 Locate bracket mounting studs through outer casing and into pump body.
45 Carefully drive outer casing fully home over pump body.
46 Fully tighten bracket mounting studs.
47 Fit spring and flow control valve to pump.
48 Fit high pressure outlet union to pump.
49 Fit Woodruff key to drive shaft spindle.
50 Reverse operations 1 to 5.

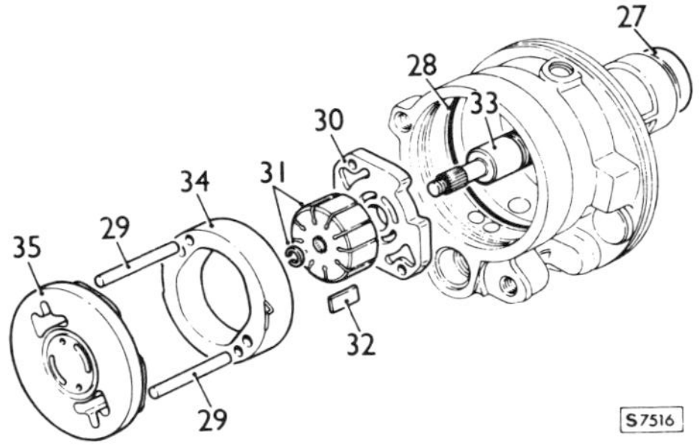

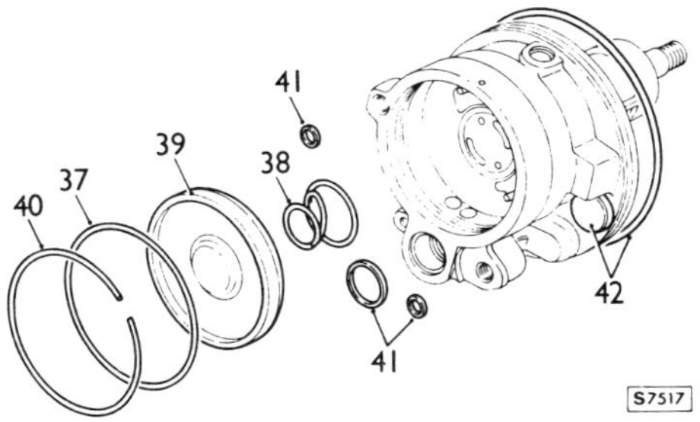

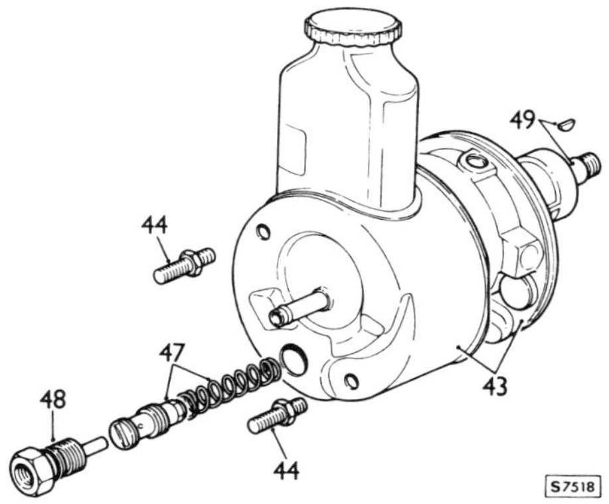

UPPER STEERING COLUMN

Remove and refit   57.40.02

Removing

1 Remove instrument module, see 88.20.01.

57–7

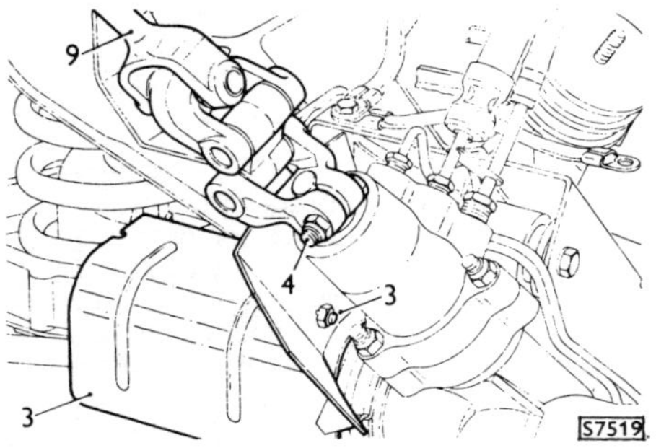

OUTER TIE-ROD BALL JOINT

Remove and refit   57.55.02

Service tool: Ball joint separator JD.24.

Removing

1. Release nut locking outer tie rod, do not turn back nut, release of nut should be sufficient to crack joint between nut and tie rod.
2. Remove nut securing outer ball joint to steering arm.
3. Using ball joint separator JD.24 separate ball joint from steering.
4. Unscrew outside rod assembly from inner tie rod.

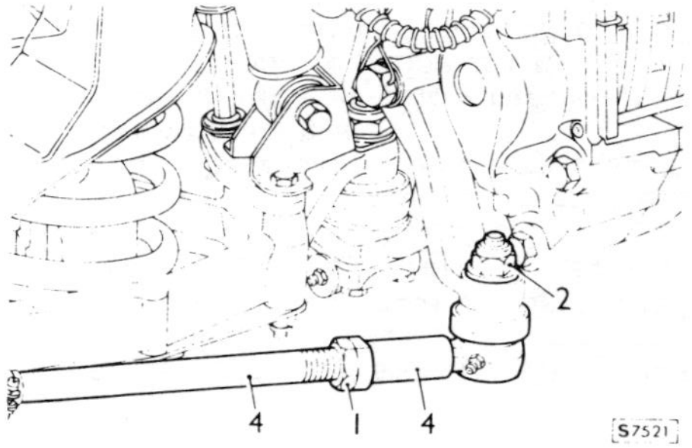

Refitting

5. Reverse operations 1 to 4.
6. Check front wheel alignment, see 57.65.01.

INNER TIE ROD AND BALL JOINT

Remove and refit   57.55.03

Removing

1. Remove outer ball joint, see 57.55.02.
2. Remove outer tie rod locking nut.
3. Remove clips securing gaiter to rack tube and tie rod.
4. Ensuring sufficient waste rag is at hand slide gaiter from tie rod.
5. Clean all grease from inner ball joint.

3. Remove bolts securing right hand gaiter heatshield to steering rack, remove heatshield.
4. Remove pinch bolt securing lower universal joint to pinion shaft.
5. From inside driver's footwell remove pinch bolts securing universal joint to lower and upper column.
6. Slide universal joint free from upper column.
7. Remove universal joint from lower column.

Refitting

8. Reverse operations 1 to 7.

STEERING COLUMN LOCK

Remove and refit   57.40.31

Removing

1. Remove upper column for access, see 57.40.02.
2. Using suitable centre punch unscrew both shear bolts from lock.
3. Remove lock.

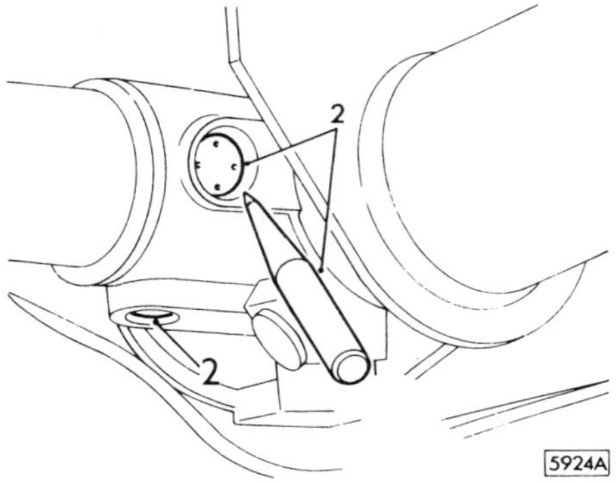

Refitting

4. Locate new lock in column.
5. Fit new shear bolts and tighten until head shears.
6. Refit upper column.

Refitting

11. Reverse operations 1 to 10. Ensure that steering wheel is in central position and that excessive force is not used when refitting lower column to upper column as damage to nylon shear plugs may occur.

STEERING COLUMN ADJUSTING CLAMP

Remove and refit   57.40.07

Removing

1. Remove steering wheel, see 57.60.01.
2. Remove bolt securing adjuster assembly to upper column.
3. Slacken nut and release grub screw securing collet adaptor to upper column, withdraw adjuster assembly from column.
4. Remove three screws securing 'U' plate to adjuster locknut, remove plate.
5. Unscrew collet adaptor from adjuster locknut.
6. Remove impact rubber washer from splined shaft.
7. Remove circlip securing adjuster locknut to splined shaft, slide locknut from shaft.
8. Slacken screw securing split collet to upper shaft, slide shaft from collet.

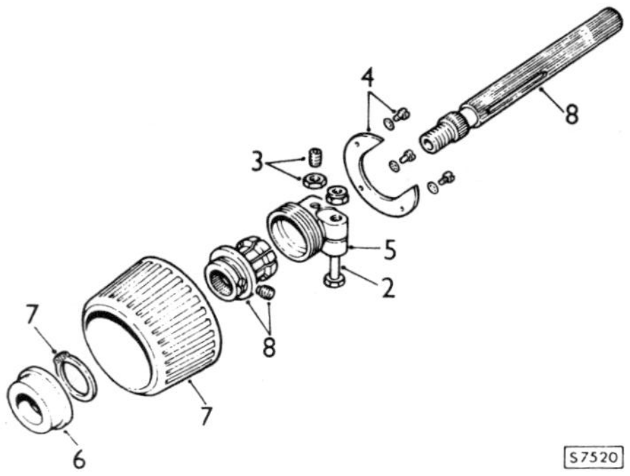

Inspection

9. Check split collet for spline wear, no radial movement between collet and splined shaft should exist.

Refitting

10. Slightly smear splined shaft with grease.
11. Fit split collet to shaft ensuring screw locating hole aligns with groove in shaft.
12. Fit locating screw to split collet. Tightening screw to full extent, turn back screw one half turn.
13. Position locking ring over split collet and secure with circlip.
14. Refit collet adaptor to locking ring.
15. Refit 'U' plate to adjuster locking ring.
16. Centralise horn slip ring needle in upper column and slide adjuster shaft into column; ensure tongue of collet adaptor aligns with groove in nylon bush.
17. Refit impact rubber and bush to shaft.
18. Reverse operations 1 to 3.

STEERING COLUMN UNIVERSAL JOINT

Remove and refit   57.40.25

Removing

1. Adjust wheels to straight ahead position.
2. Remove screws securing right hand heatshield adjacent to exhaust downpipe, remove heatshield.

57–8

6 Turn back tab washer securing ball joint locknut.
7 Release ball joint locknuts from rack, recover thrust spring and packing washer.

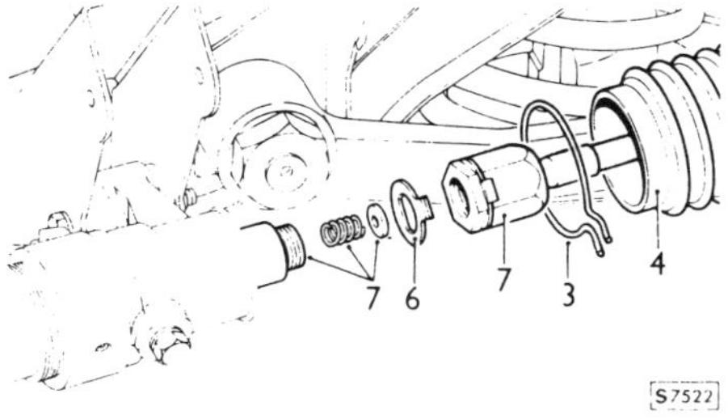

Refitting
8 Reverse operations 1 to 7 ensuring new tab washer is fitted in operation 7.
9 Coat inner ball joint with 2 oz (60 g) of recommended grease.
10 Check wheel alignment, see 57.65.01.

STEERING LEVER

Remove and refit 57.55.29

Service tool: Ball joint separator JD.24.

Removing
1 Remove nut securing outer tie rod ball joint to steering lever.
2 Using ball joint separator JD.24 detach ball joint from steering lever.
3 Remove locking wire securing steering lever mounting bolts.
4 Remove lever mounting bolts, note amount and relative position of shims fitted between lever and caliper mountings.

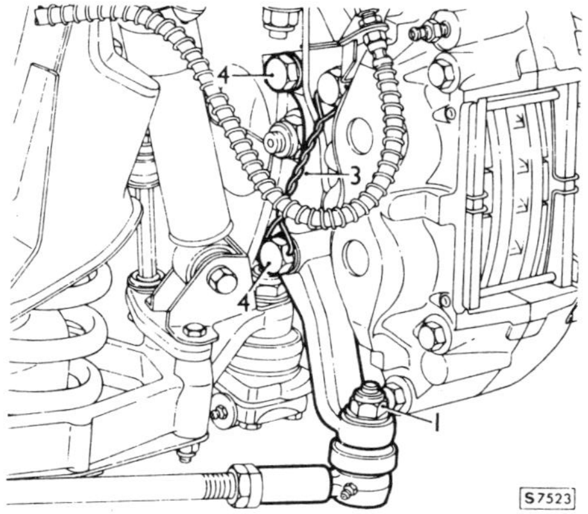

3 Remove two screws securing horn pad to steering wheel, lift pad from wheel.

Refitting
4 Reverse operations 1 to 3.

FRONT WHEEL ALIGNMENT

Check and adjust 57.65.01

Service tool: Rack centralising tool, Jaguar Part No. 12279.

Check
1 Ensure tyres are at correct pressures.
2 Adjust wheels to straight ahead position.
3 Remove grease nipple from rack adjuster nut.
4 Insert centralising tool into grease nipple locating hole, push tool on to back of rack bar.
5 Slowly turn steering wheel until centralising tool locates in recess in back of rack bar.
6 Check toe-in of wheels by using appropriate light beam equipment or an approved track setting gauge.

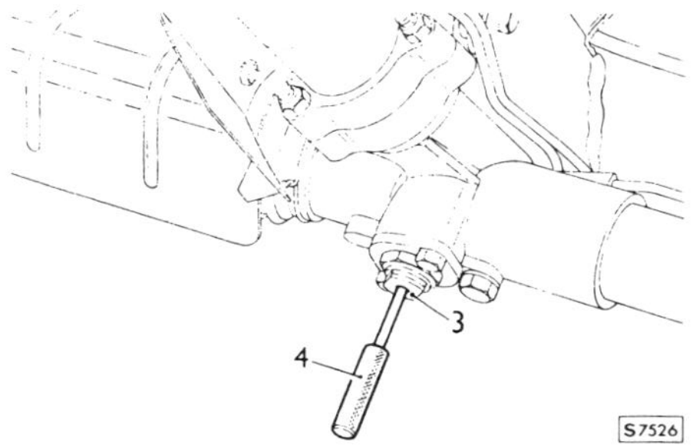

Refitting
5 Reverse operations 1 to 4, torque steering lever mounting bolts to 49 to 55 lb ft (6,8 to 7,6 kg m).
6 Check wheel alignment, see 57.65.01.

STEERING WHEEL

Remove and refit 57.60.01

Removing
1 Remove steering wheel horn pad, see 57.60.03.
2 Adjust wheels to straight ahead position, remove ignition key and lock steering wheel.
3 Remove horn contact rod from upper column.
4 Remove nut securing steering wheel to upper column.
5 Gently tap steering wheel withdrawing wheel from column, recover split collets.

Refitting
6 Reverse operations 1 to 5.

STEERING WHEEL PAD

Remove and refit 57.60.03

Removing
1 Slacken wheel adjusting locknut.
2 Pull wheel out to maximum extended position.

11 Recheck alignment.
12 Remove centralising tool and refit grease nipple.

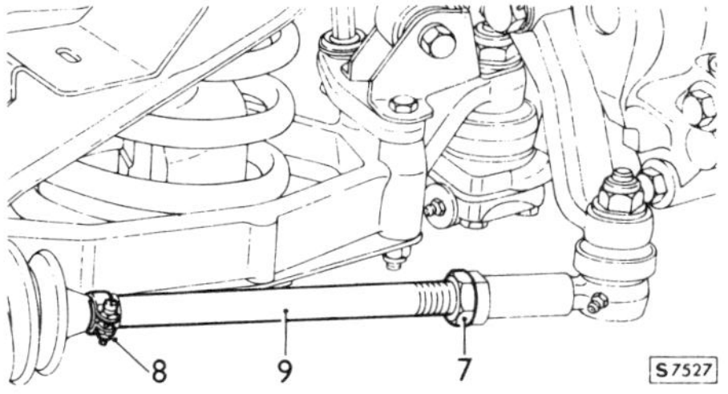

NOTE: It should be appreciated that front wheel alignment check is a routine maintenance operation and as such where toe-in exceeds the correct setting of $\frac{1}{16}$ in to $\frac{1}{8}$ in (1,6 mm to 3 mm) by more than $\frac{1}{8}$ in (3 mm) the possibility of damage to tie rods or steering levers should be investigated. Where this is necessary tie rods should be visually examined for bending and steering levers checked against the dimensions shown below.

A = 3.248 in (82,5 mm)
A = 3.252 in (82,6 mm)
B = 3.895 in (96,52 mm)
C = 0.875 in (22,23 mm)
D = 2.33 in (59,18 mm)
E = 5.136 in (136,14 mm)
F = 0.71 in (18,0 mm)
F = 0.70 in (17,8 mm)
G = 2.150 in (54,61 mm)

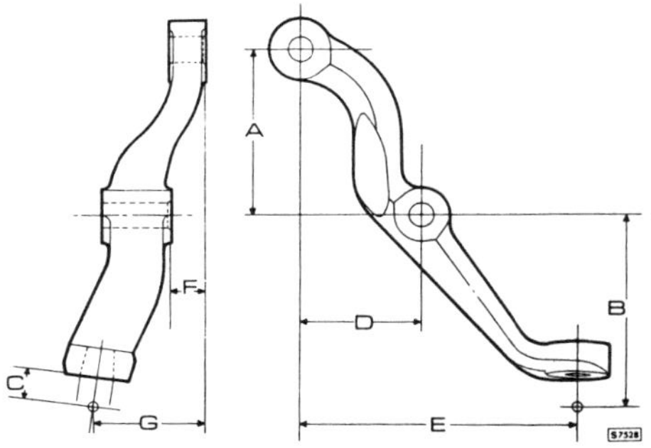

Adjust
7 Slacken locknuts at outer ends of the tie rods.
8 Slacken clips securing gaiters to tie rods.
9 Turn tie rods by equal amounts until correct alignment of $\frac{1}{16}$ in to $\frac{1}{8}$ in (1,6 mm to 3 mm) toe-in is achieved.
10 Tighten tie rod locknuts and gaiter clips.

57—9

CASTOR ANGLE

Check and adjust 57.65.04

Service tool: Suspension links JD.25.

Check

NOTE: Before checking castor angle all the suspension bushes, upper and lower wishbone ball joints and shock absorber mountings should be checked for signs of deterioration, damage or distortion.

1. Ensure car is standing on level ground and tyres are at correct pressures.
2. Manufacture two suspension setting tubes to dimensions shown.
3. Compress front suspension and insert setting tubes over rebound rubbers and rebound stops flange. Suspension is now locked in mid-laden position.

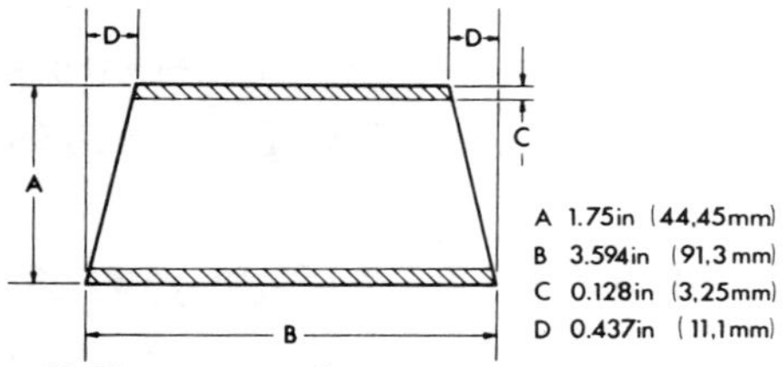

A 1.75in (44.45mm)
B 3.59in (91.3 mm)
C 0.128in (3.25mm)
D 0.437in (11.1mm)

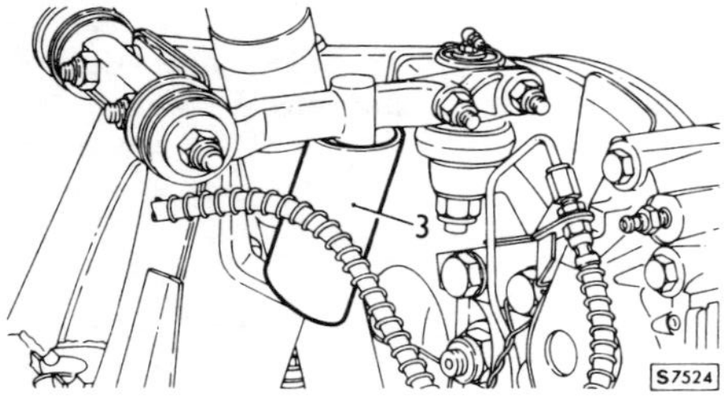

4. Compress rear suspension and hook suspension links JD.25 through lower holes of suspension mounting and over nut securing hub lower pivot pin.

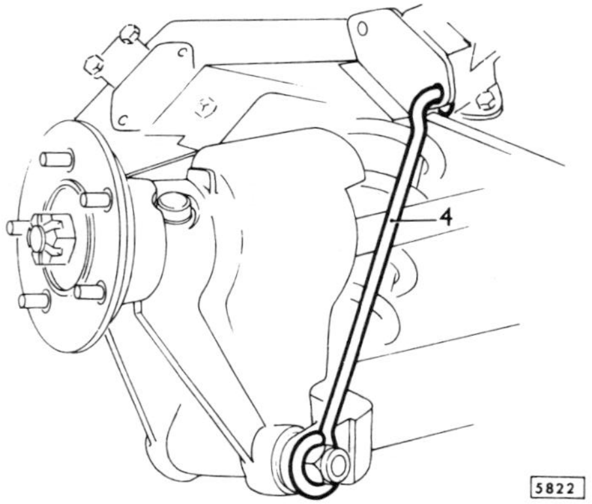

CAMBER ANGLE

Check and adjust 57.65.05

Check

1. Set car in mid laden condition, see 57.65.01 operations 1 to 4.
2. Check camber angle using approved gauge, correct camber is ½° positive ± ¼° with both front wheels being within ¼° of each other.
3. Rotate wheels through 180° and recheck camber angle.

Adjust

4. Slacken nuts securing upper wishbone to suspension turret.
5. Add or remove equal amount of shims from both shim locations.
   NOTE: Removal of shims decreases camber angle; addition of shims increases camber angle. Transposing of one $\frac{1}{16}$ in (1,6 mm) shim will alter camber angle by approximately ¼°. 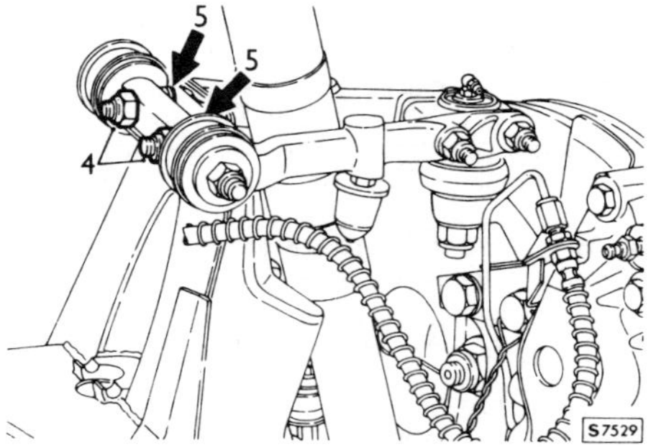
6. Reverse operation 4 and re-check camber angle.
7. If camber angle correct reverse operation 1; if angle not correct repeat operations 4 to 6.
8. Check front wheel alignment, see 57.65.01.

5. Check castor angle using approved gauge, correct castor angle is 3½° positive ± ¼°.

Adjust

6. Slacken four nuts securing upper wishbone to ball joint.
7. Remove shim from one side of ball joint, and fit to opposite side of ball joint.
   NOTE: Removal of front shim to rear position decreases castor angle, rear shim to front location increases castor angle. Transposing of one shim $\frac{1}{16}$ in (1,6 mm) thick will alter castor angle by approximately ¼°.
8. Reverse operation 6 and re-check castor angle.
9. If castor angle correct reverse operations 1 to 4, if angle not within limits repeat operations 7 and 8.
10. Check front wheel alignment, see 57.65.01.

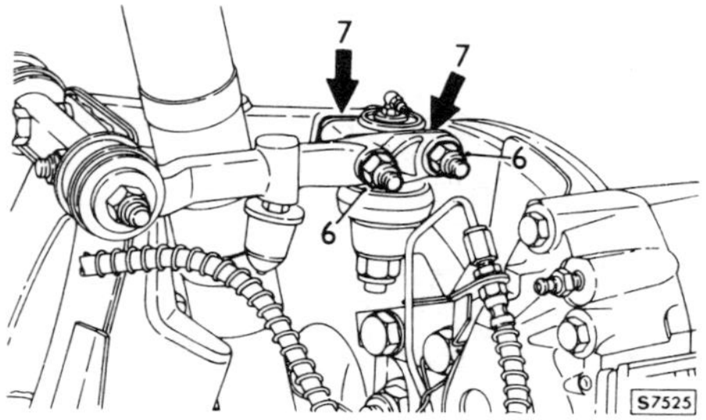

ACCIDENTAL DAMAGE

The dimensional drawings are provided to assist in assessing accidental damage. A component suspected of being damaged should be removed from the car, cleaned off and the dimensions checked and compared with those given in the appropriate illustration. Components found to be dimensionally inaccurate, or damaged in any way MUST be scrapped and NO ATTEMPT made to straighten and re-use.

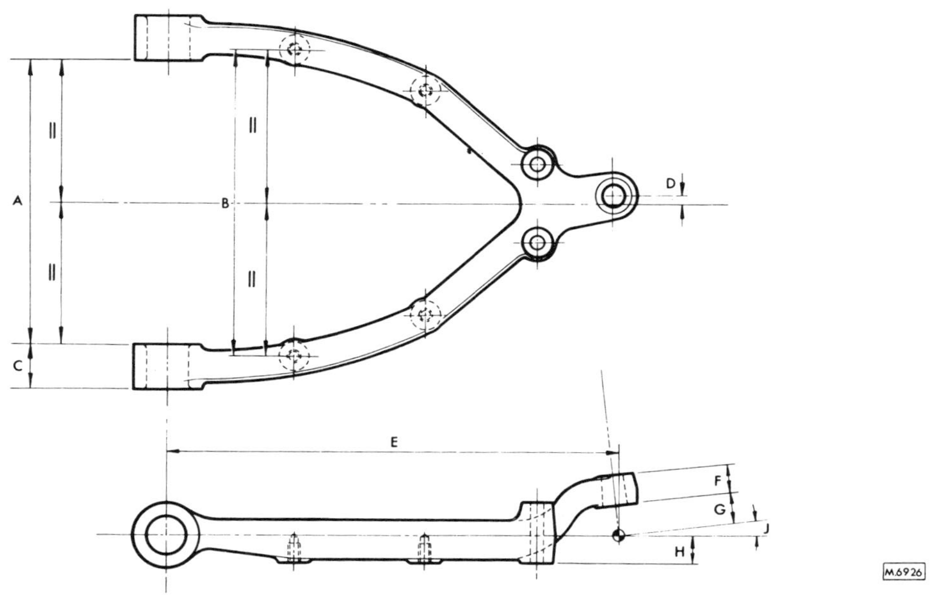

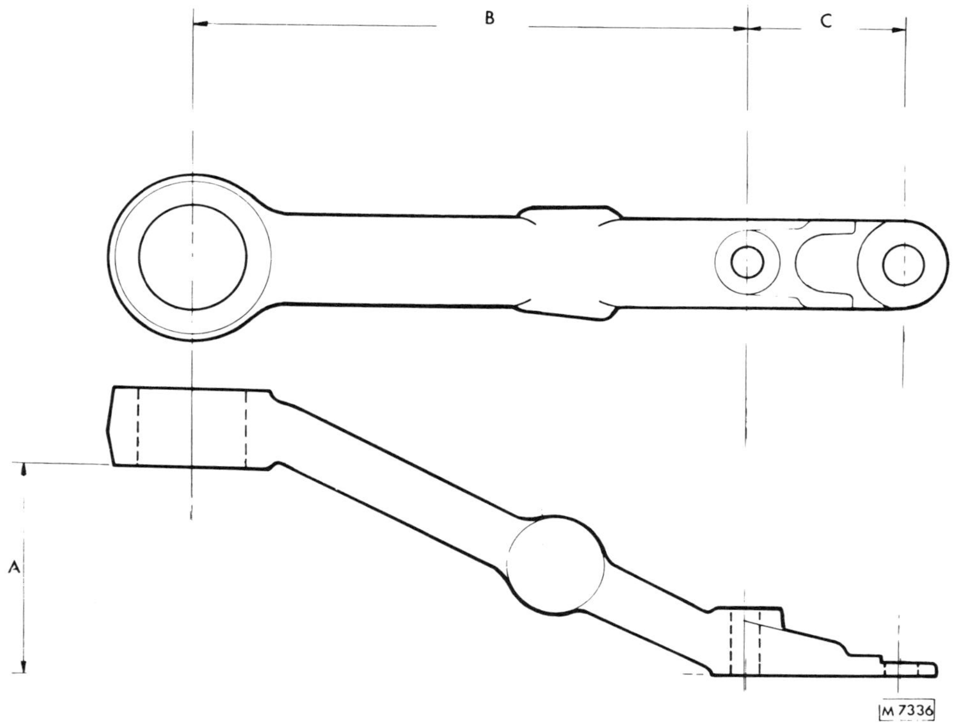

DIMENSION – UPPER WISHBONE ARM – FRONT

A 2.37 to 2.39 in (60,20 to 60,71 mm)
B 6.356 to 6.376 in (161,44 to 161,95 mm)
C 1.74 to 1.76 in (44,20 to 44,70 mm)

continued

DIMENSION – LOWER WISHBONE

A 8.87 to 8.89 in (225,30 to 225,81 mm)
B 9.62 to 9.63 in (244,35 to 244,60 mm)
C 1.365 to 1.385 in (34,67 to 35,18 mm)
D 0.23 to 0.25 in (5,84 to 6,35 mm)
E 13.93 to 13.95 in (353,82 to 354,33 mm)
F 0.90 to 0.92 in (22,86 to 23,37 mm)
G 1.05 to 1.07 in (26,67 to 27,18 mm)
H 0.83 to 0.85 in (21,08 to 21,59 mm)

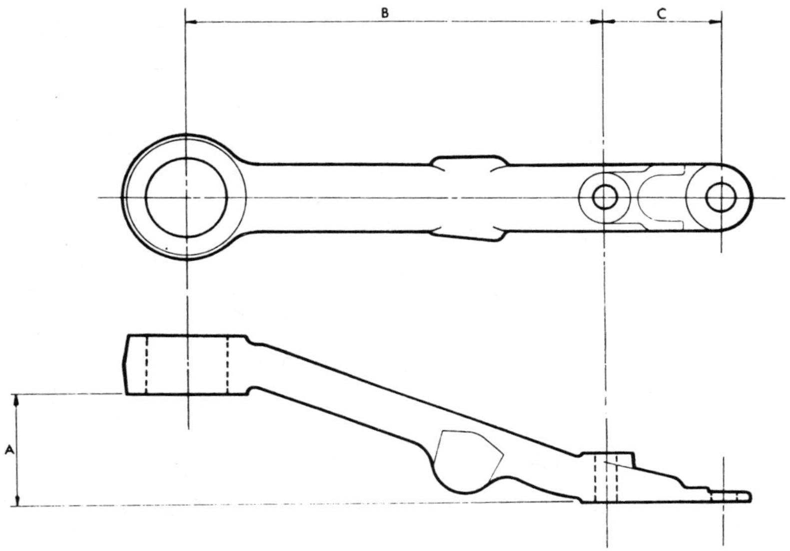

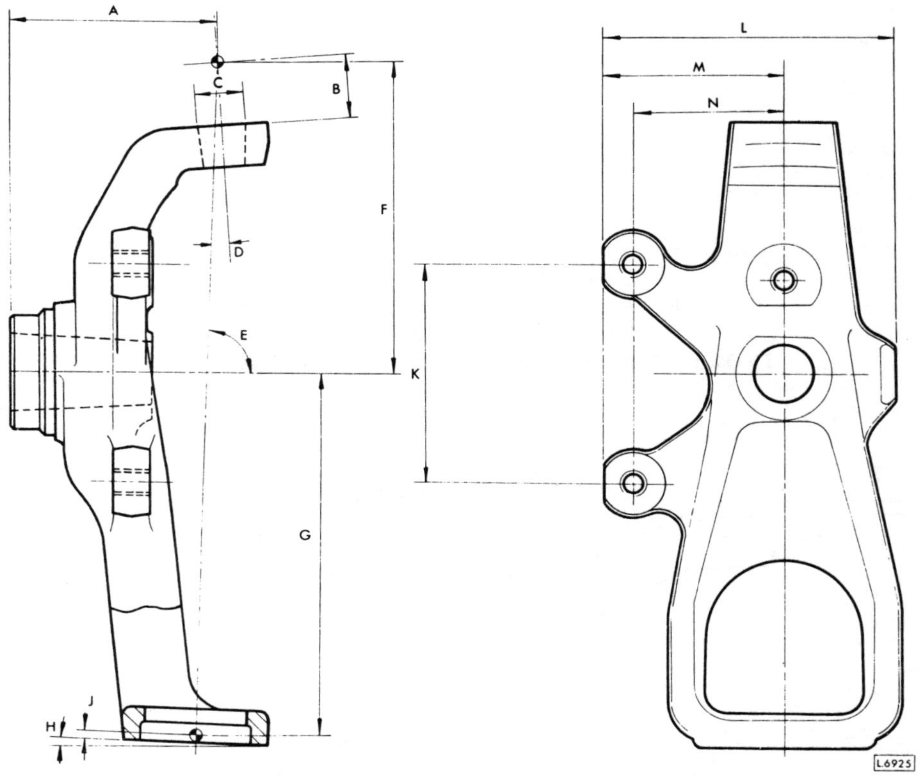

DIMENSION – STUB AXLE CARRIER

A 3.21 in (81,53 mm)
B 1.0 in (2,54 mm)
C 0.749 to 0.751 in (19,025 to 19,075 mm) dia.
D 5 degrees
E 2 degrees
F 4.42 to 4.54 in (112,3 to 115,3 mm)
G 2.80 in (71,1 mm)
H 2.328 to 2.332 in (59,13 to 59,23 mm)
J 3.499 to 3.502 in (88,88 to 88,95 mm)

DIMENSION – UPPER WISHBONE ARM – REAR

A 1.74 to 1.76 in (44,20 to 44,70 mm)
B 6.356 to 6.376 in (161,44 to 161,95 mm)
C 1.74 to 1.76 in (44,20 to 44,70 mm)

60–2

ANTI-ROLL BAR

Remove and refit 60.10.01

Service tool: Steering joint taper separator JD.24

Removing

1. Jack up front of car and place on stands. Leave jack in position under front suspension cross-member.
2. Remove both front wheels.
3. Prise out plastic drive fasteners securing vertical diaphragms in wheel arches.
4. Remove two screws and 13 plastic drive fasteners securing spoiler undertray to body and spoiler.
5. Detach nut securing L.H. steering tie rod to steering arm. Detach tie rod from arm using tool JD.24.
6. Remove self locking nuts, special washers and rubber pads securing anti-roll bar links to anti-roll bar on both sides.
7. Remove front suspension mounting bolts. Lower jack and collect washers, spacers and bush sleeves.
8. Remove nuts and bolts securing anti-roll bar to body frame and detach keeper plates.
9. Remove rubbers from anti-roll bar.

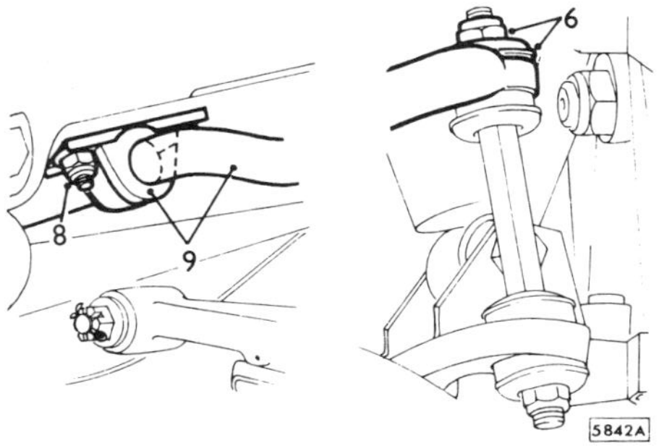

Refitting

10. Manoeuvre anti-roll bar into position through L.H. wheel arch.
11. Replace rubbers, keeper plates and rubber retainers on anti-roll bar.

NOTE: Split in bushes is to be to rear of bar. Fitting of bushes will be greatly facilitated if a proprietary rubber lubricant or a solution of twelve parts water to one part liquid soap is used.

13. Replace four bolts, from top, and four self-locking nuts attaching anti-roll bar mountings to body frame. Do not tighten nuts until full weight of car is on suspension.
14. Replace anti-roll bar on link arms, fit rubbers, special washers and nuts.
15. Replace steering tie-rod on L.H. steering arm. Tighten nut to 45 to 50 lbf. ft (6,22 to 6,91 kgf. m).
16. Raise jack to line up front attachments and replace bolts with washers, spacers and bush sleeves. Tighten nuts to 95 to 115 lbf. ft (13,13 to 15,90 kgf. m).
17. Replace spoiler undertray, using new plastic drive fasteners.
18. Replace vertical diaphragms, using new plastic drive fasteners.
19. Refit front wheels.
20. Lower car on to wheels and fully tighten anti-roll bar link nuts to 14 to 18 lbf. ft (1,94 to 2,48 kgf. m) and mounting nuts to 27 to 32 lbf. ft (3,73 to 4,42 kgf. m).

ANTI-ROLL BAR LINK

Remove and refit 60.10.02

Removing

1. Jack up front of car and rest on stands.
2. Remove self locking nut, special washer and rubber pad securing anti-roll bar link to lug on wishbone.
3. Remove self locking nut, special washer and rubber pad securing anti-roll bar link to anti-roll bar.
4. Remove upper nut on link at opposite end of anti-roll bar.
5. Lift link clear and recover two spacer tubes, two rubber pads and special washers.

Refitting

6. Check condition of rubber pads, renew if damaged in any way.
7. Pivot anti-roll bar upwards and place link in position, with inner cups and bushes in position.
8. Replace opposing upper bush, cup and nut.
9. Replace upper and lower bushes, cups and nuts.

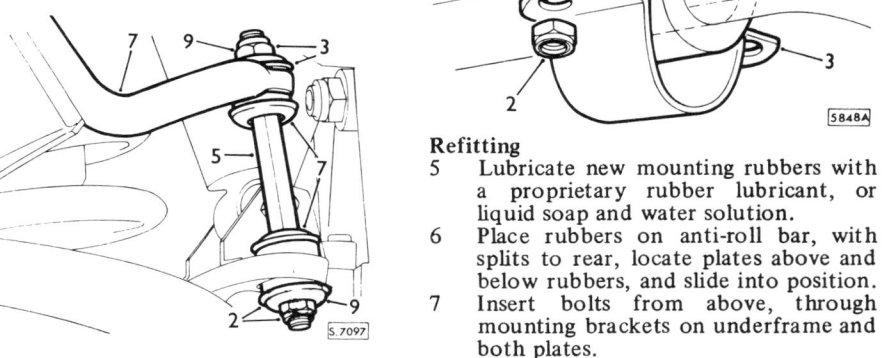

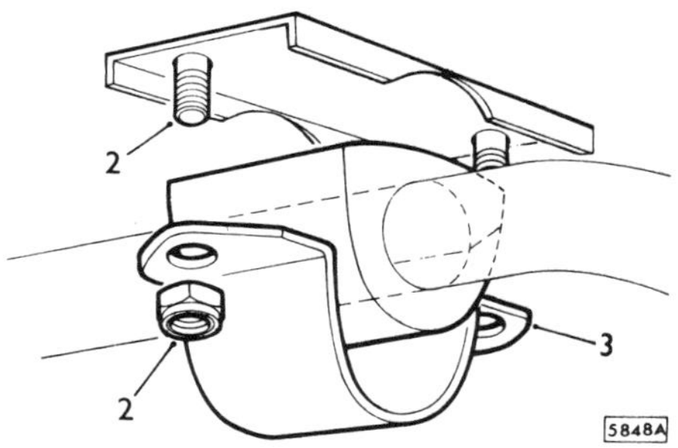

Refitting

5. Lubricate new mounting rubbers with a proprietary rubber lubricant, or liquid soap and water solution.
6. Place rubbers on anti-roll bar, with splits to rear, locate plates above and below rubbers, and slide into position.
7. Insert bolts from above, through mounting brackets on underframe and both plates.
8. Ensure that rubbers are centrally located in plates and replace nuts on bolts. Do not tighten.
9. Check that full weight of car is supported by suspension and tighten nuts to 27 to 32 lbf. ft (3,73 to 4,42 kgf. m).

10. Lower car on to wheels.
11. Fully tighten self locking nuts at top and bottom of links, torque 14 to 18 lbf. ft (1,94 to 2,48 kgf. m).

ANTI-ROLL BAR LINK BUSHES

Remove and refit 60.10.03

Removing

1. Remove anti-roll bar links, see 60.10.02.

Refitting

2. Renew rubber pads and refit anti-roll bar links.

ANTI-ROLL BAR MOUNTING RUBBERS – SET

Remove and refit 60.10.05

Removing

1. Place car on ramp or over pit. Apply handbrake.
2. Undo and remove four bolts securing anti-roll bar mountings to underframe brackets.
3. Ease plates away from rubbers.
4. Move rubbers along anti-roll bar, remove and discard.

FRONT SUSPENSION RIDING HEIGHT

Check and adjust 60.10.18

1. Check that car is full of petrol, oil and water, and that tyre pressures are correctly adjusted.
2. Position car with front wheels on slip plates.
3. Press downwards on front bumper to depress car and slowly release.
4. Measure distance between lower face of cross-beam and ground on both sides of car. Obtain values for dimension 'A', right and left hand. Correct height is 6.0 in (152 mm) minimum plus thickness of slip plates.
5. If necessary, fit or remove packing rings beneath springs to achieve this dimension; see operation 60.20.01. Packing rings are ⅛ in (3,18 mm) thick, and vary the riding height by 5⁄16 in (7,93 mm).

continued

60–3

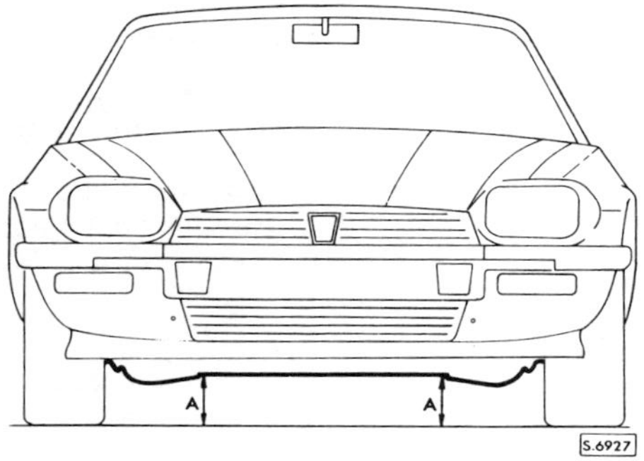

BALL JOINT – LOWER

Overhaul 60.15.13

Service tool: Steering joint taper separator JD.24

Removing

1. Jack up front of car and place on stands.
2. Remove road wheel.
3. Disconnect flexible brake hose from caliper pipe and bracket and immediately plug open ends to prevent ingress of dirt and loss of fluid.
4. Remove self locking nut and washer securing steering tie rod ball joint. Separate rod from steering arm using service tool JD.24.
5. Turn hub carrier to improve access and remove two nuts, bolts and plain washers securing ball joint to upper wishbone levers.
   NOTE: Take careful note of fitted position of packing piece and shims; these control castor angle.
6. Lower hub assembly and remove self locking nut and washer securing lower ball joint to lower wishbone.
7. Extract ball joint from stub axle carrier, using taper separator tool JD.24.

Refitting

CAUTION: Bolts securing ball joints to upper wishbones must be fitted with heads forward.

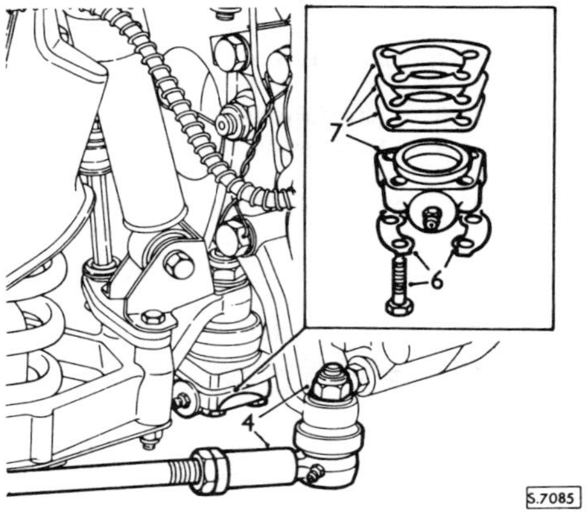

8. Clean and examine all parts for wear.
   CAUTION: In order to obtain correct adjustment of ball joint it is necessary to shim to correct clearance. Excessive wear on ball pin and sockets must not be adjusted by shims. Worn parts must be renewed.
9. Remove shims one by one until ball pin is tight in its socket with screws fully tightened to 15 to 20 lbf. ft (2,10 to 2,75 kgf. m).
   NOTE: Shims are available in .002 in (,05 mm) and .004 in (,10 mm) thickness.
10. Remove screws, ball pin cap, shims and socket. Add shims to the value of .004 in to .006 in (,10 mm to ,15 mm).
11. Lightly grease ball pin and socket. Refit socket ball pin cap and new tab washers. Refit and tighten screws; torque to 15 to 20 lbf. ft (2,1 to 2,75 kgf. m).
12. When correctly adjusted, hub and stub axle carrier can be pivoted with very slight drag.
13. Turn up tab washers and charge joint with correct grease.
14. Replace steering tie rod, tighten nut to 35 to 50 lbf. ft (4,84 to 6,91 kgf. m).
15. Remove jack from below spring pan.
16. Replace front wheel.
17. Remove car from stands.

8. Grease replacement ball joint and place in position in stub axle carrier.
9. Holding ball joint against taper fit washer and tighten retaining nut to 35 to 50 lbf. ft (4,84 to 6,91 kgf. m).
10. Clean wishbone ends and castor shims and lift into position.
11. Replace retaining bolts from the front and insert castor shims as removed in operation 5.
12. Replace washers and nuts and tighten to 26 to 32 lbf. ft (3,60 to 4,42 kgf. m).
13. Remove support wire.
14. Replace road wheel.
15. Remove car from stands.
16. Check castor and camber angles, see 57.65.04 and 57.65.05.

BALL JOINT – UPPER

Remove and refit 60.15.02

Service tool: Steering joint taper separator JD.24.

The upper wishbone ball joint cannot be dismantled and if worn, the complete assembly must be replaced.

Removing

1. Jack up car and place on stands.
2. Remove road wheel.
3. Turn steering to full lock to improve access to hub and ball joint.
4. Wire stub axle carrier to cross member turret to prevent strain on brake hose when upper joint is disconnected.
5. Remove two nuts, bolts and plain washers securing ball joint to upper wishbone levers.
   NOTE: Take careful note of fitted position of packing pieces and shims. These control castor angle.
6. Swing disc-hub assembly away from wishbone and remove self locking nut and plain washer securing ball joint to stub axle carrier.

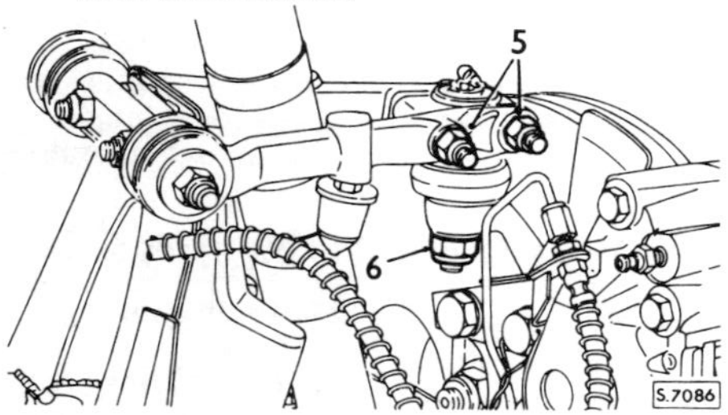

BALL JOINT – LOWER

Adjust 60.15.04

Service tool: Steering joint taper separator JD.24.

1. Jack up front of car and place on stands.
2. Remove front wheel.
3. Place jack beneath front spring seat pan and raise sufficient to relieve stub axle carrier of spring pressure.
4. Remove self locking nut and washer securing steering tie rod ball joint. Separate tie rod from steering arm using service tool JD.24.
5. Lift hub and stub axle carrier assembly to reveal any free play in lower ball joint.
6. Bend back tab washers, remove four screws securing ball pin cap to stub axle carrier.
7. Detach ball pin cap, shims and socket from stub axle carrier.

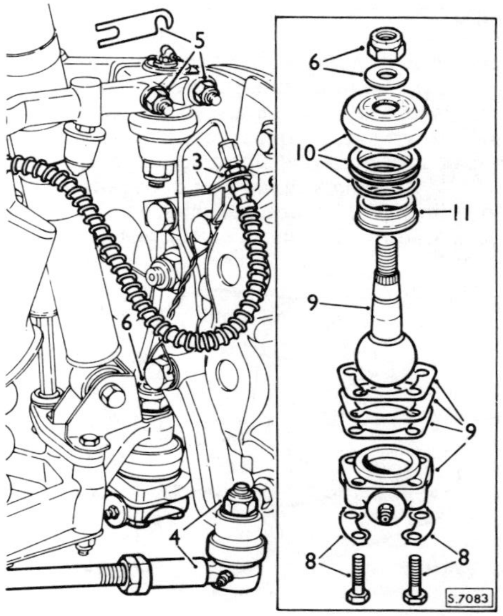

7 Separate ball joint from wishbone using service tool JD.24 and remove assembly from car.
8 Bend back tab washers and remove four screws securing ball pin cap to stub axle carrier. Discard tab washers.
9 Detach ball pin cap, shims and lower socket. Lift out ball pin.
10 Release moulded gaiter clip from upper socket and remove gaiter.
11 Remove upper socket from stub axle carrier.

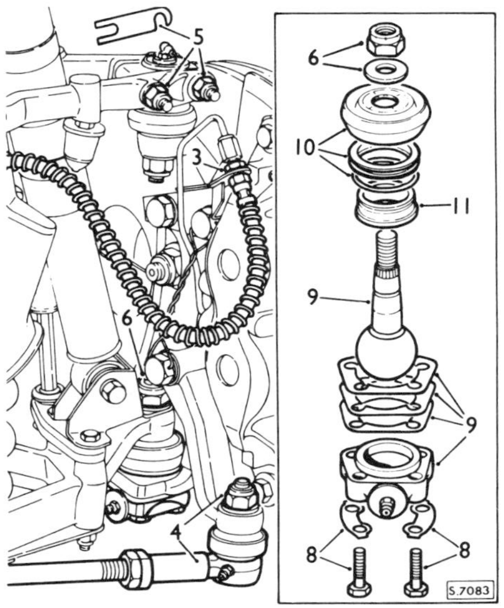

12 Clean and inspect all parts.

Overhaul
CAUTION: In order to obtain correct adjustment of ball joint it is necessary to shim to correct clearance. Excessive wear on ball pin and sockets must not be adjusted by shims. Worn parts must be renewed.

13 Fit new upper socket to stub axle carrier.
14 Engage lip of moulded gaiter clip with recess in socket. Lip must be near lower face of clip.
15 Fit new plastic gaiter to clip and secure with new plastic retaining ring.

16 Lightly grease replacement ball pin and place in position.
17 Mount ball pin cap in vice, cut away old lower socket, clean and fit new socket.
18 Refit shims removed in operation 9 and replace cap over ball.
19 Replace four setscrews with new locking tabs and tighten, checking freedom of ball pin, which should be slightly stiff in socket with setscrews tightened to 15 to 20 lbf. ft (2,1 to 2,75 kgf. m).

NOTE: If ball pin is loose in socket, remove shims which are 0.002, 0.004 and 0.010 in (0,05, 0,10 and 0,25 mm) thick. If ball pin is excessively tight add shims until ball pin is slightly stiff in socket with setscrews fully tightened.

20 Turn down locking tabs of setscrews.
21 Enter ball pin in lower wishbone, fit washer and tighten nut to 45 to 55 lbf. ft (6,23 to 7,60 kgf. m). 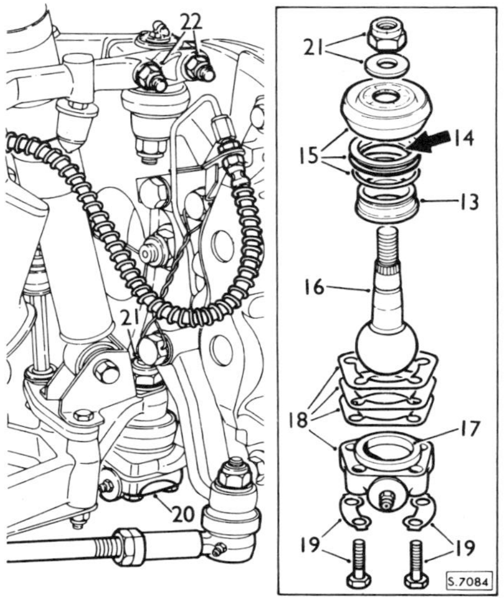
22 Line up stub axle carrier with upper wishbone and insert bolts from front.
23 Replace packing piece and shims removed in operation 5 and tighten nuts to 26 to 32 lbf. ft (3,60 to 4,42 kgf. m).

24 Refit steering tie rod to steering arm. Tighten nut to 35 to 50 lbf. ft (4,84 to 6,91 kgf. m).
25 Remove blanking plug and attach flexible hose to bracket. Tighten nut to 10 to 12 lbf. ft (1,38 to 1,66 kgf. m).
26 Reconnect caliper pipe to flexible hose. Tighten to 6.3 to 7 lbf. ft (0,89 to 0,96 kgf. m).
27 Replace road wheel.
28 Remove car from stands.
29 Bleed brakes, see 70.25.02.
30 Check castor and camber angles, see 57.65.04 to 57.65.05

FRONT SPRING

Remove and refit 60.20.01

Service tools: Spring compressor JD.6F, or JD.6D with adaptor JD.6D-1 with spring plate locating pegs JD.6E-6

Removing
1 Jack up front of car and place on stand.
2 Remove road wheel.
3 Fit spring compressor tool JD.6D with adaptor JD.6D-1 and compress spring sufficiently to relieve load on seat pan fastening

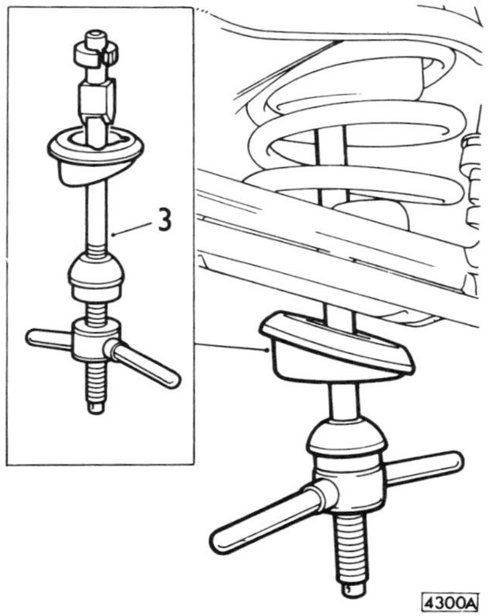

4 Remove four setscrews and washers and two nuts, bolts and washers securing spring plate to lower wishbone.
5 Slacken spring compressor tool and remove, together with spring plate, spring and packers.
NOTE: Record position of packers to assist during replacement.

Refitting
6 Fit spring plate locating pegs JD.6E-6 in two tapped holes below lower wishbone.
7 Replace packers removed in operation 5 in suspension turret.
WARNING: A MAXIMUM OF TWO PACKERS MAY BE FITTED IN SUSPENSION TURRET
8 Assemble spring, packing, spring plate and compressor tool as removed in operation 5 and lift into position in turret with locating pegs through holes in spring plate.
9 Centralise spring and compress tool.
10 Remove plate locating pegs and refit setscrews and bolts; tighten to 27 to 32 lbf. ft (3,74 to 4,42 kgf. m).
11 Remove tool and adaptor.
12 Replace road wheel.
13 Remove car from stands.
14 Check front suspension riding height, see 60.10.18.

FRONT HUB ASSEMBLY

Remove and refit 60.25.01

Removing
1 Jack up front of car and place on stands.
2 Remove road wheel.
3 Through aperture in disc shield remove five bolts and washers holding hub assembly to brake disc.
4 Remove hub grease cap, extract split pin, and remove nut and washer from stub axle.
5 Withdraw hub by hand.

Refitting
6 Pack hub with specified grease and refit to stub axle.

continued

7 Fit bearing, nut and washer to stub axle and tighten nut to give end float of .002 in to .006 in (0,05 mm to 0,15 mm).
8 Renew split pin.
9 Refit grease cap. Ensure that vent hole in grease cap is clear.
10 Replace five bolts and washers securing hub assembly to brake disc. Tighten to 30 to 36 lbf. ft (4,15 to 4,98 kgf. m). 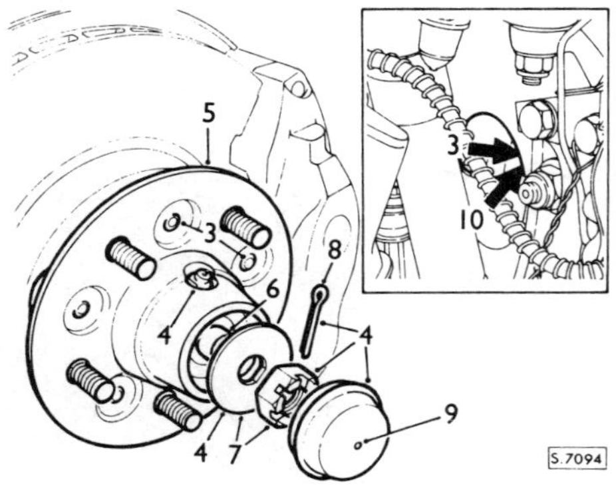
11 Refit road wheel.
12 Remove car from stands.

FRONT HUB BEARING END FLOAT 60.25.13

Check and adjust

1 Jack up front of car and place on stands.
2 Remove nave plate and grease cap. Clear vent hole in cap.
3 Remove and discard split pin.
4 Tighten nut to eliminate all end float and back off slightly to give end float of 0,002 in to 0.005 in (0,05 mm to 0,15 mm).
5 Fit new split pin.
6 Replace grease cap and nave plate.
7 Remove car from stands.

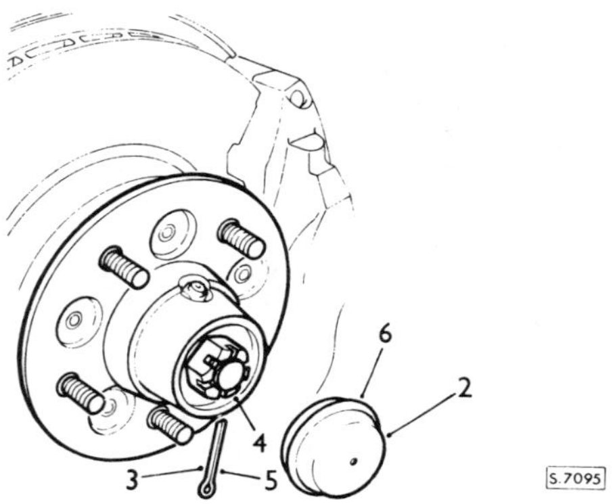

FRONT HUB GREASE SEAL 60.25.15

Remove and refit

Removing
1 Remove front hub assembly, see 60.25.01 operations 1 to 5.
2 Extract and discard grease seal.

Refitting
3 Place new seal against hub with sealing edge next to hub.
4 Tap seal squarely into position, using suitable 'bell-piece'.
5 Fit front hub assembly, see 60.25.01 operations 6 to 12.

FRONT HUB BEARINGS 60.25.14

Remove and refit

Removing
1 Remove front hub assembly, see 60.25.01 operations 1 to 5.
2 Collect outer bearing.
3 Extract grease seal and inner bearing race.
4 Drift bearing cups from hub; grooves are provided in the abutment shoulders for this purpose.
5 Clean out hub assembly.

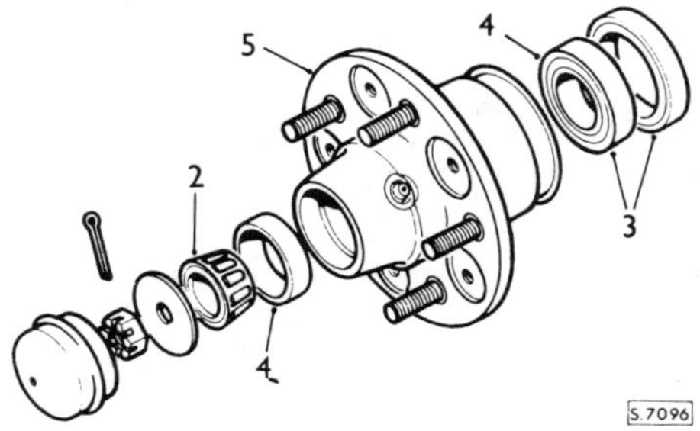

Refitting
6 Tap replacement cups into position, ensuring that they are seated square to abutments.
7 Lubricate large bearing race and fit to cup.
8 Fit new grease seal.
9 Fit front hub assembly, see 60.25.01 operations 6 to 12.

FRONT HUB STUB AXLE 60.25.22

Remove and refit

Service tool: Steering joint taper separator JD.24.

Removing
1 Jack up front of car and place on stands.
2 Remove road wheel.
3 Remove self locking nut and washer securing steering tie rod ball joint. Separate rod from steering arm using tool JD.24.
4 Disconnect flexible brake hose from caliper and immediately plug hose to prevent ingress of dirt and loss of fluid.
5 Remove two nuts, bolts and plain washers securing upper ball joint to upper wishbone levers.
   NOTE: Take careful note of fitted position of bolts and also position of packing piece and shims; these control castor angle.
6 Remove self-locking nut and washer and separate lower ball joint from wishbone; draw assembly from car.
7 Remove five bolts and washers holding hub assembly to brake disc. Bolt heads are accessible through aperture in disc shield.
8 Remove grease cap, split pin and castellated nut securing hub assembly to stub axle; draw assembly clear. Discard split pin.
9 Remove nyloc nut and plain washer securing stub axle to stub axle carrier. 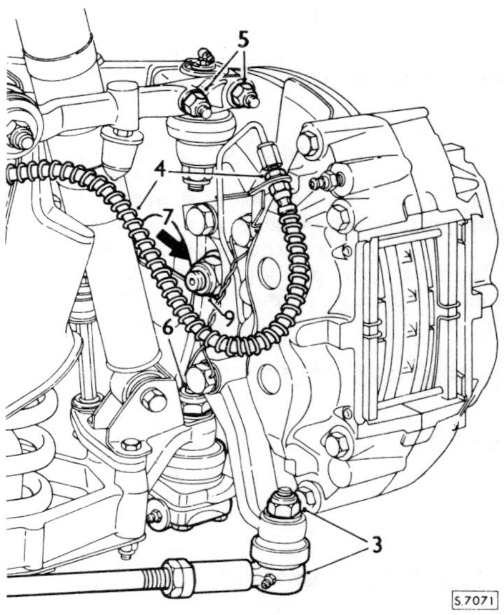
10 Support stub axle carrier and drift out stub axle.

Refitting
11 Fit stub axle to carrier. Tighten nut to 80 to 90 lbf. ft (11,1 to 12,4 kgf. m).
12 Clean and grease hub and fit to stub axle.
13 Grease and refit outer bearing. Replace washer and nut. Do not tighten.
14 Replace five bolts and washers securing hub to brake disc. Tighten to 30 to 36 lbf. ft (4,15 to 5,0 kgf. m).
15 Replace lower ball joint stem in wishbone. Tighten nut to 45 to 55 lbf. ft (6,25 to 7,60 kgf. m).
16 Align upper ball joint and replace bolts and nuts.
   CAUTION: Bolts securing upper ball joint in upper wishbone must be fitted from front of car towards rear.
17 Refit castor shims as removed in operation 5. Tighten nuts to 26 to 32 lbf. ft (3,60 to 4,42 kgf. m).
18 Replace steering tie rod joint in steering arm. Tighten nut to 35 to 50 lbf. ft (4,84 to 6,91 kgf. m).

60–6

FRONT HUB STUB AXLE CARRIER

Remove and refit   60.25.23

Service tool: Steering joint taper separator, JD.24.

Removing

1. Remove front hub and stub axle carrier, see 60.25.22, operations 1 to 6.
2. Cut locking wire, remove caliper retaining bolts and detach brake caliper, noting location of shims.
3. Remove steering arm retaining bolt and detach steering arm and brake pipe mounting bracket.
4. Undo disc shield retaining nuts and remove disc shields and clips.
5. Remove grease cap, split pin and castellated nut securing hub assembly to stub axle; draw assembly clear. Discard split pin.
6. Bend back tab washers and remove four screws securing ball pin cap to stub axle carrier. Discard tab washers.
7. Detach ball pin cap, shims and lower socket. Retain shims. Lift out ball pin.
8. Release moulded gaiter clip from upper socket and remove gaiter.
9. Remove upper socket from stub axle carrier.
10. Remove nut securing upper ball pin to stub axle carrier.
11. Extract upper ball pin from stub axle carrier using taper separator tool JD.24.
12. Remove nyloc nut and plain washer securing stub axle to stub axle carrier.
13. Support stub axle carrier and drift out stub axle. Remove weather shield.
14. Remove pads from brake caliper, mark for refitting.
15. Clean and inspect all parts.

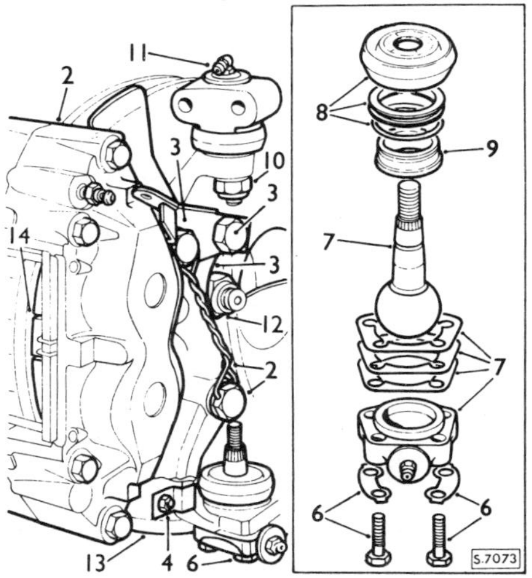

Refitting

16. Replace weather shield in stub axle carrier and refit stub axle. Tighten nut to 80 to 90 lbf. ft (11,1 to 12,4 kgf. m).
17. Grease upper ball joint and replace in stub axle carrier. Fit washer and tighten nut to 35 to 50 lbf. ft (4,84 to 6,91 kgf. m).
18. Refit lower ball joint upper socket in stub axle carrier.
19. Engage lip of moulded gaiter clip with recess in socket. Lip must be near lower face of clip.
20. Refit plastic gaiter to clip and secure with plastic retaining ring.
21. Grease ball pin and replace in socket.
22. Replace shims removed in operation 7 and refit lower socket and cap, with new locking tabs. Tighten setscrews to 15 to 20 lbf. ft (2,1 to 2,75 kgf. m). NOTE: If ball pin is loose in socket, remove enough shims to cause ball to be slightly stiff in socket with setscrews fully tightened.
23. Turn down locking tabs of setscrews.
24. Pack hub with grease and replace hub and disc assembly on stub axle.
25. Fit washer and nut to stub axle and tighten sufficiently to give end float of 0.002 in to 0.006 in (0,05 mm to 0,15 mm).
26. Fit new split pin and replace grease cap.
27. Place disc shields and steering arm in position on stub axle carrier.
28. Fit upper and lower steering arm bolts and tighten upper bolt only to 50 to 55 lbf. ft (6,92 to 7,60 kgf. m).
29. Remove lower steering arm bolt for caliper attachment.
30. Replace disc shield clips and tighten nuts to 6 to 7 lbf ft (0,83 to 0,97 kgf m).
31. Place brake caliper in position and insert bolts and shims. Tighten bolts to 50 to 60 lbf. ft (6,92 to 8,30 kgf. m).
32. Check caliper for centralisation on disc.
33. Wire up caliper bolts.
34. Replace pads in calipers as removed in operation 14. 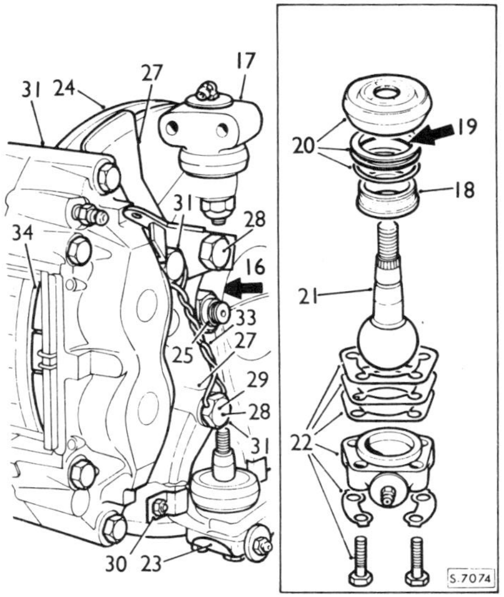
35. Refit lower ball pin in wishbone, fit washer and tighten nut to 45 to 55 lbf. ft (6,23 to 7,60 kgf. m).
36. Line up stub axle carrier with upper wishbone and insert bolts from front.
37. Replace packing piece and shims and tighten nuts to 26 to 32 lbf. ft (3,60 to 4,42 kgf. m).
38. Refit steering tie rod to steering arm. Tighten nut to 35 to 50 lbf. ft (4,84 to 6,91 kgf. m).
39. Remove blanking plugs, attach flexible hose to bracket and connect to caliper pipe. Tighten flexible hose nut to 10 to 12 lbf. ft (1,38 to 1,66 kgf. m) and caliper pipe nut to 6.3 to 7 lbf. ft (0,88 to 0,96 kgf. m).
40. Replace road wheel.
41. Remove car from stands.
42. Bleed brakes, see 70.25.02.
43. Check steering geometry, see 57.65.02.

---

19. Tighten hub nut sufficiently to give end float of 0.002 in to 0.006 in (0,05 mm to 0,15 mm).
20. Fit new split pin.
21. Replace grease cap.
22. Attach flexible hose to bracket. Tighten nut to 10 to 12 lbf. ft (1,38 to 1,66 kgf. m).
23. Reconnect caliper pipe to flexible hose. Tighten to 6.3 to 7 lbf. ft (0,88 to 0,96 kgf. m).
24. Replace road wheel.
25. Remove car from stands.
26. Bleed brakes, see 70.25.02.
27. Check steering geometry, see 57.65.02.

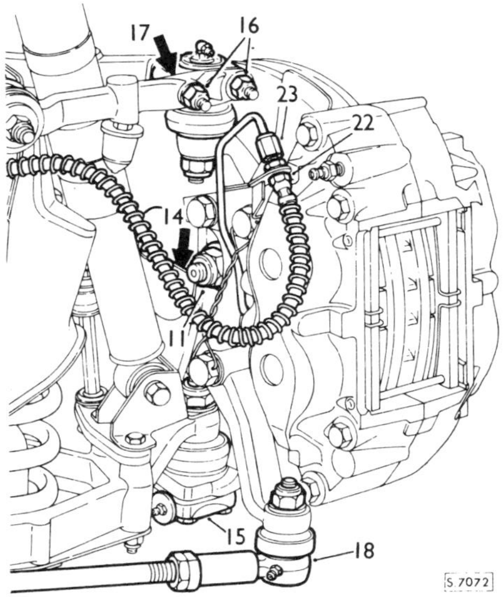

60–7

FRONT HUB STUDS

Remove and refit 60.25.29

Removing
1. Remove front hub, see 60.25.01.
2. Using power press and suitable mandrel, press stud(s) from hub.

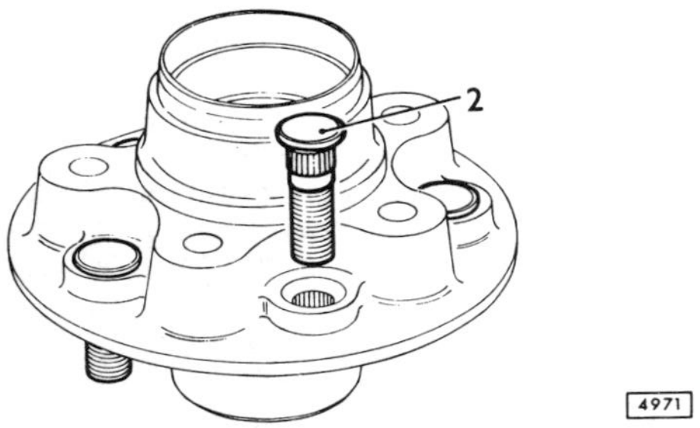

Refitting
3. Use power press and suitable mandrel to press stud(s) into hub.
4. Refit front hub, see 60.25.01.

FRONT DAMPER

Remove and refit 60.30.02

Removing
1. Beneath bonnet, remove locknut, nut, outer washer buffer and inner washer from damper front mounting.
2. Jack up front of car and place on stands.
3. Remove self locking nut and bolt from bottom mounting.
4. Draw damper from car.
   NOTE: If damper is to be refitted, wire up in compressed position.

Refitting
5. Fit lower mounting. Tighten nut to 32 to 36 lbf. ft (4,42 to 4,98 kgf. m).
6. Remove spring retaining wire clip from damper.
7. Fit upper mounting. Tighten nuts to 27 to 32 lbf. ft (3,73 to 4,42 kgf. m).

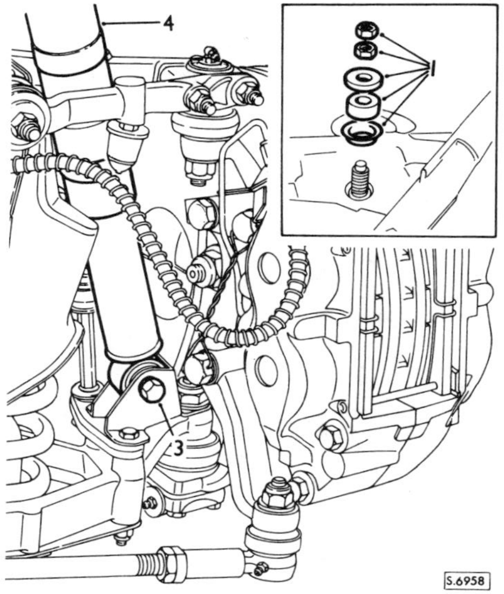

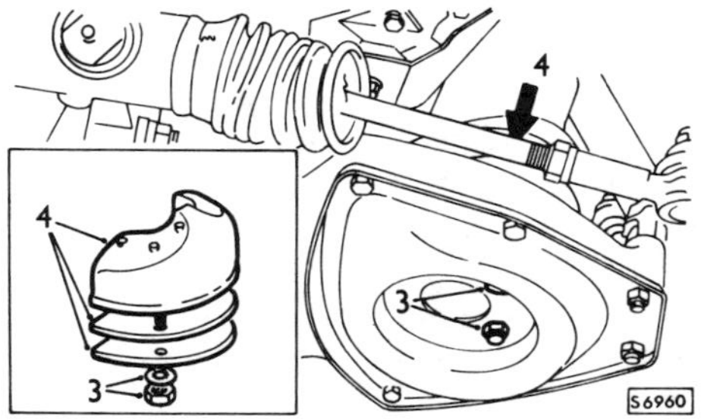

Refitting
6. Prise spring coils apart with bar and insert bump stop.
7. Place spacers on bump stop studs and insert studs through holes in spring pan.
8. Tighten nuts to 8 to 10 lbf. ft (1,11 to 1,38 kgf. m).
9. Remove car from stands.

REBOUND STOPS

Remove and refit 60.30.14

NOTE: Rebound stops must only be replaced as a pair; uneven loads will be placed on upper wishbone if this is not done.

Removing
1. Jack up front of car and place stand beneath spring seat pan.
2. Lower car on to stand.
3. Remove road wheel.
4. Unscrew rebound stops from upper wishbone.

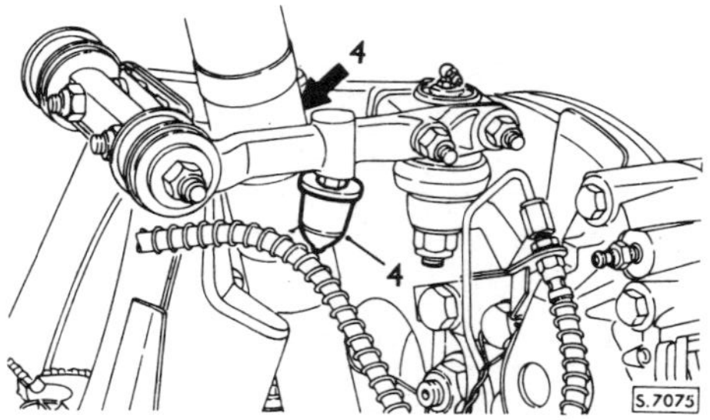

Refitting
5. Fit new stops and tighten to 8 to 10 lbf. ft (1,11 to 1,38 kgf. m).
6. Remove car from stands.

BUMP STOP

Remove and refit 60.30.10

Removing
1. Jack up front of car and place on stands.
2. Remove road wheel.
3. Remove two plain nuts and spring washers securing bump stop.
4. Swing steering to full lock to improve access and lift bump stop from location in spring pan.
5. Prise spring coils apart with bar and extract bump stop. Retain two spacers.

4. Remove two nuts, bolts and plain washers securing upper ball joint to upper wishbone levers.
   NOTE: Take careful note of fitted position of bolts and also position of packing piece and shims, these control castor angle.
5. Tie stub axle carrier to road spring turret to prevent damage to brake flexible hose.
6. Remove two bolts, special washers and nyloc nuts securing upper wishbone fulcrum shaft to road spring turret.
   NOTE: Take careful note of position of shims as these control camber angle.
7. Manoeuvre wishbone assembly clear of damper unit.

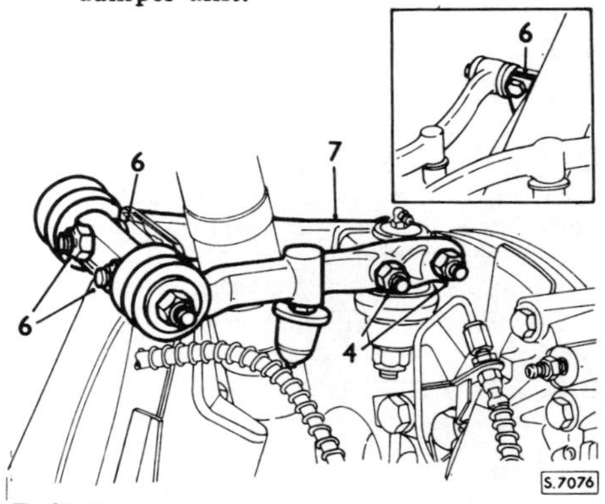

Refitting
8. Replace wishbone in position and insert bolts attaching fulcrum shaft to road spring turret, replacing spacers and special washers as removed in operation 6. No washers are fitted under nuts.
9. Refit nuts and tighten to 49 to 55 lbf. ft (6,78 to 7,60 kgf. m).
10. Remove axle carrier support and align upper ball joint with wishbone levers; insert bolts but do not tighten nuts.
    NOTE: Bolts must be inserted from front.
11. Replace shims as removed in operation 4 and tighten nuts to 26 to 32 lbf. ft (3,60 to 4,42 kgf. m).
12. Replace road wheel.
13. Remove car from stand.
14. Check steering geometry, see 57.65.02.

WISHBONE – UPPER

Remove and refit 60.35.01

Removing
1. Jack up front of car and place stand beneath spring seat pan.
2. Lower car on to stand.
3. Remove road wheel.

60–8

WISHBONE – LOWER 60.35.02

Remove and refit

Service tools: Steering joint taper separator JD.24. Spring compressor tool JD.6D and adaptor JD.6D-1. Spring plate locating pegs JD.6E-6.

Removing

1. Remove front suspension unit, see 60.35.05. Invert unit.
2. Fit spring compressor tool JD.6D with adaptor JD.6D-1 and compress spring sufficiently to relieve load on spring plate fastening.
3. Remove four setscrews and washers and two nuts, bolts and washers securing spring plate to lower wishbone.
4. Slacken spring compressor tool and remove, together with spring plate, spring and packers.
   NOTE: Record position of packers to assist during replacement.
5. Remove self locking nuts and washers securing steering tie rod ball joints; separate rod from arms using tool JD.24.
6. Remove two nuts, bolts and plain washers securing ball joint to upper and wishbone levers.
   NOTE: Take careful note of fitted position of packing piece and shims; these control castor angle.
7. Remove self locking nut and washer securing lower ball joint to lower wishbone and separate taper using joint separator JD.24.
8. Detach hub-disc assembly and lift damper and mounting bracket clear of lower wishbone.
9. Remove upper mounting bolt of rack assembly; collect washers and slacken lower bolts.
10. Pivot rack assembly about lower bolts to clear fulcrum shaft.
11. Remove split pin at fulcrum shaft nut and remove nut and plain washer. Discard split pin.
12. Drift fulcrum shaft from cross member and collect two inner spacer washers.

5. Disconnect pressure and return hoses from power steering pump. Fit blanking plugs to pipes and pump ports.
6. Remove nut securing engine to left-hand mounting.
7. Remove nuts, washers, rubber and cup securing left-hand damper to wheel arch.
8. Remove cover and element from right-hand air cleaner.
9. Repeat operations 6 and 7 on right-hand engine mounting and damper.
10. Place engine lifting hooks in front lifting eyes and fit support bracket M53A. Tighten nuts just sufficiently to take engine weight off front mountings.
11. Slacken front road wheel nuts.
12. Raise front of car and place on stands.
    WARNING: ENSURE THAT STANDS ARE CORRECTLY POSITIONED AT SPIGOTS AND APPLY HANDBRAKE SECURELY. P L A C E A U T O M A T I C TRANSMISSION CONTROL IN 'P'.
13. Remove front road wheels.
14. From inside car, slacken pinch bolts securing universal joint to upper and lower columns. Remove upper bolt.
15. Slacken clip securing draught seal to lower column.
16. Turn steering to provide access to pinch bolt securing lower universal joint to pinion shaft and remove bolt.
17. Turn steering to straight ahead position and lock by removing ignition key.
18. Detach lower column from upper universal joint and then from pinion shaft.
19. Detach brake hoses from caliper brackets and plug hoses and caliper pipes.
20. Remove setscrew securing earthing strap to L.H. body frame, forward of power steering oil cooler.
21. Remove sump attachment setscrew securing earthing strap to engine.
22. Place trolley jack under suspension cross beam and insert suitable packing securely to support cross beam.

continued

24. Refit tie rods to steering arms and tighten nuts to 45 to 50 lbf. ft (6,23 to 6,92 kgf. m).
25. Refit front suspension unit, see 60.35.05.
26. With full weight of car on suspension, tighten fulcrum shaft nut to 32 to 50 lbf. ft (4,43 to 6,91 kgf. m) and fit new split pin.
    CAUTION: Do not fully tighten fulcrum shaft nut until full weight of car is on suspension.
27. Check front wheel alignment, see 57.65.01.
28. Check castor angle, see 57.65.04.
29. Check camber angle, see 57.65.05.

FRONT SUSPENSION

Remove and refit 60.35.05

Service tools: Engine support bracket M53A. Rack centralising tool Jaguar Part No. 12297.

Removing

1. Disconnect battery earth lead in boot.
2. Disconnect air temperature sensor wire from left-hand air cleaner.
3. Remove cover and element from left-hand air cleaner.
4. Remove fluid from power steering reservoir; using tube as syphon, or suitable syringe. Discard fluid and replace cap on reservoir.
   WARNING: TAKE GREAT CARE TO ENSURE THAT TUBE OR SYRINGE ARE PERFECTLY CLEAN AND THAT NO DIRT IS A L L O W E D T O E N T E R RESERVOIR.

Refitting

13. Place wishbone in position, insert packing washers, align in riding position and refit fulcrum shaft. Do not fully tighten nut.
14. Replace damper and bracket and fit spring plate locating pegs JD.6E-6 in wishbone arms.
15. Replace packers removed in operation 4 in spring turret.
    WARNING: A MAXIMUM OF TWO PACKERS MAY BE FITTED IN SUSPENSION TURRET.
16. Assemble spring, packing, spring plate and compressor tool as removed in operation 4 and lift into position in turret, with locating pegs through holes in seat pan.
17. Centralise spring and compress tool.
18. Remove plate locating pegs and refit setscrews and bolts; tighten to 27 to 32 lbf. ft (3,74 to 4,42 kgf. m).
19. Replace lower ball joint pin in wishbone, fit washer and tighten nut to 45 to 55 lbf. ft (6,23 to 7,60 kgf. m).
20. Line up upper ball joint and insert bolts from front.
21. Replace packing piece and shims removed in operation 6 and tighten nuts to 26 to 32 lbf. ft (3.60 to 4,42 kgf. m).
22. Remove spring compressor tool and adaptor.
23. Swing steering rack back into position, replace washers and upper bolt and tighten all three nuts to 14 to 18 lbf. ft (1,94 to 2,49 kgf. m).

60–9

23 Remove nuts securing rear mounting brackets on cross beam to mounting rubbers.
24 Prise out plastic fasteners securing vertical diaphragms in wheel arches.
25 Release anti-roll bar from upper ends of links.
26 Remove two screws and 13 plastic drive fasteners securing spoiler undertray to body and spoiler.
27 Remove front suspension mounting bolts. Collect washers, spacers and bush sleeves.
28 Slowly lower jack and carefully manoeuvre suspension clear of mountings.
NOTE: IF FURTHER ATTENTION IS REQUIRED TO SUSPENSION UNIT, TRANSFER TO WORK BENCH BY USING ADEQUATE LIFTING TACKLE. DO NOT ATTEMPT TO PLACE IN POSITION BY HAND.

Refitting

29 Detach heat shield protecting gaiter on pinion side of rack.
30 Transfer suspension unit to trolley jack.
31 Manoeuvre jack and suspension unit into mounting position under car.
32 Raise jack carefully, engaging engine mountings, rear cross beam mountings and upper damper attachments.
33 Line up sleeves and spacers in front mountings and insert bolts, with washers under bolt heads.
34 Fit nuts to rear cross beam mountings and tighten to 22 to 26 lbf.ft (3,04 to 3,59 kgf. m)
35 Fit front engine mounting nuts and tighten.
36 Place cups, rubbers, washers and nuts in position on damper stems and tighten nuts to 27 to 32 lbf.ft (3,73 to 4,42 kgf. m). Tighten locknuts to same torque.
37 Fit washers and nuts to front mounting bolts and tighten to 95 to 115 lbf.ft (13,13 to 15,90 kgf. m).
38 Refit anti-roll bar links through ends of anti-roll bar, replace rubbers, cup washers and nuts and tighten to 14 to 18 lbf.ft (1,94 to 2,49 kgf. m).
39 Refit flexible brake hoses to brackets.
40 Tighten nuts to 10 to 12 lbf. ft (1,38 to 1,66 kgf. m).
41 Unplug hoses and caliper pipes and connect. Tighten union nuts to 6.3 to 7 lbf. ft (0,87 to 0,97 kgf. m).
42 Replace sump setscrew securing earthing strap and tighten to 12 to 15 lbf. ft (1,66 to 2,07 kgf. m).
43 Replace setscrew securing earthing strap to body frame.
44 Remove grease nipple from rack damper plug and insert rack centralising tool (Jaguar Part No. 12297) into position with rack centralised.
45 Engage lower universal joint with pinion shaft, replace bolt and tighten nut to 14 to 18 lbf.ft (1,94 to 2,49 kgf. m).
46 Inside car, slide upper universal joint as far as possible up upper column splines and slide lower column as far as possible into universal joint.
47 Replace upper bolt and tighten nuts on both bolts to 14 to 18 lbf. ft (1,94 to 2,49 kgf. m).
48 Replace draught seal in position and tighten its securing clip.
49 Remove rack centralising pin and replace grease nipple in rack damper plug.
50 Check that at least ⅜ in (9,5 mm) clearance exists between 'hinged' sections of lower universal joint.
51 Replace heat shield over gaiter.
52 Release nuts on engine support tool hooks and detach tool.
53 Unplug power steering hoses and ports and reconnect hoses. Tighten feed hose union nut to 18 to 20 lbf. ft (2,49 to 2,77 kgf. m).
54 Remove reservoir cap, refill with fresh fluid and bleed power steering system, see 57.15.02.
55 Replace spoiler undertray and secure with setscrews and new plastic drive fasteners.
56 Replace vertical diaphragms in wheel arches and secure with new plastic drive fasteners.
57 Replace road wheels.
58 Remove stands and lower car.
59 Refit air cleaner elements and covers.
60 Reconnect air temperature sensor wire to left-hand air cleaner.
61 Bleed brakes, see 70.25.02.

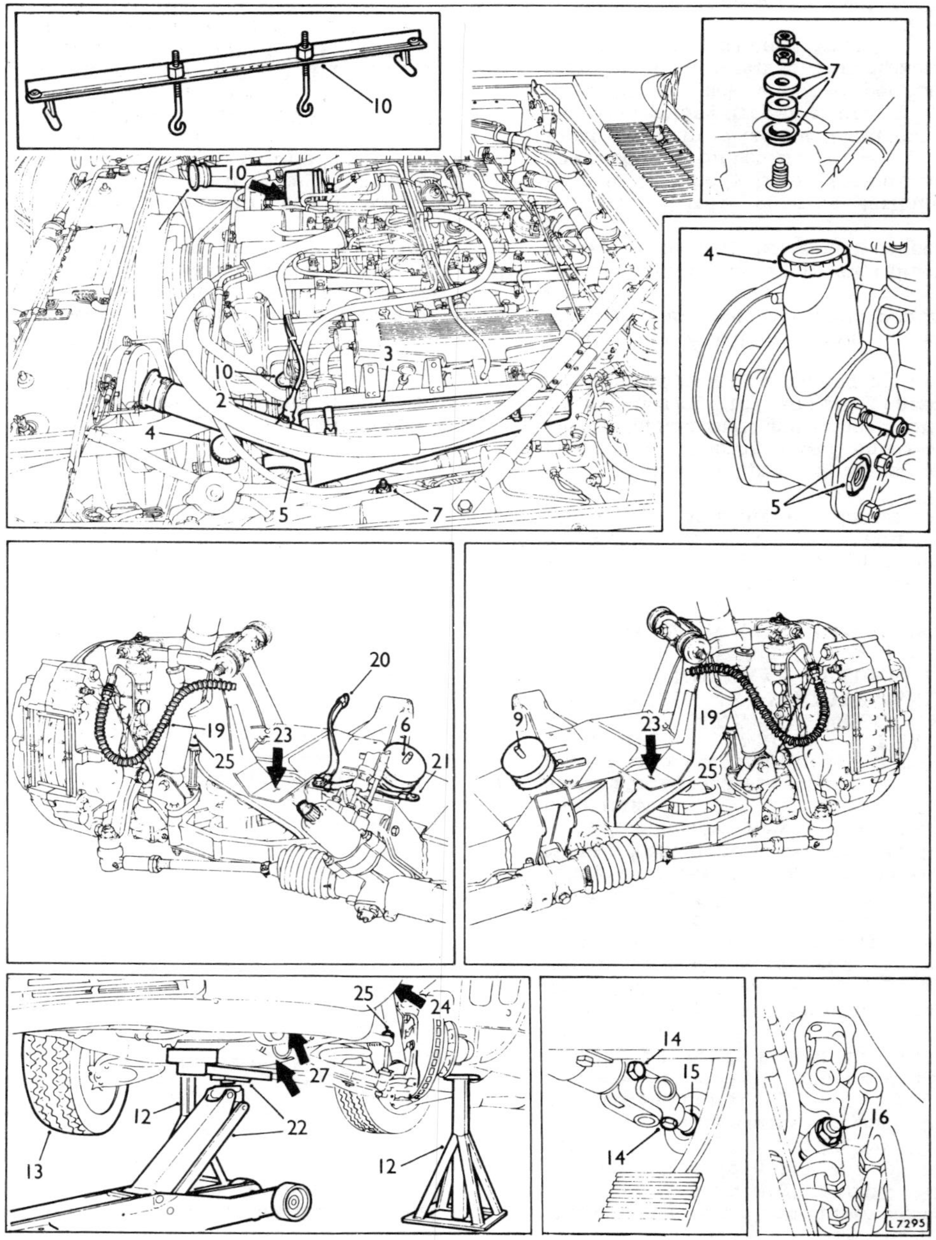

60—10

SUSPENSION UNIT MOUNTING BUSH 60.35.06

Remove and refit

NOTE: A worn or damaged bush infers that undue load has been imposed upon the opposite side bush. **Bushes must therefore be changed as a pair.**

Removing

1. Jack up front of car by trolley jack under cross beam.
2. Fit stands under spigots.
3. Remove both front wheels.
4. Adjust jack to take load off mounting bushes.
5. Remove vertical diaphragm from front of one wheel aperture.
6. Remove self locking nut securing mounting bolt.
7. Drift bolt outwards, clear of bush and bracket. Collect washers and spacers.
8. Slacken nut on clamp bolt securing mounting bush and drift out bush.
9. Remove sleeve from bush.
10. Repeat operations 5 to 9 on opposite side.

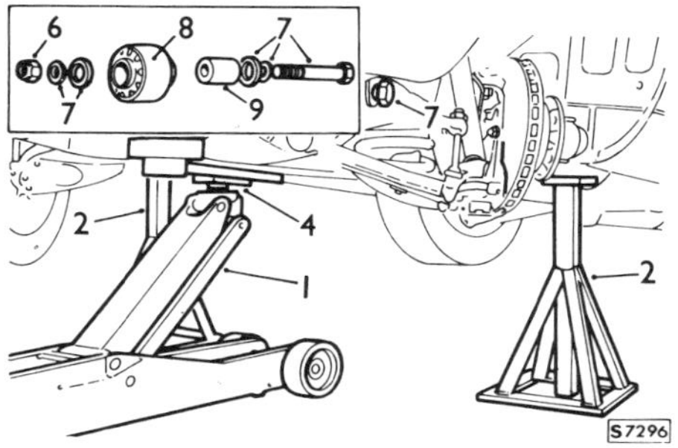

Refitting

11. Insert sleeve in replacement bush, open housing clamp and place bush in position.
12. Replace spacer between bush sleeve and bracket and insert bolt from outside of bracket. Fit washers as shown and tighten nut to 95 to 115 lbf.ft (13,13 to 15,90 kgf.m).
13. Tighten clamp bolt nut to 25 to 30 lbf.ft (3,46 to 4,15 kgf.m).
14. Refit diaphragm and secure with new plastic studs.
15. Repeat operations 11 to 15 on opposite side.
16. Reseal diaphragms to wings with underseal.
17. Replace front wheels.
18. Remove car from stands.

SUSPENSION UNIT MOUNTING – REAR

Remove and refit 60.35.07

NOTE: If one mounting is worn or damaged, excessive load has been applied to the mounting on the other side. Mountings must therefore be changed as a pair.

Removing

1. Jack up front of car and place on stands.
2. Remove front road wheels.
3. Remove air cleaners to provide access to fixings, see 19.10.01 and 02.
4. Position trolley jack to support front suspension crossmember.
5. From below car, detach nuts securing mountings to suspension cross beam.
6. Lower jack to free cross beam from mountings. Collect heat shields, if fitted.
7. Remove rear bolts securing mountings to body structure.

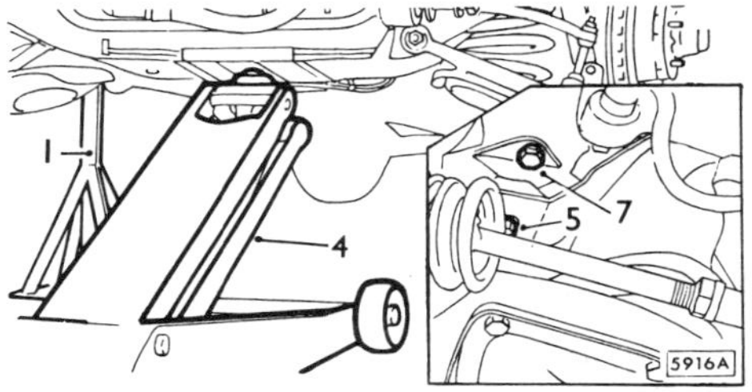

6. Remove facing washers from rubber bush, push out central sleeve and drift bush out of wishbone arm.
7. Unscrew rebound stop from arm.

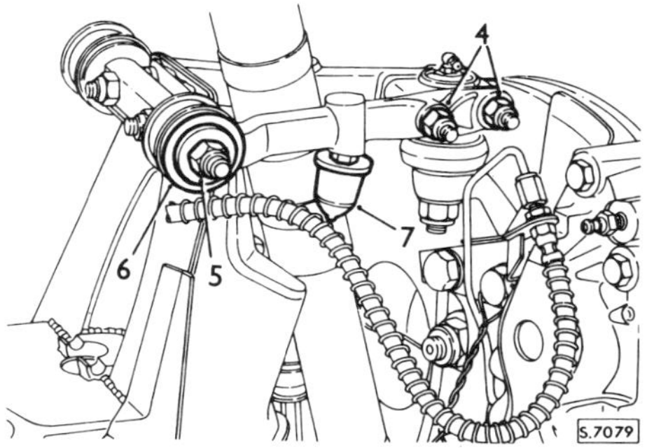

Refitting

11. Lever suspension cross beam down on one side and place mounting in position to body structure.
   NOTE: Mountings are not symmetrical, and are fitted with studs nearer to centre line of car than bolt holes.
12. Fit rear bolt and washer securing mounting to body structure but do not tighten.
13. From above, fit front bolt and washer securing mounting to body structure and tighten to 22 to 26 lbf. ft (3,04 to 3,59 kgf. m).
14. Tighten rear mounting attachment bolt to 22 to 26 lbf. ft (3,04 to 3,59 kgf. m).
15. Repeat operations 11 to 14 on opposite side.
16. Raise jack, engaging studs from mountings in suspension cross beam brackets. Replace heat shields, if fitted.
17. Fit nuts and washers securing mountings to cross beam and tighten to 14 to 18 lbf. ft (1,94 to 2,48 kgf. m).
18. Refit front wheels.
19. Remove car from stands.
20. Replace air cleaners.

WISHBONE ARM – UPPER

Overhaul 60.35.08

NOTE: If one wishbone arm is found to require replacement, check its mating arm. If one bush requires replacement, replace both bushes.

1. Jack up front of car and place on stands.
2. Remove road wheel and turn steering to full lock to improve access.
3. Wire stub axle carrier to cross member turret to prevent strain on brake hose when upper joint is disconnected.
4. Remove upper ball joint retaining bolts. Note location of shims.
5. Detach nut from fulcrum shaft and draw off wishbone arm. Collect washers.

8. Clean wishbone arm and check dimensionally against dimensions quoted on sheet 60.04.
   NOTE: No attempt may be made to correct an arm not to dimensions by bending. Arms failing inspection must be scrapped.
9. Inspect rubber bush. If not in perfect condition, repeat operations 5 to 8 on mating arm, and discard both bushes.
10. Replace rebound stop. If new stop is fitted, also renew stop in mating arm. Tighten to 8 to 10 lbf. ft (1,11 to 1,38 kgf. m).
11. Coat bush with Esso Process Oil 'L' (after removing facing washer if new bush is being fitted) and refit in wishbone arm.
12. Insert sleeve in bush and replace facing washers.
13. Replace large washer on fulcrum shaft and refit wishbone arm. Replace outer washers and nut but do not tighten.
14. Lift hub assembly to align upper ball joint and replace bolts and shims.
15. Replace washers and nuts and tighten to 26 to 32 lbf. ft (3,60 to 4,42 kgf. m).
16. Remove support wire.
17. Replace road wheel.
18. Remove car from stands.
19. Tighten fulcrum shaft nut to 45 to 55 lbf. ft (6,23 to 7,60 kgf. m).
20. Check steering geometry, see 57.65.02.

60–11

WISHBONE – LOWER

Overhaul   60.35.09

1. Remove lower wishbone, see 60.35.02 operations 1 to 12.

Dismantling

2. Using a press and suitable mandrel, remove bushes from wishbone arms.

Reassembling

3. Using a press and suitable mandrel, fit new bushes to wishbone arms, ensure each bush is central in arm.
   CAUTION: New bushes must be coated with Esso Process Oil 'L' before they are pressed in to wishbone arms.
4. Refit lower wishbone, see 60.35.02 operations 13 to 29.

WARNING

THIS CAR IS FITTED WITH A POWR-LOK DIFFERENTIAL. UNDER NO CIRCUMSTANCES MUST ENGINE BE RUN WITH CAR IN GEAR AND ONE WHEEL OFF THE GROUND. IF IT IS FOUND NECESSARY TO TURN TRANSMISSION WITH CAR IN GEAR, BOTH WHEELS MUST BE RAISED.

ACCIDENTAL DAMAGE

The dimensional drawing below is provided to assist in assessing accidental damage. A component suspected of being damaged should be removed from the car and cleaned off, the dimensions should then be checked and compared with those given in the illustration.

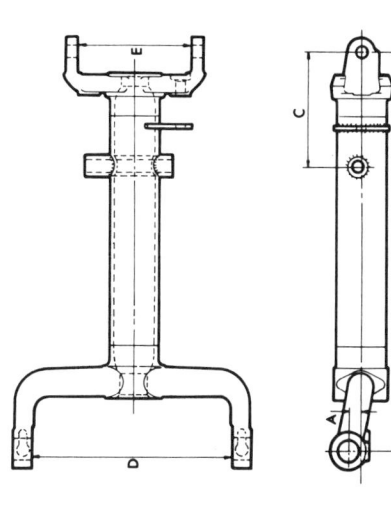

DIMENSION

A 0.62 to 0.64 in (15,75 to 16,26 mm)
B 20.45 to 20.47 in (519,43 to 519,94 mm)
C 5.93 to 5.95 in (150,62 to 151,13 mm)
D 10.632 to 10.642 in (270,05 to 270,31 mm)
E 6.12 to 6.13 in (155,45 to 155,70 mm)

REAR HUB AND CARRIER ASSEMBLY

Remove and refit 64.15.01

Service tools: Hub puller JD.1D, Dummy shaft JD.14, Thread protector JD.1C/7.

Removing

1 Raise rear of car and place on stands.
2 Remove road wheel.
3 Remove split pin, nut and washer from end of drive shaft.
4 Remove fulcrum shaft grease nipple from hub carrier.
5 Place thread protector JD.1C/7 on end of drive shaft.
6 Fit hub puller JD.1D to rear hub and secure. Withdraw hub and carrier from drive shaft and remove hub puller.
7 Remove thread protector and slide spacer off drive shaft. Inspect oil seal inner track.
8 Remove nut from fulcrum shaft, drift out shaft and insert dummy shaft JD.14. Detach hub and carrier. Collect washers and shims and secure with adhesive tape.

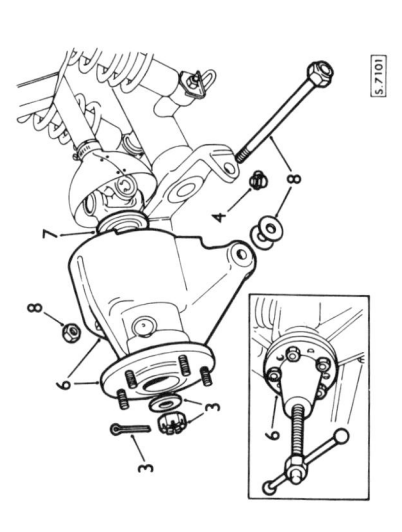

Refitting

10 Line up hub and carrier assembly, place shims and washers in position and drive in fulcrum shaft. Collect dummy shaft.
11 Tighten fulcrum shaft nut to 97 to 107 lbf. ft (13,4 to 14,8 kgf. m).
12 Apply Loctite 'Stud Lock' to outer two thirds of drive shaft spline, using a small brush.
13 Refit spacer ring to shaft.
14 Lift hub assembly and refit to drive shaft. Replace washer, tighten nut to 100 to 120 lbf. ft (13,85 to 16,6 kgf. m), and fit new split pin.
15 Replace grease nipple.

16 Replace road wheel.
17 Remove car from stands.

REAR HUB AND CARRIER ASSEMBLY

Overhaul 64.15.07

Service tools: Master spacer JD.15, Dummy shaft JD.14, Press tool JD.16C, Hand press SL.14, Press tool JD.20A, Tool JD.20A-1, Adaptor JD.16C-1.

Dismantling

1 Remove rear hub and carrier assembly, see 64.15.01, operations 1 to 9.
2 Prise out oil seal retainers from fulcrum shaft housing and remove seals, dummy shaft, bearings, distance tubes and shims.
3 Mount hub carrier in vice and drift out bearing cups from fulcrum shaft housing.
4 Transfer hub carrier to press and remove hub assembly from carrier.
5 Drift out inner hub bearing cup, with seal and bearing, from hub carrier.
6 Drift out outer bearing cup.
7 Fit hand press SL.14 with adaptors JD.16C-1 to hub and pull outer bearing from hub.
8 Remove oil seal track from hub shaft and clean and inspect all parts.
NOTE: When inspecting components, pay particular attention to oil seal tracks, when a minute score can considerably shorten oil seal life. For further details on inspection of seals and bearings refer to 'General Fitting Instructions', Group 07.

Reassembling

9 Replace outer oil seal track on hub shaft.
10 Press outer bearing cone into position on hub shaft and grease bearing with 70 cc of Retinax 'A'.
11 Press outer and inner cups of bearings into hub carrier, using tool JD.20A with adaptor JD.20A-1.
12 Drift new outer oil seal into position in hub carrier and lower carrier on to hub shaft and outer bearing.
13 Place inner bearing into position for fitting.
14 Place master spacer JD.15 in position as shown and press bearing on to hub shaft.

continued

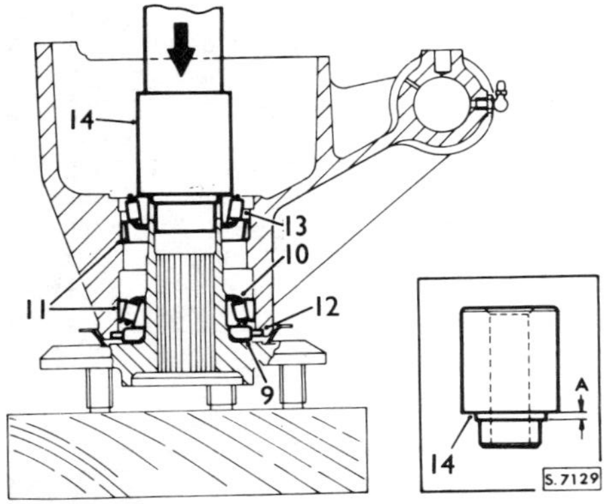

28 Replace spacer tube, pack fulcrum shaft housing with grease and replace bearing and oil seal track.
29 Push out fulcrum shaft by inserting dummy shaft and detach temporary spacers from fulcrum shaft. Check that oil seal tracks are in position.
30 Press new oil seals into fulcrum shaft housings and secure with oil seal retainers.
31 Replace rear hub and carrier assembly, see 64.15.01, operations 10 to 17.

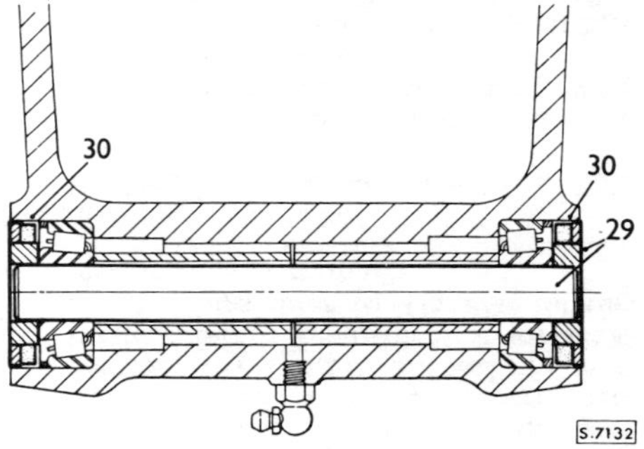

REAR HUB BEARING END-FLOAT

64.15.13

Check and adjust

NOTE: End float is controlled by a spacer located next to the universal joint on the hub shaft. Spacers are available in thickness from 0.109 in (2,77 m) to 0.151 in (3,84 mm) in 0.003 in (0,076 mm) steps. End float is normally 0.001 to 0.003 in (0,026 to 0,076 mm) and must be rectified if it exceeds 0.005 in (0,127 mm) by changing the spacer for a thicker one.

Service tools: Hub remover JD.1D. Thread protector JD.1C-7. Backlash gauge JD.13.

Checking
1 Raise car and place on stands.
2 Remove rear road wheel.
3 Tap hub towards car.
4 Clamp tool JD.13 to hub carrier web, as shown, so that stylus of dial gauge contacts hub flange.

---

22 Place a large washer (e.g. inner fork thrust washer) next to one oil seal track.
23 Cover exposed plain length of fulcrum shaft with suitable temporary spacers, fit nuts and tighten to 97 to 107 lbf.ft (13,4 to 14,8 kgf.m).
24 Apply pressure to fulcrum shaft at large washer end, turning it to settle taper rollers and using feeler gauge measure minimum distance between large washer and hub carrier.
25 Apply pressure to opposite end of fulcrum shaft and measure maximum distance between washer and hub carrier.
NOTE: End play of fulcrum shaft in hub carrier is now obtained by subtracting measurement obtained in operation 24 from that in 25. This end play must be replaced by a preload of 0.002 in (0,05 mm) by removing shims, to a total thickness of 0.002 in (0,05 mm) more than the end play, from between spacer tubes:
For example:
Assume end play found in operation 24 to be 0.010 in (0,25 mm).
Therefore shims to the value of
0.010 + 0.002 in = 0.012 in
(0,25 + 0,05 mm = 0,30 mm)
must be removed to give correct preload.

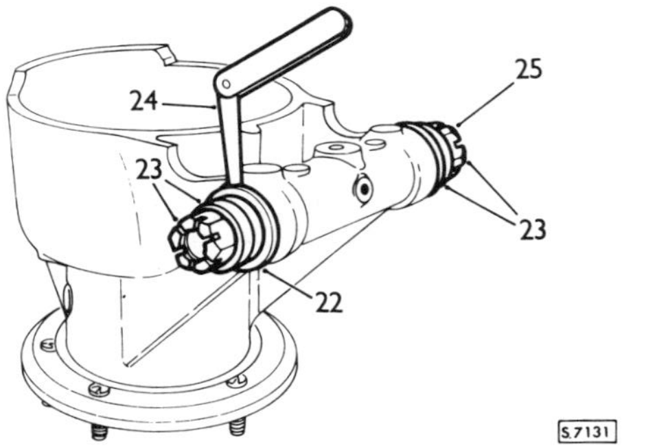

26 Release nut from large washer end of fulcrum shaft and detach spacers, washer, oil seal track and bearing.
27 Remove one spacer tube and extract shims to thickness established to give preload.

NOTE: Master spacer has a diameter of length 'A' equivalent to a spacer of 0.15 in (3,81 mm). Calculate the spacer required to give end float of .001 to .003 in (,025 to ,076 mm). Spacers are supplied in thicknesses of .109 to .151 in (2,77 to 3,84 mm) in steps of .003 in (,076 mm) and are lettered A to R (less letters I, N and O).

SPACER LETTER	THICKNESS inches mm
A	.109 2,77
B	.112 2,85
C	.115 2,92
D	.118 3,00
E	.121 3,07
F	.124 3,15
G	.127 3,23
H	.130 3,30
J	.133 3,38
K	.136 3,45
L	.139 3,53
M	.142 3,61
P	.145 3,68
Q	.148 3,76
R	.151 3,84

For example, assume end-float to be .026 in (,66 mm). Subtract required nominal end float of .002 in (,050 mm) from measured end float giving .024 in (,61 mm). Since special collar is .150 in (3,81 mm) thick, the thickness of the spacer to be fitted will be .150 in to .024 in i.e. .126 in (3,20 mm). The nearest spacer is .127 in (3,23 mm) so letter G spacer should be fitted in place of special collar.

17 Remove adaptor and fit new inner bearing oil seal to hub carrier.
18 Fit fulcrum shaft bearing cups to hub carrier and insert one bearing.
19 Secure fulcrum shaft vertically in suitably protected jaws of vice and slide bearing in hub carrier over shaft.
20 Replace distance tubes and shims as removed in operation 2, adding 0.010 in (0,25 mm) extra shims. (One extra 0.003 in [0,076 mm] shim and one extra 0.007 in [0,178 mm] shim).
21 Fit second bearing over fulcrum shaft, remove hub assembly from vice and replace oil seal tracks outside bearings.

15 Transfer hub and carrier assembly to vice, set up dial gauge and spacer JD.15 as shown and measure end float, lifting carrier by using two screwdrivers as levers.

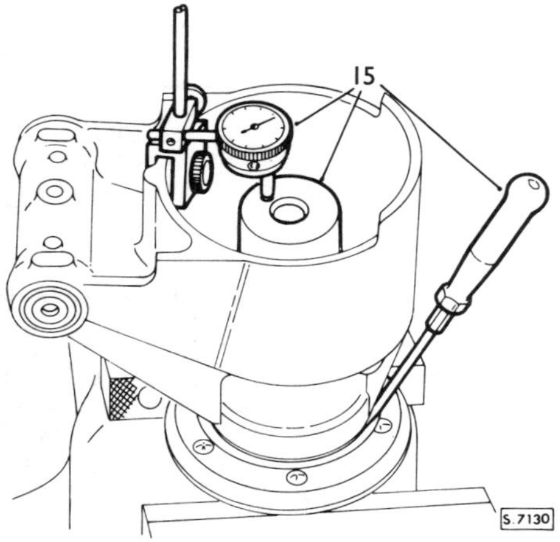

16 Select spacer to be fitted on drive shaft.

64—2

5 Note reading of dial gauge.
6 Using two levers between hub and hub carrier boss, press hub outwards. Take care not to damage water thrower. Note altered reading on dial gauge. The difference between dial gauge readings in operations 5 and 6 represents end float of hub bearings. If this exceeds 0.005 in (0,127 mm) refer to 'Adjust' below. Otherwise proceed with operation 7.

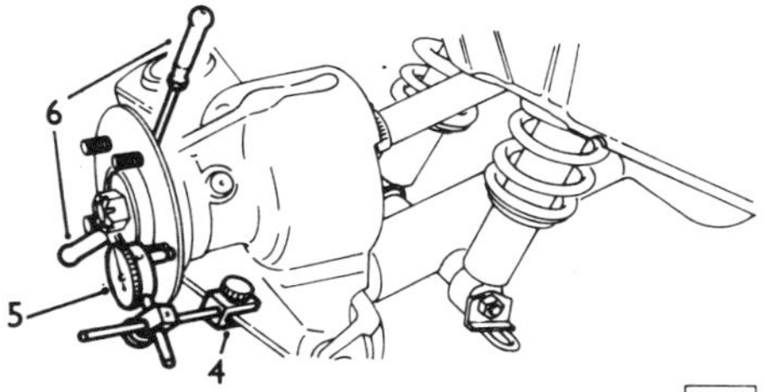

Adjusting
10 Remove split pin, nut and washer from end of drive shaft.
11 Remove fulcrum shaft grease nipple from hub carrier.
12 Place thread protector JD.1C/7 on end of drive shaft.
13 Fit hub puller JD.1D to rear hub and secure. Withdraw hub and carrier from drive shaft and remove hub puller and thread protector.
14 Remove spacer from drive shaft and measure thickness with micrometer.
NOTE: A simple calculation will now give the thickness of spacer required to reduce end float to specified 0.001 in (0,026 to 0,076 mm) i.e. If end float measured in operations 5 and 6 was 0.007 in (0,203 mm) a replacement spacer will need to be 0.005 in (0,127 mm) thicker than that removed to reduce end float to 0.002 in (0,051 mm).
As spacers are supplied in 0.003 in (0,075 mm) steps of thickness, a spacer 0.006 in (0,152 mm) thicker would be used, reducing end-float to 0.001 in (0,026 mm).

15 Clean dried 'Loctite' from drive shaft splines.
16 Place selected spacer on drive shaft.
17 Apply Loctite 'Stud Lock' to outer two thirds of drive shaft splines, using a small brush.
18 Enter drive shaft in hub and drift hub on to shaft.
19 Replace washer, tighten nut to 100 to 120 lbf. ft (13,83 to 16,6 kgf. m) and fit new split pin.
20 Replace grease nipple.
21 Repeat operations 3 to 6, to re-check end float.
22 Carry out operations 7 to 9.

REAR HUB OIL SEALS
Remove and refit   64.15.15

The degree of dismantling required to change rear hub oil seals is extensive; full rear hub overhaul procedure should therefore be carried out, see 64.15.07 and all oil seals, including outer wishbone fulcrum oil seals changed. Renew grease content of both hub and fulcrum bearing assemblies.

REAR HUB WHEEL STUDS
Remove and refit   64.15.26

Removing
1 Remove rear hub and carrier assembly, see 64.15.01.
2 Support hub carrier and press out hub using handpress and suitable mandrel.
3 Prise old oil seals from hub carrier.
4 Draw outer bearing and oil seal track from hub.
5 Use a narrow, sharp cold chisel to open peening securing water thrower. Remove thrower.
6 Support hub, and file or grind staking from faulty stud(s).
7 Unscrew stud(s) from hub flange.

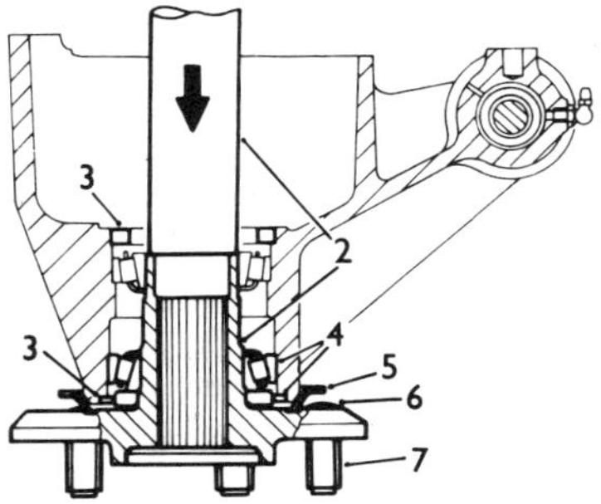

Refitting
8 Screw new stud(s) into hub and stake in four places to back of flange.
9 Fit water thrower to hub and use blunt cold chisel to peen over flange in three or four places.
10 Press oil seal track and outer bearing race on to hub.
11 Press new outer and inner oil seals into hub.
12 Fit hub into hub carrier and pack with suitable grease.
13 Locate inner bearing over hub and press into position.
14 Refit rear hub and carrier assembly.

REAR ROAD SPRINGS (ONE SIDE)
Remove and refit   64.20.01

Service tools: Handpress SL.14, Adaptor JD.11B.

Removing
NOTE: Rear springs can be removed with rear suspension fitted to car.
1 Remove rear road wheel.
2 Support rear of car on stands.
3 Remove washers and nuts securing hydraulic dampers to wishbone.
4 Drift out damper mounting pin.
5 Recover spacer and tie down bracket. Remove self-locking nut and bolt securing top of hydraulic damper to suspension unit cross-beam.
6 Withdraw hydraulic damper and road spring assembly.

7 Using tools SL.14 and JD.11B compress road spring until collets and spring seat can be removed. 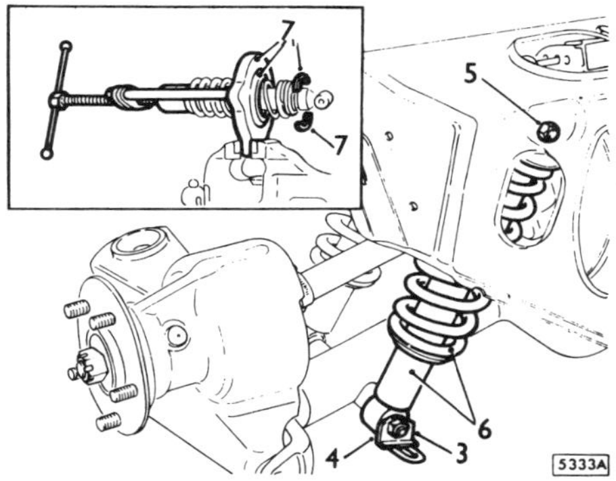
8 Release spring pressure and withdraw hydraulic damper from road spring.

Refitting
Reverse operations 1 to 9 inclusive.

REAR SUSPENSION UNIT
Remove and refit   64.25.01

Removing
1 Raise rear end of car and place on stands forward of radius arm anchor points.
2 Place jack with suitable adaptor in position under unit.
3 Remove rear road wheels.
4 Detach clips and fasteners securing rear silencers and remove silencers.

continued

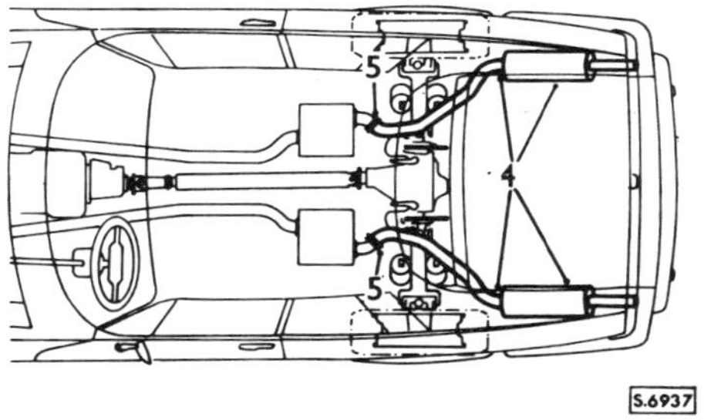

5 Release clips securing intermediate pipes to front silencers and remove clips from joints.
6 Remove locking wires and unscrew radius arm retaining bolts from body.
7 Detach forward ends of radius arms from mountings on body.
8 Remove bolts securing anti-roll bar links to radius arms and push bar clear of radius arms. 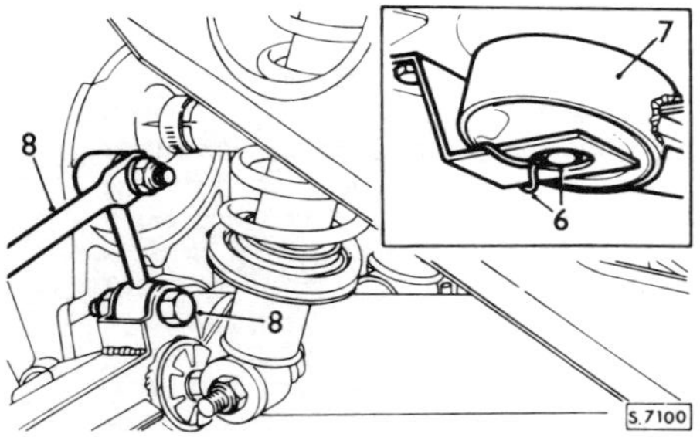
9 Remove bolts attaching propeller shaft to drive flange, lower rear end of propeller shaft.
10 Disconnect brake pipe union from body mounting bracket. Plug to prevent entry of dirt.
11 Detach flexible brake hose from bracket. Plug hose end.
12 Release handbrake adjustment (under rear carpet).
13 Release handbrake cable inner and outer members from caliper arms.
14 On cars fitted with optional overdrive, remove spring clip and detach speedometer drive from final drive unit.
15 Disconnect electrical lead from overdrive unit, if fitted.
16 Remove bolts attaching suspension mounting rubbers to body.
17 Lower jack to remove suspension unit.
18 Remove intermediate exhaust pipes.

Refitting
19 Replace intermediate exhaust pipes.
20 Jack unit into position.
21 Replace bolts attaching mountings to body and tighten to 27 to 32 lbf. ft (3,74 to 4,42 kgf. m).
22 Lift front ends of radius arms, clean and lightly grease spigots, replace bolts, tighten to 40 to 45 lbf. ft (5,6 to 6,2 kgf. m) and wire lock.
23 Refit anti-roll bar links to radius arms and tighten nuts.
24 Connect forward silencers to intermediate pipes, making joints with Fire Gum and tightening nuts to 11 to 13 lbf. ft (1,53 to 1,79 kgf. m).
25 Raise propeller shaft and couple to final drive flange. Tighten nuts to 27 to 32 lbf. ft (3,74 to 4,42 kgf. m).
26 Replace flexible brake hose.
27 Refit pipe union nut. Tighten to 6.3 to 7 lbf. ft (0,88 to 0,96 kgf. m).
28 Replace handbrake cable in caliper levers and adjust in car, see 70.35.10.
29 Refit speedometer drive to 20HU final drive unit and replace spring clip.
30 Reconnect electrical control lead to overdrive unit, if fitted.
31 Refit rear silencers, making joints with Fire Gum and tightening clip nuts to 11 to 13 lbf. ft (1,53 to 1,79 kgf. m).
32 Replace rear wheels and remove car from jack and stands.
33 Bleed brakes, see 70.25.02.

REAR SUSPENSION MOUNTING – SINGLE   64.25.02

Remove and refit

Removing
1 Place rear of car on stands, leaving jack supporting final drive unit.
2 Remove rear wheel adjacent to mountings.
3 Remove both mounting stud nuts.
4 Remove all four bolts attaching both mountings to suspension.
5 Lower suspension to allow removal of mounting.
6 Remove two bolts attaching mounting to body structure.

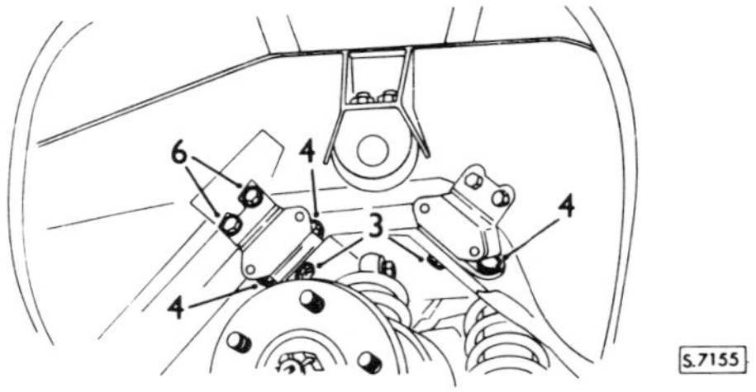

Refitting
7 Fit bolts attaching replacement mounting to body. Tighten nuts to 27 to 32 lbf. ft (3,75 to 4,42 kgf. m).
8 Raise suspension, aligning studs in cross-beam holes.
9 Fit bolts attaching mountings to cross-beam.
10 Fit nuts and tighten to 14 to 18 lbf. ft (1,94 to 2,48 kgf. m).

REAR SUSPENSION MOUNTING – PAIR   64.25.03

Remove and refit

Proceed as for 64.25.02 but remove four bolts in operation 6, detaching both mountings from body.

REAR SUSPENSION UNIT   64.25.06

Overhaul

The rear suspension unit is an assembly comprising individual units, the removal, refitting and overhaul of each being covered elsewhere in this Manual. For this reason, an overhaul procedure is not given for the rear suspension unit assembly proper, although it is advisable to check all bushes, fulcrum bearings and oil seals for damage or leakage whenever the unit is removed from the car.

REAR SUSPENSION HEIGHT   64.25.12

Check

1 Ensure radiator topped up with coolant.
2 Ensure engine sump filled to correct level with specified lubricant.
3 Ensure tyre pressure correct.

4 Note contents of fuel tank.
NOTE: Fuel tank holds a total of 20 Imperial gallons (24 U.S. gallons or 90 litres).
5 Calculate ballast weights required to represent difference between weight of fuel tank contents and weight of full tank.
NOTE: Full fuel tank totals approximately 150 lb (68 kg).
6 Place ballast weights required beneath rear window.
7 Roll car forward three lengths on perfectly level surface.
8 Measure distance between lower surface of rear cross-member and ground at both sides of car. (Dimension A must be 7.55±.25 in [190,6±6,4 mm]).
If dimension is not correct, check all bushes and bearing points of rear suspension. If no cause discovered, rear road springs must be changed. Remove all four springs and change as complete set, see operation 64.20.01.

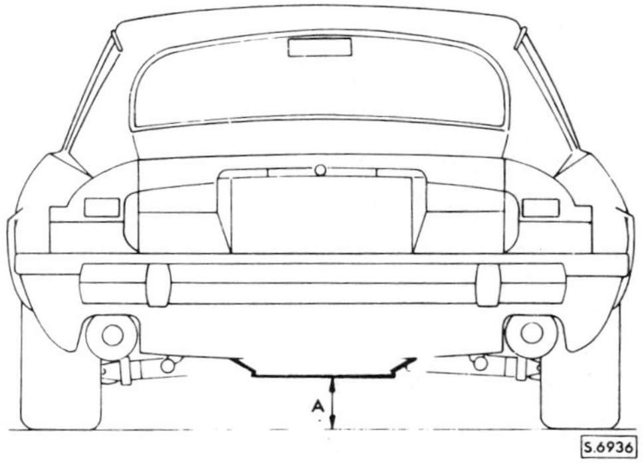

REAR SUSPENSION CAMBER ANGLE   64.25.18

Check and adjust

Service tool: Setting links JD.25

Checking
1 Set car on level surface.
2 Ensure tyre pressures correct.

64-4

3 Hook one end of setting link, tool JD.25, in lower hole of rear mounting, depress body until other end of setting link can be slid over outer wishbone fulcrum nut. Repeat on other side of car.

4 Set camber gauge against each rear tyre and read off camber angle. The correct reading should be $3/4 \pm 1/4°$ negative. If these limits are not met, note deviation and adjust camber angle, see operations 6 to 19 below inclusive. If result satisfactory continue with operation 5.

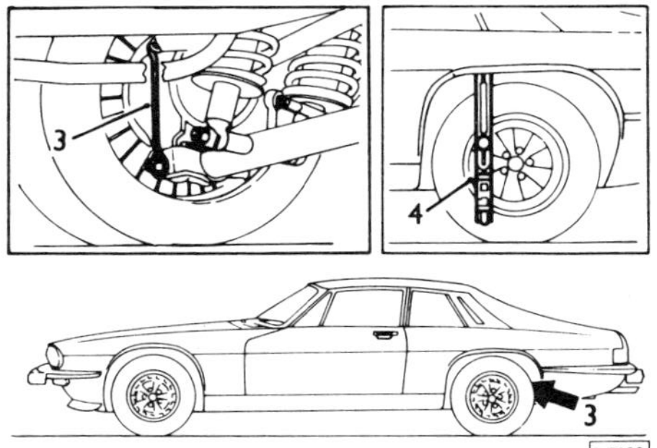

5 Remove setting links.

Adjust

6 Remove setting links.
7 Jack up rear of car and place stands to support body.
8 Remove road wheel.
9 Remove lower wishbone outer fulcrum grease nipple.
10 Remove clip securing inner universal joint cover. Slide cover clear of joint.
11 Remove four steel locknuts securing drive shaft flange to brake disc.
12 Separate drive shaft from disc to enable shims to be fitted.
NOTE: Addition of one shim .020 in (,5 mm) will alter camber position $1/4°$.

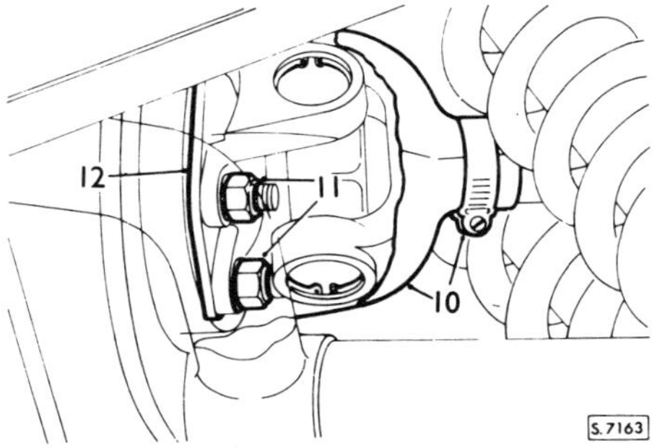

13 Add or remove shims as required.
14 Refit drive shaft flange to brake disc. Tighten nuts to 49 to 55 lbf. ft (6,78 to 7,60 kgf. m).
15 Slide inner universal joint cover back into position and tighten clip.
16 Replace grease nipple in outer fulcrum.
17 Replace road wheel and remove car from stands.
18 Replace setting links.
19 Recheck camber angle by repeating operation 4.

REAR HYDRAULIC DAMPERS

Remove and refit   64.30.01

Service tools: Handpress SL.14, Adaptor JD.11B.

Removing
NOTE: Rear hydraulic dampers can be removed with rear suspension unit fitted to car.

1 Remove rear road wheel.
2 Support rear of car on stands.
3 Remove washers and nuts securing hydraulic dampers to wishbone. Remove tie-down bracket.
4 Drift out damper mounting pin. Recover spacer at forward end of mounting pin tube.
5 Remove self-locking nut and bolt securing top of hydraulic damper to suspension unit cross beam.
6 Withdraw hydraulic damper and road spring assembly.
7 Using tools SL.14 and JD.11B compress road spring until collets and spring seat can be removed.

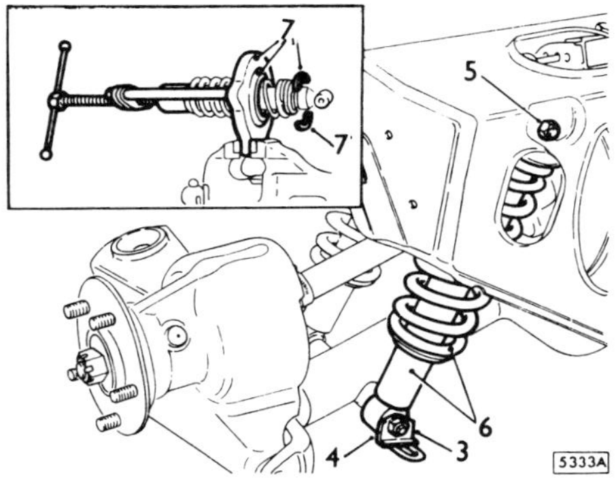

8 Release spring pressure and withdraw hydraulic damper from road spring.
9 Remove collar from upper eye of damper.

Refitting
NOTE: Hydraulic dampers fitted to this car are of the gas pressurized type and therefore need not be exercised before installation.

10 Refit collar to upper eye, align damper eyes and place spring over damper.
11 Assemble tool SL.14 and adaptors JD.11B to damper and spring.
12 Compress spring, insert spring seat and collets and detach tool.
13 Replace damper and spring assembly in cross beam, fit upper mounting bolt and tighten nut to 32 to 36 lbf. ft (4,45 to 4,95 kgf. m).
14 Drive in damper bottom mounting pin, replacing spacer removed in operation 4.
15 Fit tie-down bracket and tighten nuts to 32 to 36 lbf. ft (4,45 to 4,95 kgf. m).
16 Replace rear road wheel.
17 Remove car from stands.

Refitting
3 Place replacement bump stop in position, fit washers, and tighten nuts to 8 to 10 lbf. ft (1,11 to 1,38 kgf. m).
4 Replace rear road wheel.

ANTI-ROLL BAR

Remove and refit   64.35.08

Removing
1 Place car on ramp or over pit. Apply handbrake.
2 Raise left hand rear end of car and place on stand.
3 Remove left hand rear wheel.
4 Slacken exhaust pipe clamp nuts on right hand side, forward of cross beam.
5 Remove two bolts securing anti-roll bar to tops of links.

continued

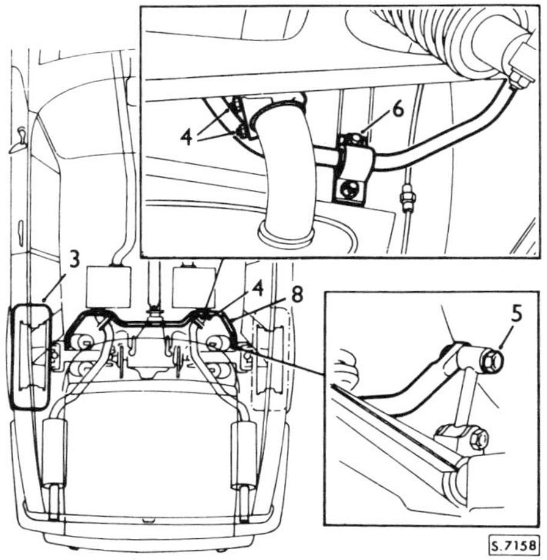

BUMP STOP

Remove and refit   64.30.15

Removing
1 Remove rear road wheel.
2 Remove two self-locking nuts and washers and detach bump stop.

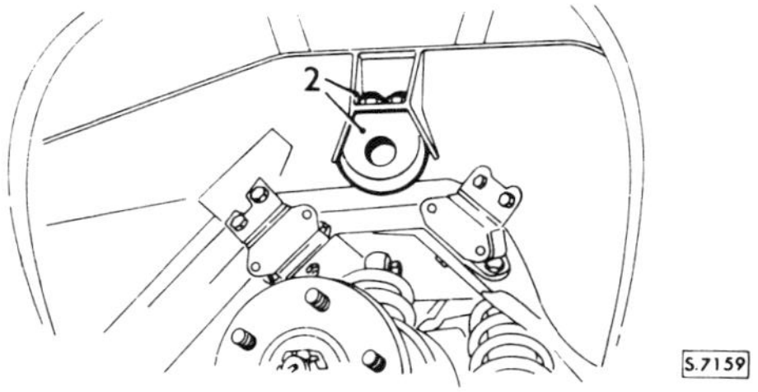

64—5

6 Remove four nuts securing anti-roll bar mountings to rear seat underpan.
7 Remove mounting plates and rubbers.
8 Separate exhaust joint slightly to clear anti-roll bar, and extricate bar through left hand wheel arch.

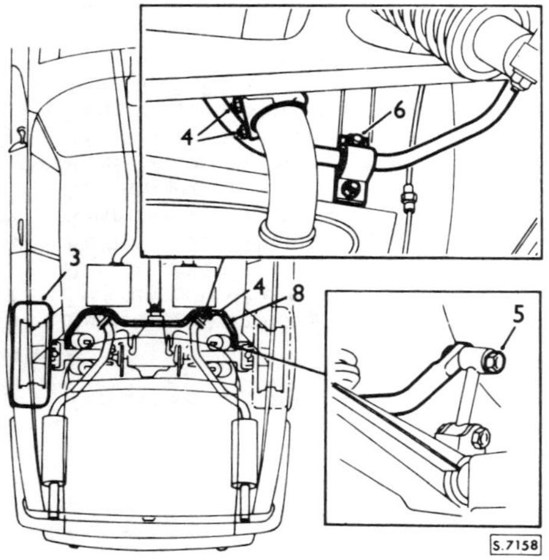

WISHBONE

Remove and refit   64.35.15

Service tool: Dummy shaft JD.14

Removing
1 Remove rear road wheel.
2 Support rear of car on stands forward of radius arms.
3 Remove one self-locking nut from outer fulcrum shaft, drift out shaft and remove.
4 Fit dummy shaft JD.14 to hub carrier assembly. Retain shims and oil seal retaining washers at each side of fulcrum with adhesive tape.
5 Raise hub and drive shaft clear of wishbone and suspend with strong wire from cross beam.
6 Remove bolt securing anti-roll bar link to radius arm and raise link clear of arm.
7 Remove locking wire and unscrew radius arm retaining bolt from body.
8 Detach forward end of radius arm from mounting on body.
9 Remove 14 bolts and setscrews securing bottom tie plate to cross-beam and inner fulcrum brackets.
10 Remove nuts and washers securing dampers to wishbone.
11 Drift out damper mounting pin. Recover spacer and tie-down bracket.
12 Remove rear nut from inner fulcrum shaft.
13 Drift fulcrum shaft forward to free wishbone from inner fulcrum. Remove wishbone and radius arm.
14 Collect oil seals, distance washers and bearings.
15 Remove bolt securing radius arm to wishbone; discard tab washer.
16 Remove grease nipples from wishbone ends. Clean and inspect all parts.

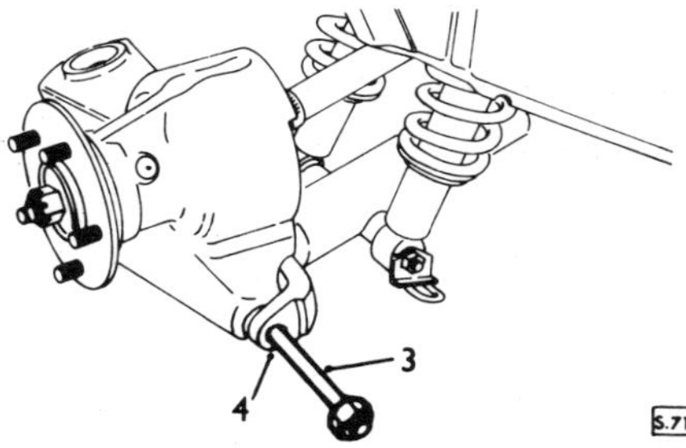

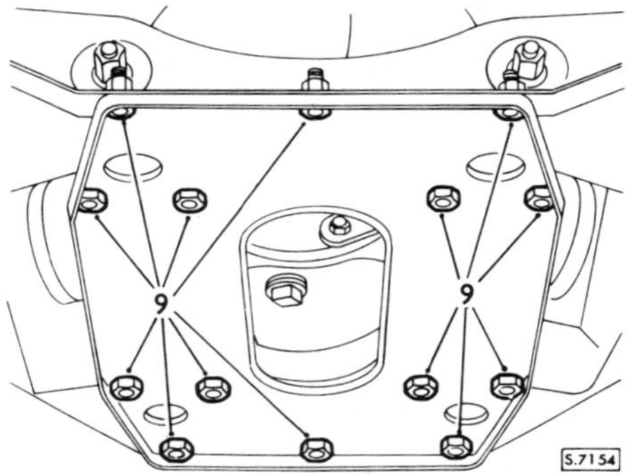

Refitting
17 Smear needle bearing cage with grease and press into wishbone inner fulcrum boss, engraved face outwards.
18 From opposite end of boss, press in second needle bearing cage, again with engraved face outwards.
19 Fit grease nipple.
20 Insert bearing tube.
21 Repeat 17, 18, 19 and 20 for other boss.
22 Attach radius arm to wishbone. Fit new tab washer and tighten nut to 60 to 70 lbf. ft (8,30 to 9,68 kgf. m).
23 Smear four outer thrust washers, inner thrust washers, new oil seals and oil seal retainers with grease and place into position on wishbone.

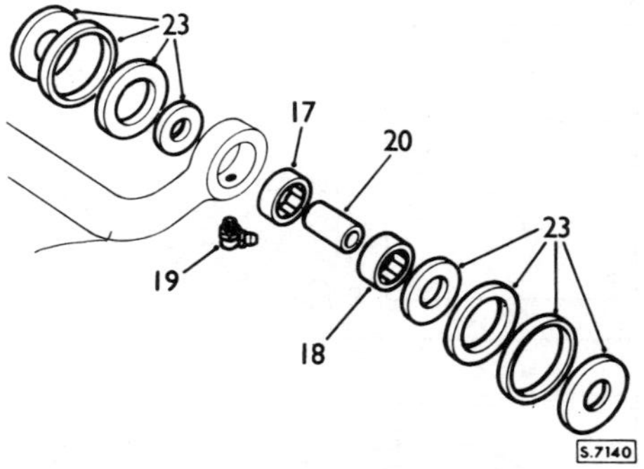

Refitting
9 Insert anti-roll bar through left hand wheel arch and manoeuvre into position.
10 Fit bolts attaching links to bar. Do not tighten at this stage.
11 Lubricate mounting rubbers with a proprietary rubber lubricant, or liquid soap and water solution, and place rubbers in position on bar.
12 Fit mounting plates over rubbers, place plates over studs and centralise rubbers in mountings. Fit nuts but do not tighten.
13 Replace road wheel and remove stand.
14 With car weight supported by suspension, tighten mounting nuts and link bolts.
15 Replace right-hand exhaust pipe to correct position and tighten nuts to 11 to 13 lbf. ft (1,52 to 1,80 kgf. m).
16 Remove car from ramp.

24 Offer up wishbone to inner fulcrum mounting bracket with radius arm bracket towards front of suspension unit.
NOTE: Take great care not to displace any of the fulcrum bearing components.
25 Carefully enter dummy shaft, tool JD.14, from each end to retain bearing assemblies and locate wishbone with mounting bracket.
26 Smear fulcrum shaft with grease and gently drift it through fulcrum to chase out dummy shafts.
NOTE: It is advisable to maintain a slight reaction pressure on dummy shafts as they emerge from fulcrum. This ensures that thrust washers are not knocked out of position. Should this happen, fulcrum shaft, dummy shaft and wishbone must be removed, and installation operations 23 to 26 repeated.
27 Fit self-locking nut to fulcrum shaft. Tighten to torque of 45 to 50 lbf. ft (6,2 to 6,9 kgf. m).

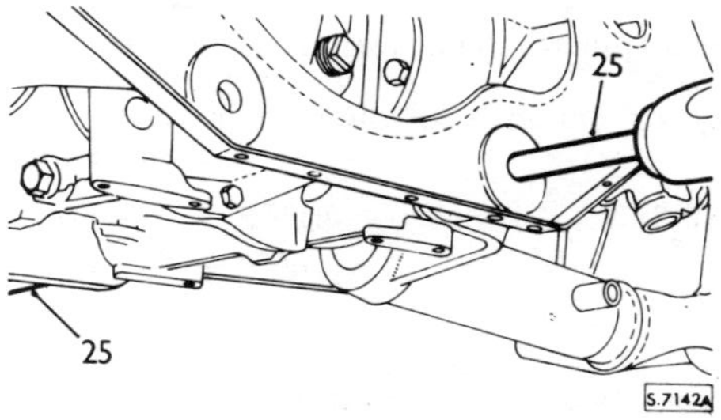

64–6

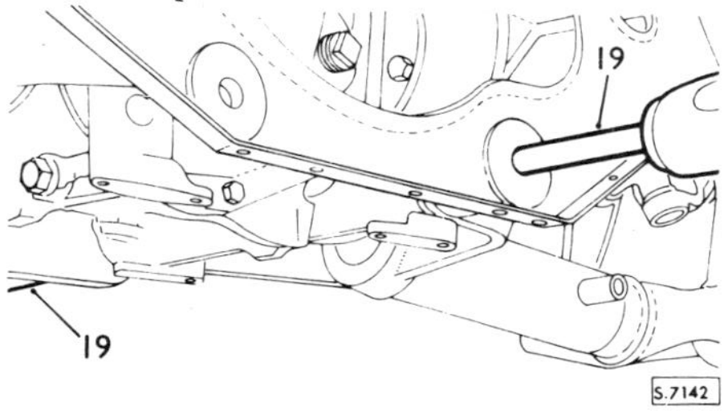

20 Drift dummy shafts from fulcrum bracket with fulcrum shaft. Restrain dummy shafts to prevent spacers or thrust washers dropping out of position.
21 Tighten fulcrum shaft nut to 45 to 50 lbf. ft (6,23 to 6,91 kgf. m).
22 Remove wire suspending hub assembly from cross beam.
23 Replace damper lower mounting shaft, refitting spacer and tie-down bracket. Tighten nuts to 32 to 36 lbf. ft (4,43 to 4,97 kgf. m).
24 Clean spigot on body, raise radius arm and replace bolt. Tighten to 40 to 45 lbf. ft (5,54 to 6,22 kgf. m) and wire lock bolt.
25 Bolt anti-roll bar link to radius arm and tighten.
26 Replace bottom tie plate and tighten bolts and setscrews to 14 to 18 lbf. ft (1,94 to 2,48 kgf. m).
27 Replace road wheel.
28 Remove car from stands.

ANTI-ROLL BAR LINK

Remove and refit   64.35.26

Removing
1 Place car on ramp or over pit. Apply handbrake.
2 Remove lower nut and bolt attaching link to radius arm and ease link clear of lugs.

continued

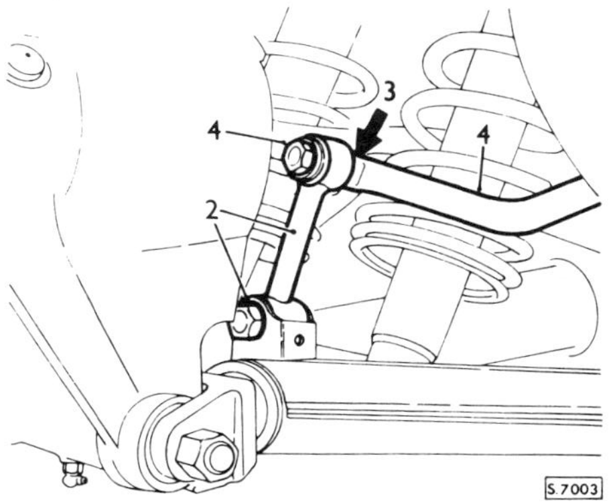

10 Drift fulcrum shaft forward to free wishbone from inner fulcrum.
11 Collect oil seals, distance washers and bearings.
12 Tap spacer tube from between lugs of fulcrum bracket.
13 Remove locking wire from two setscrews securing fulcrum bracket to final drive unit. Remove setscrews and withdraw fulcrum bracket, noting position and number of shims at each attachment point.

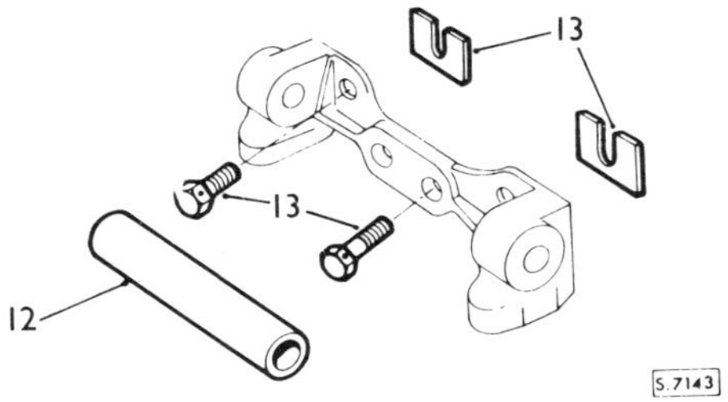

Refitting
14 Position fulcrum bracket against final drive unit and locate loosely with two setscrews.
15 Replace shims removed in operation 13 between bracket and final drive unit.
16 Tighten mounting setscrews to 60 to 65 lbf. ft (8,30 to 8,98 kgf. m) and wire lock.
17 Replace spacer tube between lugs of fulcrum bracket.
18 Clean, inspect and grease wishbone bearings, thrust washer etc. Refit with new oil seals.
19 Offer up wishbone to fulcrum bracket lugs and locate with dummy shafts, tool number JD.14. Take great care not to displace any component during this operation.

28 Raise wishbone and replace damper mounting pin, spacer and tie-down bracket. Tighten nuts to 32 to 36 lbf. ft (4,45 to 4,95 kgf. m).
29 Raise radius arm, clean and lightly grease spigot, replace bolt, tighten to 40 to 45 lbf. ft (5,5 to 6,2 kgf. m) and wire lock.
30 Remove wire suspending hub assembly from cross beam.
31 Remove adhesive tape attaching shims and washers to hub carrier, fit new seals, replace retainers and shims, and line up with wishbone.
32 Chase out dummy shaft with fulcrum shaft and tighten self locking nuts to 97 to 107 lbf. ft (13,4 to 14,8 kgf. m).
33 Refit anti-roll bar link to radius arm and tighten nut.
34 Replace bottom tie plate and tighten 14 bolts and setscrews to 14 to 18 lbf ft (1,94 to 2,48 kgf m).
35 Replace rear road wheel.
36 Remove car from stands.
37 Check rear suspension camber angle, see 64.25.18.
38 Grease wishbone bearings.

WISHBONE BEARINGS

Remove and refit   64.35.16

Operation instructions for removal and refitting of wishbone outer fulcrum bearings are given in Rear hub and Carrier assembly, remove and refit see 64.15.01 and overhaul 64.15.07.
The wishbone inner fulcrum bearings are covered in Wishbone, see remove and refit — 64.35.15.

WISHBONE OIL SEALS

Remove and refit   64.35.17

Refer to Wishbone, see remove and refit — 64.35.15.

ANTI-ROLL BAR RUBBERS

Remove and refit   64.35.18

Removing
NOTE: Always change rubbers as a pair.
1 Place car on ramp or over pit. Apply handbrake.
2 Remove four nuts securing anti-roll bar mountings to rear seat underpan.
3 Remove mounting plates and rubbers.

Refitting
4 Lubricate mounting rubbers with a proprietary rubber lubricant or liquid soap and water solution, and place rubbers in position on bar.
5 Fit mounting plates over rubbers, place plates over studs and centralise rubbers in mountings. Fit nuts and tighten.
NOTE: If car has been jacked, or raised in a wheel-free state on ramp, do not tighten nuts until full weight of car is on suspension.

INNER FULCRUM BRACKET (ONE)

Remove and refit   64.35.21

Service tools: Dummy shaft JD.14.

Removing
1 Remove adjacent rear road wheel.
2 Support rear of car on stands forward of radius arms.
3 Remove 14 bolts and setscrews securing bottom tie plate to cross beam and inner fulcrum brackets.
4 Detach forward end of radius arm from mounting on body.
5 Remove bolt securing anti-roll bar link to radius arm and raise link clear of arm.
6 Remove nuts and washers securing damper to wishbone.
7 Drift out damper mounting pin; recover spacer and tie down bracket.
8 Suspend hub and drive shaft assembly from cross beam with string wire.
9 Remove rear nut from inner fulcrum shaft.

64—7

RADIUS ARM BUSHES 64.35.29

Remove and refit

Service tool: Mandrel JD.21.

Removing

1 Remove radius arm, see 64.35.28.
2 Use mandrel tool JD.21 and press front bush from housing.

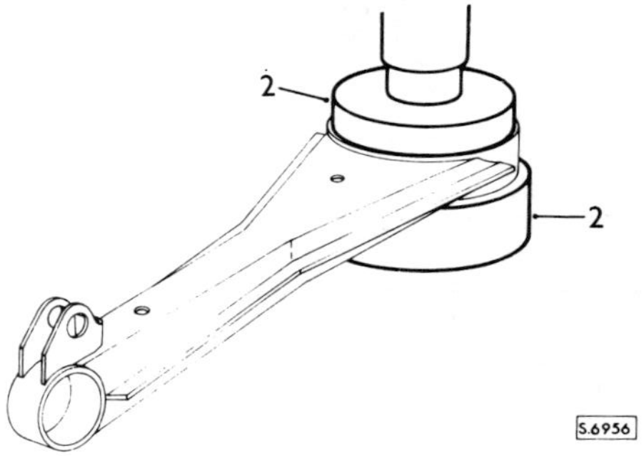

3 Use mandrel tool JD.21 and press rear bush from housing.

Refitting

4 Press new bush into rear bush housing so that bush is central in radius arm.
5 Use mandrel and press new bush into front bush housing so that holes in bush rubber are in line with centre line of radius arm. Press bush into radius arm until bush ring is flush with bush housing. When pressing bush, have small hole in bush core upwards.
6 Replace radius arm, see 64.35.28.

---

3 Remove nut from upper bolt.
4 Lever anti-roll bar end downwards and extract bolt and link clear of anti-roll bar. Remove bolt from link.

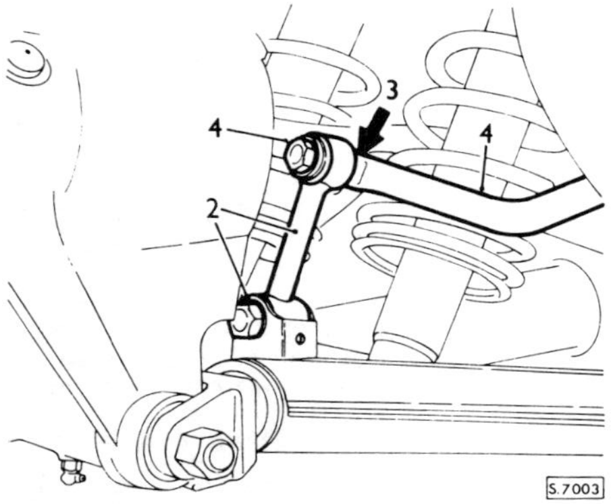

Refitting

5 Fit upper bolt to one end of link, with washer under bolt head.
6 Lever anti-roll bar end downwards, and insert bolt with link.
7 Line up lower bush with lugs on radius arm area, insert bolt, refit washer and nut and tighten.
8 Fit upper nut and tighten.
9 Remove car from ramp.

ANTI-ROLL BAR LINK

Overhaul 64.35.27

1 Remove anti-roll bar link, see 64.35.27
2 Press bush sleeve out of defective rubber bush in vice, using suitable mandrel.
3 Extract rubber bush from link.

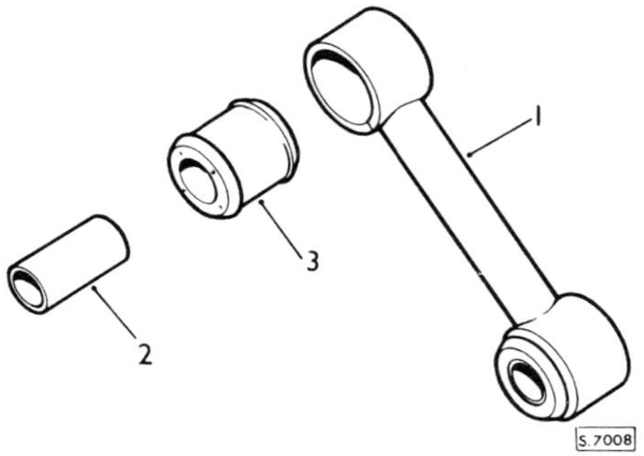

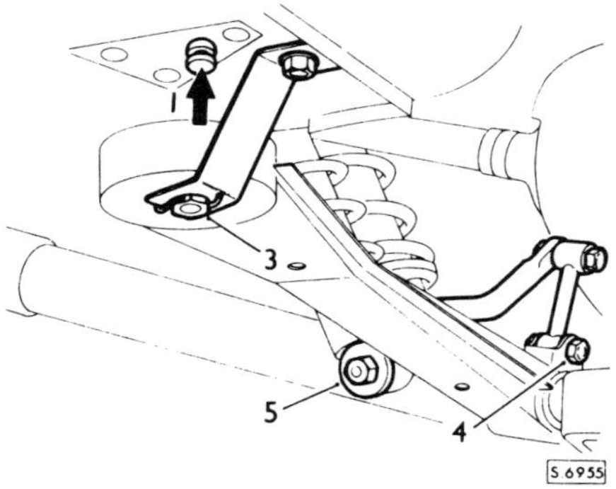

4 Remove bolt attaching anti-roll bar link to radius arm. Raise link clear of arm.
5 Remove nut from forward end of shaft carrying damper/spring unit. Detach washer.
6 Drift shaft backwards sufficiently to displace spacer. Collect spacer and swing damper/spring unit towards centre line of car.
7 Bend back tab washer and unscrew bolt attaching radius arm to wishbone. Discard tab washer.
8 Detach radius arm, clean and inspect both bushes. Refer to 64.35.29 for replacement.

NOTE: If either bush is found to require replacement, replace also the same bush in the other radius arm.

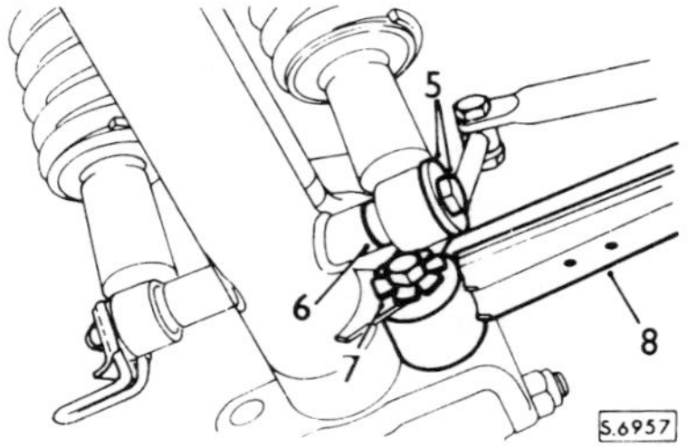

Refitting

4 Lightly lubricate replacement bush internally and externally with oil or soft soap.
5 Insert rubber bush in link and push approximately half way through bush housing.
6 Introduce bush sleeve to protruding end of rubber bush and press into place in vice using suitable spacer behind bush housing.
7 Repeat operations 2 to 6 on second bush of link.

RADIUS ARM 64.35.28

Remove and refit

Removing

1 Jack up rear of car and support on stands forward of radius arm anchor points.
2 Remove rear road wheel.
3 Remove locking wire and unscrew bolt securing radius arm to body.

Refitting

9 Place replacement radius arm in position and refit bolt attaching arm to wishbone, with new tab washer. Tighten bolt to 60 to 70 lbf. ft (8,3 to 9,68 kgf. m) and bend over tab washer.
10 Line up damper/spring unit with wishbone, hold spacer in position and tap shaft forward.
11 Fit washer and nut to damper mounting shaft and tighten to 32 to 36 lbf. ft (4,43 to 4,98 kgf. m).
12 Replace forward mounting bolt and tighten to 40 to 45 lbf. ft (5,53 to 6,22 kgf. m) and renew locking wire.
13 Replace bolt attaching anti-roll bar link to radius arm, fit washer and nut and tighten.
14 Replace road wheel.
15 Remove car from stands.

64–8

BRAKE SYSTEM

Description 70.00.00

The brake system is made up of the following main components:

1. Pedal box.
2. Servo unit.
3. Tandem master cylinder.
4. Pressure differential warning actuator (P.D.W.A.).
5. Four disc brake assemblies, front discs being ventilated.
6. Two handbrake calipers.

The above components provide the car with a dual braking system in which the front and rear caliper assemblies are totally independent of each other. Thus, in the event of a brake line fracture or a partial loss of brake fluid, one pair of brake calipers will at all times be operative. The purpose of the P.D.W.A. unit is to monitor front and rear brake line pressures and give visual warning to the driver should an undue fall in brake line pressure occur.

Operation of brake system

On application of the brake pedal the servo unit which is directly coupled to the master cylinder, transfers increased pedal pressure to the master cylinder primary piston 'A' causing the piston to move forward past the by-pass port 'P' to establish rear brake line pressure in chamber 'B'. Pressure from the primary cylinder return spring 'C' and rear brake line pressure force the secondary piston 'D' forward past the by-pass port 'P' to establish front brake line pressure in chamber 'E'.

Front and rear braking pressures enter the P.D.W.A. unit at ports 'F' and 'G' act on either end of the shuttle valve 'H' and travel to front and rear calipers via ports 'J' and 'K'. Should a fall in front or rear braking pressure occur the resultant pressure imbalance causes displacement of the shuttle valve, which in turn operates the switch 'L' and illuminates a warning light in instrument panel. In order to reset the shuttle valve the cause of fall in brake line pressure must first be established and rectified. During bleeding of the brake system which follows rectification the shuttle valve automatically resets, and extinguishes the warning light. Brake pressure entering the caliper 'M' forces the pistons 'T' out to act on the friction pads 'U' which in turn clamp the brake disc 'V'. On release of the brake pedal, brake line pressure collapses which allows the piston seals 'W' to retract the pistons into the caliper. Withdrawal of the pistons into the caliper is just sufficient for the friction pads to be in a relaxed position from the disc. This sequence provides automatic adjustment for brake pad lining wear.

Should the brake servo unit become inoperative front and rear braking systems will still operate but at a greatly reduced brake line pressure. A divided brake fluid reservoir 'R' ensures that in the event of fluid loss to front or rear brake systems one pair of brake calipers will at all times be operative. The fluid level indicator 'S' provides visual warning to the driver should the level of fluid in the reservoir fall to an unsatisfactory level.

KEY TO COMPONENT LOCATION

1. Brake fluid reservoir.
2. Master cylinder.
3. P.D.W.A. unit.
4. Brake calipers.

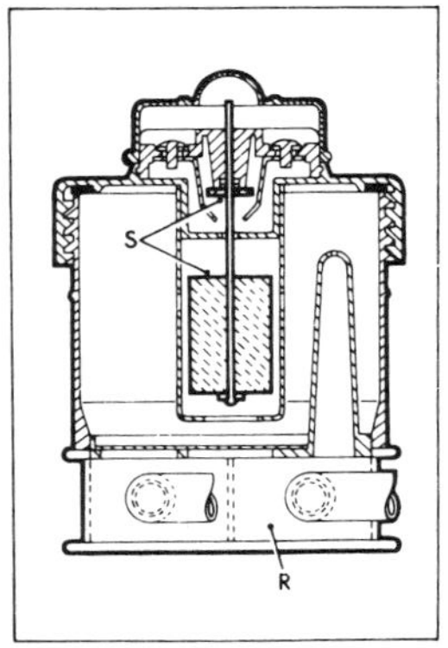

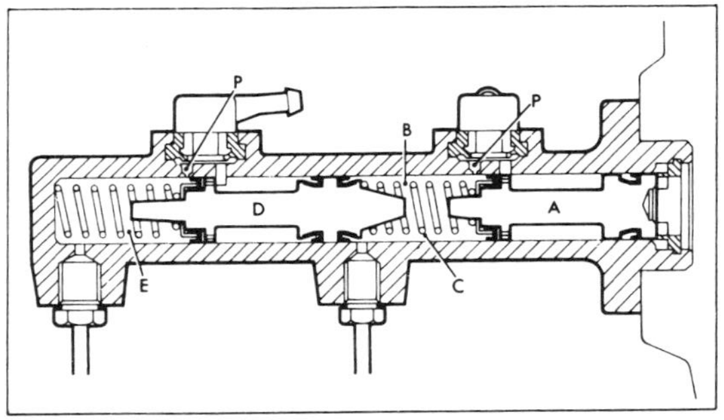

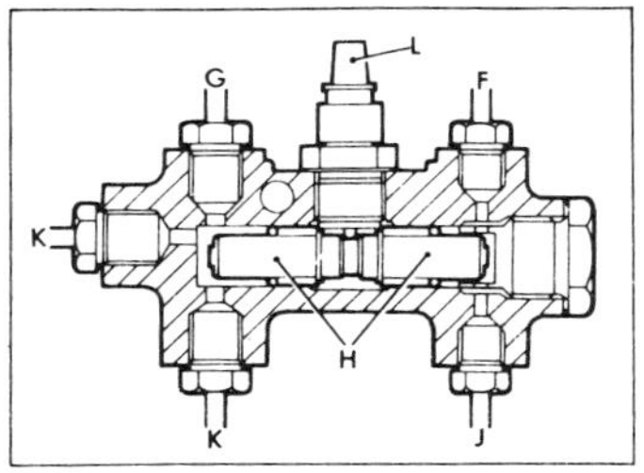

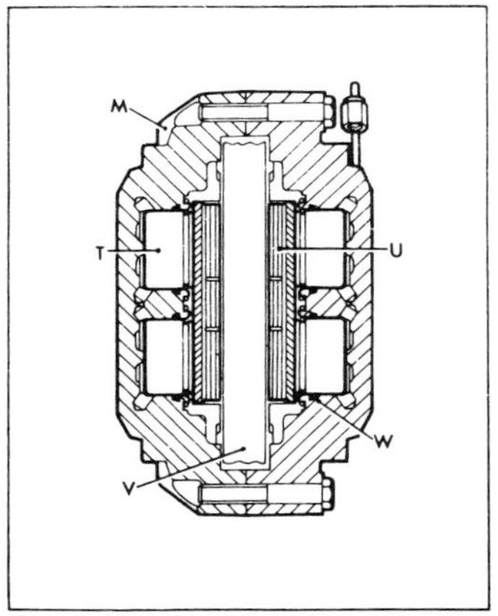

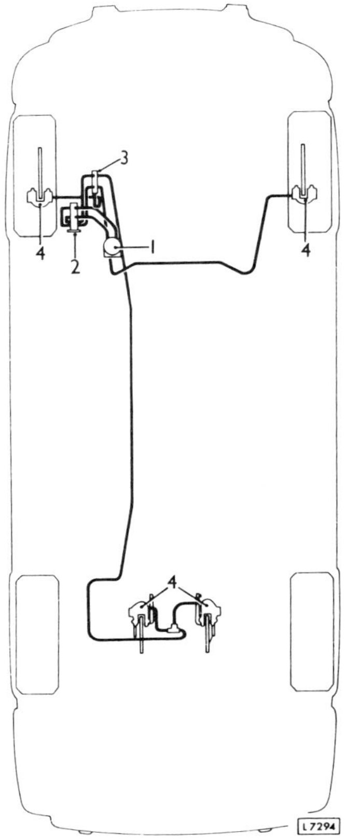

SYMPTOM AND DIAGNOSIS CHART FOR HYDRAULIC BRAKE SYSTEM

SYMPTOM	DIAGNOSIS	ACTION
FADE	Incorrect pads. Overloaded vehicle. Excessive braking. Old hydraulic fluid.	Replace the pads. Decrease vehicle load or renew hydraulic fluid as necessary.
SPONGY PEDAL	Air in system. Badly lined pads. Weak master cylinder mounting.	Check for air in the system, and bleed if necessary. Check the master cylinder mounting, pads and discs and replace as necessary.
LONG PEDAL	Discs running out pushing pads back. Distorted damping shims. Misplaced dust covers.	Check that the disc run out does not exceed 0.004 in (0,0101 mm). Rotate the disc on the hub. Check the disc/hub mounting faces.
BRAKES BINDING	Handbrake incorrectly adjusted. Seals swollen. Seized pistons. Servo faulty.	Check and adjust handbrake linkage. Check for seized pistons, repair or replace as necessary. Refer to Servo check, 70.50.03.
HARD PEDAL — POOR BRAKING	Incorrect pads. Glazed pads. Pads wet, greasy or not bedded correctly. Servo unit inoperative. Seized caliper pistons. Defective shock absorbers causing wheel bounce.	Replace the pads or if glazed, lightly rub down with rough sand-paper. Refer to Servo check, if servo is faulty. Check caliper for damage and repair as necessary. Fit new shock absorbers.
BRAKES PULLING	Seized pistons. Variation in pads. Unsuitable tyres or pressures. Defective shock absorbers. Loose brakes. Greasy pads. Faulty discs, suspension or steering.	Check tyre pressures, seized pistons, greasy pads or loose brakes; then check suspension, steering and repair or replace as necessary. Fit new shock absorbers.
FALL IN FLUID LEVEL	Worn disc pads. External leak. Leak in servo unit.	Check the pads for wear and for hydraulic fluid leakage. Refer to Servo check, 70.50.03.
DISC BRAKE SQUEAL — PAD RATTLE	Worn retaining pins. Worn discs. Worn pads.	Renew the retaining pins, or discs. Fit new pads.
UNEVEN OR EXCESSIVE PAD WEAR	Disc corroded. Disc badly scored. Incorrect friction pads.	Check the disc for corrosion, or scoring and replace if necessary. Fit new pads with correct friction material.

SYMPTOM	DIAGNOSIS	ACTION
BRAKE WARNING LIGHT ILLUMINATED (WITH HANDBRAKE RELEASED)	Fluid level low, combination valve or P.D.W.A. unit operated. Short in electrical warning circuit.	Top up reservoir. Check for leaks in system and pads for wear. Check electrical circuit.

BRAKES

IMPORTANT NOTE

Before commencing the following operations you are advised to familiarise yourself with the information in Section 07 — General Fitting Instructions. Pay especial regard to the notes on Unified and Metric unions, nuts and adaptors.

CAUTION: THROUGHOUT THE FOLLOWING OPERATIONS ABSOLUTE CLEANLINESS MUST BE OBSERVED IN ORDER TO PREVENT GRIT OR OTHER FOREIGN MATTER CONTAMINATING THE BRAKE SYSTEM. SHOULD THE SYSTEM BE FLUSHED OR COMPONENTS OVERHAULED, ONLY GIRLING BRAKE CLEANER MAY BE USED.

ALL BRAKE SYSTEM SEALS MUST BE COVERED IN CLEAN BRAKE FLUID AND ASSEMBLED USING FINGERS ONLY.

BRAKE FLUID

DURING OPERATIONS WHICH NECESSITATE HANDLING OF BRAKE FLUID, EXTREME CARE MUST BE OBSERVED, ON NO ACCOUNT MUST BRAKE FLUID CONTACT THE CAR PAINTWORK. IN INSTANCES WHERE THIS HAS OCCURRED THE CONTAMINATED AREA MUST IMMEDIATELY BE CLEANED USING A CLEAN CLOTH AND WHITE SPIRIT. THIS SHOULD BE FOLLOWED BY WASHING THE AREA WITH CLEAN WATER. METHYLATED SPIRIT MUST NOT BE USED TO CLEAN THE CONTAMINATED AREA.

FRONT DISCS

Remove and refit 70.10.10

Removing

1. Remove brake caliper friction pads, see 70.40.02.
2. Remove front hub, see 60.25.01.
3. Remove locking wire from caliper mounting bolts.
4. Remove caliper mounting bolts, recover and note position of shims located between steering arm and caliper.
5. Slacken bolt securing steering arm to hub carrier.
6. Gently easing caliper aside remove disc.

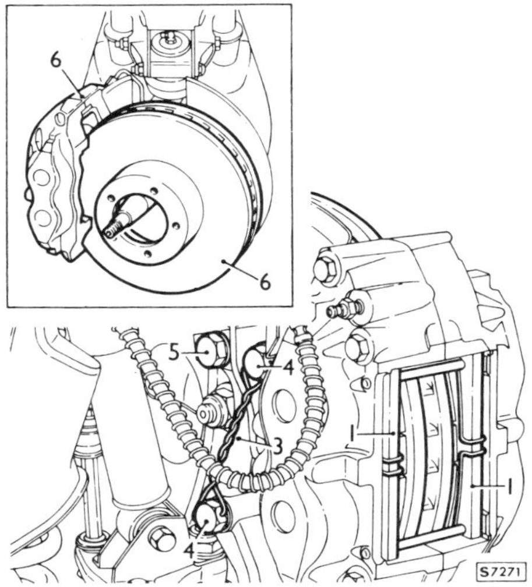

Inspection

7. Examine disc for cracks and heavy scoring, light scratches and scoring are not detrimental and may be ignored. If doubt exists a new disc should be fitted.

Refitting

8. If original disc is refitted reverse operations 1 to 6 and ensure caliper mounting bolts are torqued to 55 lb ft (7,5 kg m).
9. If new disc is fitted reverse operations 2 to 6 ensuring mounting bolts are not wire locked.
10. Check gap between caliper abutments and disc face. Gap on opposite sides of disc may differ by up to 0.010 in (0,25 mm) but gap on upper and lower abutment on same side of disc should be same.
11. If disc is not central of caliper carry out operations 12 – 14. If disc is central proceed to operation 15.
12. Remove one caliper mounting bolt and add or withdraw shim required to centralise disc, refit caliper bolt.
13. Repeat operation 12 on remaining caliper mounting bolt.
14. Repeat operation 10.
15. Torque caliper mounting bolts to 55 lb ft (7,5 kg m) wire lock bolts.
16. Refit brake friction pads.

REAR DISCS

Remove and refit 70.10.11

Removing

1. Place rear of car on stands and remove road wheel adjacent to brake disc to be removed.
2. Remove locking wire securing radius arm locking bolt, remove bolt.
3. Lever radius arm from spigot anchor point.
4. Position support blocks below hub.
5. Remove shock absorber lower fulcrum pin, recover distance piece washers, and towing bracket.
6. Slacken clip securing inner universal joint shroud, slide shroud clear of joint.
7. Remove nuts securing universal joint to brake disc.
8. Tap disc mounting bolts towards final drive unit.
9. Carefully separate universal joint from brake disc, collect camber angle shims held on disc mounting.
10. Remove brake caliper, see 70.55.03.
11. Remove brake disc from mounting bolts.
   NOTE: Do not disturb shims mounted between final drive flange and brake disc.

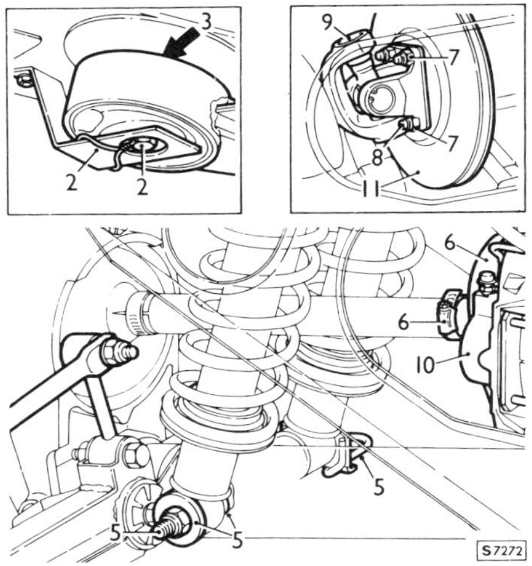

Inspection

NOTE: The condition of discs are a vital factor in efficient functioning of the brakes.

12. Examine surface of disc, which should be smooth. Scratches and light scoring are not detrimental after normal use. Should doubt exist a new disc should be fitted.

Refitting

13. Locate new disc on mounting bolts.
14. Position inner universal joint on disc mounting bolts.
15. Fit nuts securing universal joint to brake disc.
16. Offer brake caliper to mountings and secure with mounting bolts.
17. Check gap between caliper abutment and disc on both sides of caliper. If gap is not equal to within 0.010 in (0,25 mm) carry out operations 18 to 22. If disc is equidistant proceed to operation 23.
18. Remove caliper and disc adding or withdrawing shims from mounting bolts. Note thickness of shim added or withdrawn during this operation.
19. Refit caliper and disc and repeat operation 17.
20. Remove nuts securing universal joint to brake disc and separate universal joint from mounting bolts.
21. If shim was ADDED in operation 18 REMOVE shim of same size from camber angle shims. If shim was REMOVED, ADD shim to camber angle shims. This operation corrects camber angle to that prior to operation 18.
22. Refit universal joint to mounting bolts and secure with nuts.
23. Torque caliper mounting bolts to 49 – 55 lb ft (6,7 – 7,6 kg m). Wire lock bolts.
24. Reverse operation 2 to 6. Prior to fitting radius arm to spigot, wire brush spigot and smear with grease.
25. Fit brake friction pads, ensuring to check pads for wear (minimum thickness 0.125 in (3,17 mm).
26. Refit handbrake caliper.
27. Fit brake feed pipe to caliper, tighten connector at three-way union.
28. Bleed brakes, see 70.25.02.
29. Refit road wheel.
30. Check and if necessary adjust camber angle, see 64.25.18.

DISC SHIELDS – FRONT

Remove and refit 70.10.18

Removing

1. Raise front of car and place on stands.
2. Remove appropriate road wheel.
3. Swing steering onto full lock to allow access to disc shield.
4. Remove locking wire securing caliper mounting bolts.
5. Remove nuts securing lower disc shield clamps to stub axle, remove lower disc shield.
6. Remove bolt securing front section of disc shield to stub axle, lift shield from axle.
7. Release union securing caliper bridge pipe to hose, plug pipes to prevent loss of fluid.
8. Remove nut securing brake hose to mounting bracket, slide hose from bracket.

continued

70–3

9 Remove top caliper mounting bolt securing disc shield top section and caliper to stub axle carrier.
10 Remove top section of disc shield.

Refitting

11 Reverse operations 3 to 10, ensure brake hose is not twisted when securing to mounting bracket. Torque steering arm/caliper mounting bolt to 50 to 60 lb ft (6,9 to 8,3 kg m).
12 Bleed brakes, see 70.25.02.
13 Refit road wheel.

HOSES

General fitting and removal instructions   70.15.00

Removing

1 Clean unions of hose to be removed.
2 Ensure pipe sealing plugs are at hand.
3 Fully release unions securing fluid pipes to hose ends.
4 Withdraw pipe unions from hose ends, plug pipes to prevent loss of fluid and ingress of dirt.
5 Remove locknuts securing hose ends to mounting brackets.
6 Remove hose from car.

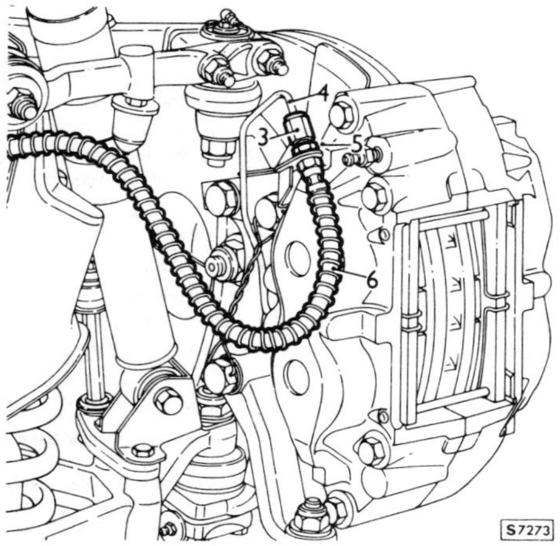

Inspection

7 After thoroughly cleaning hose examine for any signs of deterioration or damage. If doubt exists a new hose must be fitted.
8 Thoroughly clean bore of hose by feeding compressed air into one end of hose.

Refitting

9 Reverse operations 3 to 5.
10 Bleed brakes, see 70.25.02.

THREE-WAY CONNECTOR REAR

Remove and refit   70.15.34

Removing

1 Disconnect three feed pipe unions at connector, plug pipes to prevent loss of fluid and ingress of dirt.
2 Remove nut and bolt securing three-way connector to suspension unit, collect spacer and connector.

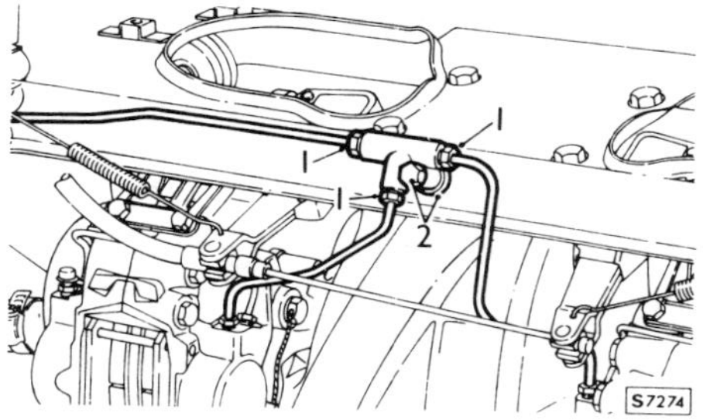

Refitting

3 Reverse operations 1 and 2.
4 Bleed brakes, see 70.25.02.

PIPES

General fitting and removal instructions   70.20.00

Removing

1 Clean unions of pipe to be removed.
2 Ensure pipe sealing plugs are at hand.
3 Fully release pipe unions.
4 Withdraw pipe from car, plug open ends of pipe remaining on car.

Inspection

5 Thoroughly clean bore of pipe by feeding compressed air into one end of pipe.
6 After thoroughly cleaning pipe examine for any sign of fracture or damage. If doubt exists as to condition exists a new pipe must be fitted.

Refitting

7 Reverse operations 3 to 6.
8 Bleed brakes, see 70.26.02.

BRAKES

Bleed   70.25.01

Bleeding the brake system is not a routine maintenance operation and should only be necessary where air has contaminated the fluid or a portion of the system has been disconnected.

Bleeding

1 Ensure fluid reservoir is topped up with fluid of correct specification.
2 Attach bleeder tube to left-hand rear bleed screw, immerse open end of tube in small jar partially filled with clean brake fluid.
3 Position gear selector in neutral and run engine at tick over.
4 Slacken left-hand rear bleed screw.
5 Operate brake pedal through full stroke until fluid issuing from tube is free of air bubbles.
NOTE: The fluid level in reservoir must be checked at regular intervals and topped up as necessary.
6 Keep pedal fully depressed and close bleed screw.
7 Repeat operations 2, 4, 5 and 6 on right-hand rear brake.
8 Continue operations 2, 4, 5 and 6 on remaining front brakes.
9 Check tighten all bleed screws and fit protective caps.
10 Top up reservoir as necessary.
11 Apply normal working load to brake pedal for several minutes, if pedal moves or feels spongy further bleeding of system is required.

12 When pedal 'feel' is satisfactory release handbrake; brake warning light should extinguish. If warning light remains illuminated carry out remaining operation.
13 Operate brake pedal applying heavy pedal pressure, warning light should extinguish; if light remains illuminated carry out P.D.W.A. Check operation, see 70.25.08.

PRESSURE DIFFERENTIAL WARNING ACTUATOR

Check and reset   70.25.08

1 With gear selector in 'N' (Neutral) or 'P' (Park) and handbrake applied, start engine, brake warning light should be illuminated. If warning light not illuminated fit new bulb.
2 Release handbrake, warning light should extinguish, if light remains illuminated carry out operation 3.
3 Check brake reservoir fluid level, top up as necessary, if warning light remains illuminated carry out remaining operation.
4 Disconnect electrical connector from P.D.W.A. switch, if warning light goes out P.D.W.A. has operated; if light remains illuminated check for 'short' in brake warning electrical circuit or a sticking reservoir fluid level switch.
NOTE: If P.D.W.A. unit has operated a major defect in the brake system is indicated.

Reset

Resetting of the P.D.W.A. unit is achieved automatically during bleeding of the brake system, which should only be carried out following rectification of defect that caused the shuttle valve displacement.

70–4

PRESSURE DIFFERENTIAL WARNING ACTUATOR (P.D.W.A.)

Remove and refit 70.25.13

Removing
1. Remove appropriate air cleaner box, see 19.10.08.
2. Disconnect electrical lead from P.D.W.A. switch.
3. Disconnect all feed pipes from P.D.W.A. plug pipes to prevent loss of fluid.
4. Remove bolt securing P.D.W.A. to wing valance. 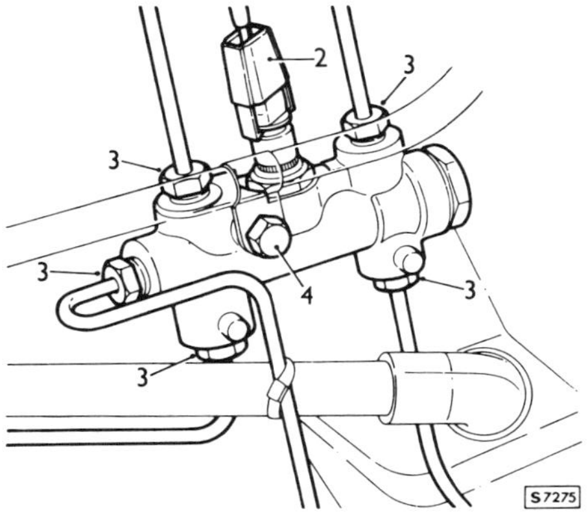
5. Lift P.D.W.A. from car.

Refitting
6. Reverse operations 1 to 5.
7. Bleed brakes, see 70.25.02.

PRESSURE DIFFERENTIAL WARNING ACTUATOR

Overhaul 70.25.14

NOTE: Overhaul of the P.D.W.A. unit is not possible and the following test should be carried out at intervals detailed in Group 10.

Operational check of P.D.W.A. unit
1. Ensure car is adequately chocked.
2. Check brake fluid level and top-up if necessary.
3. Ensure gear selector lever is in 'N' (Neutral) or 'P' (Park). Check that with ignition on and handbrake applied 'Brake Warning' light is illuminated.
4. Run engine at tickover and release handbrake.
5. Apply heavy foot pressure to brake pedal.
   NOTE: The brake pedal should be fully depressed and kept fully applied throughout operations 6 to 8.
6. Release any brake caliper bleed nipple, ensure ejecting fluid is collected in a jar or waste rag.
7. 'Brake Warning' light should illuminate.
8. Close bleed nipple.
9. Release and re-apply foot pressure to brake pedal.
10. 'Brake Warning' light should extinguish.
11. Switch off engine and apply handbrake.
12. Top-up brake fluid reservoir.
13. Should warning light fail to illuminate during operation 6, repeat operations 4 to 12. A new P.D.W.A. unit is required if warning light fails to illuminate during repeat operation.

BRAKE SYSTEM – FLUSH

Drain, flush and bleed 70.25.17

Service tool: Lever 64932392

Draining
1. Slacken all road wheel nuts.
2. Jack up front of car and place on stands.
3. Jack up rear of car and place on stands.
4. Remove all road wheels.
5. Attached bleeder tube to rear left-hand caliper bleed screw with open end of tube in suitable container.
6. Slacken bleed screw.
7. Operate brake pedal slowly through full stroke, until 'rear' brake section of fluid reservoir is drained and fluid ceases to issue from bleed tube.
8. Remove rear left-hand caliper friction pads, see 70.40.03.

WARNING: ON NO ACCOUNT OPERATE BRAKE PEDAL WHILE FRICTION PADS ARE REMOVED.

9. Using special tool 64932392, lever pistons into bores expelling remaining trapped fluid into container.
10. Replace friction pads.
    NOTE: It is not necessary to replace retaining pins and clips at this time.
11. Close bleed screw.
12. Discard expelled fluid.
13. Repeat operations 5 to 12 on right-hand rear and front calipers.

Flushing
14. Fill fluid reservoir with Castrol/Girling brake flushing fluid.
15. Attach bleeder tube to rear left-hand caliper bleed screw with open end of tube in container.
16. Slacken bleed screw.
17. Operate brake pedal slowly through full stroke, until clear flushing fluid issues from tube.
    NOTE: The fluid level in the reservoir must be checked at regular intervals and topped up as necessary.
18. Close bleed screw and operate pedal two or three times.
19. Repeat operations 15 to 18 on remaining rear and front calipers.
20. Carry out operations 5 to 10 on front brake calipers.
21. Secure rear friction pads with retaining pins and clips.
22. Repeat operations 5 to 10 on front brake calipers.
23. Secure front friction pads with retaining pins, clips and anti-chatter springs.
24. Close bleed screws on front and rear calipers.
25. Discard expelled flushing fluid.

Refilling
26. Fill brake reservoir with new brake fluid of correct specification.
27. Bleed brakes, see 70.25.02.
    NOTE: Prior to closing bleed screw during the bleeding of each caliper, check that issuing brake fluid is completely free of flushing fluid.
28. Refit road wheels to car.
29. Remove stands.

TANDEM MASTER CYLINDER

Remove and refit 70.30.08

Removing
1. Peel cover from brake fluid reservoir cap and disconnect wires from fluid level indicator.
2. Remove reservoir cap and filter.
3. Using suitable syringe remove brake fluid from reservoir.
4. Remove clips securing hoses to master cylinder adaptors, use waste rag to collect remaining fluid from feed pipes.
5. Disconnect brake pipes from master cylinder outlet unions, plug pipes to prevent loss of fluid and ingress of dirt.
6. Remove nuts securing master cylinder to servo unit, lift master cylinder from mounting studs.

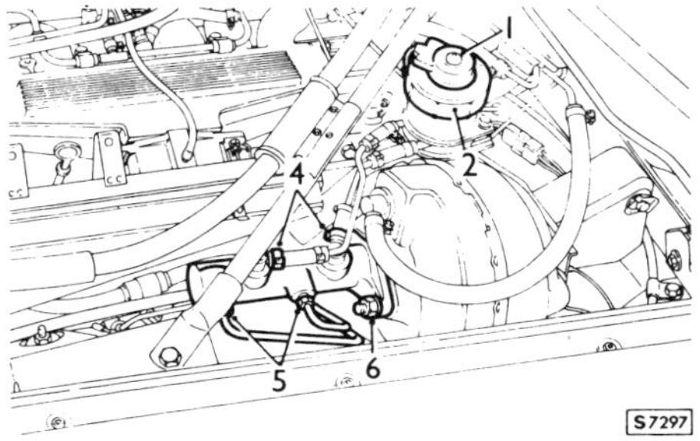

NOTE: Before a master cylinder is removed from a direct acting servo it is imperative that the brake pedal is depressed and released at least 10 times. This is to ensure that no vacuum exists to operate the servo. Operation of the servo when the master cylinder is not in place can cause its mechanism to travel past its normal limit; this can damage the servo beyond repair.

continued

70–5

Refitting
7 Reverse operations 3 to 6.
8 Bleed brakes, see 70.25.01.
9 Refit reservoir filter and sealing cap.
10 Connect leads to fluid level indicator and replace cover.

MASTER CYLINDER

Overhaul 70.30.09

1 Remove master cylinder, see 70.30.08.
NOTE: Overhaul of the master cylinder should be carried out with the work area, tools and hands in a clean condition.

Dismantling
2 Carefully prise reservoir hose adaptors from sealing grommets.
3 Using suitable screwdriver lever sealing grommets from master cylinder.
4 Press primary piston into bore of cylinder and withdraw secondary piston stop pin from forward grommet housing.
5 Remove circlip.
6 Tap flange end of cylinder on wooden block to remove primary piston and spring, secondary piston and spring. It may prove necessary to feed compressed air into cylinder front delivery port.
NOTE: Once piston assemblies are withdrawn appropriate piston and spring must be kept together. In event of springs being mixed, the secondary piston spring can be easily identified, it being slightly thicker and longer than the primary spring. 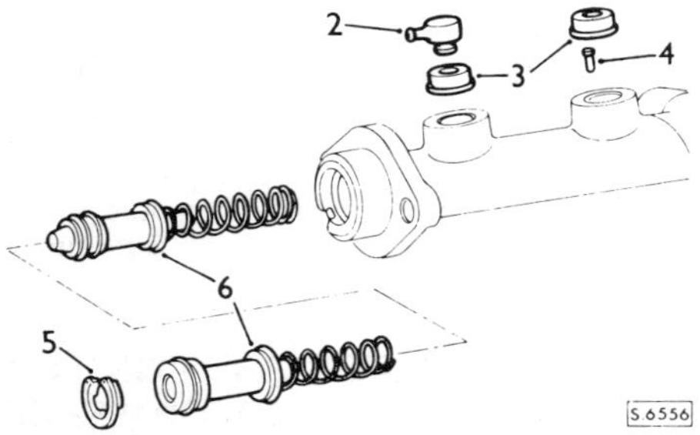
7 Remove spring, spring seat, recuperating seal and washer from secondary piston.
8 Carefully prise seals from rear of secondary piston.
9 Remove spring, spring seat, recuperating seal and washer from primary piston.
10 Carefully prise seal from rear of primary piston.

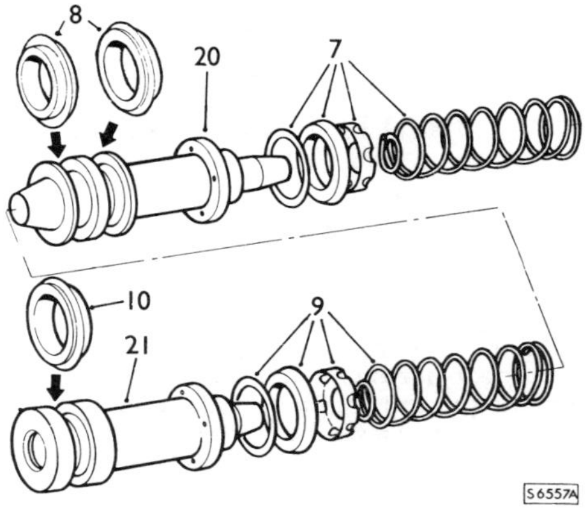

Inspection
11 Discard all old seals and associated items that will be replaced by those contained within service kit.
12 Clean all parts with Girling cleaning fluid and dry with lint-free cloth.
13 Examine piston and bore of cylinder for visible score marks and corrosion. If doubt exists as to condition of components replace suspect item.

Re-assembling
WARNING: TO HELP PREVENT DAMAGE IT IS ESSENTIAL THAT GENEROUS AMOUNTS OF CLEAN BRAKE FLUID ARE USED AT ALL STAGES OF SEAL ASSEMBLY.
14 Carefully fit inner seal of secondary piston in locating groove, ensure seal lip faces forwards.
15 Fit remaining seal in locating groove, ensure seal lip faces towards primary piston, i.e. in opposite direction of seal fitted in operation 14.
16 Fit washer, recuperating seal, spring seat and spring over forward end of secondary piston.
17 Carefully fit rear seal of primary piston in locating groove, ensure seal lip faces forward, i.e. away from circlip.
18 Fit washer, recuperating seal, spring seat and spring over forward end of primary piston.
19 Generously lubricate bore of master cylinder with clean brake fluid.
WARNING: ADHERENCE TO THE FOLLOWING INSTRUCTION IS VITALLY IMPORTANT. FAILURE TO COMPLY WILL RESULT IN DAMAGED PISTON SEALS.
20 Secure master cylinder in vice and generously lubricate piston seals in new brake fluid. Offer secondary piston assembly to cylinder till recuperating seal rests centrally in mouth of cylinder. Ensuring seal is not trapped, slowly rotate and rock piston assembly whilst GENTLY introducing piston into cylinder bore. Once recuperating seal enters bore of cylinder SLOWLY push piston into bore in one continuous movement.
21 Repeat operations 19 and 20 with primary piston and spring. 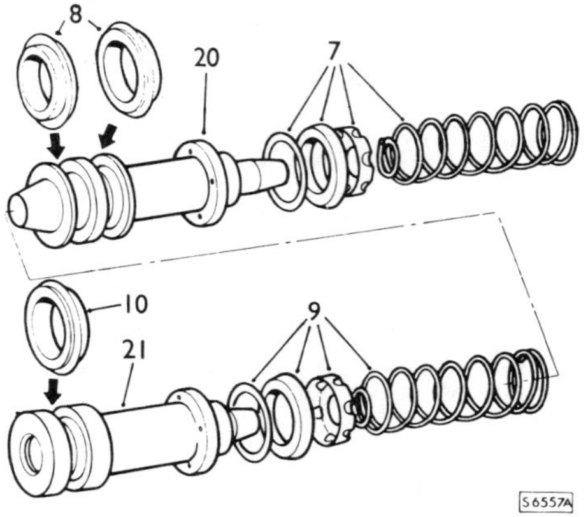
22 Pressing piston into bore of cylinder, fit circlip.
23 Press primary piston into bore of cylinder to full extent, fit secondary piston stop pin.
24 Fit sealing grommets to master cylinder.
25 Lubricate hose adaptors with brake fluid and press into sealing grommets.

FLUID RESERVOIR – MASTER CYLINDER

Remove and refit 70.30.16

Removing
1 Restraining fluid level indicator remove reservoir cap (cap turns independent of fluid level indicator). Position cap and indicator to one side.
2 Remove reservoir filter.
3 Using suitable syringe remove fluid from reservoir.
4 Remove clips securing hoses to reservoir.
5 Remove nuts, bolts and flat washers securing reservoir to mounting bracket.
6 Slide reservoir free from hoses and lift from car. Plug hose ends to prevent loss of fluid and ingress of dirt.

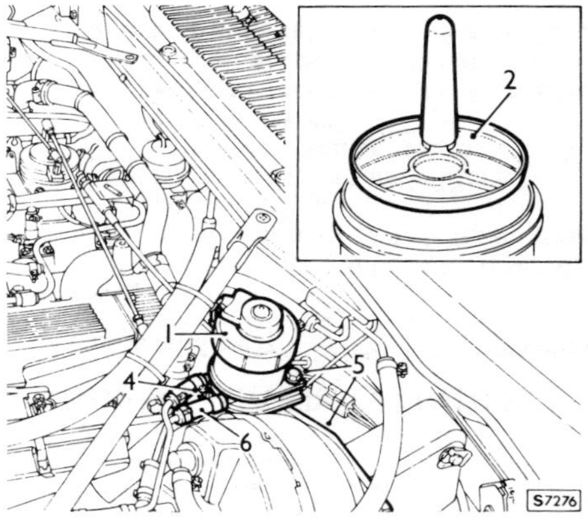

Refitting
7 Reverse operations 4 to 6.
8 Bleed brakes, see 70.25.02.

70–6

PEDAL BOX

Remove and refit 70.35.03

Removing

1. Remove bolt securing stay bar to wing valance, swing bar clear of wing.
2. Prise vacuum pipe adaptor from servo unit, swing pipe and adaptor clear of servo unit.
3. Remove nuts securing master cylinder to servo unit.

Right-hand drive cars only

4. Disconnect leads from throttle kick-down switch.
5. Slacken locknuts securing throttle cable to pedestal.
6. Rotate pedestal turntable and release throttle cable nipple from locating recess.
7. Position cable clear of servo unit.

All cars

8. Remove bolts securing fluid reservoir to mounting bracket, rest reservoir on engine.

Manual transmission cars only

9. Release unions securing pipe to clutch master cylinder and hose connection, remove pipe from car. To prevent ingress of dirt plug ends of master cylinder union and hose connection.

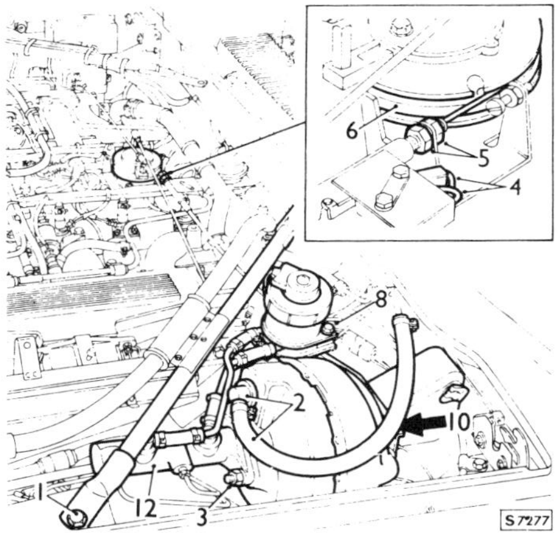

All cars

10. Slacken nuts securing reservoir mounting bracket to pedal box, swing bracket into vertical position.
11. Remove four nuts securing pedal box assembly to bulkhead.
12. Ease master cylinder from mounting bolts and clear of servo.
13. Manoeuvre pedal box and servo unit clear of bulkhead and lift from car.
14. Remove spring clip securing servo push rod clevis pin.
15. Remove clevis pin and spring washer securing push rod to brake pedal lever.
16. Remove four nuts securing pedal box to servo unit, recover reservoir mounting bracket and lay aside.
17. Detach pedal box from servo unit.

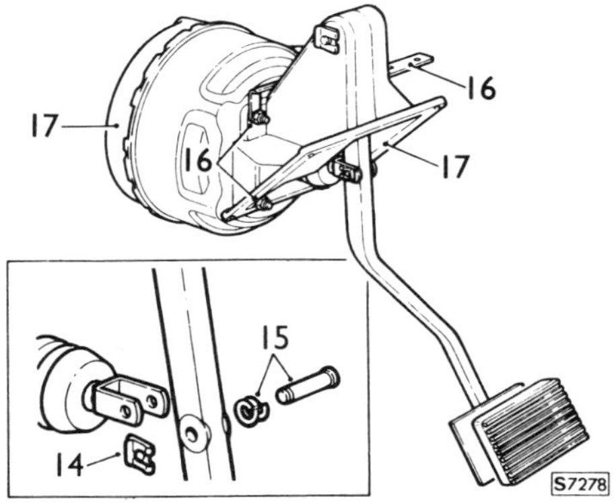

Inspection

4. Clean pedal lever components.
5. Examine pivot pin and lever bush for wear. Should doubt as to condition exist a new component should be fitted.

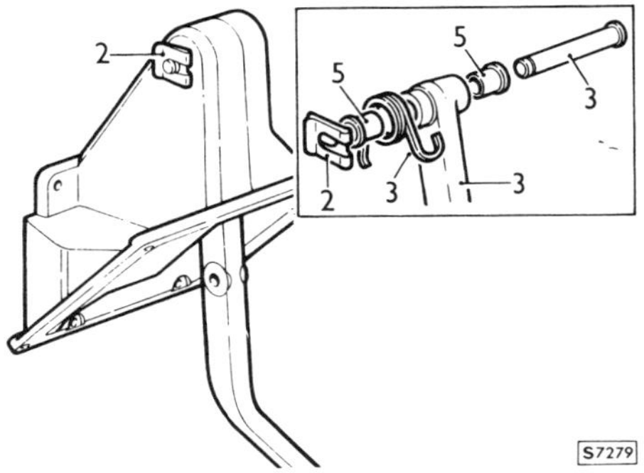

Re-assembling

6. Position pedal lever return spring so hook of spring locates over lever.
7. Locate pedal lever and return spring in pedal box and align lever boss with pivot pin hole.
8. Enter pivot pin into pedal box and drive fully home.
9. Reverse operations 1 and 2.

Refitting

18. Reverse operations 1 to 17.
19. Bleed clutch on manual transmission cars.

PEDAL BOX

Overhaul 70.35.04

1. Remove pedal box, see 70.35.03.

Dismantling

2. Remove spring clip securing pedal lever pivot pin.
3. Remove pivot pin from pedal box, withdraw pedal lever and return spring.

5. Disconnect feed and earth wire from brake warning switch.
6. Raise carpet on sill for access to cable adjusting nuts.
7. Slacken adjusting and lock nuts on handbrake cable.
8. Remove adjustable nipple from handbrake cable.
9. Remove bolts securing handbrake assembly and earth lead to sill, recover distance pieces located between handbrake and sill.
10. Remove pivot bolt securing handbrake to sill, recover distance piece.

continued

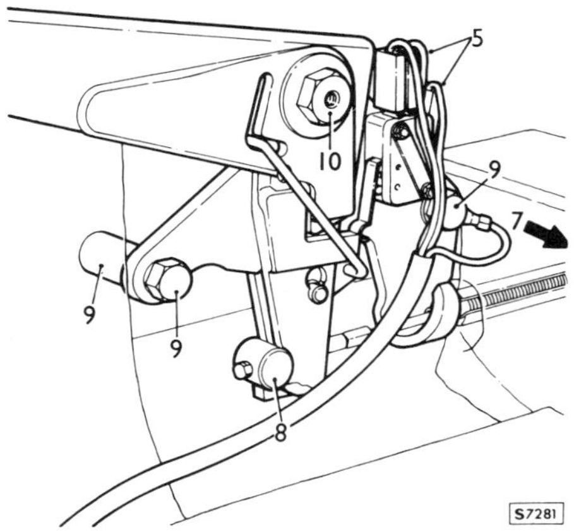

HANDBRAKE LEVER ASSEMBLY

Remove and refit 70.35.08

Removing

1. Disconnect battery, see 86.15.20.
2. Remove front seat, see 76.70.01.
3. Remove bolt securing handbrake mechanism cover.
4. Raise handbrake lever and slide cover from lever.

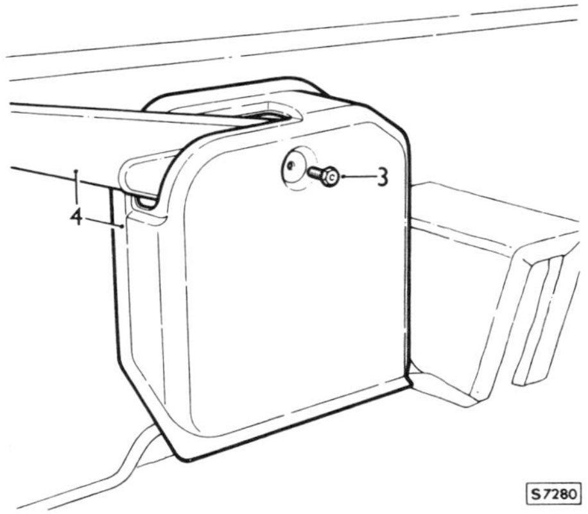

70-7

11 Disengage handbrake from cable, lift handbrake from car.
NOTE: If new handbrake assembly is to be fitted remove warning light switch from old handbrake. Fit and adjust switch to new handbrake, see 86.65.46.

Refitting
12 Reverse operations 5 to 11.
13 Adjust handbrake cable, see 70.35.10.

HANDBRAKE CABLE

Adjust   70.35.10

Adjusting
1 Raise sill carpet adjacent to rear of driver's seat for access to cable adjusting nuts.
2 Turn back cable locknut.
3 Tighten adjusting nut to a point when handbrake is fully off a slight amount of slack is apparent within the cable.
NOTE: Should cable be adjusted so that all slack is removed, binding of handbrake caliper may result. 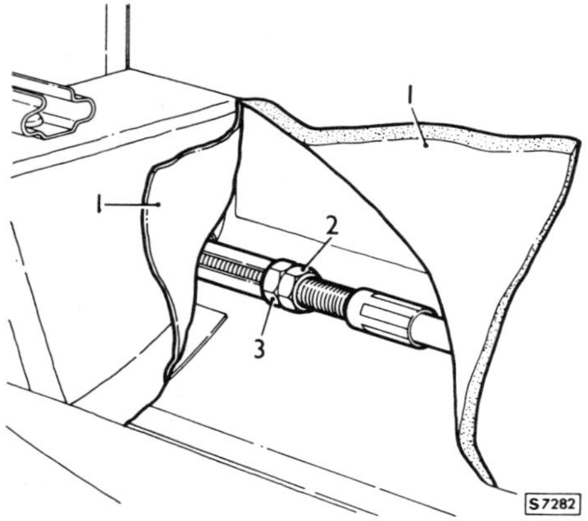
4 Tighten cable locknut.
5 Replace sill carpet.

HANDBRAKE CABLE ASSEMBLY

Remove and refit   70.35.14

Removing
1 Remove front seat, see 76.70.01.
2 Remove bolt securing handbrake mechanism cover.
3 Raise handbrake lever and slide cover from lever.
4 Raise sill carpet to allow access to cable adjusting nut.
5 Slacken cable lock and adjusting nuts.
6 Remove nipple from cable end.
7 Ease cable free from handbrake mechanism.
8 Disengage nipple and cable from operating arm of handbrake caliper.
9 Disconnect outer cable from handbrake caliper operating arm.
10 Cut clips securing protective sleeve to outer cable, slide sleeve from cable.
11 Remove protective cover from outer cable.
12 From handbrake lever mechanism in car pull cable free from body guide tube. 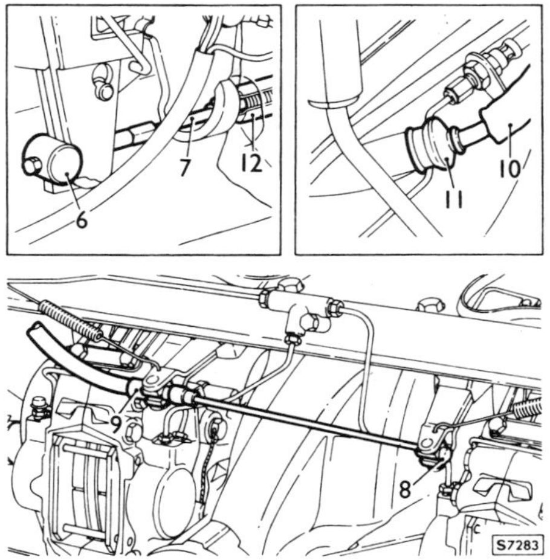
13 Lift cable from car.
14 Remove adjusting nut, lock nut and guide tube from cable.

Refitting
15 Reverse operations 6 to 14.
16 Adjust handbrake cable, see 70.35.10.
17 Reverse operations 1 to 4.

BRAKE PADS – FRONT

Remove and refit   70.40.02

Service tool: Lever 64932392

Removing
1 Remove road wheel.
2 Remove clips securing pad retaining pins.
3 Remove upper pin.
4 Remove anti-chatter springs.
5 Remove lower retaining pin.
6 Withdraw brake pads.

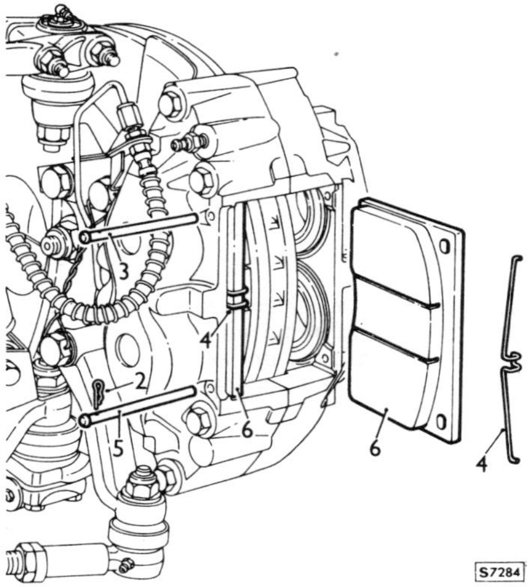

Inspection
7 Examine pads for wear; if thickness of friction pad is less than 0.125 in (3,17 mm) new pads must be fitted.

Refitting
NOTE: It is advisable to reduce the level of brake fluid in reservoir before fitting new pads.
8 Lever caliper pistons into cylinder bores using service tool 64932392.
9 Locate brake pads in caliper.
10 Reverse operations 1 to 5.
11 Top-up fluid reservoir.
12 Run engine and apply brake pedal several times until pedal feels solid.

BRAKE PADS – REAR

Remove and refit   70.40.03

Service tool: Lever 64932392

Removing
1 Jack up rear of car and place on stands.
2 Remove road wheels.
3 Remove clips securing pad mounting pins.
4 Remove mounting pins.
5 Withdraw friction pad.

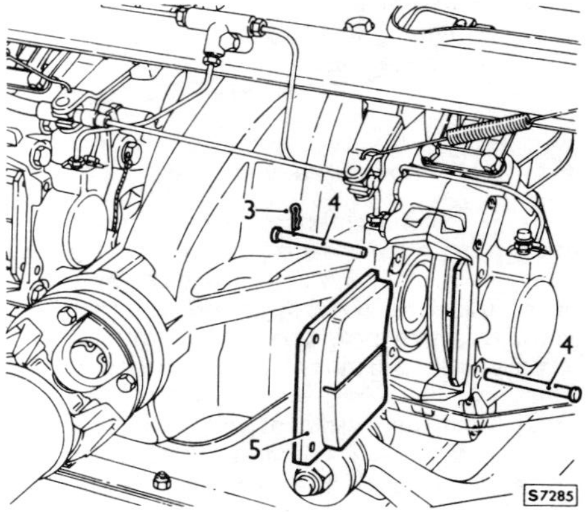

Inspection
6 Examine pads for wear, if thickness of friction pad is less than 0.125 in (3,17 mm) new pads must be fitted.

Refitting
NOTE: Prior to fitting new brake pads it is advisable to reduce level of brake fluid in reservoir.
7 Using service tool 64932392 lever pistons into cylinder bores.
8 Locate brake pads in caliper.
9 Fit mounting pins ensuring upper pin enters caliper from centre line of car and lower pin enters from wheel side of car.
10 Fit retaining clips to mounting pins.
11 Top up fluid reservoir.
12 Reverse operations 1 and 2.
13 Run engine and apply brake pedal several times until pedal feels solid.

70–8

HANDBRAKE PADS

Remove and refit 70.40.04

Removing
1. Remove handbrake caliper, see 70.55.04.
2. Remove nut and spring washer securing pads to brake pad carriers, remove pads.

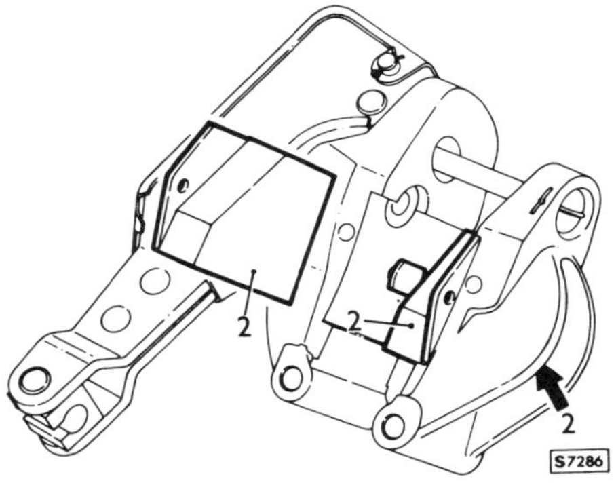

Refitting
3. Holding one pad carrier wind remaining one out two or three turns.
4. Fit new brake pads to pad carriers.
5. Refit handbrake caliper.
6. Operate handbrake several times to adjust pads to correct clearance.

SERVO ASSEMBLY

Remove and refit 70.50.01

Removing
NOTE: Removal of servo assembly is completely covered by operation 70.35.03, Pedal box removal.

BRAKE SERVO – CHECK AND TEST PROCEDURE 70.50.03

The following test and concluded symptoms are only relevant when carried out with the hydraulic braking system in a satisfactory condition.

Servo test and check
1. Jack up front of car and confirm one wheel turns freely. Start engine, allow vacuum to build up and apply brake pedal several times. It should be possible to rotate wheel immediately pedal is released. If brakes bind, a major defect within the servo unit is indicated.
2. With engine running apply brake pedal several times and check operation of pedal. If response is sluggish, check condition of vacuum hoses and servo unit air filter.
3. Allow vacuum to build up, switch off engine and operate brake pedal. Approximately two or three applications should be vacuum assisted, any less indicates a leaking vacuum system or inoperative non-return valve.
4. Switch off engine and operate brake pedal several times to evacuate vacuum in system. Hold a light foot pressure on pedal and start engine. If servo is operating correctly, pedal will fall under existing foot pressure. If pedal remains stationary a leaking vacuum system is indicated.

RESERVAC TANK

Remove and refit 70.50.04

Removing
1. Remove right-hand front wheel.
2. Remove screws securing lower edge of diaphragm panel to front skirt panel.
3. Prise free plastic studs securing top of diaphragm panel to inner wing.
4. Remove diaphragm panel from wheel arch.
5. Remove bolt securing clamp around waist of reservac.
6. Slacken clip securing vacuum hose to tank.
7. Ease reservac tank clear of clamp and disengage from vacuum pipe.
8. Lift reservac from car.

Refitting
9. Reverse operations 1 to 8.

SERVO ASSEMBLY

Overhaul 70.50.06

The servo assembly is a sealed unit and overhaul is not possible, should operation of the servo unit deteriorate to an extent where braking efficiency is affected, see 70.50.03, a new unit must be fitted.

NON-RETURN VALVE

Remove and refit 70.50.15

Removing
1. Slacken clip securing vacuum hose to non-return valve.

Right-hand valves only
2. Detach heater vacuum hose from manifold stub pipe.

All valves
3. Remove valve from manifold, recover washer.

Refitting
4. Blow through valve in both directions checking operation of non-return action.
5. Reverse operations 1 to 3.

FRONT CALIPER

Remove and refit 70.55.02

Removing
1. Disconnect feed pipe at support bracket, plug pipes to prevent loss of fluid and ingress of dirt.
2. Remove locking wire from caliper mounting bolts.
NOTE: Under no circumstances should four setbolts which secure both halves of caliper be removed.
3. Remove caliper mounting bolts, recover and note position of shims located between steering arm and caliper.
4. Withdraw caliper from disc.

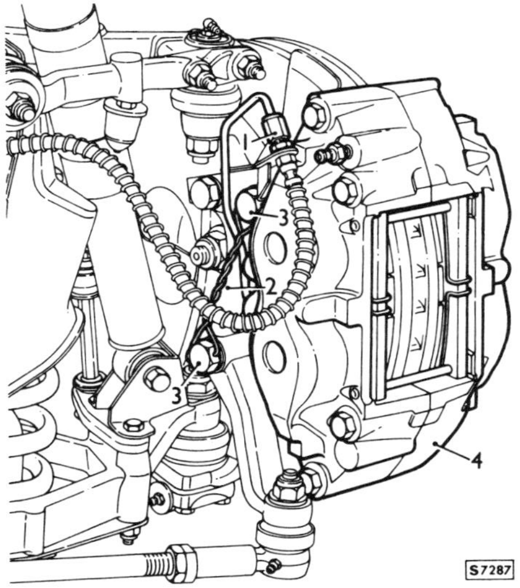

Refitting
5. If new caliper is to be fitted reverse operations 3 and 4 and carry out caliper/disc centralisation. Check and adjust, see 70.10.10. If no adjustment is necessary reverse operations 1 to 4.
6. Where original caliper is refitted reverse operations 1 to 5 ensuring that caliper mounting bolts are torqued to 55 lb ft (7,5 kg m).
7. Bleed brakes, see 70.25.01.

REAR CALIPER

Remove and refit 70.55.03

Removing

1. Remove handbrake caliper, see 70.55.04.
2. Slacken caliper feed pipe union at three-way connector.
3. Disconnect feed pipe at caliper, swing pipe clear of caliper, plug pipes to prevent loss of fluid and ingress of dirt.
4. Remove locking wire securing caliper mounting bolts.

CAUTION: Under no circumstances should caliper be split by removing set bolts securing two halves of caliper.

5. Remove caliper mounting bolts.
6. Slide caliper around brake disc and withdraw through gap exposed by removal of tie plate.
7. Release springs from caliper arms and position handbrake cable clear of calipers.
8. If new caliper is fitted reverse operations 5 and 6 and carry out caliper/disc centralisation. Check and adjust, see 70.10.11. If no adjustment is necessary reverse operations 1 to 4. Where original caliper is refitted reverse operations 1 to 6 ensuring caliper mounting bolts are torqued to 55 lb ft (7,6 kg m).
9. Fit brake friction pads to caliper.
10. Bleed brakes, see 70.25.02.
11. Check camber angle, and adjust if necessary, see 64.25.18.

HANDBRAKE CALIPER

Remove and refit 70.55.04

Removing

1. Raise carpet adjacent to handbrake mounting for access to cable adjusting nuts.
2. Ensure handbrake is released; turn back cable adjusting nuts.
3. Remove nuts and bolts securing tie plate to suspension unit, remove tie plate.
4. Using suitable lever, move caliper arm to centre line of car and disengage cable from arm.
5. Slide rubber cover from outer cable at caliper arm. 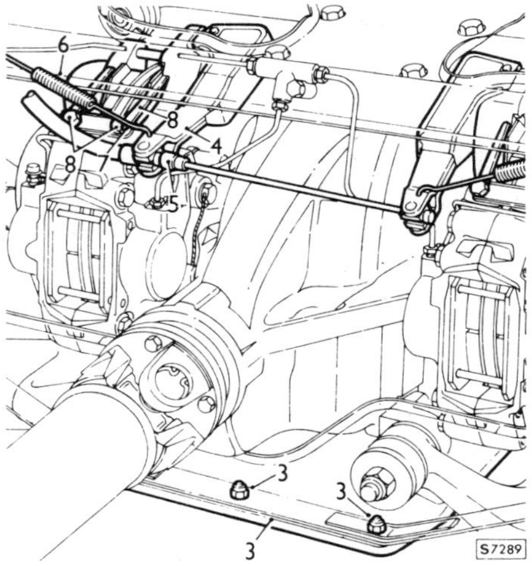
6. Turn down locking tabs securing caliper mounting bolts.
7. Remove mounting bolts, tab washer and retraction plate.
8. Slide caliper around brake disc and withdraw through gap exposed by removal of tie plate.

Refitting

9. Examine pad linings for wear, if lining is less than 0.125 in (3,2 mm) thick, new pad must be fitted, see 70.40.04.
10. Operate caliper arm until adjuster ratchet ceases to click, this adjusts pads to correct clearance.
11. Reverse operations 1 to 6.

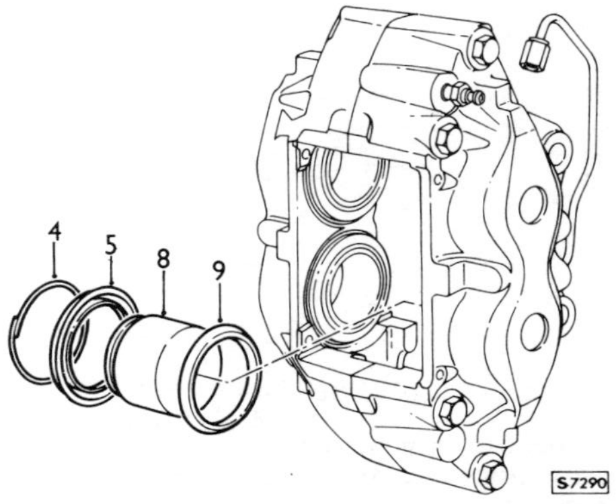

Inspection

10. Using Girling brake cleaner thoroughly clean piston, cylinder bore and seal groove.
11. Examine piston and cylinder bore for signs of corrosion or scratches. Should doubt exist as to condition a new component must be fitted.

Assembling

12. Coat new seals in Girling disc brake lubricant.
13. Using fingers ONLY fit new seals to recess in cylinder bore.
14. Coat piston in clean disc brake lubricant.

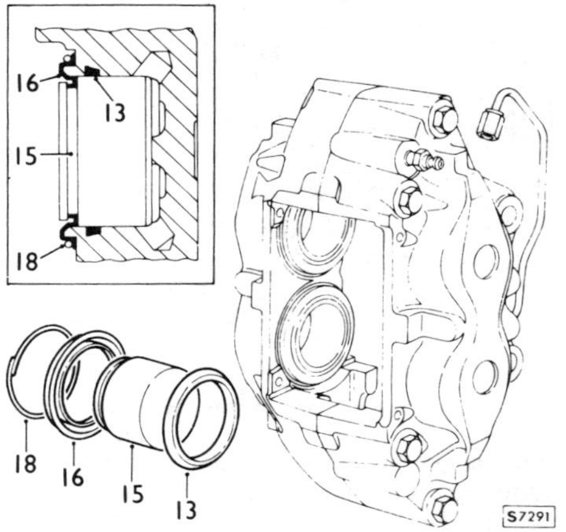

BRAKE CALIPER – FRONT

Overhaul 70.55.13

Service tool: 18G 672 Piston clamp.

1. Remove front friction pads, see 70.40.02.
2. Remove front caliper, see 70.55.02.
3. Thoroughly clean caliper with Girling brake cleaner.

Dismantling

CAUTION: Under no circumstances must caliper halves be separated.

4. Remove spring clips securing piston dust covers.
5. Remove covers from pistons.
6. Fit piston clamp to any half of caliper.
7. To expel pistons carefully feed compressed air into caliper fluid inlet port.
8. Remove pistons from caliper.

WARNING: EXTREME CARE MUST BE TAKEN NOT TO DAMAGE CYLINDER BORE WHEN EXTRACTING SEALS.

9. Carefully prise seals from recess in cylinder wall.

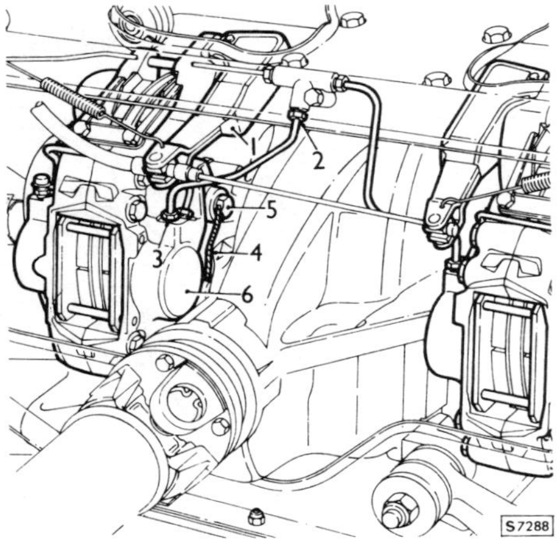

Refitting

NOTE: Prior to refitting handbrake and main caliper friction pads examine pad linings for wear. Where lining is less than 0.125 in (3,2 mm) thick on main caliper and handbrake caliper fitment of new linings is necessary.

70–10

15 Enter pistons into cylinder bores.
16 Fit new dust covers over pistons.
17 Push pistons fully home.
18 Locate dust cover over rim in caliper, secure with spring clips.

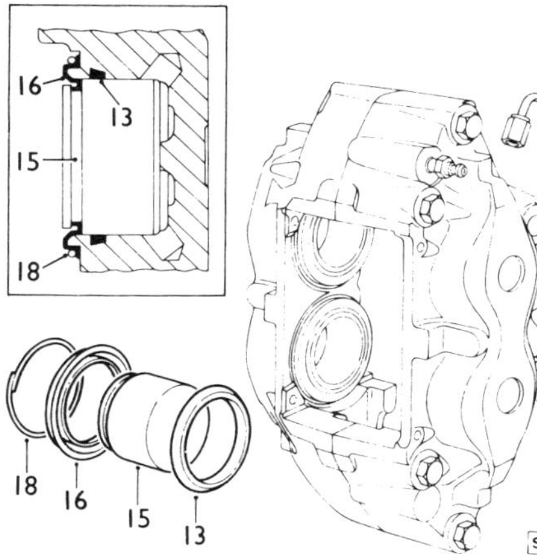

5 Remove dust seal from caliper cylinder bore.
WARNING: EXTREME CARE MUST BE TAKEN NOT TO DAMAGE THE CALIPER CYLINDER BORE WHEN EXTRACTING SEAL.
6 Carefully prise seal from recess in cylinder bore.

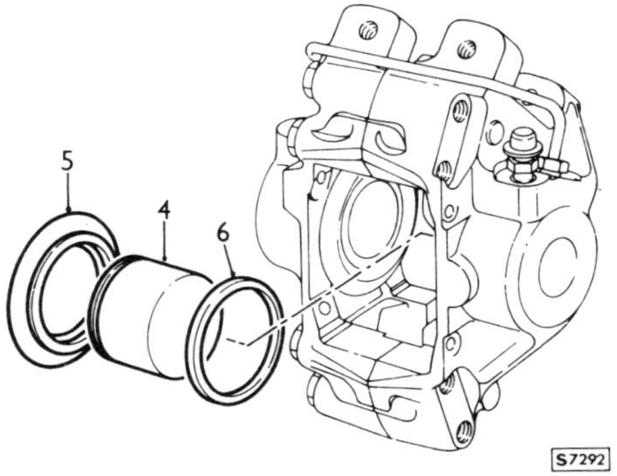

Inspection
7 Using Girling brake cleaner thoroughly clean piston, cylinder bore and seal recess.
8 Examine piston and cylinder for signs of corrosion or scratches. Should doubt in condition exist, a new component must be fitted.

Assembly
9 Coat new seal with Girling disc brake lubricant.
10 Using fingers ONLY fit new seal to recess in cylinder bore.
11 Locate dust cover in outer groove in cylinder bore.
12 Coat piston in clean disc brake lubricant.
13 Enter piston into cylinder bore through dust seal.
14 Locate dust seal into groove in piston.
15 Release piston clamp and fit to opposite side of caliper to press 'serviced' piston fully home.
16 Repeat operations 4 to 15 on remaining cylinder piston.
17 Remove piston clamp.
18 Refit rear brake caliper to car.

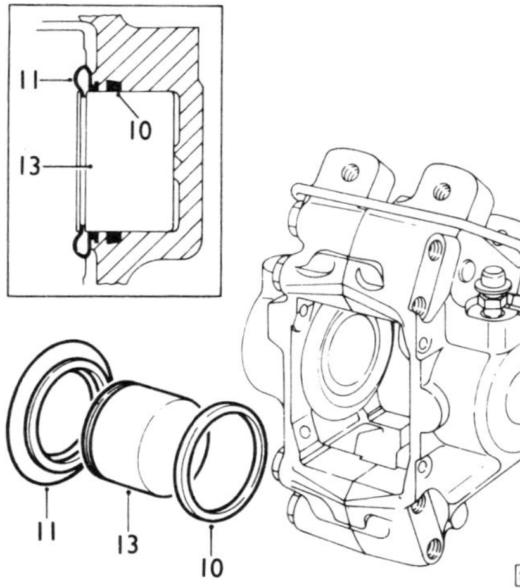

19 Release piston clamp and fit to opposite half of caliper.
20 Repeat operations 5 to 18 on remaining two pistons.
21 Refit caliper to car.

BRAKE CALIPER – REAR

Overhaul   70.55.14

Service tool: 18G 672 Piston clamp.
1 Remove rear brake caliper, see 70.55.03.
2 Thoroughly clean caliper using Girling cleaning fluid.

Dismantling
CAUTION: Under no circumstances must the caliper halves be separated.
3 Fit piston clamp to retain one piston in location.
4 Carefully feed compressed air into caliper fluid inlet port expelling on piston, disengage dust seal and remove one piston.

70–11

FUEL FILLER FLAP

Remove and refit   76.10.25

Removing
1. Open fuel filler flap.
2. Remove bolts securing flap and hinge mechanism to body, remove flap assembly.

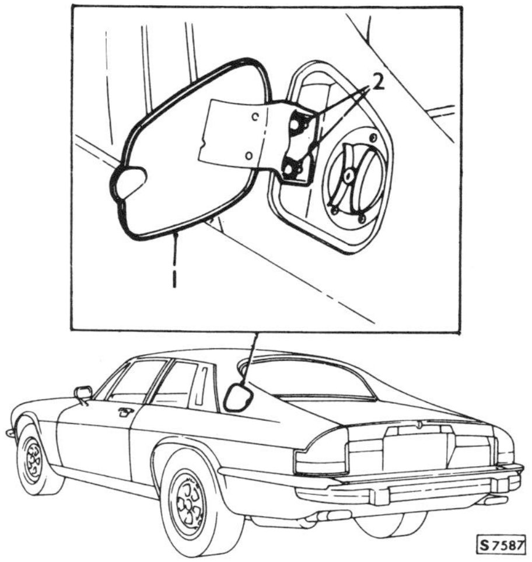

Refitting
3. Refit flap and hinge assembly, do not fully tighten securing bolts.
4. Adjust flap to correct position.
5. Fully tighten securing bolts.

FRONT SPOILER

Remove and refit   76.10.46

Removing
1. Place vehicle on ramp and raise ramp.
2. Remove two screws retaining oil cooler grille and detach grille.
3. Prise out and discard plastic drive fasteners securing spoiler undertray to body and spoiler.
4. Remove screws and detach spoiler undertray.
5. Remove screws securing spoiler.

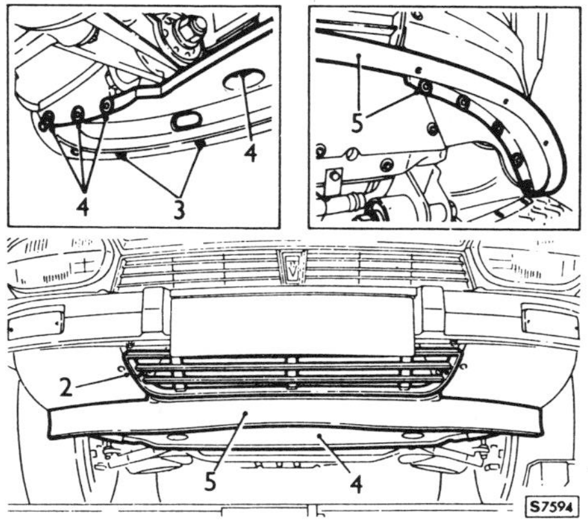

Refitting
6. Offer up spoiler and align holes.
7. Replace and tighten screws securing spoiler.
8. Offer up spoiler undertray and secure with new drive fasteners.
9. Replace and tighten screws securing spoiler undertray.
10. Fit oil cooler grille and secure with two screws.
11. Remove car from ramp.

'A' POST TRIM PAD

Remove and refit   76.13.07

Removing
1. Unclip section of crash roll adjacent to 'A' post trim pad.
2. Remove screws securing trim pad to 'A' post, lift trim pad from car.

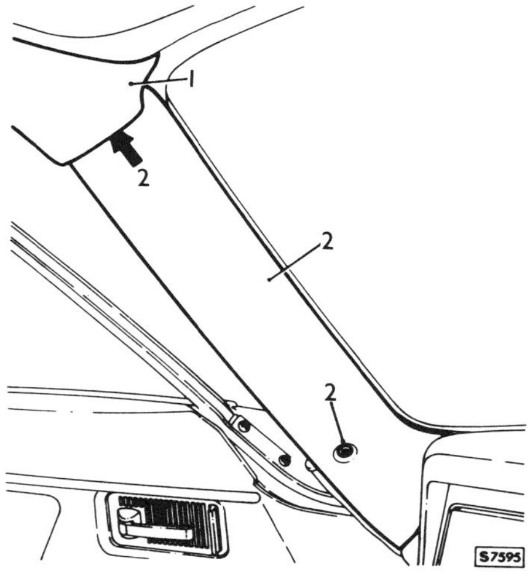

Refitting
3. Reverse operations 1 and 2.

'B' POST TRIM PAD

Remove and refit   76.13.08

Removing
1. Press seat belt upper fixing cover upwards to release from belt lug.
2. Pull cover clear of lug and remove bolt securing seat belt to 'B' post, recover washers and spacer.
3. Manoeuvre 'B' post trim pad free from crash roll.

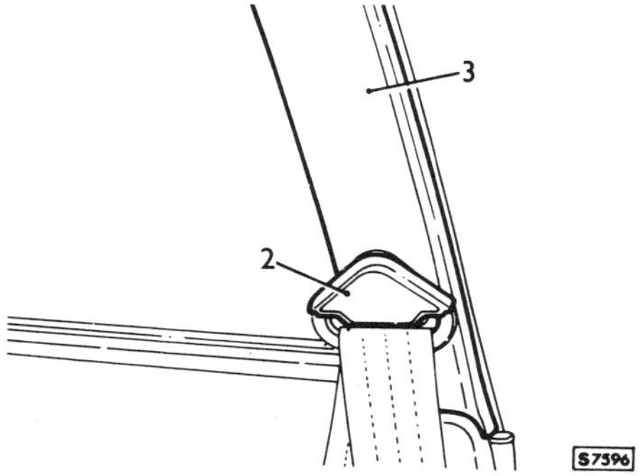

4. Lift trim pad from car.

Refitting
5. Reverse operations 1 to 4.

REAR QUARTER TRIM PAD – LOWER

Remove and refit   76.13.12

Removing
1. Remove rear seat cushion, see 76.70.37.
2. Remove two screws at base of rear seat back and lift out seat back.
3. Lift off rear arm rest adjacent to panel.
4. Remove cover from front seat belt mounting on 'B' post, detach bolt and remove seat belt, mounting bolt, spacers and spring washers.
5. Detach trim pad from 'B' post and remove four screws securing side trim panel.
6. Unscrew and remove bolt attaching front seat harness to inner sill. Lift out belt end fitting and collect spacer and spring.
7. Remove trim pad, threading belt and fittings through hole in pad as it is withdrawn.

continued

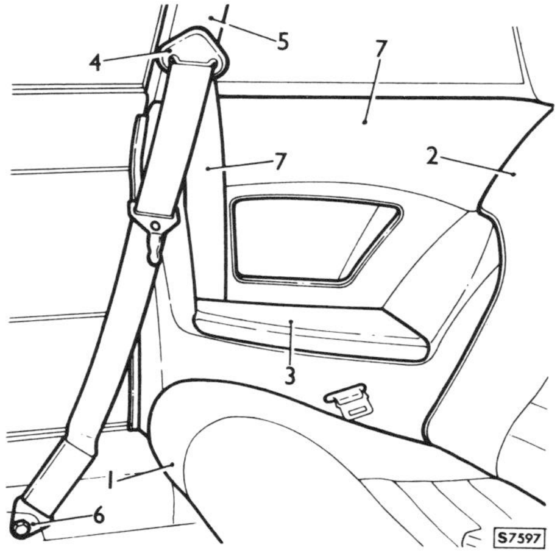

76–1

Refitting

8 Place trim pad near its location in car and thread seat belt and fittings through hole.
9 Fit spring to belt end fitting, position spacer and replace bolt. Insert bolt in inner sill and tighten.
10 Replace trim panel and secure with four screws.
11 Replace trim on 'B' post.
12 Raise front seat belt lug to mounting on 'B' post, insert bolt in lug, place spring washer and spacers on bolt, and tighten bolt in mounting. Replace plastic cover.
13 Push arm rest back into location.
14 Replace seat back and secure with two screws.
15 Replace seat cushion, see 76.70.37.

REAR QUARTER TRIM PAD – UPPER

Remove and refit 76.13.13

Removing

1 Remove rear seat squab, see 76.70.38.
2 Unclip and remove rear parcel shelf.
3 Pull cantrail crash roll fastener away from cantrail near trim pad.
4 Remove two screws securing trim pad. Unclip and ease trim pad from location.
5 Disconnect leads from interior light and remove trim pad.
6 Detach interior light from pad.

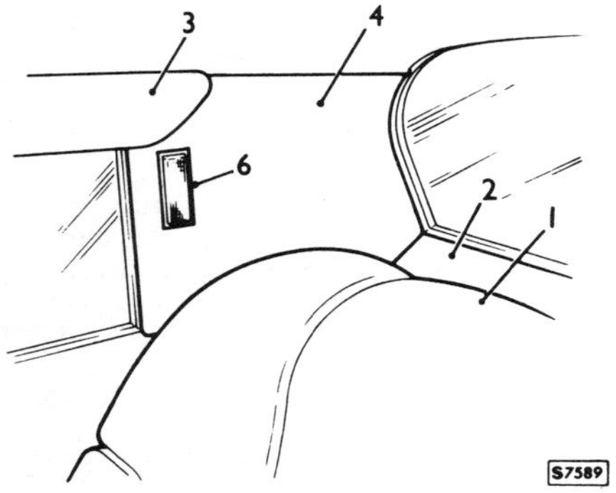

Refitting

7 Fit new clip to replacement trim pad and transfer interior light to it.
8 Offer up pad and connect leads.
9 Place panel in correct location and secure with two screws. Tighten screws.
10 Press cantrail crash roll fastener back into position.
11 Replace rear parcel shelf.
12 Refit rear seat squab, see 76.70.38.

BONNET

Remove and refit 76.16.01

Removing

1 Remove bonnet stay, see 76.16.14.
2 Remove radiator grille, see 76.55.03.
3 Remove four bolts securing bonnet to hinges.
4 Lift bonnet from car.

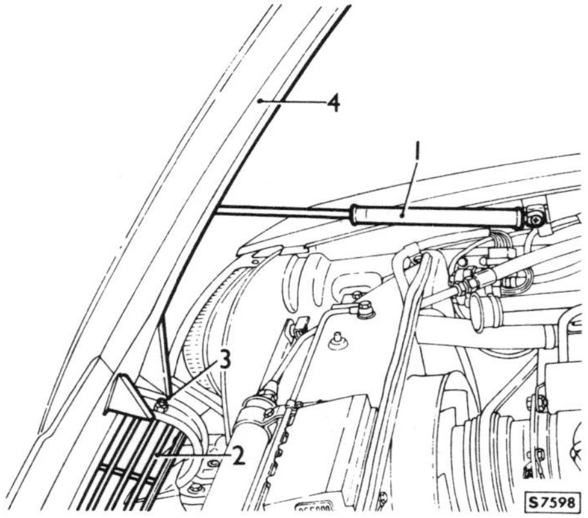

Refitting

5 Fit bonnet to hinges, do not fully tighten securing bolts.
6 Close bonnet and if necessary adjust position to centralise between wing valances.
7 Open bonnet and fully tighten securing bolts.
8 Reverse operations 1 and 2.

BONNET HINGES – SET

Remove and refit 76.16.12

Removing

1 Remove bonnet assembly, see 76.16.01.
2 Mark location of hinges to crossmember.
3 Remove four bolts securing hinges to crossmember; remove bonnet hinges.

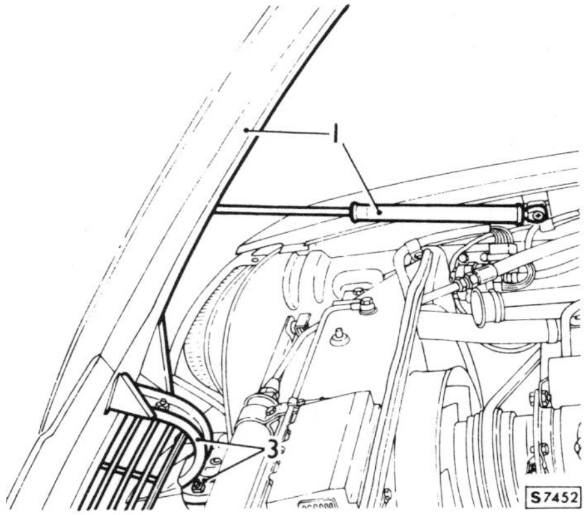

Refitting

4 Locate new hinges to crossmember where marked in operation 7 above.
5 Fit and tighten four bolts securing hinges to crossmember.
6 Locate bonnet assembly to hinges; fit but do not tighten four securing bolts.
7 Align bonnet assembly; tighten four securing bolts.
8 Reverse operations 1 and 2.

BONNET STAY

Remove and refit 76.16.14

Removing

1 Remove screws securing stay to bonnet and wing valance.
2 Remove stay, recovering back plates and seating blocks.

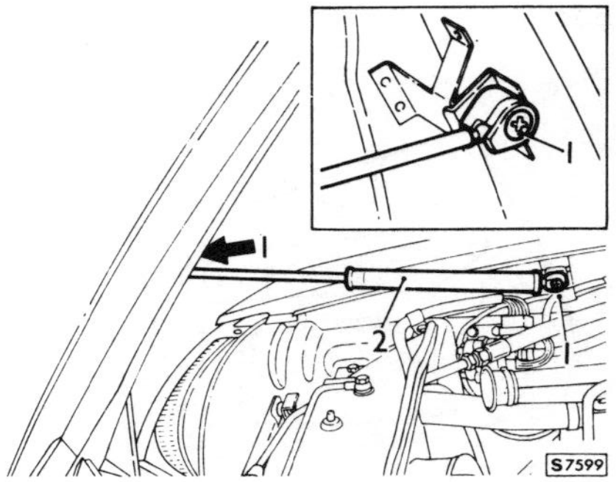

Refitting

3 Reverse operations 1 and 2.

BONNET LOCK

Adjust 76.16.20

1 Slacken but do not remove two setscrews securing bonnet lock to back plate.
2 Raise or lower lock as required (screw locating holes are elongated).
3 Pinch up setscrews to secure lock.
4 Close bonnet; check adjustment by operating bonnet release handle.
5 Raise bonnet, re-adjust as necessary.
6 Fully tighten securing setscrews.

76–2

BONNET LOCK – LEFT HAND

Remove and refit 76.16.21

Removing

1 Slacken bolt securing control cable to bonnet release handle nipple; release cable.
2 Slacken bolt securing control cable to lock pivot; release cable.
3 Remove four screws securing lock assembly to bulkhead.
4 Remove lock assembly complete with lock to release handle control cable.
5 Remove two bolts securing lock to backplate; remove backplate.

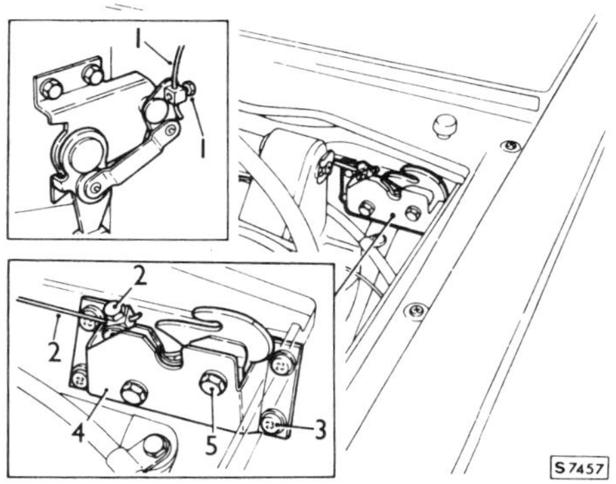

Refitting

7 Reverse operations 1 to 6 lightly greasing control cables.
8 Adjust bonnet lock, see 76.16.20.
9 Adjust bonnet lock control cable, see 76.16.28.

BONNET LOCK – RIGHT HAND

Remove and refit 76.16.26

Removing

1 Remove two screws securing cold start relay cover to valance; remove cover.
2 Remove four screws securing lock assembly to bulkhead.
3 Slacken bolt securing control cable to left-hand bonnet lock pivot; release cable.
4 Remove lock assembly complete with control cable.
5 Remove control cable.
6 Remove two bolts securing lock to backplate; remove backplate.

Refitting

7 Reverse operations 1 to 6 lightly greasing control cable.
8 Adjust bonnet lock, see 76.16.20.
9 Adjust bonnet lock control cable, see 76.16.28.

BONNET LOCK CONTROL CABLE

Adjust 76.16.28

1 Slacken bolt securing control cable at left-hand bonnet lock. 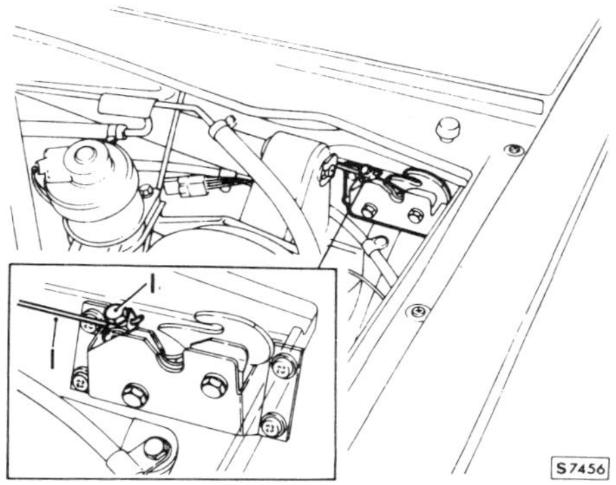
2 Adjust control cable as required by tightening or slackening in nipple.
3 Tighten bolt.
4 Close bonnet and check control cable adjustment by operating release handle.
5 Re-adjust as necessary.

BONNET LOCK TRIGGER/RELEASE HANDLE

Remove and refit 76.16.30

Removing

1 Remove driver's footwell side trim pad, see 76.13.01.
2 Move release handle to bonnet open position.
3 Slacken control cable clamping screw; release control cable.
4 Move release handle to bonnet closed position.
5 Remove three bolts and washers securing release handle mounting bracket to body. 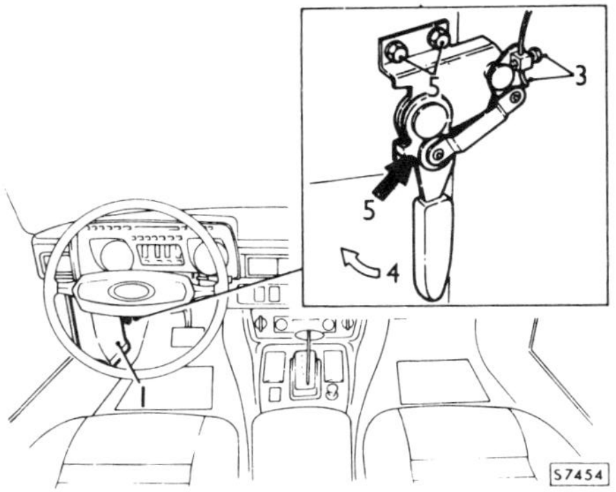
6 Remove release handle assembly.

Refitting

7 Reverse operations 1 to 6.
8 Check bonnet alignment.
9 Adjust control cable, see 76.16.28.

BOOT LID

Remove and refit 76.19.01

Removing

1 Peel back boot carpet to gain access to electrical harness.
2 Disconnect harness at block connector and single bullet connector.
3 Remove plastic clips securing harness to right hand boot lid hinge.
4 Remove four bolts and washers securing boot lid to hinges.
5 Remove boot lid assembly.

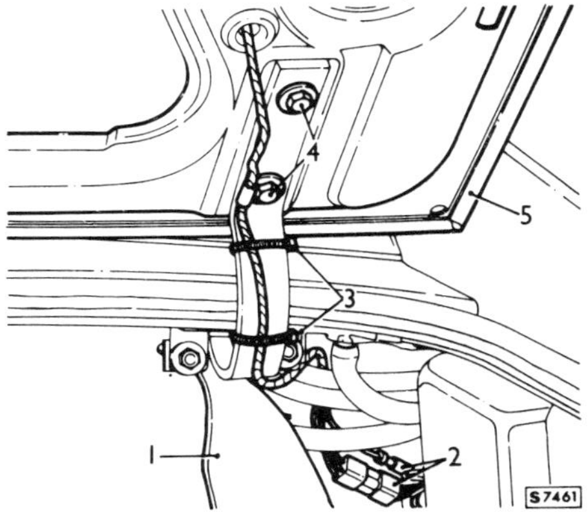

Refitting

6 Locate boot lid to hinges; fit but do not fully tighten four bolts and washers.
7 Close boot lid and check alignment, adjust by moving boot lid as required on hinges elongated bolt holes.
8 Fully tighten hinge securing bolts.
9 Reverse operations 1 to 3.

BOOT LID SEAL

Remove and refit 76.19.06

Removing

1 Remove screws securing finisher panel to boot aperture valance, lift panel from boot.
2 Peel boot seal from rail mounting.

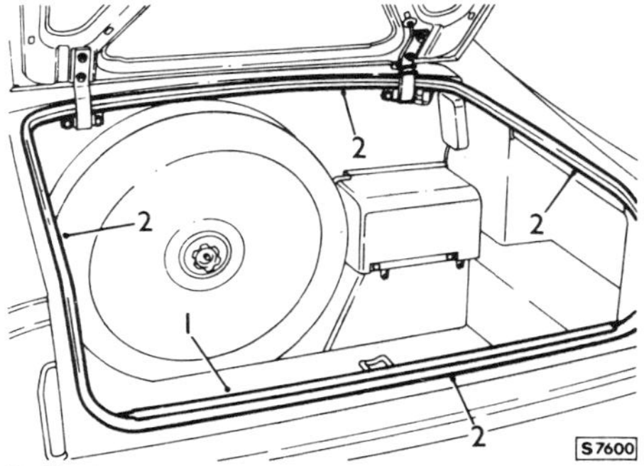

Refitting

3 Reverse operations 1 and 2.

76–3

BOOT LID HINGES

Remove and refit 76.19.07

Removing
1. Remove boot lid, see 76.19.01.
2. Remove rear parcel tray shelf, see 76.67.06.
3. Remove bolt and washer securing boot hinges to rear parcel tray shelf plate.
4. Remove two bolts and washers securing hinges to mounting brackets.
5. Remove hinge assemblies.

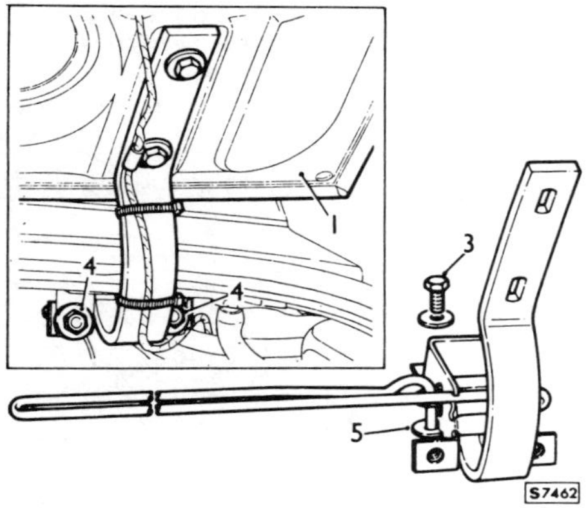

Refitting
Reverse operations 1 to 5.

BOOT LID LOCK

Remove and refit 76.19.11

Removing
1. Remove clip securing boot lid handle operating lever to boot lid lock operating rod; disconnect rod from lever.
2. Remove three bolts and washers securing lock assembly to boot lid; remove lock assembly.
3. Release spring clip securing operating rod to lock operating lever; remove operating rod.

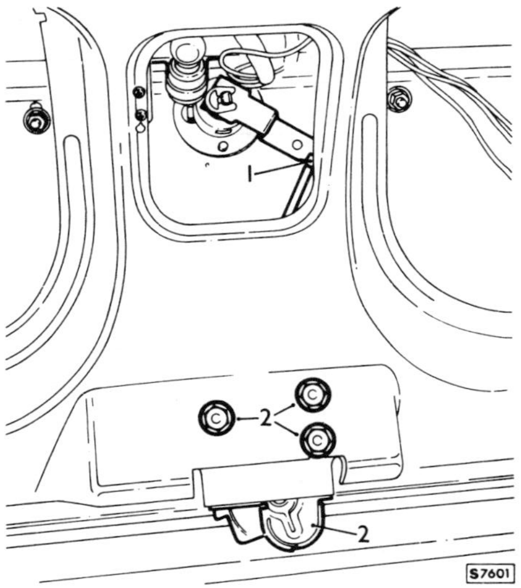

Refitting
4. Smear lock operating mechanism with suitable grease.
5. Refit operating rod to lock operating lever using new spring clip.
6. Locate lock assembly to boot lid, fit but do not tighten three securing bolts and washers.
7. Refit operating rod to boot lid handle operating lever using new spring clip.
8. Close boot lid; check adjustment of lock; adjust as necessary.
9. Fully tighten three bolts securing lock assembly to boot lid.

BOOT LID LOCK STRIKER

Remove and refit 76.19.12

Removing
1. Remove screws securing finisher panel to boot aperture sill, lift panel from boot.
2. Mark striker legs along top face of clamp plate for reference when refitting.
3. Slacken bolts securing striker clamp plate.
4. Slide striker free from clamp.

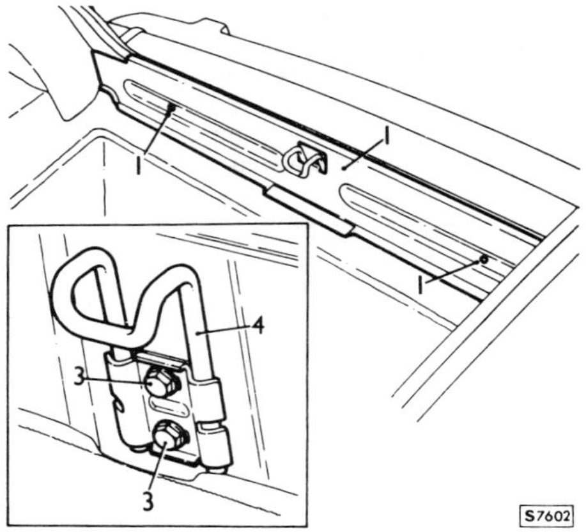

Refitting
5. Locate striker behind clamp plate aligning scribed marks to top face of plate.
6. Tighten clamp securing bolts.
7. Close boot lid, adjust position of striker should more than push effort be required.
8. Refit finisher panel to boot aperture sill.

BOOT LID HANDLE/LOCK ASSEMBLY

Remove and refit 76.19.17

Removing
1. Remove number plate and reverse lamp assemblies, see 86.40.87.
2. Remove spring clip securing boot lid handle operating lever to boot lid lock operating rod; disconnect rod from lever.
3. Remove two screws and spring washers securing handle/lock assembly to boot lid.
4. Remove plate from rear of assembly.
5. Remove handle/lock assembly by manoeuvring through boot lid aperture.

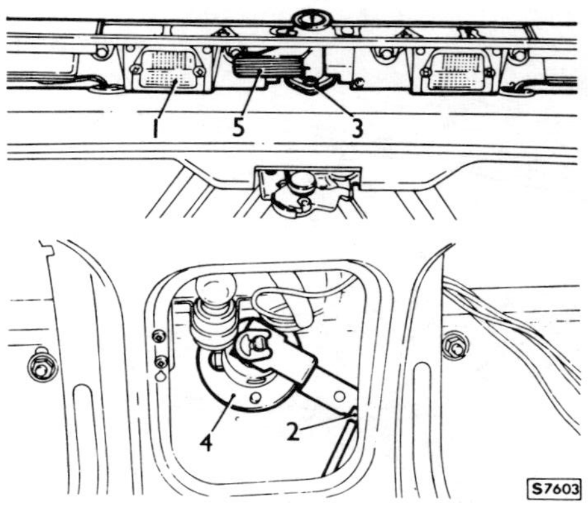

FRONT BUMPER CENTRE SECTION

Remove and refit 76.22.11

Removing
1. Remove six screws securing radiator grille, remove grille.
2. Remove seven nuts, bolts and washers securing centre section to beam.
3. Remove centre section from beam.

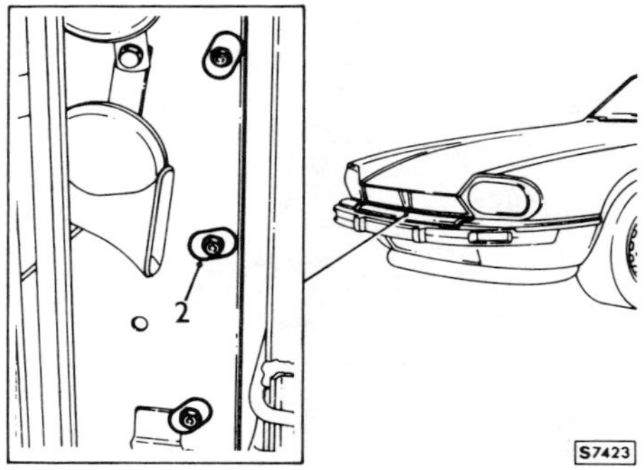

Refitting
Reverse operations 1 to 3.

76—4

REAR BUMPER CENTRE SECTION

Remove and refit   76.22.12

Removing
1. Remove five nuts and oval washers securing centre section to beam.
2. Remove centre section from beam.

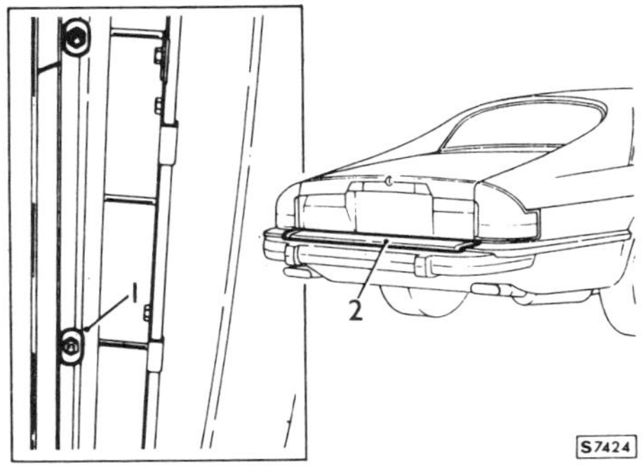

Refitting
Reverse operations 1 and 2.

REAR BUMPER SIDE SECTION

Remove and refit   76.22.13

Removing
1. Ease back luggage compartment side carpet.
2. Remove three nuts and washers securing side section to body.
3. Remove side section.

Refitting
Reverse operations 1 to 3.

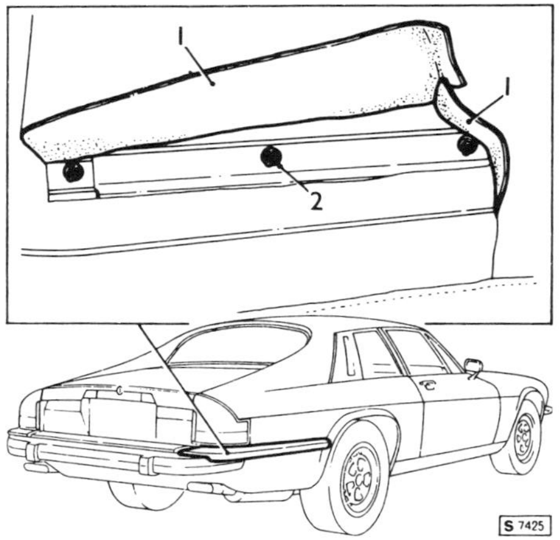

ENERGY ABSORBING BEAM – FRONT

Remove and refit   76.22.26

Removing
1. Remove front bumper centre section, see 76.22.11.
2. Remove front energy absorbing beam cover, see 76.22.28.
3. Remove two nuts, bolts and washers securing beam to struts.
4. Remove lower spacers from beam.
5. Remove energy absorbing beam.
6. Remove four screws securing spacer blocks to beam; remove spacer blocks.
7. Remove flasher lamp securing screw spire nuts.

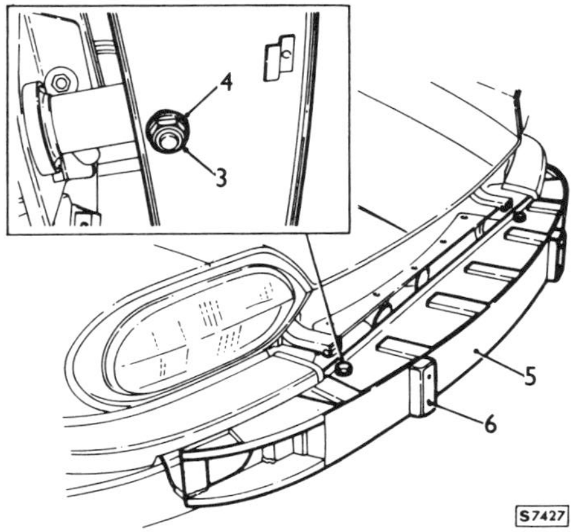

Refitting
Reverse operations 1 to 7.

FRONT BUMPER QUARTER SECTION

Remove and refit   76.22.16

Removing
1. Remove four bolts and oval washers securing quarter section to body.
2. Remove quarter section.

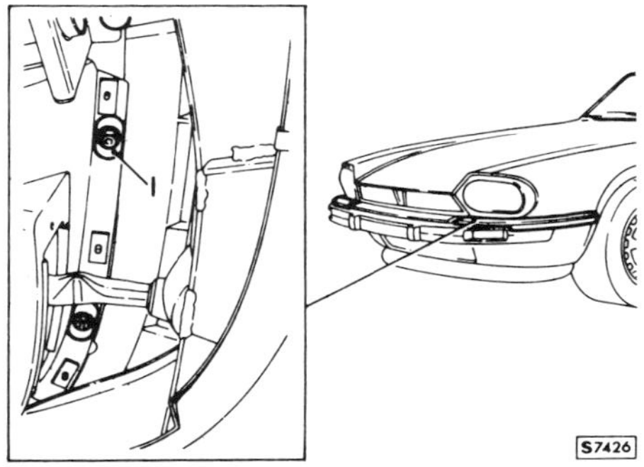

Refitting
Reverse operations 1 and 2.

4. Remove lower spacers from beam.
5. Remove energy absorbing beam.

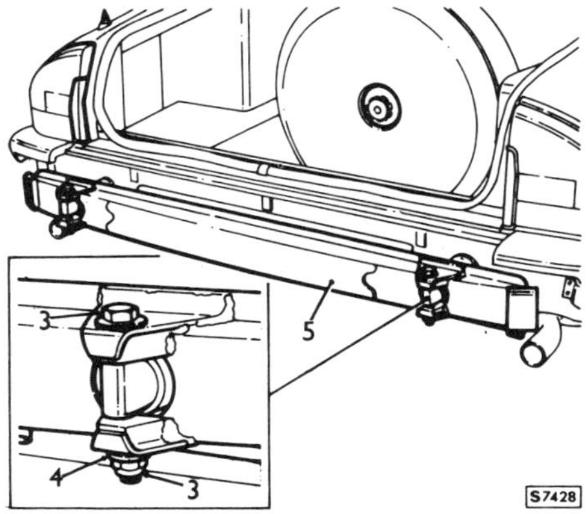

Refitting
Reverse operations 1 to 5.

ENERGY ABSORBING BEAM COVER – FRONT

Remove and refit   76.22.28

Removing
1. Remove front bumper centre section, see 76.22.11.
2. Remove front flasher lamp assemblies, see 86.40.42.
3. Remove five screws securing cover to beam.

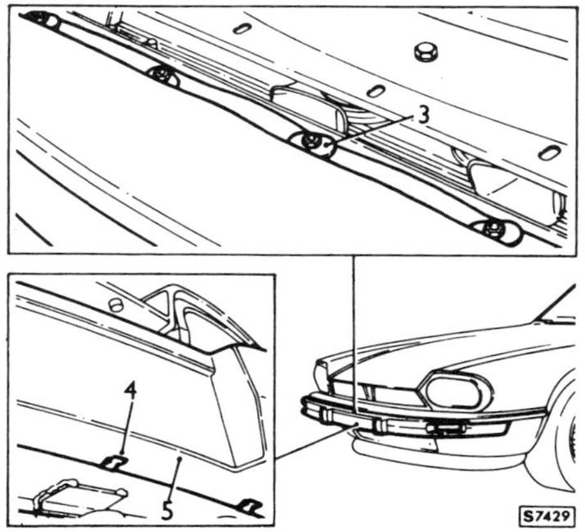

ENERGY ABSORBING BEAM – REAR

Remove and refit   76.22.27

Removing
1. Remove rear bumper centre section, see 76.22.12.
2. Remove rear energy absorbing beam cover, see 76.22.29.
3. Remove two nuts, bolts and washers securing beam to struts.

76–5

4 Disengage locating clips from beam cover.
5 Pull cover from beam.

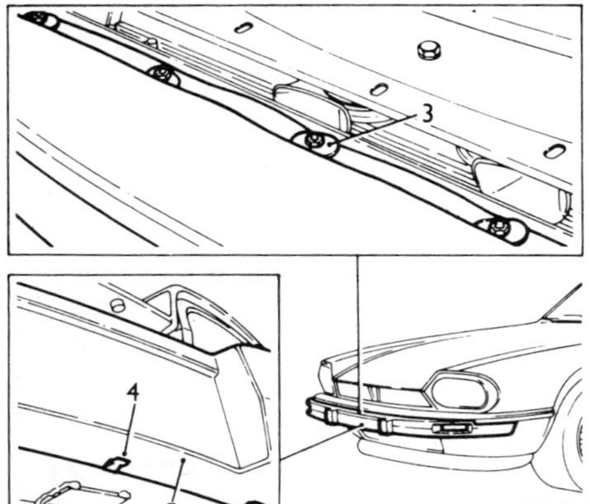

3 Slacken seven screws securing lower clamps, disengage cover.
4 Release cover from wing mounting and remove cover.

Refitting
Reverse operations 1 to 4.

ENERGY ABSORBING STRUT – REAR

Remove and refit 76.22.32

Removing
1 Remove rear energy absorbing beam, see 76.22.32.
2 Peel back carpet to gain access to strut securing nut.
3 Remove nut and washer securing strut to body.
4 Remove strut assembly.
5 Remove strut mounting rubber.

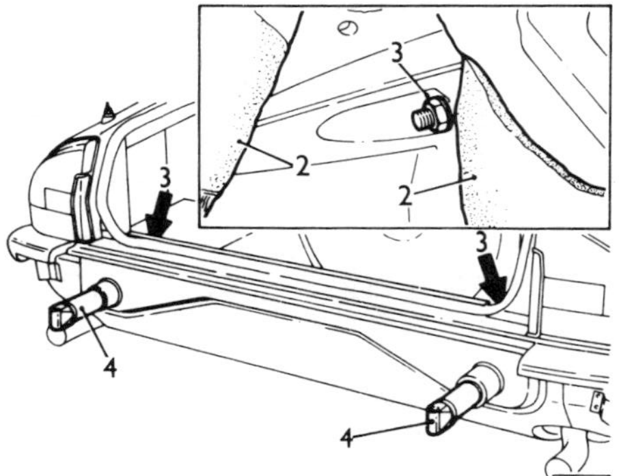

Refitting
Reverse operations 1 to 5.

ENERGY ABSORBING BEAM COVER – REAR

Remove and refit 76.22.29

Removing
1 Remove rear bumper centre section, see 76.22.12.
2 Remove four screws securing upper clamps.

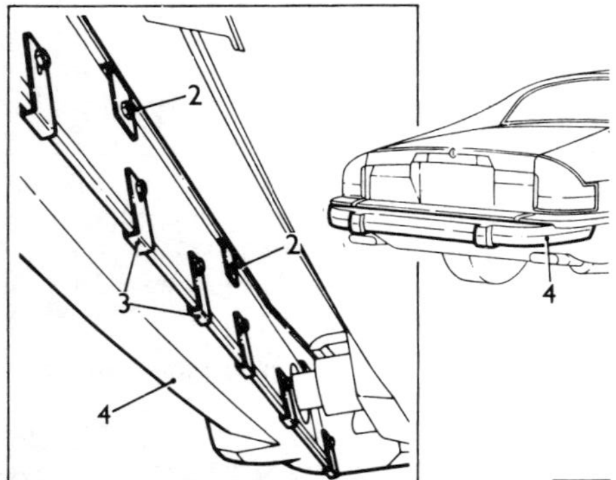

3 Remove four screws securing side air vent grilles and forward side panels and grilles.
4 Unscrew gear control knob.
5 Raise stowage lid to improve access to central screw, and remove three screws securing switch panel to console.
6 Ease switch panel clear of console, noting position of wiring.
7 Disconnect 16 wiring connectors and remove switch panel.
8 Remove two screws from under top edge of control panel location, securing front end of console.
9 Move seats fully forward, tilt seat backs forward and turn down carpet on tunnel near bottom rear corners of console. Remove two screws exposed under carpet.
10 Remove two screws securing stop light failure sensor (forward of stowage) and ease sensor and wiring aside.

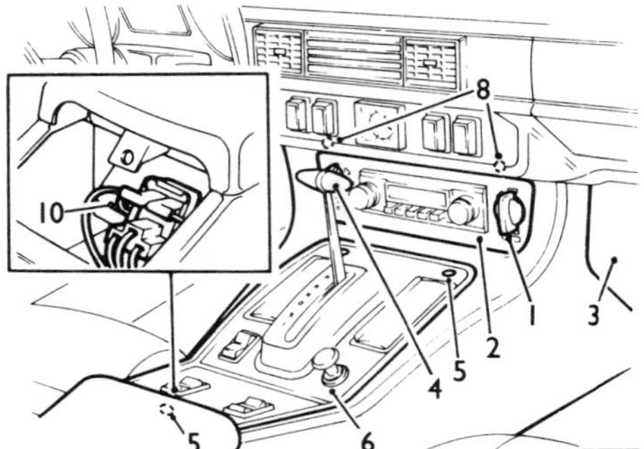

11 Remove console assembly.
12 Remove screws securing stowage lid hinge and check strap to console and detach stowage lid.
13 Remove three further screws securing stowage trough to console, and detach stowage.
14 Remove four screws securing air vent box to console, below stowage trough, and detach air box.
15 Invert console, bend back lugs of rear ash tray and detach ash tray and surround panel.

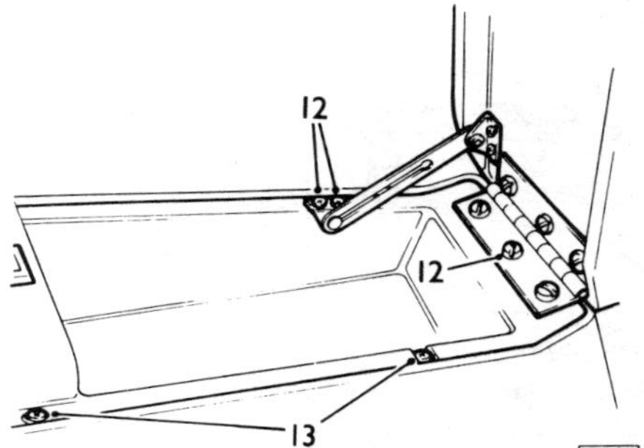

Refitting
16 Place ash tray surround panel, followed by ash tray, in position on replacement console and secure by bending lugs.
17 Place air vent box in position and secure with four screws.
18 Place stowage trough in position and insert three screws.
19 Replace stowage lid by fixing hinge to console with three screws.
20 Secure check strip with two screws and close lid.
21 Place console in car and pass temperature and fan control panel through console aperture.
22 Connect air tubes to vent box and insert four console securing screws.
23 Tighten screws and replace carpet.
24 Open stowage lid and replace stop light failure sensor. Fit and tighten two screws.
25 Refit temperature and fan control panel, tighten brass fasteners and replace knobs.
26 Reconnect 16 wiring connectors to switch panel, replace panel and secure with three screws. Close stowage lid.
27 Refit gear control knob.
28 Replace air vent grilles and forward side panels and secure with four screws.
29 Return seats to normal position.

CONSOLE ASSEMBLY – FRONT

Remove and refit 76.25.01

Removing
1 Pull off control knobs from temperature and fan controls.
2 Unscrew two brass fasteners below control knobs and detach control panel.

76-6

AUTOMATIC TRANSMISSION SELECTOR QUADRANT

Remove and refit   76.25.08

Removing
1. Unscrew left and right hand sections of selector lever handle.
2. Remove screws securing console finisher panel to console.
3. Raise panel from console. 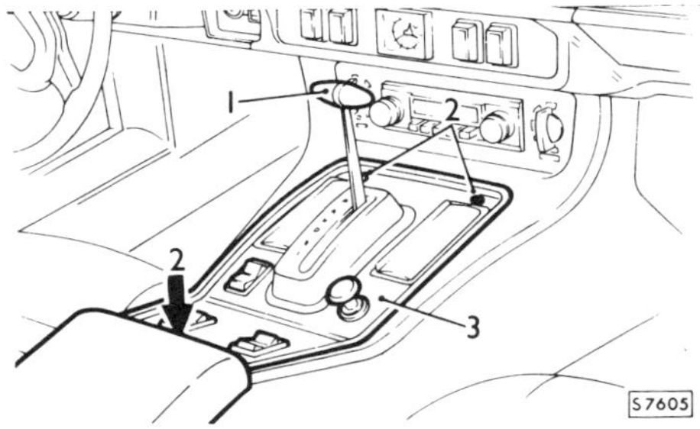
4. Disconnect electrical leads from cigar lighter.
5. Ensuring leads to window lift and door lock switches are not detached lift panel over selector lever, lay panel to one side.
6. Remove nuts securing quadrant cover to mounting plate, lift earth lead from mounting stud.
7. Remove cover from quadrant. 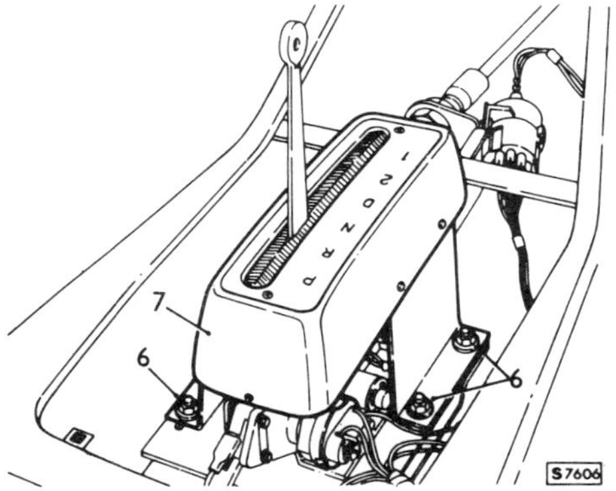
8. Noting location detach electrical leads from reverse switch and inhibitor switch.
9. Remove split pin and washer securing transmission selector cable to quadrant lever, detach cable from mounting pin.
10. Remove bolts securing opticell unit and selector cable mounting bracket to quadrant.
    NOTE: Position of quadrant bracket on mounting studs should be marked for reference when refitting.
11. Remove four bolts securing quadrant assembly to transmission tunnel.
12. Lift quadrant from car.

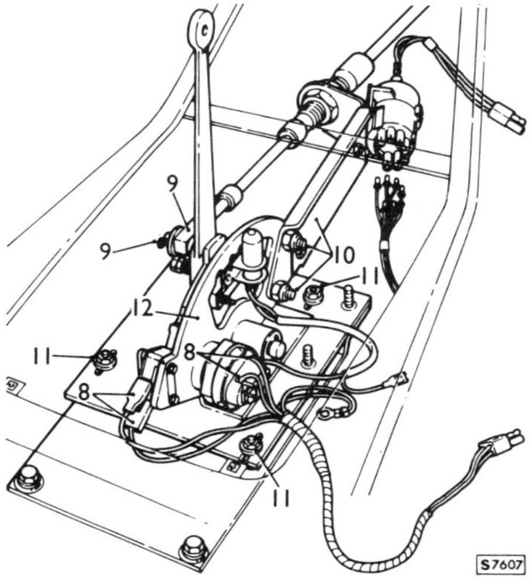

Refitting
13. Reverse operations 6 to 12.
14. Check gear selector cable adjustment, see 44.15.08.
15. Reverse operations 1 to 5.

DOOR

Remove and refit   76.28.01

Removing
NOTE: Door **MUST** be adequately supported in the half open position during this operation.
1. Remove inertia switch cover (driver's door only).
2. Release inertia switch from spring clips; swing switch aside (driver's door only).
3. Remove two screws securing footwell side trim pad.
4. Remove three screws securing underscuttle casing.
5. Remove underscuttle casing.
6. Remove footwell side trim pad.
7. Remove two screws securing door lock solenoid relay to body; ease relay aside.
8. Disconnect six door harness electrical connectors and earth from door lock solenoid relay mounting bracket. 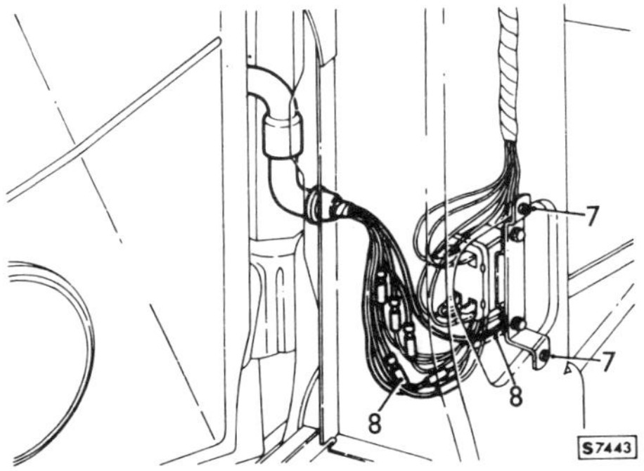
9. Disconnect radio cable connector (if fitted).
10. Remove six nuts and washers securing hinge plates to body; remove hinge plates.
11. With assistance, remove door assembly.
12. Transfer components to new door unit as required.

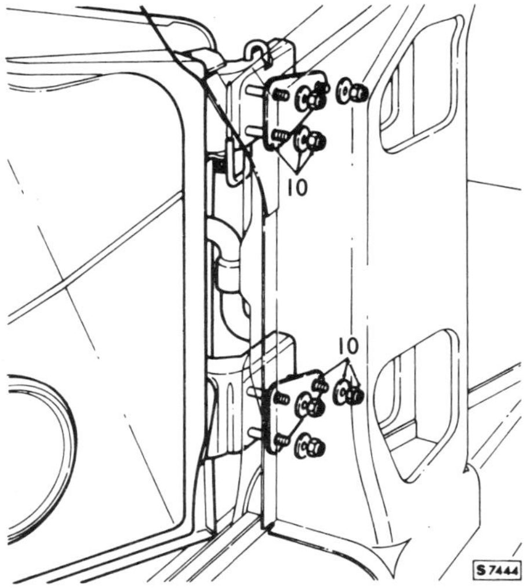

Refitting
13. Locate new door assembly and hinges to body; secure using one nut and washer on each hinge, do not tighten fully.
14. Remove support.
15. Close door, check alignment; re-align door as required.
16. Fit remaining nuts and washers and tighten fully.
17. Reverse operations 1 to 9. Ensure button on inertia switch is depressed.

DOOR HINGE

Remove and refit   76.28.42

Removing
NOTE: Door **MUST** be adequately supported in the half open position during this operation.
1. Remove door trim pad, see 76.34.01.
2. Remove inertia switch cover (driver's door only).
3. Release inertia switch from spring clips; swing switch aside (driver's door only).
4. Remove two screws securing footwell side trim pad.

continued

76–7

3 Using suitable tool prise rear quarter glass and rubber from body.
4 Remove seal from glass.

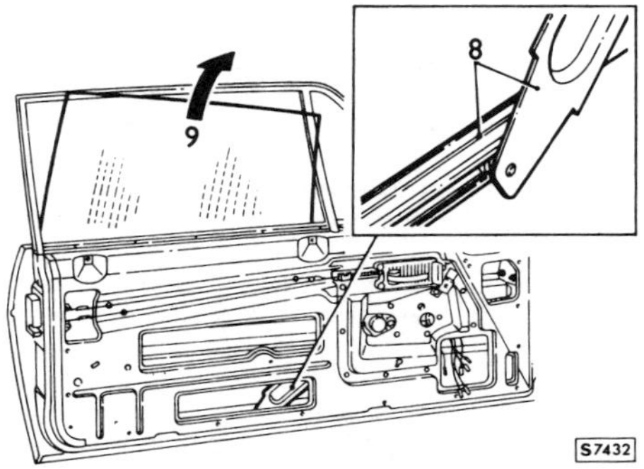

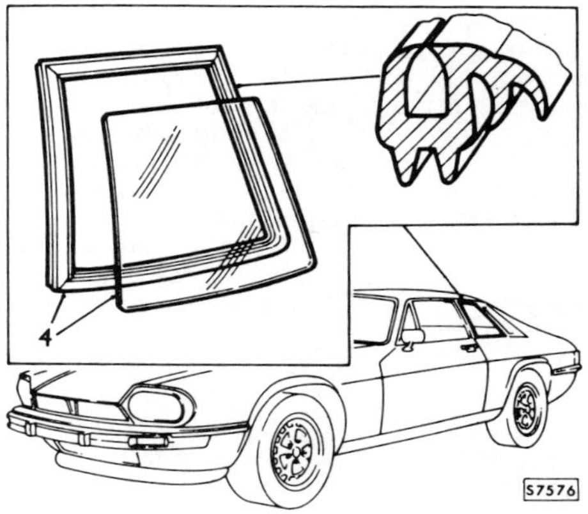

8 Disconnect window lift remote control mechanism from window glass by tilting glass forward.
9 With glass still in the tilted position, carefully lift clear of door.

Refitting
Reverse operations 1 to 9.

QUARTERLIGHT – FRONT

Remove and refit   76.31.29

Removing
1 Remove two screws securing quarterlight from the door panel.
2 Lower door glass.
3 Ease aside door glass rubber.
4 Remove two screws securing quarterlight to door glass frame.
5 Remove quarterlight assembly.
6 Remove quarterlight glass and rubber from frame.

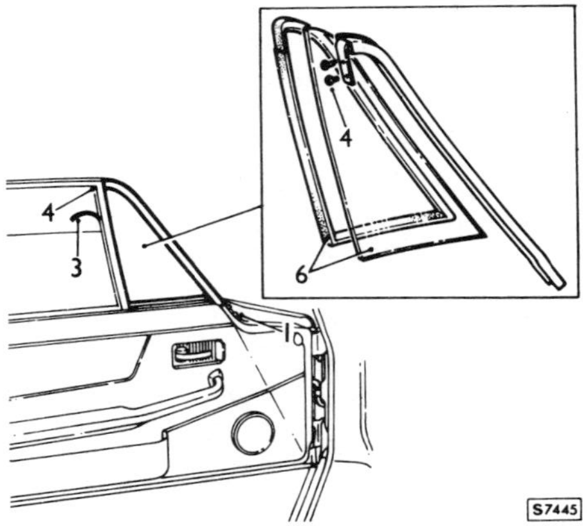

Refitting
5 Apply sealant into locating grooves of new seal.
6 Fit glass to seal.
7 Fit glass and seal to quarterlight location.

DOOR GLASS WEATHERSTRIP

Remove and refit   76.31.50

Removing
1 Remove front quarterlight, see 76.31.29.

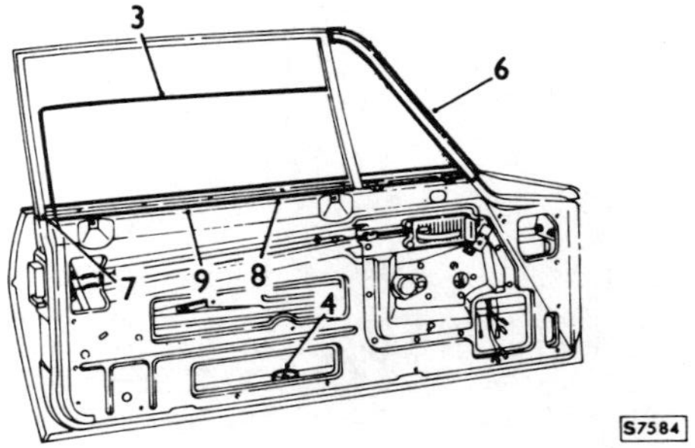

DOOR GLASS – SIDE FRONT – EACH

Remove and refit   76.31.01

Removing
1 Remove door trim pad, see 76.34.01.
2 With glass in raised position remove two screws and washers securing window channel to door.
3 Ease rubber clear of channel.
4 Remove window channel.
5 Remove two bolts, plain and shakeproof washers securing window stop to base of door.
6 Remove window stop.
7 Lower door glass.
NOTE: When lowering window glass care should be taken to guide glass by hand.

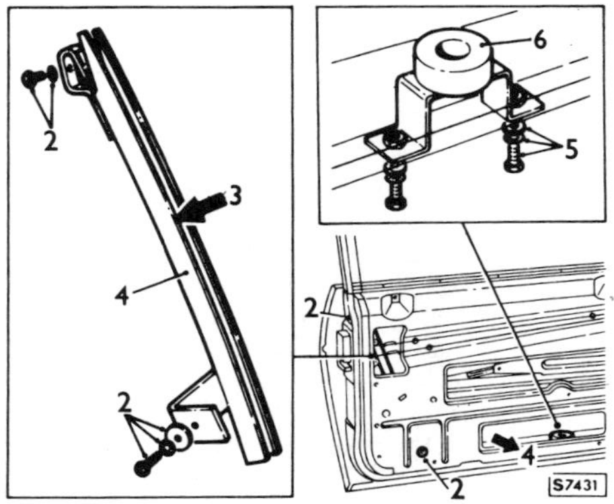

QUARTERLIGHT – REAR

Remove and refit   76.31.31

Removing
1 Remove rear quarter upper trim pad, see 76.13.13.
2 Remove rear quarter lower trim pad, see 76.13.12.

Refitting
Reverse operations 1 to 6.

5 Remove three screws securing underscuttle casing.
6 Remove underscuttle casing.
7 Remove footwell side trim pad.
8 Remove two screws securing door lock solenoid relay to body; ease relay aside.
9 Remove sponge rubber packing from access aperture (upper hinge only).
10 Remove three nuts and washers securing hinge plate to body; remove hinge plate.
11 Remove four nuts and washers (three on lower hinge) securing hinge plate to door; remove hinge plate.
12 Remove hinge.

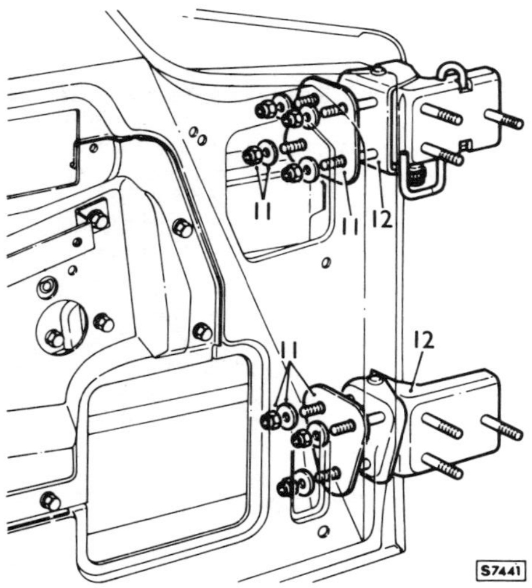

Refitting
13 Locate new hinge and hinge plate to body; secure using one nut and washer only, do not tighten fully.
14 Locate new hinge and hinge plate to door; secure using one nut and washer only, do not tighten fully.
15 Remove support.
16 Close door, check alignment; re-align door as required.
17 Fit remaining nuts and washers and tighten fully.
18 Reverse operations 1 to 9.

76–8

2 Remove door trim pad, see 76.34.01.
3 With door glass in raised position remove two bolts, plain and shakeproof washer securing window stop to base of door.
4 Remove window stop.
5 Lower door glass.
6 Remove three screws securing quarterlight channel to door; remove channel.
7 Release clip securing door finisher at rear edge of door.
8 Remove eight screws securing door finisher to door; remove door finisher.
9 Remove weather strip from door finisher.

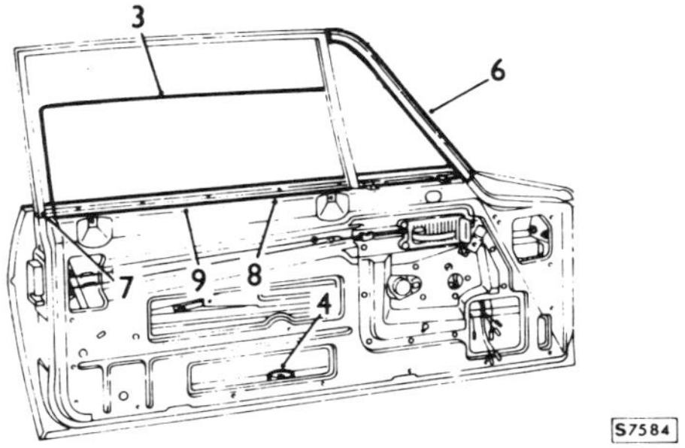

Refitting

Reverse operations 1 to 9.

DOOR TRIM PAD 76.34.01.

Remove and refit

Removing

1 Release arm rest finisher from spring clip.
2 Remove screw securing arm rest to door trim pad.
3 Slacken two screws securing interior light switch plate to door; pull pad clear of striker plate.
4 Remove screw securing bottom edge of pad to door.
5 Unclip door trim pad from door.
6 Disconnect radio speaker electrical connectors.
7 Remove door trim pad assembly.

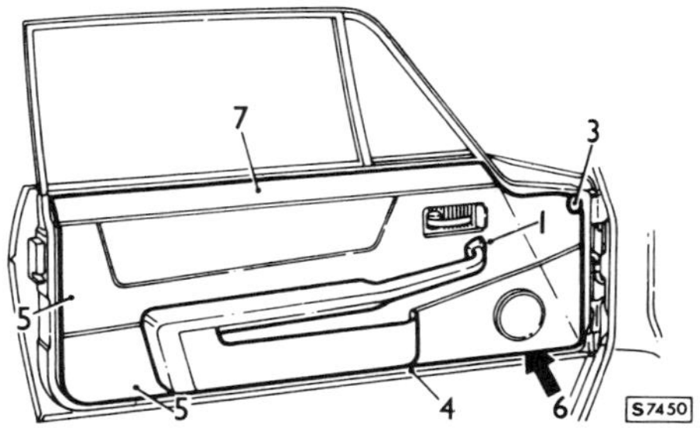

DOOR LOCK – ADJUST

Adjust 76.37.01

WARNING: IF ANY OF THE FOLLOWING SYMPTOMS BECOME EVIDENT, IMMEDIATE REMEDIAL ACTION MUST BE TAKEN AS OUTLINED BELOW:

a Door fails to fully close.
b Door fails to open on operation of inside handle.
c Door opens upon initial movement of inside handle.
d Door fails to lock upon operation of inside lock lever.

Adjusting

1 Remove door trim pad, see 76.34.01.
   NOTE: When symptoms a, b or c are evident proceed as follows.
2 Squeeze inside handle link rod spring connector and slightly operate handle, release spring connector. Close door and check for evidence of a, b or c.
3 Continue operation 2, adjusting link rod to left or right of spring connector until door opens when handle is half operated and door fully opens and closes.
   NOTE: If symptom d is evident proceed as follows.
4 Squeeze spring connector joining lock lever link rods, slightly operate lock lever and release spring connector. Close door and check for evidence of d.

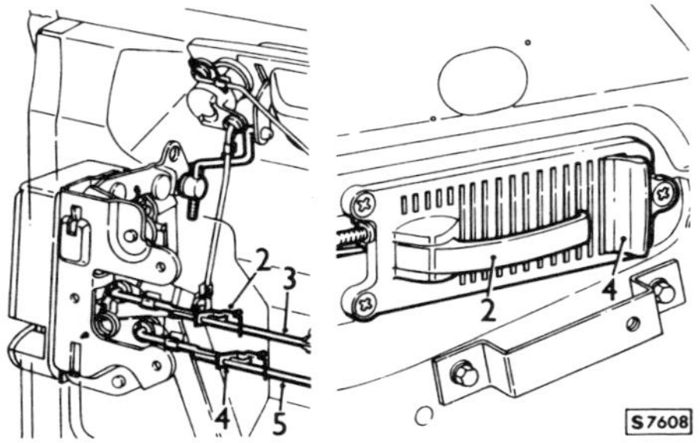

5 Continue operation 4, adjusting link rod to left or right of spring connector until door locks with lever in fully rear position.
6 Refit door trim casing.

DOOR ARMREST 76.34.22

Remove and refit

Removing

1 Remove door trim pad, see 76.34.01:
2 Remove two screws securing retaining bracket to door trim pad; remove retaining bracket.
3 Straighten six arm rest securing lugs.
4 Remove arm rest and pocket assembly.

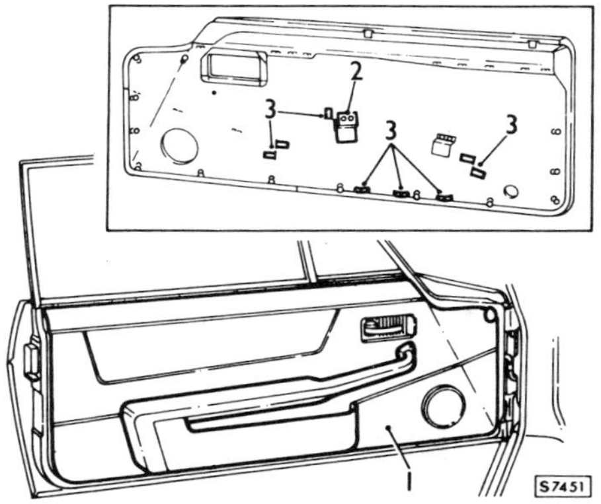

Refitting

Reverse operations 1 to 4.

DOOR LOCK

Remove and refit 76.37.12

Removing

1 Remove door trim pad, see 76.34.01.
2 Disconnect inner handle to lock connecting rods from anti-rattle and connecting clips.
3 Ease plastic sheeting to one side.
4 Disconnect outer handle to lock connecting rods.
5 Remove four screws securing outer lock unit to door.
6 Remove outer lock unit.
7 Remove inner lock unit.

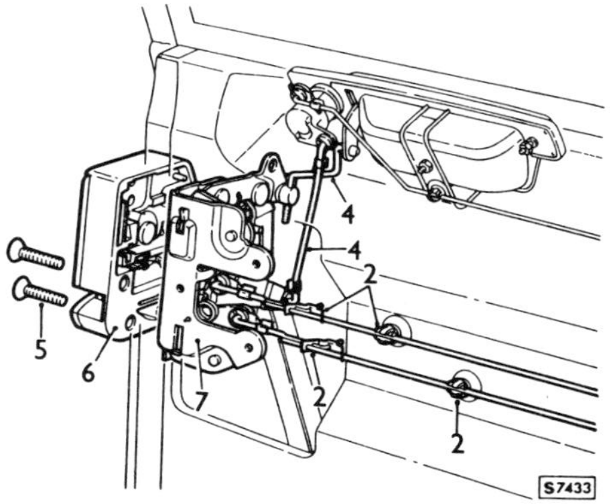

Refitting

Reverse operations 1 to 7.

76–9

DOOR LOCK STRIKER PLATE

Remove and refit 76.37.23

Removing
1. Remove rear seat cushion, see 76.70.37.
2. Remove two screws at base of rear seat back and lift out seat back.
3. Lift off rear arm rest adjacent to striker plate.
4. Remove cover from front seat belt mounting on 'B' post, detach bolt and remove seat belt, mounting bolt, spacers and spring washer.
5. Detach trim pad from 'B' post and remove four screws securing side trim panel.
6. Move trim panel aside and remove striker plate attachment screws.
7. Detach striker plate.

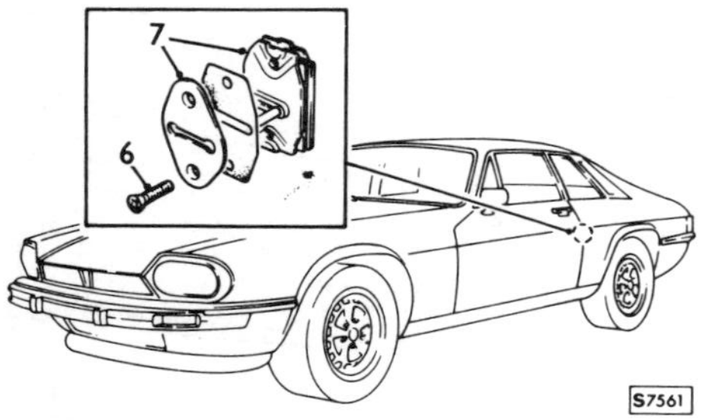

Refitting
8. Fit striker plate and secure with two screws.
9. Replace trim panel and secure with four screws.
10. Refit 'B' post trim pad.
11. Replace front seat belt mounting on 'B' post and refit bolt with spring washer and two spacers. Fit plastic cover.
12. Push arm rest back into location.
13. Replace seat back and secure with two screws.
14. Replace seat cushion, see 76.70.37.
15. Check adjustment of striker plate.

DOOR LOCK STRIKER – ADJUST

Adjust 76.37.27

Adjusting
1. Slacken setscrews securing striker plate to body.
2. Reposition striker and tighten setscrews.
3. Close door, door should close with minimum push effort.
4. Continue operation 2 until door closes as described in operation 3.

FACIA PANEL

Remove and refit 76.46.01

Removing
1. Slacken steering wheel adjuster and pull wheel out to full extent.
2. Remove screws securing upper and lower shroud to steering column, lift shrouds from column.
3. Remove pinch bolt securing steering wheel and adjuster mechanism to upper column.
4. Slacken locknut and Allen screw adjacent to pinch bolt.
5. Withdraw steering wheel and adjuster assembly from upper column.
6. Remove 'A' post trim pads, see 76.13.07.
7. Remove instrument module, see 88.20.01.
8. Remove driver's underscuttle casing, see 76.46.11.
9. Remove interior and map lamp mounted in facia, see 86.45.04.
10. Remove passenger underscuttle casing, see 76.46.15.
11. Remove facia switch panel, see 76.46.16.
12. Remove glove compartment liner, see 76.52.03.
13. Pull air conditioning control knobs from spindles.
14. Remove radio or locking rings securing radio blanking panel to console, remove panel.
15. Remove two screws securing facia and console to air conditioning unit.
16. Slacken nuts securing top rear portion of facia to bulkhead.
17. Remove bolts securing sides of facia to bulkhead, access to these bolts is through either underscuttle aperture.
18. Remove nut securing main light switch to mounting, ease switch from mounting.
19. Carefully remove facia from car.

NOTE: If facia is removed for access only proceed to refitting operation, if new facia is to be fitted carry out following operations.

20. Remove glove box lid, see 76.52.02.
21. Push centre air vent retaining clips towards facia and disengage clips from air vent.
22. Remove centre air vent from facia.
23. Remove nut securing centre air vent trim to facia, lift trim from facia.
24. Remove facia side air vents.
25. Remove screws securing glove box lid catch to facia, remove catch.
26. Prise heat sensor tubing from facia. 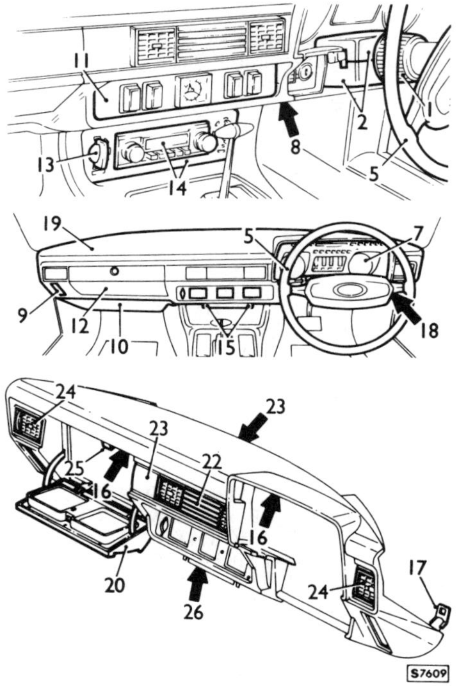
27. Remove ignition and light switch bracket clamp screws, remove switch brackets.
28. Remove underscuttle casing securing clips from lip of facia.

Refitting
29. Reverse operations 20 to 28.
30. Relocate facia to bulkhead.
31. Reverse operations 16 and 17 ensuring earth lead is fastened between near side mounting and securing bolt.
32. Feed wires for facia lights through rear cover of light mountings.
33. Reconnect light units to wires and refit to facia.
34. Reverse operations 13 to 15.
35. Refit facia switch panel.
36. Refit instrument module to facia.
37. Refit glove box liner.
38. Fit light switch surround around switch assembly.
39. Reconnect fibre optic to surround, tighten surround securing screw.
40. Reverse operations 1 to 6.
41. Refit driver's and passenger's underscuttle casing.

DRIVER'S UNDERSCUTTLE CASING

Remove and refit 76.46.11

Removing
1. Remove screws securing casing to facia and scuttle.
2. Carefully lower casing sufficiently to gain access to rheostat and speedometer trip.

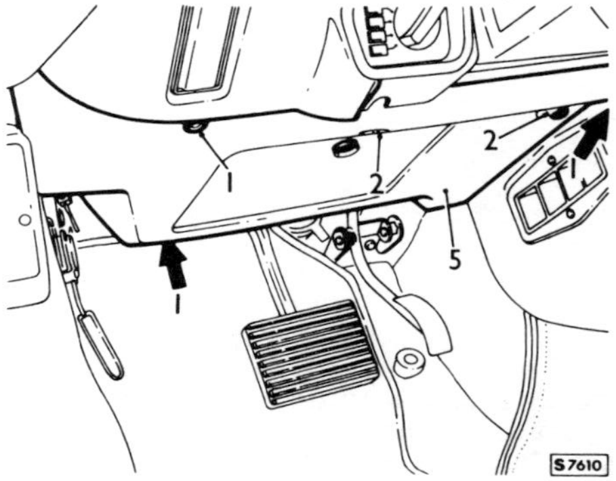

76–10

3   Detach wires from rheostat.
4   Disconnect speedometer trip from casing.
5   Lift casing from car.

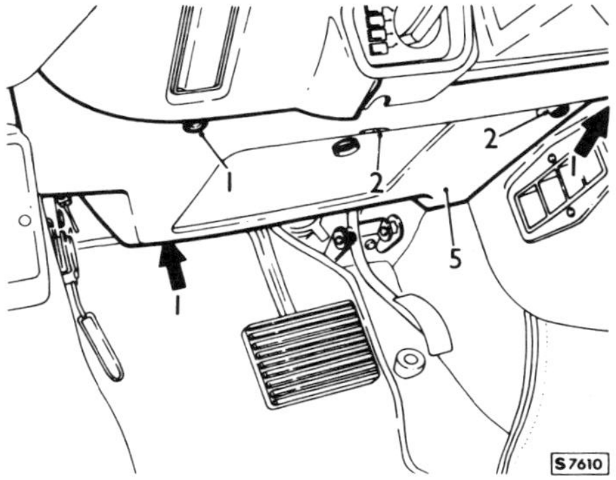

Refitting
Reverse operations 1 to 5.

PASSENGER UNDERSCUTTLE CASING

Remove and refit   76.46.15

Removing
1   Remove screws securing casing to facia panel and underscuttle.
2   Manoeuvre casing from location.

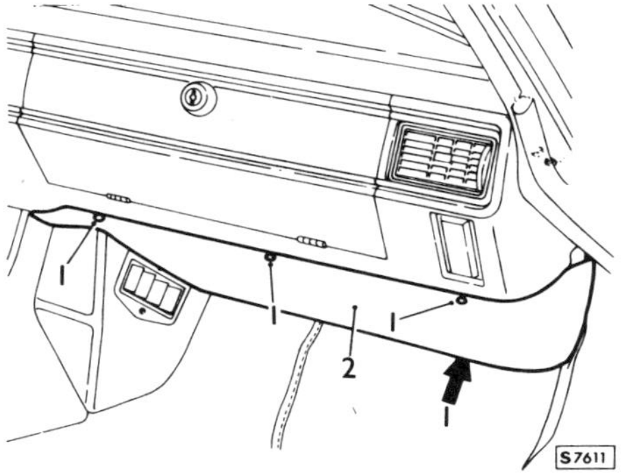

Refitting
Reverse operations 1 and 2.

FACIA SWITCH PANEL

Remove and refit   76.46.16

Removing
1   Ease clock from its aperture in panel, disconnect leads and remove bulb holder. Remove clock.
2   Working through clock aperture, prise switch panel away from its six press-in attachments.
3   Ease panel clear of facia, manoeuvring hazard warning switch wires past sharp edge of facia aperture. 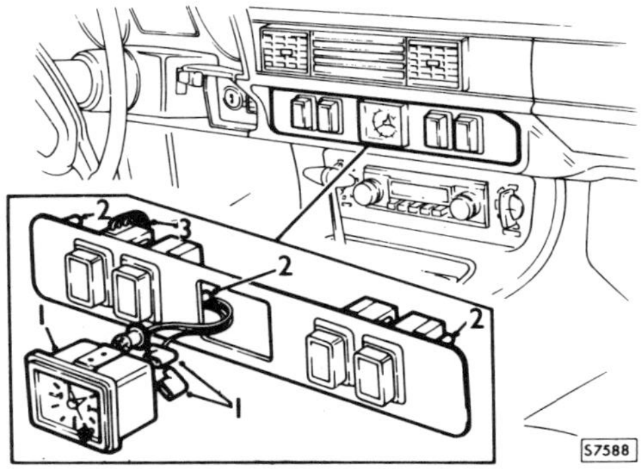
4   Disconnect wiring at switches and bulb holders.
5   Remove all switches and bulb holders from panel.

Refitting
6   Fit bulb holders and switches to new panel.
7   Offer up panel to facia and connect wiring.
8   Manoeuvre wiring through facia aperture and press panel fasteners home.
9   Offer up clock, reconnect wires, replace bulb holder and press clock into its aperture in panel.

GLOVE BOX LID

Remove and refit   76.52.02

Removing
1   Remove screws securing vanity mirror and tray to glove box lid, lift assembly from lid.
2   Remove nuts securing lid check straps, detach straps from mounting studs.
3   Remove screws securing glove box lid to hinges, lift lid from car.
4   Remove nut securing lid lock, remove lock from lid.
5   Remove fasteners securing metal trim to lid, remove trim.

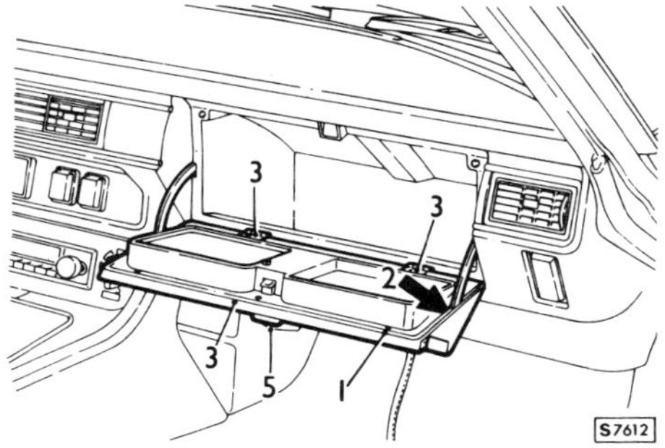

Refitting
Reverse operations 1 to 5.

GLOVE BOX LINER

Remove and refit   76.52.03

Removing
1   Remove screws securing liner to facia.
2   Carefully prising liner around latch withdraw liner from facia.

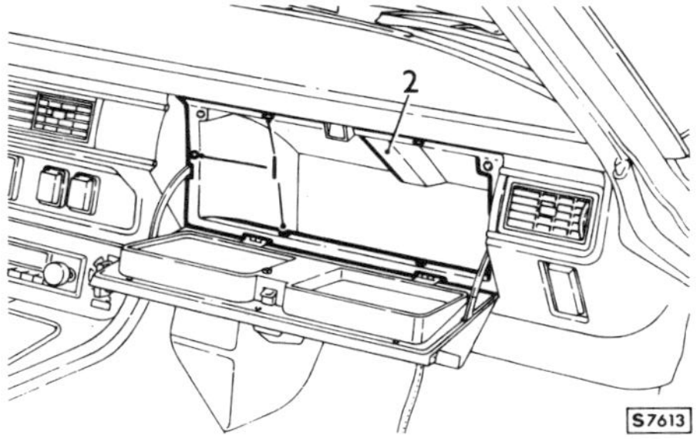

Refitting
Reverse operations 1 and 2.

RADIATOR GRILLE

Remove and refit   76.55.03

Removing
1   Remove screws securing grille to bonnet, recover washers and distance pieces.

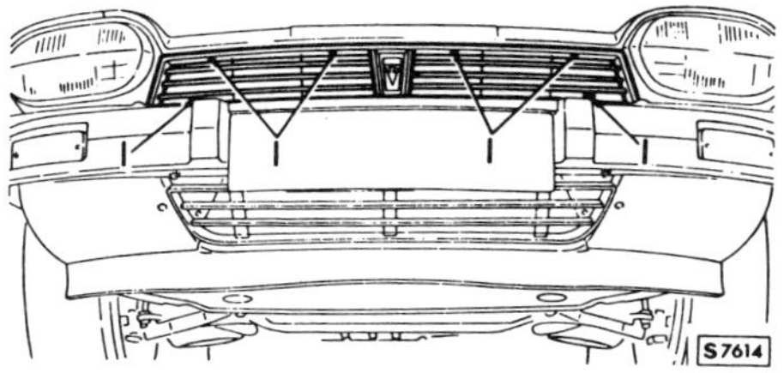

Refitting
2   Refit grille to bonnet, ensure distance pieces are fitted between grille upper mountings.

RADIATOR LOWER GRILLE

Remove and refit   76.55.06

Removing
1   Remove screws securing grille to wings.
2   Remove grille.

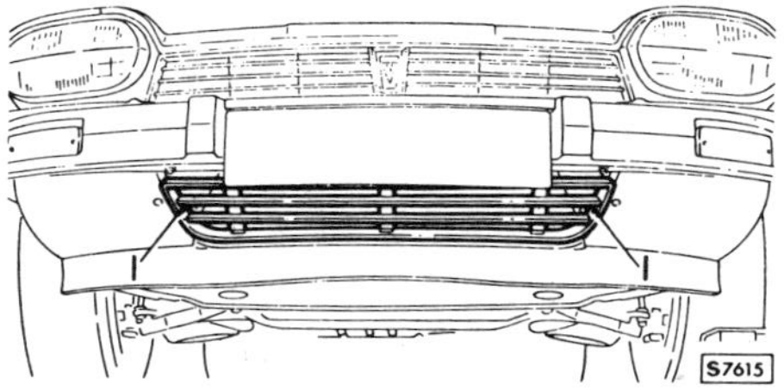

Refitting
Reverse operations 1 and 2.

REAR PARCEL SHELF

Remove and refit   76.67.06

Removing
1. Remove rear seat squab, see 76.30.38.
2. Carefully prise shelf from panel.

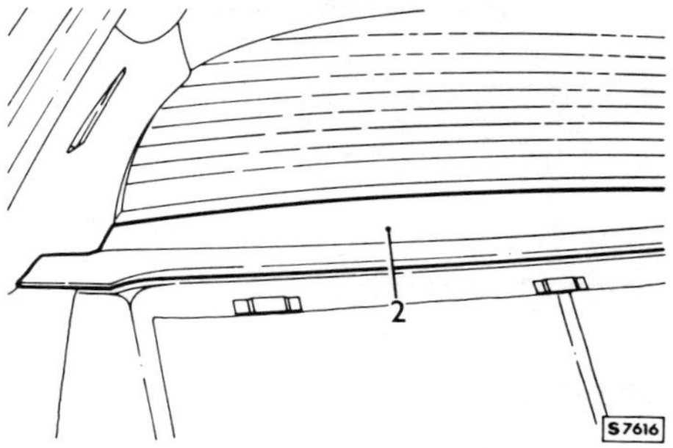

Refitting
Reverse operations 1 and 2.

FRONT ASHTRAY

Remove and refit   76.67.13

Removing
1. Open ashtray cover and withdraw ash container.
2. Remove screws securing ash container holder to console.
3. Withdraw holder and securing bracket from console.

continued

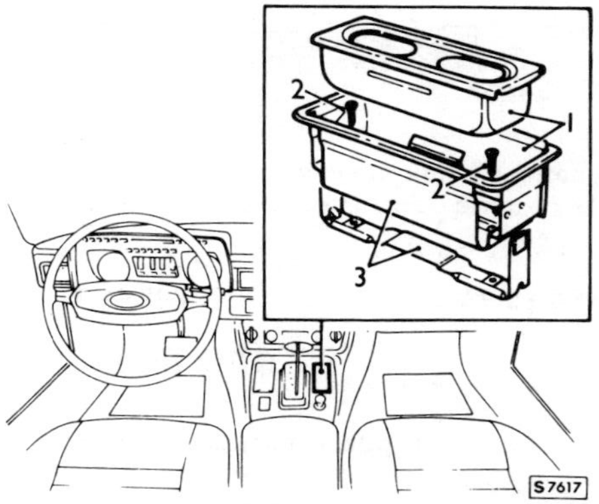

9. Bend back six tabs securing headlining.
10. Remove complete headlining assembly via passenger door.

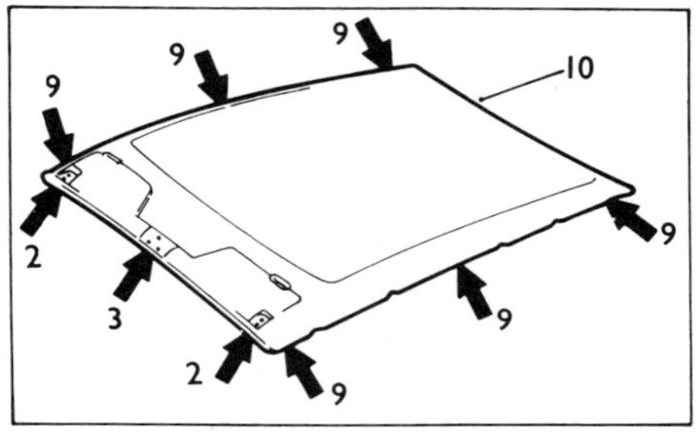

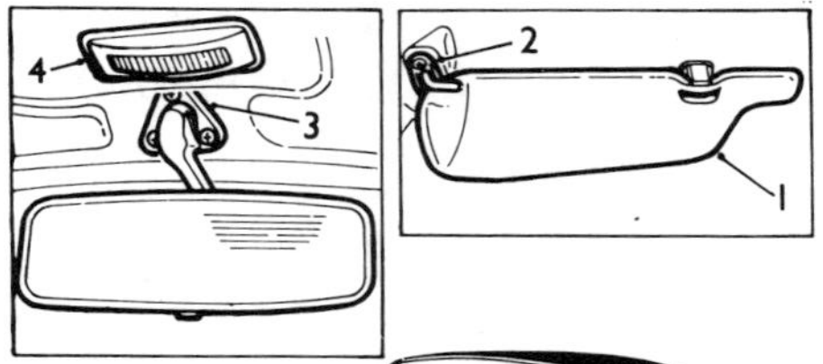

Refitting
11. Introduce replacement headlining through passenger door, raise into position and secure with clips.
12. Ease top forward edge of rear window surround over rear edge of headlining.
13. Refit rear quarter upper trim pads, see 76.13.13.
14. Replace rear shelf, see 76.67.06.
15. Refit cantrail crash rolls, see 76.13.10.
16. Reconnect leads to interior light and refit light.
17. Refit interior mirror, see 76.10.51.
18. Refit both sun visors, see 76.10.47.
19. Refit sun visor retaining brackets and tighten screws.
20. Replace rear seat squab, see 76.70.38.

HEADLINING

Remove and refit   76.64.01

WARNING: THIS OPERATION SHOULD NOT BE ATTEMPTED BY PERSONS KNOWN TO BE ALLERGIC TO GLASS FIBRE (FIBREGLASS). SHOULD SKIN AREAS DEVELOP A RASH OR IF ITCHING OCCURS, WASH AFFECTED AREA WITH WATER AND SEEK MEDICAL ADVICE IMMEDIATELY. ALWAYS WEAR GLOVES, FACE MASK AND GOGGLES WHEN HANDLING HEADLINING.

Removing
1. Remove both sun visors.
2. Remove two screws securing sun visor retaining brackets.
3. Remove interior mirror.
4. Displace front interior light, disconnect leads and remove light.
5. Remove cantrail crash rolls from both sides, see 76.13.10.
6. Remove rear seat squab to provide access to rear shelf, see 76.70.38.
7. Remove rear shelf to provide access to quarter trim pads, see 76.67.06.
8. Remove rear quarter upper trim pads from both sides, see 76.13.13.

DOOR OUTSIDE HANDLE

Remove and refit   76.58.01

Removing
1. Remove door trim pad, see 76.34.01.
2. Remove door glass, see 76.31.01.
3. Remove three clips securing control arms to outside handle.
4. Disconnect control arms from outside handle.
5. Remove two nuts and washers securing door handle.
6. Remove securing bracket.
7. Remove outside handle assembly.
8. Remove and discard gasket.

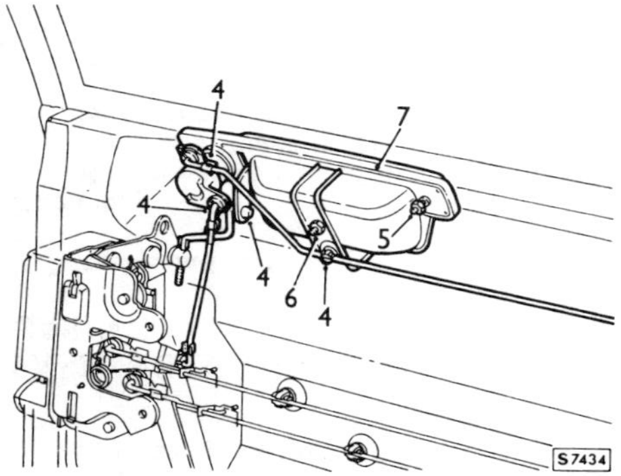

Refitting
Reverse operations 1 to 8 using new gasket and control arm clips.

DOOR INSIDE HANDLE

Remove and refit   76.58.18

Removing
1. Remove door trim, see 76.34.01.
2. Remove three screws securing handle to door.
3. Disconnect control arms from anti-rattle clips and lock clips.
4. Ease handle clear of door panel.
5. Release lock control arm from clip.
6. Release handle control arm from plastic clip.
7. Remove interior handle assembly.

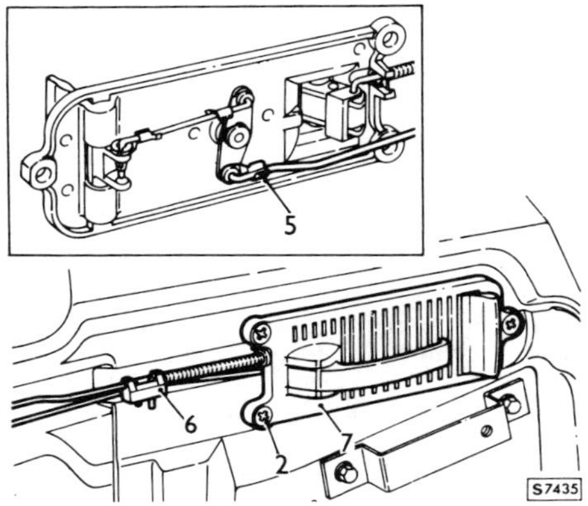

Refitting
Reverse operations 1 to 7 using new control arm clips.

Refitting

4 Slightly secure bracket with one screw to holder unit.
5 Fit holder and bracket to console.
6 Align unsecured portion of bracket with hole in holder.
7 Fit remaining bracket securing screw, fully tighten both screws.
8 Fit ash container to holder.

REAR ASHTRAY

Remove and refit   76.67.14

Removing

1 Remove screws securing console stowage lid check strap.
2 Remove screws securing stowage lid hinge, lift lid from console.
3 Remove screws securing stowage liner to console, prise liner lock catch from console and remove liner. 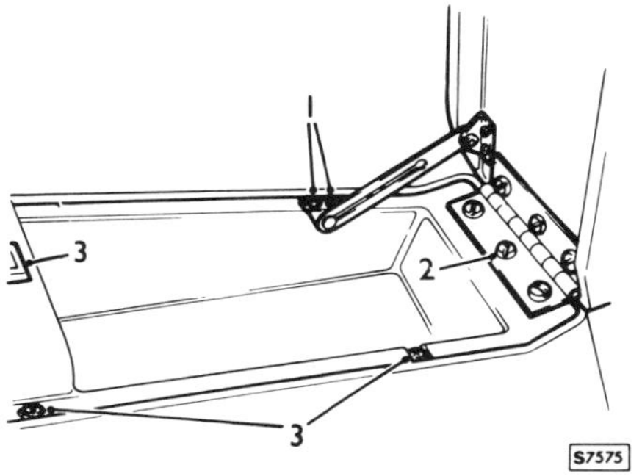
4 Open ashtray, push down and remove ash container from holder.
5 Straighten upper clips securing holder to console.
6 Lift holder and bezel surround from console.

Refitting

Reverse operations 1 to 6.

FRONT SEAT

Remove and refit   76.70.01

Removing

1 Operate seat release lever and push seat to full rear position.
2 Remove two Allen head screws securing seat runner assembly to floor mountings.
3 Operate seat release lever and push seat to full forward position.
4 Remove two Allen head screws securing rear of runner assembly to floor mountings.

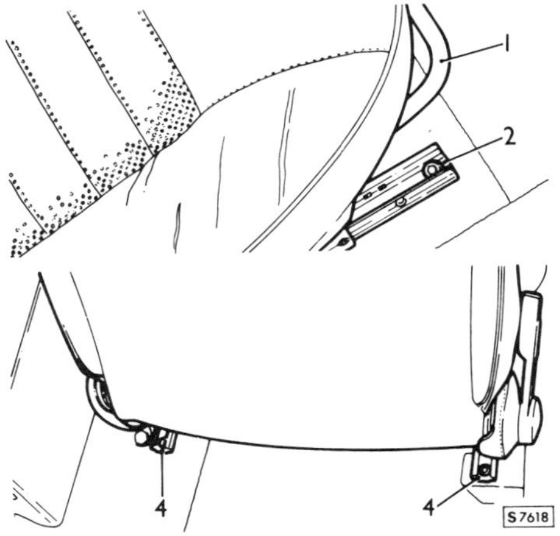

5 Lift seat assembly from car.
6 Recover distance pieces from front mountings.

Refitting

Reverse operations 1 to 6.

FRONT SEAT CUSHION

Remove and refit   76.70.02

Removing

1 Remove front seat and adjuster assembly, see 76.70.24.
2 Remove screw securing rake angle adjustment lever, lift lever and plastic disc from seat.
3 Carefully prise trim plate from seat.
4 Remove screw securing trim plate to side of seat, prise plate from seat.

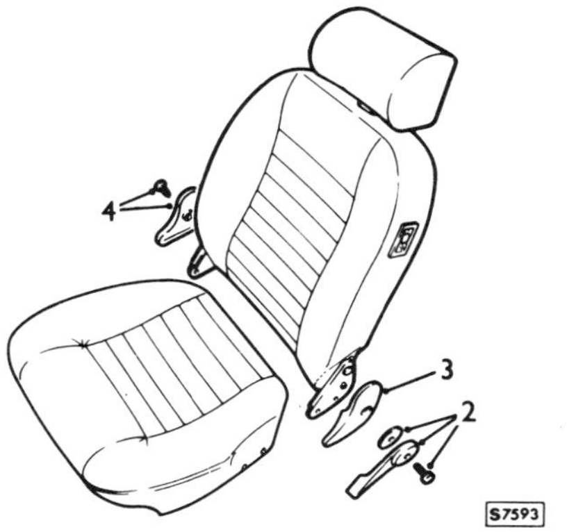

5 Remove bolts securing seat cushion to back rest, separate cushion from back rest.

Refitting

Reverse operations 1 to 5.

FRONT SEAT RUNNER AND ADJUSTER ASSEMBLY

Remove and refit   76.70.24

Removing

1 Remove front seat, see 76.70.01. 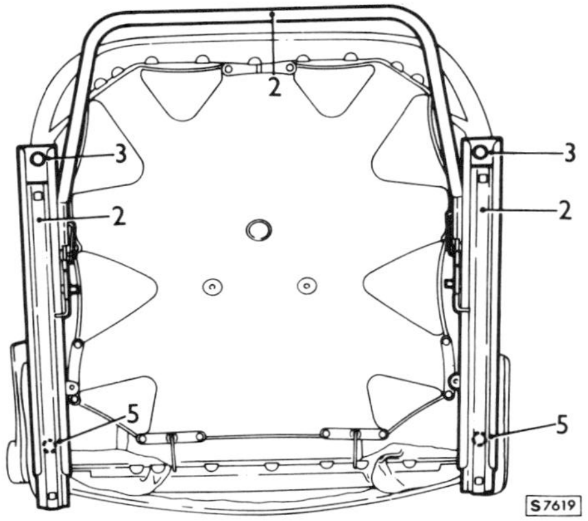
2 Operate runner adjusting lever and push runners to rear exposing screws securing runners to front of seat.
3 Remove securing screws.
4 Adjust runners to full forward position.
5 Remove screws securing runners to rear of seat.
6 Lift runner and adjuster assembly from seat, recover washers.

Refitting

Reverse operations 1 to 6.

REAR SEAT CUSHION

Remove and refit   76.70.37

Removing

1 Move both front seats fully forward and tilt backrests forward.
2 Remove two screws below front edge of rear seat cushion.
3 Remove screw retaining belt anchorage surround in centre of rear seat cushion. Lift off surround panel and seat cushion.

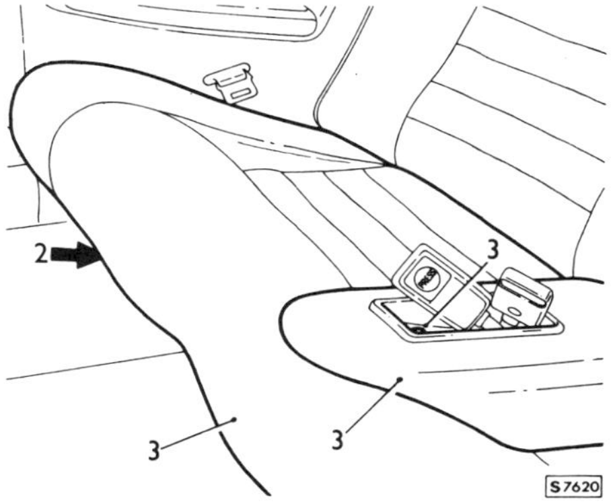

Refitting

4 Replace seat cushion and fit two screws.
5 Replace belt anchorage surround and fit retaining screw.
6 Return front seats to their original positions.

76—13

REAR SEAT SQUAB

Remove and refit 76.70.38

Removing
1. Remove rear seat cushion, see 76.70.37.
2. Remove screws securing seat squab to seat pan, lift squab from car.

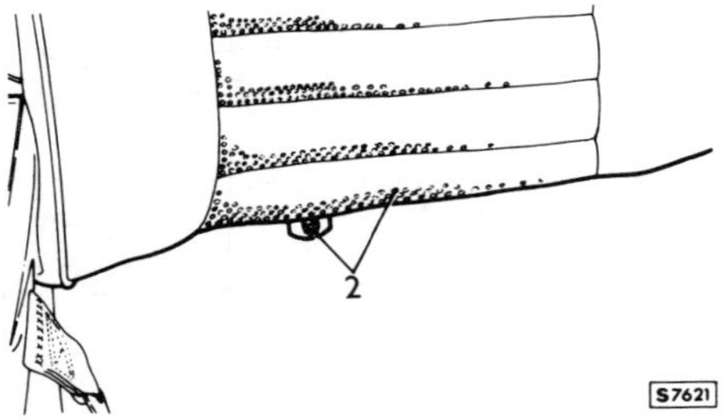

Refitting
Reverse operations 1 and 2.

FRONT SEAT BELT

Remove and refit 76.73.10

Removing
1. Position seat in fully forward position. 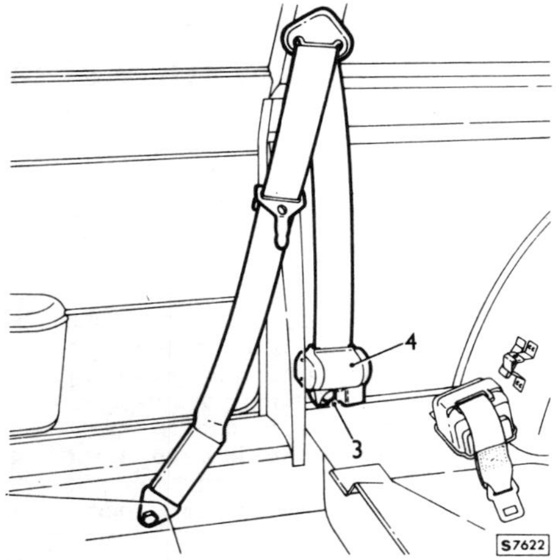
2. Remove rear quarter lower trim pad, see 76.13.12.
3. Remove bolt securing belt reel to body.
4. Lift belt reel from car.

Refitting
Reverse operations 1 to 4.

REAR SEAT BELT

Remove and refit 76.73.18

Removing
1. Remove rear quarter lower trim pad, see 76.13.12.
2. Lift off cover and remove bolt securing belt reel to body.
3. Lift reel from car.
4. Remove bolt securing seat belt buckle to tunnel, lift buckle from car.

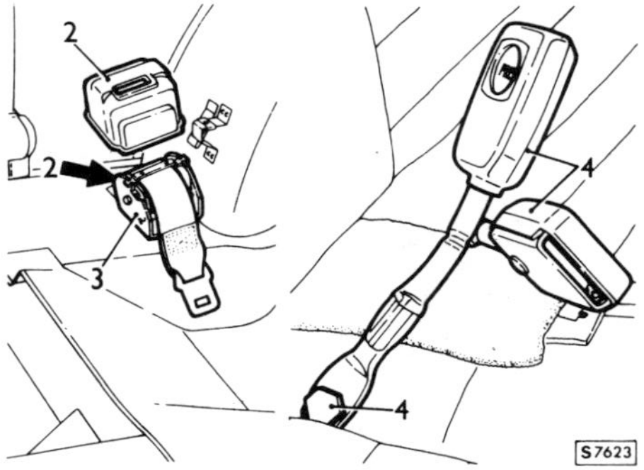

Refitting
Reverse operations 1 to 4.

76–14

AIR CONDITIONING SYSTEM

System description   82.00.00

The air conditioning system is comprised of the following components:

1. Compressor with magnetic clutch.
2. Condenser.
3. Combined receiver/drier unit.
4. Combined evaporator coil, expansion valve, heater coil and control system.
5. Blower motors.
6. Temperature sensing devices (thermistors) to compare in car and ambient temperatures.

The refrigeration cycle is best described with reference to figures 1 and 2. The refrigerant used is to specification R.12 (refrigerant 12) which is a halogenated hydrocarbon (dichlorodifluoromethane).

The heart of the automobile refrigerant system is the compressor. Its purpose is two fold; to raise the pressure of the refrigerant vapour and correspondingly raise its temperature.

The suction side of the compressor, point 1, pulls in superheated refrigerant vapour. The compression cycle occurs between 1 and 2 of figure 1, work being done on the vapour to raise its pressure and add heat.

The fact that heat is added is shown by point 2 on figure 2 being to the right of point 1. The pressure difference is given by the vertical axis of figure 2. The high pressure, high temperature vapour is delivered to a fin and tube construction condenser located in front of the engine coolant radiator where heat flow takes place from the high temperature vapour to the surrounding air. As the refrigerant passes through the condenser, heat transfer and a reduction in temperature takes place, the gas giving up its latent heat and condensing to a cool liquid. However, the length of the condenser is selected so that further heat loss takes place, sub-cooling the refrigerant liquid to ensure complete condensation. In figure 2 these conditions are shown between points 3 and 5. At point 3 condensation commences, passes through a wet vapour state, and is complete at point 4. The sub-cooling is shown taking place between points 4 and 5.

The high pressure liquid now passes to the receiver/drier, a reservoir for the liquid content of the system. The receiver/drier incorporates a filter and a limited capacity dehydrating element to remove traces of moisture from the refrigerant.

From the receiver/drier, liquid refrigerant passes to the expansion valve and evaporator unit point 6 of figure 1.

The expansion valve is the dividing point in the system, a step change from a high pressure area into a low through a small metering orifice. The metering orifice is protected by a gauze filter in the inlet union of the expansion valve. The orifice size is controlled by the temperature at the outlet from the evaporator unit and by the inlet pressure to the expansion valve. A quantity of liquid refrigerant passes the expansion valve orifice and expands suddenly as it enters the low pressure area. As the liquid passes through the coils of the evaporator unit heat transfer takes place from the car interior air to the liquid causing it to boil.

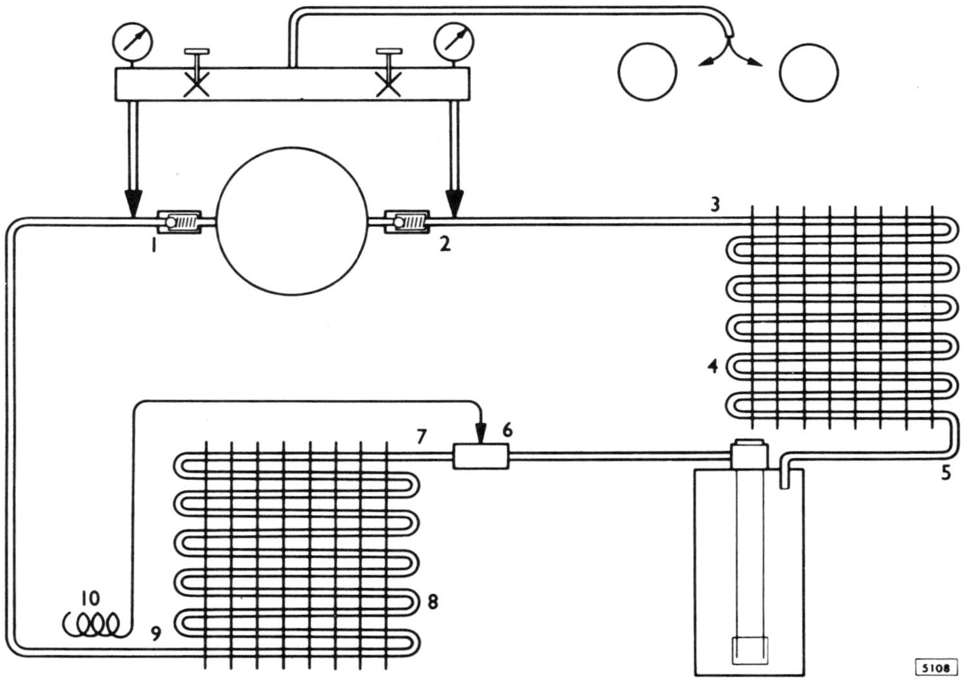

Fig. 1

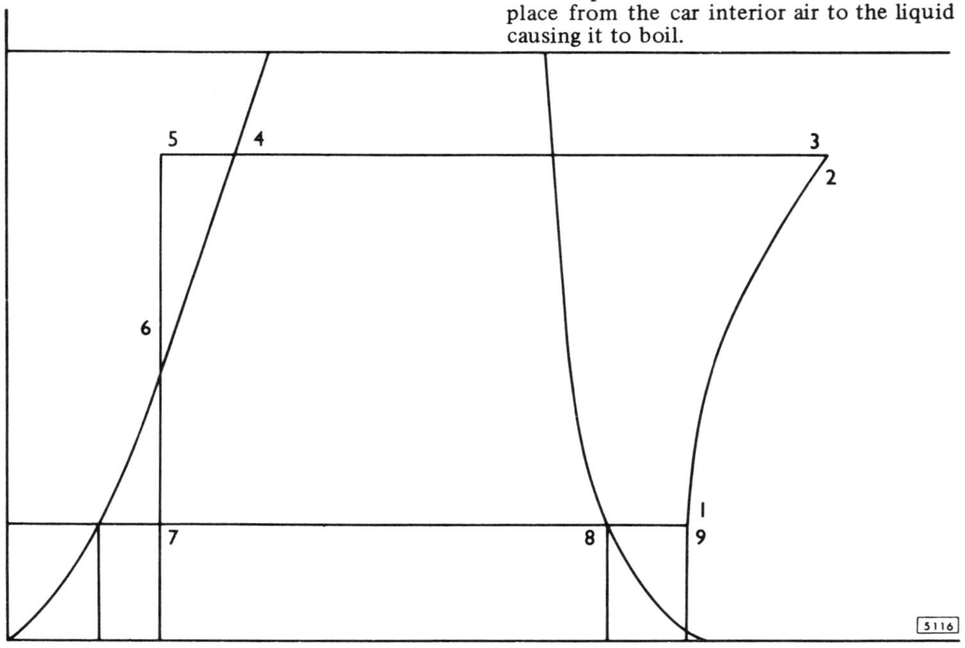

Fig. 2

82—1

The length of the evaporator coil is so chosen that the liquid refrigerant has completely vapourised at approximately three quarters through, the remaining length serving to super-heat the vapour. This ensures that no liquid refrigerant reaches the compressor, and that as much heat as possible is absorbed from the car interior. The expansion valve is fitted at the outlet of the evaporator. The capillary senses the outlet gas temperature and sets the expansion valve to meter the supply of cold liquid to the input of the evaporator.

For example, should the outlet pipe temperature fall, the expansion valve closes, cutting off the flow of liquid refrigerant until the temperature rises to the preset level. The super-heated vapour is then drawn to the suction side of the compressor, point 1 of figure 1, and the cycle continues.

Moisture from air passing over the matrix of the evaporator unit condenses on to the cool fins and drains from the evaporator via rubber tubes. Use of the air conditioning system can therefore result in a pool of water beneath the vehicle after parking. This is completely normal and does not indicate malfunction of the system.

It can be seen that ice formation is possible upon the fins of the evaporator unit. Anti-icing of the coil, and control of the vehicle interior temperature is provided by the thermostat. The capillary senses the temperature of the evaporator coil and, via a switch, de-energises the compressor clutch when the temperature falls to the preset level. This stops the flow of refrigerant and allows the coil to heat up until the thermostat switch re-closes. The compressor clutch then engages to re-start the cycle.

Component description

Compressor

The compressor utilizes three double acting pistons disposed axially around the compressor drive shaft. The pistons are actuated by a swash plate pressed on to the shaft. A magnetic clutch is used to drive the compressor shaft. When current passes through the clutch coil, the armature clutch plate assembly, keyed to the compressor shaft, is drawn rearwards against the belt driven pulley that is free wheeling upon the same shaft. This locks pulley and armature plate together to drive the compressor. When current ceases to flow, springs in the armature plate draw the clutch face from the pulley. The compressor comes to rest and the pulley continues to free wheel.

Condenser

The condenser is a copper tube and aluminium fin heat transfer unit, fitted forward of the engine coolant radiator. The condenser transfers heat from the refrigerant flowing through it to the airstream drawn through it by the engine cooling fans.

Receiver/Drier

The receiver/drier is a cylindrical tank that serves as a reservoir for the refrigerant. The liquid refrigerant is fed in, and is drawn out, via a screen and filter. The outlet tube has a sight glass fitted through which it can be determined by observing the refrigerant condition, whether there is adequate refrigerant in the system. A cloth sac filled with moisture absorbing granules is located inside the tank. This retains any traces of moisture that may be present in the refrigerant.

Expansion Valve

The expansion valve controls the flow of refrigerant through the evaporator coil to achieve optimum cooling efficiency. To do this, the valve senses the outlet pipe temperature and inlet pipe pressure and increases or decreases the flow of refrigerant liquid to maintain the outlet temperature constant.

Evaporator

The evaporator is a tube and fin heat exchanger into which the liquid refrigerant is metered. The air content of the vehicle, when forced over the fins by the recirculating blower fans, gives up its heat to boil the refrigerant.

Air conditioning heater/cooler unit and Automatic temperature sensing and control system

Automatic control is achieved by comparing car interior temperature and the temperature selected. This comparison provides an error signal to the air conditioning control unit, demanding an increase or decrease in car interior temperature. When the selected temperature is reached, the control unit will maintain it.

The error signal is detected across a Wheatstone Bridge circuit; two arms of which are fixed resistors, one arm contains the in-car thermistor and the fourth arm the temperature selection potentiometer. An error signal will be detected if car interior temperature is above or below that set on the temperature selection potentiometer. The error signal is amplified, and via relays, switches the servo motor to run clockwise or anti-clockwise. The position of the servo motor cam shaft directly determines the heating or cooling effect of the air conditioning system. Full heating and full cooling, are at opposite extremes of cam shaft travel.

The ambient thermistor in the Wheatstone Bridge circuit modifies the effect of the in-car thermistor. The result is a slightly colder interior temperature on hot days, and vice versa. A potentiometer driven by the servo motor is connected into the bridge circuit, modifying dynamic response. This provides control system damping, preventing excessive fluctuations in discharge air temperature.

The servo motor shaft mounted cams control seven functions:—

1 Air discharge temperature — The cam shaft moves blend flaps to vary air flow progressively from full cold to full heat. The cams are set to provide cooler air at head level, than to foot level, when the unit is in the low-medium heating mode. This prevents stuffiness at head level.

2 Fan speeds — The cam shaft alters fan speed progressively to increase air flow at full cold or full heat positions. Four fan speeds are available on cooling, three on heating. On low heating or cooling the cam shaft selects a low fan speed, preventing noise and excessive air movement.

3 Mode — The cam shaft controls a vacuum switch so that the distribution of air in the car is automatically controlled by a vacuum operated flap. Cold air is distributed from the face level vents, and hot air is distributed mainly from foot level vents with a bleed of air from screen vents.

4 Fresh/Recirculated Air — To improve performance the cam shaft selects recirculated air on maximum cooling. Fresh air is selected for all other requirements.

5 Water Valve — On maximum cold, the cam shaft controls a second vacuum switch to switch off the water valve controlling flow through the heater block.

6 Water Thermostat — A thermostat is fitted to prevent the system operating until engine water is hot enough to produce warm air. When on cooling mode the cam shaft overrides this switch, and allows the system to operate immediately.

7 Evaporator thermostat — A thermostat is fitted to prevent icing of the evaporator. Under conditions where icing would be impossible and maximum cooling performance is required, the thermostat is overriden by the camshaft.

continued

Manual Controls

1 Temperature selection potentiometer

This provides for the selection of interior temperature from 65°F to 85°F approximately.

B. Mode Switch –

1 Off – The system is switch 'off' and the fresh air intake is closed.

2 Auto – The system operates automatically.

3 Hi Auto and Lo Auto – These switch positions select a fixed high or low fan speed independently from that selected by the automatic control.

4 Defrost – This switch directs 90% air flow to the screen by closing the lower heater flap and opening the bleed flap to the screen outlets. At the same time an additional resistor is switched into the Wheatstone Bridge circuit to ensure that the servo motor cam shaft runs to the full heat position.

Method of Temperature Variation

All air passes through the fans and evaporator matrix. In the evaporator air is cooled and dehumidified. After leaving the evaporator, four blend flaps control the degree of heat added by the heater matrix. On maximum cold all air by-passes the heater matrix, and on maximum heat all air passes through the heater. Intermediate temperatures are obtained by varying the mixtures of cooled and heated air. One pair of blend flaps control the air to the car interior upper level, and another pair control the temperature to the lower level.

GENERAL SECTION

This section contains safety precautions, general information, good practice and standards that must be followed when working upon the air conditioning system. A fault finding and rectification section is included.

SAFETY PRECAUTIONS

The air conditioning equipment is manufactured for use only with Refrigerant 12 (dichlorodifluoromethane) and **extreme** care must be taken **NEVER** to use a methyl chloride refrigerant.

The chemical reaction between methylchloride and the aluminium parts of the compressor will result in the formation of products which burn spontaneously on exposure to air, or decompose with violence in the presence of moisture. The suitable refrigerant is supplied under the following trade names:

FREON 12
ARCTON 12
ISCEON 12

or any refrigerant to specification 12.

Goggles and gloves must be worn while working with the refrigerant.

WARNING: EXTREME CARE SHOULD BE EXERCISED IN HANDLING THE REFRIGERANT. LIQUID REFRIGERANT AT ATMOSPHERIC PRESSURE BOILS AT −29°C (−20°F) SERIOUS DAMAGE OR BLINDNESS MAY OCCUR IF REFRIGERANT IS ALLOWED TO CONTACT THE EYES.

FIRST AID: IF REFRIGERANT SHOULD CONTACT THE EYES OR SKIN, SPLASH THE EYES, OR AFFECTED AREA WITH COLD WATER FOR SEVERAL MINUTES. DO NOT RUB. AS SOON AS POSSIBLE THEREAFTER, OBTAIN TREATMENT FROM A DOCTOR OR EYE SPECIALIST.

GENERAL INFORMATION

Torque levels to be used when tightening all connections are as follows:

Torque loading

		lb ft.	kg. m	
1	Compressor/Condenser	30 to 35	4,15 to 4,84	
2	Condenser/Compressor	21 to 27	2,90 to 3,73	
3	Condensor/Receiver/Drier	15 to 20	2,10 to 2,76	
4	Receiver/Drier/Condenser	11 to 13	1,52 to 1,80	(Aluminium tank)
		30 to 35	4,15 to 4,84	(Steel tank)
5	Receiver/Drier/Evaporator	11 to 13	1,52 to 1,80	(Aluminium tank)
		30 to 35	4,15 to 4,84	(Steel tank)
6	Evaporator/Receiver/Drier	11 to 13	1,52 to 1,80	
7	Expansion valve/Evaporator	15 to 20	2,10 to 2,76	
8	Evaporator/Compressor	21 to 27	2,90 to 3,73	
9	Compressor/Evaporator	30 to 35	4,15 to 4,84	

GOOD PRACTICE

1 The protective sealing plugs must remain in position on all replacement components and hoses until immediately before assembly.

2 Any part arriving for assembly without sealing plugs in position must be returned to the supplier as defective.

3 It is essential that a second backing spanner is always used when tightening all joints. This minimises distortion and strain on components or connecting pipes.

4 Components must not be lifted by connecting pipes, hoses or capillary tubes.

5 Care must be taken not to damage fins on condenser or evaporator matrices. Any damage must be rectified by the use of fin combs.

6 Before assembly of tube and hose joints, use a small amount of clean new refrigerant oil on the sealing seat.

7 Refrigerant oil for any purpose must be kept very clean and capped at all times. This will prevent the oil absorbing moisture.

8 Before assembly the condition of joints and flares must be examined. Dirt and even minor damage can cause leaks at the high pressures encountered in the system.

9 Dirty end fittings can only be cleaned using a clean cloth wetted with alcohol.

10 After removing sealing plugs and immediately before assembly, visually check the bore of pipes and components. Where ANY dirt or moisture is discovered, the part must be rejected.

11 All components must be allowed to reach room temperature before sealing plugs are removed. This prevents condensation should the components be cold initially.

12 Before finally tightening hose connections ensure that the hose lies in the correct position, is not kinked or twisted, and will not be trapped by subsequent operations e.g. closing bonnet, refitting battery.

13 Check hose is correctly fitted in clips or strapped to subframe members.

14 The Frigidaire compressor must be stored horizontally and sump down. It must not be rotated before fitting and charging. Do not remove the shipping plate until immediately before assembly. Always use new 'O' ring seals beneath union housing plate.

15 Components or hoses removed must be sealed immediately after removal.

16 AFTER A SYSTEM HAS BEEN OPENED TWICE THE RECEIVER/DRIER MUST BE RENEWED.

82–3

FAULT FINDING CHART 1.

Direct access to the air conditioning unit is very limited. The following checks are designed to be carried out with the minimum of test equipment and confined to those areas that are readily accessible. The sequence must be strictly adhered to, each check is based on the satisfactory result of the previous test. The standard cable colour code is used to identify leads and terminal connections referred to in the tests.

Before commencing checks, run engine until normal running temperature is reached, this also ensures that sufficient vacuum is available for tests. For cooling tests the engine must be running for compressor clutch to operate.

Note 1: The voltages quoted in the chart are nominal, assuming an actual 12 volts in the system.

2: This procedure applies only for ambient temperatures between 65°F (18.3°C) – 85°F (29.4°C) which is the controlled range of the unit. At higher temperatures it will not be possible to set a heating mode.

3: Actuators have been tested for operations down to 8 in. (20.32 cm.) of mercury. Check vacuum level of system, and upstream of suspect valves, before rejecting valves.

Equipment Required
1. Voltmeter, capable of covering 0–13 volts D.C.
2. Continuity tester.
3. Ohmmeter capable of covering 0–20K ohms.
4. Vacuum gauge, (not essential) to check vacuum level.

TEST 1:

AIR CONDITIONING CONTROL KNOB POSITIONS — L.H. 75° — R.H. 'OFF'

ITEM	CORRECT OPERATION	FAULT SYMPTOM	CHECK	RESULT	ACTION
1–1	BLOWERS 'OFF'	AIR FLOW AT OUTLETS	1–1–1 Single relay NY terminals for voltage	(a) 'O' Volts (b) 12 Volts	Change switch 'C' Change relay
1–2	FRESH AIR FLAPS CLOSE. RECIRCULATION FLAPS OPEN	RECIRCULATION FLAP CLOSED	1–2–1 Disconnect vacuum line at bottom of solenoid valve	(a) No vacuum (b) Vacuum	Blockage in supply line 1–2–2
			1–2–2 Disconnect vacuum line at solenoid valve outlet	(a) No Vacuum (b) Vacuum	1–2–3 1–2–4
			1–2–3 Check voltage at 'K' terminal of solenoid	(a) 12 Volts (b) 'O' Volts	Change solenoid Trace circuit from solenoid to switch 'C'
			1–2–4 Pull open flap by hand	(a) Not possible (b) Possible	Mechanical link fault 1–2–5
		ONE SIDE CLOSED	1–2–5 Check vacuum at actuator inlet	(a) No Vacuum (b) Vacuum	Blockage in line to solenoid valve Actuator flap linkage fault

TEST 2:

CONTROL KNOB POSITIONS — L.H. 75° — R.H. 'DEFROST'

2–1	BLOWERS TO HIGH SPEED	BLOWERS 'OFF'	2–1–1 Check voltages at fan motor connectors	(a) Voltage NOT 'O' Volts (b) 'O' Volts	Blower electrical fault-change 2–1–2
			2–1–2 Check voltages at triple relay, GS lead and terminal	(a) 12 Volts (Break in GS line(s) relay to blower connections) (b) 10 Volts (c) 5 Volts (d) 'O' Volts	Switching O.K. 2–1–10 2–1–7 2–1–3
			2–1–3 Check voltage at NW lead on triple relay	(a) 'O' Volts (b) 12 Volts	Electrical fault in main supply 2–1–4

82–5

ITEM	CORRECT OPERATION	FAULT SYMPTOM	CHECK	RESULT		ACTION
			2-1-4 Check voltage at NY lead on single relay	(a)	'O' Volts	Break in NY or switch fault
				(b)	12 Volts	2-1-5
			2-1-5 Check earth resistance at single relay	(a)	High resistance	Earth contact loose
				(b)	Low resistance	2-1-6
			2-1-6 Check Y terminal on single relay	(a)	'O' Volts	Replace single relay
				(b)	12 Volts	Resistor assembly fault. Check Y & GS connections
		BLOWERS AT LOW SPEED ONLY (2-1-2c)	2-1-7 Check earth resistance at B.U. lead on triple relay	(a)	High resistance	Break in B.U. or Switch fault in switch D, or poor earth at 'D'
				(b)	Low resistance	2-1-8
			2-1-8 Check voltage at BG lead on triple relay	(a)	'O' Volts	Break in BG lead to switch 'A'
				(b)	12 Volts	2-1-9
			2-1-9 Check voltage at relay terminal 'R'	(a)	'O' Volts	Replace relay
				(b)	12 Volts	Fault in resistor assembly or connection to harness lead 'R'
		BLOWERS AT MED 2 SPEED (2-1-2b)	2-1-10 Check voltage at BW lead, triple relay	(a)	'O' Volts	Break in BW line to hand control
				(b)	12 Volts	Replace relay
2-2	HOT AIR TO SCREEN	NO AIR TO SCREEN (AIR TO FOOT LEVEL ONE/BOTH SIDES)	2-2-1 Check vacuum at 'T' piece under crash roll (green line)	(a)	Vacuum	Vacuum switch fault on hand control
				(b)	No vacuum	See Test 5
		COLD AIR FOOT LEVEL, COLD AIR SCREEN	2-2-2 Check air at outlets	(a)	Servo has not traversed	See Test 6-4
				(b)	Water valve failure	
		COLD AIR FOOT LEVEL OR COLD AIR SCREEN	2-2-3 Check air at outlets			Mechanical link or flap fault See Test 6

TEST 3: L.H. HEATING MODE (HIGHER THAN AMBIENT) / R.H. AUTO-HI

CONTROL KNOB POSITIONS

ITEM	CORRECT OPERATION	FAULT SYMPTOM	CHECK	RESULT	ACTION
3–1	BLOWER SPEED HIGH	BLOWERS OFF	3–1–1 Listen to blowers		Water temperature switch faulty or harness break in N & NY lines
3–2	SCREEN FLAPS CLOSE AIR TO FOOT LEVEL	HIGH FLOW AT SCREEN OR COLD AIR TO FOOT LEVEL	3–2–1 Check vacuum at 'T' piece under crash roll (green line)	(a) Vacuum	Actuator fault or flap mechanical failure
				(b) No Vacuum	Faulty vacuum switch 'C' or line blocked

TEST 4: L.H. 85° / R.H. AUTO-LO

CONTROL KNOB POSITIONS

ITEM	CORRECT OPERATION	FAULT SYMPTOM	CHECK	RESULT	ACTION
4–1	BLOWERS TO LOW SPEED	BLOWERS OFF	4–1–1 Check voltage at NY lead on single relay	(a) '0' Volts	Break in NY lead
				(b) 12 Volts	4–1–2
			4–1–2 Check earth resistance at single relay	(a) High resistance	Earth contact loose
				(b) Low resistance	4–1–3
			4–1–3 Check voltage at Y terminal on single relay	(a) '0' Volts	Replace relay
				(b) 12 Volts	Resistor assembly fault
		BLOWERS AT MED 1 (8.5 V)			Switch 'D' faulty

TEST 5: AMPLIFIER/SERVO RESPONSE

ITEM	CORRECT OPERATION	CHECK	RESULT	ACTION
5–1	SERVO TRAVERSE MOVE LEFT HAND KNOB BOTH WAYS	5–1–1 Check that servo traverses (listen for servo motor)	(a) Movement	Test 6
			(b) No movement	5–1–2
		5–1–2 Check voltages at P & PR connector and move L.H. control	(a) Voltage changes ± 12 Volts	Servo motor, condenser, wiring, disconnected
			(b) Both voltages at 12 Volts	Replace amplifier

82–7

ITEM	CORRECT OPERATION	FAULT SYMPTOM	CHECK	RESULT		ACTION	
				(c)	Both leads at earth potential		5-1-3
				(d)	One or both leads open circuit		Line break in RP leads
				(e)	PR Terminal at constant 12 Volts (Continuous heating)		Switch off unit to to 5-1-5
			5-1-3 Check amplifier supply at fuse (N line)	(a)	'O' Volts		Harness break N line
				(b)	12 Volts		5-1-4
			5-1-4 Disconnect ambient sensor lead 'U' and check resistance of servo socket at No. 7 (U) to No. 4 (W)	(a)	Open circuit		Temperature selector disconnected
				(b)	0–2K		Amplifier
			5-1-5 Switch OFF unit. Check resistance between amplifier fuse and IN-CAR sensor lead WR	(a)	High resistance		Switch A fault
				(b)	Low resistance		5-1-6
			5-1-6 Check resistance between amplifier socket No. 3 (N) and No. 7 (U)	(a)	Open circuit		Break in harness or sensor fault
				(b)	14K ohms		5-1-7
		R Terminal at constant 12V (Constant cooling)	5-1-7 Check resistance between amplifier socket No. 7 (U) and No. 4 (W). Rotate L.H. control	(a)	No variation		Temperature selector faulty
				(b)	Variation		5-1-8
			5-1-8 Check resistance between amplifier socket No. 5 (G) and 8(Y) and between No. 8(Y) and 9(0)	(a)	0 or Open circuit		Feed back potentiometer or harness fault
				(b)	Up to 2K ohms		Amplifier fault

TEST 6:

AUTOMATIC FUNCTIONS

| 6-1 | Limit cooling & heating | Servo will not traverse | | | | | Servo could be damaged See Test 5 |

ITEM	CORRECT OPERATION	FAULT SYMPTOM	CHECK	RESULT	ACTION
6–2	REFRIGERATION SYSTEM	OUTLET TEMPERATURE INCREASES	6–2–1 Check voltage and fuse in NY lead	(a) 'O' Volts (b) 12 Volts	Break in NY lead Fuse blown
			6–2–2 Check voltage at clutch terminal	(a) 'O' Volts (b) 12 Volts	6–2–2 Servo switch lead break GN Clutch fault loss of refrigerant
	Full cooling	LOW FLOW RATE DESPITE HIGH SPEED	6–2–3 Check air flow at outlets	(a) Evaporator blockage (b) Icing in matrix	Ranco thermostat fault
6–3	FRESH AIR/HIGH SPEED Full cooling	FANS NOT AT HIGH SPEED & RECIRCULATION FLAPS CLOSED	6–3–1 Check voltage in servo WG lead	(a) 'O' Volts (b) 12 Volts	Servo switch fault Test 2–1
		FANS NOT AT HIGH SPEED & RECIRCULATION FLAPS OPEN	6–3–2 Test 2–1	(a) Fault condition (b) No fault	Rectify at 2–1 Harness break WG lead
		FANS AT HIGH SPEED AND RECIRCULATION FLAPS CLOSED	6–3–3 Test 1–2	(a) Fault condition (b) No fault	Rectify as 1–2 Break in Servo branch of K lead
6–4	WATER VALVE Full cooling	WATER PIPES FEEL HOT (Valve open)	6–4–1 Check vacuum at actuator	(a) Vacuum (b) No vacuum	Actuator or valve fault Servo valve fault
	Full heating	AIR NOT HOT (Valve closed)	6–4–2 Check vacuum at actuator	(a) Vacuum (b) No vacuum	Servo valve fault Actuator or valve sticking
6–5	WATER THERMOSTAT BYPASS (Disconnect Thermostat leads) Full cooling	BLOWERS OFF			Harness break N or NY Servo switch fault
	Heating mode	BLOWERS ON			Servo switch fault
6–6	BLOWERS MED 1 and MED 2 Full heating	BLOWERS NOT ON MED 1	6–6–1 Check voltage at triple relay BY lead	(a) 12 Volts (b) 'O' Volts	6–6–2 6–6–3
			6–6–2 Check voltage at triple relay U lead	(a) 'O' Volts (b) 12 Volts	Relay fault Lead or resistance unit fault

ITEM	CORRECT OPERATION	FAULT SYMPTOM	CHECK	RESULT	ACTION
	Transverse full heating to full cooling	NO INTERMEDIATE FAN SPEEDS — MED 1 and MED 2 OFF	6–6–3		NY Lead break, or Servo switch fault
		MED 2 MISSING	6–6–4		BE lead break in servo limb or MED 2 switch fault
6–7	FACE FLAP (MODE)— Full cooling	NO AIR TO FACE LEVEL	6–7–1 Check vacuum at servo vacuum valve S9	(a) Vacuum (b) No vacuum	Blockage or actuator fault Servo vacuum valve fault
	Full heating	AIR TO FACE LEVEL	6–7–2 Check vacuum at servo vacuum valve S9	(a) Vacuum (b) No vacuum	Servo valve fault Blockage or actuator fault
6–8	INCORRECT AIR TEMPERATURES AND AIR DISTRIBUTION				
			6–8–1 If above tests are satisfactory 6–8–2 Full cooling 6–8–3 Full heating		Blend flap fault Ensure flaps are seated — readjust linkage as necessary

FAULT FINDING CHART 2.

Procedures to resolve the causes given are included in the following section.

CAUSE		(a) Unusually low reading of compound gauge	(b) Unusually high reading of compound gauge	(c) Unusually low reading of high pressure gauge	(d) Unusually high reading of high pressure gauge
Air or non condensable gases in system	(m)				X
Unusually hot running engine	(l)				X
Insufficient air over condenser	(k)				X
Restriction in high pressure side	(j)				X
Loose capillary tube connection at evaporator coil outlet	(i)		X		
Very high heat load	(h)		X		X
Excessive refrigerant charge	(g)		X		X
Evaporator coil blocked with ice	(f)	X		X	
Defective expansion valve	(e)	X	X	X	
Low capillary charge in expansion valve	(d)	X		X	
Partial restriction in expansion valve	(c)	X		X	
Very light heat load	(b)	X		X	
Low refrigerant charge	(a)	X		X	

SYMPTOM

82−11

FAULT FINDING AND RECTIFICATION

Carry out preliminary tests – 82.30.16, before continuing with relevant check.

Low refrigerant charge

1. Run engine at 1000 – 1200 r.p.m.
2. Switch on air conditioning system to high delivery setting.
3. Set thermostat control to mid range position.
4. Fit charging manifold as described in 82.30.06, first ensuring both charging manifold valves closed.
5. With the following ambient air temperatures, the corresponding delivery pressure gauge readings should be as given.

Ambient Temp.	Pressure gauge lb/sq. in	kg/cm²
60°F (16°C)	100 – 150	7,03 – 10,2
80°F (27°C)	140 – 190	9,84 – 13,36
100°F (38°C)	180 – 225	11,65 – 15,8
110°F (43.5°C)	215 – 250	15,1 – 17,56

NOTE: The higher pressure gauge readings for each temperature reading would only occur if very little air is flowing over condenser. Readings will decrease as the vehicle cools down from extended air conditioning system use.

Check

Check receiver/drier sight glass for excessive foaming.
Check expansion valve for intermittent hissing.

Correction

Test for leaks – 82.30.09, and rectify as necessary.

IMPORTANT: The system MUST be depressurised before any brazing repair is attempted.

Recharge the system – 82.30.08 on completion of repair.

Very light heat load

Check

Low outside ambient.
Evaporator blower intake for obstruction.
Outlet louvres for obstruction.
Remove blower motor assembly and check fan for damage.
Evaporator coils for ice formation.

Correction

Clear or remove any obstruction.
Renew or repair fan blades.
If ice formation present, set thermostat control to higher position.

Partial restriction of expansion valve

Check

Feel expansion valve and connection hoses. If a restriction is present ice can form on the valve body and on the pipe from valve to matrix. If the valve is functioning correctly, the intake hose will feel warm.

Correction

Remove valve – 82.25.01. Clean filter in inlet connection. If this filter is not obstructed fit replacement valve.

Low capillary charge in expansion valve

Check

Observe suction side pressure gauge, and enclose capillary sensing coil in the hand. The unusual heat should cause the expansion valve to flood and show sudden rise in suction side pressure. If the pressure does not rise, it can be assumed that capillary charge is weak.

Correction

Replace expansion valve – 82.25.01.

Defective expansion valve

Check

Carry out checks as for partial restriction of expansion valve.

Correction

As for partial restriction of expansion valve.

Evaporator coil blocked with ice

Check

Thermostat sensor not in contact with fins of evaporator.
Thermostat setting.
Expansion valve.

Correction

Reposition sensor to touch fins.
Reposition thermostat to lower level.
Renew expansion valve.

Excessive refrigerant charge

Check

Observe charging manifold gauge readings. If both suction and discharge pressure readings are unusually high for the prevailing ambient temperature this indicates the possibility of the system being overfilled. Ensure centre hose of charging manifold pointing in safe direction, and slowly open pressure side valve to bleed off some refrigerant.
If the gauge pressures fall, this confirms excessive refrigerant charge.

Correction

Continue to slowly bleed off refrigerant until both gauge readings are approximately normal.

CAUTION: Keep close watch on receiver/drier sight glass during this operation. Any bubbles appearing will indicate that too much refrigerant has been removed.

Very high heat load

Check

Very hot or very humid day.
Vehicle heater is switched off.
All windows and doors are closed.

Correction

Ensure heater is switched off and that all air vents, windows, and doors are closed.

Loose capillary tube connection at evaporator coil outlet

Check

Observe suction side of compressor connector union housing for severe frosting or icing.
Note: A slight degree of frosting is normal.

Correction

If icing is heavy, check that expansion valve capillary coil is in correct contact and clamped to evaporator outlet pipe. Tighten clip as necessary. Do not overtighten to damage capillary coil.
Replace expansion valve – 82.25.01.

Restriction in high pressure side

Check

For restriction in condenser, receiver/drier, and hoses connecting these units. Any restriction or partial blockage will create a drop in temperature at the point of restriction. This temperature drop will be obvious to the touch, and in some cases frost or sweating may occur at that point.

Correction

Replace component or hose affected.

Insufficient air over condenser

Check

That the condenser matrix is not damaged or obstructed.
Vehicle radiator matrix is not damaged.
Cooling fan runs continually while air conditioning system is on.
Direction of rotation of fan.

Correction

Blow out matrices with compressed air. Use hose if necessary to soften caked mud.
Dress deformed finning.
Check wiring to fan and replace fan if wiring correct.

82–12

Unusually hot running engine

Check
Engine cooling system.
Ensure radiator blind (if fitted) is not in operation.

Air or non-condensable gas in system

Check
If all other methods fail to reduce head pressure to a satisfactory level, check for air in the system.
Pour cold water over condenser to accelerate condensing action.
If there is excessive refrigerant in system the pressure will momentarily fall.
Air in the system will not condense. The pressure will therefore remain high.

Correction
Sweep (purge) the system – 82.30.07.
Change receiver/drier unit.

COMPRESSOR DRIVE BELT

Adjust 82.10.01

NOTE: For cars to North American Federal specification see operation 17.25.13.

1 Slacken nut and bolt securing auxiliary pulley mounting arm.
2 Slacken adjusting link locknut.
3 Adjust belt tension by means of adjusting link nut; correct tension is as follows:—
A load of 6.4 lb (2,9 kg) must give a total belt deflection of 0.22 in (5,6 mm) when applied at mid point of belt.
4 Reverse operations 1 and 2.

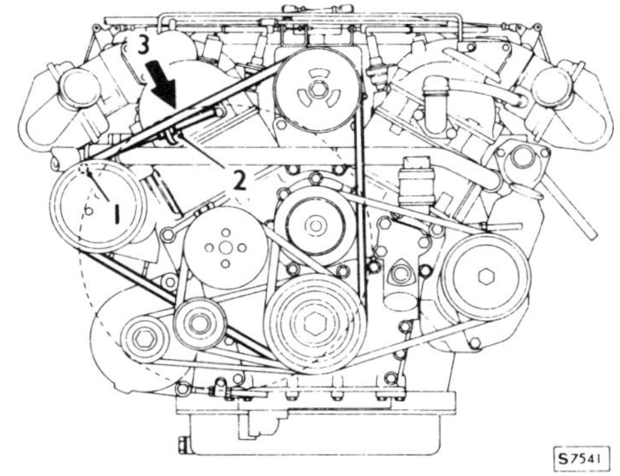

Refitting
9 Reverse operations 1 to 8.
10 Adjust compressor drive belt, see 82.10.01.

COMPRESSOR OIL LEVEL

Check 82.10.14

The oil level cannot be checked while the compressor is installed but the level must not change throughout normal service. However, owing to the fact that oil is normally held in suspension and comes to rest whenever the system is shut down, that oil within a component removed is lost to the total quantity.

In order to compensate for this loss, a specific quantity of oil is to be added for each component replaced; the quantity relating to each component is given below.

If oil has been lost from the system owing to accident damage depressurisation or incorrect depressurisation procedure, the compressor must be removed from the installation to check oil level, see 82.10.20.

The following amounts of oil should be added to the system whenever any of the components below are changed.

1 As a new compressor contains 10–11 fl oz (285–310 cm^3) of 525 viscosity refrigerant oil, the DRAINING of some oil will be necessary. Prior to fitment of a new compressor it is necessary to drain and weigh the oil from the old compressor, the difference between this amount and 11 fl oz (310 cm^3) is the quantity of oil that should be drained from the new compressor. NOTE: It is not sufficiently accurate to estimate the amount of oil remaining in the old compressor by weighing the individual compressor assemblies, the oil must be drained and weighed. Where circumstances do not permit this procedure, 5 fl oz (141 cm^3) should be drained from the new compressor.

2 If a new evaporator is fitted 3 fl oz (85 cm^3) of oil should be added to the system.
3 If a new condenser is fitted 2 fl oz (51 cm^3) of oil should be added.
4 If a new receiver/drier is fitted 1 fl oz (28 cm^3) of oil should be added.
5 If new hoses are fitted additional oil will not be required.
NOTE: The addition of oil to the system can be either, directly into the compressor sump or into the compressor charging port.

It is again emphasised that opening of the system at any union must IMMEDIATELY be followed by sealing of the open ends, failure to do so may cause a highly corrosive mixture to develop within the system which could render permanent damage to the compressor.

Operators should ensure that whenever a component is changed, or the system opened, a new receiver/drier is fitted.

COMPRESSOR DRIVE BELT

Remove and refit 82.10.02

NOTE: For cars to North American Federal specification see operation 17.25.15.

Removing
1 Remove fan belt, see 26.20.07.
2 Remove power assisted steering/water pump drive belt, see 57.20.02.
3 Slacken nut and bolt securing auxiliary pulley mounting arm.
4 Slacken adjusting link locknut.
5 Slacken nut on adjusting link securing bolt.
6 Slacken adjusting link pivot bolt.
7 Pivot pulley toward engine; disengage belt.
8 Remove compressor drive belt.

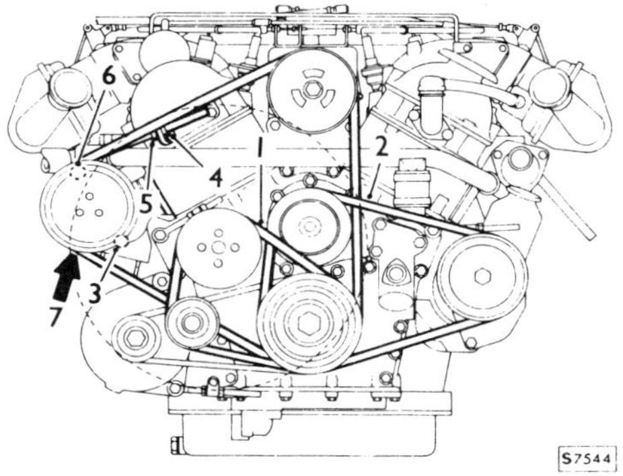

82–13

COMPRESSOR

Remove and refit   82.10.20

W A R N I N G : B E F O R E COMMENCING WORK ON THIS OPERATION REFER TO GENERAL SECTION – 82.00.00. DO NOT OPERATE COMPRESSOR UNTIL SYSTEM IS COMPLETELY CHARGED.

NOTE: Ensure that suitable clean, dry male and female sealing caps are to hand.

Removing

1. Disconnect battery earth lead.
2. Depressurise air conditioning system, see 82.30.05.
3. Remove bolt, plain and spring washer securing fuel cooler bracket to front of compressor.
4. Remove bolt and spring washer securing valve and union assembly to rear of compressor.
5. Remove clamp plate.
   NOTE: **Immediately** seal all connection orifices using clean, dry caps. Discard 'O' ring seals.
6. Swing fuel cooler unit clear of compressor.
7. Remove bolt and spring washer securing silencer unit bracket to front of compressor, swing silencer unit clear of compressor.
8. Disconnect electrical connectors from clutch. 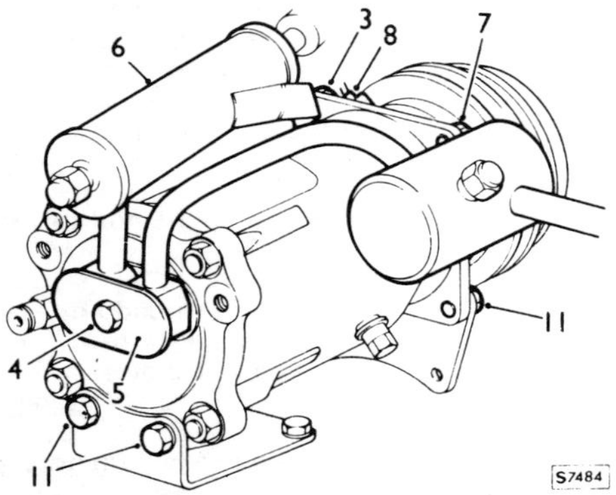
9. Slacken compressor drive belt, see 82.10.01.
10. Remove two nuts and washers securing ignition amplifier to mounting bracket, swing amplifier unit clear.
11. Remove four bolts, plain and shakeproof washers securing compressor mounting flanges to mounting brackets.
12. Keeping compressor horizontal, sump down, lift from engine compartment.
13. Drain oil from compressor sump into a suitable clean container and accurately measure quantity.
    CAUTION: If oil shows any sign of contamination system must be swept, see 82.50.07 and receiver/drier unit replaced, see 82.17.01.
14. Drain oil if any from replacement compressor.
    NOTE: Following manufacturers instructions transfer replacement parts from defective compressor to replacement.
15. If replacement compressor contained oil on receipt, refill with clean refrigerant oil equal in quantity to that removed in operation 14. If replacement compressor contained no oil, fill with clean refrigerant oil equal in quantity to that removed in operation 14, **PLUS** 1 oz (29 g).
    NOTE: This compensates for oil lost by retention in freshly drained unit, see 82.10.14.

Refitting

16. Place compressor in position, sump down.
17. Fit but do not tighten two bolts and washers securing compressor front mounting flange and earth strap to mounting bracket.
18. Fit but do not tighten two bolts, plain and spring washers securing compressor rear mounting flange to mounting bracket.
19. Fully tighten four securing bolts.
20. Position ignition amplifier on to mounting bracket, secure using two nuts and washers.
21. Refit compressor drive belt, adjust tension, see 82.10.01.
22. Remove shipping plate from rear of compressor, transfer to defective compressor.
23. Locate silencer and fuel cooler units to mounting position.
24. Fit silencer and fuel cooler unit brackets to front of compressor using two bolts and washers.
25. Using new 'O' ring seals clamp silencer and fuel cooler units to rear of compressor, secure using clamp plate bolt and spring washer to torque of 10 lb. ft – 25 lb. ft (4,5 kg. m – 11,3 kg. m).
    NOTE: Ensure 'O' ring seals are not displaced.
26. Reconnect electrical connections to clutch.
27. Reconnect battery.
28. Charge air conditioning system, see 82.30.08.
    CAUTION: After charging, cycle clutch in and out 10 times by selecting 'OFF-LO' 'AUTO-OFF' on mode selector switch with engine running. This ensures that pulley faces and clutch plate are correctly bedded in before a high demand is made upon them.

CONDENSER UNIT

Remove and refit   82.15.07

Removing

Before commencing this operation ensure that suitable clean, dry sealing plugs and caps are to hand.

1. Disconnect battery earth lead.
2. Depressurise air conditioning system, see 82.30.05.
3. Partially drain engine coolant.
4. Remove three nuts and washers securing fan cowl to radiator top rail; pull fan cowl clear of mounting studs.
5. Remove banjo bolt securing coolant bleed pipe to radiator, retrieve spacer from beneath banjo connections.
6. Remove four nuts and two washers securing receiver/drier unit mounting clamps and coolant bleed pipe securing clips to radiator top rail.
7. Slacken two nuts and bolts securing receiver/drier clamps, swing mounting clamps clear of radiator top rail.
8. Release union nut securing condenser pipe to receiver/drier; fit cap to receiver/drier.
9. Remove two nuts and washers securing condenser unit mounting brackets to radiator top rail.
10. Release union nut securing condenser pipe to main feed pipe. Fit plug to main feed pipe.
11. Remove four bolts and washers securing radiator top rail to wing valances.
12. Ease radiator top rail clear of condenser mounting brackets.
13. Lift condenser unit from car.

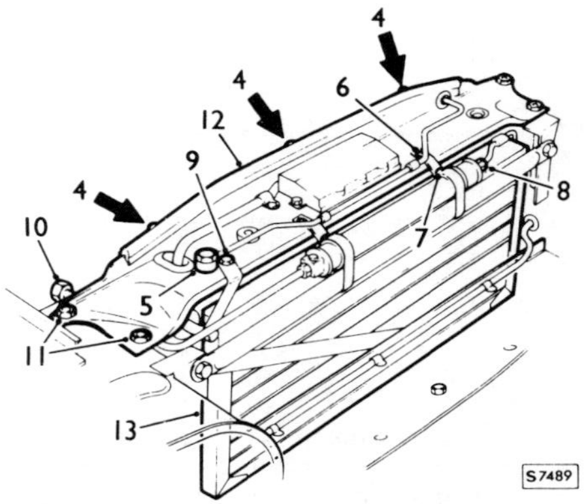

Refitting

14. Reverse operations 1 to 13.
15. Charge air conditioning system, see 82.30.08.

CONDENSER UNIT

Leak testing

The condenser unit can be tested for leaks whilst removed from car.

1. Seal off outlet pipe union with suitable cap, nut and sealing disc.
2. Connect a small refrigerant container to condenser inlet union with flexible hose.
3. Open container valve and allow a quantity of refrigerant to enter condenser.
4. Pass leak detector around condenser tubes. Pay particular attention to U bends to each end of condenser unit.
5. Replace defective unit with new unit.

82–14

RECEIVER/DRIER UNIT

Servicing

If the drying agent in the receiver/drier unit becomes completely saturated with water the unit must be replaced. If the system is allowed to remain open for a long period, or even a short time in very humid conditions, the drier unit must be replaced. DO NOT REMOVE the protective sealing caps from a new unit until it has been fitted and is ready for coupling to pipe unions.

CAUTION: It is of the utmost importance that water is not allowed in the system. Refrigerant '12' is a hydrocarbon containing chlorine and fluorine halogens. The hydrogen in water, can under certain temperature conditions, hydrolize with chlorine or fluorine to produce hydrochloric and hydrofluoric acids. These acids will attack the copper tubing and polished surfaces within the air conditioning system.

RECEIVER/DRIER UNIT

Remove and refit   82.17.01

Removing

1. Evacuate air conditioning system, see 82.30.06.
2. Disconnect union nuts securing pipe lines.

NOTE: Blank off unions and pipe lines using clean, dry caps.

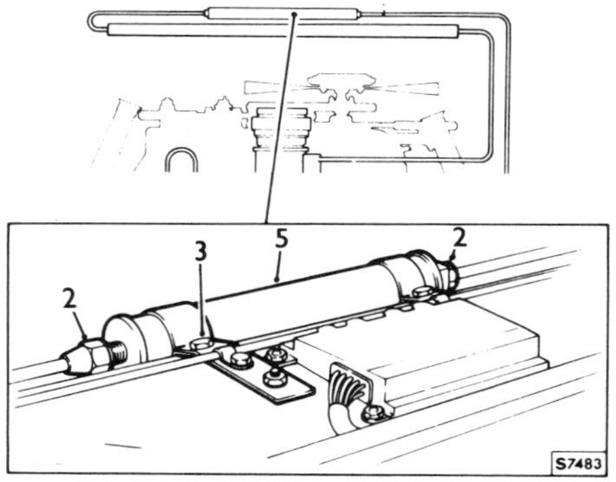

3. Slacken the nuts and bolts securing receiver/drier clamps.
4. Ease feed pipe clear of drier unit.
5. Manoeuvre receiver/drier unit clear of securing clamps.

Refitting

6. Reverse operations 1 to 5.
7. Charge air conditioning system, see 82.30.06

AMBIENT TEMPERATURE SENSOR

Remove and refit   82.20.02

Removing

1. Disconnect battery earth lead.
2. Remove underscuttle casing, see 76.46.15.
3. Note position of and disconnect electrical connectors from sensor.
4. Remove sensor from blower motor assembly.

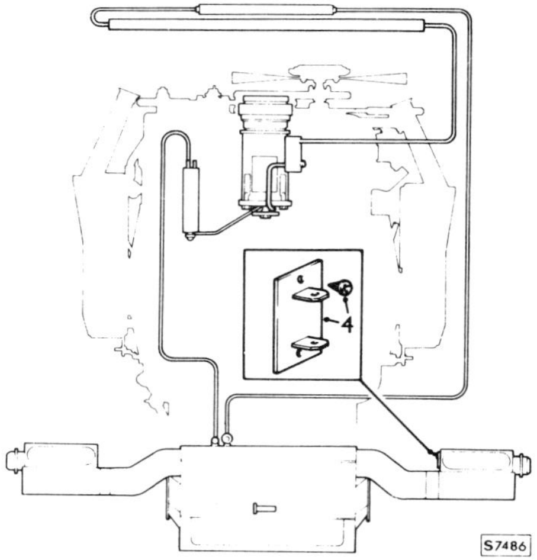

BLOWER MOTOR POWER RELAY

Remove and refit   82.20.06

Removing

1. Disconnect battery earth lead.
2. Remove left hand cheek and air vent assembly from centre console, see 76.25.02.
3. Disconnect relay from mounting block.

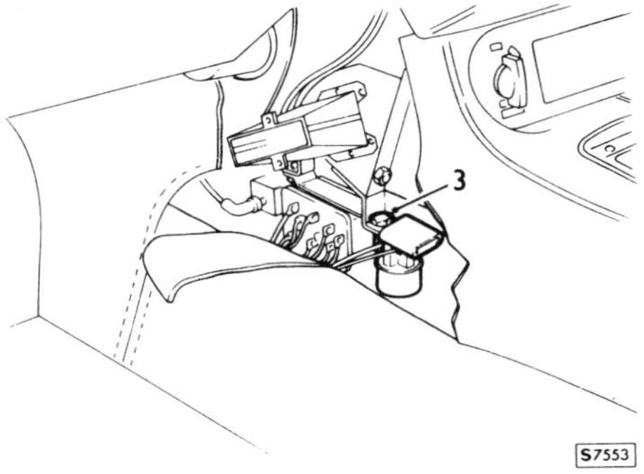

Refitting

Reverse operations 1 to 4.

IN CAR TEMPERATURE SENSOR

Remove and refit   82.20.03

Removing

1. Disconnect battery earth lead.
2. Remove facia panel assembly, see 76.46.01.
3. Remove air tube.
4. Disconnect cables from sensor terminals.
5. Release two plastic studs securing sensor to mounting bracket, remove sensor from bracket. Sensor is fitted to the crash roll on later cars.

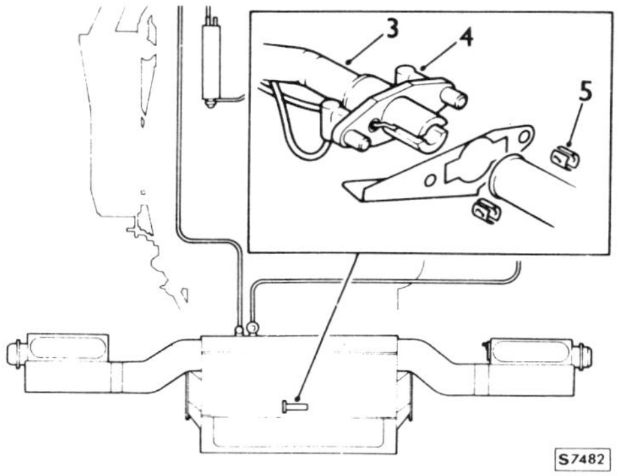

Refitting

Reverse operations 1 to 5.

BLOWER MOTOR POWER RELAY

Refitting

Reverse operations 1 to 3.

This relay and the adjacent relay are replaced by a four-way unit on later cars — see illustration.

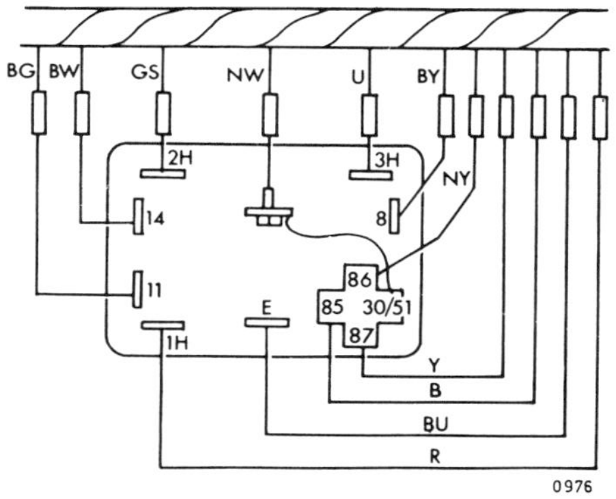

Refitting

Reverse operations 1 to 3.

The in-car sensor is fitted to the crash roll on later cars.

82–15

2 Place link system in **FULL COLD** position by rotating servo-motor clockwise. Apply slight pressure to main link ensuring bottom heater flap only is sealed. (Check for correct sealing by applying slight anti-clockwise pressure to defrost lever — there should be no resultant movement.)

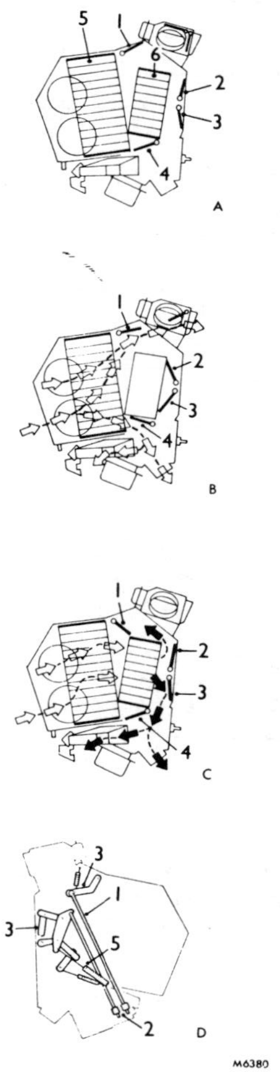

TEMPERATURE SELECTOR

Remove and refit 82.20.10

Removing
1 Disconnect battery earth lead.
2 Remove console assembly, see 76.25.01.
3 Peel back carpet at bottom end of right hand air conditioning unit stay.
4 Remove bolt and washer securing stay to transmission tunnel; ease stay aside to give access to switch panel and switch cover at right hand side.
5 Remove two screws securing switch cover to unit at right hand side.
6 Remove nut and washer securing switch cover to unit at lower, right hand side, release harness 'P' clip from stud.
7 Remove switch cover.
8 Note position of and disconnect vacuum pipes from vacuum switch.
9 Disconnect earth lead from lower micro-switch.
10 Remove nut and washer securing switch panel at upper right hand side.
11 Peel back carpet at bottom end of left hand air conditioning unit stay.
12 Remove bolt and washer securing stay aside to give access to switch panel.
13 Remove two nuts and washers securing switch panel to unit at left hand side.
14 Release switch panel from mounting studs.
15 Note colour of and disconnect temperature selector cables at snap connectors.
16 Remove two screws securing selector assembly to switch panel, remove selector assembly.

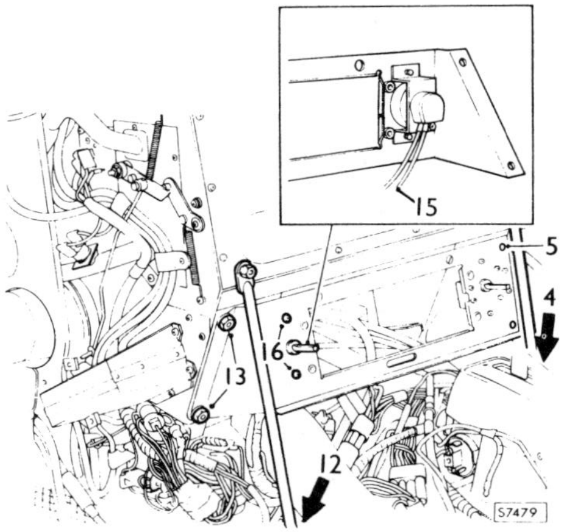

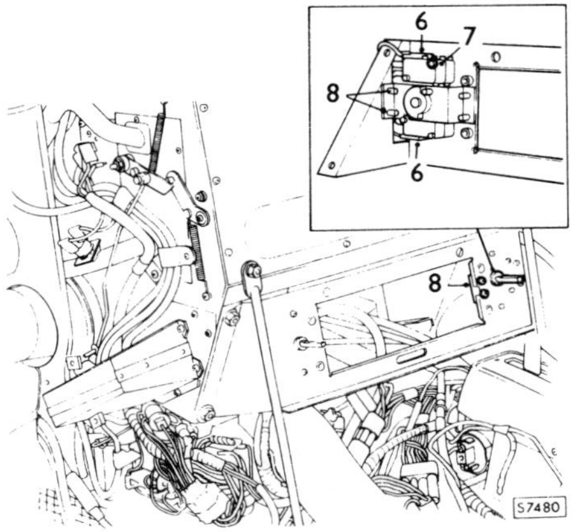

9 Remove circlip securing cam assembly, remove cam assembly.
NOTE: If replacement items are fitted, care must be taken to ensure that only correct replacements are used and that individual items are replaced in correct position.

Refitting
10 Insert new selector unit into mounting bracket. Ensure locating washer is in position.
11 Reverse operations 1 to 7.
NOTE: When refitting cams ensure that vacuum switch operating rod is pressed back to allow camshaft into position.

Refitting
Reverse operations 1 to 16.

MODE SELECTOR

Remove and refit 82.20.11

Removing
1 Carry out operations 1 to 15 of 'Temperature Selector – Remove and refit', see 82.20.10.
2 Disconnect opticell fibre elements by pulling each from opticell lens hood.
3 Disconnect mode selector cables at block connector.
4 Remove switch panel and radio aperture escutcheon assemblies.
5 Remove radio aperture escutcheon from switch panel.
6 Note position of and disconnect cables from mode selector micro-switches.
7 Note positions of micro-switches, remove two screws and nuts securing each pair of switches. Retain distance pieces for refit operation.
8 Remove two crosshead screws securing vacuum switch mounting bracket, remove vacuum switch assembly.

PROCEDURE FOR FLAP LINK ADJUSTMENT 82.20.17

NOTE: When adjustment procedure is in progress the main system tension spring '**Must not be connected**'. Only bottom by-pass spring may be connected during adjustment.

1 Ensure top heater flap adjuster, and bottom by-pass adjustable link are free to move.

82–16

3   When in FULL COLD position apply pressure to top heater flap in clockwise direction ensuring that flap seals. Lock top heater flap adjuster. Check that both heater flaps close together by moving main link.
4   Place link system in 'full heat' position by rotating servo motor anti-clockwise. Top heater flap should just make contact with casing. Maintain this position, seal bottom by-pass flap with link, and lock.
5   With main linkage system now adjusted, rotate main link to ensure flaps seal at each end of travel, and check that link system moves freely.
6   Attach main system tension spring to linkage, check that bottom by-pass flap seals correctly.

KEY TO DIAGRAMS

A   FLAP LOCATION
1   Top By-Pass Flap.
2   Top Heater Flap.
3   Bottom Heater Flap.
4   Bottom By-Pass Flap.
5   Evaporator Matrix.
6   Heater matrix.

B   FULL COLD CONDITION
1   Flap Open (Cool Air).
2
3 } Flaps Closed (No Hot Air).
4   Flap Open (Cool Air).

C   FULL HEAT CONDITION
1   Flap Closed (No Cool Air)
2
3 } Flaps Open (Heated Air)
4   Flap Closed (No Cool Air)

D   FLAP OPERATING RODS
1   Adjusting Rods.
2   Adjusters (On Servo Unit).
3   Operating Linkage For Flaps.
4   Main Tension Spring.

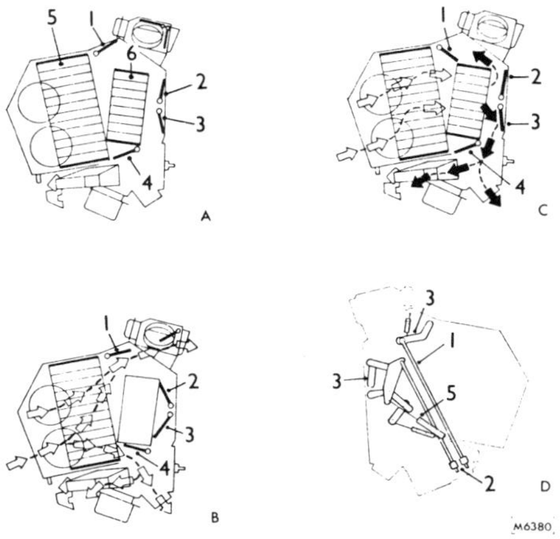

Refitting
Reverse operations 1 to 6.
NOTE: If a replacement unit is fitted, ensure capillary is formed to exact dimensions of unserviceable unit. When refitting capillary probe section ensure that it makes contact with evaporator matrix.

THERMOSTAT

Adjust   82.20.19

1   Remove clip from rear of thermostat and locate adjusting screw. Note that one turn clockwise = +3°C.
2   Refit clip.

DATA
Cut-in temperature = 2 ± 1°C.
Cut-out temperature = 1 ± 1°C.

THERMOSTAT

Remove and refit   82.20.18

Removing
1   Disconnect battery earth lead.
2   Remove right hand underscuttle casing, see 76.46.11/15.
3   Remove right hand console side casing, see 76.25.02.
4   Remove two screws securing thermostat to support bracket.
5   Note position of and disconnect cables at lucar connections.
6   Carefully remove thermostat by pulling capillary tube from heater/cooler unit.

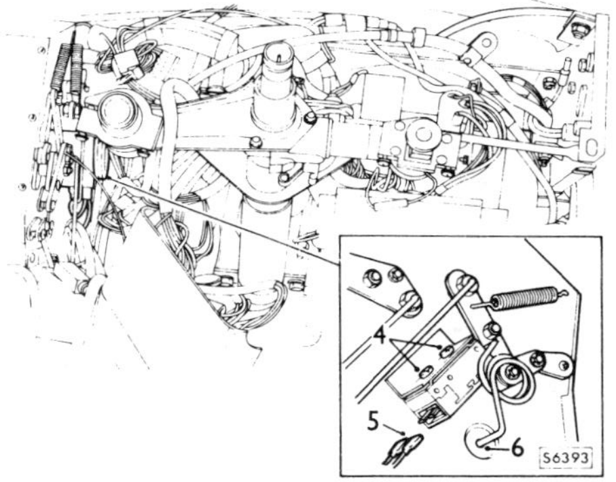

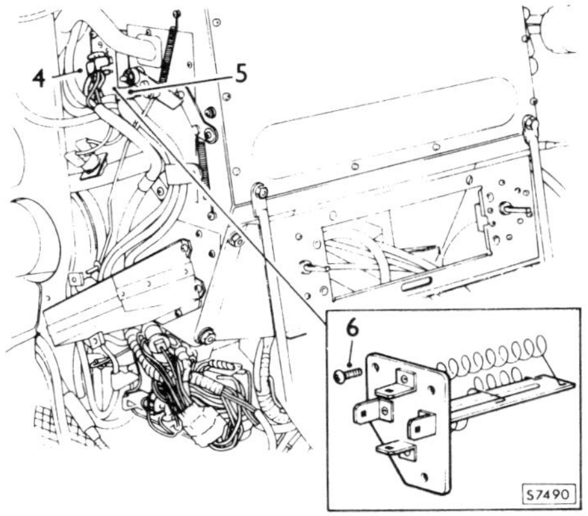

6   Remove three crosshead screws securing resistance unit to heater/cooler unit, remove resistance unit.

Refitting
Reverse operations 1 to 6.

BLOWER MOTOR RESISTOR UNIT

Remove and refit — Right hand drive   82.20.26

Removing
1   Disconnect battery earth lead.
2   Remove glove box compartment liner, see 76.53.03.
3   Note position of and disconnect cables from resistance unit lucar connections.
4   Remove screw securing vacuum hose 'P' clip, move hose assembly to one side.
5   Remove three crosshead screws securing resistance unit to heater/cooler unit, remove resistance unit.

Refitting
Reverse operations 1 to 5.

BLOWER MOTOR RESISTOR UNIT

Remove and refit — Left hand drive   82.20.26

Removing
1   Disconnect battery earth lead.
2   Remove drivers side underscuttle casing, see 76.46.11.
3   Remove drivers console side casing, see 76.25.02.
4   Note position of and disconnect cables from resistance unit lucar connections.
5   Remove screw securing vacuum hose 'P' clip, remove hose assembly to one side.

82—17

BLOWER MOTOR RELAY

Remove and refit   82.20.27

Removing
1. Disconnect battery earth lead.
2. Remove left hand console side casing, see 76.25.02.
3. Remove four screws securing footwell air outlet duct, remove duct.
4. Peel back carpet to gain access to relay.
5. Note position of and disconnect electrical connectors at relay.
6. Remove screw securing brown/white cables to relay terminal, disconnect cables.
7. Remove nut securing relay to rear mounting stud.
8. Remove two nuts securing earth cables, vacuum solenoid and relay to front mounting stud.
9. Remove blower motor relay.

Refitting
Reverse operations 1 to 9.

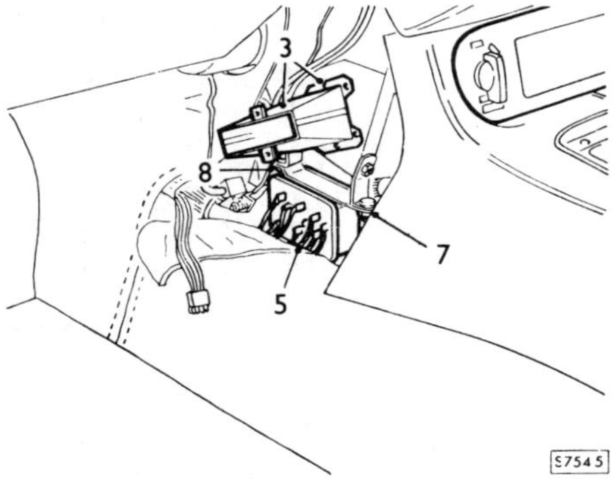

This relay and the adjacent relay are replaced by a four-way unit on later cars. See diagram, 82.20.03.

WATER VALVE TEMPERATURE SWITCH

Remove and refit   82.20.29

Removing
1. Disconnect battery earth lead.
2. Remove left hand underscuttle casing, see 76.46.11 (L.H. drive cars).
3. Remove left hand console side casing, see 76.25.02 (L.H. drive cars).
4. Remove glovebox liner, see 76.52.03 (R.H. drive cars).
5. Disconnect cables at lucar connectors on switch.
6. Withdraw two securing screws and remove switch.

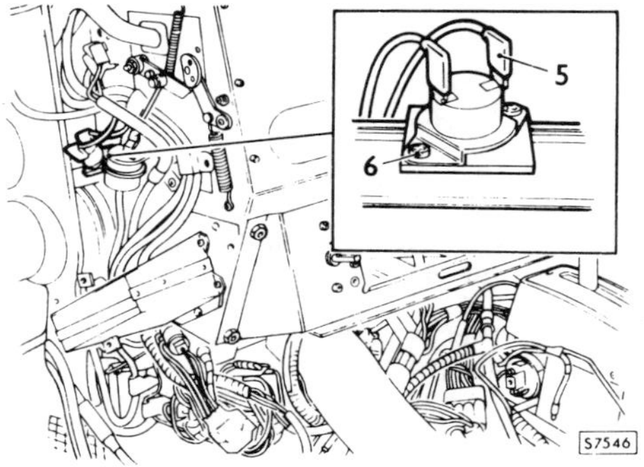

Refitting
Reverse operations 1 to 5.

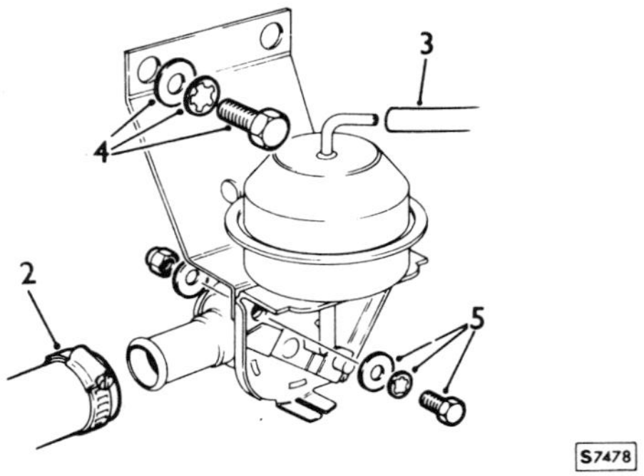

4. Remove two bolts securing valve and bracket assembly to bulkhead, remove water valve assembly.
5. Remove two bolts securing valve to bracket, remove valve from bracket.

Refitting
Reverse operations 1 to 5.

VENTILATOR(S)

Remove and refit – Centre   82.20.38
                Right hand   82.20.39
                Left hand   82.20.40

Removing
1. Remove facia, see 76.46.01.
2. Depress clips securing ventilator to facia.
3. Remove ventilator.

Refitting
Reverse operations 1 to 5.

WATER VALVE

Remove and refit   82.20.33

Removing
1. Drain coolant, see 26.10.01.
2. Release clips securing hoses to water valve; disconnect hoses.
3. Disconnect vacuum hose from top of valve.

Refitting
Reverse operations 1 to 6.

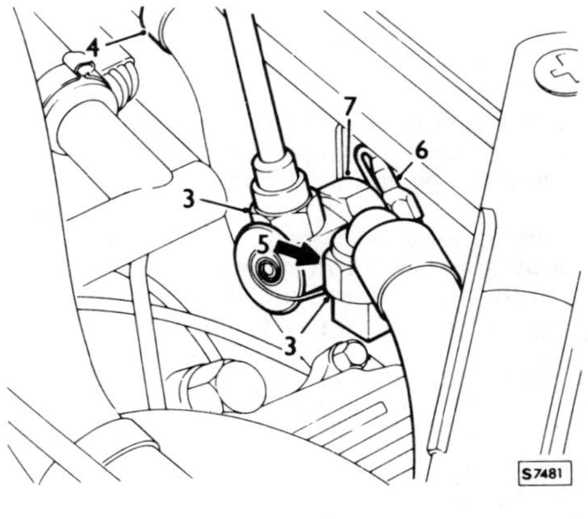

Refitting
11. Reverse operations 1 to 10 using new gauge filter.
12. Charge system, see 82.30.08.

BLOWER ASSEMBLY

Remove and refit – R.H. Unit   82.25.13

The blower fans are heavy duty motors with impellors attached. Speed is varied by controlled switching of resistances in series with the motors. The right hand unit has the ambient temperature sensor mounted in the inlet duct. Air flow control flaps are operated by a vacuum actuator situated in the side of the inlet duct.

Removing
1. Disconnect battery earth lead.
2. Remove dash liner, see 76.46.11.
3. Remove console side casing, see 76.25.02.

EXPANSION VALVE

Remove and refit   82.25.01

Removing
1. Depressurise air conditioning system, see 82.30.05.
2. Partially drain engine coolant.
3. Disconnect hose unions and seal with clean blanking caps.
4. Release clip securing water valve hose, disconnect hose from valve and move clear of expansion valve.
5. Remove asbestos padding from capillary tube.
6. Disconnect capillary tube at union.
7. Release valve by unscrewing union nut.
   NOTE: To avoid straining joint or pipe ensure that valve is held firmly as union is unscrewed.
8. Slacken two screws securing capillary tube clamp to valve.
9. Remove valve assembly manoeuvring capillary tube clear of clamp.
10. Retrieve gauge filter from valve and discard.

82–18

Refitting
Reverse operations 1 to 17.

L.H. Drive cars only
4 Remove glove box liner.
5 Remove nuts securing component mounting plate to blower assembly and lower plate. Secure plate clear of work area.

All cars
6 Disconnect 'in car' sensor hose from blower assembly duct.
7 Disconnect pliable duct from stub pipes at side of air conditioning unit.
8 Disconnect blower harness at block connector.
9 Disconnect at vacuum pipe from flap operating screw on blower assembly.

R.H. Drive cars only
10 Remove nuts securing bottom of fuse box mounting plate.
11 Slacken nuts securing top of fuse box mounting plate.
12 Ease fuse box clear of blower assembly and secure clear of work area.

All cars
13 Open recirculation flap in base of blower assembly and hold open with a suitable wedge.
14 Remove bolts securing blower assembly to mounting brackets.
15 Ease blower assembly from location.
16 Remove tape securing ducting to assembly.
17 Remove ducting from assembly.

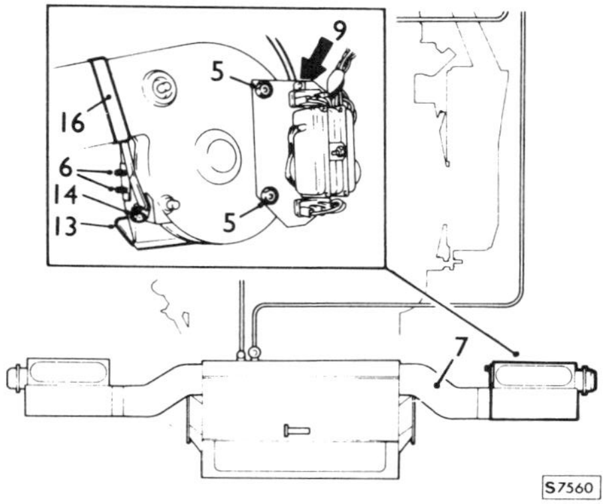

BLOWER ASSEMBLY

Remove and refit – L.H. Unit 82.25.14

Removing
1 Disconnect battery earth lead.
2 Remove dash liner, see 76.46.11.
3 Remove console side casing, see 76.25.02.

R.H. Drive cars only
4 Remove glove box liner.
5 Remove nuts securing component mounting plate to blower assembly, lower plate and secure clear of work area.

L.H. Drive cars only
6 Remove nuts securing bottom of fuse box mounting plate to blower assembly.
7 Slacken nuts securing top of fuse box mounting plate.
8 Ease fuse box clear of blower assembly and secure clear of work area.

All cars
9 Disconnect pliable ducting, fan to stub pipes, at side of air conditioning unit.
10 Disconnect blower harness at block connector.
11 Disconnect vacuum pipe from flap operating servo on blower assembly.
12 Open recirculation flap in base of blower assembly and hold open with suitable wedge.
13 Remove bolts securing blower assembly to mounting brackets.
14 Ease blower assembly from location.
15 Remove tape securing ducting to assembly.
16 Remove ducting from assembly.

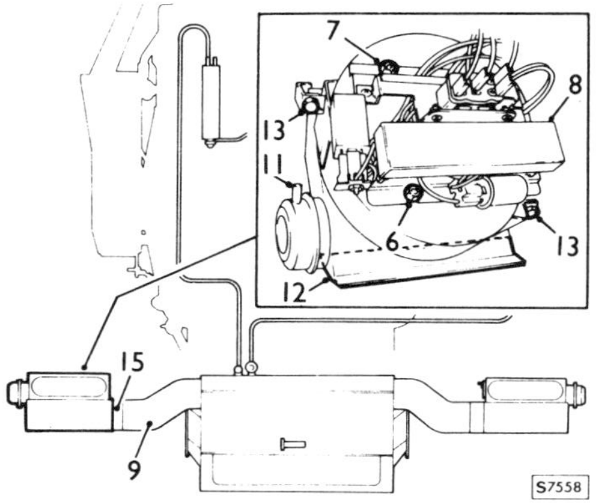

Refitting
Reverse operations 1 to 16.

HEATER MATRIX

Remove and refit 82.25.19

Removing
1 Remove air conditioning unit, see 82.25.21.
   IMPORTANT: Note the positions of all operating rods and levers before commencing the following operations. It is recommended that positions are marked with a scriber or similar method.
2 Disconnect cables from blower motor resistance unit.
3 Disconnect cables from water valve temperature switch.
4 Remove screws securing harness clip and bracket.
5 Remove screws securing vacuum pipe clips.
6 Disengage return springs from operating levers.
7 Remove screw securing lower flap operating lever to flap hinge and remove lever.
8 Slacken screw securing operating rod from vacuum servo to flap operating lever on matrix cover.
9 Release lever from rod.

10 Remove screws securing matrix cover plate to unit.
11 Remove screws securing heater matrix pipes retaining bracket to unit and remove bracket.
12 Remove pipe clips.
13 Ease matrix clear of unit using a straight pull.
14 Remove asbestos sleeve from top pipe.
15 Remove cover plate.
16 Remove water valve temperature switch from lower pipe.

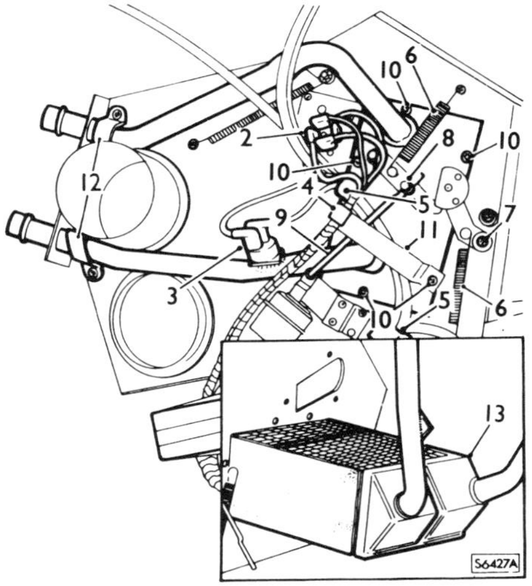

Refitting
Reverse operations 1 to 16.
NOTE: Ensure internal sponge pads and seals are refitted correctly, and replaced if necessary.

82–19

EVAPORATOR

Remove and refit 82.25.20

Removing

1. Remove air conditioning unit, see 82.25.21.
2. Remove screws securing heater matrix pipe retaining bracket to unit, and remove bracket.
3. Remove screws securing back plate to unit.
4. Ease rubber pad from back plate and remove screws securing expansion valve mounting plate to back plate.
5. Ease back plate over expansion valve. **NOTE:** Take great care to prevent damage to capillary tube.
6. Disconnect cables from thermostat on side of unit.
7. Remove screws securing unit to mounting bracket.
8. Ease thermostat capillary clear of evaporator.
9. Ease evaporator from housing.

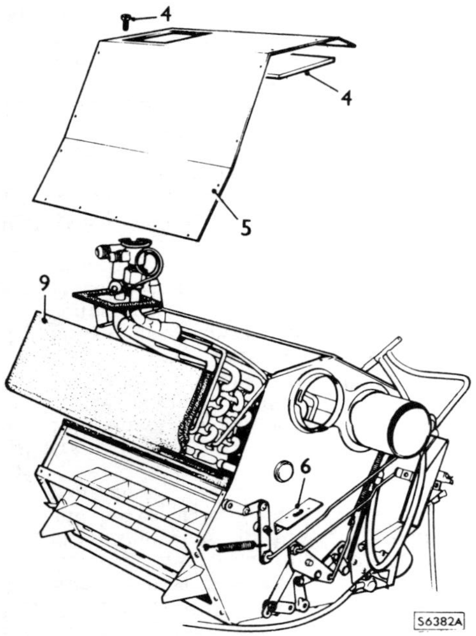

Refitting

Reverse operations 1 to 9.
NOTE: Ensure internal sponge pads seals are refitted or replaced as necessary.

AIR CONDITIONING UNIT

Remove and refit 82.25.21

NOTE: Water formed by condensation is discharged from the unit via drain tubes situated in the transmission tunnel. The condensate may collect beneath the car when parked, if the unit has been in use. This is a normal occurrence and requires no investigation.

Removing

1. Disconnect battery earth lead.
2. Depressurise air conditioning system, see 82.30.05.
3. Drain cooling system, see 26.10.01.
4. Remove facia, see 76.25.01.
5. Remove centre console, see 76.46.01.
6. Disconnect air conditioning hoses at bulkhead, expansion valve, connectors in engine compartment.
7. Disconnect heater hoses at bulkhead connectors to unit strut pipes.
8. Remove nuts securing unit to bulkhead in engine compartment.
9. Unclip main harness from securing clips on screen rail.
10. Remove bolts securing demist duct support rail to body mounting points.
11. Remove support rail.
12. Disconnect pliable ducting between unit and blower motors from stub pipes unit.
13. Remove rear compartment ducts.
14. Remove nuts and bolts securing unit support stays, recover stays.
15. Remove automatic gearbox selector quadrant cover.
16. Remove bolts securing upper steering column to mounting bracket, recover spacers and packing washers.
17. Remove bolts securing earth leads, and support stays to steering column mounting bracket. Recover washers.
18. Remove bolt securing mounting bracket to screen rail and recover bracket.
    NOTE: To facilitate refitting it is advised that the positions of ALL electrical multi pin plugs and connectors are noted. The positions and routes of all vacuum pipes noted, and where necessary mark mating ends of both electrical and vacuum connectors with identifying tape.
19. Disconnect blower motor flap vacuum pipes at 'T' piece.
20. Disconnect demister duct vacuum pipe at servo.
21. Disconnect main panel harness electrical connectors.
22. Remove harness securing clips.
23. Unclip 'in car' sensor from Left hand bulkhead support.
24. Remove nuts securing switch mounting panel to unit.
25. Remove screws securing mode switch protector plate, recover plate.
26. Disconnect vacuum pipes from mode switch assembly.
27. Disconnect earth wire and motor harness multi pin plug at air conditioning main harness.
28. Disconnect mode switch wires at snap connectors and recover switch mounting assembly.
29. Disconnect remaining vacuum pipes beneath unit.
30. Disconnect remaining block connectors and 'in car' sensor cables at snap connectors.
31. Disconnect feed wires from ambient sensor in right hand blower assembly.
32. Disconnect multi pin connector of windscreen wiper motor harness at bulkhead.
33. Ease drain tubes clear of grommets in transmission tunnel.
34. Ease main panel harness clear of unit.
35. Ease demist duct vane securing studs from screen rail and recover demist duct assembly.

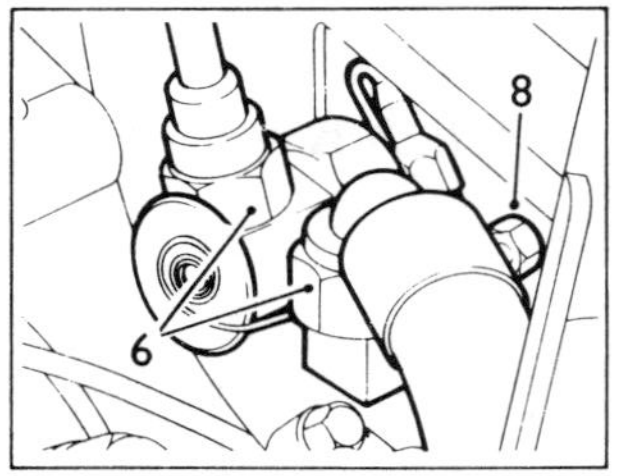

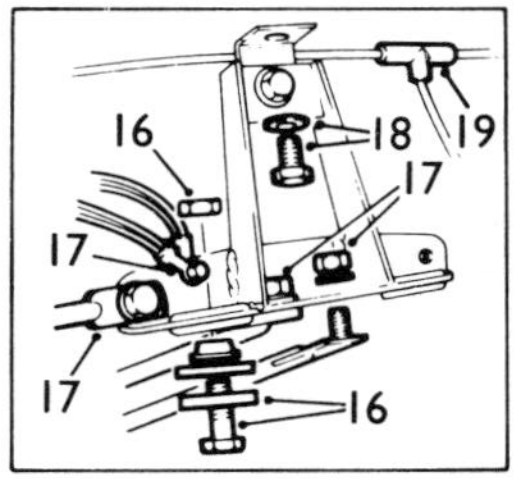

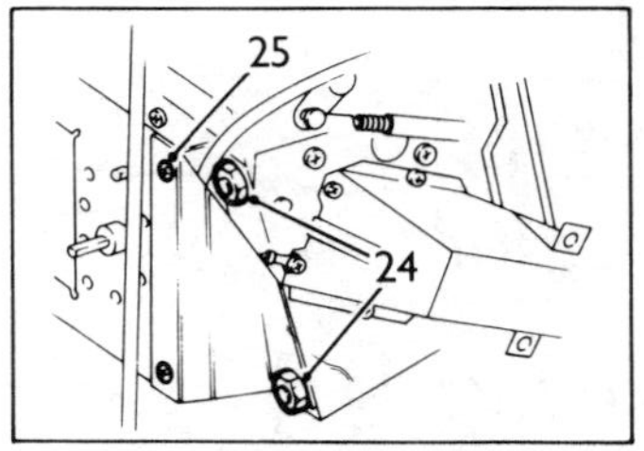

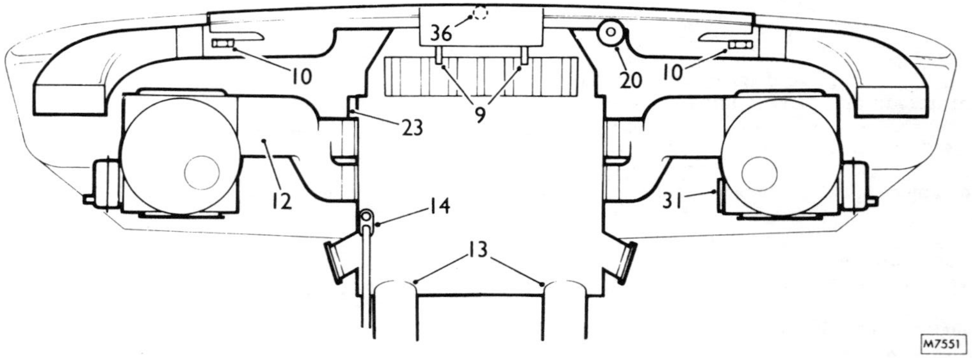

82–20

6 Note position of vacuum hoses to mode control switch assembly and disconnect.
7 Disconnect earth lead.
8 Remove control switch mounting panel securing nuts.
9 Remove L.H. support stay lower securing bolt.
10 Remove switch mounting panel to one side after disconnecting L.H. control switch cables.
11 Remove screws securing R.H. footwell vent to unit and remove vent.
12 Mark position of servo controlled operating rods and slacken clamp screws.
13 Note and disconnect vacuum pipes from actuators.
14 Disconnect electrical harness multi pin connectors.
15 Remove screws securing assembly to air conditioner housing and ease assembly clear.
16 Remove control unit cones.

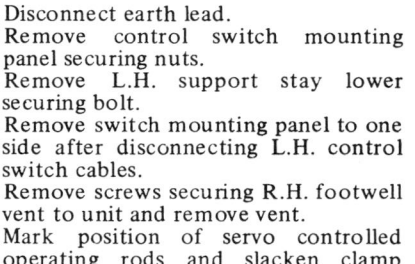

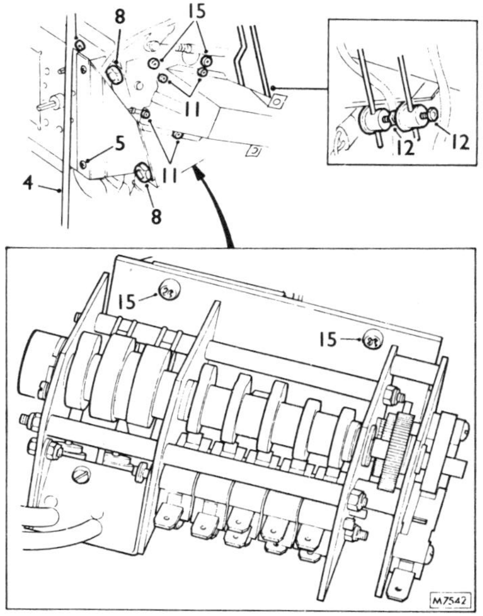

Refitting
Reverse operations 1 to 16.

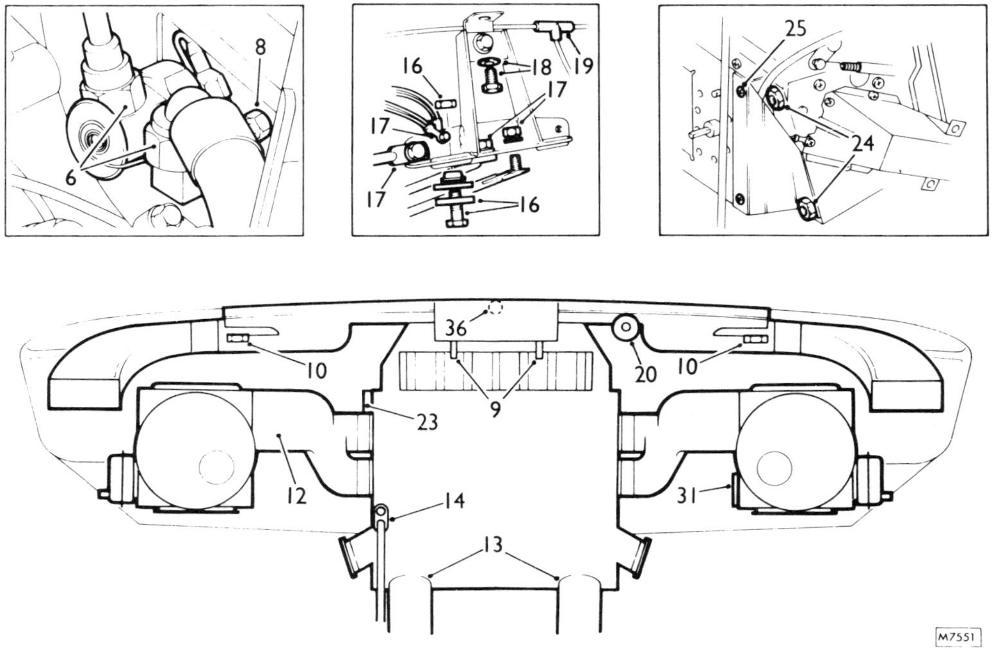

VACUUM SOLENOID

Remove and refit 82.25.23

Removing
1 Remove console L.H. side facing.
2 Remove screws securing footwell outlet vent to air conditioning unit.
3 Ease carpet clear of work area.
4 Remove nut securing earth leads to solenoid and blower relay mounting bolt.
5 Remove nut securing solenoid to mounting bolt.
6 Disconnect vacuum hoses from unit.
7 Disconnect electrical connectors.
8 Recover unit.

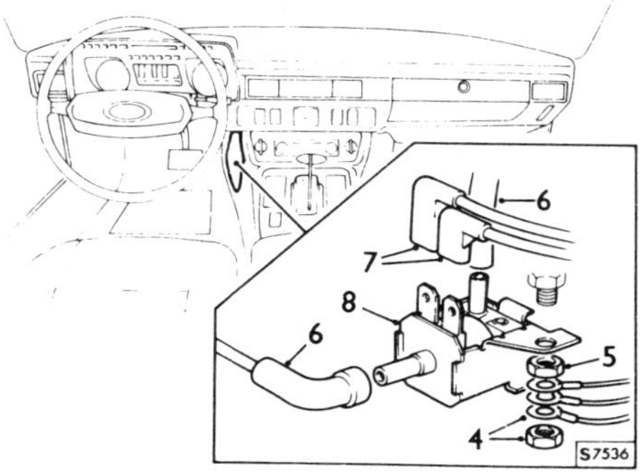

Refitting
Reverse operations 1 to 8.

SERVO AND CONTROL UNIT

Remove and refit 82.25.24

Removing
1 Disconnect battery earth lead.
2 Remove centre console, see 76.25.01.
3 Remove rear footwell ducting.
4 Remove lower bolt securing support stay at R.H. side of air conditioning unit.
5 Remove R.H. mode control switch assembly protective cover and release harness 'P' clip.

36 Remove screw securing air conditioning unit to top rail.
37 Manoeuvre unit from location. Take great care to prevent damage to unit or surrounding components.
38 Remove unit to work bench.
39 Remove face level vent and brackets from unit.
40 Remove demist duct assembly from unit.

Refitting
41 Replace all insulating pads. Reverse operations 1 to 40.

82—21

AMPLIFIER UNIT

Remove and refit   82.25.29

Removing
1. Disconnect battery, see 86.15.20.
2. Remove left hand console cheek and air vent.
3. Withdraw four retaining screws from air outlet duct and remove duct.
4. Disconnect cables from relay box, note cable positions to assist in refitting operation.
5. Remove nut retaining vacuum solenoid and relay box.
6. Slacken other nut retaining relay box.
7. Swing relay box clear of access to amplifier unit. 
8. Release amplifier retaining clip with flat bladed screwdriver.
9. Manipulate amplifier from beneath heater/cooler unit.
10. Disconnect amplifier cable harness at multi-pin plug/socket and earth connector.

Refitting
Reverse operations 1 to 10.

BLOWER ASSEMBLY

Overhaul   82.25.30

1. Disconnect battery, see 86.15.20.
2. Remove blower assembly. Right hand, see 82.25.13. Left hand, see 82.25.14.

Dismantling
3. Withdraw three self tapping screws from air inlet casing.
4. Part air inlet casing and motor assembly and disconnect electrical connections at lucars.
   NOTE: It is recommended at this stage that positions of various components are marked either with paint or scriber. One cable lucar has a raised projection which matches the aperture in motor casing. This assists in ensuring connections are replaced correctly and direction of rotation of motor is not altered.
5. Unscrew three bolts securing fan motor mounting bracket to outlet duct. 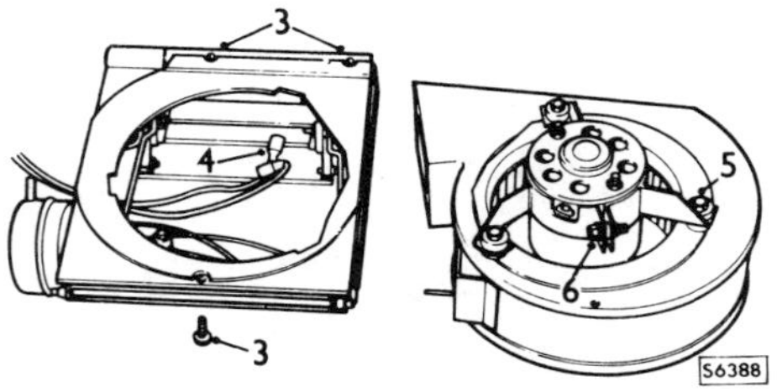
6. Release nut and bolt securing motor mounting clip and remove motor fan bracket.
7. Using appropriate Allen key remove impellor from motor shaft, note position on shaft for reassembling.
   NOTE: Blowers and impellor are handed for rotation and impellor operation, ensure correct motor and impellor are fitted if replacement is necessary.

Reassembling
Reverse operations 1 to 7.
NOTE: Ensure that impellor does not foul outlet duct, and that motor mounting clip does not obscure lucar connector spades.

CHARGING AND TESTING EQUIPMENT

Fit and remove   82.30.01

The charging and testing equipment consists of a charging manifold A fitted with two stop valves B and C and two pressure gauges D and E, a vacuum pump F, and a supply of refrigerant gas, G. One gauge is a compound type, reading both vacuum and positive pressure, and is connected to the suction side of the union housing; the other is a high pressure gauge and is connected to the delivery side.

WARNING: FOR SAFETY REASONS, THE ACCURACY OF BOTH GAUGES MUST BE CHECKED AT FREQUENT INTERVALS.

The stop valves enable either suction or delivery hoses, or both, to be connected to the centre port of the manifold. The centre hose H can be connected either to a vacuum pump or to a supply of refrigerant. Two hose connectors must be fitted with depressers to operate the Schrader valves. The equipment should be fitted with a means of accurately weighing the refrigerant container during the charging process.

Fitting
1. To fit the charging manifold to the air conditioning system, remove the protective sealing caps from the Schrader valves on fuel cooler and silencer.
2. Ensure both manifold stop valves are fully closed (screwed in).
3. Quickly fit hose connectors to correct Schrader valves. The gauges will display system pressures on suction and delivery sides.

Removing
4. To remove the equipment from the vehicle, quickly unscrew each connector in turn. This ensures that the Schrader valves are held open for the shortest possible time. Refit the valve sealing caps.

AIR CONDITIONING SYSTEM

Depressurise   82.30.05

Depressurising the system means that the system is vented until the refrigerant remaining is at atmospheric pressure. The system is then resealed to prevent air contaminating the components.

This procedure **MUST** be carried out before any connection is released.

It is very important that the method used is **EXACTLY** as described. Too rapid venting entrains the compressor lubricating oil and necessitates refilling to the correct level.

WARNING: NO SMOKING. POINT THE VENT HOSE IN A SAFE DIRECTION.

1. Remove protective cap from the discharge Schrader valve.
2. Using a piece of hose approximately 36 in (910 mm) long, fitted with a suitable connector, vent the system by **SLOWLY** screwing the hose connector on to the discharge Schrader valve.
3. If oil is seen escaping or if the vented gas becomes dense and white, **IMMEDIATELY** slow the flow rate by unscrewing the hose.
4. As the flow rate falls, the hose connector can be screwed further on to the union.
5. When no further gas escapes, and the hose connector is fully home, **IMMEDIATELY** unscrew it as quickly as possible.
6. Refit protective sealing cap.

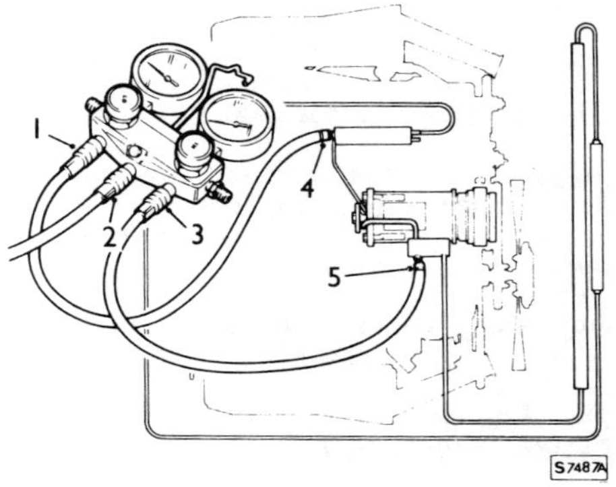

AIR CONDITIONING SYSTEM

Evacuate 82.30.06

The system is evacuated by removing all residual gas or air after depressurisation and/or repair using a vacuum pump. Evacuation must be carried out before charging, as the ability of the system to hold a high vacuum is a measure of its tightness: the vacuum also assists in drawing in the charge of refrigerant. The evacuation process serves to boil off any moisture in the system if ambient temperature is high enough. In conditions of low ambient temperature the purging method of system cleansing must be used before charging.

Any sign of a rapid fall in vacuum indicates a serious leak. This must be found and rectified **IMMEDIATELY**, as air is being drawn in through the leak.

Follow the procedure given under leak test before proceeding. The procedure given here refers to the evacuation of an old system. The method for a new system is similar, but no decrease in vacuum is permitted.

1. Remove both protective caps from the Schrader valves.
2. Ensure both manifold stop valves are fully closed (screwed in).
3. Quickly fit hose connectors to correct Schrader valves.
4. Fit centre hose of charging manifold to vacuum pump connection.
5. Fully open both valves of charging manifold.
6. Start vacuum pump.
7. Wait until a vacuum of 28 in Hg (50.8 Torr) has been drawn, or when the maximum that can be achieved with the prevailing barometric conditions is obtained.
8. Close both valves on charging manifold.
9. Switch off vacuum pump and wait 20 minutes.
10. A very slight pressure rise may occur due to the slow evaporation of liquid refrigerant or moisture entrained in the compressor oil.
11. If the vacuum holds satisfactorily, switch on vacuum pump, open both charging manifold valves and allow the pump to pull on the system for a further 20 minutes.
12. Fully close both charging manifold valves.
13. Switch off vacuum pump.
14. Disconnect centre hose from vacuum pump connection.

The air conditioning system is now ready for charging.

AIR CONDITIONING SYSTEM

Sweep (Purge) 82.30.07

The sweeping, or purging, operation given below may be used in addition to, and following, evacuation as a method of removing the last traces of moisture if ambient temperature is low.

The operation must be carried out if the system has inadvertently been left open for longer than a few minutes on a humid day. The operation must also be carried out if moisture is suspected in the system following the diagnosis of a fault.

The receiver/drier **MUST** be replaced immediately after the purging operation and before the final evacuation operation commences.

1. Evacuate system, see 82.30.06.
2. Disconnect vacuum connection from pump and connect to refrigerant supply.
3. Open refrigerant supply valve.
4. To purge length of hose, slightly crack centre connector at charging manifold; retighten connection.
5. Slowly open the suction side valve on charging manifold and allow ½ lb to 1 lb (0,23 kg to 0,45 kg) of gas to enter system.
6. Close suction side valve on charging manifold.
7. Close refrigerant valve.
8. Leave for 10 minutes.
9. Disconnect hose from refrigerant supply.
10. Open both valves on charging manifold to allow refrigerant to escape slowly from system **IMMEDIATELY** flow stops, reconnect hose to refrigerant supply.
11. Close suction side valve on charging manifold.
12. Loosen hose connection at suction side of charging manifold.
13. Open refrigerant supply valve slowly and allow gas to pass through entire system and escape at charging manifold for about 5 seconds.
14. Close refrigerant supply valve.
15. Close valve at pressure side of charging manifold.
16. Tighten suction side hose connection while gas still flows.
17. If system is being purged to remove excess moisture, change receiver/drier before proceeding to evacuation and charging. If system is being purged owing to low ambient temperature, evacuate and charge immediately.

AIR CONDITIONING SYSTEM

Charge 82.30.08

Charging the air conditioning system is the process of adding a specific quantity of refrigerant to the circuit. Before attempting the charging operation the system **MUST** have been evacuated and, if necessary, swept (purged) immediately beforehand. No delay between evacuation and charging procedures is permissible. Great care must be taken to charge correctly, as under charging will result in very inefficient operation, and over charging will result in very high pressures and possible damage to components.

NOTE: The **minimum** charge weight to fully charge the system is independent of temperature (be it engine or ambient).

The **maximum** charge that can be safely induced into the system to suit all climates and high ambient temperatures is 2.75 lb.

1. Evacuate the system, see 82.30.06.
2. Connect centre hose of charging manifold to supply of refrigerant. The supply available must be at least 7.2 lb (3,3 kg) weight.
3. Open refrigerant supply valve.
4. Purge centre hose by momentarily cracking connection at manifold block; retighten connector.
5. Record weight of refrigerant supply source.
6. Open both valves on charging manifold and allow refrigerant source pressure to fill vacuum in system.
7. Between ½ lb to 1 lb (0,23 kg and 0,45 kg) weight will enter the system. Record quantity.
   NOTE: The quantity drawn in will vary with ambient temperature.
8. Close pressure side valve on manifold block.
9. Ensure all clear and start vehicle engine. Run engine at 1500 r.p.m.
   Set air conditioning system blower speed control to 'FAST'.
   NOTE: This engages compressor clutch to start system circulation, and runs blower motors at fast speed to heat evaporator coil. Vapour will be turned to liquid in the condenser and stored in the receiver/drier.
10. Control flow of refrigerant with suction side valve on charging manifold, and allow a total weight (including operation 7) of 2½ lb ± 2 oz (1,13 kg ± 0,057 kg) refrigerant to enter system.
11. Close suction side valve.
    NOTE: Alternatively, observe sight glass on receiver/drier until sight glass clears, and no bubbles or foam are visible. Close suction valve.
    Reopen suction valve for 2 to 5 minutes (2 minutes if ambient temperature low, 5 minutes if high).
    This may allow an additional ¼ lb (0,11 kg) of refrigerant to enter the system.
12. Run system for 5 minutes, observing sight glass.

continued

14. If foaming very slight, switch off engine.
    NOTE: It is normal for there to be slight foaming if ambient air temperature is 70°F (21°C) or below.
15. Close refrigerant supply valve; disconnect hose.
16. Quickly disconnect hoses from Schrader valves on union block.
17. Fit protective sealing caps.
18. Switch on engine and check function of air conditioning system.
19. Switch off engine; flush engine compartment and interior of vehicle with shop compressed air line.
20. Conduct a leak test on installation.

AIR CONDITIONING SYSTEM

Leak test 82.30.09

The system shall show no leaks when tested by a detector with high sensitivity ideally of 1 lb (0,45 kg) in 32 years. Exceptions are the receiver/drier sight glass and uncapped Schrader valves, which must show no leakage when tested by a detector with sensitivity of 1 lb (0,45 kg) in 15 years.
Do not smoke while conducting the leak tests.
For safety reasons the discharge pressure gauge on the charging manifold must be checked at frequent intervals.
The testing area must be well ventilated, but free from draughts.
The system must be operated at high pressure before leak testing. As compressor discharge pressures are variable with ambient temperature the following procedure must be used.

Pressurising

1. Remove Schrader valve protective sealing cap from discharge connector.
2. Ensure both valves on charging manifold closed.
3. Quickly screw pressure hose from charging manifold block on to union.
4. If necessary blank off condenser.
5. Set heater control to full hot, fast fan.
6. Set air conditioning controls to full cold.
7. Start engine and allow discharge pressure to reach 225 lb/sq in (15,76 bars).
   Under no circumstances allow the pressure to rise above 250 lb/sq in (17,58 bars).
   CAUTION: Do not allow engine to overheat.
8. When pressure reaches 225 lb/sq in (15,76 bars) turn off engine.
9. Continue with Testing.

Testing

1. All joints and fittings shall be free of excess oil to eliminate the possibility of false readings caused by refrigerant absorption in the oil. For this reason any joint tightened to eliminate leakage should be cleared with compressed air to remove refrigerant vapour.
2. Since refrigerant vapour is heavier than air, the detector probe must be moved in the area below the joint tested.
3. The detector probe must be held for at least 3 seconds closer than .250 in (6,4 mm) to the joint tested.
4. The detector should be cleared with uncontaminated air before each usage.
5. False readings may occur if the detectors are used in atmospheres where solvents or volatile compounds containing halides (Fluorine, Bromine, Chlorine or Iodine) are present e.g. Trichlorethylene.
   Cigarette smoke and exhaust fumes may also cause false readings.
6. If the exact location of a leak is in doubt, liquid soap solution should be brushed on to the area and the position of the bubble observed.
7. The detector probe should be held at the air conditioning outlets with the system off and the fans turned on and off quickly to flow a small quantity of air. This procedure will find any leaks in the evaporator coil. The car body must be cleared of refrigerant before this test.

On some cars in which the servo action is fairly fast the separation of the fan speeds may not be discernable.

Aspiration and intermediate position check
Remove the heat source from the sensor. Within ten minutes, depending upon ambient conditions, the unit should shift off recirculation and the blowers will drop to one of the intermediate speeds. This test can be carried out on the road since thermistor aspiration will be better and hence the test will be performed more quickly. In certain high ambient conditions the system will be reluctant to come off recirculation, in which case the intermediate modes can be checked by inching the servo through these positions. This is done by turning the L.H. knob slightly clockwise until the servo motor is heard to operate, and then returning it to a lower position to stop the servo motor at the desired position.

Defrost and Fan vibration check
Turn the R.H. knob to defrost. The centre outlet flap should close and the screen outlets open. Air to the footwells should be cut off leaving air to the upper ducts only. The fans should shift to maximum speed and hot air should issue from the upper ducts. Fan vibration is best assessed under these conditions. Tests in accordance with the defrost schedule can be carried out at this point if desired.

Outlet Vent valve check
Check that air can be cut off from the outer face level vents by rotating the wheels beneath the outlets.

Settled mid range and Hi speed override check
Set the R.H. knob to auto. Set the L.H. knob to 75° and wait for the unit to settle. The fans should now be on low speed. Turn the R.H. knob to 'HI'. Maximum fan speeds should now be engaged.

AIR CONDITIONING – TEST OPERATION 82.30.11

NOTE: During the following tests windows should be closed and footwell fresh air vents shut 'off'.

Warm up and check operation of Thermostatic cut out and low speed override. R.H. contol to 'auto'.
With the engine cold turn the L.H. knob to full heat. Start the engine and run at 1000 r.p.m. If after any previous running the camshaft has turned to the cold position the servo will operate for a few seconds and then shut down. As the water temperature reaches 40°C the system will start up, the centre outlet and the fans will slip up to speed 2. This can be checked by turning the R.H. knob to low when a drop in speed should be noticed.

Sequence of operation check. R.H. control to 'auto'.

With the engine warm turn the L.H. knob to 65°. Operate the cigar lighter or other heat source and hold the heated unit about 1" below the sensor inlet hole, which is situated below the centre parcel shelf. The unit should then go through the following sequence in approximately 20 seconds.

1. Blower speeds will drop to low.
2. Temperatures will decrease, the upper temperature dropping more quickly than the footwell temperature.
3. After approximately 10 seconds the centre outlet flap will open.
4. Approximately 1 second after this the fan speeds will shift up to a medium 1.
5. A further 1 second later the fan speed will shift up to medium 2.
6. Another 1 second later the fan speed will shift to maximum, at the same time the recirculating flaps will close and the rush of air into the air boxes will be felt along the bottom edge of the lower trim panels. Turn the R.H. knob to 'LO' which should cause the fan speeds to drop. Return R.H. knob to auto setting.

82–24

AIR CONDITIONING EQUIPMENT

Preliminary tests 82.30.16

The following check must be carried out to ensure that the system is basically functional. These checks may also be used to ensure satisfactory operation after any rectification has been done. If the system proves unsatisfactory in any way refer to fault finding.

1. Check blower fans are giving air flow expected in relation to control switch position. Check that air delivered is equal at both outlets.

2. Check that compressor clutch is operating correctly, engaging and releasing immediately control switch is set to an 'ON' position.
   NOTE: The engine must be running and the thermostat control set fully cool.

3. Check radiator cooling fan starts operating when compressor clutch engages.
   NOTE: The engine must be running for this check.

4. Check that the compressor drive belt is correctly adjusted and is not slipping at higher engine speeds, at idle speed, or on sudden acceleration of the engine, with the compressor clutch engaged.

5. Observe sight glass on receiver/drier and check for frothing or bubbles with engine running at 1000 r.p.m. Slowly increase engine speed and repeat check at 1800 r.p.m.
   NOTE: It is normal for there to be slight foaming if ambient air temperature is below 70°F (21°C).

6. Check for frosting on connector union housing, the region around the suction part is normally cold, and **slight** frosting is permissible.

7. Check by feel along pipe lines for sudden temperature changes that would indicate blockage at that point.

Conclusions

If the above minimum requirements are met then it can be assumed that:—
a The thermostats are opening correctly.
b The water valve is opening fully.
c The flaps and linkages are correctly adjusted for the heating mode.
d The fans give adequate airflow at maximum speed.

If the above criteria are not met the causes may be related to:—

Thermostats

The water temperature gauge will not achieve 'Normal' position within 7 minutes and the air outlet temperature remains low. The thermostat(s) must be removed and checked for sticking open.

Water Valve

The temperature gauge reads 'Normal' but the air outlet temperature remains low. Check that the vacuum operated water valve is subjected to at least 8½" (21,6 cm) Hg of vacuum. If the valve is under adequate vacuum, change the valve. However if the vacuum is low, check that vacuum is being supplied to the whole system, that the water valve vacuum actuator, is operational and that the water valve vacuum switch is operational (See that the supply from the switch to valve is not pinched or trapped).

Flaps and Linkages

Inadequate flap sealing will result in low air velocity at the screen outlets. Check that the centre facia flap closes fully on 'Defrost' and that only a small air bleed to the footwells occurs. These leaks can be detected by hand and may be rectified by adjusting the linkage. Excessive airflow from the screen outlets in the heat mode may be caused by the demist control flap sticking open.

Blowers

If following flap inspection the air flow is still low, investigations should be carried out into the blower assemblies. Check that full voltage is being received on maximum speed and that the units are correctly wired for rotation. If all is correct the only remaining procedure is to change the fan assembly.

DEFROST AND DEMIST TESTS 82.30.15

Purpose

To ascertain that the heating/air conditioning system is functioning correctly in the 'Defrost' mode, and that adequate airflow is maintained in the heat mode to ensure that the windscreen remains mist-free.

Method

a Set the Control to '85°'.
b Set R.H. Control to 'Defrost'.
c Close end of dash outlets.
d Start engine and run for 7 minutes at 1500 r.p.m.
e During the running period measure the airflow from each screen outlet using checking ducts and velometer. Ensure that the centre dash outlet is closed and that it seals satisfactorily. The velocity from the screen outlets should be:—

    1550 ft/min

Also during the running period turn the R.H. control to 'HI' and open end of dash outlets. Using the screen outlet and end of dash checking ducts measure the resulting air velocity. This should be:

Minimum Velocity (ft/min)
Screen   End of dash
500   850

f At the end of 7 minutes running at 1500 rev/min check that the water temperature gauge indicates 'Normal'. Using mercury in glass thermometers check that the following minimum screen outlet temperatures are achieved.

Plenum Inlet °C	Screen Outlet °C (minimum)
10	54
12	55
14	55.5
16	56.5
18	57
20	58
22	58.5
24	59.5

CHARGING VALVE CORE 82.30.12

Remove and refit

A possible reason for very slow charging is a bent or damaged Schrader valve depressor. Do not attempt to straighten. The valve core must be replaced.
If excessive leakage is detected from the Schrader valve cores at the rear of the compressor, use a soap solution to ensure that the valve core itself is at fault. If the valve core is leaking replace it by following this procedure. Ensure replacement clean dry valve core is to hand before commencing operation.

Removing

1. Depressurise the system, see 82.20.05.
2. Remove valve core using a Schrader removing tool.

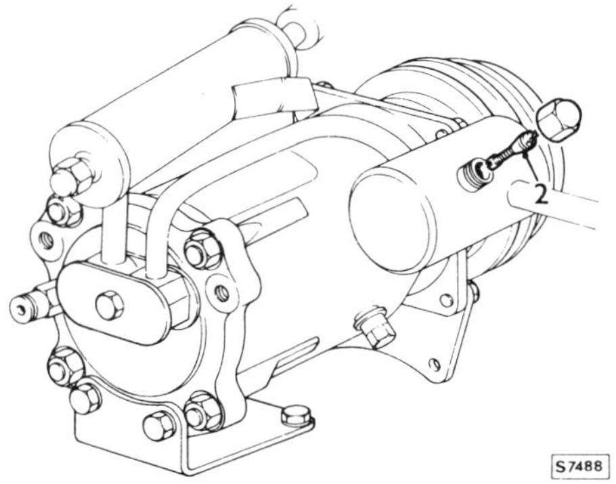

Refitting

3. Insert new valve into union and ensuring threads not crossed, screw home.
   NOTE: Do not overtighten.
4. Charge the system, see 82.30.08.

82–25

8 Place a thermometer in the air outlet louvres. Run the vehicle on the road and note drop in temperature with air conditioning system switched on or off.

9 Ensure condenser matrix is free of mud, road dirt, leaves or insects that would prevent free air flow. If necessary clear the matrix.

10 If the foregoing checks are not met satisfactorily, refer to rectification and fault finding procedures.

Description

The windscreen wiper motor and gearbox assembly is a two speed self parking unit driving two wiper arm wheel boxes via a flexible drive. The motor is a two pole permanent magnet type, the field assembly comprising two ceramic magnets housed in a cylindrical yoke. A worm gear formed on the extended armature shaft drives a moulded gear within the gearbox. The rotary motion of the motor being converted to a linear movement by a connecting rod actuated by a crankpin carried on the gear. The gearbox incorporates the self park mechanism which automatically parks the wiper blades at the end of the wiping cycle. This is achieved by extending the movement of the connecting rod via a slider block which operates a limit switch which is part of the terminal assembly unit. Two speed operation is obtained by switching the positive feed to the third brush when the higher speed is selected.

CAUTION: Whenever the slider block or limit switch and terminal assembly are dismantled it is essential on reassembly to maintain the correct operating relationship between the cam of the slider block and the switch operating plunger. To ensure correct parking sequence the slider block reassembly must be an exact reversal of the dismantling procedure. It is advisable to note relative positions of main gear, crankpin and slider block before dismantling commences.

The washer system consists of a reservoir and permanent magnet motor driving a pump. The motor is remotely controlled from a push button switch on the steering column, and is energised while the button is depressed.

Operation and maintenance

The windscreen wiper and washer are controlled from a single, steering column mounted, operating stalk. To operate windscreen wipers and washer proceed as follows.

1 SLOW (normal) operation. Move switch upwards to SLOW position. This position should be selected for moderate rain and heavy snow conditions.

2 FAST — move switch to next upwards position from SLOW. This position should be selected for heavy rain only.

3 SINGLE WIPE. Lift stalk against spring pressure and release, this will operate the windscreen wipers once only.

4 WASHER. Press button mounted in the end of the operating stalk. The washer will only operate when button is depressed.

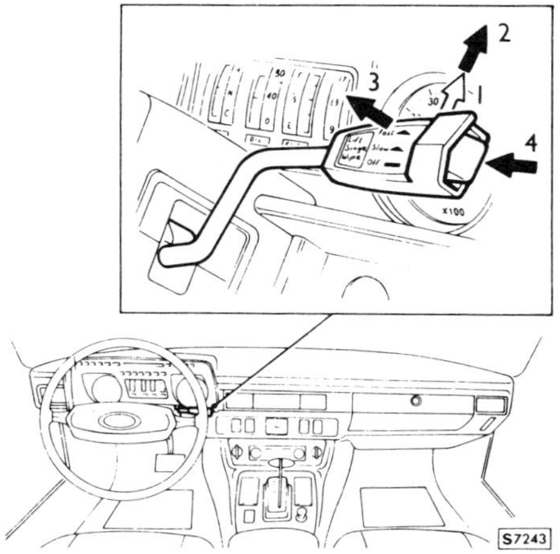

The washer reservoir should be filled with soft water where possible. If soft water is not available and continued use of hard water is necessary occasional attention should be paid to washer jet outlet holes. It is permissible to clear deposits from outlets with thin wire when necessary. The washer bottle should also be cleaned out, and the filter flushed occasionally. Windscreen washer additives should be confined to proprietary brands such as Trico or Holts.

CAUTION: Denatured alcohol or methylated spirit must NOT be used.

Washer jets can be adjusted with a screwdriver as illustrated, the jet should be adjusted to strike the top of the windscreen.

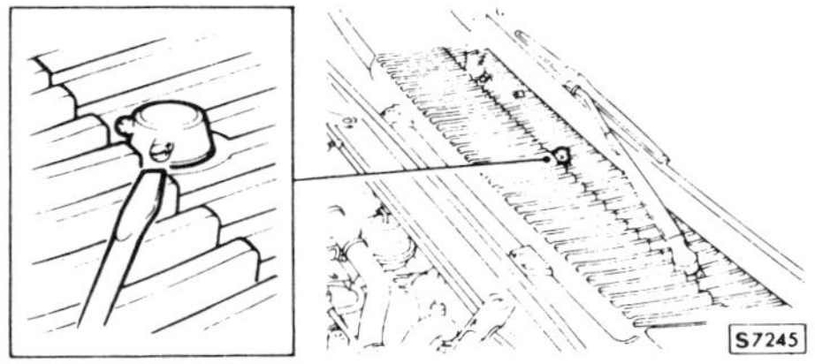

WASHER RESERVOIR

Remove and refit   84.10.01

Removing
1 Release and raise washer reservoir lid.
2 Disconnect pick up pipe from elbow connector lid.
3 Remove pump feed pipe, and elbow from lid.
4 Slacken nuts securing mounting bracket.
5 Remove reservoir.
6 Remove reservoir bracket retaining nuts, bolts and spring washers.
7 Recover bracket.

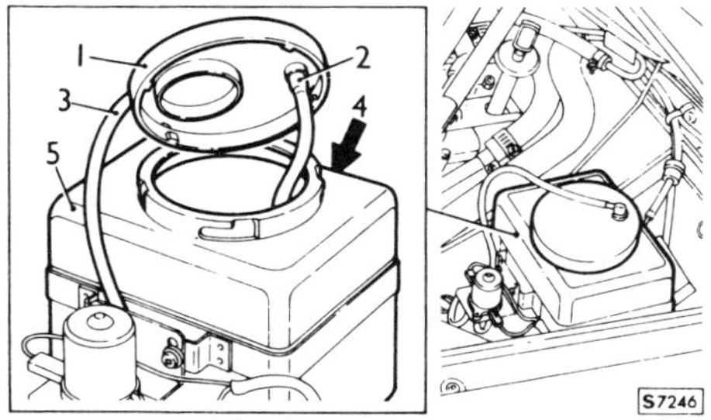

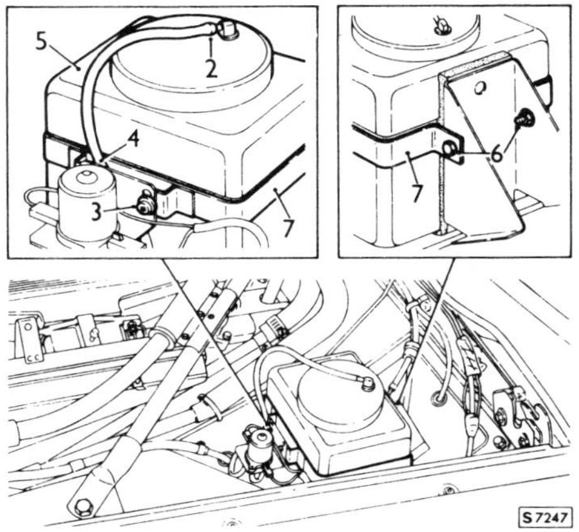

Refitting
Reverse operations 1 to 6.

WASHER JETS

Remove and refit   84.10.09

Removing
1 Disconnect battery earth lead.
2 Remove wiper arms, see 84.15.01.
3 Remove air inlet grille retaining nuts, bolts and washers.
4 Raise grille and motor assembly.
5 Disconnect washer pump to jet tube at jet.
6 Remove jet assembly retaining nut and shakeproof washer.
7 Remove jet assembly.

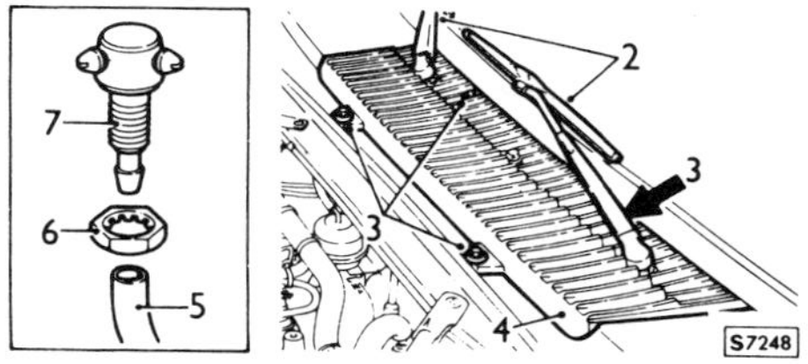

Refitting
Reverse operations 1 to 7.

WASHER RESERVOIR BRACKET

Remove and refit   84.10.02

Removing
1 Disconnect battery earth lead.
2 Disconnect washer pump feed pipe at reservoir lid connector.
3 Remove washer pump retaining screws, recover nuts and spring washers.
4 Withdraw feed pipe from bracket.
5 Remove washer reservoir.

84–1

WASHER PUMP

Remove and refit   84.10.21

Removing
1. Disconnect battery earth lead.
2. Disconnect supply cables at washer pump lucar connectors.
3. Disconnect washer feed pipes.
4. Withdraw screws securing pump to bracket.
5. Recover shakeproof washer and nuts.
6. Recover pump.

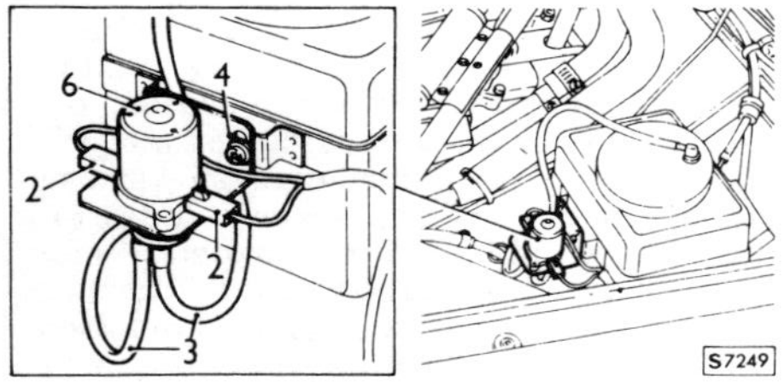

Refitting
Reverse operations 1 to 6.

WINDSCREEN WIPER ARMS

Remove and refit   84.15.01

Removing
1. Disconnect battery earth lead.
2. Raise plastic cover from spindle nut.
3. Remove arm retaining nut from spindle.
4. Remove arm assembly from spindle.
   NOTE: The position of arm in relationship to spline should be noted at this point.
5. Depress blade retaining clip and remove blade.

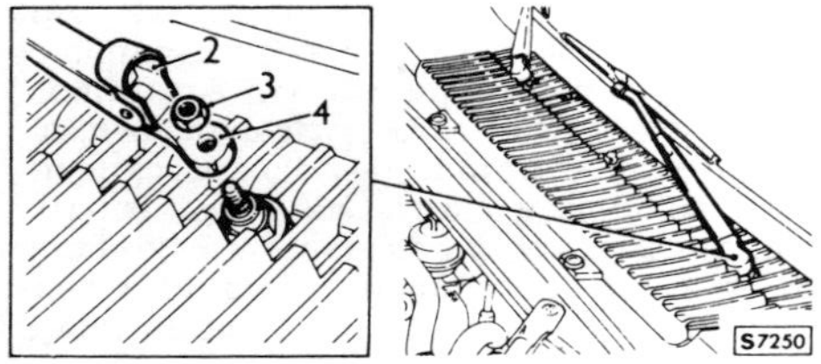

Refitting
6. Reverse operations 4 and 5.
   NOTE: Ensure arm assembly is replaced in same position as noted at 4 above.
7. Reverse operations 1 to 3.
8. Wet screen and check operation of wipers, check sweep area of arms and if necessary adjust.

WINDSCREEN WIPER BLADES

Remove and refit   84.15.05

Removing
1. Raise wiper arm from screen.
2. Depress blade retaining clip.
3. Remove blade.

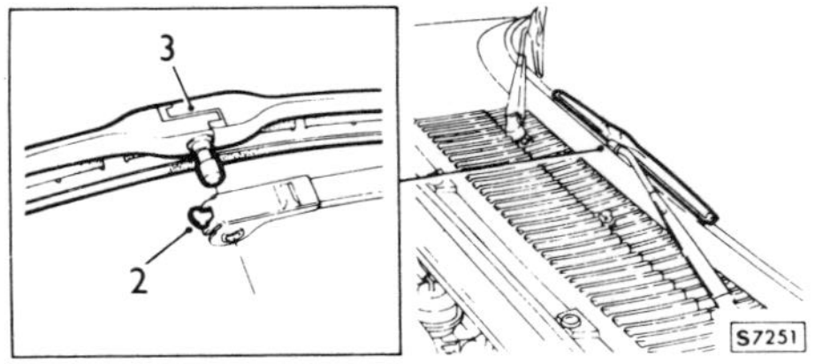

Refitting
Reverse operations 1 to 3.

WIPER MOTOR

Remove and refit   84.15.12

Removing
1. Disconnect battery earth lead.
2. Remove wiper arms, see 84.15.01.
3. Remove air inlet grille retaining nuts, bolts and washers.
4. Raise grille and wiper assembly.
5. Disconnect washer tube at jet.
6. Disconnect multi plug connector at bulkhead.
7. Remove grille and wiper assembly.
8. Remove nuts securing motor mounting bracket to grille.
9. Remove bolts securing bracket to motor.
10. Remove brackets.
11. Invert grille/motor assembly and remove wheel box spindle nuts.
12. Invert grille/motor assembly and remove motor and drive rack assembly from grille.
13. Remove rack cover plate retaining bolts and lift off cover.
14. Remove rack assembly.
15. Remove motor shroud, disconnect harness at gearbox connector and lay both aside.

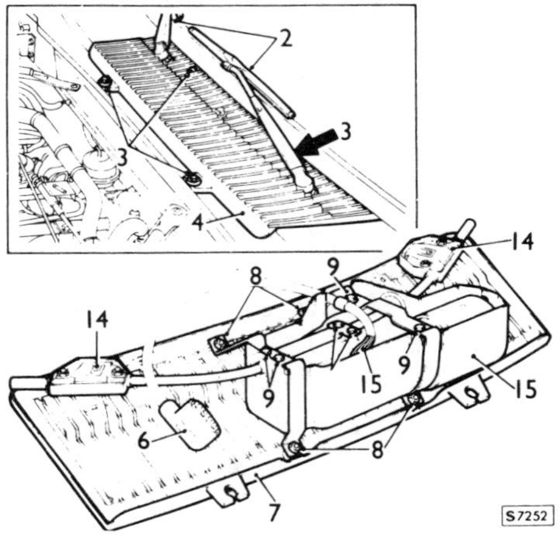

Refitting
Reverse operations 1 to 15.

WINDSCREEN WIPER MOTOR GEAR ASSEMBLY

Remove and refit   84.15.14

Removing
1. Remove windscreen wiper motor, see 84.15.12.
2. Remove screws securing plastic switch cover.
3. Remove gearbox cover plate.
   NOTE: At this point you are advised to mark the relative positions of driven gear, crankpin and slider block.
4. Remove circlip washer retaining gear assembly.
5. Remove gear and crankpin assembly. Note position of belled washer.

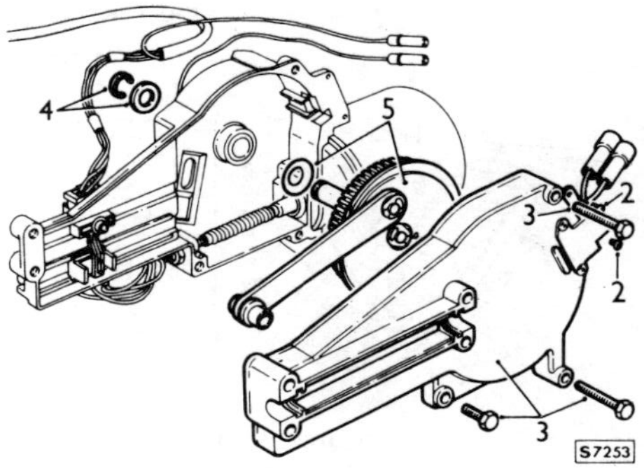

Refitting
Reverse operations 1 to 5.
NOTE: Ensure new gear assembly is fitted as noted in 3 above.

WINDSCREEN WIPER MOTOR RACK DRIVE

Remove and refit

Removing
1. Carry out operation 84.15.12 items 1 to 14.
2. Remove wheelbox back plate nuts.
3. Remove wheelbox back plate and shroud.
4. Pull wheelbox clear of rack drive.
5. Remove bundy tube from rack.

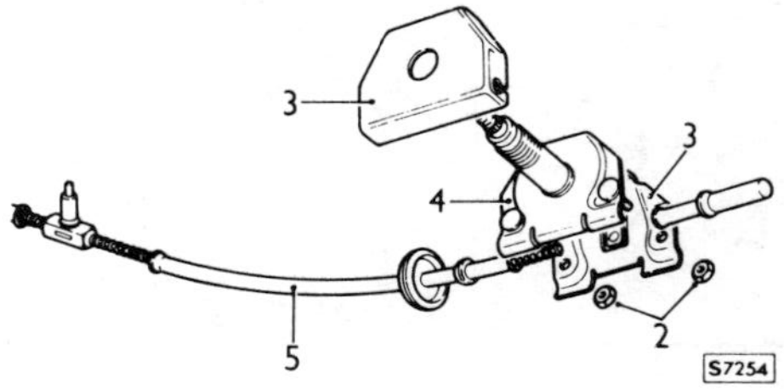

Refitting
Reverse operations 1 to 5.

84.15.24

WINDSCREEN WIPER WHEEL BOXES

Remove and refit
Left hand   84.15.28
Right hand   84.15.29

Removing
1. Carry out operation 84.15.24 items 1 to 4.

Refitting
Reverse items 1 to 4 of operation 84.15.24.

WINDSCREEN WIPER/WASHER CONTROL SWITCH

Remove and refit   84.15.34

Removing
1. Disconnect battery earth lead.
2. Slacken steering wheel adjustment ring and extend to maximum travel.
3. Remove screws securing steering column lower shroud and remove shroud.
4. Remove steering wheel, see 57.60.01.
5. Remove screw securing upper shroud to bracket on steering column.
6. Slacken pinch screw securing switch assembly to steering column. 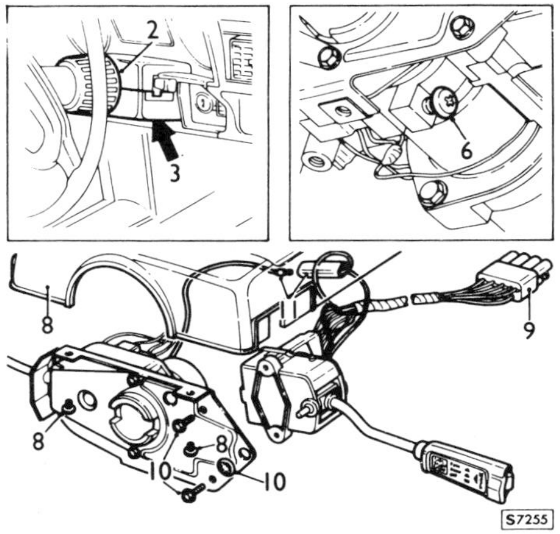
7. Ease switch assembly and upper shroud off steering column.
8. Remove shroud from switch assembly.
9. Disconnect harness at block connectors.
10. Remove spice nut and screws securing switch mounting plate.
11. Disconnect earth cable at snap connector.

Refitting
Reverse operations 1 to 11.

WINDSCREEN WIPER MOTOR HARNESS – BULKHEAD CONNECTOR

Remove and refit   84.15.35

Removing
1. Disconnect battery earth lead.
2. Remove wiper arms, see 84.15.01.
3. Remove fresh air intake, see 80.15.29.
4. Disconnect harness at multi pin connector on bulkhead.
5. Disconnect washer tube.
6. Remove bulkhead connector retaining screws.
7. Ease connector from location.
8. Disconnect panel harness connector and secure harness.

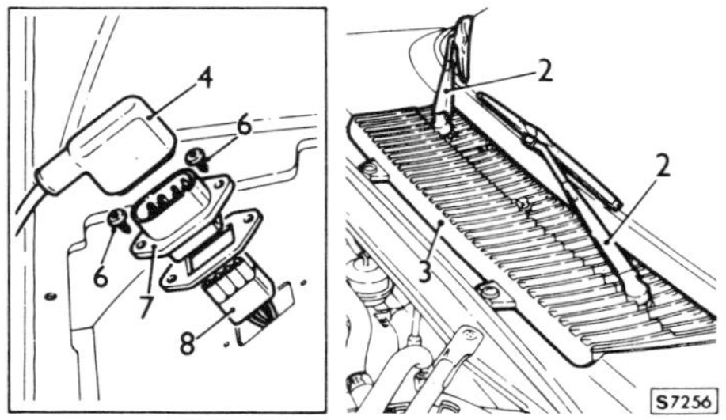

Refitting
Reverse operations 1 to 8.

84–3

ALTERNATOR

Description 86.10.00

The Lucas alternators, types 20ACR or 25ACR, fitted to specification of car, are high-output three-phase machines which produce current at idling speed.

The heatsink — rectifier, terminal block assembly can be removed complete. There are six silicon diodes connected to form a full-wave rectifier bridge circuit, and three silicon diodes which supply current to the rotor winding. Individual diodes cannot be removed from the heatsink assemblies. Regulation is by a Lucas type 8TRD control unit mounted in the slip-ring end bracket. There is no provision for adjustment in service.

Individual connectors are used to connect external wiring to the alternator. The alternators' main negative terminals are connected internally to the body of the machine. Provision is made for the connection of external negative wiring if required.

Surge protection device

The surge protection device is a special avalanche diode, fitted to the outer face of the slip-ring end bracket (not to be confused with a suppression capacitor, similarly fitted in the end bracket). The avalanche diode is connected between terminal 'IND' and frame and its purpose is to protect the regulator from damage by absorbing high transient voltages which occur in the charging system due to faulty cable connections, or if the cables are temporarily disconnected at the battery whilst the engine is running. (The surge protection device is intended to provide limited protection for the regulator under normal working conditions and therefore the service precaution not to disconnect any of the charging system cables, particularly those at the battery, while the engine is running, should still be observed.)

An alternative high output alternator, the Motorola 9AR 2512P, is fitted to some later cars; instructions for its overhaul, which differ in some details from those for Lucas alternators, are given in the appropriate sections of the manual.

CAUTION: No part of the charging circuit should be connected or disconnected while the engine is running.

When using electric-arc welding equipment in the vicinity of the engine take the following precautions to avoid damage to the semi-conductor devices used in the alternator and control box, and also the ignition system.

Disconnect battery earthed lead.
Disconnect alternator output cables.
Disconnect ignition and amplifier unit.

ALTERNATOR

Test (in situ) Lucas alternators only 86.10.01

(a) Output test

Equipment required: A moving coil ammeter or multi-range test meter on range 0–75 amperes.

This test should be carried out with the alternator at normal temperature. Run cold engine at 3,000 rev/min for 3 to 4 minutes.

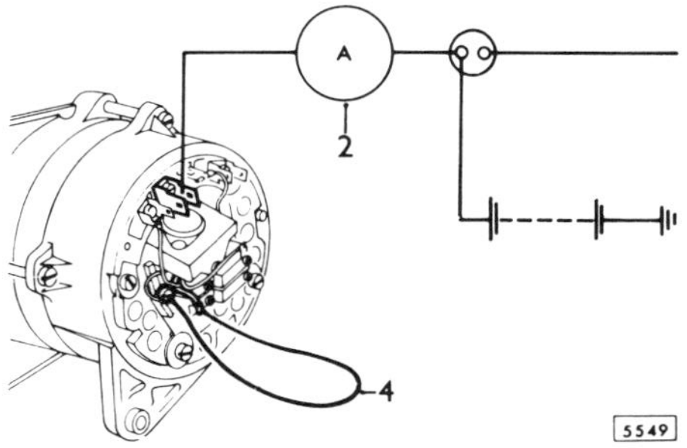

1 Disconnect battery earth lead.
2 Connect ammeter in series with alternator main output cable and starter solenoid.
3 Remove connectors from alternator. Remove moulded end cover and remake connectors.

4 Connect jumper lead to short out the 'F' and '−' terminals of the control unit.
   (This makes regulator inoperative by effectively linking 'F' green lead to alternator frame.)
5 Re-connect battery earth lead.
6 Switch on all vehicle lighting, headlights on main beam. Switch on ignition and check warning light is on.
7 Start engine, slowly increase speed to 3,000 rev/min. Ammeter reading should equal maximum rated output of 66 amperes.

(b) Voltage drop test (in situ)

Equipment required: A moving coil voltmeter multi-range test meter on 0–30 volt range.

To check for high resistance in the charging circuit.

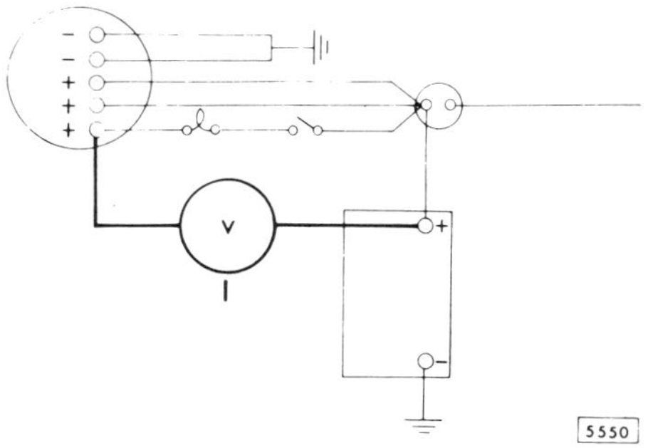

1 Connect voltmeter between battery +ve terminal and alternator main output terminal.

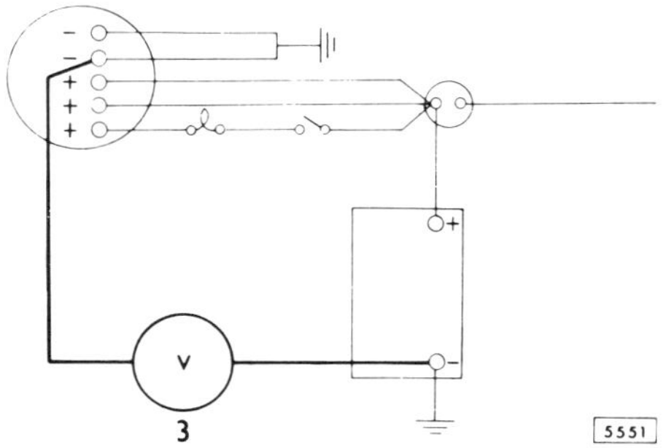

2 Switch on all vehicle lighting, headlights on main beam.
   Start engine and run at 3,000 rev/min. Note voltmeter reading. Stop engine.
3 Transfer voltmeter connections to battery earth and alternator negative terminal.
4 Repeat operation 2. Note voltmeter reading.
5 Voltage should not exceed 0.5 volts for positive side. Higher readings indicate high resistance in the circuit.

(c) Control unit test

Equipment required: A moving coil ammeter and moving coil voltmeter or multi-range test meters.

Circuit wiring must be in good condition, and all connections clean and secure. The battery must be in a well charged condition or be temporarily replaced by a charged unit.

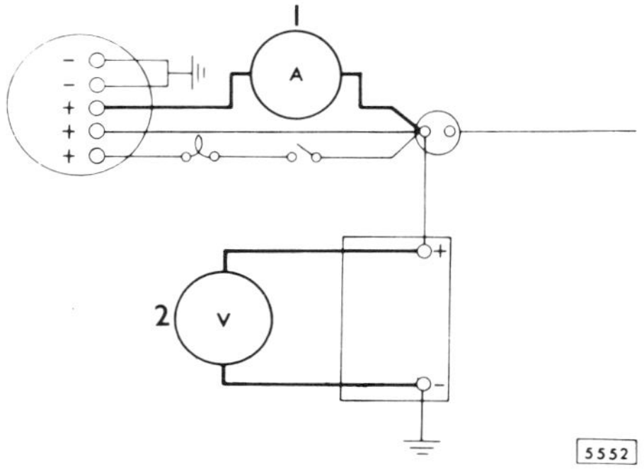

1 Connect ammeter in series with starter solenoid and alternator main output cable.
2 Connect voltmeter between battery terminals.
3 Start engine and run at 3,000 rev/min until the ammeter reads less than 10 amperes. Voltmeter reading should be between 13.6 volts and 14.4 volts.
4 An unstable reading or a reading outside the specified limits indicates a faulty control unit.

86–1

ALTERNATOR

Test (in situ) Motorola alternators 86.10.01

Equipment required: Voltmeter and ammeter, field rheostat.

NOTE: Before commencing tests ensure that battery is fully charged. If not, disconnect battery before recharging it.

Never disconnect battery, alternator or regulator with engine running.

Do not earth field winding (terminal marked 'EX', connected to regulator by green lead).

On cars fitted with air conditioning it is advisable to remove alternator from vehicle before carrying out tests 1 and 3 and to substitute bench tests for tests 4, 5 and 6.

Always disconnect battery when removing or refitting alternator.

Test 1
Ignition switched OFF. Check of stator windings. Check voltage on one of the three phases of stator windings, accessible to a probe from voltmeter passing through ventilation hole as shown.
Connect voltmeter first between phase and earth, then between winding and positive terminal, observing correct polarity.
Indication of any reading other than zero on voltmeter shows defective positive rectifier diode, necessitating changing of diode bridge, see 86.10.08.

Test 2
Ignition switched OFF. Check of battery connections. Check voltage at B+ terminal on alternator and at battery positive terminal. Voltage should be the same at both points. If voltage at B+ terminal is lower than battery voltage, or fluctuates, check for broken wires, faulty connections or corroded terminals.

Test 3
Ignition ON, engine not running. Check of field circuit.
Check voltage at slip ring, by touching probe of voltmeter on field terminal 'EX' with regulator attachment screws removed. If voltmeter reading is higher than 2 volts, field circuit is defective, remove brush holder by detaching green regulator lead from field terminal 'EX' and remove two setscrews, with washers, securing brush holder to alternator. Check that brushes are free to slide, undamaged and not excessively worn; new brushes protrude by approximately 0.35 in (9 mm) from the brush holder, and complete brush holder must be renewed if either brush protrudes by less than 0.15 in (4 mm). Ensure that brush leads are not frayed and are securely attached to brushes, and that slip-rings are clean. If in doubt, refer to 86.10.14 and check brushes electrically.
If voltmeter reads zero, check connections to regulator, ignition switch and ignition indicator lamp.
Also check regulator circuit by detaching its green lead from field 'EX' terminal and measuring voltage across field windings, which should not exceed 2 volts. If this voltage is between 8 and 12 volts, alternator is defective. If correct, proceed to test 5.

Test 4
Ignition ON, engine running faster than idle. Further check of field circuit.
If incorrect readings were obtained in Test 3, re-test field circuit by disconnecting regulator from field terminal 'EX' and connecting ammeter between this terminal and output terminal B+. If meter indicates current less than 1 amp, re-check brushes, leads and slip-rings.
CAUTION: Use a field rheostat in series with ammeter, so that excessive current which could flow if field is shorted will not damage ammeter.

Test 5
Ignition ON, engine running faster than idle.
Check of output voltage.

Check voltage both at output terminal (B+) and at positive terminal of battery. Correct voltage at both points is 14.2 volts ± 0.5 volts, at 25°C (77°F).
If difference between battery voltage and voltage at B+ terminal is more than 0.3 volts, check wiring and terminals for corrosion or breaks.

Test 6
Field lead disconnected, regulator disconnected, output terminal shorted to field terminal, ignition on, engine running at fast idle. Regulator and diodes check.
With alternator connected as specified above and shown in diagram, check voltage between output terminal B+ and earth. If voltage rises to 14 to 16 volts in this test, regulator is defective. Replace by new regulator.
If output voltage does not rise, and field circuit has been found satisfactory in tests 3 or 4, then either alternator stator or rectifier diodes are defective. Remove alternator for further attention, see 86.10.02.

ALTERNATOR

Remove and refit 86.10.02

Removing
1. Disconnect battery, see 86.15.20.
2. Remove right-hand air cleaner, see 19.10.01.
3. Remove air pump if fitted, see 17.25.07.
4. Slacken nut securing alternator mounting bolt.
5. Slacken nut securing alternator adjusting link.
6. Slacken trunnion block to mounting bracket securing bolt.
7. Withdraw electrical connector from alternator.
8. Slacken alternator adjusting link locknuts.
9. Ease alternator drive belt off pulley.
10. Withdraw bolts 4 and 5.
11. Remove alternator complete with pulley, from engine compartment.

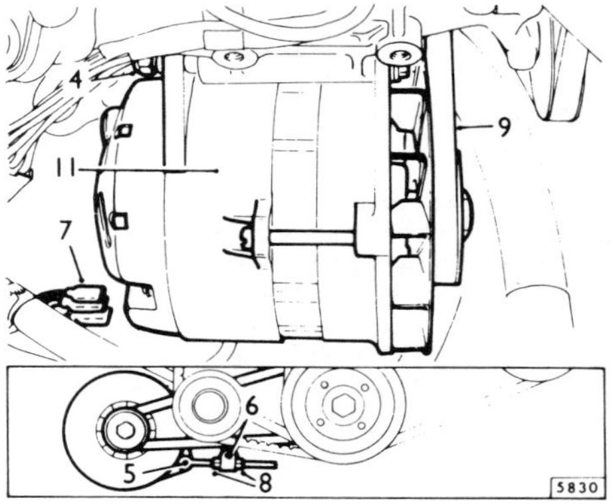

Refitting
Reverse operations 1 to 11.

86–2

ALTERNATOR – DRIVE BELT

Remove and refit 86.10.03

Removing
1. Remove right-hand air cleaner, see 19.10.01.
2. Remove fan belt, see 26.20.07.
3. Remove water pump/power assisted steering pump belt, see 57.20.02.
4. Remove air delivery pump/air conditioning compressor belt, see 17.25.15.
5. Slacken alternator mounting bolt securing nut.
6. Slacken alternator adjusting link securing nut.
7. Slacken trunnion block to mounting bracket securing bolt.
8. Slacken adjusting nuts.
9. Remove alternator belt.

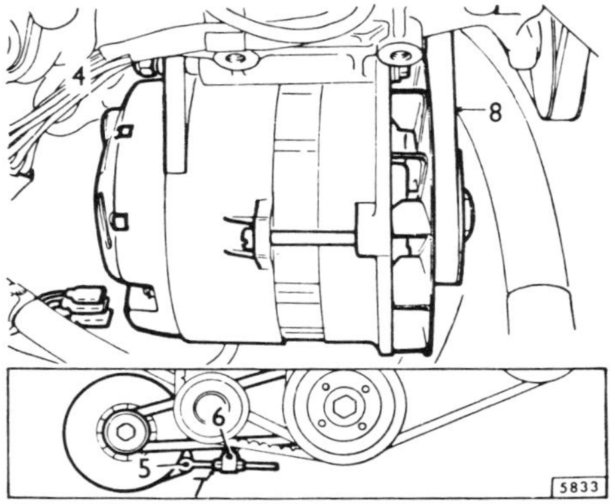

Refitting
Reverse operations 1 to 9.

ALTERNATOR PULLEY

Remove and refit 86.10.04

Removing
1. Slacken securing nut on pulley.
2. Slacken alternator drive belt, see 86.10.05.
3. Slip drive belt from pulley.
4. Remove nut and washer from alternator pulley.
5. Withdraw pulley from alternator drive shaft.

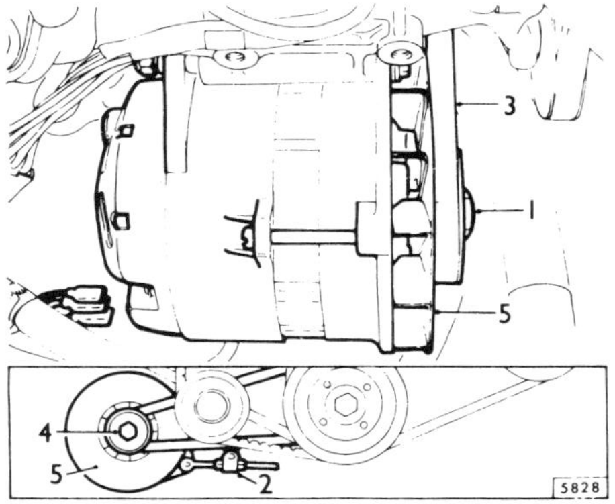

Refitting
Reverse operations 1 to 5.

ALTERNATOR DRIVE BELT

Adjustment 86.10.05

NOTE: Alternator drive belt adjustment can only be carried out successfully from underneath the car.

1. Slacken nut securing alternator mounting bolt.
2. Slacken nut securing adjusting link to alternator.
3. Slacken bolt securing trunnion block to mounting bracket.
4. Slacken adjusting link locknut.
5. Adjust belt by means of adjusting link nut. Correct tension is as follows:
   A load of 3.2 lb (1.5 kg) must give a total belt deflection of 0.17 in (4,4 mm) when applied at mid point of belt.
6. Reverse operations 1 to 4.

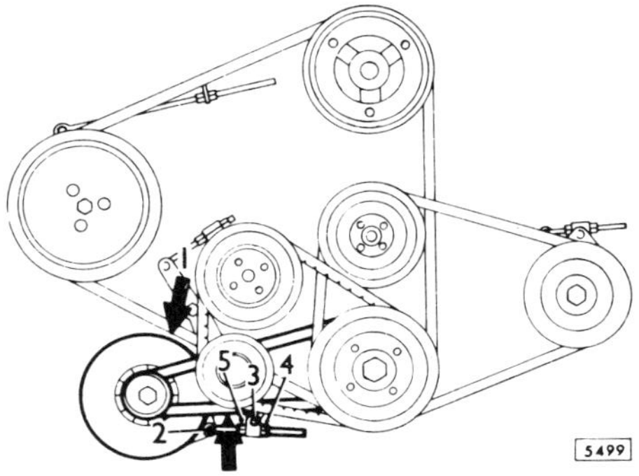

ALTERNATOR – MOTOROLA 9AR 2533 P

Overhaul 86.10.08

Dismantling
1. Detach nut, shakeproof washer and connector blade from B+ terminal at end cover.
2. Remove setscrew and washer securing capacitor to alternator case, separate Lucars and detach capacitor.
3. Withdraw three screws and remove moulded rear cover.
4. Remove two setscrews and washers, separate two Lucars and detach regulator.
5. Remove two setscrews and washers and lift out brush holder.
6. Clamp pulley, unscrew pulley nut and remove small washer, pulley, fan and large washer from alternator spindle.
7. Extract Woodruff key from spindle and remove spacer.
8. Remove four through-bolts; collect washers and square trapped nuts.
9. If casing halves do not readily separate, clamp alternator spindle in protected jaws of vice and draw off rear housing, with stator and diode bridge. Rear bearing will remain on spindle.
   CAUTION: Take care to avoid damage to stator and windings by rotor.
10. Remove alternator spindle from vice and draw off front housing. Collect short spacer adjacent to rotor.
11. If necessary, remove front bearing from housing by withdrawing three screws securing retaining plate and pressing out bearing.
12. If necessary, draw rear bearing off alternator spindle, using an extractor reacting against spindle end.
13. Mark position of stator ring in rear housing to ensure that it is correctly replaced.
14. Unsolder leads of three-phase windings and D+ (red) lead from diode bridge.
   CAUTION: Avoid transmitting excessive heat to diodes by incorporating a thermal shunt by using long-nosed pliers to grip each terminal as wire is unsoldered.

86–3

15 Withdraw two setscrews and lift out diode bridge. Collect washers.
16 Lift housing off stator, detach two terminals from housing and remove D+ lead complete.
17 Extract 'O' ring from bearing housing.
 NOTE: To remove diode bridge with a minimum of dismantling, carry out operations 1, 2, 3, 14 and 15 only. Refer to 86.10.14 for details on inspection and bench testing.

Reassembling

18 Fit new 'O' ring into recess in rear bearing housing.
19 Replace D+ (red) lead assembly in rear housing, securing it with two setscrews and washers at Lucar carrier and bolt and nut at D+ terminal. Thread loose end of lead through hole below D+ terminal.
20 Place stator and coils in marked position with three leads passing back through housing. Rest stator, with housing on top of it, on non-abrasive surface.
21 Lower diode bridge, with terminals and capacitor fitted, into position in housing, with three leads passing through gaps between fins. Secure with two setscrews and washers, trapping capacitor connector under R.H. setscrew.
22 Using long-nosed pliers (as a thermal shunt) to grip each terminal in turn and prevent excess heat reaching diode, solder three-phase winding leads and D+ lead to diode bridge. Do not overheat diode bridge.
23 If required, press new bearing on to rear end of rotor spindle.
24 Press spindle and bearing into position in rear housing.
25 Place short spacer over front end of spindle, ensuring that its larger inside diameter is next to rotor.
26 If necessary, press new front bearing into front housing and secure with retaining plate; apply Loctite to screw threads and to tapped holes in plate.
27 Press front housing into position and insert four through-bolts with plain washers under heads.
28 Coat threads of through-bolts and trapped nuts with Loctite and tighten to 3.6 lbf ft (0,5 kgf m).
29 Place plain spacer over spindle, insert Woodruff key and replace large washer, fan, pulley, small washer and nut on spindle.
30 Tighten nut to 29 lbf ft (4,0 kgf m).
31 Reverse operations 1 to 5.

ALTERNATOR

Inspection and testing (Lucas alternators)   86.10.14

Brush gear and slip-ring inspection
The serviceability of the brushes is gauged by the length protruding beyond the brush moulding in the free position. This amount should not exceed 0.3 in (8 mm). If renewal is necessary care must be taken to replace the leaf spring at the side of the inner brush. The surface of the slip-rings should be smooth and free from grease or dirt. Servicing is confined to cleaning with a petrol moistened cloth or finest grade glass-paper.

NOTE: Emery cloth or similar abrasive must not be used. The slip rings cannot be machined.

(a) Brush Replacement
1 Remove the small screws securing the brush retaining plates and regulator cables.
2 Replace brushes with new units and refit brush retaining plates and regulator cables.
3 Brush spring pressure should be checked with a push type spring tension gauge. This should indicate 9 to 13 oz (255 to 368 grammes) when brush face is flush with the moulding.

(b) ROTOR Testing

Equipment required: An ohmmeter, or a 12-volt battery and ammeter. A 110-volt a.c. supply and a 15-watt test lamp.

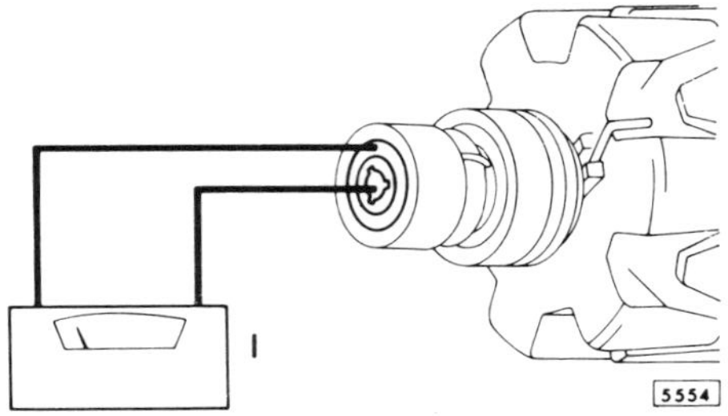

1 Connect the ohmmeter between the slip-rings. Resistance should be 3.2 ohms at 20°C.

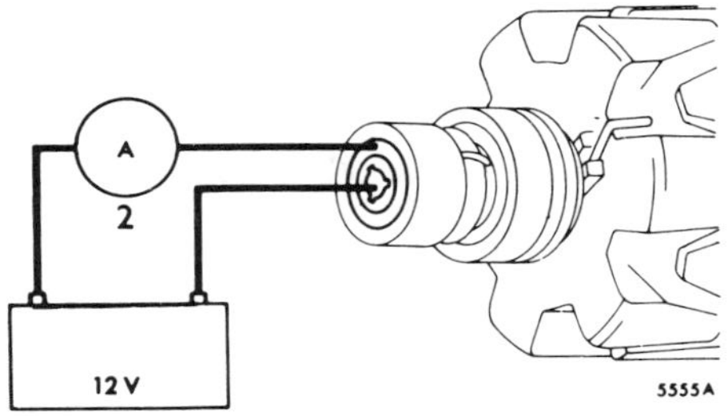

2 Alternatively connect ammeter and battery slip-rings. The ammeter should read approximately 3 amperes.

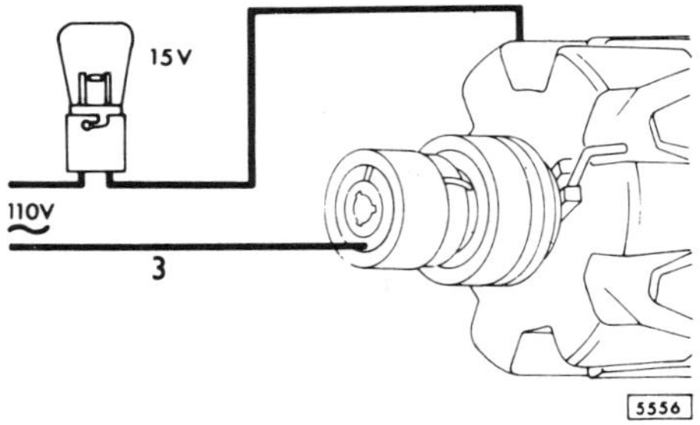

3 To test for defective insulation between slip-rings and rotor poles, connect the 110-volt supply and 15-watt test lamp between slip-rings and rotor poles in turn. If the lamp lights, the coil is earthed to the rotor core. A replacement rotor, slip-ring assembly should be fitted.

(c) STATOR Testing

Equipment required: A 12-volt battery and 36-watt test lamp. A 110-volt a.c. supply and a 15-watt test lamp.

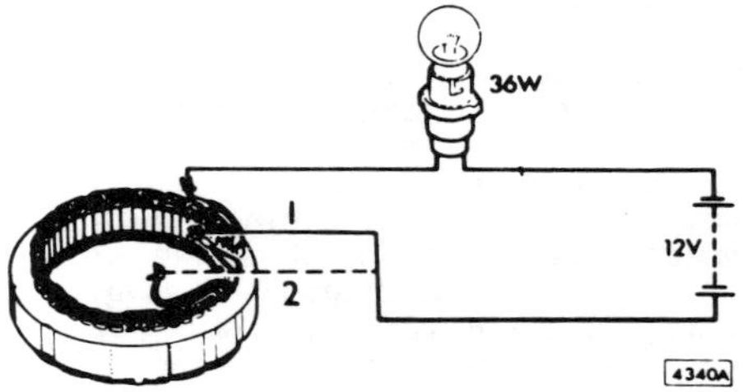

1 Check continuity of stator windings between any pair of wires by connecting in series a 12-volt battery and test lamp of not less than 36 watts. Failure of the test lamp to light means that part of the stator winding is open-circuit and a replacement stator must be fitted.

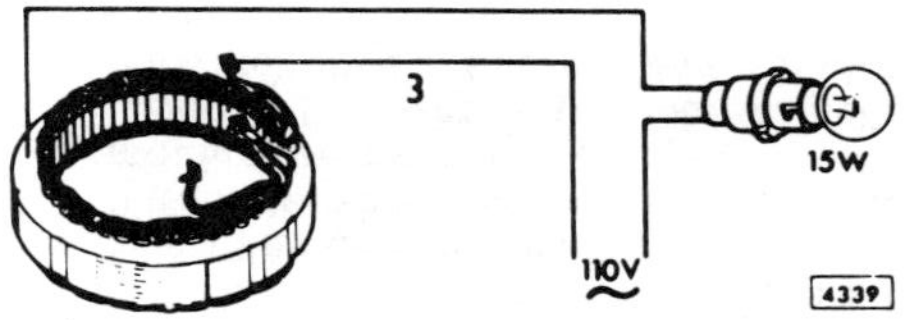

2 Test stator insulation with 110-volt test lamp. Connect test leads to laminated yoke and any one of the three stator cables. If the lamp lights, the stator coils are earthed. A replacement stator must be fitted.

(d) DIODES
Testing

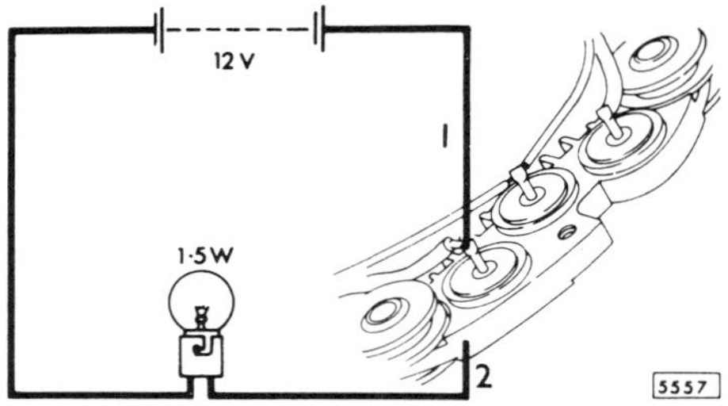

1 Connect one battery terminal to the heatsink under test.
2 Connect the other battery terminal in series with the test lamp and each diode pin in turn.
3 Reverse connections to heatsinks and diode pins. The lamp should light in one direction only. Should the lamp light in both tests, or not light at all the diode is defective and a new rectifier heatsink assembly must be fitted.
4 To prevent damage to diode assemblies during soldering operations it is important that a thermal shunt is used.

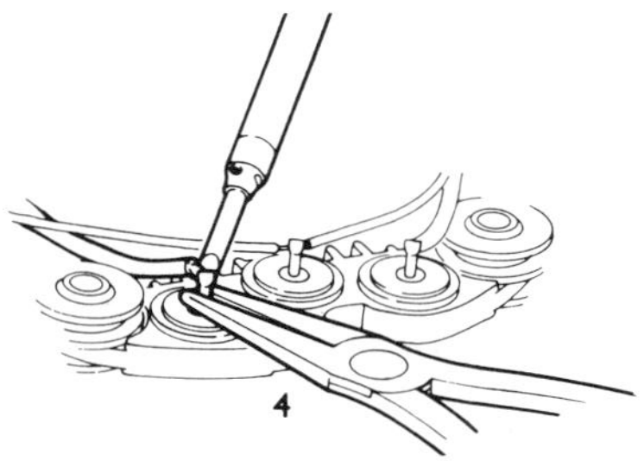

NOTE: Only 'M' grade 45–55 tin-lead solder should be used.

INSPECTION AND TESTING (MOTOROLA 9 AR 2533 ALTERNATOR)

Brush gear and slip-ring inspection

Equipment required: Compressor spring testing gauge, range up to 250 g (½ lb) 12-volt supply, test probes and indicator bulb, ohmmeter.

1 Remove alternator from car, see 86.10.02.
2 Remove brush holder by carrying out operations 4 and 5 of Alternator overhaul, dismantling, see 86.10.08.
3 Measure length of brushes protruding from housing; if either brush measures less than 0.15 in (4 mm) complete brush holder must be renewed.
4 Using push-type spring tension gauge, measure load required to press each brush face in turn flush with housing. If either reading is less than 4¼ oz (120 g) complete brush holder must be renewed.
5 Using 12-volt supply, bulb and test probes, touch field Lucar terminal (EX) with one probe and the adjacent brush with the other. Bulb should light and remain lit without flickering when brush is moved in holder.
6 Repeat operation 5 on second brush and negative terminal plate. Again, bulb should light and remain lit without flickering when brush is moved.
7 Touch each brush with one probe; bulb must not light or even flicker when brushes are moved.

NOTE: If failure is recorded in any of the tests made in operations 3 to 7 above, complete brush holder assembly must be renewed. Separate brushes and springs, etc., are not supplied as spares.

ROTOR

Testing

8 Connect ohmmeter in series with probes and touch each slip-ring with one probe. Ohmmeter should record a resistance of between 3.8 and 5.2 ohms.
9 Touch alternator casing with one probe and each slip-ring in turn with the other probe. Ohmmeter should indicate infinite resistance.

NOTE: Incorrect reading in operations 8 and 9 indicate defective wiring in field coils or between coils and slip-rings; a low reading in operation 8 would imply shorting of the field coils, a high reading a wire breakage. Indication of any current flow in operation 9 implies a breakdown of insulation. Rotor assembly must be removed for further inspection. Refer to operations 6 to 10 in 86.10.08 for removal, 10 and 11 below for further inspection and 24 to 29 in 86.10.08 for replacement of rotor in alternator casing.

10 Clean slip-ring with a lint-free petrol-soaked cloth; wipe off any petrol adhering to rotor.
11 Using micrometer, measure diameter of both slip-rings. Limits of diameter of newly fitted rings are 1.244 to 1.240 in (31,6 to 31,5 mm) for both rings, and rotor may not be refitted if diameter of either ring is less than 1.226 in (31,15 mm).

NOTE: Replacement of worn slip-rings on Motorola alternators is possible but it is most strongly urged that this work should be entrusted to specialists equipped with the special tools required for the fitting and machining of slip-rings.

STATOR

Testing

12 Visually inspect the portion of stator coils which can be seen through regulator housing for signs of damage due to overheating.
13 Check stator insulation by use of ohmmeter, touching one probe on alternator casing and the other on to each phase winding in turn. The phase windings are accessible to a probe through ventilation holes in rear moulded cover. If any current is shown to flow the stator insulation is defective and it must be removed – 86.10.08 operations 6 to 17.

DIODES

Testing

Equipment required: 12-volt supply, test probes and indicator bulb.

14 Remove diode bridge, see 86.10.08, operations 1, 2, 3, 14 and 15.
15 Check positive diodes by connecting probe from indicator bulb to each phase terminal in turn, the second probe being in contact with terminal B+. Then reverse probes. Indicator bulb should only light in one direction of circuit; if it lights in both, diode is shorted, and if it does not light in either direction, diode is in open circuit. Complete diode bridge must then be renewed. Individual diodes are not supplied as spares.

CAUTION: Complete circuits for shortest possible time to avoid damage to diodes.

16 Check negative diodes similarly, but touching second probe on diode bridge, and again reversing probes. Indications should be as for operation 15.
17 Check field diodes by holding one probe in contact with D+ terminal and second probe on each phase terminal, in turn. Then reverse probes and repeat. Indications should be as in operation 15.

CONTROL UNIT

Remove and refit (Lucas alternators only) 86.10.26

Removing
1. Disconnect battery earth lead.
2. Remove the alternator, see 86.10.02.
3. Withdraw two retaining screws and remove cover.
4. Disconnect control box connector leads.
   NOTE: Take note of cable positions.
5. Withdraw control box securing screw. Recover solid connector from control box to brush assembly.
6. Remove control box — recover insulating spacer washer from beneath control box fixing point.

Refitting
Reverse operations 1 to 6.
CAUTION: ENSURE INSULATING PIECE IS RE-FITTED BENEATH CONTROL BOX FIXING POINT AND THAT THE ALUMINIUM CASING OF THE CONTROL BOX DOES NOT CONTACT ANY PART OF THE ALTERNATOR FRAME AS THIS WILL CAUSE THE FIELD CIRCUIT TO BE FULLY SWITCHED ON AND THE ALTERNATOR TO SUPPLY MAXIMUM OUTPUT REGARDLESS OF BATTERY CONDITION.

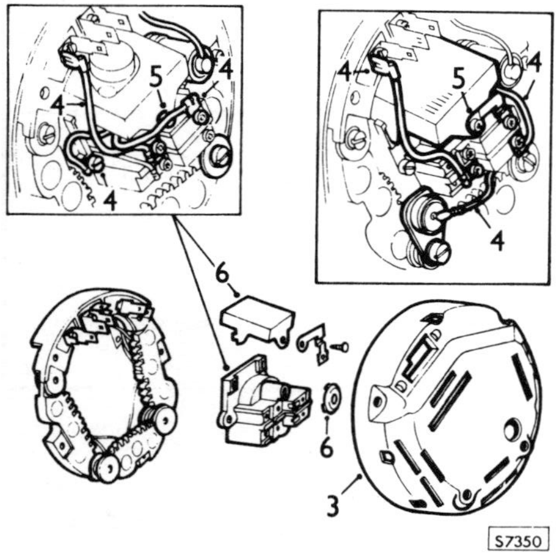

BATTERY 86.15.00

WARNING: THE BATTERY FITTED TO THIS VEHICLE HAS SPECIAL TOPPING-UP FACILITIES. WHEN BATTERY CHARGING IS CARRIED OUT THE VENT COVER SHOULD BE LEFT IN POSITION ALLOWING GAS TO ESCAPE OR FLOODING OF ELECTROLYTE WILL RESULT.

Description
The battery is a special high-performance type and is located in the luggage compartment.

Data
Battery type: Lucas 12-volt 68Ah, Pacemaker CP13/11.

BATTERY

Remove and refit 86.15.01

Removing
1. Remove battery cover by releasing two Dzus fasteners and easing cover from beneath retaining fold on battery clamp.
2. Ease back battery terminal cover from terminals, slacken clamp bolts and disconnect leads from battery. 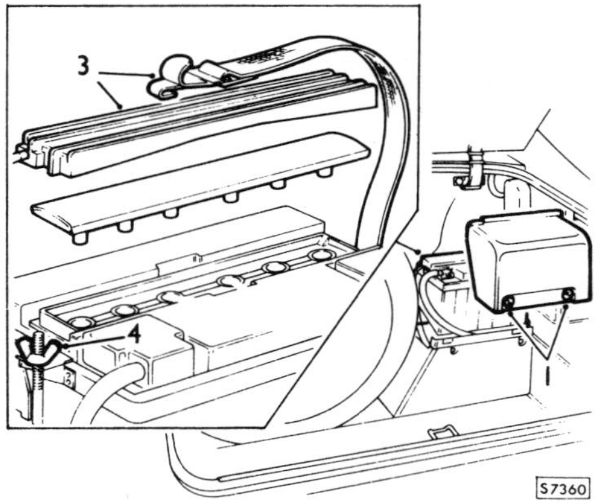
3. Release battery filler cover securing strap and remove filler cover.
4. Release battery clamp securing nuts, remove clamp.
5. Recover battery from mounting tray.

Refitting
Reverse operations 1 to 5. Smear terminals with petroleum jelly before re-connecting battery leads.
NOTE: The battery must be kept level at all times to prevent spillage of electrolyte, consequent damage to vehicle finish, and possible personal injury.

BATTERY

Test 86.15.02

It is not possible to test this battery with a high-rate discharge meter, due to the location of the intercell connectors. The battery top must not be drilled in an attempt to locate the connectors. Check the specific gravity of the electrolyte in each cell using an hydrometer. A variation of more than 40 points (0.040) in any cell reading means that the battery is suspect and should be removed for testing by a battery agent. If possible prove the battery by substitution.

State of charge — S.G. readings
1. Lift and tilt the battery vent cover to one side.
2. Insert the hydrometer into each cell through the filling tube and note the readings.

STATE OF CHARGE	SPECIFIC GRAVITY READINGS CORRECTED TO 60° F (15° C)	
	CLIMATES NORMALLY Below 77° F (25° C)	CLIMATES NORMALLY Above 77° F (25° C)
FULLY CHARGED	1.270 – 1.290	1.210 – 1.230
70% CHARGED	1.230 – 1.250	1.170 – 1.190
DISCHARGED	1.100 – 1.120	1.050 – 1.070

Electrolyte temperature correction
For every 18° F (10° C) below 60° F (15° C) subtract 0.007.
For every 18° F (10° C) above 60° F (15° C) add 0.007.

BATTERY TRAY

Remove and refit   86.15.11

Removing
1. Remove battery, see 86.15.01.
2. Remove spare wheel.
3. Withdraw two screws securing fuel pipe protection plate to battery tray.
4. Lift carpet and remove securing clip from plate.
5. Remove protective plate.
6. Remove bolts securing fuel expansion tank to mounting.
7. Ease tank clear to give access to battery tray securing nuts.
8. Remove four nuts and two set bolts securing battery tray.
9. Disconnect drain tube from tray.
10. Remove battery tray from boot.

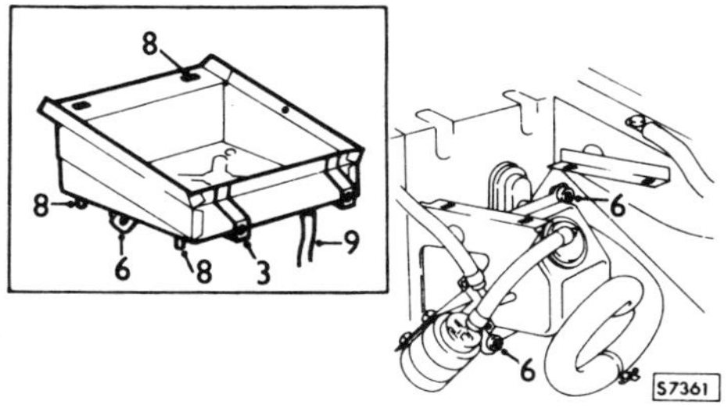

Refitting
Reverse operations 1 to 10.

BATTERY LEAD – NEGATIVE

Remove and refit   86.15.19

Removing
1. Release DZUS fasteners and remove battery cover.
2. Ease back cover of negative terminal, slacken clamp bolt and disconnect lead from negative terminal.
3. Remove bolt securing earth lead to earth point in boot (trunk).
4. Recover earth lead.

Refitting
Reverse operations 1 to 4.

BATTERY TERMINALS

Disconnect   86.15.20

1. Remove battery cover.
2. Ease back terminal insulated cover.
3. Slacken clamp bolt.
4. Lift connector from terminal post.

Re-connect
Reverse operations 1 to 4.

ELECTRICALLY OPERATED WINDOWS AND DOOR LOCKS

Description   86.25.00

The electrically operated window lift system comprises two motors, driver's and passenger's control panel, relay and thermal overload cut-out. The electrically operated door lock circuit comprises a solenoid for each door lock, thermal overload relay, two relays and a selector switch mounted in the centre console escutcheon.

Operation

With the window lift master switch ON, operation of either of the centre 'off' two pole switches will cause the associated window lift motor to run in the selected direction. Selections can only be made on one switch at a time, the driver's switch over-riding the other panel switch. The circuit is arranged so that the operation of each switch isolates the subsequent circuit, thus preventing operation of more than one motor at a time and protecting the circuit from an overload condition.

Fault conditions, i.e. sticking windows, or overload will result in excessive current consumption causing the thermal cut-out to operate. The cut-out will re-set after a short interval allowing normal operation of window lift motors to be resumed. If the condition persists a detailed examination of the system is required.

The door lock selector switch is of the centre 'off' rocker variety, operation in one direction locks the doors from inside, preventing access from outside when the car is occupied. An opposite selection on the switch will open the door locks. The door lock solenoids will remain in the last selected position when the switch is returned to the off position. Manual operation of the conventional door handles from inside the car will over-ride the door lock solenoids.

NOTE: Rapidly repeated operation of the door locks will result in an overload condition, causing the thermal cut-out to operate, isolating the door lock solenoid circuit. A short wait is necessary before the thermal cut-out automatically re-sets.

WINDOW LIFT MOTOR

Remove and refit   86.25.05

Removing
1. Disconnect battery earth lead.
2. Remove door trim pad, see 76.34.01.
3. Remove inner door handle securing screws.
4. Disengage remote control levers from handle.
5. Lower window to full extent.
6. Ease waterproof sheet clear of motor connections.
7. Disconnect motor electrical connections.
8. Remove from motor securing bolts.
9. Ease window lift assembly towards rear of door, disengaging regulator from locating channel.
10. Raise glass normally and wedge at highest point.

continued

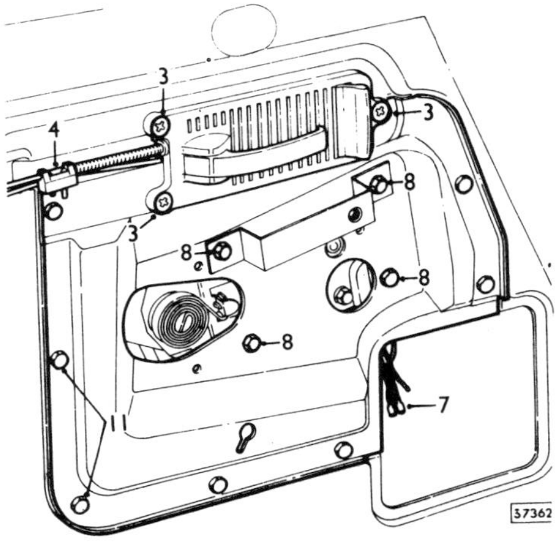

86–7

11 Withdraw seven securing screws and remove mounting plate. 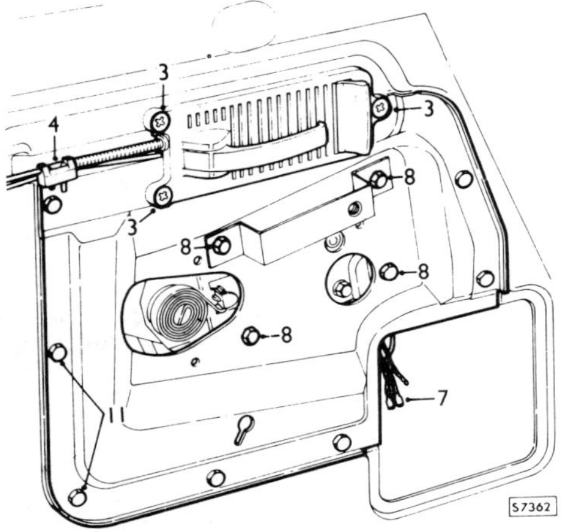
12 Remove motor and quadrant assembly from door.
13 Before disengaging regulator from motor ensure that lifting arm and quadrant of regulator are clamped in a vice.
   NOTE: This prevents spring disengaging suddenly, causing possible damage.
14 Remove bolts securing regulator to motor.

Refitting
15 Fit motor to regulator.
16 Remove from vice.
17 Locate assembly in door.
18 Remove wedge and lower glass to lowest position.
19 Engage regulator arm in channel and replace waterproof sheet.
20 Offer mounting plate to motor, locate and replace motor securing bolts.
21 Refit mounting plate and secure in position.
22 Reverse operations 1 to 4.

WINDOW LIFT SWITCH(ES)   86.25.13
Remove and refit

Removing
1 Disconnect battery earth lead.
2 Withdraw three screws securing centre panel in console.
3 Lift centre console for access to rear of switches.
4 Note position of cables and orientation of switch, disconnect cables.
5 Depress securing tags and push switch through panel.

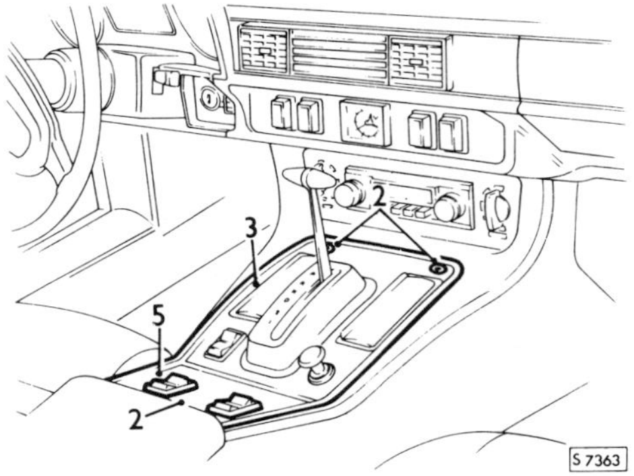

Refitting
Reverse operations 1 to 5.

DOOR LOCK SELECTOR SWITCH   86.25.14
Remove and refit
Refer to operation 86.25.13.

WINDOW LIFT RELAY   86.25.28
Remove and refit

Removing
1 Disconnect battery earth lead.
2 Remove passenger side dash liner, see 76.46.11.
3 Remove nuts securing mounting plate to fan motor unit.

4 Remove nuts, bolts and shakeproof washers securing relay to plate.
   NOTE: European specification cars – Relay is mounted back to back with low coolant control unit.
   North American specification cars – Relay is mounted on rear of plate.
5 Disconnect electrical connectors after noting positions.

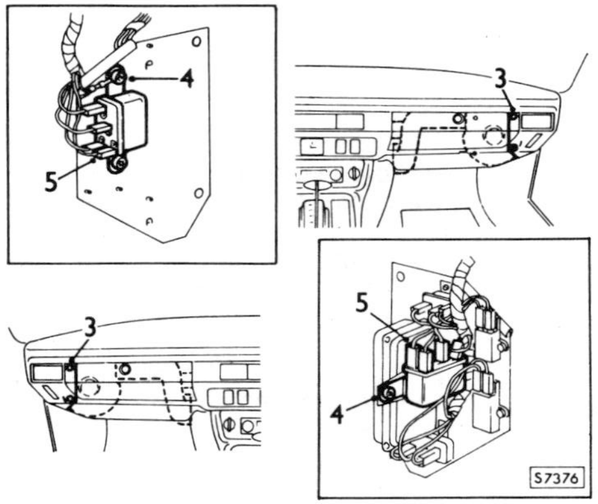

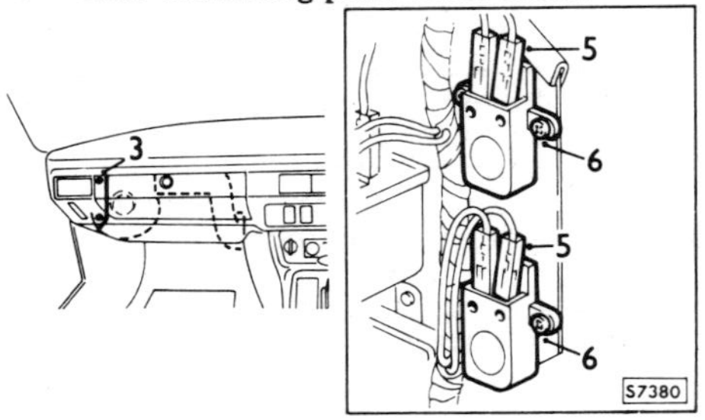

5 Note position of cables and disconnect at Lucars on relevant circuit breaker.
6 Withdraw two screws securing unit to mounting plate.

Refitting
Reverse operations 1 to 6.

DOOR LOCK SOLENOIDS   86.25.32
Remove and refit

Removing
1 Raise window to fully closed.
2 Disconnect battery, see 86.15.20.
3 Remove arm rests and door trim, see 76.34.01.
4 Remove two setscrews securing solenoid to door stretcher.
5 Disconnect cables at snap connectors.

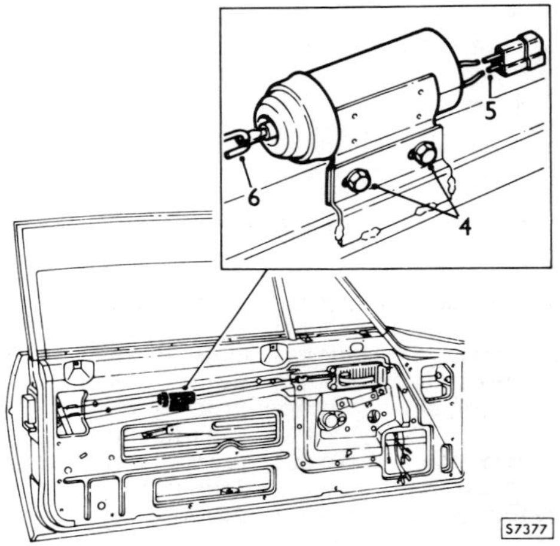

CIRCUIT BREAKERS   86.25.31
Remove and refit

Removing
1 Disconnect battery, see 86.15.20.
2 Remove passenger side dash liner, see 76.46.11.
3 Remove nuts and shakeproof washers securing mounting plate to fan motor.
4 Ease mounting plate from studs.

6 Unhook solenoid operating piston from door lock push rod and remove from door.

Refitting
Reverse operations 1 to 6.

86–8

DOOR LOCK SOLENOID RELAY

Remove and refit 86.25.33

Removing
1 Disconnect battery earth lead.
2 Remove passenger side dash liner, see 76.46.11.
3 Remove securing screws from footwell side casing.
4 Remove passenger screws from footwell side casing.
5 Remove screws securing relay mounting bracket to body side panel. 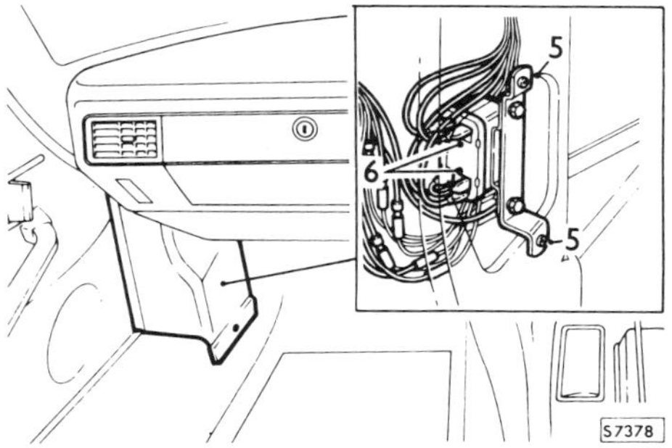
6 Note position of harness connector leads to relay and disconnect.
7 Remove nuts, bolts and washers securing relay to bracket. Recover relay and lay bracket aside.

Refitting
Reverse operations 1 to 7.

HORNS

Description 86.30.00

Twin horns are fitted, mounted on front lower cross-member behind and beneath front bumper. Both horns operate simultaneously and are energised by a relay. The relay is connected to the battery through the ignition and switch so that the horns will only operate with ignition switched 'ON'.

HORN PUSH

Remove and refit 86.30.01

Removing
1 Disconnect battery, see 86.15.20.
2 Slacken steering wheel adjustment ring and extend column to full extent.
3 Withdraw two screws from behind push and lift push from steering wheel.
4 Withdraw four screws from push backplate. 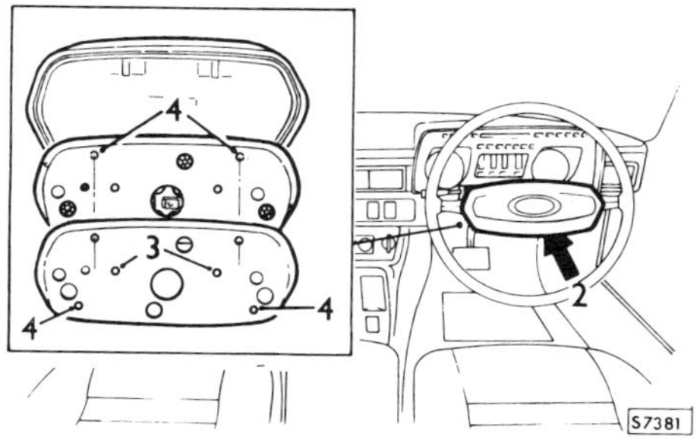
5 Ease trim pad from push and recover push contact unit.

Refitting
Reverse operations 1 to 5.

HORNS

Remove and refit 86.30.09

Removing
1 Disconnect battery earth lead.
2 Remove nut securing horns to mounting beneath front bumper apron.
3 Lower horns and recover distance pieces and washers.
4 Disconnect supply leads at lucar connectors.
5 Recover horns.

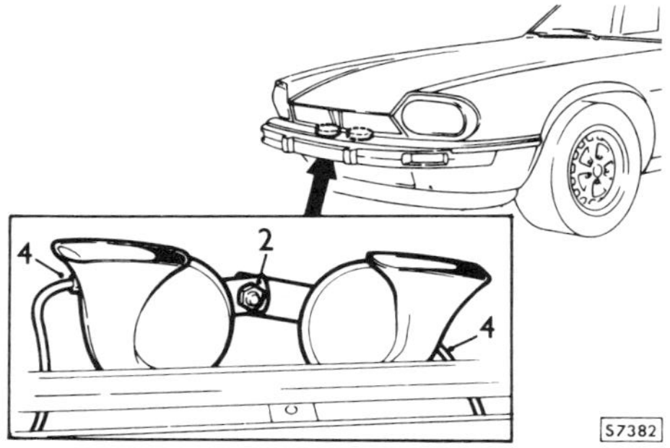

Refitting
Reverse operations 1 to 5.

HORNS

Adjust 86.30.08

1 Carry out operation 86.30.09 items 1 to 3.
2 Adjust note of horn using diaphragm tension screw on top of horn.
3 Switch ignition 'ON' and check operation and note of horn. Adjust until satisfactory.
4 Repeat for second horn.
5 Refit horns as detailed in 86.30.09.

HORN RELAY

Remove and refit 86.30.18

Removing
1 Disconnect battery earth cable.
2 Note position and colour of cables to headlight fuse boxes.
3 Disconnect cables from headlight fuse boxes.
4 Withdraw screws securing relay cover to wing valance and remove cover.
5 Note position and colour of cables and disconnect from relay.
6 Remove securing nuts and shakeproof washers from horn relay.

NOTE: A revised relay (round) is supplied to replace the one fitted to early cars. The revised wiring is as follows:

W1 to 85   C1 to 87
W2 to 86   C2 to 30/51
            C3 to 87A

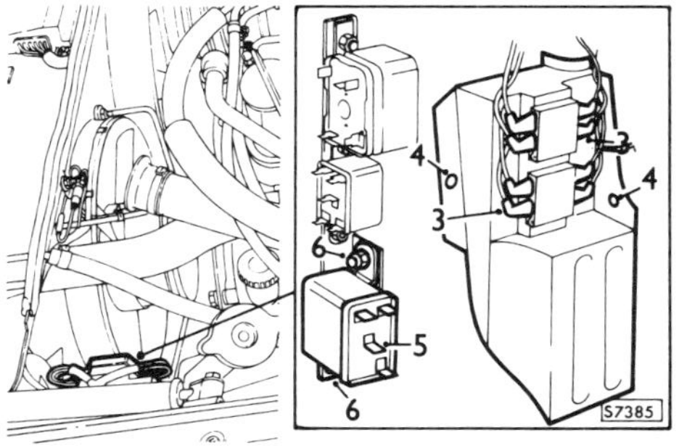

7 Remove relay from location.

Refitting
Reverse operations 1 to 7.

THE IGNITION SYSTEM (12 CYLINDER CARS)

Description 86.35.00

The 'OPUS' Electronic ignition system, which is fitted, comprises the following:—
1 DISTRIBUTOR – Model 36 DE 12.
2 AMPLIFIER UNIT – Model AB3.
3 BALLAST RESISTANCE UNIT – Model 9BR.
4 IGNITION COIL – Model 13C 12.
5 WIRE TO TACHOMETER.

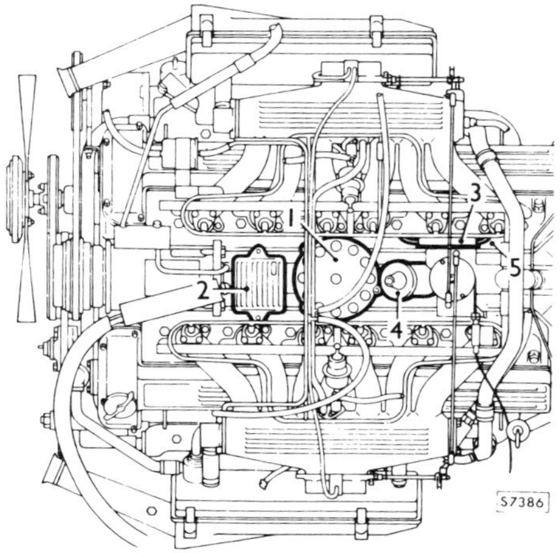

produced at the H.T. output terminal of the ignition coil.

DISTRIBUTOR CAP

Remove and refit   86.35.10

Removing
1. Check all H.T. sparking plug leads and ensure each is adequately identified.
2. Detach leads from sparking plugs.
3. Detach H.T. lead from ignition coil.
4. Unscrew three captive screws and detach distributor cap.
5. Check carbon contact is in good condition and moves freely on spring.

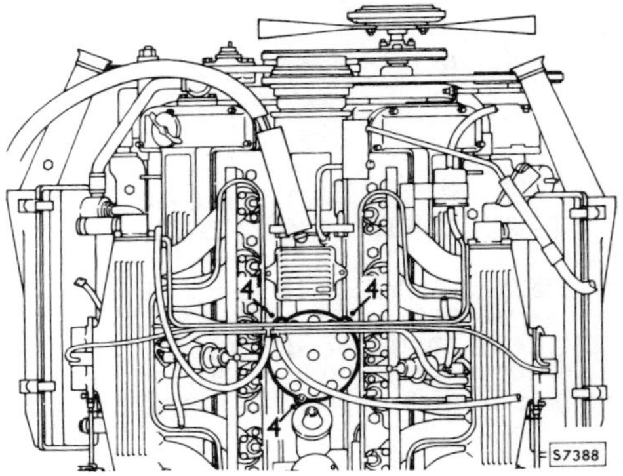

Refitting
Reverse operations 2 to 4.

DISTRIBUTOR LEADS

Remove and refit   86.35.11

Removing
1. Pull faulty lead from distributor cap and related sparking plug.
2. Fit replacement H.T. lead.
3. Transfer identification sleeve or renew.
4. Fit termination.

NOTE: If ready made lead is used ensure it is of correct length and carries correct identification sleeves.

Refitting
5. Fit new lead to sparking plug and to distributor cap.

Distributor
The 'OPUS' distributor comprises:—
1. The centrifugal auto-advance mechanism.
2. Vacuum unit (Advance or Retard).
3. High tension rotor.
4. Electronic timing rotor.
5. Pick-up module assembly.
6. Trigger unit.

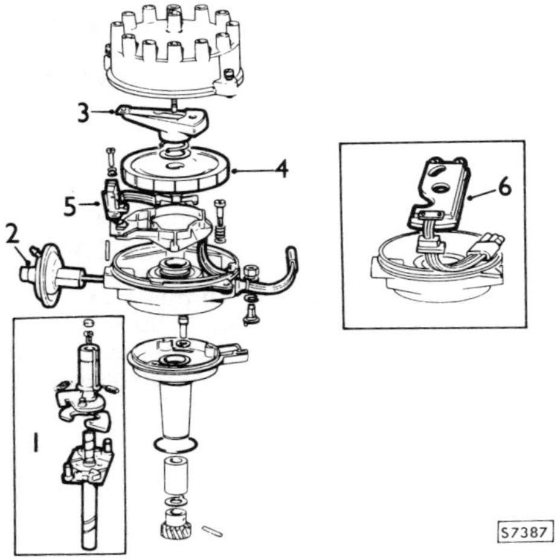

The timing rotor and pick-up module, working in conjunction with a separate amplifier unit, replace the contact breaker and cam of a conventional distributor.

The timing rotor is a glass-filled nylon disc with small ferrite rods embedded into its outer edge, the number and spacing of the rods corresponding with the number of cylinders and firing angles of the engine. An air gap (adjustable to specified limits) exists between the rotor and the ferrite core of the stationary pick-up module. The pick-up module assembly comprises a magnetically-balanced small transformer, the primary (input) and secondary (output) windings.

The trigger unit is located within the distributor. It consists of two reed switches mounted radially and 180 degrees apart, parallel to the plane of the high tension rotor arm. The rotor arm has a magnet fitted beneath the tail which closes each switch alternately, one each revolution of the crankshaft. The trigger unit switches initiate the start of the fuel injection period and also give the ECU an engine speed reference depending upon the frequency at which the successive switches are operated.

CAUTION: Magnetic balancing of the pick-up module. This unit is balanced during manufacture and the setting cannot alter in service. The sealed ferrite adjusting screw must not be disturbed.

Automatic control of the ignition timing is provided by the vacuum unit which varies the static timing position of the pick-up module in relation to the ferrite rods in timing rotor.

The distributor timing rotor and pick-up module generate an electronic timing signal, which is fed to the amplifier unit via external cables.

CAUTION: The length of this triple-core extruded type cable must not be altered and the cables must not be separated or replaced by loose individual cables.

Amplifier Unit — Item 1
This interprets the timing signals from the distributor. The power transistor incorporated in the printed circuit then functions as an electronic switch in the primary circuit of the ignition coil. The unit is connected to the ignition coil via a ballast resistance unit and external cables.

Ballast Resistance Unit — Item 2
An encapsulated assembly comprising three resistors in an aluminium heat sink fixing bracket. External wiring connects two of the resistors in series with the ignition coil primary winding.

The third resistor unit is associated with the function of one of the transistors in the amplifier unit.

Ignition Coil — Item 3
A specially designed fluid-cooled, high-performance, ballast-ignition ignition coil.

The coil terminals are marked '+' and '—', and have different types of Lucar connector to prevent incorrect cable connection.

CAUTION: The 'OPUS' coil is NOT interchangeable with any other type.

Operation
Normally when the engine is stationary, the distributor timing rotor will be in a position where none of the ferrite rods will be in close proximity with the ferrite core of the pick-up module.

When the ignition is switched on, a power transistor in the amplifier unit is in a conductive state and the ignition coil primary winding circuit is complete via the emitter/collector electrodes of the power transistor.

Simultaneously, a sinusoidal (pulsating a.c.) voltage is applied by the amplifier unit to the distributor pick-up module primary windings and a small residual a.c. voltage is produced at the pick-up secondary windings which at this stage is magnetically balanced.

The voltage at the pick-up module secondary terminals is applied to the amplifier unit, but the residual voltage at this stage is insufficient to have any effect on transistor circuits which control the switching off of the power transistor in the output stage of the amplifier unit.

When the engine is cranked, one of the ferrite rods in the rotor, now brought into close proximity with the ferrite core of the module causes 'magnetic unbalancing' of the module core, resulting in an increase in the voltage at the module output terminals.

This 'unbalancing' and voltage increases to maximum as the rotor rod traverses the centre and upper limbs of the module 'E' shaped core.

Maximum voltage is then applied to the amplifier unit, where it is rectified, the resulting direct (d.c.) current is then used to operate the transistor circuits which control the switching off of the power transistor in the output stage. With the power transistor switched off, its emitter/collector electrodes ceases to conduct and the coil primary winding is disconnected which causes a rapid collapse of the primary winding magnetic field through the secondary windings of the ignition coil, resulting in a high-tension (H.T.) voltage being

86–10

ROTOR ARM

Remove and refit 86.35.16

Removing
1 Remove three captive screws and detach distributor cap 86.35.10.
2 Withdraw H.T. rotor.

NOTE: Care must be taken not to damage the trigger unit glass case.

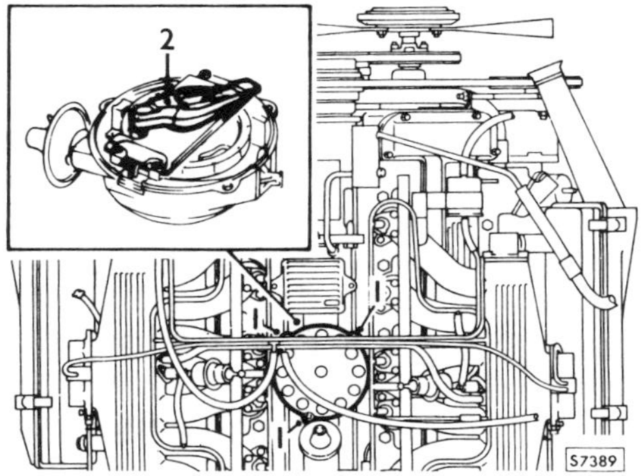

Refitting
3 Place two or three drops of clean engine oil on rotor carrier shaft oil pad.
4 Replace H.T. rotor. Ensure keyway engaged, and rotor pushed fully home.
5 Refit distributor cover and tighten three screws.

ELECTRONIC TIMING ROTOR

Remove and refit 86.35.17

Removing
1 Remove the rotor arm, see 86.35.16.
2 Remove trigger unit, see 19.22.26.
3 Remove circlip and wave washer.
4 Detach electronic timing rotor.

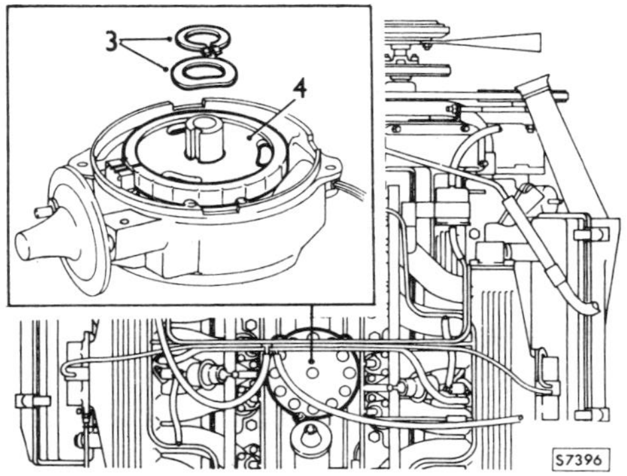

Refitting
Reverse operations 1 to 3.

PICK UP MODULE

Remove and refit 86.35.18

Removing
1 Remove electronic timing rotor, see 86.35.17.
2 Remove two cheese head screws securing pick up module to pick up arm and recover two spring and two plain washers.
3 Prise cable grommet inwards from body of distributor.
4 Feed cable and cable connector in through hole and lift pick up module clear.

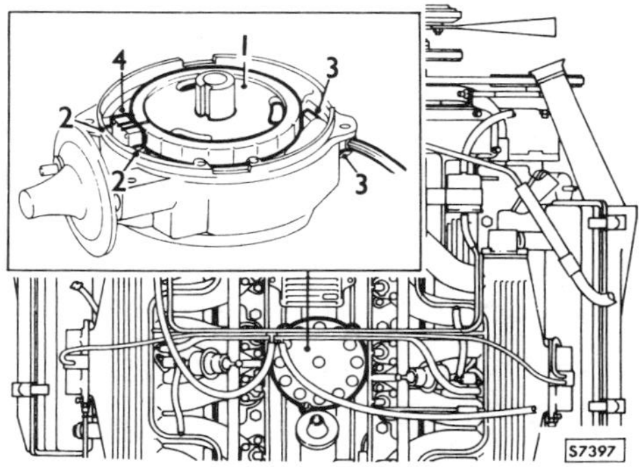

Refitting
5 Feed cable connector out through hole in distributor body. Fit grommet into hole in body, wide end first.
6 Locate pick up module on pick up arm, pick up core towards distributor shaft. Loosely secure using two cheese head screws, plain and spring washers.
7 Fit electronic timing rotor and secure using a wave washer and circlip.
8 Use feeler gauges to set distance between pick up module E core faces and timing rotor outer edge to 0.020 in to 0.022 in (0,50 mm to 0,55 mm).
9 Tighten both pick up securing screws.
10 Fit rotor arm.

DISTRIBUTOR

Overhaul 86.35.26

Dismantling
1 Remove distributor, see 86.35.20, electronic timing rotor, see 86.35.17 and trigger unit, see 19.22.26.
2 Lift vacuum operating rod from peg on pick up arm.
3 Prise cable grommet from body of distributor.
4 Remove pick up arm bearing spring. Slide pick up arm sideways to disengage it from bearing. Lift from micro housing, drawing cable in through hole. Detach pick-up module.
5 Use a pin punch 0.073 in (1,85 mm) to tap out roll pin securing vacuum unit in micro housing.
6 Withdraw vacuum unit from micro housing.
7 Remove three spring loaded screws and lift micro housing from distributor body.
8 Extract felt pad from top of rotor carrier shaft and release screw.
9 Release control springs from fixing posts.
10 Lift rotor carrier shaft from distributor shaft. Collect centrifugal weights.

Inspection
11 Check control springs for correct length.
12 Check pivot holes in centrifugal weights for wear or deformation.
13 Check distributor shaft for undue play.
NOTE: If any part of the distributor body assembly is found to be defective, the complete assembly must be renewed.

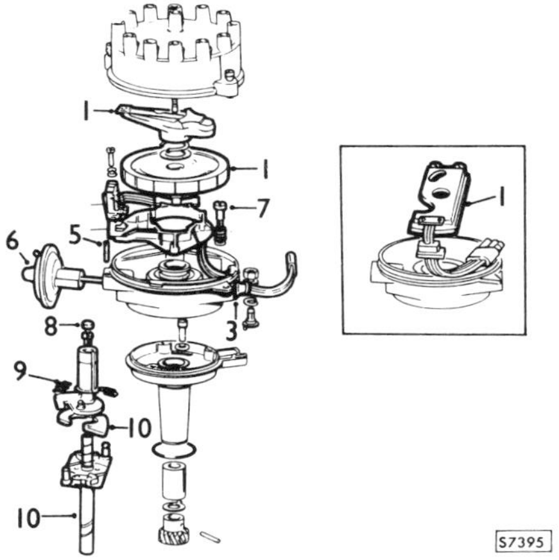

Assembly
14 Smear centrifugal weights and rotor carrier pivot posts with either Rocol grease No. 30863 or Mobilgrease No. 2. Assemble weights to pivot posts.
15 Lubricate bore of rotor carrier shaft with clean engine oil and fit to distributor shaft. Retain with round headed screw. Fit oil pad.
16 Fit control springs.
NOTE: Ensure three socket headed screws and plain washers in place through slots in distributor body base.
17 Liberally smear auto advance mechanism with grease previously specified.
18 Fit micro housing to distributor body, ensuring micro adjustment eccentric peg engages in slot.
19 Secure micro housing to body using screws, plain washers and springs. Tighten screws to just short of coil binding.
20 Loosely secure pick up module to pick up arm using two cheese head screws, plain and spring washers.

86—11

21 Pass pick up module connector and cable out through hole in micro housing and locate pick up arm on rotor carrier shaft.
22 Fit bearing spring.
23 Engage wide part of cable grommet in hole and prise into position.
24 Place vacuum unit in position and secure with a new roll pin.
25 Fit vacuum operating rod to peg on pick up arm.
26 Fit electronic timing rotor and secure using a wave washer and circlip.
27 Use feeler gauges to set distance between pick up module E core faces and timing rotor outer edge to 0.020 in to 0.022 in (0,50 mm to 0,55 mm).
28 Tighten both pick up module securing screws.
29 Fit trigger unit.
30 Fit rotor arm.
31 Fit distributor.

DISTRIBUTOR

Remove and refit – Engine dismantling and reassembling only 86.35.20

Removing
1 Remove three captive screws and detach distributor cover.
2 Disconnect cable at connecting plug.
3 Disconnect pipe from vacuum retard unit.
4 Release three Allen screws, accessible through slots in micro housing, withdraw distributor.

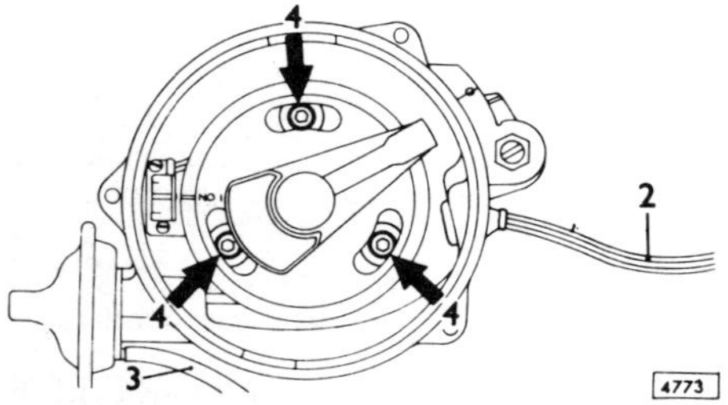

Refitting
5 Rotate engine until mark 'A' etched on crankshaft damper is in line with 10° B.T.D.C. mark on timing plate.
   CAUTION: No. 1 piston 'A' bank must be on firing stroke. Both inlet and exhaust valves in cylinder will be closed and removal of sparking plug will enable an observation to be made to ascertain that this is so. DO NOT rotate engine backwards.
6 Rotate distributor until No. 1 cylinder mark on timing rotor is in alignment with mark on pick up module.
7 Reverse operations 1 to 4 ensuring that marks on timing rotor and pick up module do not move out of alignment.
8 Check ignition timing, see 86.35.29/7.

DISTRIBUTOR

Remove and refit – Service replacement only 86.35.20/1

Removing
1 Remove three captive screws and detach distributor cover.
2 Rotate engine until No. 1 cylinder mark on timing rotor is in alignment with mark on pick up module.
3 Disconnect cable assembly at connector plug.
4 Disconnect pipe from vacuum retard unit.
5 Release three Allen screws, accessible through slots in micro-housing, and withdraw distributor. DO NOT rotate engine.

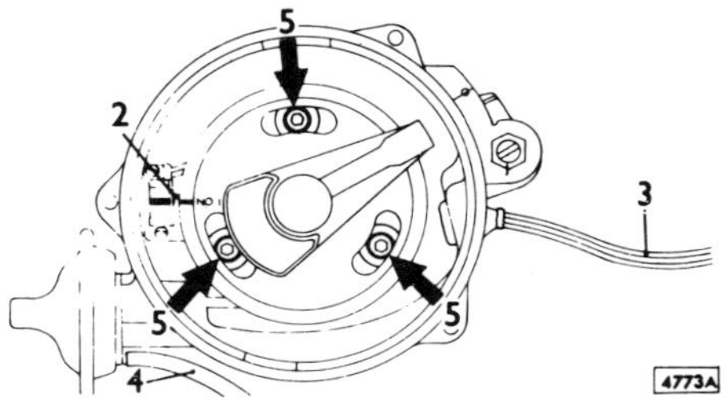

Refitting
6 Reverse operations 3 to 5.
7 Check that No. 1 cylinder mark on timing rotor is in alignment with mark on pick up module.
8 Refit distributor cap.
9 Check ignition timing, see 86.35.29/7.

'OPUS' IGNITION SYSTEM

Checking 86.35.29

EQUIPMENT REQUIRED:
DC Moving coil voltmeter–0–20V scale.
Hydrometer.
Ohmmeter.
H.T. Jumper lead.

Preliminary procedure

(a) BATTERY 86.35.29/1

Test
It is NOT possible to test this battery with a high rate discharge meter, due to the location of the intercell connectors. The battery top must not be drilled in an attempt to locate the connectors. Check the specific gravity of the electrolyte in each cell using an hydrometer. A variation of more than 40 points (0.040) in any cell reading means that the battery is suspect and should be removed for testing by a battery agent. If possible prove the battery by substitution.

State of Charge – S.G. Readings
1 Lift and tilt the battery vent cover to one side.
2 Insert the hydrometer into each cell through the filling tube and note the readings.

STATE OF CHARGE	SPECIFIC GRAVITY READINGS CORRECTED TO 60° F (15° C)		CLIMATES NORMALLY Above 77° F (25° C)
	CLIMATES NORMALLY Below 77° F (25° C)	CLIMATES NORMALLY (25° C)	
FULLY CHARGED	1.270 – 1.290	1.230 – 1.250	1.210 – 1.230
70% CHARGED	1.230 – 1.250		1.170 – 1.190
DISCHARGED	1.100 – 1.120		1.050 – 1.070

Electrolyte temperature correction
For every 18°F (10°C) below 60°F (15°C) subtract 0.007.
For every 18°F (10°C) above 60°F (15°C) add 0.007.

b Ensure that spark plugs are checked and that fuel is available before commencing tests on the OPUS ignition system.
c Check all electrical connections for security and cleanliness.

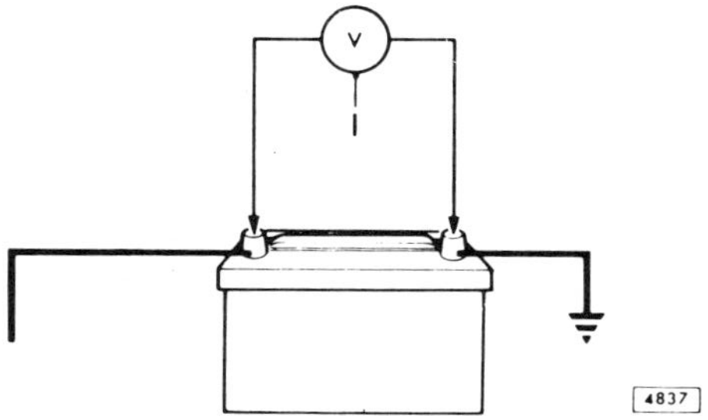

Circuit test

1 Check battery voltage. Full 12 volts should be registered.
2 Disconnect cable from coil L.T. terminal marked '−'.
3 Connect voltmeter between battery earth and frame.
4 Operate starter, check voltmeter reading which should not exceed 0.5 volts. Refit coil cable.
5 If more than 0.5 volts is registered rectify faulty connection between frame and battery.

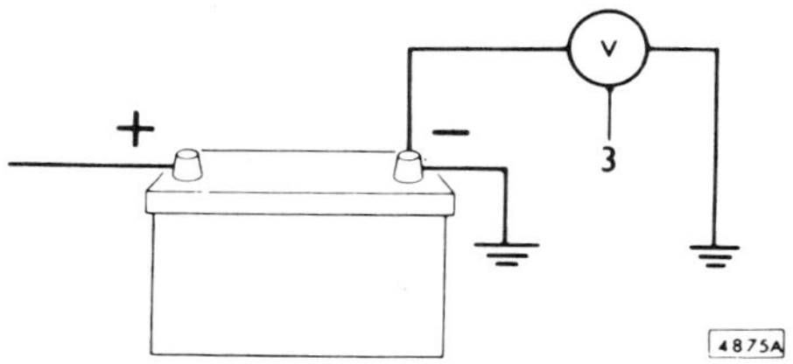

Check for sparking   86.35.29/2

1 Disconnect H.T. lead from distributor cover and hold free end approx. ¼ in (6 mm) from unpainted part of engine block.
Ignition 'ON', crank engine. Regular sparking should occur.
NOTE: If no sparking occurs proceed to 86.35.29/3. If sparking occurs, check the following:
a Distributor cover for cleanliness and cracks.
b H.T. cables.
c Rotor arm.
d Spark plugs.
e Fuel supply.

Ballast resistor check   86.35.29/3

1 Withdraw socket from amplifier side of ballast resistor.
2 Connect voltmeter between battery earth and each terminal of ballast resistor amplifier output in turn.
CAUTION: **Voltmeter test lead must not come into contact with ballast resistor housing.**
3 With ignition 'ON' meter should read battery voltage.
NOTE: If satisfactory proceed to 86.35.29/4.
If no reading is obtained check supply to SW terminal, replace ballast resistor if battery voltage is registered. If no reading at SW terminal trace circuit back through ignition switch.

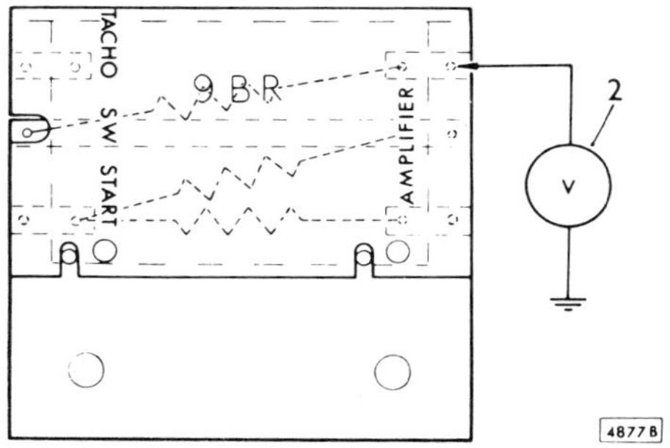

Coil voltage check   86.35.29/4

1 Reconnect socket to amplifier side of ballast resistor.
2 Disconnect amplifier/distributor socket.
3 Connect voltmeter between battery earth and '+' terminal on coil.
4 Ignition 'ON' reading should be 4–6 volts.
NOTE: High reading indicates a faulty coil or amplifier.
No reading – check supply to '+' terminal or coil at Green/White cable with voltmeter, reading should be battery volts.
If below 4 volts is registered, check value of each resistor in ballast unit with ohmmeter.
Readings obtained should be:
Resistor 1 – 7.6 – 9.2 ohms.
Resistor 2 – 0.72 – 0.80 ohms.
Resistor 3 – 0.9 – 1.0 ohms.
Renew resistor unit if outside these limits. Check battery supply voltage to unit.
Connection 4 – Straight through internal connection.
Connection 5 – Tachometer terminal 100 ohms.
5 If above 6 volts is registered, remove distributor cover, crank engine until two of the timing rotor rods (1) are equi-distant either side of the pick up module core (2) and re-check.
6 Check ignition coil for continuity and resistance of primary winding between '+' and '−' terminals.
Renew coil if outside limits of 0.8 – 1.0 ohms, or open circuited.

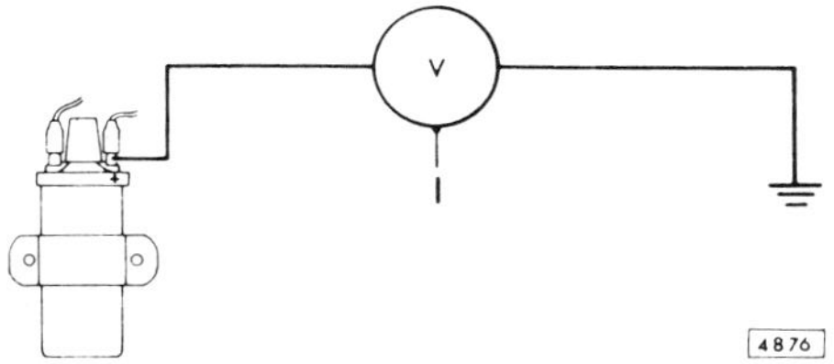

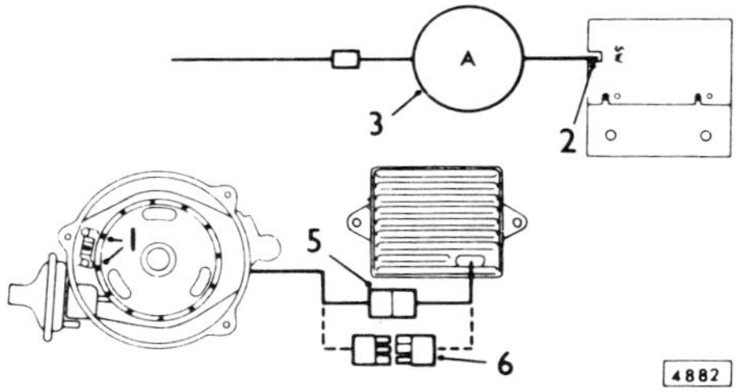

continued

Data

1. Nominal voltage .... (measured at 'SW' terminal 12V (Negative Earth)
2. Stall current (measured at 'SW' terminal of ballast resistance unit) .... 5.0 – 6.5A
3. Ignition coil primary winding resistance (measured between L.T. terminals '+' and '–') .... 0.8 – 1.0 ohm at 20° C
4. Distributor Pick-up Module
   (a) Primary (input) winding resistance (measured between centre terminal and outer terminal with red cable) .... 2.5 ohms nominal at 20° C
   (b) Secondary (output) winding resistance (measured between centre terminal and outer terminal with black cable) .... 0.9 ohm nominal at 20° C
   (c) Gap between pick-up module 'E' core faces and timing rotor-outer edge .... 0.020 in – 0.022 in (0,50 mm – 0,55 mm)
5. Centrifugal Auto Advance Details
   Run up to 100 distributor rev/min and set gauge to read zero degrees. Check at following speeds.

   Distributor R.P.M.   **Distributor advance degrees**
Distributor R.P.M.	Distributor advance degrees
100	No advance
400	No advance
600	½ – 2½
1000	6.0 – 8.0
1300	8.0 – 10.0
2600	11.0 – 13.0
3500	12.0 – 14.0

6. Distributor advance
   (a) No vacuum advance below 6 in (15,2 cm) Hg
   (b) Maximum vacuum advance at 10 in (25,4 cm) Hg .... 4 degrees

Rotor arm insulation check   86.35.29/8

1. Hold free end of jumper lead approx. ⅛ in (3 mm) from rotor arm electrode.
2. Ignition 'ON', crank engine. No spark should occur.
   NOTE: If there is H.T. sparking, replace rotor arm.

On completion of checks, refit distributor cover, remove test meter and jumper leads. Visually check for cleanliness and security of cables, connectors, and spark plugs.

Pick up module check   86.35.29/9

1. Withdraw three retaining screws and remove distributor cap.
2. Carry out visual check of pick up module, ensure faces of 'E' core are in line and parallel with edge of timing rotor. 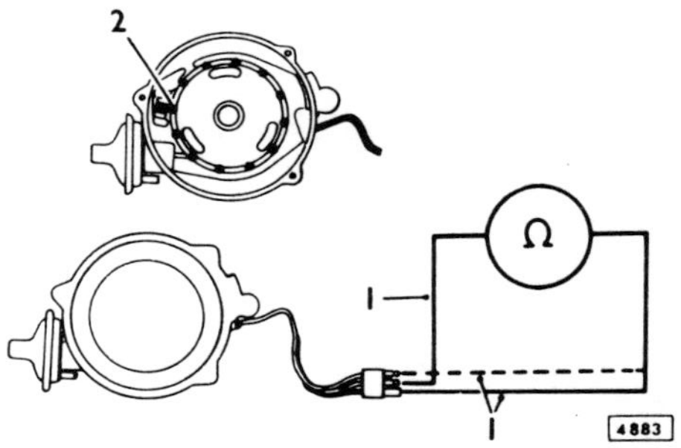
3. Check air gap between module core and timing rotor is within limits 0.020 – 0.022 in (0,50 – 0,55 mm).
4. Check resistance values between centre and each outer terminal at distributor connector. Centre terminal to red cable 2.5 ohms ± 10%, centre terminal to black cable 0.9 ohms ± 10%.
   NOTE: If core faces are out of line, or ohmmeter reading incorrect pick up module should be replaced.

Coil primary winding check   86.35.29/5

1. Disconnect lead at '–' terminal on coil.
2. Connect voltmeter between battery earth and coil '–' terminal. Ignition 'ON', meter should read battery voltage.
3. **NOTE:** No reading, replace coil.

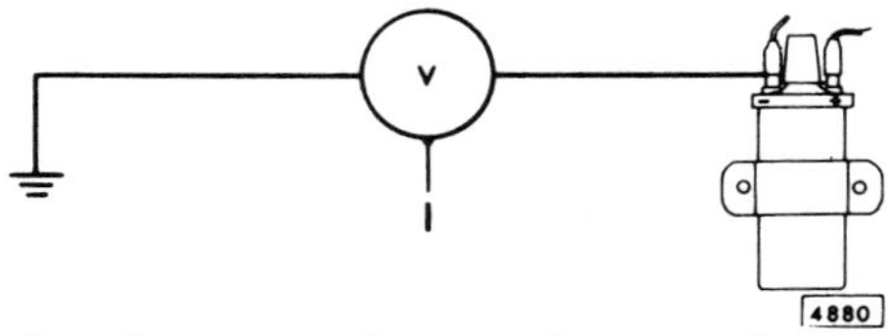

Amplifier volts drop   86.35.29/6

1. Reconnect lead to '–' terminal on coil.
2. Connect voltmeter between coil '–' terminal and battery earth. Ignition 'ON', meter should read 0-2 volts.
   NOTE: High reading, battery voltage, replace amplifier. High reading between 2 volts and battery voltage check for earth fault.

IGNITION TIMING

Check   86.35.29/7

1. Disconnect vacuum pipe from vacuum unit.
   NOTE: For 1977-78 Californian and Australian cars check with stroboscope that (with vacuum pipe connected) the ignition timing is 4° A.T.D.C. at idling speed (750 rev/min).
2. Slacken locknut of micro adjustment control and set vernier at zero.
3. Set engine idling speed at 500 to 600 rev/min.
4. Check timing with a stroboscope and adjust vernier until timing is 10° B.T.D.C. (0° on cars for Sweden).
5. Tighten locknut, refit vacuum pipe.
6. Reset engine idling speed at 650 to 750 rev/min.

86–14

Coil H.T. and amplifier switching
86.35.29/10

1. Connect voltmeter between battery earth and '−' terminal of coil.
2. Connect H.T. jumper lead to H.T. connection on coil.
3. Hold free end of jumper lead ¼ in (6 mm) from engine block.
4. Ignition 'ON', crank engine. Meter should read 3-4 volts, fluctuate with H.T. sparking.
   NOTE: If meter reads 3-4 volts with no H.T. fluctuation, replace coil. If meter reads 0-2 volts, or battery voltage with no H.T. spark, replace amplifier.
   Readings obtained should be:–
   Resistor 1 – 7.6 – 9.2 ohms.
   Resistor 2 – 0.72 – 0.80 ohms.
   Resistor 3 – 0.9 – 1.0 ohms.
   Renew resistor unit if outside these limits. Check battery supply voltage to unit.
   Connection 4 – Straight through internal connection.
   Connection 5 – Tachometer terminal 100 ohms.
5. If above 6 volts is registered, remove distributor cover, crank engine until two of the timing rotor rods (1) are equi-distant either side of the pick-up module core (2) and re-check.
6. Check ignition coil for continuity and resistance of primary winding between '+' and '−' terminals.
   Renew coil if outside limits of 0.8 – 1.0 ohms, or open circuited.

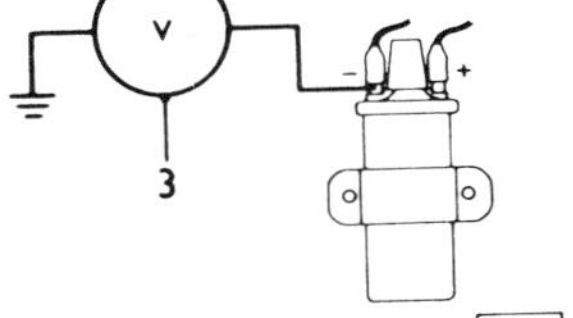

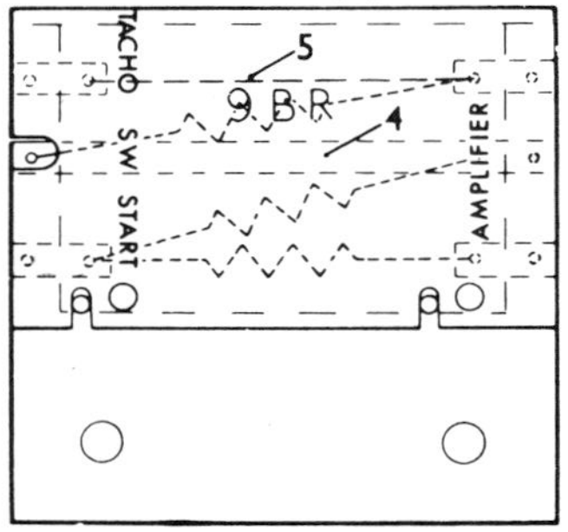

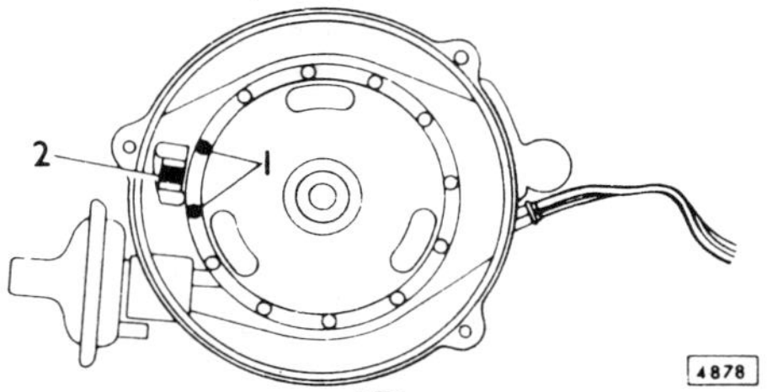

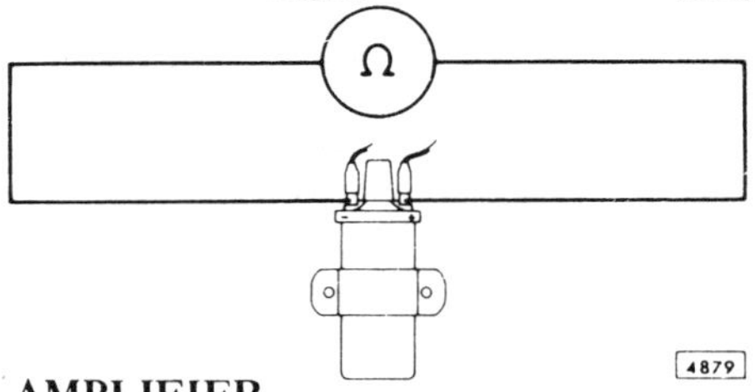

AMPLIFIER
Remove and refit **86.35.30**

Removing
1. Disconnect battery earth lead.
2. Disconnect amplifier to ballast resistor harness at resistor connector and coil.
3. Manoeuvre connector and harness clear of plug leads.
4. Lift amplifier to distributor harness connector clear of amplifier cooling vanes and disconnect at plug and socket.
5. Remove nuts and shakeproof washers securing amplifier to bracket and manoeuvre amplifier clear.

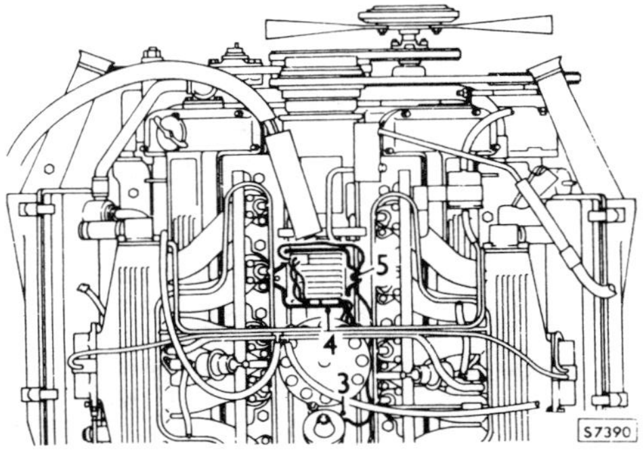

Refitting
Reverse operations 1 to 5.

IGNITION COIL
Remove and refit **86.35.32**

Removing
1. Disconnect battery earth lead.
2. Disconnect H.T. lead.
3. Disconnect L.T. leads at lucars on coil.
4. Remove bolts and shakeproof washers securing coil to throttle pedestal.
5. Remove coil from locations.

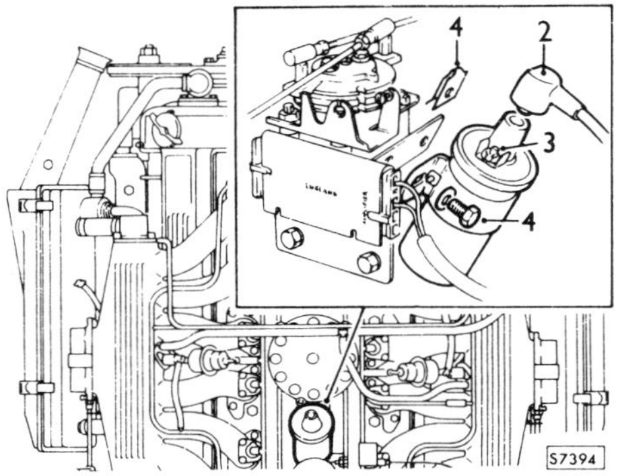

Refitting
Reverse operations 1 to 5.

BALLAST RESISTOR
Remove and refit **86.35.33**

Removing
1. Disconnect battery earth lead.
2. Disconnect block connectors at ballast resistor.
3. Disconnect No. 5 injector electrical connector (for access only).
4. Disconnect throttle operating rod at bell crank on induction manifold and swing aside.
5. Remove bolts and shakeproof washers securing ballast resistor to throttle pedestal.
6. Manoeuvre resistor clear of location.

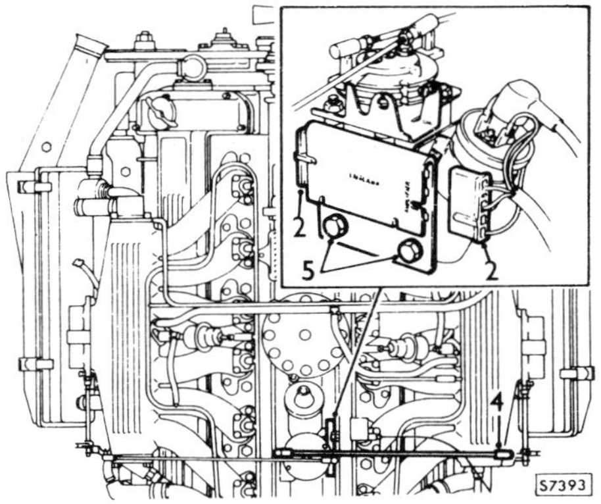

Refitting
Reverse operations 1 to 6.

HEADLAMP RIM FINISHER
Remove and refit **86.40.01**

Removing
1. Raise bonnet.
   EUROPEAN CARS ONLY.
2. Remove screw securing top of finisher to top of lamp housing.
3. Ease top of finisher away from lamp unit and lift bottom locating spigots from housing.

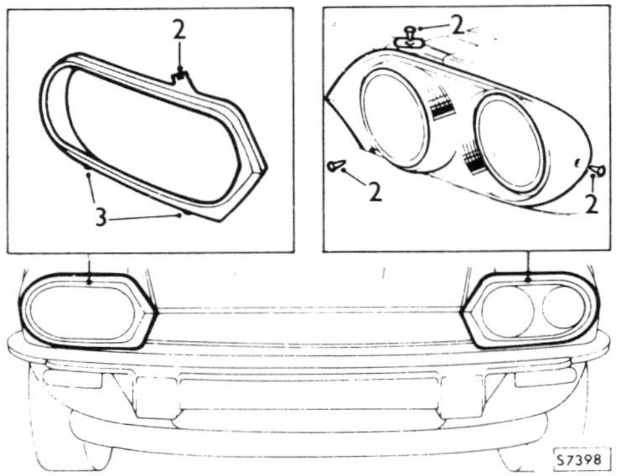

86–15

NORTH AMERICAN SPECIFICATION CARS ONLY.
2 Remove three screws securing finisher.
NOTE: To locate bottom screw it may be necessary to ease bumper sleeve away from lamp unit finisher.
3 Remove finisher.

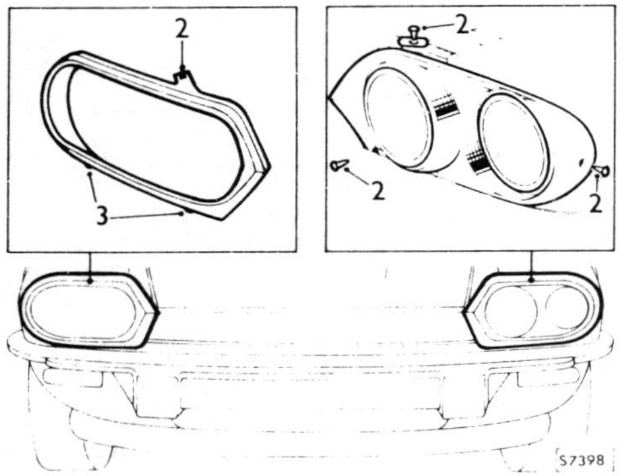

Refitting
Reverse operations 1 to 3.

HEADLAMP ASSEMBLY 86.40.02

Remove and refit

Removing
1 Remove headlamp rim finisher, see 86.40.01.

EUROPEAN CARS ONLY.
2 Depress nylon securing tabs retaining lamp unit.
3 Withdraw unit from housing and disconnect harness at block connectors behind unit.

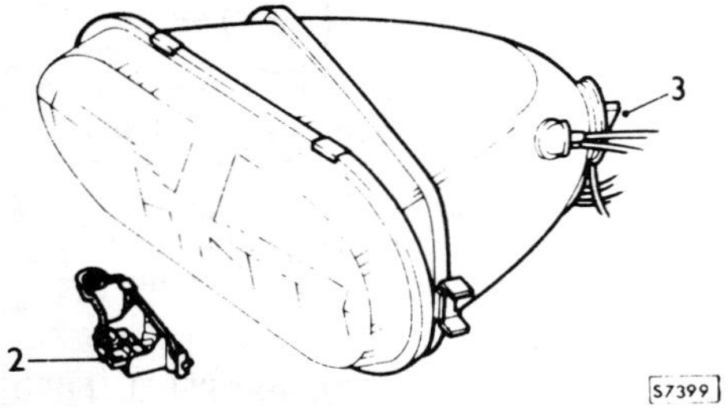

NORTH AMERICAN SPECIFICATION CARS ONLY.
2 Remove four screws securing assembly to housing.
3 Disconnect harness at block connectors on rear of sealed beam units.

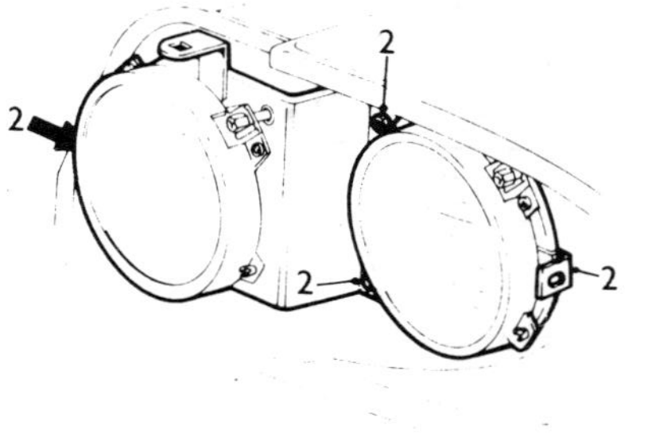

Refitting
Reverse operations 1 to 3.

HEADLAMP SEALED BEAM UNIT/BULB 86.40.09

Remove and refit

Removing
ALL CARS
1 Disconnect battery earth lead.
2 Remove headlamp rim finisher, see 84.40.01.

EUROPEAN SPECIFICATION CARS ONLY
3 Remove headlamp assembly, see 86.40.02.
4 Release clips securing defective bulb.
5 Withdraw bulb.

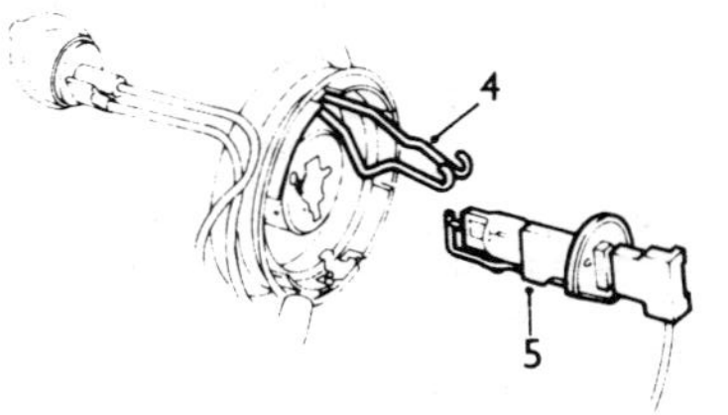

NORTH AMERICAN SPECIFICATION CARS ONLY
3 Slacken three screws securing beam unit retaining ring.
NOTE: Do not disturb beam setting screws.
4 Remove retainer by turning until it releases from locating slots.
5 Withdraw beam unit and disconnect harness at block connector.

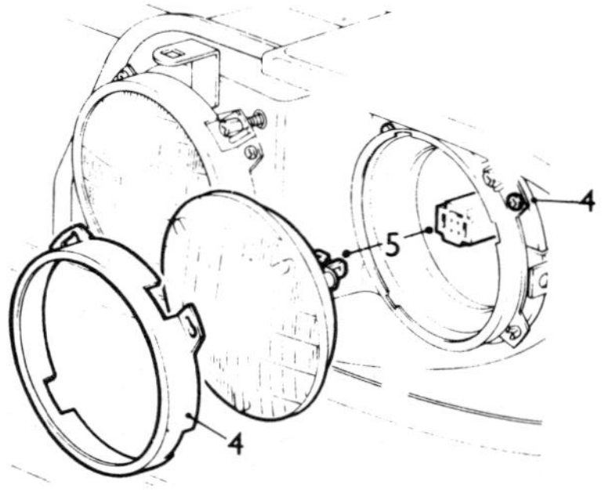

Refitting
Reverse operations 1 to 5.

HEADLAMP PILOT BULB 86.40.11

EUROPEAN SPECIFICATION CARS ONLY

Remove and refit

Removing
1 Disconnect battery earth lead.
2 Remove headlamp assembly, see 86.40.02.
3 Withdraw pilot bulb holder from rubber mountings in reflectors.
4 Remove bulb from holder.

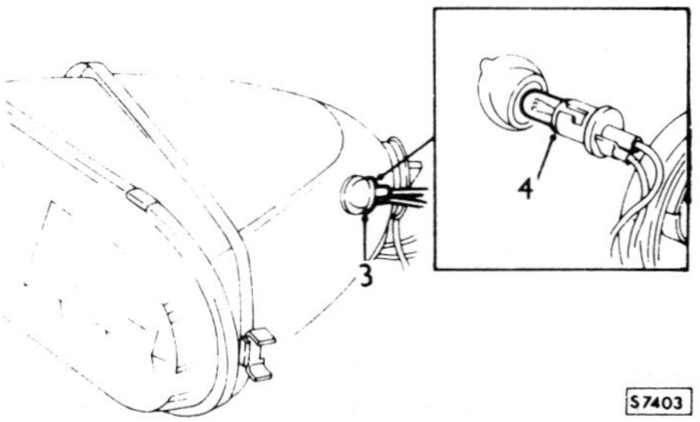

Refitting
Reverse operations 1 to 4.

HEADLAMP ALIGNMENT 86.40.18

Headlamp beam setting should only be carried out by qualified personnel, and with approved beam setting apparatus.

Adjustment
The adjustment screws are set diagonally opposite each other. The upper screw is for vertical alignment, the lower screw for horizontal alignments. Operations 2 and 3 refer.

1 Remove headlamp rim finisher, see 86.40.01.
2 Turn top screw anti-clockwise to lower the beam, clockwise to raise the beam.
3 Turn side screw anti-clockwise to move beam to left, clockwise to move beam right.

FRONT FLASHER LAMP LENS 86.40.40

Remove and refit

Removing
1 Remove screws securing lens to lamp assembly.
2 Remove lens.

Refitting
Reverse operations 1 and 2.

FRONT FLASHER LAMP BULB 86.40.41

Remove and refit

Removing
1 Remove lens, see 86.40.40.
2 Remove bulb.

Refitting
Reverse operations 1 and 2.

SIDE MARKER ASSEMBLY

Remove and refit   Front, 86.40.59

Removing
1. Disconnect battery, see 86.15.20.
2. Remove retaining nuts and lockwashers from captive retaining bolts inside wheel arch.
3. Disconnect cables from snap connectors.
4. Check condition of seals while assembly is removed from car.

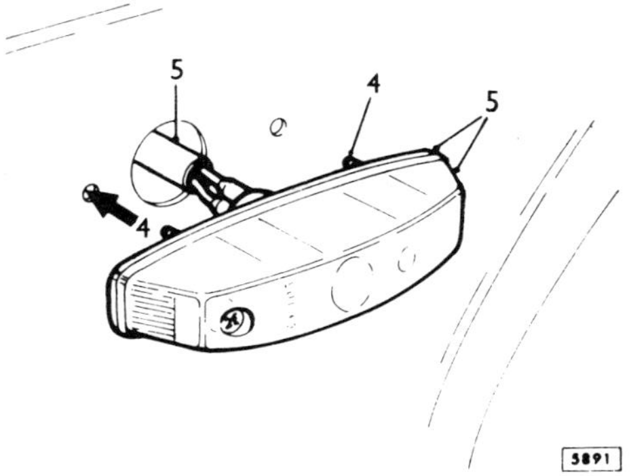

5. Disconnect cables from snap connectors. Check condition of seals while assembly is removed from car.

Refitting
Reverse operations 1 to 6.

SIDE MARKER LENS   Front, 86.40.57
                                   Rear, 86.40.62

Remove and refit

Removing
1. Withdraw one cross-head retaining screw.
2. Remove lens, note retaining clip.

Refitting
Reverse operations 1 and 2.

SIDE MARKER BULB   Front, 86.40.58
                                      Rear, 86.40.63

Remove and refit

Removing
1. Remove lens, see 86.40.57.
2. Withdraw bulb.

Refitting
Reverse operations 1 and 2.

TAIL, STOP AND FLASHER ASSEMBLY LENS

Remove and refit   86.40.68

Removing
1. Remove retaining screws from lens.
2. Remove lens.

Refitting
Reverse operations 1 and 2.

FRONT FLASHER LAMP ASSEMBLY

Remove and refit   86.40.42

Removing
1. Disconnect battery earth lead.
2. Remove bulb, see 86.40.41.
3. Remove nut and bolt securing assembly outer edge.
4. Remove drive screw securing inner edge.
5. Ease assembly clear of energy absorbing beam and disconnect at harness block connector.
6. Remove assembly.

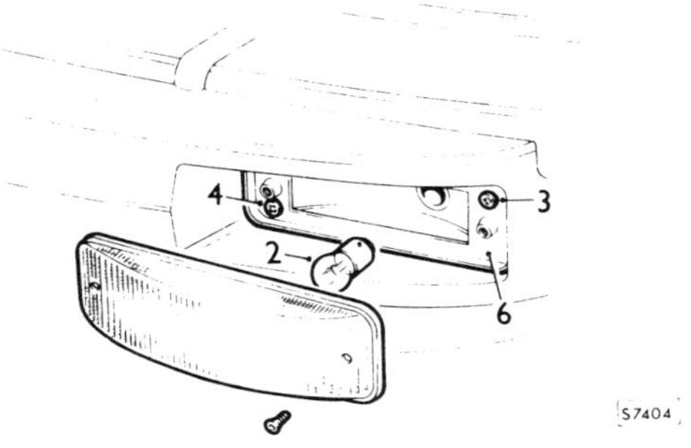

Refitting
Reverse operations 1 to 6.

FRONT FLASHER REPEATER LENS (If fitted)

Remove and refit   86.40.51

Removing
1. Withdraw one cross-head securing screw.
2. Remove lens.

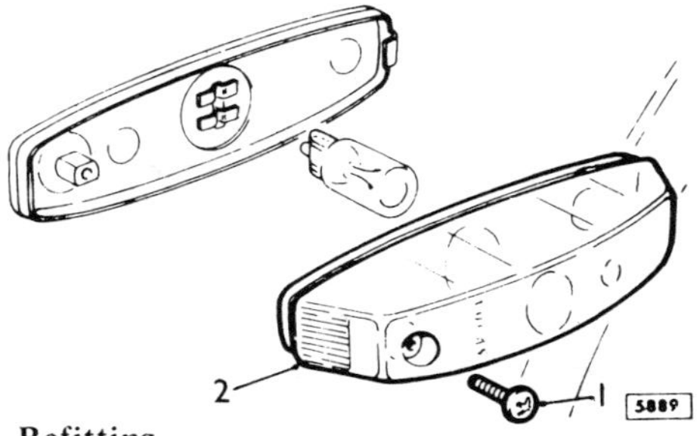

FRONT FLASHER REPEATER BULB (If fitted)

Remove and refit   86.40.52

Removing
1. Remove lens, see 86.40.51.
2. Withdraw bulb.

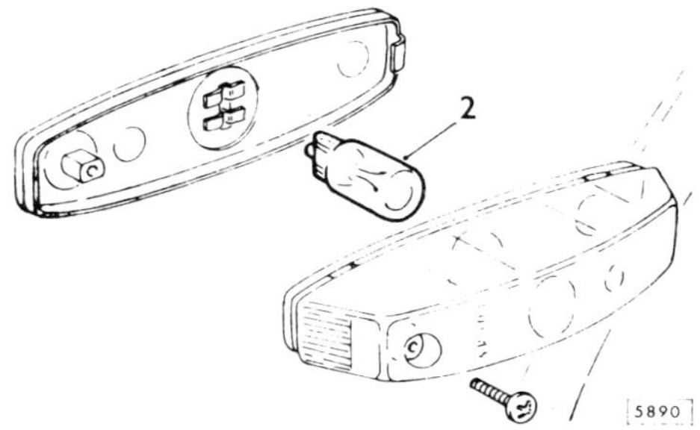

Refitting
Reverse operations 1 and 2.

FRONT FLASHER REPEATER ASSEMBLY (If fitted)

Remove and refit   86.40.53

Removing
1. Disconnect battery earth lead, see 86.15.20.
2. Remove lens, see 86.40.51.
3. Remove bulb.
4. Remove two nuts and lock washer from captive retaining bolts.

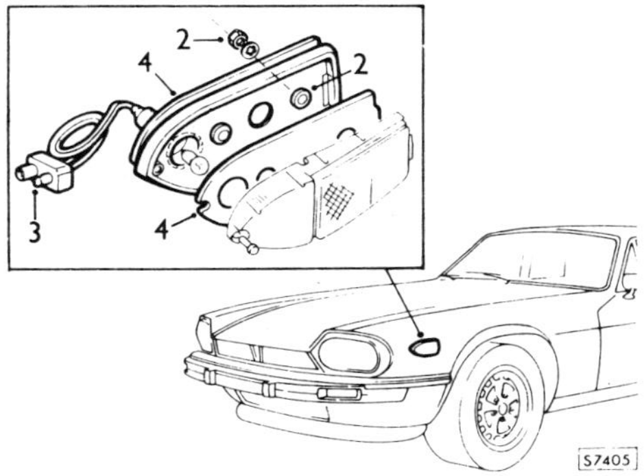

Refitting
Reverse operations 1 and 2.

ROOF LAMP BULB   86.45.01

Remove and refit

Removing
1 Disconnect battery earth lead.
2 Prise lamp assembly from mounting in head lining and clear of aperture.
3 Remove shroud from rear of assembly.
4 Remove bulb.

Refitting
Reverse operations 1 to 4.

ROOF LAMP ASSEMBLY   86.45.02

Remove and refit

Removing
1 Disconnect battery earth lead.
2 Prise lamp assembly from mounting in head lining and clear of aperture.
3 Disconnect electrical connectors from lamp terminals.
4 Remove assembly.

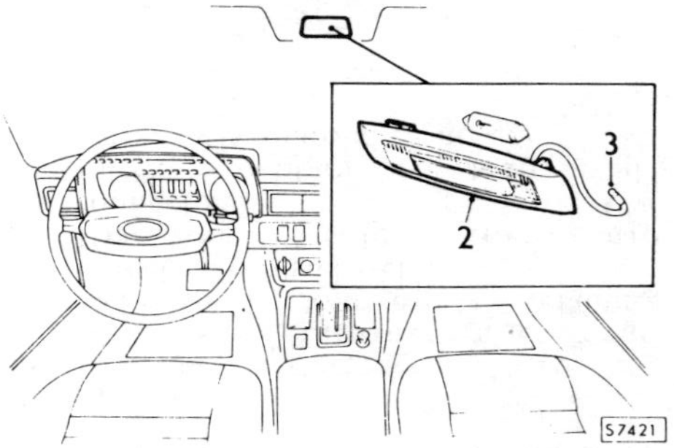

Refitting
Reverse operations 1 to 4.

REVERSE LAMP LENS   86.40.89

Remove and refit

Removing
1 Withdraw two screws retaining lens.
2 Remove lens, examine seal for condition.

Refitting
Reverse operations 1 and 2.

REVERSE LAMP BULB   86.40.90

Remove and refit

Removing
1 Remove lens, see 86.40.89.
2 Remove bulb.

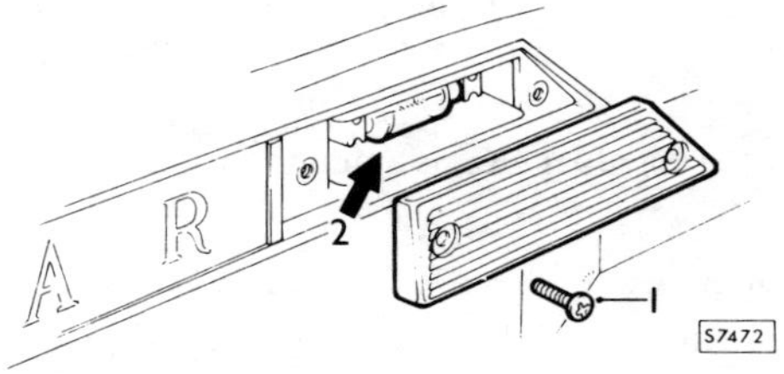

TAIL, STOP AND FLASHER ASSEMBLY BULB(S)   86.40.69

Remove and refit

Removing
1 Remove lens, see 86.40.68.
2 Remove bulb(s) from holder.

Refitting
Reverse operations 1 and 2.

TAIL, STOP AND FLASHER ASSEMBLY   86.40.70

Remove and refit

Removing
1 Disconnect battery earth lead.
2 Open boot (trunk) for access to rear of assembly.
3 Remove three retaining nuts and washers from rear of assembly.
4 Withdraw assembly sufficient to gain access to harness connector.
5 Disconnect harness at block connector.
6 Remove assembly.

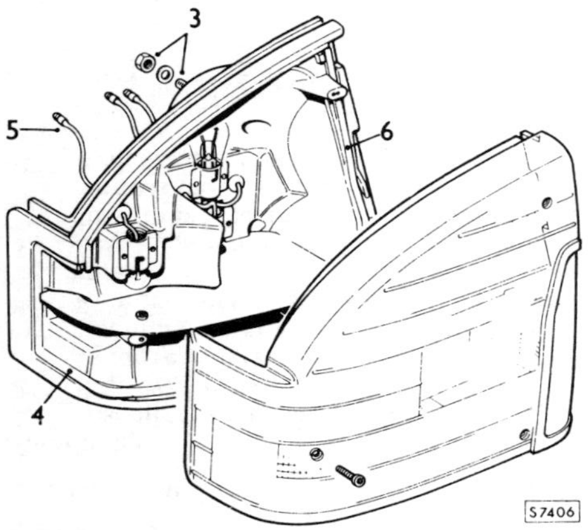

5 Remove bolts securing assembly to boot lid.
6 Ease lamp assembly and harness clear of boot lid.

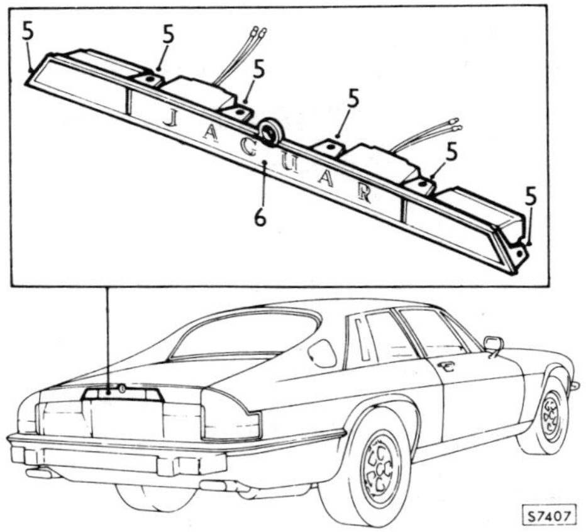

Refitting
Reverse operations 1 to 6.

NUMBER PLATE LAMP LENS   86.40.84

Remove and refit

Removing
1 Raise boot lid for access.
2 Remove screws securing lens to assembly.
3 Lower lens and remove bulb.
4 Ease bulb holder clear of lens.
5 Remove spire nuts from spigots on lens holder. 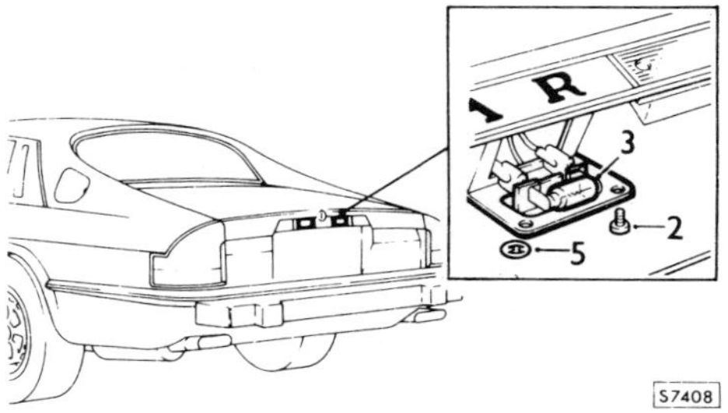
6 Recover lens from location.

Refitting
Reverse operations 1 to 6.

NUMBER PLATE LAMP BULB   86.40.85

Remove and refit

Removing
1 Operation 86.40.84 items 1 to 3 refer.

Refitting
Reverse operations 1 to 3.

NUMBER PLATE AND REVERSE LAMP ASSEMBLY   86.40.87

Remove and refit

Removing
1 Disconnect battery earth lead.
2 Raise boot lid for access.
3 Disconnect block connectors.
4 Displace grommets from boot lid and feed harness and connectors through them.

Refitting
Reverse operations 1 to 6.

INTERIOR AND MAP LAMP BULB
Remove and refit   86.45.03

Removing
1. Disconnect battery earth lead.
2. Prise interior lamp from location point and clear of aperture.
3. Remove bulb from holder.

Refitting
Reverse operations 1 to 3.

INTERIOR AND MAP LAMP ASSEMBLIES
Remove and refit   86.45.04

Removing
1. Disconnect battery earth lead.
2. Prise assembly from location point and clear of aperture.
3. Disconnect electrical connectors at rear of assembly.
    NOTE: Passenger side interior lamp, located in fascia side casing, doubles as map lamp and has more electrical connectors than other interior lamps.
4. Remove assembly.

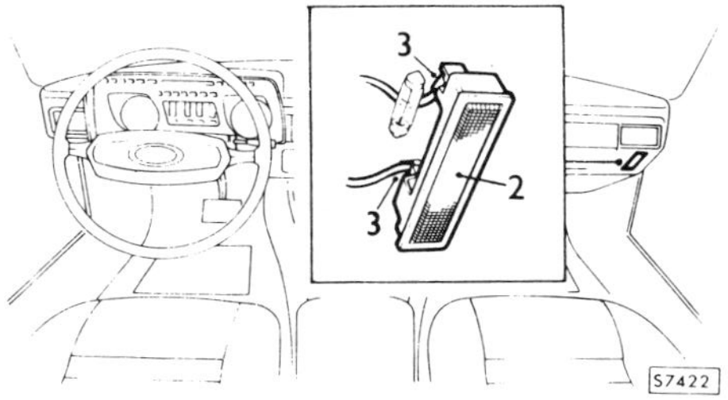

Refitting
Reverse operations 1 to 4.

LUGGAGE COMPARTMENT LIGHT BULB
Remove and refit   86.45.15

Removing
1. Open luggage compartment.
2. Access to bulb is through aperture in luggage compartment lid.
3. Remove bulb.

LUGGAGE COMPARTMENT LAMP ASSEMBLY
Remove and refit   86.45.16

Removing
1. Disconnect battery earth lead, see 86.15.20.
2. Withdraw two screws retaining bracket.
3. Disconnect cable from snap connectors.

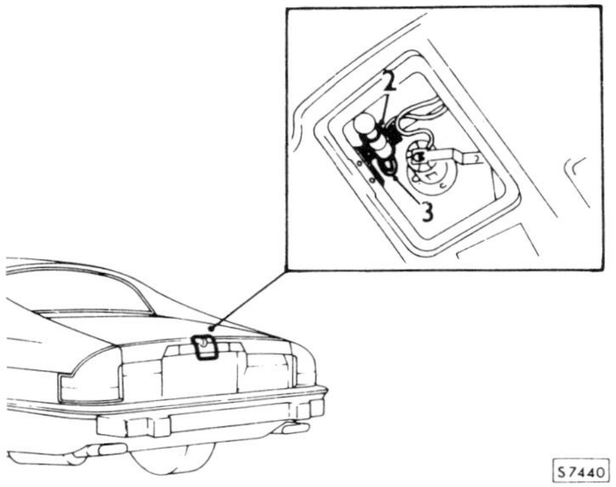

Refitting
Reverse operations 1 to 3.

OPTICELL UNIT
Remove and refit   86.45.27

Removing
1. Disconnect battery earth lead.
2. Raise centre glove box lid.
3. Withdraw three screws securing console centre panel.
4. Lift centre panel clear of console.
5. Note positions of cables on door lock selector switch and disconnect.
6. Lift centre panel clear of gear/transmission selector lever.
7. Disconnect feed cable to cigar lighter at snap connector if necessary to allow full movement of panel
8. Remove four nuts securing gear selector quadrant cover to tunnel.
9. Lift cover clear of quadrant.
10. Remove nuts securing opticell mounting bracket and lift assembly clear.
11. Disconnect fibre elements and opticell feed cable.
12. Withdraw two screws retaining opticell unit and remove unit from bracket.

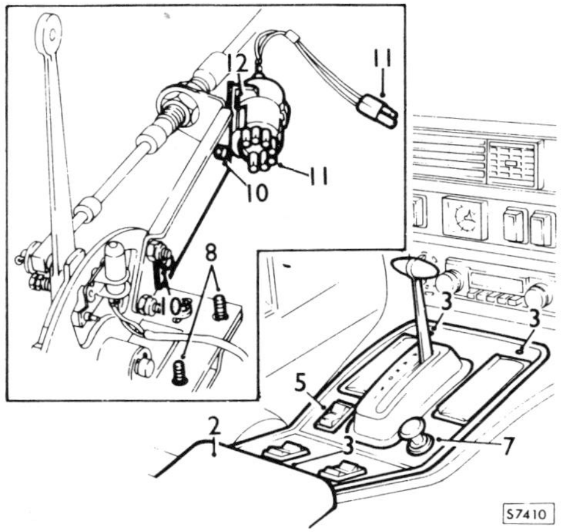

Refitting
Reverse operations 1 to 12.

OPTICELL BULB
Remove and refit   86.45.28

Removing
1. Disconnect battery earth lead.
2. Raise centre glove box lid.
3. Withdraw three screws securing console centre panel.
4. Lift centre panel clear of console for access.
5. Prise opticell unit bulb holder from rear of unit.
6. Remove bulb.

Refitting
Reverse operations 1 to 6.

PANEL SWITCH ILLUMINATION BULB
Remove and refit   86.45.31

Removing
1. Disconnect battery earth lead.
2. Prise switch mounting panel from fascia.
3. Pull bulb holder from diffuser.
4. Remove bulb from holder.

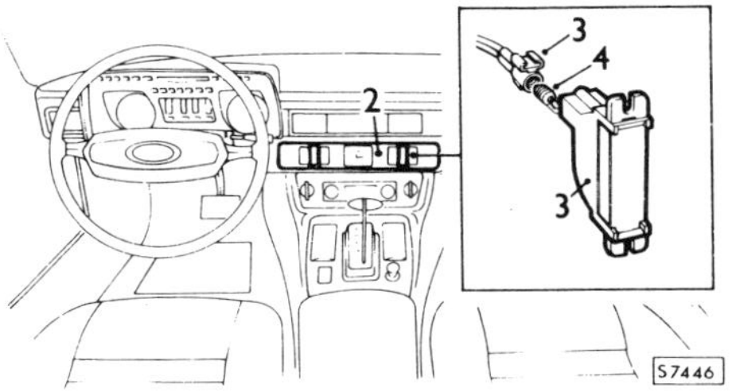

Refitting
Reverse operations 1 to 4.

AUTOMATIC TRANSMISSION SELECTOR LAMP BULB
Remove and refit   86.45.40

Removing
1. Raise centre console glove box lid.
2. Remove three screws securing centre panel to console.
3. Raise centre panel and disconnect cables from door lock switch and cigar lighter.
   NOTE: Take note of position of cables to facilitate correct refitting.

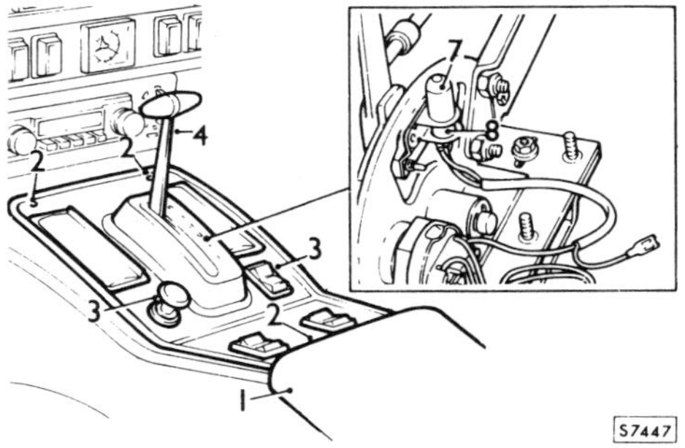

86–19

4 Raise centre panel over selector and remove for access.
5 Remove nuts securing selector quadrant cover.
6 Raise cover clear of quadrant.
7 Remove shroud from selector lamp.
8 Remove bulb.

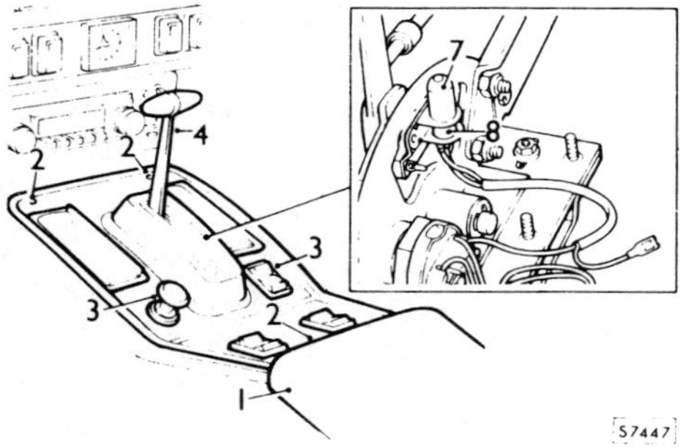

Refitting
Reverse operations 1 to 8.

INSTRUMENT ILLUMINATION BULB

Remove and refit 86.45.48

This operation serves for the following operations:

Speedometer Illumination Bulb 86.45.49
Oil Gauge Illumination Bulb 86.45.50
Temperature Gauge Illumination Bulb 86.45.51
Fuel Gauge Illumination Bulb 86.45.52
Tachometer Illumination Bulb 86.55.53
Battery Condition Indicator Bulb 86.55.56

Removing
1 Disconnect battery earth lead.
2 Remove instrument panel module, see 88.20.01.
3 Remove bulb holder.
4 Remove bulb from holder.

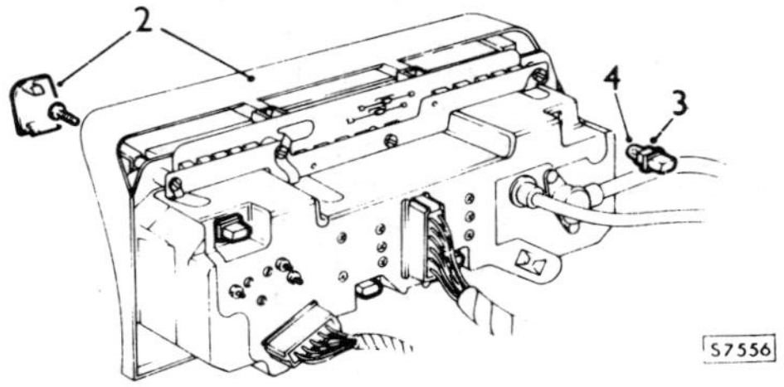

5 Remove nut and shakeproof washer securing earth terminal and relay to mounting plate.
6 Note positions of electrical connectors and disconnect from lucars on base of relay.
7 Recover relay.

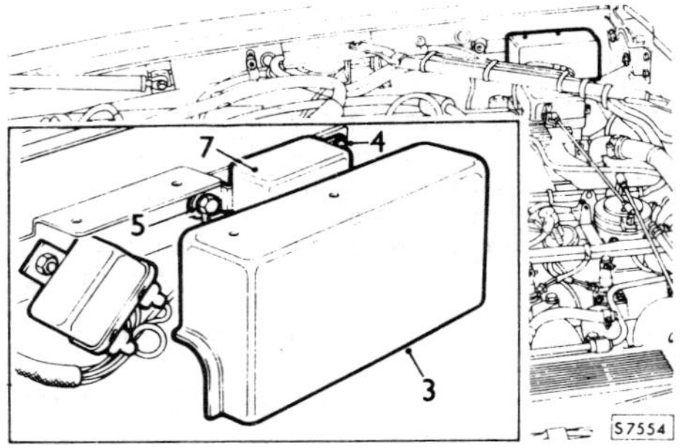

Refitting
Reverse operations 1 to 7.

CLOCK ILLUMINATION BULB

Remove and refit 86.45.54

Removing
1 Prise mounting panel from fascia.
2 Pull bulb holder from back of clock.
3 Remove bulb from holder.

Refitting
Reverse operations 1 to 3.

WARNING LAMP BULB(S)

Remove and refit 86.45.61

Removing
1 Slacken steering wheel adjustment ring and raise steering wheel to full extent.
2 Prise out covers from each end of warning light strip.
3 Remove screws securing warning lamp lens cover to instrument module.
4 Remove cover.
5 Remove defective bulb.

Refitting
Reverse operations 1 to 5.

STARTER RELAY

Remove and refit 86.55.05

Removing
1 Disconnect battery earth lead.
2 Remove screws securing relay cover to mounting plate.
3 Remove cover.
4 Remove screw securing relay and mounting plate to wing valance.

Refitting
Reverse operations 1 to 3.

STEERING LOCK/SAFETY BELT AUDIO UNIT (If fitted)

Remove and refit 86.55.13

Removing
1 Disconnect battery earth lead.
2 Remove passenger side dash liner, see 76.46.11.
3 Remove nuts and shakeproof washers securing mounting plate to fan motor case.
4 Ease mounting plate clear of studs.
5 Remove nuts, bolts, and shakeproof washers securing unit to mounting plate.
6 Disconnect block connector from unit.
7 Recover unit.

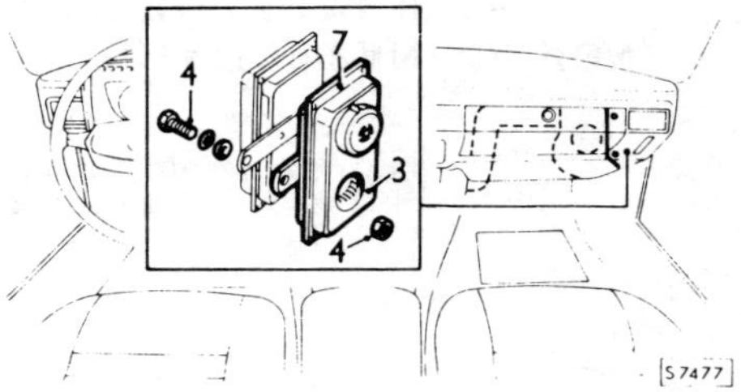

Refitting
Reverse operations 1 to 7.

HAZARD FLASHER/TURN SIGNAL UNIT

Remove and refit 86.55.12

Removing
1 Disconnect battery earth lead.
2 Remove fuse box access cover from driver's dash liner.
3 Locate unit, above fuse block and push upwards to remove from multi pin adaptor.

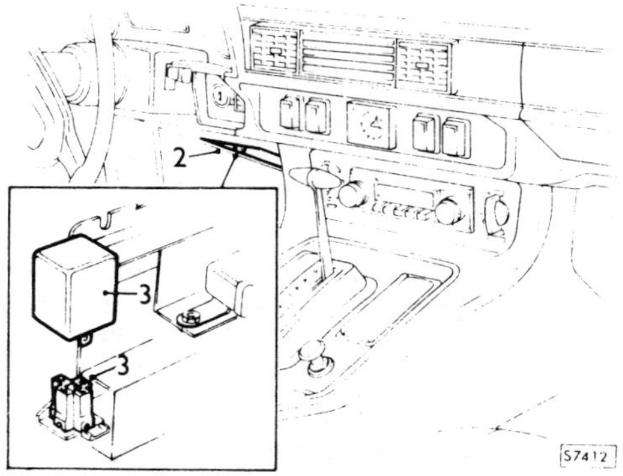

Refitting
Reverse operations 1 to 7.

86—20

HEADLAMP RELAY 86.55.17

Remove and refit

Removing
1. Disconnect battery earth lead.
2. Disconnect electrical connectors from headlamp fuses located on relay cover, left hand wing valance.
3. Remove screws securing relay cover to wing valance and remove cover.
4. Disconnect electrical connectors from relay.
   NOTE: Take note of positions before disconnection takes place.
5. Remove nuts securing relay to mounting plate, recover washers.
6. Ease horn relay from mounting stud to facilitate removal of headlamp relay.
7. Remove headlamp relay.

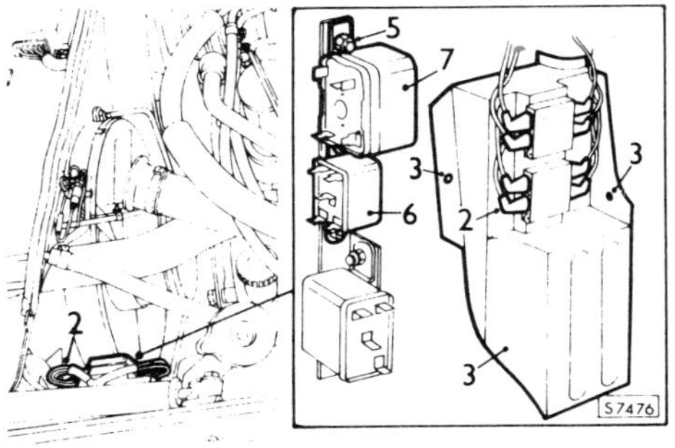

Refitting
Reverse operations 1 to 7.

PARK LAMP FAILURE WARNING SENSOR 86.55.22

Remove and refit

Removing
ALL CARS
1. Disconnect battery earth lead.

EUROPEAN SPECIFICATION CARS–RIGHT HAND SIDE FRONT AND REAR.
NORTH AMERICAN SPECIFICATION CARS–RIGHT HAND FRONT ONLY.

2. Remove passenger side dash liner, see 76.46.11.
3. Remove nuts securing mounting plate to fan motor, ease mounting plate clear of location.
4. Remove electrical connectors and drive screw.
5. Remove sensor from locating piece on mounting plate.

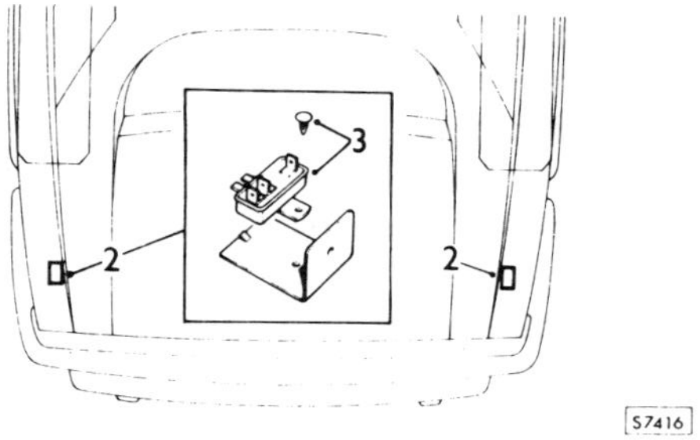

EUROPEAN SPECIFICATION CARS – LEFT HAND SIDE FRONT AND REAR.
NORTH AMERICAN SPECIFICATION CARS – LEFT HAND FRONT ONLY.

2. Remove passenger side dash liner, see 76.46.11.
3. Remove electrical connectors.
4. Remove drive screw securing sensor to mounting plate.
5. Remove sensor from locating piece on mounting plate.

NORTH AMERICAN SPECIFICATION CARS ONLY – LEFT AND RIGHT REAR.

2. Open boot and locate sensor related to left or right hand rear park light beneath rear body side panel forward of lamp unit assemblies.
3. Pull sensor from mounting bracket, located and secured by plastic spigot.
4. Disconnect electrical connectors.
5. Remove sensor.

Refitting
Reverse operations 1 to 5 relevant to model specification and component location.

IGNITION/STARTER CONTROLLED RELAY 86.55.28

Remove and refit

Removing
1. Disconnect battery earth lead.
2. Remove fuse block access cover from driver's side dash liner.
3. Remove screws, nuts and shakeproof washers securing relay to mounting bracket above fuse block.
4. Note position of electrical connectors and disconnect.
5. Recover relay.

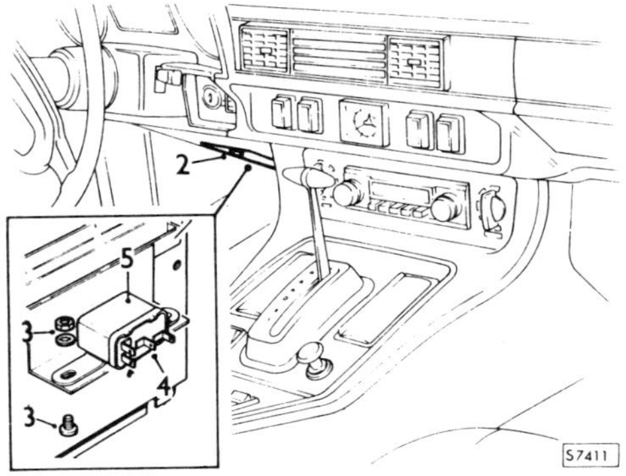

Refitting
Reverse operations 1 to 5.

STOP LIGHT FAILURE SENSOR 86.55.34

Remove and refit

Removing
1. Disconnect battery earth lead.
2. Raise centre console glove box lid for access to centre panel rear securing screw.
3. Remove screws securing centre panel.
4. Raise centre panel, note position of electrical connectors to door lock operating switch and disconnect.
5. Lift centre panel over gear/transmission selector handle.

continued

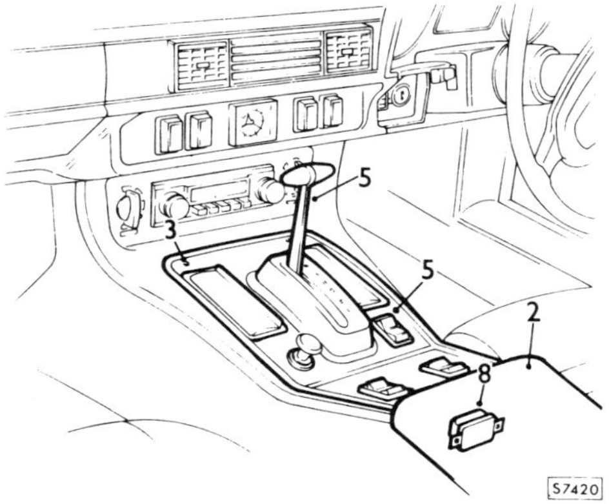

LOW COOLANT CONTROL UNIT 86.55.33

Remove and refit

Removing
1. Disconnect battery earth lead.
2. Remove passenger side dash liner, see 76.46.11.
3. Remove nuts and shakeproof washers securing mounting plate to fan motor.
4. Ease mounting plate clear of studs.
5. Remove nuts, bolts and shakeproof washers securing unit to mounting plate.
6. Disconnect block connector from main harness.
7. Recover unit.

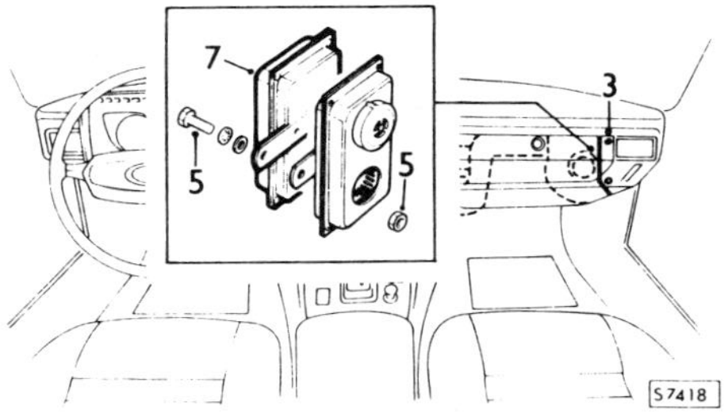

86–21

6 Disconnect sensor harness at block connector and lucars.
7 Remove screws securing sensor to mounting bracket.
8 Recover sensor.

Refitting
Reverse operations 1 to 8.

WARNING BUZZER/CONTROL UNIT

Remove and refit 86.57.01

Removing
1 Disconnect battery earth lead.
2 Remove passenger side dash liner, see 76.46.11.
3 Disconnect multi-pin plug from unit.
4 Remove front screw, nut and washer securing unit to mounting bracket.
5 Disconnect earth lead.
6 Remove rear securing screw, nut and washer.
7 Recover warning buzzer/control unit.

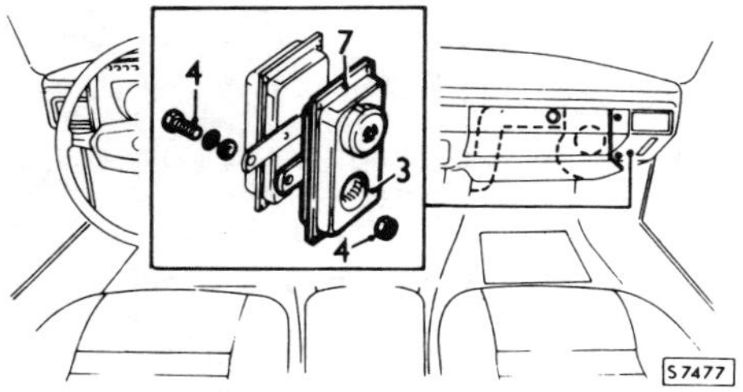

5 Disconnect switch unit block connector.
6 Recover belt/switch unit.

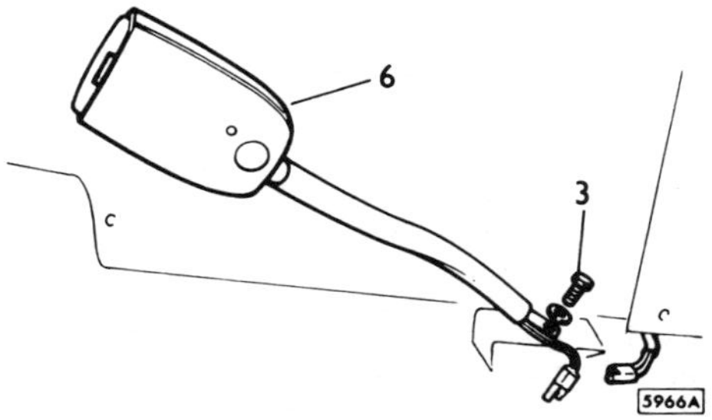

Refitting
Reverse operations 1 to 6.

STARTER MOTOR

Remove and refit 86.60.01

Removing
1 Disconnect battery earth lead.
2 Drain power steering system.
3 Remove pinch bolt securing lower column universal joint to steering column.
4 Remove pinch bolt securing universal joint to lower column.
5 Remove bolts securing steering gaiter heatshield to mounting bracket, remove heatshield.
6 Remove bolt securing lower column to steering unit.
7 In car – disengage lower column upper universal joint from bottom of steering column.
8 Slacken draught exclusion collar securing clip and remove universal joint from lower column.
9 Disengage lower column from steering unit and ease back through bulkhead.
10 Release clip securing power steering feed and return pipes to bridge piece and rack tube.
11 Disconnect feed and return pipes from steering unit. Support pipes clear of rack.
NOTE: Fit plugs to pipes and steering unit.

12 Remove bolts securing exhaust down pipe to intermediate pipe (on pinion side of engine).
13 Disengage intermediate pipe from down pipe and remove olive.
14 Remove screws securing front exhaust heatshield to underfloor, remove heatshield.
15 Slacken lower bolts securing steering rack to front suspension unit.
16 Remove bolt securing rack and heatshield mounting bracket to suspension.
17 Lower rack carefully, ease past sump.
18 Retrieve distance/packing washers from rack top mounting joint.
19 Remove nuts securing exhaust down pipe to manifold, remove down pipe.

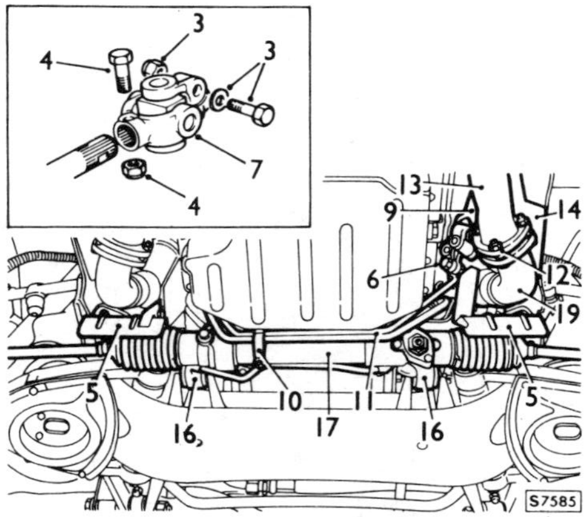

20 Remove nuts securing starter motor heatshield and cable retaining clip. Recover heatshield.
21 Remove nut securing feed cable from alternator at starter motor and disconnect cable.
22 Disconnect solenoid to terminal post connector at terminal post. Secure cable with string to wing stay to facilitate refitting. Disconnect cable at bulkhead lucar.
23 Remove bolts securing starter to bell housing.
24 Detach cable attached to wing stay, ensure that string is long enough to pull through engine bay.

25 Manoeuvre starter motor clear of location. Detach string from solenoid cable.

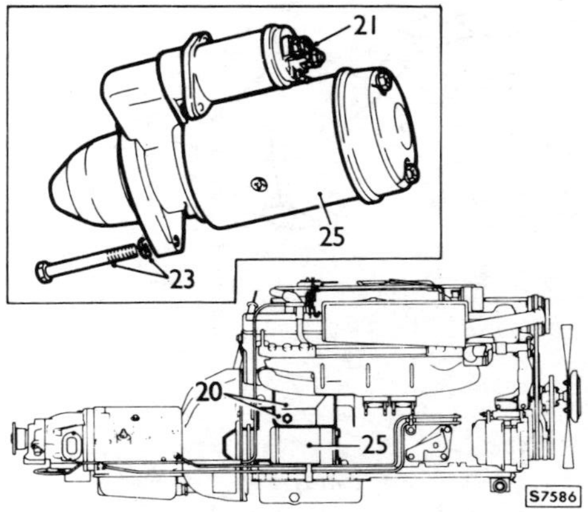

26 Remove heatshield bracket and distance piece from motor.

Refitting
Reverse operations 1 to 26.
NOTE: Power steering system must be topped up and bled.

STARTER MOTOR ROLLER CLUTCH DRIVE UNIT

Remove and refit 86.60.05

Removing
1 Remove starter motor, see 86.60.01.
2 Remove solenoid complete with bridge strap (copper link).
3 Remove solenoid unit from drive end fixing bracket.
4 Remove engagement lever pivot pin.

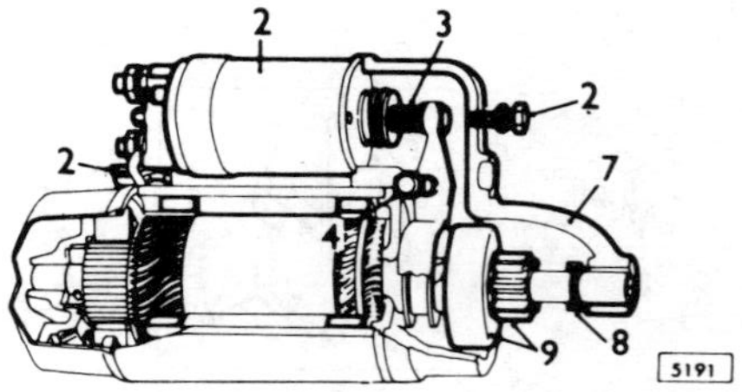

BELT SWITCH(ES)

Remove and refit 86.57.25

Removing
1 Disconnect battery earth lead.
2 Push seat forward to full extent of travel.
3 Remove bolt securing belt unit to floor pan.
4 Raise belt and ease connector leads clear of carpet.

86–22

5 Withdraw through bolts, but do not remove end bracket or commutator end cover.
6 Mount starter motor vertically in a vice (drive end uppermost).
7 Withdraw drive end fixing bracket.
8 Remove jump ring from groove on drive shaft.
9 Remove collar and drive unit from shaft.

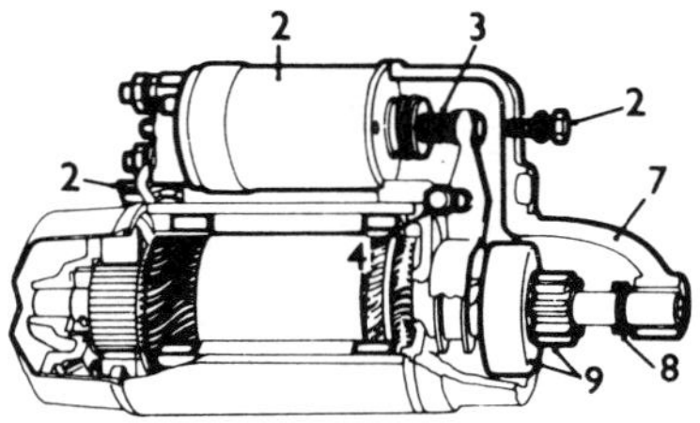

Refitting

10 Reverse operations 3 to 9.
   Tightening torques:
   Through bolts 8.0 lb ft (1,1 kg m).
   Solenoid unit fixing bolts 4.5 lb ft (0,62 kg m).
   a Smear all moving parts of drive unit liberally with grease. Use Retinax 'A' (Home and cold climate countries) Shell SB.2628 (Hot climate countries).
11 Connect solenoid terminal 'STA' to starter motor casing.
12 Connect a 6 volt supply between solenoid operating 'Lucar' terminal and starter motor casing.
13 With solenoid energised and drive assembly now in engaged position, press pinion lightly back towards armature to take up any slack in drive operating mechanism and then set position of eccentric pivot pin to obtain 0.005 in to 0.015 in (0,127 mm to 0,381 mm) clearance between pinion and thrust collar.
14 Apply sealing compound and tighten locknut.
15 Refit bridge strap (copper link).
16 Refit starter motor, see 86.60.01.

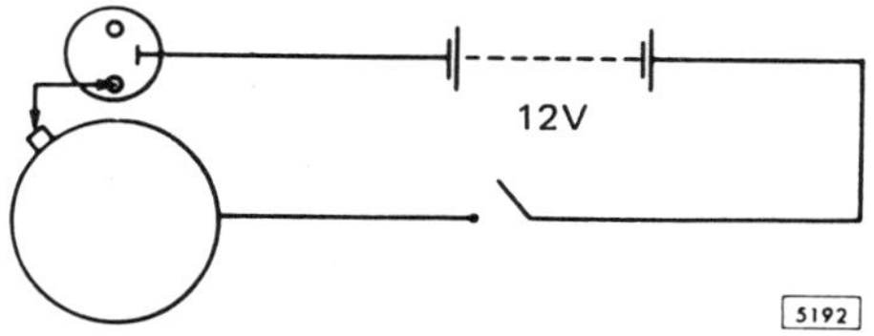

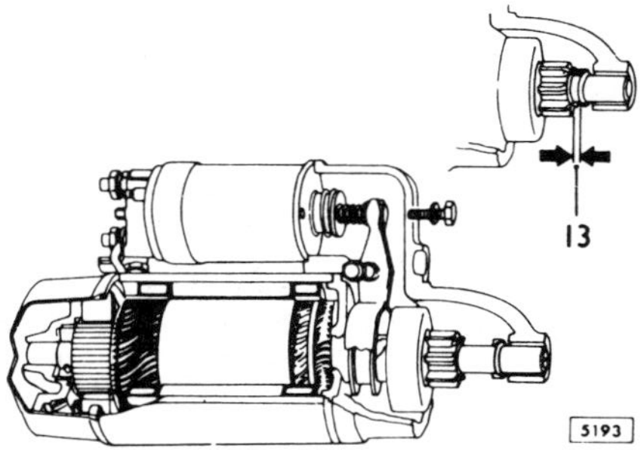

STARTER MOTOR SOLENOID UNIT

Remove and refit   86.60.08

Removing
1 Remove starter motor, see 86.60.01.
2 Remove link connecting solenoid to yoke terminal.
3 Remove two fixings, withdraw solenoid from bracket. Collect gasket.
4 Release plunger from top of drive engagement lever.

Refitting
Reverse operations 1 to 4 inclusive.

STARTER MOTOR SOLENOID UNIT

Test   86.60.09

The following checks assume that the pinion travel has been correctly set.
1 Remove bridge strap connecting solenoid to motor.
2 Connect a 12 volt d.c. supply, with switch between solenoid 'Lucar' and large terminal 'STA'. **DO NOT CLOSE SWITCH.**
3 Connect a separately energised 60 watt test lamp across solenoid main terminals.
4 Close switch. Solenoid should be heard to operate, and lamp should light with full brilliance.
5 Open switch. Lamp should go out.

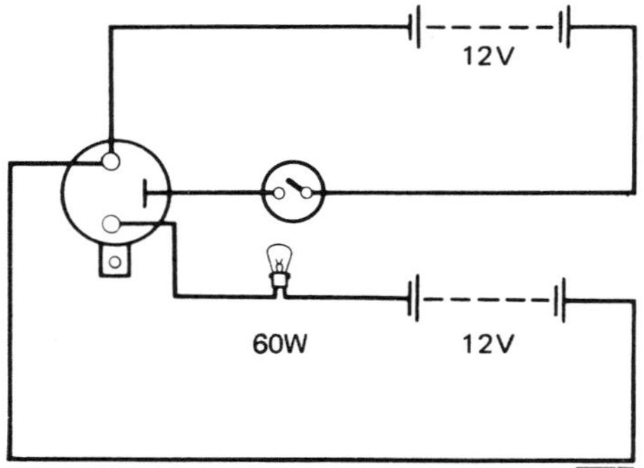

14 Remove intermediate bracket.
15 Check individual components as detailed under relevant sub-headings.

Re-assembling
16 Reverse operations 1 to 15. Tighten the through bolts to a torque of 8.0 lb ft (1,1 kg m).

Brush replacement
17 Renew brushes when worn to 0.313 in (8 mm) length.
   NOTE: The insulated brush connectors are hot pressed to the free ends of the field coils. To replace, cut off the worn brush connectors approximately 0.125 in (3 mm) from the joint.
   Open out and tin the loop of the replacement brush. Place the tinned loop over the stub of the brush connector, squeeze up and solder.

Brush box insulation test
18 Connect a 110 volt a.c. 13 watt test lamp between a clean part of the end bracket and each of the two insulated brush boxes.
   If the lamp lights, renew the commutator end bracket assembly.

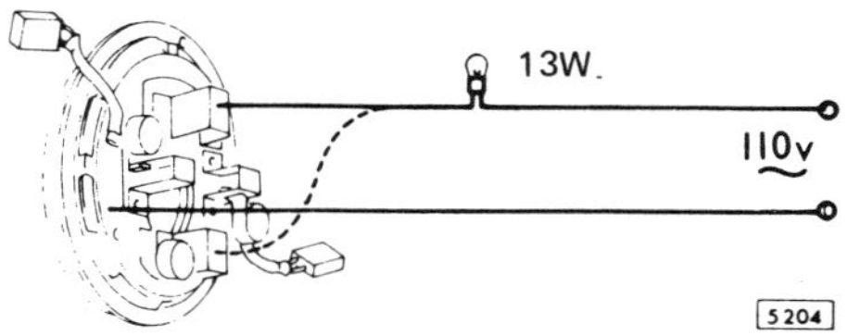

STARTER MOTOR

Overhaul   86.60.13

Dismantling
1 Remove link connecting solenoid to yoke terminal.
2 Remove two fixings, withdraw solenoid from bracket. Collect gasket.
3 Release plunger from top of drive engagement lever.
4 Remove commutator-end sealing cover.
5 Withdraw through bolts.
6 Lift brushes from boxes and detach commutator-end bracket from yoke.
7 Withdraw yoke and field coil assembly from armature and intermediate bracket. Collect sealing ring fitted to intermediate bracket.
8 Slacken locknut and remove drive engagement pivot pin from fixing bracket.
9 Remove drive end fixing bracket.
10 Remove drive engagement lever.
11 Using suitable piece of tubing drift the thrust collar away from the jump ring on armature shaft.
12 Remove jump ring.
13 Withdraw roller clutch drive assembly.

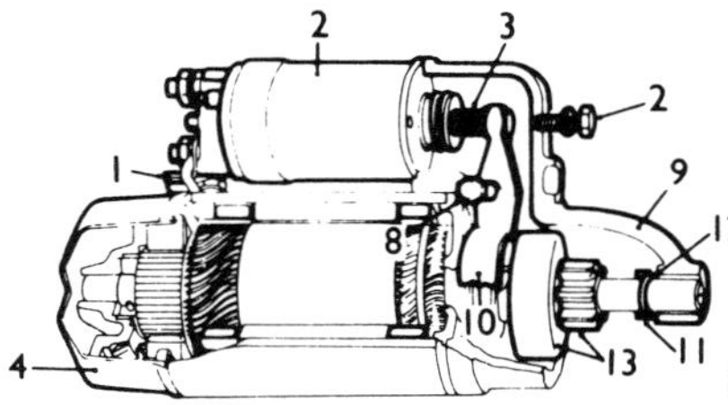

86—23

Commutator cleaning

19 Clean the commutator if not scored with a petrol moistened cloth.
Worn commutators should be cleaned with fine glass paper or mounted in a lathe and a fine cut taken with a sharp tool. Finally polish with very fine glass paper. DO NOT UNDER CUT INSULATORS BETWEEN SEGMENTS.
NOTE: Armatures must not be skimmed below a minimum diameter of 1.5 in (38 mm). Replace if below this limit.

Armature – checking

20 Armature conductors lifted from risers indicate overspeeding. Carefully resolder conductors or replace armature. Check clutch operation. Armatures showing signs of fouling indicate worn bearings or un-true shaft.
Renew armature or bearings as required.
No attempt should be made to machine an untrue shaft.

Armature insulation test

21 Connect a 110-volt a.c. 15 watt test lamp between any one of the commutator segments of the shaft.
If lamp lights renew armature.

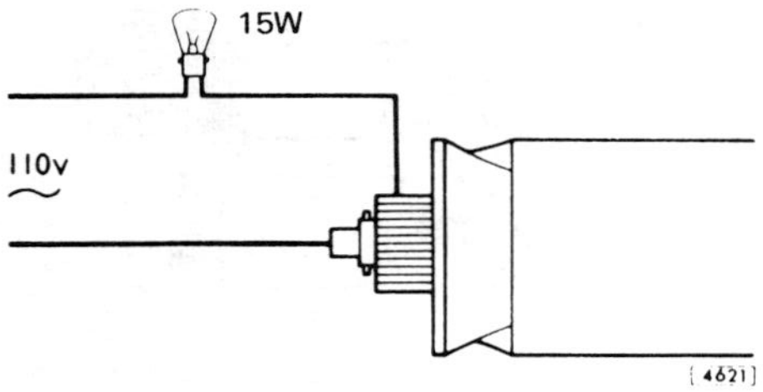

Field coil – test

22 Check continuity of winding by connecting a 12 volt test lamp and battery between the terminal post and each brush (with the armature removed). An open circuit is indicated if lamp does not light.
Replace faulty coils.

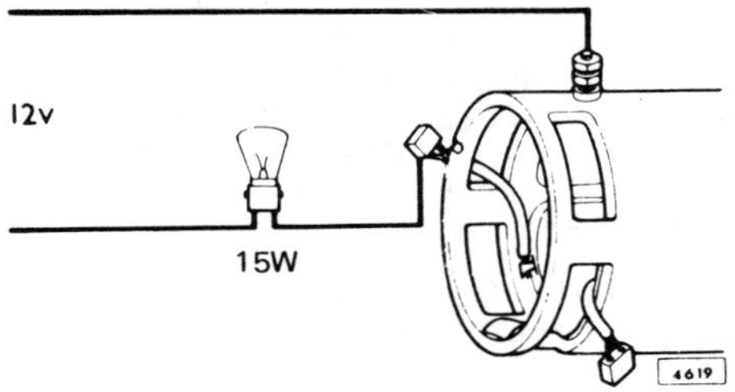

NOTE: Porous bronze bushes must not be reamed out after fitting.

Roller clutch drive – checking

26 Check that pinion is free to move on shaft splines, and clutch assembly operates correctly. Replace faulty or sticking units.

Pinion movement – setting

27 After re-assembly of the starter (cranking) motor pinion movement must be reset as follows:
Connect the 'Lucar' solenoid terminal in series with a switch to a 10 volt battery.
Connect other battery terminal to starter yoke.
Close switch. (This throws the drive assembly forward into the engage position.) Measure the distance between pinion and thrust washer on armature shaft extension.
NOTE: Pinion should be pressed lightly towards armature to take up any slack in engaging linkage. Correct setting should be 0.005 in to 0.015 in (0,127 mm to 0,381 mm).
To adjust, slacken the eccentric pin securing nut and turn pin until correct setting is obtained.
NOTE: Arc of adjustment is 180° and the head of the arrow on the pivot pin should be set only between the arrow heads on the drive end casting.
Tighten securing nut to retain pin position after setting.

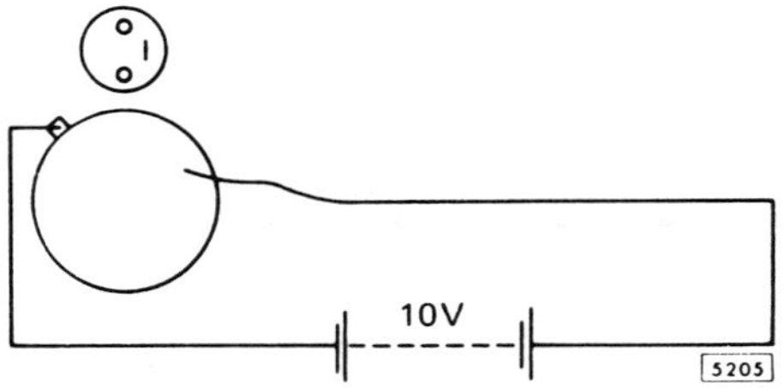

23 Check coil insulation with a 110 volt a.c. 15 watt test lamp connected between the terminal post and a clean part of the yoke.
Renew field coils if bulb is illuminated.

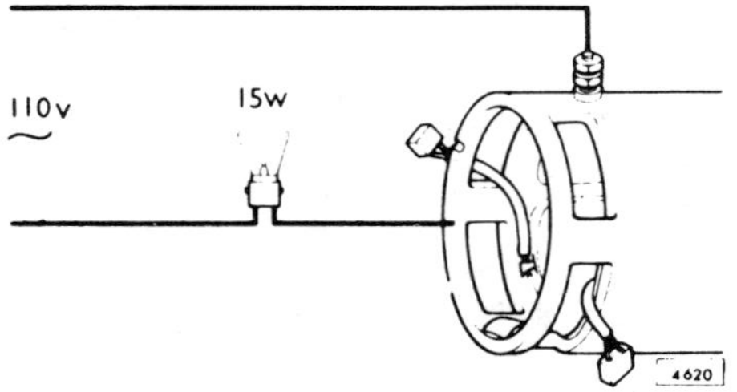

24 To replace field coils, unscrew the four poleshoe retaining screws using a wheel-operated screwdriver. Remove coils, pole shoes and insulation pieces.
Fit new coils over shoes, and replace in yoke, taking care that the taping around the coils is not trapped between the shoes and yoke.
Locate shoes by lightly tightening the screws, fit insulation pieces, and finally tighten screws with wheel-screwdriver.

Bearing – replacement

25 Replace bearings if excessive side play of shaft is evident.
Bushes in intermediate and drive end brackets should be pressed out, commutator end bracket bush must be withdrawn with a withdrawal tool. Soak bushes in clean engine oil for 24 hours before refitting. Refit by using a shouldered polished mandrel, 0.0005 in (0,013 mm) greater in diameter than shaft.

STARTER MOTOR

Bench testing   86.60.14

The following bench tests will determine if the fault is with the motor or solenoid unit.

1 Clamp motor in vice.
2 Connect a 12 volt battery, using heavy duty cables, to the motor frame and motor terminal.
3 Check that motor operates under light running conditions. If necessary equipment is available check light running current and speed against figures stated under 'Performance Data'.
4 If starter motor fails test, dismantle for overhaul.
If starter operates check or replace solenoid unit as follows:
5 Transfer cable from motor terminal to main solenoid terminal.
6 Fit jumper lead and touch to Lucar solenoid connector.
7 If motor does not operate, solenoid or solenoid contacts are faulty. Check and replace as necessary.

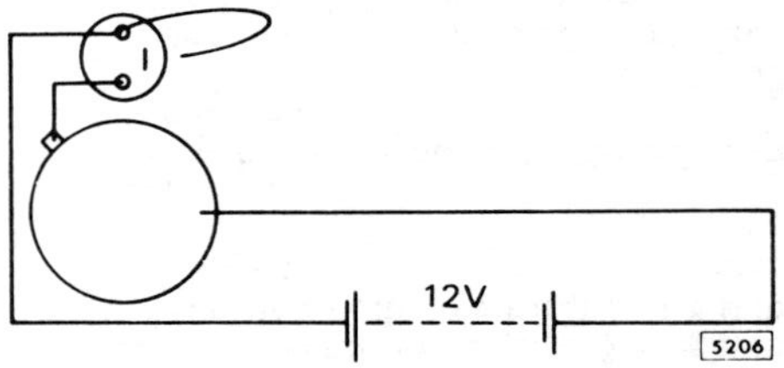

86–24

IGNITION SWITCH

Remove and refit   86.65.03

Removing
1. Disconnect battery earth lead.
2. Remove driver's side dash casing, see 76.46.11.
3. Remove screw securing shroud to fascia, and instrument module surround.
4. Slacken screws securing shroud to mounting bracket.
5. Ease shroud clear of location to gain access to grub screw securing switch.
6. Slacken grub screw, ease switch and associated harness clear of location.
7. Disconnect block connector after removing rubber retaining collar.
8. Remove switch unit.

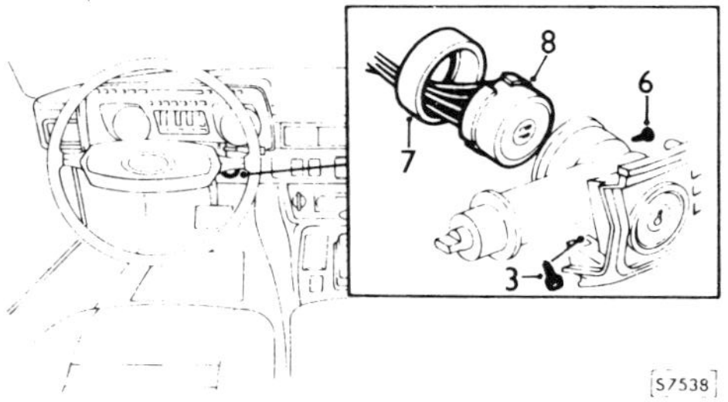

Refitting
Reverse operations 1 to 8.

PANEL SWITCHES

Remove and refit   86.65.06

NOTE: This operation applies to the following:

Interior Light Switch	86.65.13
Map Light Switch	86.65.24
Backlight Heater Switch	86.65.36
Hazard Warning Switch	86.65.50

Removing
1. Disconnect battery earth lead.
2. Prise switch panel from fascia.
3. Disconnect block connector from switch unit.
4. Depress retaining clips on top and bottom of switch.
5. Push switch through panel.

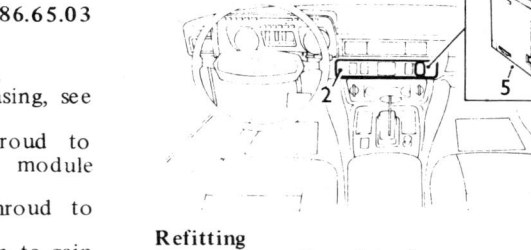

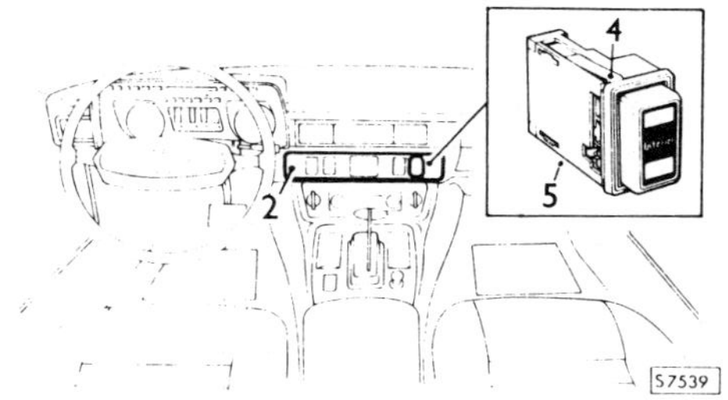

PANEL LIGHT RHEOSTAT

Remove and refit   86.65.07

Removing
1. Disconnect battery earth lead.
2. Remove driver's side dash liner.
3. Remove grommet from rheostat aperture in driver's side of dash liner.
4. Remove rheostat control knob lug depressing spring loaded pin and pulling knob from shaft.
5. Remove knurled nut securing rheostat to mounting bracket.
6. Remove rheostat from mounting bracket.

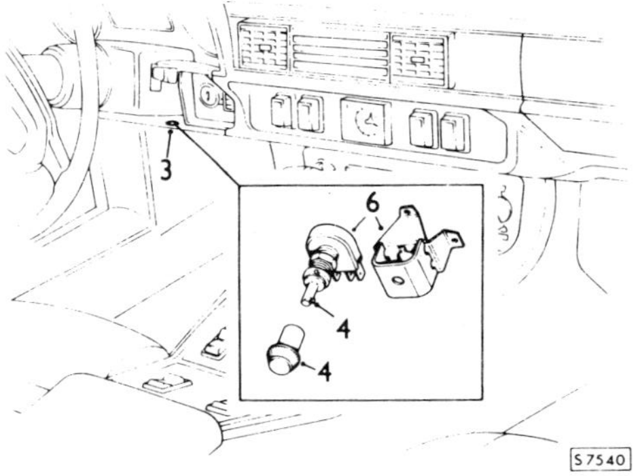

Refitting
Reverse operations 1 to 6.

MASTER LIGHTING SWITCH

Remove and refit   86.65.09

Removing
1. Disconnect battery earth lead.
2. Remove driver's side dash liner, see 76.46.11.
3. Remove screw securing switch shroud to fascia, and instrument module surround.
4. Slacken screws securing shroud to lower mounting bracket.
5. Ease shroud clear to give access to spring loaded pin retaining operating knob.
6. Depress pin and remove knob.
7. Remove shroud from location.
8. Remove nut securing switch to mounting bracket. 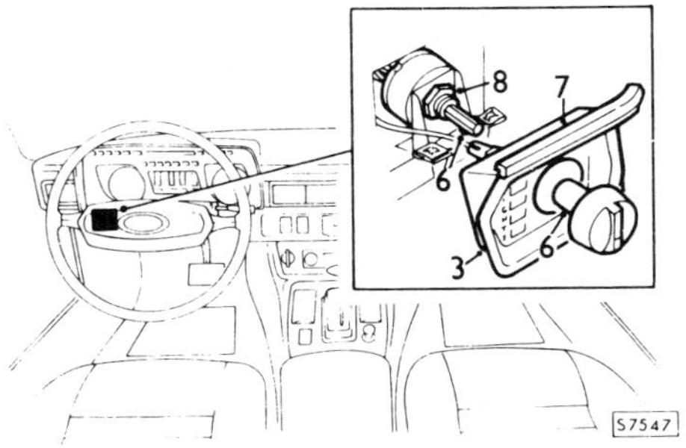
9. Remove switch from bracket and disconnect harness at block connector.

Refitting
Reverse operations 1 to 9.

DOOR PILLAR SWITCH

Remove and refit   86.65.15

Removing
1. Disconnect battery earth lead.
2. Remove screw securing switch to door pillar.
3. Withdraw switch.
4. Disconnect cable at Lucar.

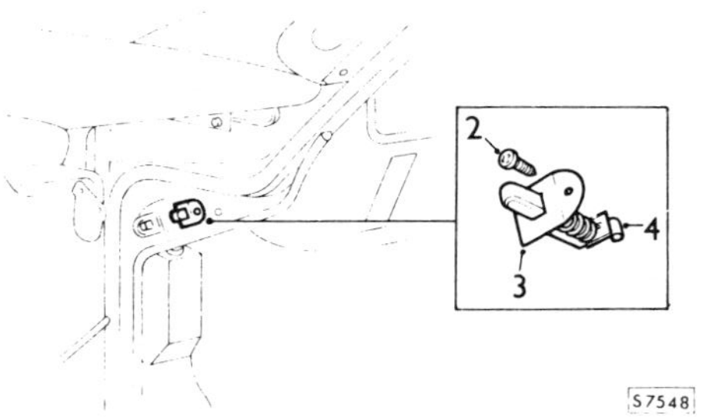

Refitting
Reverse operations 1 to 4.

REVERSE LIGHT SWITCH

Remove and refit   86.65.20

Removing
1. Disconnect battery earth lead.
2. Raise centre console glove box lid for access to rear fixing screw.
3. Remove console centre panel fixing screws.
4. Raise centre panel, note position of cables at switches and disconnect at lucars on back of switches.
5. Remove centre panel.
6. Remove nuts securing transmission selector mechanism cover.
7. Remove selector mechanism cover.
8. Note position of cables on reverse light switch and disconnect.
9. Unscrew switch from location.

continued

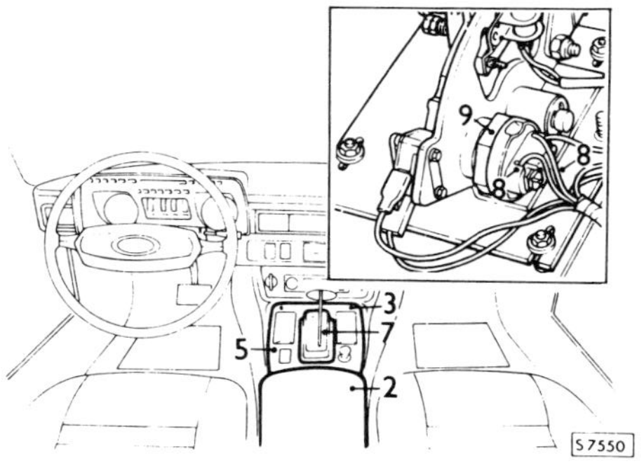

86–25

Refitting
Reverse operations 1 to 9.
NOTE: When refitting ensure that operating plunger of reverse light switch protrudes through mounting plate by an amount sufficient for car to operate switch when reverse is selected.

LUGGAGE COMPARTMENT LIGHT SWITCH 86.65.22

Remove and refit

The switching device for the luggage compartment light is an integral part of the lamp assembly. Removal and refitting is detailed under 86.45.16. It is not possible to service this unit, defective units must be replaced.

DOOR SWITCH, KEY ALARM 86.65.27

Remove and refit

Removing
1. Disconnect battery earth lead.
2. Remove screw securing switch to door pillar.
3. Withdraw switch.
4. Disconnect cables at Lucars.

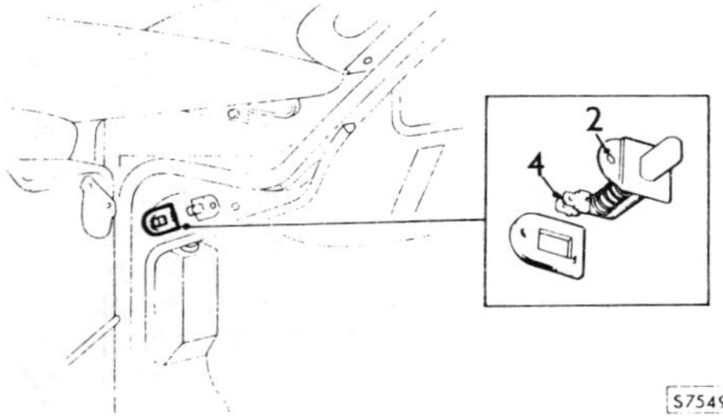

Refitting
Reverse operations 1 to 4.

OIL PRESSURE SWITCH 86.65.30

Remove and refit
Refer to operation 88.25.08.

COMBINED WINDSCREEN WIPER/WASHER SWITCH 86.65.41

Remove and refit
Refer to operation 84.15.34.

HAND BRAKE WARNING SWITCH 86.65.45

Remove and refit

Removing
1. Disconnect battery earth lead.
2. Remove screw securing hand brake mechanism cover.
3. Slide cover clear of mechanism.
4. Disconnect electrical harness from lucars on switch.
5. Remove set bolts securing switch to hand brake assembly.
6. Remove bolt and spacer, recover switch.

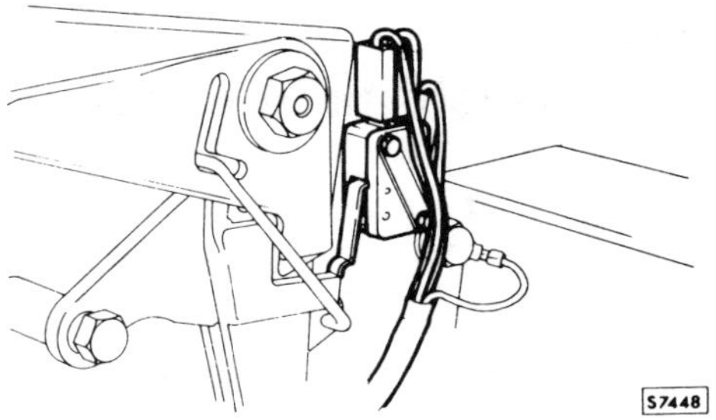

4. Adjust switch with hand brake in OFF position until warning light just goes out.
5. Tighten securing bolts.
6. Check light comes on with hand brake applied and goes off when hand brake is released. Re-adjust as necessary.
7. Refit cover.

COMBINED DIRECTION INDICATOR/HEADLIGHT/FLASHER/DIP SWITCH 86.65.55

Remove and refit

Removing
1. Disconnect battery earth lead.
2. Slacken steering wheel adjustment ring and extend to maximum travel.
3. Remove screws securing steering column lower shroud and remove shroud.
4. Remove steering wheel, see 57.60.01.
5. Remove screw securing upper shroud to bracket on steering column.
6. Slacken pinch screw securing switch assembly to steering column.
7. Ease switch assembly and upper shroud from switch assembly.
8. Remove shroud from switch assembly.
9. Disconnect harness at block connectors.
10. Remove spire nuts and screws securing switch mounting plate.
11. Disconnect earth cable at snap connector.
12. Remove wiper/washer switch from assembly.
CAUTION: Do not attempt to separate Direction/Headlight/Flasher switch from mounting bracket, if the switch is faulty a complete switch and bracket assembly must be fitted.

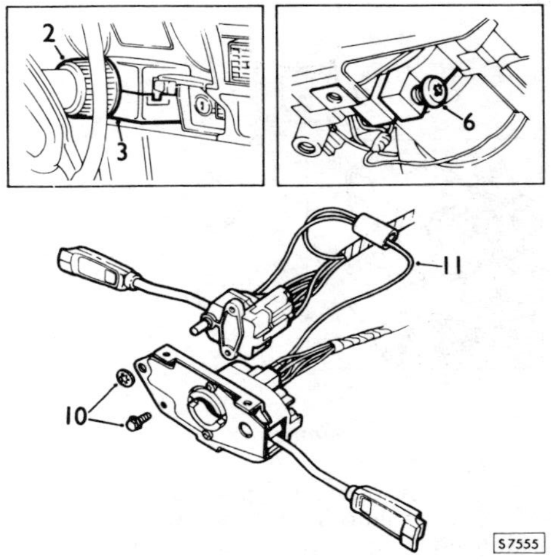

Refitting
Reverse operations 1 to 12.

HAND BRAKE WARNING SWITCH 86.65.46

Adjust
1. Remove screws securing hand brake mechanism cover.
2. Slide cover clear of mechanism.
3. Slacken set bolts securing switch.

STOP LIGHT SWITCH 86.65.51

Remove and refit

Removing
1. Disconnect battery earth lead.
2. Disconnect electrical connectors from lucars on switch.
3. Remove switch securing bolt.
4. Remove threads of keep plate and switch.

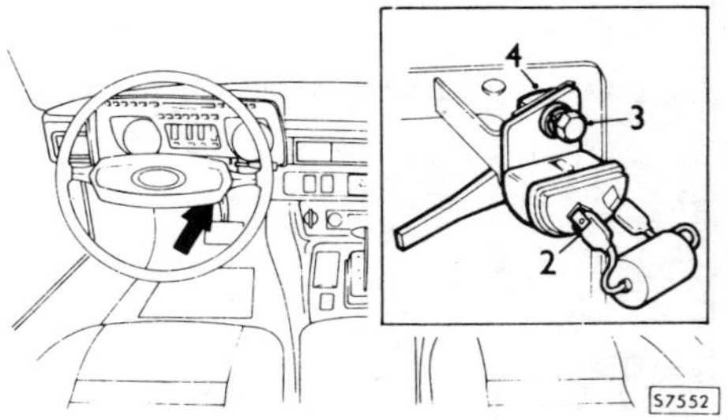

Refitting
Reverse operations 1 to 4, adjust switch for correct operation.

86–26

STOP LIGHT SWITCH

Adjust 86.65.56

1. Slacken switch securing bolt.
2. Adjust switch position until stop lights operate when brake pedal is depressed, and are OFF when pedal is fully released.
   NOTE: Ignition switch must be on while adjustments are carried out.
3. Tighten securing bolt.

FUEL CUT-OFF INERTIA SWITCH

Remove and refit 86.65.58

Removing
1. Disconnect battery earth lead.
2. Pull protective cover from inertia switch location.
3. Remove inertia switch from retaining clips.
4. Disconnect electrical connectors from lucars. 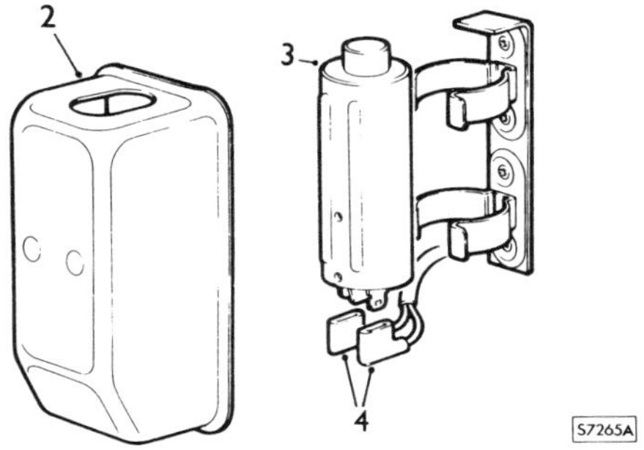
5. Recover switch.

Refitting
Reverse operations 1 to 5.

FUEL CUT-OFF INERTIA SWITCH

Reset 86.65.59

The inertia switch is fitted in the electrical supply to the fuel pumps. Should the car be subjected to heavy impact forces, the switch will operate, isolating the fuel pumps and ensuring fuel is not pumped to a potentially dangerous area. The switch is located on the side of the fascia on the driver's side 'A' post. Press the button mounted on top of the switch to re-set after operation.

CIGAR LIGHTER

Remove and refit 86.65.60

Removing
1. Disconnect battery earth lead.
2. Remove screws securing console centre panel.
   NOTE: Raise console glove box lid for access to rear screw.
3. Raise centre panel for access to cigar lighter assembly.
4. Disconnect electrical connectors from cigar lighter.
5. Remove cigar lighter lamp holder by depressing sides and unclipping from assembly.
6. Unscrew and remove cigar lighter lower sleeve.
7. Remove cigar lighter from panel.

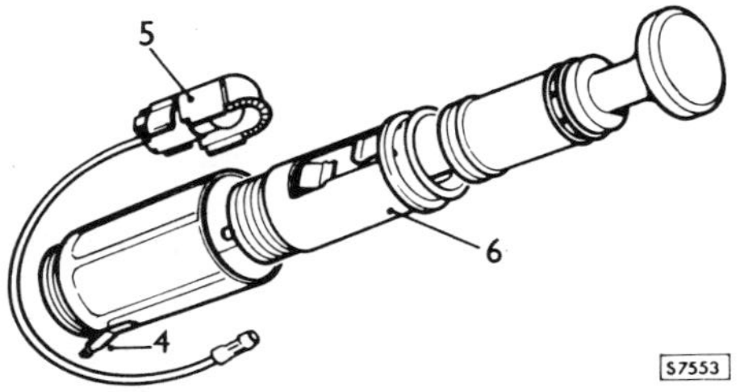

Refitting
Reverse operations 1 to 7.

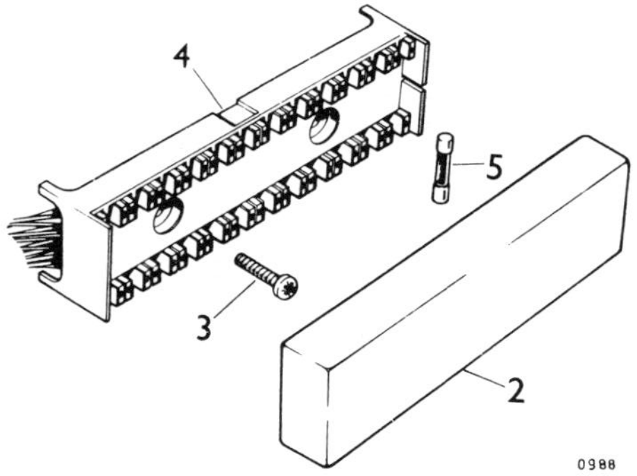

FUSE BOX

Remove and refit 86.70.01

1. Remove driver's underscuttle casing, see 76.46.01.
2. Remove fuse box cover.
3. Remove retaining screws.
4. Release fuse box.
   The fuse box is part of the harness but, should the box need replacing, a fuse box with short length of wire is supplied as a spares item.

Refitting
5. Reverse operations 1 to 4. Transfer fuses (5).

Fuse No.	Amps
1	25
2	15
3	35
4	10
5	15
6	10
7	10
8	3
9	3
10	50
11	35
12	35

The following pages contain details of:

COMPONENT LOCATION
WIRING DIAGRAMS

To assist in identification and location, the symbols and cable colour codes are given below. A master key to location and wiring diagrams is given with the component location chart. Extracts from this master key are given with the appropriate systems diagram.

CABLE COLOUR CODE

N. Brown	**P.** Purple	**W.** White	
U. Blue	**G.** Green	**Y.** Yellow	
R. Red	**L.** Light	**B.** Black	
K. Pink	**S.** Slate	**O.** Orange	

When a cable has two colour code letters, the first denotes the Main Colour and the second the Tracer Colour.

LAMP BULBS

LAMP FUNCTION	PART NUMBER	WATTS	TYPE
Outer Headlight (Main and dipped beam)	—	See local Dealer	Sealed Beam
Inboard main beam only	—	See local Dealer	Sealed Beam
Front flasher lamps	GLB 380	21	Bayonet
Tail flasher lamps	GLB 382	21	Bayonet
Tail lamp	GLB 207	5	Bayonet
Stop lamp	GLB 382	21	Festoon
Reverse lamp	11740	6	Festoon
Number plate lamp	GLB 254	6	Festoon
Side marker lamp	GLB 233	5	Bayonet
Map/interior lamps	12273	10	Festoon
Luggage compartment lamp	GLB 989	5	Bayonet
Warning lamps	C38966	1.2	Miniature capless
Instrument lamps	C30309	2.2	Capless
Cigar lighter			
Automatic transmission selector illumination	GLB 643	22	Miniature bayonet
Fibre optic light source	GLB 988	5	
Catalyst/EGR warning lamp	GLB 989	5	
	GLB 281	2	Bayonet

FUSES

FUSE No.	PROTECTED CIRCUITS	PART NUMBER	CURRENT CAPACITY
1	Fog lamps (if fitted)	GFS 425	25A
2	Hazard warning	GFS 415	15A
3	Map/interior lamps, cigar lighter, electric aerial (if fitted), clock, seat belt warning lamp, boot lamps	GFS 435	35A
4	Panel lights	GFS 410	10A
5	Direction indicators, stop lamp kickdown	GFS 415	15A
6	Reversing lights	GFS 410	10A
7	Panel switches, cigar lighter illumination, number plate lamp, luggage compartment lamp, fibre optic unit, gear selector illumination	GFS 410	10A
8	Side/tail lamps (LH)	GFS 43	3A
9	Side/tail lamps (RH)	GFS 43	3A
10	Air conditioning motor	GFS 450	50A
11	Windscreen wipers, air conditioning relay and clutch, windscreen washer, horn relay winding, cooling fan relay winding	GFS 435	35A
12	Heated back light	GFS 435	35A
	Headlamp (main beam)	GFS 425	25A
	Headlamp (dipped beam)	GFS 410	10A

SYMBOLS USED

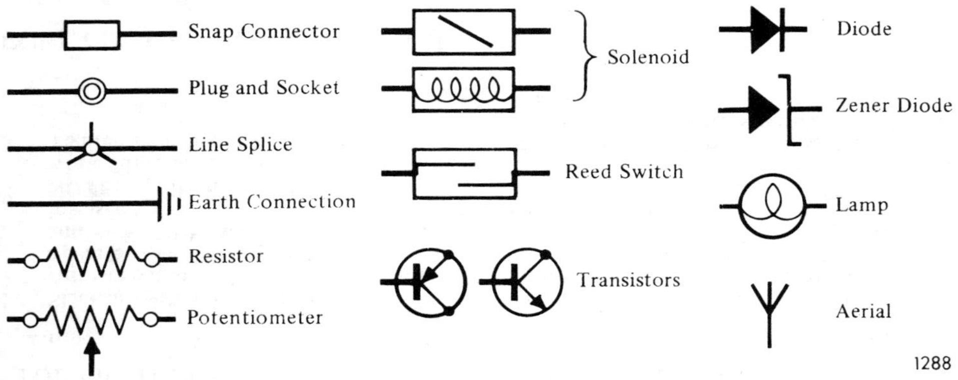

- Snap Connector
- Plug and Socket
- Line Splice
- Earth Connection
- Resistor
- Potentiometer
- Solenoid
- Reed Switch
- Transistors
- Diode
- Zener Diode
- Lamp
- Aerial

86—28

1288

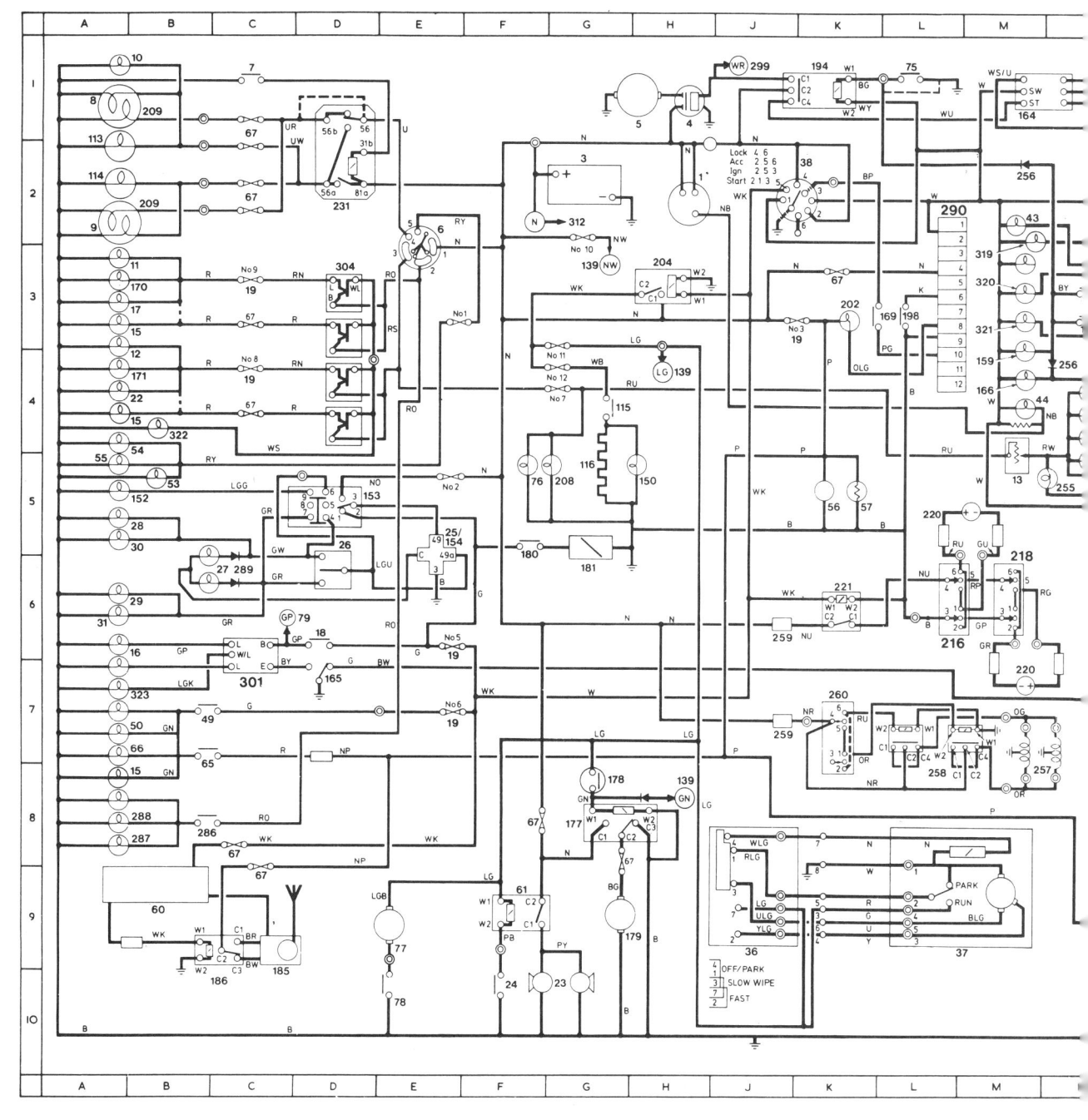

COMPONENT	No.	COMPONENT	No.	COMPONENT	No.
Alternator		Boot light	66	Fibre optics illumination lamp	255
Aerial motor	185	Boot light switch	65	Flasher unit (part of 154)	25
Aerial motor relay	186	Brake differential pressure switch	160	Flasher lamp RH front	28
Air conditioning ambient sensor	265	Brake failure warning light	323	Flasher lamp LH front	29
Air conditioning amplifier	261	Brake fluid level switch	182	Flasher lamp RH rear	30
Air conditioning blower	33	Brake fluid level warning light	159	Flasher lamp LH rear	31
Air conditioning blower relay	189	Buzzer alarm	168	Fog lamp RH	54
Air conditioning blower resistor	188	Cigar lighter	57	Fog lamp LH	55
Air conditioning compressor clutch	190	Cigar lighter illumination	208	Fog lamp warning light	53
Air conditioning control switch	192	Clock	56	Fuel gauge	34
Air conditioning/heater (to)	139	Cold start injectors	300	Fuel gauge tank unit	35
Air conditioning in-car sensor	264	Cold start relay	299	Fuel injection amplifier	292
Air conditioning servo	262	Coolant temperature sensor	305	Fuel injection control unit (E.C.U.)	293
Air conditioning temperature selector	327	Direction indicator switch	26	Fuel injection main relay	312
Air conditioning thermostat	191	Direction indicator warning light	27	Fuel level warning light	319
Air conditioning vacuum valve	263	Distributor	40	Fuel pump	41
Air conditioning water temperature sensor	47	Door lock solenoid	257	Fuel pump relay	314
Air temperature sensor	297	Door lock solenoid relay	258	Fuel tank change-over switch	140
Auto gearbox kickdown solenoid	181	Door lock switch	260	Full throttle switch (Australia only)	326
Auto gearbox kickdown pressure switch	180	Door switch	21	Fuse box	19
Auto gearbox selector lamp	76	Door switch (buzzer alarm)	169	Handbrake switch	165
Auto gearbox start inhibit switch	75	EGR control unit	291	Handbrake warning lamp	166
Ballast resistor	164	EGR thermo switch	308	Hazard warning flasher unit (includes 25)	154
Battery	3	EGR valve	307	Hazard warning lamp	152
Battery condition indicator	146	EGR warning switch } See Service	277	Hazard warning switch	153
Blocking diode — brake warning	256	EGR warning lamp } Interval	278	Headlamp dip switch	7
Blocking diode — direction indicators	289	EGR diode } Counter	284	Headlamp dip beam	209
Blocking diode — inhibit incorrect — polarity	315			Headlamp inner RH	113

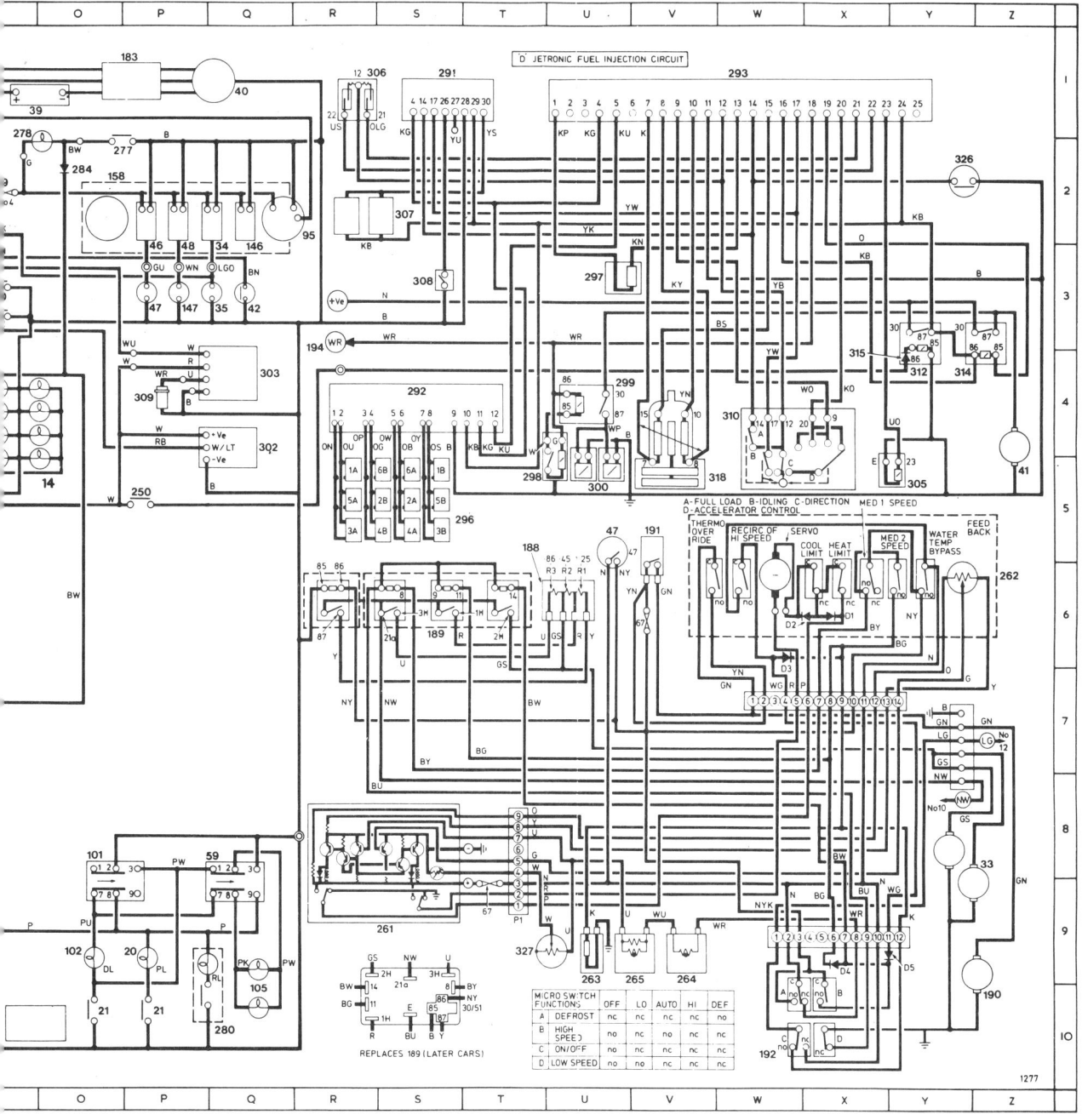

COMPONENT	No.	COMPONENT	No.	COMPONENT	No.
eadlamp inner LH	114	Oil pressure switch (for 48 or 295)	42	Side lamp LH or (Headlamp pilot lamp)	12
eadlamp outer RH	8	Oil pressure transmitter (for 48)	147	Service Interval Counter Switch	277
eadlamp outer LH	9	Oil pressure warning lamp	43	Service Interval Counter warning light *	278
eadlamp pilot lamp (see sidelamp)	11	Overvoltage CU	302	Starter motor	5
eadlamp relay	231	Overvoltage W/L	321	Starter solenoid	4
orn	23	Panel lamps	14	Starter solenoid/ballast resistor relay	194
orn push	24	Panel lamps rheostat	13	Stop lamps	16
orn relay	61	Park lamp failure sensor	304	Stop lamps failure sensor	301
gnition amplifier	183	Park lamp failure warning light	322	Stop lamp switch	18
gnition coil	39	Printed circuit (instrument panel)	158	Tail lamp RH	17
gnition protection relay	204	Radiator cooling fan motor	179	Tail lamp LH	22
gnition switch	38	Radiator cooling fan relay	177	Thermal circuit breaker	259
gnition warning lamp	44	Radiator cooling fan thermostat	178	Thermotime switch	298
nertia switch	250	Radio	60	Throttle switch	310
njectors	296	Rear fog guard lamp	288	Trigger unit	306
nterior light	20	Rear fog guard switch	286	Trailer socket	79
nterior light rear	111	Rear fog guard warning lamp	287	Water temperature gauge	46
nterior light switch	59	Rear window demist switch	115	Water temperature transmitter for gauge	47
ine fuse	67	Rear window demist unit	116	Window lift motor	220
ow coolant control unit	303	Rear window demist warning lamp	150	Window lift safety relay	221
ow coolant sensor	309	Reverse lamps	50	Window lift switch RH front	216
ow coolant warning light	320	Reverse lamps switch	49	Window lift switch LH front	218
ain beam warning light	10	Revolution counter	95	Windscreen washer pump	77
ain light switch	6	Roof light	280	Windscreen washer switch	78
ap light	102	Seat belt switch — driver	198	Windscreen wiper motor	220
ap light switch	101	Seat belt warning control unit	290	Windscreen wiper switch	36
umber plate lamp	15	Seat belt warning lamp	202		
il pressure gauge	48	Side lamp RH or (Headlamp pilot lamp)	11	* SIC operates EGR/Catalyst warning light	

86-29

COMPONENT LOCATION

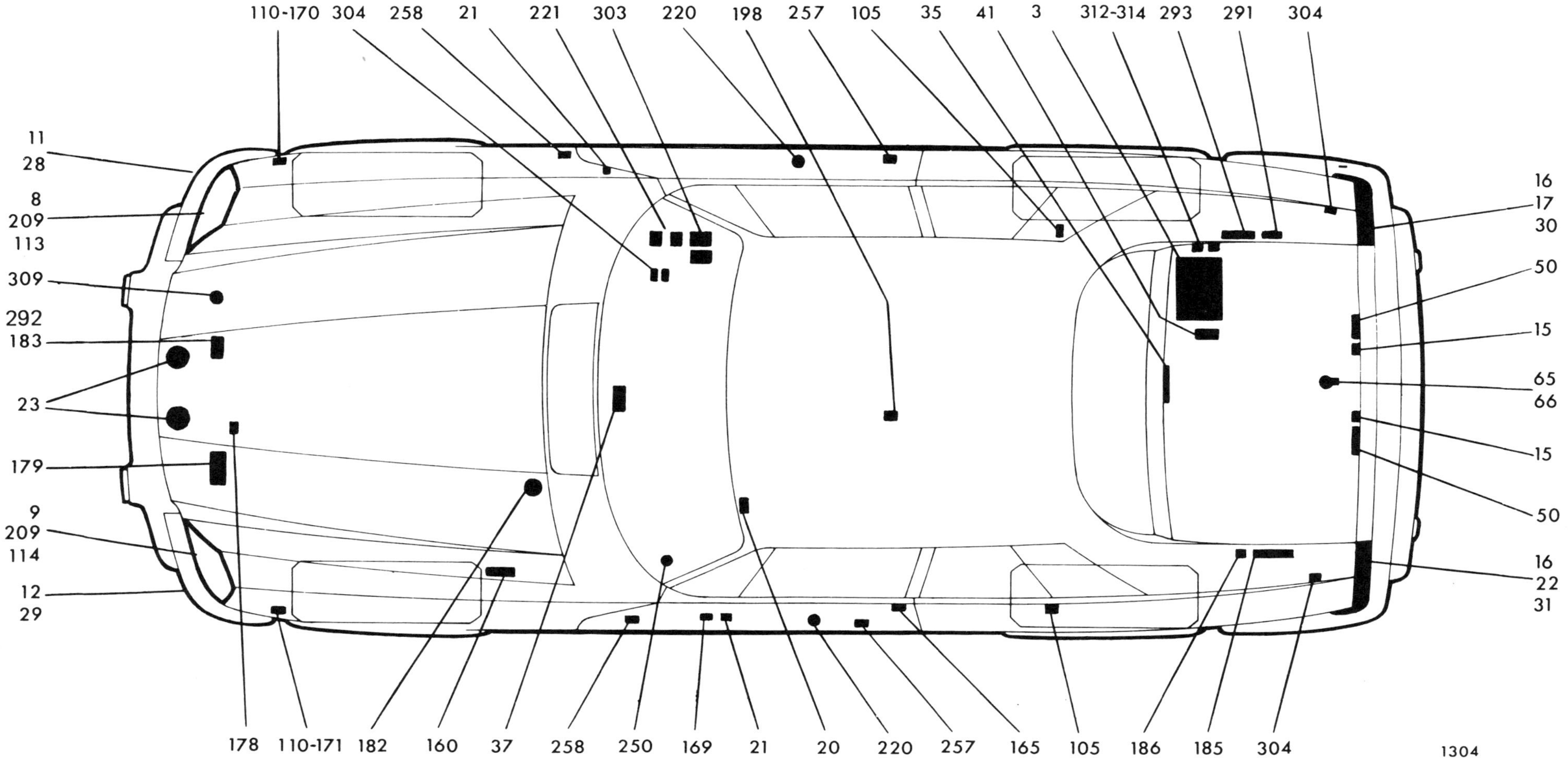

COMPONENT LOCATION

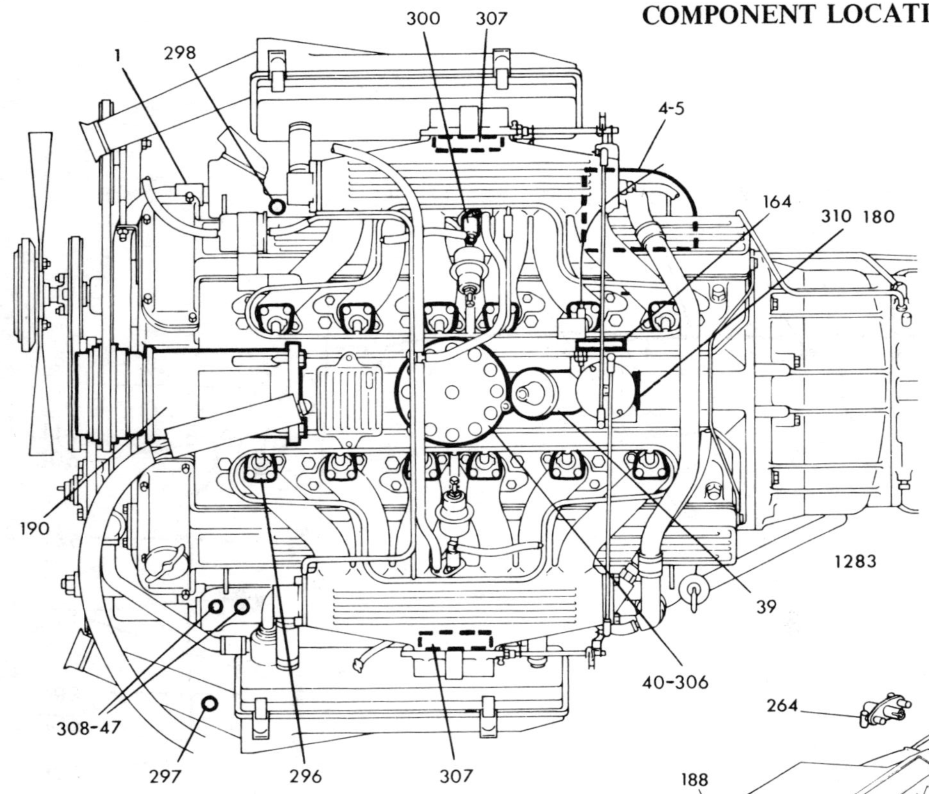

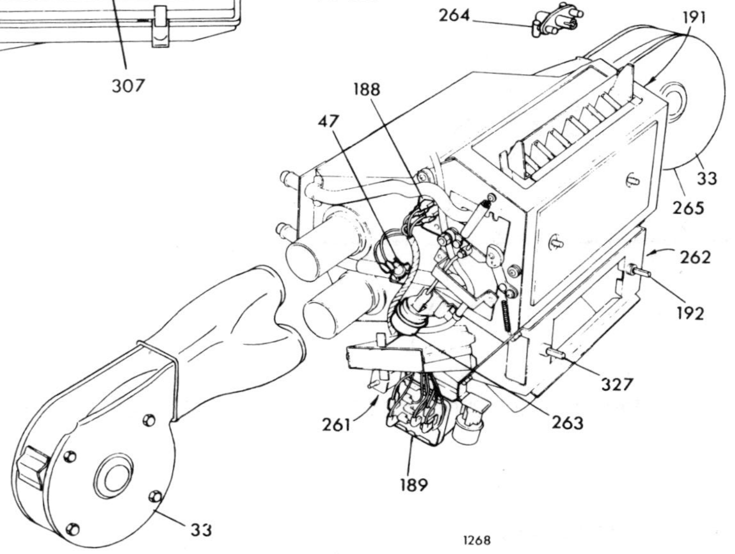

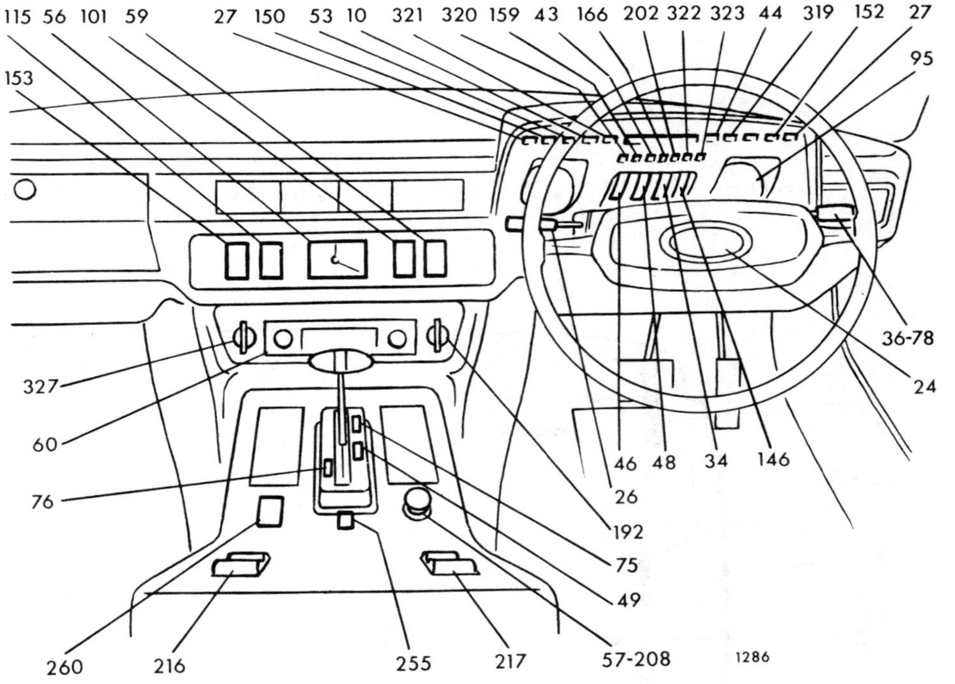

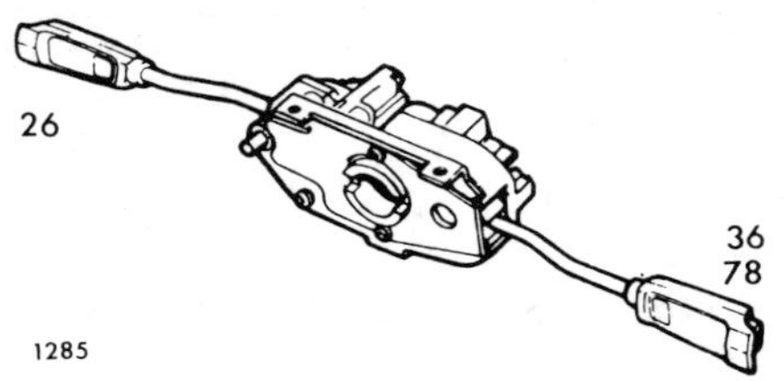

86-32

BATTERY CONDITION INDICATOR

Remove and refit   88.10.07

Removing
1. Disconnect battery earth lead.
2. Remove instrument module, see 88.20.01.
3. Remove instrument panel lens, see 88.20.17.
4. Remove screws connecting printed circuit to instrument from rear of module. 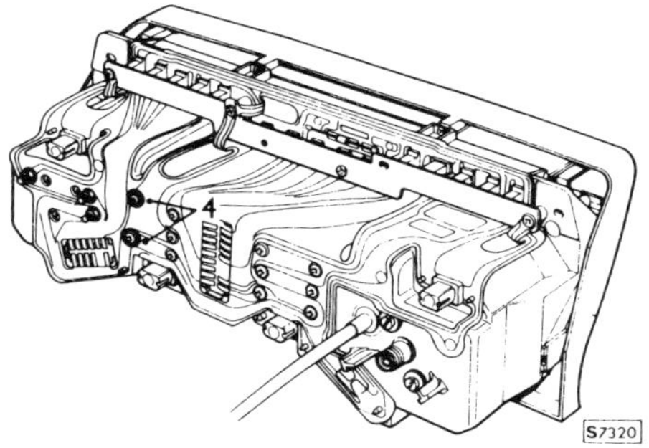
5. Prise off spire washers securing instrument, from spigots.
6. Recover instrument.

Refitting
Reverse operations 1 to 6.
NOTE: New spire washers should be used when refitting instruments.

CLOCK

Remove and refit   88.15.07

Removing
1. Disconnect battery earth lead.
2. Prise clock from switch panel.
3. Pull clock illumination bulb holder from casing.
4. Disconnect supply leads at lucars on rear of clock.
5. Recover clock.

Refitting
Reverse operations 1 to 5.
NOTE: To reset clock, set hands by depressing reset knob in right hand lower corner of clock face and turn hands to required position. To adjust, prise clock from switch panel and adjust by turning adjustment screw on rear of clock in appropriate direction.

INSTRUMENT PANEL LENS ASSEMBLY

Remove and refit   88.20.17

Removing
1. Remove instrument panel, see 88.20.01.
2. Remove seven screws retaining lens assembly.
3. Recover assembly.

Refitting
Reverse operations 1 to 3.

INSTRUMENT PANEL (MODULE)

Remove and refit   88.20.01

Removing
1. Disconnect battery earth lead.
2. Remove drivers side dash casing, see 76.46.11.
3. Remove centre securing strip and screw from instrument module surround.
4. Remove screws securing side pieces of surround and recover surround.
5. Prise off covers from securing screw apertures.
6. Remove screws securing instrument panel module to fascia.
7. Disconnect speedometer cable from angle drive at rear of speedometer. 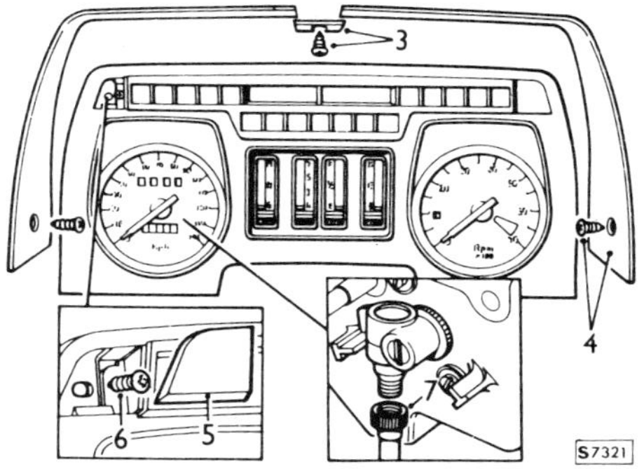
8. Ease module forward and disconnect cable harness at block connectors.
9. Slacken steering wheel adjustment ring and extend to maximum travel.
10. Manoeuvre instrument panel module clear of housing.

Refitting
Reverse operations 1 to 10.

PRINTED CIRCUIT

Remove and refit   88.20.19

Removing
1. Remove instrument panel module, see 88.20.01.
2. Remove four instrument illumination bulb holders from rear of assembly.
3. Note positions of all fixing screws, connectors etc.
4. Remove warning lamp retaining bar/earth strip.
5. Remove three nuts connecting printed circuit to tachometer.
6. Remove seventeen warning lamp holders. 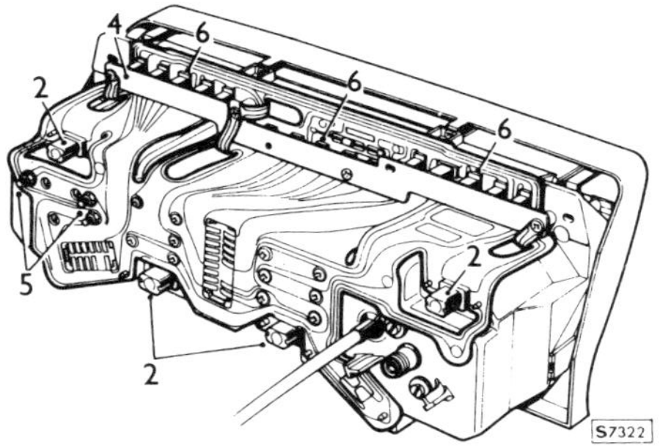
7. Remove screws connecting printed circuit to centre gauges.
8. Ease printed circuit clear of locating spigots.
9. Recover printed circuit.

Refitting
Reverse operations 1 to 9.
NOTE: Care should be taken to ensure printed circuits are not torn or deformed, and that all connecting tags are correctly positioned under terminals of components.

OIL GAUGE   88.25.01

COOLANT TEMPERATURE GAUGE   88.25.14

FUEL GAUGE   88.25.26

Remove and refit
Refer to operation 88.10.07.

OIL PRESSURE TRANSMITTER

Remove and refit   88.25.07

Removing
1. Disconnect battery earth lead.
2. Disconnect supply lead from lucar on transmitter.
3. Withdraw transmitter from manifold.

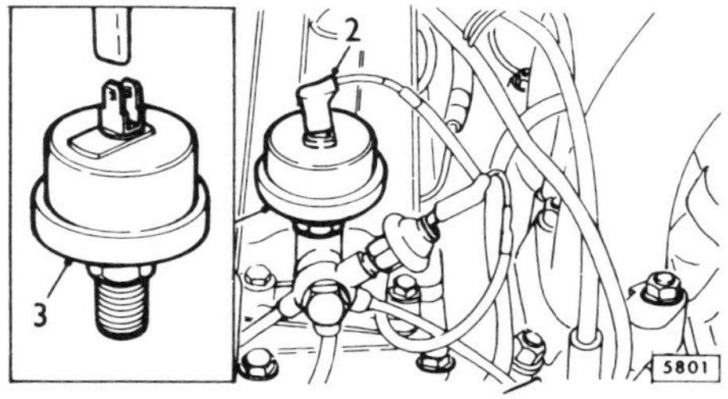

Refitting
Reverse operations 1 to 3.

OIL PRESSURE WARNING SWITCH

Remove and refit 88.25.08

Removing
1. Disconnect battery earth lead.
2. Disconnect supply lead from lucar on switch.
3. Withdraw switch from manifold.

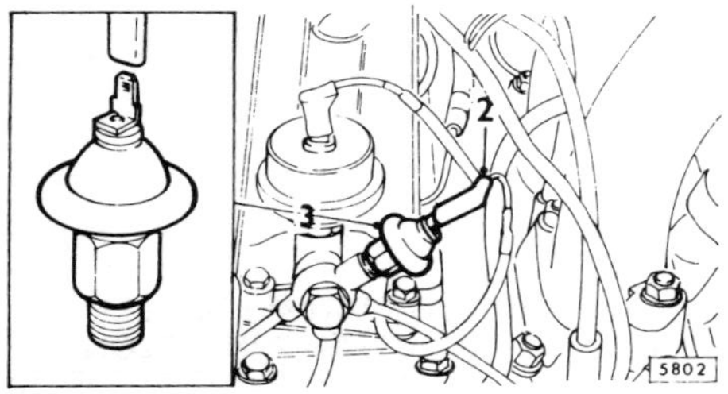

Refitting
Reverse operations 1 to 3.

COOLANT TEMPERATURE TRANSMITTER

Remove and refit 88.25.20

Removing
1. Disconnect battery earth lead.
2. Remove header tank cap to depressurise cooling system.
3. Disconnect supply lead from lucar on transmitter. 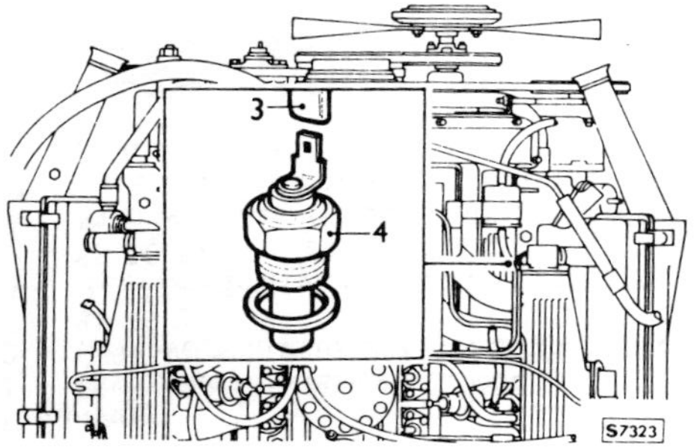
4. Withdraw switch.

Refitting
Reverse operations 1 to 4.

FUEL GAUGE TANK UNIT

Remove and refit 88.25.32

Removing
1. Disconnect battery earth lead.
2. Drain fuel tank, see 19.55.02.
3. Knock out and remove tank unit locking ring.
   NOTE: Discard locking ring.
4. Turn unit through 180° and remove from tank.
5. Remove sealing ring from tank unit and discard.

Refitting
Reverse operations 1 to 5.

SPEEDOMETER

Remove and refit 88.30.01

Removing
1. Disconnect battery earth lead.
2. Remove instrument panel and module, see 88.20.01.
3. Remove instrument panel lens assembly, see 88.20.17.
4. Remove speedometer angle drive gearbox from rear of instrument.
5. Remove screws securing speedometer mechanism to housing.
6. Recover speedometer mechanism.

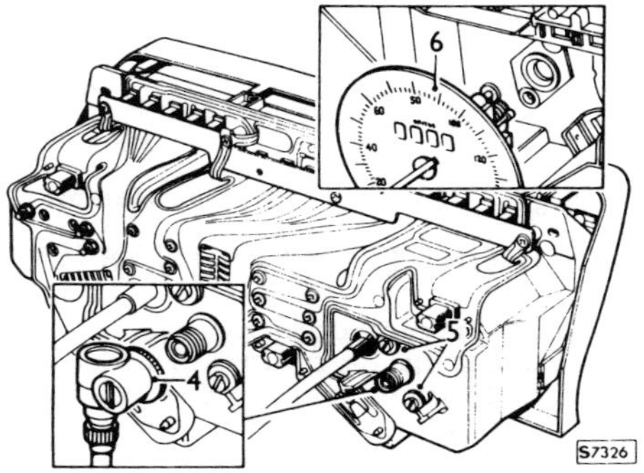

SPEEDOMETER TRIP RESET

Remove and refit 88.30.02

Removing
1. Remove speedometer, see 88.30.01.
2. Depress retaining tangs and remove trip reset.

Refitting
Reverse operations 1 and 2.

SPEEDOMETER CABLE ASSEMBLY

Not U.S.A. Specification

Remove and refit 88.30.06

Removing
1. Raise car on ramp.
2. Disconnect cable from gearbox angle drive.
3. Disconnect cable from instrument angle drive. Operation 88.30.07 refers.
4. Manoeuvre cable clear of air conditioning unit for removal.
5. Displace grommet from gearbox tunnel aperture. 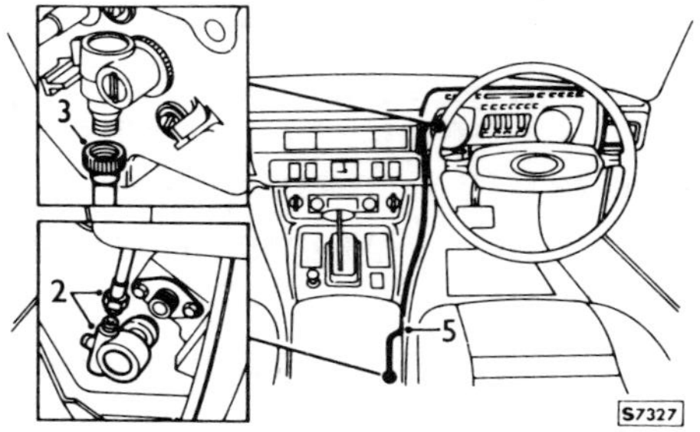
6. Feed cable into car and remove grommet.

Refitting
Reverse operations 1 to 6.

SPEEDOMETER CABLE INNER

Remove and refit 88.30.07

Removing
1. Remove passenger side dash casing, see 76.46.11.
2. Disconnect cable from angle drive at rear of speedometer.
3. Withdraw inner cable from outer.

Refitting
Reverse operations 1 to 3, lubricate cable sparingly before refitting. Use only TSD 119 or equivalent.

SPEEDOMETER ANGLE DRIVE, INSTRUMENT

Remove and refit 88.30.15

Removing
1. Remove passenger side dash casing, see 76.46.11.
2. Disconnect speedometer cable at angle drive.
3. Unscrew knurled retaining ring and remove angle drive from rear of instrument.

Refitting
Reverse operations 1 to 3.

88–2

SPEEDOMETER ANGLE DRIVE, GEARBOX

Remove and refit 88.30.16

Removing
1. Raise car on ramp.
2. Slacken nut securing speedometer cable to angle drive.
3. Place cable to allow access to angle drive.
4. Unscrew knurled retaining ring and remove angle drive.

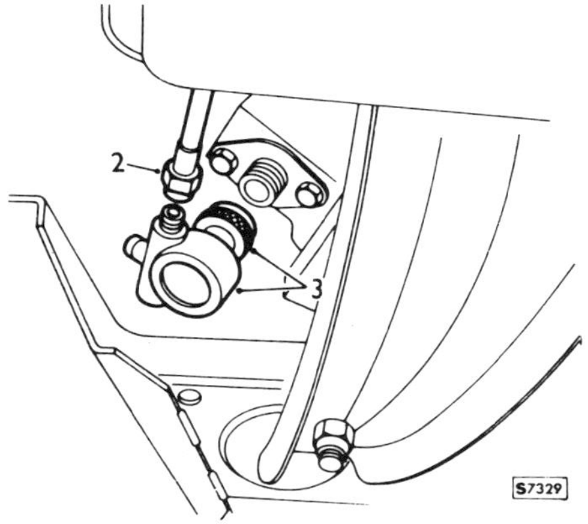

Refitting
Reverse operations 1 to 4.

TACHOMETER

Remove and refit 88.30.21

Removing
1. Disconnect battery earth lead.
2. Remove instrument panel module, see 88.20.01.
3. Remove instrument panel lens assembly, see 88.20.17.
4. Remove nuts connecting printed circuit to tachometer.
5. Remove securing screws and recover tachometer mechanism from housing.

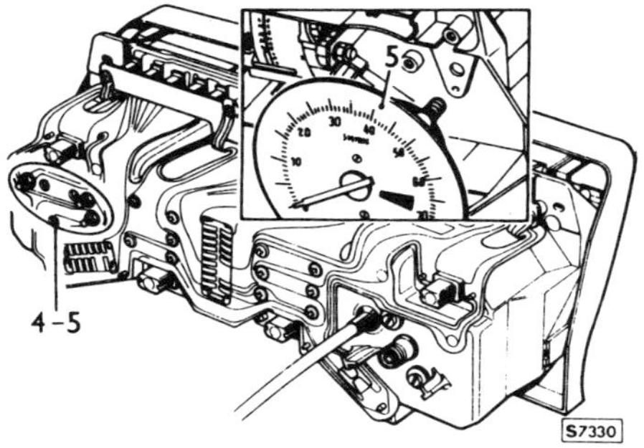

Refitting
Reverse operations 1 to 5.

88—3

SERVICE TOOLS

SECTION 12

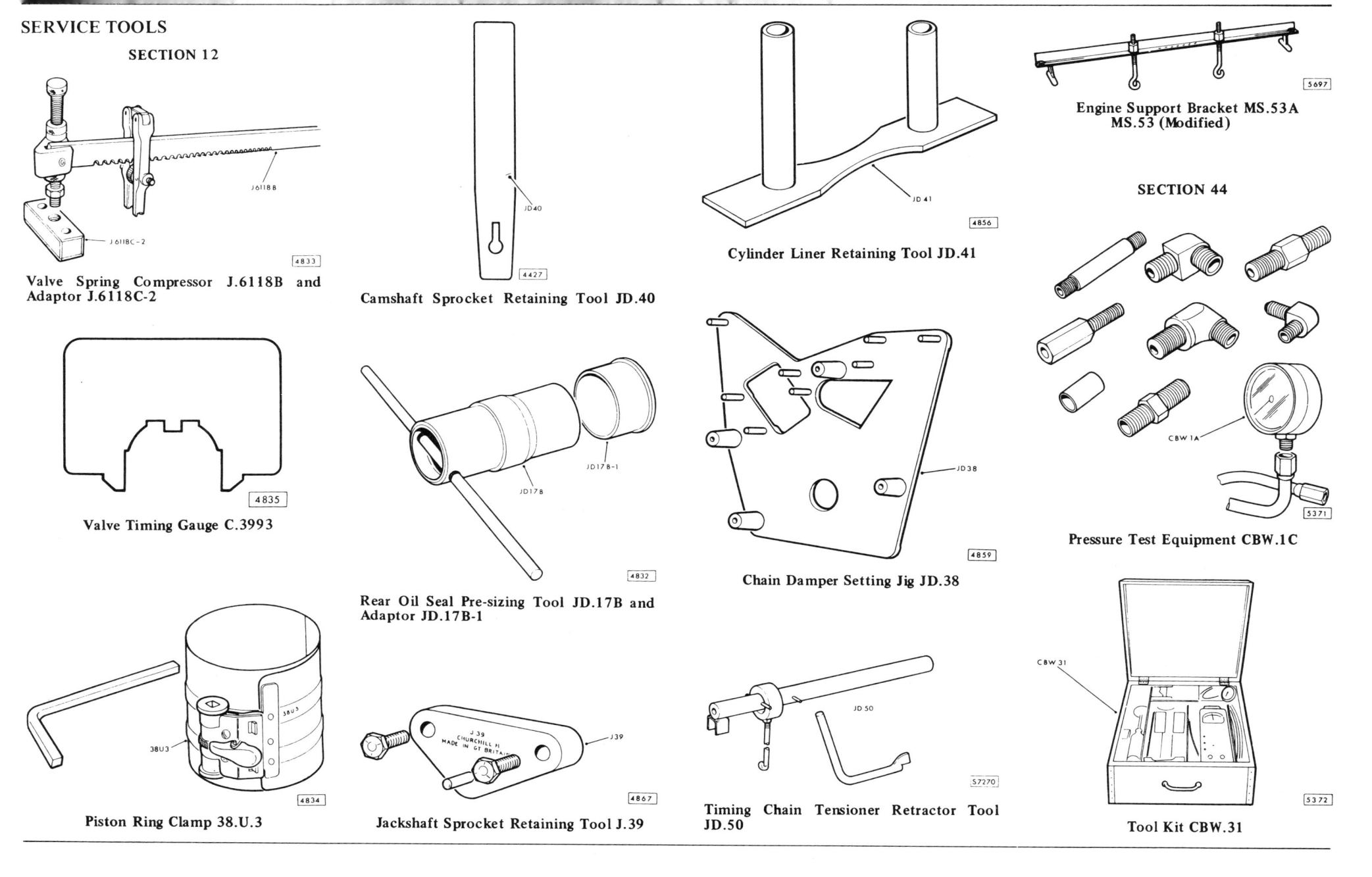

Valve Spring Compressor J.6118B and Adaptor J.6118C-2

Camshaft Sprocket Retaining Tool JD.40

Valve Timing Gauge C.3993

Rear Oil Seal Pre-sizing Tool JD.17B and Adaptor JD.17B-1

Piston Ring Clamp 38.U.3

Engine Support Bracket MS.53A MS.53 (Modified)

SECTION 44

Pressure Test Equipment CBW.1C

Tool Kit CBW.31

Cylinder Liner Retaining Tool JD.41

Chain Damper Setting Jig JD.38

Timing Chain Tensioner Retractor Tool JD.50

Jackshaft Sprocket Retaining Tool J.39

99—1

SECTION 44 (continued)

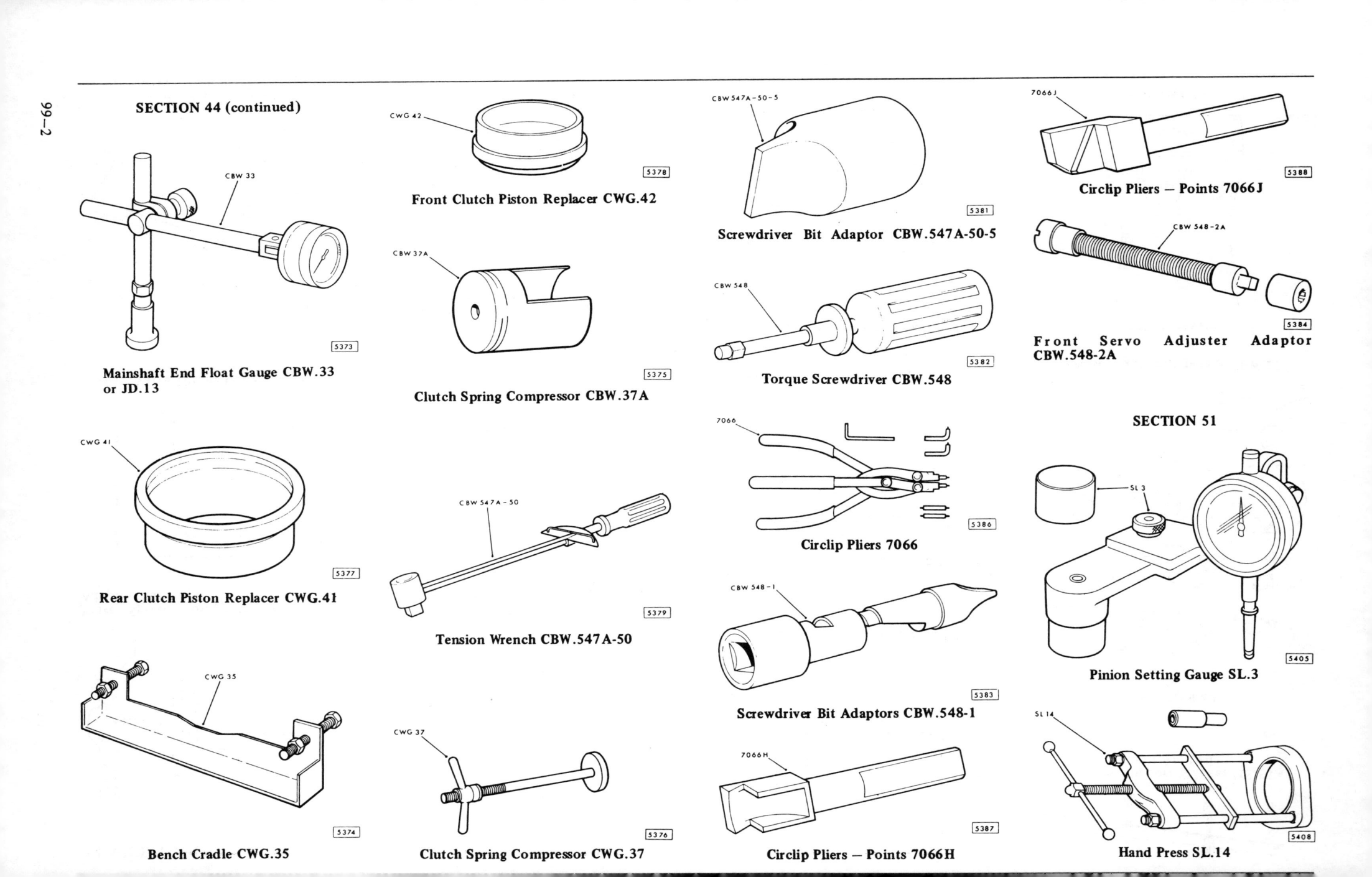

- Mainshaft End Float Gauge CBW.33 or JD.13
- Front Clutch Piston Replacer CWG.42
- Screwdriver Bit Adaptor CBW.547A-50-5
- Circlip Pliers – Points 7066J
- Rear Clutch Piston Replacer CWG.41
- Clutch Spring Compressor CBW.37A
- Torque Screwdriver CBW.548
- Front Servo Adjuster Adaptor CBW.548-2A
- Bench Cradle CWG.35
- Tension Wrench CBW.547A-50
- Circlip Pliers 7066

SECTION 51

- Clutch Spring Compressor CWG.37
- Screwdriver Bit Adaptors CBW.548-1
- Pinion Setting Gauge SL.3
- Circlip Pliers – Points 7066H
- Hand Press SL.14

99—2

SECTION 51 (continued)

Pinion Bearing Cone Remover/Replacer SL. 47-1

Rear Hub Bearing Cone Remover SL.14-7

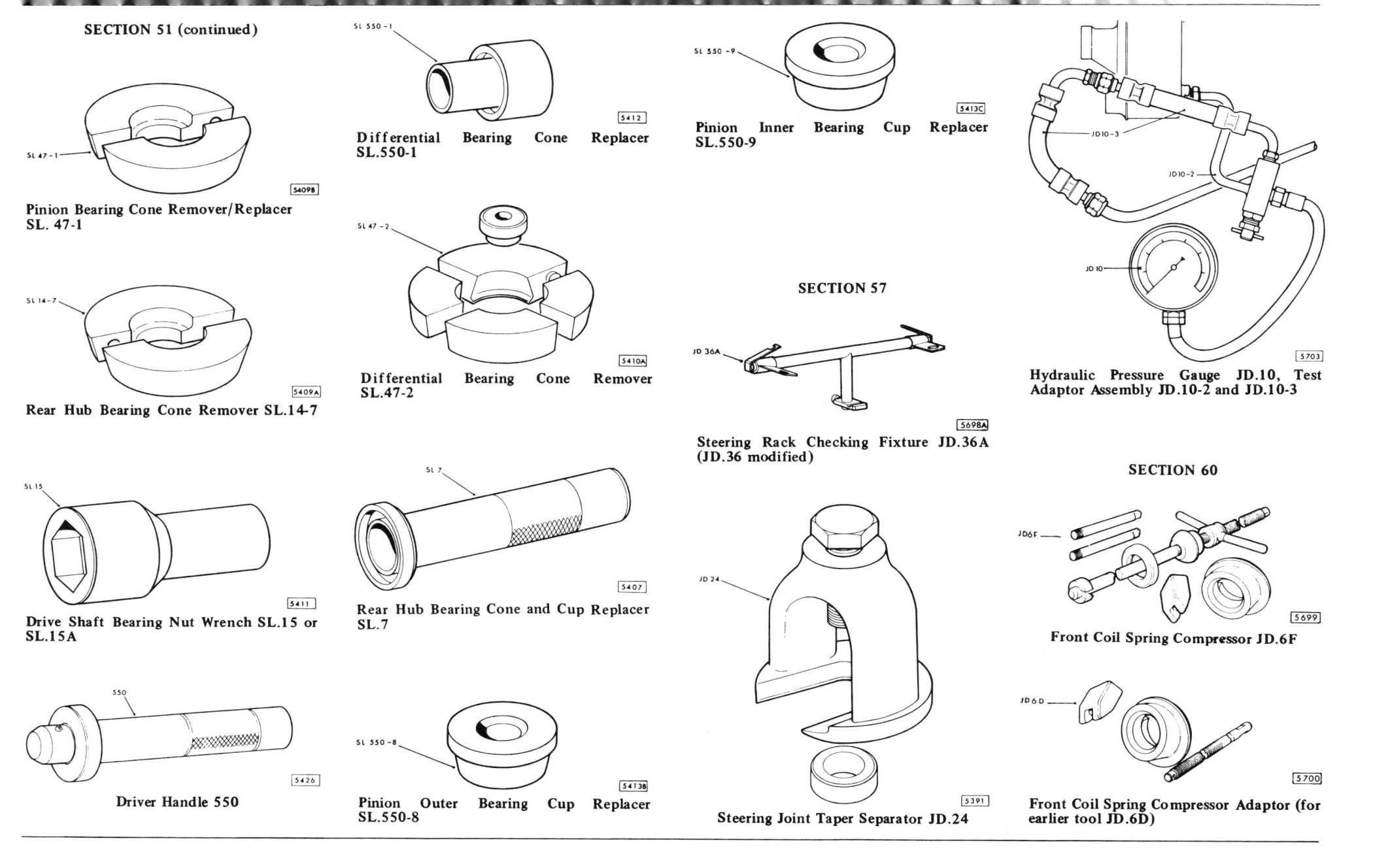

Differential Bearing Cone Replacer SL.550-1

Differential Bearing Cone Remover SL.47-2

Rear Hub Bearing Cone and Cup Replacer SL.7

Drive Shaft Bearing Nut Wrench SL.15 or SL.15A

Driver Handle 550

Pinion Inner Bearing Cup Replacer SL.550-9

Pinion Outer Bearing Cup Replacer SL.550-8

SECTION 57

Steering Rack Checking Fixture JD.36A (JD.36 modified)

Steering Joint Taper Separator JD.24

Hydraulic Pressure Gauge JD.10, Test Adaptor Assembly JD.10-2 and JD.10-3

SECTION 60

Front Coil Spring Compressor JD.6F

Front Coil Spring Compressor Adaptor (for earlier tool JD.6D)

SECTION 76

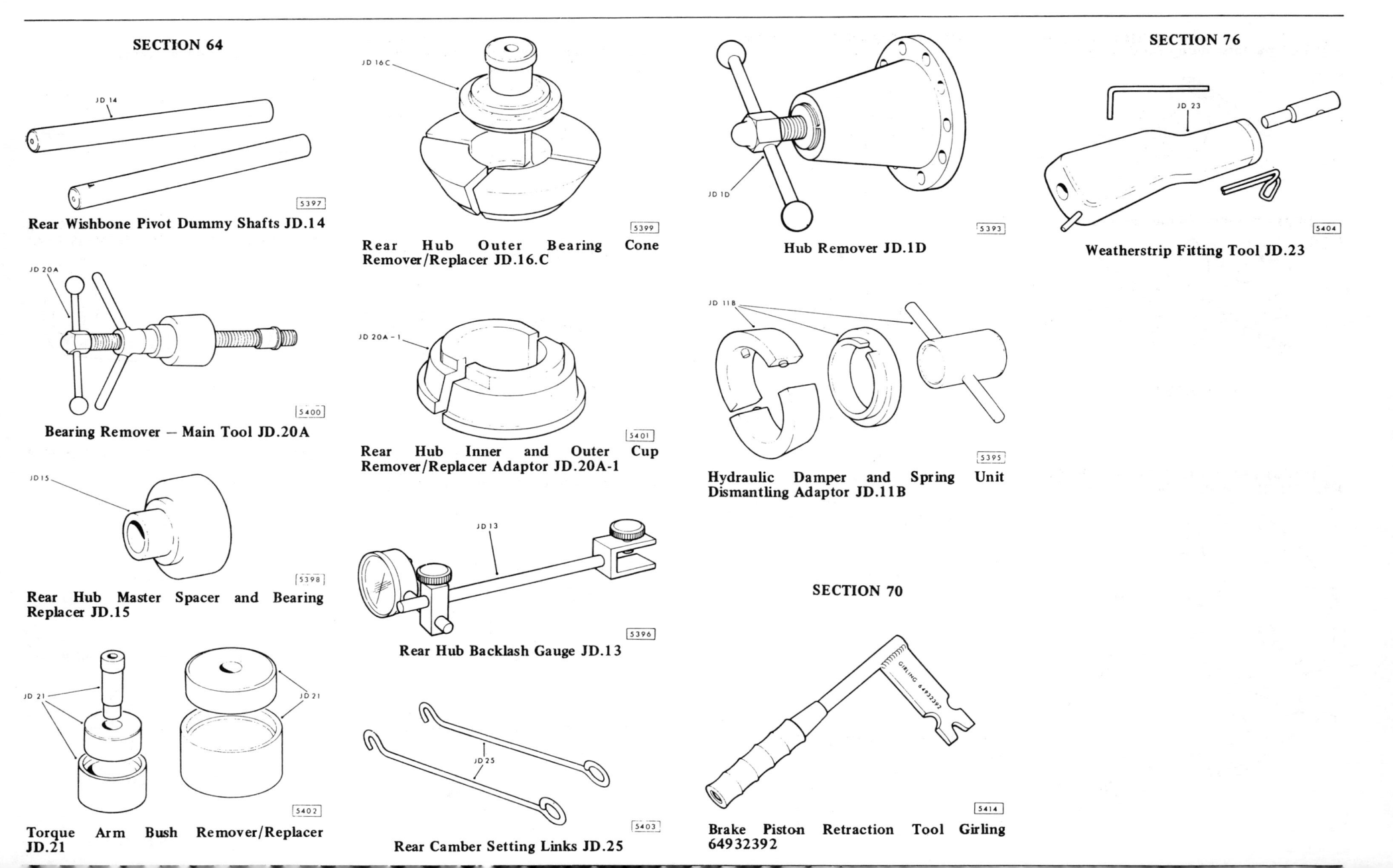

Weatherstrip Fitting Tool JD.23

SECTION 64

Rear Wishbone Pivot Dummy Shafts JD.14

Hub Remover JD.1D

Hydraulic Damper and Spring Unit Dismantling Adaptor JD.11B

Rear Hub Outer Bearing Cone Remover/Replacer JD.16.C

Rear Hub Inner and Outer Cup Remover/Replacer Adaptor JD.20A-1

Bearing Remover — Main Tool JD.20A

Rear Hub Master Spacer and Bearing Replacer JD.15

Rear Hub Backlash Gauge JD.13

SECTION 70

Brake Piston Retraction Tool Girling 64932392

Rear Camber Setting Links JD.25

Torque Arm Bush Remover/Replacer JD.21

BY APPOINTMENT
TO H M QUEEN ELIZABETH
THE QUEEN MOTHER
B L CARS LIMITED
MANUFACTURERS OF DAIMLER JAGUAR
ROVER CARS AND LAND ROVERS

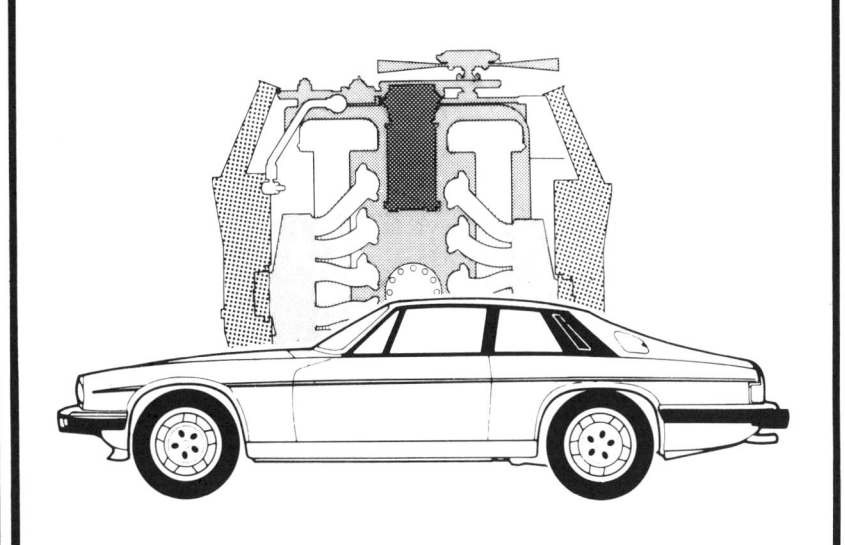

JAGUAR
XJ-S H.E.

Supplement to the Repair Operation Manual

SUPPLEMENT A • XJ-S H.E. 1979-1984

INTRODUCTION

This supplement is to be used in conjunction with the current XJS Repair Operation Manual, part number AKM 3455.

It covers the changes introduced since the above Repair Operation Manual was last published.

By using the appropriate service tools and carrying out the procedures as detailed a skilled technician will be able to complete the operation within the time stated in the Supplement to the Repair Operation Times.

Due to the fact that the electrical systems of the XJS-HE are sophisticated and of necessary complexity, section 86 of this supplement has been devoted entirely to Electrical Fault Diagnosis, by dividing the circuits into more easily understood subsystems.

The correct and speedy diagnosis of faults together with the Location and Access charts and diagrams will enable the isolation of faulty components. With each system diagram is a description of the circuit and test procedures to ensure the fast, accurate repair of electrical faults.

Some of the tests described require the use of specialised equipment and it is in the interest of efficiency and safety that it is used.

SPECIFICATION

Users are advised that the specification details set out in this book apply to a range of vehicles and not to any one. For the specification of a particular vehicle purchasers should consult their Dealer.

Jaguar Cars Ltd., reserve the right to vary their specifications with or without notice and at such times and in such manner as they think fit. Major as well as minor changes may be involved in accordance with the Manufacturer's policy of constant product improvement.

Whilst every effort is made to ensure the accuracy of the particulars contained in this book, neither Jaguar Cars Ltd., nor the Dealer, by whom this book is supplied, shall in any circumstances be held liable for any inaccuracy or the consequences thereof.

COPYRIGHT

© Jaguar Cars Ltd 1984

All rights reserved. No part of this publication may be reproduced, stored in a retrieval system or transmitted in any form, electronic, mechanical, photocopying, recording or other means without prior written permission of Jaguar Cars Ltd., Service Department, Radford, Coventry CV6 3GB.

CONTENTS

	Operation No.	Page No.
General Specification Data	—	04—1
Engine Tuning and Data	—	05—1
Recommended Lubricants, Fuel and Fluids, Capacities	—	09—1
Maintenance Summary	—	10—1

Engine

	Operation No.	Page No.
Camshaft cover gasket — LH — Remove and refit	12.29.40	12—2
— RH — Remove and refit	12.29.41	12—2
Crankshaft front oil seal — Remove and refit	12.21.14	12—1
Cylinder head gasket — LH — Remove and refit	12.29.02	12—1
— RH — Remove and refit	12.29.03	12—1
Description	12.00.00	12—1

Emission Control

	Operation No.	Page No.
Access Chart	—	17—3
Delay valve — Remove and refit	17.55.04	17—3
Dump valve — Remove and refit	17.55.01	17—3
Ignition vacuum advance system — Description	17.00.00	17—1
Solenoid air switch — Remove and refit	17.55.05	17—3
Solenoid valve — 2 way — Remove and refit	17.55.02	17—3
— 3 way — Remove and refit	17.55.03	17—3
Time delay unit — Remove and refit	17.55.07	17—3
Vacuum regulator — Remove and refit	17.55.06	17—3
Water temperature switch — Remove and refit	17.55.08	17—3
Vacuum system 1983 on	—	17—4

Fuel Injection System

	Operation No.	Page No.
Access Chart	—	19—7
Cold start System — Description	—	19—4
Cranking Enrichment — Description	—	19—4
Description	—	19—1
Engine Load Sensing — Description	—	19—4
Extra Air Valve — Description	—	19—5
Flooding Protection — Description	—	19—5
Fuel Control System — Description	—	19—4
Fuel Cooler — Remove and refit	19.40.40	19—7
Fuel Filter — Remove and refit	19.25.02	19—7
Fuel Rail — LH — Remove and refit	19.60.05	19—8
— RH — Remove and refit	19.60.04	19—8
Inertia Switch — Remove and refit	19.22.09	19—7
Injectors — LH — Remove and refit	19.60.03	19—8
— RH — Remove and refit	19.60.01	19—8
Lambda Sensor	—	19—5
Power Resistor — Remove and refit	19.22.44	19—7
Pressure Regulator — Remove and refit	19.44.11	19—7
Relays — Description	—	19—5
Temperature Sensor — Description	—	19—5
Throttle Potentiometer — Description	—	19—5
— Adjuster	—	19—6

Cooling System

	Operation No.	Page No.
Atmospheric catchment tank — Remove and refit	26.15.03	26—2
Coolant — Drain and refill	26.10.01	26—2
Description	26.00.00	26—1
Hose, expansion tank to atmospheric catchment tank — Remove and refit	26.30.62	26—2

Manifold and Exhaust System

	Operation No.	Page No.
Induction manifold gaskets — LH — Remove and refit	30.15.24	30—1
— RH — Remove and refit	30.15.25	30—1

Steering

	Operation No.	Page No.
1983 M.Y. on changes	—	57—1

Brakes

	Operation No.	Page No.
1983 M.Y. on changes	—	70—1

Body	**Operation No.**	**Page No.**
Bumper — front — centre section/blade — Remove and refit	76.22.37	76—1
— rear — centre section/blade — Remove and refit	76.22.12	76—1
Companion box liner — rear quarter lower — Remove and refit	76.13.51	76—1
Door trim pad — Remove and refit	76.34.01	76—1
— side front, veneer panel — Remove and refit	76.34.21	76—2
Energy absorbing beam — front — Remove and refit	76.22.26	76—1
— rear — Remove and refit	76.22.27	76—1
Facing trim pad — companion box, rear quarter panel lower — Remove and refit	76.13.55	76—1
Fascia switch panel — veneer panel — Remove and refit	76.46.30	76—2
Fascia veneer panel — passenger side — Remove and refit	76.46.25	76—2
— driver's side — Remove and refit	76.46.26	76—2
— centre — Remove and refit	76.46.29	76—2
Glove box lid — veneer panel — Remove and refit	76.52.12	76—2
Quarter bumper — front — Remove and refit	76.22.16	76—1
— rear — Remove and refit	76.22.17	76—1
— rear, rubber buffer — Remove and refit	76.22.63	76—1
Rear quarter trim casing — lower — Remove and refit	76.13.12	76—1
Rear seat arm rest — Remove and refit	76.70.39	76—2

Electrical system

Alternator Testing	—	86—7
Battery Testing	—	86—6
Bulb Chart	—	86—5
Bulb Failure Units	—	86—22
Cable Colour Code	—	86—6
Component Access	—	86—3
Component Location	—	86—1
Cooling Fan	—	86—26
Door Locks	—	86—23
Door Mirrors	—	86—24
Flasher Lamps	—	86—19
Fuse Charts	—	86—5
Head and Fog Lamps	—	86—11
Headlamp Wash/Wipe	—	86—12
Heated Rear Screen and Door Mirror	—	86—16
Horns	—	86—27
Ignition System Testing	—	86—7
Instruments	—	86—18
Interior Lamps	—	86—21
Low Coolant Control	—	86—23
Overcharge Warning Control	—	86—25
Parking Lamps	—	86—10
Radio, Clock and Luggage Compartment Lamps	—	86—17
Seat Belt Logic Control	—	86—26
Speed Control	—	86—13
Starter Circuit	—	86—7
Symbols	—	86—6
Window Lift	—	86—20
Windscreen Wipers	—	86—15
Vacuum Time Delay Unit	—	86—25
Revised Specification 1983/4	—	86—30
Trip Computer	—	86—28
Windscreen Wipers	—	86—28
Lucas MF3 Battery	—	86—30

GENERAL SPECIFICATION DATA

Engine

Number of cylinders	12	
Stroke	70 mm	2.756 in
Bore	90 mm	3.543 in
Cubic capacity	5343 cm³	326.0 in³
Compression ratio	'S' or 'L', Non HE	HE only 12.5:1

Cylinder Block

Material (cylinder block)	Aluminium alloy	
Angle of cylinders	60° Vee	
Type of cylinder liner	Slip fit, wet liner	
Material (liners)	Cast iron	
Normal size of bore after honing:		
Grade 'A' — Red	89.98 mm	3.543 in
Grade 'B' — Green	90.01 mm	3.544 in
Outside diameter of liner — both grades	97,99m + 0.22 mm − 0.00 mm	3.858 in + 0.001 in −0.00 in
Main line bore for main bearings	80,429 to 80,434 mm	3.1665 to 3.1667 in

Cylinder Heads

Material	Aluminium Alloy	
Valve seat angle: Inlet	44½°	
Exhaust	44½°	

Material	Manganese molybdenum steel	
Number of main bearings	7	
Main bearing type	Vandervell V.P.3	
Journal diameter	76,218 to 76,231 mm	3.0007 to 3.0012 in
Journal length: Front	29,72 to 29,97 mm	1.170 to 1.180 in
Centre	36,20 to 36,22 mm	1.425 to 1.426 in
Intermediate	30,43 to 30,53 mm	1.198 to 1.202 in
Rear	36,20 to 36,22 mm	1.452 to 1.426 in
Thrust taken	Centre bearing thrust washers	
Thrust washer thickness	2,57 to 2,62 mm	0.101 to 0.103 in
Permissible end-float	0,10 to 0,15 mm	0.004 to 0.006 in
Width of main bearing: Front	24,40 to 24,65 mm	0.963 to 0.973 in
Centre	30,2 to 30,5 mm	1.190 to 1.200 in
Intermediate	24,40 to 24,65 mm	0.963 to 0.973 in
Rear	30,2 to 30,5 mm	1,190 to 1.200 in
Diametrical clearance: all bearings	0,04 to 0,07 mm	0.0015 to 0.003 in
Crankpin diameter	58,40 to 58,42 mm	2.2994 to 2.3000 in
Crankpin length	43,15 to 43,20 mm	1.699 to 1.701 in

Connecting Rods

Length between centres	151,4 mm + 0,12 mm − 0,00 mm	5.96 in + 0.005 in − 0.000 in
Big-end bearing material	VP2C	
Bore for big-end bearing	62,0 mm + 0,15 mm − 0,00 mm	2.441 in + 0.006 in − 0.000 in
Width of big-end bearing	18,3 to 18,5 mm	0.720 to 0.730 in
Big-end diametrical clearance	0,04 to 0,09 mm	0.0015 to 0.0034 in
Big-end side clearance	0,17 to 0,33 mm	0.007 to 0.013 in
Small-end bush material	VP.10	
Bore for small-end bush	26,98 mm + 0,025 mm − 0,00 mm	1.062 in + 0.001 in − 0.000 in
Width of small-end bush	26,2 to 26,7 mm	1.03 to 1.05 in
Bore diameter of small-end bush	23,813 to 23,818 mm	0.9375 to 0.9377 in

Pistons

Type	Solid skirt	
Skirt clearance (measured midway down bore across bottom of piston skirt)	0,03 to 0,04 mm	0.0012 to 0.017 in

04—1

GENERAL SPECIFICATION DATA

Piston Rings

Number of compression rings	2	
Number of oil control rings	1	
Top compression ring thickness	3,81 to 4,06 mm	0.150 to 0.160 in
Second compression ring thickness	3,81 to 4,06 mm	0.150 to 0.160 in
Oil control ring width	Self expanding	
Width of oil control ring rails	2,62 ± 0,07 mm	0.103 ± 0,003 in
Top compression ring width	1,58 to 1,60 mm	0.062 to 0.063 in
Second compression ring width	1,96 to 1,98 mm	0.077 to 0.078 in
Side clearance of top compression ring in groove	0,07 mm	0.0029 in
Side clearance of second compression ring in groove	0,09 mm	0.0034 in
Side clearance of oil control rings in groove	0,14 to 0,17 mm	0.0055 to 0.0065 in
Top compression ring gap in bore	0,36 to 0,51 mm	0.14 to 0.020 in
Second compression ring gap in bore	0,25 to 0,38 mm	0.010 to 0.015 in
Gap of oil control ring rails in bore	0,38 to 1,14 mm	0.015 to 0.045 in

Gudgeon Pins

Type	Fully floating	
Length	79,25 to 79,38 mm	3.120 to 3.125 in
Outside diameter: Grade 'A' Red	23,81 mm	0.9375 in
Grade 'B' Green	23,76 mm	0.9373 in

Camshafts

Number of journals	7 per shaft	
Number of bearings	7 per shaft (14 half bearings)	
Type of bearings	Aluminium alloy — camshafts run direct in caps and tappet block.	
Journal diameter: All journals	26,93 mm + 0.013 mm − 0,000 mm	1.0615 in + 0.0005 in − 0,000 in
Diametrical clearance	0,03 to 0,07 mm	0.001 to 0.003 in
Thrust taken	Front end of shafts	

Jackshaft

Number of bearings	3	
Diametrical clearance in block	0,013 to 0,076 mm	0.0005 to 0.0003 in
Thrust taken	Front end of shaft	
Permissible end-float	0,13 mm	0.005 in
Line bore of front bearing	31,78 to 31,80 mm	1.251 to 1.252 in
Line bore of centre and rear bearing	30,23 to 30,25 mm	1.190 to 1.191 in

Valves and Valve Springs

Inlet valve material	Silico chrome steel	
Exhaust valve material	Austenitic steel	
Inlet valve head diameter (except HE)	41,22 to 41,32 mm	1.623 to 1.627 in
Inlet valve head diameter HE	41,15 to 41,40 mm	1.620 to 1.630 in
Exhaust valve head diameter (except HE)	34,5 to 34,6 mm	1.358 to 1.362 in
Exhaust valve head diameter HE	34,32 to 34,6 mm	1.355 to 1.362 in
Valve stem diameter: Inlet and exhaust	7,854 to 7,866 mm	0.3092 to 0.3093 in
Valve lift	9,5 mm	0.375 in
Inlet valve clearance (except HE)	0,305 to 0,356 mm	0.012 to 0.014 in
Inlet valve clearance HE	0,254 to 0,305 mm	0.010 to 0.012 in
Exhaust valve clearance (except HE)	0,305 to 0,356 mm	0.012 to 0.014 in
Exhaust valve clearance HE	0,254 to 0,305 mm	0.010 to 0.012 in
Outer valve spring free length	53,4 mm	2.108 in
Inner valve spring free length	44,0 mm	1.734 in

Valve Guides and Seats

Valve guide material	Cast iron	
Inlet valve guide length	48,5 mm	1.910 in
Exhaust valve guide length (except HE)	54,0 mm	2.125 in
Exhaust valve guide length HE	43,02 mm	1.725 in
Inlet valve guide outside diameter	As exhaust valve guide	
Exhaust valve guide outside diameter:		
Standard	12,75 to 12,72 mm	0.502 to 0.501 in
First oversize (2 grooves)	12,88 to 12,85 mm	0.507 to 0.506 in
Second oversize (3 grooves)	13,01 to 12,98 mm	0.512 to 0.511 in
Inlet valve guide finished bore	7,90 to 7,92 mm	0.311 to 0.312 in
Exhaust valve guide finished bore	7,90 to 7,92 mm	0.311 to 0.312 in
Maximum clearance between valve stem and guide	0,05 to 0,06 mm	0.0020 to 0.0023 in
Interference fit in cylinder head	0,05 to 0,15 mm	0.002 to 0.006 in
Valve seat insert material	Sintered iron	

GENERAL SPECIFICATION DATA

Service Replacements	Inlet valve seat insert outside diameter (except HE)	43,30 mm $^{+0,01}_{-0,00}$ mm	1.744 in $^{+0.0005}_{-0.0000}$ in
	Inlet valve seat insert HE diameter	42,93 mm $^{+0,01}_{-0,00}$ mm	1.690 in $^{+0.0005}_{-0.0000}$ in
	Exhaust valve seat insert outside diameter	38,17 mm $^{+0,01}_{-0,00}$ mm	1.503 in $^{+0.0005}_{-0.0000}$ in
	Inlet valve seat inside diameter (except HE)	35,56 mm $^{+0,17}_{-0,00}$ mm to 39,74 mm $^{+0,25}_{-0,00}$ mm	1.400 in $^{+0.0030}_{-0.0000}$ in to 1.565 in $^{+0.0100}_{-0.0000}$ in
	Inlet valve seat inside diameter HE	35,56 mm $^{+0,17}_{-0,00}$ mm to 39,95 mm $^{+0,25}_{-0,00}$ mm	1.400 in $^{+0.0005}_{-0.0000}$ in to 1.573 in $^{+0.0100}_{-0.0000}$ in
	Exhaust valve seat inside diameter (except HE)	30,10 mm $^{+0,07}_{-0,00}$ mm to 33,40 mm $^{+0,12}_{-0,00}$ mm	1.185 in $^{+0.0030}_{-0.0000}$ in to 1.315 in $^{+0.0050}_{-0.0000}$ in
	Exhaust valve seat inside diameter HE	30,45 mm $^{+0,07}_{-0,00}$ mm to 33,51 mm $^{+0,25}_{-0,00}$ mm	1.199 in $^{+0.0030}_{-0.0000}$ in to 1.280 in $^{+0.0100}_{-0.0000}$ in
Tappets and Tappet Guides	Tappet material	Cast iron (chilled)	
	Outside diameter of tappet	34,87 to 34,90 mm	1.373 to 1.374 in
	Diametrical clearance	0,02 to 0,04 mm	0.001 to 0.002 in
Lubricating System	Oil pump	Epicyclic gear type	
	Oil pump gears:		
	Driving gear outside diameter:		
	Diametrical clearance	0,127 mm to 0,305 mm	0.005 to 0.012 in
	Radial clearance	0,065 to 0,152 mm	0.0052 to 0.006 in
	Driven gear outside diameter:		
	Diametrical clearance	0,178 to 0,254 mm	0.007 to 0.010 in
	Radial clearance	0,09 to 0,13 mm	0.0035 to 0.005 in
	Driven gear internal diameter:		
	Diametrical clearance	0,28 to 0,46 mm	0.011 to 0.018 in
	Radial clearance	0,14 to 0,23 mm	0.0055 to 0.009 in
	Side clearance: driving and driven gear	0,115 to 0,165 mm	0.0045 to 0.0065 in
	Oil filter type	Full flow, disposable canister	
Timing Chain and Sprockets	Type of chain	Duplex endless	
	Pitch	9,5 mm	0.375 in
	Number of pitches	180	
	Camshaft sprockets: Number of teeth (each)	42	
	Crankshaft sprocket: Number of teeth	21	
	Jackshaft sprocket: Number of teeth	21	
Braking System	Front brakes: Make and type	Girling; ventilated discs, bridge-type calipers	
	Rear brakes: Make and type	Girling; damped discs, bridge-type calipers incorporating handbrake friction pads	
	Handbrake: Type	Mechanical, operating on rear discs	
	Disc diameter: Front	284 mm	11.18 in
	Rear	263,5 mm	10.375 in
	Disc thickness: Front	24,13 mm	0.95 in
	Rear	12,7 mm	0.50 in
	Master cylinder bore diameter	22,28 mm	0.875 in
	Brake operation	Hydraulic	
	Hydraulic fluid	Castrol/Girling Universal Brake and Clutch Fluid—exceeding specification S.A.E. J.1703/D	
	Main brake friction pad material	Ferodo 2430 slotted	
	Hand brake friction pad material	Mintex M68/1	
	Servo unit refs.: L.H.D.	Girling 64049748	
	R.H.D.	Girling 64049747	
Front Suspension	Type	Independent, coil spring	
	Castor angle	$3\frac{1}{2}° \pm \frac{1}{4}°$ positive	
	Camber angle	$\frac{1}{2}° \pm \frac{1}{4}°$ negative	
	Front wheel alignment	0 to 1,6 mm toe in	0 to $\frac{1}{16}$ in toe in
	Dampers	Telescopic, gas filled	
Rear Suspension	Type	Independent, coil springs, co-axial with dampers	
	Camber angle	$\frac{3}{4}° \pm \frac{1}{4}°$ negative	
	Rear wheel alingment	Parallel ± 0,08 mm	Parallel ± $\frac{1}{32}$ in
	Dampers	Telescopic gas filled	

GENERAL SPECIFICATION DATA

Final Drive Unit
- Type .... Hypoid with Power Lok differential
- Ratio 2.88:1 (49/17)

Automatic Gearbox
- Make and type .... General Motors GM 400
- Ratios: First gear .... 2.48 : 1
- Second gear .... 1.48 : 1
- Third gear .... 1.00 : 1
- Reverse .... 2.07 : 1
- Torque converter .... 2.00 : 1

Cooling System
- Water pump. Type .... Centrifugal, with two outlets
- Drive .... Belt driven from crankshaft
- No. of cooling fans .... Two (one 12-bladed, belt-driven through Holset coupling, plus 14-bladed electrically driven, thermostatically controlled)
- Cooling system control .... 2 thermostats
- Thermostat opening temperature .... 82°C / 180°F
- Thermostat fully open temperature .... 93.5° to 96°C / 200° to 205°F
- Engine filler cap pressure rating .... 1,41 kg/cm^2 / 20 lb/in^2
- Header tank cap pressure rating .... 1,05 kg/cm^2 / 15 lb/in^2

Fuel Injection Equipment
- Make and type:
  - NAS, UK and European markets .... Lucas Digital 'P', pressure sensing

	'A' Emissions N.A.S. only	'B' Emissions U.K./Europe
Injector reference no.	Lucas 73178A	Lucas 73178A
Cold start injector — reference no.	Lucas 73180A	Lucas 73180A
Pressure regulator — reference no.	Lucas 73177A	Lucas 73177A
Throttle switch — reference no.	Lucas type 193SA	Lucas type 193SA
Water temperature sensor — reference no.	Lucas 73198A	Lucas 73198A
Air temperature sensor — reference no.	Lucas 73197A	Lucas 73197A
Thermotime switch — reference no.	Lucas 33704A	Lucas 33704A
Extra air valve — reference no.	Lucas 73225A	Lucas 73192A
Deceleration valve — reference no.	Lucas 73224A	Lucas 73224A
Supplementary air valve — reference no.	—	Tecalemit TDA832
Full throttle micro-switch — reference no.	Burgess VBFYR1	—
Electrical control unit — reference no.	Lucas 83606A	Lucas 83605A
Lambda sensors — reference no.	Lucas 73211A	—
Power resistor — reference no.	Lucas 73196A	Lucas 73196A

Fuel System Pump
- Make and type .... Lucas 73202A — Electrical roller cell pump with integral relief valve and non-return valve
- Pressure .... 2,5 bar / 36 lbf/in^2

Power Assisted Steering
- Type .... Rack and pinion
- Number of turns lock to lock 2.75
- Turning circle, kerb to kerb .... 12,6m / 41 ft

Electrical Equipment
Battery
- Make and type .... Lucas CP13/11/8, GBY 218
- Voltage 12V
- Number of plates per cell 13
- Capacity at 10-hour rate .... 60 Ah
- Capacity at 12-hour rate .... 70 Ah

GENERAL SPECIFICATION DATA

Alternator	Make and type	early cars — Lucas 25ACR (Later cars — Lucas A133)
	Nominal voltage	12V
	Cut-in-voltage	13.5V at 1500 rev/min
	Polarity	Negative earth
	Maximum output	66A (75A)
	Maximum operative speed	15,000 rev/min
	Rotor winding resistance	3.6 ohms at 20°C (2.43 ohms)
	Brush spring pressure	255 to 369 gf — 9 to 13 ozf (130 to 270 gf — 4.7 to 9.8 ozf)
Starter Motor	Make and type	Lucas M45 pre-engaged
	Lock torque (at 940 amps)	4,01 kgf m — 29 lbf ft
	Torque at 1000 rev/min (at 535 amps)	1,80 kgf m — 13 lbf ft
	Light running current	100A at 5000 to 6000 rev/min
Distributor	UK Europe and NAS — Digital E.F.I. cars	Lucas 36DM12
Wiper Motor	Make and type	Lucas 16W
	Light running speed, rack disconnected (after 60 seconds from cold)	Normal: 46 to 52 rev/min, high: 60 to 70 rev/min
	Light running current (after 60 seconds from cold)	Normal: 1.5A, high: 2.0A
Tyre Data	Fitted as complete sets only	
	Type:	Pirelli Cinturato P5 215/70VR15, or Dunlop 215/70 VR15 SP Sport Super D7

Pressure

	Front	Rear
For speeds above 160 km/hr (100 mph) with driver, passenger and 27,2 kg (60 lb) luggage	2,20 Bar 2,25 kgf/cm² 32 lbf/in²	2,06 Bar 2,11 kgf/cm² 30 lbf/in²
For speeds above 160 km/hr (100 mph) with full load (including luggage) of 326 kg (720 lb)	2,20 Bar 2,25 kgf/cm² 32 lbf/in²	2,20 Bar 2,25 kgf/cm² 32 lbf/in²

For maximum comfort in countries where speeds are not in excees of 160 km/hr (100 mph), the above inflation pressure may be reduced by 0,42 kg/cm² (6 lb/in², 0,41 Bar) on front and rear tyres.

Tyre Replacement and Wheel Interchanging

When replacement of tyres is necessary, it is preferable to fit a complete car set. Should either front or rear tyres only show a necessity for replacement, new tyres must be fitted to replace the worn ones. No attempt must be made to interchange tyres from front to rear or vice-versa as tyre wear produces characteristic patterns depending upon their position and if such position is changed after wear has occurred, the performance of the tyre will be adversely affected. It should be remembered that new tyres require to be balanced.

The radial-ply tyres specified above are designed to meet the high-speed performance of which this car is capable.

Only tyres of identical specification as shown under 'Tyre Data' must be fitted as replacements and, if to different tread pattern, should not be fitted in mixed form.

UNDER NO CIRCUMSTANCES SHOULD CROSS-PLY TYRES BE FITTED.

Recommended Snow Tyre

The following information relates to the only tyre recommended for Jaguar and Daimler Cars. The use of snow tyres fitted with studs is not permitted in certain countries.

TYRE DESIGNATION	RECOMMENDED FITMENT	ROAD SPEED AND TYRE PRESSURES	REMARKS
Dunlop Weathermaster 185SR 15 SP M & S (Mud and Slush)	Complete sets only	Up to 137 km/h (85 mph): FRONT — REAR 1,86 bar — 1,79 bar 1,90 kgf/cm² — 1,83 kgf/cm² 27 lbf/in² — 26 lbf/in² From 137 km/h (85 mph) up to a maximum of 161 km/h (100 mph) FRONT — REAR 2,41 bar — 2,35 bar 2,46 kgf/cm² — 2,39 kgf/cm² 35 lbf/in² — 34 lbf/in²	1. Snow chains may be fitted to rear wheels only. 2. Tyres may be fitted with studs provided maximum speed does not exceed 121 km/h (75 mph) 3. Inner tubes with the wording 'Weathermaster only' are available and MUST be fitted when using 185SR 15 SP M& S Dunlop Weathermaster tyres.

ENGINE TUNING & DATA — 5.3 LITRE

ENGINE TUNING & DATA — 5.3 LITRE

General Data

Ignition timing: PI (Digital)
'A' Emission spec. (USA) pre 1983	25°-27° B.T.D.C.	
'A' Emission spec. (USA) 1983 on	18° B.T.D.C.	at 3000 rev/min. engine hot, vac. pipe disconnected
'B' Emission spec. (UK & Europe) Not HE	24° B.T.D.C.	
'B' Emission spec. (UK & Europe) HE	18° B.T.D.C.	

Initial static setting, to start engine only.
- 'A' Emission spec. ..... 9° ± 1° B.T.D.C.
- 'B' Emission spec. ..... 4° ± 1° B.T.D.C.

Cranking pressure:
- S ..... 11,6 kgf/cm² ..... 165 lbf/in²
- L ..... 9,5 kgf/cm² ..... 135 lbf/in²
- HE ..... 14,1 to 16,9 kgf/cm² ..... 200 to 240 lbf/in² (guide only)

Idling speed ..... 750 rev/min ± 25 rev/min

Firing order: 'A' Bank — Right-hand; 'B' Bank — Left-hand ..... 1A, 6B, 5A, 2B, 3A, 4B, 6A, 1B, 2A, 5B, 4A, 3B (Cylinders numbered from front of engine)

Timing marks ..... T.D.C. mark on crank damper rim and degree scale on bracket bolted to sump

Spark Plugs

- Make/type except HE ..... Champion N10Y or Unipart GSP151
- Make/type HE ..... Chamption BN5 or Unipart GSP360
- Gap except HE ..... 0,9 mm ..... 0.035 in
- Gap HE ..... 0,64 mm ..... 0.025 in

Ignition Coil

- Make/type except HE ..... Lucas 22C. 12 (Japan and Australia only) / Lucas 54402269 (all other markets)
- Make/type HE ..... Lucas 35C6 — 2 off
- Primary resistance at 20°C (68°F) ..... 0.9 to 1.1 ohms
- Consumption: stationary ..... 5.0 to 6.5 amps
- running ..... 2.5 to 3.0 amps
- Ballast resistance ..... 0.9 to 1.0 ohms

Distributor

- Make/type except HE ..... Lucas Opus 36 DE 12
- Make/type HE ..... Lucas constant energy 36DM12
- Rotation of rotor (looking down at rotor) ..... Anti-clockwise
- Pick-up module to timing rotor gap: except HE ..... 0,51 to 0,55 mm ..... 0.020 to 0.022 in
- HE ..... 0,15 to 0,20 mm ..... 0.006 to 0.008 in

Centrifugal advance

Crankshaft degrees and rev/min

	Except HE		HE only
Decelerating check, vacuum pipe disconnected	34° to 38° at	6200	20° to 24°
	29° to 33° at	4000	18° to 22°
	22° to 26° at	2000	12° to 16°
	No advance below	900	No advance below

Vacuum advance (crankshaft degrees)

- Maximum: except HE ..... 4° at 254 mmHg ..... 10 inHg
- HE ..... 12° at 381 mmHg ..... 15 inHg
- Starts: except HE ..... 152 mmHG ..... 6 inHg
- HE ..... 127 mmHg ..... 5 inHg

Fuel Injection Equipment

- Make/type: except HE and Digital ..... Bosch-Lucas electronically controlled
- Digital and HE ..... Lucas Digital F.I.

Exhaust Emission

Exhaust gas analyser reading at engine idle speed ..... 1 to 2% max CO at 750 rev/min without air injection

RECOMMENDED LUBRICANTS, FUEL AND FLUIDS, CAPACITIES

RECOMMENDED LUBRICANTS, FLUIDS, CAPACITIES AND DIMENSIONS

Engine Oil — Recommended S.A.E. Viscosity Range/Ambient Temperature Scale

Component — Model	Temperature Range	Specification	S.A.E. Viscosity Rating	Approved Brands Available in U.K. for Temperatures Above −10°C (14°F)
Engine Distributor Oil Can	Above −10°C (14°F) −20°C to 10°C (−4°F to 50°F) Below −10°C (14°F)	BLSO OL.02 or MIL-L-2104 B or A.P.1 SE	10W/50, 15W/50, 20W/40, 20W/50 10W/30, 10W/40 10W/50 5W/20, 5W/30	Unipart Super Multigrade, BP Super Viscostatic, Castrol GTX, Duckhams Q Motor Oil, Esso Uniflow, Fina Super Grade, Mobiloil Super, Shell Super Oil, Texaco Havoline
Powr-Lok Differential — Initial Fill — Refill	All All	Use only approved brands of fluid specially formulated for Powr-Lok	90 90	Shell Spirax Super 90 Shell Spirax Super 90 BP Gear Oil 1453, BP Limslip Gear Oil 90/1, Castrol G722, Castrol Hypoy LS, Duckhams Hypoid 90 DL, Texaco 3450 Gear Oil Veedol Multigear Limited Slip S.A.E. 90
Drain and Top-Up — Top-up only if above oil not available	All	MIL-L-2105B	EP 90	BP Gear Oil S.A.E. 90 EP, Castrol Hypoy, Duckhams Hypoid 90, Esso Gear Oil GX 90/140 Mobilube HDGO, Shell Spirax HD 90 Texaco Multigear Lubricant EP 90
GM 400 Automatic Transmission	All	Dexron 2D		BP Autran DX, Castrol TQ Dexron, Esso ATF Dexron, Mobil ATF 220 Dexron Shell ATF Dexron, Texaco Texamatic Fluid 6673
Power Assisted Steering	All	Above Specification or Dexron 2D	—	BP Autran DX, Castrol TQ Dexron Esso ATF Dexron, Mobil ATF 220 Dexron Shell ATF Dexron, Texaco Texamatic Fluid 6673
Grease Points	All	Multipurpose Lithium Grease, N.L.C.I. Consistency No. 2	—	BP Energrease L8, Castrol LM, Duckhams LB10, Esso Multipurpose H, Fina Marson HTL2, Mobilgrease MP, Shell Retinax A, Texaco Marfak

09—1

RECOMMENDED LUBRICANTS, FUEL AND FLUIDS, CAPACITIES

COOLING SYSTEM

Additive	Barr's Leak inhibitor, 2 sachets per car
Anti-freeze	BP Type HS25, Union Carbide UT 184 Unipart Universal, or Bluecol 'U' If these are not available, anti-freeze conforming to specification B.S. 3150 or 3152 may be used Concentration — U.K. only 40% S.G. 1.065 All other markets 55% S.G. 1.074
In territories where anti-freeze is unnecessary the cooling system must be filled with a solution of Marston Corrosion Inhibitor Concentrate SQ36. Always top-up the cooling system with recommended strength of anti-freeze or Corrosion Inhibitor, NEVER with water only.	

CAPACITIES

	Litres	Imperial	U.S.
Engine refill (including filter)	10,7	19 pt	22,8 pt
Automatic transmission unit	9,1	16 pt	19,2 pt
Final drive unit	1,6	2,75 pt	3,25 pt
Cooling system, including reservoir and heater or air conditioning			
Not HE	21,2	37,5 pt	45 pt
HE	19,5	35 pt	42 pt
Fuel tank	90,9	20 gal	24 gal

DIMENSIONS

Wheelbase	2591 mm	102 in
Track: Front	1482 mm	58.4 in
Rear	1497 mm	58.9 in
Overall length	4743 mm	186.7 in
Overal width	1793 mm	70.6 in
Overall height	1264 mm	49.7 in
Turning circle: between walls	12.6 m	41 ft
Ground clearance: kerb condition	140 mm	5.5 in

WEIGHTS

UK and European Models	Kg	lb
Kerb weight	1755	3869
Gross vehicle weight (GVW)	2110	4651
*Gross combination weight (GCW)	3610	7959
Maximum permitted front axle load (FAW)	1065	2348
Maximum permitted rear axle load (RAW)	1070	2359
Federal Models	**Kg**	**lb**
Gross vehicle weight rating (GVWR)	2132	4700
Gross axle weight rating — Front (GAWR — Front)	1078	2376
Gross axle weight rating — Rear (GAWR — Rear)	1054	2324

* Gross combination weight is the gross vehicle weight plus maximum trailer weight.

RECOMMENDED LUBRICANTS, FUEL AND FLUIDS, CAPACITIES

FUEL REQUIREMENTS

In U.S.A. use unleaded fuel with minimum octane rating of 91 RON.

In the United Kingdom use '4 STAR' fuel, 97 octane

If, of necessity, the car has to be operated on lower octane fuel, do not use full throttle otherwise detonation may occur with resultant piston trouble.

RECOMMENDED HYDRAULIC FLUID

Braking System

Castrol-Girling Universal Brake and Clutch fluid. This fluid exceeds S.A.E. J1703/D specification.
NOTE: Check all pipes in the brake system at the start and finish of each winter period for possible corrosion due to salt and grit used on the roads.

MAINTENANCE SUMMARY

MAINTENANCE SUMMARY—UK & EUROPE

OPERATION	Interval in Kilometres x 1000 Interval in Miles x 1000	1.6 1	12 7.5	24 15
PASSENGER COMPARTMENT				
Fit protection kit		X	X	X
Check condition and security of seats and seat belts		X	X	X
Check operation of seat belt warning system		X		
Check footbrake operation		X	X	X
Drive on lift; stop engine		X	X	X
Check operation of lamps		X		
Check operation of horns		X		
Check operation of warning indicators		X		
Check operation of windscreen wipers		X		
Check operation of windscreen washers		X		
Check security of handbrake — release fully after checking		X	X	X
Check rear-view mirrors for security and function		X		
Check operation of boot lamp		X		
EXTERIOR				
Open bonnet — fit wing covers		X	X	X
Raise lift to convenient working height with wheels free to rotate		X	X	X
Mark stud to wheel relationship			X	X
Remove front wheels			X	
Remove road wheels — front and rear				X
Check that tyres are the correct size and type		X	X	X
Check tyre tread depth		X	X	X
Check tyres visually for external lumps, bulges and uneven wear		X	X	X
Check tyres visually for external exposure of ply or cord		X	X	X
Check/adjust tyre pressure		X	X	X
Inspect brake pads for wear and discs for condition			X	X
Adjust front hub bearing end-float				X
Grease hubs				X
Check for oil leaks from steering and fluid leaks from suspension system		X	X	X
Check condition and security of steering unit joints and gaiters		X	X	X
Refit road wheels in original position			X	X
Check tightness of road wheel fastenings		X	X	X
UNDERBODY				
Raise lift to convenient height		X	X	X
Drain engine oil		X	X	X
Check/top up gearbox oil		X	X	X
Grease all points excluding hubs			X	X
Check/top up rear axle/final drive oil		X	X	X
Check visually hydraulic hoses, pipes and unions for chafing, cracks, leaks and corrosion		X	X	X
Check exhaust system for leakage and security		X	X	X
Lubricate handbrake mechanical linkage and cables		X	X	X
Check condition of handbrake pads				X
Lubricate automatic gearbox exposed selector linkage		X	X	X
Check tightness of propshaft coupling bolts		X		X
Check security of accessible engine mountings		X		
Check condition and security of steering unit, joints and gaiters		X	X	X
Check security and condition of suspension fixings		X	X	X
Check steering rack for oil leaks		X	X	X
Check power steering for leaks, hydraulic pipes and unions for chafing, corrosion and security		X	X	X
Check shock absorbers for fluid leaks		X	X	X
Renew engine oil filter element			X	X
Refit engine drain plug		X	X	X
Check for oil leaks — engine and transmission		X	X	X
Lower lift		X	X	X

10—1

MAINTENANCE SUMMARY—UK & EUROPE

OPERATION	Interval in Kilometres x 1000	1.6	12	24
	Interval in Miles x 1000	1	7.5	15
ENGINE COMPARTMENT				
Fit exhaust extractor pipe		X	X	X
Fill engine with oil		X	X	X
Lubricate accelerator control linkage and pedal pivot		X		
Renew air cleaner element(s)				X
Check security of accessible engine mountings		X		
Check driving belts; adjust or renew		X		X
Check and adjust spark plugs			X	
Renew spark plugs				X
Check/top-up battery electrolyte		X	X	X
Clean and grease battery connections		X	X	X
Check/top-up brake fluid reservoir		X	X	X
Check brake servo hose(s) for security and condition		X	X	X
Check/top-up windscreen washer reservoir		X		
Check cooling and heater system for leaks and hoses for security and condition		X	X	X
Check/top-up cooling system		X		
Renew fuel filter				X
Clean engine breather filter (where applicable)				X
Check crankcase breathing system for leaks, hoses for security and condition		X		X
Check/top-up fluid in power steering reservoir; check security and condition of oil pressure hose at oil filter		X	X	X
Run engine and check for sealing of oil filter; stop engine			X	X
Check/top-up engine oil			X	X
Check electronic instruments and check data		X		X
Lubricate distributor (not cam wiping pad) at 22,500 mile intervals — run engine		X		
Disconnect vacuum pipe, check dwell angle, adjust as necessary		X		X
Check ignition timing		X	X	X
Check distributor automatic advance		X		X
Check advance increases as vacuum pipe is reconnected		X		X
DOOR AND WINDOW MECHANISMS				
Lubricate all locks, hinges and door check mechanisms (not steering lock)		X		X
Check operation of all door, bonnet and boot locks		X		
Check operation of window controls		X		
Check and if necessary renew windscreen wiper blades			X	X
UNDER BONNET				
Check power steering system for leaks, hydraulic pipes and unions for chafing and corrosion		X	X	X
Check for oil leaks from engine and transmission		X	X	X
Check/top-up automatic gearbox fluid		X	X	
Re-check tension if driving belt has been renewed		X		X
Remove wing covers		X	X	X
Fill in details and fix Leycare and other appropriate Unipart underbonnet stickers		X	X	X
Close bonnet		X	X	X
Remove exhaust extractor pipe		X	X	X
WHEELS AND TYRES				
Remove spare wheel		X	X	X
Check that tyre complies with manufacturer's specification		X	X	X
Check tyre tread depth		X	X	X

10—2

MAINTENANCE SUMMARY

MAINTENANCE SUMMARY—UK & EUROPE

OPERATION	Interval in Kilometres x 1000 Interval in Miles x 1000	1.6 1	12 7.5	24 15
WHEELS AND TYRES cont.				
Check tyres visually for external exposure of cord or ply		X	X	X
Check tyres visually for external lumps or bulges		X	X	X
Check/adjust tyre pressure		X	X	X
Refit spare wheel		X	X	X
MISCELLANEOUS				
Check/adjust headlamp alignment		X		X
Check/adjust front wheel alignment		X		X
Drive off lift		X	X	X
Carry out road or roller test		X	X	X
Check operation of seat belt inertia mechanism		X	X	X
Ensure cleanliness of controls, door handles, steering wheel, etc		X	X	X
Remove protection kit		X	X	X
Report additional work required		X	X	X

At 48 000 km (38 000 mile) intervals:
  Change final drive oil
  Change brake fluid
  Change coolant ensuring that 40% UK and 55% all other markets anti-freeze content is present upon replenishment
  Renew automatic transmission filter (GM 400 only)
  Renew automatic transmission fluid

At 96 000 km (60 000 mile) intervals
  Renew all fluid seals in hydraulic system; examine and renew if necessary all flexible hoses
  Examine working surfaces of master cylinder and calipers. Renew if necessary

OPTIONAL SERVICES

OPERATION	Interval in Kilometres x 1000 Interval in Miles x 1000	12 7.5	24 15
Check operation of lamps			X
Check operation of horns			X
Check operation of warning indicators			X
Check operation of windscreen wipers			X
Check operation of windscreen washers			X
Check operation of window controls		X	X
Check sunroof and controls for correct operation (if fitted)			X
Check operation of headlamp wipe/wash (if fitted)			X
Check rear view mirrors for security and function			X
Check operation of boot lamp			X
Check/top-up windscreen washer reservoir			X
Check/top-up cooling system		X	X
Check operation of all door, bonnet and boot locks		X	X
Check operation of cruise control (if fitted)			X
Lubricate all locks, hinges and door check mechanisms (not steering lock)		X	
Check/adjust headlamp alignment		X	
Check/adjust front wheel alignment		X	

MAINTENANCE SUMMARY — NORTH AMERICAN MARKETS

Service Code Letter	Distance x 1000 in miles and kilometres — The period between services should not exceed 12 months							
A Km	1.5							
A Miles	1							
B Km		12		36		60		92
B Miles		7.5		22.5		37.5		52.5
C Km			24				72	
C Miles			15				45	
D Km					48			96
D Miles					30			60

OPERATION DESCRIPTION	SERVICE			
	A	B	C	D
LUBRICATION				
Lubricate all grease points		X	X	X
Renew engine oil and engine oil filter	X	X	X	X
Check/top-up brake fluid reservoir	X	X	X	X
Check/top-up automatic transmission fluid	X	X	X	X
Check battery condition	X	X	X	X
Check/top-up cooling system	X	X	X	X
Check/top-up rear axle oil	X	X	X	X
Renew Automatic Transmission fluid and filter				X
Lubricate all locks and hinges (not steering lock)	X	X	X	X
Check/top-up power steering reservoir	X	X	X	X
ENGINE				
Check all driving belts — adjust				X
Renew air cleaner element(s)				X
Check security of engine mountings	X			
Check for oil leaks	X	X	X	X
IGNITION				
Renew spark plugs				X
Lubricate distributor				X
FUEL AND EXHAUST SYSTEMS				
Check fuel system for leaks, pipes and unions for chafing and corrosion	X	X	X	X
Check exhaust system for leaks and security	X	X	X	X
Renew oxygen sensor				X
Renew fuel filter				52.5 only
TRANSMISSION, BRAKES, STEERING AND SUSPENSION				
Check condition and security of steering unit, joints and gaiters		X	X	X
Inspect brake pads for wear and discs for condition		X	X	X
Check brake servo hoses for security and condition	X	X	X	X
Check/adjust front wheel alignment	X	X		
Check foot and hand brakes	X			
Check visually brake hydraulic pipes and unions for cracks, chafing, leaks and corrosion	X	X	X	X
Check/adjust front hub bearing end float			X	X
Check tightness of propeller shaft coupling bolts			X	X

10—4

MAINTENANCE SUMMARY

MAINTENANCE SUMMARY — NORTH AMERICAN MARKETS

OPERATION DESCRIPTION	SERVICE			
	A	B	C	D
WHEELS AND TYRES				
Check tyres for tread depth and visually for external cuts in fabric, exposure of ply or cord structure, lumps or bulges	X	X	X	X
Check that tyres comply with manufacturer's specification	X	X	X	X
Check/adjust tyre pressure, including spare wheel	X	X	X	X
Check tightness of road wheel fastenings	X	X	X	X
ELECTRICAL				
Check/adjust operation of washers and top up reservoir	X	X	X	X
Check function of original equipment, i.e. lamps, horns, wipers and all warning indicators	X	X	X	X
Check wiper blades and arms; renew if necessary		X	X	X
Check/adjust headlamp alignment	X	X	X	X
BODY				
Check operation, security and operation of seats and seat belts	X	X	X	X
Check operation of all door, bonnet and luggage compartment locks	X	X	X	X
Check operation of window controls	X	X	X	X
GENERAL				
Road/roller test and check function of all instrumentation	X	X	X	X
Report additional work required	X	X	X	X

ENGINE

V12 HE ENGINE

Description 12.00.00

The HE engine features a new type of combustion chamber design, which gives more efficient combustion by controlling the movement of the fuel/air mixture, it also allows the use of a higher compression ratio, increased from 10:1 to 12.5:1.

The combustion chamber consists of two areas, a small one beneath the inlet valve with a narrow guide channel leading to a larger area beneath the exhaust valve, where the spark plug is located.

As the piston commences its compression stroke, the mixture is compressed in the inlet valve area and swirls along the guide channel into the larger combustion area where it is ignited.

The strong turbulence and 'squish' effect, produced by the compression and swirling of the mixture from the small to the large combustion chamber areas, gives better fuel/air mixing and controlled even burning. It also allows the use of a very lean mixture.

The new inlet valve has a flat face and forms the ceiling of the small compression area, the exhaust valve face remains slightly dished.

The pistons are new and have a dished crown instead of the previous flat face.

CRANKSHAFT FRONT OIL SEAL

Remove and refit 12.21.14

Removing

Drive the vehicle onto a ramp and disconnect the battery.

Open the bonnet and remove both air cleaner elements and covers.

Slacken the air conditioning compressor drive belt; position the auxiliary pulley away from its mounting position.

Remove the fan 'Torquatrol' fan assembly, from the auxiliary pulley.

Slacken the fan belt. Remove the nuts and bolts securing the idler pulley housing to the timing cover; remove the pulley housing.

Slacken the power steering pump drive belt. Remove all the above drive belts from the crank pulley.

Remove the crankshaft outer pulley securing bolts, dognut and spacers; remove the crankshaft outer pulley.

Raise the ramp and remove the alternator drive belt.

Undo the crank damper securing bolt and remove the crank damper, damper mounting cone and woodruff key.

Remove and discard the seal. Remove the oil seal collar.

Refitting

To refit, reverse the above procedure, ensure complete cleanliness, fit a new oil seal. Correctly tension all drive belts.

CYLINDER HEAD GASKET — LEFT-HAND

Remove and refit 12.29.02

Removing

Open the bonnet, depressurise the fuel system and drain the coolant.
Disconnet the battery.
Remove both air cleaner elements and covers.
Remove the ignition amplifier. Remove both inlet manifolds and gaskets; fit suitable blanking plugs to the inlet ports.
Remove both camshaft covers and gaskets.
Remove the LH wing stay.
Disconnect the fuel return hose from the fuel cooler; carefully secure the fuel cooler assembly away from the cylinder head. Plug the pipe and hose.
Disconnect the HT leads from the spark plugs.
Disconnect the LH top hose from the thermostat housing, and the LH expansion hose from the water rail.
Slacken the clip securing the coolant cross pipe hose to the LH water rail. Remove the bolts securing the water rail.
Remove the engine lifting eyes, disconnect the water rail and gaskets from the coolant crosspipe; remove the water rail.
Disconnect the camshaft oil feed pipe from the LH cylinder head. Remove the exhaust manifold heat shield, engine dipstick and tube assembly.
Jack up the front of the vehicle and place on stands.
Remove the LH down pipe heat shields and disconnect the downpipe from the exhaust manifolds. Remove the rear manifold lower securing nuts.
Lower the vehicle; remove the rear LH exhaust manifold. Slacken the fan belt and remove the fan/'Torquatrol' assembly from the pulley and position away from the engine. Disconnect the expansion hose from the coolant filler pipe neck.
Rotate the engine using the crankshaft damper nut, to position the camshaft 180° B.T.D.C. Bend back the tab washers and slacken the camshaft sprocket retaining bolts. Rotate the engine until the valve timing gauge C.3993 can be fitted. Bend back the tab washers and slacken the remaining camshaft sprocket retaining bolts.
Remove the rubber grommet from the front timing cover, and using JD 50 release the tension on the timing chain.
Remove the camshaft sprocket retaining bolts and tab washers. Displace the sprocket onto the chain guide and retain using JD 40. Do not rotate the engine again until the camshaft sprocket is refitted to the camshaft.
Starting in the centre and working outwards, remove the nuts securing the cylinder head to the cylinder block and timing cover.
Remove the cylinder head, discard the head gasket and downpipe sealing rings.

Refitting

Fit a new cylinder head gasket and downpipe sealing rings. Ensure that no wires or pipes can be trapped and refit the cylinder head to the cylinder block and timing cover.
Fit and tighten the cylinder head nuts, starting at the centre and working outwards.
Remove service tool JD 40 and refit the camshaft sprocket to the camshaft, fit new tab washers, refit but do not tighten the accessible bolts.
Release the timing chain tensioner and rotate the engine using the crankshaft pulley damper nut, to gain access to fit the remaining camshaft sprocket bolts.
Tighten all the camshaft sprocket bolts and secure with the tab washers. Refit the tensioner access plug.
Refit the fan/'Torquatrol' assembly. Refit and correctly tension the fan belt. Reconnect and secure the expansion hose to the filler pipe neck.
Using new gaskets fit and secure the exhaust manifolds, use upper securing nuts only.
Refit the exhaust downpipe to the exhaust manifolds, fit but do not tighten the centre securing nuts. Refit the heat shield, fit but do not tighten the heat shield front securing nuts.
Jack up the front of the vehicle and place on stands. Refit the heat shield rear securing nut; tighten all downpipe and exhaust manifold securing nuts. Lower the vehicle.
Refit the engine dipstick assembly. Refit the exhaust manifold downpipe.
Refit the camshaft oil feed pipe to the LH cylinder head, using new copper washers.
Refit, using new gaskets, the LH water rail assembly. Reconnect the top hose and expansion tank hose to the water rail.
Reconnect the HT leads to the spark plugs. Release the fuel cooler from its secured position and reconnect the fuel return hose.
Refit the LH wing stay. Refit the camshaft covers and gaskets. Remove the blanking plugs from the inlet ports and refit the induction manifolds, use new gaskets.
Refit the amplifier, air cleaner elements and covers. Reconnect the battery.
Refill the cooling system with coolant of the correct specification.

CYLINDER HEAD GASKET — RIGHT-HAND

Remove and refit 12.29.03

Removing

Open the bonnet; depressurise the fuel system and drain the coolant.
Disconnect the battery.
Remove the RH air cleaner cover and element.
Remove the RH induction manifold and gaskets. Fit suitable blanking plugs to the inlet ports.
Slacken the air conditioning compressor drive belt, and displace the auxiliary pulley bracket from the cylinder head.
Slacken the clip securing the coolant cross pipe to the RH water rail. Disconnect the thermotime switch and coolant temperature transmitter feed wires.

12—1

ENGINE

Disconnect the top hose from the RH thermostat cover.
Remove the air conditioning compressor securing bolts, and chock the compressor clear of the cylinder head.
Disconnect the HT leads from the spark plugs. Unclip the harness and throttle cable from the RH wing stay and remove the wing stay.
Remove the harnesses from the air valve crosspipe. Disconnect the crosspipe from the LH manifold. Disconnect the ECU, brake and other minor vacuum pipes, after the first noting fitted positions, from the crosspipe, and remove the pipe.
Disconnect the water valve hose from the RH water rail.
Remove the automatic transmission dipstick tube assembly, lifting eye bracket and the RH water rail assembly.
Disconnect the RH camshaft oil feed pipe, discard the copper washers, and remove the camshaft cover.
Disconnect the exhaust manifold heat shield and disconnect the downpipe from the manifold. Remove the front exhaust manifold and the upper and front lower nuts securing the rear exhaust manifold.
Raise the front of the vehicle and place on stands. Remove the starter motor heat shield and heat shield spacer.
Remove the lower rear manifold securing nut. Lower the vehicle.
Remove the remaining rear exhaust manifold securing nuts.
Remove the RH engine mounting upper securing nut.
Using a suitable block and jack, raise the engine sufficiently to enable the removal of the rear exhaust manifold. Lower the engine on the jack.
Slacken the fan belt and remove the fan/'Torquatrol' assembly from the pulley, position away from the engine.
Disconnect the expansion hose from the coolant filler neck.
Rotate the engine using the crankshaft damper nut, to position the camshaft 180° B.T.D.C. Bend back the tab washers and slacken the camshaft sprocket retaining bolts.
Rotate the engine until the valve timing gauge C.3993 can be fitted. Bend back the tab washers and slacken the remaining camshaft sprocket retaining bolts.
Remove the rubber grommet from the front timing cover, and using JD 50 release the tension on the timing chain.
Remove the camshaft sprocket retaining bolts and tab washers. Displace the sprocket onto the chain guide and retain using JD 40. Do not rotate the engine again until the camshaft sprocket is refitted to the camshaft.
Starting in the centre and working outwards, remove the nuts securing the cylinder head to the cylinder block and timing cover.
Remove the cylinder head, discard the head gasket and downpipe sealing rings.

Refitting

Fit a new cylinder head gasket and downpipe sealing rings. Ensure that no wires or pipes can be trapped and refit the cylinder head to the cylinder block and timing cover.

Fit and tighten the cylinder head nuts, starting at the centre and working outwards.
Remove service tool JD 40 and refit the camshaft sprocket to the camshaft, fit new tab washers, refit but do not tighten the accessible bolts.
Release the timing chain tensioner and rotate the engine using the crankshaft damper nut, to gain access to fit the remaining camshaft sprocket bolts. Tighten all the camshaft sprocket bolts and secure with the tab washers. Refit the tensioner access plug.
Jack up the RH side of the engine, fit rear exhaust manifold and new gasket, retain with one nut. Lower the engine and refit the RH engine mounting. Fit and tighten all the rear exhaust manifold securing nuts except the lower rear one.
Raise the front of the vehicle and place on stands.
Fit and tighten the remaining rear exhaust manifold nut. Refit the starter motor heat shield and spacer. Lower the vehicle.
Refit the front exhaust manifold and new gasket. Connect and secure the downpipe to the exhaust manifolds. Refit the exhaust manifold heat shield.
Refit the fan/'Torquatrol' assembly. Refit and correctly tension the fan belt.
Refit the expansion hose to the coolant filler neck.
Using a new gasket and half moon seal, refit and secure the RH camshaft cover.
Reconnect the RH camshaft oil feed pipe, using new sealing washers.
Refit the water rail, use new gaskets. Refit the transmission dipstick assembly, and the engine lifting eyes. Connect the coolant hose from the heater valve to the water rail.
Refit the air valve crosspipe, connect to the LH manifold.
Reconnect all the vacuum pipes to the crosspipe assembly and secure the harnesses.
Refit the RH wing stay. Reposition the harness and throttle cable to the wing stay.
Reconnect the HT leads to the spark plugs.
Release the chock from the air conditioning compressor and refit the compressor to the engine, ensure that the earth lead is fitted to the front LH bolt.
Refit the electric relays to their mounting brackets.
Refit the top hose and coolant crosspipe to the RH water rail. Reconnect the harness to the thermotime switch and the coolant temperature transmitter.
Refit the auxiliary pulley to the cylinder head.
Remove the blanking plugs from the inlet ports and using a new gasket, refit the RH induction manifold. Refit the air cleaner element and cover. Reconnect the battery. Refill the cooling system with coolant of the correct specification.

CAMSHAFT COVER GASKET — LEFT-HAND

Remove and refit   12.29.40

Removing

Open the bonnet.
Depressurise the fuel system.
Disconnect the battery, and remove the LH air cleaner cover and element.
Remove the ignition amplifier.
Remove the LH inlet manifold and gaskets. Fit suitable blanking plugs to the inlet ports.
Remove the nuts and bolts securing the camshaft cover.
Remove the harness clip from the front stud and remove the camshaft cover.
Remove and discard the camshaft cover gasket and the half moon seal from the rear of the cylinder head.

Refitting

Ensure that the camshaft cover gasket is correctly aligned.
Use new gaskets.
Reverse the above procedure.

CAMSHAFT COVER GASKET — RIGHT-HAND

Remove and refit   12.29.41

Removing

Open the bonnet.
Depressurise the fuel system.
Disconnect the battery.
Remove the RH air cleaner cover and element.
Remove the RH inlet manifold and gaskets, fit suitable blanking plugs to the inlet ports.
Remove the nuts and bolts securing the camshaft cover. Remove the bolt securing the air conditioning compressor silencer to the front of the compressor.
Carefully remove the camshaft cover from underneath the air conditioning pipe and the rear coolant hose. Discard the gasket and the half moon seal from the rear of the cylinder head.

Refitting

Ensure that the camshaft cover gasket is correctly aligned.
Use new gaskets.
Reverse the above procedure.

EMISSION CONTROL

IGNITION VACUUM ADVANCE SYSTEM
'B' EMISSION V12 HE ENGINES

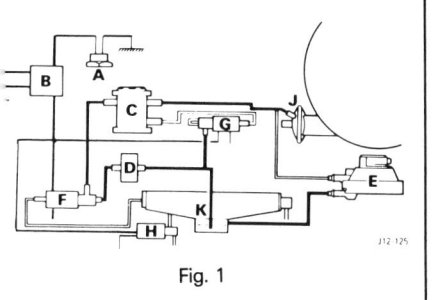

Fig. 1

Vacuum Advance System Component Location

A Water temperature switch
B Time delay module
C Vacuum regulator
D Vacuum delay valve
E Vacuum dump valve
F 3-way solenoid valve
G 2-way solenoid valve
H Solenoid air switch
J Distributor advance capsule
K Throttle edging tapping

The Vacuum System has FIVE modes

1. Idle — cold (below 38°C)
2. Light throttle — cold (below 38°C) below 4000 rpm
3. 'Vacuum Dumping' Operation
4. Idle — hot (above 45°C)
5. Light throttle — hot (above 45°C) below 4000 rpm

Ignition advance operation with engine coolant temperature below 38°C and below 4000 rpm

The water temperature sensor (A) energises the 3-way solenoid (F), the 2-way solenoid (G) via the time delay module (B) which 'holds' for fifteen minutes.

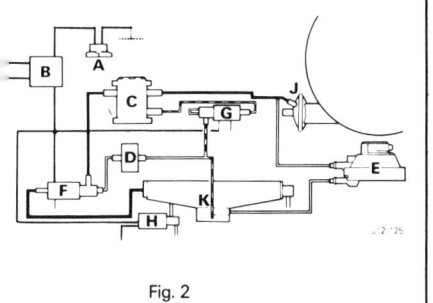

Fig. 2

1. Idle cold (Fig. 2)

During idle no vacuum is available at the throttle edge tapping (K) therefore there is no vacuum advance at idle when cold. To compensate for this, the solenoid air switch (H) is energised (open) for fifteen minutes to give a steady idle speed during this warm-up period, by drawing air directly from the air cleaner into inlet manifold. 'B' bank inlet manifold is fed by the engine balance pipe.

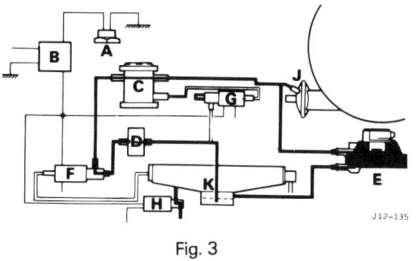

Fig. 3

2. Light throttle below 38°C below 4000 rev/min (Fig. 3)

The throttle edge tapping (K) sends a vacuum signal through the delay valve (D), the 3-way solenoid (F) and the regulator (C) to the distributor advance capsule (J) whilst the throttle is open. The lower port in regulator (C) is closed so that the regulator is inoperative. In this condition full advance is available and the vacuum is equal either side of the dump valve which remains closed.

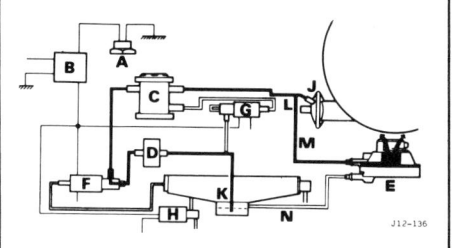

Fig. 4

3. Vacuum dumping, rapid acceleration and high engine speeds over 4000 rpm coolant below 38°C (Fig. 4)

The dump valve (E) has two vacuum connections, one to the throttle housing (N) on the engine side of the butterfly which acts on a diaphragm inside the valve, the other at the 'T' piece (L) between the regulator and the distributor vacuum capsule. This line (M) under certain circumstances will be vented to atmosphere.

When a vacuum is present at the throttle housing (N) the vacuum signal received at the dump valve (E) acts on the diaphragm which closes the valve to atmosphere. When the vacuum signal disappears the valve opens to atmosphere. Therefore, on rapid throttle openings, or high engine speed operations the vacuum signal is dumped to prevent the possibility of detonation or ignition knock.

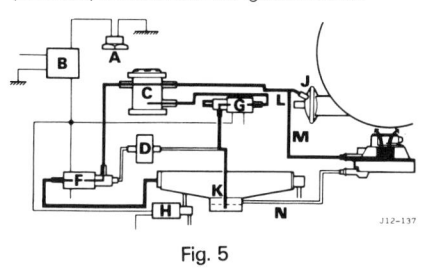

Fig. 5

Engine temperature above 45°C (Fig. 5)

The water temperature switch (A) de-energises the 3-way solenoid (F) and 2-way solenoid (G) and the solenoid air switch (H); the vacuum signal is now taken directly from the inlet manifold via the 3-way solenoid (F) and the regulator valve (C) to the advance capsule (J).

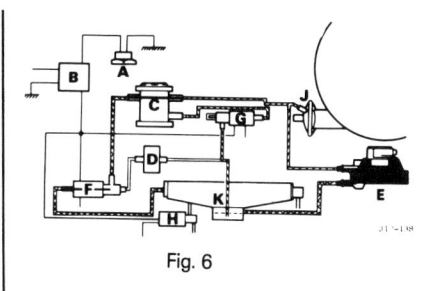

Fig. 6

4. Idle hot above 45°C (Fig. 6)

Atmospheric pressure is acting on the lower part of the regulator valve (C) via the 2-way solenoid (G) and through the throttle edge tapping (K). This allows the regulator to limit the depression on the vacuum advance capsule to 318 mmHg (12.5 Hg) to achieve a smooth idle condition by venting to atmosphere via (K) any vacuum signal in excess of 318 mmHg (12.5 Hg).

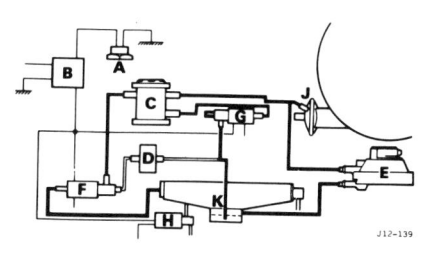

Fig. 7

5. Light throttle hot above 45°C and below 4000 rev/min (Fig. 7)

Throttle edge tapping (K) supplies a vacuum signal rather than atmospheric pressure to the lower part of the regulator. This signal removes the restriction from the regulator thus allowing full manifold depression to act upon the vacuum capsule (J).

At this point the vacuum is equal either side of the dump valve and so it remains closed.

17—1

EMISSION CONTROL

IGNITION VACUUM ADVANCE SYSTEM A AND C EMISSION V12 HE ENGINES — PRE 1983 M.Y.

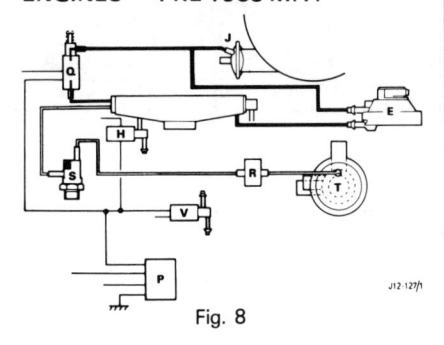

Fig. 8

- E   Dump valve
- J   Dist. advance capsule
- P   Timer/relay
- Q   Solenoid vacuum valve
- R   Delay valve
- S   Thermal valve
- T   Air switch valve
- V   Purge control valve

Under normal operating conditions (Fig. 8) the distributor advance vacuum signal is fed from the inlet manifold via solenoid vacuum valve (Q) to the distributor advance capsule (J).

However to obtain faster catalytic 'light off' this vacuum signal is inhibited for the first 45 seconds after the engine has started. This function is obtained by means of a timer/relay (P) and solenoid vacuum valve (Q).

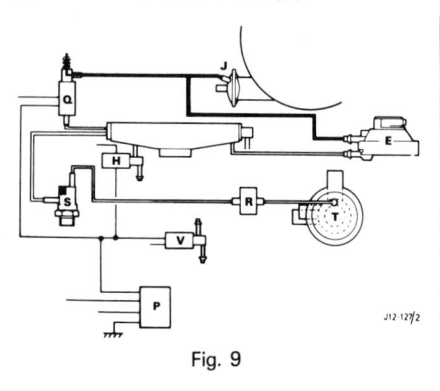

Fig. 9

When the solenoid vacuum valve (Q) is energised the distributor vacuum capsule (J) is ported to atmosphere for 45 seconds (Fig. 9) after which the timer/relay (P) cuts out, de-energises the solenoid vacuum valve (Q) and ports the distributor advance capsule to the inlet manifold. The timer/relay is reset and activated on each occasion that the starter motor is energised regardless of engine temperature.

The solenoid vacuum valve causes a restriction which can delay the destruction of the vacuum signal at the distributor when the throttle is opened rapidly. This condition may cause detonation. To overcome this problem a vacuum dump valve (E) is introduced into the system.

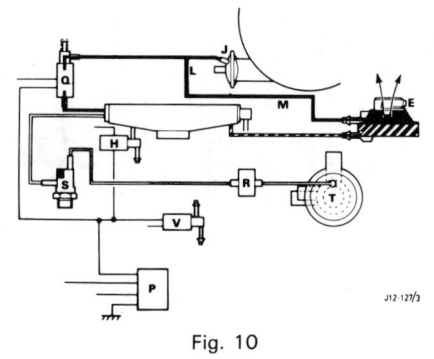

Fig. 10

Operation of dump valve (Fig. 10)

The dump valve (E) has two vacuum connections, one to the inlet manifold which acts on a diaphragm inside the valve, the other at the 'T' piece (L) between the solenoid vacuum valve (Q) and distributor vacuum capsule (J). This line (M) under certain circumstances is vented to atmosphere. When a vacuum is present at the inlet manifold the vacuum signal received at the dump valve (E) acts on the diaphragm which closes the valve to atmosphere. When the vacuum signal disappears the valve opens to atmosphere. Therefore on rapid throttle openings the vacuum signal is 'dumped' to prevent the possibility of detonation or ignition knock.

NOTE: Ignition timing

The ignition is 18° ± 2° B.T.D.C. at 3000 rpm with the vacuum pipe disconnected at the distributor and the engine at normal operating temperature.

Air injection switching system (Fig. 11)

Secondary air is supplied to the exhaust ports only during the engine warm up phase. When the engine coolant temperature reaches a pre-determined level, delivery of secondary air is switched from the exhaust ports to atmosphere. This function is achieved by means of a vacuum operated air switching valve, and the vacuum source is controlled by a thermal vacuum valve, located in the right-hand rear coolant outlet rail.

When the water temperature is below 38°C the thermal vacuum valve (S) allows air injection to the cylinder head exhaust ports via the air switching valve (T) when the water temperature reaches 45°C the thermal vacuum valve (S) closes. This causes the air switching valve (T) to switch secondary air from the exhaust ports to atmosphere (air cleaner).

A delay valve (R) is fitted between the thermal vacuum valve (S) and the air switching valve (T), to retain air injection to the exhaust ports during acceleration phases by inhibiting the loss of vacuum from the air switching valve (T).

Vacuum Component Location — Key to Fig. 11

- E   Dump valve
- J   Vacuum advance unit
- Q   Solenoid vacuum valve
- R   Delay valve
- S   Thermal valve
- T   Air switching valve
- V   Charcoal canister
- W   Purge control valve
- X   Pressure/vacuum relief valve
- Y   Pressure regulator (fuel system)
- Z   Vacuum switch (air conditioning)

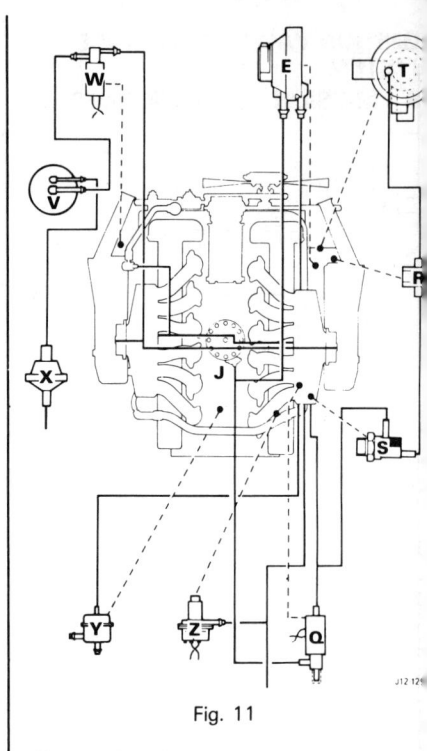

Fig. 11

Charcoal canister

A canister containing activated charcoal located in the right hand front wheel arch, is used to store hydrocarbon emmissions from the fuel tanks.
Filter pads are fitted above and below the charcoal to prevent the ingress of foreign matter into the charcoal or the passage of charcoal into the purge line. Emissions from the fuel tank enter the canister at the bottom and the purging air enters the canister at the top passing through the charcoal to the purge outlet also at the bottom.

Canister purging

Canister purging is obtained by connecting the purge pipe to a vacuum source and drawing a controlled quantity of air through the charcoal in the canister.

The purge depression is taken from ports in both throttle housings close to and upstream of the throttle butterflies. A restrictor fitted at the point where the purge pipe divides is used to control the maximum purge air flow rate. In order to inhibit canister purging until the catalytic convertors are 'alight' a purge control valve (W) is mounted in the purge line. The purge control valve is operated by the timer/relay such that when it is energised (closed), canister purging does not occur for the first 45 seconds after the engine has started.

Pressure relief valve

The pressure relief valve is used to control the transfer of vapour from the fuel vapour separators to the charcoal canister. The valve is designed to prevent the flow from the tanks until a pre-set pressure of 0,14 kgf/cm² (2 lbf/in²) is exceeded.

17—2

EMISSION CONTROL

ACCESS CHART

OPERATION NUMBER	DESCRIPTION	LOCATION	Fig. No.
17.55.01	Dump Valve — Remove and refit	Attached to RH Thermostat Housing	E, Fig. 1
17.55.02	Two Way Solenoid Valve — Remove and refit	Bolted to RH Throttle Housing	G, Fig. 1
17.55.03	Three Way Solenoid Valve — Remove and refit	Bolted below RH Inlet Manifold adjacent to Overrun valve.	F, Fig. 1
17.55.04	Delay Valve — Remove and refit	Located underneath RH Inlet Manifold	D, Fig. 1
17.55.05	Solenoid Air Switch — Remove and refit	Bolted to top of RH Inlet Manifold	H, Fig. 1
17.55.06	Vacuum Regulator — Remove and refit	Located below RH Inlet Manifold	C, Fig. 1
17.55.07	Time Delay Unit — Remove and refit	Located passenger side component board	B, Fig. 1
17.55.08	Water Temperature Switch — Remove and refit	In RH Rear Water Rail	A, Fig. 1

DUMP VALVE

Remove and refit   17.55.01

Removing

Carefully displace the dump valve from the bracket on the right-hand thermostat housing.
Note the position and disconnect the valve vacuum pipes.
Remove the valve.
On refitting ensure the correct connections of vacuum pipes.

TWO WAY SOLENOID VALVE

Remove and refit   17.55.02

Removing

Remove the bolt securing the valve to the right-hand throttle housing and displace the valve for access.
Note the position of the vacuum pipes and cable connections.
Disconnect and remove the valve.
On refitting ensure the pipes and cables are correctly connected.

THREE WAY SOLENOID VALVE

Remove and refit   17.55.03

Removing

Remove the right-hand air cleaner assembly.
Remove the bolt securing the valve and displace the valve for access.
Note the position of the vacuum pipes and cable connections.
Disconnect and remove the valve.
On refitting ensure the pipes and cables are correctly connected.

DELAY VALVE

Remove and refit   17.55.04

Removing

Remove the right-hand air cleaner cover and element note the position the vacuum pipes are fitted.
Disconnect the pipes and remove the valve.
On refitting ensure the vacuum pipes are fitted the correct way round.

SOLENOID AIR SWITCH

Remove and refit   17.55.05

Removing

Remove the bolt securing the air switch to the right-hand induction manifold.
Note the position of the vacuum hose and electrical connections.
Disconnect and remove the switch.
On refitting ensure the electrical connections are correct.

VACUUM REGULATOR

Remove and refit   17.55.06

Removing

Remove the right-hand air cleaner assembly.
Remove the securing bolt and displace the three-way solenoid valve.
Note and disconnect the vacuum pipes.
Remove the vacuum regulator.
On refitting ensure the correct connections of the vacuum pipes.

TIME DELAY UNIT

Remove and refit   17.55.07

Removing

Disconnect the battery.
Remove the passenger side dash liner.
Remove the nut securing the delay unit to the component panel.
Disconnect the block connector and remove the unit.
On refitting ensure the connector is secure.

WATER TEMPERATURE SWITCH

Remove and refit   17.55.08

Removing

Remove the right-hand air cleaner assembly.
Carefully remove and refit the system pressure cap to release any residual pressure from the cooling system.
Disconnect the electrical connections and remove the switch.
On refitting ensure the switch is tightened down.
Replenish the cooling system if necessary

EMISSION CONTROL

'A' & 'C' EMISSION — FEDERAL, CANADIAN & JAPANESE SPECIFICATION 1983 ON

The vacuum system for these markets was modified to incorporate a vacuum regulator similar to that previously used on vehicles to European specification. It must be emphasised that these regulators are NOT interchangeable.

Ignition Vacuum Advance System

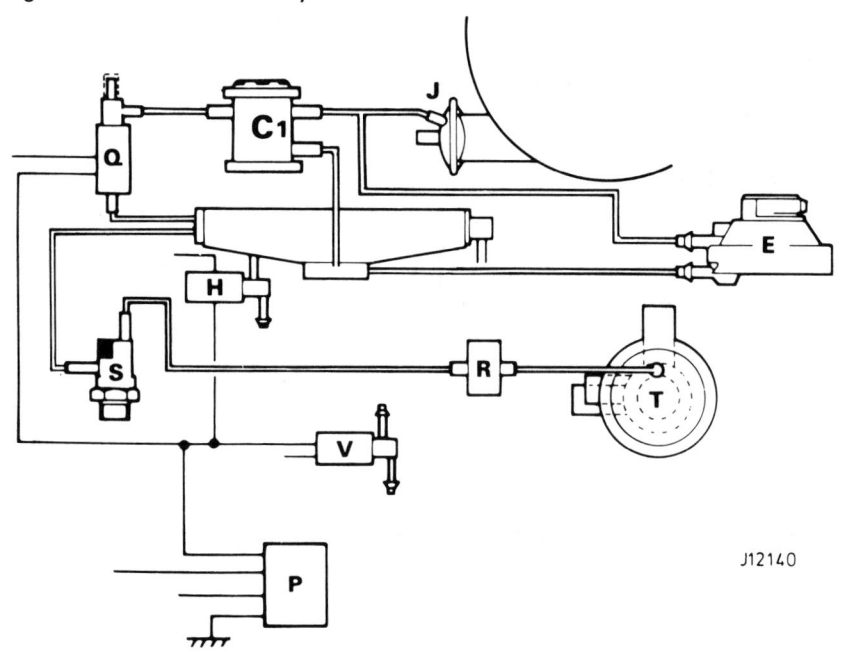

'A' & 'C' Emission Vacuum Advance System

E	Dump Valve
J	Distributor Advance Capsule
P	Timer Relay
Q	Solenoid Vacuum Valve
R	Delay Unit
S	Thermal Valve
T	Air Switching Valve
V	Purge Control Valve
C1	Vacuum Regulator

The vacuum advance system has three modes of operation:

1. For the first 45 seconds after starting.
2. Idle — after first 45 seconds.
3. Vacuum dumping.

Fig. 12

First 45 Seconds After Starting (Refer to Fig. 13)

To obtain faster catalytic "light off" the vacuum signal is inhibited for the first 45 seconds after the engine has started. This function is obtained by means of a timer/relay (P) and solenoid vacuum valve (Q).
When the solenoid vacuum valve (Q) is energised the distributor vacuum capsule (J) is ported to atmosphere for 45 seconds (Fig. 2) after which the timer/relay (P) cuts out, de-energises the solenoid vacuum valve (Q) and ports the distributor advance capsule to the inlet manifold. The timer/relay is reset and activated on each occasion that the starter motor is energised regardless of engine temperature.

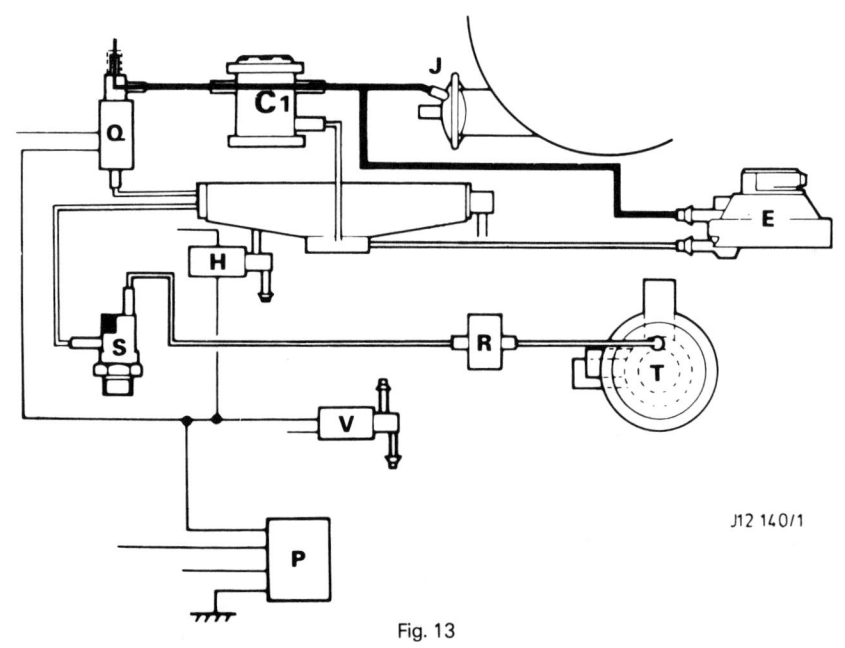

Fig. 13

17—4

EMISSION CONTROL

Idle — After First 45 Seconds (Refer to Fig. 14)

After the time relay (P) de-energises the solenoid vacuum valve (Q). The distributor capsule (J) is ported to the inlet manifold to give vacuum advance. However, to obtain a more efficient idle condition a vacuum regulator (C1) has been introduced into the system, located between the solenoid vacuum valve (Q) and the distributor advance capsule (J). The lower port of the vacuum regulator (C1) being sourced to the air cleaner side of the throttle butterfly. Under idle conditions atmospheric pressure operates on the vacuum regulator (C1) thereby limiting the vacuum available to the distributor advance capsule (J) to 0.372 ± 0.033 bars (11.0" ± 1" Hg).

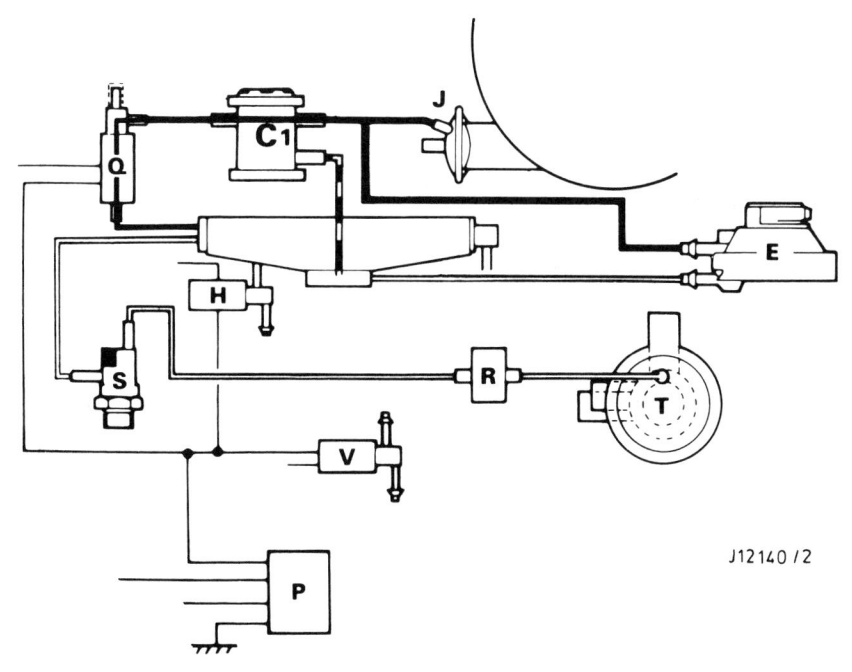

Fig. 14

Vacuum Dumping

The vacuum regulator (C1) and solenoid vacuum valve (Q) cause a restriction in the system which can delay the destruction of the vacuum signal when the throttle is opened rapidly. This condition can cause detonation. To overcome this problem a vacuum dump valve (E) is introduced into the system.

EMISSION CONTROL

Operation of Dump Valve (Refer to Fig. 15)

The dump valve (E) has two vacuum connections, one to the inlet manifold which acts on a diaphragm inside the valve the other at the 'T' piece (L) between the vacuum regulator (C1) and the distributor vacuum capsule (J). This line (M) under certain circumstances is vented to atmosphere. When a vacuum is present at the inlet manifold the vacuum signal received at the dump valve (E) acts on the diaphragm which closes the valve to atmosphere. When the vacuum signal disappears the valve opens to atmosphere. Therefore on rapid throttle openings the vacuum signal is 'dumped' to prevent the possibility of detonation.

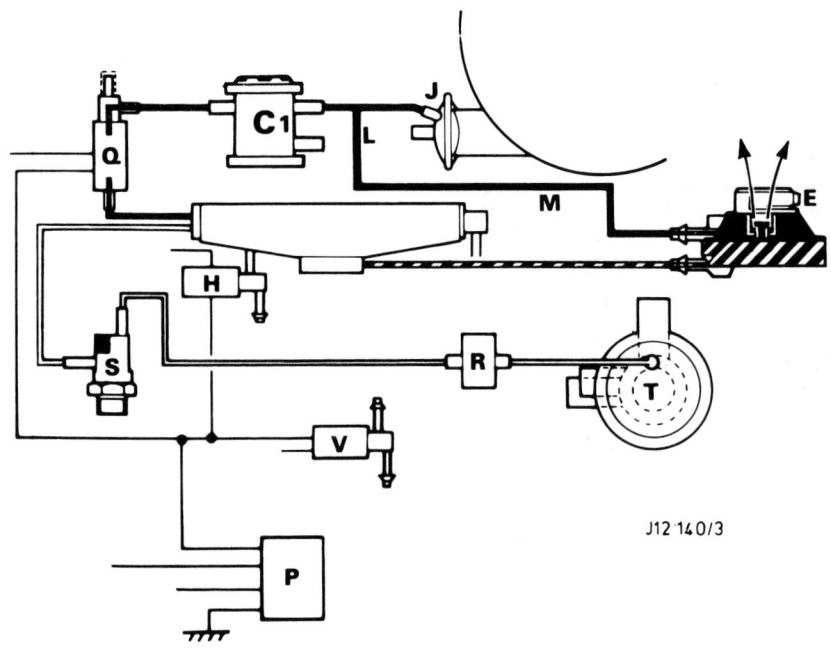

Fig. 15

Air Injection Switching System

Secondary air is supplied to the exhaust ports only during the engine warm up phase. When the engine coolant temperature reaches a pre-determined level, delivery of secondary air is switched from the exhaust ports to atmosphere. This function is achieved by means of a vacuum operated air switching valve, and the vacuum source is controlled by a thermal vacuum valve, located in the right hand rear coolant outlet rail.

EMISSION CONTROL

'D' EMISSION — AUSTRALIAN SPECIFICATION

Ignition Vacuum Advance System

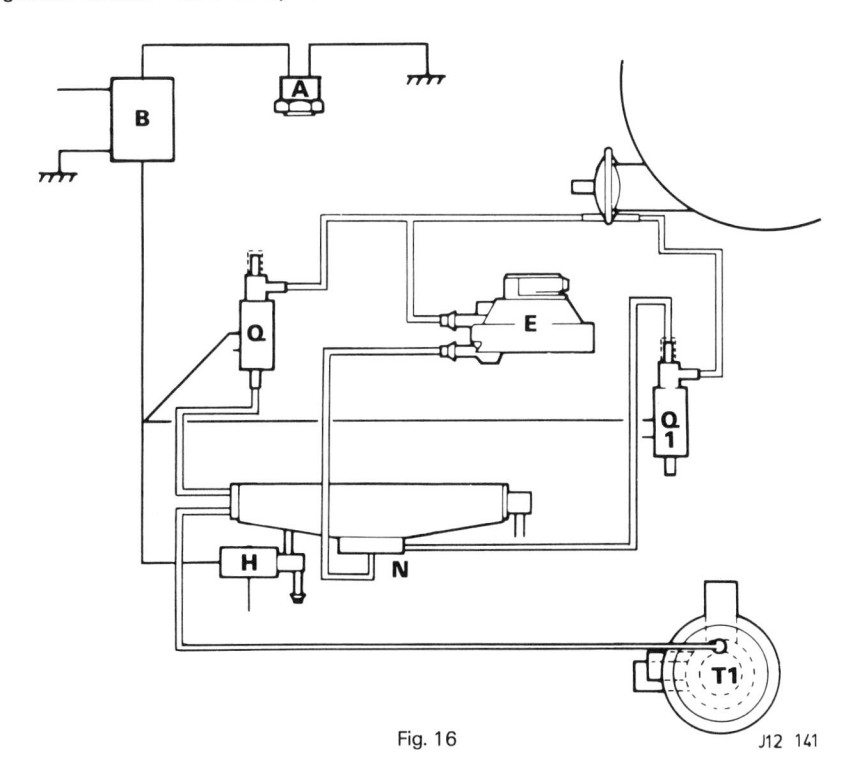

Fig. 16

'D' Emission Vacuum Advance System

- A   Water Temp. Switch
- B   Time Delay Module
- E   Vacuum Dump Valve
- Q   2 Way Solenoid Valve
- H   Solenoid Air Switch
- Q1   2 Way Solenoid Valve
- T1   Diverter Valve

The 'D' emission vacuum system has three modes of operation.

1. Idle — cold (below 38°C)
2. Idle — hot (above 45°C)
3. Vacuum dumping

Idle Cold — Below 38°C and 4000 rev/min (Refer to Fig. 17)

The water temperature sensor (A) energises both two-way solenoids (Q) and (Q1) via the timer relay (B) which holds for fifteen minutes. The two-way solenoid (Q), vents the distributor advance capsule to atmosphere so inhibiting vacuum advance and two-way solenoid (Q1), directing the inlet manifold vacuum to react on the retard side of the distributor advance/retard capsule giving approximately 6 degrees of retardation (i.e. 6 degrees ATDC).

As a consequence of the ignition setting and low CO level, sufficient air has to be introduced into the engine to maintain a satisfactory idle. A solenoid air switch (H) achieves this by directing air from the air cleaner into the inlet manifold, bypassing the throttle butterfly.

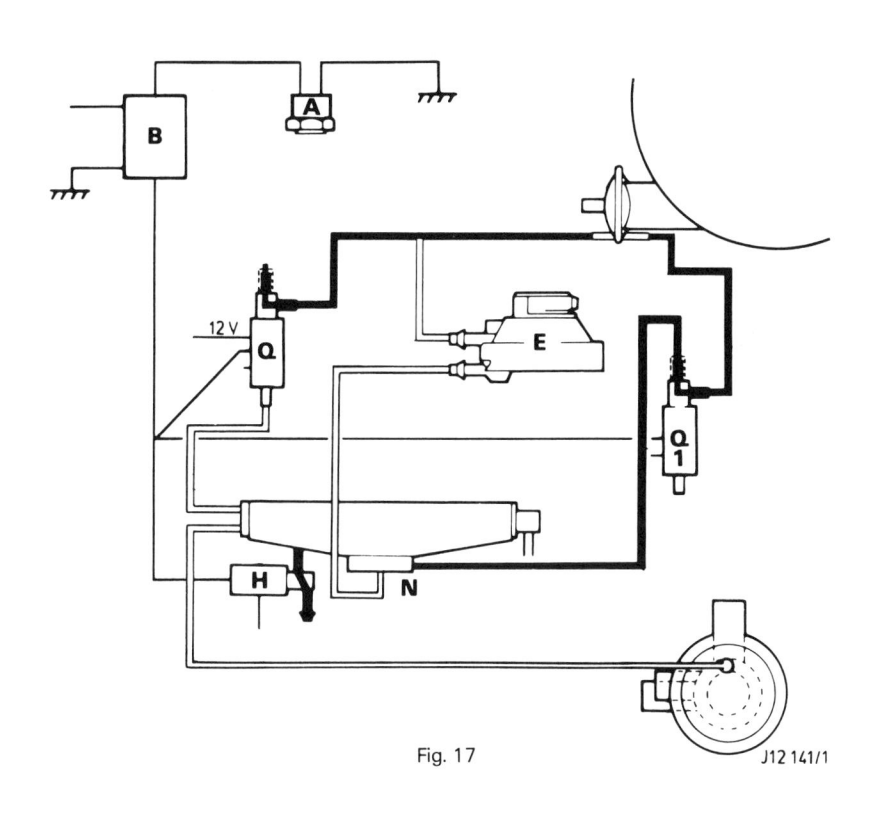

Fig. 17

17—7

EMISSION CONTROL

Idle Hot — Above 45°C (Refer to Fig. 18)

After fifteen minutes the timer/relay (B) de-energises the solenoids (Q) and (Q1) this removes the vacuum retard and allows manifold vacuum to react on the advance side of the distributor vacuum capsule. Additional air is no longer required therefore the solenoid air switch (H) is also de-energised by the timer/relay (B).

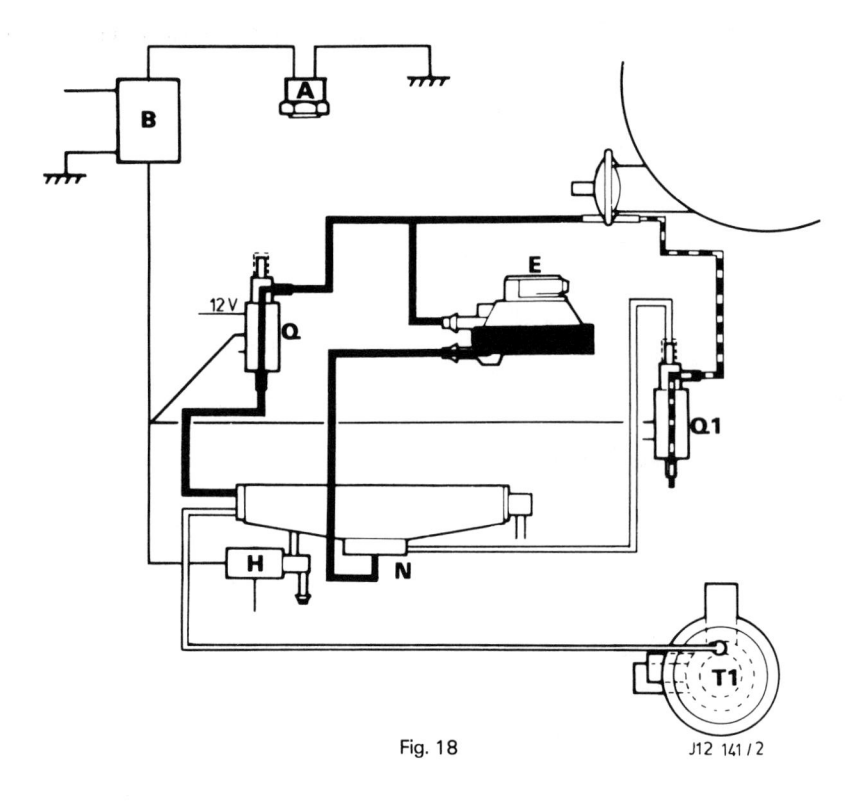

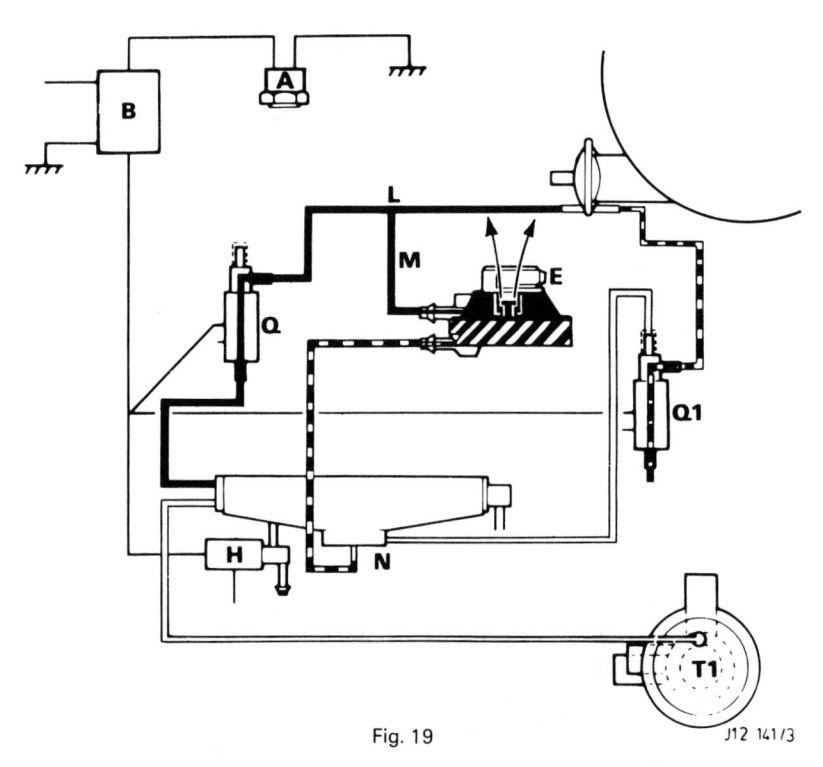

Fig. 18

Vacuum Dumping — Rapid Acceleration & High Engine Speeds Over 4000 rev/min (Refer to Fig. 19)

The dump valve (E) has two connections one to the throttle housing on the engine side of the butterfly which acts on a diaphragm inside the valve, the second at the 'T' piece (L) between the solenoid (Q) and the distributor advance/retard capsule.

This line (M), under certain conditions, will be vented to atmosphere.

When a vacuum is present in the throttle housing (N) the vacuum signal received at the dump valve (E) acts on a diaphragm which closes the valve to atmosphere. When the vacuum signal disappears the valve opens to atmosphere.

Therefore on rapid throttle openings or high engine speeds the vacuum signal is dumped to prevent the possibility of detonation.

Air Injection

If air injection was continuous at all times, then particularly during the warm up period with the engine at high rev/min on the overrun it would backfire through the exhaust system.

To prevent this, a vacuum sensed diverter valve (T1) is incorporated in the system. This allows the air, normally used for injection, to be diverted to atmosphere under conditions of high vacuum.

Fig. 19

17—8

EMISSION CONTROL

ADDITIONAL ENGINE — FAULT DIAGNOSIS 1983 ON

Should a malfunction with the distributor or vacuum system be suspected, e.g. poor performance or poor idle, check against the following values.

	Condition	Specification	Ignition Setting
1.	3000 rev/min Vac. pipe disconnected	All Models	18 degrees BTDC
2.	500 rev/min ± 25 rev/min Vac. pipe disconnected	All Models	1.5 degrees ATDC to 3.5 degrees BTDC

3. If the above setting (2) is not recorded, the mechanical distributor advance mechanism is suspect.

	Check vacuum at distributor capsule at idle	USA, Canadian and Japanese models	0.372 ±0.033 Bars (11" ± 1" Hg) Taken after 45 seconds engine running time
		European models	0.422 ± 0.033 Bars (12.5" ± 1" Hg) Taken after 15 minutes engine running time
		Australian models	Not applicable vacuum regulator not fitted

If high — suspect:
(a) Vacuum regulator
(b) No vacuum bleed to atmosphere from lower part of vacuum regulator

If low — suspect:
(c) Leaking vacuum regulator
(d) Vacuum leak

If correct — proceed to check 4.

4.	500 rev/min Vac. pipe connected	USA, Canadian and Japanese models	19 degrees BTDC ± 7 degrees After 45 seconds engine running time
		European models	21.5 degrees BTDC ± 6.5 degrees After 15 minutes running time
		Australian models	17.5 degrees BTDC ± 4.5 degrees After 15 minutes running time

The tolerances in (4) are theoretically based on those achievable in checks (2) and (3), if not correct, suspect distributor vacuum capsule.

FUEL INJECTION SYSTEM

THE ELECTRONIC FUEL INJECTION SYSTEM

Digital 'P' Pressure Sensing Type

Description

The electronic fuel injection system is divided into two sub-systems interconnected only at the injectors.
The systems are:
1. The fuel system delivering to the injectors a constant supply of fuel at the correct pressure, 2,5 bar (38 lbf in).
2. The electronic sensing and control system which monitors engine operating conditions of load, speed, temperature (coolant; and induction air) and throttle movement. The control system then produces electrical pulses of appropriate width to hold open the injector solenoid valves and allows the correct quantity of fuel to flow through the nozzle for each engine cycle.

As fuel is held constant, varying the pulse width increases or decreases the amount of fuel passed through the injector to comply with the engine requirements.
Pulse width and therefore fuel quantity is also modified to provide enrichment during starting and warming-up, at closed throttle, full throttle and while the throttle is actually opening.
The injectors are operated by the Electronic Control Unit (ECU) in two groups of six. Each is further broken into two sub-groups of three, although each pair of sub-groups is operated simultaneously to make up the two groups of six twice per engine cycle.

Injection in two groups of six:

A Bank			B Bank		
1A	3A	5A	1B	3B	5B
2A	4A	6A	2B	4B	6B

Firing order:
1A 6B 5A 2B 3A 4B 6A 1B 2A 5B 4A 3B

Cylinders numbered from front of engine.

The induction system is basically the same as that on a carburetted engine: tuned ram pipe, air cleaners, plenum chambers and induction ports. Air is drawn through paper element cleaners to a butterfly valve for each bank and to individual ports for each cylinder leading off the plenum chamber. The injectors are positioned at the cylinder head end of each port so that fuel is directed at the back of each inlet valve.

	A	(1)	(2)	(3)	(4)	(5)	(6)
Front							
	B	(1)	(2)	(3)	(4)	(5)	(6)

(1) Inlet opens 13° B.T.D.C.
(2) Inlet closes 55° A.B.D.C.
(3) Ignition with engine at normal operating temperature
  U.K./Europe 24° B.T.D.C.
  All others 25—27° B.T.D.C.
(4) Exhaust opens 55° B.B.D.C.
(5) Exhaust closes 13° A.T.D.C.

North American — Emission A   Fig. 1

19—1

FUEL INJECTION SYSTEM

Fig. 2

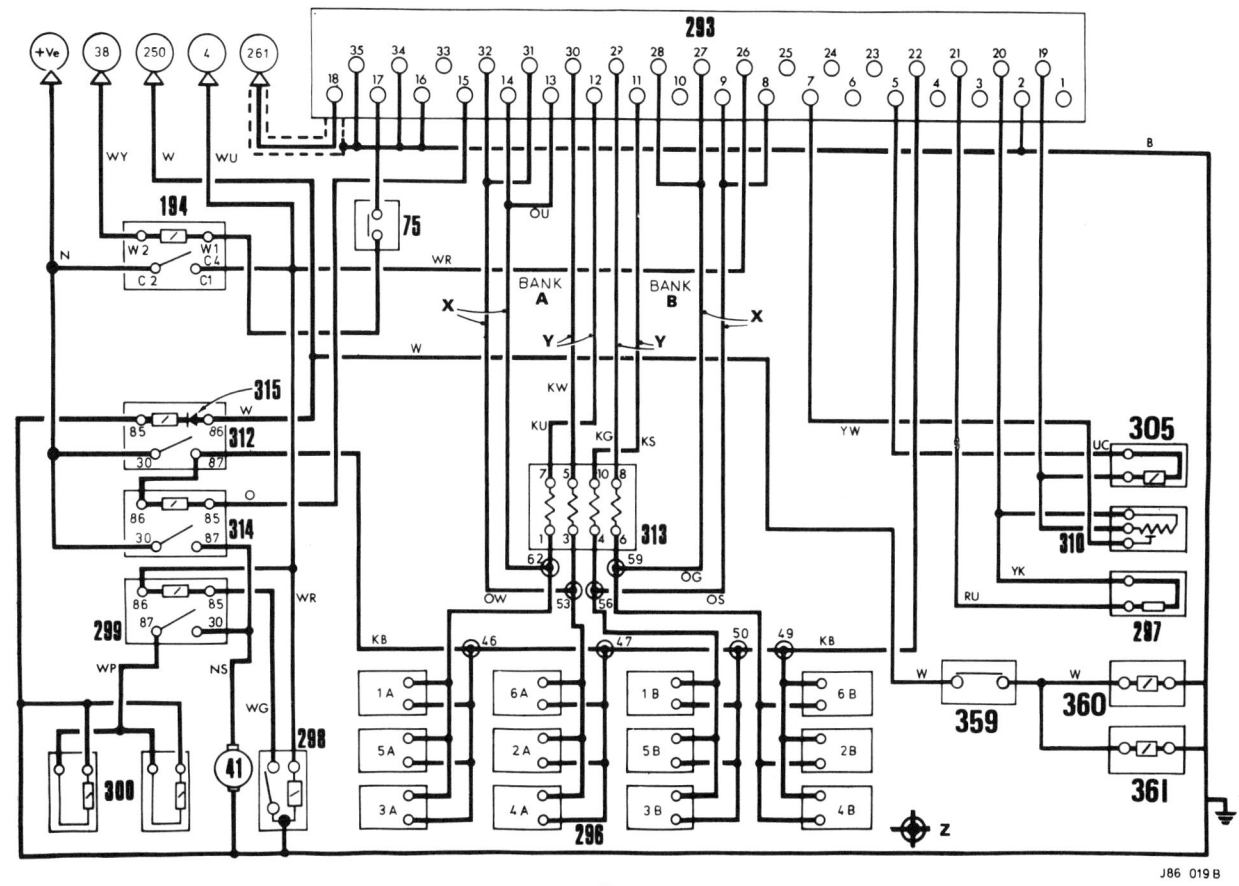

U.K. & Europe — Emission B

MASTER KEY FOR CIRCUITS EMISSION A AND B

	4.	Starter solenoid
	38.	Ignition switch pin 1
	38.	Ignition switch pin 3
	39.	Amplifier
	41.	Fuel pump
	75.	Start inhibit switch
	140.	Fuel change-over switch
	261.	Amplifier
	194.	Starter relay
	250.	Inertia switch
A	293.	E.C.U.
&	296.	Injectors
B	297.	Air temperature sensor
	298.	Thermotime switch
	299.	Cold start relay
	300.	Cold start injectors
	305.	Coolant sensor
	310.	Throttle potentiometer
	312.	Main relay
	313.	Power resistors
	314.	Pump relay
	315.	Block diode

A	316.	(Oxygen) sensors
A	326.	Vacuum switch
A	349.	Micro-switch
A	353.	Feed-back monitor socket
A	354.	Disable socket
	355.	Feedback monitor relay
B	359.	Oil temperature switch
B	360.	Vacuum changeover switch
B	361.	Supplementary air valve
X		Turn on
Y		Hold on
Z		Part of 10 way engine harness

19—2

FUEL INJECTION SYSTEM

ELECTRONIC FUEL INJECTION

Electronic control unit (E.C.U.)

PARAMETERS CONTROLLED Stoichiometric (air/fuel) ratio

	PARAMETERS SENSED	SOURCE	RESULT
Primary (Digital)	1. Inlet manifold absolute pressure 2. Engine speed	Sensor in the E.C.U. Igntion coil	
Secondary (Analogue)	3. Throttle movement and position 4. 'Closed loop' correction 5. Supply voltage 6. Oxygen partial pressure in exhaust	Potentiometer E.C.U. Oxygen sensors	Triggers injector pulses
	7. Full load	Vacuum switch	Opens 'closed loop'
	8. Engine coolant temperature 9. Starter signal 10. Intake air temperature	Sensor Starter relay Sensor	Triggers cold start pulses

The system contains an integrated circuit for the dedicated fuel injection control chip and an analogue/digital converter for the intake manifold depression signal.

The fuelling information is stored in a (ROM) Read Only Memory (1288-bit words) so that for any combination of manifold pressure and speed the memory gives a number proportional to the amount of fuel required by the engine. The injectors will be energised for a time proportional to the number computed, plus the constant of the proportionality varied according to the secondary parameters.

To ensure a fast opening of the injectors the full battery voltage is applied but, to reduce power dissipation when they are open, drive is maintained via a resistor pack external to the ECU.

Fuel Supply

Fuel is drawn from the small sealed sump tank (1) at rear of the car by a fuel pump (2) via a non-return valve (3) to a fuel rail through an in-line filter (4) and a pressure regulator (5). Fuel is controlled so that the pressure drop across the injector nozzle is maintained at a constant 2,5 bar (36.25 lbf/in²). Excess fuel is returned to the tank via a fuel cooler (6). The twelve fuel injectors (7) are connected to the fuel rail (8) and are electro-mechanically operated to inject into each inlet port. Fuel is also supplied to a cold start injector (9) which is only operated during the starting of a cold engine.

Fuel Control

The metering of fuel is controlled by regulating the time that the injectors are open during each engine cycle. The main criteria that govern the fuel requirements of the engine are the air consumption which reflects the engine load, and the engine speed. Other operating conditions such as engine coolant and air intake temperatures modify fuel requirements. An Electronic Control Unit (ECU) receives information from sensors placed about the engine and computes the quantity of fuel required and thus the time that the injectors remain open. The ignition LT circuit switch triggers a group of injectors at every third spark, alternating between engine banks, each time.

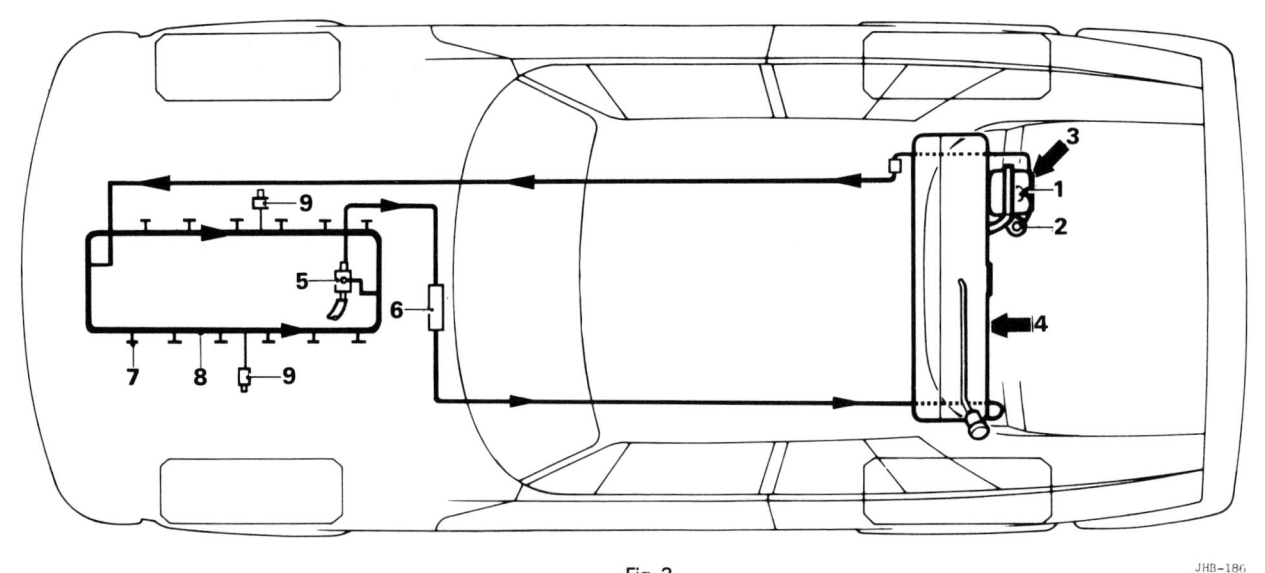

Fig. 3

Key to Fig. 3
1. Sump pump
2. Fuel pump
3. Non-return valve
4. Fuel filter
5. Pressure regulator
6. Fuel cooler
7. Injectors
8. Fuel rail
9. Cold start injector

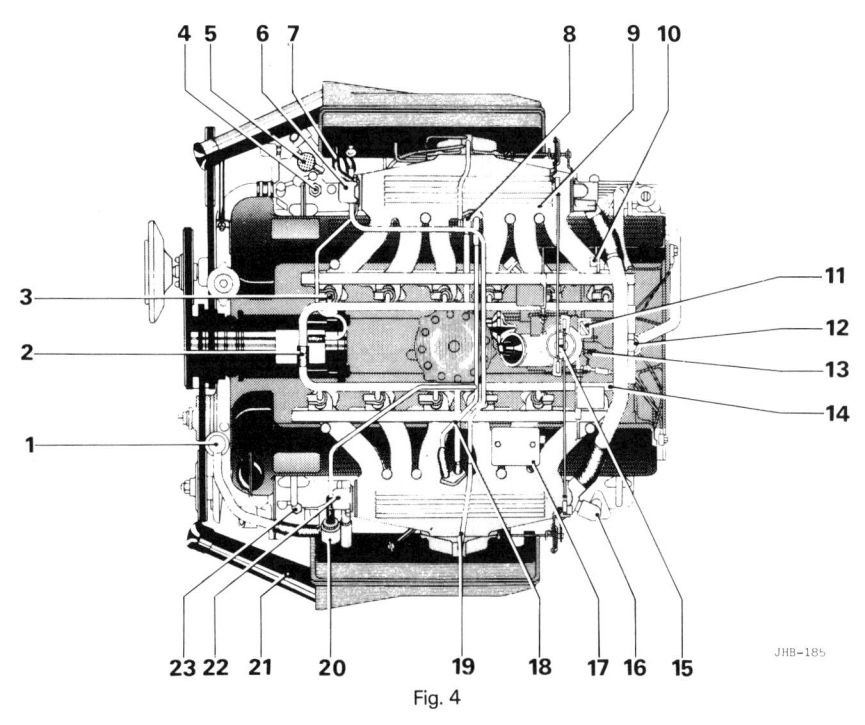

FUEL INJECTION SYSTEM

Key to Fig. 4

1. Crankcase breather
2. Fuel rail
3. Fuel injector
4. Thermo-time switch
5. Over-run valve
6. Vacuum dump valve
7. Supplementary air valve
8. Cold start injectors
9. Solenoid vacuum valve
10. Vacuum switch
11. Throttle position switch
12. Check valve (NAS cars only)
13. Fuel pressure regulator
14. Air injection distributor rail (NAS cars only)
15. Throttle potentiometer (below throttle pulley)
16. Extra air valve
17. Ignition system amplifier
18. Crankcase vent
19. Canister purge (NAS cars only)
20. PCV valve
21. Air temperature sensor
22. Over-run valve
23. Coolant temperature sensor

Fig. 4

Fuel Control System

The pulse duration is varied by the electronic control unit (1, Fig. 5) according to inputs from engine and chassis mounted sensors. The control parameters sensed fall into two groups. The first being the intake manifold absolute pressure and the engine speed. The second consists of the coolant temperature, intake air temperature, throttle movement and postion, exhaust gas oxygen content, battery voltage, and the starter signal.

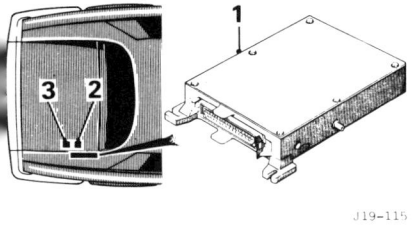

Fig. 5

Cranking Enrichment

The ECU increases the pulse width during engine cranking, in addition to any enrichment due to the coolant temperature sensor or the cold start injectors, this increase reduces slightly when cranking stops, but falls to normal after a few seconds. This temporary enrichment sustains the engine during initial running.

Flooding Protection System

When the ignition is switched on, but the engine is not cranking, the fuel pump will run for two seconds to raise the pressure in the fuel rail; it is then automatically switched off by the ECU. Only after cranking has started is the fuel pump switched on again. Switching control is built into the ECU circuitry. This system prevents flooding if any injectors become faulty (remain open) when the ignition is left on.

Engine Load Sensing

The driver controls engine power output by varying the throttle opening and therefore the flow of air into the engine. The air flow determines the pressure that exists within the plenum chamber, this pressure therefore is a measure of the demand upon the engine. The pressure is also used to provide the principle control of fuel quantity, being converted by the manifold pressure sensor in the ECU into an electrical signal. This signal varies the width of the injector operating pulse as appropriate. The pressure sensor is fitted with a separate diaphragm system that compensates for ambient barometric variations. The Manifold Pressure Sensor is located in the ECU and connected by a pipe to the inlet manifold balance pipe.

Full Load Fuelling — Cars to USA

To obtain maximum engine power it is necessary to inhibit the 'closed loop' system and simultaneously increase the fuelling level. This is obtained by using a vacuum operated electrical switch (1, Fig. 6) sensing inlet manifold depression, and a micro-switch (2, Fig. 6) operated by the throttle pulley spindle. These two switches are wired in parallel, so that either or both can signal the need for full load fuelling.

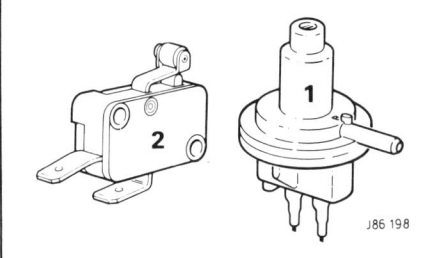

Fig. 6

Vacuum Switch

The contacts of this switch are operated by a spring loaded diaphragm in a chamber. This senses inlet manifold vacuum such that, when the depression falls below a certain value e.g. when the engine is operating at low speed, with part throttle near full load condition then the contacts close.

The closing of the contacts causes the fuel system to go 'Open loop' and simultaneously introduce a fuel enrichment of 12%.

Cold Start System

For cold starting, additional fuel is injected into the inlet manifolds by two cold start injectors (8, Fig. 4). These are controlled by the cold start relay (1, Fig. 7) and the Thermotime switch (4, Fig. 4). The Thermotime switch senses coolant temperature and, depending on what temperature, makes or breaks the ground circuit of the relay.

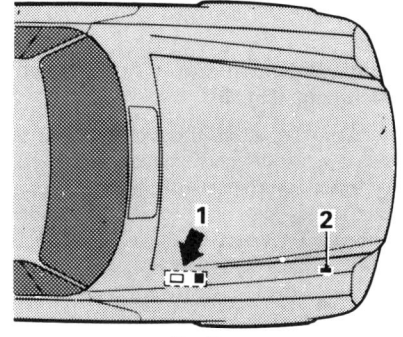

Fig. 7

When the starter switch is operated, the cold start relay is energised with its circuit completed via the Thermotime switch, which also limits the time for which the relay is energised, to a maximum of twelve seconds under extreme cold conditions.

The enrichment is in addition to that provided by the coolant temperature sensor.

If the temperature is above the rated value 35°C (95°F) of the Thermotime switch it will not operate at all, no starting enrichment being required.

19—4

FUEL INJECTION SYSTEM

Temperature Sensors

The temperature of the air taken into the engine through the inlet manifold and the temperature of the coolant in the cylinder block are constantly monitored. The information is fed directly to the ECU. The air temperature sensor (21, Fig. 4) has a small effect on the injector pulse width, and should be regarded as a trimming rather than a control device. It ensures that the fuel supplied is directly related to the weight of air drawn in by the engine. Therefore, as the weight (density) of the air charge increases with a falling temperature, so the amount of fuel supplied is also increased to maintain optimum fuel/air ratio.

The coolant temperature sensor (23, Fig. 4) has a greater control although it functions mainly while the engine is initially warming-up. The sensor operates in conjunction with the cold start system and the auxiliary air valve to form a completely automatic equivalent of the carburetter choke.

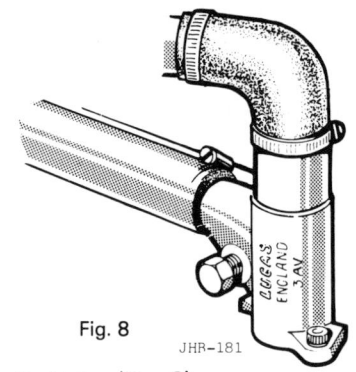

Fig. 8

Extra Air Valve (Fig. 8)

The Extra Air Valve is controlled by coolant temperature. To prevent stalling at cold start and cold idle conditions due to the increased drag of the engine, the valve opens to allow air to by-pass the throttles and so increase engine speed.

In addition to the main coolant temperature regulated air passage, the auxiliary air valve has a by-pass controlled by an adjusting screw. The screw controls the idle speed by regulating the airflow.

Throttle Potentiometer — Acceleration Enrichment (Fig. 9)

To ensure the vehicle road performance is satisfactory, with good throttle response, acceleration enrichment is necessary. Appropriate signals are provided by the throttle potentiometer, which is mounted on the throttle spindle and indicates the throttle position to the ECU.

(a) By opening the throttle very quickly all the injectors are simultaneously energised for one pulse. This ensures that there is enough fuel available at the inlet ports for the air admitted by the sudden opening of the throttle, and gives a smooth quick response to the throttle. The duration of this extra pulse is controlled by the engine temperature signals.

(b) By lengthening the normal injection pulse. This is done in proportion to the rate at which the throttle is moved, and it takes a short time to decay when the throttle movement stops. Enrichment this way is also varied according to the engine temperature.

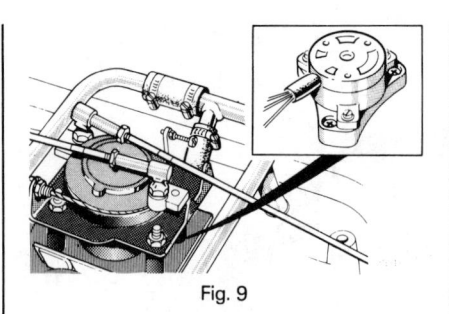

Fig. 9

Throttle Position Switch

This micro-switch is mounted such that its contacts are closed when the throttle is opened beyond a certain position; this condition has the same effect as the closing of the contacts of the full load vacuum switch. The throttle position switch is required to provide full load fuelling when the vehicle is operating under high speed full load running conditions and the mainfold depression is unable to operate the vacuum switch.

Lambda Sensors (Oxygen Sensors) — Cars to USA

The Lambda sensors (one fitted in each exhaust manifold) measure the oxygen concentration in the exhaust system. Excessive free oxygen over a certain proportion indicates a weak mixture, whereas insufficient free oxygen indicates a rich mixture. A signal is fed to the ECU to compensate for these variations by revising the applied injector pulse width. The Lambda sensors are initially set using the Lucas EFI Feedback Monitor Unit. A service interval counter in conjunction with a warning lamp indicates when the Lambda sensors must be replaced.

Fuel Pressure Regulator

The fuel pressure regulator (13, Fig. 4) maintains a constant pressure drop across the injector nozzles. It is connected to mainifod depression which operates against a spring loaded diaphragm.

Fuel Recirculation System

A regulator maintains the correct fuel pressure in the fuel rail. Fuel in excess of the required quantity is returned to the fuel tank.

Fuel Cut-Off Inertia Switch

The electrical supply to the fuel pump is passed through an inertia sensitive switch.

Should the car be subjected to heavy impact forces, the switch opens, isolating the pump and ensuring that fuel is not spilled in a potentially dangerous situation.

The switch is located in the drivers footwell on the front door pillar and can be reset after operation by pressing in the button on top of the switch.

Relays (Fig. 10)

Four relays are used in the electrical control system. They are the main, fuel pump, cold start, and the feed back monitor relays. Details of wiring connections are shown in the circuit diagram (Fig. 10).

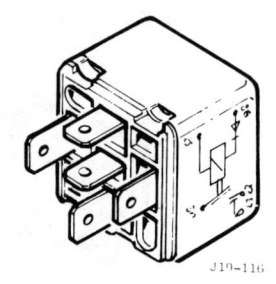

Fig. 10

Good Practice

The following instructions must be strictly observed:

1. Always disconnect the battery before removing any components.

2. Always depressurise the fuel system before disconnecting any fuel pipe (see 19.50.02).

3. When removing fuelling components always clamp fuel pipes approximately 38 mm (1.5 in) from the unit being removed. Do not overtighten clamp.

4. Ensure that a supply of cloth is available to absorb any spillage that may occur.

5. When reconnecting electrical components, always ensure that good contact is made by the connector before fitting covers.

6. Always ensure that ground connections are made to clean, bare metal, and are tightly fastened, using correct screws and washers.

Maintenance

There is no routine maintenance laid down for the electronic fuel injection system other than that at all service intervals, the electrical connections must be checked for security. The fuel filter must be discarded and a replacement fitted at 50,000 miles (80.000 km) intervals, when the air cleaner must also be renewed.

WARNING:

1. Do not run the engine with the battery disconnected.

2. Do not use a high-speed battery charger as a starting aid.

3. When using a high-speed battery charger to charge the battery, the battery MUST be disconnected from the vehicle's electrical system.

4. Ensure that the battery is connected with correct polarity.

5. No battery larger than 12V may be used.

FUEL INJECTION SYSTEM

LAMBDA SENSORS
Test Procedure

Check the ignition timing and idle speed.
Run the engine until it reaches normal operating temperature.
Disconnect the Lambda plug from the harness socket (1, Fig. 11) otherwise the Lambda sensors will not function while the vehicle is in 'Neutral' or 'Park'.

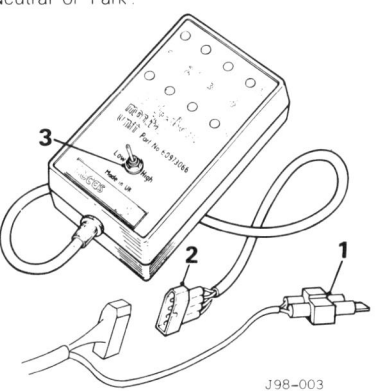

Fig. 11

Connect the Feedback Monitor Unit to the fuel setting diagnostic plug (2, Fig. 11).
Position the Monitor Unit switch (3, Fig. 11) to 'Low'.
With the engine running, bulb 2 in either row, together with bulb 2 or 3, in the other row, should glow.

NOTE: If the readings are incorrect adjust the fuel settings to comply with the vehicle emission regulations.

Disconnect the pressure regulator vacuum pipe and temporarily seal the vacuum take-off in the manifold.
With the engine running, the monitor unit indicator bulbs, should show that a 'Richening up' of the exhaust gases, has taken place i.e. bulb(s) 2 should extinguish and bulb(s) 1 glow, and/or bulb 3 extinguishes and bulb 1 or 2 glows.
If the Monitor unit indicator bulbs do not change, check the electrical circuit between the Lambda sensors and the ECU. If satisfactory, suspect the Lambda sensor(s) and/or ECU. Re-connect the pressure regulator vacuum pipe.
Disconnect the Feedback Monitor Unit and make good all connections.

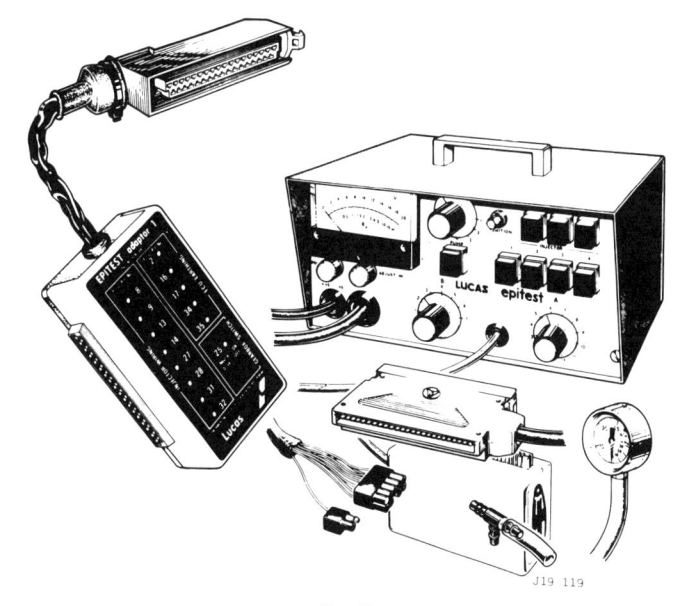

Fig. 13

LUCAS POTENTIOMETER ADJUSTER

Remove the throttle linkage from the engine.
Slacken the potentiometer fixing screws (1, Fig. 12).

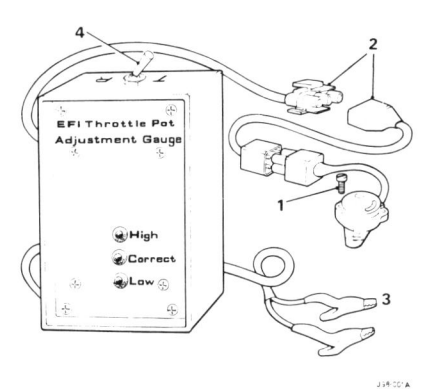

Fig. 12

Connect the adjuster to the potentiometer (2, Fig. 12).
Connect the potentiometer gauge to the battery (3, Fig. 12).
Move the switch to 'T' (4, Fig. 12).
Adjust by rotating the potentiometer to right or left until CORRECT lamp is lit.
Re-tighten the fixing screws.
Refit the throttle linkage.

Test Equipment

The 'Epitest' (1, Fig. 13) and the 'Epitest Adaptor' (2, Fig. 13) test equipment has been developed to enable quick location of faults to be carried out on the vehicle. It is supplied complete with the necessary multiplugs, fuel pressure gauge and the operating instructors.

FUEL INJECTION SYSTEM

ACCESS CHART

Operation No.	Description	Access	Fixing	Location	Diagrams 1 and 2
19.10.01	Air Cleaner (Left-hand)	Engine	Clips and Bolts		
19.10.02	Air Cleaner (Right-hand)	Engine	Clips and Bolts		
19.10.08	Air Cleaner Element	Engine	Clips		
19.20.01	Throttle Pedal	Foot Well	Bolts		
19.20.03	Throttle Pedestal	Engine	Nuts		
19.20.06	Throttle Cable	Engine	Clips		
19.20.16	Auxiliary Air Valve	Engine	Screw	16, Fig. 4	
19.20.22	Over Run Valve	Engine	Screws	22, Fig. 4	
19.22.09	Inertia Switch	Drivers Side Dash Liner	Screws		250
19.22.18	Coolant Temperature Sensor	Engine	Screw In	23, Fig. 4	305
19.22.20	Thermotime Switch	Engine	Screw In	4, Fig. 4	298
19.22.22	Air Temperature Sensor	Engine	Screw In	21, Fig. 4	297
19.22.31	Cold Start Relay	Engine Compartment	Securing Tag	1, Fig. 7	299
19.22.34	Electronic Control Unit	Luggage Compartment	Securing Screws	1, Fig. 5	293
19.22.35	Throttle Potentiometer	Engine	Securing Screws	15, Fig 4	310
19.22.39	Pump Relay	Luggage Compartment	Securing Tag	3, Fig. 5	34
19.22.44.	Power Resistors	Engine Compartment	Nuts and Screws	2, Fig. 7	313
19.25.02/01	Fuel Filter	Luggage Compartment	Bracket	4, Fig. 3	
19.45.11	Fuel Pressure Regulator	Engine	Bracket	13, Fig. 4	
19.45.08	Fuel Pump	Luggage Compartment	Screws	2 Fig. 3	41
19.55.01/01	Fuel Tank	Luggage Compartment	Securing Strap		
19.60.01/01	Injectors	Engine	Nuts And Clamps	3 Fig. 4	312
19.22.38	Main Relay	Luggage Compartment	Securing Tag	2 Fig 5	312

INERTIA SWITCH

Remove and Refit 19.22.09

Removing

Disconnect the battery.
Remove the trip reset cable rubber knob and trim panel securing screws.
Disconnect the trip reset cable from the bracket and carefully displace the drivers side dash liner (1, Fig. 4).

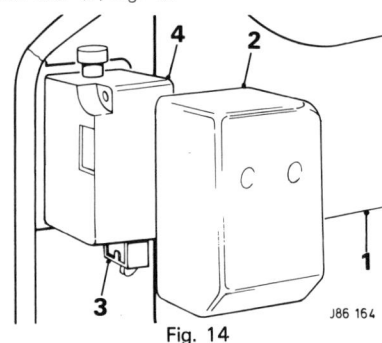

Fig. 14

Remove the switch cover (2, Fig. 14).
Disconnect the switch block connector (3, Fig. 14).
Remove the two screws securing the switch and remove the switch (4, Fig. 14).
On refitting the switch ensure that it is correctly reset.

POWER RESISTOR

Remove and refit 19.22.44

Removing

Disconnect the battery.
Disconnect the power resistor box block connector.
Remove the nuts and bolts securing the power resistor box to the right hand inner wing.

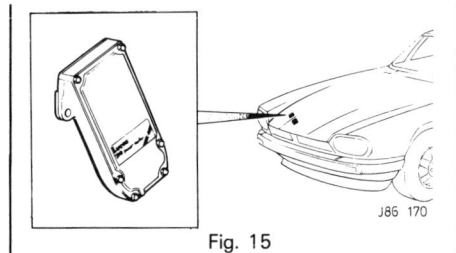

Fig. 15

Remove the power resistor box (Fig. 15). on refitting ensure the block connector is fully connected and secure.

FUEL MAIN FILTER

Remove and refit 19.25.02

Removing

Depressurise the fuel system.
Disconnect the battery (1, Fig. 16).
Remove the spare wheel cover and the spare wheel.
Fit pipe clamps to the outlet and inlet hoses.
Disconnect the filter hoses (2, Fig. 16).

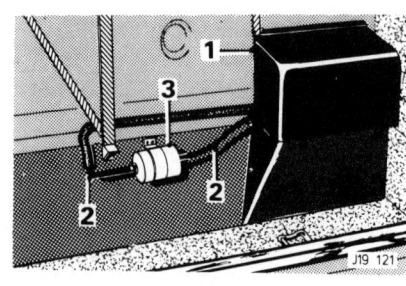

Fig. 16

Fit plugs to the disconnected hoses and to the filter (3, Fig. 16).
Slacken the filter clamp screw and remove the filter.
On refitting ensure the hose connections are secure.

FUEL COOLER

Remove and refit 19.40.40

Removing

De-gas the air conditioning system.
Depressurise the fuel system.
Disconnect the battery.
Disconnect the fuel hoses (1, Fig. 17) from the cooler and fit plugs to the unions to prevent ingress of dirt.

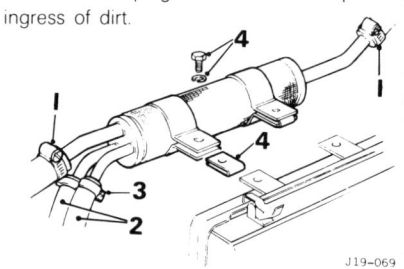

Fig. 17

Disconnect the air conditioning hoses (2, Fig. 17) from the fuel cooler and plug the hoses (3, Fig. 17) to prevent contamination of the air conditioning system.
Remove the nuts and bolts (4, Fig. 17) securing the fuel cooler to the left hand air cleaner back plate.
Remove the fuel cooler.
On refitting ensure all connections are secure and tight.
Fit new 'O' rings to the air conditioning unions.

PRESSURE REGULATOR

Remove and refit 19.45.11

Removing

Depressurise the fuel system.
Disconnect the fuel cooler to regulator hose from the regulator (1, Fig. 18).
Plug the hose and the regulator.
Remove the nut securing the regulator to the mounting bracket.

19—7

FUEL INJECTION SYSTEM

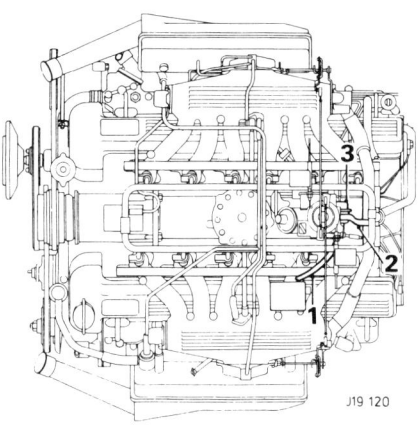

Fig. 18

Disconnect the regulator to the fuel rail hose from the fuel rail (2, Fig. 18).
Plug the hose and the rail.
Displace the regulator from the mounting bracket for access.
Disconnect the vacuum pipe from the regulator and remove the regulator (3, Fig. 18).
On refitting ensure that all fuel and vacuum pipe connections are secure.

INJECTORS

Right-hand Bank   19.60.01
Remove and refit

Removing

Depressurise the fuel system.
Disconnect the battery
Remove the right-hand fuel rail (1, Fig. 19)

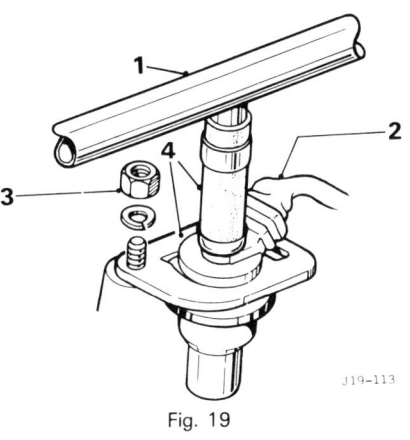

Fig. 19

Disconnect the block connectors (2, Fig. 19)
Remove the securing nuts and clamps (3, Fig. 19).
Remove the injectors (4, Fig. 19).
On refitting ensure that all pipes and electrical connections are secure.

INJECTORS

Left-hand Bank
Remove and refit   19.60.03

Removing

As for right-hand bank of injectors removing left-hand fuel rail instead of the right-hand bank.

FUEL RAIL

Right-hand Side
Removing   19.60.04

Depressurise the fuel system.
Disconnect the battery.
Disconnect the right-hand throttle rod (1, Fig. 20) and the throttle cable (2, Fig. 20) from the throttle pedestal.
Disconnect the manifold cross pipe from the right-hand manifold (3, Fig. 20).
Disconnect the cold start injector hose from the fuel rail (4, Fig. 20).

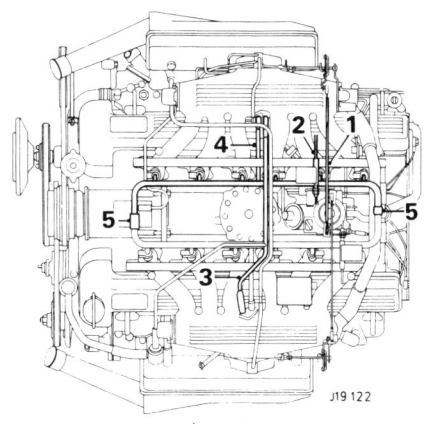

Fig. 20

Plug the hose and rail.
Slacken the hose clips and disconnect the fuel rail halves (5, Fig. 20).
Plug the hoses and fuel rail.
Slacken the injector hose clips and remove the fuel rail.
On refitting ensure all fuel hose connections are secure.

FUEL RAIL

Left-hand Side
Remove and refit   19.60.05

Removing

Depressurise the fuel system.
Disconnect the battery.
Disconnect the left-hand throttle rod (1, Fig. 21) from the throttle pedestal
Disconnect the regulator to fuel rail hose from the fuel rail (2, Fig. 21).
Plug the fuel rail and the hose.
Disconnect the manifold crosspipe from the left-hand inlet manifold (3, Fig. 21).

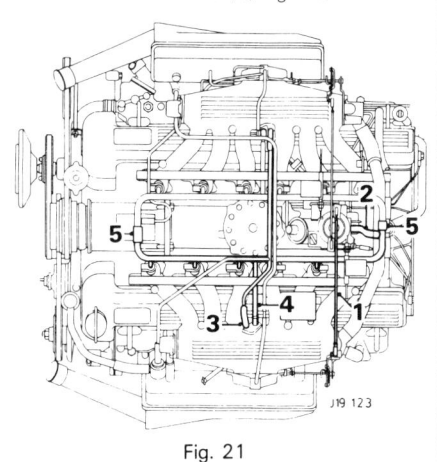

Fig. 21

Disconnect the cold start injector hose from the fuel rail (4, Fig. 21).
Plug the fuel rail and the hose.
Slacken the hose clips and disconnect the fuel rail halves (5, Fig. 21).
Plug the fuel rail and hoses.
Slacken the injector hose clips and remove the fuel rail.
On refitting ensure the hose connections are secure.

NOTE — 1983 M.Y. ON

To improve the method of assembly, and reduce the number of underbonnet fuel hose connections, a new fuel rail was introduced at the following VIN's:

    XJS
    112050 A Emission
    111999 B Emission
    111980 D Emission

The new fuel rail incorporated reversed fir tree connections and can be identified by the direct union connection of the pressure regulator to the fuel rail.

19—8

COOLING SYSTEM

Fig. 1

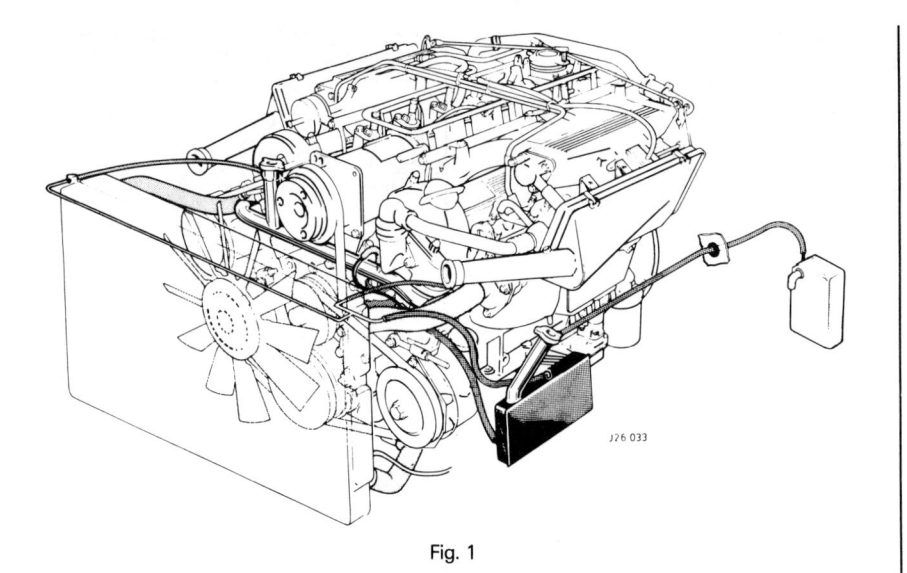

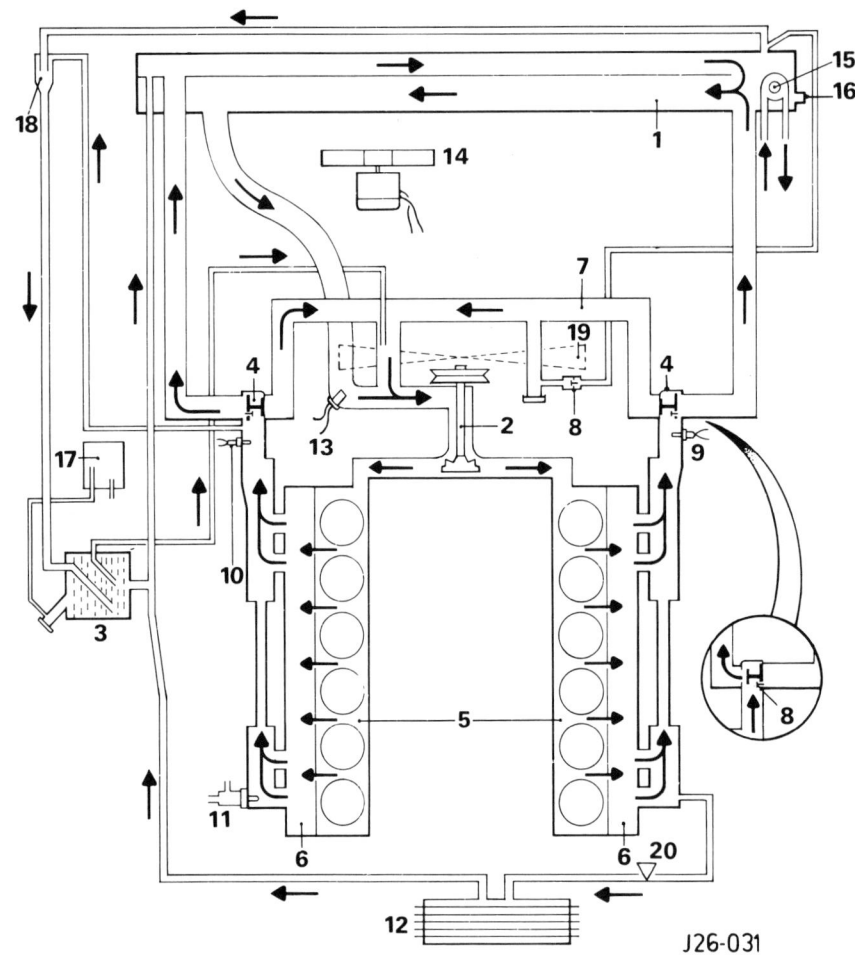

Fig. 2

COOLING SYSTEM (Fig. 2)

The cooling system consists of a radiator matrix (1), a water pump (2), belt driven by the engine crankshaft; and a remote header tank (3). Two thermostatic valves (4) are fitted, one to each cylinder tank, to ensure rapid warm up from cold.

Under start conditions coolant is forced by the water pump equally through each cylinder block and cylinder head (5 and 6) to the thermostatic valve housings. The valves are closed and coolant is therefore returned via the engine cross-pipe (7) to the water pump inlet. During this period the radiator is under pump suction and air is bled by jiggle pins (8) in each thermostatic valve.

Note. When fitting a replacement thermostat the thermostat MUST be fitted with the jiggle pin at the top of the housing.

The engine contains air pockets which have to be purged before effective cooling is possible. The air entrained by the coolant rises to the highest point on each side of the engine, the thermostat housings, then through the jiggle pins to the top of the radiator.

During this phase the thermotime switch (9), the coolant temperature sensor (10) and the auxiliary air valve (11) function as an automatic choke and warm up the system. Full pump suction draws coolant from the base of the radiator and starts the full cooling circuit.

At this time pump suction also appears at the heater matrix (12), and the remote header tanks carry out an air separation function in addition to providing a reservoir of coolant.

When coolant temperature rises to a pre-determined level the thermostatic valves open and allow coolant to flow into the top of the radiator.

A thermostatic fan switch (13) is fitted in the water pump suction elbow. The switch starts the radiator electric cooling fan (14) should the temperature of the coolant leaving the radiator rise above a pre-determined level.

A cooling tube coil (15) is included in the fabrication of the right-hand end tank of the radiator, and is connected in series with the automatic transmission hydraulic fluid circulation.

A drain tap (16) is located in the base of the right-hand end tank.

Key to 12 Cylinder HE cooling system

1. Radiator matrix
2. Water pump
3. Remote header tank
4. Thermostat
5. Cylinder block
6. Cylinder head
7. Engine crosspipe
8. Jiggle pins
9. Thermotime switch
10. Coolant temperature sensor
11. Auxiliary air valve
12. Heater matrix
13. Thermostatic fan switch
14. Radiator electric cooling fan
15. Automatic transmission fluid cooling tube coil
16. Radiator drain tap
17. Atmospheric catch tank
18. Venting jet
19. Engine driven fan
20. Heater water control valve

COOLING SYSTEM

Filler Caps

The cooling system has 2 filler caps, the Engine Filler Cap (Fig. 3) and the Header Tank Cap (Fig. 4).

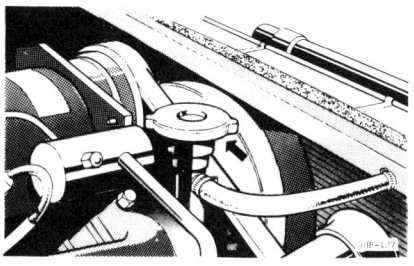

Fig. 3

The engine filler pipe cap which is set to 1.41 kg/cm² (20 lb/in²) and ensures a reliable seal for the radiator bleed pipe.

Fig. 4

The header tank cap controls the system pressure to 1,05 kg/cm² (15 lb/in²) and is retained by a chain to ensure that it is not incorrectly fitted to the engine filler pipe.

CAUTION: On no account must the two filler caps be interchanged.

Checking the coolant level

COLD

Remove the cap from the engine filler pipe (Fig. 3). If the coolant level is below 7 cm (3 in) from the seal face of the neck, top up with coolant of the correct concentration. See Coolant Concentration Chart. Refit Cap.

HOT

WARNING: CHECKING THE LEVEL OF COOLANT WHEN HOT IS NOT RECOMMENDED DUE TO THE DANGER OF SCALDING.

If it is necessary to check the coolant level when hot use a cloth or gloves to protect the hands. Carefully remove the cap on the header tank (Fig. 4) this will release the pressure in the system. The engine filler pipe cap (Fig. 3) may then be removed and the system filled with the correct coolant. See Coolant Concentration Chart. Refit both caps.

COOLANT

Drain and refill 26.10.01

Remove the engine filler and header tank caps, open the radiator drain tap. Allow the coolant to drain from the system. This will allow approximately 14 litres (25 pints) to drain without disconnecting hoses. If the system is being drained to refill with new anti-freeze (recommended every 2 years) then the following procedure should be carried out.

Close the radiator drain tap. Fill the system with plain water using the engine filler pipe, fit both caps. Set the air conditioning/heater to 'Defrost'. Run the engine at 1500 rev/min for 5 minutes. Open both filler caps and turn the radiator drain tap 'on' to drain the water as before.

Close the radiator drain tap and put in the correct amount of anti-freeze/corrosion inhibitor (see Chart) Fill the system with water through the engine filler pipe until the header tank is full. Replace the cap on the header tank and continue to fill the system, replace the cap when the system level is full and stabilised. Run the engine for a few minutes to mix the concentration.

Recommended antifreezes — B.P. HS25, Bluecol 'U', Union Carbide UT 184, and Unipart Universal.

The correct concentration for UK market vehicles is 40% all other markets is 55%.

COOLANT CONCENTRATION CHART

Capacity	Metric Litres	Imperial Pints	U.S. Pints	Specific Gravity
Cooling System Complete	19.5	35	42	
40% concentration	7.75	14	17	1.065
55% concentration	10.75	19.25	23	1.074

ATMOSPHERIC CATCHMENT TANK

Remove and refit 26.15.03

Removing

Jack up and place the front of the vehicle on stands.
Remove the front left-hand road wheel.
Remove the screws securing the wheel arch rear panel and remove the panel.
Unclip the catchment tank.
Unclip the overflow pipe from the body and displace the tank from its mounting position.
Disconnect and remove the catchment tank from the joining hose.
On refitting apply new body sealant to the wheel arch panel seal.

HOSE EXPANSION TANK TO ATMOSPHERIC CATCHMENT TANK

Remove and refit 26.30.62

Removing

Jack up and place the front of the vehicle on stands.
Remove the left-hand front wheel.
Remove the screws securing the wheel arch rear panel and remove the panel.
Unclip the catchment tank and overflow pipe from the body.
Displace the tank from its mounting position.
Slacken the hose clip and remove the catchment tank complete with the joining hose from the vehicle.
Remove the hose clip from the hose.
Displace the hose from its mounting position and remove the hose from the header tank.
On refitting apply new body sealant to the wheel arch seal.

26—2

MANIFOLD AND EXHAUST SYSTEM

INDUCTION MANIFOLD GASKETS — LEFT-HAND

Remove and refit   30.15.24

Removing

Open the bonnet, depressurise the fuel system.
Remove the LH air cleaner cover and element.
Disconnect the battery.
Remove the bolts securing the fuel cooler to the LH air cleaner back plate.
Disconnect the fuel regulator to the fuel cooler hose, from the fuel cooler. Fit suitable plugs to the hose and cooler.
Renove the ignition amplifier.
Remove the clip securing the harness to the fuel rail. Displace the regulator hose from the injector mounting position. Disconnect all the LH injectors and the cold start injector.
Disconnect the LH throttle rod from the throttle pedestal.
Disconnect the harness from the coolant temperature transmitter.
Disconnect the hoses from the PC valve in the air cleaner back plate. Disconnect the manifold cross pipe from the LH inlet manifold and move to one side.
Remove the bolt securing the ignition amplifier harness to the inlet manifold.
Disconnect the rear balance pipe hose and the auxiliary air valve hose, from the inlet manifold.
Disconnect the auxiliary air valve hose from the air cleaner back plate.
Remove the throttle return spring.
Disconnect the fuel regulator to LH fuel rail hose. Slacken the clips and separate the fuel rail halves, plug hoses and fuel rail ends.
Remove the nuts securing the LH inlet manifold, and remove inlet manifold. Discard all gaskets.

Refitting

Reverse the above procedure, fit new gaskets, ensure that all fuel hose connections are tight and all electrical connections are secure.

INDUCTION MANIFOLD GASKETS — RIGHT-HAND

Remove and refit   30.15.25

Removing

Open the bonnet, depressurise the fuel system.
Disconnect the battery.
Remove the RH air cleaner cover and element.
Disconnect the fuel feed hose from the engine bundy pipe; fit suitable blanking plugs to the hose and pipe ends.
Remove the dump valve from the mounting bracket; disconnect the coolant temperature and two way solenoid valve feed wires.
Remove the manifold cross pipe from the RH inlet manifold.
Disconnect all the RH injectors and the cold start injector.
Remove the clips securing the harness to the RH fuel rail.
Disconnect the throttle cable and RH throttle rod from the throttle pedestal. Disconnect the vacuum pipe from the distributor vacuum capsule.
Remove the screw securing the solenoid air switch to the RH inlet manifold and remove the earth wires.
Remove the nuts and bolts securing the brake fluid reservoir, disconnect the electrical leads, and carefully displace the reservoir to one side.
Remove the overrun valve securing screws, disconnect the valve from the RH inlet manifold. Remove the throttle linkage.
Disconnect the vacuum pipe from the three way solenoid valve and disconnect the electrical leads. Disconnect the electrical leads from the water temperature switch. Remove the throttle return spring.
Slacken the clips and separate the fuel rail halves, plug hoses and fuel rail ends.
Remove the nuts securing the RH manifold and remove the inlet manifold. Discard all gaskets.

Refitting

Reverse the above procedure, fit new gaskets; ensure that all fuel hose connections are tight and all electrical connections are secure.

STEERING

STEERING 1983 M.Y. ON

Steering Tie Rod Ends (Track Rod Ends) All Models

Commencing at the Vehicle Identification Numbers listed below, an alternative design of tie rod was introduced. It should be noted that the new tie rod ends are sealed and require no lubrication. A feature of the new tie rod is that there is NO SPRING LOADED FREE MOVEMENT. To avoid any confusion over the identification of these tie rod ends, the old and new type are listed below:

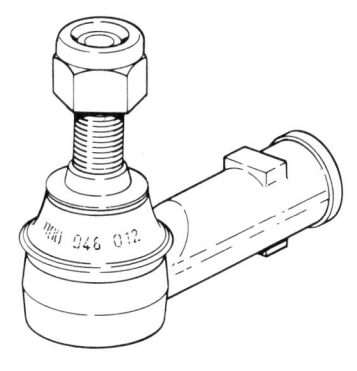

NEW TYPE **OLD TYPE**

JSI 187

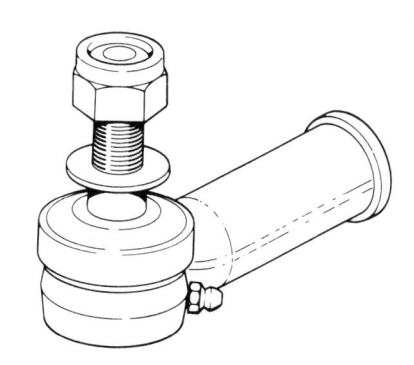

Vehicle Identification Number
107365   XJS

BRAKES

BRAKES 1983 M.Y. ON

Brakes — All Models

1. The brake system pressure differential warning actuator (PDWA) unit has been deleted.

NOTE: This deletion in no way affects the performance of the braking system as the conventional split system is retained.

2. All steel brake pipes from 1983 M.Y. on model year cars are plastic coated. This improves the corrosion resistance of the pipe work.

Brake Pad Material Change All Models

A semi metallic brake pad lining was introduced on Jaguar vehicles from:

VIN 109447 XJS

Identification of semi metallic pads is by the friction material code FER 3401 printed on the rear face of the material adjacent to the pad batch number.

Semi metallic pads may be used in vehicle sets as a retrospective fit on Jaguar vehicles with 4 pot caliper front brakes.

Under no circumstances should semi metallic and non semi metallic brake pads be mixed.

Brake pads must be used in vehicle sets only.

It is therefore necessary to check lining specification **on the complete vehicle** before replacing brake pads in axle sets to ensure that mixing does not occur.

BODY

REAR QUARTER TRIM CASING (Lower)

Remove and refit   76.13.12

Removing

Remove the rear seat cushion and squab.
Remove the upper 'B' post trim pad.
Remove the front seat belt lower anchorage bolt, spacing washers and spring.
Displace the 'Furplex' trim from the door flange at the trim position.
Carefully displace the trim casing flap from the door flange and remove the casing front securing screws.
Remove the remaining casing securing screws and displace the casing from the mounting position for access to the rear speaker cables.
Disconnect the rear speaker cables.
Reposition the seat belt through the casing and remove the casing assembly.
Remove the screws securing the arm rest and remove the arm rest.
Remove the companion box lower securing rivets.
Remove the trim clips and displace the upper securing tabs.
Remove the companion box assembly.
Replace the tabs securing the rear seat belt blank or surround (if fitted) and the front seat belt guides.
Remove the guides.

Refitting

On refitting ensure the speaker cables are re-connected correctly and the trim is securely adhered to the body flange.

COMPANION BOX LINER — REAR QUARTER LOWER

Remove and refit   76.13.51

Removing

Remove the rear seat cushion and squab.
Remove the 'B' post trim pad.
Remove the rear quarter lower trim casing assembly and remove the rear speaker.
Remove the speaker grille and by releasing the companion box securing tabs remove the companion box liner.

FACING TRIM PAD — COMPANION BOX — REAR QUARTER (Lower)

Remove and refit   76.13.55

Removing

Remove the rear seat cushion and squab.
Remove the upper 'B' post trim pad.
Remove the rear quarter lower trim casing assembly.
Remove the companion box assembly and arm rest.
Remove the trim pad lower securing rivets and clips.
Displace the upper securing tabs and remove the trim pad.

BUMPER — REAR CENTRE SECTION — BLADE

Remove and refit   76.22.12

Removing

Displace the boot side carpeting for access to quarter bumper securing bolts.
Remove the quarter bumper securing bolts, and nuts securing the centre blade to the body flange.
Remove the blade, quarter bumper and rubber buffer assembly.

NOTE: It is advisable to carry out this operation with two men so as to avoid damage to paintwork.

Remove nuts and bolts securing the quarter bumpers to the blade.
Remove the quarter bumpers and rubber joint finishers.

QUARTER BUMPER — FRONT

Remove and refit   76.22.16

Removing

Remove the quarter bumper trim finisher.
Remove the quarter bumper blade securing nuts.
Remove the quarter bumper blade to the centre blade securing nuts and bolts.
Remove the quarter bumper blade and rubber joint finisher.

QUARTER BUMPER — REAR

Remove and refit   76.22.17

Removing

Remove the rear centre blade assembly.
Remove the nuts and bolts securing the quarter bumper to the centre blade.
Remove the quarter bumper and the rubber joint finisher.

ENERGY ABSORBING BEAM 'FRONT'

Remove and refit   76.22.26

Removing

Disconnect the battery.
Disconnect the front flasher lamps.
Remove the nuts securing the beam mounting brackets to body struts.
Remove the beam, cover and lamp assembly.
Remove the nuts securing the brackets to the beam and remove the brackets.
Remove the flasher lamp securing screws and remove the lamps with the sealing rubbers.
Remove the bolts and clips securing the beam cover to the beam and remove the cover.
Remove the flasher lamp securing cage nuts.
On refitting ensure that the beam assembly is aligned before all the mounting bracket securing nuts are fully tightened.

ENERGY ABSORBING BEAM — REAR

Remove and refit   76.22.27

Removing

Remove the nuts securing the fog guard lamps to the rear beam and displace the lamps from the mounting position.
Slacken the beam bumper blade to body securing nuts.
Remove the beam to the bracket securing nuts.
Carefully remove the beam assembly ensuring the rear fog guard lamps are not damaged.
Remove the clips securing the beam cover and remove the cover.

FRONT BUMPER — CENTRE SECTION — BLADE

Remove and refit   76.22.37

Removing

Remove the upper radiator grille.
Remove the nuts and bolts securing the bumper blade to the quarter bumpers.
Remove the nuts and bolts securing the finisher to the body.
Remove the blade and the finisher assembly.
Drill out the pop rivets securing the finisher to the blade and remove the finisher.
On refitting ensure the beam is aligned before the mounting nuts are fully tightened. Also align the fog lamps before the securing nuts are fully tightened.

QUARTER BUMPER — REAR — RUBBER BUFFER

Remove and refit   76.22.63

Removing

Remove the rear bumper centre blade assembly.
Remove the nuts securing the rubber buffer and remove the buffer.

DOOR TRIM PAD

Remove and refit   76.34.01

Removing

Disconnect the battery.
Remove the door mirror switch surround panel.
To avoid damage to the arm rest leather and stitching apply a suitable masking to the area adjacent to the chrome finisher.
Displace the chrome finisher and remove the arm rest upper securing screw.
Remove the backing plate.
Remove the arm rest lower securing screw and the door pillar switch plate securing screws.

(continued)

BODY

Remove the switch plate.
Unclip the trim pad from the door and lift the trim pad assembly to release the locating lugs.
Displace the trim for access to disconnect the speaker and the puddle lamp cables.
Remove the trim pad assembly.
On refitting ensure the speaker is connected correctly and on reconnecting the battery the puddle lamps operate satisfactory.

DOOR TRIM PAD — SIDE FRONT — VENEER PANEL

Remove and refit   76.34.21

Removing

Disconnect the battery.
Remove the door mirror switch mounting panel and the door trim pad assembly.
Remove the nuts securing the veneer panel to the trim pad and remove the veneer panel.

FASCIA VENEER PANEL — PASSENGER SIDE   76.46.25
FASCIA VENEER PANEL — DRIVERS SIDE   74.46.26
FASCIA VENEER PANEL — CENTRE   76.46.29

Remove and refit

Removing

For the centre and passenger side veneer panels open the glove box for access.
By using a suitable long thin bladed instrument carefully release the veneer panel retaining clips and remove the appropriate panel.

FASCIA SWITCH PANEL — VENEER PANEL

Remove and refit   76.46.30

Removing

Disconnect the battery.
Remove the electric clock.
By using a long thin bladed instrument carefully unclip and displace the switch panel.
Note and disconnect the switch block connectors and the switch illuminator bulb holders.
Remove the panel assembly.
Unclip and remove the switches from the panel.
Remove the spire nuts securing the bulb holders to the panel and remove the bulb holders.
On refitting use new spire nuts to secure the bulb holders.
Ensure the electrical connectors are secure and the correct operation of components.

GLOVE BOX LID — VENEER PANEL

Remove and refit   76.52.12

Removing

Remove the screws securing the vanity mirror moulding to the rear of the glove box lid.
Remove the assembly.
Remove the screws securing the veneer panel to the glove box lid and remove the panel.
On refitting ensure the veneer panel is correctly aligned before fully tightening the securing screws.

REAR SEAT ARM REST

Remove and refit   76.70.39

Removing

Remove the rear seat cushion and squab.
Remove the 'B' post upper trim pad.
Remove the rear quarter lower trim casing assembly.
Remove the screws securing the arm rest to the trim casing and remove the arm rest.

ELECTRICAL FAULT DIAGNOSIS

COMPONENT LOCATION — ENGINE COMPARTMENT

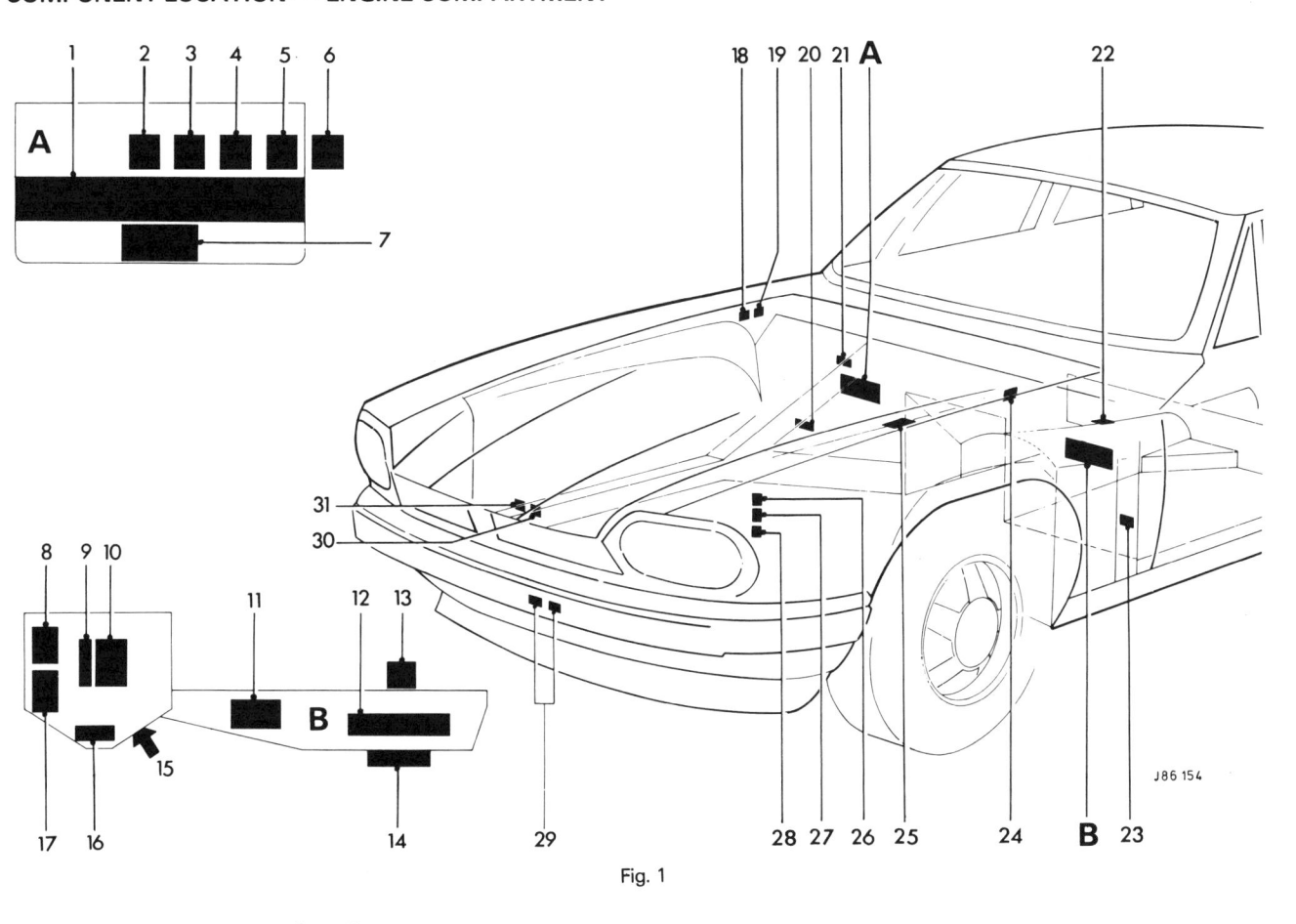

Fig. 1

COMPONENT LOCATION KEY (Fig. 1)

1. Main fuse box
2. Heated rear screen delay unit
3. Headlamp inhibit relay
4. Ignition protection relay
5. Warning light bulb check unit
6. Direction indicator flasher unit
7. Voltage over-charge warning light unit

} Component Panel-A Drivers Side

8. Door lock thermal circuit breaker
9. Low coolant warning light control unit
10. Vacuum delay unit
11. Interior light delay unit
12. Auxiliary fuse box
13. Windscreen wiper delay unit
14. Speed control unit (if fitted)
15. Window lift relay
16. Front parking lamp bulb failure unit
17. Window lift thermal circuit breaker

} Component Panel-B Passengers Side

18. Cold start relay
19. Starter relay
20. Door lock relay
21. Main fuse box
22. Stop lamp bulb failure unit
23. Door unlock relay
24. Cooling fan diode pack
25. Headlamp fuse box
26. Headlamp relay
27. Horn relay
28. Cooling fan relay
29. Horns
30. Headlamp wash/wipe relay
31. Headlamp wash/wipe diode pack

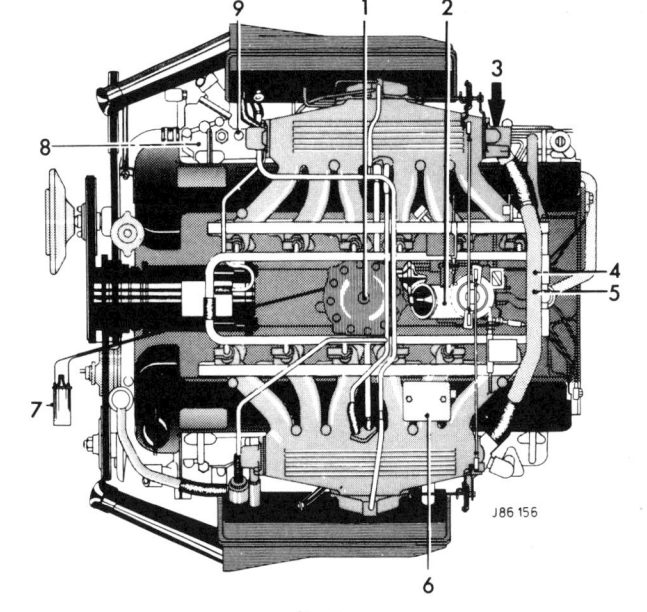

Fig. 2

COMPONENT LOCATION KEY (Fig. 2)

1. Distributor
2. Main HT coil
3. Starter motor
4. Oil pressure transmitter
5. Oil pressure warning switch
6. Amplifier
7. Auxilliary HT coil
8. Alternator
9. Coolant temperature transmitter

86—1

ELECTRICAL FAULT DIAGNOSIS

COMPONENT LOCATION — LUGGAGE COMPARTMENT

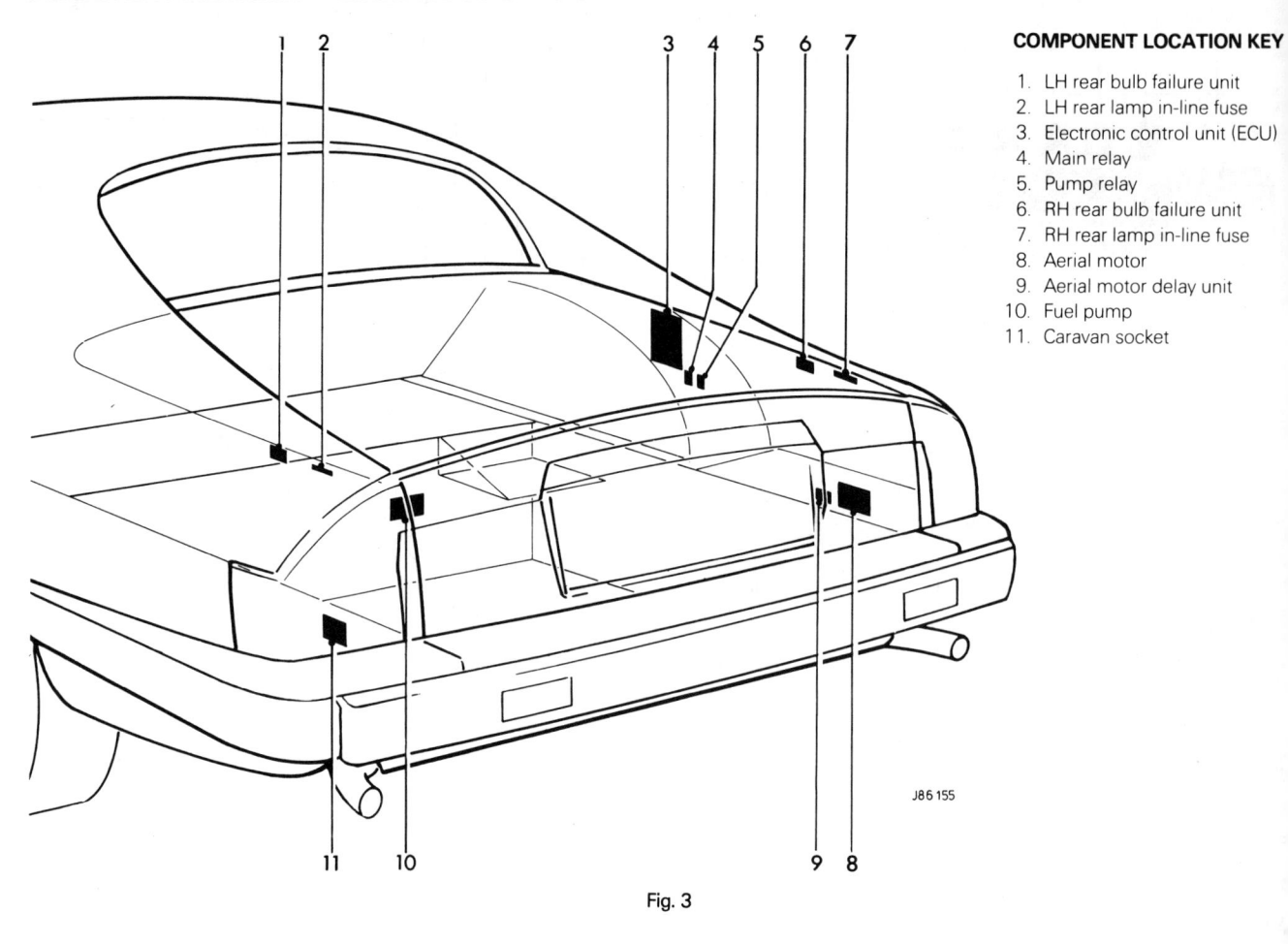

COMPONENT LOCATION KEY

1. LH rear bulb failure unit
2. LH rear lamp in-line fuse
3. Electronic control unit (ECU)
4. Main relay
5. Pump relay
6. RH rear bulb failure unit
7. RH rear lamp in-line fuse
8. Aerial motor
9. Aerial motor delay unit
10. Fuel pump
11. Caravan socket

Fig. 3

COMPONENT LOCATION — DRIVING COMPARTMENT

COMPONENT LOCATION KEY

1. Inertia switch
2. Wiper wheel box
3. Wiper motor
4. Washer jets
5. Door lock solenoid
6. Wiper wheel box
7. Rear speaker
8. Window lift motor
9. Front speaker
10. Door pillar switches
11. Washer reservoir
12. Washer motor

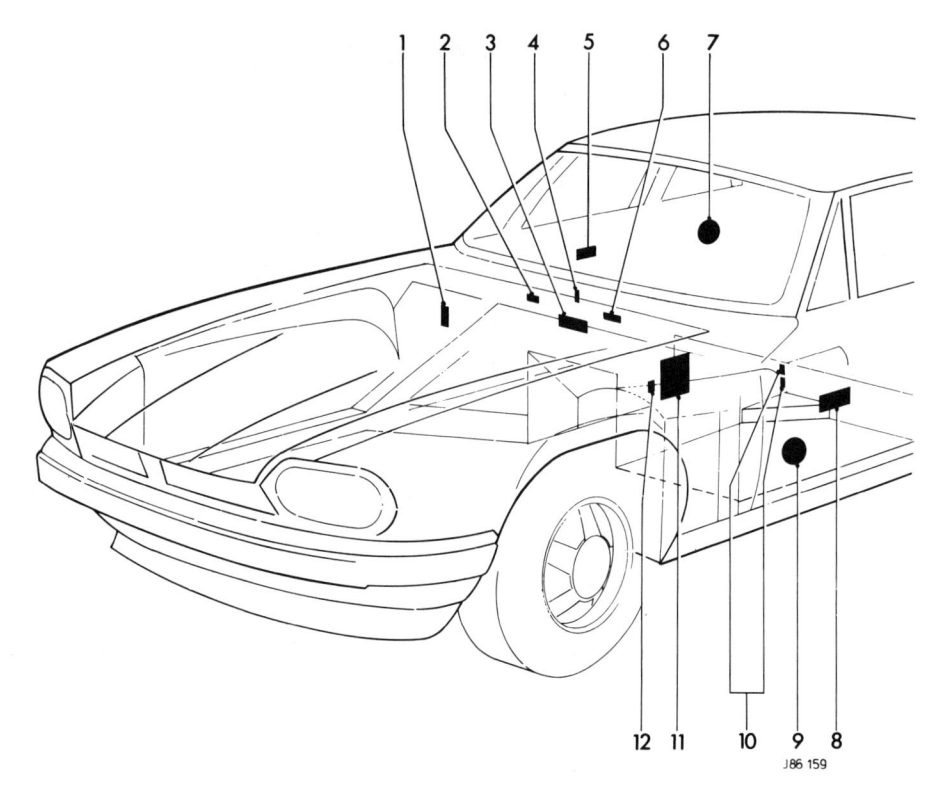

Fig. 4

ELECTRICAL FAULT DIAGNOSIS

COMPONENT LOCATION — FACIA

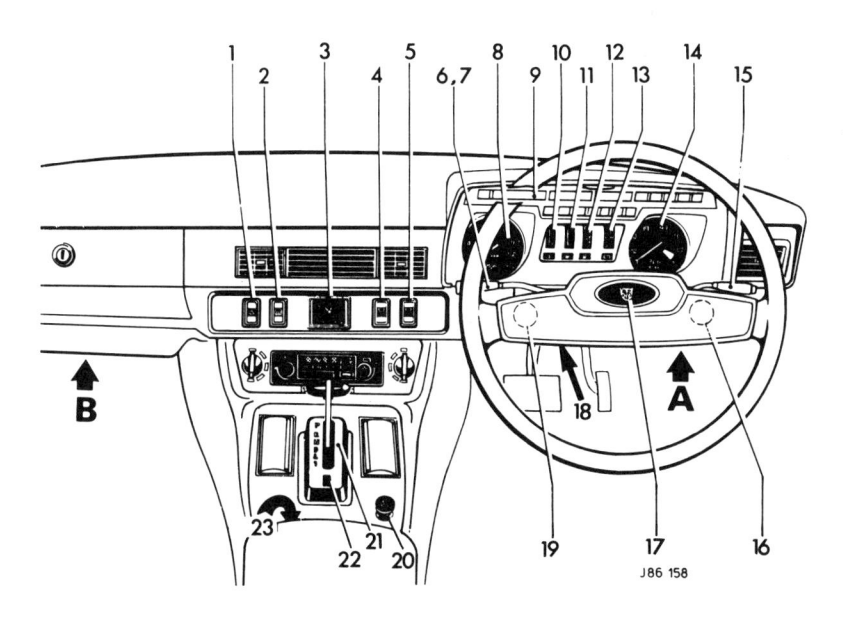

Fig. 5

COMPONENT LOCATION KEY
1. Hazard lamp switch
2. Heated rear screen
3. Clock
4. Map lamp switch
5. Interior light switch
6. Speed control set switch
7. Flasher lamp switch
8. Speedometer
9. Warning lamps
10. Temperature gauge
11. Oil pressure gauge
12. Fuel gauge
13. Battery condition gauge
14. Tachometer
15. Window wash/wiper switch
16. Ignition switch
17. Horn push
18. Stop lamp switch
19. Master light switch
20. Cigar lighter
21. Optiflex unit
22. Cruise control
23. Window lift switches.
A — RH component panel
B — LH component panel

COMPONENT ACCESS

The electrical component access list contains the details of the component location, access, fixings and also the Component Location Fig. number and the Wiring Circuit Fig. number.

Description	Access	Fixing	Location Diagram	Circuit Diagram
Aerial Motor Relay	Luggage Compartment	Securing Tab	9, Fig. 3	5, Fig. 6
Alternator	Right-hand Engine	Nuts and Bolts	8, Fig. 2	2, Fig. 6
Alternator Drive Belt	Right-hand Engine	Nuts and Bolts		
Amplifier	Engine	Securing Bolts	6, Fig. 2	12, Fig. 8
Auxiliary Ignition Coil	Front of Radiator	Securing Bolts	7, Fig. 2	10, Fig. 8
Battery	Luggage Compartment	Bracket		1, Fig. 6
Battery Condition Indicator	Instrument Panel	Spire Washers	13, Fig. 5	5, Fig. 4
Battery Lead	Luggage Compartment	Nuts and Bolts		
Battery Tray	Luggage Compartment	Securing Bolts		
Cigar Lighter	Centre Console	Screw	20, Fig. 5	
Clock	Switch Panel	Clips	3, Fig. 5	1, Fig. 6
Coolant Temperature Gauge	Instrument Panel	Spire Washers	10, Fig. 5	2, Fig. 7
Coolant Temperature Transmitter	Engine Water Rail	Screw	9, Fig. 2	12, Fig. 7
Courtesy Light Delay Unit	Component Panel B	Securing Tab	11, Fig. 1	5, Fig. 21
Distributor	Engine	Allen Screws	1, Fig. 2	13, Fig. 8
Distributor Cap	Engine	Screws	1, Fig. 2	11, Fig. 8
Distributor Rotor Arm	Engine	Push-on Fit	1, Fig. 2	11, Fig. 8
Distributor Vacuum Unit	Engine	Screw	1, Fig. 2	13, Fig. 8
Door Guard Lamp	Door Trim Pad	Screws		Fig. 21
Door Guard Lamp Lens	Door	Screws		Fig. 21
Door Lock Circuit Breaker	Component Panel B	Screws	8, Fig. 1	5, Fig. 23
Door Lock or Unlock Relay	Left-hand or Right-hand Footwell Side Casing	Screws	23, Fig. 1	3, Fig. 23
Door Lock Solenoid	Doors	Screws	5, Fig. 4	Fig. 23
Door Switch	Door Pillar	Screws	10, Fig. 4	6, Fig. 21
Door Switch Key Alarm	Door Pillar	Screws	10, Fig. 4	6, Fig. 29
Electric Mirror Control Switch	Drivers Door	Screw		1-6, Fig. 24
Electric Mirror Switch Panel	Drivers	Screw		
Front Flasher Lamp	Front Bumper	Nuts and Bolts		Fig. 19
Front Speaker	Front Doors	Screws	9, Fig. 4	
Fuel Gauge	Instrument Panel	Spire Washer	12, Fig. 5	4, Fig. 17
Fuse Boxes	Component Panel A and B	Screws	A-B, Fig. 5	
Handbrake Warning Switch	Handbrake Cover	Screws		9, Fig. 22
Hazard Warning Flasher Unit	Component Panel A	Securing Tab	6, Fig. 1	2, Fig. 19
Headlamp Assembly		Nylon Tabs		Fig. 10
Headlamp Inhibit Relay	Component Panel A	Securing Tab	3, Fig. 1	2, Fig. 10
Headlamp Pilot Bulb		Push-in Fit		Fig. 9
Headlamp Relay	Engine Compartment	Screws	26, Fig. 1	4, Fig. 10
Headlamp Rim Finisher		Screws		
Heated Rear Screen Relay	Component Panel A	Securing Tab	2, Fig. 1	1, Fig. 15

86—3

ELECTRICAL FAULT DIAGNOSIS

COMPONENT ACCESS

Description	Access	Fixing	Location Diagram	Circuit Diagram
Horn Push	Steering Column	Screws	17, Fig. 5	1, Fig. 30
Horn Relay	Engine Compartment	Securing Tab	27, Fig. 1	2, Fig. 30
Horns	Front Bumper Apron	Nuts	29, Fig. 1	4, Fig. 30
Ignition Protection Relay	Component Panel A	Securing Tab	4, Fig. 1	7, Fig. 6
Ignition Switch	Driver Side Dash Casing	Grub Screws	16, Fig. 5	4, Fig. 6
Indicator, Headlamp Dip Switch	Steering Column	Clamp	7, Fig. 5	3, Fig. 19
Inertia Switch	Door Pillar	Clips	1, Fig. 4	
Instrument Panel	Facia Panel	Retaining Screws	Fig. 5	Fig. 17
Instrument Panel Lens Assembly	Facia	Retaining Screws	Fig. 5	
Interior and Map Lamps	Facia Side Casing	Securing Tabs		Fig. 21
Low Coolant Warning Unit	Component Panel B	Screw	9, Fig. 1	1, Fig. 25
Luggage Compartment Lamp	Luggage Compartment	Screws		Fig. 6
Main Ignition Coil	Engine	Securing Bolts	2, Fig. 2	9, Fig. 8
Master Light Switch	Driver Side Dash Liner	Securing Nut	19, Fig. 5	1, Fig. 10
Number Plate and reverse lamp	Luggage Compartment	Securing Nuts		Fig. 9
Oil Gauge	Instrument Panel	Spire Washers	11, Fig. 5	3, Fig. 17
Oil Pressure Switch	Rear of Engine	Screw	5, Fig. 2	11, Fig. 17
Oil Pressure Transmitter	Rear of Engine	Screw in	4, Fig. 2	11, Fig. 17
Oil Pressure Warning Switch	Rear of Engine	Screw in	5, Fig. 2	
Opticell Unit	Centre Console	Securing Nuts	21, Fig. 5	3, Fig. 9
Panel Light Rheostat	Driver Side Dash Liner	Securing Nut	Fig. 5	
Panel Switches	Switch Panel	Clips	Fig. 5	
Pick-up Module	Distributor	Screws	1, Fig. 2	13, Fig. 8
Printed Circuit	Instrument Panel	Nuts and Screws	Fig. 5	
Rearlamp Cluster	Luggage Compartment	Securing Nuts		Fig. 9
Rearlamp Failure Warning Sensor	Luggage Compartment	Screws	1-6, Fig. 3	18-19, Fig. 9
Rear Speaker	Rear Companion Box	Screws	7, Fig. 4	
Repeater Lamp	Front Wing	Securing Nuts		Fig. 19
Reverse Light Switch	Centre Console	Screws		
Roof Lamp	Roof Lining	Securing Tab		Fig. 21
Sidelamp Failure Warning Sensor	Component Panel	Screws	16, Fig. 1	20, Fig. 9
Side Marker Lamp	Front Wing	Securing Nuts		30, Fig. 9
Starter Motor	Engine	Securing Bolts	3, Fig. 2	15, Fig. 6
Starter Motor Relay	Engine Compartment	Screws	19, Fig. 1	11, Fig. 6
Stoplight Failure Warning Sensor	Centre Console	Screws	22, Fig. 1	13, Fig. 22
Stoplight Switch	Footbrake	Screw	18, Fig. 5	10, Fig. 22
Tachometer	Instrument Panel	Nuts and Screws	14, Fig. 5	6, Fig. 17
Tank Unit	Fuel Tank	Locking Ring		10, Fig. 17
Voltage Overcharge Unit	Component Panel A	Screw	7, Fig. 1	1, Fig. 27
Washer Jets	Air Intake Grille	Retaining Nut	4, Fig. 4	
Washer Pump	Engine Compartment	Securing Screws	12, Fig. 4	3, Fig. 11
Washer Reservoir	Engine Compartment	Bracket	11, Fig. 4	
Window Lift Circuit Breaker	Component Panel B	Screws	17, Fig. 2	5, Fig. 20
Window Lift Motor	Doors	Securing Bolts	8, Fig. 4	1, Fig. 20
Window Lift Relay	Component Panel B	Securing Tab	15, Fig. 2	6, Fig. 20
Window Lift Switches	Centre Console	Securing Tabs	23, Fig. 5	3, Fig. 20
Wiper Arms		Securing Nut		
Wiper Blades		Securing Clip		
Wiper Motor	Air Intake Grille	Bracket	3, Fig. 4	3, Fig. 14
Wiper Motor Delay Unit	Component Panel B	Securing Tab	13, Fig. 1	2, Fig. 14
Wiper Motor Gear Assembly	Air Intake Grille	Circlip	Fig. 4	
Wiper Motor Harness	Air Intake Grille	Screws	Fig. 4	Fig. 14
Wiper Wash Switch	Steering Column	Clamp	15, Fig. 5	1, Fig. 14
Wiper Wheel Boxes	Air Intake Grille	Retaining Nut	6, Fig. 4	

ELECTRICAL FAULT DIAGNOSIS

FUSES

MAIN FUSE BOX

Fuse No.	Protected Circuit	Fuse Capacity	Unipart Number
1.	Front Fog Lights	20A	GFS420
2.	Hazard Warning, Seat Belt Logic	15A	GFS415
3.	Clock, Aerial, Caravan Boot Lamp	35A	GFS435
4.	Panel Instruments, Reverse Light	10A	GFS410
5.	Direction Indicators, Stop Lamps, Auto Kick Down Switch	15A	GFS415
6.	Fog Rear Guard	10A	GFS410
7.	Panel/Cigar Lighter/Selector Illumination	10A	GFS410
8.	Door Locks, Electric Mirrors	3A	GFS43
9.	Wipers	35A	GFS435
10.	Air Conditioning Motors	50A	GFS450
11.	Air Conditioning Controls, Horn, Washers, Radiator Cooling Fan	35A	GFS435
12.	Heated Rear Screen, Heated Mirrors	35A	GFS435

AUXILIARY FUSE BOX

Fuse No.	Protected Circuit	Fuse Capacity	Unipart Number
13.	Interior and Map Lights	10A	GFS410
14.	L.H. Side Lights	3A	GFS43
15.	R.H. Side Lights	3A	GFS43
16.	Cigar Lighter	20A	GFS420
17.	Speed Control	3A	GFS43

HEADLAMP FUSE BOX

Fuse No.	Protected Circuit	Fuse Capacity	Unipart Number
1.	Radiator Auxiliary Cooling Fan Motor Relay	25A	GFS425
2.	L.H. Main Beam	25A	GFS425
3.	L.H. Dip Beam	10A	GFS410
4.	R.H. Main Beam	25A	GFS425
5.	R.H. Dip Beam	10A	GFS410

BULBS

BULB CHART

	Watts	Lucas Part No.	Unipart No.	Notes
Headlamps — outer	37.5/60			Tungsten sealed beam light unit
— inner	50			Tungsten sealed beam light unit
Front Flasher and Side Lamp	21/5	380		
Stop Lamp	21	382	GLB 382	
Tail Lamp	5	207	GLB 207	
Rear Flasher	21	382	GLB382	
Reversing Light	21	273	GLB 273	
Number Plate Lamp	6	254	GLB 254	Festoon bulb
Sidemarker	4	233	GLB 233	
Flasher Side Repeaters—where fitted	4	233	GLB 233	
Interior/Map Lamps	6	254	GLB 254	Festoon bulb
Roof Lamp	10			Festoon bulb
Boot Lamp	5	239	GLB 239	Festoon bulb
Fibre Optic Light Source	6	254	GLB 254	
Instrument Illumination	2.2	987	GLB 987	
Warning Lights	1.2	286	GLB 286	
Automatic Selector Illumination	2.2	987	GLB 987	
Cigar Lighter Illuminator	2.2	987	GLB 987	

ELECTRICAL FAULT DIAGNOSIS

SYMBOLS AND CABLE COLOUR CODE

CABLE COLOUR CODE When a cable has two colour code letters, the first denotes the Main Colour and the second the Tracer Colour.

N. Brown — Positive Cable

B. Black — Negative Cable

W. White
K. Pink } Ignition switch controlled
G. Green

R. Red   Y. Yellow   O. Orange   S. Slate   L. Light

SYMBOLS USED

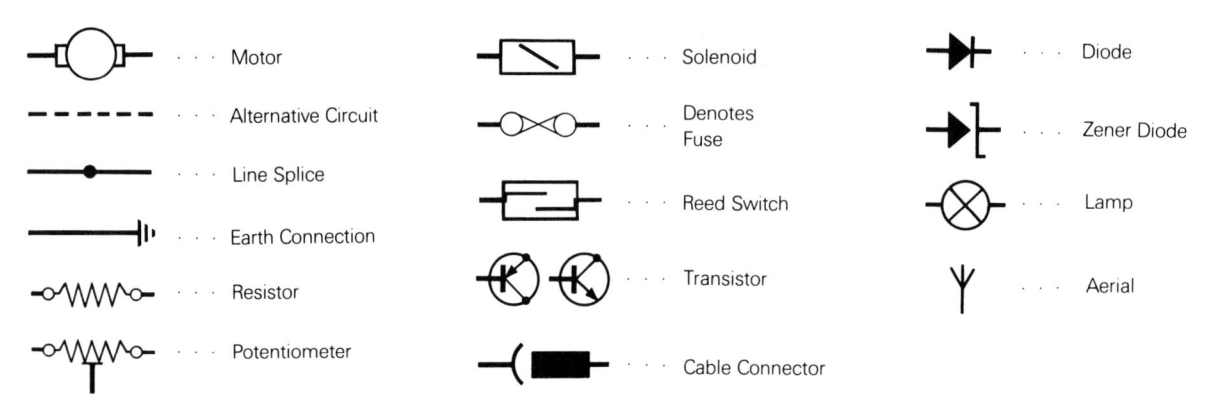

Symbol	Name	Symbol	Name	Symbol	Name
—	Motor	—	Solenoid	—	Diode
— — — —	Alternative Circuit	—	Denotes Fuse	—	Zener Diode
—	Line Splice	—	Reed Switch	—	Lamp
—	Earth Connection	—	Transistor	—	Aerial
—	Resistor	—	Cable Connector		
—	Potentiometer				

BATTERY

Testing — Hydrometer

Testing should commence at the source of supply; the battery itself. If the battery is discharged or unserviceable, the readings in the other tests will be affected.

There is a relationship between the state of battery charge and the strength of the electrolyte. As the battery becomes discharged, the specific gravity (SG) of the electrolyte becomes lower. The SG of the electrolyte is measured by means of a hydrometer. This instrument consists of a glass tube, with a rubber bulb fitted on one end. Inside the tube, there is a float which is calibrated from 1.130 to 1.300.

When the end of the hydrometer is inserted in the battery cell, and the rubber bulb is pressed and then released, a small quantity of the electrolyte is drawn into the tube. The position of the float is determined by the specific gravity of the electrolyte. When the specific gravity is high, the float maintains a high position inside the tube, and if the specific gravity is low the float sinks to a lower position.

From the specific gravity (SG) readings, a fairly accurate indication of the battery state of charge can be obtained.

Electrolyte Temperature Correction

For every 10°C (18°F) below 15°C (60°F) subtract 0.007.
For every 10°C (18°F) above 15°C (60°F) add 0.007.

The hydrometer gives an accurate indication of the battery condition. If there is a variation of more than 40 points (0.040) between any cells, the battery is suspect and should be thoroughly checked.

If the battery is less than 70% charged, it should be recharged from an external source.

Battery Testing — Heavy Discharge Test

This test should be carried out as a further check of the battery condition. A heavy discharge tester should be applied to the battery terminals.

The test ensures that the battery is capable of supplying the heavy currents required by the starter at the moment of starting the engine.

The tester should be set to discharge the battery at 3 times the ampere hour rate (20 hour rate) for 15 seconds. (Example: If the battery has a capacity of 50 ampere hours (20 hour rate), the tester should be set to 150 amps on the ammeter). Observe the voltmeter during the battery discharge. If the voltmeter reading is above 9.6 volts, the battery is considered satisfactory. If the voltage falls below 9.6 volts, the battery is suspect and should be removed for further testing.

Specific Gravity Readings

State of Charge	Climates normally below 25°C (77°F)	Climates normally above 25°C (77°F)
Fully charged	1.270 — 1.290	1.210 — 1.230
70° charged	1.230 — 1.250	1.170 — 1.190
Discharged	1.110 — 1.130	1.050 — 1.070

ELECTRICAL FAULT DIAGNOSIS

IGNITION, ALTERNATOR AND STARTER SYSTEMS

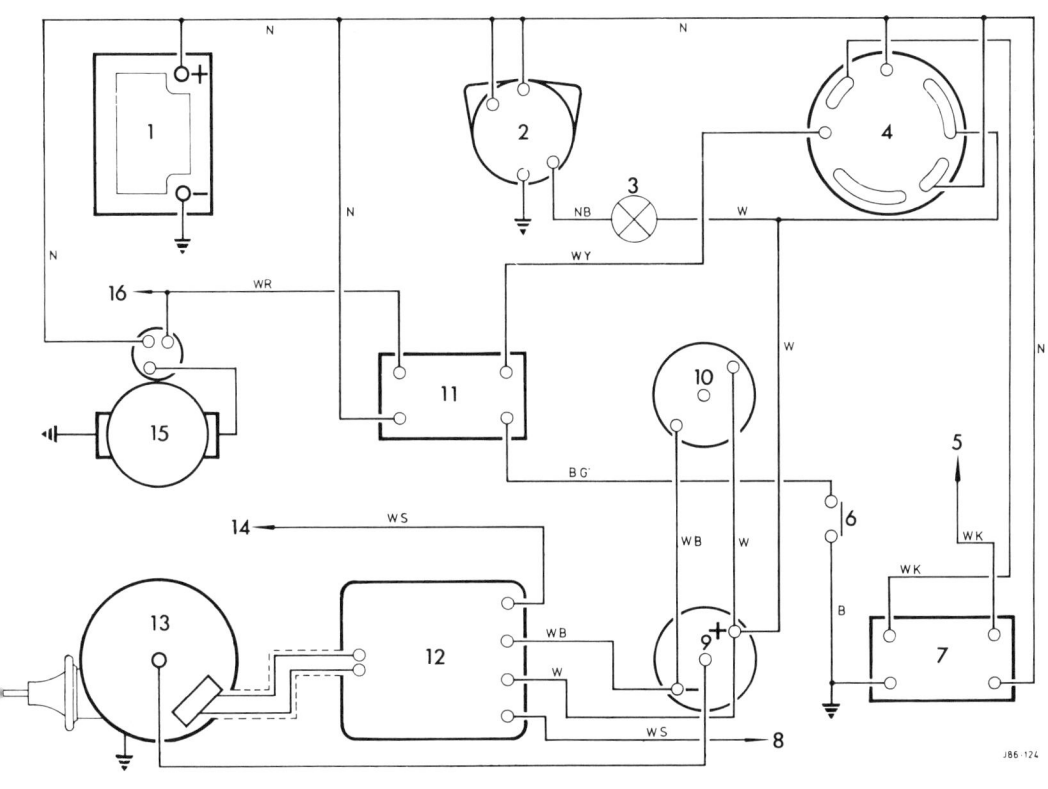

KEY TO DIAGRAM
1. Battery
2. Alternator
3. Warning light
4. Ignition switch
5. To fuse No. 12
6. Automatic gearbox safety switch
7. Ignition protection relay
8. To tachometer
9. Main HT coil
10. Auxiliary coil
11. Starter relay
12. Amplifier
13. Distributor
14. To ECU
15. Starter motor
16. To ECU

Checking for Excessive Voltage Drop in the Starter Circuit

If the previous test has proved that the battery and the battery connections are satisfactory, a moving coil voltmeter (0 — 20 volt range) should be used to determine whether there is excessive voltage drop in the circuit.

NOTE: During the voltmeter checks, the starter should crank the engine, without starting it.

Petrol engines: The low-tension circuit of the ignition coil should be disconnected between the coil and distributor.

Test 1

Checking the Battery Terminal Voltage Under Load Conditions

Connect the voltmeter across the battery terminals and operate the starter switch. The reading should be about 10.0 volts, and for diesel engines (12 volt systems), 9.0 volts. Proceed to Test 2.

A low voltage reading would indicate excessive current flow in the circuit. The starter should be removed for bench testing.

NOTE: If the solenoid operates intermittently during the test or the engine is cranked at a low or irregular speed, there is insufficient voltage at the solenoid operating winding terminal or the solenoid is faulty.

To check the switching circuit for high resistance, connect the voltmeter between the solenoid operating winding terminal and earth (commutator end bracket).

When the switch contacts are closed the reading on the voltmeter should be slightly less than the reading in Test 1. A satisfactory reading will indicate that there is a negligable voltage drop in the circuit and that the fault is in the solenoid.

If the reading is appreciably lower than in Test 1, check the switching circuit for high resistance or faulty connections.

Test 2

Checking the Starter Terminal Voltage Under Load Conditions

Having ascertained the battery voltage under load, the voltage across the starter is checked with the voltmeter connected between the starter input terminal and earth (commutator end bracket). When the operating switch is closed, the reading should be not more than 0.5 volt below that obtained in Test 1.

If the reading is within this limit, the starter circuit is satisfactory. If there is a low reading across the starter, but the voltage at the battery is satisfactory, it indicates a high resistance in the cable or at the solenoid contacts. Proceed to Test 3.

Test 3

Checking the Voltage Drop on the Insulated Line

The voltage drop on the insulated line is then checked with the voltmeter connected between the starter input terminal and the battery (insulated) terminal.

When the operating switch is open, the voltmeter should register battery voltage. When the operating switch is closed, the voltmeter reading should be practically zero.

A high voltmeter reading indicates a high resistance in the starter circuit. All insulated connections at the battery, solenoid and starter should be checked. Proceed to Test 4.

Test 4

Checking the Voltage Drop Across the Solenoid Contacts

To check the voltage drop across the solenoid contacts, connect the voltmeter across the two main solenoid terminals. Crank the engine.

A zero or fractional reading on the voltmeter indicates that the high resistance deduced in Test 3 must be due either to high resistance starter cables or soldered connections.

A high reading (similar to that in Test 3) indicates a faulty solenoid or connections.

86—7

ELECTRICAL FAULT DIAGNOSIS

IGNITION, ALTERNATOR AND STARTER SYSTEMS

Test 5

Checking the Voltage Drop on the Earth Line

Finally, check the voltage drop on the earth line. Connect the voltmeter between the battery earth terminal and the starter earth (commutator end bracket) as shown in Fig. 66. When the operating switch is closed, the voltmeter reading should be practically zero.

NOTE: The total voltage drop in the starting circuit (i.e. insulated line and earth line) must not exceed 0.5 volt.

If meter reading is high, clean and tighten all earth connections and check bonding strap.
The bonding strap must make good electrical contact with the chassis and engine block. If the bonding strap is frayed, it will have a serious effect on the performance of the starter. It may even immobilise the vehicle.

ALTERNATOR

Drive Belt Tension and Condition

There should be approximately ½ inch play when moderate pressure is applied to the longest run of the alternator drive belt.
If the drive belt is oily, worn or cracked, it must be replaced.

Test 1

Remove connections from the alternator.
Switch ignition on.
Do not start the engine.
Connect the Voltmeter between a good earth and each of the disconnected leads in turn.
The voltmeter should indicate battery voltage.

If the voltmeter reads zero when connected to the 'IND' terminal and earth, check the warning light bulb, and all connections to the warning light.
If the voltmeter reads zero when connected to the 'MAIN OUTPUT' terminal and earth, check the wiring and connections to the starter solenoid and battery.
If the voltmeter reads zero when connected to the 'S' terminal and earth, check the wiring and connections to the battery. The alternator output is based directly on battery condition, if the battery sensor circuit is broken there will be no output from the alternator.

NOTE: On alternators with internal battery condition sensing, the 'S' terminal is used for physical retention of the plug only.

Test 2

Replace connections.
Switch ignition on.
Connect the voltmeter between a good earth and the 'IND' terminal.
The voltmeter should indicate 1.5 to 2 volts.
If the reading is zero, check the surge-protection diode.

NOTE: The cover must not be removed if the alternator is within the warranty period.

If the reading is 12 volts proceed to test 3.

Test 3

Connect the voltmeter between a good earth and the 'F' or green lead, or the case of the 14 TR.
Switch ignition on.
The voltmeter should indicate approximately .5 volt.
If the reading is 12 volts, the control box is faulty.
If the reading is .5 volt but on test 2 the reading was 12 volts, check the brushes and rotor.

Test 4

Connect the voltmeter between the battery insulated terminal and the alternator main output terminal.
Start and run the engine at a constant 3000 rpm, approximately.
The voltmeter reading should not exceed .5 volt.
If the voltmeter reading is higher than .5 volt, check the wiring from alternator to battery for loose or dirty connections.

Test 5

Start and run the engine at a constant 3000 rpm approximately. Check the voltage at alternator main output terminal and the 'IND' terminal.
The difference should not exceed .5 volt.
If there is a difference of .5 volt, change the rectifier pack.

Test 6

Disconnect the earth cable.
Connect the ammeter between the starter solenoid and the alternator main output cable.
Connect the voltmeter across the battery terminals.
Reconnect the earth cable.

Start and run the engine at a constant 3000 rpm approximately until the ammeter is reading 10 amps or less.
The voltmeter reading should be within 13.6 to 14.4 volts.
If the reading is outside these limits, change the control box.

IGNITION SYSTEM

Spark Plugs

The V12 HE engine uses Champion BN5 or Unipart GSP 360 plugs, these plugs have a taper shoulder and must not be over tightened or seizure within the cylinder head could result. The torque figure is 0.968 to 1.24 kgf-m (7 to 9 lbf ft).
Attention should be paid in cleaning the area around the plugs before removing.

Ignition protection relay

The ignition protection relay protects the ignition switch against overloading when all the auxiliary units are in operation such as the heated rear screen, wiper motors, etc.

ELECTRICAL FAULT DIAGNOSIS

IGNITION, ALTERNATOR AND STARTER SYSTEMS

CONSTANT ENERGY IGNITION

Operation

A voltage signal generated by the reluctor and pick up assembly is interpreted by the amplifier which switches on and off the current flowing in the primary winding of the HT coils. When a reluctor tooth passes across the pick up limb, the magnetic field strength around the pick up winding is intensified creating a voltage in the winding. The rise and fall of this voltage is sensed by the amplifier and is used to trigger the output of the transistorised amplifier.

Two HT coils are incorporated on the V12 HE engines. The main coil primary winding is connected in parallel with the primary winding of the auxiliary coil. The HT section of the auxiliary coil is not used and the HT outlet is sealed.

The auxiliary coil enables the ignition system to achieve the required performance at high engine speeds under load.

The coils used in this system are a nominal 6 volts. There is no separate ballast resistor in the circuit. The amplifier controls the maximum current flowing in the primary circuit.

The fuel injection lead to the ECU and the lead to the tachometer are taken from the amplifier. The distributor incorporates the standard automatic advance system. The traditional cam and contact breaker are replaced by an anti-flash shield, reluctor, and pick up assembly. The reluctor is a gear like component with as many teeth as there are cylinders to the engine and is mounted on th distributor shaft in place of the cam. The pick up consists of a winding around a pole piece attached to a permanent magnet, and is prewired with two leads terminating in a moulded two pin inhibited connector. During normal service the air gap between the reluctor and the pick-up does not alter and will only require re-setting if it has been tampered with. The normal setting is 0.20 to 0.35 mm (0.008 to .014 in).

The pick-up resistance should be 2 to 5 K ohms.

DISTRIBUTOR

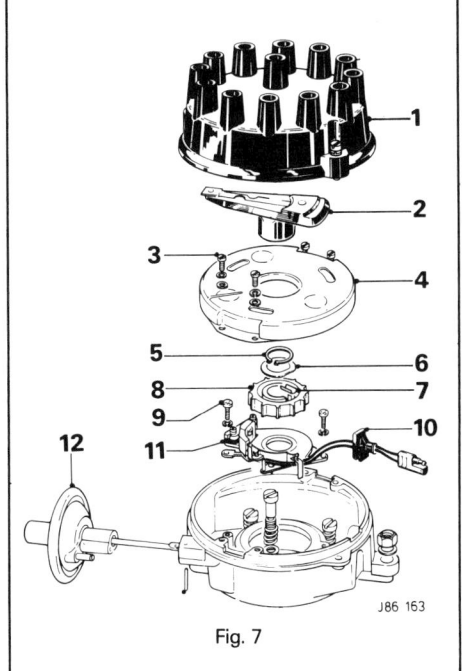

Fig. 7

KEY TO DISTRIBUTOR
1. Distributor Cap
2. Rotor
3. Flash Shield Screw
4. Flash Shield
5. Circlip
6. Flat Washer
7. Reluctor Key
8. Reluctor
9. Pick-up Screw
10. Grommet
11. Pick-up
12. Vacuum Unit

FAULT FINDING

Test 1

Check battery. A heavy discharge test applied to the battery terminals will determine whether the battery is capable of supplying the heavy currents required by the starter motor.
Check the specific gravity of the electrolyte in each cell. A variation of 0.040 in any cell means the battery is suspect.

Test 2

Check for HT spark. Remove the HT lead from the centre of the distributor cap 'A' and position the end of the lead approximately 6 mm (0.25 in) from a good earth on the engine. Crank the engine and if a spark is obtained, check the HT leads, spark plugs, distributor cover, and the rotor.

Test 3

With the ignition switched on. The voltage at the positive terminal of the MAIN coil 'C' should be 12 volts. If the voltage is below 11 volts check the wiring to/from the ignition switch 'G'.

Test 4

Disconnect the leads from the negative terminal of the MAIN coil 'C'. With the igniton switched on a 12 volt reading should be obtained from the negative terminal. A zero reading would indicate a faulty MAIN coil. If a 12 volt reading is obtained, reconnect the disconnected leads to the main coil and repeat Test 3 and 4 at the AUXILIARY coil 'D'.

Test 5

Disconnect the distributor pick-up leads from the amplifier 'B'. Measure the resistance of the distributor pick-up 'A'. It should be 2.2 to 4.8 K ohms. An incorrect reading indicates a faulty pick-up coil.

Test 6

Connect a voltmeter between the positive terminal of the battery and the negative terminal of the main coil 'C'. Switch on the ignition and the voltmeter should indicate a zero reading. Crank the engine and the voltmeter reading should rise to between 2 and 3 volts. If the voltmeter remains at zero the amplifier 'B' is suspect.

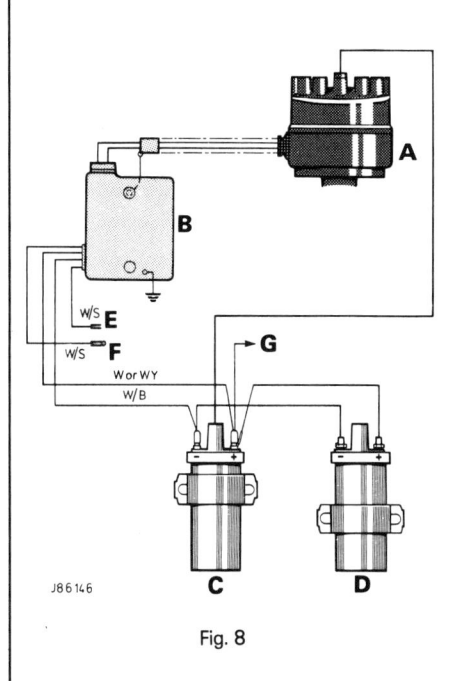

Fig. 8

86—9

ELECTRICAL FAULT DIAGNOSIS

FRONT AND REAR PARKING LAMPS

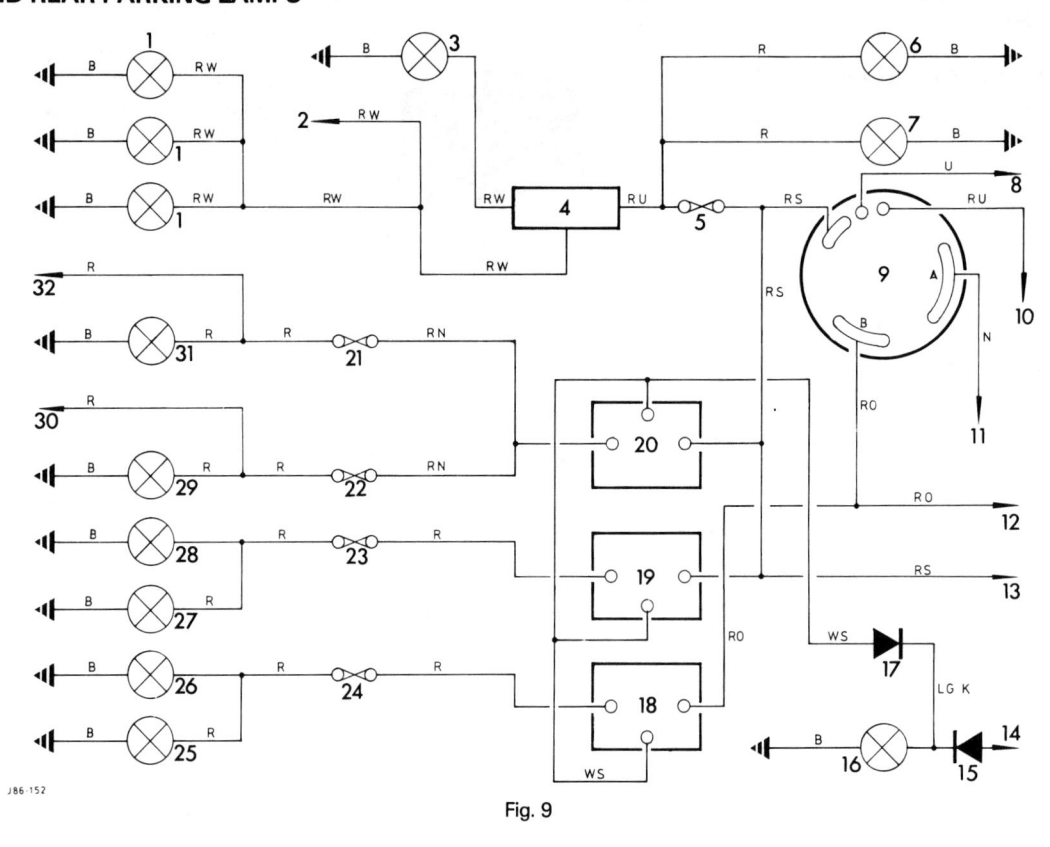

Fig. 9

KEY TO DIAGRAM

1. Switch and clock lamps
2. To panel lamps
3. Fibre optic lamp
4. Panel light rheostat
5. Fuse No 7
6. Cigar lighter lamp
7. Gearbox selector lamp
8. To headlamp relay
9. Master light switch
10. To headlamp inhibit relay
11. Terminal post
12. To caravan socket
13. To caravan socket
14. To stoplamp bulb failure unit
15. Warning lamp blocking diode
16. Warning lamp
17. Warning lamp blocking diode
18. RH rear bulb failure unit
19. LH rear bulb failure unit
20. Front parking lamp bulb failure unit
21. Fuse No 15
22. Fuse No 14.
23. Rear lamp in-line fuse
24. Rear lamp in-line fuse
25. RH Number plate lamp
26. RH rear lamp
27. LH Number plate lamp
28. LH rear lamp
29. LH front parking lamp
30. To side marker lamp (if fitted)
31. RH front parking lamp
32. To side marker lamp (if fitted)

FRONT AND REAR PARKING LAMPS

With the master light switch in the parking lamp on position current flows to the lamp via the switch contacts A and B, the bulb failure units and fuses. The current flowing through the bulb failure units will cause the bulb failure warning lamp to glow for 15 to 30 seconds. If the warning lamp fails to go out, then there is a bulb failure or a circuit fault in the front parking lamp, rear lamps, or number-plate lamps.

Fault Finding

Check the fuses and all connections, ensuring the earth connections are clean and tight.

With the master light switch in the parking lamp on position, battery voltage should be obtained at the B and the L terminals of the bulb failure unit.

If battery voltage is obtained at the B terminal but a zero reading at the L terminal replace the bulb failure unit.

ELECTRICAL FAULT DIAGNOSIS

HEAD AND FOG LAMPS

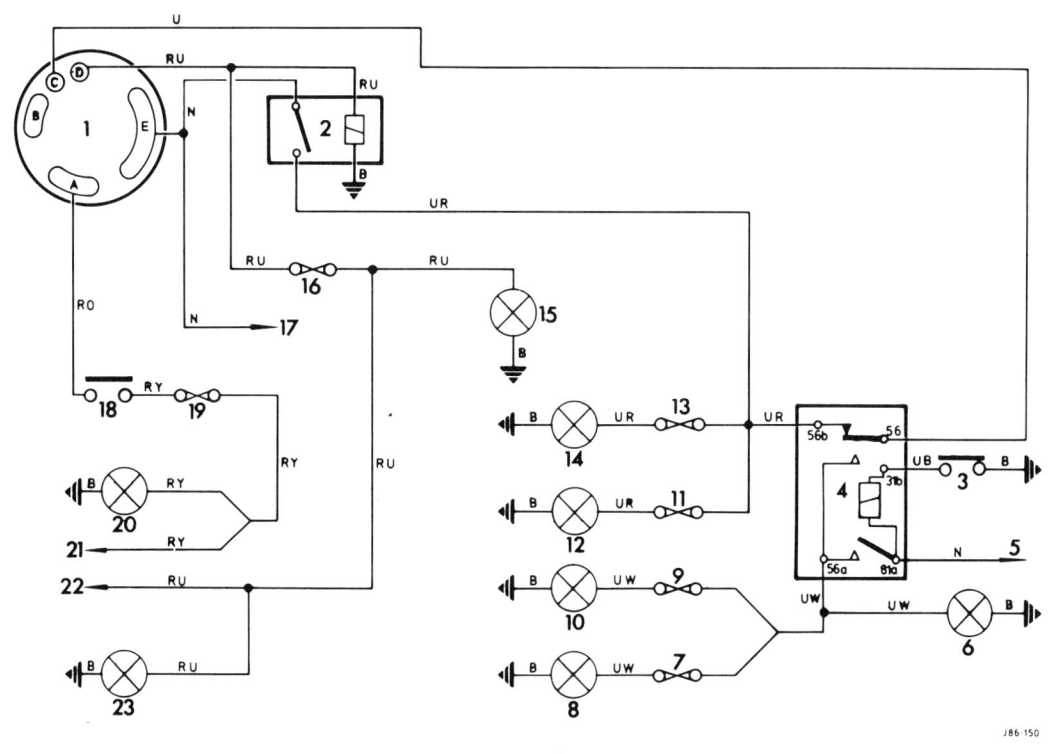

Fig. 10

KEY TO DIAGRAM

1. Master light switch
2. Headlamp inhibit relay
3. Headlamp flash switch
4. Headlamp relay
5. Terminal post
6. Main beam warning lamp
7. Fuse No 2
8. LH main beam
9. Fuse No 4
10. RH main beam
11. Fuse No 3
12. LH dip beam
13. Fuse No 5
14. RH dip beam
15. Fog lamp warning lamp
16. Fuse No 6
17. To terminal post
18. Fog lamp switch
19. Fuse No 1
20. Front fog lamp
21. RH fog lamp (if fitted)
22. LH rear fog lamp (if fitted)
23. Rear fog lamp

HEAD AND FOG LAMPS

With the master light switch in the headlamp position the contacts A, B, C and E are connected together in the main lighting switch to supply power to the headlamp relay. The headlamp flash switch activates the headlamp relay which in turn selects the main or dip headlamp filaments.

When the rear fog guard lamps are selected contacts A, B, D and E are connected together supplying power to the rear fog guard lamps. The headlamp inhibit relay is also energised supplying current to the headlamp dip filaments only.

The headlamp flash switch activates the headlamp relay, supplying current to the headlamp main filaments or dip filaments.

Fault Finding

Check the fuses and all connections. The earth connections should be clean and tight.

With the master light switch in the off position battery voltage should be obtained at terminal 81a of the headlamp relay. Short terminal 31b to earth and battery voltage should then be obtained at terminal 56a the headlamp main beam terminal. Should a zero reading be obtained replace the relay.

With the master light switch in the headlamp on position battery voltage should be obtained at terminal 56. By shorting terminal 31b on and off to earth, the relay should switch battery voltage alternately to terminals 56a and 56b. The main and dip beam headlamp terminals.

With master light switch in the fog-lamp position battery voltage should be obtained at terminals 85, 87 and 30/51 of the headlamp inhibit relay. If battery voltage is present at terminals 85 and 30/51 but not at terminal 87 replace the relay.

86—11

ELECTRICAL FAULT DIAGNOSIS

HEADLAMP WASH WIPE

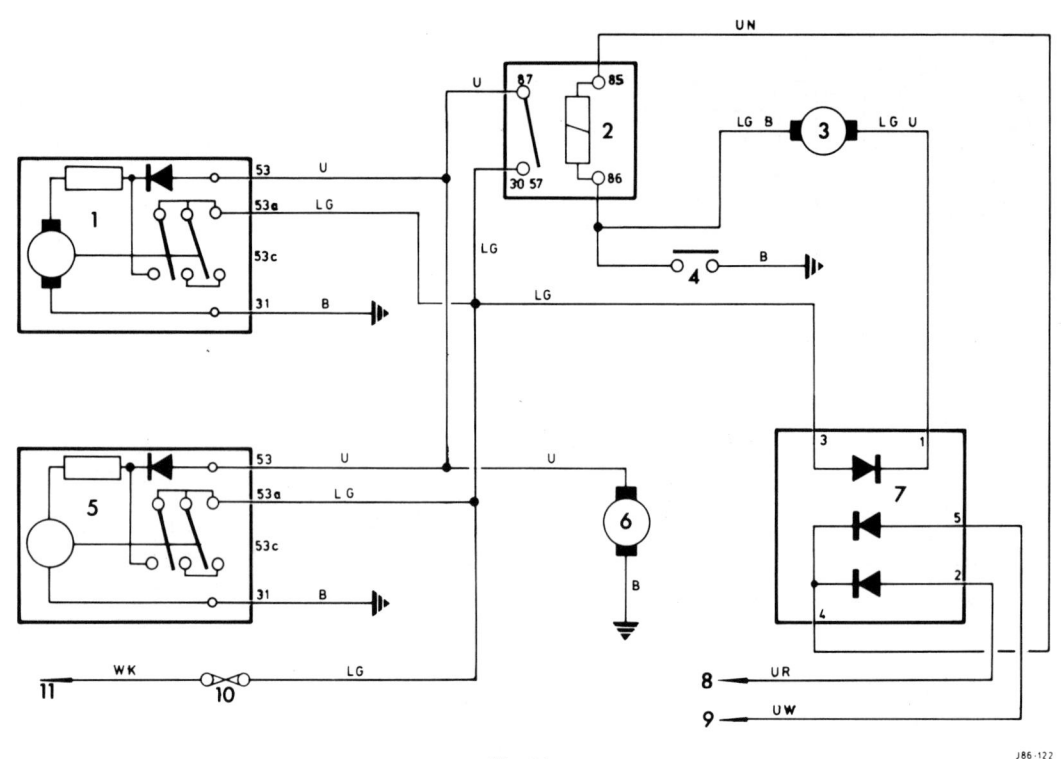

Fig. 11

KEY TO DIAGRAM

1. RH headlamp wiper motor
2. Wash wipe relay
3. Windscreen washer pump
4. Windscreen washer switch
5. LH headlamp wiper motor
6. Headlamp washer pump
7. Diode pack
8. To headlamp dip beam
9. To headlamp main beam
10. Fuse No 11

HEADLAMP WASH WIPE

Description

The headlamp wash wipe circuits will only be activated with headlamps switched to main or dip beam. When the headlamps are switched on power is applied to the wash wipe relay via a diode in the wash wipe diode unit. Power is also supplied to the windscreen washer motor via a diode in the wash wipe diode unit. When the windscreen washer switch is operated the circuit is completed to earth energising the wash wipe relay and the headlamp washer motor. Power is now supplied to the headlamp wiper motors via the relay contacts. When the windscreen washer switch is released the relay is de-energised thus switching off the headlamp washer motor. The headlamp wiper motors will continue to operate via power being applied to terminal 53a on the wipers until the wiper internal switch contacts open.

Fault Finding

Check that all connections are clean and tight. Check the fuse.
Ensure the earth connections are clean and tight.
With the ignition switched on. Battery voltage should be obtained at terminals 1 and 3 of the diode module. If battery voltage is obtained at terminal 3 but a zero reading at terminal 1 a faulty diode is indicated in the diode module.
With the headlamps switched to main beam. Battery voltage should be obtained at terminals 5 and 4 of the diode module., With the headlamps switched to dip beam. Battery voltage should be obtained at terminals 2 and 4 of the diode module. A zero reading at terminal 4 in either main or dip beam position indicates a faulty diode in the module.
With the ignition switched on. Battery voltage should be obtained at terminal 30/51 of the wipe/wash relay. With headlamps switched on. Battery voltage should be obtained at terminals 85 and 86 of the relay. When the windscreen washer is operated the terminal 86 of the relay should drop to zero, and battery voltage should then be obtained at terminal 87 of the relay. If the voltage remains at 12 volts at terminal 86 of the relay. Check the wash/wipe switch, and wiring. If the terminal voltage drops to zero voltage replace the relay.
With the ignition switched on, battery voltage should be obtained at terminal 53a of the headlamp wiper motor. With headlamps switched on, battery voltage should be obtained at terminal 53 of the wiper motor. Terminal 31 of the motor should be earthed. If the voltage reading and the earth are satisfactory remove the wiper motor and bench check.

ELECTRICAL FAULT DIAGNOSIS

SPEED CONTROL SYSTEM

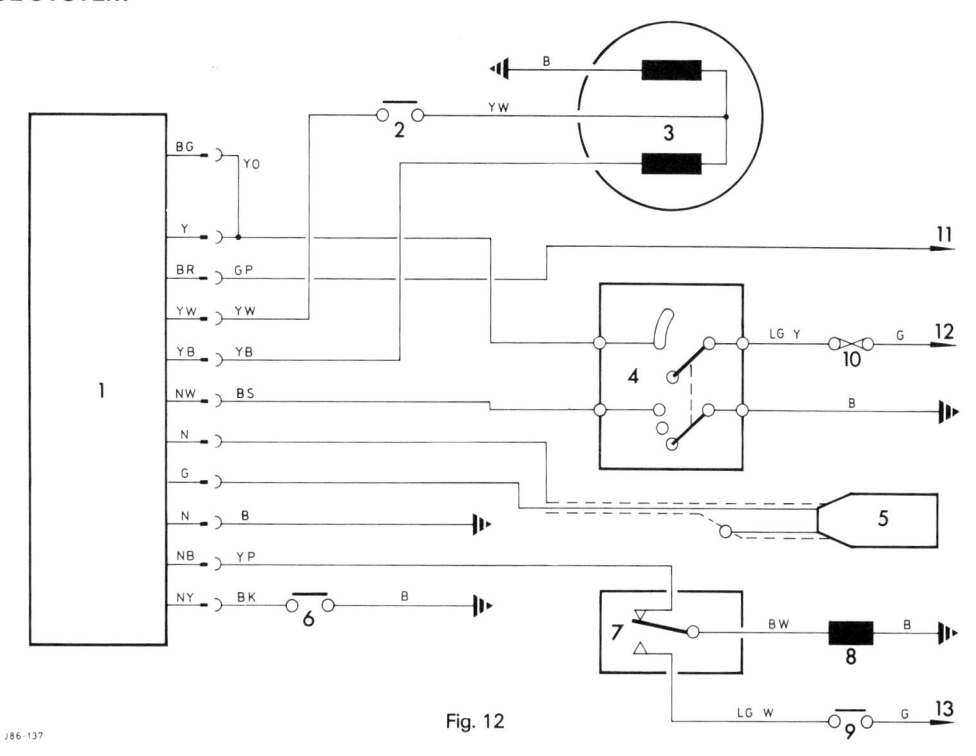

Fig. 12

KEY TO DIAGRAM

1. Speed control unit
2. Brake operated switch
3. Actuator
4. Control switch
5. Magnetic pickup
6. Set switch
7. Inhibit switch
8. Kick down solenoid
9. Kick down switch
10. Fuse No 17
11. To stoplamp switch
12. Hazard switch
13. To fuse No 5

Throttle Actuator

A vacuum operated Throttle Actuator is located under the bonnet, directly in front of the distributor.

The unit contains a control solenoid valve, which controls the vacuum allowed in the unit, a dump solenoid valve, which seals the unit when the system is engaged, and a set of bellows held extended by an internal spring.

One end of the bellows is fixed to the unit body, containing the valves, whilst the other is connected to the accelerator pedal mechanism, via the actuator cable.

An electrical signal transmitted by the speed control unit, triggers the solenoids, and as manifold vacuum evacuates the unit it overcomes the spring pressure, and the bellows compress so opening the throttle.

Speed Control Unit

The Electronic Speed Control Unit is situated on the passenger side under scuttle area, adjacent to the blower motor.

The unit receives signals from the speed transducer and sends out signals to the actuator solenoids, altering the vacuum in the actuator unit to open or close the throttle as required.

The control unit is engaged and the memory recorded when the set button is pressed. After an override the memorised speed may be recalled by operation of the 'RESUME' switch. The memory is cancelled when the control switch is moved to 'OFF'.

When the brake pedal is touched the unit signals the throttle actuator to dump its vacuum, causing the throttle to close.

Adjustment of the 'Set Speed' is possible via an access hole in the speed control unit.

Speed Sensing

The magnetic speed transducer is mounted to the rear suspension cradle, adjacent to the propshaft flange. The transducer transmits a signal to the control unit indicating the speed of the vehicle.

Two magnets mounted diametrically opposite each other on the differential/propshaft flange, generate a signal which is picked up by the transducer and sent to the speed control unit. The air gap between the magnets and transducer should be 7 mm ± 1 mm (0.275 ± 0.040 in).

Master ON/OFF Resume Switch

The Master switch is positioned at the rear of the gear selector cover. This switch has three positions, 'OFF', 'ON' and 'RESUME'.

The 'RESUME' mode is spring biased to the 'ON' position. The 'OFF' position ensures that the control unit is isolated from its supply voltage, and consequently cancels the control system memory.

Set Switch

The set switch is located in the end of the indicator stalk. The switch is used to trigger the speed control unit to bring the system into operation. A single push on the button will cause the vehicle to cruise within ± 1 mph of the speed at which the set switch was pressed. If the button is constantly depressed the vehicle will accelerate until the button is released.

Inhibit Switch

The inhibit switch, comprising a cam and micro-switch, is mounted on the gear selector mechanism, and inhibits the cruise control from operating in any selector position other than 'D'.

Brake Operated Switches

The existing brake light switch is utilised for cancelling the cruise mode when the brakes are applied. As an additional safety feature, a second brake pedal operated switch, which makes and breaks the direct feed signal from the control unit to actuator, is also fitted.

This second switch is mounted in tandem with the brake light switch, but operates fractionally later. When the switch is operated, the current to the actuator is cut.

This switch functions completely independantly to the brake light cancellation or control unit commands.

86—13

ELECTRICAL FAULT DIAGNOSIS

SPEED CONTROL SYSTEM

STATIC TEST

These tests are carried out using the ECONOCRUISE Installation Tester (Fig 13).

1. Disconnect the main harness multiplug connector from the electronic speed control unit.
2. Connect the Test Unit leads to the control unit and main harness plug.
3. Switch on the ignition, move the gear selector to the 'D' position; position the master switch to 'ON' and the Test Unit to the 'STATIC' position.
   Lamps 1, 2, 5, 7 and 9, should illuminate indicating correct continuity of the wiring.
4. To check the Neutral Gear position override, move the gear lever to 'N', lamp No 8 should illuminate.
5. The set switch can be checked, by pressing it, lamp No 3 should illuminate.
6. To check the 'RESUME' position of the main function switch; select 'RESUME' and lamp No 4 should illuminate.

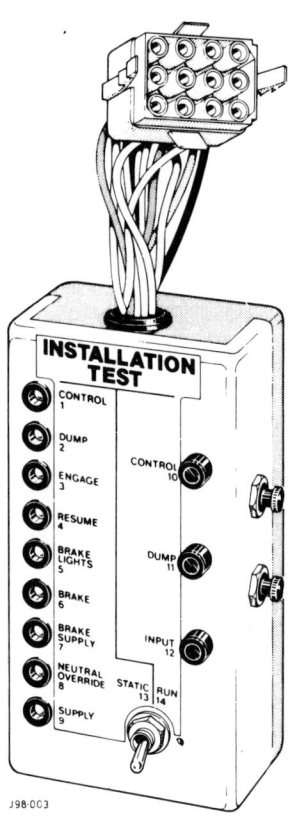

Fig. 13

7. To check the Brake Cancellation and Safety switches, press the brake pedal, lamps 1, 2 and 5, should extinguish and No 6 illuminate.
   If both lamps 5 and 6 illuminate together, check for an open circuit in the Brake Light Switch/Brake Light Circuit. If this circuit is not continuous, then the speed control system will not function.

Running Test

1. Switch the Test Unit to the 'RUN' position.
2. Start the engine, press and hold the dump and control buttons on the Test Unit, the engine revolutions should increase, release buttons. This test indicates that the actuator functions.
   Engage gear and drive off. The lower indicator lamp, on the run side of the Test Unit, should flicker, indicating the presence of an impulse from the Transducer Unit, mounted at the propshaft flange.
3. At approximately 80 km/h, 50 m/h press 'Set' button and release, the middle lamp marked 'Dump', should illuminate and the top light marked 'Control', should flicker. Press brake and both lamps should go out.

Checking the Set Speed

This check is the main dynamic test which ensures that there is no surge or drop off of the set speed when the engage command is given. Before carrying out this test, ensure that:

(i) The actuator cable is adjusted so that the free play at the actuator does not exceed 1 mm (0.040 in).
(ii) The air gap at the speed transducer is 7 ± 1 mm (0.275 ± 0.040 in).
(iii) The inhibit switch only operates in the 'D' position.

To check the set speed, proceed as follows:

1. Switch system on.
2. Drive at approximately 80 km/h 50 m/h on a quiet, flat road.
3. Engage the cruise control and remove foot from the accelerator pedal.
4. Record the speed at which the system is cruising.
5. Press and release the set button; allow the system to settle to the cruise speed again.
6. Note the new cruise speed. If the system is correctly set, then there should not be any increase or decrease in the noted cruising speed. If there is a change, then the Speed Control Unit will require adjustment.
7. Remove the rubber grommet from the side of the Control Unit and adjust the set speed potentiometer, with a suitable screwdriver, clockwise to increase or anti-clockwise to decrease the cruise speed.

Repeat the above procedure until the system is set correctly.

Fault Diagnosis

A. SYSTEM DOES NOT ENGAGE — at any speed above Low Speed Lock Out.

Causes:

1. Control unit malfunction.
2. Engage switch failure.
3. Inhibit switch failure.
4. Inhibit switch incorrectly adjusted.
5. Back-up switch failure.
6. Brake light switch, incorrectly adjusted, or failed. This switch must operate before the back-up switch.
7. Master switch malfunction.
8. Main fuse (No 12) blown.
9. Stop lamp fuse (No 11) blown.
10. Transducer air gap too large.
11. Transducer unit knocked out of alignment.
12. Loss of magnetic tab at transducer pick-up.
13. No vacuum supply to throttle actuator.

B. SYSTEM DOES NOT FUNCTION AT LOW SPEED

Causes:

1. Vehicle speed below low speed lock out. System not designed to function below 22 to 25 mph.
2. Loss of magnetic tab at transducer pick-up.

C. SYSTEM WILL NOT RESUME

Causes:

1. Speed control unit malfunction.
2. Low speed lock out speed too high — loss of magnetic tab at transducer pick-up (Vehicle also will not engage at low speeds).

D. SYSTEM HUNTS AT LOW SPEED

Causes:

1. Air gap at transducer too large. Reset to 7 ± 1 mm (0.275 ± 0.040 in).
2. Actuator cable too slack.
3. Control unit malfunction.

ELECTRICAL FAULT DIAGNOSIS

WINDSCREEN WIPER CIRCUIT

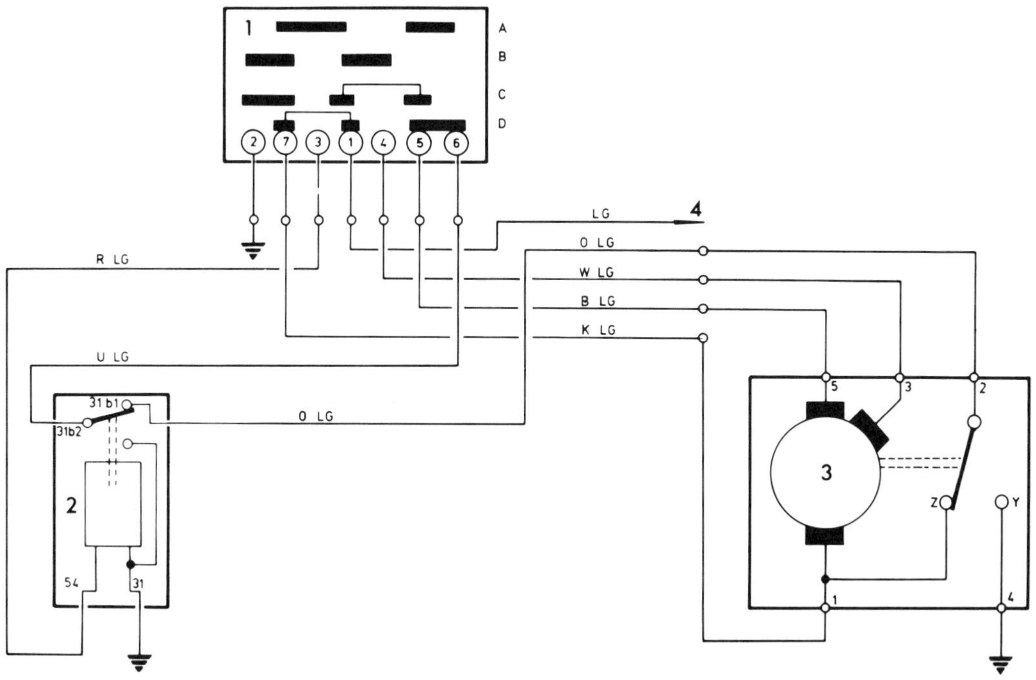

KEY TO DIAGRAM

1. Control switch
2. Delay unit
3. Wiper motor
4. To fuse No 9

Switch Positions

A Intermediate/Flick wipe
B Fast wipe
C Slow wipe
D Off

Park Switch Position

Z Park
Y Run

Fig. 14

WINDSCREEN WIPERS

DESCRIPTION

The following description of the Windscreen Wiper Operation should be studied with the accompanying circuit diagram.

OFF The Load Relay is energised by selection of 'IGN ON' and applies 12 volts through fuse 7 to the terminal 1 on the wiper switch. 12 volts is internally fed via the wiper switch contacts to terminal 7 and from there passes to terminal 1 on the Wiper Motor. This applies 12 volts to one side of the motor winding and, via the closed 'PARK' switch to the closed contacts of the Wiper Timer at terminal 31 b1. Terminal 31 b2 of the timer passes the 12 volts via terminal 6 of the switch and internal connection to terminal 5. From terminal 5 the 12 volt supply is applied to terminal 5 on the motor and thereby the opposite brush to terminal 1, completely stalling the motor.

SLOW The 12 volts on terminal 1 of the switch is applied to terminal 5 and to terminal 5 on the wiper motor. An earth connection to terminal 2 on the switch is internally applied to terminal 7 and thereby to terminal 1 on the motor. The motor then runs in a forward direction at a slow speed.

FAST The earth connection detailed in 'SLOW' is maintained but the 12 volt supply is moved via terminal 4 on the switch and terminal 3 on the motor to the high speed brush. The motor then runs in a forward direction at a high speed.

OFF When 'OFF' is selected the 'RUN' position of the Park/Run contact in the motor applies an earth at connection 4 of the motor via connection 2 and the timer contacts to position 6 on the switch. Then via the internal connection to 5 on the switch applying the earth to terminal 5 on the motor and the brush previously supplied with 12 volts. The motor therefore stops. Meanwhile 12 volts is applied to the previously earthed motor contact 1 via switch contacts 1 and 7; the motor therefore immediately stops the sweep and runs in a reverse direction to the limit of its travel. At that point the internal Park/Run switch moves to the 'PARK' position, removes the earth at 4 and applies 12 volts via the timer contact to the opposite brush. The motor instantly stalls.

Single Sweep Operation

When this position is selected and released, 12 volts are applied both to the coil of the timer unit, operating its contact, and to the terminal 1 on the motor. An earth is applied to the opposite brush of the motor, terminal 5 via the timer contact 31 b2 and an internal connection to terminal 31. The motor therefore starts and runs in a reverse direction. As the supply to the timer is applied then instantly removed the contacts operate then relax to the 'AT REST' position. The motor earth on the brush of terminal 5 is then achieved through the Park/Run contacts at the 'RUN' position until the motor reaches the end of the sweep. The Park/Run switch then returns to the PARK position, applying 12 volts to brush 5 as previously described. The motor therefore stalls.

Intermittent Operation

When position 'D' is selected 12 volts are applied to the operating coil of the timer and simultaneously to terminal 1 of the wiper motor. Terminal 5 of the wiper motor is connected via terminal 5 and 6 of the switch to the timer contact 31 b2(1). When 12 volts are applied to the timer its operation consists of an instantaneous 'flip flop' then a delay of approximately 5 seconds and another 'flip flop'.

This action applies instantaneous earth to terminal 5 of the motor, starting it running in a reverse direction. The motor then obtains its own earth via the Park/Run contact at 'RUN' to sustain it for one sweep. As the motor is running in a reverse direction, the 'PARK' switch closes at the end of the sweep and applies 12 volts via the new closed contacts of the Timer to the opposite brush to stall the motor in the 'PARK' position.

After approximately 5 seconds the Timer again 'flip flops' removing the 12 volts from the motor brush at 5 and replaces it with an earth. The cycle already described then repeats itself until the Wiper Switch is restored to the 'OFF' position cancelling the operation of the timer contacts. The motor then stalls as previously described.

Fault Finding

Check the fuse and all connections ensuring the earth connections are clean and tight.
With the ignition switched on battery voltage should be obtained at terminals 1, 2 and 5 of the wiper motor. Battery voltage should also be obtained at terminals 31 b1 and 31 b2 on the delay unit. With the wiper switch in the slow run position and the ignition switched on battery voltage should be obtained at terminal 5 of the wiper motor. The wiper motor earth circuit is via terminals 7 and 2 of the wiper switch. In the fast run position battery voltage should be obtained at terminal 3 of the wiper motor. The earth circuit is the same as the slow run position.
In the intermittent wipe position battery voltage should be obtained at terminal 1 of the wiper motor and terminal 54 on the delay unit. The earth is switched intermittently via the delay unit and the Park/Run switch in the wiper motor.

86—15

ELECTRICAL FAULT DIAGNOSIS

HEATED REAR WINDOW

Description

The timer unit switches off the heated rear screen circuit approximately ten minutes after the heated rear screen circuit has been switched on. The timer resets whenever the circuit is switched off or when the ignition is switched off.

A small relay, and the electronic components are mounted on a circuit board inside a yellow thermoplastic cover.

The unit is polarity conscious but reverse polarity connection will not result in damage.

The terminal 2 is connected to a positive supply via fuse 12, terminal 3 is connected to earth. When terminal 1 is switched to earth by the control switch the relay is energised connecting terminals 2 and 4 together, thus supplying power to the heated rear screen, heated door mirrors and the warning light. When the timing cycle is completed the relay is de-energised and the relay contacts open.

Fault Finding

Check the fuse and all connections ensuring the earth connections are clean and tight.

With the ignition switched on and a voltmeter connected between a good earth and terminal 2 of the relay, the voltmeter should indicate battery voltage. With the ignition switched on, the heated rear screen switched on and the voltmeter connected between terminals 1 and 2 of the relay, the voltmeter should indicate battery voltage. If a satisfactory reading was obtained in the first check but a zero reading is obtained in the second check, check the wiring to/from the control switch and the control switch itself. With the ignition and the heated rear screen switched on battery voltage should be obtained at terminal 4 of the relay. If terminal 4 gives a zero reading and the previous tests are satisfactory replace the relay.

HEATED REAR WINDOW AND DOOR MIRRORS

KEY TO DIAGRAM

1. Delay unit
2. Control switch
3. Fuse No. 12
4. To ignition switch protection relay
5. Heated rear window
6. RH door mirror
7. Warning lamp
8. LH door mirror

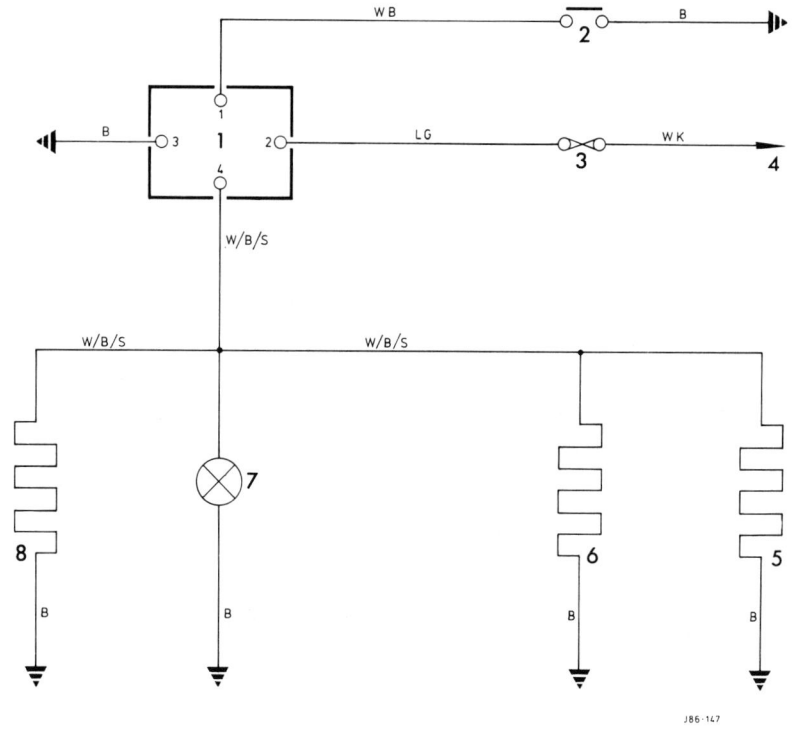

Fig. 15

ELECTRICAL FAULT DIAGNOSIS

RADIO, CLOCK AND LUGGAGE COMPARTMENT LAMPS

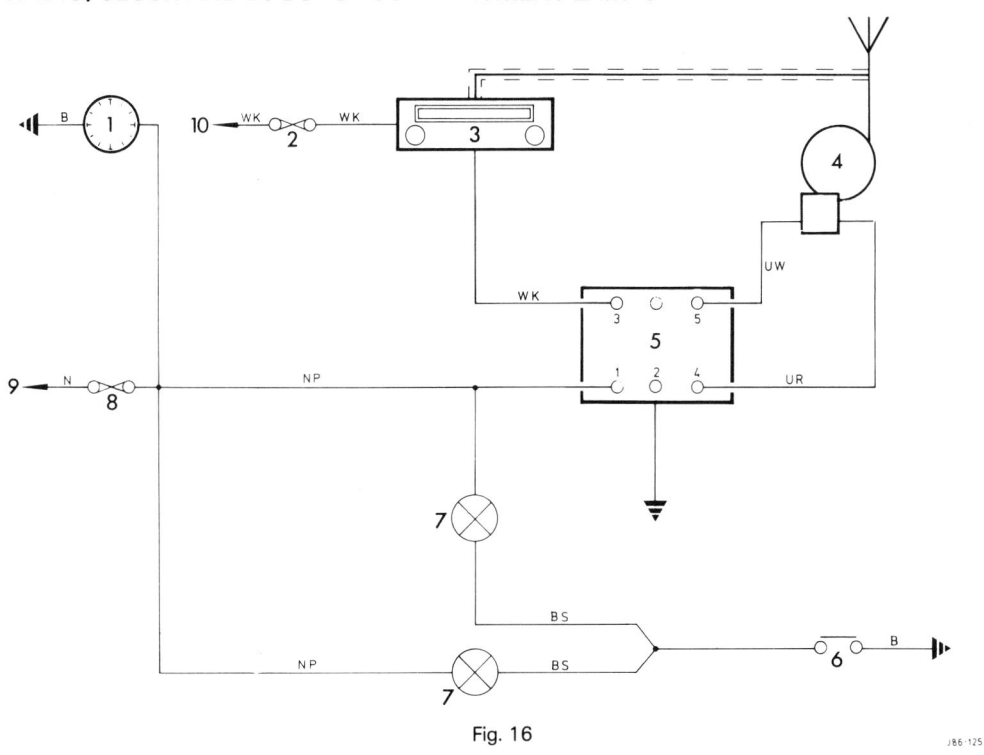

KEY TO DIAGRAM

1. Clock
2. Line fuse
3. Radio
4. Aerial motor
5. Delay relay
6. Boot lamps switch
7. Boot lamps
8. Fuse No 3
9. Terminal post
10. To ignition switch

Fig. 16

AERIAL

Circuit

The aerial maintains its extended position for a period of 10 seconds after the radio has been switched off. The aerial will then retract. If the radio is switched on before the 10 seconds has elapsed, the aerial will remain in the extended position.

With terminal 1 of the relay connected to a positive supply via fuse 5 and the terminal 2 connected to earth. The positive supply from the radio to terminal 3 of the relay turns on a transistor which charges a capacitor in a timing circuit. It also activates a relay which connects terminals 1 and 5. This allows current to flow to the aerial motor and thus drive the aerial to the extended position. On switching the aerial off the aerial will remain extended for a period of 10 seconds then the relay will become de-activated the contact between terminals 1 and 5 will become disconnected, but the contact will be made between terminals 1 and 4. This will allow current to flow to the aerial motor to retract the aerial.

Fault Finding

Should the aerial fail to extend when the radio is switched on.

Check the fuse and all connections ensuring the earth connections are clean and tight. Battery voltage should be obtained at terminals 1 and 3 of the delay unit. If the battery voltage is obtained bridge the leads from terminals 1 and 5 together. Should the aerial now extend replace the delay unit.

86—17

ELECTRICAL FAULT DIAGNOSIS

INSTRUMENT CIRCUIT

KEY TO DIAGRAM

1. To fuse No. 4
2. Temperature gauge
3. Oil pressure gauge
4. Fuel gauge
5. Battery gauge
6. Tachometer
7. To amplifier
8. To ignition switch
9. Low fuel warning lamp
10. Fuel tank unit
11. Oil pressure transmitter
12. Coolant temperature transmitter

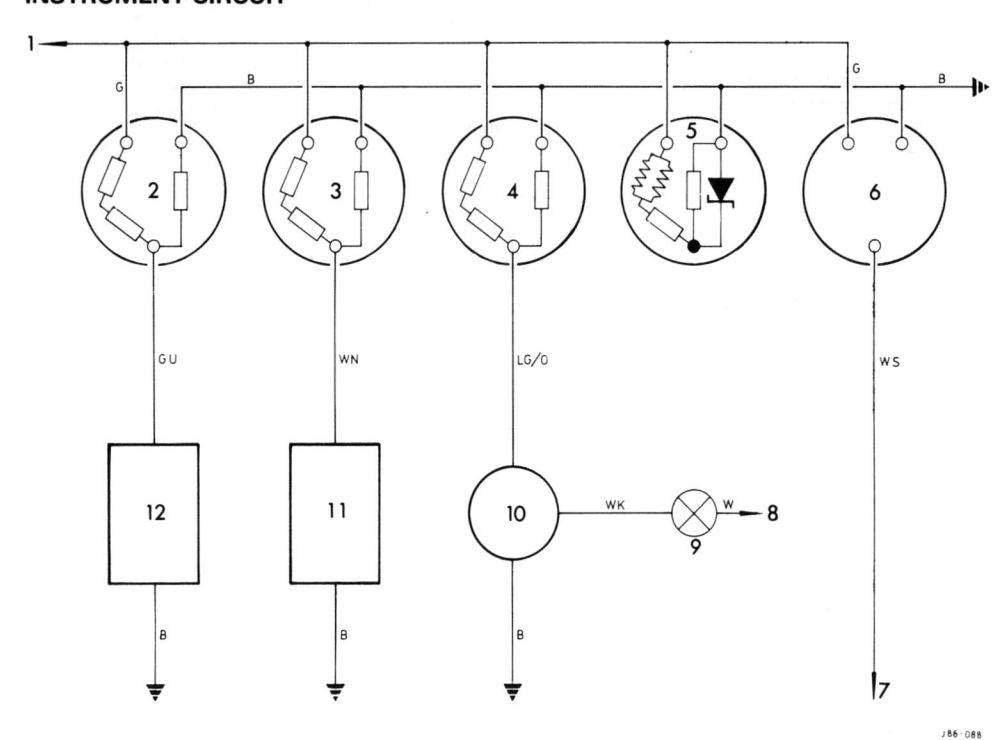

Fig. 17

INSTRUMENTS

Description

The fuel gauge, temperature gauge, oil gauge and battery gauge, are all air cored instruments. An air cored instrument can be considered as a magnet in a magnetic field. A bar magnet pivoted in its centre is mounted in the centre of three coils. To cause the magnet and therefore the pointer to move, the current flow in one of the three coils is changed, which in turn alters the magnetic strength in that coil. The transmitter or sensors are all variable resistors, and are connected across or in parallel with one of the coils. As the resistance varies in the transmitter the current flow through the coil will also vary. This will alter the magnetic field strength causing the pointer to move.

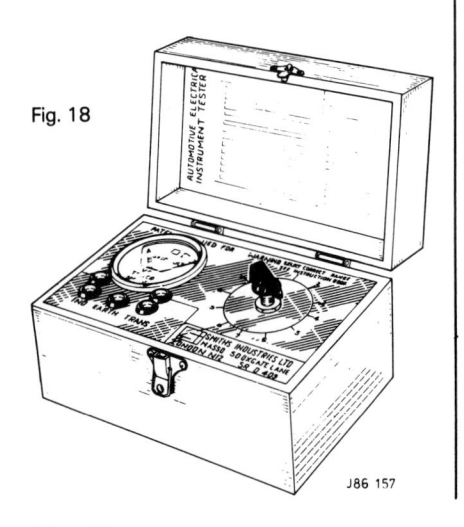

Testing

Instrument tester SR/409 (Fig. 18) is recommended for testing the instruments on the vehicle.
Plug the black lead into the earth socket of the instrument tester and the red lead into the A range socket. Move the switch to position 5. Disconnect the lead from the tank unit, temperature transmitter, or oil pressure transducer which ever is appropriate, and connect the disconnected lead to the red lead from the tester. Connect the black lead from the tester to earth.
Switch the ignition on and the appropriate gauge should move to the lower calibration marks. The calibration marks are two small dots which appear above or below the instrument scale.
Move the tester switch to position 6, the needle of the gauge should now move to high calibration marks.

Fig. 18

86—18

ELECTRICAL FAULT DIAGNOSIS

FLASHER INDICATORS

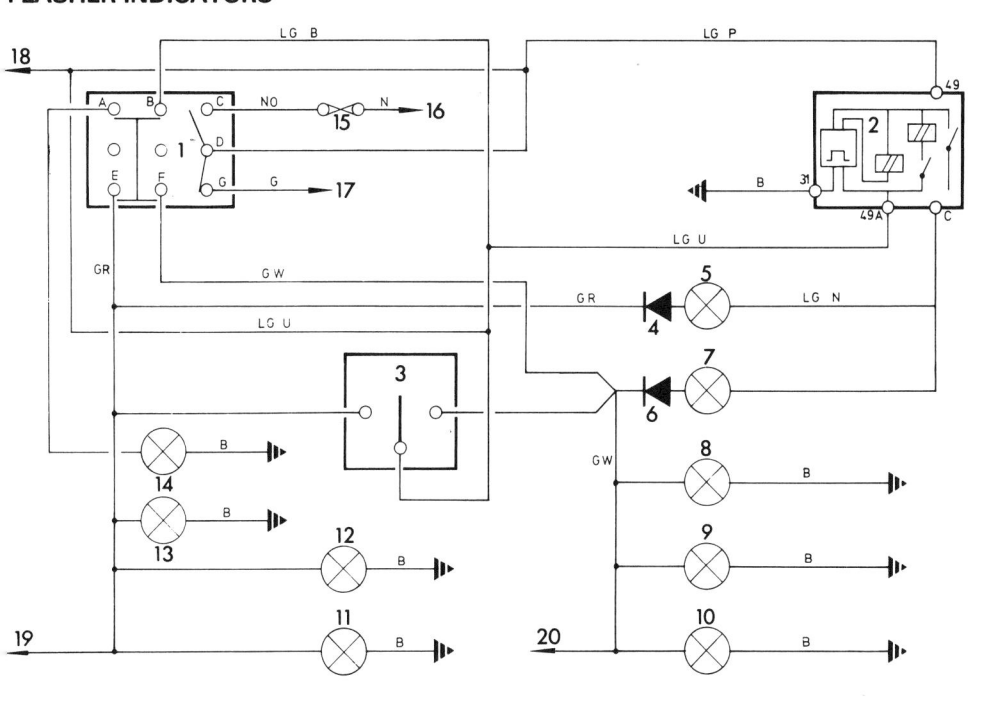

Fig. 19

KEY TO DIAGRAM

1. Hazard switch
2. Flasher unit
3. Flasher switch
4. Blocking diode
5. LH warning lamp
6. Blocking diode
7. RH warning lamp
8. RH repeater lamp
9. RH front indicator lamp
10. RH rear indicator lamp
11. LH rear indicator lamp
12. LH front indicator lamp
13. LH repeater lamp
14. Hazard warning lamp
15. Fuse No 2
16. To terminal post
17. To fuse No 5
18. To acoustic unit
    Australia Only
19. Caravan socket
20. Caravan socket

FLASHER LAMPS

Description

With the ignition switched on and the left hand flasher lamps selected, current flows at the appropriate flash rate from fuse No 5 to the flasher lamps via the hazard switch, flasher unit and the flasher switch. The warning light is supplied with flashing signal via the C terminal on the flasher unit. The circuit to earth for the warning light is diode 6 and the right hand flasher lamps.

When the hazard lamps are selected terminals C and D in the hazard switch are connected. The terminals D and G are disconnected. Terminals A, B, E and F are connected together. This allows current to flow to all the flasher lamps via fuse No 2, the hazard switch and the flasher unit.

Fault Finding

Check the fuse and all connections ensuring the earth connections are clean and tight.

With the ignition switched on battery voltage should be obtained at terminal 49 on the flasher unit. If a zero reading is obtained check the wiring to/from the hazard lamp switch and the hazard lamp switch itself.

With the ignition switched on and the flasher lamps switched on to either right or left hand. Bridge terminals 49 and 49A together with an ammeter. If the lamps now illuminate and the ammeter registers approximately 3.5 amps replace the flasher unit.

Should the lamps still fail to illuminate check the flasher switch and wiring.

86—19

ELECTRICAL FAULT DIAGNOSIS

WINDOW LIFT CIRCUIT

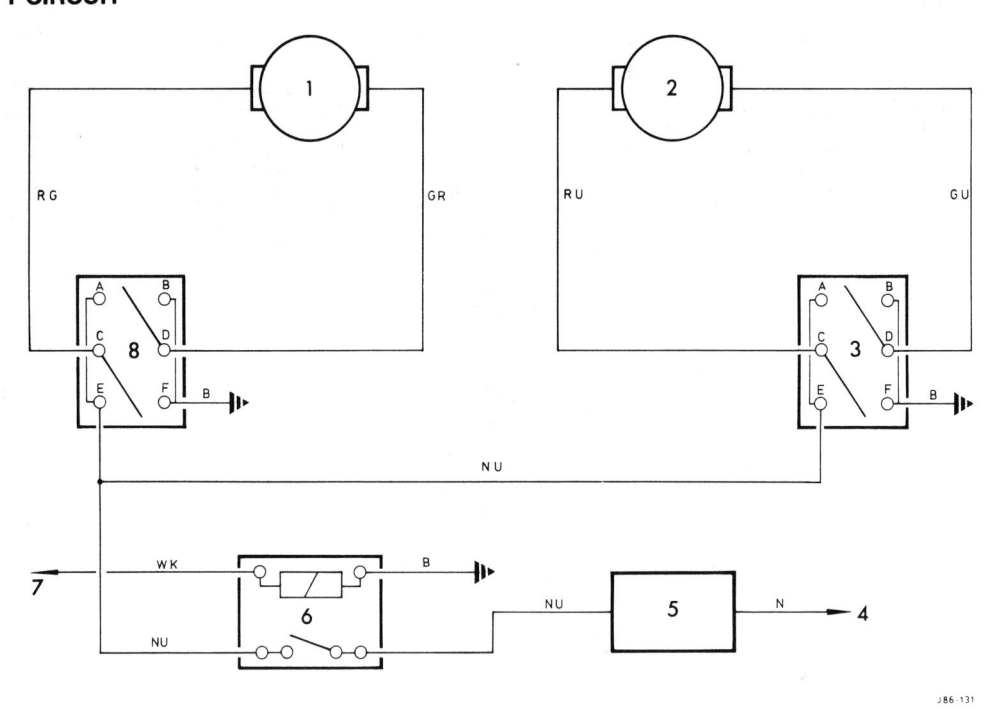

Fig. 20

KEY TO DIAGRAM

1. LH window lift motor
2. RH window lift motor
3. LH switch
4. RH switch
5. Thermal circuit breaker
6. Window lift relay
7. To ignition switch

ELECTRICALLY OPERATED WINDOWS

Description

Power is supplied to the relay contacts from the main battery supply via a thermal circuit breaker. The window lift relay is activated when the ignition is switched on. When the control switch is operated to wind the windows down, the contacts AD and CF are connected. This allows current to flow to the motor via contacts AD and the circuit to earth is completed via contacts CF When the control switch is operated to wind the window up, the contacts BD and CE are connected. This allows the current to flow through the motor in the opposite direction via contacts CE and the circuit to earth is completed via contacts BD

Fault Finding

Check the fuse and all connections, ensuring the earth connections are clean and tight.
Check the thermal circuit breaker by joining the circuit breaker leads together. Switch on the ignition and operate the window lift switches.

Should the windows operate satisfactorily replace the circuit breaker.
With the ignition switched off battery voltage should be obtained at terminal C1 of the window lift relay. With the ignition switched on battery voltage should also be obtained at terminals W1 and C2 of the relay. If battery voltage is obtained at terminals W1 and C1 but a zero reading at terminal C2 replace the relay.
With the ignition switched on battery voltage should be obtained at the brown and blue lead terminal of the left hand window lift switch. Operate the switch. Battery voltage should now be obtained at the red and green lead terminal when the switch is operated in one direction, or the green and red lead terminal when the switch is operated in the opposite direction. Should a zero reading be at either test point replace the switch.
The same checks apply for the right hand window lift switch. Noting the switch cable colours are red and blue for one direction. Green and blue for the reverse direction.
If the checks prove satisfactory, check the window lift motor wiring for continuity. Should the wiring prove satisfactory remove the window lift motor for bench checks.

86—20

ELECTRICAL FAULT DIAGNOSIS

INTERIOR LAMPS

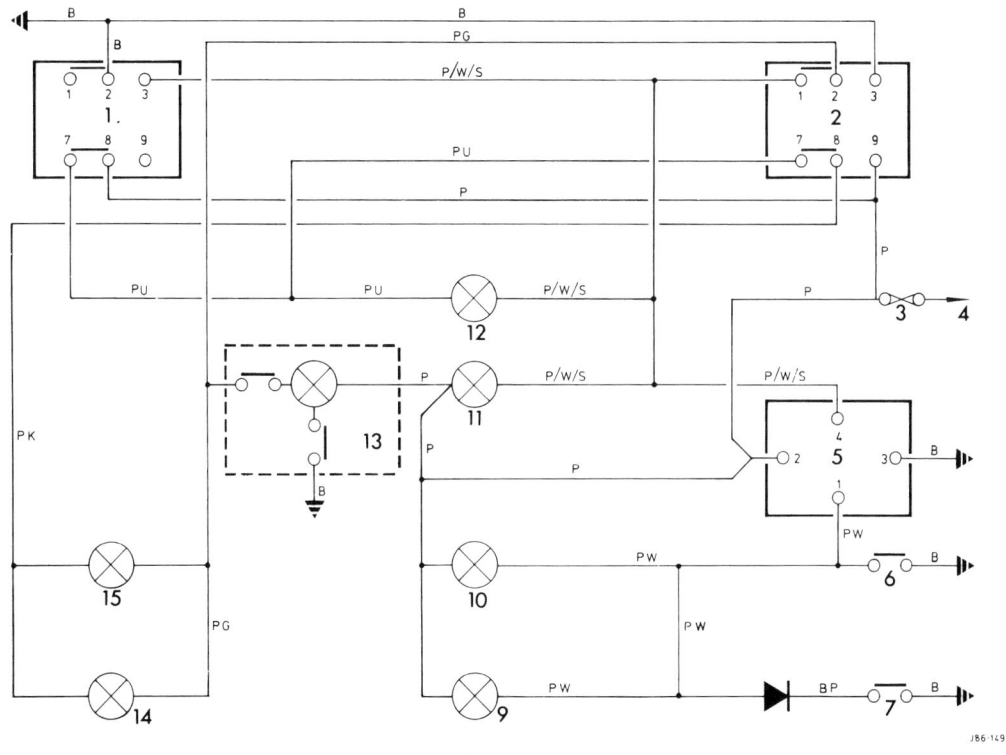

Fig. 21

KEY TO DIAGRAM

1. Map light switch
2. Interior light switch
3. Fuse No 13
4. To terminal post
5. Delay unit
6. LH door switch
7. RH door switch
8. Blocking diode
9. Puddle lamp
10. Puddle lamp
11. Front passenger lamp
12. Drivers lamp
13. Roof lamp
14. Rear passenger lamp
15. Rear passenger lamp

INTERIOR LAMPS

Description

The courtesy lamp delay unit controls the operation of the vehicle interior lamps so that they remain on for approximately 10 seconds after the doors are closed. The puddle lamps are not affected by the delay unit and will switch on and off as the doors are opened and closed.

The delay switch is polarity conscious, but a reverse polarity connection will not result in damage to the unit.

With terminal 2 connected to a positive supply via the fuse No 13 and terminal 3 connected to earth. When terminal 1 is earthed via a door switch a transistor charges a capacitor in a timing circuit which joins terminals 3 and 4 together via an internal relay. When terminal 1, earth circuit, is broken (a door closed) the capacitor commences to discharge turning off the relay at the end of the prescribed period which in turn switches off the interior lamps.

Power is supplied to the rear passenger lamps via the map and interior lamp switches. Power supplied to the drivers lamp via the map lamp switch. The roof lamp, front passenger lamp and the puddle lamps are supplied with power from the fuse. The circuit to earth for the rear passenger lamps and the roof lamp is through the interior lamp switch and the delay unit. The circuit to earth for the driver and passenger lamps is through the delay unit. With the interior lamps switched on by the panel switch the delay unit is by-passed and therefore the delay unit will not operate.

Fault Finding

Check the fuse and all connections ensuring the earth connections are clean and tight.

Battery voltage should be obtained at terminal 2 of the delay unit. If the voltage is satisfactory bridge terminal 1 of the delay unit to earth. If the interior lamps now illuminate, check the door switches, and the wiring to the door switches. If the lamps still operate unsatisfactorily bridge terminals 3 and 4 together on the delay unit. Should the lamps now illuminate replace the delay unit.

ELECTRICAL FAULT DIAGNOSIS

WARNING LIGHT FAILURE UNIT AND STOP LAMP BULB FAILURE UNIT

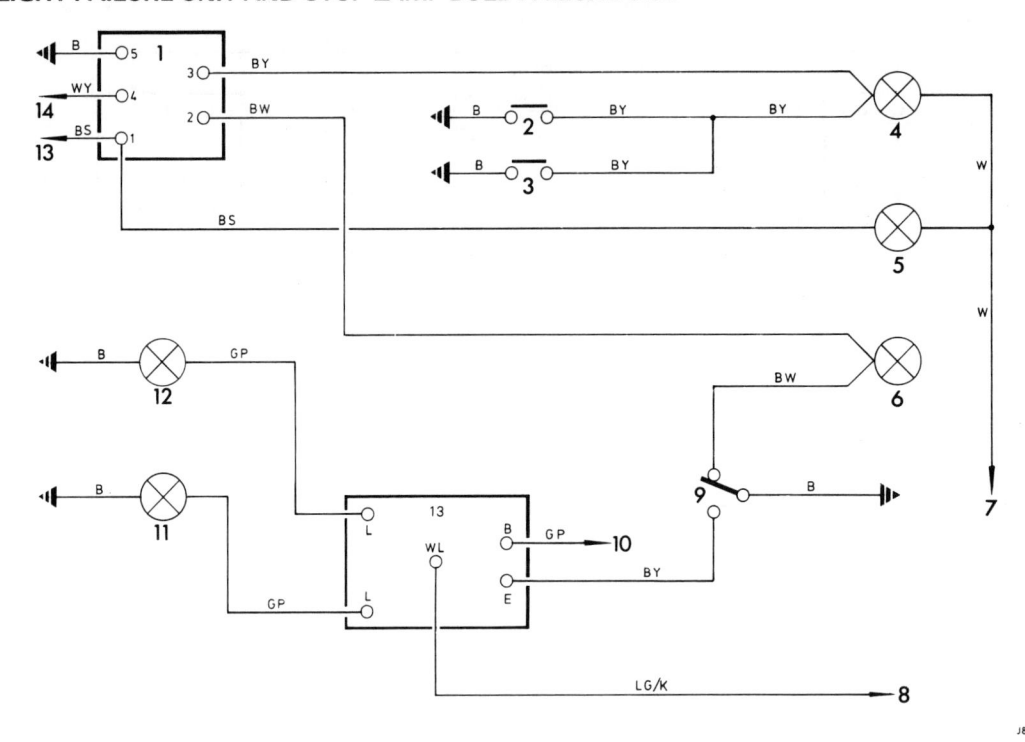

Fig. 22

KEY TO DIAGRAM

1. Warning light failure unit
2. Reservoir brake switch
3. PDWA
4. Brake warning lamp
5. Oxygen warning lamp
6. Handbrake warning lamp
7. To ignition switch
8. To bulb failure warning lamp via blocking diode
9. Handbrake switch
10. To footbrake switch
11. LH stop lamp
12. RH stop lamp
13. Stop lamp bulb failure unit

WARNING LIGHT FAILURE UNIT

Description

When the starter motor is activated a voltage supply to the warning light failure unit switches a transistor circuit to earth which completes the circuit of the oxygen warning lamp, brake warning lamp, and the handbrake warning lamp, causing the lamps to glow. This indicates that the warning lamps are operating satisfactorily. A failure of a warning lamp should be investigated immediately.

Fault Finding

To test the warning light failure unit, switch the ignition on and short terminal 1 of the brake light failure unit to earth. The oxygen sensor warning lamp should glow. Should the warning lamp fail to glow check the warning lamp bulb, and the warning lamp supply voltage.
Repeat the test at terminal 2 for the handbrake warning lamp, and terminal 3 for the brake warning lamp. If the above checks prove satisfactory replace the failure unit.

STOP LAMP BULB FAILURE UNIT

Description

The stop lamps are supplied with current via a bulb failure unit. Should the bulb failure warning lamp glow with the master light switch off, the ignition switched on, the handbrake released and the foot brake depressed, a circuit fault or a faulty bulb is indicated.

Fault Finding

To test the stop lamp failure unit switch on the parking lamps. The bulb failure warning lamp should glow for 30 seconds this proves the warning lamp is satisfactory. Switch off the parking lamps and switch the ignition on. Remove a stop lamp bulb, release the handbrake, and depress the foot brake. The warning lamp should glow. Should the warning lamp fail to glow replace the bulb failure unit.

ELECTRICAL FAULT DIAGNOSIS

ELECTRICALLY OPERATED DOOR LOCKS

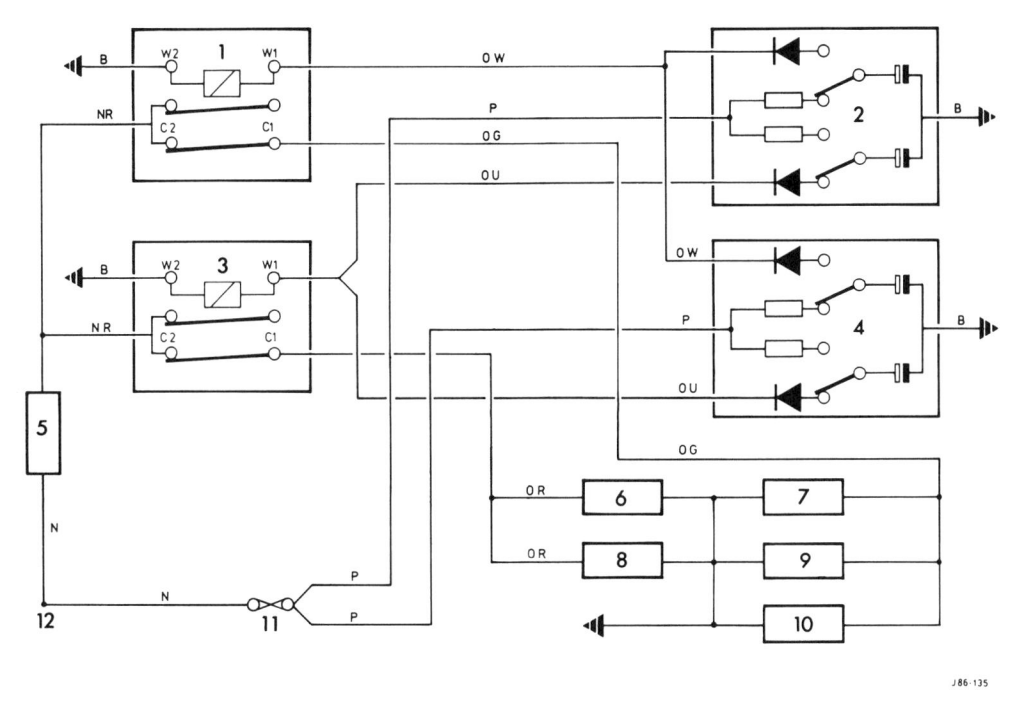

Fig. 23

KEY TO DIAGRAM

1. Lock relay
2. Trigger unit
3. Unlock relay
4. Trigger unit
5. Thermal circuit breaker
6. LH unlock solenoid
7. LH lock solenoid
8. RH unlock solenoid
9. RH lock solenoid
10. Boot lock solenoid
11. Fuse No 8
12. Terminal post

ELECTRICALLY OPERATED DOOR LOCKS

Description

The electric door lock circuit is activated from either of the two doors if the key is turned in either door lock, the two doors and the boot lock will activate into the lock position, or the doors will unlock leaving the boot compartment solenoid in the locked position.

The two interior door lock control levers will also operate the two doors and the boot solenoid into the locked position, and the doors into the unlocked position. The 12 volt supply is taken from the terminal post through a thermal circuit breaker to the lock and unlock relays. 12 volts is also taken from fuse No 8 to the resistors in the right and left hand trigger units. The trigger units contain 2 diodes, 2 resistors, 2 electrolytic capacitors and 2 micro switches. The capacitors are charged to the battery potential via the resistors.

If one of the trigger units is turned to the unlocked position (by the door key or by the interior levers) the micro switches are moved to the unlock position, allowing the capacitor to discharge through a diode a surge of current to the unlock relay coil. This action closes the relay contacts thus completing the circuit to the door solenoids. After the initial discharge from the capacitor the unlock relay will de-activate reverting to its normal contacts open position. The second trigger unit will be moved to the unlocked position mechanically by the action of the solenoid to which it is attached. While the trigger units are kept in the unlocked position the door lock capacitors are kept in a charged state.

The lock and unlock circuits are protected from each other by the diodes.

Fault Finding

Should the door locks not operate electrically check that all the connections are secure ensuring that the earth connections are clean and tight. Check the fuse.

Check the thermal circuit breaker by joining the thermal breaker leads together. If the door locks now operate replace the thermal circuit breaker.

To check the lock or unlock relay, battery voltage should be obtained at the C1 terminal of the appropriate relay.

Disconnect the lead from the W1 terminal and connect a voltmeter between the disconnected lead and a good earth. Operate the locks and battery voltage will be obtained from the disconnected lead which will gradually drop. If a zero reading is obtained at the disconnected lead check that battery voltage is obtained on the purple lead of the trigger unit, if satisfactory replace the door lock solenoid assembly.

86—23

ELECTRICAL FAULT DIAGNOSIS

ELECTRICALLY OPERATED DOOR MIRRORS

KEY TO DIAGRAM

1. RH mirror switch
2. RH mirror vertical motor
3. RH mirror horizontal motor
4. LH mirror horizontal motor
5. LH mirror vertical motor
6. LH mirror switch
7. To fuse No 8
8. To fuse No 8

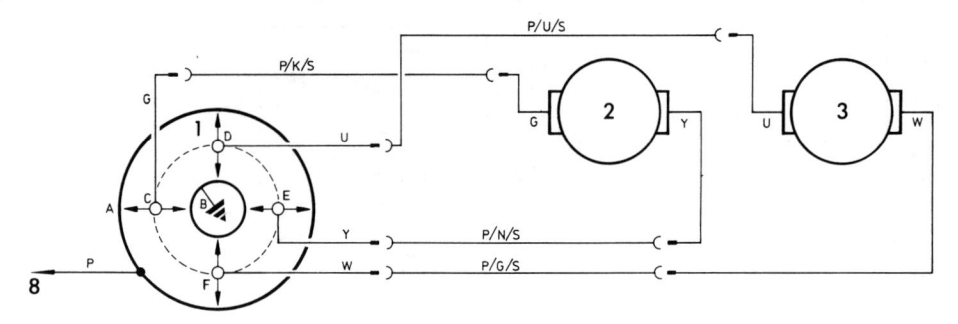

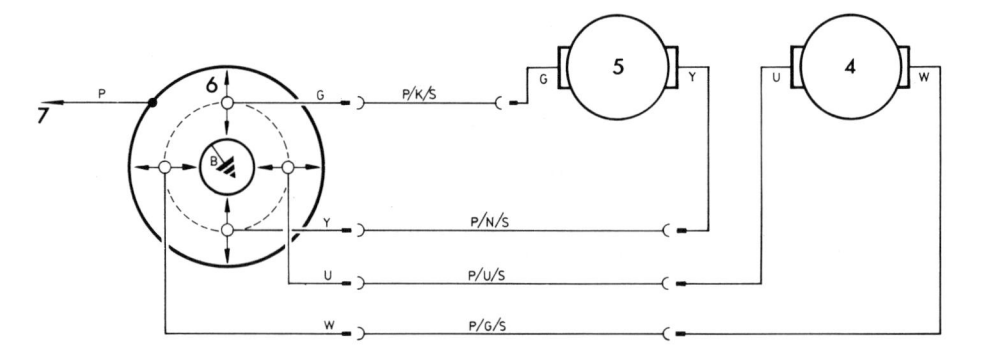

Fig. 24

ELECTRIC DOOR MIRRORS

Description

The right and left hand mirror controls are located on the drivers door. The horizontal and the vertical motors are operated from one control switch. With the control switch in the up position the current flow to motor 3 is via the outer ring A and contact D. The earth circuit is via contact F and the inner ring B.

With the switch in the down position the circuit is completed in the opposite direction via the outer ring A and contact F. The circuit to earth is completed via the contact D and the inner ring B. Similar current flow takes place when the control is operated in a horizontal position using the contact C, E and the motor 2.

LOW COOLANT CONTROL UNIT

Description

With a positive supply to the white wire on the control unit and the black wire earthed, the sensor will partially earth through the coolant and the warning lamp will glow for a few seconds then go out. If the coolant falls below the sensor level the partial earth circuit will be broken and the warning lamp will flash on and off.

LOW COOLANT WARNING LAMP CIRCUIT

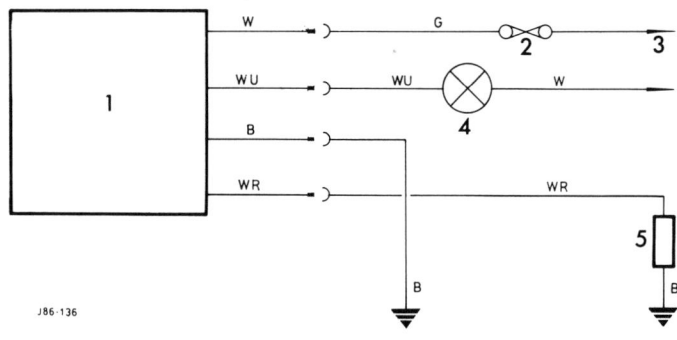

Fig. 25

Fault Finding

Should the warning lamp fail to glow when the ignition is switched on, check all connections and the fuse, ensuring the earth connections are clean and tight.

With the ignition switched on battery voltage should be obtained from the white lead on the control unit. With the white and blue lead from the warning lamp shorted to earth, switch the ignition on and the low coolant warning lamp should glow. Should the lamp fail to glow check the warning lamp bulb and wiring.

With the white and red lead from the sensor unit disconnected, the ignition switch on the warning lamp should flash on and off. Should the lamp fail to flash replace the control unit.

KEY TO DIAGRAM

1. Warning light control unit
2. Fuse No 4
3. To ignition switch
4. Coolant temperature sensor

86—24

ELECTRICAL FAULT DIAGNOSIS

VACUUM DELAY

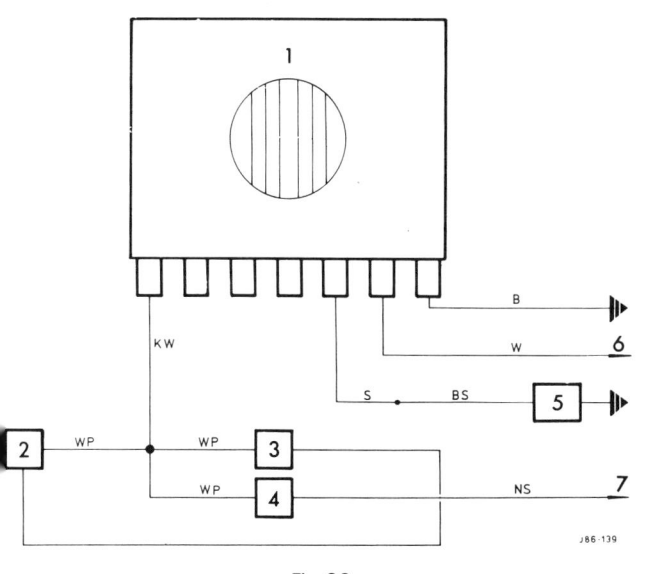

Fig. 26

KEY TO DIAGRAM

1. Delay unit
2. 3 way solenoid valve
3. 2 way solenoid valve
4. Solenoid air valve
5. Coolant temperature switch
6. To pump relay
7. To positive supply

VACUUM DELAY

Description

A positive supply is via the white cable from the ignition switch and earthed via the black cable. When the coolant temperature is below 38°C the coolant temperature sensor switches the vacuum delay unit to earth via the black and slate cable. This starts a timing circuit which switches the three solenoids to earth and energises them for a period of 15 minutes.

At temperatures above 45°C the coolant temperature sensor is not switched to earth so the delay unit is inoperative.

OVERCHARGE WARNING LIGHT CONTROL UNIT

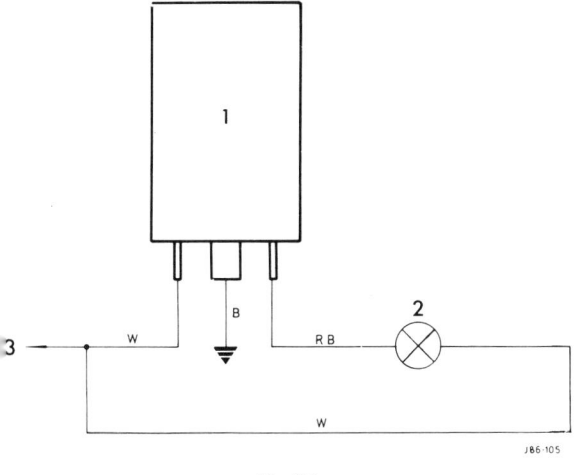

Fig. 27

KEY TO DIAGRAM

1. Warning light control unit
2. Warning lamp
3. To ignition switch

OVERCHARGE WARNING LIGHT CONTROL UNIT

Description

With the ignition switched on the earth circuit of the overcharge voltage lamp is completed via the internal circuit of the overcharge voltage control unit.

The control unit circuit will illuminate the warning light to indicate a malfunction in the alternator. The alternator control box will be suspect by not controlling the output of the alternator and cause the alternator to overcharge the battery.

86—25

ELECTRICAL FAULT DIAGNOSIS

COOLING FAN

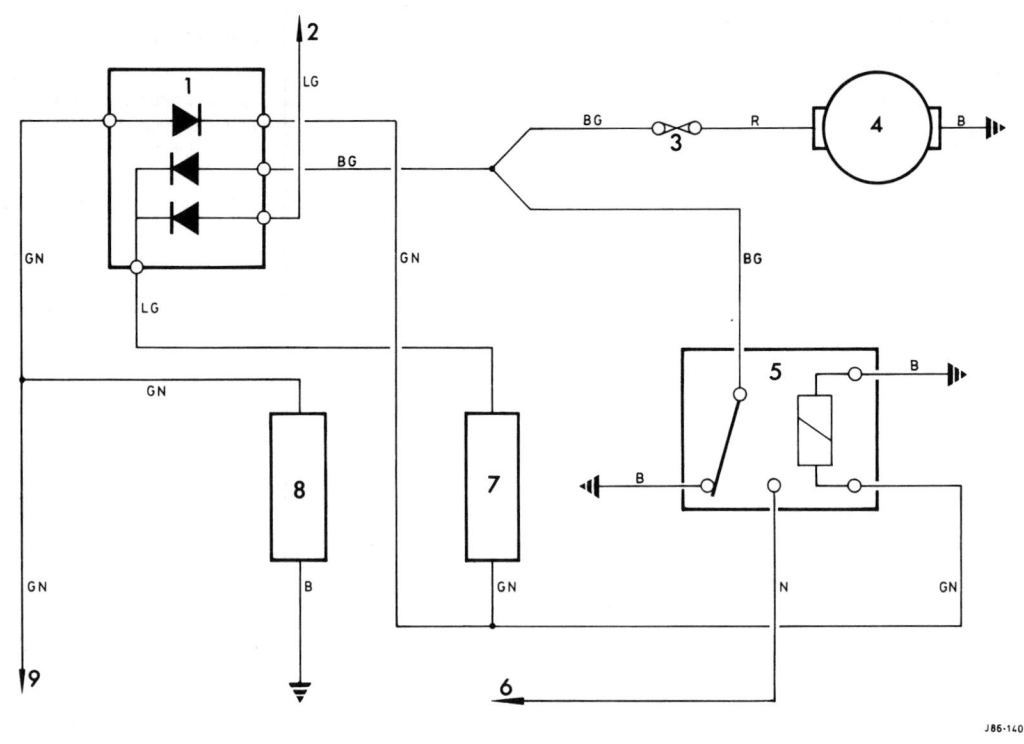

Fig. 28

KEY TO DIAGRAM

1. Diode pack
2. Fuse No 11
3. In-line fuse
4. Cooling fan
5. Fan relay
6. To fuse No 1 (Headlamp fusebox)
7. Radiator fan thermostat
8. Compressor clutch
9. To air conditioning

COOLING FAN

Description

With the coolant temperature cool the thermostat contacts are open which prevents the cooling fan relay from being energised, this in turn prevents the cooling fan from being activated.

With the engine running and the coolant warm the thermostat contacts close. This allows the relay coil circuit to be completed via a diode in the diode pack and the thermosat contacts. The cooling fan circuit is now completed via the relay contacts.

When the ignition is switched off but the the coolant temperature is still hot the cooling fan will still be activated via the relay contacts. The relay will remain energised via a diode in the diode pack, the thermostat contacts, and the relay contacts.

When the coolant temperature cools sufficiently the thermostat contacts will open and the relay will become de-energised. The relay contacts will then switch an earth to the positive brush of the cooling fan motor.

SEAT BELT LOGIC SYSTEM

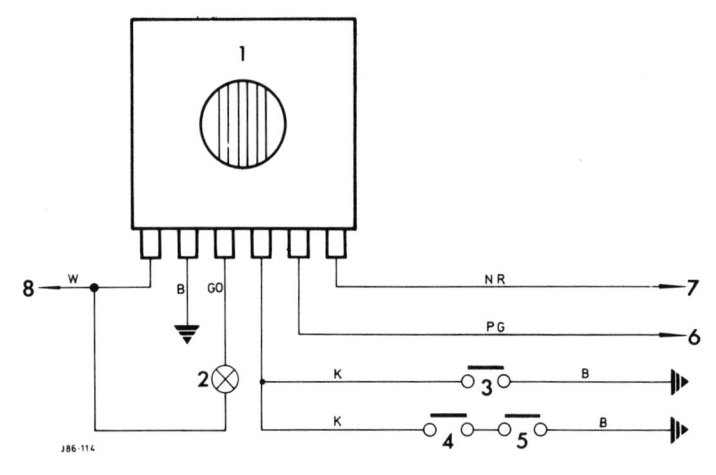

Fig. 29

KEY TO DIAGRAM

1. Logic unit
2. Warning lamp
3. Drivers seat belt switch
4. Passenger seat belt switch
5. Passenger seat switch
6. To door switch
7. To fuse No 2
8. To ignition switch

SEAT BELT WARNING SYSTEM

Description

The seat belt logic unit operates an audible signal for approximately 10 seconds giving a warning to the driver that the seat belts are not fastened.

The control unit operates two timing circuits, one is for the warning lamp which will glow for 10 seconds even when the seat belts are in use. The second timing circuit controls the audible signal for when the seat belts are not in use.

The seat switch, and the seat belt switches completes the earth circuit of the logic control unit. This completed circuit starts a timing cycle for the audible warning which will cease to operate after the 10 seconds have elapsed or when the seat belt is fastened thus breaking the circuit to earth.

86—26

ELECTRICAL FAULT DIAGNOSIS

HORN CIRCUIT

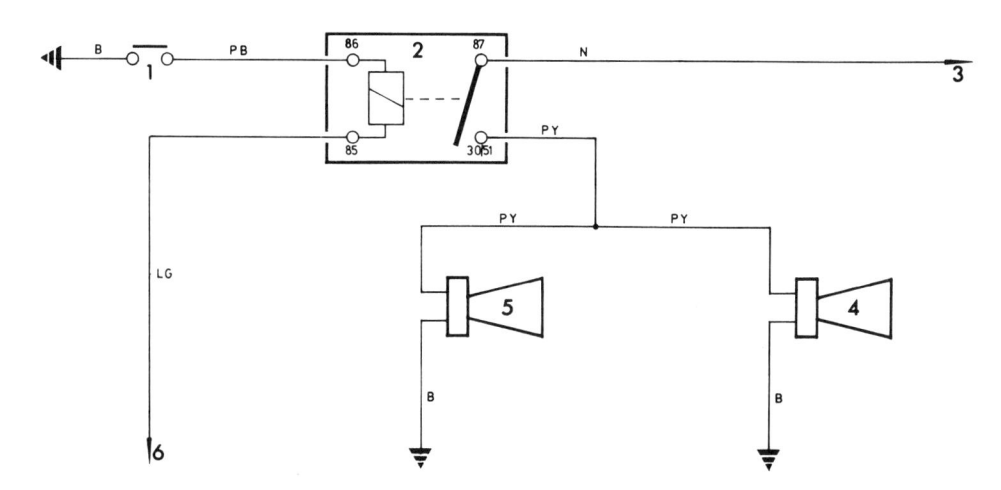

Fig. 30

KEY TO DIAGRAM

1. Horn push
2. Horn relay
3. To terminal post
4. Horn
5. Horn
6. Fuse No 11

HORN

Circuit

The current to the horn relay coil is via fuse No. 11 and the circuit to earth is via the horn push. With the coil energised the current flow to the horns is via the relay contacts.

Fault Finding

With the ignition switched on, battery voltage should be obtained at the terminals 85, 86 and 87 of the relay. If battery voltage is obtained at terminal 85 but not at terminal 86 the relay is faulty.

When the horn push is operated the terminal 86 voltage should drop to zero. If terminal 86 remains at battery voltage check the horn push and the wiring to the horn push. With the horn push pressed and the relay terminal 86 voltage at zero, battery voltage should be obtained at the terminal 30/51 of the relay. A zero reading at terminal 30/51 indicates a faulty relay.

Should the above tests prove satisfactory check the wiring to the horns and both horns.

ELECTRICAL

WINDSCREEN WIPERS 1983 M.Y. ON

Windscreen Wiper Arm/Blades

From the following VIN

XJS   112100

Modified windscreen wiper arms and blades were introduced.

The new design is a top fixing hook type known as 'Trico Series 10'.

These new assemblies may be fitted in CAR SETS ONLY.

To ensure that windscreen wipers perform correctly it is important to ensure that the new type arms and blades are fitted and positioned in accordance with the dimensions.

The inset (B) XJS indicate inner blade support from where measurements should be taken.

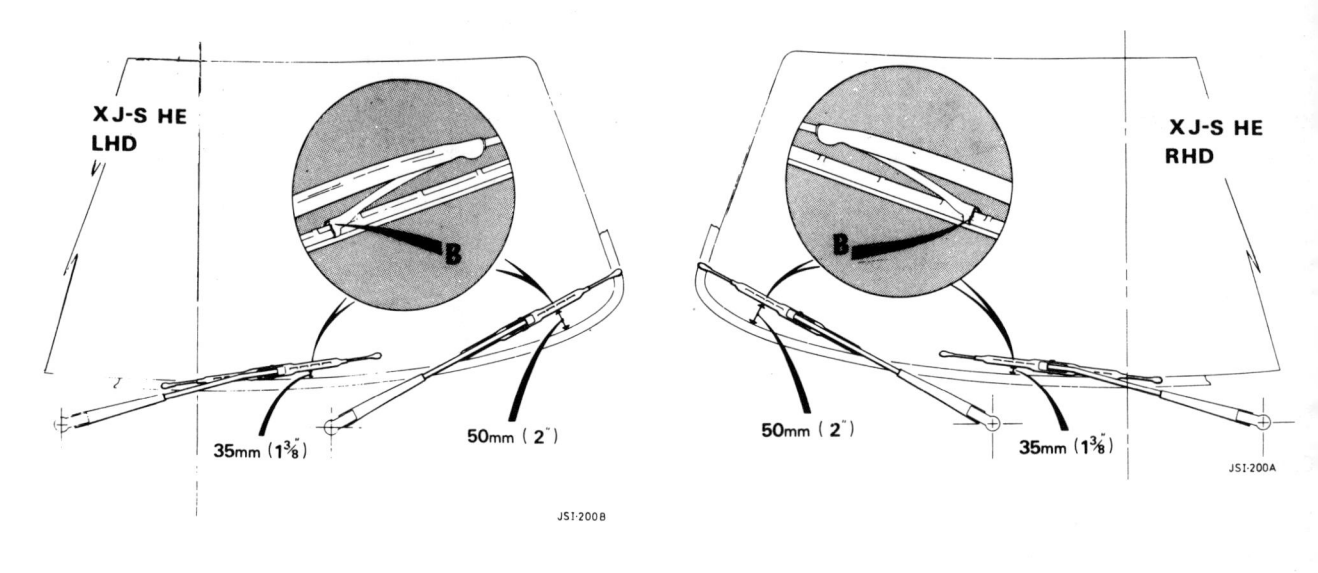

TRIP COMPUTER 1984 M.Y. ON

Description

The trip computer records fuel usage, time and distance. By storing the three sets of information and relating one to another it computes fuel consumption, both average consumption for the journey or a current consumption figure updated every three seconds. The information may be displayed either in miles and gallons or litres and kilometres.

Computer Controls

There are nine controls on the computer face, the use of each is described followed by examples of their use.

mls km — Use this switch to display metric or imperial/US units.

RESET — Press for 5 seconds to switch all function displays to zero.

DISP — Press to switch display off (function updating continues).

TIME — Press to display time of day — press again to display elapsed time since reset — after 6 seconds, display will revert to time of day.

AV.SPD — Press to display average speed since reset.

DIST — Press to display distance travelled since reset.

AVE — Press to display average fuel consumption since reset.

INST — Press to display the fuel consumption at that time.

FUEL — Press to display fuel consumed since reset.

To show which function is on display the relevant button will be illuminated. When the vehicle lights are switched on the computer illumination is dimmed but the legend plate is illuminated.

Time of Day Clock

The digital clock is displayed on the computer when the ignition is on, it is not zero'd as the computer functions are by the RESET button.

To set time: Press the RESET and TIME buttons together to set the hours display and the RESET and DISP buttons together to set minutes display.

ELECTRICAL

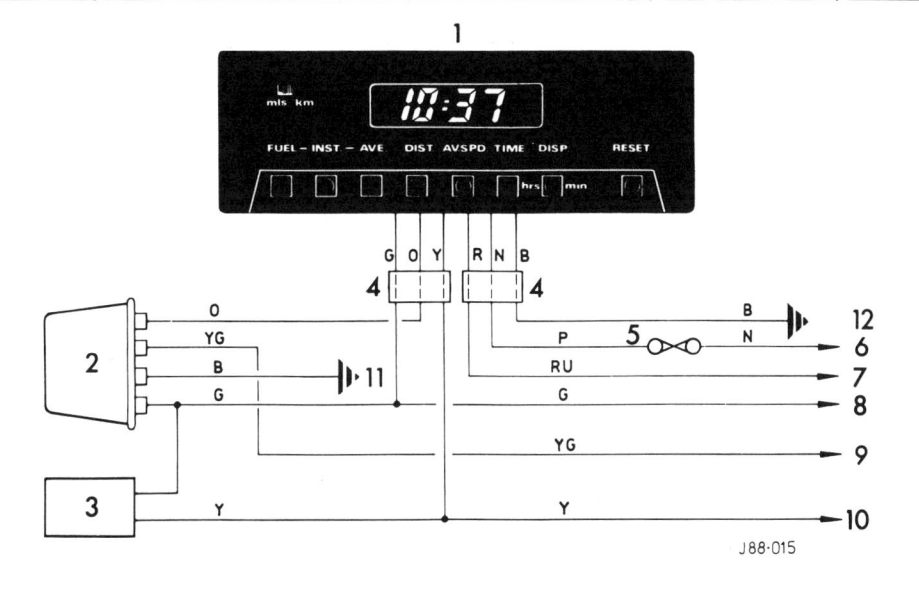

Fig. 9

KEY TO DIAGRAM

1. Trip Computer
2. Interface Unit
3. Speed Transducer
4. Connector Blocks
5. Fuse No. 16 (2 amps)
6. Terminal Post
7. To Sidelamps
8. To Fuse No. 4
9. To ECU
10. To Speedometer

The signals required to operate the trip computer are picked up from the ECU via the interface unit (2, Fig. 9) and the pulse generator (3, Fig. 9). A 12 volt supply is via the fuse (5, Fig 9). This supply voltage enables the clock to function and for the computer to retain information it has received. A second 12 volt supply is via fuse (8, Fig. 9) this supply enables the computer to display information when the ignition is switched on. The third 12 volt supply is via the red and blue lead (4, Fig. 9). This supply voltage enables the display and the buttons to dim when the sidelamps are switched on. The legend strip is also illuminated.

Fault Diagnosis

Check the fuses and all connections. Ensure the earth connections are clean and tight. With the ignition switched off 12 volts should be obtained on the purple lead to the trip computer.

The voltmeter should give the following readings with the ignition switched on:

12 volts at the green lead to the trip computer, the green lead to the pulse generator, the green lead and the yellow/green lead to the interface unit.

With the engine running a voltage should be obtained at the orange lead to the computer. A zero reading indicates a faulty interface unit or lack of continuity in the wiring between the computer and the interface unit. Recheck at the interface unit located in the luggage compartment.

With the rear of the vehicle jacked up, and on stands. Start the engine and put the vehicle into drive. A voltage should then be obtained at the yellow lead to the computer. A zero reading indicates a faulty pulse generator or lack of continuity in the wiring between the pulse generator and the computer.

86—29

ELECTRICAL

Fault	Action
Computer inoperative Screen blank All voltages correct	Replace computer
Time of day displayed Average speed and distance displayed Fuel characteristics zero Speedometer operating All battery voltages correct Zero voltage on orange lead	Replace interface unit
Time of day displayed All other functions zero Speedometer not operating All battery voltages correct Zero voltage on yellow lead	Replace the pulse generator (Speed transducer)
Trip computer does not dim with the sidelamps switched on Battery voltage at the red/blue cable connection	Trip computer faulty
More than one light emitting diode illuminated at the same time All battery voltages correct	Trip computer faulty

HALOGEN HEADLAMPS 1984 SPECIFICATION

XJS HE U.S.A.

All XJS models for U.S.A./Canada markets are equipped with halogen sealed beam headlight assemblies.

Outer 37.5W (Dip)/60W Main
Inner 50W

Speed E.C.U.

A new speed Electronic Control Unit (E.C.U.) has been introduced at:

Vehicle Identification Number
346688 Saloon   108321 XJS HE

The new E.C.U. incorporates revised circuitry and harness termination due to the speed input signal being derived from the speed transducer located in the transmission unit and not the propeller shaft.

NOTE: The new speed E.C.U. is NOT interchangeable with the one previously fitted.

LUCAS 'MF 3' BATTERY

From vehicle identification number VIN 112450, all Jaguar XJS 'HE' Models are equipped with a new Lucas low maintenance battery known as the 'Lucas MF.3'.
Interchangeability with the previous Lucas battery is not affected.

General Information

Battery Type — QOCP 13/11.8
Plate Capacity — 13
AMP Hour Rating — 68 amp

Topping up Intervals

For the 'Lucas MF.3' are 3 years or 80,000 kms (50,000 miles) and therefore for convenience, this work may be carried out at the 72,000 kms (45,000 miles) service.
Distilled water should be used to maintain the electrolyte level between the max/min level markings on the battery container.

Charging Procedures:

(a) Temperature
 (i) Charging should only be carried out if the electrolyte temperature is between 3°C and 40°C. During charging the temperature should not be allowed to rise too high; if it reaches 45°C the charge rate should be halved and if it reaches 50°C the charge must be interrupted until the temperature has dropped to below 40°C. If the temperature is below 3°C it should be warmed to at least 3°C before commencing charging.
 (ii) Should 'topping up' be required, this should be carried out during the latter stages of charging. However, in cases where the tops of the plates are exposed, distilled water should be added to just cover the plates before commencing charging.

(b) Charging of Low Maintenance Batteries
Charging should be contined until the battery gases freely for at least two hours. The fully charged condition can be detected when the electrolyte density remains stable over three consecutive hourly readings at the end of charge. If any cells fail to gas after prolonged charging, then it is likely that these cells are faulty.

(c) Monitoring of Battery Condition
The state of charge of a battery can be determined in a number of ways and the two most convenient ways are by measuring, accurately, the open circuit voltage or by measuring the density of the electrolyte.
 (i) Open Circuit Voltage
 This only gives a reasonable indication of state of charge provided:
 1. The battery has stood open circuit for at least 12 hours since it was last charged or discharged.
 2. The measuring instrument (voltmeter) is itself accurate and has a sensitivity of at least 10,000 ohms per volt.
 The following table gives approximate state of charge v/s voltage:

Open Circuit Volts	State of Charge %
12.60	95
12.50	85
12.40	75
12.30	65
12.20	55
12.10	45
12.00	35
11.90	25
11.80	15
11.70	5

 (iii) Electrolyte Density
 This measurement is potentially more accurate but less convenient than open circuit voltage; it will give good indication of state of charge provided:
 1. The battery has not been topped up with distilled water since it was recharged.
 2. The denisty measuring device (usually a hydrometer) is itself accurate.
 The following table gives state of charge v/s electrolyte density:

Density kg/m^3	State of Charge %
1280	100
1260	90
1240	80
1220	70
1200	60
1180	50
1140	30
1100	10
1080	0 (flat)

BY APPOINTMENT
TO H M QUEEN ELIZABETH
THE QUEEN MOTHER
B L CARS LIMITED
MANUFACTURERS OF DAIMLER JAGUAR
ROVER CARS AND LAND ROVERS

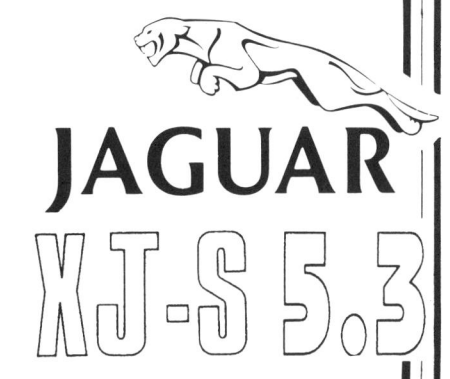

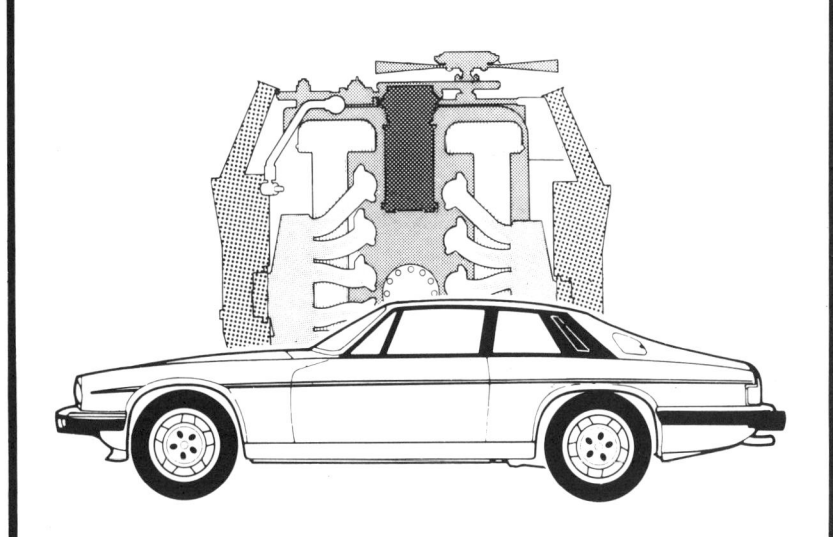

JAGUAR
XJ-S 5.3

Supplement to the
Repair Operation Manual

SUPPLEMENT B • XJ-S 5·3 1984-1988 1/2

INTRODUCTION

This supplement is to be used in conjunction with XJS Repair Operation Manual, part number AKM 3455 and Supplement A - XJS HE. It covers the changes introduced since the above supplement was published in 1984. By using the appropriate service tools and carrying out the procedures as detailed a skilled technician will be able to complete the operation within the time stated in the Supplement to the Repair Operation Times.

SPECIFICATION

Users are advised that the specification details set out in this book apply to a range of vehicles and not to any one. For the specification of a particular vehicle purchasers should consult their Dealer.

Jaguar Cars Ltd, reserve the right to vary their specifications with or without notice and at such times and in such manner as they think fit. Major as well as minor changes may be involved in accordance with the Manufacturer's policy of constant product improvement. Whilst every effort is made to ensure the accuracy of the particulars contained in this book, neither Jaguar Cars Ltd, nor the Dealer, by whom this book is supplied, shall in any circumstances be held liable for any inaccuracy or the consequences thereof.

COPYRIGHT

All nights reserved. No part of this publication may be reproduced, stored in a retrieval system or transmitted in any form, electronic, mechanical, photocopying, recording or other means without prior written permission of Jaguar Cars Ltd .

CONTENTS

Section 1 - Lucas `P` Type Fuel Injection System.

		Page			Page
1.1	System Description	4	1.20	Air Temperature Sensor, Renew	11
1.2	Ignition System Fault Finding	6	1.21	Cold Start Relay, Renew	11
1.3	Air Cleaner Assembly, Renew	7	1.22	Cold Start System, Test	11
1.4	Air Cleaner Element, Renew	7	1.23	Power Resistor, Renew	12
1.5	Throttle Pedal, Renew	7	1.24	Throttle Potentiometer, Renew	12
1.6	Throttle Linkage, Check And Adjust	7	1.25	Electronic Control Unit (ECU), Renew	12
1.7	Throttle Cable, Renew	8	1.26	Main Relay, Renew	13
1.8	Throttle Butterfly Valves, Adjust	8	1.27	Pump Relay, Test	13
1.9	Auxiliary Air Valve, Test	9	1.28	Pump Relay, Renew	13
1.10	Auxiliary Air Valve, Renew	9	1.29	Main Fuel Filter, Renew	13
1.11	Idle Speed, Adjust	9	1.30	Fuel Cooler, Renew	14
1.12	Overrun Valve, Test	9	1.31	Fuel Pump, Renew	14
1.13	Overrun Valve, Renew	9	1.32	Fuel System, Depressurise	14
1.14	Inertia Switch, Renew	10	1.33	Fuel System, Pressure Test	15
1.15	Coolant Temperature Sensor, Test	10	1.34	Pressure Regulator, Renew	15
1.16	Coolant Temperature Sensor, Renew	10	1.35	Injectors, Renew	15
1.17	Thermotime Switch, Test	10	1.36	Fuel Rail (Right Hand Side), Renew	15
1.18	Thermotime Switch, Renew	11	1.37	Fuel Rail (Left Hand Side), Renew	16
1.19	Air Temperature Sensor, Test	11	1.38	Fuel Tank, Renew	16

Section - 2 Body.

2.1	Cabriolet Hood, Renew	17	2.13	Rear Stowage Compartment Lid, Renew	20
2.2	Cabriolet Hood Latches, Renew	17	2.14	Rear Stowage Compartment Luggage Rail, Renew	20
2.3	Cabriolet Hood Seal, Renew	17	2.15	Rear Stowage Compartment Lock Barrel, Renew	20
2.4	Cabriolet Hood Guides, Renew	18	2.16	Console Glove Box Lid, Renew	21
2.5	Cabriolet Targa Top Seal, Renew	18	2.17	Console Glove Box, Renew	21
2.6	Cabriolet Roll Bar Trim Pad, Renew	18	2.18	Door Lock, Renew	21
2.7	Cabriolet Front Headlining Trim, Renew	19	2.19	Passive Restraint Seat Belts, General & Testing	22
2.8	Cabriolet Cantrail Trim Pad, Renew	19	2.20	Passive Restraint Motor And Drive Assembly, Renew	25
2.9	Cabriolet Targa Top Handle Finisher, Renew	19	2.21	Passive Restraint Inertia Reel, Renew	26
2.10	Rear Stowage Compartment Strut, Renew	19	2.22	Passive Restraint Lap Belt Inertia Reel, Renew	26
2.11	Rear Quarter Oddments Tray, Renew	20	2.23	Passive Restraint ECU, Renew	27
2.12	Rear Stowage Compartment, Renew	20	2.24	Passive Restraint Relay, Renew	28

General Specifications.

Wiring Diagrams

1 LUCAS `P` TYPE FUEL INJECTION SYSTEM.

WARNING:

FIRE PRECAUTIONS:

WHEN WORK IS BEING CARRIED OUT ON THE FUEL SYSTEM, FUEL AND FUEL VAPOUR MAY BE PRESENT WHICH IS EXTREMELY FLAMMABLE. GREAT CARE MUST BE TAKEN AND THE FOLLOWING PRECAUTIONS MUST BE STRICTLY ADHERED TO:

1. **SMOKING MUST NOT BE ALLOWED NEAR THE AREA:**
2. **"NO SMOKING" WARNING SIGNS MUST BE POSTED ROUND THE AREA:**
3. **A CO^2 FIRE EXTINGUISHER MUST BE CLOSE AT HAND:**
4. **DRY SAND MUST BE AVAILABLE TO SOAK UP ANY SPILLAGE:**
5. **ENSURE THE FUEL IS EMPTIED FROM THE TANK USING SUITABLE FIREPROOF EQUIPMENT:**
6. **FUEL MUST NOT BE EMPTIED INTO A PIT:**
7. **THE WORKING AREA MUST BE WELL VENTILATED:**
8. **ENSURE THE FUEL IS EMPTIED INTO AN AUTHORISED EXPLOSION PROOF CONTAINER:**
9. **THE DISCARDED TANK MUST NOT BE DISPOSED OF UNTIL IT IS RENDERED SAFE FROM EXPLOSION:**

TEST EQUIPMENT

1. Epitest
2. Epitest adapter
3. Fuel pressure gauge
4. EFI Throttle potentiometer adjustment gauge
5. EFI Feedback monitor unit
6. Infra red CO meter
7. Multi test meter
8. Idle mixture adjustment key Lucas No 60730551

GOOD PRACTICE

The following instructions must be strictly observed:

1. Always disconnect the battery earth lead before removing any components.
2. Always depressurise the fuel system before disconnecting any fuel pipes.
3. When removing fuelling components always clamp the fuel pipes approximately 38 mm (1.5 in) from the unit being removed. Do not over tighten clamp.
4. Ensure rags are available to absorb any spillage that may occur. Observe all fire precautions.
5. When reconnecting electrical components always ensure that good contact is made by the connector before fitting the rubber cover. Always ensure the ground connections are made on to clean bare metal, and are tightly fastened using the correct screws and washers.

Key to Fig. 1, Component Location Diagram

1. Crankcase breather
2. Fuel rail
3. Fuel injector
4. Thermotime switch
5. Overrun valve
6. Vacuum dump valve
7. Supplementary air valve
8. Cold start injector
9. Solenoid vacuum valve
10. Vacuum switch
11. Throttle position switch
12. Check valve (N. America only)
15. Throttle potentiometer (below pulley)
16. Extra air valve
17. Ignition amplifier
18. Crankcase vent
19. Canister purge
20. PCV valve
21. Air temperature sensor
22. Overrun valve
23. Coolant temperature sensor

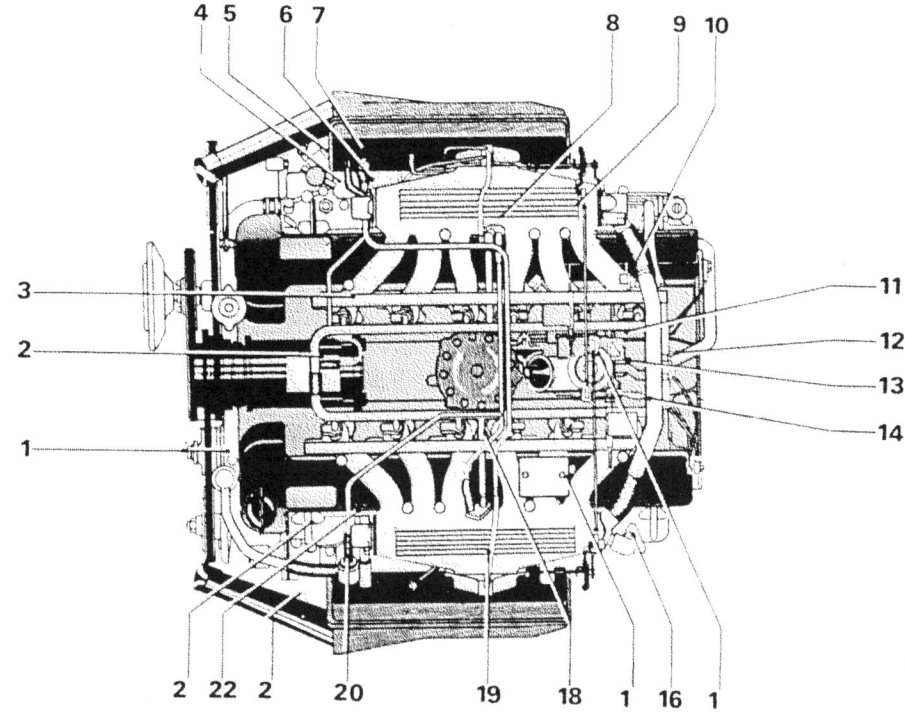

Fig. 1

3

WARNING

1. **DO NOT LET THE ENGINE RUN WITHOUT THE BATTERY BEING CONNECTED.**
2. **DO NOT USE A HIGH-SPEED BATTERY CHARGER AS A STARTING AID.**
3. **WHEN USING A HIGH-SPEED CHARGER TO CHARGE THE BATTERY, THE BATTERY MUST BE DISCONNECTED FROM THE REST OF THE VEHICLE'S ELECTRICAL SYSTEM.**
4. **WHEN INSTALLING, ENSURE THAT THE BATTERY IS CONNECTED TO THE CORRECT POLARITY,**
5. **ENSURE THE BATTERY IS NO LARGER THAN THE NOMINAL 12 VOLTS.**
6. **TEST ALL COMPONENT REMOVAL THAT RESULTS IN FUEL OR FUEL VAPOUR BEING PRESENT —IT IS IMPERATIVE TO OBSERVE ALL FIRE PRECAUTIONS AS DETAILED ABOVE.**

1.1 SYSTEM DESCRIPTION

This is an indirect injection system which incorporates solenoid-operated low pressure injectors (Fig. 2) intermittently spraying fuel into the inlet port of each cylinder. A digital ECU (Fig. 3) with an integral manifold pressure sensor, governs the amount of fuel injected by varying the spray duration to suit a set of parameters laid down by a series of sensors. These are:

Manifold pressure sensor
speed sensor (derived from ignition pulses)
engine temperature sensor (Fig. 4)
inlet air temperature sensor (Fig. 5)
throttle position potentiometer (Fig. 6).

Calculations derived from the above, together with the prevalent battery voltage, ensure that the correct fuelling mixture is maintained across all running conditions.

During cold start conditions, to maintain a good idle speed, a temperature-sensitive auxiliary air valve (Fig. 7) supplies extra air (above that normally supplied by the idle bypass). The valve is mounted on the cylinder head.

The fuel pressure regulator (Fig. 8) is attached to the fuel rail, and bleeds fuel back to the fuel tank while the pump is running, thereby maintaining system pressure at 3 bar.

The ignition system is of the constant energy type, using a distributor (Fig. 9) incorporating a cover (1), rotor arm (2), anti-flash shield (3), reluctor (4), and a pick-up assembly (5). When a tooth of the reluctor passes across the pick-up limb, the magnetic field strength is intensified creating a voltage in the wind-

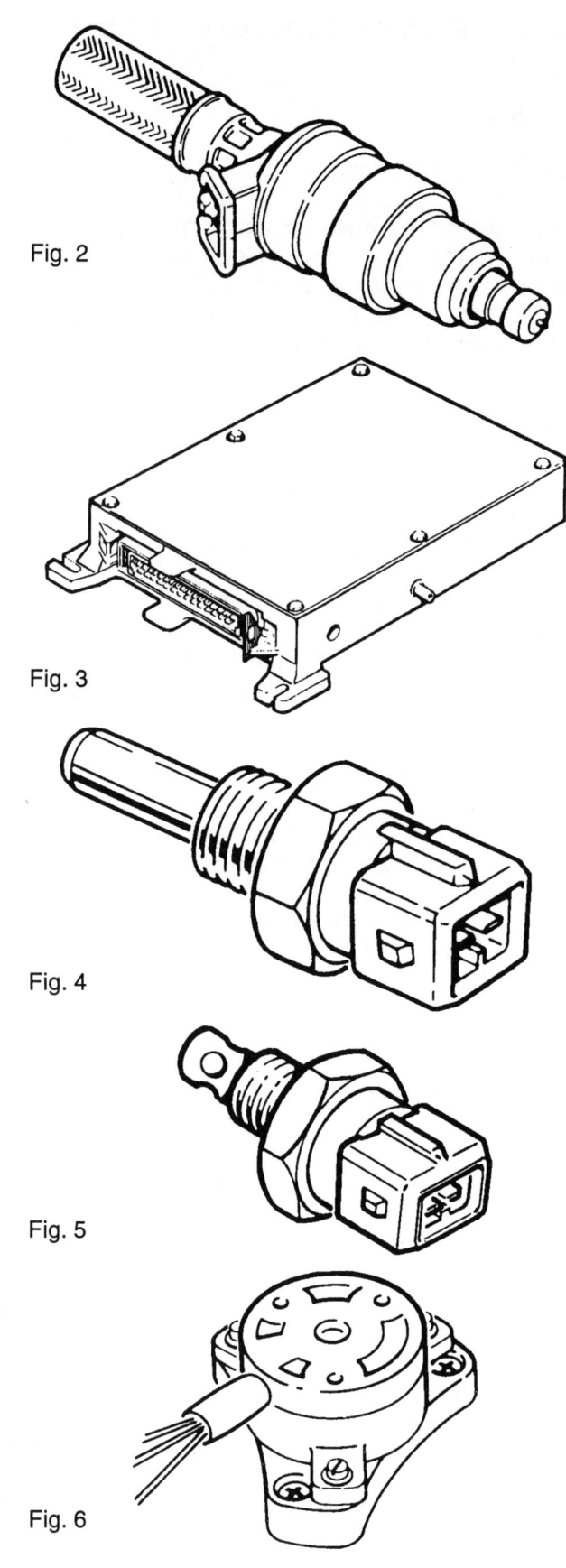

Fig. 2

Fig. 3

Fig. 4

Fig. 5

Fig. 6

ing. The rise of this voltage is sensed by an amplifier (Fig. 10) causing it to switch the current flowing in the primary winding of the ignition coil on and off. Two coils are employed.

WARNING: DO NOT TAMPER WITH THE AMPLIFIER, AS IT CONTAINS BERYLIA, AN EXTREMELY DANGEROUS SUBSTANCE IF HANDLED.

Fig.7

Fig.8

Fig.9

Fig.10

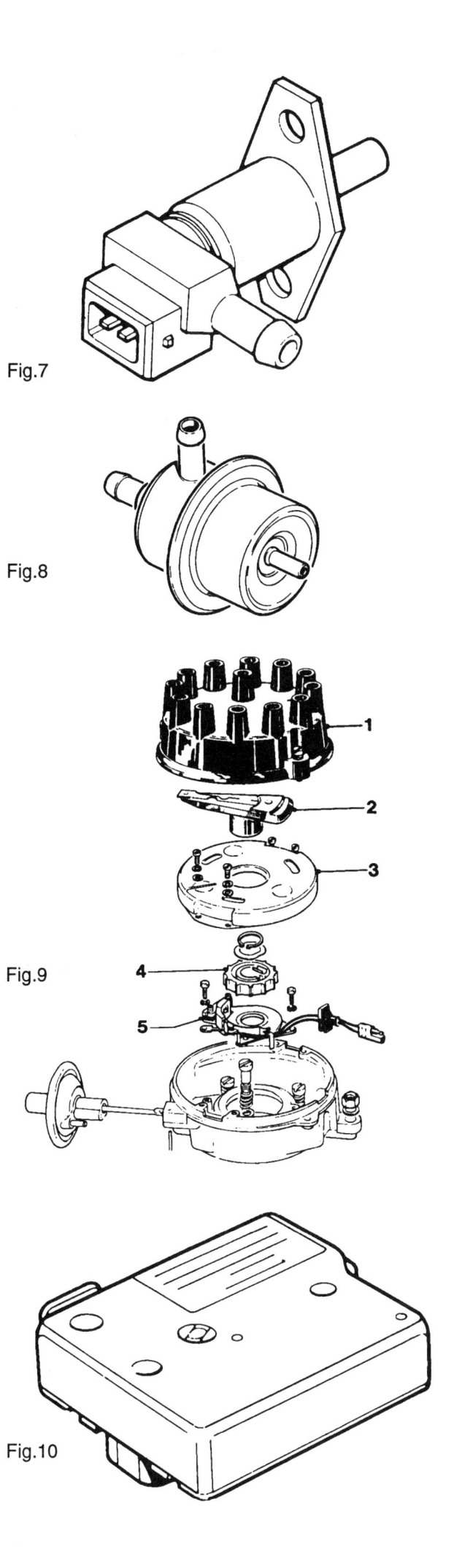

Relays

Four relays are used in the fuel injection system. They are the main relay which supplies battery voltage to the system, the pump relay which controls the fuel pump, the idle speed relay (certain markets only) and the lambda sensor relay (certain markets only).

Firing order with the cylinders numbered from the front:
1A 6B 5A 2B 3A 4B 6A 1B 2A 5B 4A 3B

The induction system is basically the same as that on carburetted engine: tuned ram pipe, air cleaners, plenum chambers and induction ports. Air is drawn through paper element cleaners to a butterfly valve for each bank and to individual ports for each cylinder leading off the plenum chamber. The injectors are positioned at the cylinder head end of each port so that fuel is directed at the back of each inlet valve.

```
              A (1 ) (2) (3) (4) \5) (6)
    Front
              B (1 ) (2) (3) (4) (5) (6)
```

(1) Inlet opens   13 deg. BTDC
(2) Inlet closes   55 deg. ABDC
(3) Ignition with engine at normal operating temperature
    UK/Europe   24 deg. BTDC
    All others   25-27 deg. BTDC
(4) Exhaust opens   55 deg. BBDC
(5) Exhaust closes   13 deg. ATDC

Fuel supply (Fig. 11)

Fuel is drawn from a small tank (1) at the rear of the car by a fuel pump (2) via a non-return valve (3) to a fuel rail through an in-line filter (4) and a pressure regulator (5). Fuel is controlled so that the pressure drop across the injector nozzle is maintained at a constant 2.5 bar (36.25 lb/in2). Excess fuel is returned to the fuel tank via a fuel cooler (6). The twelve fuel injectors (7) are connected to the fuel rail (8) and are electro-mechanically operated to inject fuel into each inlet port. Fuel is also supplied to cold start injectors (9) which only operate while starting a cold engine.

Full load fuelling (USA)

To obtain maximum engine power it is necessary to inhibit the "closed loop" system and simultaneously increase the fuelling level. This is obtained by using a vacuum operated switch, sensing inlet manifold depression and a micro switch operated by the throttle pulley spindle. These two switches wired in parallel, so that either or both can signal the need for full load fuelling.

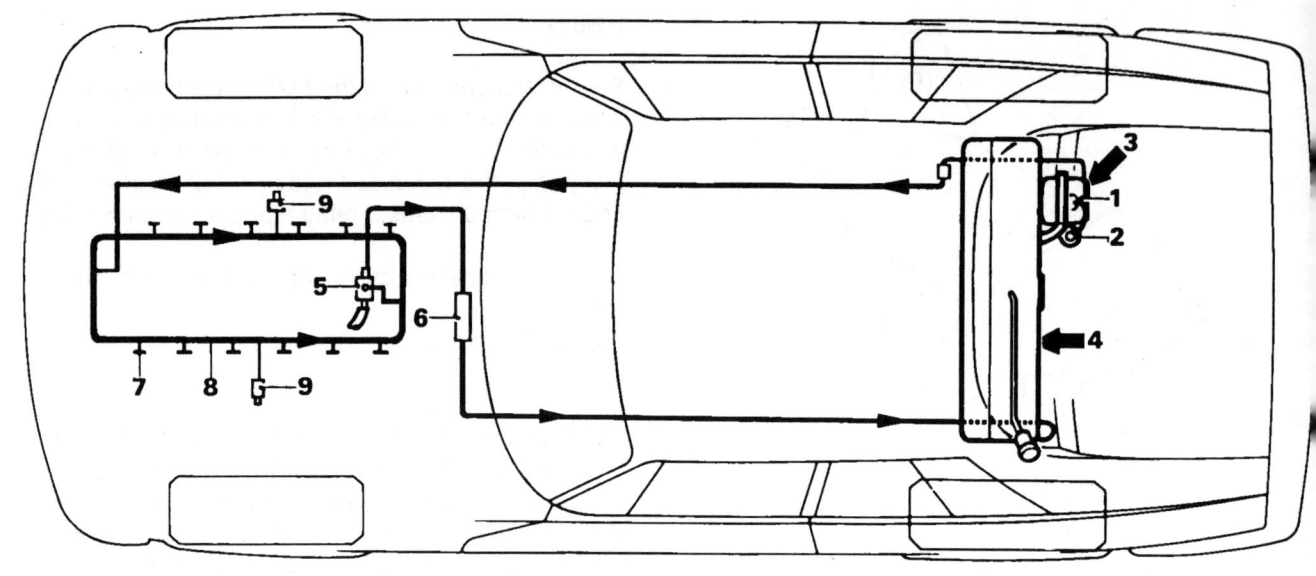

Fig.11

Vacuum switch (Fig. 12)

The contacts of this switch are operated by a spring loaded diaphragm in a chamber.

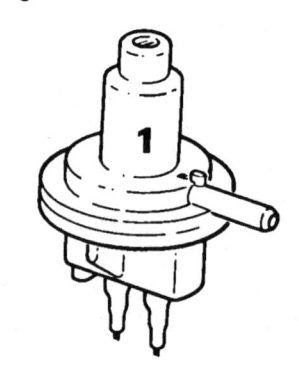

Fig.12

This senses inlet manifold vacuum such that, when the depression falls below a certain value e.g. when the engine is operating at low speed, with part throttle near full load condition then the contacts close. The closing contacts causes the fuel system to go `Open loop` and simultaneously introduce a fuel enrichment of 12%.

1.2 IGNITION SYSTEM FAULT FINDING

Test 1

Check battery. A heavy discharge test applied to the battery terminals will determine whether the battery is capable of supplying the heavy currents required by the starter motor.
Check the specific gravity of the electrolyte in each cell. A variation of 0.040 in any cell means the battery is suspect.

Test 2

Check for HT spark. Remove the HT lead from the centre of the distributor cap 'A' and position the end of the lead approximately 6 mm (0.25 in.) from a good earth point on the engine. Crank the engine and if a spark is obtained, check the HT leads, spark plugs, distributor cap and the rotor. If test is satisfactory reconnect the HT lead.

Test 3

With the ignition switched on. the voltage at the positive terminal of the 'MAIN' HT coil 'C' should be battery voltage. If the voltage is more than a volt below battery voltage check the wiring to/from the ignition switch, check the switch and wiring to/from battery to switch.

Test 4

Disconnect the leads from the negative terminal of the 'MAIN' HT coil. With the ignition switched on a battery voltage reading should be obtained from the HT coil negative terminal. A zero reading would indicate a faulty 'MAIN' coil. If a battery voltage reading is obtained, reconnect the disconnected leads to the 'MAIN' coil and repeat tests 3 and 4 at the auxiliary coil.
If tests are satisfactory reconnect all disconnected leads and switch ignition off.

Test 5

Disconnect the distributor pick-up leads from the amplifier 'B'. Measure the resistance of the distributor pick-up 'A' with an ohm meter. The resistance should be 2.2 to 4.8 K ohms. An incorrect reading indicates a faulty pick-up coil, If test is satisfactory reconnect the pick-up leads.

Connect a voltmeter between the positive terminal of the battery and the negative terminal of the 'MAIN' HT coil. Switch the ignition on, the voltmeter should indicate zero volts. Crank the engine, the voltmeter reading should rise between 2 to 3 volts. If the voltmeter reading remains at zero the amplifier is suspect

1.3 AIR CLEANER ASSEMBLY, RENEW

Disconnect the battery earth lead.
Disconnect the air temperature sensor lead from the sensor (1 Fig. 13) (left hand only).
Unclip the air cleaner element cover and displace the cover from the element (2 Fig. 13).
Remove the element (3 Fig. 13) and the cover.
Disconnect the crankcase breather pipe and auxiliary air valve pipes (left hand cleaner only) (4 Fig. 13) from the air cleaner.
Remove the four bolts securing the air cleaner to the throttle butterfly housing (5 Fig. 13) and remove the air cleaner.
Remove the air temperature sensor.
On refitting, ensure all mating surfaces are clean and smeared with silicon sealant.

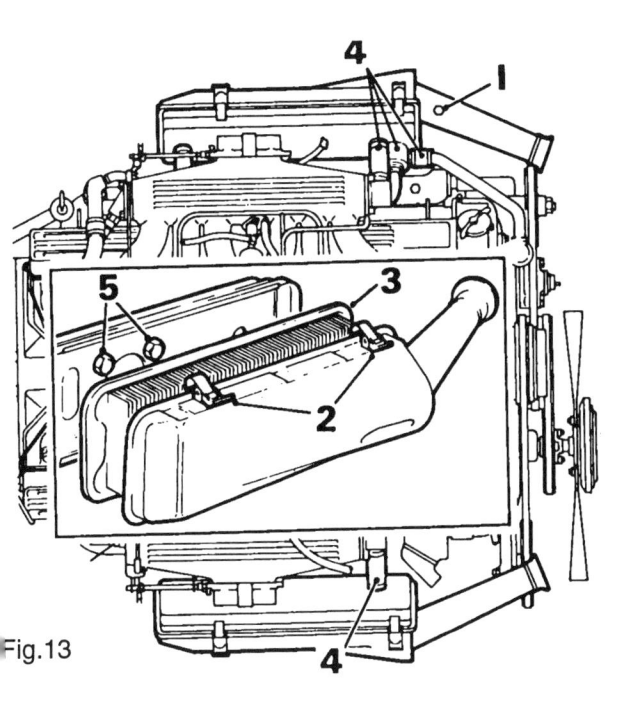

Fig.13

1.4 AIR CLEANER ELEMENT, RENEW

Disconnect the air temperature sensor lead from the sensor (1 Fig. 13) (left hand only).
Unclip the air cleaner element cover and displace the cover from the element.
Remove the element (2 Fig. 14) and the cover.
On fitting new element ensure that it is correctly oriented, with the metal plate opposite the throttle housing.

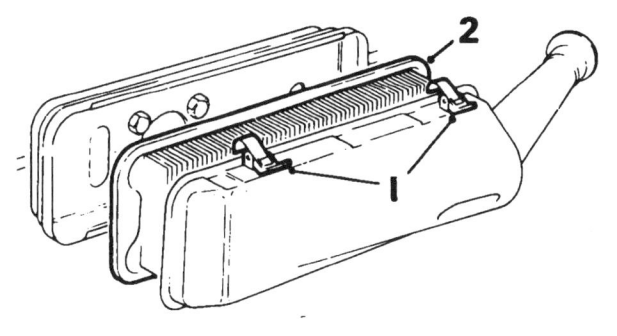

Fig.14

1.5 THROTTLE PEDAL, RENEW

Fold the carpet away from the base of the throttle pedal.
Remove the nuts and washers securing the base of the pedal (1 Fig. 15).
Pull the base of the pedal away from mounting plate and disengage the spring from the pedal (2 Fig. 15).
Refitting a new pedal is the reversal of the removal procedure.

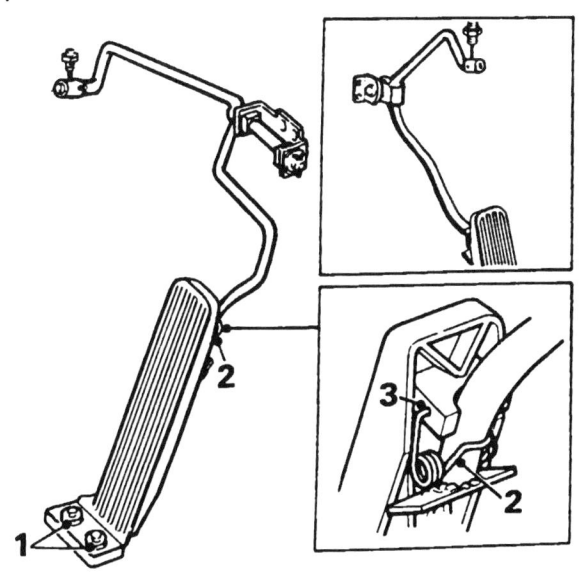

Fig.15

1.6 THROTTLE LINKAGE, CHECK AND ADJUST

Check.

Ensure throttle return springs are correctly secured that throttle pulley moves freely, resting against closed stop when released (1 Fig. 16).
Ensure that throttle butterfly closed stop screw has not been moved. If signs of tampering are present, check and if necessary, adjust (2 Fig. 16).

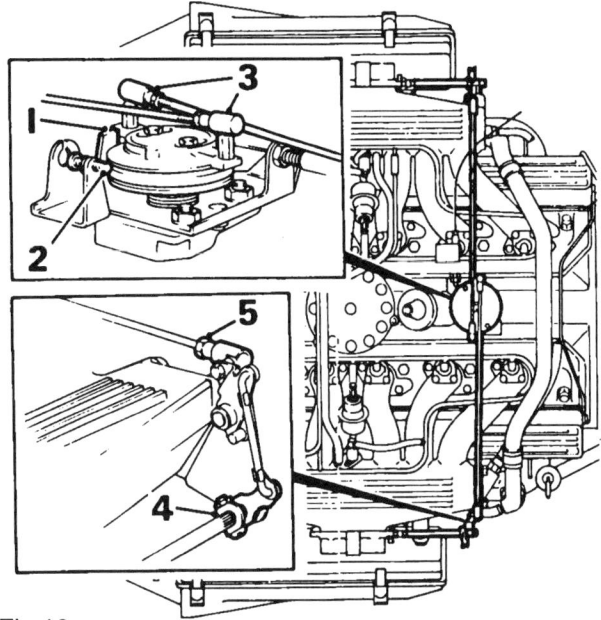

Fig.16

Ensure that throttle pulley can be rotated to touch fully open stop and that throttle butterfly valve stop arm is touching throttle housing.
If conditions of above operations are not satisfied, proceed with operation 'Adjust'.

Adjust

If throttle butterfly closed stop has been moved, adjust the stop.
Check for worn pivots, damaged rods or linkage and trace of any stiffness.
Renew items as necessary.
Release throttle cross-rods from throttle pulley (3 Fig. 16)
Slacken clamps securing levers to rear of throttle shafts (4 Fig. 16).
With butterfly valve against closed stop, bellcrank against stop, and play in coupling taken up in opening direction, tighten clamp to lock lever to throttle shaft. Repeat for other side of engine.
Offer cross-rods to ball connectors on pulley; rods must locate without moving pulley or linkage.
Slacken locknuts on cross-rods and adjust length of rods to locate pulley ball connector while bellcrank against closed stop (5 Fig. 16).
Tighten locknuts and ensure ball joints remain free.
Slacken locknut on throttle pulley, fully open stop and wind back adjustment screw.
Hold throttle pulley fully open and ensure that both throttle butterfly stop arms are against throttle housing.
Set fully open stop to just touch throttle pulley and tighten locknut.
Check operation of throttle switch.
Check kickdown switch adjustment.

1.7 THROTTLE CABLE, RENEW

Disconnect battery earth lead. Disconnect electrical connectors from kickdown switch (1 Fig. 17).
Slacken locknut and disconnect throttle cable from pulley (2 Fig. 17).
Disconnect cable assembly from bulkhead tube. (3 Fig. 17).
Remove and discard split pin securing throttle cable to throttle pedal, pull off sleeve and release cable from pedal. (4 Fig. 17).
Pull cable through to engine compartment.

Note: On right-hand drive cars the cable is secured to the bulkhead by a nut above the driver's footwell.

Release cable from pedestal.
Slacken throttle cable locknut at pedestal bracket, remove throttle cable assembly.
Examine bulkhead grommet for damage, renew as necessary.
Fit grommet to cable.
Refitting is the reversal of the above operations, apply suitable sealing compound around cable and

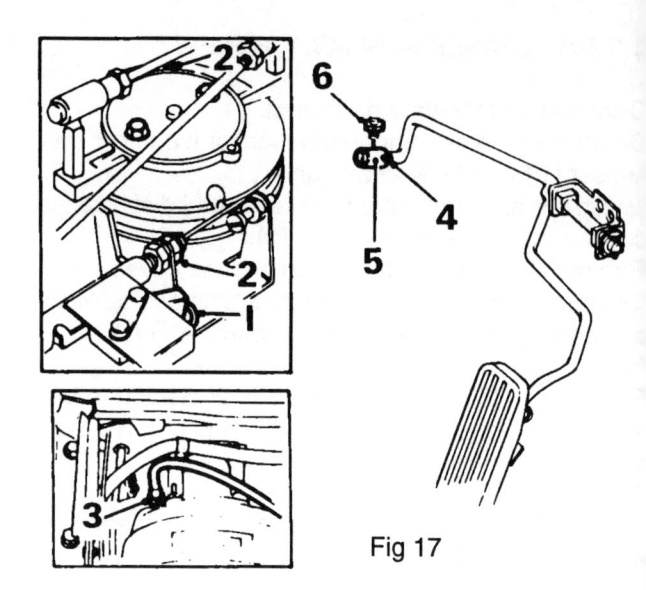

Fig 17

grommet at bulkhead (right-hand drive cars).
Check throttle linkage adjustment.
Check kickdown switch adjustment.
Re-connect battery earth lead.

1.8 THROTTLE BUTTERFLY VALVES, ADJUST

CAUTION: Any adjustment must be carried out on both butterfly valves. It is NOT permitted to adjust one valve only.

Remove both air cleaners.
Slacken locknut on throttle butterfly stop screw. Wind back screw.
Ensure throttle butterfly valve closes fully.
Insert a 0.002 in. (0,05 mm) feeler gauge between top of valve and housing to hold valve open. (1 Fig. 18).
Adjust stop screw to just touch stop arm and tighten locknut with feeler in position. (2 Fig. 18).
Withdraw feeler.
Press stop arm against stop screw and repeat on other side of engine.
Seal threads of adjusting screws and locknuts using a blob of paint.
Refit air-cleaners.

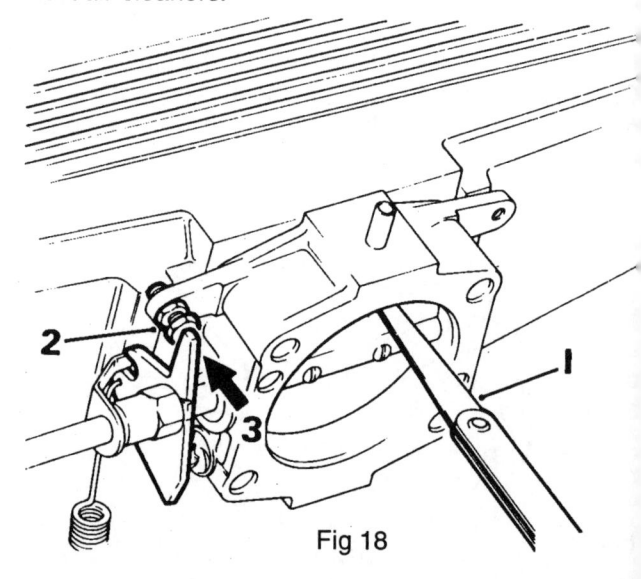

Fig 18

Check throttle linkage adjustment.
Check operation of throttle switch.
Check kickdown switch adjustment.

1.9 AUXILIARY AIR VALVE, TEST

Remove left-hand air cleaner element. Remove auxiliary air valve.
Fully close idle speed adjustment screw.
Immerse auxiliary air valve bulb in a container of boiling water and observe valve head through side port.
Valve should move smoothly to closed position.
Quickly blow through side port; no air should pass.
Allow valve bulb to cool. Valve head should move smoothly back to open main air passage.
If valve performance is satisfactory reset idle speed adjustment screw and refit valve.
If valve performance is not satisfactory, fit replacement component.

1.10 AUXILIARY AIR VALVE, RENEW

CAUTION: This procedure MUST ONLY be carried out on a cold or cool engine.

Carefully remove pressure cap from remote header tank to release any cooling system residual pressure.
Replace cap tightly.
Slacken the clips securing the air hoses to the auxiliary air valve and pull the hoses clear (1 Fig. 19).
Remove two screws and washers securing auxiliary air valve to coolant pipe, lift valve clear (2 Fig. 19).
Clean all traces of gasket from coolant pipe, taking care not to damage seating area.
Scribe a line on idle speed adjusting screws and note number of turns required to screw fully in (3 Fig. 19).

On refitting set idle speed adjustment screw of replacement valve to the number of turns open as noted previously.
Coat the new gasket with suitable non-hardening sealing compound locate the valve on the coolant pipe and secure using the two screws and washers.
Fit and secure the hoses.
Reconnect the battery earth lead.
Check coolant level at remote header tank, if necessary top up.
Check, if necessary adjust idle speed.

1.11 IDLE SPEED, ADJUST

Ensure engine is at normal operating temperature.
Check throttle linkage for correct operation, check that return springs are secure and effective.
Start engine, run for two to three minutes.
Set idle speed adjustment screw to achieve 750 rev/min..

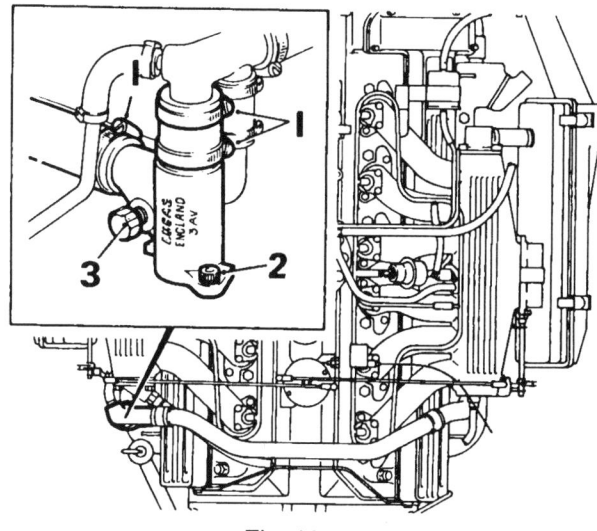

Fig. 19

Note: If it proves impossible to reduce idle speed to specified level, check ALL pipes and hoses to inlet manifolds for security and condition, also check operation of overrun valves.

Check security of injectors and cold start injectors.
Ensure that all joints are tight and that inlet manifold to cylinder head fastenings are tight.
Ensure that both throttle butterfly closed stops show no signs of tampering; if they do, adjust throttle butterfly valves.
Check operation of overrun valves.
If the above operations do not reduce idle speed, check operation of auxiliary air valve.

1.12 OVERRUN VALVE, TEST

Remove both air cleaners.
Pull the inlet pipes clear. Block the overrun valve inlet pipes.
Start engine. If idle speed is correct, stop engine, unblock one valve only.
Restart engine. If idle speed still correct, stop engine and unblock second valve.
Restart engine. If idle speed too fast renew relevant valve.

1.13 OVERRUN VALVE, RENEW

Remove three screws securing overrun valve to inlet manifold, retrieve spacer.
Remove valve and body by releasing air cleaner hose.
Remove and discard gasket, remove all traces of gasket, ensuring mating surfaces are not damaged.
Remove valve from body.
Reverse the above operations ensuring new gaskets are fitted.

1.14 INERTIA SWITCH, RENEW

Disconnect battery earth lead.
Remove the trip reset cable rubber knob and trim panel securing screws.
Disconnect the trip reset cable from the bracket and carefully displace the drivers side dash liner (1 Fig. 20).
Remove the switch cover (2 Fig. 20).
Disconnect switch block connector (3 Fig. 20).
Remove the two screws securing the switch and remove the switch (4 Fig. 20).

On refitting a new switch ensure it is correctly reset.

1.15 COOLANT TEMPERATURE SENSOR, TEST

Disconnect battery earth lead.
Pull electrical connector from coolant temperature sensor.
Connect suitable ohmmeter between terminals, note resistance reading. The reading is subject to change according to temperature and should closely approximate to the relevant resistance value given in the table.
Disconnect ohmmeter.
Check resistance between each terminal in turn and body of sensor. A very high resistance reading (open circuit) must be obtained.
Refit electrical connector to temperature sensor.
Re-connect battery earth lead.

Coolant Temperature (Deg. C)	Resistance (Kilohms)
-10	9.2
0	5.9
+10	5.9
+20	2.5
+30	1.7
+40	1.18
+50	0.84
+60	0.60
+70	0.435
+80	0.325
+90	0.250
+100	0.190

1.16 COOLANT TEMPERATURE SENSOR, RENEW

Note: This procedure MUST ONLY be carried out on a cold or cool engine.

Disconnect battery earth lead.
Pull electrical connector from coolant temperature sensor.
Carefully remove pressure cap from remote header tank to release any cooling system residual pressure.
Replace cap tightly.

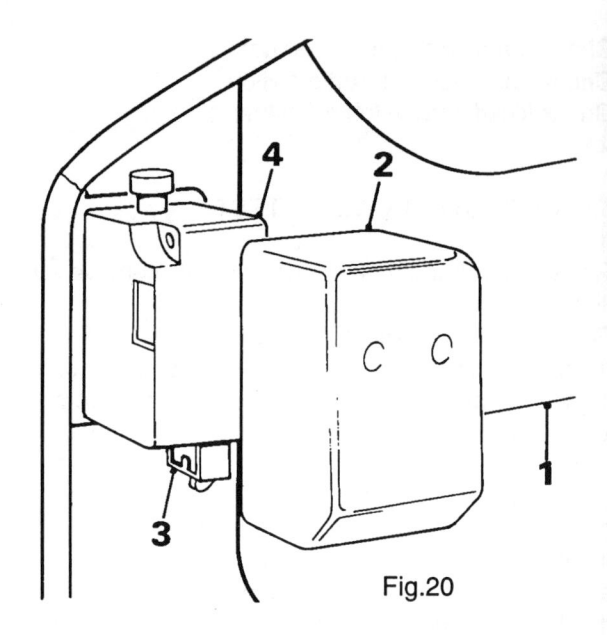

Fig.20

Note: The replacement component is prepared at this point and the transfer made as quickly as possible.

Ensure sealing washer located on replacement temperature sensor and coat threads with suitable sealing compound.
Remove existing temperature sensor from thermostat housing.
Screw replacement temperature sensor into thermostat housing.
Refit electrical connector to temperature sensor.
Re-connect battery earth lead.
Check coolant level at remote header bank. Top up as necessary.

1.17 THERMOTIME SWITCH, TEST

Equipment required: Stop watch, ohmmeter, single pole switch, jump lead for connecting switch to battery and thermotime switch and a thermometer.

Note: Check coolant temperature with thermometer and note reading before carrying out procedures detailed below. Check rated value of thermotime switch (stamped on body flat).

Disconnect battery earth lead. Pull electrical connector from thermotime switch.
'A' coolant temperature higher than switch rated value: Connect ohmmeter between terminal 'W' and earth. A very high resistance reading (open circuit) should be obtained.
Renew switch if a very low resistance reading (short circuit) is obtained.
'B' coolant temperature lower than switch rated value: Connect ohmmeter between terminal 'W' and earth. A very low resistance reading (short circuit) should be obtained.
Connect 12V supply via isolating switch to terminal 'G' of thermotime switch.

Using stop watch, check time delay between making isolating switch and indication on ohmmeter changing from low to high resistance. Delay period must closely approximate to time indicated in table for specific coolant temperature.
Renew thermotime switch if necessary.
Reconnect battery earth lead.

Coolant Temperature (deg. C)	Delay 15 deg.) C Switch (seconds)	Delay 35 deg. C Switch (seconds)
-20	8	8
-10	5.7	6.5
0	3.5	1.2
+10	1.2	3.5
+15	0	2.7
+20	-	2.7
+30	-	0.5
+35	-	0

1.18 THERMOTIME SWITCH, RENEW

Note: This procedure MUST ONLY be carried out on a cold or cool engine.

Disconnect battery earth lead.
Pull electrical connector (1 Fig 21) from thermotime switch (2 Fig 21).
Carefully remove and refit pressure cap from remote header tank to release any cooling system residual pressure.
Ensure that a new sealing washer is located on the replacement unit and the threads are coated with suitable sealing compound.
Remove thermotime switch from the thermostat housing .
Fit replacement thermotime switch.
Check coolant level at remote header tank, top up as necessary .

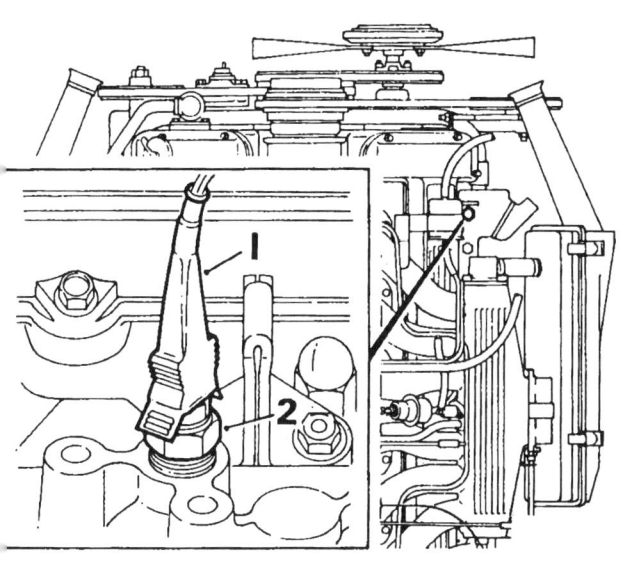

Fig 21

1.19 AIR TEMPERATURE SENSOR, TEST

Disconnect battery earth lead. Pull electrical connector from air temperature sensor.
Connect suitable ohmmeter between terminals, note resistance reading given in the table.
Disconnect ohmmeter.

Check resistance between each terminal in turn and body of sensor. A very high resistance reading (open circuit) must be obtained.
Refit electrical connector to temperature sensor.
Reconnect battery earth lead.

Ambient Air Temperature (deg.C)	Resistance (ohms)
-10	960
0	640
+10	435
+20	300
+30	210
+40	150
+50	180

1.20 AIR TEMPERATURE SENSOR, RENEW

Disconnect battery earth lead.
Pull electrical connector from air temperature sensor.
Remove sensor from air cleaner ram pipe.

Refitting is the reversal of the above operations.

1.21 COLD START RELAY, RENEW

Disconnect battery earth lead.
Remove two screws securing cover to relay mounting bracket.
Remove the cover.
Displace relay from mounting tag.
Remove the relay from the cable harness block connector.

Refitting is the reversal of the above operations.

1.22 COLD START SYSTEM, TEST (refer to Fig. 22)

WARNING: THIS TEST RESULTS IN FUEL VAPOUR BEING PRESENT IN THE ENGINE COMPARTMENT. IT IS THEREFORE IMPERATIVE THAT ALL DUE PRECAUTIONS ARE TAKEN AGAINST FIRE AND EXPLOSION

1. Remove two setscrews and washers securing cold start injectors to inlet manifolds. Remove cold start injectors.

2. Arrange containers to collect sprayed fuel.

Fig. 22

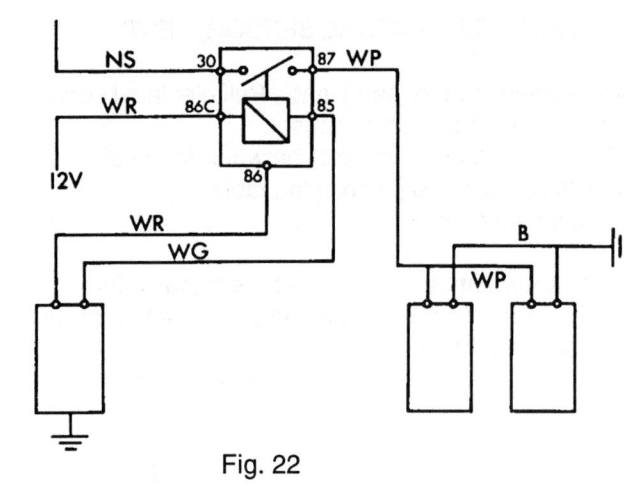

3. Disconnect negative terminal from ignition coil.

Note: If engine is cold (below 15 deg. C), proceed with operations 4 and 5. If engine is hot (above 1 5 deg. C) proceed with operations 7 to 11.

4. Switch on ignition and observe cold start injectors for fuel leakage. Crank engine one or two revolutions. Injectors should spray while engine cranks. Note: Do not operate starter motor longer than necessary to make this, and the following observations.

5. If injectors do not spray, carry out the following tests.
   a. Crank engine. Check for battery voltage at cold start injector supply (White/Pink cable). If yes: check all electrical plug and earth connections to cold start injectors. If satisfactory, cold-start injectors are suspect. If no: Proceed to test (b).
   b. Crank engine. Check for battery voltage at terminal 87. If yes: Check cables between cold start relay and cold start injectors. If no: Proceed to test (c).
   c. Crank engine. Check for battery voltage at terminal 30. If yes: Cold start relay not energised or contacts faulty. Proceed to test (d). If no: Check supply from pump relay.
   d. Crank engine. Check for battery voltage at terminal 86. If yes: Cold start relay not energised or contacts faulty. Proceed to test (e). Disconnect lead at terminal 85.

6. Check for battery voltage at terminal 85. If yes: Bridge terminal 85 to earth. Cold start relay should energise. Check voltage at terminal 87. If 0 volt, replace relay. If satisfactory, check cables and connections to thermotime switch. If satisfactory, thermotime switch is suspect. If no: Replace cold start relay.

7. Replace cold start injectors and all cables and connectors removed.

8. Crank engine. Voltage at terminal 87 of cold start relay must be 0 volt.

9. If battery voltage present, disconnect terminal 85. Crank engine and re-check voltage at terminal 87.

10. If voltage at terminal 87 is 0 volt, thermotime switch is faulty and should be replaced.

11. If battery voltage is present, replace cold start relay.

12. If cold start injectors pass fuel while voltage atterminal 87 is 0 volt, injector(s) must be renewed.

1.23 POWER RESISTOR, RENEW

Disconnect battery earth lead
Disconnect the power resistor block connector.
Remove the nuts and bolts securing the power resistor to the right hand inner wing.
Remove the power resistor (Fig. 23).
Refitting a new power resistor is the reversal of the above procedure ensuring the block connector is fully connected and secure.

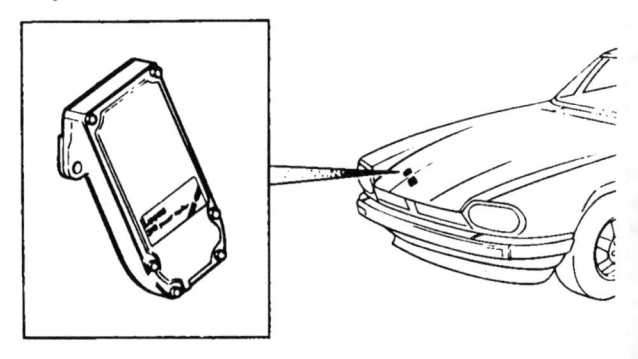

Fig. 23

1.24 THROTTLE POTENTIOMETER, RENEW

Disconnect the battery earth lead.
Remove the throttle cross-rods from the throttle pulley.
Disconnect the throttle potentiometer in-line cable connector.
Remove the nuts and washers securing the throttle pulley plate to the throttle pedestal and lift the plate clear.
Remove the screws securing the potentiometer and lift the potentiometer clear.
To refit new potentiometer, reverse the above operations.

1.25 ELECTRONIC CONTROL UNIT (ECU), RENEW

Disconnect battery earth lead.
Remove right-hand luggage compartment trim panel.
Remove screw securing ECU harness clamp, move clamp aside.
Unclip ECU and cover.

Fig. 24

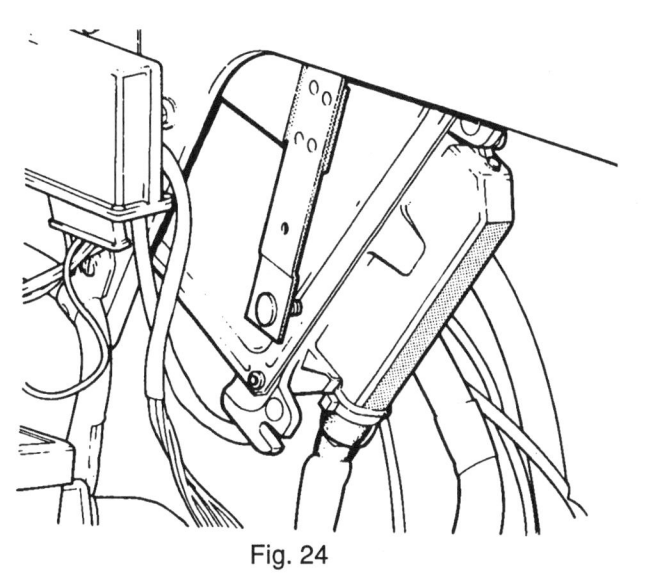

Locate handle on harness plug and withdraw plug.
Pull ECU (Fig. 24) clear of mounting bracket.

Caution: The idle fueling potentiometer is pre-set and must not be moved.

Refitting is the reversal of the above operations.
Check idle CO level.

1.26 MAIN RELAY, RENEW

Disconnect battery earth lead.
Remove right-hand luggage compartment trim panel.
Displace the relay from the securing tag.
Remove the main relay from the cable harness block connector.

Refitting is the reversal of the above operations.

1.27 PUMP RELAY, TEST (refer to Fig. 25)

Switch on ignition. Pump should run for one to two seconds, then stop.
If pump does not run, or does not stop, check systematically as follows:
Check inertia switch cut-out button pressed in.
Detach inertia switch cover and ensure both cables are secure.
Pull electrical connectors from switch and check continuity across terminals.
Pull button out and check open circuit.
Remove ohmmeter, replace electrical connectors, reset button and refit cover.
If inertia switch satisfactory, earth (ground) pump relay terminal 85, switch on ignition and check circuit systematically as detailed below.

a. Check for battery voltage at terminal 86 of main relay. If yes: Proceed to test b. If no: Check battery supply from ignition switch via inertia switch.

b. Check for battery voltage at terminal 87 of main

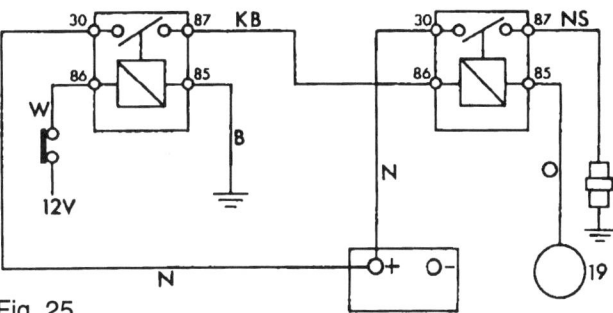

Fig. 25

relay. If yes: Proceed to test c. If no: Check for battery voltage at earth (ground) lead and connection from terminal 85 of main relay. If satisfactory renew main relay.

c. Check for battery voltage at terminal 86 of pump relay. If yes: Proceed to test d. If no: Open circuit between terminals 87 of main relay and 86 of pump relay. If satisfactory proceed to test d.

d. Check for battery voltage at terminal 87 of pump relay. If yes: Proceed to test e. If no: Check for battery voltage to earth (ground) lead and connections from terminal 85 of pump relay. If satisfactory, renew pump relay.

e. Check for battery voltage at supply lead (NS) and connections to fuel pump. If yes: Faulty pump or earth (ground) connections. If no: Open circuit between terminal 87 of pump relay and supply lead connection to fuel pump.

1.28 PUMP RELAY, RENEW

Disconnect battery earth lead.
Remove right-hand luggage compartment trim panel.
Displace the relay from the securing tag.
Remove the main relay from the cable harness block connector (Fig. 26).
Refitting is the reversal of the above operations.

1.29 MAIN FUEL FILTER, RENEW

Depressurise the fuel system.
Disconnect the battery earth lead.

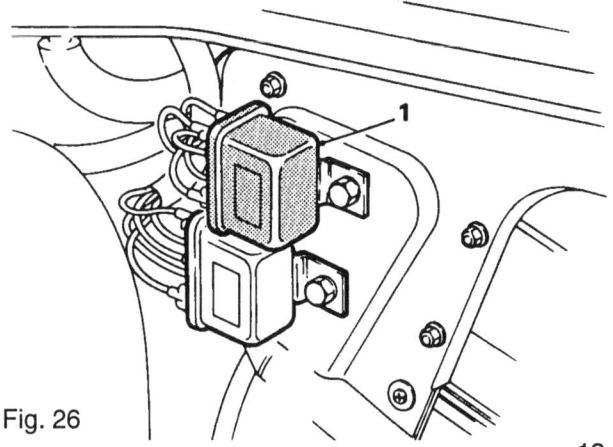

Fig. 26

Fig. 27

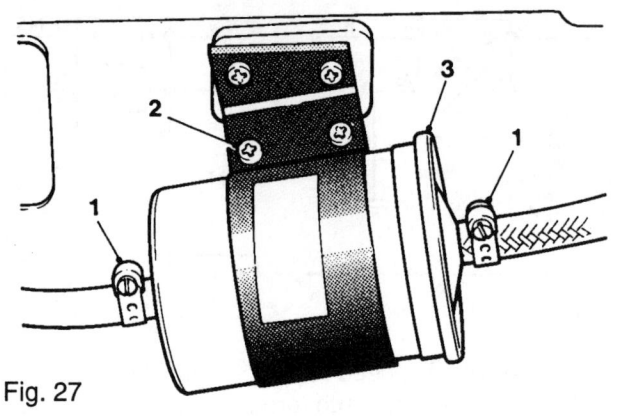

Remove the spare wheel cover and the spare wheel.
Fit clamps to the filter inlet and outlet hoses.
Disconnect the fuel filter hoses (1 Fig. 27).
Fit plugs to the disconnected hoses and to the filter.
Slacken the filter securing clamp screw (2 Fig. 27) and remove the filter (3 Fig. 27).

Refitting a new filter is the reversal of the above operations ensuring the hoses are secure and do not leak.

1.30 FUEL COOLER, RENEW

Depressurise the air conditioning system.
Depressurise the fuel system.
Disconnect the battery earth lead.
Disconnect the fuel hoses (1 Fig. 28) from the fuel cooler and fit plugs to the hoses and fuel cooler to prevent ingress of dirt.
Disconnect the air conditioning hoses from the fuel cooler and plug the hoses (2 Fig. 28) to prevent contamination of the air conditioning system.
Remove the nuts and bolts securing the fuel cooler (4 Fig. 28).
Remove the fuel cooler.

Refitting a new fuel cooler is the reversal of the above operations ensuring all connection are secure and new 'O' rings are fitted to the air-conditioning unions.

1.31 FUEL PUMP, RENEW

Depressurise fuel system.
Disconnect battery earth lead.

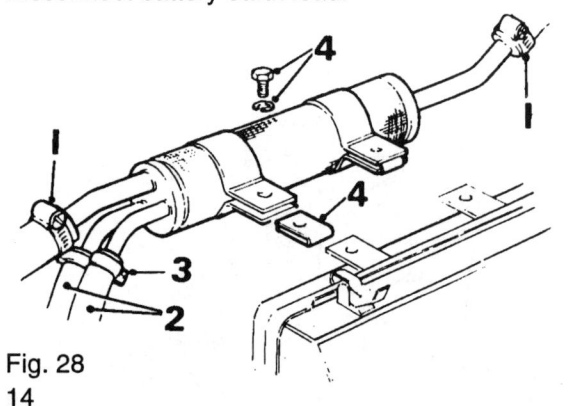

Fig. 28

Remove spare wheel.
Remove two screws securing fuel pump cover to battery tray.
Peel back floor carpet, release fuel pump cover from floor clips.
Fit clamps to inlet and outlet hoses (1 Fig. 29).
Remove two Nyloc nuts and plain washers securing fuel pump retaining band to mounting, retrieve spacer washers and remove fuel pump assembly (2 Fig. 29).
Remove electrical connector from fuel pump socket (3 Fig. 29).

CAUTION: Place a suitable receptacle beneath car to collect spilled fuel.

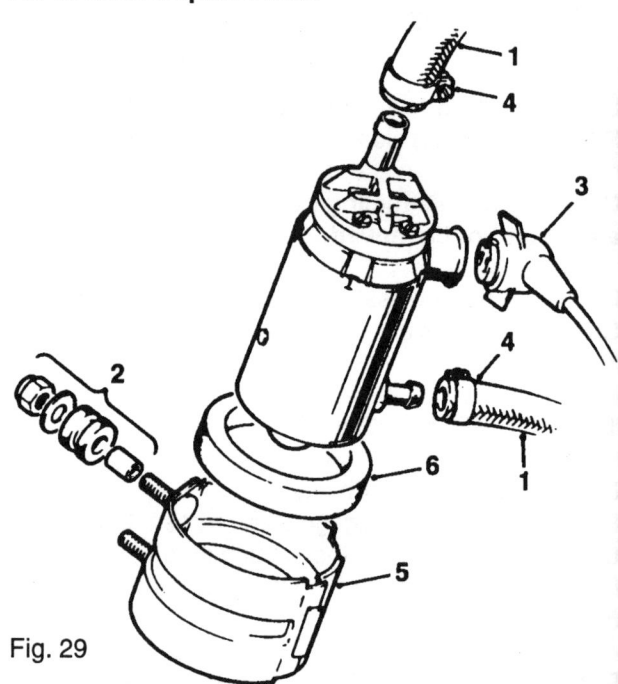

Fig. 29

Release two hose clips securing inlet and outlet hoses to fuel pump, disconnect hoses (4 Fig. 29).
Separate and remove two halves of fuel pump retainer band (5 Fig. 29).
Remove two foam rubber insulation bands (6 Fig. 29). Refitting is the reversal of the above operations.

1.32 FUEL SYSTEM, DEPRESSURISE

CAUTION: The fuel system must always be depressurised before disconnecting any fuelling component.

Remove right-hand luggage compartment trim.
Remove fuel pump relay (Fig. 30).
Switch ignition on and crank engine for a few seconds.

Note: The fuel system will now be depressurised.

On completion of operations to fuelling components and prior to starting the engine, proceed as follows:
Refit fuel pump relay.
Fit right-hand luggage compartment trim.

Fig. 30

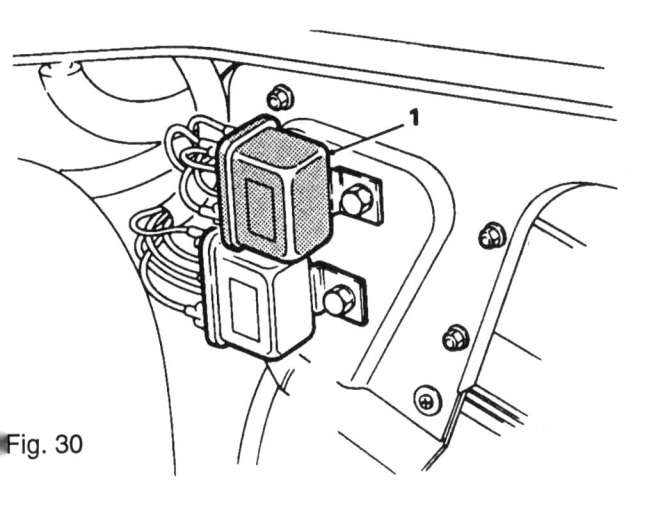

1.33 FUEL SYSTEM, PRESSURE TEST

Depressurise fuel systems.
Slacken clip securing left-hand cold start injector hose to fuel rail, disconnect hose.
Connect pressure gauge pipe to fuel rail and secure using hose clip.
Pull positive low tension lead from ignition coil.
Connect terminal 85 of pump relay to earth (ground).
Switch on ignition and check pressure gauge reading. Reading must be between 28.5 and 30.8 lbf/in^2 (2,0-2,2 kgf/cm^2).
If reading high, check for blockage in return line.
If reading low, check for blockage or choked filter in supply line or pump suction pipes.
Switch off ignition, depressurise fuel system and remove pressure gauge.
Switch on ignition and check for leaks.
Remove earth (ground) connection from terminal 85 of pump relay.
Re-connect low tension lead at ignition coil.

1.34 PRESSURE REGULATOR, RENEW

Depressurise the fuel system.
Disconnect the fuel cooler to regulator hose from the regulator.
Plug the hose and the regulator to prevent contamination.
Remove the nut securing the regulator to the mounting bracket.
Disconnect regulator to fuel rail hose from the fuel rail.
Plug the hose and fuel rail.
Displace the regulator from the mounting bracket for access.
Disconnect the vacuum pipe from the regulator and remove the regulator.

Refitting a new regulator is the reversal of the above operations ensuring all fuel and vacuum pipe connections are secure and tight.

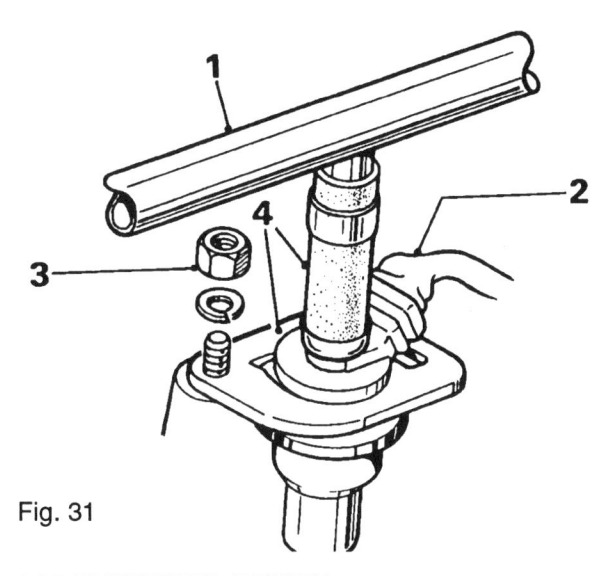

Fig. 31

1.35 INJECTORS, RENEW

Depressurise the fuel system.
Disconnect the battery earth lead.
Remove the right hand fuel rail (1 Fig. 31).
Disconnect the injector harness block connectors (2 Fig. 31).
Remove the securing nuts and clamps (3 Fig. 31).
Remove the injectors (4 Fig. 31).
Refitting new injectors is the reversal of the above procedure ensuring all pipes and block connectors are secure.

1.36 FUEL RAIL (Right hand side), RENEW

Depressurise the fuel system.
Disconnect the battery earth lead.
Disconnect the right hand throttle rod (1 Fig. 32) and the throttle cable (2 Fig. 32) from the throttle pedestal.
Disconnect the manifold cross pipe from the right hand manifold (3 Fig. 32).
Disconnect the cold start injector hose from the fuel rail (4 Fig. 32).
Plug the hose and rail.
Slacken the hose clips and disconnect the fuel rail

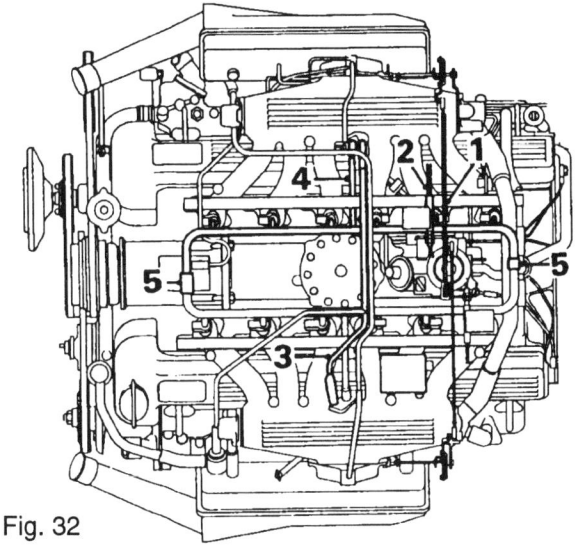

Fig. 32

halves (5 Fig. 32).
Plug the hoses and fuel rail.
Slacken the injector hose clips and remove the fuel rail.
Refitting a new fuel rail is the reversal of the above operations ensuring all hose connections secure and tight.

1.37 FUEL RAIL (Left hand side), RENEW

Depressurise the fuel system.
Disconnect the battery earth lead.
Disconnect the left hand throttle rod from the throttle pedestal (1 Fig. 33) .
Disconnect the regulator to fuel rail hose from the fuel rail (2 Fig. 33).
Disconnect the manifold cross pipe from the left hand inlet manifold (3 Fig. 33).
Disconnect the cold start injector hose from the fuel rail (4 Fig. 33).
Plug the hose and rail.
Slacken the hose clips and disconnect the fuel rail halves (5 Fig. 33).
Plug the hoses and fuel rail.
Slacken the injector hose clips and remove the fuel rail.
Refitting a new fuel rail is the reversal of the above operations ensuring all hose connections secure and tight .

1.38 FUEL TANK, RENEW

WARNING: THIS OPERATION RESULTS IN FUEL AND FUEL VAPOUR BEING PRESENT WHICH IS EXTREMELY FLAMMABLE, GREAT CARE MUST BE TAKEN WHEN WORK IS BEING CARRIED OUT ON THE FUEL SYSTEM AND THE FOLLOWING PRECAUTIONS MUST BE STRICTLY ADHERED TO:
SMOKING MUST NOT BE ALLOWED NEAR THE AREA :
`NO SMOKING` WARNING SIGNS MUST BE POSTED ROUND THE AREA:
A CO^2 FIRE EXTINGUISHER MUST BE CLOSE AT HAND:
DRY SAND MUST BE AVAILABLE TO SOAK UP ANY SPILLAGE:
ENSURE THE FUEL IS EMPTIED FROM THE TANK USING SUITABLE FIREPROOF EQUIPMENT:
FUEL MUST NOT BE EMPTIED INTO A PIT:
THE WORKING AREA MUST BE WELL VENTILATED:
ENSURE THE FUEL IS EMPTIED INTO AN AUTHORISED EXPLOSION PROOF CONTAINER:
THE DISCARDED TANK MUST NOT BE DISPOSED OF UNTIL IT IS RENDERED SAFE FROM EXPLOSION:

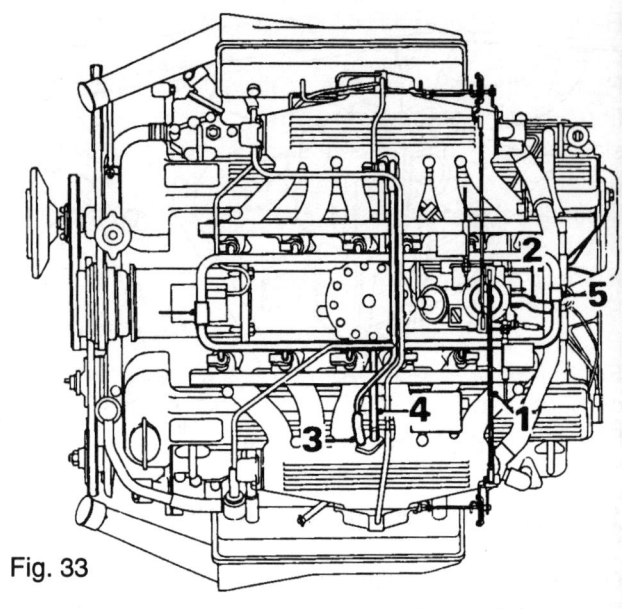

Fig. 33

Remove the luggage compartment RH side panel.
Remove the pump relay from its bracket and remove the relay from the multi-plug connector.
Crank the engine to depressurise the fuel system.
Disconnect and remove the battery.
Remove the spare wheel cover and the spare wheel.
Empty the fuel tank into an explosion-proof container using suitable fireproof equipment.
Clamp the fuel return hose and disconnect the union nut at the fuel tank (1 Fig. 34).
Release the hose clip securing the expansion tank supply hose to pipe, and disconnect the hose (2 Fig. 34).
Release the union nut securing the expansion tank return pipe to the fuel tank (3 Fig. 34).
Release the three hose clips securing the vent hoses to the fuel tank vent pipes.
Note the position of and disconnect the hoses (4 Fig. 34)
Note the position of and disconnect the electrical connectors from the fuel gauge tank unit (5 Fig. 34).
Release the hose clips securing the fuel filler assembly to the body, and remove the assembly (6 Fig. 34).
Remove the two bolts and release the fuel tank securing straps (7 Fig. 34).

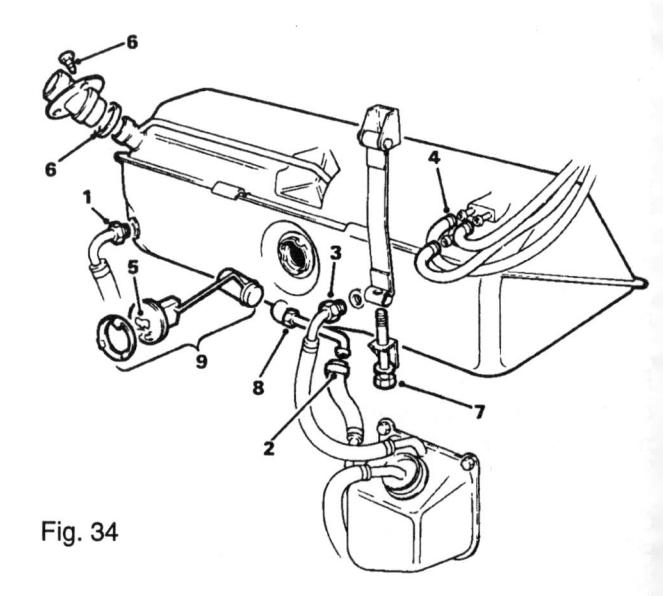

Fig. 34

Remove the LH and RH luggage compartment trim pads.
Remove the seven screws, plain washers and four fibre washers (each side) securing the LH and RH side luggage compartment side panel to body lower panel.
Disconnect the earth leads from the relay bracket, and lower the panel.
Remove the expansion tank supply pipe (8 Fig. 34).
Remove the fuel tank.
Remove the tank unit (9 Fig. 34) as required.
Fitting a new tank is a reversal of the removal procedure, ensuring that all earth and electrical connections are secure, and checking for fuel leaks.

2 BODY

2.1 CABRIOLET HOOD, RENEW

Open the door.
Release the hood locking catch.
Release the headlining-to-`D` post velcro fastening.
Fold the hood to the rearward position (Fig. 35).
Release the outer hood velcro fastening.
Undo and remove the hood-to-body securing screws.
Push the hood upwards for access.
Release the head lining to rear shelf securing screws.
Undo and remove the hinge plate to `D` post securing screws.
Displace and remove the hood assembly.

Fit the new hood to the body.
Fit the rubber spacers.
Tighten the headlining-to-shelf securing screws.
Fit new `O` ring seals to all screws.
Fit and tighten the securing screws
Raise the hood to align the hinge plates to the `D` post.
Fit but do not fully tighten the securing screws.
Fully seat the hood and secure the catches.
Tighten the hinge plate securing screws.
Release the hood locking catches.
Fold the hood rearwards.
Tighten the hinge plate lower securing screws.
Seat the hood and secure the top catches on the roof rail.
Lock the hood catch.
Position the headlining on the `D` post and secure the velcro fastening.
Position the headlining on the rear shelf and fasten with velcro.
Secure the outer cover with the velcro fastening.
Close the door.

2.2 CABRIOLET HOOD LATCHES, RENEW

Open the door. Release the hood locking catches.
Release the headlining trim to `D` post velcro

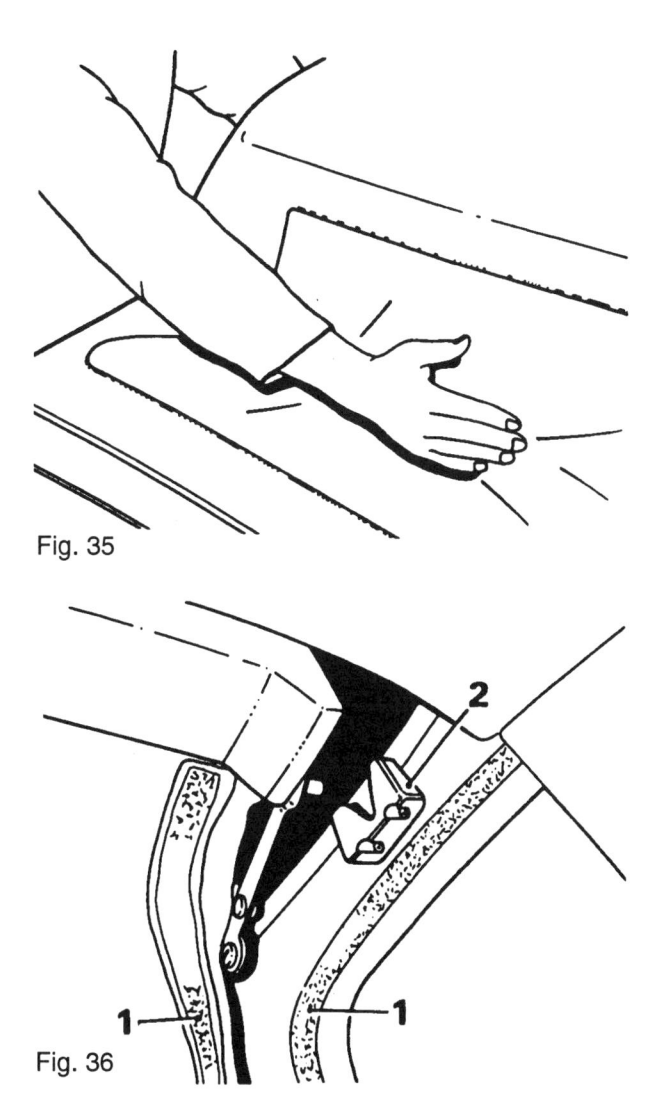

Fig. 35

Fig. 36

fastening (1 Fig. 36).
Fold the hood to the rearward position.
Undo and remove the canopy latch securing screws.
Displace and remove the latches (2 Fig. 36).

Fit new latches to the roll bar position.
Fit but do not fully tighten the securing screws.
Raise and fully seat the hood.
Secure the hood locking catch.
Final-position the latches and tighten the securing screws.
Position the headlining to the `D` post and secure the velcro fastening.

2.3 CABRIOLET HOOD SEAL, RENEW

Open the door.
Release the hood locking catches.
Release the headlining trim to `D` post velcro fastening (1 Fig. 37).
Fold the hood to the rearward position.
Release the outer cover velcro fastening.
Undo and remove the three outer hood securing screws each side.
Displace the hood for access.
Undo and remove the seal retaining clamp plate (latches) securing screws.
Remove the clamp plates (latches) (2 Fig. 37).

Fig. 37

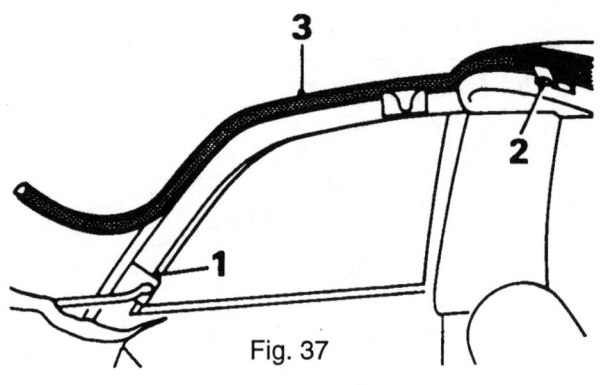

Displace and remove the seal (3 Fig. 37).
Clean the seating surface.
Fit the new seal to the roll bar.
Locate and fully seat the seal to the corners.
Fully seat the seal to the side members, taking care with the seal to avoid splitting.
Fit the seal retaining clamps.
Fit and tighten the securing screws.
Position the hood on the body.
Fit new `O` ring seals to all screws.
Fit and tighten the securing screws.
Secure the velcro fastening.
Raise the hood to the roll bar position.
Secure the hood catches to the roll bar mountings.
Lock the hood catch.
Position the headlining on the `D` post and secure with the velcro fastening.

2.4 CABRIOLET HOOD GUIDES, RENEW

Open the door. Release the hood locking catches.
Release the headlining trim to `D` post velcro fastening (1 Fig. 38).
Fold the hood to the rearward position.
Undo and remove the canopy guide securing screws.
Displace and remove the canopy guides (2 Fig. 38).

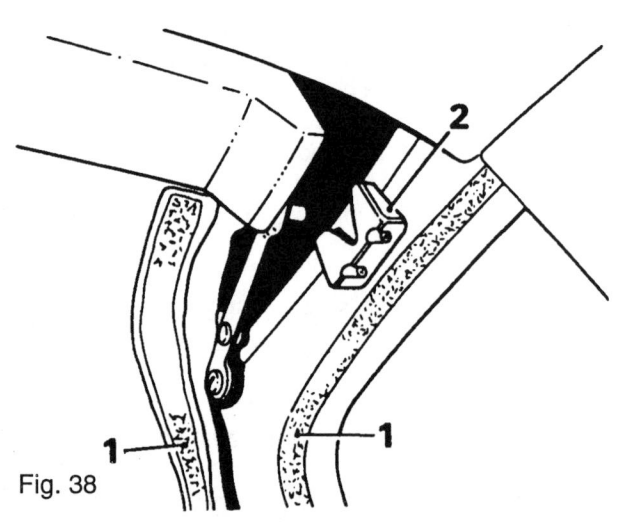

Fig. 38

Fit the new canopy guides.
Fit but do not fully tighten the securing screws.
Seat the hood and secure the top catches to the roll bar mountings.
Lock the hood catch.
Position the guides to the hood guide pins, and tighten the securing screws.
Position the headlining on the `D` post and secure with the velcro fastening.

2.5 CABRIOLET TARGA TOP SEAL, RENEW

Open the door. Remove the Targa top (1 Fig. 39).
Undo and remove the Targa top latch securing screws.
Displace and remove the latches.
Displace and remove the seal (2 Fig. 39).
Clean the seal seating surface.
Fit a new seal to the aperture.

Note: The red marker aligns to the front and centre.

Fully seat the seal to the `A` post - avoid splitting the seal.
Seat the seal to the rear corner and locate on the cantrail trim.
Fully seat the remaining corner to the `A` post.
Seat the seal to the top rail.
Fully seat the seal to the roll bar and cantrail and locate over the roll bar trim.
Fit the latches to the top rail.
Fit but do not tighten the securing screws.
Refit the Targa tops. Tighten the securing screws.

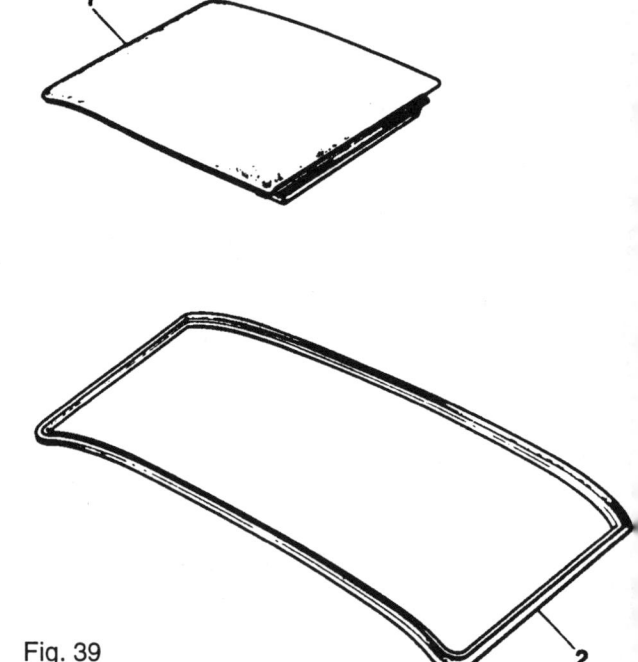

Fig. 39

2.6 CABRIOLET ROLL BAR TRIM PAD, RENEW

Remove the left and right hand Targa tops.
Move the seat fully forwards.
Incline the seat squab.
Remove the seat belt upper anchorage bolt finisher.
Remove the securing bolt.
Displace the seat belt.
Remove the spacer.
Remove the coat hook.
Displace and remove the B post trim pad.

Release the headlining trim to `D` post velcro fastening.
Release the hood locking catch.
Fold the hood to the rearward position.
Undo and remove the roll bar latch securing screws.
Displace and remove the latches.
Displace the Targa top-to-roll bar seal.
Carefully displace the hood canopy to roll bar seal.
Displace and remove the trim pad (1 Fig. 40).

Place the new trim pad in position and insert the trim under the `B` post trim.
Position the Targa top seal on the roll bar and fully seat it.
Position the hood canopy seal on the roll bar and fully seat it.
Fit the B post trim.
Refit the latches to the roll bar.
Fit and tighten the securing screws.
Raise the hood to align the guides.
Seat the hood to the roll bar.
Lock the hood catch.
Position the headlining on the `D` post and secure with the velcro fastening.
Refit the targa tops.

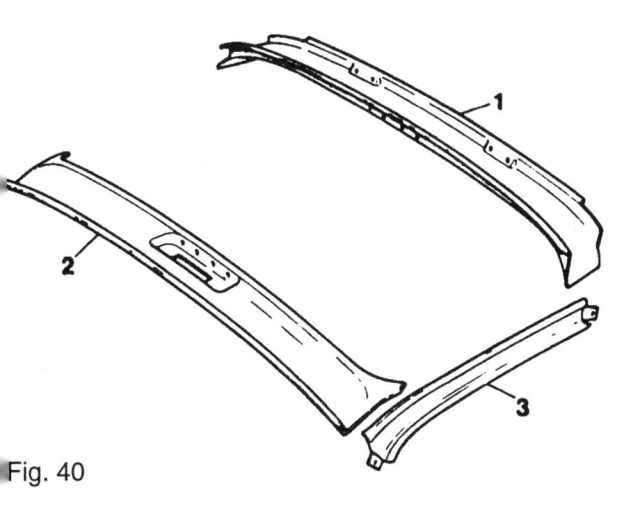

Fig. 40

2.7 CABRIOLET FRONT HEADLINING TRIM, RENEW

Open the door.
Remove the Targa tops.
Remove both left and right hand `A` post trim pads.
Remove the sun visors and the interior mirror.
Remove the roof lamp assembly.
Displace the Targa top seal at the forward edge.
Undo and remove the front Targa latch securing screws.
Remove the latches.
Undo and remove the left and right hand cantrail trim forward securing screws.
Displace the cantrail trim panels for access.
Undo and remove the front headlining trim securing screws.
Remove the trim (2 Fig. 40).
Remove the courtesy light shield.
Fit the courtesy light shield to the new trim panel.
Position the trim panel and align the fixing holes.

Fit and tighten the securing screws.
Position the cantrail trim panels.
Fit and tighten the securing screws.
Position the Targa top seal and fully seat over the trim panel edges.
Fit the Targa top latches.
Fit and tighten the securing screws.
Refit the courtesy lamp.
Refit the interior mirror.
Fit the sun visors.
Refit the `A` post trim finishers.
Fit the Targa tops.

2.8 CABRIOLET CANTRAIL TRIM PAD, RENEW

Open the door.
Remove the Targa tops.
Fold down the hood canopy.
Remove the `A` post trim pad.
Remove the `B` post trim pad.
Remove the roll bar finisher.
Undo and remove the cantrail trim securing screws.
Displace and remove the trim pad (3 Fig. 40).

Place the new trim in position and align the fixing holes.
Fit and tighten the securing screws.
Refit the roll bar finisher.
Fit the `B` post trim.
Fit the `A` post trim.
Refit the hood canopy.
Fit the targa tops.

2.9 CABRIOLET TARGA TOP HANDLE FINISHER, RENEW

Open the door.
Remove the Targa tops.
Undo and remove the handle finisher screws.
Displace and remove the finisher.

Fit the new finisher.
Fit and tighten the securing screws.
Refit the Targa tops.

2.10 REAR STOWAGE COMPARTMENT STRUT, RENEW

Open the door.
Move the front seat forwards for access.
Open the stowage compartment lid.
Undo and remove the strut securing screws.
Remove the strut (Fig. 41).

Fit the new strut.
Fit and tighten the securing screws.

19

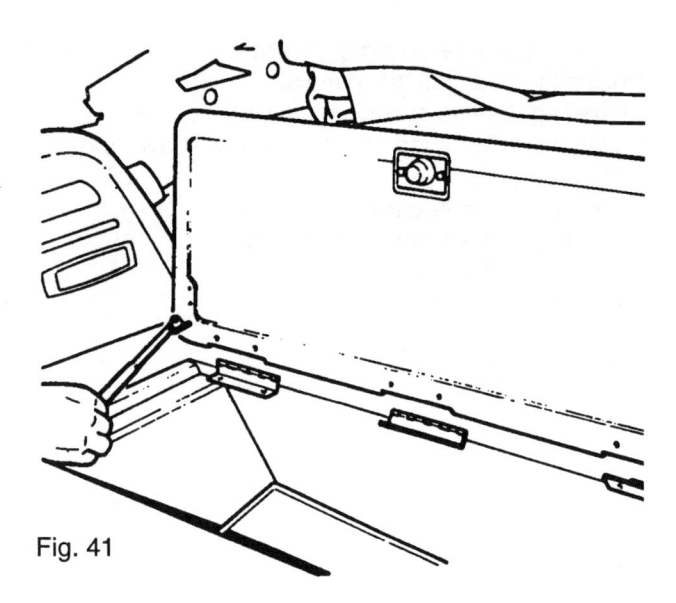

Fig. 41

2.11 REAR QUARTER ODDMENTS TRAY, RENEW

Open the door.
Move and incline the seat fully forwards.
Open the rear stowage box lid.
Undo and remove the oddments tray securing screws.
Displace and remove the oddments tray.
Transfer the mounting brackets to the new component.
Remove oddments tray.

Fit the oddments tray and fit and tighten the securing screws.

2.12 REAR STOWAGE COMPARTMENT, RENEW

Open the door.
Move and incline the seats fully forwards.
Open the rear stowage box lid (1 Fig. 42).
Undo and remove the oddments tray securing screws.
Remove the oddments tray.
Remove the stowage compartment carpets.
Undo and remove the stowage compartment-to-transmission tunnel securing screw.
Displace the carpets inside the stowage compartment.
Remove the stowage compartment securing screws.
Displace the rear floor carpets for access.
Remove the seat belt slider bars.
Remove the spacers.
Remove the stowage compartment taking care with the seat belt slider bars (2 Fig. 42).
Place the unit on a bench.

Transfer the hinges, brackets, luggage rail and map pockets to the new stowage compartment.
Fit the new assembly to the vehicle.
Align and fit the seat belt slider bars.
Fit and tighten the unit securing screws.
Position the inside carpets, rear floor carpets and the stowage carpets.

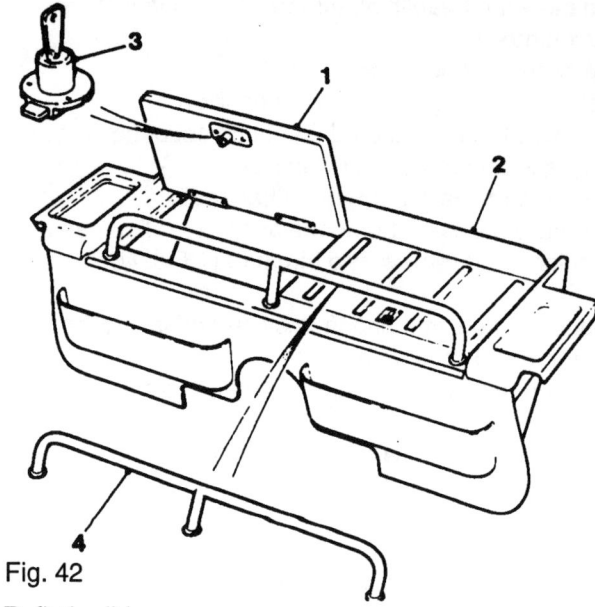

Fig. 42

Refit the lids.
Refit the oddments tray.

2.13 REAR STOWAGE COMPARTMENT LID, RENEW

Open the door.
Move and incline the seats fully forwards.
Open the rear stowage box lid.
Undo and remove the lid to hinge securing screws.
Remove the lid assembly (1 Fig. 42).
Transfer the lifting handle, lock assembly and lock finisher to the new lid.

Fit the new lid assembly in position and align the hinges.
Fit and tighten the securing screws.

2.14 REAR STOWAGE COMPARTMENTLUGGAGE RAIL, RENEW

Open the door.
Move and incline the seats fully forwards.
Open the rear stowage box lid.
Undo and remove the luggage rail securing bolts.
Remove the rail (4 Fig. 42).

Fit the new rail and secure with the bolts.

2.15 REAR STOWAGE COMPARTMENT LOCK BARREL, RENEW

Open the door.
Move and incline the seats fully forwards.
Open the rear stowage box lid.
Displace and remove the lock lever securing clip.
Displace and remove the lock lever. Withdraw the lock barrel (3 Fig. 42).
Insert the key into the new lock.
Fit the assembly to the compartment lid.
Fit the lock lever.

Fit the securing clip.
Close the lid. Turn the key to operate the lock.
Withdraw the key.

2.16 CONSOLE GLOVE BOX LID, RENEW

Open the door.
Open the console glove box lid.
Undo and remove the lid hinge (3 Fig. 43) securing screws.
Partially close the lid and disengage the link from the bracket (4 Fig. 43). Remove the lid (1 Fig.43).
Transfer the link and hinge to the new lid.
Fit the assembly to the console and engage the link to the bracket.
Fit and tighten the hinge securing screws.

Fig. 43

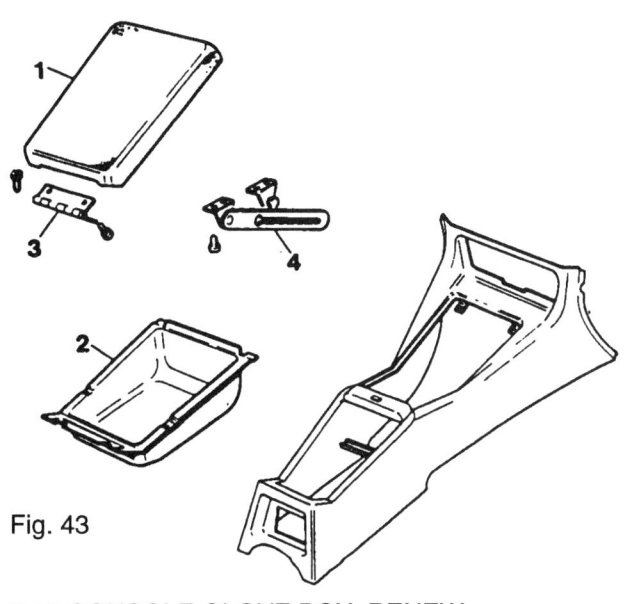

2.17 CONSOLE GLOVE BOX, RENEW

Open the door.
Open the console glove box lid.
Undo and remove the lid hinge (3 Fig. 43) securing screws.
Partially close the lid and disengage the link from the bracket (4 Fig. 43).
Remove the lid (1 Fig. 43).
Undo and remove the glove box pocket screws.
Remove the glove box lid link bracket.
Displace and remove the pocket (2 Fig. 43).
Fit the new glove box pocket.
Fit the glove box link bracket.
Fit and tighten the screws.
Fit the lid assembly to the glovebox, engaging the link to the bracket.
Fit and tighten the hinge securing screws.

2.18 DOOR LOCK, RENEW

Open the door.
Slide the armrest escutcheon plate from position (1 Fig. 44).
Undo and remove the armrest upper securing screw.
Undo and remove the door pocket securing screw (2 Fig. 44).

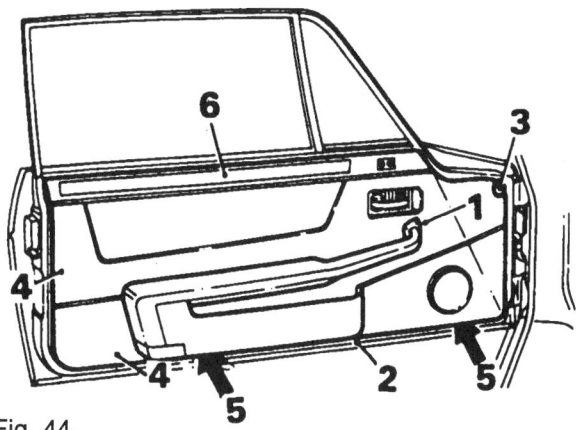

Fig. 44

Undo and remove the interior light striker plate securing screws (3 Fig. 44).
Remove the striker plate.
Remove the electric door mirror control switch mounting panel.
Displace the trim pad securing clips (4 Fig. 44)
Lift the trim pad to disengage the locating tangs.
Displace the trim pad from the door.
Note and disconnect the speaker wires (5 Fig. 44).
Note and disconnect the door guard lamp wires (5 Fig. 44).
Remove the trim pad (6 Fig. 44).

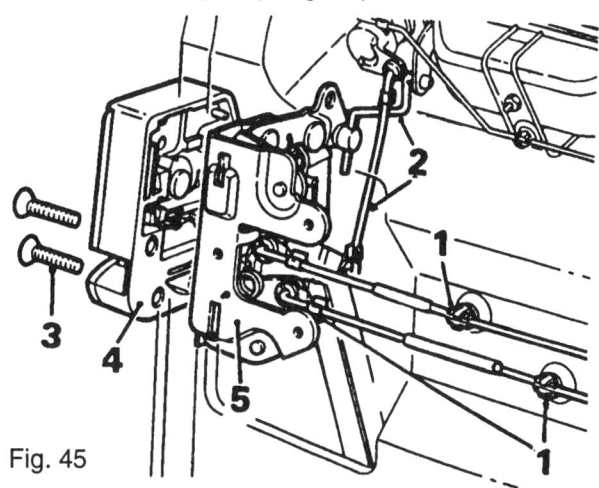

Fig. 45

Disconnect the inner handle to lock operating rods from the anti-rattle and connecting clips (1 Fig. 45).
Displace the water curtain for access.
Disconnect the outer handle to lock operating rods (2 Fig. 45).
Remove the lock unit securing screws (3 Fig. 45).
Remove the outer and inner units (4 & 5 Fig. 45).

Fit the new inner and outer lock units into position.
Fit and tighten the securing screws.
Reconnect the operating rods, fit the connecting clips.
Secure the rods in the anti-rattle clips.
Refit the trim pad to the door, correctly routing the electric mirror harness.
Connect the speaker wires.
Connect the door guard lamp wires.
Locate the door trim pad to the door, fully seat the retaining tangs.

Key to Fig. 46

1. Passive restraint 'B' post anchorage
2. Passive restraint diagonal belt and emergency buckle
3. Passive restraint runner/tongue-assembly
4. Lap belt inertia reel
5. Passive restraint diagonal belt inertia reel
6. Lap belt buckle
7. Lap belt and tongue

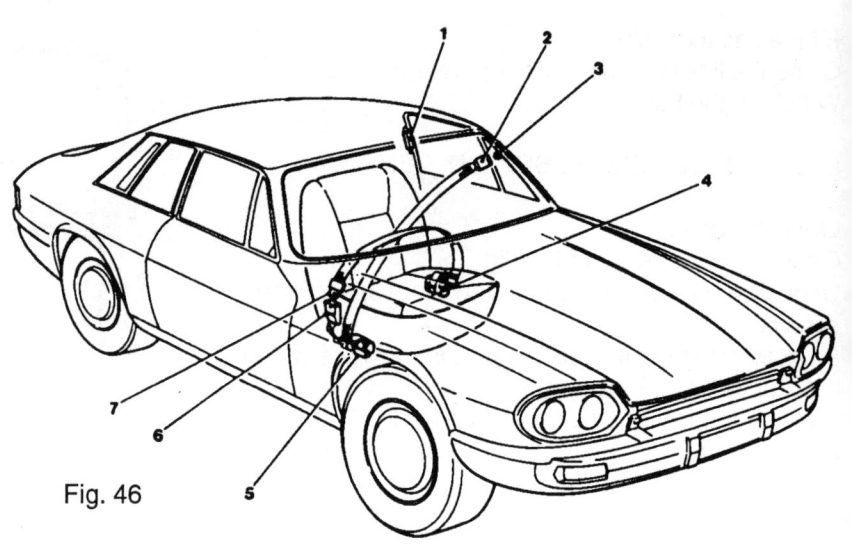

Fig. 46

Fully seat the retaining clips.
Fit the interior light striker plate.
Fit and tighten the securing screws.
Fit and tighten the door pocket securing screws.
Fully seat the escutcheon.

2.19 PASSIVE RESTRAINT SEAT BELTS (U.S.A. ONLY)

In order to comply with vehicle legislation in the United States of America, passive restraint belts were introduced to the front seats of all U.S.A. Market Jaguar XJS vehicles built after 2nd February 1987 (U.S.A. 1988 Model Year).

DESCRIPTION

The passive belt is a single diagonal seat belt, fed from an inertia reel assembly mounted on the seat slide (centre console side) to a motorised runner mounted into a guide rail, attached to the inner cantrail. A quick-release buckle for emergency use is located at the cantrail end of the belt. An independent second belt (inertia reel lap type) to be used in conjunction with the passive belt is fitted and must be manually secured by the occupant.

The passive belt system is driven by a motor and winch assembly mounted behind the rear lower quarter trim panel - one to the L.H. and one to the R.H. side of the vehicle.

A warning light will illuminate on the instrument pack to indicate that the passive restraint belt is not located in the driver's side anchorage or to the runner/tongue assembly. In addition there is a six second audible warning.

OPERATION

The system is actuated when the weight switch is operated on the driver's seat cushion sensor pad, the doors are closed and the ignition key is inserted into the ignition starter/switch.

The diagonal seat belt deploys automatically to the non-restraining position when the ignition key is removed or the adjacent door is opened. However, if reverse gear is selected, a relay is energised to prevent the driver's diagonal seat belt travelling to the non-restraining position if the driver's door is open. This will help safeguard the driver from being caught by the moving belt webbing should the occupant wish to lean out with the door open whilst reversing.

When the driver's seat is occupied (with doors closed, key in ignition) but the front passenger seat is vacant, the passenger diagonal belt will also deploy to the restraining position so as not to obscure the door mirror.

A warning light will illuminate on the instrument pack when the driver's seat is occupied (with doors closed and the ignition key turned to position 2) until the passive restraint belt is located in the driver's side anchorage with the belt in the runner/tongue assembly. Additionally, an audible warning will sound for approximately six seconds after the ignition key has been turned to position 2, if the belt is not located in the restraining position.

Note: An audible warning also sounds continuously if the driver's door is left open with the key in the ignition.

To release a diagonal belt in an emergency, the buckle at the shoulder end of the belt may be released by operating the red button marked 'Press'. If the belt is released for any reason it MUST be reconnected before the vehicle is driven.

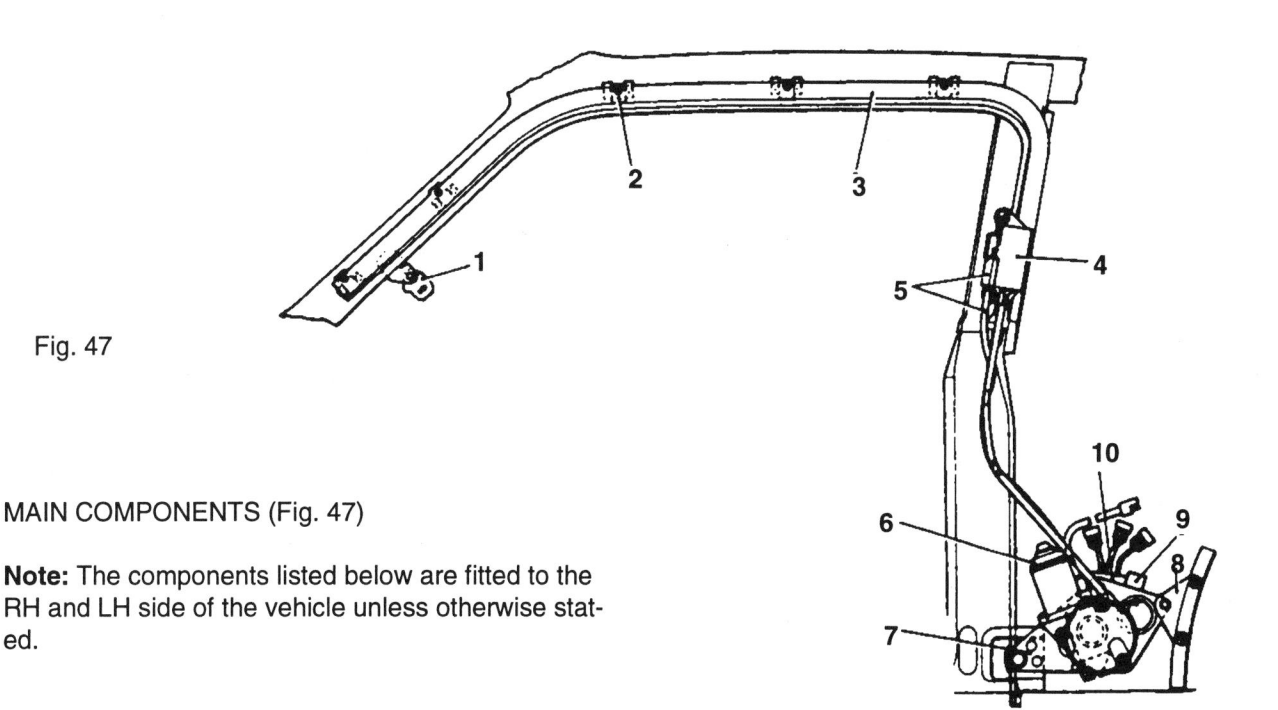

Fig. 47

MAIN COMPONENTS (Fig. 47)

Note: The components listed below are fitted to the RH and LH side of the vehicle unless otherwise stated.

Component	Location
1. Assembly runner and tongue	End of cable in the runner rail
2. Brackets, runner rail to cantrail	On 'A' post and cantrail
3. Runner Rail	On 'A' post, cantrail and 'B' post
4. Anchorage	On 'B' post
5. Rear dock switch with proximity switch for warning light	On 'B' post, part of anchorage
6. Motor and winch	Behind rear quarter trim panel
7. Motor and winch mounting plate	Behind rear quarter trim panel
8. Motor and winch mounting bracket	Behind rear quarter trim panel
9. Passive restraint reverse inhibit relay	Mounted on passive restraint electronic control unit bracket
10. Passive restraint electronic control unit with 3 gothic type connectors (Common, RH and LH side channels)	Behind drivers side rear quarter trim panel

SECONDARY COMPONENTS

Component	Location
Cantrail crash-roll	On cantrail and 'A' post
Trim finisher outer	On 'B' post
Trim finisher inner	On 'B' post
Passive restraint inertia reel	Under front seat
Lap belt inertia reel	Under front seat
Emergency tongue kit	In glovebox
Electronic control unit mounting bracket assembly	Behind driver's side rear quarter trim panel
Rear dock and proximity switch connector	Behind rear quarter trim panel
Motor and drive connector	Behind rear quarter trim panel
Weight sensor switch	Under driver's seat cushion
Door switch	On 'A' post, combined with courtesy light switch
Power pick-up points: two 15 amp fuses Nos. 17 and 18	In main fuse box (driver's side under-scuttle)

USE ON EARLY VEHICLES

Incorporation of the passive restraint system necessitated modifications at the body-in-white stage. Fitment of the passive restraint system is therefore not possible to vehicles manufactured before 2nd February 1987.

FAULT DIAGNOSIS

WARNING: SHOULD THE SEATS OR SEAT SLIDES BE REMOVED FOR ANY REASON, UPON REASSEMBLY CHECK FOR CORRECT SEAT BELT OPERATION.

Fault	Cause
Belt inoperative	Belt twisted
Faulty lap belt operation	Inadequate retention or damage to inertia reel. Check integrity of fasteners and that reel is correctly set on bracket.
Lap belt becomes difficult to find	Chrome guide loop insecure
Webbing wear/fray	Damage to chrome guide loop, webbing guide, running loop or reel shield
Seat slide jam	Inertia reel running loop insecure
Runner/tongue loose	Damage/break in cable
Runner/tongue stalls	Damage or obstructions to runner rail, trim finisher, motor/winch assembly
Runner/tongue fails to engage in 'B' post anchorage	Damage/obstructions or inadequate retention of anchorage

SYSTEM CHECKS

The following checks must be carried out at every 15,000 miles (24,000 km) service interval (the period between services must not exceed 12 months).

INERTIA SWITCH

1 The inertia switch is located in the driver's footwell beside the 'A' post. Sit in the driving position. Ensure the gear lever is not in reverse. Close the driver's door and insert the ignition key. The passive diagonal belts should move to the restraining position. Switch on the ignition (position 2). Raise the test/reset button which protrudes through the top of the inertia switch cover.

2. Open the doors: the diagonal belts should not move from the restraining position. Remove the ignition key: the diagonal belt should not move from the restraining position. To exit from the vehicle first operate the emergency release buckle button on the cantrail end of the belt. Once outside the vehicle observe that the runner/tongue assemblies remain in the restraining position.

3. Re-enter the vehicle, leave the door open, reset the inertia switch by depressing the button. Observe that the runners move to the non-restraining position. Re-attach the belt buckle to the runner/tongue assembly with the red 'press' button facing inboard.

4. If the passive restraint fails to comply with the above, locate and correct the fault. Suspect: inertia switch, passive restraint ECU common connector pin No.7.

SEAT BELT WARNING

1. Sit in the driver's seat with the key in the ignition and both doors closed. The diagonal belt should automatically deploy to the rear dock (restraining position). Turn on the ignition (position 2): the warning light should not be illuminated.

2. Remove diagonal seat belt from runner/tongue assembly. The seat belt warning light should illuminate. If it does not, locate and correct the fault. Suspect: rear dock proximity switch, passive restraint ECU common connector pin No.2, LH side connector pin No.7.

REVERSE INHIBIT

1. Sit in the driver's seat, close both doors, insert the key in the ignition and turn it to position 2. The diagonal belt should automatically deploy to therear dock (restraining position) and the warning light should not be illuminated.

2. Select reverse gear and open the driver's door.The diagonal seat belt should remain in the restraining position. If the seat belt travels to the non-restraining position, locate and correct the fault.

3 Open the driver's door, and ensuring body or limbs are not in the path of deploying seat belt, disengage reverse gear. The seat belt should travel to the front dock (non-restraining position). If it does not, locate and correct the fault. Suspect the reverse inhibit relay.

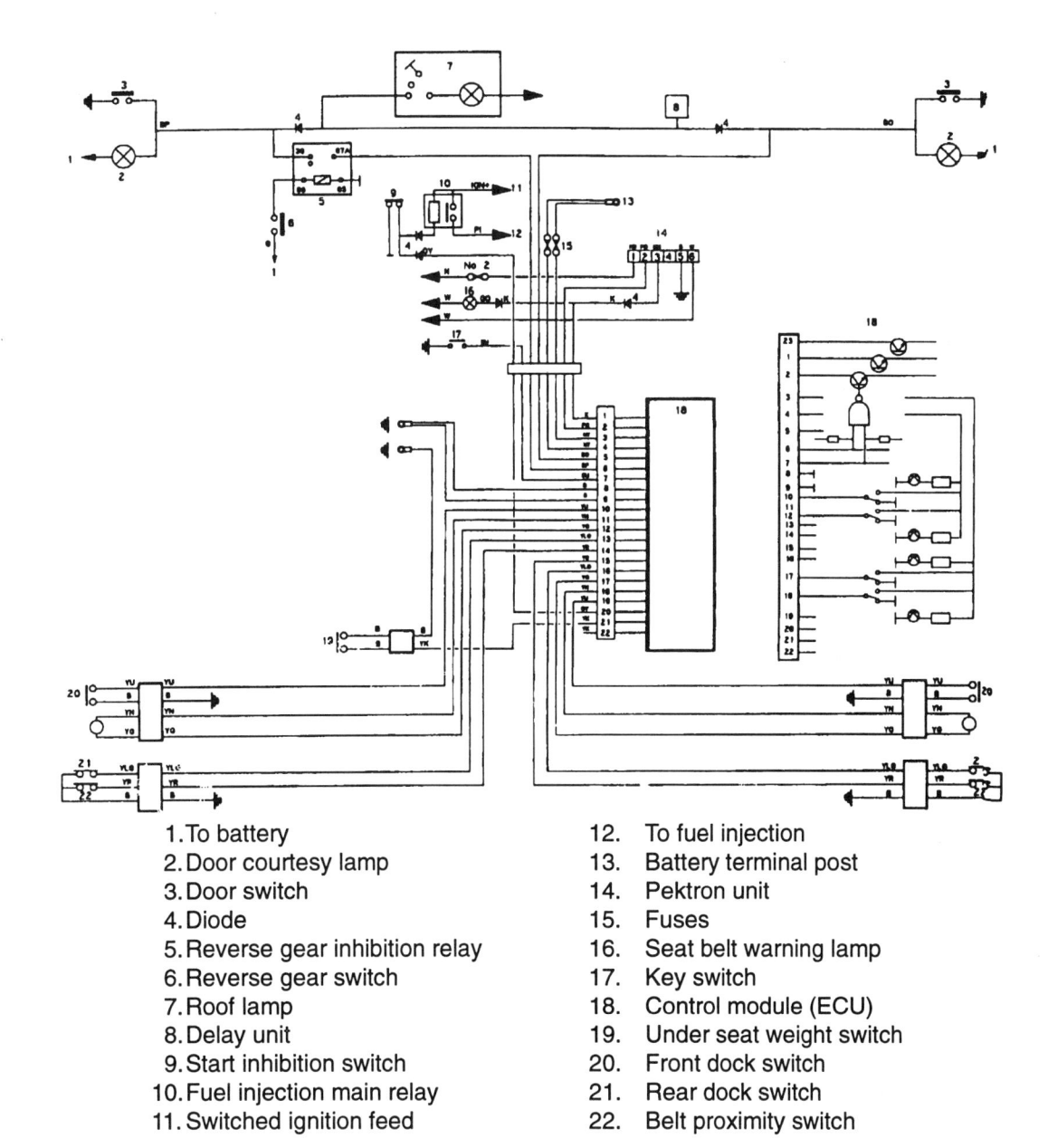

1. To battery
2. Door courtesy lamp
3. Door switch
4. Diode
5. Reverse gear inhibition relay
6. Reverse gear switch
7. Roof lamp
8. Delay unit
9. Start inhibition switch
10. Fuel injection main relay
11. Switched ignition feed
12. To fuel injection
13. Battery terminal post
14. Pektron unit
15. Fuses
16. Seat belt warning lamp
17. Key switch
18. Control module (ECU)
19. Under seat weight switch
20. Front dock switch
21. Rear dock switch
22. Belt proximity switch

Fig. 48 Passive Restraint System Circuit Diagram

2.20 PASSIVE RESTRAINT MOTOR AND DRIVE ASSEMBLY, RENEW

Disconnect the passive restraint belt buckle.
Remove the front seat squab.
Remove the rear seat cushion and squab.
Remove the front seat belt upper and lower anchorage bolts, spacers and springs (2 Fig.49).
Displace the 'Furflex' trim from the door flange (3 Fig. 49).
Carefully displace the trim pad flap from the door flange, remove the front securing screws.
Remove the remaining trim pad screws and displace the pad for access to the rear speaker wires.
Disconnect the speaker feed wires.
Reposition the seat belt through the trim pad.
Remove the pad (4 Fig. 49).
Remove the `B` post trim pad and cantrail crash roll.
Remove the motor and drive securing bolts/nuts.
Disconnect the multi plugs. Remove the drive to `B` post securing bolts.

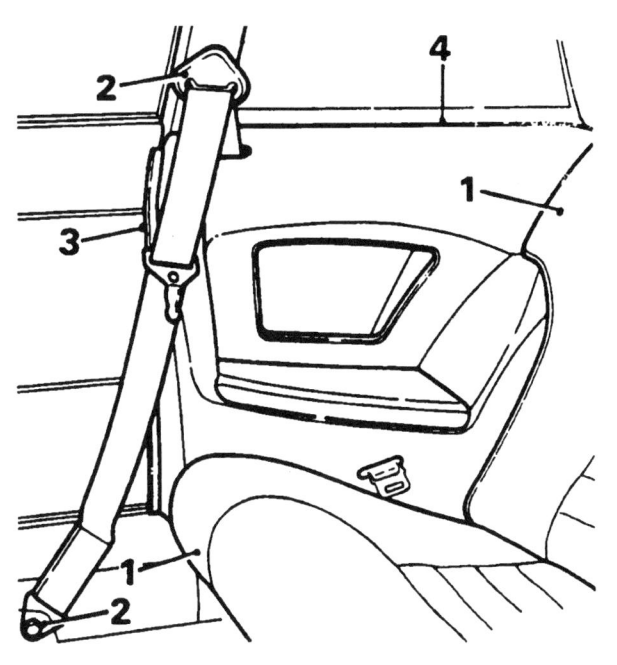

Fig. 49

Fig. 50

Remove the keyhole bracket.
Displace and remove the motor (Fig. 50).
Withdraw the drive cable from the run channel.

Fit the new assembly to the vehicle, feed the drive into the run channel.
Locate, but do not fully tighten, the drive upper securing bolt.
Fit the keyhole bracket, fit and tighten the lower securing bolt.
Fully tighten the upper securing bolt.
Locate the motor mounting bracket.
Fit and tighten the securing bolts/nuts.
Reconnect the multi plugs.
Refit the `B` post trim pad and cantrail crash roll.
Feed the seat belt through the trim pad and fit the pad in position.
Reconnect the speaker wires, fit the trim pad, fit and tighten the screws.
Position the flap to the door flange, fit and tighten the screws.
Fit the 'Furflex' trim.
Fit and tighten the seat belt upper and lower anchorage bolts.
Refit the rear seat cushion and squab.
Fit the front seat and reconnect the passive seat belt.

2.21 PASSIVE RESTRAINT INERTIA REEL, RENEW

Disconnect the passive restraint belt buckle.
Remove the front seat squab.
Place the seat assembly on a workbench.
Operate the seat slide adjuster and move the seat slide fully rearwards for access to one seat slide securing bolt.
Remove the bolt and spacer.
Operate the seat slide adjuster and move the seat slide fully forwards for access to the remaining bolt.
Remove the bolt and spacer.
Remove the slide and reel assembly from the seat.

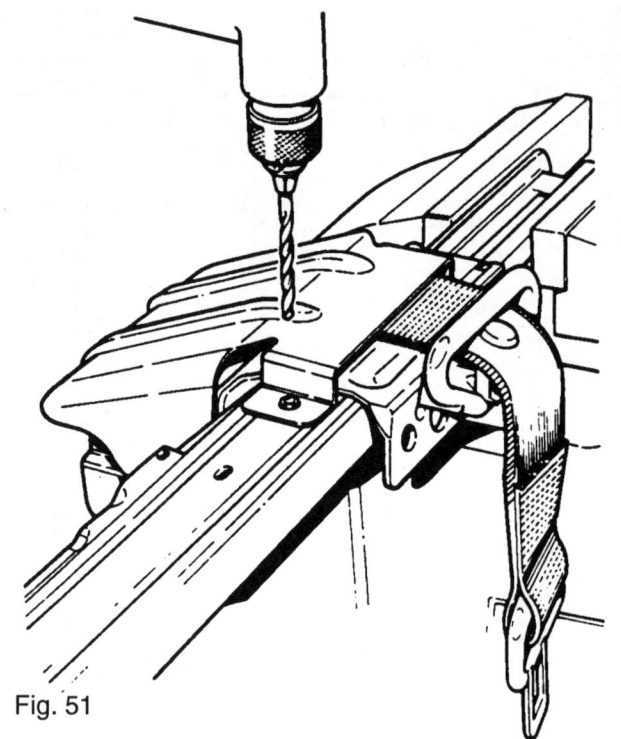

Fig. 51

Secure the assembly in a vice.
Remove the lap belt buckle.
Drill out the rivets securing the reel cover (Fig. 51), remove the cover.
Remove the belt guide bolt finisher and remove the bolt.
Displace the belt and guide from the runner.
Remove the reel to runner securing bolt/nut (1 Fig. 52); remove the reel.

Fit the new reel - hold the reel level to enable the belt to be withdrawn.
Fit and tighten the securing nut.
Align the guide to the runner.
Fit and tighten the securing bolt.
Fit the cover to the reel, secure with pop rivets.
Fit the runner assembly to the seat - align the seat slide adjuster bar to the runner.
Fit the spacer.
Fit but do not tighten the securing bolt.
Operate the slider bar. Fit the remaining spacer. Fit the remaining securing bolt.
Fully tighten the securing bolts.
Invert the seat.
Withdraw the belt from the reel.
Fit the guide to the seat. Align the belt to the guide.
Fit and tighten the securing screws.
Refit the seat to the vehicle.
Reconnect the passive restraint belt buckle.

2.22 PASSIVE RESTRAINT LAP BELT INERTIA REEL, RENEW

Disconnect the passive restraint belt buckle.
Remove the front seat squab.
Place the seat assembly on a workbench.
Operate the seat slide adjuster and move the seat slide fully rearwards for access to one seat slide

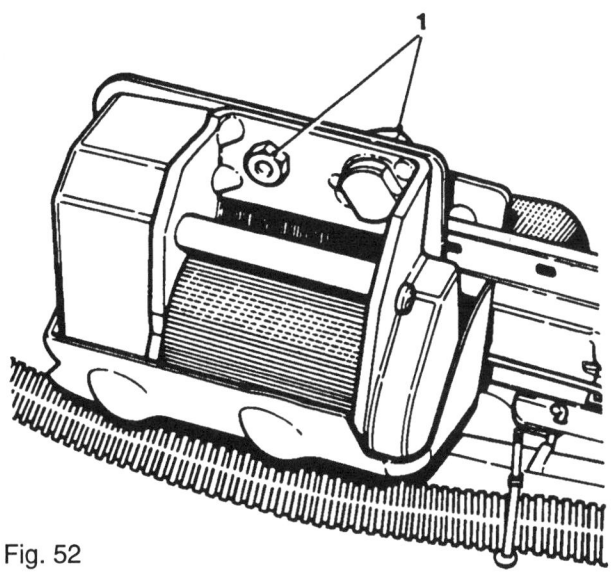

Fig. 52

securing bolt.
Remove the bolt and spacer.
Operate the seat slide adjuster and move the seat slide fully forwards for access to the remaining bolt.
Remove the bolt and spacer.
Remove the slide and reel assembly from the seat.
Secure the assembly in a vice.
Remove the lap belt buckle.
Drill out the rivets securing the reel cover (Fig. 51), remove the cover.
Remove the belt guide bolt finisher and remove the bolt.
Displace the belt and guide from the runner.
Remove the reel to runner securing bolt & nut (1 Fig. 52).
Remove the reel.

Fit the new reel - hold the reel level to enable the belt to be withdrawn.
Fit and tighten the securing nut. Align the guide to the runner.
Fit and tighten the securing bolt.
Fit the cover to the reel, secure with pop rivets.
Fit the runner assembly to the seat - align the seat slide adjuster bar to the runner.
Fit the spacer.
Fit but do not tighten the securing bolt.
Operate the slider bar. Fit the remaining spacer. Fit the remaining securing bolt.
Fully tighten the securing bolts.
Invert the seat. Withdraw the belt from the reel.
Fit the guide to the seat. Align the belt to the guide.
Fit and tighten the securing screws.
Refit the seat to the vehicle.
Reconnect the passive restraint belt buckle.

2.23 PASSIVE RESTRAINT E.C.U., RENEW

Disconnect the passive restraint belt buckle.
Remove the front seat squab.
Remove the rear seat cushion and squab.
Remove the front seat belt upper and lower anchorage bolts, spacers and springs (2 Fig. 49).
Displace the 'Furflex' trim from the door flange (3 Fig. 49).
Carefully displace the trim pad flap from the door flange.
Remove the front securing screws.
Remove the remaining trim pad screws and displace the pad for access to the rear speaker wires.
Disconnect the speaker feed wires.
Reposition the seat belt through the trim pad.
Remove the pad (4 Fig. 49).
Remove the `B` post trim pad and cantrail crash roll.
Remove the motor to mounting bracket securing bolts/nut.
Displace the motor for access.
Disconnect the E.C.U. multi plugs.
Remove the relay base from the mounting bracket (1 Fig. 53).

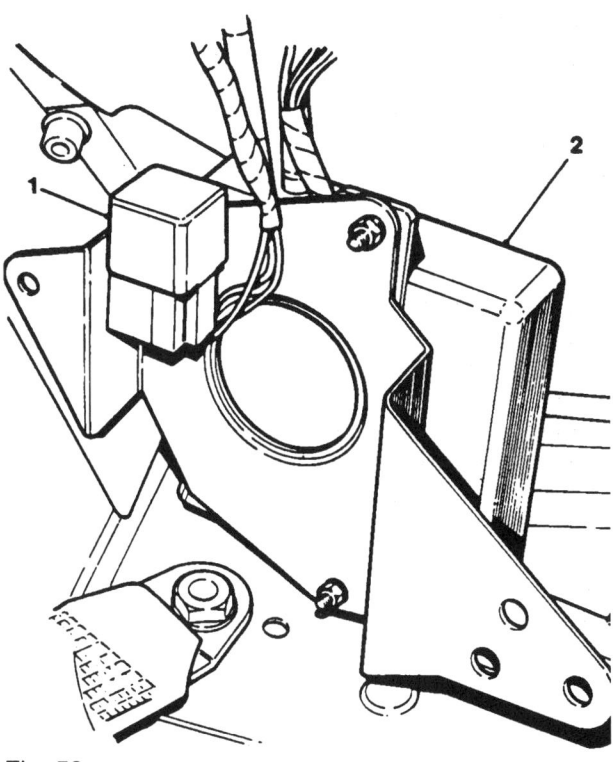

Fig. 53

Remove the E.C.U. assembly (2 Fig. 53).
Remove the E.C.U. from the bracket.

Fit the new E.C.U. to the mounting bracket; fit and tighten the securing nuts.
Fit the assembly to the car and connect the relay base.
Fit the motor/drive assembly to the mounting bracket and fit and tighten the securing nuts.
Reconnect the multi plugs.
Refit the `B` post trim pad and cantrail crash roll.
Feed the seat belt through the trim pad and fit the pad to position.
Reconnect the speaker wires, fit the trim pad, fit and tighten the screws.
Position the flap to the door flange; fit and tighten the screws.
Fit the 'Furflex' trim.

Fit and tighten the seat belt upper and lower anchorage bolts.
Refit the rear seat cushion and squab.
Fit the front seat and reconnect the passive seat belt.

2.24 PASSIVE RESTRAINT RELAY, RENEW

Disconnect the passive restraint belt buckle.
Remove the front seat squab.
Remove the rear seat cushion and squab.
Remove the front seat belt upper and lower anchorage bolts, spacers and springs (2 Fig. 49).
Displace the 'Furflex' trim from the door flange (3 Fig. 49).
Carefully displace the trim pad flap from the door flange; remove the front securing screws.
Remove the remaining trim pad screws and displace the pad for access to the rear speaker wires.
Disconnect the speaker feed wires.
Reposition the seat belt through the trim pad.
Remove the pad (4 Fig. 49).
Remove the `B` post trim pad and cantrail crash roll.
Displace and remove the relay (1 Fig.53) from the mounting bracket (2 Fig. 53).

Fit a new relay to the mounting bracket.
Refit the `B` post trim pad and cantrail crash roll.
Feed the seat belt through the trim pad and fit the pad in position.
Reconnect the speaker wires, fit the trim pad, fit and tighten the screws.
Position the flap to the door flange,.
Fit and tighten the screws.
Fit the 'Furflex' trim.
Fit and tighten the seat belt upper and lower anchorage bolts.
Refit the rear seat cushion and squab.
Fit the front seat and reconnect the passive seat belt.

3 General Specification

Fuses

Main Fuse Box - R.H. Steering

Fuse No.	Value	Protected Circuit
1	20A	Cigar lighter
2	15A	Hazard warning, seat belt logic
3	35A	Clock, aerial, caravan, boot lamp
4	10A	Panel instruments, reverse light
5	15A	Direction indicators, stop lamps, kickdown switch (where fitted)
6	10A	Fog rear guard (where fitted)
7	10A	Panel/cigar lighter and selector illumination
8	3A	Door locks, electric mirrors.
9	35A	Windscreen wipers
10	50A	Air conditioning motors
11	35A	Air conditioning controls, horn, windscreen washers, radiator cooling fan
12	35A	Heated rear screen, heated door mirrors

Main Fuse Box - L.H. Steering

Fuse No.	Value	Protected Circuit
1	20A	Front fog lamps
2	15A	Hazard warning, seat belt logic
3	35A	Clock, aerial, caravan, boot lamp
4	10A	Panel instruments, reverse light
5	15A	Direction indicators, stop lamps, kickdown switch (where fitted)
6	10A	Fog rear guard (where fitted)
7	10A	Panel/cigar lighter and selector illumination
8	3A	Door locks, electric mirrors.
9	35A	Windscreen wipers
10	50A	Air conditioning motors
11	35A	Air conditioning controls, horn, windscreen washers, radiator cooling fan
12	35A	Heated rear screen, heated door mirrors

Auxiliary Fuse Box - R.H. Steering

Fuse No.	Value	Protected Circuit
13	20A	Interior & map lamps
14	3A	Left-hand front side lights
15	3A	Right-hand front side lights
16	20A	Front fog lamps (where fitted)
17	3A	Speed control

Auxiliary Fuse Box - L.H. Steering

14	2A	Left-hand front side lights
15	2A	Right-hand front side lights.
16	10A	Cigar lighter
17	25A	Speed control

Headlamp Fuse Box

Fuse No.	Value	Protected Circuit
1	25A	Radiator auxiliary cooling fan motor relay
2	25A	Left-hand main beam headlamp
3	10A	Left-hand dip beam headlamp
4	25A	Right-hand main beam headlamp
5	10A	Right-hand dip beam headlamp

WIRING DIAGRAMS

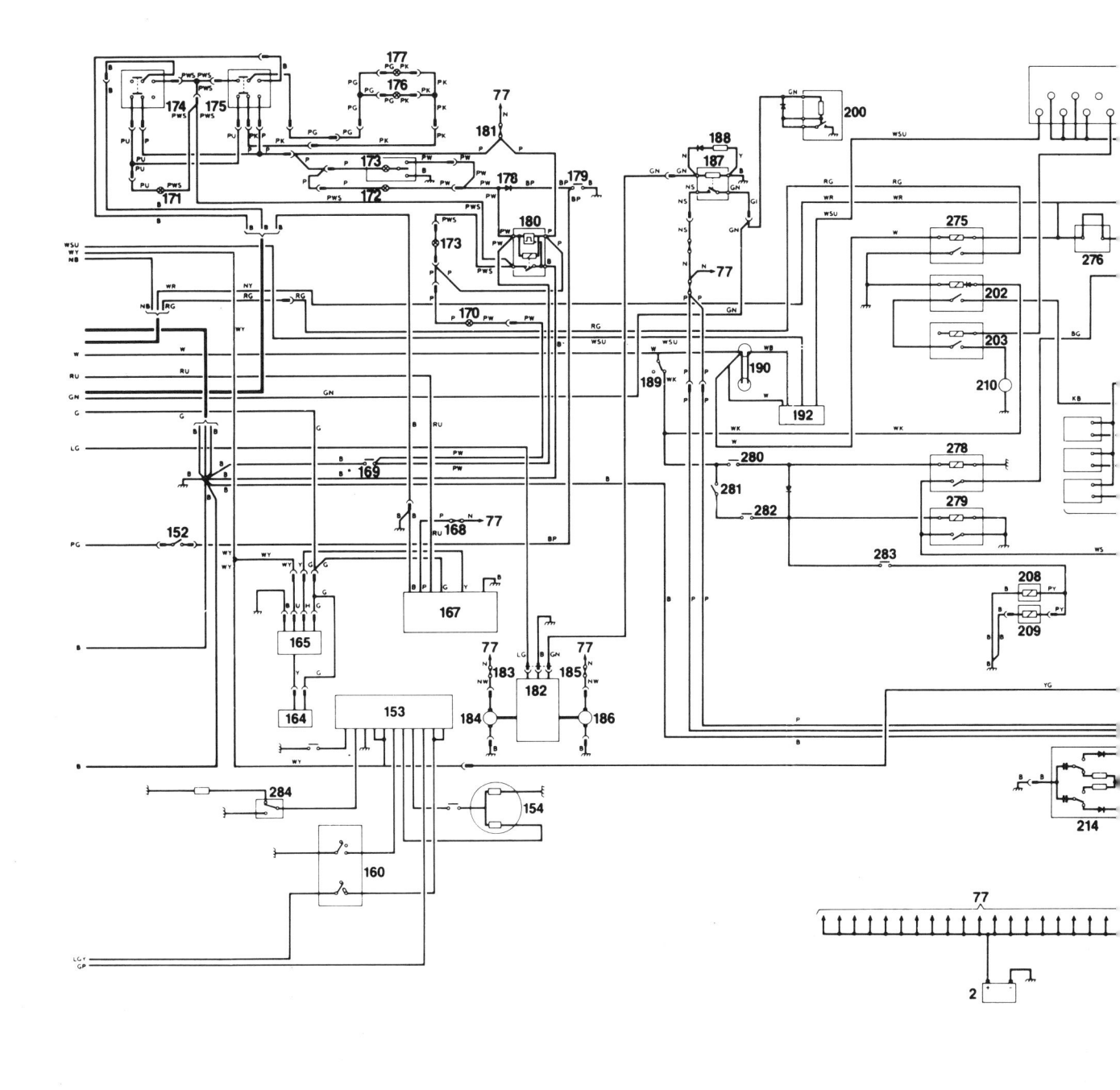

Air Conditioning Diagram

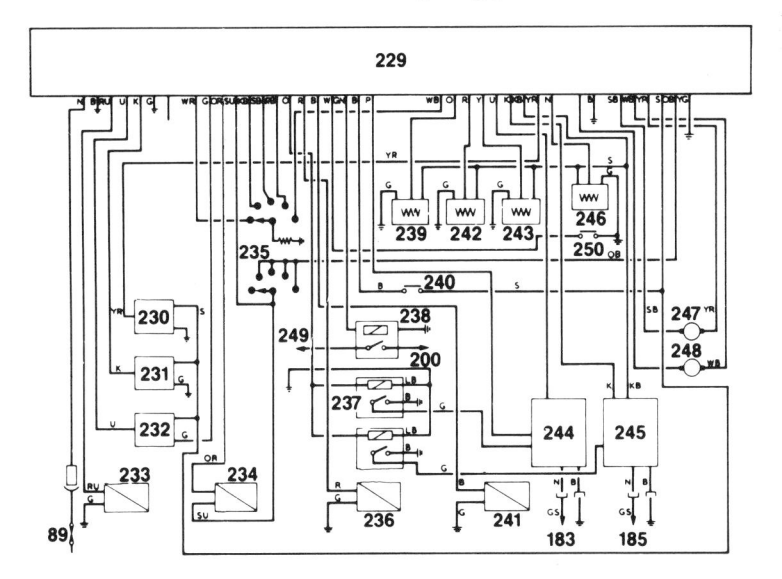

AIR CONDITIONING DIAGRAM

- 229 Control module
- 230 Ambient temperature sensor
- 231 Evaporator temperature sensor
- 232 In-car temperature sensor
- 233 Recirculation vacuum solenoid
- 234 Defrost vacuum solenoid
- 235 Blower motor control switch
- 236 Water valve vacuum solenoid
- 237 High speed relays
- 238 Compressor clutch relay
- 239 Differential control
- 240 Coolant temperature switch
- 241 Centre vent vacuum solenoid
- 242 Lower servo feedback potentiometer
- 243 Upper servo feedback potentiometer
- 244 LH Blower motor assembly
- 245 RH Blower motor assembly
- 246 Temperature demand control
- 247 Upper servo
- 248 Lower servo
- 249 V12 in line fuse. 3.6 fuse No. 16
- 250 Auto/manual switch

PASSIVE RESTRAINT

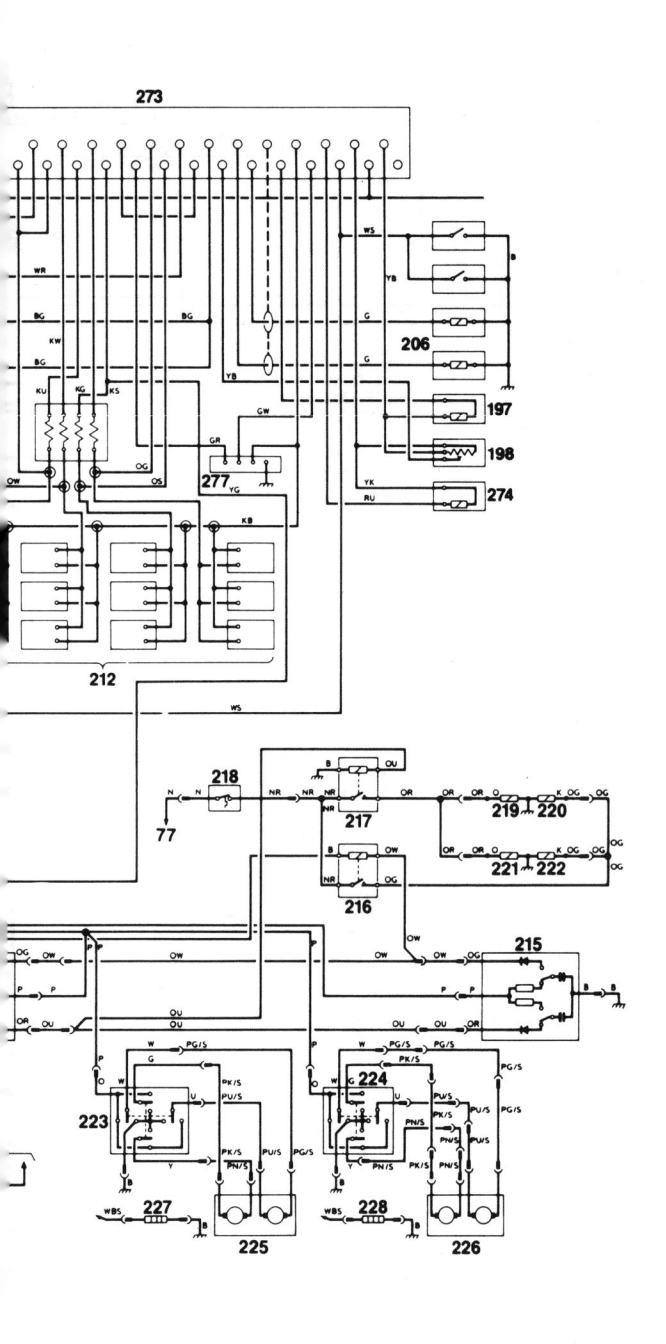

251 Diode
252 Reverse gear inhibition relay
253 Reverse gear switch
254 Delay unit
255 Start inhibition switch
256 Fuel injection main relay
257 Switched ignition feed
258 To fuel injection
259 Battery terminal post
260 Pectron unit
261 Fuses
262 Seat belt warning lamp
263 Key switch
264 Control module
265 Under seat weight switch
266 Front door switch
267 Rear dock switch
268 Belt proximity switch

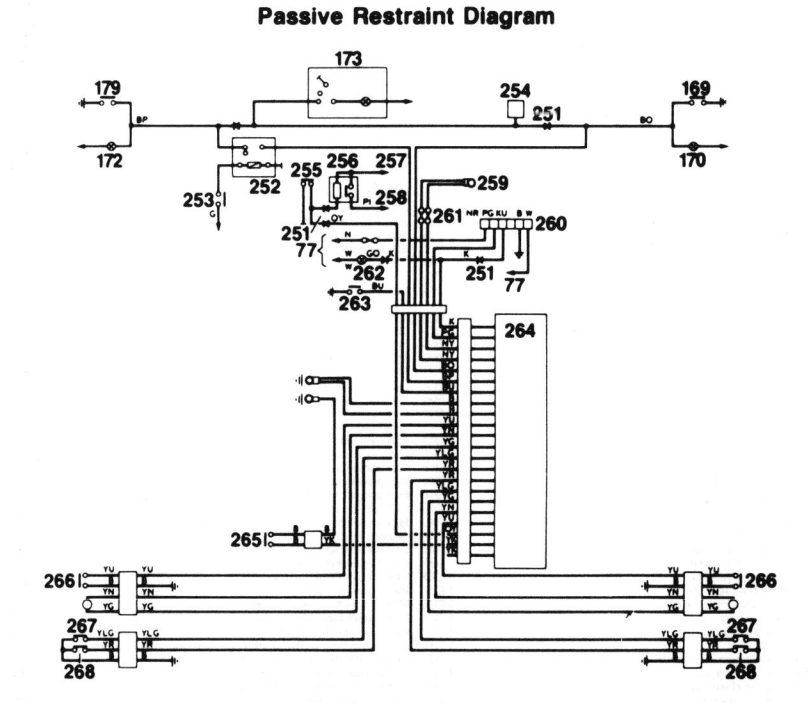

Passive Restraint Diagram

JAGUAR
XJ-S 5.3 & 6.0

BY APPOINTMENT
TO H M QUEEN ELIZABETH
THE QUEEN MOTHER
B L CARS LIMITED
MANUFACTURERS OF DAIMLER JAGUAR
ROVER CARS AND LAND ROVERS

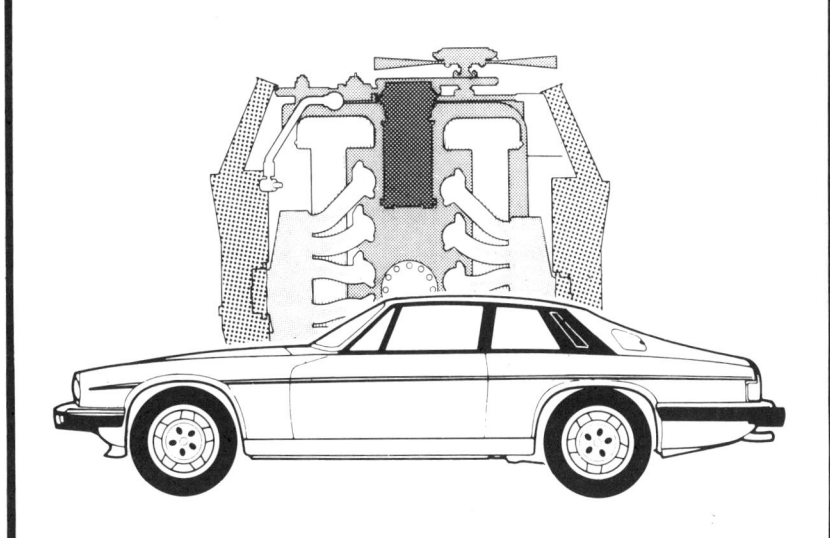

Supplement to the
Repair Operation Manual

SUPPLEMENT C • XJ-S 5·3 & 6·0 1988½ -

INTRODUCTION

This supplement is to be used in conjunction with XJS Repair Operation Manual Part No. AKM 3455 and its **supplements A and B sections**.

It covers later vehicles, mainly dealing with those modifications brought about in line with the introduction of the 6 Litre engine and four-speed electronically-controlled transmission, but also covering ABS brake systems, body modifications, and more. Please note that, due to the ever increasing sophistication of the later versions of these vehicles, it is necessary to cover Electrical Circuits in-depth in a separate publication. A limited selection of circuits are printed at the rear of this Supplement, but full coverage will only be available by reference to the 1992 Model Year-on Service Manual (Part Number JJM 10 04 06 / 02) and the relevant Model Year's 'Electrical Guide'.

SPECIFICATION

Users are advised that the specification details set out in this book apply to a range of vehicles and not to any one. For the specification of a particular vehicle purchasers should consult their Dealer.

Jaguar Cars Ltd. reserve the right to vary their specifications with or without notice and at such times and in such a manner as they think fit. Major as well as minor changes may be involved in accordance with the Manufacturer's policy of constant product improvement. Whilst every effort is made to ensure the accuracy of the particulars contained in this supplement, neither Jaguar Cars Ltd., nor the Dealer, by whom this book is supplied, shall in any circumstances be held liable for any inaccuracy or the consequences thereof.

COPYRIGHT

All rights reserved. No part of this publication may be reproduced, stored in a retrieval system or transmitted in any form, electronic, mechanical, photocopying, recording or other means without prior written permission of Jaguar Cars Ltd.

CONTENTS

Chapter		Page	Chapter		Page
1	6.0 Litre Engine	3	9	Exhaust System (6.0 Litre)	49
2	Emission Control System	5	10	Automatic Transmission (6.0 Litre)	52
	(5.3 Litre 1992 MY on)		11	Front Suspension (1993.5 MY on)	61
3	Emission Control System (6.0 Litre)	8	12	Rear Suspension (1993.5 MY on)	63
4	Engine Management System	13	13	Anti-Lock Braking Systems	64
	(5.3 Litre 1992 MY on)		14	Body (Including SRS Systems)(1992 MY on)	99
5	Engine Management System (6.0 Litre)	22	15	Air Conditioning System (1994 MY on)	124
6	Fuel System (5.3 Litre 1992 MY on)	33	16	Electrical System (1992 MY on)	131
7	Fuel System (6.0 Litre)	46	17	Instruments (1992 MY on)	138
8	Cooling System (1993 MY on)	48		Wiring Diagrams, ECU, Relay and Fuses etc.	145

GENERAL SPECIFICATIONS

Engine - 6.0 Litre

Stroke	78.5 mm
Bore	90.0 mm
Capacity	5994 cc
Compression ratio	11.0:1
Maximum power (with air injection)	301 BHP @ 5350 RPM
Maximum torque (with air injection)	350 lbf / ft @ 2850 RPM
Maximum power (without air injection)	304 BHP @ 5350 RPM
Maximum torque (without air injection)	355 lbf / ft @ 2850 RPM

TRANSMISSION - FOUR SPEED

Transmission - Four Speed

Make / Model	Powertrain 4L80E
Gear ratios:	
1st	2.482:1
2nd	1.482:1
3rd	1.000:1
4th	0.750:1
Reverse	2.077:1
Transmission fluid	Dexron II E
Fluid capacity	12.8 Litres
Weight	105 kg dry (118 kg wet)

Chapter 1

6.0 LITRE ENGINE (1993.25 MODEL YEAR ON)

INTRODUCTION

Items listed in this section are those which are changed relative to the 5.3 Litre engine. The engine is available with two different exhaust systems to comply with the emission regulations in particular countries.

Repair operations are generally similar whether the 5.3 or 6.0 litre variant is fitted.

Engine performance with each system fitted is as follows:

Twin Catalyst Without Air Injection (EEC Markets)

Maximum Power 296 bhp (300 PS, 220 kW) at 5350 RPM
Maximum Torque 347 lbf ft (471 Nm) at 2850 RPM

90% of maximum torque available from idle to 4800 RPM

Triple Catalyst With Air Injection (Rest of World Markets including Japan*)

Maximum Power 285 bhp (289 PS, 212 kW) at 5400 RPM
Maximum Torque 341 lbf ft (462 Nm) at 2850 RPM

90% of maximum torque available from idle to 4500 RPM

* Japanese Market vehicles also carry grass shields and over-temperature sensors.

CYLINDER BLOCK

Cylinder Block Casting:

The mounting flange for the transmission has been modified to accept the Hydra-Matic 4L80E four speed automatic transmission. The engine oil dipstick tube has been repositioned to the rear of B-Bank.

Crankshaft:

A new crankshaft made from forged steel has an increased crank-throw to provide the longer stroke of 78.5 mm instead of 70 mm. To enhance refinement, increased balance masses are incorporated. Minor modifications have been made to the windage tray to accommodate the increased crank-throw.

Cylinder Liner:

This is of a similar construction to that of the 5.3 Litre engine, but is 2 mm shorter at its lower end to provide clearance for the connecting rod due to the increased crank-throw.

Bearing Shells:

Shells for the main bearings and for the connecting rods are as fitted to the 5.3 Litre engine, but graded for size to reduce the maximum bearing clearances by 20%.

Crankcase Breather System:

An improved breather system has been incorporated to maintain a more consistent depression within the crankcase throughout the speed range.

PISTON ASSEMBLIES

Pistons:

The gudgeon (wrist) pin to piston crown height has been reduced to allow for the increased stroke and to provide the desired compression ratio of 11:1. The shallow bowl in the piston crown and the solid skirt features have been retained from the 5.3 Litre engine.

Piston Rings:

These have been designed with lower tangential loading, coupled with reduced thickness to provide better sealing properties. This reduces piston blow-by and gives major improvements in oil consumption. Nominal thicknesses of the new rings are:

Top - 1.5 mm.
Second - 1.5 mm.
Oil control - 3.0 mm.

CYLINDER HEADS

Combustion Chamber:

Combustion chamber volume is increased to provide the revised compression ratio of the engine.

Valves:

The inlet valve is shorter (by 0.6 mm) than the 5.3 Litre version to accommodate the modifications made to the combustion chamber. Otherwise the valves are identical to those on the 5.3 Litre engine.

Camshafts:

The cam profile has been modified to reduce valve train operating noise at mid to high engine speed. The changes are to the ramp angles and a reduction of 0.025 mm to the maximum lift.

ENGINE ANCILLARY UNITS

Canister Purge and Air Injection:

To improve the operation of these systems and to delete the vacuum switches which controlled them they are now controlled directly from the engine management control module.

Engine Rear Mounting:

A new, single spring rear mounting is now fitted to the gearbox extension housing.

Engine Top Cover:

This cover (1 Fig. 1) and its retaining clips (2 Fig. 1) are identical to that fitted to the 6 litre saloon.

COLD AIR INTAKE

The new cold air intake system improves the performance of the engine by drawing in ambient temperature air instead of air that has been warmed by passing through the radiator core. The cooler, denser air has a higher proportion of oxygen which improves combustion. Air flow to the engine is also improved due to the re-siting of the air intakes.

The air cleaner covers (3 Fig. 1) are a new design in moulded plastic which connect to the intake ducts (5 Fig. 1) with silicone rubber hoses (4 Fig. 1).

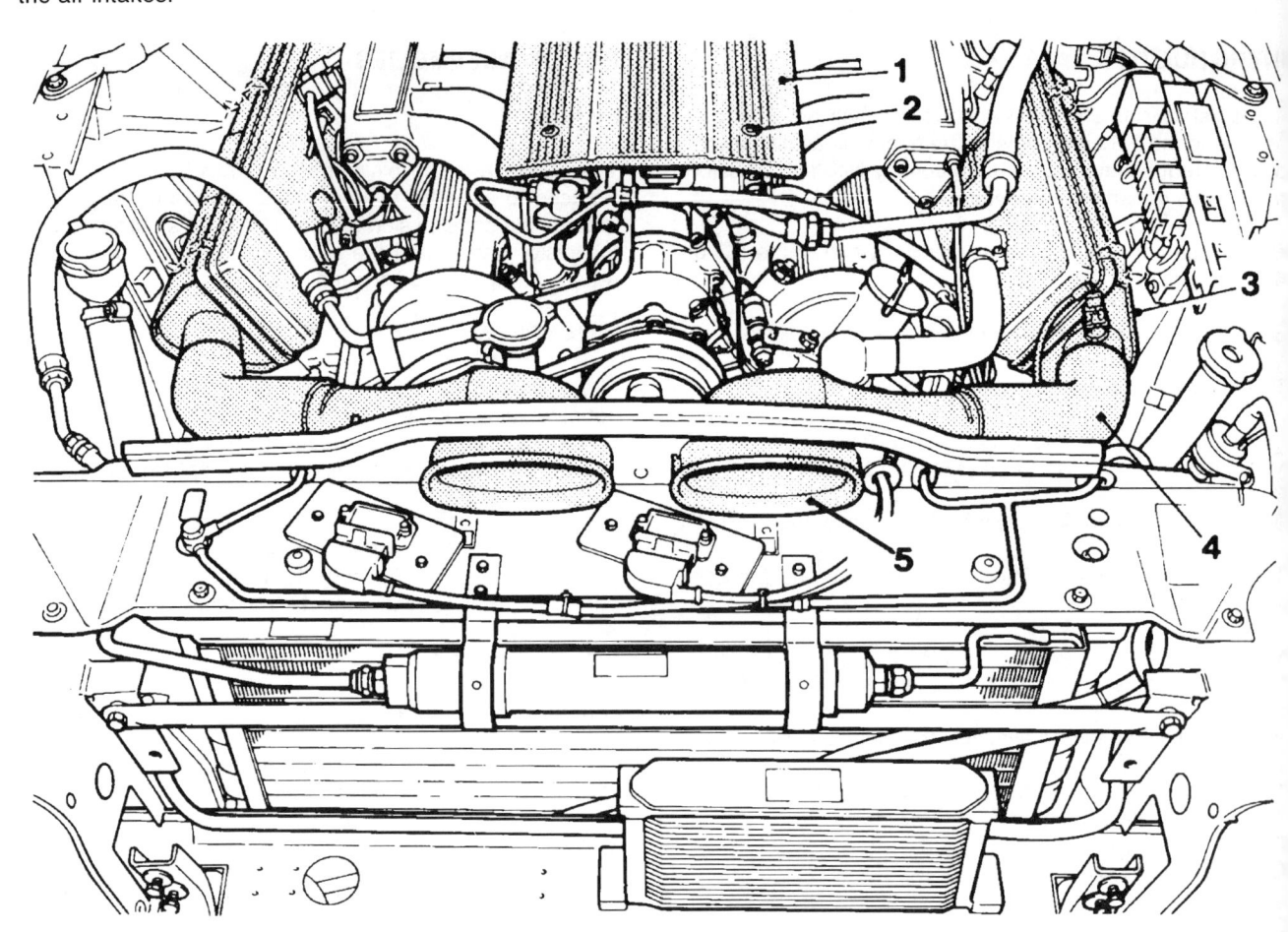

Fig. 1 Cold Air Intake and Engine Cover

Chapter 2

EMISSION CONTROL SYSTEM (5.3 LITRE 1992 MODEL YEAR ON)

AIR INJECTION, DESCRIPTION

To reduce emissions of carbon monoxide and hydrocarbons during the engine warm up period, secondary air is delivered to the exhaust manifolds via air rails in order to aid oxidisation.

Air is supplied to the air rails from the air delivery pump (8 Fig. 1) through the diverter valve outlet (10 Fig. 1) and twin check valves (11 Fig. 1). These check valves prevent exhaust gases from entering the pump should the pump drive belt fail or the exhaust system become blocked.
Upon start-up, a solenoid valve (2 Fig. 1) is electrically activated by the timer (1 Fig. 1) for 45 seconds, opening the vacuum line between the right-hand inlet manifold (4 Fig. 1) and the diverter valve (7 Fig. 1) which is attached to the air delivery pump. This 45 second period of air injection allows time for the catalytic converters to light up.

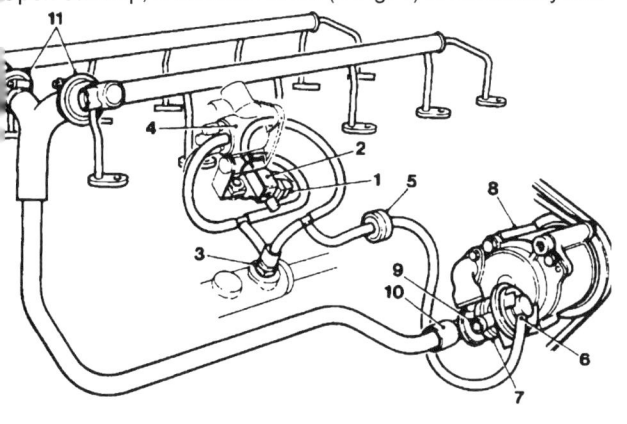

The vacuum controlled diverter valve (7 Fig. 1) on the air delivery pump, ensures that the secondary air is only supplied to the exhaust manifold ports during the engine warm up period.

A thermal vacuum valve (3 Fig. 1) is situated in the water rail which when a pre-determined coolant temperature of 38°C is reached, switches the diverter valve so that it redirects or dumps the air-flow back to the air filter from the port (9 Fig. 1).

In the vacuum line between the thermal valve and the diverter valve vacuum inlet (6 Fig. 1) is a delay valve (5 Fig. 1); this valve ensures that the vacuum from the inlet manifold is maintained at a constant level through any full throttle period.

AIR DELIVERY PUMP, RENEW

Note: No servicing or overhaul is possible. In the event of failure, a service exchange unit must be fitted.

Remove the RH air cleaner cover.
Slacken the adjuster rod trunnion bolt.
Remove the adjuster rod locknut (6 Fig. 2).
Remove the drive belt.
Remove the three air pump pulley securing bolts (2 Fig. 2).

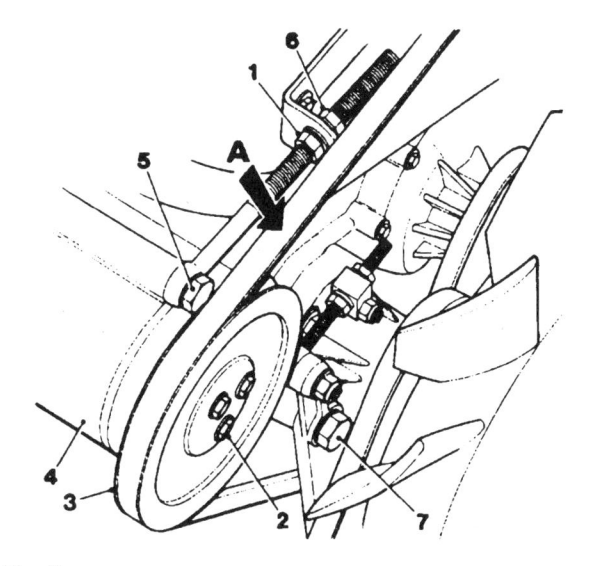

Fig. 2

CAUTION:

Do not use a screwdriver or wedge to prise off the pulley or extensive damage to the filter element will result.

Remove the pump pulley (3 Fig. 2).
Pivot the pump away from the engine.
Support the pump, remove the mounting bolt, and retrieve the spacers and washers.
Disconnect the vacuum hose from the top of the diverter valve.
Release the air rail and air filter return hoses from the diverter valve.
Discard the sealing ring.
Remove the air delivery pump (4 Fig. 2).
Remove the diverter valve from the air delivery pump.
Fit the valve to the new pump.
Fit the pump in the reverse order of the removal procedure.
Adjust the belt.

AIR DELIVERY PUMP DRIVE BELT, ADJUST

Remove the RH air cleaner cover.
Slacken the adjuster rod bolt (5 Fig. 2).
Slacken the mounting bolt (7 Fig. 2).
Adjust the belt tension by turning the adjuster nut (1 Fig. 2) as necessary.
The tension is checked and adjusted as follows:

Apply a load of 6.4 lb (2.9 kg) to the belt at point 'A' (Fig. 2).
Adjust to give a total belt deflection of 0.22 in. (5.6 mm).

Tighten all fixings and refit the air cleaner cover.

EVAPORATIVE LOSS CONTROL SYSTEM, DESCRIPTION

Charcoal Canister
A canister containing activated charcoal granules, located in the right hand front wheel arch, is used to store hydrocarbon emissions from the fuel tank.

Filter pads are fitted above and below the charcoal to keep foreign matter out of the purge line. Vapour emissions from the fuel tank enter the canister via a purge control (Rochester) valve which is vacuum operated through hoses connected to the inlet manifold balance pipe. The fuel vapour is adsorbed in the canister by being drawn through the charcoal.

Purging air enters the canister at the side, passing through the charcoal to the two vacuum operated purge valves at the bottom of the canister.

Canister purging
The charcoal canister is purged by a twin stage purging system. The two vacuum operated purge valves are connected by hoses to twin tappings on the right-hand throttle body of the inlet manifold. When the engine is idling there is no purging but as the throttle is increased the butterfly blade starts to open. First it exposes a port which goes through a thermal valve and providing the temperature is above 45°C, a signal is produced which opens the first purge valve on the canister. This allows the canister to purge into the inlet manifolds via a restrictor. As the butterfly continues to open it exposes a second port, which goes through the same thermal valve, opening the second purge valve. However, in this second route there is a delay valve which holds back the purge action, necessitating further throttle opening, thus preventing stalling on drive-away.

CHARCOAL CANISTER, RENEW
Remove the left-hand front road wheel. Remove the under-arch access cover.

Note the positions of the hoses that connect to the two purge valves (1 Fig. 3) the Rochester valve (2 Fig. 3) and

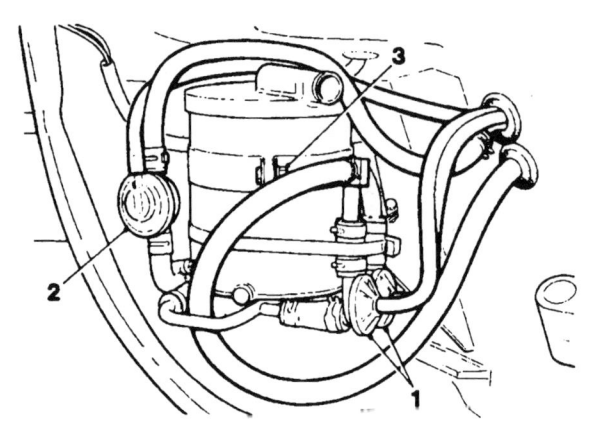

Fig. 3

the hoses to the canister stub pipes and remove them by slackening the hose clips.
Undo the canister clamp securing nut (3 Fig. 3) and remove.
Prise open the clamp and remove the canister.
Fit the new canister within the clamp.
Fit and tighten the clamp securing nut.
Reconnect the hoses and tighten the hose clips.
Refit the access cover.
Fit the road wheel.

VACUUM PURGE CONTROL (ROCHESTER) VALVE, RENEW

Slacken and slide back the two hose clips above the valve and similarly the one below and remove the valve (2 Fig. 3).
Fit a new valve; reposition and tighten the three hose clips.

TWIN PURGE CONTROL VALVES, RENEW

Slacken and slide back the hose clips that connect the hoses to the right hand purge control valve.
Carefully disconnect the hoses and remove the purge valve (1 Fig. 3).
Fit a new purge valve and reconnect the hoses.
Reposition and tighten the hose clips.

To remove and renew the left-hand purge valve the procedure is the same as for the right-hand.

PURGE CONTROL THERMAL VALVE, RENEW

Open the bonnet and fit wing protection.
Unscrew the purge control thermal valve from beneath the right-hand inlet manifold.
Fit a new valve.
Remove wing protection and close the bonnet.

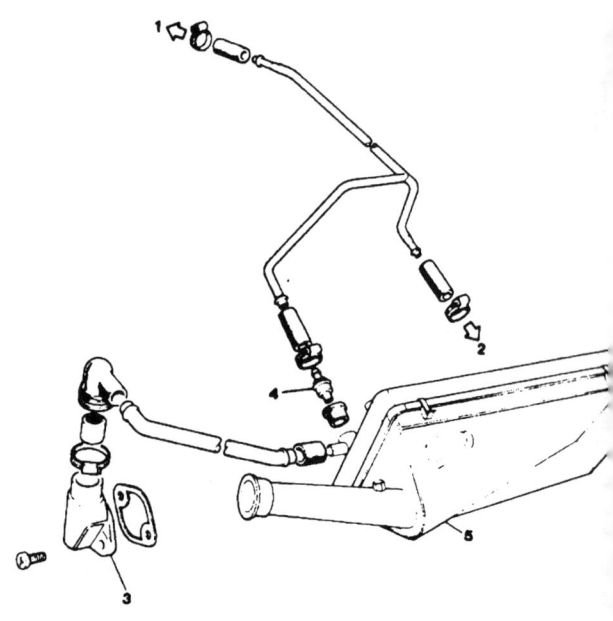

Fig. 4
1. To right manifold
2. To left manifold
3. Engine breather filter
4. Control valve
5. Air cleaner

CRANKCASE VENTILATION SYSTEM, DESCRIPTION

The crankcase ventilation system is a means of purging unwanted gases from the engine. The gases from the crankcase are drawn by vacuum from the left-hand and right-hand manifold through a filter in the engine breather and via pipework are discharged into the engine's combustion chamber

ENGINE BREATHER FILTER, RENEW

Remove the hose clip securing the rubber cover to the breather housing.
Disconnect the breather pipe from the rubber cover.
Remove the rubber cover.
Lift out the filter (Fig. 5).

Fit the new filter.
Refit the rubber cover.
Reconnect the breather pipe and secure with the hose clip.

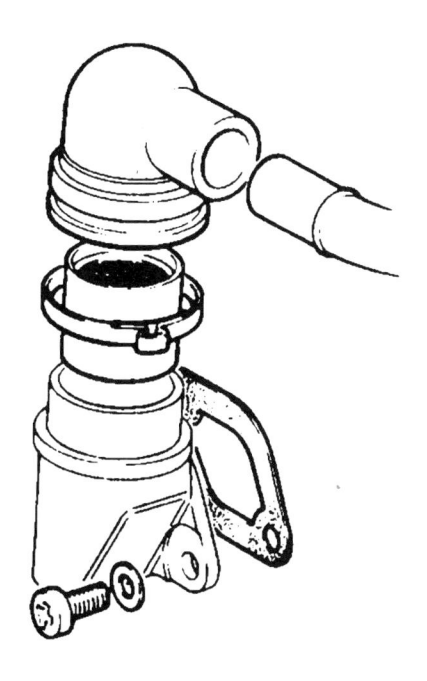

Fig. 5

OXYGEN SENSORS, RENEW

The oxygen sensors (1 Fig. 6) are located in the left-hand and right-hand exhaust down-pipes, just in front of the under-floor catalytic converters (2 Fig. 6).

Open bonnet and fit wing protection.
Disconnect oxygen sensor harness multi-plug.

Remove harness grommet from inner wing.
Raise vehicle front end and support on two stands.
Detach sensor harness from retaining clips from under wheel arch.
Unscrew and remove oxygen sensor.
Fit new sensor into exhaust downpipe and tighten.
Route new sensor harness under wheel arch and secure in retaining clips.

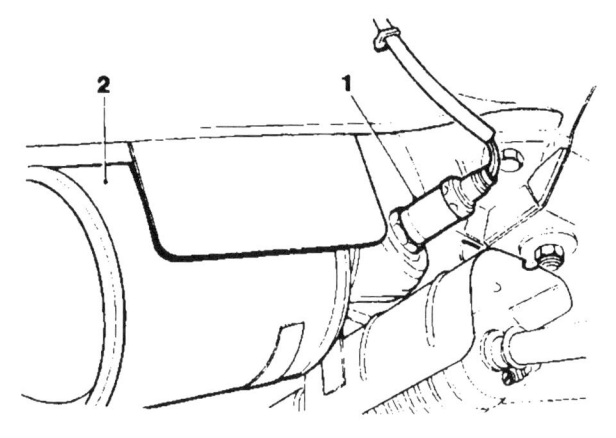

Fig. 6

Reposition harness multi-plugs through inner wing.
Fully seat body grommet.
Lower vehicle from stands.
Reconnect sensor harness multi-plugs.
Remove wing protection and close bonnet.

Torque Figures
Oxygen sensor to catalyst 50 - 60 Nm

CATALYTIC CONVERTERS, DESCRIPTION

Catalytic converters are fitted into the exhaust system to reduce emission of carbon monoxide, hydrocarbon and oxides of nitrogen.

In each left-hand and right-hand down-pipe, close to the engine, there is a catalytic converter unit. This is integral with the down-pipe and is only serviced by renewing the complete down-pipe assembly.

In the under-floor exhaust system there is a further catalytic converter fitted in the left-hand and right-hand exhaust pipes. These under-floor catalytic converters are renewable by disconnecting the exhaust pipe system.

Chapter 3

EMISSION CONTROL SYSTEM 6 LITRE

INTRODUCTION

Component Identification

Traditional Name	New Name (To SAE J1930)	Acronym
Air Pump	Secondary Air Injection Pump	AIRP
Air Injection System	Secondary Air Injection System	AIR
Air Pump Clutch	Secondary Air Injection Magnetic Clutch	AIRPC
Non Return Valve	Secondary Air Injection Check Valve	AIRC
Air Switching Valve	Secondary Air Injection Switching Valve	AIRS
Delay Valve	Secondary Air Injection Delay Valve	AIRD
Solenoid Vacuum	Valve Secondary Air Injection Solenoid Vacuum Valve	AIRV
Air Injection Relay	Secondary Air Injection Relay	AIRR
Evaporative Emission Control	Evaporative Emission Control System	EVAP
Purge Control Valve	Evap Purge Control Valve	EVAPP
Lambda Sensor	Heated Oxygen Sensor	HO2S
Fuelling Failure	Diagnostic Trouble Code	DTC
Injection ECM	Powertrain Control Module Fuel	PCMF
Ignition ECM	Powertrain Control Module Ignition	PCMI
Check Engine Light	Malfunction Indicator Lamp	MIL
On Board Diagnostics	On Board Diagnostic System	OBD
Manifold Pressure Sensor	Manifold Absolute Pressure Sensor	MAPS
Coolant Temperature Sensor	Engine Coolant Temperature Sensor	ECTS
Air Temperature Sensor	Intake Air Temperature Sensor	IATS
Throttle Potentiometer	Throttle Position Sensor	TPS
Crankcase Ventilation System	Crankcase Ventilation System	CV

Covered in this section are the functions and systems that are added to the vehicle for emission control reasons, but in some cases also offering additional vehicle refinement e.g. crankcase breather and evaporative emission systems.

Except for the crankcase ventilation system, the systems are all microprocessor controlled by the Powertrain Control Module Fuel (PCMF).

The following systems are covered:

- Crankcase Ventilation System (CV).
- Evaporative Emission Control System (EVAP).
- Heated Oxygen Sensor (HO2S).
- On Board Diagnostic System (OBD).
- Secondary Air Injection System (AIR).

Note: Not all systems are used in all markets.

SECONDARY AIR INJECTION SYSTEM (AIR), DESCRIPTION

AIR will be used for a period of time each time the engine is started. After a cold start, a few minutes; hot start, a few seconds.

Air is pumped into the exhaust ports through the air rails (9 Fig. 1). Mixing and oxidation takes place, the heat generated reduces the time for the catalysts to reach operating temperature.

The mechanically driven rotary vane air pump (AIRP. 4 Fig. 1) has a magnetic clutch (AIRPC. 5 Fig. 1) mounted upon its drive shaft. The clutch enables the pump to be engaged or disengaged under microprocessor control from the PCMF via the Secondary Air Injection Relay (AIRR. 8 Fig. 1). The relay also drives the Secondary Air Injection Solenoid Vacuum Valve (AIRV. 6 Fig. 1). This valve controls a vacuum signal to the Secondary Air Injection Switching Valve (AIRS. 2 Fig. 1), backing up the Secondary Air Injection Check Valves (AIRC. 1 Fig. 1). It prevents back flow of exhaust gas to the AIRP when the AIRP is not operating. It also ensures that air is not drawn into the exhaust system by the exhaust pressure pulsations.

The AIRS vacuum supply is routed from the inlet manifold (7 Fig. 1) through a Delay Valve (AIRD. 3 Fig. 1) and the AIRV to the AIRS. The AIRD retains sufficient vacuum signal during short periods of high engine load operation to keep the AIRS open.

The AIR system will be turned off (by de-energizing the relay) if the engine speed exceeds 2500 rpm during AIR operation. This prevents damage to the pump due to excessive back pressure or speed.

Operation of the AIR system is monitored by the On Board Diagnostic System (OBD). With closed loop feedback control operational and the engine at idle (throttle closed), the AIR system is turned on. The Heated Oxygen Sensor (HO2S) signal responds to the air pumped into the exhaust system. It adds extra fuel (the feedback integrator voltage increases) to maintain the mean Air Fuel Ratio at the HO2S at 14.7:1.

If the integrator voltage changes by less than 1.0V the flag will be set and Diagnostic Trouble Code (DTC) 67 will be stored.

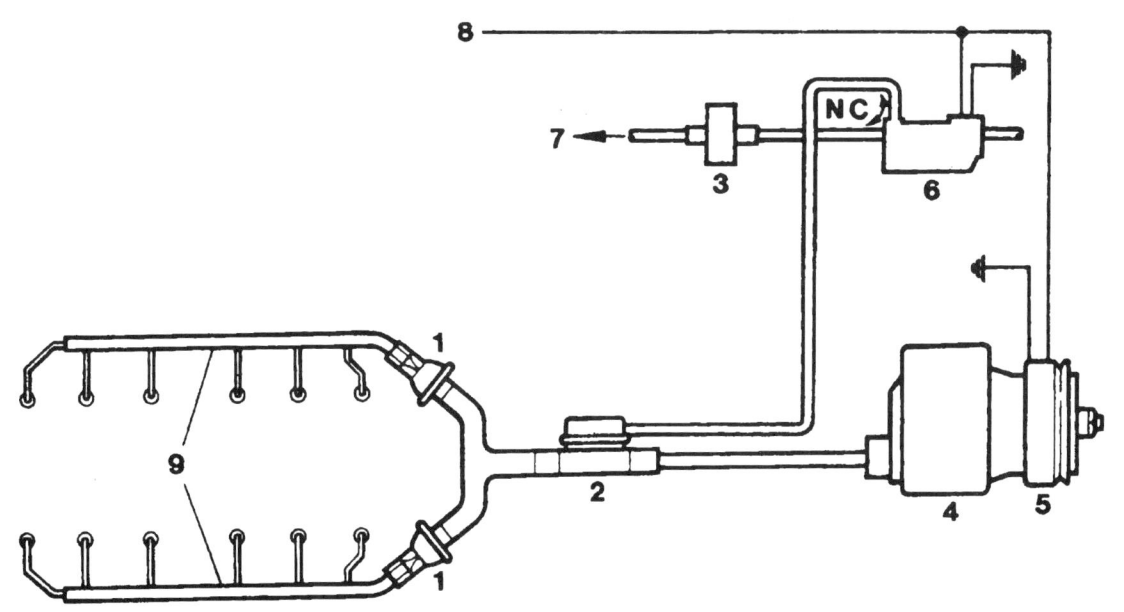

Fig. 1 Secondary Air Injection System (AIR).
1. Secondary Air Injection Check Valves (AIRC)
2. Secondary Air Injection Switching Valve (AIRS)
3. Secondary Air Injection Delay Valve (AIRD)
4. Secondary Air Injection Pump (AIRP)
5. Secondary Air Injection Magnetic Clutch (AIRPC)
6. Secondary Air Injection Solenoid Vacuum Valve (AIRV)
7. Manifold Vacuum
8. Secondary Air Injection Relay (AIRR) output
9. Air Rails

During engine set up using JDS, the system will be tested by turning on the AIR and measuring the HO2S integrator feedback deflection.

There is one On Board Diagnostic (OBD) code associated with the Secondary Air Injection System:

DTC FF67 Secondary Air Injection System (AIR).

At programmed intervals the PCMF turns on the AIR system at idle, when the engine is at normal operating temperature. Charcoal Canister purge will be turned off for this test to take place. The HO2S feedback system will be operating and the air pumped into the exhaust ports will cause a deflection of the feedback integrator voltage.
If the integrator voltage does not rise by the required amount, on both engine banks, the code will be set.
Failure of the test indicates that the air flow from the AIR system into the engine is less than it should be.
JDS can be used to trace faults in the Secondary Air Injection System.

CRANKCASE VENTILATION SYSTEM (CV), DESCRIPTION

To ensure that piston blow-by gases do not escape from the engine crankcase to atmosphere, a depression is maintained in the crankcase under all operating conditions. This is achieved by scavenging from the front of 'B' bank tappet block (1 Fig. 2), and from the rear of the jack shaft cover (2 Fig. 2) below the throttle position sensor (TPS).
The crankcase emissions are then fed into the engine intake manifolds. At part throttle (3 Fig. 2) through restrictors located in the outlet pipes of the oil separator, and into the air cleaner back plate (4 Fig. 2) from the tappet block housing at full throttle.
During part throttle operation, air is drawn from the air cleaner tube, along the full throttle breather hose in the direction marked (5 Fig. 2). Fresh air is drawn into the engine through the full load oil separator (1 Fig. 2).
Blow-by gases are removed from the crankcase through the part load oil separator (7 Fig. 2). this is mounted below the jack shaft cover. The separator contains two knit mesh oil screens. Each of the part load oil separator tubes has a restrictor fitted. This ensures that the blow-by gases are consumed equally by each bank of the engine, in proportion to manifold vacuum.
Under full throttle conditions crankcase gases are drawn into the inlet manifold through the full throttle breather hose (6 Fig. 2), air cleaner back plate and the throttle ('B' bank). The part load system will also vent a small proportion of the gases directly into the inlet manifolds.
Icing in cold weather conditions is prevented by the physical location of the part load ventilation components on the engine.

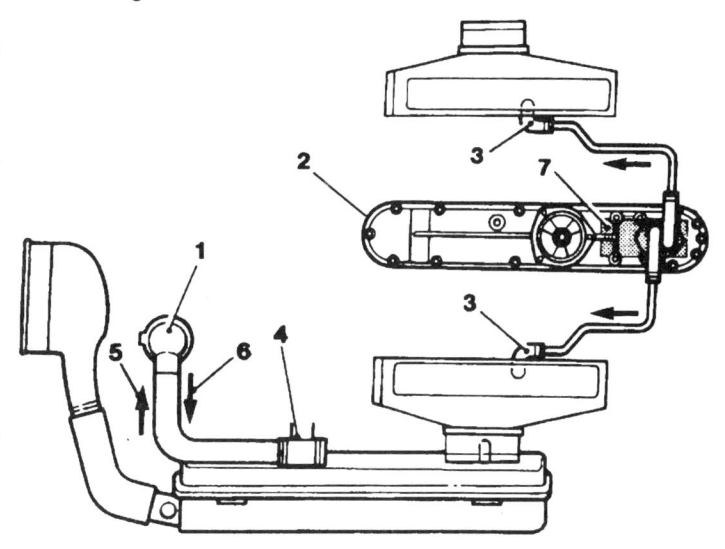

Fig. 2 Crankcase Ventilation System
1. Full Load Oil Separator
2. Jackshaft Cover
3. Part Throttle Breather Pipes
4. Full Throttle Breather Hose
5. Part Load Air Flow
6. Full Throttle Gas Flow
7. Part Throttle Oil Separator

EVAPORATIVE EMISSION CONTROL SYSTEM (EVAP), DESCRIPTION

The maximum fuel level in the tank is limited by fuel covering the vapour vent pipe. This is necessary to accommodate up to 7% fuel expansion, due to fuel volume increasing as the fuel temperature increases. The vent pipe is located at the maximum fuel level. It is linked to the filler neck via a tube on the top of the tank (4 Fig. 3). During filling, fuel rises in the filler tube. When the vent pipe is covered by fuel, it causes the fuel delivery nozzle to shut off.

In vehicle use the fuel tank is vented through a liquid/vapour separator and vapour pipe, to an activated charcoal canister (1 Fig. 3). The charcoal canister is located under the left-hand front wheel arch. The flow of vapour to the canister is controlled by a Tank Pressure Control Valve (2 Fig. 3).

The tank pressure control valve is a three-function valve, as follows:

(1) When the engine is not running, but due to increasing ambient or vehicle temperature the fuel temperature is rising, vapour is generated and the valve functions as a pressure relief valve. It limits the tank pressure to about 1.0 lbf/in^2.

(2) If the fuel in the tank is cooling down no vapour will be formed. The fuel tank will draw air/vapour back from the charcoal canister through a non-return valve in the tank pressure control valve (negative tank pressure could collapse the fuel tank).

(3) With the engine running a manifold vacuum signal is connected to the valve diaphragm, opening the valve and reducing the fuel tank pressure to near zero.

A further fuel tank over pressure safeguard is provided by the fuel cap.

The vapour that leaves the fuel tank enters the charcoal canister and is absorbed by the activated charcoal. The canister is air purged by engine vacuum via two electrically (mark/space ratio) operated Purge Control Valves (EVAPP) (3 Fig. 3) controlled by the Powertrain Control Module Fuel (PCMF).

The EVAPP valves are connected one to each inlet manifold (5 Fig. 3). This method ensures that vapour is delivered proportional to manifold vacuum to each bank of the engine.

The rate of purge flow is dependent upon engine speed, engine load and charcoal canister loading. The maximum flow is limited by a 3/32 in. dia. restrictor fitted in the hose between the EVAPP valve and the inlet manifold.

With normal canister loading the full mapped purge rate will be used. If the vapour load becomes excessive it will be detected by the HO2S feedback signal. The PCMF will reduce the purge rate at lower engine speeds until the vapour load reduces and the full purge rate can be reinstated.

This method allows aggressive purge rates to be used without causing driveability problems.

Charcoal canister purging commences a short time after HO2S closed loop operation is initiated.

There are no OBD codes associated with the charcoal canister purge system.

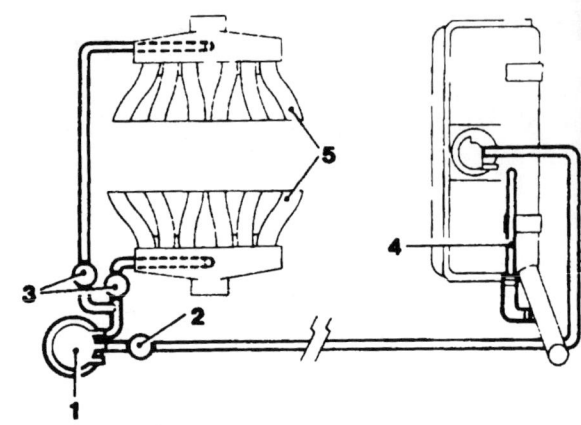

Fig. 3 Evaporative Emission Control System (EVAP).
1. Charcoal Canister
2. Tank Pressure Control Valve
3. Electrical Purge Control Valve (EVAPP)
4. Fuel Tank Limited Fill Vapour Pipe
5. Inlet Manifold

HEATED OXYGEN SENSOR (HO2S), DESCRIPTION

A pair of Heated Oxygen Sensors (HO2S) (Fig. 4) are mounted one in each exhaust downpipe in front of the downpipe catalyst. The signal from the HO2S is received by the PCMF and modifies the Fuel Injector (FI) 'on time' Lengthening or shortening the mapped pulse duration to achieve an air fuel ratio (AFR) of lambda 1.0 (14.7 : 1) The FI pulse duration is controlled separately on each engine bank.

One part of the ceramic body of the HO2S is located in the path of the exhaust gas, and the other is in contact with ambient air. The surface of the ceramic body (zirconium dioxide) is provided with electrodes made of a thin gas-permeable layer.

The ceramic material used in the sensor begins to conduct oxygen ions at a temperature of approximately 300°C. An internal heater is used to bring the sensor up to temperature quickly.

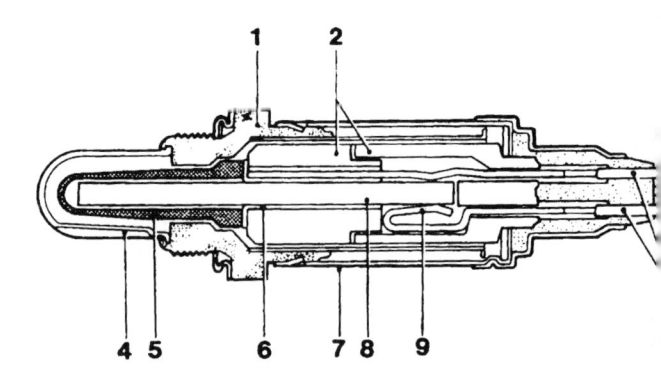

Fig. 4 Heated Oxygen Sensor (HO2S)
1. Sensor Housing
2. Protective Ceramic Tube
3. Connection Cable
4. Protective Tube with Slots
5. Active Sensor Ceramic
6. Contact Element
7. Protective Sleeve
8. Heater Element
9. Clamp Terminals for Heater

The sensor output varies according to the oxygen levels at the ambient air and exhaust sides of the sensor.

When the AFR is richer than lambda 1.0 the voltage output is high (about 800mV) (1 Fig. 5). When the AFR is leaner than lambda 1.0 the output voltage will be low (about 200mV) (2 Fig. 5).

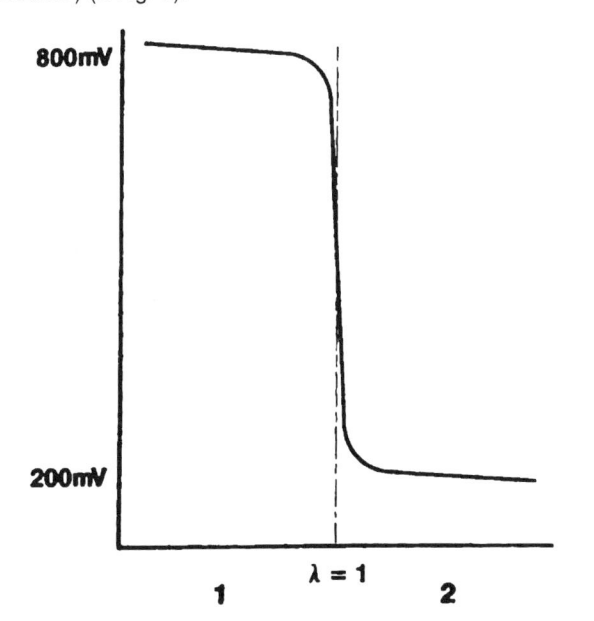

Fig. 5

The PCMF receives the HO2S signal and continually switches the fuelling rich, lean, rich in response to the HO2S signal changing. The speed with which the pulse duration is changed is mapped with a 4 X 4 load / speed map. This function is also used to bias the fuelling slightly rich or lean of lambda 1.0. By this means the best use of catalyst conversion is achieved. Lean bias improves oxidation of carbon monoxide (CO) and hydrocarbons (HC). Rich bias reduces oxides of nitrogen (NOx).

The switching signal of the HO2S on each bank is averaged as a correction to the main load / speed fuel map. This correction is known as the integrator voltage. The range is 0 to 5V, the mid point 2.5V being zero correction on the fuel map. Voltages greater than 2.5V are adding fuel (lengthening pulse duration) and voltages less than 2.5V are removing fuel (shortening pulse duration). The total correction range is approximately +/- 20%.

Catalytic convertors are fitted into the exhaust system to oxidize carbon monoxide (CO) and hydrocarbons (HC) and reduce oxides of nitrogen (NOx) emissions. Very close AFR control is required to optimise emissions.

Unleaded fuel must be used on catalyst equipped vehicles and labels are displayed to show this. A filler neck designed only to fit unleaded fuel pump nozzles is used in most markets.

There are five OBD codes associated with the HO2S or using the HO2S signal; they are:

DTC FF23 Fuel Supply.
DTC FF34 Fuel Injector (FI) 'A' Bank.
DTC FF36 Fuel Injector (FI) 'B' Bank.

With the engine hot at idle and feedback operating. If the integrator voltage for one or both engine banks is on the rich or lean clamp (0 or 5V). The system will add to or remove fuel from the main fuel map (25 - 37.5 - 50%). When both sensors were on clamp and they respond to the fuel change DTC FF23 Fuel Supply will be set. If one sensor only was on clamp and responds to the fuel change leaving clamp, DTC FF34 or 36 will be set.

Default: The integrator voltage for the faulty bank will be clamped to mid point, i.e. main fuel map.

DTC FF44 Heated Oxygen Sensor Circuit (HO2S) 'A' Bank.
DTC FF45 Heated Oxygen Sensor Circuit (HO2S) 'B' Bank.

With the engine hot at idle and feedback operating. If the integrator voltage for one engine bank is on the rich or lean clamp (0 or 5V). The system will add to or remove fuel from the main fuel map (25 - 37.5 - 50%). If the sensor fails to respond and remains on clamp, DTC FF44 or 45 will be set.

Default: The integrator voltage for the faulty bank will be clamped to mid point, i.e. main fuel map.

JDS and/or the flow charts can be used to trace faults of this type.

The HO2S heater circuit and signal screen earth to the engine are particularly important to the correct operation of the system.

ON BOARD DIAGNOSTIC SYSTEM (OBD), DESCRIPTION

In the event of an engine management fault being detected by the OBD system, the Malfunction Indicator Lamp (MIL) will be illuminated and display 'Check Engine'. If a 'Limp Home' default is available, the default condition will be imposed for the remainder of the journey or until the fault disappears.

When a fault has been signalled, the likely area responsible for the malfunction can be shown when the vehicle is stationary. Switch off the ignition, wait at least 5 seconds, then turn the ignition switch to position 'II' (do not start the engine). The OBD code will be displayed on the trip computer display as FF + XX (where XX is a number that identifies the fault code).

All Diagnostic Trouble Codes (DTC) are, when triggered, stored in the PCMF memory. In addition, the MIL lamp is illuminated. The DTC codes are displayed on the trip computer in order of priority, one at a time. The next code is only displayed when the preceding one is rectified or cleared. JDS can be used to see all codes stored by the PCMF at the same time.

The MIL will be extinguished when the fault is rectified and the PCMF memory is cleared by JDS.

Vehicles using secondary air injection as part of the emission control system will have the MIL latched. i.e. the MIL will be illuminated on each journey once a fault code is stored. Vehicles without secondary air injection will not have the MIL latched, it will only be illuminated for the remainder of the journey during which the fault occurred.

DIAGNOSTIC TROUBLE CODES (DTC), DESCRIPTION

Definitions of DTC codes flagged by the PCMF and displayed on the Trip Computer Screen are as follows:

DTC	Description	Function
13	PCMF Manifold Absolute Pressure Sensor (MAPS) or signal pipe signal pipe	Looks for fluctuating MAPS vacuum signal and vacuum 'v' throttle position
14	ECTS or associated wiring	Looks for ECTS resistance out of range or static during engine warm-up
16	IATSF or associated wiring	Looks for IATSF resistance out of range
17	TPS or associated wiring	Looks for TPS resistance out of range
18	TPS / MAPS calibration	Looks for low TPS signal at high load
19	TPS / MAPS calibration	Looks for high TPS signal at low load
23	Fuel Supply	Looks for poor HO2S feedback control on both A and B banks of the engine
34	A Bank FI	Looks for uneven HO2S feedback signal due to FI flow or wiring problem on one bank only
36	B Bank FI	Looks for uneven HO2S feedback signal due to FI flow or wiring problem on one bank only
44	A Bank HO2S	Looks for correct HO2S output
45	B Bank HO2S	Looks for correct HO2S output
49	Ballast Resistor	Looks for presence of Ballast Resistor
67	AIR	Looks for AIR to be detected by the HO2S
77	Speed Signal	Looks for engine speed signal from PCM

Chapter 4

ENGINE MANAGEMENT SYSTEM (5.3 LITRE 1992 MODEL YEAR ON)

DIGITAL IGNITION SYSTEM, DESCRIPTION

The Marelli Digital Ignition System is a fully digital microprocessor controlled system comprising:

- An Electronic Control Unit (ECU).
- A Distributor.
- Two HT coils (A and B banks).
- Two Amplifier units (A and B banks).
- A Crankshaft (TDC) sensor.
- A Flywheel sensor.
- A Coolant temperature sensor.
- An Air temperature switch.
- A Throttle idle switch.
- One 6K8 resistor.
- An Ignition map select link.

The microprocessor, located within the ECU (Electronic Control Unit), controls the timing of the low tension voltage, via the two amplifier units located on the front cross member, to two ignition coils which are located between the two cylinder banks on brackets either side of the distributor. The coils in turn supply the high tension voltage via the distributor to the sparking plugs.

The front coil supplies 'B' bank and the rear coil supplies 'A' bank.

The ECU, which is located in the front passenger footwell, controls the ignition timing according to the inputs from sensors mounted on the engine. The control parameters fall into two groups;
Primary, consisting of intake manifold vacuum and engine speed.

Secondary, consisting of throttle position, engine coolant temperature and intake air temperature.

Engine Starting

The ignition timing during cranking and start-up is dependent on engine speed and engine coolant temperature. The engine speed is sensed by an inductive pick up sensor (located at the rear of the engine), the speed signal is obtained by the rotation of the starter gear ring with segments, inducing a signal frequency proportional to the engine speed.

The engine coolant temperature is sensed by a sensor located on the front of 'A' bank engine coolant rail.

The timing information is stored in memory and permits an early establishment of a good idling condition.

Hot Starting

To aid restarting a hot engine, a 'Hot start system' is fitted which works in conjunction with the fuel system.

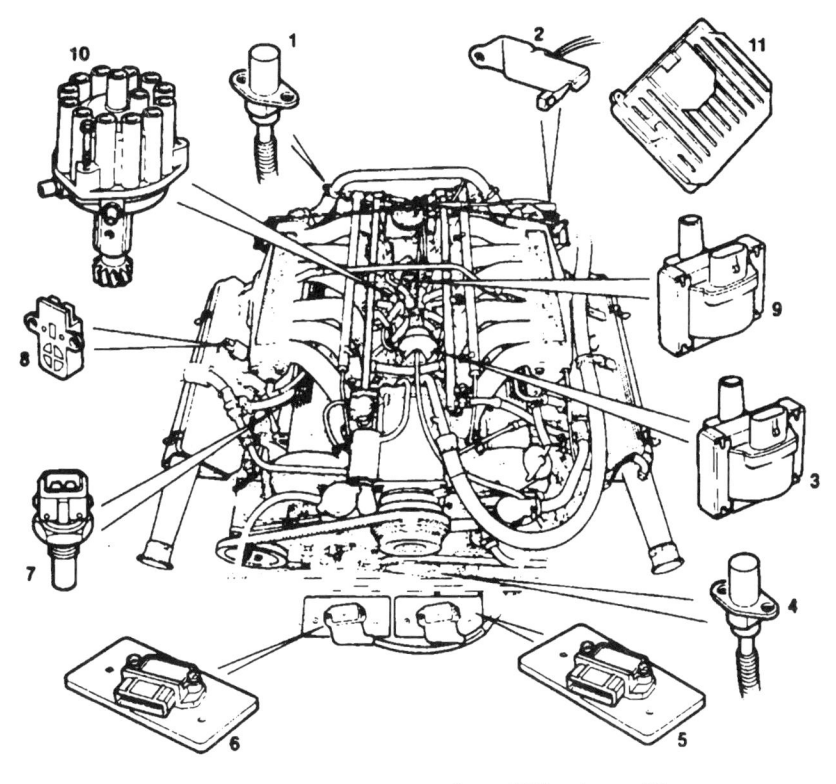

Fig. 1
1. Flywheel sensor
2. Throttle idle switch
3. Ignition coil
4. Crankshaft sensor
5. 'B' bank amplifier
6. 'A' bank amplifier
7. Coolant temperature sensor
8. Air temperature switch
9. Ignition coil (rear)
10. Distributor
11. Electronic control unit (in front passenger footwell)

An electrical temperature switch is fitted to the underside of the right hand fuel rail. When the fuel temperature exceeds 70°C at hot starting, the switch commands an increase of fuel pressure from the pump to the fuel rails via the fuel pressure regulator, to compensate for evaporation.

The pressure increase is limited to a 45 second period by a timer module which is located on the passenger under-scuttle component panel.

A solenoid vacuum valve controls the vacuum signal commanded by the fuel rail temperature switch and the 45 second timer.

Closed Throttle Running

The ignition timing for closed throttle running, idling and overrun conditions is programmed independently according to engine speed and engine coolant temperature, the timing information being stored in memory. This function is controlled by a throttle mounted switch located on the end of the 'B' bank bell-crank lever rod.

Open Throttle Running

The ignition timing for open throttle running is controlled primarily by intake manifold pressure and engine speed. The information for the ignition requirements is stored in a pre-programmed memory, so that for any given manifold pressure and engine speed the memory will generate a number which relates to the timing required. The timing thus computed is then subject to adjustment dependent on prevailing engine coolant and inlet air temperature.

The inlet air temperature is determined by a sensor located on the back plate of the 'A' bank air cleaner. Intake manifold pressure is sensed by an absolute pressure aneroid capsule located in the ECU and ported to the intake manifold. The crankshaft position is determined by an inductive pick up sensor sensing the position of a three toothed wheel mounted at the front of the crankshaft.

MANIFOLD PRESSURE CONTROL SYSTEM

Cold Engine

With the engine coolant below 45°C, the manifold pressure signal to the ECU is switched by means of a solenoid vacuum valve, via a delay valve, to a port in the 'A' bank intake throttle housing. This port is located so that manifold pressure is not available to the ECU when the throttles are closed. As the throttles are opened from the idle position, exposing the port to manifold pressure, the pressure signal is supplied to the ECU via a delay valve and solenoid valve located under the rear of 'A' bank manifold. This reduces the rate of ignition advance on acceleration following a cold start and so effects a reduction in exhaust emissions under these conditions.

Hot Engine

When the engine coolant exceeds 45°C, the manifold pressure signal to the ECU is switched by means of a coolant temperature switch (located at the rear of the 'A' bank coolant rail) and the solenoid vacuum valve from the throttle edge port to the intake manifold. This allows full manifold pressure to the ECU under all conditions and when required a maximum ignition advance.

Manifold Pressure Control System (Ignition) Low Loss Catalyst Vehicles

Each time the engine is started from cold or hot, the system starts a counter which controls the ignition off idle through the engine control unit (ECU), above full load. Effectively, there is no vacuum advance at part load until the counter ends its cycle, which is approximately 45 seconds. The ignition then works normally with the manifold pressure signal connected directly to the ignition ECU.

USA Specification Catalyst Vehicles

With these vehicles, whether starting from hot or cold, the manifold pressure signals connected directly to the ignition ECU at all times. There are no temporary inhibitions or functions effecting the ignition timing other than normal cranking, idle and temperature functions.

PRIMARY INPUTS

The engine will not start without these inputs.

- **TDC Sensor**

The TDC sensor is mounted so that it is triggered by a 3 toothed wheel attached to the crankshaft pulley. This gives an 8° ATDC reference for each cylinder on 'A' bank.
The correct sensor air gap is 0.018 - 0.042 in.
The sensor signal can only be checked using an oscilloscope (Fig 2). Connect the 'scope positive to ECU pin 1 or the sensor blue lead.

Note: Fig. 2 'A' = 160 teeth per engine revolution.

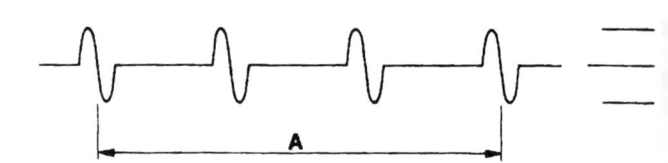

Fig. 2

Connect the 'scope negative to ECU pin 2 or the sensor screen.

- **Flywheel sensor**

The flywheel sensor is mounted so that it is triggered by each flywheel ring gear tooth to give an engine speed signal. The correct air gap is 0.018 - 0.042 in. The sensor signal can only be checked using an oscilloscope (Fig 3). Connect the 'scope positive to ECU pin 16 or the sensor red lead. Connect the 'scope negative to ECU pin 3 or the sensor screen.

Note: Fig. 3 'A' = 160 teeth per engine revolution.

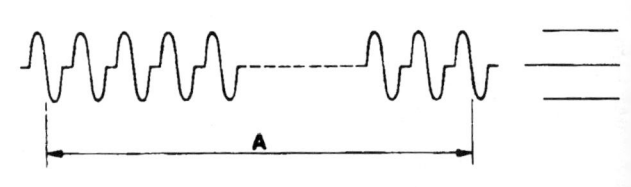

Fig. 3

SECONDARY INPUTS

These inputs will not prevent the engine from starting; however they do perform important functions and must not be neglected.

Coolant temperature sensor

This sensor gives a variable resistance with respect to temperature. This allows the ECU to trim the ignition timing dependent on coolant temperature. The table below gives approximate resistance (in Ohms) against temperature.

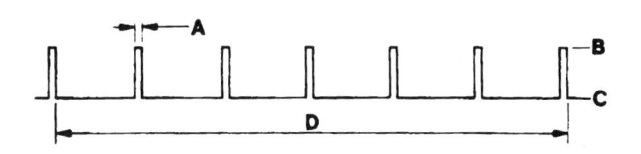

Fig. 4

A - 400 micro seconds
B - Approximately 1 volt less than battery voltage
C - Zero volts
D - 1 engine revolution

Temperature (°C)	10	20	30	40	50	60	70	80	90
Resistance (Ohms)	4K4	2K5	1K6	1K2	840	630	470	320	210

Air temperature switch

This switch is a bi-metal unit (located on the back plate of the 'A' bank air cleaner) to switch on at 75°C this perform the function of retarding the ignition timing at high air inlet temperatures.

Check the output using an oscilloscope.

Connect the positive to ECU pin 24 or the under-bonnet connection between the ignition harness and the main injection harness (screened lead). Connect the negative to the chassis or engine earth. The ignition ECU also triggers the signal to the ignition amplifier units.

EMISSION	FUEL OCTANE	CATALYST	HARNESS LINK
A	95	twin	in
B	95	none	in
E	91	none	out
F	95	twin low-loss	in
H	91 or 95 dependent on territory	twin low-loss	in or out
J	95	twin low-loss	in

Throttle idle switch

This switch enables the ignition ECU to sense when the throttle is in the closed position, the engine is idling or in a overrun condition. It is therefore important that the contacts of this switch are 'CLOSED' when the throttle is in the closed position.

Map select link

This link permits the selection of either of the two ignition maps contained in memory. If the link is in place the ignition map suitable for 95 octane fuel is selected and if the link is removed the ignition map suitable for 91 octane fuel is selected.

OUTPUTS

The ignition ECU generates the speed signal to the injection ECU. The frequency of the signal is 6 pulses per engine revolution.

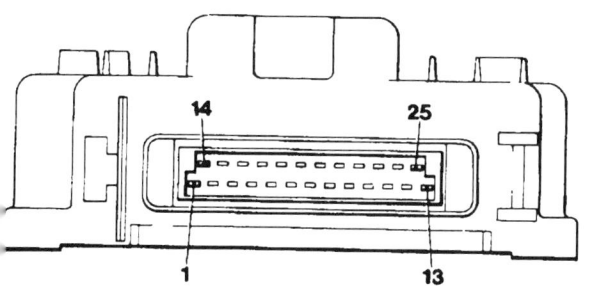

ECU PIN CONNECTIONS

Refer to Fig. 5 for pin positions.

Pin No.	Connection
1	TDC sensor
2	TDC sensor earth
3	Flywheel sensor earth
4	Earth
5	Switch earths
6	Water temperature sensor earth
7	Not connected
8	Not connected
9	LH amplifier unit (6)
10	LH amplifier unit (3)
11	Earth
12	Earth
13	Ignition switch
14	RH amplifier unit (6)
15	RH amplifier unit (3)
16	Flywheel sensor
17	Map selection
18	Idle switch
19	Water temperature sensor
20	Diagnostic request input
21	Serial output (diagnostics)
22	Gearbox control (1)
23	Air temperature switch
24	Engine speed output
25	Gearbox control (2)

Note: Pin connections refer to non-catalyst, single-catalyst and full-catalyst ECUs.

Fig. 5

ELECTRONIC CONTROL UNIT (ECU), RENEW

Open the boot and disconnect the battery earth lead.
Open the passenger door and remove the footwell carpet.
Remove cover from the fuel cut off inertia switch.
Remove the draught welting from the 'A' post.
Remove the 'A' post lower trim pad (1 Fig. 6).

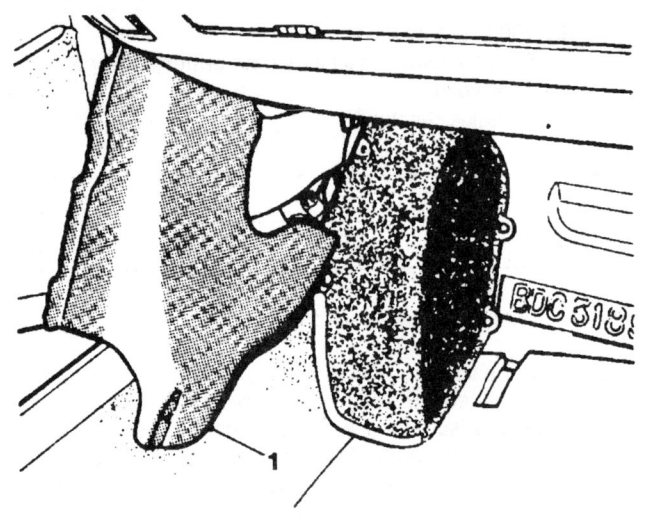

Fig. 6

Remove ECU harness 'P' clip.
Remove the nuts/bolts securing the ECU cover and remove the cover.

Disconnect the ECU multi-plug connector (1 Fig. 7).
Disconnect the ECU vacuum pipe (2 Fig. 7).
Remove the ECU / cover assembly (3 Fig. 7).

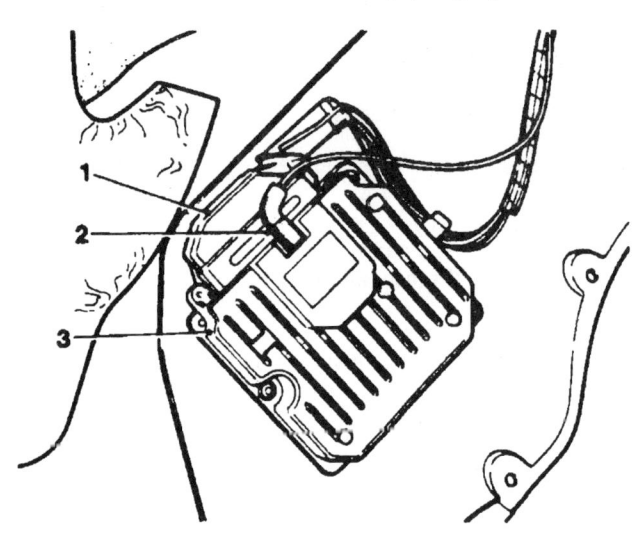

Fig. 7

Remove the nuts securing the ECU to the bracket.

Remove the ECU.
Fit the new ECU to the bracket.
Refit the cover.
Connect the vacuum pipe.
Connect the multi-plug.
Fit the ECU / cover into position.
Fit and tighten the nuts / bolts.
Refit the trim pad and draught welting.
Refit cover for the fuel cut off inertia switch.
Reposition the carpet.
Connect the battery earth lead.

HOT START SYSTEM, DESCRIPTION

To aid restarting a hot engine, a 'Hot start system' is fitted which works in conjunction with the fuel system.

An electrical temperature switch is fitted to the underside of the right hand fuel rail (1 Fig. 8). When the fuel temperature exceeds 70°C at hot starting, the switch commands an increase of fuel pressure from the pump to the fuel rails via the fuel pressure regulator (2 Fig. 8) to compensate for evaporation.

The pressure increase is limited to a 45 second period by a timer module (6 Fig. 8) which is located on the passenger under-scuttle component panel.

A solenoid vacuum valve (3 Fig. 8) controls the vacuum signal commanded by the fuel rail temperature switch and the 45 second timer.

Note: On non-catalyst vehicles the 45 second timer displaces the 15 minute timer module. The 15 minute timer module is relocated in the luggage compartment right-hand side near the fuel pump relay and the main relay.

A vacuum delay switch (4 Fig. 8) controls the way the extra pressure is applied at hot starting, to give a gradual reduction in fuel pressure over the 45 second period from the maximum, down to normal pressure.

CAUTION:

When disconnecting and reconnecting the rubber vacuum hoses between the inlet manifold, pressure regulating valve and associated units, ensure that the ends are plugged to keep dirt out.

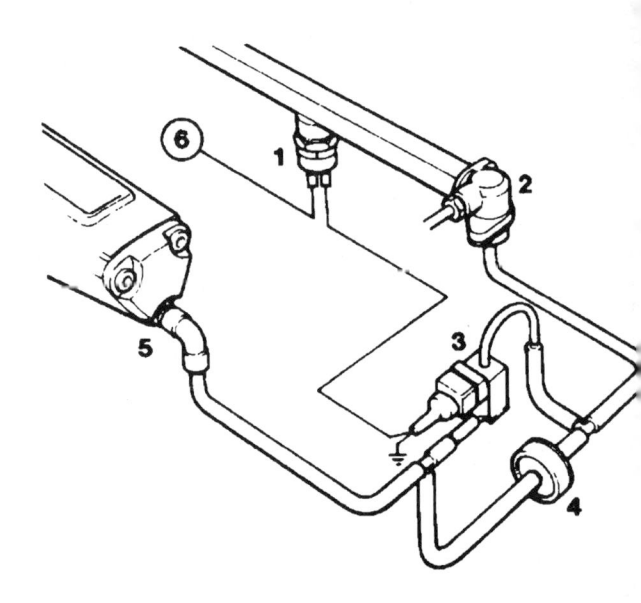

Fig. 8
1. Electrical temperature switch.
2. Fuel pressure regulator.
3. Solenoid vacuum valve.
4. Vacuum delay valve.
5. Inlet manifold connector.
6. Second timer.

RENEW OPERATIONS

> **WARNING:**
>
> SOME OF THE FOLLOWING OPERATIONS RESULT IN FUEL VAPOUR BEING PRESENT WHICH IS EXTREMELY FLAMMABLE. GREAT CARE MUST BE TAKEN WHEN WORK IS BEING CARRIED OUT ON THE ELECTRICAL TEMPERATURE SWITCH OR THE PRESSURE REGULATOR VALVE. ADHERE STRICTLY TO THE FOLLOWING PRECAUTIONS:
>
> SMOKING MUST NOT BE ALLOWED NEAR THE AREA.
>
> NO SMOKING WARNING SIGNS MUST BE POSTED AROUND THE AREA.
>
> A CO2 FIRE EXTINGUISHER MUST BE CLOSE AT HAND.
>
> DRY SAND MUST BE AVAILABLE TO SOAK UP ANY SPILLAGE.
>
> DRAINED OFF FUEL MUST NOT BE EMPTIED INTO A PIT.
>
> THE WORK AREA MUST BE WELL VENTILATED.
>
> ENSURE THAT ANY DRAINED OFF FUEL IS EMPTIED INTO AN AUTHORISED EXPLOSION PROOF CONTAINER.

ELECTRICAL TEMPERATURE SWITCH, RENEW

Open the luggage compartment.
Disconnect battery earth lead.
Open the bonnet and fit wing protection.
Disconnect the electrical leads from the two spade lugs at the base of the temperature switch (1 Fig. 9).
Unscrew the brass locknut securing the switch body (2 Fig. 9).

Position suitable cloth to catch possible fuel residue.
Unscrew and remove the switch.
Screw in a new switch and tighten.
Re-tighten the locknut to 11 to 14 Nm and reconnect the two leads.
Remove wing protection and close bonnet.
Reconnect the battery earth lead.
Close luggage compartment.

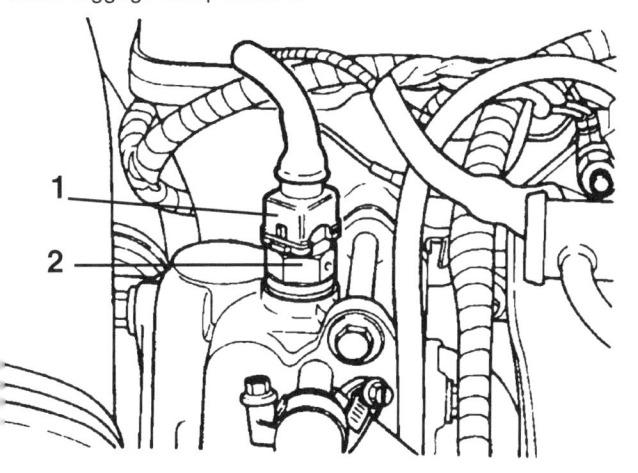

Fig. 9

PRESSURE REGULATOR VALVE, RENEW

Depressurise the fuel system prior to disconnecting any fuel pipe connections or removing components as follows:
Open the luggage compartment lid and disconnect the battery earth lead.
Remove the luggage compartment right-hand trim panel.
Displace the relays from the securing tags at the back of the hockey stick panel adjacent to the fuelling ECU (Fig. 10).

Identify the fuel pump relay; silver relay on a yellow-black base connector.

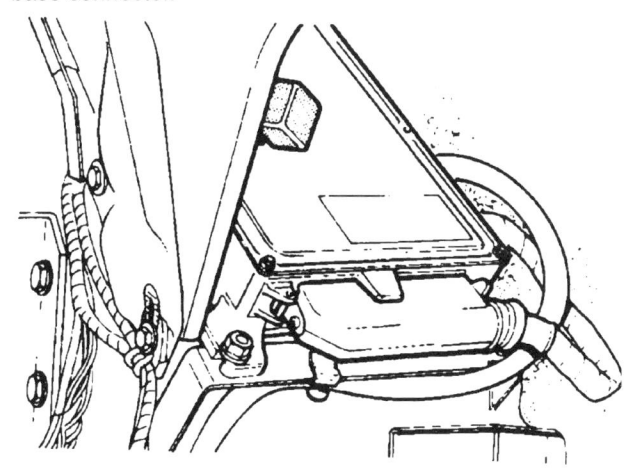

Fig. 10

Reconnect the battery earth lead.
Open the driver's door and crank the engine for 10 seconds to depressurise the fuel system.
Close the driver's door.
Disconnect the battery earth lead, refit the relay and reposition the base to the securing tag.
Reposition the trim panel.
Reconnect the battery earth lead.

Open the bonnet and fit wing protection.
Disconnect vacuum hose (1 Fig. 11) from the base of the pressure regulator valve.

Position suitable cloth to catch possible fuel residue.
Undo and release the fuel return pipe mounting from 'B' bank camshaft cover.
Undo the return pipe union (2 Fig. 11) from the regulator valve.
Withdraw the pipe. Remove and discard the seal.
Fit blanking plugs to pipe and valve.
Remove the two bolts (3 Fig. 11) securing the regulator valve to the fuel rail.

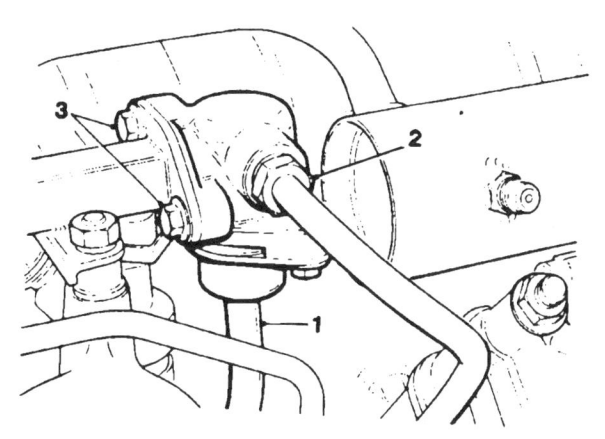

Fig. 11

Remove the valve and discard the seal.
Clean seal faces.
Lubricate and position new seal ring.
Position and seat new regulator valve, fit and tighten securing bolts.
Remove blanking plugs, fit a new seal to the return pipe.
Locate and tighten pipe (10 to 12 Nm) to regulator.
Remove cloth.
Connect vacuum hose to the valve.
Reposition and secure return pipe mounting bracket.
Remove wing protection and close bonnet.

VACUUM DELAY VALVE, RENEW

Open the bonnet and fit wing protection.
Carefully pull the rubber vacuum hoses off their connectors from each side of the vacuum delay valve (4 Fig. 12).

Fit a new vacuum delay valve with the red disc side AWAY from the inlet manifold connection.
Re-connect the rubber vacuum hoses to each side of the new valve.
Remove the wing protection and close the bonnet.

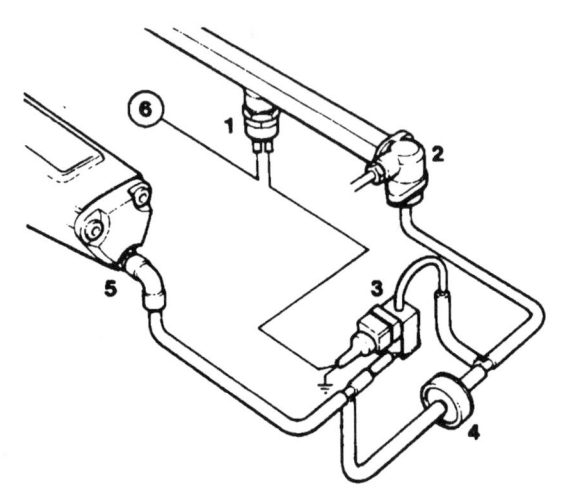

Fig. 12

FLYWHEEL SENSOR, RENEW

Open the boot and disconnect the battery earth lead.
Open the bonnet.
Disconnect the sensor multi-plug cable connector.
Fit a draw string to the sensor lead.
Jack up the vehicle and support with axle stands.
Remove the heat shield.
Remove the sensor securing bolts and displace the sensor from its position.
Displace the lead from its retaining clips.
Withdraw the sensor and draw string downwards.
Remove the drawstring and the sensor (Fig. 13).
Fit the new sensor to the drawstring.
Pull up the sensor lead from above.
Fit the new sensor into position.
Fit and tighten the securing bolts.
Secure the lead in the retaining clips.
Remove the drawstring.
Connect the sensor multi-plug.
Connect the battery earth lead.

AIR TEMPERATURE SWITCH, RENEW

Open the bonnet.
Disconnect the two switch lucar connectors.

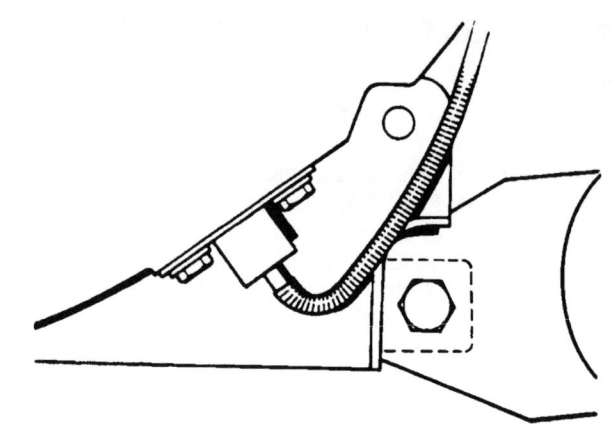

Fig. 13

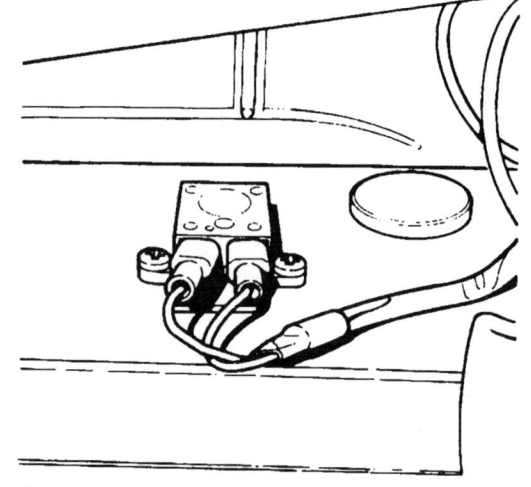

Fig. 14

Remove the screws securing the switch to the air cleaner back plate.
Displace and remove the switch (Fig. 14).
Fitting a new switch is the reversal of the removal procedure.

THROTTLE IDLE SWITCH, RENEW

Open the bonnet.
Remove the bolts (1 Fig. 15) securing the switch to the manifold.
Displace the switch from its location.
Remove the spacer from the manifold.
Disconnect the switch multi-plug connector.
Remove the switch (2 Fig. 15).
Fit the new switch into position.
Fit the spacer, fit and tighten the securing bolts.
Connect the multi-plug.
Adjust the switch (as follows).

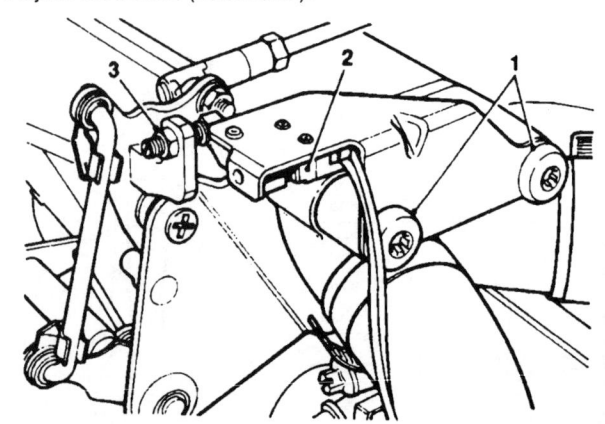

Fig. 15

THROTTLE IDLE SWITCH, ADJUST

Open the bonnet.
Slacken the locknut on the adjusting screw (3 Fig. 15).
Turn the adjusting screw until the micro switch emits and audible click as the throttle just closes.
Tighten the lock nut.
Reconnect the multi-plug.

VACUUM SOLENOID VALVE, RENEW

Open the bonnet.
Remove the air cleaner.
Remove the screw securing the vacuum valve support bracket.
Displace the valve rearwards.
Disconnect the hoses (2, 3 Fig. 16).
Disconnect the multi-plug connector (1 Fig. 16).
Remove the valve assembly.
Fit the new valve into position.
Fit the hoses and multi-plug.
Fit the valve to the support bracket.
Fit and tighten the screws.
Refit the air cleaner.

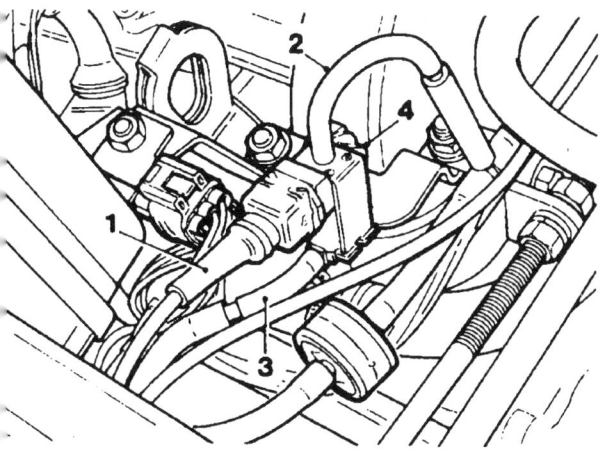

Fig. 16

CRANKSHAFT SENSOR, RENEW

Open the bonnet.
Jack up the vehicle and support with axle stands.
Remove the bolts securing the sensor (1 Fig. 17).
Remove the timing bracket.
Displace the sensor (2 Fig. 17).
Displace the sensor cable harness from the lower 'P' clip.
From above, disconnect the sensor cable harness multi-plug connector.
Fit a draw string to the sensor cable harness connector.
Displace the sensor harness from the upper 'P' clip.
From below, carefully pull the sensor and cable harness from its location leaving the draw string in place.
Remove the sensor assembly from the draw string.
Fit the new sensor to the drawstring.
Pull up the sensor lead from above.
Fit the new sensor into position.
Fit and tighten the securing bolts.
Secure the lead in the retaining clips.
Remove the drawstring.
Connect the sensor multi-plug.
Connect the battery earth lead.

FUEL INJECTORS, RENEW

Depressurise the fuel system prior to disconnecting any fuel pipe connections or removing components as follows:

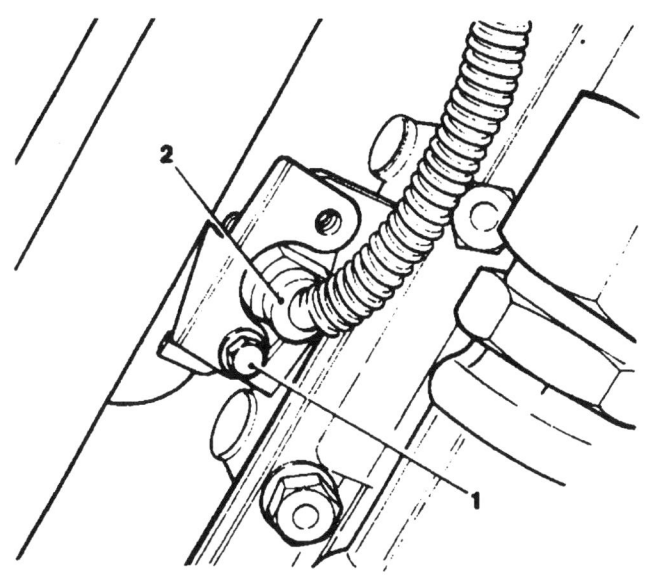

Fig. 17

Open the luggage compartment lid and disconnect the battery earth lead.
Remove the luggage compartment right-hand trim panel.
Displace the relays from the securing tags at the back of the hockey stick panel adjacent to the fuelling ECU (Fig. 18).

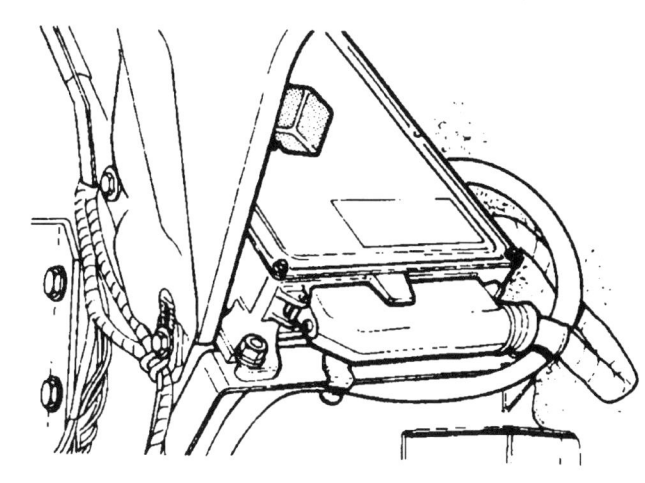

Fig. 18

Identify the fuel pump relay; silver relay on a yellow-black base connector.
Reconnect the battery earth lead.

Open the driver's door and crank the engine for 10 seconds to depressurise the fuel system.
Close the driver's door.
Disconnect the battery earth lead, refit the relay and reposition the base to the securing tag.
Reposition the trim panel.

Open the bonnet and fit wing protection.
Disconnect throttle breather pipe hose from manifold stub pipes.
Remove the nut securing the fuel return pipe bracket on the camshaft cover (1 Fig. 19).
Remove the bracket.
Place a suitable cloth below the fuel regulator return pipe union.
Undo the fuel regulator union nut (2 Fig. 19).
Detach the regulator pipe from the regulator.
Remove and discard the seal.
Fit blanking plugs.
Remove cloth.

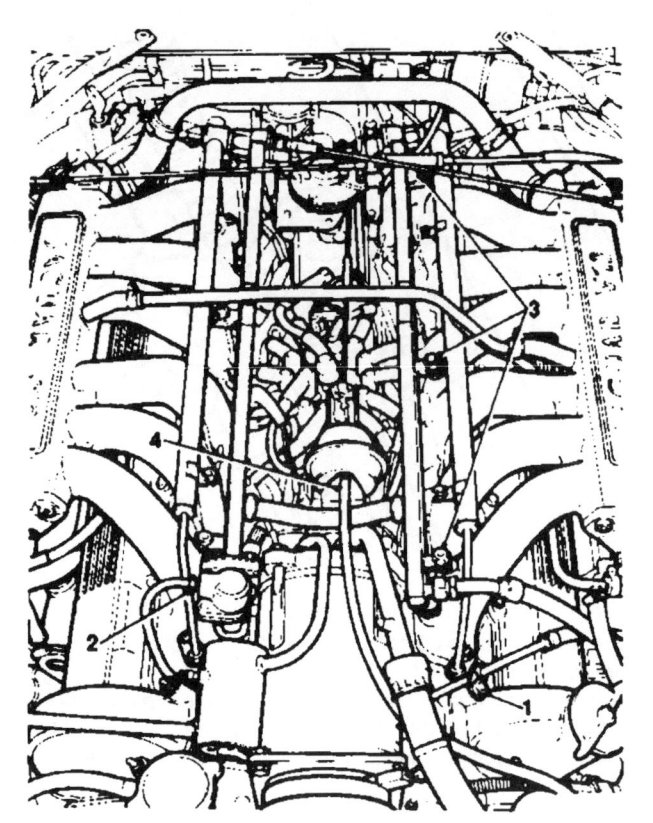

Fig. 19

Remove the multi-plug from the failed injector.
Undo the three nuts securing the fuel rail to the inlet manifold (3 Fig 19).

Lift off fuel rail and injector assembly (1 Fig 20).
Remove injector retaining clip (2 Fig 20).
Remove and discard injector 'O' rings.
Remove and discard remaining injector manifold seals.
Lubricate and fit new injector seal on the fuel rail.
Remove plugs from fuel feed pipe and fuel rail.
Fit new seal to fuel rail.
Fit feed pipe to fuel rail and tighten union nut.
Fit throttle return spring bracket and secure with bolt.
Hook return spring onto bracket.

Fit throttle cable to the pedestal bracket.
Engage throttle cable into throttle cam.
Tighten throttle cable lock nut finger tight.
Adjust cable as necessary then fully tighten lock nut.
Fit throttle cross rods to the bell crank levers at each end of rods.
Connect rods to pedestal.
Remove wing protection.
Close bonnet.
Reconnect the battery earth lead.
Close luggage compartment.

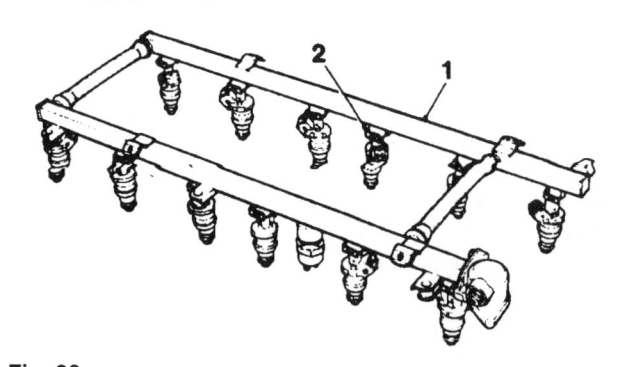

Fig. 20

IDLE SPEED, ADJUST

Ensure that engine is at normal operating temperature.
Check throttle linkage for correct operation.
Check that return springs are secure and effective.

Start engine and run for two or three minutes.
Check idle speed; if it is outside + 50 of an idle speed of 750 rev/min then it needs to be adjusted.
Set idle speed adjustment screw on auxiliary air valve to achieve 750 rev/min + 25.

Note: If it proves impossible to reduce idle speed to specified level, check ALL pipes and hoses to inlet manifolds for security and condition.

Check security of injectors.
Ensure all joints are tight and that inlet manifold to cylinder head fastenings are tight.
Ensure that both throttle butterfly closed stops show no signs of tampering; if they do, adjust throttle butterfly valves.
If the above operations do not reduce idle speed, check operation of auxiliary air valve.

COOLANT TEMPERATURE SENSOR, RENEW

Note: This procedure must only be carried out on a cool or cold engine.

Disconnect battery earth lead.
Pull electrical connector from coolant temperature sensor.
Carefully remove pressure cap from remote header tank to release any cooling system residual pressure.
Replace cap tightly.

Note: The new component should be prepared at this point so that the renewal can be made as quickly as possible.

Coat threads of sealing washer on new sensor with a suitable sealing compound.
Remove existing temperature sensor from thermostat housing.
Screw new temperature sensor into thermostat housing.
Plug in electrical connector to temperature sensor.
Re-connect battery earth lead.
Check coolant level at remote header tank. If necessary, top up.

COOLANT TEMPERATURE SENSOR, TEST

Disconnect battery earth lead.
Pull electrical connector from coolant temperature sensor.
Connect suitable ohmmeter between terminals, note resistance reading. The reading is subject to change according to temperature and should closely approximate to the relevant resistance value given in the table below.
Disconnect the ohmmeter.
Check resistance between each terminal in turn and body of sensor.
A very high resistance reading (open circuit) must be obtained.

Refit electrical connector to temperature sensor.
Re-connect battery earth lead.

Coolant Temperature (°C)	Resistance (kilohms)
-10	9.2
0	5.9
10	5.9
20	2.5
30	1.7
40	1.18
50	0.84
60	0.60
70	0.435
80	0.325
90	0.250
100	0.190

THROTTLE POTENTIOMETER, RENEW

Disconnect the battery earth lead.
Remove the throttle cross rods and cruise control valve actuating rod from the throttle pulley.
Disconnect the throttle potentiometer in-line cable multi-plug connector.
Remove the bolts securing the throttle pulley plate to the throttle pedestal and lift the plate clear.
To fit a new potentiometer, reverse the above operations.

MAIN RELAY, RENEW

Open the luggage compartment lid and disconnect the battery earth lead.
Remove the luggage compartment right-hand trim panel.
Displace the relays from the securing tags at the back of the hockey stick panel adjacent to the fuelling ECU (Fig. 21).
Identify the main relay; silver relay with red stripe on a red base connector.
Fit the new relay and reposition the base to the securing tag.
Reposition the trim panel.
Reconnect the battery earth lead.

PUMP RELAY, RENEW

Open the luggage compartment lid and disconnect the battery earth lead.
Remove the luggage compartment right-hand trim panel.
Displace the relays from the securing tags at the back of the hockey stick panel adjacent to the fuelling ECU (Fig. 21).
Identify the fuel pump relay; silver relay on a yellow-black base connector.
Fit the new relay and reposition the base to the securing tag.
Reposition the trim panel.
Reconnect the battery earth lead.

PUMP RELAY, TEST

Switch on ignition. Pump should run for one or two seconds, then stop.
Note: If pump does not run, or does not stop, check systematically as follows:

Check that inertia switch cut-out button is pressed in.
Detach inertia switch cover and ensure both cables are secure.
Pull electrical connectors from switch and check continuity across terminals with ohmmeter.
Pull button out and check open circuit.
Remove ohmmeter, replace electrical connectors, reset button and refit cover.
If inertia switch is satisfactory, earth (ground) pump relay terminal 85, switch on ignition and check circuit systematically as detailed below.

a   Check for battery voltage at terminal 86 of main relay. If yes: proceed to test b. If no: check battery supply from ignition switch via inertia switch.

b   Check for battery voltage at terminal 87 of main relay. If yes: Proceed to test c. If no: check for battery voltage at earth (ground) lead and connection from terminal 85 of main relay. If satisfactory, renew main relay.

c   Check for battery voltage at terminal 86 of pump relay. If yes: proceed to test d. If no: open circuit between terminals 87 of main relay and 86 of pump relay. If satisfactory proceed to test d.

d   Check for battery voltage at terminal 87 of pump relay. If yes: Proceed to test e. If no: check for battery voltage to earth (ground) lead and connections from terminal 85 of pump relay. If satisfactory, renew pump relay.

e   Check for battery voltage at supply lead (NS) and connections to fuel pump. If yes: faulty pump or earth (ground) connections. If no: open circuit between terminal 87 of pump relay and supply-lead connections to pump.

SPARKING PLUGS, RENEW

Open the bonnet.
Note the positions of the twelve leads and then disconnect from the plugs.
Undo and remove the plugs.
Fit new plugs, tightening to 23 to 28 Nm and reconnect plug leads.

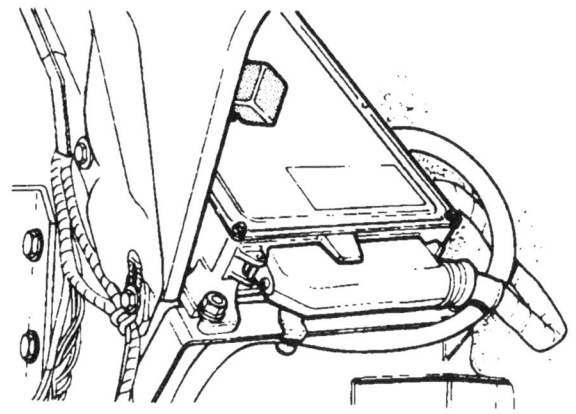

Fig. 21

Chapter 5

ENGINE MANAGEMENT SYSTEM (6 Litre)

IGNITION CONTROL SYSTEM (IC)

Description

The Ignition System (IC) is controlled by the Powertrain Control Module Ignition (PCMI). It is a microprocessor based system and uses discrete components and analogue to digital circuits to interface between the microprocessor and the input sensors and output devices.

Software programmed into the microprocessor is divided into Control Code and Data (engine calibration). The control code will be common to all specifications. Data will be specific to the requirements of each specification or market.

The sensor inputs to the PCMI are continuously scanned and the ignition timing calculated according to engine conditions. It provides switching signals to the two Ignition Modules (IMs) and in turn the IMs control the ignition coil. Low tension circuits (LT), dwell and coil current control is provided by the IMs.

The two IM units are located on top of the radiator cross member. The ignition coils are located in the vee between the cylinder heads. The front coil supplies 'B' bank (left-hand side) and the rear coil supplies 'A' bank (right-hand side).

The coils supply the high tension (HT) voltage via the Distributor Ignition (DI) to the sparking plugs.

In addition the PCMI provides a speed output signal for use by other vehicle systems, these are;

Powertrain Control Module Fuel (PCMF), Transmission Control Module (TCM), Fuel Pump Control Module and the Driver's Tachometer.

PCMI Inputs And Outputs.

Inputs	Outputs
Manifold Absolute Pressure	A Bank Coil Drive
Crankshaft Position	B Bank Coil Drive
Crankshaft Speed Signal	Engine Speed
Ignition Switch 12V	Gear shift Acknowledge
Closed throttle/idle switch	Gear shift Request
	Air Temperature

The control parameters fall into two groups;

Primary: intake manifold absolute pressure (MAP), engine speed and position.
Secondary: throttle position, air intake temperature, retard request and acknowledgement signals.

Manifold Absolute Pressure Control.

The Manifold Absolute Pressure Sensor (MAPS) is mounted within the Powertrain Control Module Ignition (PCMI) and is directly connected to the rear of 'A' bank inlet manifold. There are no temporary inhibitions or functions affecting the ignition timing other than normal cranking, idle and temperature functions. Air must be able to flow freely between the inlet manifold and the MAP sensor.

SOFTWARE FUNCTIONS

Main Map, Open Throttle Running.

The ignition timing for open throttle running is controlled primarily by the main map, which is based on Manifold Absolute Pressure (MAP) and engine speed.

The information for the ignition requirements is stored in a look-up table in memory. Up to 16 MAP and 16 speed sites may be used. So that for any given MAP and engine speed the memory will generate a number that relates to the timing required. When the engine load and/or speed is between sites, a function known as two dimensional interpolation is used to calculate an ignition value from the surrounding sites.

The timing computed is then subject to adjustment dependant upon intake air temperature and Transmission Control Module (TCM) retard request signal.

The intake air temperature signal is determined by an NTC thermistor located in the 'A' bank air inlet adjacent to the throttle body.

The Intake Air Temperature Sensor Ignition (IATSI) software function, applies ignition retard to the map value dependent upon the air temperature measured to prevent detonation with high air intake temperatures.

The ignition timing is also retarded to reduce engine power in response to a retard request signal from the TCM. The retard function will operate in reverse gear at wider throttle openings and during forward gear shifts to improve gear shift quality. The ignition timing will be retarded up to 20° crank with a limit of 8° ATDC. The retard is applied for a maximum of 1.2 seconds, or less if the retard request signal is removed. The ignition timing ramps back to the load/speed map over 0.5 second.

The PCMI sends a retard acknowledge signal back to the TCM. If the acknowledge signal is not received by the TCM the transmission will default into a protected mode.

Engine Position and Speed.

A variable reluctance device known as the Crankshaft Position Sensor (CKPS) is used to recognise engine position. It is triggered by a 3 toothed wheel attached to the crankshaft pulley. The voltage induced in the sensor by the passing teeth provides an 8° ATDC reference for each cylinder on 'A' bank. The signal is decoded by the PCMI. The PCMI generates the 'B' bank position in software from this information. A similar Crankshaft Flywheel Sensor uses the starter ring gear teeth to provide a signal from which the PCMI calculates engine speed.

Engine Cranking and Starting.

Engine cranking and starting is recognised by decoding the speed signal. The ignition timing during cranking and start up is dependent on engine speed, intake air temperature and throttle position.

The ignition timing moves from the crank to run function at 350 rpm as the speed increases, providing a fast speed rise to the idle speed.

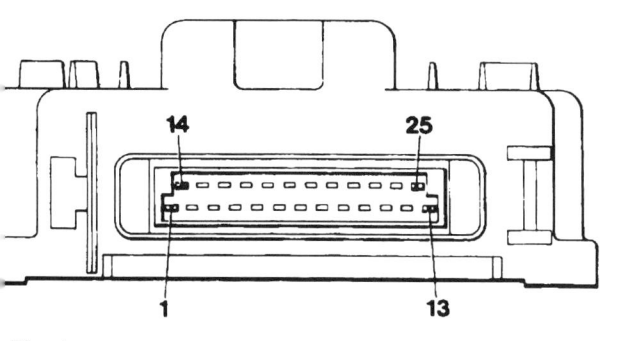

Fig. 1

Pin No.	Connection	Pin No.	Connection
1.	CKPS Signal	14.	Bank IM (6)
2.	CKPS Earth	15.	A Bank IM (3)
3.	CKFS Earth	16.	CKFS Signal
4.	Earth	17.	Retard Link
5.	Switch Earths	18.	Closed Throttle Switch
6.	IATSI Earth	19.	IATSI Signal
7.	Not Connected	20.	Diagnostic Request (not used)
8.	Not Connected	21.	Gearshift Acknowledge Signal
9.	B Bank IM (6)	22.	Not Connected
10.	B Bank IM (3)	23.	2° Retard Link
11.	Earth	24.	Engine Speed Output
12.	Earth	25.	Gearshift Request
13.	Ignition Switch 12V Supply		

2 Component Location - Ignition Control System (IC)

Closed Throttle Running.

A Closed Throttle Switch operated by the throttle pulley provides a closed throttle signal to the PCMI. The ignition for idle and overrun is programmed separately from the main map. It is also modified by engine speed and intake air temperature.

CRANKSHAFT POSITION SENSOR (CKPS), DESCRIPTION

The CKPS (Fig. 3) is a variable reluctance device consisting of a bobbin coil with a magnetic core.

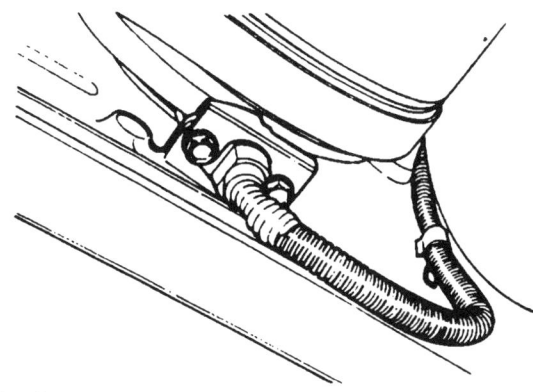

Fig. 3

It is mounted below the crankshaft damper on the timing cover. The 3-toothed timing wheel on the back of the damper pulley assembly, induces a signal in the coil as each tooth passes. This signal is received by the PCMI.
The three teeth give an 8° ATDC reference for each cylinder on 'A' bank on its firing stroke.
The correct sensor air gap is 0.2 to 0.8 mm.
The sensor signal (Fig. 4) can only be checked using an oscilloscope.

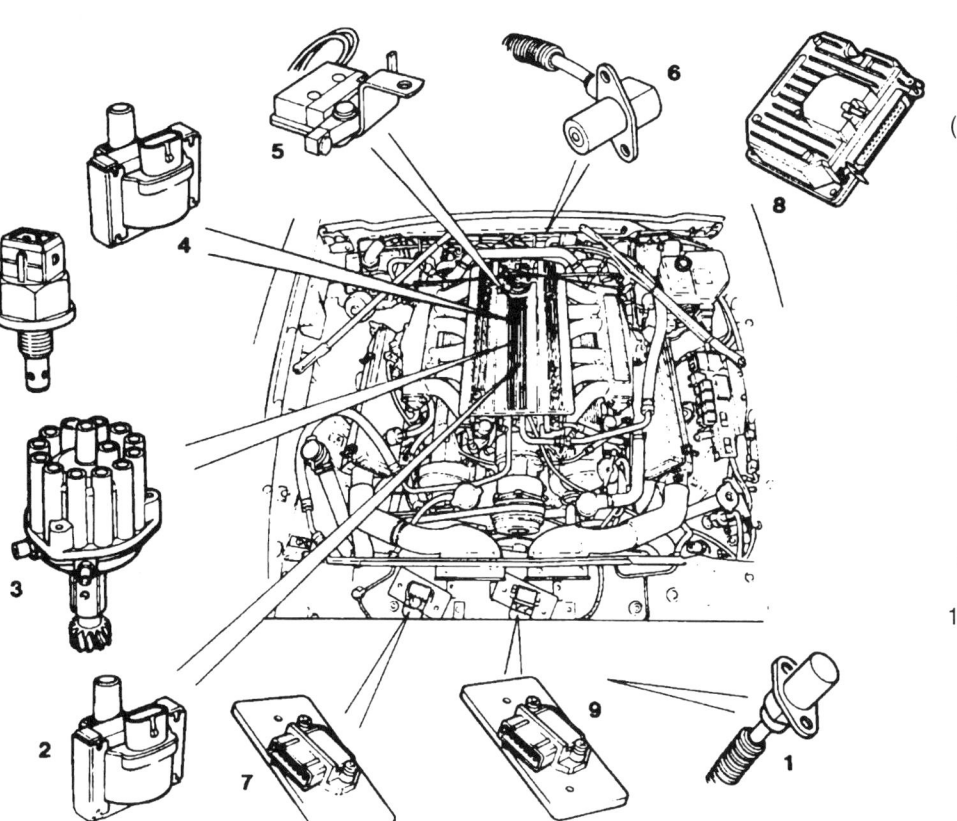

(key to Fig. 2)
1. Crankshaft Position Sensor (CKP)
2. Ignition Coil ('B' Bank)
3. Distributor Ignition (DI)
4. Ignition Coil ('A' Bank)
5. Closed Throttle Switch
6. Crankshaft Flywheel Sensor (CKFS)
7. LH 'B' Bank Ignition Module (IM)
8. Powertrain Control Module Ignition (PCMI) (located at side of front passenger footwell)
9. RH 'A' Bank Ignition Module (IM)
10. Intake Air Temperature Sensor (IATSI)

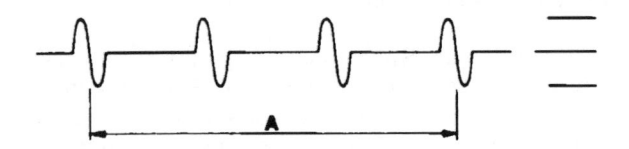

Fig. 4

Connect the oscilloscope positive to PCMI pin 1 or the sensor blue lead.
Connect the oscilloscope negative to PCMI pin 2 or the sensor screen.

Note: Fig. 4 'A' = 1 engine revolution

Crankshaft Flywheel Sensor (CKFS), Description

The CKFS (Fig. 5) is a variable reluctance device consisting of a bobbin coil with a magnetic core.

It is mounted on the starter ring gear lower cover. The 160 starter ring gear teeth induce a signal in the coil as each tooth passes the pole piece of the sensor.
The signal is used to generate engine speed reference.

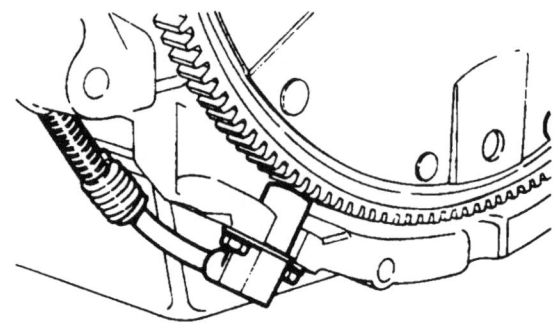

Fig. 5

The correct sensor air gap is 0.2 to 0.8 mm.
The sensor signal (Fig. 6) can only be checked using an oscilloscope.

Connect the oscilloscope positive to PCMI pin 16 or the sensor red lead.
Connect the oscilloscope negative to PCMI pin 3 or the sensor screen.

Note: Fig. 6 'A' = 160 teeth per engine revolution.

Intake Air Temperature Sensor Ignition (IATSI),

Description

The IATSI (Fig. 7) is an NTC thermistor, i.e. its resistance decreases as the temperature increases. It is located in the 'A' bank air inlet near to the throttle body. The temperature signal is received by the PCMI and ignition retard is used to avoid detonation at high inlet air temperatures. Two temperature maps are programmed with retard laws. One of the maps is used for the main load / speed maps. Ignition retard is used above 25°C air inlet temperature.

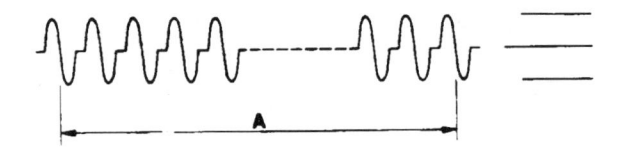

Fig. 6

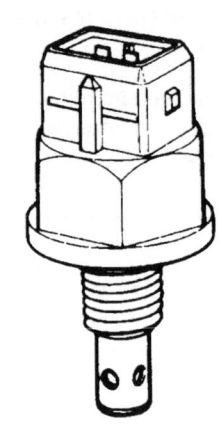

Fig. 7

A separate law is used to retard cranking ignition at air inlet temperatures above 75°C.

Component Diagnostic Check

Remove the IATSI from the 'A' bank air intake tube. The IATSI should be tested in air.

Air temperature and component temperature should be allowed to stabilize. Measure the temperature with an accurately calibrated thermometer (e.g. mercury/glass).

Using a DMM set to ohms, measure the resistance across the component blades, note the resistance and temperature. The recorded readings should fall within the band on the graph (Fig. 8), if not, recheck temperature and resistance measurement.

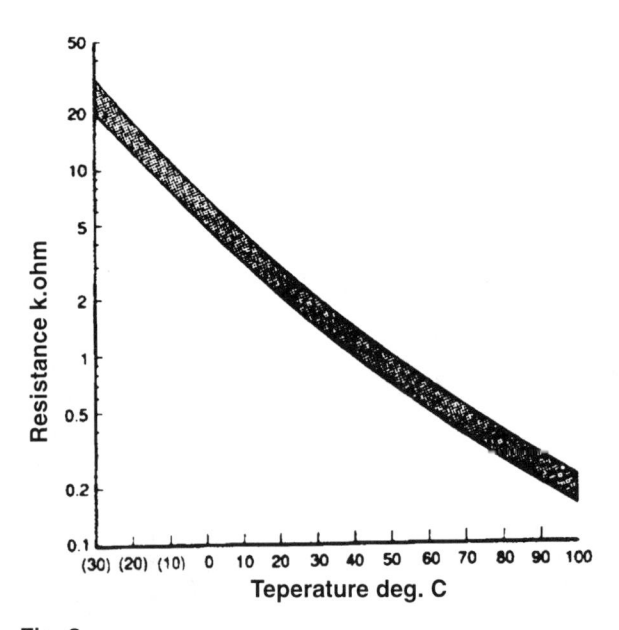

Fig. 8

Check each connector pin to the metal case, the reading should be high (infinity). Reject components that fall outside the tolerance band.

Engine Speed Signal

The PCMI generates a speed signal (Fig. 9) which is used by the Powertrain Control Module Fuel (PCMF), Transmission Control Module (TCM), Fuel Pump Module Controller and the Driver's Tachometer. The frequency of the signal is 6 pulses per engine revolution. Check the output using an oscilloscope.

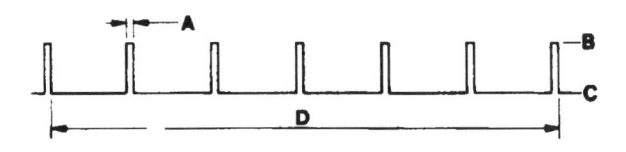

Fig. 9
A = 400 micro seconds
B = Approximately battery voltage -1V
C = <1V
D = 1 engine revolution

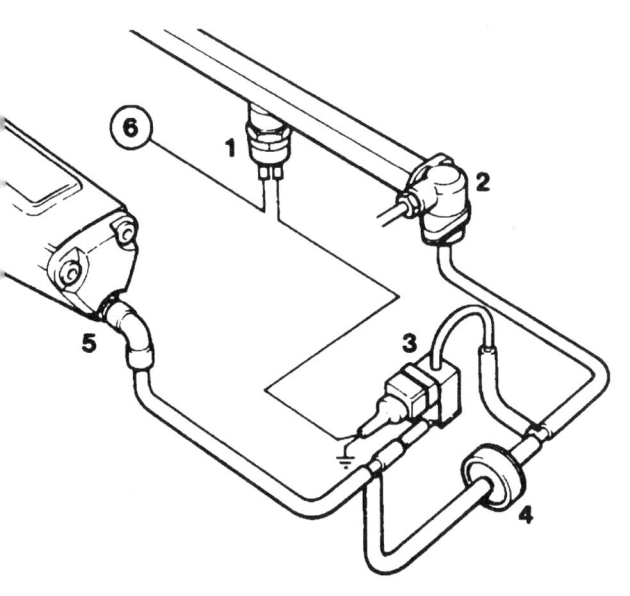

Fig. 10
1. Electrical temperature switch
2. Fuel Pressure Regulator
3. Solenoid Vacuum Valve
4. Vacuum Delay Valve
5. Inlet Manifold Connector
6. Electrical Supply

Connect the 'scope positive to ignition PCMI pin 24.
Connect the 'scope negative (ground) to the chassis or engine earth.

Hot Start System, Description

To aid restarting a hot engine, a 'Hot Start System' is fitted that works together with the fuel pressure control system.

An electrical temperature switch (1 Fig. 10) is fitted to the underside of the right-hand fuel rail. When the fuel temperature exceeds 70°C at hot starting, the switch is closed and energises a solenoid valve (3 Fig. 10). This valve closes the vacuum signal line from the inlet manifold to the fuel pressure regulator diaphragm chamber (2 Fig. 10). Full regulated fuel pressure (3.0 Bar) is applied to the injectors.

The higher fuel pressure at idle and part load will suppress vapour generated in the injectors by the high injector body temperatures.

The vacuum signal by-passes the solenoid valve through the delay valve (4 Fig. 10). This allows the vacuum signal to be gradually returned to the fuel pressure regulator over approximately 45 seconds.

The Powertrain Control Module Fuel (PCMF) secondary air injection control function provides the electrical supply for the temperature switch and solenoid valve (including non AIR vehicles). This control ensures that the hot start system will be disabled at all times other than immediately following an engine start.

CAUTION:

When disconnecting and reconnecting the rubber vacuum hoses between the inlet manifold, fuel pressure regulator and associated units, ensure that the ends are plugged to keep dirt out.

When it is necessary to check the fuel rail switch operation. Great care must be taken to change its temperature slowly when approaching the switching points. The switch closes on rising temperature at 70 +/- 4°C and opens on falling temperature at 60 +/- 3°C.

Powertrain Control Module Fuel (PCMF) System, Description

The Powertrain Control Module Fuel (PCMF) is a microprocessor based system. It maintains optimum engine performance over the engine's full load and speed range, by metering the fuel delivered to each cylinder by the Fuel Injector (FI). It uses discrete components and analogue to digital circuits to interface between the microprocessor and the input sensors and output devices.

Software programmed into a ROM is divided into control code and data (engine calibration). The control code will be common to all specifications. Data will be specific to the requirements of each specification or market.

PCMF Inputs and Outputs

The control parameters fall into two groups:

Primary: consisting of intake manifold absolute pressure (MAP) and engine speed.

Inputs
Engine Coolant Temperature Sensor
Oxygen Sensor A Bank
Oxygen Sensor B Bank
Throttle Position Sensor
Manifold Absolute Pressure
Speed Signal
Fuel Level
Intake Air Temperature Sensor
Serial Communication
Ignition on 12V
Permanent 12V

Outputs
Fuel Used
Diagnostic Trouble Codes
A Bank Fuel Injectors
B Bank Fuel Injectors
Purge Valve Drive
Secondary Air Injection Relay
Malfunction Indicator Lamp
Serial Communication
Oxygen Sensor Heater Relay

Secondary: consisting of throttle position/movement, engine coolant temperature, intake air temperature, oxygen sensor signal and supply voltage.

Fuel Metering

A pair of electrically driven Fuel Pump Modules (FPM) are mounted in the fuel tank one behind the other. They deliver fuel to the solenoid operated Fuel Injectors (FI) via the fuel rail. The PCMF runs the pump for about one second after the Ignition Key is turned on, to pressurise the fuel rail. When the engine speed signal is received from the Powertrain Control Module Ignition (PCMI). The fuel pump will continue to run until either, the Ignition is turned off, or, until approximately one second after there is no speed signal.

At engine speeds up to 2840 rpm, the first FPM (rear module) is used. At engine speeds higher than 2840 rpm, the second FPM is switched in by the fuel pump module controller. On falling engine speed, the system reverts to single pump operation at 2000 rpm.

Fuel pressure is controlled by a fuel pressure regulator. Fuel pressure fluctuates between 2.4 and 3.0 Bar depending on the manifold vacuum acting on the regulator diaphragm. The differential pressure across the FI nozzle therefore remains constant and the quantity of fuel injected for a given FI pulse duration is maintained constant.

The PCMF provides engine over speed control and imposes an engine speed limit of 6000 rpm by cutting Fuel Injector (FI) operation. The FI begin operating again as the speed falls to 5900 rpm.

Software Functions.

Main Map.

The primary control functions of the system are Manifold Absolute Pressure (MAP) (engine load) and Engine speed. The fuel requirements are programmed into a look-up table of 8 engine loads X 16 engine speeds.

Exhaust Catalysts are used in all markets. Apart from full load operation, the fully warm engine will be running at Lambda 1 (14.7:1 Air Fuel Ratio (AFR)). Typical engines are mapped over the full engine load and speed range on the engine test bed. Full load operation, decided by throttle position and engine speed uses a richer mixture to achieve maximum engine power. The feedback system is inhibited and the AFR will be approximately 12 : 1.

Further refinement of the fuel map is carried out during emission development and road testing, over a range of climates representative of conditions experienced by vehicles in service.

Usually, the load and speed at which the engine is running will be between load and speed sites. A function known as two dimensional interpolation is used with the surrounding sites to calculate the correct FI pulse duration for the between - sites condition.

Cranking and Starting

The PCMF receives the engine speed signal from the PCMI (six pulses per engine revolution). Cranking is decided from the speed signal. The PCMF will fire the FI three times per engine revolution. The FI pulse duration during cranking is dependent upon engine coolant temperature. At 250 - 450 rpm depending upon engine coolant temperature, the FI frequency changes to the normal once per revolution. Half the engines' fuel requirement for the combustion cycle is delivered at each injection.

After Start Enrichment and Warm-up Law

Once the engine speed crosses the crank-to-run threshold (250 - 450 rpm) after start enrichment is applied by the PCMF. The amount is programmed against engine coolant temperature. The after start enrichment is decremented against engine revolutions down to the warm up law. This law is also programmed against engine coolant temperature. It provides the necessary additional enrichment to sustain the engine during the cold and part warm phase after starting. The normal load/speed fuel map provides the basic fuelling on which these functions work.

Intake Air Temperature

The PCMF receives an air temperature signal from the Intake Air Temperature Sensor Fuel (IATSF). This function is programmed to increase the FI pulse duration by approximately 2% / 10°C as intake air temperature falls to maintain the AFR constant.

Supply Voltage Corrections

The PCMF monitors the electrical system voltage, i.e. the battery voltage. FI opening time is affected by voltage and results in a corresponding change in the quantity of fuel delivered. The PCMF battery voltage compensation function adjusts the FI pulse duration accordingly.

Throttle Demand Corrections

The PCMF receives a throttle position and movement signal from the Throttle Position Sensor (TPS). The signal is used to decide the requirement for closed throttle (idle, overrun and fuel cut-off), opening and closing throttle (acceleration and deceleration) and wide throttle (full power demands).

Closed throttle recognition for the idle function is adaptive in the range 220 - 750 mV. It is not necessary to readjust the TPS idle setting unless it fails the JDS check. This is carried out as a background task during engine set up.

The overrun fuel cut-off function operates on closed throttle. With the engine fully warm, the FI will be inhibited when the throttle closes with engine speed higher than 1500 rpm. The FI will be reinstated at 1100 rpm. At lower engine coolant temperatures higher 'cut-out' and 'cut-in' speeds are used to avoid driveability problems.

The rate of throttle opening and engine coolant temperature is used to determine the requirement for acceleration enrichment. With moderate rates of throttle opening the regular FI pulse duration is lengthened for a short period to provide good engine response. Faster rates of throttle opening will cause an extra injection pulse to be generated, in addition to the lengthening of the normal FI pulses.

Falling TPS voltage identifies throttle closing. It will be detected in the same way and the FI pulse duration will be momentarily shortened to avoid a rich mixture, which has an adverse effect on exhaust emissions.

The TPS voltage and engine speed are used to identify the full load enrichment requirement. The enrichment is added to the pulse duration obtained from the load/speed map. When full load enrichment is enabled, the HO2S feedback function is inhibited.

Air Fuel Ratio (AFR) Corrections

On catalyst equipped vehicles exhaust pollutants are reduced to a minimum using a closed loop system to monitor the oxygen content of the exhaust gas. Air/fuel mixture is corrected to maintain an intake ratio of approximately 14.7:1 (Lambda 1.0).

This is achieved by the PCMF in response to an input from a HO2S mounted in the exhaust downpipe. Each time the AFR crosses lambda 1.0 the fuel trim is reversed, lean to rich, rich to lean etc. The rate at which fuel trim change is applied varies with load and speed using a 4 X 4 map. These fuel trim characteristics are known as the feedback integrator functions. Each bank of the engine is controlled separately. The average integrator voltage provides an overall trim for the main fuel map.

Idle Mixture Adjustment

The PCMF adjusts idle mixture using an adaptive function on catalyst (feedback) equipped vehicles. Feedback integrator average voltage is monitored at idle. The adaptive function adds to, or subtracts from, the FI pulse duration to centralize the integrator voltage. The correction (plus or minus) is applied to the whole load/speed map. Adaption is applied separately to each bank of the engine and stored in memory. Re-adaption is carried out as necessary during each journey.

Secondary Air Injection Control (AIR) System

The PCMF is software programmed to control the AIR system. It controls the operation of the Secondary Air Injection Relay (AIRR). The AIRR in turn switches the engine driven Secondary Air Injection Pump Clutch (AIRPC) and the Secondary Air Injection Solenoid Valve (AIRV). The AIRV vacuum circuit controls the operation of the

Secondary Air Injection Switching Valve (AIRS).

Air Injection is used following each engine start to promote a reaction in the exhaust down pipe catalyst. This reaction generates heat and reduces the time required to bring the catalyst up to working temperature.

AIR operation is also checked by the PCMF OBD function using feedback integrator deflection to check the system operation. The AIR system is turned off if the engine speed exceeds 2500 rpm during operation. This prevents damage to the pump due to excessive back pressure or speed.

Evaporative Emission Control System (EVAP)

Fuel vapour from the vehicle fuel tank is stored in the charcoal canister and purged by engine inlet manifold vacuum, when the vehicle is in use. The conditions during which vapour is purged, and the purge rate, are programmed to suit the vehicle under all environmental conditions.

Purge rate has to be sufficiently high to prevent the charcoal canister overflowing and venting vapour to atmosphere. Purge rates set too high would cause driveability problems.

Vapour will be purged once the HO2S feedback system is operational. The rate of purge is determined from a load / speed map. A small amount of purge is used at idle, the purge valve operation can be detected by holding or touching the valves. They will be pulsing in response to the mark/space ratio signal from the PCMF.

The valves are normally open, although when the ignition is turned off the valves are held closed for a short period. This prevents vapour being drawn into the engine that could cause 'running on'. The valve is also energised to close it with ignition on, but the engine not running.

On Board Diagnostic System (OBD)

The OBD function of the PCMF monitors the system sensors and output devices for correct operation while the vehicle is in use. Once a fault is identified, the Malfunction Indicator Lamp (MIL) is illuminated to inform the driver that a problem has been identified. The MIL will remain on for the remainder of the journey, i.e. until the ignition is turned off. In certain markets the MIL will not illuminate on successive journeys if the fault was temporary and not present when the engine is restarted. A fuel level signal is used as an input to the OBD system. It is used to prevent false flagging of Diagnostic Trouble Codes (DTC) if the fuel level is low enough to cause momentary loss of fuel pressure.

In all cases, the fault code will be stored for retrieval when required.

The area in which the fault occurred will be identified by turning on the ignition but not starting the engine. The Diagnostic Trouble Code (DTC) code will be displayed on the trip computer display as FF + XX (where XX is a number that identifies the fault code).

If more than one fault is present, the highest priority code will be displayed. JDS can be used to look at all codes stored at once. This will be helpful in some instances. Certain types of fault may trigger more than one code. When a fault is detected in a sensor or input circuit, a default value will be substituted to allow continued use of the vehicle. Some degradation in performance may be observed. Where faults are detected in output devices, there will be no default. Continued use of the vehicle will depend upon the exact nature of the fault.
There are two OBD codes which are directly associated with the PCMF inputs, they are:

FF13 Manifold Absolute Pressure Sensor (MAPS) or signal pipe.

This OBD function carries out three checks on the MAPS signal pipe and vacuum signal.

1. As the engine starts, and runs up to speed through cranking, the MAP signal is monitored to detect the change in manifold vacuum that confirms that the MAPS is receiving the engine manifold vacuum signal.

2. While the engine is running, if the MAP signal is less than 80 Torr when the engine speed is less than 1500 rpm. The code will be set (check inhibited below 350 rpm).

3. While the engine is running, the MAP signal must be less than 550 Torr when the engine speed is less than 1000 rpm. The code will be set (check inhibited below 350 rpm).

Default: When DTC FF13 is set, the MAPS function will not be used. The fuelling will be controlled by throttle angle (TPS signal) and engine speed. HO2S feedback will be disabled.

FF77 Speed Signal input from PCMI.

This OBD function checks for the presence of the speed signal received from the PCMI.

Temporary loss of the signal may cause stalling or starting difficulties. The OBD code is for information only. Complete loss of the signal will prevent the engine running.

Default: There is no default for this DTC.

OTHER SYSTEMS ASSOCIATED WITH THE ENGINE MANAGEMENT SYSTEM.

Idle Air Supply

Auxiliary Air Valve, Description (see Fig. 11)

The Auxiliary Air Valve is mounted in the 'B' bank rear water rail. It uses a wax motor responding to water temperature. It drives a cup shaped sleeve against a spring inside a tube, to close an orifice in the side of the tube. The orifice is shaped to provide the correct amount of air at idle to the engine during warm up.
A bypass with a control screw is provided for hot idle speed setting. The top outlet tube is connected to the rear of 'B' bank inlet manifold using a rubber tube. Inlet air to the valve is provided from the back of 'B' bank air cleaner through a suitable tube.

Test

Remove the auxiliary air valve from the water rail.
Place a piece of tape over the bypass hole in the side port, or fully close using the idle speed adjustment screw.
Immerse auxiliary air valve bulb in boiling water and observe valve head through side port.
Valve should move smoothly to closed position.
Quickly blow through side port, no air should pass.
Allow valve to cool. Valve head should move smoothly back to open main air passage.
If valve performance is satisfactory remove tape and refit valve.
If valve performance is unsatisfactory, fit replacement component.
Check, if necessary, adjust idle speed.

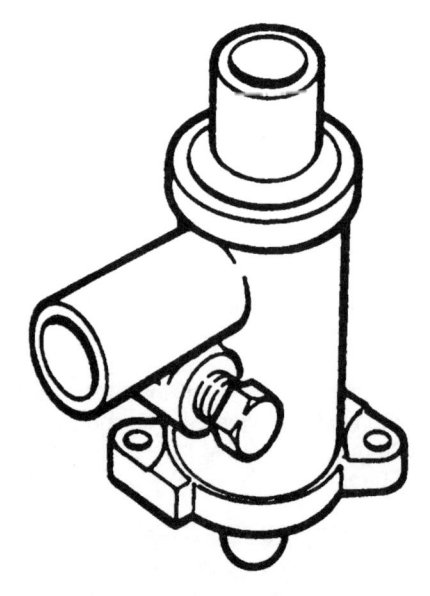

Fig. 11

Supplementary Air Valve, Description (see Fig. 12)

The Supplementary Air Valve is mounted in the back of 'A' bank Air Cleaner. It is connected via a rubber hose to 'A' bank inlet manifold, inboard of the throttle. The valve is energised when a transmission gear (e.g. Drive or Reverse) is selected and the Air Conditioning Compressor Clutch is engaged. The air supplied by the valve compensates for the extra engine load, and maintains idle speed to prevent engine stalling.

Intake Air Temperature Sensor Fuel (IATSF), Description

The IATSF (Fig. 15) is an NTC thermistor, i.e. its resistance decreases as the temperature increases. It is located in the 'B' bank air filter back plate. The temperature signal is

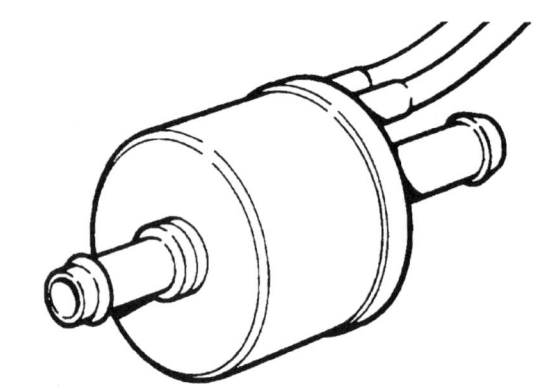

Fig. 12

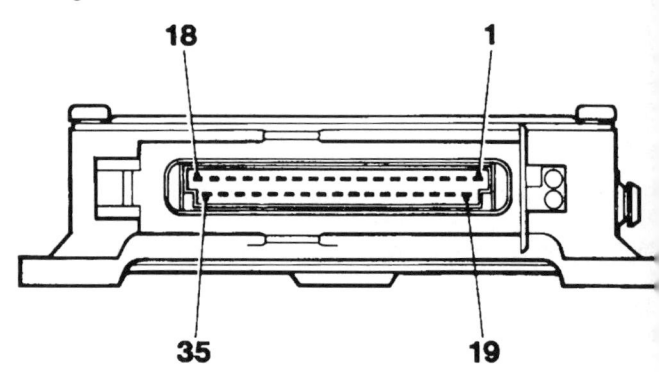

Fig. 13 PCMF Pin Connections

Pin No.	Connection	Pin No.	Connection
1.	Low current earth	19.	Low fuel level
2.	Fuel used / DTC	20.	TPS +ve
3.	Fused 12V supply	21.	IATS signal
4.	Serial in (L)	22.	Ignition 12V
5.	ECTS signal	23.	HO2S earth
6.	HO2S Bank A signal	24.	HO2S Bank B signal
7.	TPS signal	25.	EVAPP signal
8.	FI Bank B turn on	26.	AIRR signal
9.	FI Bank B turn on	27.	FI Bank B turn on
10.	Serial out (K)	28.	FI Bank B turn on
11.	FI Bank B hold on	29.	FI Bank B hold on
12.	FI Bank A hold on	30.	FI Bank A hold on
13.	FI Bank A turn on	31.	FI Bank A turn on
14.	FI Bank A turn on	32.	FI Bank A turn on
15.	HO2S Heater Relay	33.	MIL
16.	Earth	34.	Earth
17.	Earth	35.	Earth
18.	Speed signal from PCMI		

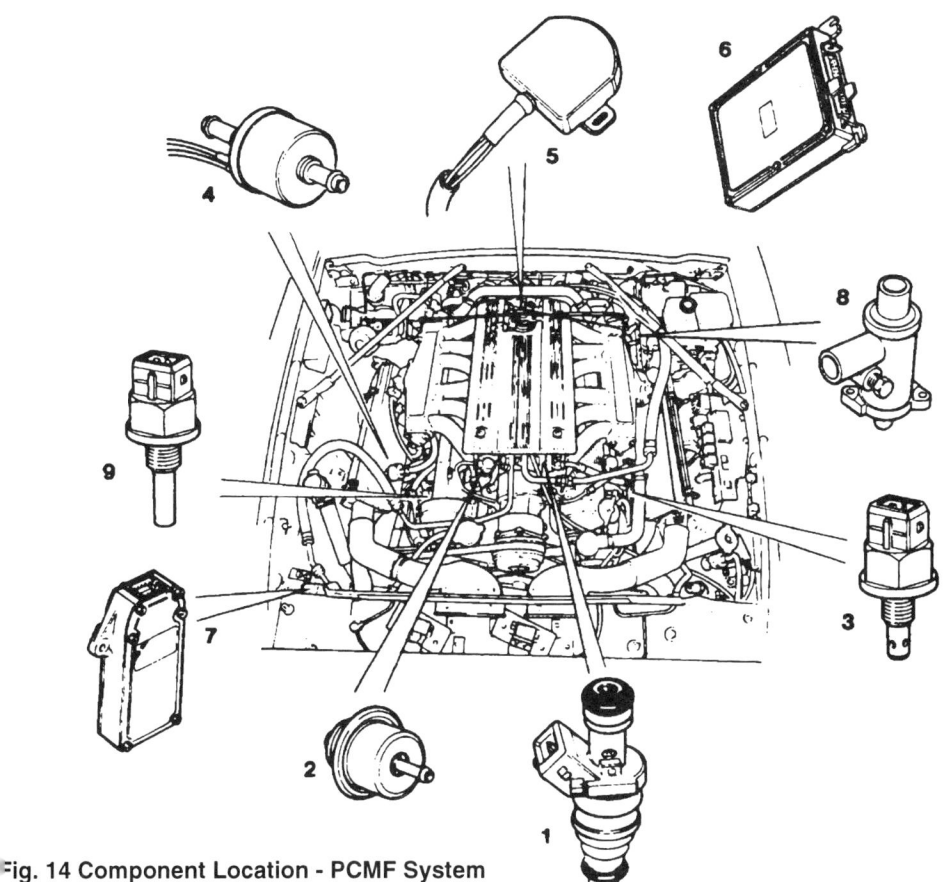

(key to Fig. 14)

1. Fuel Injector (FI)
2. Fuel Pressure Regulator
3. Intake Air Temperature Sensor (IATS)
4. Supplementary Air Valve
5. Throttle Position Sensor (TPS)
6. Powertrain Control Module Fuel (PCMF)
7. Power Resistor
8. Auxiliary Air Valve
9. Engine Coolant Temperature Sensor (ECTS)

Fig. 14 Component Location - PCMF System

received by the PCMF and the FI pulse duration is increased by approximately 2% per 10°C as intake air temperature falls to maintain AFR constant.

Component Diagnostic Check

Remove the IATSF from the 'B' bank air filter back plate. The IATSF should be tested in air.
Air temperature and component temperature should be allowed to stabilise. Measure the temperature with an accurately calibrated thermometer (e.g. mercury / glass).
Using a DMM set to ohms, measure the resistance across the component blades. Note the resistance and temperature. The recorded readings should fall within the band on the graph (Fig. 16). If not, recheck temperature and resistance measurement.

Check each connector pin to the metal case. The reading should be high (infinity). Reject components that fall outside the tolerance band.

The OBD code associated with the IATSF is:

DTC FF16 Intake Air Temperature Sensor Fuel (IATSF) Circuit.

The PCMF monitors the IATSF voltage. If it falls outside the range 0.1 to 4.9V a fault is flagged after several out of range readings.
Open or short circuit faults are the likely cause.

Default:

When DTC FF16 is recognised, the air temperature law defaults to a trim value of 30°C. JDS can be used to trace faults in this circuit.

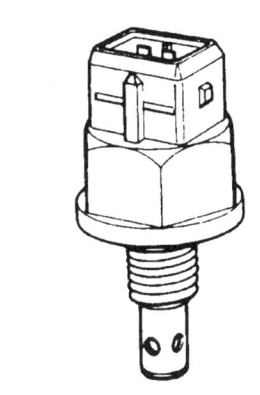

Fig. 15

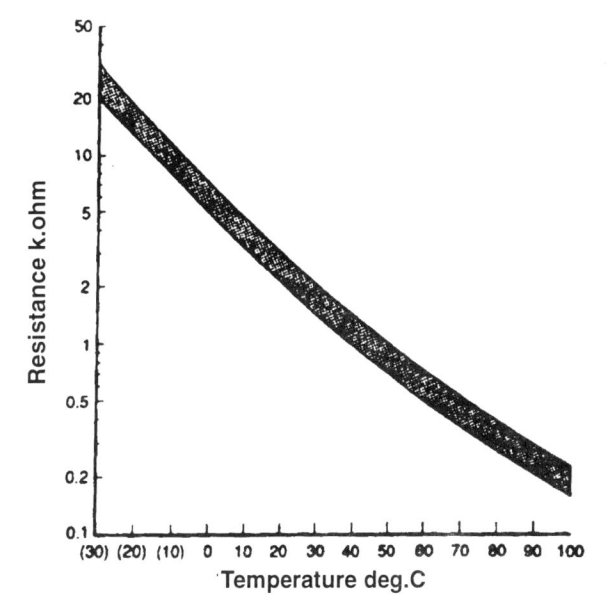

Fig. 16

ENGINE COOLANT TEMPERATURE SENSOR FUEL (ECTSF),

DESCRIPTION

The ECTSF (Fig. 17) is an NTC thermistor. Its resistance decreases as its temperature increases.
The PCMF converts the voltage from the sensor into a digital number that relates to the coolant temperature.

The temperature is used for a number of software functions. They include start and warm up, acceleration enrichment, secondary air injection control and evaporative purge control.

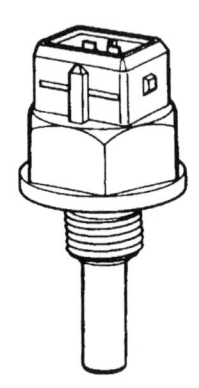

Fig. 17

Component Diagnostic Check

Remove the ECTSF from the water thermostat housing.
The ECTSF should be tested in hot water.
The component temperature should be allowed to stabilise. Measure the temperature with an accurately calibrated thermometer (e.g. mercury/glass).
Using a DMM set to ohms, measure the resistance across the component blades. Note the resistance and temperature.
The recorded readings should fall within the band on the graph (Fig. 18). If not, recheck temperature and resistance measurement.
Check each connector pin to the metal case. The reading should be high (infinity). Reject components that fall outside the tolerance band.

The OBD code associated with the ECTSF is:

DTC FF14 Engine Coolant Temperature Sensor Fuel (ECTSF) Circuit.

The ECTSF circuit is monitored in three ways.

a). From a cold start the indicated temperature has to rise above 53°C. within 15360 engine revolutions (approximately eighteen minutes). Failure to achieve this would suggest either a high resistance sensor/wiring or a faulty or missing engine coolant thermostat.
b). With the engine running the amount that temperature indication is allowed to decrease is also monitored. From a fully warm engine, temperature indication falling below 53°C. will flag a fault. This function is designed to catch intermittent high impedance faults in the sensor or wiring.
c). If the ECTSF signal voltage falls outside the range 0.1 to 4.9V for several consecutive readings a fault will be flagged. Open or short circuit harness/component faults are the most likely cause.

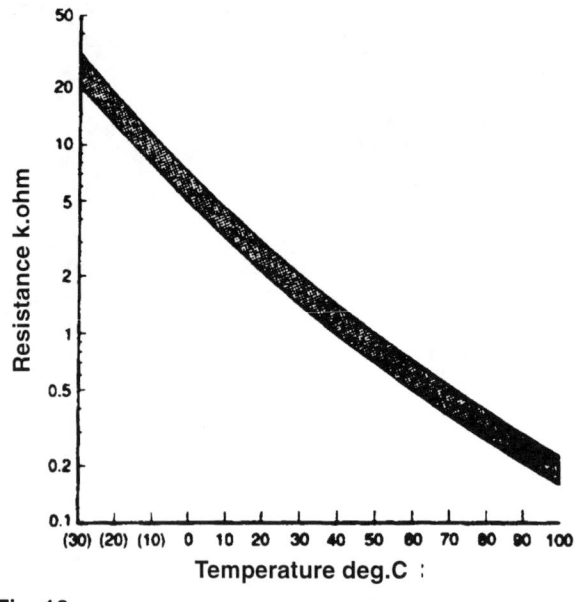

Fig. 18
Default:

When DTC FF14 is recognised, the warm up law defaults to a trim value of 26°C. Suitable start and after start values are used. The IATSF is taken as the temperature indication for starting. This enables the engine to be started over a wide temperature range.
JDS can be used to trace faults in this circuit.
Fuel Injector (FI), Description
The FIs (Fig. 19) are solenoid valves, using a pintle valve the tip of which is a waisted shape passing through the orifice at the bottom of the FI body.

Electrical pulses (pulse duration) applied to the FI coil by the PCMF output stage, overcome spring pressure and attract the head of the pintle to the coil armature. Fuel is allowed to flow through the annulus between the pintle tip and the orifice.

The waisted shape of the pintle tip and orifice cause the fuel expelled from the FI at 3.0 Bar pressure to form a cone shaped spray of small fuel particles. Valve lift is approximately 0.15 mm (0.006 in.) fully open, this is reached in about 1.0 millisecond.

Current drawn by the FI is controlled by the PCMF to reduce power consumption. A high current 'turn-on' signal is followed by a 'hold on' signal using the power resistor circuit to control the injector current.

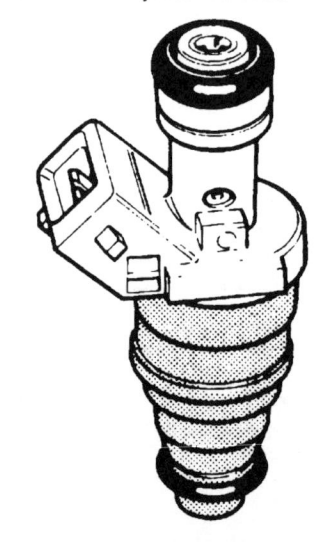

Fig. 19

The fuel inlet at the top of the FI, carries a small thimble filter to trap any debris that may be present in the fuel.

O' Rings are used to mount the FI top and bottom into the fuel rail and inlet manifold respectively. Care should be exercised whenever FIs are refitted that the 'O' rings are undamaged.

Plugged FI 'down on flow' is a fairly common problem. It is usually caused by running the vehicle on fuel that does not include the correct detergents to keep the FI nozzle clean. Deposits bake onto the tip during hot soak conditions and the build up of deposits gradually reduces the fuel flow.

It is corrected in terms of AFR (catalyst vehicles) by the feedback adding to the pulse duration. There is a limit to the range available (20%) and eventually driveability problems will result or OBD codes will be flagged.

There are four OBD codes associated with the injectors and drive circuit, they are:

DTC FF23 Fuel Supply.

The PCMF monitors the HO2S integrator feedback correction. If the maximum fuelling correction is being applied to both banks of the engine, a fuel change is applied. If the feedback then responds and leaves the integrator upper or lower limit on both engine banks the code will be set.

Default:

The integrator voltage will be clamped to mid point, i.e. main fuel map.

> **DTC FF34 Fuel Injectors (FI) 'A' Bank.**
> **DTC FF36 Fuel Injectors (FI) 'B' Bank.**

The PCMF monitors the HO2S integrator feedback correction. If the maximum fuelling correction is being applied to one bank of the engine, a fuel change is applied to that bank only. If the feedback then responds and leaves the integrator upper or lower limit the code for that bank will be set.

Default:

The integrator voltage for the faulty bank will be clamped to mid point, i.e. main fuel map.

> **DTC FF49 Ballast Resistor Circuit.**

The PCMF monitors the four FI output stages and will detect the absence of the ballast resistor in any one of the four circuits. When the fault is detected, the code will be set.

Default:

The PCMF will limit the FI pulse duration to a value that will avoid damaging the output stage. The limit on pulse duration will cause loss of vehicle performance at high load. But can be driven safely at lower speeds.

JDS can be used to trace faults in the injector and ballast resistor circuits.

FUEL PRESSURE REGULATOR, DESCRIPTION

The fuel pressure regulator is placed as near the fuel rail as possible to achieve good dynamic control of fuel pressure. The requirements being to have the same pressure across each FI, and deliver a similar quantity of fuel to each of the twelve cylinders.

The main component of the pressure regulator is the diaphragm, below which a steel valve is mounted on a spherical bearing. The upper side of the diaphragm has a spring retaining cup fitted to it. The spring fitted in the cup between the diaphragm and the upper body of the regulator is set during manufacture, by deforming the top cover.

The fuel pressure at which the diaphragm lifts at atmospheric pressure is therefore set (3,0 Bar). When the valve lifts off the seat fuel is spilled down the return line back to the fuel tank, maintaining the fuel pressure at the control value.

The tube on the top of the upper case is connected to manifold vacuum. When the engine is running the vacuum 'assists' the fuel under pressure to lift the diaphragm against the spring.

The fuel pressure across the FI nozzle will be maintained at 3.0 Bar.

The range of fuel pressure measured with a gauge will be 2.3 to 3.0 Bar, usually about 2.5 Bar at idle.

Testing the Fuel Pressure Regulator.

Use care and observe the normal safety precautions when breaking fuel line connections.

The fuel pressure can be measured between the fuel delivery pipe and the fuel rail.

Turn on the Ignition several times to prime the fuel rail. The fuel pressure should be close to 3.0 Bar. The pressure should be maintained when the fuel pump stops.

Only if the pressure drops quickly to zero (less than 15 to 20 minutes) should the regulator be changed.

Note: If pressure loss is observed, ensure that it is being lost through the regulator and not through FI or FP non-return valve.

Immediate and complete loss of fuel pressure during hot soak (engine off) conditions, can contribute to long cranking times when the engine is restarted.

THROTTLE POSITION SENSOR (TPS), DESCRIPTION

The TPS (Fig. 20) contains two separate potentiometer tracks with wipers driven by a common spindle connected to the throttle pulley drive shaft.

The output of one potentiometer is used by the Transmission Control Module (TCM) to measure throttle position, and the other potentiometer is used by the PCMF. The two potentiometers have similar resistance 'V' angle of rotation characteristics.

The PCMF uses the TPS signal for several purposes, they are:

Full Throttle.
  Full Load Enrichment
  HO2S feedback inhibited

Part Throttle.
  Main Fuel Map
  Charcoal Canister Purge enabled

Throttle Closed.
  Adaptive Idle Fuel Trim
  Overrun Fuel Cut-off

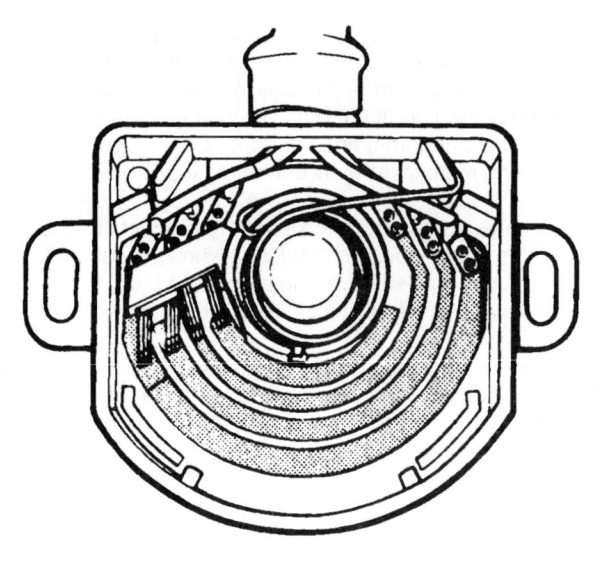

Fig. 20

Charcoal Canister Purge, which operates when lambda feedback is enabled and idle adaption is completed.

Opening Throttle.
Acceleration Enrichment
1. Ordinary FI pulse duration lengthened for a short period.
2. Extra FI pulse used with fast throttle opening.

Closing Throttle.
Deceleration Enleanment
Reduction of FI pulse duration for a short period as the throttle closes.

The TPS output relative to the throttle pulley is set with the throttles and pulley closed using JDS. The software function in the PCMF recognising closed throttle is adaptive and will recognise voltage in the range 0.22 to 0.75V as closed throttle. Small changes in voltage at closed throttle will therefore have no effect. Other functions are adjusted accordingly, relative to closed throttle.

Testing the TPS and Circuit.
The resistance of the TPS potentiometer tracks can be measured using a suitable high impedance Digital Multi-Meter (DMM).
PCMF track 4.0 to 6.0 Kohms.
TCM track 4.0 to 6.0 Kohms.
Setting the TPS position should be carried out using JDS. Resetting an existing unit will only be necessary if it fails the JDS check. Carried out as a background task during the engine set-up procedure.
There are three OBD codes associated with the TPS they are:

DTC FF17 Throttle Position Sensor (TPS) Circuit.

TPS voltage out of range. The PCMF monitors the wiper voltage. If the signal falls below 166 mV, i.e. below minimum recognition of closed throttle. The DTC will be set. Loss of the wiper signal or 5V supply will be recognised, as would a worn drive mechanism.

Note: Loss of the ground signal to pin 1 of the PCMF connector will not be recognised by this code. But the engine will not restart. The signal voltage on the wiper being greater than that required to trigger the 'clear flood' function which inhibits the FIs during cranking.

Default:

When DTC FF17 is flagged, the default is to a fixed part throttle voltage. None of the normal TPS functions will operate. i.e. no Overrun Fuel Cut-off, no Full Load Enrichment, no acceleration transient fuel enrichment o closed throttle re-adaption.

DTC FF18 Throttle Position Sensor (TPS) / MAP Calibration.

TPS and MAP out of normal range compatibility. The Manifold Absolute Pressure (MAP) is less than 150 Torr while the TPS voltage is greater than 3V. The DTC will be set.

Default:

When DTC FF18 is flagged, the TPS input within the PCMF is clamped to a part throttle value. There will be no Full Load Enrichment.

DTC FF19 Throttle Position Sensor (TPS) / Map Calibration.

TPS and MAP out of normal range compatibility. The Manifold Absolute Pressure (MAP) is greater than 550 Torr while the TPS voltage is less than 800 mV. The DTC will be set.

Default:

When DTC FF19 is flagged, there will be no closed throttle functions. The engine will run on the main fuel map only.

Note: Both FF18 & 19 are inhibited if the engine speed falls below 350 rpm, this prevents the faults being flagged if the engine stalls.

JDS can be used to trace faults in this circuit.

Chapter 6

FUEL SYSTEM (5.3 Litre 1992 Model year on)

SAFETY PRECAUTIONS

> **WARNINGS:**
>
> OPERATIONS INVOLVING THE FUEL SYSTEM RESULT IN FUEL AND FUEL VAPOUR BEING PRESENT IN THE ATMOSPHERE.
>
> FUEL VAPOUR IS EXTREMELY FLAMMABLE, THEREFORE GREAT CARE MUST BE TAKEN WHEN WORK IS CARRIED OUT ON THE FUEL SYSTEM.
>
> THE FOLLOWING PRECAUTIONS MUST BE STRICTLY ADHERED TO:
>
> 1. SMOKING MUST NOT BE ALLOWED NEAR THE AREA.
> 2. 'NO SMOKING' WARNING SIGNS MUST BE POSTED ROUND THE AREA.
> 3. A CO_2 FIRE EXTINGUISHER MUST BE CLOSE AT HAND.
> 4. DRY SAND MUST BE AVAILABLE TO SOAK UP ANY SPILLAGE.
> 5. EMPTY THE FUEL FROM THE TANK INTO AN AUTHORISED EXPLOSION PROOF CONTAINER USING SUITABLE FIREPROOF EQUIPMENT.
> 6. FUEL MUST NOT BE EMPTIED INTO A PIT.
> 7. THE WORKING AREA MUST BE WELL VENTILATED.
> 8. THE BATTERY MUST BE DISCONNECTED BEFORE CARRYING OUT WORK ON THE FUEL SYSTEM.
> 9. THE DISCARDED TANK MUST NOT BE DISPOSED OF UNTIL IT IS RENDERED SAFE FROM EXPLOSION.
> 10. MAINTENANCE PERSONNEL SHOULD HAVE UNDERTAKEN SPECIALIST TRAINING BEFORE BEING ALLOWED TO REPAIR COMPONENTS ASSOCIATED WITH THE FUEL SYSTEM.
> 11. DEPRESSURISE THE FUEL SYSTEM PRIOR TO DISCONNECTING ANY FUEL PIPE CONNECTIONS OR REMOVING COMPONENTS.

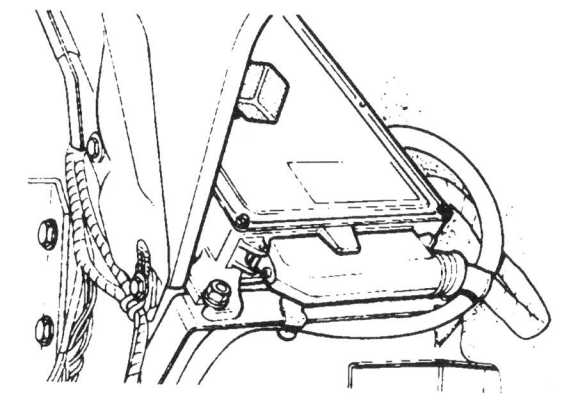

Fig. 1

DEPRESSURIZING THE FUEL SYSTEM

Depressurise the fuel system prior to disconnecting any fuel pipe connections or removing components as follows; Open the luggage compartment and remove the right-hand trim liner.

Locate the relays mounted at the rear of the right-hand 'hockey stick' panel adjacent to the fuelling ECU.

Displace relays from their mounting and identify the fuel pump relay (silver relay on a black/yellow base). (Fig. 1). Disconnect the relay from the base.

Crank the engine to depressurise the system.
Reposition the relay and trim liner, close the luggage compartment.

FUEL SYSTEM, DESCRIPTION

The fuel system used is of the recirculatory type. The fuel tank is mounted across the car behind the rear passenger seats.

Fuel is drawn from the fuel tank into the in-tank fuel module by the fuel pump, with the fuel passing through a venturi on the fuel pump pressure side.

A 70 micron filter is incorporated at the inlet port of the module to prevent ingress of particle contaminants. A further 400 micron rock filter is included at the inlet port of the fuel pump to prevent the passage of smaller participle contaminants into the fuel pump and the rest of the fuel system.

Fuel is pumped from the base of the fuel tank through a flexible hose to an in-line filter. From the filter, fuel flows through under-floor pipework to the front of the vehicle where a flexible hose connects the under-floor feed pipe to the inlet port of the fuel rail.

The fuel rail is mounted on the engine and has a fuel pressure regulator attached at the return port. The fuel pressure regulator is used to control fuel line pressure and maintains a constant delivery of fuel to the fuel injectors across the fuel rail. This is accomplished by applying inlet manifold pressure to a diaphragm within the pressure regulator.

Pressure is kept constant within the range 241.30 - 310.25 kN/m^2 (35 - 45 lbf/in^2).

Unused fuel from the engine is returned to the in-tank module through return hoses and under-floor piping. A 70 micron filter is used in the return inlet of the module to capture debris that may be present in the fuel rail or return line at initial start-up and so prevent it from entering the system.

Both the feed and return ports of the module use non-return valves for safety.

Electronic Fuel Injection

A digital type electronic control unit (ECU) with an integral manifold pressure sensor, governs the amount of fuel injected. The manifold pressure and speed signal derived from ignition pulses, provide the main control for the fuel injected. Additional sensors are used to monitor engine temperature, inlet air temperature and throttle position, thereby ensuring optimum fuelling is computed for all engine operating conditions.

Fuel Control System

The metering of fuel is controlled by regulating the time that the injectors are open during each engine cycle. The frequency of the injectors is dependent on engine speed and conditions. The basic pulse length is mapped against speed and intake manifold pressure which is sensed by a transducer located in the electronic control unit and linked to the intake manifold by a pipe. Information on engine speed is derived from the ignition trigger pulses in the ignition system.

Each cylinder bank has six injectors which are electrically connected in parallel. The injectors of both cylinder banks are fired six times per engine cycle, this operation being triggered from the output of the ignition ECU.

The injectors are energized for a time proportional to the figure given on the base map plus a constant of proportionality which is varied according to secondary control parameters, i.e. engine coolant temperature, inlet air temperature, throttle movement and position, and battery voltage.

The fuel pump is energised only when the ECU senses an engine cranking signal or an engine running signal. This prevents the engine being flooded with fuel in the event of an injector sticking open.

Cranking Fuelling

Cranking fuelling provides six injections per engine cycle instead of the normal two. This reduces after a set number of injections.

Coolant Temperature Enrichment

Temperature enrichment is provided during starting and warm-up. This is achieved by increasing the injector 'on' time above that of basic requirements and is implemented by the ECU in response to an input from the coolant temperature sensor. The enrichment is reduced with increasing engine speed and load.

After-start Enrichment

After-start enrichment is provided to supply added fuel during warm-up. The enrichment is coolant temperature dependant (the colder the temperature, the more fuel is supplied). This is achieved by the ECU which increases the injector 'on' time above that of basic requirements and then decreases the amount of additional fuel supplied at a fixed rate over a number of engine revolutions.

Temperature Sensors

The temperature of the air taken into the engine through the inlet manifold and the temperature of the coolant in the cylinder block are constantly monitored. The information is fed directly to the ECU.

The air temperature sensor has a small effect on the injector pulse width, and should be regarded as a trimming rather than a control device. It ensures that the fuel supplied is directly related to the weight of air drawn in by the engine. Therefore, as the weight (density) of the air charge increases with a falling temperature, so the amount of fuel supplied is also increased to maintain optimum fuel/air ratio. The coolant temperature sensor has a greater control although it functions mainly while the engine is initially warming-up.

Full Load Fuelling

To obtain maximum engine power it is necessary to inhibit the 'closed loop' system and simultaneously increase the fuelling level. This is determined by throttle position and engine speed.

Flooding Protection System

When the ignition is switched on, but the engine is not cranking, the fuel pump will run for two seconds to raise the pressure in the fuel rail; it is then automatically switched off by the ECU. Only after cranking has started is the fuel pump switched on again. Switching control is built into the ECU circuitry. This system prevents flooding if any injectors become faulty (remain open) when the ignition is left on.

Engine Load Sensing

The driver controls engine power output by varying the throttle opening and therefore the flow of air into the engine. The airflow determines the pressure that exists within the polonium chamber, this pressure therefore is a measure of the demand upon the engine. The pressure is also used to provide the principle control of fuel quantity, being converted by the manifold pressure sensor in the ECU into an electrical signal. This signal varies the width of the injector operating pulse as appropriate. The pressure sensor is fitted with a separate diaphragm system that compensates for ambient barometric variations. The manifold pressure sensor is located in the ECU and is connected by a pipe to the inlet manifold balance pipe.

Vacuum Switch

The contacts of this switch are operated by a spring loaded diaphragm in a chamber. This senses inlet manifold vacuum such that, when the depression falls below a certain value, e.g. when the engine is operating at low speed, with part throttle near full load condition then the contacts close.

The closing of the contacts causes the fuel system to go 'open loop' and simultaneously introduce a fuel enrichment sufficient to achieve the maximum power available for that throttle opening.

Fuel Pump / In-tank Module

The fuel pump is fitted inside the in-tank module (1 Fig. 2), which is fixed to the centre base of the fuel tank, via a rubber cradle (2 Fig. 2) and steel bracket (6 Fig. 2). The rubber cradle behaves as an acoustic baffle to deaden pump vibration and noise. There are three large tabs (3 Fig. 2) and one small tab on the module casing, which locate into slots in the centre recess of the rubber mounting (4 Fig. 2), this ensures that the pump can only be mounted in the correct orientation. The periphery of the rubber mounting incorporates four protrusions each having a locating slot (5 Fig. 2), three slots are large and one is small, which locate onto tabs on the steel support bracket (7 Fig. 2).

It is important to ensure correct orientation and fitting of fuel pump module with rubber mounting and steel support bracket.

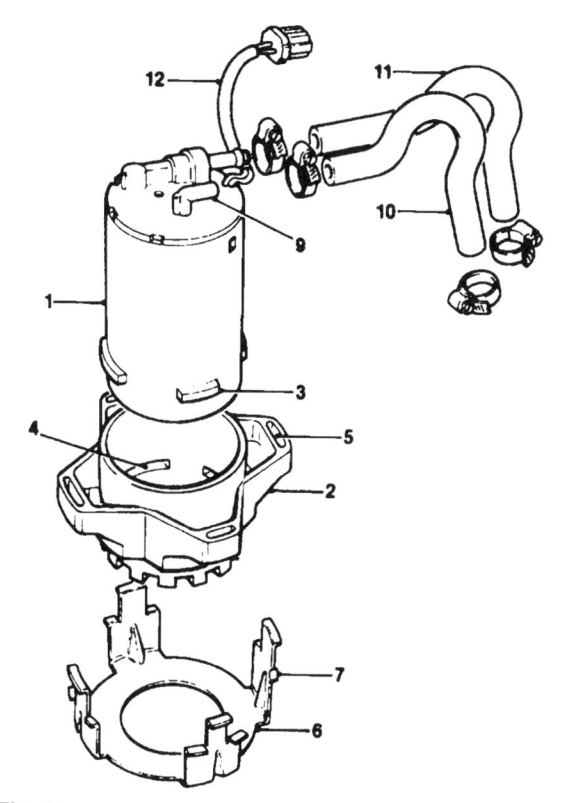

Fig. 2

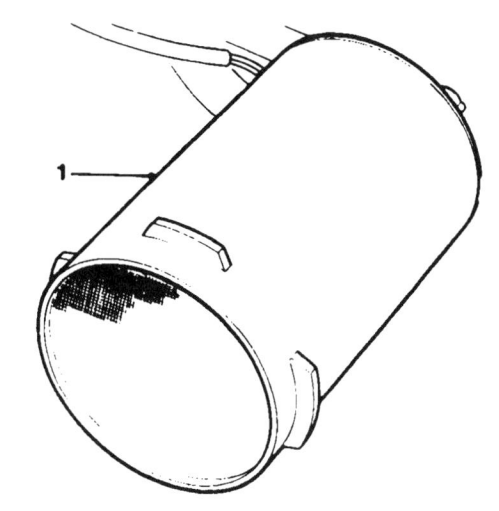

Fig. 3

The fuel pump draws fuel into the module from the fuel tank through a filter in the base of the module (1 Fig. 3).

Fuel is pumped out of the module through a port at the top of the module (9 Fig. 2). There are two rubber moulded hoses connected to the top of the module, the short hose (10 Fig. 2) connects from the module's outlet port to the under-floor fuel feed pipe at the base of the fuel tank. The long hose (11 Fig. 2) connects the fuel return pipe to the module's return port.

Connections between the hoses and solid pipes are underneath the car. The only direct connection between pump module and the tank is at the evaporative loss flange via a two wire electrical connection (12 Fig. 2).

The pump delivers approximately 120 Litres/min. @ 3 Bar.

The complete fuel pump module is replaced when renewing the fuel pump.

Fuel Filter

The main in-line fuel filter is mounted under-floor on the left-hand side of the vehicle forward of the axle carrier. The unit is made from stainless steel.

> **CAUTION:**
>
> Directions for fuel flow are shown on the filter, it is important that the filter is connected the correct way round.

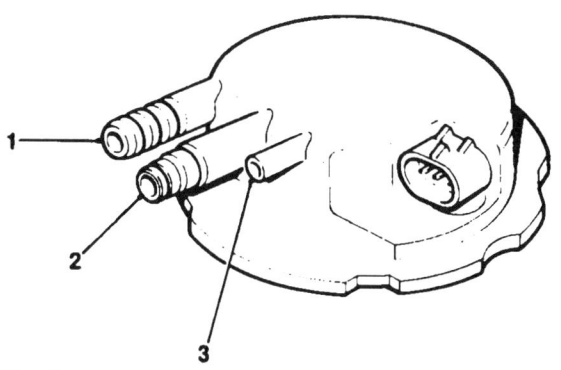

Fig. 4

Evaporative Loss Flange

The flange is mounted to the tank by a seal and locking ring, it has three external outlet ports and one electrical connection.

The large port (1 Fig. 4) allows the tank to vent vapour during refuelling and is directly connected to the fuel filler neck.

The intermediate port (2 Fig. 4) has a 'Legris' quick-fit connector moulding and allows for over pressurization; it is connected to atmosphere by a nylon pipe the outlet of which is underneath the vehicle.

The small outlet port (3 Fig. 4) is used for vehicles incorporating evaporative loss control and uses a nylon pipe to connect the tank to the charcoal canister located at the front of the vehicle. For cars not fitted with evaporative loss control, the pipe is vented directly to atmosphere via a hose passing underneath the car. In both evaporative loss control vehicles and non-evaporative loss control vehicles, a Rochester valve is included in the line, controlled by manifold pressure.

Connectors and Hoses

Legris quick-fit connectors are used throughout the fuel system (Fig. 5). At connections between hoses and the under-floor pipes, the pipes have a ridge or paint line to indicate how far the connector should be pushed onto the pipe end.

There are two sizes of connector, 8 mm and 10 mm, this is to ensure that the feed pipe is not connected as the return pipe and vice versa. The 10 mm connector is used for the feed line.

To release connectors, push and hold the locking ring (1 Fig. 5) towards the connector fitting while moving the connector and pipe end apart.

Always ensure pipe connectors are pushed fully home when refitting by having the pipe shoulder (2 Fig. 5) abut the locking ring.

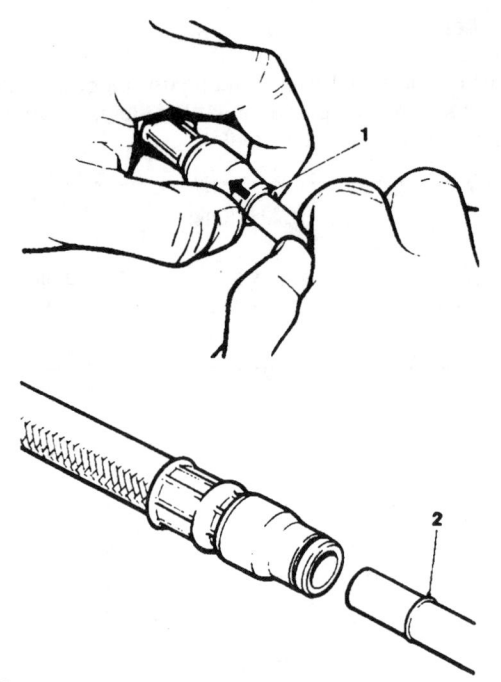

Fig. 5

> **CAUTION:**
>
> Pipes and component stubs must be fully inserted into connector ends to ensure proper fit; failure to do so could cause fuel leakage.

In-tank fuel hoses

Service tool JD 175 is available for the tightening of jubilee clips used for attaching feed and return hoses to the in-tank module.

Throttle Position Switch

This micro-switch (1 Fig. 6) is mounted such that its contacts are closed when the throttle is opened beyond a certain position; this condition has the same effect as the closing of the contacts of the full load vacuum switch. The throttle position switch is required to provide full load fuelling when the vehicle is operated under high speed, full load running conditions and the presence of intake manifold depression prevents the operation of the vacuum switch.

Throttle Potentiometer

To ensure the vehicle road performance is satisfactory, with good throttle response, acceleration enrichment is necessary. Signals are provided by the throttle potentiometer (1 Fig. 7), which is mounted on the throttle spindle and

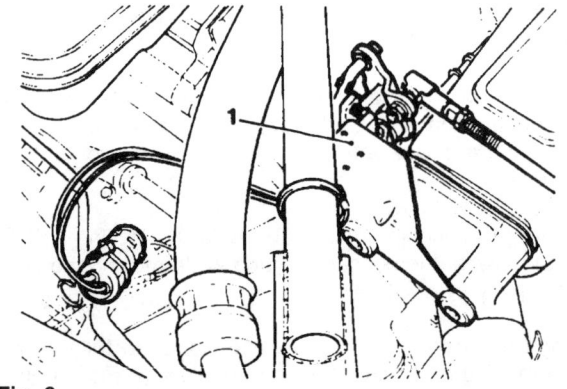

Fig. 6

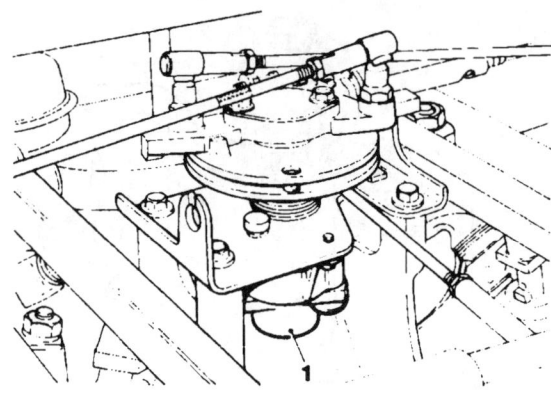

Fig. 7

indicates the throttle position to the ECU. When the throttle is opened the fuelling is enriched and when closed the fuelling is weakened. When the throttle is opened very quickly, all the injectors are simultaneously energized for one pulse.

This ensures that there is enough fuel available at the inlet ports for the air admitted by the sudden opening of the throttle. The duration of this extra pulse is controlled by the engine temperature signal and is longer with a cold engine. Lengthening the normal injection pulses is done in proportion to the rate at which the throttle is opened and it takes a short time to decay when the throttle movement stops. Enrichment in this way is also varied according to the engine temperature. The fuel cut off function is controlled by the throttle potentiometer and the conditions under which it occurs are programmed into the ECU memory.

Auxiliary Air Valve

In order to maintain idling speed during cold start and warm up, more air than that supplied by the normal idle bypass is required.

This extra air is supplied by the auxiliary air valve (1 Fig. 8) which is temperature sensitive so that the idle speed can be controlled throughout the warm-up phase. The auxiliary air valve consists of a variable orifice, controlled by an expansion element. It is mounted on the engine cylinder head where it is responsive to coolant temperatures. By adjusting the profile of the variable orifice (2 Fig. 8) according to coolant temperature the engine idle speed can be controlled.

Injectors

Each fuel injector (Fig. 9) consists of a solenoid operated needle valve with the movable plunger rigidly attached to the nozzle needle. In the closed position, a helical compression spring holds the needle against the valve seat.

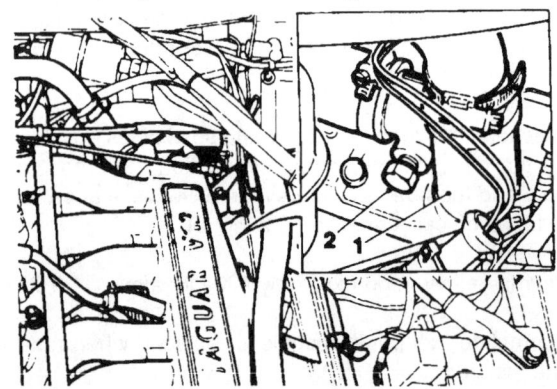

Fig. 8

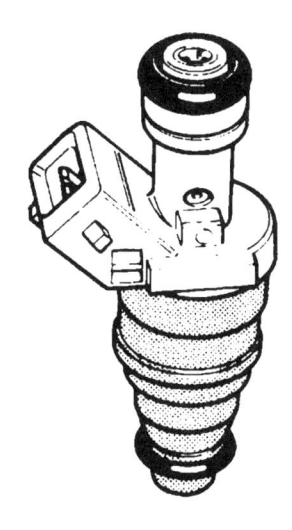

Fig. 9

The injectors have a solenoid winding mounted in the rear section of the valve body, with a guide for the nozzle needle in the front section.

The injectors are operated in two stages, initially they are operated via a pull-in circuit, then when the injectors are open a change to a hold on circuit is made via current limiting resistors for the remainder of the injection period as determined by the ECU. In this way the heating effect on the output transistors of the ECU is reduced. It also ensures a rapid response from the injectors.

To open the injectors at the speeds required by the engine a fairly high current is needed. The ECU has an output stage to deliver this current but to protect the output transistors of the ECU from injector faults a power resistor is wired in series with each three injectors. These resistors will limit current to a safe value, thus protecting the ECU. The power resistors (one for each group of injectors) are housed in a single unit to the right side of the engine valence by two screws.

The injectors are operated in two groups of six, which is further broken down into two sub-groups of three, although each pair of sub-groups is operated simultaneously to make up the two groups. The injectors of both cylinder banks are fired six times per engine cycle, this operation being triggered from the output of the ignition ECU.

A Bank	B Bank
1A - 3A - 5A	1B - 3B - 5B
2A - 4A - 6A	2B - 4B - 6B

The metering of fuel is controlled by regulating the time that the injectors are open during each engine cycle. The frequency of the injectors is dependent on engine speed and conditions. The basic pulse length is mapped against speed and intake manifold pressure which is sensed by a transducer located in the electronic control unit and linked to the intake manifold by a pipe. Information on engine speed is derived from the ignition trigger pulses in the ignition system.

The injectors are energized for a time proportional to the figure given on the base map plus a constant of proportionality which is varied according to secondary control parameters, i.e. engine coolant temperature, inlet air temperature, throttle movement and position and battery voltage.

Fuel Pressure Regulator

The fuel pressure regulator (1 Fig. 10) is attached to the main fuel rail and consists of a metal housing containing a spring loaded diaphragm. When the pressure setting of the regulator is exceeded, the diaphragm moves, exposing an opening to an overflow duct which allows excess fuel to return to the fuel tank, causing a drop in fuel pressure. The reduced fuel pressure allows the diaphragm to move back to its original position, thereby closing the fuel return outlet. This sequence of events is repeated as long as the pump is running. In this way, fuel pressure is held constant as fuel demand varies. The pressure setting is adjusted to the correct value during production when the outer spring housing is compressed until the correct spring load is obtained. This is not adjustable in service.

The spring housing of the regulator is sealed and connected to the engine inlet manifold by a small bore pipe. By allowing it to sense inlet manifold depression, the pressure drop across the injector nozzle remains constant because the fuel pressure will alter as manifold depression alters. This arrangement ensures that the amount of fuel injected is only dependent on the duration of injector open time.

System fuel pressure is 2.5 bar (31.2 lb/in^2) above manifold pressure.

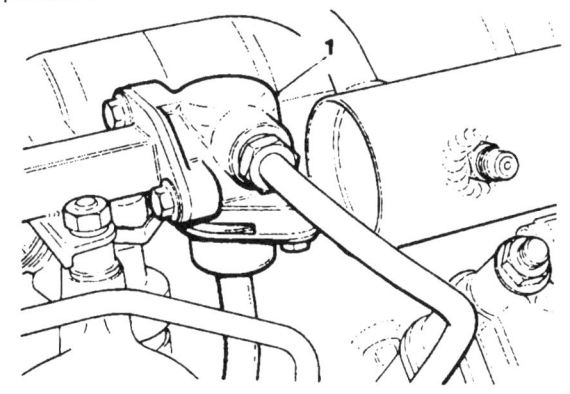

Fig. 10

Fuel Rail

The fuel rail (1 Fig. 11) is mounted and secured to the inlet manifold castings. The 12 injectors (2 Fig. 11) are directly fitted to the rail via 'O' ring seals and secured by retaining clips.

The rail is configured into two separate rails joined by two hoses (3 Fig. 11). Fuel is fed into the left-hand rail, flow pressure across all injectors is controlled by the pressure regulator valve (4 Fig. 11) mounted at the front of the right-hand rail. An electrical thermal switch (5 Fig. 11) is mounted to the underside of the right-hand rail.

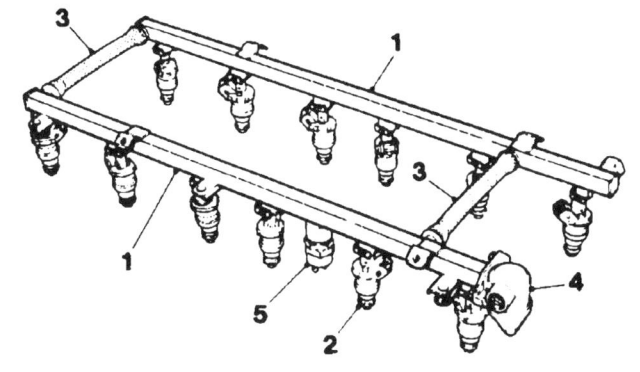

Fig. 11

37

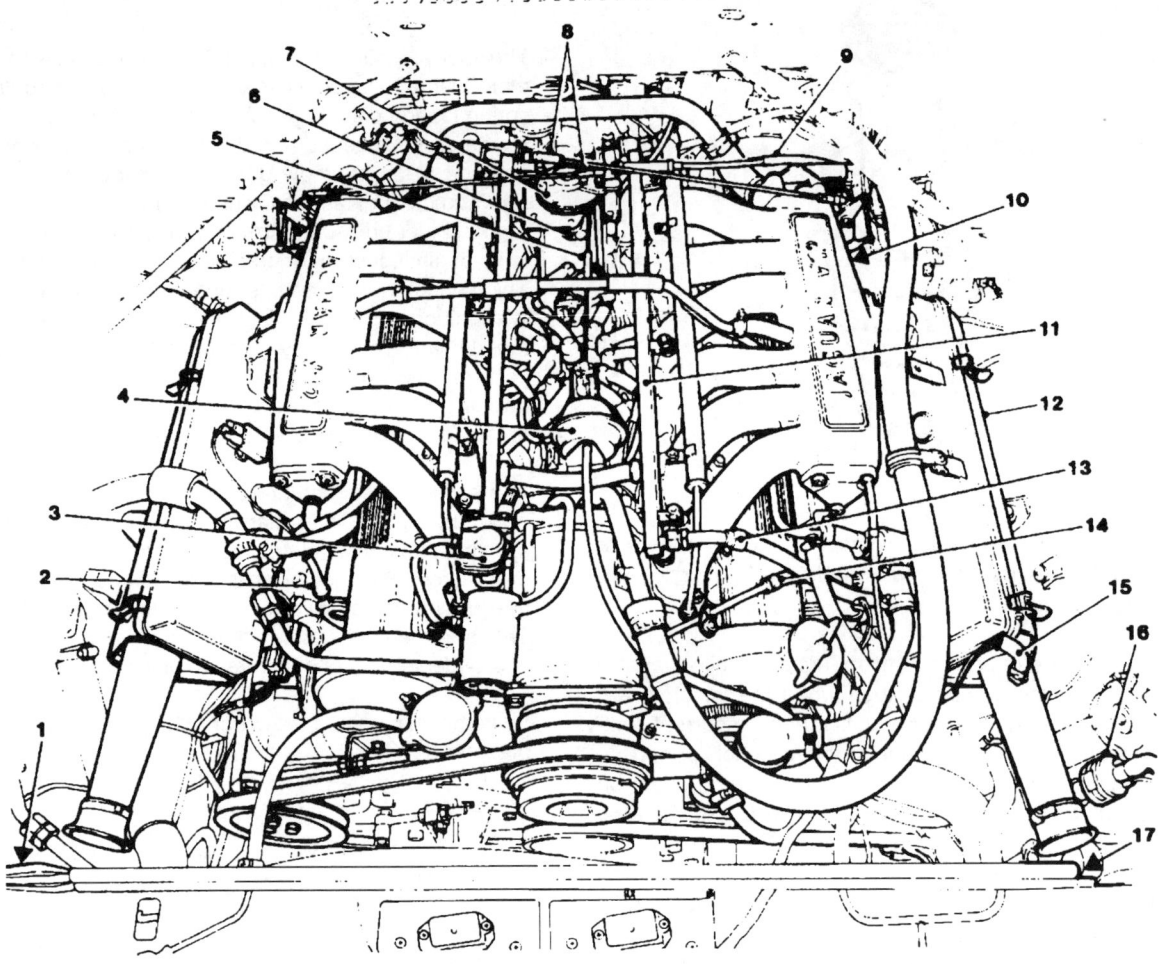

Fig. 12   Engine Compartment Component Location

(Key to Fig. 12)

1. Power resistor
2. Thermotime switch
3. Pressure regulator
4. Throttle actuator
5. Actuator rod
6. Throttle potentiometer
7. Throttle pedestal assembly
8. Throttle linkage rods
9. Throttle cable
10. Auxiliary air valve
11. Fuel rail assembly
12. Air cleaner assembly
13. Fuel feed pipe
14. Fuel return pipe
15. Air temperature sensor
16. Vacuum dump valve
17. Vacuum pump

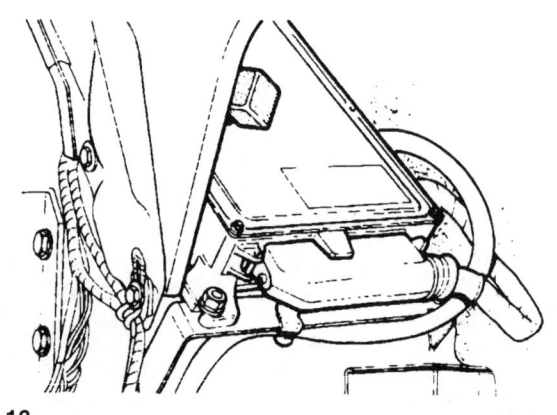

WARNING:

THIS OPERATION RESULTS IN FUEL AND FUEL VAPOUR BEING PRESENT WHICH IS EXTREMELY FLAMMABLE, GREAT CARE MUST BE TAKEN WHEN WORK IS BEING CARRIED OUT ON THE FUEL SYSTEM AND THE FOLLOWING PRECAUTIONS MUST BE STRICTLY ADHERED TO:

SMOKING MUST NOT BE ALLOWED NEAR THE AREA. NO SMOKING WARNING SIGNS MUST BE POSTED ROUND THE AREA.

A CO^2 FIRE EXTINGUISHER MUST BE CLOSE AT HAND.

DRY SAND MUST BE AVAILABLE TO SOAK UP ANY SPILLAGE.

ENSURE THE FUEL IS EMPTIED FROM THE TANK USING SUITABLE FIREPROOF EQUIPMENT. FUEL MUST NOT BE EMPTIED INTO A PIT, ONLY INTO AN AUTHORISED EXPLOSION PROOF CONTAINER.

THE WORKING AREA MUST BE WELL VENTILATED.

THE DISCARDED TANK MUST NOT BE DISPOSED OF UNTIL IT IS RENDERED SAFE FROM EXPLOSION.

Fig. 13

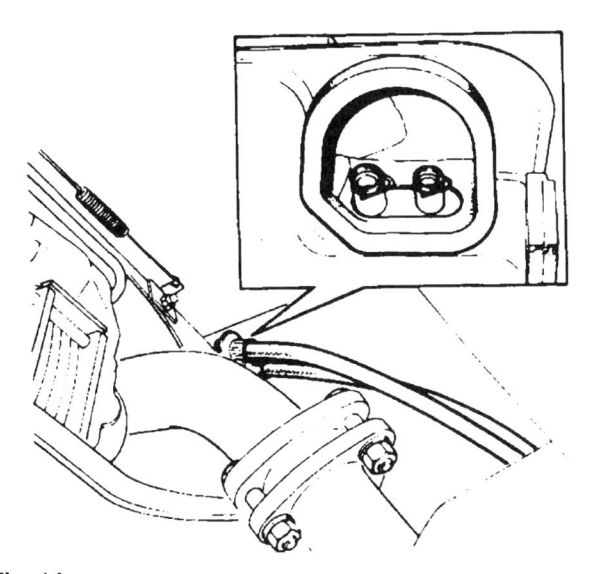

Fig. 14

FUEL TANK, RENEW

Open the luggage compartment and displace the right-hand front side liner.

Displace the fuel pump relay (silver relay on black/yellow base) from mounting at rear of hockey stick panel and remove the relay (Fig. 13).
Crank the engine to depressurise the fuel system.
Disconnect and remove the battery.
Remove the spare wheel.
Empty the fuel tank into a explosion proof container using suitable fireproof equipment.
Raise the vehicle on a ramp.
Remove the clips securing the fuel supply and return hoses (Fig. 14).
Release clip and disconnect breather hose (1 Fig. 15) from evaporative loss flange.

> **WARNING:**
>
> OPERATIONS WITHIN FUEL TANK ASSEMBLY MUST ONLY BE PERFORMED WITH USE OF SCREWDRIVER JD 175.

Using tool JD 174, displace and remove the evaporative loss flange retaining ring (2 Fig. 2).

Reposition the flange (3 Fig. 15) for access to the fuel pump harness multi-plug, disconnect the multiplug.

Using screwdriver JD 175 release the internal hose (4 Fig. 15) securing clips and disconnect hoses from the pump assembly.

Displace and remove the pump assembly (5 Fig. 15).
Displace and remove the pump rubber mounting (6 Fig. 15).

Using tool 18G 1001 displace and remove the fuel gauge unit securing ring. Remove the gauge unit and seal (7 Fig. 2).

Fitting a replacement tank is the reversal of the removal procedure.
Refit the components from the original tank.
Fitting of pump assembly and evaporative loss flange are covered elsewhere in this Section.
Ensure new 'O' rings and seals are fitted.

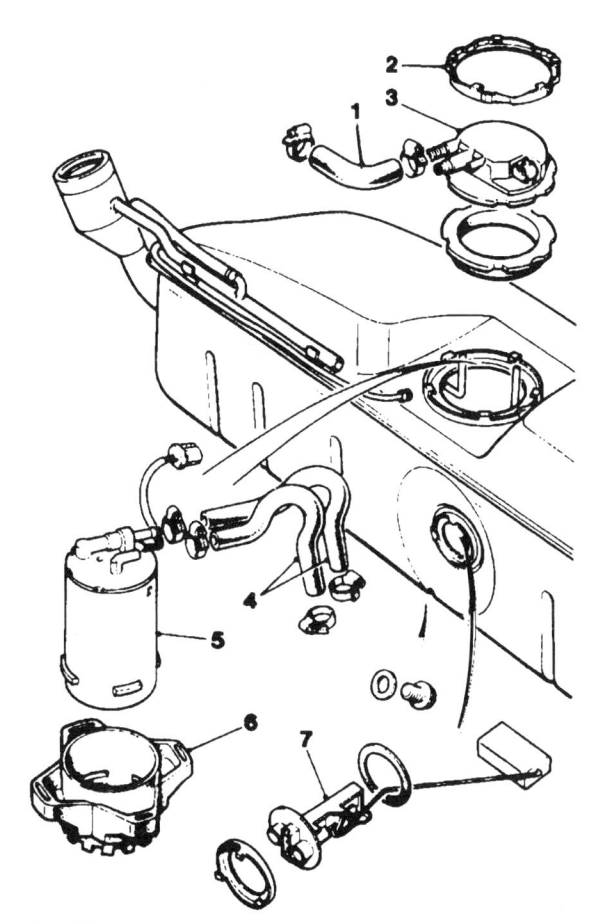

Fig. 15

Ensure all electrical connections are clean and secure.
Check all flange, hose and pipe connections for security and signs of leakage.

Service Tools

Evaporative Loss Flange Lock Ring Wrench JD 174
Hose Clip to Fuel Pump Screwdriver JD 175
Tank Sender Unit Wrench 18G 1001

Torque Figures

Fuel sender unit locking ring	5.5 - 13 Nm
Fuel filler neck hose clips	3.5 - 4.5 Nm
Tank breather hose clips	1.5 - 2 Nm
Fuel filler neck retaining screws	3.5 - 4.5 Nm
Internal hose securing clips	1.5 - 2 Nm

FUEL PUMP, RENEW

Remove the fuel tank assembly.

> **WARNING:**
>
> OPERATIONS WITHIN FUEL TANK ASSEMBLY MUST ONLY BE PERFORMED WITH USE OF SCREWDRIVER JD 175.

Release clip and disconnect breather hose from the evaporative loss flange.
Using tool JD 174, displace and remove the evaporative loss flange retaining ring.
Reposition the flange for access to the fuel pump harness multi-plug, disconnect the multi-plug. Using screwdriver JD 175 release the internal hose securing clips (1 Fig. 16) and disconnect hoses from the pump assembly.

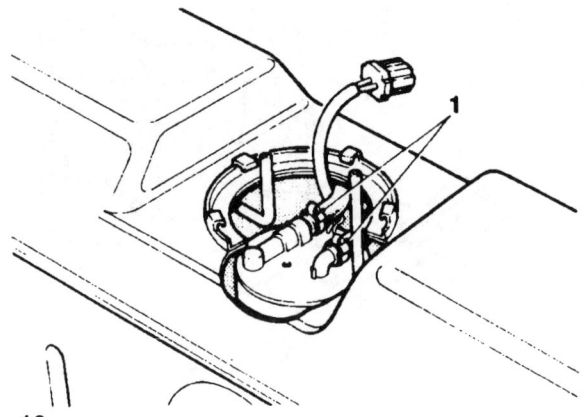

Fig. 16

Release tie wrap from around the mounting rubber and using a twisting action displace and remove the pump assembly.

Fit and fully seat the replacement fuel pump assembly to the mounting, engaging the master lug. Secure with new tie wrap.
Reposition hoses to the pump and tighten securing clips.
Fit and seat new evaporative flange seal.
Connect evaporative loss flange to pump multiplug.
Seat evaporative loss flange to seal.
Using tool JD 174 secure flange with retaining ring.
Connect and secure breather hose to flange stub pipe.

Reposition the tank assembly to the vehicle.
Refit the tank by reversing the removal operations.
Ensure new 'O' rings and seals are fitted.
Ensure all electrical connections are clean and secure.
Check all flange, hose and pipe connections for security and signs of leakage.

Service Tools

Evaporative Loss Flange Lock Ring Wrench JD 174
Hose Clip to Fuel Pump Screwdriver JD 175

Torque Figures

Tank strap securing nut	40 - 50 Nm
Fuel sender unit locking ring	5.5 - 13 Nm
Internal hose securing clips	1.5 - 2 Nm
Fuel filler neck hose clips	3.5 - 4.5 Nm
Tank breather hose clips	1.5 - 2 Nm
Fuel filler neck retaining bolts	3.5 - 4.5 Nm

MAIN FUEL FILTER, RENEW

Open the luggage compartment and displace the right-hand front side liner.
Displace the fuel pump relay (silver relay on black/yellow base) from mounting at rear of hockey stick panel and remove the relay (Fig. 17).

Crank the engine to depressurise the fuel system.
Reposition the relay and disconnect the battery earth lead.
Empty the fuel tank into an explosive proof container using suitable fireproof equipment.
Raise the vehicle on a ramp.
Filter assembly is located under-body, LHS, forward of the axle carrier.
Undo and remove the filter to fuel tank feed pipe union (1 Fig. 18), remove and discard 'O' ring seal.

Fit blanking plugs to feed pipe and filter.
Undo and remove the fuel feed pipe to filter union (2 Fig. 18), remove and discard 'O' ring seal.

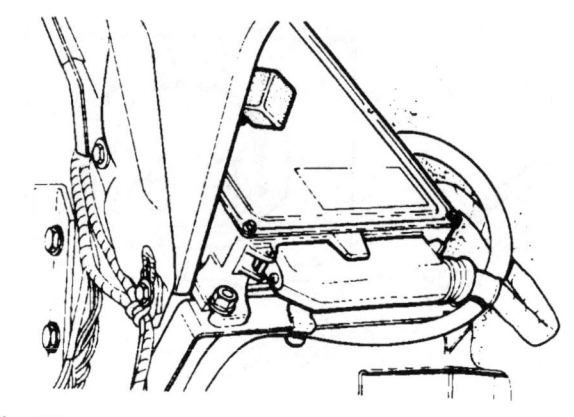

Fig. 17

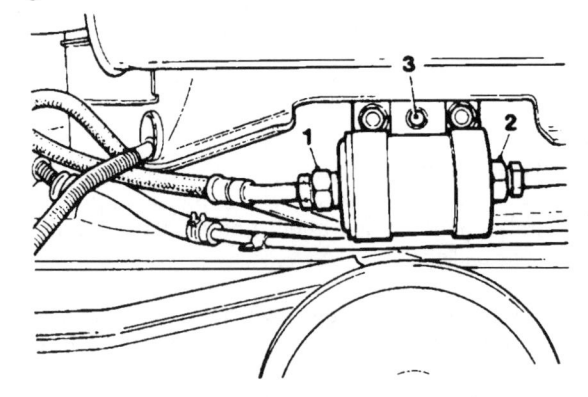

Fig. 18

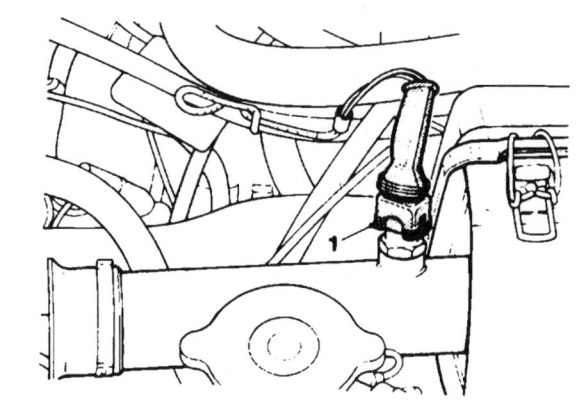

Fig. 19

Fit blanking plugs to feed pipe and filter.
Remove the filter mounting clamp bolt (3 Fig. 18) and remove the assembly.
Reverse the above operations to fit the replacement filter assembly.
Use new 'O' ring seals and check pipe connections for signs of leakage.
Lower the vehicle, reposition the trim panel and reconnect the battery.
Refill the fuel tank.

AIR CLEANER ASSEMBLY, RENEW

Disconnect the battery earth lead.
Disconnect the air temperature sensor lead from the sensor (1 Fig. 19) (left-hand only).

Unclip the air cleaner element cover (1 Fig. 20) and displace the cover from the element (2 Fig. 20).

Remove the element and the cover.
Disconnect the crankcase breather pipe and auxiliary air valve pipes (left-hand cleaner only) from the air cleaner.

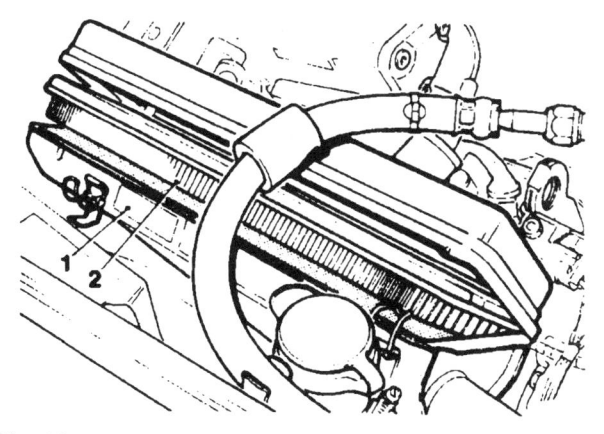

Fig. 20

Release the P clip securing the air conditioning hose to the air cleaner (left-hand only).
Remove the four bolts securing the air cleaner to the throttle butterfly housing and remove the air cleaner.
Remove the air temperature sensor.
On refitting, ensure all mating surfaces are clean and smeared with arborsil sealant and a new gasket is used.

AIR CLEANER ELEMENT, RENEW

Disconnect the air temperature sensor lead from the sensor (1 Fig. 19) (left-hand only).
Unclip the air cleaner element cover (1 Fig. 20) and displace the cover from the element (2 Fig. 20).
Remove the element and the cover.
On fitting new element ensure that it is correctly orientated, with the metal plate opposite the throttle housing.

ELECTRONIC CONTROL UNIT (ECU), RENEW

Disconnect battery earth lead.
Remove the luggage compartment right-hand trim panel.
Release the harness connector block (1 Fig. 21) and withdraw from the ECU.

Displace the vacuum hose (2 Fig. 21) from the edge clip.
Remove the two lower securing nuts and washers (3 Fig. 21).
Displace and remove the ECU from its mounted position.
Disconnect the vacuum hose from the side of the unit.
Refitting is the reversal of the above operations.

AIR TEMPERATURE SENSOR, RENEW

Disconnect battery earth lead.
Pull electrical connector from air temperature sensor (1 Fig. 22).

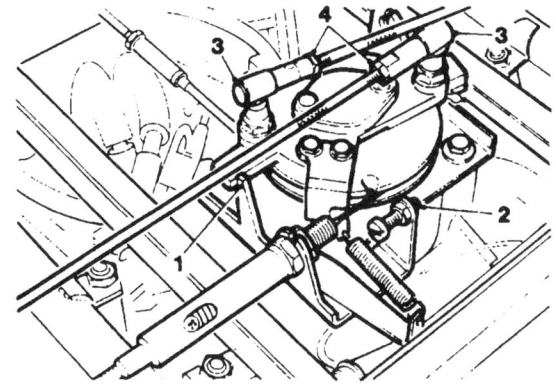

Fig. 21

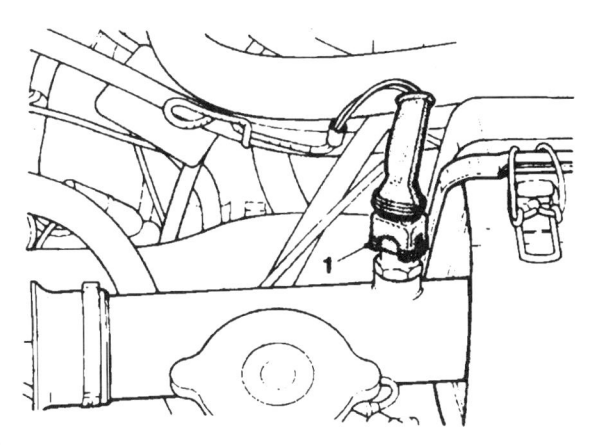

Fig. 22

Remove sensor from air cleaner ram pipe.
Refitting is the reversal of the above operations.

AIR TEMPERATURE SENSOR, TEST

Disconnect battery earth lead.
Pull electrical connector from air temperature sensor.
Connect suitable ohmmeter between terminals, note resistance reading.
Disconnect ohmmeter.

Check resistance between each terminal in turn and body of sensor. A very high resistance reading (open circuit) must be obtained. Refit electrical connector to temperature sensor. Reconnect battery earth lead.

Coolant Temperature	Delay 15°C Switch	Delay 35°C Switch
-20°C	8 secs	8 secs
-10°C	5.7 secs	6.5 secs
0°C	3.5 secs	5 secs
+10°C	1.2 secs	3.5 secs
+15°C	0 secs	2.7 secs
+20°C		2.7 secs
+30°C		0.5 secs
+35°C		0 secs

Ambient Air Temperature (°C)	Resistance (Ohms)
-10	960
0	640
+10	435
+20	300
+30	210
+40	150
+50	180

THERMOTIME SWITCH, RENEW

This procedure **MUST ONLY** be carried out on a cold or cool engine

Disconnect battery earth lead.
Pull electrical connector from thermotime switch.
Carefully remove and refit pressure cap from remote header tank to release any cooling system residual pressure.
Remove thermotime switch.

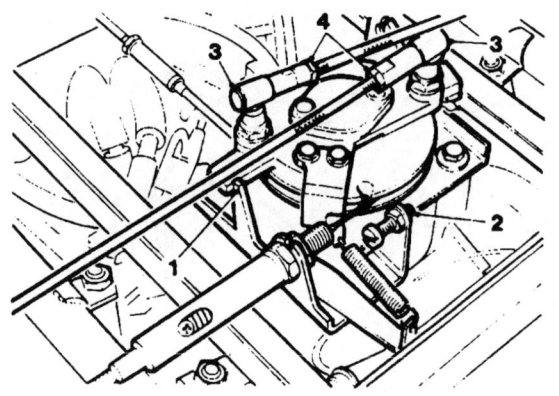

Fig. 23

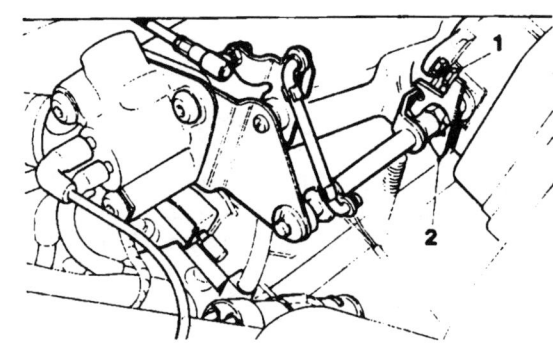

Fig. 24

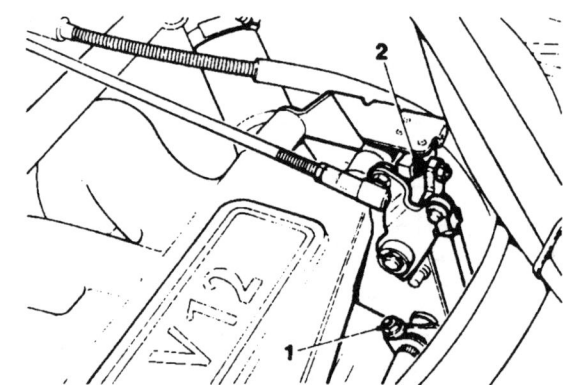

Fig. 25

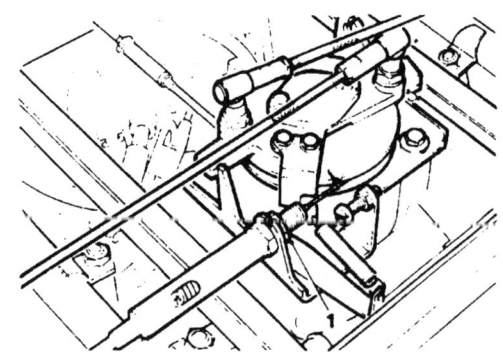

Fig. 26

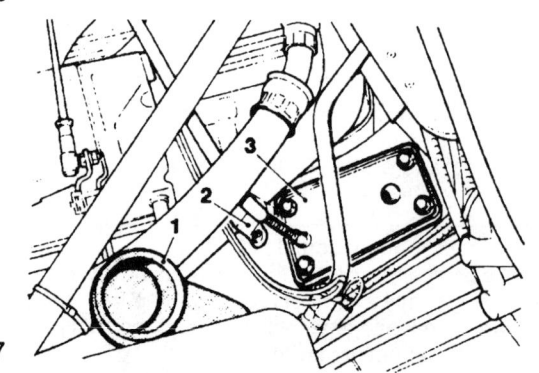

Fig. 27

Ensure that a new sealing washer is located on the replacement unit and the threads are coated with suitable sealing compound.
Locate and tighten the replacement thermotime switch.
Check coolant level at remote header tank, top up if necessary.

THROTTLE LINKAGE, CHECK AND ADJUST

Check

Ensure throttle return springs correctly secured and that throttle pulley moves freely, resting against closed stop when released (1 Fig. 23).

Ensure that throttle butterfly closed stop screw has not been moved (1 Fig. 24). If signs of tampering are present, check and if necessary, adjust.

Ensure that throttle pulley can be rotated to touch fully open stop (2 Fig. 23) and that throttle butterfly valve stop arm (2 Fig. 24) is touching throttle housing. If conditions of above operations are not satisfied, proceed with operation 'Adjust'.

Adjust

If throttle butterfly closed stop has been moved, adjust the stop.
Check for worn pivots, damaged rods or linkage and traces of any stiffness.
Renew items as necessary. Release throttle cross-rods from throttle pulley (3 Fig. 23).
Slacken clamps securing levers to rear of throttle shafts (1 Fig. 25).

With butterfly valve against closed stop, bell-crank against stop, and play in coupling taken up in opening direction, tighten clamp to lock lever to throttle shaft.
Repeat for other side of engine.

Offer cross-rods to ball connectors on pulley; rods must locate without moving pulley or linkage.
Slacken locknuts on cross-rods (4 Fig. 23) and adjust length of rods to locate pulley ball connector while bell-crank against closed stop.
Tighten locknuts and ensure ball joints remain free.
Slacken locknut on throttle pulley, fully open stop and wind back adjustment screw.
Hold throttle pulley fully open and ensure that both throttle butterfly stop arms are against throttle housing.
Set fully open stop to just touch throttle pulley and tighten locknut.
Check operation of throttle switch (2 Fig. 25).

THROTTLE CABLE, RENEW

LHD Vehicles

Disconnect the inner cable from the throttle pedestal.
Slacken and remove the outer cable locknut (1 Fig. 26) and withdraw the cable assembly from the pedestal.

Displace the tunnel carpet from around the throttle pedal.
Remove the split pin, withdraw the clevis pin and displace the inner cable from the pedal.
Displace the plenum chamber drain tube for access (1 Fig. 27).

Undo the water pipe 'P' clip (2 Fig. 27) and reposition the pipe.
Remove the bulkhead plate securing bolts (lower LH position has retaining nut).

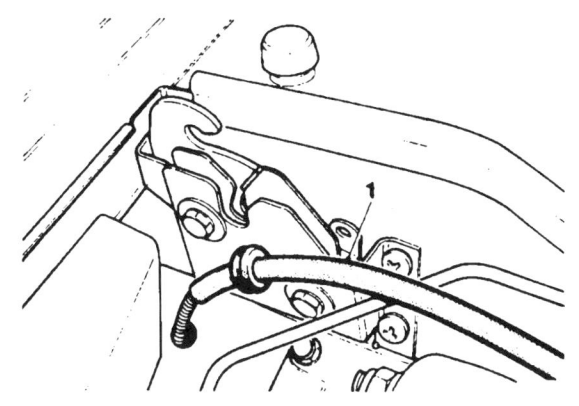

Fig. 28

Displace the cable and plate assembly (3 Fig. 27) from the bulkhead.
Remove the split pin and displace the cable from the lever arm.
Withdraw the cable from the plate.
Fit the replacement cable to the plate, connect to the lever arm and insert new split pin.
Clean the gasket faces and apply new gasket sealant.
Fit and align assembly to the bulkhead ensuring connecting rod passes into footwell.
Secure plate to bulkhead.
Reconnect rod to the pedal assembly.
Position the cable to the throttle pedestal, refit the outer cable locknut and insert the cable end to the pedestal.
Remove the slack from the inner cable, adjust outer cable position and tighten locknuts.
Check throttle linkage adjustment.

RHD Vehicles

Disconnect the inner cable from the throttle pedestal.
Slacken and remove the outer cable locknut and withdraw the cable assembly from the pedestal.
Displace the tunnel carpet from around the throttle pedal.
Remove the split pin from the cable retaining sleeve.
Slide the sleeve from position and disconnect the inner cable from the pedal.
Withdraw the outer cable from the bracket adjacent to the pedal.
Withdraw the cable (1 Fig. 28) through the bulkhead grommet and into the engine compartment.

Fit the replacement cable in position.
Connect to the pedal assembly.
Position the cable to the throttle pedestal, refit the outer cable locknut and insert the cable end to the pedestal.
Remove the slack from the inner cable, adjust outer cable position and tighten locknuts. Check throttle linkage adjustment.
Seal the bulkhead grommet with silicone sealant.

THROTTLE BUTTERFLY VALVE, ADJUST

Remove both air cleaners.
Slacken locknut on throttle butterfly stop screw.
Wind back screw (1 Fig. 29).

Ensure throttle butterfly valve closes fully.

> **CAUTION:**
>
> Any adjustment must be carried out on both butterfly valves. It is NOT permitted to adjust one valve only.

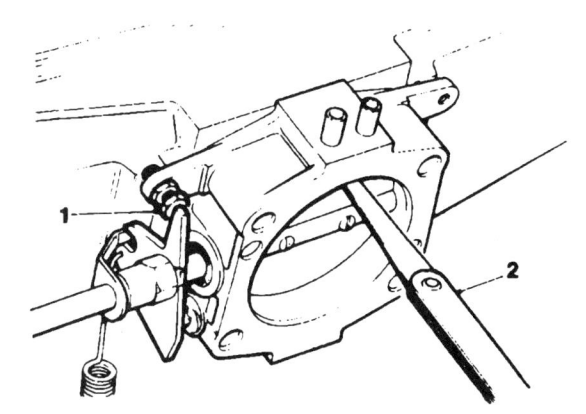

Fig. 29

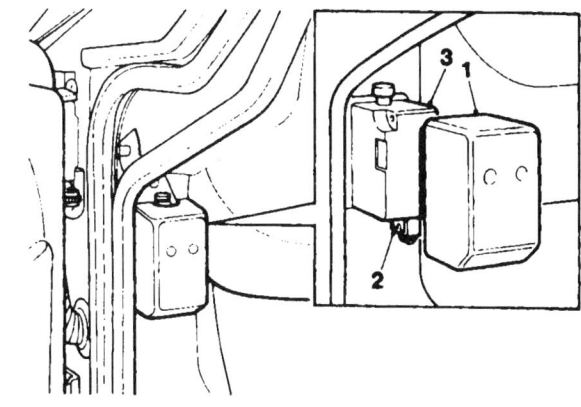

Fig. 30

Insert a 0.002 in. (0.05 mm) feeler gauge between top of valve and housing to hold valve open (2 Fig. 29).
Adjust stop screw to just touch stop arm and tighten locknut with feeler in position.
Withdraw feeler.
Press stop arm against stop screw and repeat on other side of engine.
Seal threads of adjusting screws and locknuts using a blob of paint.
Refit air-cleaners.
Check throttle linkage adjustment.

Check operation of throttle switch.

INERTIA SWITCH, RENEW

Disconnect battery earth lead. Remove the inertia switch trim cover (1 Fig. 30).

Release the switch block connector (2 Fig. 30) retaining catch and withdraw the connector.
Remove the passenger side dash liner.
Remove the two screws securing the switch and remove the switch (3 Fig. 30).
On refitting a new switch ensure it is correctly reset.

AUXILIARY AIR VALVE, RENEW

Carefully remove pressure cap from remote header tank to release any cooling system residual pressure.
Replace cap tightly.
Slacken the clips (1 Fig. 31) securing the hoses to the auxiliary air valve (2 Fig. 31) and pull the hoses clear.

> **CAUTION:**
>
> This procedure MUST ONLY be carried out on a cold or cool engine.

43

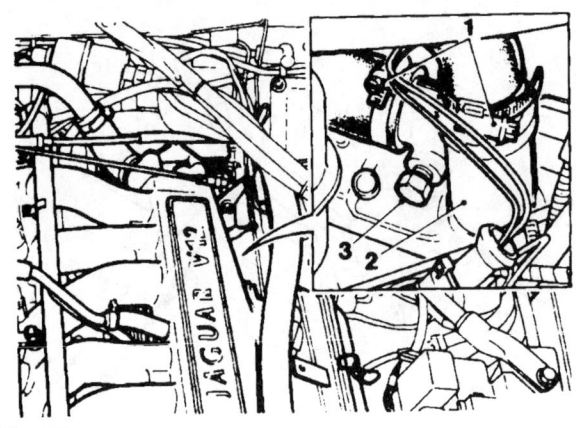

Fig. 31

Remove two screws and washers securing auxiliary air valve to coolant pipe, lift valve clear.
Clean all traces of gasket from coolant pipe, taking care not to damage seating area.
Scribe a line on idle speed adjusting screw (3 Fig. 31) and note number of turns required to screw fully in.

On refitting set idle speed adjustment screw of replacement valve to the number of turns open as noted previously.
Coat the new gasket with suitable non-hardening sealing compound.
Locate the valve on the coolant pipe and secure using the two screws and washers.
Fit and secure the hoses.
Check coolant level at remote header tank, if necessary top up.
Check, if necessary adjust idle speed.

AUXILIARY AIR VALVE, TEST

Remove auxiliary air valve.
Fully close idle speed adjustment screw.
Immerse auxiliary air valve bulb in a container of boiling water and observe valve head through side port.
Valve should move smoothly to closed position.
Quickly blow through side port; no air should pass.
Allow valve bulb to cool.
Valve head should move smoothly back to open main air passage.
If valve performance is satisfactory reset idle speed adjustment screw and refit valve.
If valve performance is not satisfactory, fit replacement component.

SPEED CONTROL SYSTEM, DESCRIPTION

The speed control system maintains the vehicle at a drivers selected speed over normal driving conditions, without the need for constant surveillance of the speedometer and continuous driver use of the throttle pedal.
Once engaged and set the speed control system will control the position of the throttle over a wide range of movement from idle to full throttle to maintain the speed set under different conditions.

System Components

Throttle actuator

A vacuum operated device containing bellows attached to the throttle by a mechanical linkage.

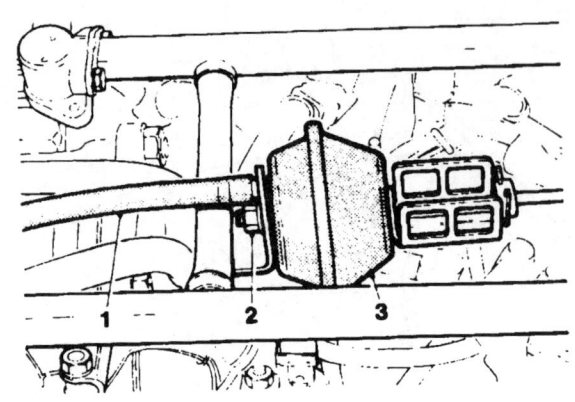

Fig. 32

Electronic Control Unit (ECU)

A device containing a microprocessor and associated electronic circuits which evaluates the signal provided by the drivers controls, the pedal switches and road speed sensors. It then transmits signals to a vacuum control unit in order to control the vehicle speed. It also retains in memory the speed setting.

Vacuum pump and regulator valve

An assembly which increases and decreases the vacuum system pressure depending on signals received from the ECU.

Dump valve

A device which vents the vacuum system pressure to atmosphere when braking or the throttle returns to idle.

Operating switches

The system is controlled by a centre console mounted master switch and a resume button on the end of the combination switch stalk.
A brake switch disengages the system.

THROTTLE ACTUATOR, RENEW

Open the bonnet and fit wing protection.
Disconnect the vacuum hose (1 Fig. 32) from the front of the throttle actuator.

Slacken the actuator rod locknut.
Release the rod from the actuator linkage.
Remove the actuator to mounting bracket securing nut (2 Fig. 32).
Displace and remove the actuator and rod assembly (3 Fig. 32).

Fit and align the replacement actuator assembly to the mounting bracket.
Fit and tighten securing nut.
Connect vacuum hose to the front of the actuator.
Tighten rod to linkage and adjust linkage to remove free play, tighten locknut.
Remove wing protection and close bonnet.

THROTTLE ACTUATOR LINKAGE ROD, ADJUST

Open the bonnet and fit wing protection.
Slacken the actuator linkage rod locknut.
Adjust rod to remove free play, tighten locknut.
Remove wing protection and close bonnet.

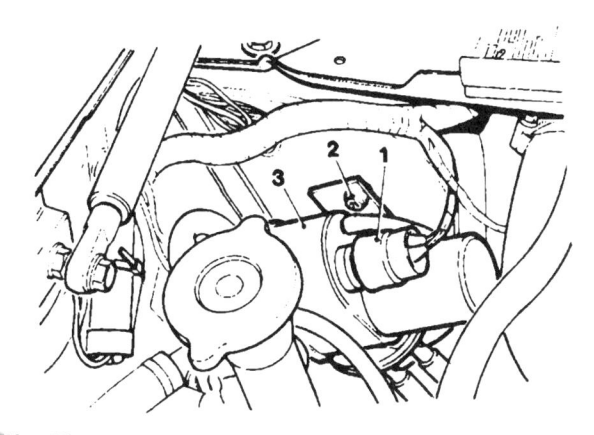

Fig. 33

VACUUM PUMP, RENEW

Open the bonnet and fit wing protection.
Disconnect the air temperature sensor connector from the left-hand air cleaner.
Release retaining clips, displace air cleaner cover and withdraw element.
Remove air cleaner cover.
Disconnect the vacuum pump harness multi-plug (1 Fig. 33).
Remove the vacuum pump mounting screws (2 Fig. 33).
Disconnect and remove the pump from the vacuum hoses (3 Fig. 33).

Fit vacuum hoses to the replacement vacuum pump.
Fit and tighten vacuum pump securing screws.
Connect vacuum pump harness multi-plug.
Refit the air cleaner element and reposition cover.
Connect the air temperature sensor multi-plug.
Remove wing covers and close bonnet.

VACUUM DUMP VALVE, RENEW

Open the bonnet and fit wing protection.
Disconnect the air temperature sensor connector from the left-hand air cleaner.
Release retaining clips, displace air cleaner cover and withdraw element.
Remove air cleaner cover.
Disconnect the vacuum hose (1 Fig. 34) from the dump valve.
Disconnect the harness multi-plug (2 Fig. 34).
Remove the valve securing bolts (3 Fig. 34) and remove the valve assembly.

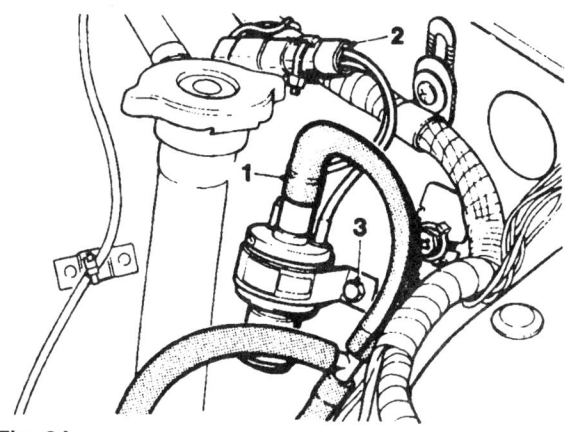

Fig. 34

Position and secure the replacement valve assembly with bolts.
Connect the harness multi-plug and refit the vacuum hose.
Refit the air cleaner element and reposition cover.
Connect the air temperature sensor multi-plug.
Remove wing covers and close bonnet.

VACUUM SUPPLY HOSE, RENEW

Open the bonnet and fit wing protection.
Disconnect the vacuum hose from the front of the actuator.
Disconnect the hose from the vacuum dump valve.
Disconnect the hose from the vacuum pump.
Remove the hose assembly and disconnect hoses from 'T' piece.

Connect new hoses to the 'T' piece and fit the assembly, routing the longest hose to the actuator.
Remove wing covers and close bonnet.

SPEED CONTROL ECU, RENEW

Reposition the passenger footwell carpet for access to the ECU panel securing screws.
Undo and remove the floor panel screws.
Displace the panel and disconnect the ECU harness multi-plug.
Release the retaining clip and displace and remove the unit.

Align the new ECU, secure with retaining clip and connect the harness multi-plug.
Reposition and secure the panel, reposition carpet.

Chapter 7

FUEL SYSTEM (6 Litre)

Introduction

The fuel systems differs in few respects compared to that fitted to the previous version.

SAFETY PRECAUTIONS

> **WARNING:**
>
> OPERATIONS INVOLVING THE FUEL SYSTEM RESULT IN FUEL AND FUEL VAPOUR BEING PRESENT IN THE ATMOSPHERE.
> FUEL VAPOUR IS EXTREMELY FLAMMABLE.
> GREAT CARE MUST BE TAKEN WHEN WORK IS CARRIED OUT ON THE FUEL SYSTEM.
> THE FOLLOWING PRECAUTIONS MUST BE STRICTLY ADHERED TO:
>
> 1. NO SMOKING IN THE WORK AREA.
> 2. 'NO SMOKING' WARNING SIGNS MUST BE POSTED ROUND THE AREA.
> 3. A CO^2 FIRE EXTINGUISHER MUST BE NEARBY.
> 4. DRY SAND MUST BE AVAILABLE TO SOAK UP ANY FUEL SPILLAGE.
> 5. EMPTY THE FUEL FROM THE TANK INTO AN AUTHORIZED EXPLOSION-PROOF CONTAINER USING SUITABLE FIREPROOF EQUIPMENT.
> 6. FUEL MUST NOT BE EMPTIED INTO A PIT.
> 7. THE WORK AREA MUST BE WELL VENTILATED.
> 8. THE BATTERY MUST BE DISCONNECTED BEFORE CARRYING OUT WORK ON THE FUEL SYSTEM.
> 9. THE OLD TANK MUST NOT BE DISPOSED OF UNTIL IT IS RENDERED SAFE FROM EXPLOSION.
> 10. MAINTENANCE PERSONNEL SHOULD HAVE UNDERTAKEN SPECIALIST TRAINING BEFORE BEING ALLOWED TO REPAIR COMPONENTS ASSOCIATED WITH THE FUEL SYSTEM.

THROTTLE LINKAGE, CHECK AND ADJUST

Check

Ensure that all throttle return springs are correctly fitted. Check that throttle pulley, linkage and throttles operate freely. The throttle pulley should close to the closed throttle stop. It should reach the full throttle stop, just before the outboard full throttle stops touch the pads on the inlet manifolds.
If the above conditions are not met, proceed to 'adjust'.

Adjust

Disconnect throttle rods from bell crank ball pins.
Slacken nut/bolt securing bottom lever to shaft.
Ensure butterfly levers are on closed throttle stops.
Lift bell crank levers to stop and tighten the nut / bolt securing the bottom lever.

Reconnect throttle rods to bell crank ball pins. Slacken throttle rod adjustment locknuts.
Individually adjust throttle rod length so that with the butterfly levers held against their stops, the pulley can be opened 0.508 to 0.762 mm (0.020 to 0.030 in.).
Tighten throttle rod locknuts and check operation.
Slacken pulley full throttle stop locknut. Open pulley to full throttle. Adjust full throttle stop screw so that pulley is on stop, just before first outboard butterfly lever touches the inlet manifold pad.
Tighten pulley full throttle stop locknut. Check throttle action including operation of ignition closed throttle switch.

FUEL FILTER

The fuel filter (Fig. 1) is mounted under the left hand front wheel arch.

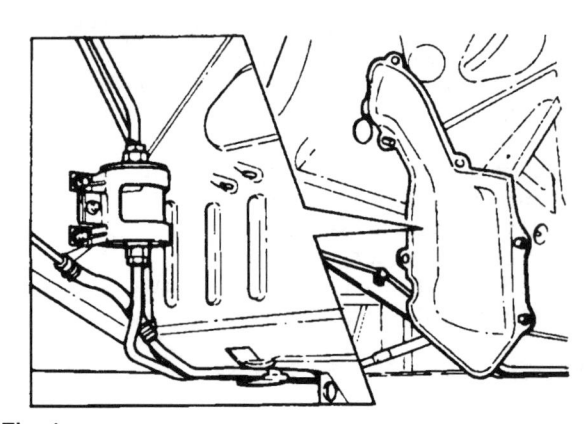

Fig. 1

> **CAUTION:**
>
> Direction of fuel flow is shown on the filter casing; it is important that the filter is fitted the correct way round.

Connectors and Hoses

Quick fit connectors are used on the fuel system (Fig. 2). At connections between hoses and the under floor pipes.

> **CAUTION:**
>
> Pipes and component stubs must be fully inserted into connector ends; failure to do so could cause fuel leakage.

The pipes have a ridge or paint line to show how far the connector should be pushed onto the pipe end.

There are two sizes of connector, 8 mm and 10 mm. This is to ensure that the feed pipe is not connected as the return pipe and vice versa. The 10 mm connector is used for the feed line.

Service tool JD 175 is available for the tightening of jubilee clips used for attaching feed and return hoses to the FPM and pipes within the tank.

Fig. 2

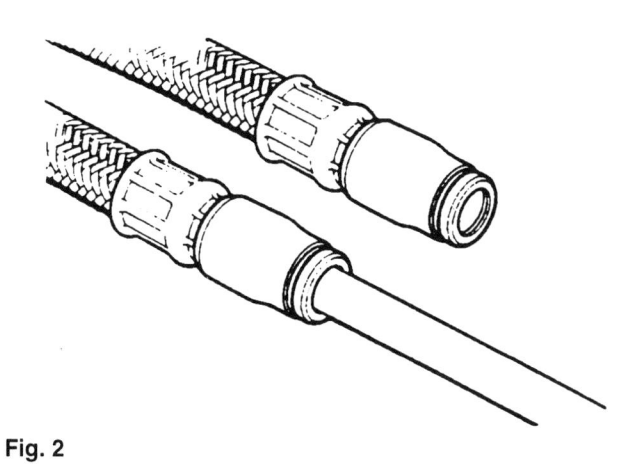

CAUTION:

It is important that the Tank Pressure Control Valve is fitted the correct way round. The valve has 'Tank' printed on one port and 'Can' (or 'Engine') printed on the other port for identification. Failure to fit the valve in the correct orientation could cause tank collapse. When the engine is running the valve is opened by the inlet manifold vacuum signal, fuel vapour generated by rising fuel temperature flows freely to the charcoal canister, keeping the fuel tank pressure near zero.

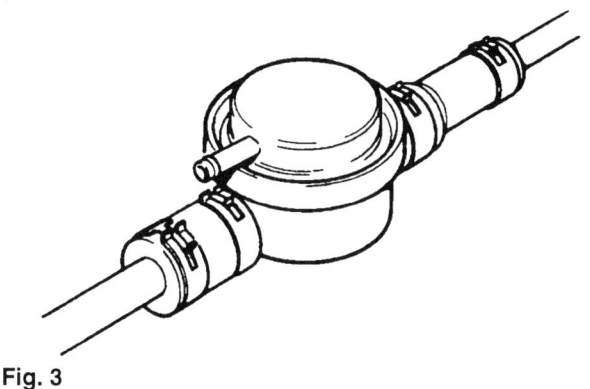

Fig. 3

TANK PRESSURE CONTROL VALVE (FIG. 3)

When the engine is not running the Tank Pressure Control Valve permits the vapour pressure in the fuel tank to rise, as a result of heat input from the vehicle and/or environment, to a maximum of 1.0 to 1.3 lbf / in^2. At which point the valve opens and allows vapour to flow from the tank to the charcoal canister.

The Tank Pressure Control Valve also contains a non-return valve. When the fuel and fuel vapour cools the pressure in the tank falls. The non return valve opens to allow flow from the charcoal canister to the fuel tank. This prevents negative pressures in the fuel tank which could cause the tank to collapse.

If the 1.25 lbf / in^2 valve fails to work for any reason, the fuel filler cap will relieve pressure at 2.0 to 3.5 lbf / in^2.

Chapter 8

COOLING SYSTEM - 93 MODEL YEAR ON

RADIATOR COOLING FAN, description

A new cooling fan is used. The fan has improved cooling performance and full radio interference suppression. It is mounted on the rear of the radiator cowl (Fig. 1).

ENGINE OIL COOLER, description

European models have a new 28 mm full width oil cooler (2 Fig. 2). The cooler is mounted on the bonnet hinge panel and the pipework routed under the radiator (Fig. 2).
The condenser is three-quarter depth and 'O' ring connections are now standard to reduce the possibility of leaks.
USA models retain the smaller type oil cooler (1 Fig. 2). All other features are as European models.

Fig. 1

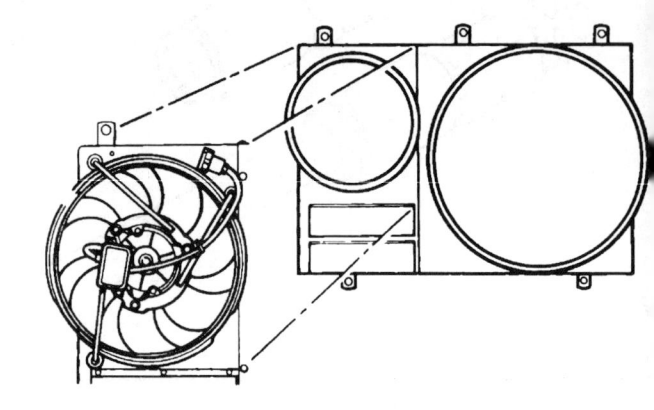

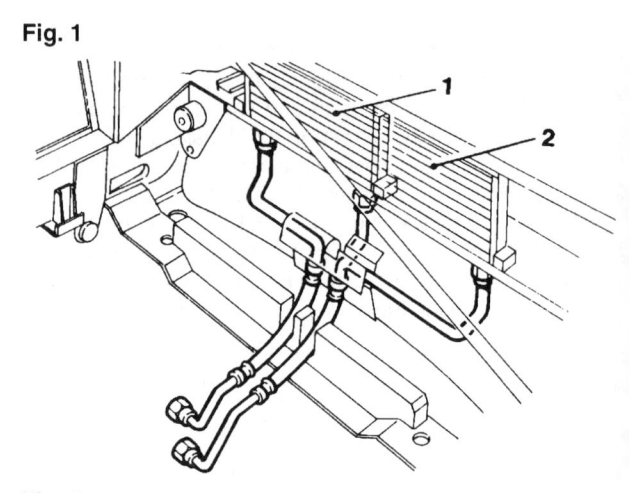

Fig. 2

Chapter 9

EXHAUST SYSTEM - 6 LITRE

INTRODUCTION

The exhaust system for 6.0 litre vehicles, introduced at 93.5 MY, is different from that used on the previous 5.3 litre models. The variations are as follows:

Note: Fig. 1 shows the exhaust arrangement for Rest of the World (ROW) vehicles.

From the front of the system:

The down pipe has an added, secondary, catalyst (1 Fig. 1).

The spherical joints between the downpipe to underfloor pipe and intermediate pipe to over-axle pipe are replaced with integral 'Torca' clamps (2 Fig. 1). These reduce the possibility of leaks, rattles and knocks.

The intermediate pipe assemblies (3 Fig. 1) now have an additional mounting (4 Fig. 1) toward the rear of the propshaft tunnel. This gives additional support, aids assembly and also helps locate the system square in the vehicle.

Larger diameter over-axle pipes (5 Fig. 1) are re-routed to give good clearance to the rear suspension cradle.

Smaller diameter rear silencers (6 Fig. 1) improve body-to-system clearance.
New tailpipe trims (7 Fig. 1) have a square profile to match the bumper aperture.

EEC Countries
The EEC version has downpipe catalysts only. There are no catalysts in the underfloor region.
The secondary catalysts in the downpipes ensure the vehicle complies with EEC emission regulations.
ROW
This version (Fig.1) has downpipe and underfloor catalysts and is specified everywhere except Japan and the EEC.
Japan
The Japanese version is the same as the ROW, except that grass shields and overheat sensors are added to the downpipes.

DOWNPIPE CATALYTIC CONVERTER, RENEW

Note: The procedure for renewal of either downpipe catalytic converter is similar. Where differences do occur, they will be shown by reference to right or left hand side only.

Disconnect the air temperature multiplug (left hand only).
Release the air box cover securing clips.
Disconnect the rubber hose from the crossmember intake duct.
Reposition the air box cover and remove the air cleaner element. Remove the air box cover.
Cut and remove the lambda sensor harness securing tie strap. Disconnect the sensor multiplugs.
Unscrew and remove the downpipe/heatshield front flange securing nut (the heatshield is on the right hand downpipe only).
Raise the vehicle on a ramp.

Slacken the intermediate pipe to downpipe clamp nut/bolt (1 Fig. 2). Slacken the intermediate pipe to overaxle pipe clamp nut/bolt (1 Fig. 3). Remove the intermediate pipe to forward mounting securing nut/bolt.
Disconnect the intermediate pipe from the overaxle pipe.
Remove the intermediate pipe from the downpipe.

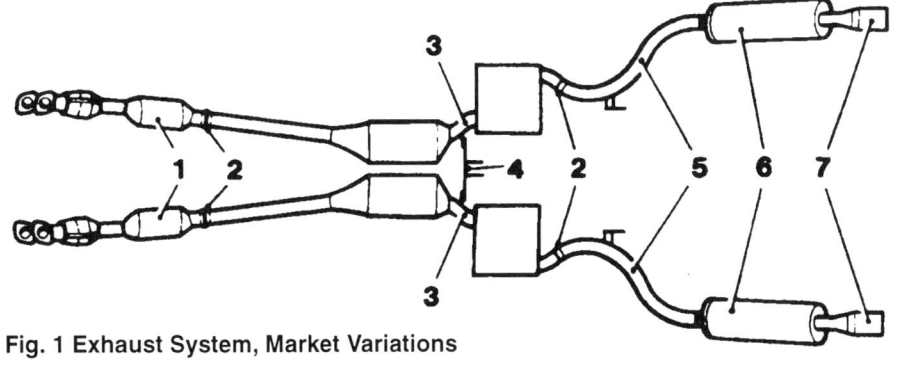

1. Secondary catalysts
2. 'Torca' clamp fastenings
3. Intermediate pipe assemblies
4. Additional mounting point, rear tunnel
5. Over-axle pipes
6. Rear silencers
7. Tailpipe trims

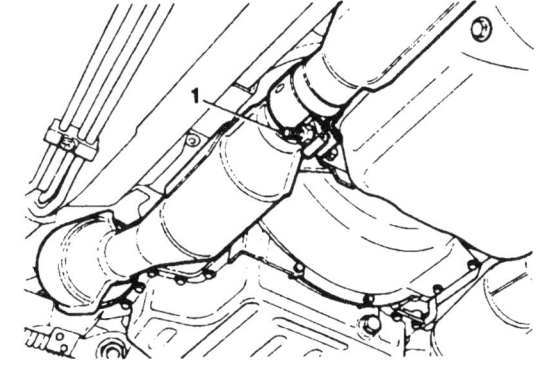

Fig. 1 Exhaust System, Market Variations

Fig. 2

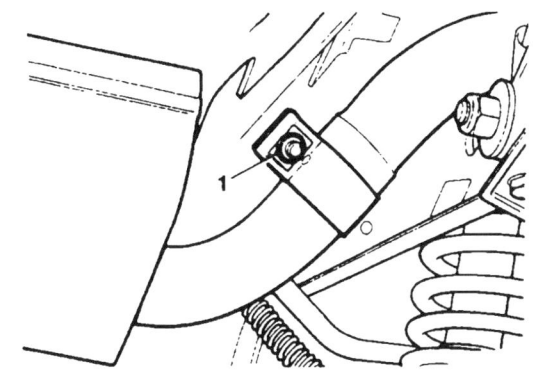

Fig. 3

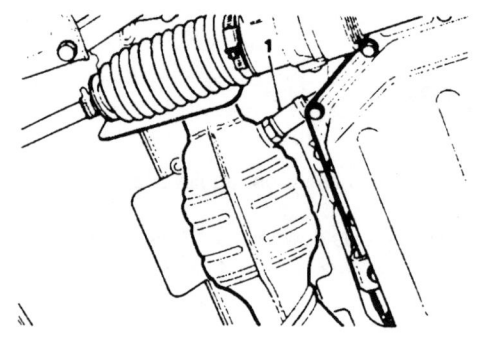

Fig. 4

Remove the steering rack heatshield securing bolts.
Retrieve the lambda sensor harness from between the engine and beam. Unscrew and remove the lambda sensor (1 Fig. 4, RH sensor only illustrated).

Unscrew and remove the remaining downpipe securing nuts.
Reposition the heatshield (right hand only).
Remove the downpipe assembly. Remove the downpipe sealing rings.
Clean all pipe / joint mating faces prior to reassembly.
Fit and seat new sealing rings to a new downpipe.
Fit and align the downpipe and position the pipe securing flanges.
Fit, but do not tighten, the downpipe inner securing nut and position the heatshield (right hand only).
Fit, but do not tighten, the downpipe inner securing nuts.
Fit, but do not tighten, the downpipe outer securing nuts.
Fully tighten all downpipe securing nuts.

Note: Ensure there is adequate clearance between the pipe and the gearbox sump (left hand only).

Fit and tighten the lambda sensor. Route the lambda sensor harness between the engine and beam.
Align the steering rack heatshield and fit and tighten the heatshield securing bolts.
Apply exhaust sealant to the intermediate pipe mating faces. Connect the intermediate pipe to the downpipe. Connect the opposite end of the pipe to the over-axle pipe. Position the intermediate pipe to the front mounting. Fit and tighten the pipe to mounting securing nut/bolt.
Tighten the intermediate pipe clamp securing nuts / bolts.
Lower the ramp.
From in the engine bay:
Connect the lambda sensor multiplugs. Fit a new lambda sensor harness tie strap.
Align the air box cover to the air box and fit a new air cleaner element. Final position the air box cover and secure with clips.
Connect the rubber hose to the air intake duct.
Connect the air temperature sensor multiplug (left hand only).
Close the bonnet.

Torque Figures

Intermediate pipes, mounting bracket to vehicle floor 20.5-27.5 Nm
Intermediate silencer to tie bar 22-28 Nm
Slip joint fastenings - 'Torca' clamps 44-60 Nm

Oils / Greases / Sealants

Exhaust sealant TIVOLI KAY ADHESIVES No. 5696

INTERMEDIATE PIPE ASSEMBLY, RENEW

Note: The intermediate pipe assembly includes the underfloor catalytic converter.
Raise the vehicle on a ramp.

Slacken the intermediate pipe to downpipe clamp nut/bolt (1 Fig. 2). Slacken the intermediate pipe to overaxle pipe clamp nut/bolt (1 Fig. 3). Remove the intermediate pipe to forward mounting securing nut/bolt.
Disconnect the intermediate pipe assembly from the downpipe. Disconnect and remove the intermediate pipe assembly from the over-axle pipe.
Clean all pipe / joint mating faces prior to reassembly.
Apply exhaust sealant to the new intermediate pipe mating faces. Connect the intermediate pipe to the downpipe. Connect the opposite end of the pipe to the over-axle pipe. Position the intermediate pipe to the front mounting. Fit and tighten the pipe to mounting securing nut/bolt.
Tighten the intermediate pipe clamp securing nuts/bolts.
Lower the ramp.

Torque Figures

Intermediate pipes, mounting bracket to vehicle floor 20.5-27.5 Nm
Intermediate silencer to tie bar 22-28 Nm
Slip joint fastenings - 'Torca' clamps 44-60 Nm

Oils / Greases / Sealants

Exhaust sealant TIVOLI KAY ADHESIVES No. 5696

OVER-AXLE PIPE, (REAR INTERMEDIATE PIPE), RENEW

Raise the vehicle on a ramp.
Slacken the rear silencer to over-axle pipe clamp securing nut/bolt (1 Fig. 5).

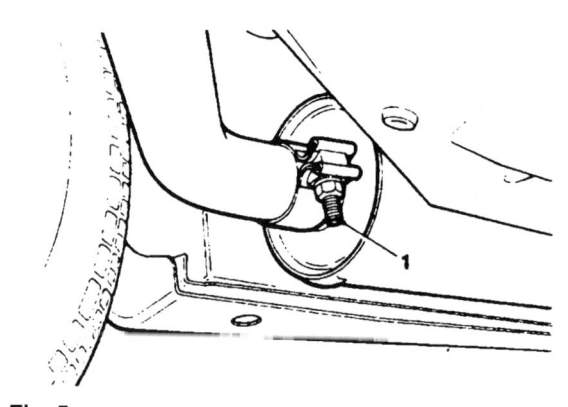

Fig. 5

Disconnect the rear silencer from the over-axle pipe.
Unscrew and remove the forward mounting to intermediate pipe securing nut/bolt.
Slacken the intermediate pipe to downpipe clamp securing nut/bolt (1 Fig. 2). Disconnect the intermediate pipe assembly from the downpipe. Slacken the intermediate pipe to overaxle pipe clamp nut/bolt (1 Fig. 3). Disconnect and remove the intermediate pipe assembly from the over-axle pipe.
Displace the over-axle pipe from the axle mounting. Remove the over-axle pipe from the rear beam.
Unscrew and remove the over-axle pipe mounting pin securing nut. Remove the mounting pin forward insulation bush. Remove the mounting pin and rear insulation bush.

Clean all pipe/joint mating faces prior to reassembly.
Fit a new forward bush and mounting pin. Fit a new rear bush to the mounting pin. Ensure they are aligned correctly. Fit and tighten the mounting pin securing nut.
Carefully fit and align the new over-axle pipe to the rear beam mounting.
Apply exhaust sealant to the pipe mating faces.
Connect the intermediate pipe to the new over-axle pipe.
Connect the intermediate pipe to the downpipe. Fit and tighten the forward mounting bracket to the pipe ring nut/bolt.
Tighten the over-axle pipe to intermediate pipe clamp securing nut/bolt. Tighten the intermediate pipe to downpipe clamp securing nut/bolt.
Apply exhaust sealant to the rear silencer joint.
Connect the rear silencer to the over-axle pipe.
Tighten the silencer clamp securing nut/bolt.
Lower the ramp.

Torque Figures

Intermediate pipes, mounting bracket to vehicle floor 20.5-27.5 Nm
Intermediate silencer to tie bar 22-28 Nm
Slip joint fastenings - 'Torca' clamps 44-60 Nm

Oils / Greases / Sealants

Exhaust sealant TIVOLI KAY ADHESIVES No. 5696

TAIL PIPE AND SILENCER, RENEW

Raise the rear end of the vehicle and support on two axle stands.
Slacken the tail pipe trim clamp securing grub screw. Remove the tail pipe trim.
Slacken the rear silencer to over-axle pipe clamp securing nut/bolt (1 Fig. 5). Disconnect the rear silencer from the over-axle pipe and remove the silencer from the rear mounting.
Clean the over-axle pipe and rear silencer mating faces.
Fit and align a new rear silencer to the rear mounting. Apply exhaust sealant to the over-axle pipe and rear silencer mating faces. Fit the new rear silencer to the over-axle pipe and tighten the clamp securing nut/bolt.
Fit and align the tail pipe trim securing clamp and the tail pipe trim. Position the clamp to the trim. Tighten the clamp securing grub screw.
Repeat the procedure on the opposite side of the vehicle to renew the vehicle set.
Lower the vehicle from the stands.

Torque Figures

Rear silencer mounting to luggage compartment floor 18.5-25.5 Nm
Slip joint fastenings - 'Torca' clamps 44-60 Nm
Tail pipe finisher grub screw 3.5-4.5 Nm

Oils / Greases / Sealants

Exhaust sealant TIVOLI KAY ADHESIVES No. 5696

FRONT MOUNTING RUBBER, RENEW

Raise the vehicle on a ramp.
Unscrew and remove the front mounting rubber to underside body securing bolts. Remove the mounting rubber.
Fit a new front mounting rubber. Tighten the mounting rubber securing bolts.
Lower the ramp.

Torque Figures

Intermediate pipes, mounting bracket to vehicle floor 20.5 - 27.5 Nm.

Chapter 10

AUTOMATIC TRANSMISSION - 6 LITRE

GENERAL DESCRIPTION

The Hydra-Matic 4L80-E is a four-speed, high torque capacity, electronically controlled automatic transmission. It consists basically of a torque converter and three planetary gear sets. There are five multiple disc clutches, one intermediate sprag clutch assembly, two roller clutch assemblies and two band assemblies to provide the drive elements necessary for correct sequential gear engagement and operation.

The torque converter containing a pump, turbine (rotor), a stator assembly, and a clutch pressure plate splined to the turbine, acts as a fluid coupling for smooth torque transmission from the engine. The converter also supplies additional torque multiplication when necessary, and the torque converter clutch (TCC) pressure plate provides a mechanical direct drive or 'lock-up' above a certain speed in top gear for greater fuel economy.

Gear shift operations are controlled from the Transmission Control Module (TCM), which governs the electronically controlled valve body situated within the transmission.

Three planetary gear sets provide reverse and the four forward ratios - the changing of which is fully automatic in relation to load, vehicle speed and throttle opening.

The transmission control module (TCM) receives and integrates various vehicle sensor input signals, and transmits operating signals to the solenoids located in the control valve assembly. These solenoids govern the transmission operating pressures, up-shift and down-shift gear selection patterns, and also the torque converter clutch operation from a pulse width modulated solenoid control.

Gear Ranges

Selectable gear positions are:

P - Park,
R - Reverse,
N - Neutral,
D - Drive, 3, 2.

P - Park position of the selector lever provides a mechanical locking of the output shaft of the transmission, and as such, must only be engaged when the the vehicle is stationary. In addition, and for extra safety, the handbrake should also be applied. It is necessary to have the ignition on and the footbrake applied to move the selector lever from the Park position. For ignition key removal the selector lever must be in the Park position. The engine can be started in the Park position.

R - Reverse enables the vehicle to be operated in a rearwards direction. The engine cannot be started in the Reverse position.

N - Neutral position enables the engine to be started and operated without driving the vehicle. It also allows the vehicle to be moved manually for access, ie. for removal of the propeller shaft.

D - Drive position allows the automatic selection of all four forward gear ratios during normal driving conditions for maximum efficiency and fuel economy. On acceleration, down-shifts are obtained by depressing the accelerator pedal or by manual selection. The engine cannot be started in this position.

3 - Manual third position allows automatic operation of the three lower gear ratios but inhibits selection of the fourth ratio. This position is used for towing a trailer or negotiating hilly terrain when greater engine braking control is required. The engine cannot be started in this position.

2 - Manual second position allows automatic operation of the two lower gear ratios but inhibits selection of the third and fourth ratios. This position is used for heavy traffic congestion or negotiating hilly terrain when even greater engine braking control is required than is provided by manual third. This ratio may be selected at any vehicle speed - even if the transmission is in third or fourth ratio, the transmission will immediately down-shift to second gear provided the vehicle speed is below 137 km/h (85 mile/h). The engine cannot be started in this position.

Note: With the performance mode switch in the 'normal' position, the vehicle will pull away in second gear. However, if more than 75% of throttle is applied when the vehicle speed is between zero and 13 km/h (8 mile/h), then first gear will be selected. From 13 to 61 km/h (8 to 38 mile/h) first gear is obtainable by 'kickdown'. In 'sport' mode the vehicle pulls away in first gear and the transmission operates fully in all four forward gears.

Basic Specifications

Transmission Type: Hydra-Matic 4L80-E
High torque capacity, Automatic overdrive with torque converter clutch assembly.

Explanation of Designation

Gear Ratios:		
	1st	2.482
	2nd	1.402
	3rd	1.000
	4th	0.750
	Reverse	2.077

Maximum Engine Torque (at turbine shaft)	597 Nm (440 lbf ft)
Maximum Gearbox Torque (into converter)	1200 Nm (885 lbf ft)
Maximum Shift Speed : (all upshifts):	6000 RPM
Torque Converter:	310 mm diameter
Stall Torque Ratio:	2.6 : 1
Transmission Fluid Capacity:	
Drain and refill only	7,3 litre (15.4 Imp. pints)
Complete fill	12,8 litre (27 Imp. pints)
Transmission Fluid:	Dexron II E
Transmission Weight: Empty	105 kg (232 lb)
Complete with fluid	117 kg (258 lb)

Fig. 1

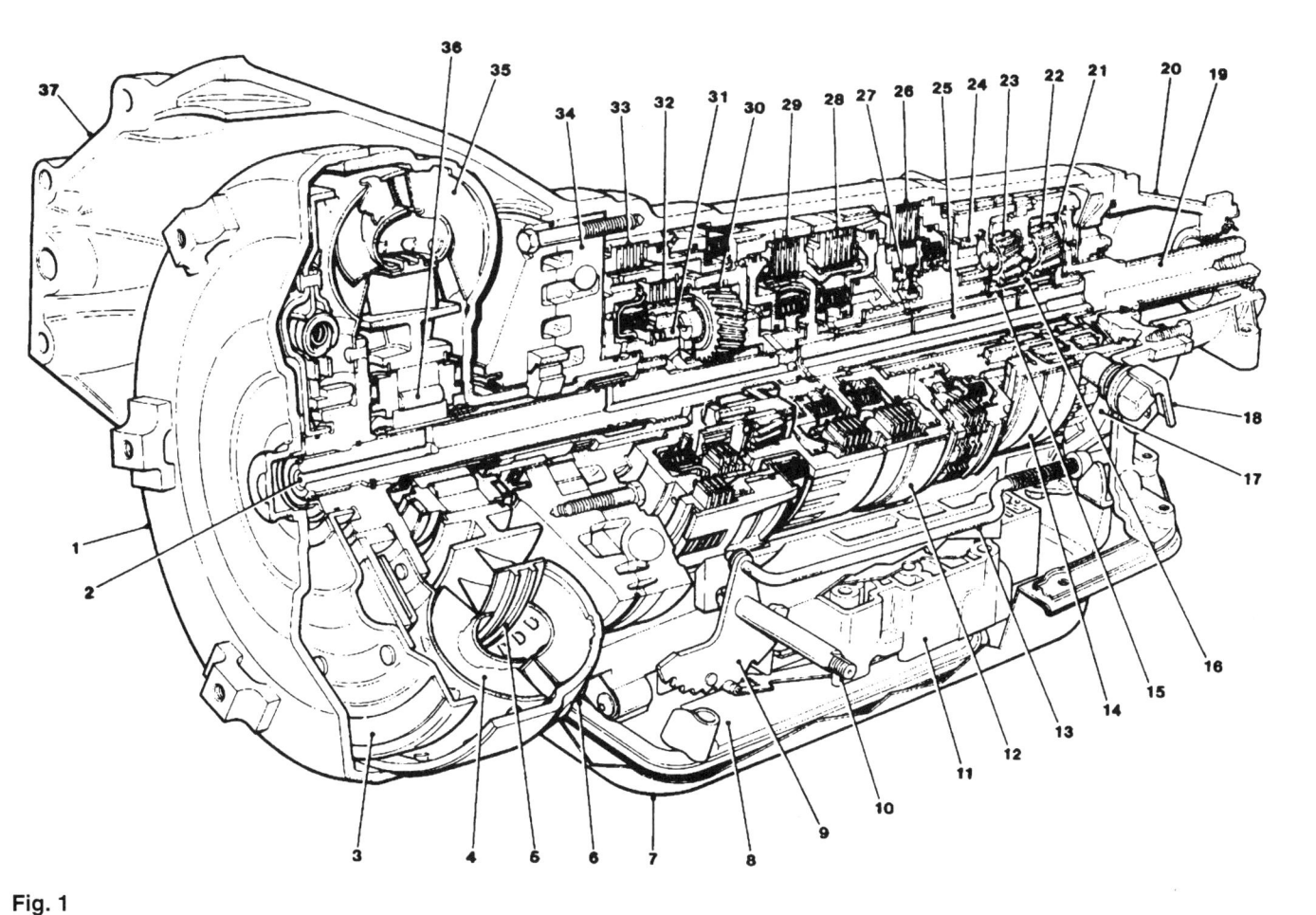

Key to Fig. 1

1. Torque converter
2. Turbine shaft
3. Pressure plate
4. Converter turbine
5. Converter stator
6. Variable force motor solenoid
7. Sump pan
8. Filter
9. Interior detent lever
10. Manual shaft
11. Control valve
12. Front band
13. Parking lock actuator
14. Rear band
15. Sun gear shaft
16. Sun gear
17. Parking lock pawl
18. Parking lock switch and activator
19. Output shaft
20. Rear extension housing
21. Rear internal gear
22. Output planetary carrier assembly
23. Reaction planetary carrier assembly
24. 'LO' roller clutch
25. Main shaft
26. Intermediate clutch
27. Intermediate sprag clutch
28. Direct clutch
29. Forward clutch
30. Overdrive planetary carrier assembly
31. Overdrive roller clutch
32. Overrun clutch
33. Fourth clutch
34. Pump assembly
35. Converter pump
36. Stator roller clutch
37. Transmission case

Lubrication flow

Transmission oil leaves the transmission unit, passes through the oil cooler and returns to the transmission unit via a connector in the case, and into the valve body and so into the lubrication pipe. The fluid is then routed to the rear of the transmission to lubricate the rear case and the rear extension housing bearing.

Lubrication fluid is also routed through the pump assembly, and into the overrun clutch housing where it is then routed to the various apply components to cool and lubricate the transmission clutches and gear sets.

Cooler Circuit

With the torque converter clutch (TCC) released, transmission fluid returning from the torque converter is routed through the TCC shift valve into the cooler circuit, ie, through the cooler feed pipe to the transmission cooler in the radiator and then back to the transmission unit.

When the TCC is applied, ie, the shift valve in the apply position, regulated converter fluid is passed through the valve and into the cooler circuit as above.

Torque Converter

The torque converter is a three element (single stator) unit which acts as a fluid coupling to connect the engine power smoothly to the transmission gear train, and as a torque multiplier to provide extra power when pulling away from rest. The three elements of the converter are the pump (impeller), the turbine, and the stator planetary gear sets (see fig. 2).

There are three planetary gear sets in this transmission, 'overdrive', 'reaction', and 'output'.

Each gear set comprises a centre or sun gear, an annulus or internal gear, and a planetary carrier assembly which contains the smaller planet gears.

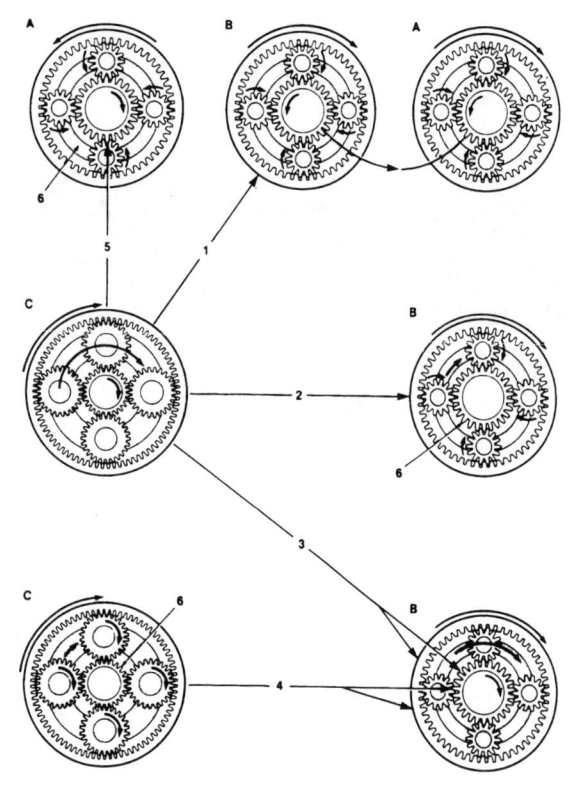

Fig. 2

1. First gear
2. Second gear
3. Third gear
4. Fourth gear
5. Reverse
6. Component held stationary
7. Reaction gear set
8. Ouput gear set
9. Overdrive gear set

Direct drive in a planetary gear set is obtained when any two parts of a gear set rotate in the same direction at the same speed, so driving the third component also at that same speed. The planetary gears in this case act as wedges to drive the entire gear set as one unit, and so the output speed of the transmission is the same as the input speed from the torque converter.

Conversely a planetary gear set reverses the direction of power flow when a carrier assembly is held stationary and power is applied to the sun gear so causing the planetary gears to act as idler gears to drive the internal gear in the opposite direction.

In first, second and third gears, the 'overdrive' roller clutch retains the 'overdrive' sun gear and carrier assembly together - so driving the internal gear at the same speed.

In first gear, the 'output' internal gear drives the 'output' carrier planet gears clockwise, which causes the sun gear to rotate anti-clockwise. As the 'output' sun gear and the 'reaction' sun gear are common, the 'reaction' carrier planet gears rotate clockwise. The 'reaction' carrier, being held stationary by the 'LO' roller clutch, then causes the 'reaction' carrier planet gears to drive the 'reaction' internal gear and the output shaft, ie. first gear.

Second gear is obtained when the 'output'/'reaction' sun gear is held stationary by the intermediate clutch, and therefore when the 'output' carrier planet gears are driven clockwise by the rear internal gear, the planet gears rotate clockwise round the stationary sun gear. The 'output' carrier planet gears drive the 'output' carrier assembly and 'output' shaft clockwise, ie. second gear.

Third gear is obtained when the direct clutch is applied; the power flow from the 'overdrive' planet gears and the forward clutch housing is then transferred to both the sun gear and the 'output' internal gear. With the power flow through the 'overdrive' planetary gear set being a direct drive, and with both the sun gear and the internal gears of the 'output' planet gear set driving at converter turbine speed, the 'output' carrier planetary gears act as wedges and drive the 'output' carrier assembly and output shaft together, ie. direct drive, third gear.

In fourth gear, the 'overdrive' sun gear is held stationary by the fourth clutch being applied, then the 'overdrive' carrier being driven clockwise, the 'overdrive' planetary gears rotate clockwise also on their axes around the stationary sun gear. This causes the planetary gears to drive the 'overdrive' internal gear clockwise and so an 'overdrive' ratio is obtained through the 'overdrive' planetary gear set.

In reverse gear, the rear brake band is applied to hold the 'reaction' carrier stationary while the direct clutch is applied to supply clockwise power flow to the sun gear. Power flow through the 'overdrive' planetary gear set being a direct drive, the sun gear drives the 'reaction' planetary gears anti-clockwise which drives the 'reaction' internal gear ('output' carrier assembly) anti-clockwise, ie. reverse gear.

Torque Converter Clutch (TCC)

When a pre-determined speed is achieved in fourth gear, the torque converter clutch is enabled by the transmission control module through the pulse width modulated solenoid and a clutch plate provides a direct drive so reducing fuel consumption. Pulsing the solenoid causes the TCC valve to modulate pressure against the TCC. This pulsing or modulated pressure allows the TCC to slip slightly so providing a smooth apply and release of the TCC.

If the transmission fluid exceeds a temperature of 125°F) the TCC will also apply in second and third gears to reduce friction generated in the torque converter.

Pulse Width Modulated (PWM) Solenoid

This solenoid provides the gradual application and release of the torque converter clutch for increased shift quality, see TCC.

Input and output speed sensors (TISS) and (TOSS)

The two speed sensors are of the magnetic induction type, and are mounted on the left hand side of the transmission - the input sensor forward of the transmission centre, and the output sensor close to the rear extension housing. The induced voltage in the input sensor is generated from machined serrations on the forward clutch housing, and in the output speed sensor by serrations on the rear carrier assembly. The information from these two sensors is used by the transmission control module (TCM) to determine, whether the engine is running, vehicle speed, the gear ratio, the TCC slip, and the turbine speed.

Shift Solenoids 'A' and 'B'

Both solenoids are attached to the valve body and are 'normally open exhaust valves'. The TCM activates the solenoid by earthing through an internal 'quad driver', activates the shift valve and so controls shift pattern.

For solenoid operation see 'active components for each gear ratio'. Solenoid 'A' is usually grey in colour, and solenoid 'B' is usually green.

Variable Force Solenoid / Motor (VFS)

The force motor is attached to the valve body and controls shift quality, dependant upon engine load, based on information received from the TCM.

Transmission Control Module (TCM)

This unit is an electronic module which controls gear shift points and shift quality. Using data received from various sensors, the TCM is 'updated' every 25 milliseconds, and adapts to changes in engine load, altitude, and other conditions. Electrical signals are then transmitted to the shift solenoids which activate the shift valves for precise shift control.

The luggage compartment mounted TCM has also a diagnostic capability such that it continually monitors the transmission's conditions and stores the performance data in its memory. This information can then be down-loaded and read using the Jaguar Diagnostic System.

Pressure Switch Manifold (PSM)

This is a gear range sensing device used by the TCM to sense the gear range that has been selected by the driver. The PSM, located on the valve body, contains five normally open pressure switches which under the various fluid pressures fed from the manual valve, provide system signals to the PCM which determines the gear range that the transmission is operating in.

Barometric Pressure Sensor

This sensor provides the TCM with altitude/atmospheric information. It is mounted adjacent to the TCM in the luggage compartment.

Transmission Temperature Sensor

The temperature of the transmission fluid is monitored by this sensor which relays information to the TCM. If the temperature rises above a certain figure, then the torque converter clutch is applied, see Torque Converter Clutch. This sensor is contained in the internal transmission harness.

Accumulators

The accumulators act as shock absorbers to cushion the engagement of the transmission clutches. The clutch 'apply' fluid pressure on one side of the accumulator piston acts against the accumulator spring pressure and the accumulator fluid pressure on the opposite side of the piston. This sequence is damped by controlling the exhaust rate of the accumulator fluid. There are accumulators for the second third and fourth clutches.

Oil Pump And Internal Valves

The oil pump is mounted in the transmission assembly situated between the torque converter and the transmission gear's casing. The pump is a constant mesh spur gear type and the drive gear, being keyed to the torque converter hub, is driven at engine speed whenever the enine is running. The oil pump contains valves for pressure regulation, torque converter clutch (TCC) enable and shift, converter limiting, and reverse boost.

Pressure Regulator Valve Train

A pressure regulating valve maintains the transmission fluid at a constant pressure, to ensure correct operation of the transmission, directing fluid into the converter limit valve, and the pump suction circuit to regulate the pump output.

Reverse boost valve

The boost valve is moved towards the pressure regulating valve when activated by torque signal fluid pressure from the variable force motor, the regulating valve is then moved against the fluid supply from the pump, and so boosts the line pressures in relation to engine torque. When reverse gear has been selected, the reverse fluid pressure forces the reverse boost valve to move towards the pressure regulator valve to boost the line pressure.

Torque Converter Clutch Enable Valve

To retain the TCC in a released condition, the TCC shift valve is held by regulated converter feed fluid which passes through the enable valve to the TCC enable circuit.

Torque Converter Clutch Shift Valve

The TCC shift valve, being held in the release position by spring force and TCC enable fluid, permits regulated converter feed fluid to pass through the valve and enter the TCC release circuit. To apply the TCC, enable fluid exhausts and the valve is shifted by signal fluid pressure

Converter Limiting Valve

This valve permits converter feed pressure from the pressure regulator valve to enter the regulated converter feed circuit. Excess converter feed pressure causes the limit valve to move against its spring pressure, and so the converter feed fluid is opened to exhaust.

Valves In The Valve Body

The valve body contains: the accumulator valve, the actuator limit valve, the TCC apply valve, the manual valve, the 3 - 4 shift valve, the 2 - 3 shift valve, and the 1 - 2 shift valve.

Torque Signal Compensatory Valve

This valve is located in the accumulator housing, and torque signal fluid pressure is fed to it in each gear range. The spring in the valve dampens any pressure irregularities in the torque signal fluid pressure that are caused by the operation of the force motor.

Checkball Valves

There are eleven checkball valves, all in the transmission case except where shown otherwise, see Valve Body And Accumulator Assembly, Removal. Their applications are as follows:

#1. -	Overrun clutch
#2. -	Second accumulator
#3. -	Front band play
#4. -	Second clutch
#5. -	Third accumulator
#6. -	Fourth accumulator
#7. -	Lo / third in the control valve body
#8. -	Third clutch
#9. -	Reverse
#10. -	Fourth clutch
#11. -	Third clutch/reverse in the control valve body

TRANSMISSION ASSEMBLY, RENEW, (INCLUDING REMOVE FOR ACCESS)

Disconnect the battery.
Remove the transmission dipstick, and the engine dipstick.
Remove the upper dipstick tube and lifting eye bracket securing bolt and remove the tube; refit the bolt.
Fit Service Tool MS 53C engine support tool across the wing channels and engage the hooks in the lifting eyes (Fig. 3). Tension the hooks to support the engine.

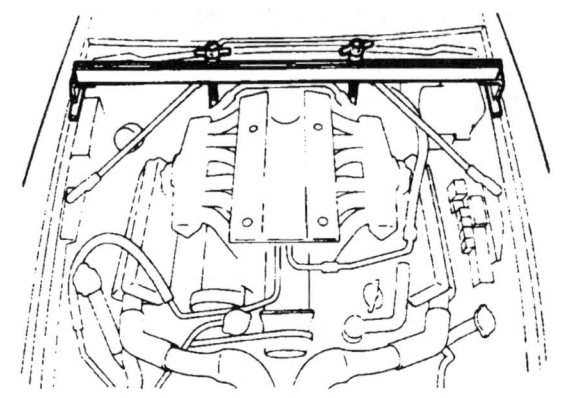

Fig. 3

Remove the fan cowl securing clips, remove the lower cowl securing nuts and position the cowl onto the engine fan to allow for engine movement.
Remove oil cooler pipe nuts, and disconnect the oil cooler pipes. Remove and discard the oil cooler pipe 'O' ring seals. Plug outlets to prevent oil loss/ingress of dirt.
Remove the downpipe catalytic converters.
Remove the underfloor catalytic converters and intermediate pipes.
Drain the transmission.
Remove the rear engine mounting.
Disconnect the propeller shaft from the transmission drive flange, and allow the shaft to lie on the exhaust heat-shield.

CAUTION:

The propeller shaft contains a sliding joint, care must be taken to ensure the shaft does not hang down.

Slowly release the engine support hooks to lower the rear of the transmission. Care must be taken to avoid damage to the camshaft oil feed pipe and the steering rack.
Remove the gear selector lever from the gearbox lever, remove the gear selector cable from the transmission housing and allow to hang clear.
Remove the transmission tunnel insulation foam.
Disconnect the harness from the transmission multi plug, and from the input and output speed sensors.
Release the harness from the clips and from behind the the speed sensor brackets and allow to hang clear.
Remove the transmission oil cooler pipes retaining clip and disconnect the pipes from the transmission housing unions, and plug the outlets.
Remove the dipstick tube and seal from the transmission housing, and plug the outlet.
Remove the flywheel sensor from its mounting bracket and torque converter stone guard.
Remove the torque converter stone guard.
Remove the flywheel sensor mounting bracket (bolts and roll pins).

Working through the access torque converter hole, remove the drive plate to torque converter retaining bolts - turning the crankshaft for access, (see Fig. 4).

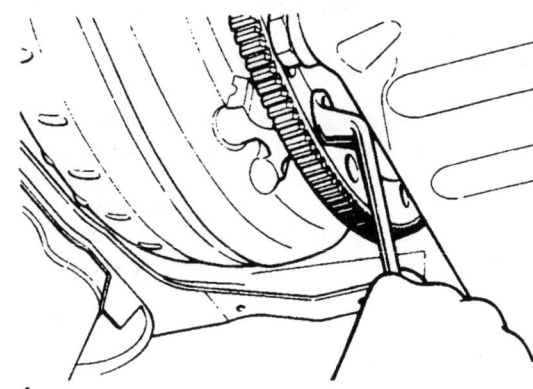

Fig. 4

Note: Use a suitable 'jam' to prevent the flywheel rotating.

Remove the starter motor retaining bolts and move the motor forward to clear the aperture.
Remove the engine to transmission securing bolt which also retains the air injection pipe/starter motor harness support bracket.
Remove six more retaining bolts, but leave the upper two.
Support the transmission with a suitable lifting device, and secure in position (Fig. 5).

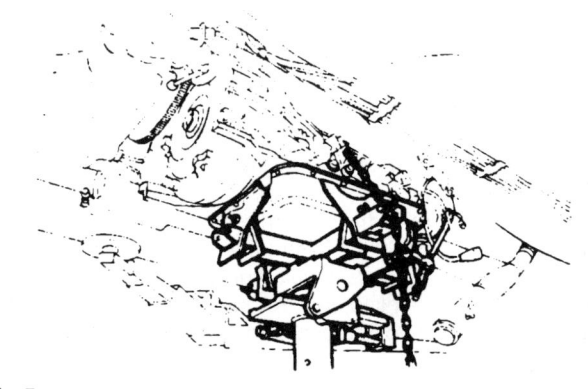

Fig.5

Remove the two remaining transmission to engine bolts, and separate the unit from the engine.
Lower the lift to remove the transmission.

Note: This point is the finish of 'Remove for Access'.

Remove the breather hose (1 Fig. 6) from its retaining clip (2 Fig. 6), and disconnect the hose from the transmission stub pipe.
Remove the hose retaining clip (2 Fig. 6) from the rear extension housing, and refit the bolt.
Remove the transmission harness from the mounting bracket, and remove the gear selector lever from the transmission shaft.
Remove the rear engine mounting spring locating cup.
Remove the oil cooler adaptor unions from the transmission housing, and plug the outlets.
Remove the propeller shaft drive flange from the transmission shaft, and fit a transit bung to the extension housing.
Remove the transmission unit from the lifting device.
Clean and inspect all components and renew all damaged components.

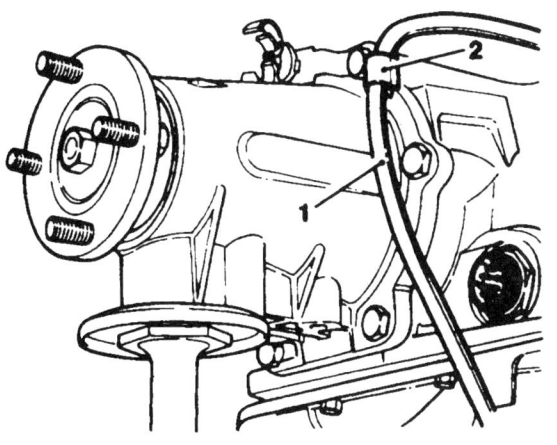

Fig. 6

Fit and secure the new transmission unit to the lifting device, and remove the transit strap and transfer it to the old transmission unit.
Remove the transit bung from the rear extension housing, and fit the propeller shaft drive flange to the transmission shaft, and secure.
Remove the blanking plugs from the transmission unit oil cooler outlets and fit the adaptor unions..
Fit the rear engine mounting spring locating cup.
Fit the gear selector lever to the transmission shaft, and the transmission harness to the mounting bracket.
Fit the breather hose retaining clip (2 Fig. 6) to the rear extension housing, and refit the bolt.
Connect the breather pipe to the transmission stub pipe.

Note: This point is the start of 'Refit for Access'.

Clean all mating faces, and raise the unit into position. Fit and secure four retaining bolts - two at the top and two at the bottom.
Remove the lifting device.
Fit the remaining bolts including the one to retain the air injection pipe/starter motor harness support bracket - lever the transmission unit to one side for access.
Re-align and secure the starter motor in position.
Align one bolt hole in the drive plate with the access aperture, then align the torque converter and fit (but do not tighten) one retaining bolt.
Turn the drive plate/torque converter and fit the remaining bolts. When all six bolts are fitted, only then secure (see Fig. 7). It may be necessary to jam the drive plate during final tightening.

Refit the rubber plug to the access aperture.
Fit the flywheel sensor bracket to the cylinder block, and refit the torque converter stone guard.

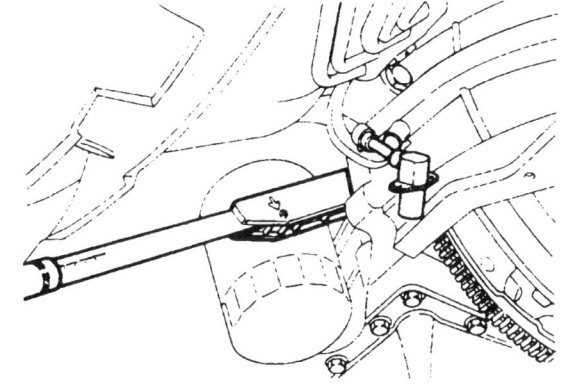

Fig. 7

Fit the flywheel sensor to its mounting bracket, and then fit the heat shield.
Remove the blanking plug and reseat the dipstick tube in position with a new seal.
Route the transmission harness speed sensor plugs behind the protection brackets, and connect the input and output speed sensors.
Connect the transmission harness multi plug.
Secure the harness to the extension housing clip.
Secure the gear selector cable to the transmission casing, and connect the cable ball pin to the selector lever.
Refit the transmission tunnel foam insulation.
Tighten the engine support hooks to raise the transmission into position.
Reconnect the propeller shaft.
Refit the rear engine mounting.
Realign the fan cowl and secure the lower mounting bolts.
Reconnect the engine oil cooler pipes using new 'O' ring seals.
Fully release the engine support hooks, and remove the support - Service Tool MS 53C.
Secure the fan cowl to its upper mountings.
Remove the RH rear lifting eye securing bolt, fit the dipstick tube in position, and refit the bolt.
Refit the transmission and engine oil dipsticks.
Check and adjust the selector cable assembly.
Refit the downpipe catalytic converters.
Reconnect the battery.

WARNING:

If contamination is suspected, the transmission oil cooler and feed/return pipes must be flushed out and thoroughly dried before connecting to the transmission unit.

Refit the RH underfloor catalytic converter, (removed to allow flushing equipment to be connected).
Align and connect the oil cooler feed and return pipes to the transmission unions, and fit to the engine sump retaining clip.
Fill the transmission unit with fluid.

Torque Figures

Detent shaft lever nut	20 - 27 Nm
Drive plate to torque converter	58.5 - 71.5 Nm
Fluid cooler connector	35 - 40.5 Nm
Transmission unit to cylinder block	41 - 49 Nm

TRANSMISSION FLUID, RENEW

Wipe clean around the transmission sump pan drain plug, remove the plug (1 Fig. 8), and allow the fluid to drain into a suitable container.

Inspect the drain plug for foreign material, wipe clean and also the sump pan drain plug face.
Remove the transmission dipstick, fit a suitable funnel to the dipstick tube and initially fill with 4.5 litres of oil.
Remove the funnel and refit the dipstick.
Ensure that the parking brake is applied, and start the engine.
Apply the footbrake and slowly operate the gear selector through all the gear positions, finally leaving the selector lever in the 'P' - park position.

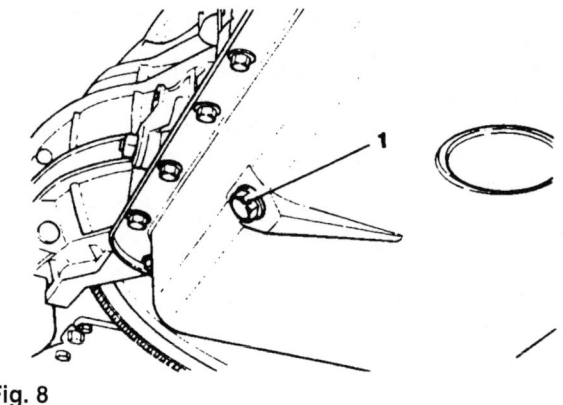

Fig. 8

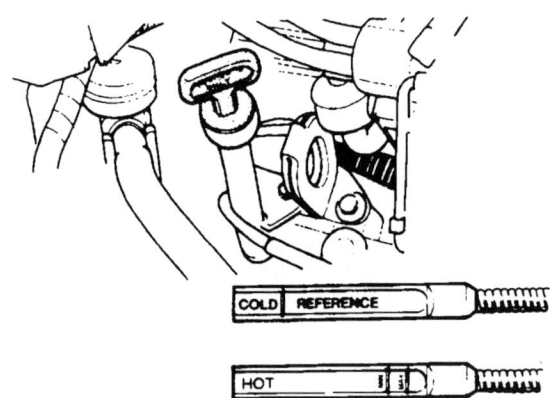

Fig. 9

Withdraw the dipstick and wipe clean. Insert the dipstick fully, withdraw it and note the fluid level. If the fluid is to the correct 'cold' mark - see Fig. 9 - replace the dipstick.

If the fluid level is low, fit the funnel and add fluid - checking frequently with the dipstick to avoid over-filling.
Remove the funnel, refit the dipstick and switch off the engine.

WARNING:

On no account must the vehicle be driven if the fluid is BELOW the cold level and the system has not been primed.

Carry out a road test until the transmission is at full operating temperature.
If the above procedure has been carried out correctly, the fluid level should be between the MIN / MAX marks on the dipstick (Fig. 9), and there should be no reason to add more fluid.

Torque Figures

Sump drain plug   20 - 27 Nm

SUMP PAN, GASKET AND FILTER, RENEW

Drain the transmission fluid.
Remove the RH underfloor catalytic converter/intermediate pipe.
Slacken the RH downpipe catalytic converter to manifold nuts, and re-position the pipe to one side by inserting a suitable piece of wood between the downpipe and the engine.
Slacken the LH underfloor pipe to downpipe catalytic converter.

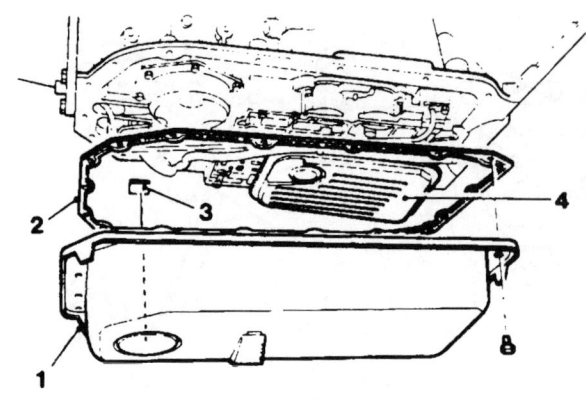

Fig. 10

Drain the residual fluid from the sump (1 Fig. 10)

Remove and inspect the gasket (2 Fig. 10) and if damaged then renew, but it may be re-used if in good condition.
Remove the magnet (3 Fig. 10) from the pan, clean and place to one side.
Place the sump aside if it is being renewed.
Clean the sump pan including the gasket surface of the transmission case.
Renew the filter (4 Fig. 10) if required by twisting and pulling downwards.
Fit the magnet to the recess in the pan.
Fit the sump pan into position with the gasket and secure.
Refit the RH underfloor catalytic converter/intermediate pipe.
Remove the block of wood, re-align the RH downpipe, and secure the manifold flange nuts.
Refill the transmission with fluid.

Torque Figures

Sump pan to case   20 - 27 Nm

GEARSHIFT INTERLOCK SOLENOID, ADJUST

Disconnect the battery.
Remove the centre console veneer panel.
Remove the selector indicator housing, and engage the gear selector to 'P' - park position to check engagement with the interlock lever.
Slacken the interlock solenoid retaining nuts, (1 Fig. 11).
Support the gear selector lever mid-way between 'P' park and 'R' reverse notches - assistance will be required, retain the gear shift interlock solenoid rearwards against spring pressure to eliminate free play, and secure the solenoid retaining nuts (1 Fig. 11).

Note: With the gear selector lever held mid-way between

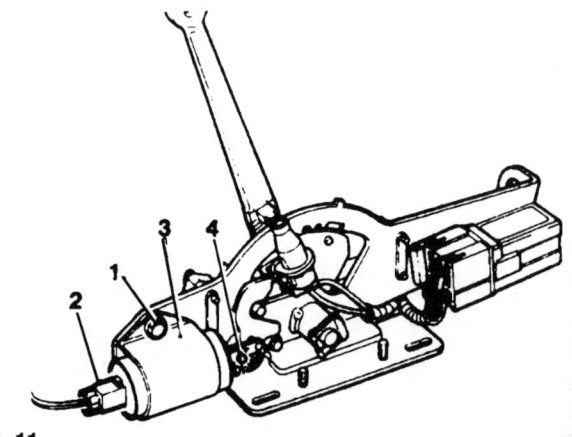

Fig. 11

58

the 'P' and 'R' positions, there should be no free play at the interlock lever.

Operate the selector lever to 'P' park position to check engagement with the interlock lever.
Refit the gear selector indicator housing.
Refit the centre console veneer panel.
Reconnect the battery.

Torque Figures

Solenoid retaining nut   6 - 8 Nm

GEARSHIFT INTERLOCK SOLENOID, RENEW

Disconnect the battery.
Remove the centre console veneer panel.
Remove the selector indicator housing.
Remove the gearshift interlock solenoid retaining nuts (1 Fig. 11), disconnect the multi plug (2 Fig. 11), and remove the interlock solenoid and plunger assembly (3 Fig. 11).
Fit the new solenoid in position, ensure the hole in the plunger engages with the interlock lever peg (4 Fig. 11).
Fit the nuts but do not secure.
Remove and discard the transit clip from the solenoid / plunger assembly (3 Fig. 11), and adjust the solenoid.
Refit the gear selector surround.
Refit the centre console veneer panel.
Reconnect the battery.

TRANSMISSION CONTROL MODULE (TCM), RENEW

Disconnect the battery.
Remove the rear seat squab.
Remove the rear quarter trim panel.
Remove the rear quarter lower trim pad.
Remove the TCM to upper mounting bracket securing nuts (1 Fig. 12) and cover (2 Fig.12).

Remove the TCM and mounting bracket assembly (3 Fig. 12) from the body mounting bracket.
Disconnect the TCM multi plug (1 Fig. 13).

Remove the lower mounting bracket to TCM securing nuts (2 Fig. 13), and remove the bracket (3 Fig. 13).
Fit and align the lower mounting bracket (3 Fig. 13) to the new TCM, and secure with the nuts (2 Fig. 13).
Connect the harness multi plug (1 Fig. 2) to the TCM and fit the TCM and mounting bracket assembly to the body mounting bracket; ensure the lower bracket slot is fully seated.
Fit the upper mounting bracket and secure with the nuts (1 Fig. 12).

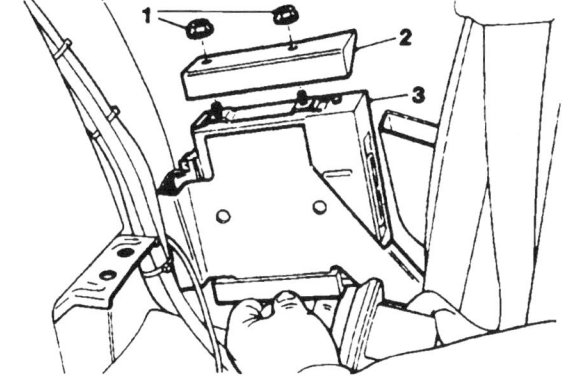

Fig. 12

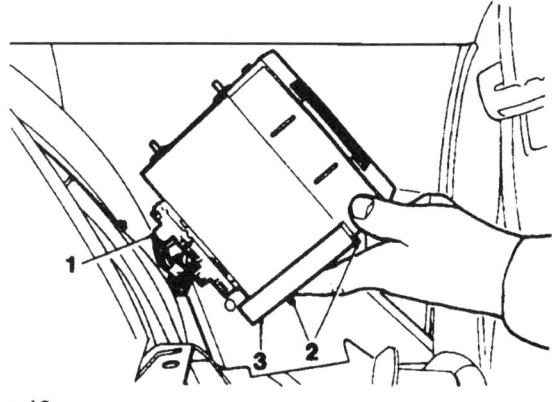

Fig. 13

Refit the rear quarter lower trim pad.
Refit the rear quarter trim panel.
Refit the rear seat squab.
Reconnect the battery.

Torque Figures

Nut - TCM to mounting bracket   2 - 3 Nm

BAROMETRIC PRESSURE SENSOR, RENEW

Disconnect the battery.
Remove the two cover retaining screws (1 Fig. 14), and remove the cover (2 Fig. 14).

Disconnect the harness multi plug (3 Fig. 14) from the barometric sensor (4 Fig. 14).
Remove the two nuts (5 Fig. 14) and remove the sensor.
Fit the new sensor into position, and secure with the two nuts.
Connect the multi plug. and refit the cover with the two screws.
Reconnect the battery.

Torque Figures

Cover screw   2 - 3 Nm
Sensor mounting nut   2 - 3 Nm

KICKDOWN SWITCH, ADJUST

Place the gear selector lever in 'P' park position.
Position the Jaguar Diagnostic System (JDS) unit alongside the vehicle, insert JDS disc 'A' in the disc drive, and switch on the unit.

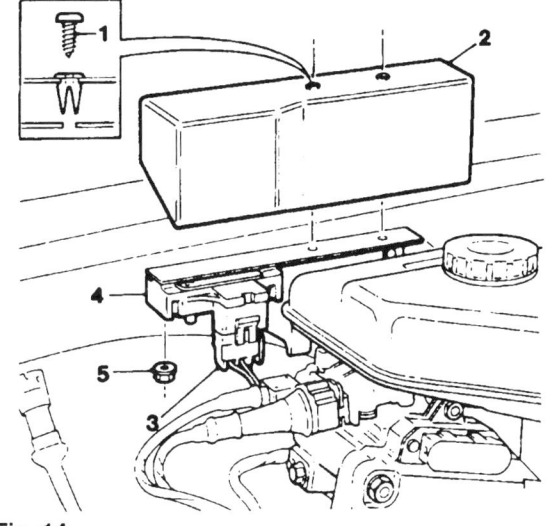

Fig. 14

Allow the JDS to go through its testing sequence, then press function '2' on the keypad.
Remove the battery cover and connect the JDS reference cables to the battery posts.
As the system is self explanatory, it is comparatively easy to follow through the set-up sequence.
All connection and adjustment instructions to the vehicle are fully designated, and must be adhered to.

Note: Access to the kickdown switch multi plug is obtained by removal of the driver's footwell carpet. The kickdown switch and relative positions of the throttle pedal is shown at Fig. 15.

When the setting-up procedure has been carried out, switch off the ignition, disconnect all the JDS cables from the vehicle including those to the battery terminals, switch off the power to the JDS unit, and remove the disc.
Refit the battery cover.

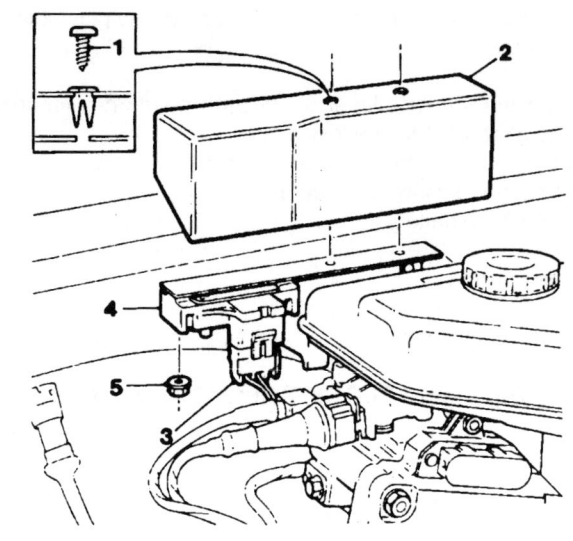

Fig. 15

KICKDOWN SWITCH, RENEW

Remove the driver's footwell carpet.
Disconnect the multi plug, and remove the kickdown switch, (see Fig. 15).
Fit, but do not tighten, the new switch.
Adjust the kickdown switch.
Refit the driver's footwell carpet.

Chapter 11

FRONT SUSPENSION (93.5 MODEL YEAR)

SUSPENSION UNIT, RENEW

Place the vehicle on a ramp.
Disconnect the battery.
Remove the left and right hand air cleaner elements.
Remove the left and right hand, upper engine mounting nuts.
Using a syringe, clear all fluid from the power steering reservoir.
Take the hooks from the Engine Support Tool, Service Tool MS 53A, and position them in the engine front lifting eyes.
Place the tool MS 53A in position over the hooks and locate in the wing channels (Fig. 1).
Remove the left hand shock absorber thread protector.
Remove the shock absorber securing nuts (1 Fig. 2).
Remove the washer, micron buffer and cup washer (2, 3 and 4 Fig. 2).
Repeat the procedure on the right hand shock absorber.

Jack up the front of the vehicle and support on two stands.
Remove the wheel and tyre assemblies.

From the wheel arch:
Unscrew the caliper brake pipe to brake hose securing union nut (1 Fig. 3). Fit a blanking plug to the pipe.

Remove the brake hose lock nut (3 Fig. 3) at the caliper abutment bracket.
Reposition the hose (4 Fig. 3) from the abutment bracket.
Remove the left hand anti-roll bar to anti-roll bar link securing nut (1 Fig. 4).

Remove the cup washer and mounting rubber.

From the engine bay:
Disconnect the left hand ABS sensor harness connector.
Remove the sensor harness 'P' clip securing bolt.
Remove the 'P' clip.
Remove the harness grommet from position in the inner wing.
From the wheel arch:
Displace the sensor harness from the retaining bracket.
Repeat the procedure on the right hand sensor harness, brake pipe / hose and anti-roll bar.
Raise the vehicle on the ramp.

Note: On convertible models only, remove the front cruciform bracing assembly.

Displace the anti-roll bar from the anti-roll bar links and position aside.
Unscrew the securing bolts and remove the right hand steering rack heat-shield.
Remove the right hand down-pipe catalytic converter.
Turn the steering to allow access to the lower column to steering rack pinion securing bolt (1 Fig. 5).

Remove the securing bolt.
Lower the ramp.

From inside the vehicle:
Remove the lower steering column to upper universal joint securing bolt (1 Fig. 6).

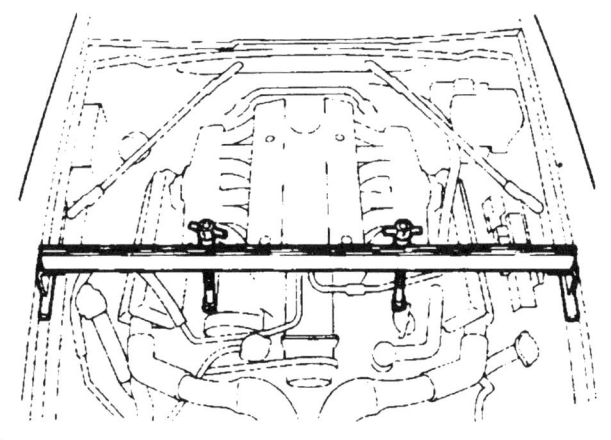

Fig. 1

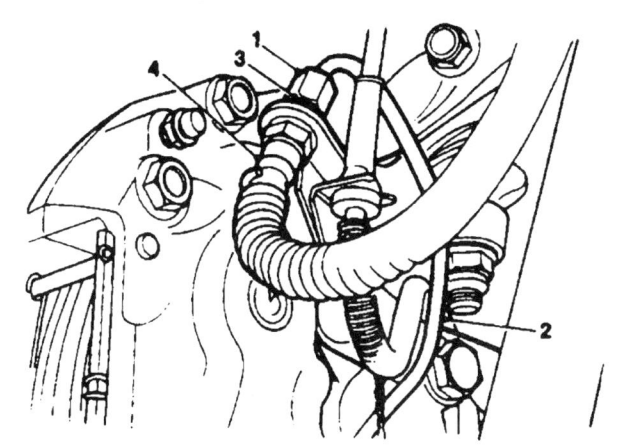

Fig. 3

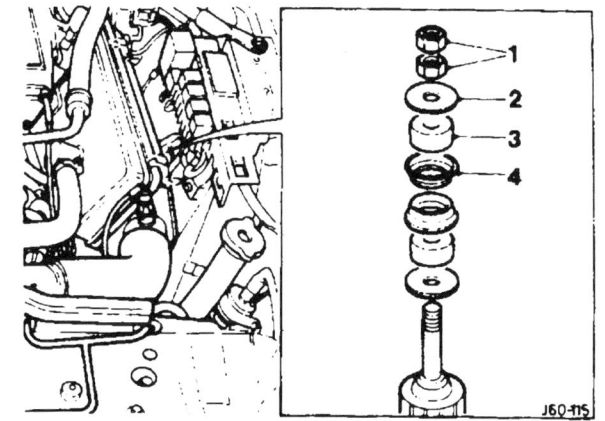

Fig. 2

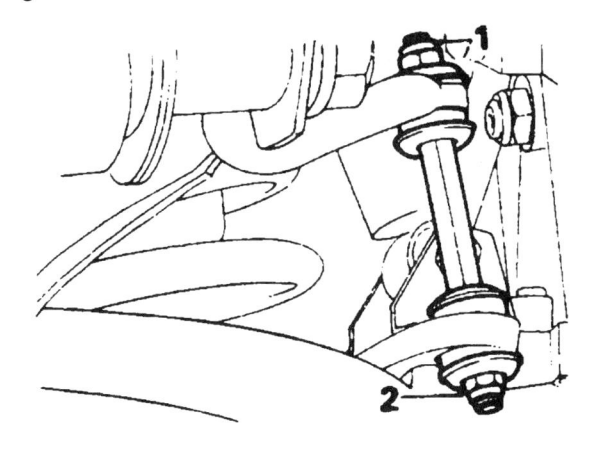

Fig. 4

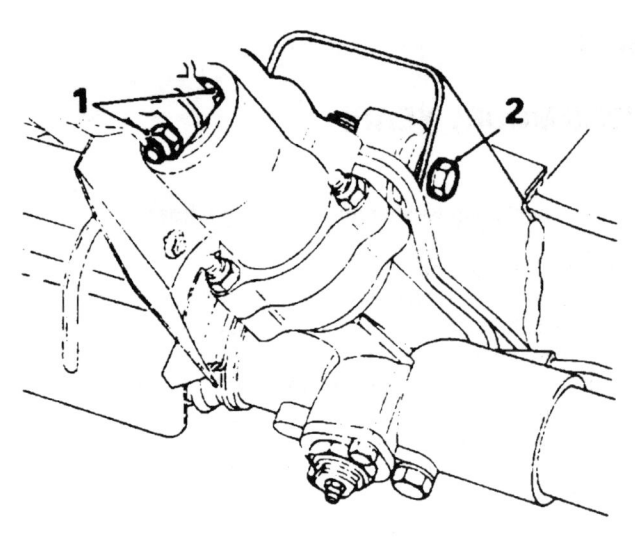

Fig. 5

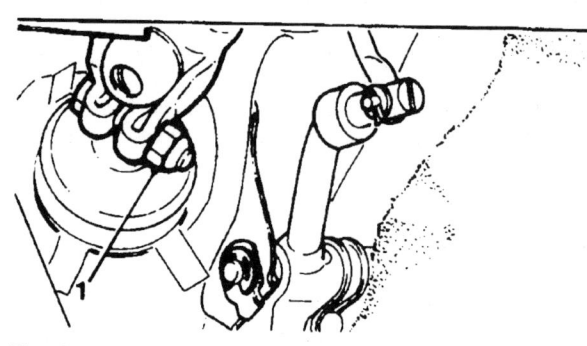

Fig. 6

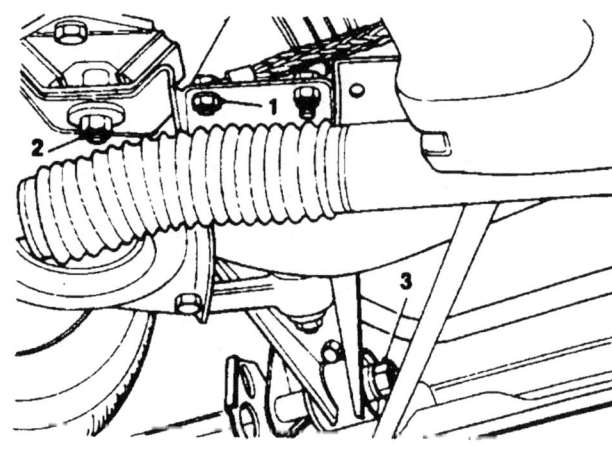

Fig. 7

Rotate the steering to give access to, and remove, the remaining universal joint securing bolt.
Ease the upper universal column clear from the bottom of the upper column.
Pull the lower column up into the vehicle and clear of the pinion.
Raise the vehicle on the ramp.
Remove all the screws securing the under-tray and remove the under-tray.
Remove the power steering feed hose union nut from the steering pump.
Reposition the hose.
Slacken the power steering return hose to pump securing clip.
Reposition the return hose and remove the hose clip.
Ensure the disconnected hoses are fitted with blanking plugs.
Remove the engine and chassis earth strap from the front beam.
Align a jack to the suspension unit and jack up the unit.
With the jack in place, remove the suspension forward mounting securing nuts/bolts (3 Fig. 7).

Note: Fig. 7 shows the under-tray removed.

Remove the spacers between the mounting and body.
Remove the rear beam mounting securing nuts. Carefully lower the jack with suspension unit from the vehicle.
Position a pulley block over the suspension unit and attach a lifting chain from the unit to the block.
Raise the lifting chain to clear the suspension unit from the jack.
Swing the pulley block clear of the jack and lower the suspension unit safely to one side.
Remove the lifting chain from the suspension unit and pulley block.
Fitting a new suspension unit is a reversal of the removal procedure.
Upon completion, refill the power steering reservoir to the correct level.
Reconnect the battery.
Bleed the ABS hydraulic system.
Remove the vehicle from the ramp.

Special Tools

Engine support tool MS 53A

Oils / Greases / Sealants

Brake fluid to a minimum DOT 4 specification

Chapter 12

REAR SUSPENSION (93.5 MODEL YEAR)

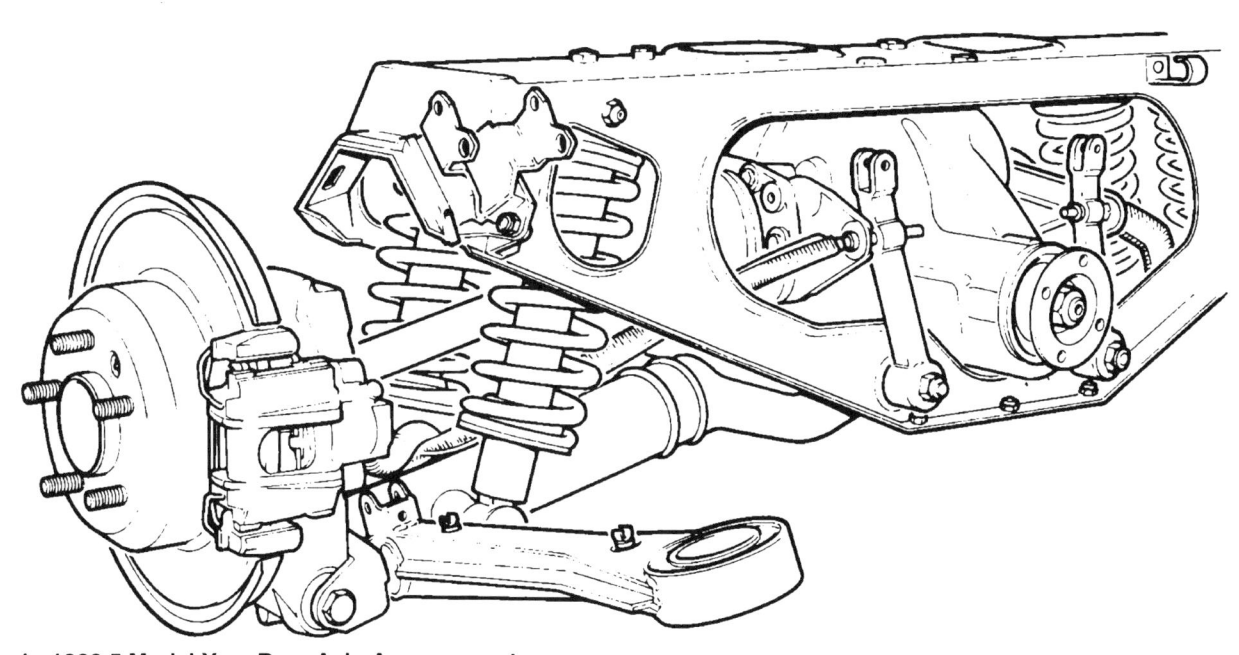

Fig. 1 1993.5 Model Year Rear Axle Arrangement

REAR HUB BEARING, RENEW

REAR HUB OIL SEAL, RENEW

REAR HUB CARRIER FULCRUM BEARING, RENEW

These three Renew Operations are covered in full within the **BRAKES** Section of this Supplement.

Chapter 13

ANTI-LOCK BRAKING SYSTEM

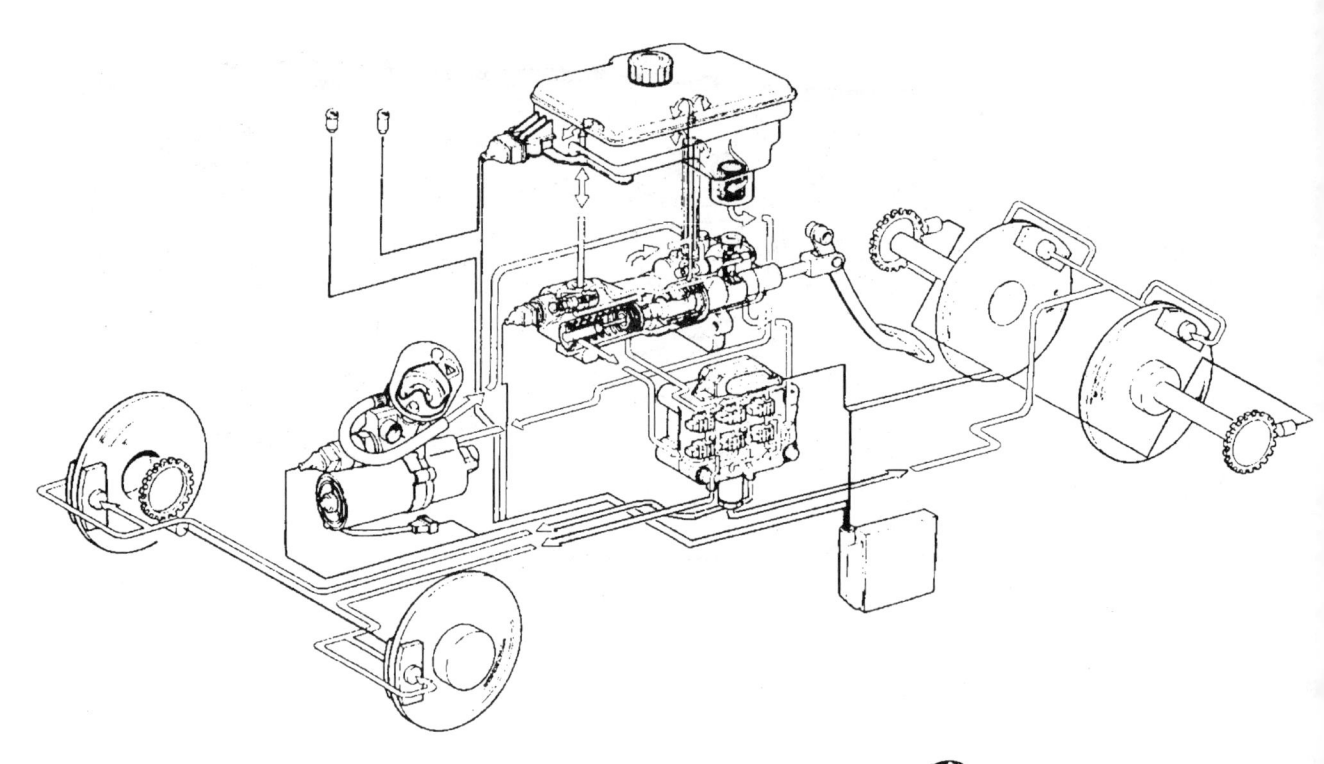

DESCRIPTION

The hydraulic components necessary for the anti-lock control are integrated with a servo-hydraulic booster and its operating components; brake fluid is used as the single working fluid. An energy source consisting of an electric motor, a pump, a pressure switch and an accumulator provides high pressure fluid to operate the booster during normal braking as well as providing a means of increasing brake line pressure when required during anti-lock braking (ABS). Also contained in this assembly are the master cylinder the solenoid valves and the reservoir.

During installation the front hydraulic circuit is operated conventionally by the master cylinder, assisted by the booster. The rear circuit is operated directly by the controlled, pressurized fluid in the booster. The front circuit is static and the rear circuit dynamic.

However when the anti-lock control is required, the front also becomes dynamic. Then inlet and outlet solenoid valves operate in each of the three circuits to control the pressure as required to prevent wheel-lock.

Sensors are installed at each wheel. Their wheel speed related signals are processed by an electronic control module (ECM) which operates the solenoid valves in the hydraulic system. The front wheels are controlled individually and the rear wheels, which are on a single hydraulic circuit, are controlled together on the select low principle. Therefore a tendency for one wheel to lock results in control of both wheels according to the need of the 'locking' wheel.

The state of the anti-lock system is continuously monitored by the ECM, which automatically switches off the system if a failure is identified, illuminating a warning lamp and leaving full, boosted braking to all wheels. Warning lamps also indicate low accumulator pressure or low fluid level in the reservoir.

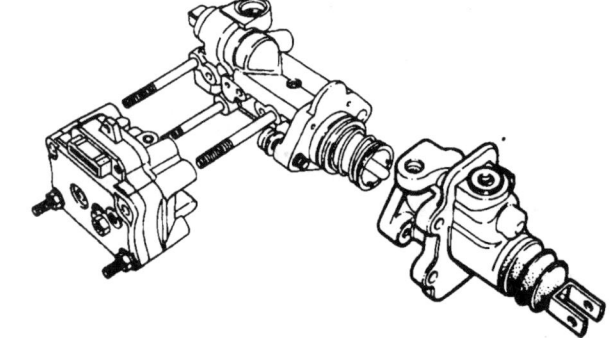

Fig. 1

If the front hydraulic circuit fails, pedal TRAVEL will increase.
If the rear circuit fails, pedal EFFORT will increase.

Hydraulic Brake Booster

One important aspect of the ABS system is the hydraulic booster (Fig. 1) which boosts the pedal force by means of hydraulic pressure. The dynamic circuit of the rear brakes is supplied from the hydraulic accumulator via a control valve in the booster.
The pressure in the booster and the rear brake circuit is proportional to the pedal force i.e.:

Low pedal force- low pressure
High pedal force- high pressure

The booster comprises an actuating piston (1 Fig. 2) and a booster piston (2 Fig. 2). The movable mechanical connection between the control valve (3 Fig. 2) and the two pistons is made by means of a scissor-lever mechanism (4 Fig. 2). The control valve (3 Fig. 2) opens the unpressurized booster chamber to the reservoir (9 Fig. 2); simultaneously the channel from the hydraulic accumulator is closed. The accumulator is constantly maintained in an operating pressure range between 140 to 180 bar.

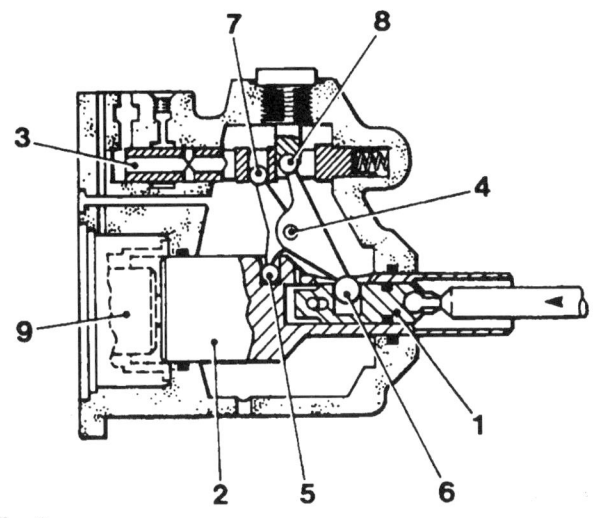

Fig. 2

As force is applied to the brake pedal, the actuating piston (1 Fig. 2) with the scissor (4 Fig. 2) moves forward. The two lower articulated balls (5 & 6 Fig. 2) move towards one another while the upper balls (7 & 8 Fig. 2) move apart. Due to this movement, the control valve (3 Fig. 2) opens the intake channel from the accumulator just after it closes the return flow. In the brake booster, a pressure is built up which is transmitted to the rear brakes and which acting simultaneously on the booster piston (2 Fig. 2) boosting the actuating force on the master cylinder piston. At the same time the pressure acts between the booster piston (2 Fig. 2) and the actuating piston (1 Fig. 2) separating the two parts. The lower articulated balls (5 & 6 Fig. 2) move apart whilst the upper balls move towards one another, this movement closes the intake by means of the control valve; the return flow remains closed.

The control valve (3 Fig. 2) is closed when the pressure acting on the actuating plate (1 Fig. 2) causes a force which is equal to the preset pedal force, i.e. when there is a balance of forces. The pressure acting on the annular circle of the booster piston (2 Fig. 2) increases the pedal force. The pedal force is increased in the ratio 1: 4, the booster ratio depends on the ratio of actuating piston area (1 Fig. 2) to booster piston area (2 Fig. 2).

The pressure in the booster is proportional to the pedal force. In maximum braking position, the control valve is open completely, the entire accumulator pressure of 180 bar acts on the booster piston. The maximum possible brake boosting is utilized. The brake pressure to the front wheel brakes can only be increased when the pedal force is increased. The brake pressure in rear wheel brakes cannot exceed 180 bar even in the case of the pedal force increase.

Electronic Control Module (ECM)

The ECM processes the signals from the four wheel sensors, converts their frequency information into values which correspond to wheel speed and then, with the data received, controls the solenoid valves during braking under ABS control.

The ECM checks the input and output signals in order to indicate any ABS disturbances. A self-monitoring function is integrated in the ECM which, in the case of a system failure, illuminates the ABS warning lamp.
During ABS control the respective solenoid valves in the wheel circuit concerned are controlled. The main valve is controlled when a front wheel is under ABS control. In order to enable the wheel to transmit the optimum brake force under all road conditions, the control of the valves must be operated very rapidly, up to 6 times a second.

When the ignition is switched on, battery voltage is supplied to pin 2 of the ECM which is fed internally to pin 8. From pin 8, the main relay coil is activated closing the relay contacts thereby allowing Battery Voltage to be applied to pins 3 and 20. which switches on the ECM thus starting the test routines. The module is protected by a 30A fuse.

During the test routines, the ABS warning lamp is illuminated by being switched to earth via pin 1. The time the lamp is illuminated depends on the charging of the hydraulic accumulator. If the system test routines prove satisfactory the electronic control module opens the earth circuit between pins 1 and 27, the warning light is then extinguished. The warning lamp is supplied with battery voltage from the ignition switch when the ignition is switched on.

ABS and Brake Warning Lamp

The ABS warning lamp warns the driver in the case of a malfunction in the ABS system i.e. loss of pressure and indicates that the ECM has switched off the ABS system. The consequence is the loss of the anti-lock function, the conventional brake with brake boosting is, however maintained.

In the case of a failure due to faulty plug connections, broken cables or defective components, such as sensors and solenoid valves, the voltage supply to the main relay is switched off by the ECM via terminal 8. The relay de-energises disconnecting the supply to pins 3 and 20.

Voltage (Refer to circuit Fig. 3) is available to the ABS warning lamp from the ignition switch. The earth circuit for the warning lamp is via the diode and contacts 30/87a of the main relay.
If the ECM is not connected the warning lamp will illuminate because it is connected to earth via the main relay.

Should the ABS and the brake warning lamp illuminate simultaneously brake pedal feels normal but the fluid is low.

The brake warning is illuminated because an earth connection for the warning lamp is made via the switch contacts located in the reservoir.

The ABS warning lamp is illuminated because fluid level switch is open and the circuit between terminals 9/10 of the ECM is broken. During the system self test, the ECM will

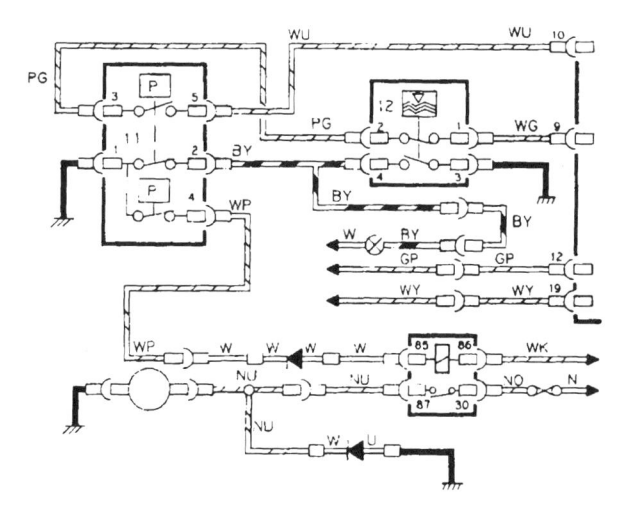

Fig. 3

recognize this failure and switch terminal 27 from the warning lamp to earth, at the same time, the system will be partially inhibited.

Should the pedal feel hard (after some brake applications) i.e. in the case of accumulator pressure below 85 bar there is no boost braking.

The brake warning lamp illuminates because the pressure is below 105 bar. The warning lamp switch contacts in the combined pressure switch close so a circuit is made to earth via terminals 1 and 2.

The ABS warning lamp illuminates simultaneously because the warning switch contacts open breaking the circuit between terminals 9 and 10 of the ECM. During the system self test, the ECM recognizes this failure and switches terminal 27 from the warning lamp to earth, at the same time, the system will be partially inhibited.

Pump Motor Operation

With the pressure in the hydraulic accumulator below 140 bar the motor will switch on.

When the ignition is switched on battery power is supplied to the relay coil, if the pressure in the hydraulic accumulator is below 140 bar, the coil will be earthed.

Battery power is then supplied to the pump motor via the closed contacts of the relay and a 30 amp fuse.

The motor operates until a pressure of 180 bar is achieved. Having reached a pressure of 180 bar the pressure switch contacts open de-activating the relay thus switching off the pump motor. When the pressure drops to 140 bar the pressure switch closes, switching on the pump motor so that the system pressure is maintained at between 140 to 180 bar.

Motor Pump Unit

With the motor pump unit, an independent energy supply is obtained by means of a motor and a pump which generates hydraulic energy, and accumulates the pressure energy in a hydraulic accumulator. From the hydraulic accumulator the pressure supply for the dynamic circuit of the rear wheel brakes, the hydraulic brake booster and the static circuit of the front wheel brakes is provided during a braking with ABS control.

The motor (1 Fig. 4) drives via a coupling (2 Fig. 4) a rotor (3 Fig. 4) which includes two pistons (4 Fig. 4) and two balls (5 Fig. 4) which move in an eccentric ring (6 Fig. 4). Brake fluid is drawn via the suction channel (7 Fig. 4), a filter (8 Fig. 4), through the control shaft (9 Fig. 4) on the upper side of the lower piston. The rotation causes a reduction of space due to the eccentric ring and ball. The piston is moved towards the control shaft and a pressure is generated. The pressure opens a check valve (10 Fig. 4),is transmitted to the accumulator and to the annular chamber of the control piston in the booster. Simultaneously the pressure acts on the tappet (11 Fig. 4) of the combined pressure warning switch (12 Fig. 4) and moves it till the system pressure of 180 bar is achieved, the pressure switch in the combined pressure switch switches the motor off.

The hydraulic system is protected against damage by the pressure control valve which releases pressure at 210 bar.

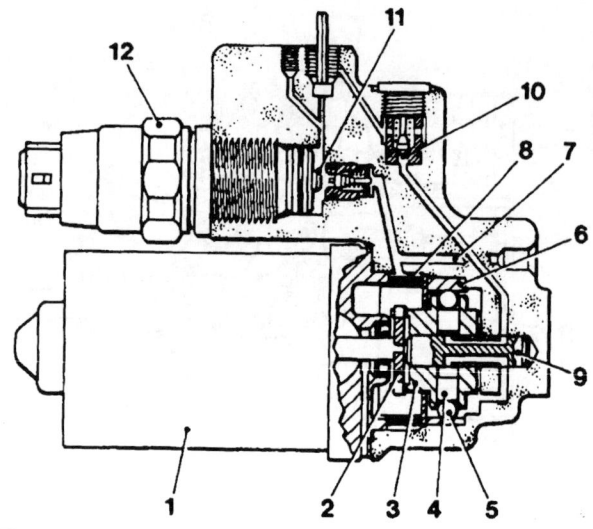

Fig. 4
Hydraulic Accumulator

The hydraulic accumulator (Fig. 5) accumulates the hydraulic pressure and makes it available to the hydraulic system for the booster and rear wheel brakes.

Owing to the fact that fluids are almost incompressible they cannot be used for an accumulation of energy. The additional use of compressible nitrogen gas in the hydraulic accumulator allows an energy accumulation. The gas and fluid must be separated by means of a membrane (1 Fig. 5).

The hydraulic accumulator has a reservoir which is divided into two chambers by a membrane (1 Fig. 5). The upper chamber (2 Fig. 5) is filled with nitrogen gas to an initial pressure of 84 bar. The lower chamber (3 Fig. 5) is filled with brake fluid supplied by the pump.

With an increasing amount of brake fluid the pressure in the system also increases, the nitrogen gas is compressed and the gas pressure is increased i.e. the nitrogen volume becomes smaller. Equally the accumulator volume for the brake fluid increases up to the cut-off pressure of 180 bar. The pressure in the accumulator is maintained by means of a check valve which is integrated in the motor pump unit and is available up to the annular chamber of the control piston.

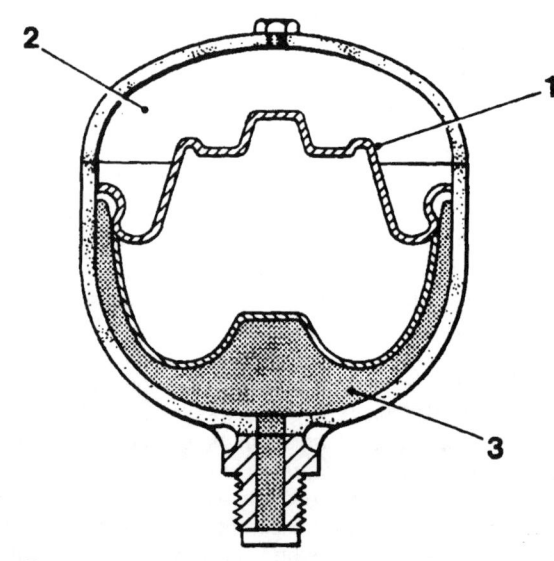

Fig. 5

Combined Pressure Warning Switch

The pump generates a pressure in the system which acts on the tappet (1 Fig. 6). The tappet moves against a spring (2 Fig. 6). At a pressure of 130 bar the warning switch (3 Fig. 6) is actuated.

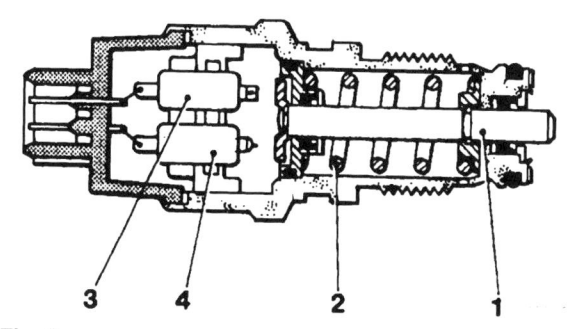

Fig. 6

First, the brake warning light will be extinguished followed a short time after by the ABS warning light. The tappet (1 Fig. 6) is moved further by pressure and opens the pressure switch (4 Fig. 6) at a pressure of 180 bar to open the pump relay coil circuit so stopping the pump motor.

If, after a number of brake applications or normal leak down, the system pressure falls below 140 bar, the pressure switch (5 Fig. 6) closes and the pump will operate until the cut-off pressure of 180 bar is reached. In this way, the system pressure is maintained within the range of 140 to 180 bar. If due to a hydraulic failure or a electrical failure, a pressure drop below 105 bar occurs, the brake warning switch (4 Fig. 6) contacts close. The brake warning lamp lights, simultaneously the ABS warning switch (3 Fig. 6) contacts open and the circuit between terminals 9 and 10 of the ECM is broken. The ECM is partially inhibited, ABS control for the front wheel brakes is switched off and the ABS warning light is illuminated.

Wheel Sensors

At each road wheel a speed sensor and a toothed rotor is fitted. By means of a magnetic core (1 Fig. 7) in a coil (2 Fig. 7) a sinusoidal alternating voltage signal is generated by a toothed rotor (3 Fig. 7) breaking the magnetic field and inducing voltage in the coil. The frequency of which is dependent on the wheel speed. The voltage is transmitted to the ECM via screened cables. The gap between the sensor and the toothed wheel is very important i.e. a large gap will generate a low voltage, and a small gap a high voltage.

Valve Block

The valve block contains three pairs of solenoid valves, one pair for each of the front brakes and one pair for the rear brakes. Each pair contain an outlet and an inlet valve. The valves are electronically operated by signals from the ECM. During braking with ABS control, the ECM provides a voltage to the inlet and outlet solenoid valves, which influence the hydraulic pressure to the brakes. The control to the rear brakes is determined by the wheel which first shows a tendency to lock. The brake pressure at both rear wheels is thus determined by the wheel having the lowest friction coefficient.

This ensures that in the case of braking on a surface with low friction coefficient neither of the rear wheels will lock.

All the valves with the exemption of the main valve have a common earth point which is connected to pin 11 of the ECM.

The earth connection pin 11 is known as the reference earth, via this connection the ECM receives test pulses for the valves.

During normal braking the anti-lock system will not be activated. However, if the braking force applied is sufficient to overcome tyre/road adhesion the anti-lock system will automatically be activated preventing the road wheel from locking.

With no current flowing through the inlet solenoids the valves are open so that during braking the brake pressure can be applied direct to the wheel brakes.

With no current flowing through the outlet solenoids the valves are closed and disconnect the wheel brakes from the reservoir.

To maintain pressure the outlet valve (1 Fig. 8) stays closed, and the inlet valve (2 Fig. 8) to a wheel with a tendency to lock' closes, ensuring the brake pressure to that wheel cannot be increased. To decrease pressure the inlet

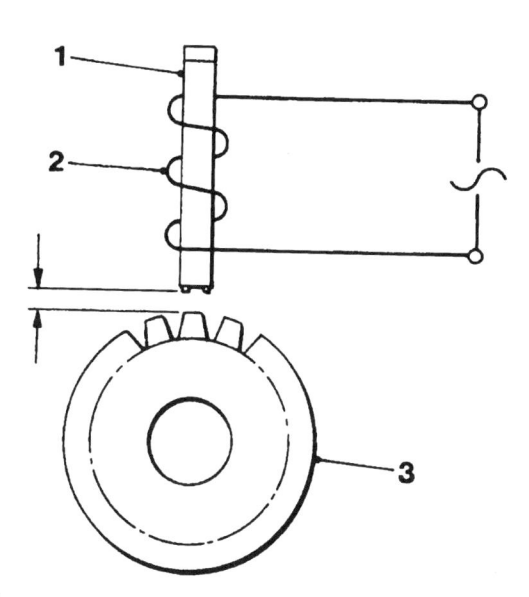

Fig. 7

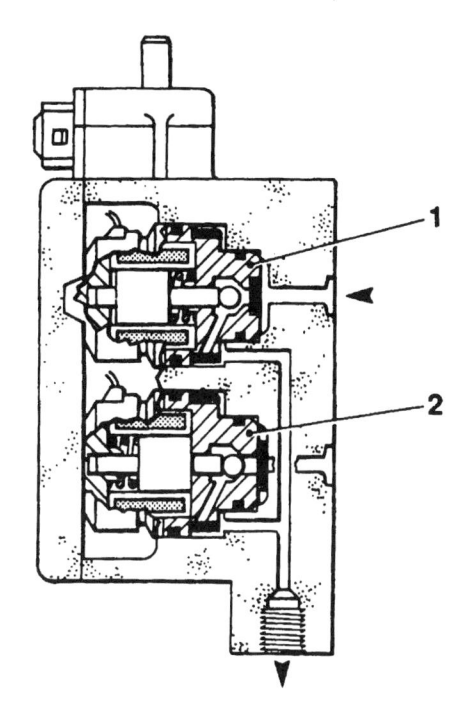

Fig. 8

valve closes and the outlet valve opens. The brake pressure to the wheel is decreased.

To increase the pressure to the wheel the inlet valve opens and the outlet valve closes and the brake pressure to the wheel is again increased almost up to the locking pressure limit.

These phases are repeated up to six times a second until the tendency for the wheel to lock is eliminated.

Main valve

The main valve is a solenoid operated valve which during ABS controls supplies the front wheel brakes with pressure and causes a push back of the pedal by applying a pressure to the positioning sleeve. In a braking position without ABS control the solenoid valve (1 Fig. 9) is de-energized and connects the reservoir (2 Fig. 9) with the master cylinder (3 Fig. 9).
The connection between the booster, the master cylinder and the return flow is closed.
At the beginning of ABS control the main solenoid valve is energized and connects the dynamic circuit of the booster with the static circuit of the master cylinder.

The connection between the master cylinder and the reservoir is closed.

Master cylinder

The master cylinder (Fig. 10) acts exclusively on the front wheel brakes, the static circuit. The pressure in the master cylinder is generated by means of force acting from the booster piston to static fluid column.

In addition, the positioning sleeve and the main valve are integrated in the master cylinder.

The brake pressure to the rear wheel brakes is supplied by the hydraulic accumulator through the control valve.

As pressure is applied to the brake pedal, the master cylinder piston (1 Fig. 10) is moved forward by the booster piston (5 Fig. 10). The control valve (2 Fig. 10) closes and a pressure is built up in the front wheel brakes. Simultaneously to the forward movement of the booster piston, the positioning sleeve (3 Fig. 10) is pulled to the left. The main valve (4 Fig. 10) is in an inoperative position. The connection between the master cylinder and the reservoir is open, the connection between the booster and the master cylinder is closed.
If there is a wheel locking tendency during braking at one or more front wheels, the main valve (4 Fig. 10) is controlled by the ECM. The main valve closes the connection between the static circuit of the master cylinder and the reservoir.

Simultaneously, the connection between the dynamic circuit in the rear wheel brakes, the booster, and the static circuit in the front wheel brakes is made.

The dynamic pressure is applied to the positioning sleeve (3 Fig. 10) which is moved to its stopping point. This movement causes the push back of the booster piston and of the pedal. At the same time, the dynamic flow-in over the primary seal (6 Fig. 10) of the master cylinder piston (1 Fig. 10) and thus the direct pressure supply in the wheels occurs.

The pressure on the right side of the primary seal is higher than on the left side and due to this the primary seal is pushed forward.

Due to the pressure compensation, the master cylinder piston (1 Fig. 10) is pushed to the booster piston (5 Fig. 10).

The direct flow into the front wheel brakes and the push back of the pedal, avoids a pedal pulsation during braking under ABS control and guarantees at the same time that sufficient reserve is available in case of rear circuit failure.

Positioning sleeve

The positioning sleeve is a safety feature which ensures that in the case of a circuit failure in the rear wheel brakes there is sufficient master cylinder stroke left during ABS controlled braking. In its inoperative position the sleeve is at its stopping point. The chamber left of the positioning sleeve is connected with the reservoir via the main valve. The chamber right of the sleeve is connected direct with the reservoir.

Braking without ABS control, pressure is applied to the brake pedal, the booster piston moves to the left pushing the positioning sleeve to the left against the spring tension.

Braking on surfaces with a high friction coefficient under ABS control, pressure is applied from the dynamic circuit by opening the main valve. The positioning sleeve moves to the right to its stopping point and pushes back the booster piston as well as the brake pedal.

At the beginning of ABS control the brake pedal has a hard pedal feel.

During braking on surfaces with a low friction coefficient the booster piston and the brake pedal moves gradually to the stopping point of the positioning sleeve. Therefore, no push back movement of the pedal can be felt.

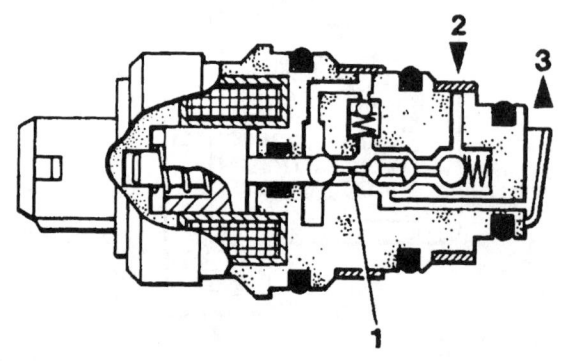

Fig. 9

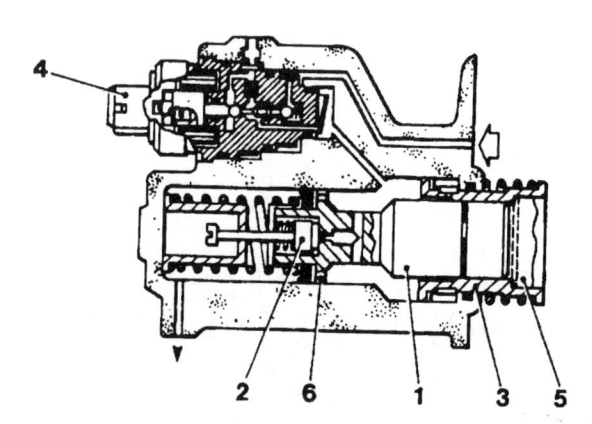

Fig. 10

Reservoir

The reservoir is made up of separate chambers. The most important chambers deliver supply to:

1. The front wheel brakes, (1 Fig. 11) i.e. for the master cylinder via the main valve.
2. The motor pump unit, i.e. for the booster and rear wheel brakes (2 Fig. 11).
3. For the return flow from the booster (3 Fig. 11), the rear brakes (2 Fig. 11), the valve block and the position sleeve (4 Fig. 11).

If due to a leakage, e.g. in the front wheel brakes, brake fluid is lost, the necessary amount of fluid for operating the rear wheel brakes is still available. The ECM is partially inhibited, the ABS for the front wheel brakes is switched off and the ABS warning light is illuminated.

In the case of a leakage in the rear wheel brakes, a residual amount of fluid is available for operating the front wheel brakes to decelerate the vehicle. The ECM is partially inhibited, the ABS for the front wheel brakes is switched off and the ABS warning light is illuminated.

The fluid level is monitored by two reed contacts on different levels inside the reservoir. The contacts are switched by contact plates (1 Fig. 12) on a float stem.

With fluid loss, the lower reed contact (2 Fig. 12) switches on the brake warning lamp, the same as a conventional brake system.

With further loss of brake fluid the upper reed contact (3 Fig. 12) is opened. The circuit between terminals 9 and 10 is broken. The ECM senses the open circuit during its continuous tests and is partially inhibited. The ABS for the front wheel brakes is switched off and the ABS warning lamp is illuminated.

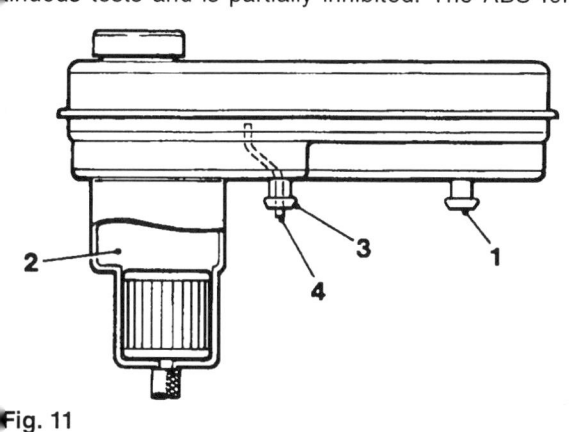

WARNING LAMP INDICATIONS WITHOUT ERROR CODE OUTPUT

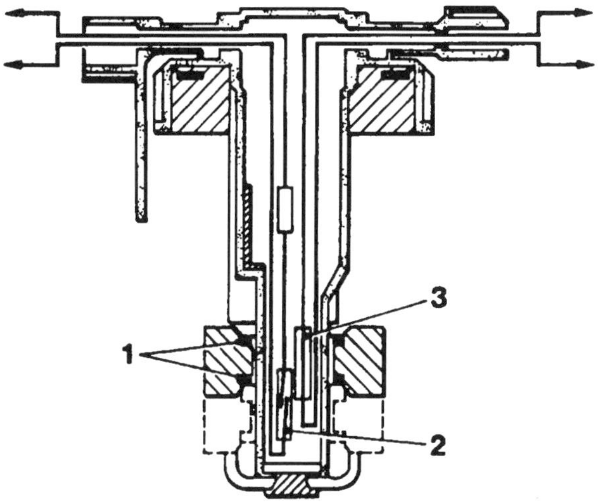

The on-board diagnosis can only monitor errors that generate electrical signals. The error code information is triggered by the diagnosis trigger input, and displayed by the warning lamp.

TEST CYCLE FOR WARNING SWITCH PATH:

After the ignition is switched 'ON' (providing the brake pressure warning light is 'OFF'), the warning lamp (WL) remains 'ON' for approximately 1.7 seconds. Then it flickers for approximately one second to test the reservoir and pressure switch path. If the warning lamp flickers continuously, this path is open or short circuited to ground potential.

IMPROPER INSTALLATION:

If the main connector is not installed in the ECM (or if the connector is loose), and the ignition is switched 'ON' (position 2), the main relay remains 'resting' and the warning lamp is switched 'ON' by the 'resting' contact of the main relay.

FAILURES OF THE ECM

FAILURES DETECTED BY INTERNAL TIME-OUT:

Certain hardware faults cause the ECM to be switched off by internal time-out. Any hardware fault will cause the warning lamp to light continuously and, since main power is cut off, the ECM is no longer capable of storing/outputting failure codes.

SHORT CIRCUIT AT THE DIAGNOSIS TRIGGER INPUT:

If the diagnosis trigger input is shorted to ground potential, the ECM goes into the diagnosis output mode when the ignition is switched on and if a failure is stored in the continuous memory. As the car accelerates and reaches 8 km/hr (5 mph), the short to ground on the diagnosis trigger unit still exists, the ECM is switched off, and the warning lamp lights continuously.

WARNING LAMP PATH FAULTS:

Short circuiting the warning lamp wire to ground potential will activate the warning lamp, but will not affect the anti-lock braking facility. The ECM cannot recognise this short circuit.

DEFECTIVE WARNING LAMP DRIVER:

If there is a defect in the warning lamp driver inside the ECM, either the warning lamp will remain continuously 'OFF', or will stay continuously 'ON', depending on the internal failure cause.

MISCELLANEOUS WARNING LAMP DISPLAYS/CONDITIONS:

In the case of intermittent V defective contacts or leads in the warning lamp driving path, the warning lamp may flicker 'ON/OFF' for undefined periods.
If the warning lamp is 'blown-out' or otherwise damaged/destroyed, no information about the status of the ECM is possible.
Note: The driver will realise that the ABS warning lamp circuitry is faulty, because the lamp will not illuminate on the ignition cycle.

Fig. 11

Fig. 12

FAULT DIAGNOSIS

FAILURE	EFFECT	RESULT	INDICATOR
Brake fluid low	Requires topping up	Brake fluid low at level 1	Brake warning lamp on
Broken sensor			No ABS
Partial intermittent failure on front axle		No ABS on rears only	ABS warning lamp on
Partial intermittent failure on front axle above 40 km/hr (25 mph)		No ABS	ABS warning lamp on
Partial intermittent failure on front axle above 20 km/hr (12 mph)		No ABS	ABS warning lamp on
Partial intermittent failure on front axle above 20 km/hr (12 mph)		No ABS	ABS warning lamp on
Pressure Switch connection broken	Accumulator will not charge	Loss of power assistance Unboosted front brakes only. No ABS	ABS warning lamp on
30 A main fuse blown (pump motor)	Accumulator will not charge	Loss of power assistance. Unboosted front brakes only No ABS	ABS and brake warning lamps on
30 A ABS fuse blown		No ABS	ABS warning lamp on
Pump connection broken	Accumulator will not charge	Loss of power assistance. Unboosted front brakes only. No ABS	ABS and brake warning lamps on when pressure drops.
Brake fluid low at level 2		Boosted brakes. ABS on rear only	ABS and brake warning lamps on
Failed front hydraulic circuit	Loss of fluid to level 2	Boosted rears with ABS only.	ABS and brake warning lamps on
Failed rear hydraulic circuit*	Loss of fluid	Unboosted front. brakes only. No ABS	ABS and brake warning lamps on

*Note: If the front hydraulic circuit fails, pedal TRAVEL will increase. If the rear circuit fails, pedal **EFFORT** will increase.

BRAKE WARNING LIGHT SWITCH

During warning light switch renewal/adjustment, the operator must ensure that the brake pedal is fully returned against its stop PRIOR TO SETTING THE SWITCH. Failure to do this may result in a no-warning lights condition.

BRAKE FLUID LEVEL

Correct brake fluid level is essential for the efficient operation of the brake system.

There are two 'MAX' marks on the reservoir. The brake fluid level must be at the highest 'MAX' level on the reservoir (Fig. 13).

Note: In some cases the fluid may be above the 'MAX' mark, this is dependent upon the charged state of the hydraulic unit. Therefore the following procedure for checking or topping up the hydraulic brake fluid level must be followed.

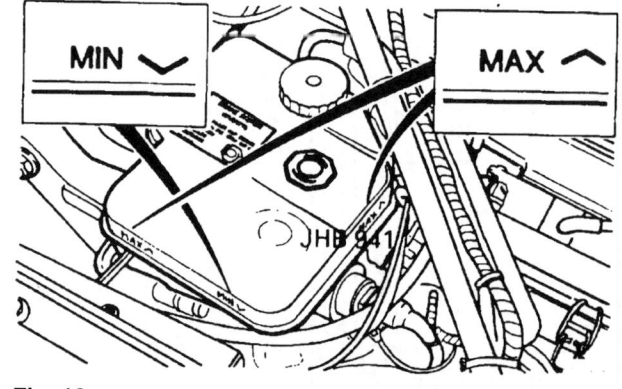

Fig. 13

> **WARNING:**
>
> THE ANTI-LOCK BRAKING SYSTEM OPERATES UNDER HIGH HYDRAULIC PRESSURE, AND GREAT CARE MUST BE EXERCISED WHEN SERVICING OR REPAIRING THE SYSTEM.
> AVOID SKIN/EYE CONTACT OR INGESTION OF BRAKE FLUID. IF SKIN OR EYES ARE ACCIDENTALLY SPLASHED WITH BRAKE FLUID, RINSE THE AFFECTED AREA IMMEDIATELY WITH PLENTY OF WATER. OBTAIN MEDICAL ATTENTION. IF BRAKE FLUID IS INGESTED, OBTAIN IMMEDIATE MEDICAL ATTENTION.

> **CAUTION:**
>
> Fluid must not be allowed to contact the vehicle paintwork. Remove any spilt fluid from the paintwork by rinsing away with running water.

1. Ensure the vehicle is on a level surface.
2. With the ignition switched 'OFF' pump the brake pedal at least 20 times, or until pedal travel becomes hard.
3. Switch the ignition 'ON'.
4. Wait for the pump to stop running.
5. Check brake fluid level. Fluid level must be at the highest 'MAX' level on the reservoir (Fig. 13). Top up using the recommended brake fluid.

The efficiency of the brakes may be impaired if fluid is used which does not meet JAGUAR specifications. Use ONLY Jaguar Brake and Clutch Fluid or Castrol Girling Universal to/or exceeding DOT 4 specification.

Also do not use brake fluid that has been exposed to atmosphere for any length of time. Moisture absorbed from the atmosphere dilutes the fluid and impairs its efficiency.

BRAKE HYDRAULIC SYSTEM - ABS, BLEEDING

> **WARNING:**
>
> THE ANTI-LOCK BRAKING SYSTEM OPERATES UNDER HIGH HYDRAULIC PRESSURE, AND GREAT CARE MUST BE EXERCISED WHEN SERVICING THE SYSTEM.
>
> AVOID SKIN/EYE CONTACT OR INGESTION OF BRAKE FLUID. IF SKIN OR EYES ARE ACCIDENTALLY SPLASHED WITH BRAKE FLUID, RINSE THE AFFECTED AREA IMMEDIATELY WITH PLENTY OF WATER. OBTAIN MEDICAL ATTENTION. IF BRAKE FLUID IS INGESTED, OBTAIN IMMEDIATE MEDICAL ATTENTION.

> **CAUTION:**
>
> Fluid must not be allowed to contact the vehicle paintwork. Remove any spilt fluid from the paintwork by rinsing away with running water. Methylated spirit (Denatured alcohol) MUST NOT be used to clean the contaminated area.
> Use only Castrol/Girling brake cleaning fluid.

> **CAUTION:**
>
> Never use methylated spirit (Denatured alcohol) for component cleaning purposes.

> **CAUTION:**
>
> Throughout the following maintenance/service operations, absolute cleanliness must be observed to prevent grit or other foreign matter contaminating the brake system.

> **CAUTION:**
>
> The following notes MUST BE read before commencing bleed procedure.

Bleeding the ABS brake system is not a routine maintenance operation and should only be necessary when either, part of the system has been disconnected or, the fluid is contaminated.

During the bleeding procedure, it is important that the level of fluid in the reservoir is maintained at approximately 2 mm below the bottom of the filler neck.

Note: The motor/pump unit cannot charge the accumulator if air, and not fluid, is standing on the low pressure side of the pump.

Therefore, if the motor/pump unit, the fluid intake hose, or the hydraulic unit of the ABS system have been renewed/disconnected, the FLUID INTAKE HOSE bleeding procedure must be used.

For all other maintenance work, use the FRONT BRAKES/REAR BRAKES bleeding procedure(s).

> **CAUTION:**
>
> DO NOT allow the pump motor to run for more than two minutes. If the motor does exceed the two minute time limit, immediately switch off the ignition (POSITION 'O'). Allow the motor to cool for at least ten minutes before continuing with the bleed procedure.

BLEED PROCEDURE (FRONT BRAKES) - FOR ALL MAINTENANCE WORK, EXCLUDING: THE MOTOR/PUMP UNIT, FLUID INTAKE HOSE, OR ACTUATION UNIT.

Ensure that the vehicle is standing level. Switch the ignition off. Discharge the hydraulic accumulator by operating the brake pedal (approximately 20 times) until the pedal travel goes hard. Top up the reservoir to approximately 2 mm below the bottom of the filler neck (Fig. 14).

Bleed the caliper furthest away from the actuation unit first, i.e., the left hand caliper on RHD cars, the right hand caliper on LHD cars.
Note: The following procedure must be done at a rate not faster than 3 seconds per cycle.

Open the bleed nipple (1 Fig. 15), fully depress the brake pedal, and hold down. Close the bleed nipple after a minimal one second. Release the pedal slowly.
Repeat until the fluid flows air free.

Note: If more than 20 cycles are required, the reservoir must be topped up.

Bleed the remaining caliper using the same procedure.

BLEED PROCEDURE (REAR BRAKES) - FOR ALL MAINTENANCE WORK, EXCLUDING: THE MOTOR/PUMP UNIT, FLUID INTAKE HOSE, OR ACTUATION UNIT.

Ensure that the vehicle is standing level. Switch the ignition off. Discharge the hydraulic accumulator by operating the brake pedal (approximately 20 times) until the pedal travel goes hard. Top up the reservoir to approximately 2 mm below the bottom of the filler neck (Fig. 14).
Open one bleed nipple (Fig. 16), fully depress the brake pedal, and hold down. Switch on the ignition and wait (A MINIMUM OF 15 SECONDS) until the fluid flows air free. Close the bleed nipple. Release the pedal slowly. Switch off the ignition. Bleed the remaining caliper using the same procedure.

Note: Ensure that the fluid level does not drop more than 10 mm below the maximum mark during the above procedure.

BLEED PROCEDURE (FLUID INTAKE HOSE) - FOR ALL MAINTENANCE WORK INVOLVING: THE MOTOR/PUMP UNIT, FLUID INTAKE HOSE, OR ACTUATION UNIT.

Ensure that the vehicle is standing level.
Switch the ignition off. Top up the reservoir to approximately 2 mm below the bottom of the filler neck. Disconnect the

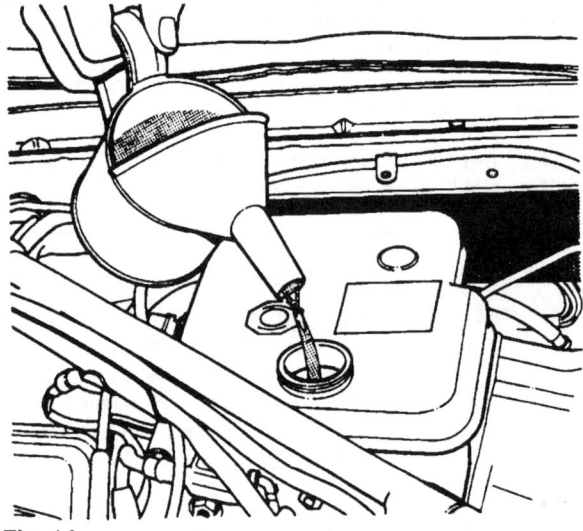

Fig. 14

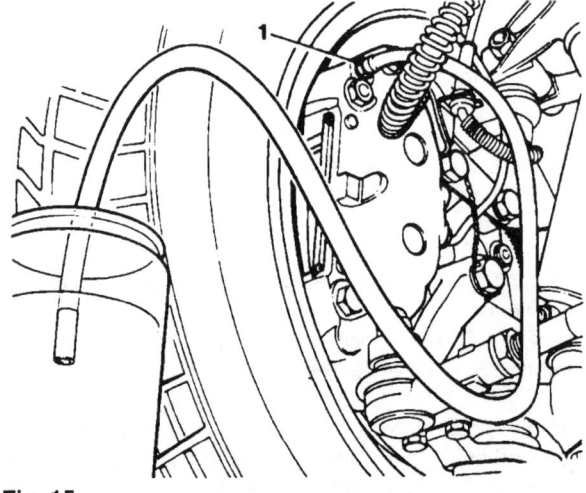

Fig. 15

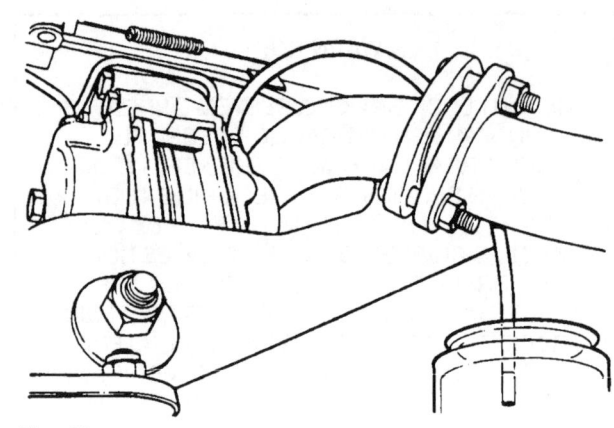

Fig. 16

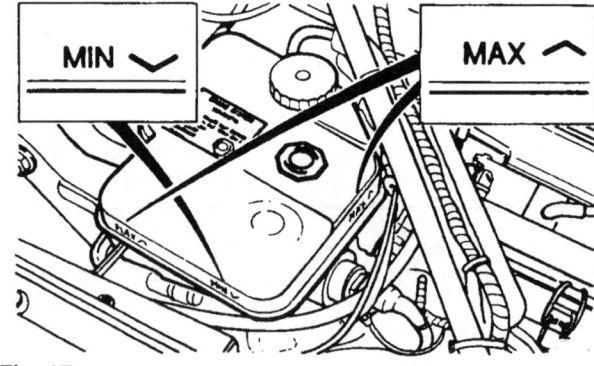

Fig. 17

fluid intake hose at the pump and allow the fluid to flow into a container until it is air free. Check that the plastic elbow '0' ring is not damaged, and reconnect the hose WHILE FLUID IS FLOWING.

Switch the ignition on and operate the brake pedal several times. If the motor/pump unit is charging, the fluid level in the reservoir will decrease. The upper cut-out point should be reached in less than 60 seconds.

ON COMPLETION OF THE BLEED PROCESS: Switch on the ignition.
Wait until the accumulator is charged to the upper cut-out point, and top up the reservoir to the maximum level (Fig. 17).

Note: If the fluid level is too high, any excess fluid must be drawn off. This can be done by using the procedure for bleeding the rear brakes.

Tighten the bleed screw to 7-13 Nm

FLUID RESERVOIR, RENEW

Note: Read the notes relating to the ABS system bleed procedure before commencing this operation.

Discharge the hydraulic accumulator by operating the brake pedal (approximately 20 times) until the pedal travel goes hard.
Open the bonnet.
Place suitable absorbent material around the reservoir area to catch any spilt fluid.

Remove the low pressure hose to pump securing clip (1 Fig. 18), disconnect the hose, remove and discard the '0' ring.
Route the end of the hose into a container. Drain the fluid.
Disconnect the reservoir multi plug.

Remove the reservoir retaining clip bolt (1 Fig. 19).

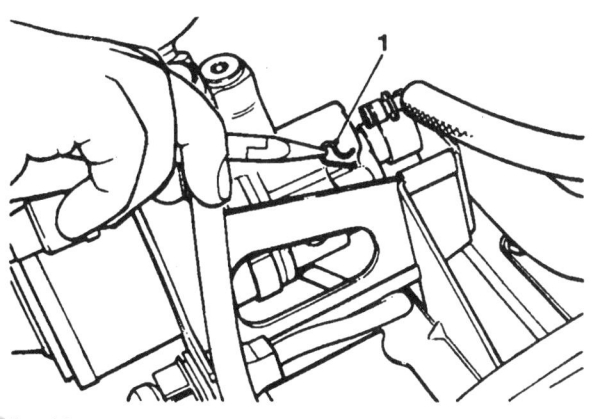

Fig. 18

Fig. 19

Displace the reservoir from position, disconnect the low pressure hose, and remove the reservoir taking care not to spill any fluid.
Remove the retaining clip, remove and discard the 'O' rings/sleeve from the actuation unit.
Fit the retaining clip to the new reservoir.
Fit new 'O' rings/sleeve to the actuation unit.
Place the reservoir in position. Connect the low pressure hose.
Fit the reservoir to the actuation unit ensuring correct location to the sleeve and 'O' rings.
Fit and tighten the retaining clip bolt (1 Fig. 19).
Reconnect the multi plug.
Connect the low pressure hose to the pump; fit the retaining clip (1 Fig. 18).
Bleed the ABS hydraulic system.
Switch on the ignition.
Wait until the accumulator is charged to the upper cut-out point, and top up the reservoir to the maximum level (Fig. 17)

ACTUATION UNIT, RENEW

Note: Read the notes relating to the ABS system bleed procedure before commencing this operation.

Discharge the hydraulic accumulator by operating the brake pedal (approximately 20 times) until the pedal travel goes hard.
Open the bonnet.
Remove the harness to RH wing stay ratchet straps, displace the harness and remove the wing stay.
Place suitable absorbent material around the reservoir area to catch any spilt fluid.
Remove the low pressure hose to pump securing clip (1 Fig. 18), disconnect the hose and remove and discard the 'O' ring.
Route the end of the hose into a container. Drain the fluid.
Disconnect the reservoir multi plug.

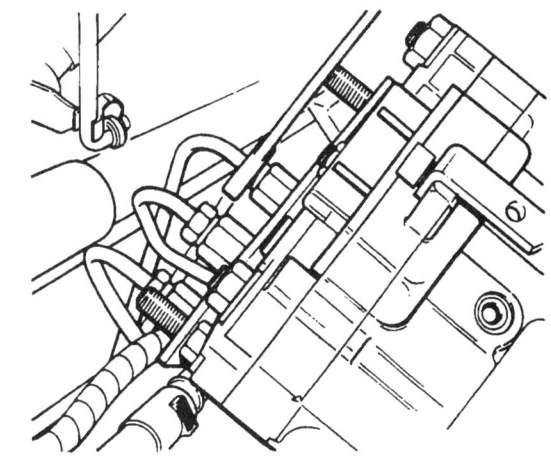

Fig. 20

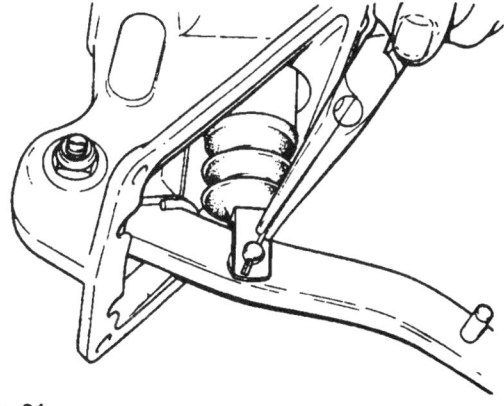

Fig. 21

Remove the reservoir retaining clip bolt (Fig. 19).
Displace the reservoir, disconnect the low pressure hose, and remove the reservoir taking care not to spill any fluid.
Disconnect the actuation unit multi plug. Disconnect the earth lead.
Place suitable absorbent material below the valve block unions to catch any spilt fluid.
Disconnect the pipes from the valve block (Fig. 20).
Disconnect the actuation unit harness
Disconnect the high pressure pipe.

Remove the pedal box to bulkhead securing bolts, displace and remove the actuation unit/pedal box assembly.
Remove the actuation unit to pedal split pin and clevis pin (Fig. 21). Remove the actuation unit securing bolts, displace and remove the unit.
Remove the high pressure pipe union. Fit blanking plugs to all ports to prevent fluid leakage.
Remove all blanking plugs from the new actuation unit.
Fit the high pressure pipe union to the unit.
Fit the actuation unit to the pedal box, fit and tighten the securing bolts.
Connect the brake pedal, fit the clevis and split pins.
Fit the assembly to the vehicle, position the valve block pipes, fully seat the assembly.
Fit and tighten the valve block pipe unions.
Fit and tighten the pedal box securing bolts.
Reconnect the high pressure pipe.
Reconnect the actuation unit harness.
Reconnect the earth lead and multi plug.

Fit new 'O' rings and sleeve (Fig. 22) to the actuation unit.
Place the reservoir in position. Connect the low pressure hose.
Fit the reservoir to the actuation unit, ensuring correct location to the sleeve and 'O' rings. Fit and tighten the retaining clip bolt.

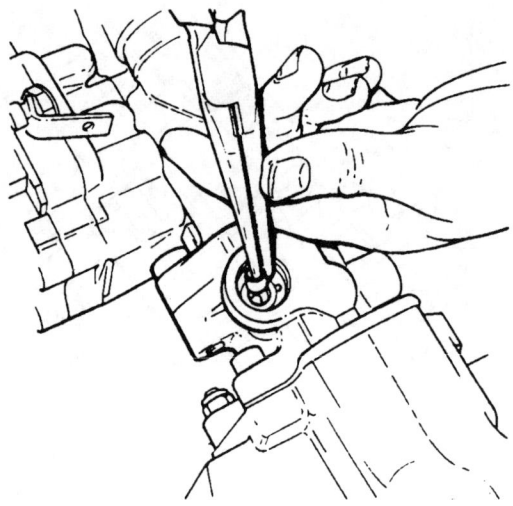

Fig. 22

Reconnect the multi plug.
Connect the low pressure hose to the pump and fit the retaining clip.
Refit the wing stay, securing the harness with new ratchet straps.
Bleed the ABS hydraulic system.
Switch on the ignition.
Wait until the accumulator is charged to the upper cut-out point, and top up the reservoir to the maximum level (Fig. 17).

Torque Figures

Bleed screw to caliper	7 - 13 Nm
High pressure pipe union	7 - 12 Nm
Pedal box to body (5/16 UNF bolt)	7 - 18 Nm
Pedal box to body (5/16 UNF nut)	1 - 18 Nm
Reservoir retaining clip bolt	6 - 8 Nm

ACCUMULATOR - ABS, RENEW

WARNING:

THE ACCUMULATOR CONTAINS PRESSURISED NITROGEN GAS. DO NOT PUNCTURE OR INCINERATE.

Note: Read the notes relating to the ABS system bleed procedure before commencing this operation.

Discharge the hydraulic accumulator by operating the brake pedal (approximately 20 times) until the pedal travel goes hard.

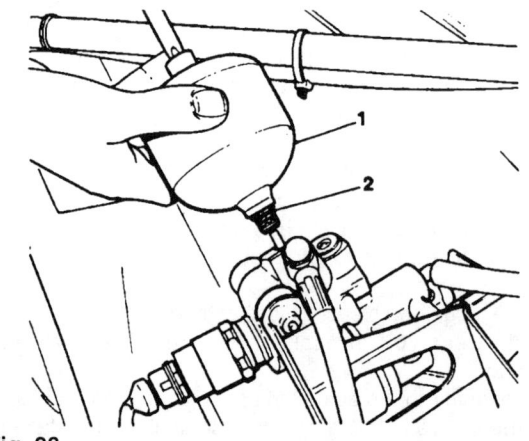

Fig. 23

Place suitable absorbent material around the accumulator area to catch any spilt fluid. Slacken off and remove the accumulator (1 Fig. 23).
Remove and discard the 'O' ring seal (2 Fig. 23). Clean the surrounding area.
Fit a new 'O' ring to the new accumulator.
Fit and tighten the accumulator.
Switch on the ignition.
Wait until the accumulator is charged to the upper cut-out point, and the ABS warning light is extinguished.
Switch off the ignition.

Torque Figures

Accumulator to pump   40 - 46 Nm

MOTOR AND PUMP UNIT, RENEW

Note: Read the notes relating to the ABS system bleed procedure before commencing this operation.

Discharge the hydraulic accumulator by operating the brake pedal (approximately 20 times) until the pedal travel goes hard.
Open the bonnet. Place suitable absorbent material around the reservoir area to catch any spilt fluid.
Remove the low pressure hose to pump securing clip (1 Fig. 18), disconnect the hose, remove and discard the 'O' ring.
Route the end of the hose into a container.
Drain the fluid.

Disconnect the pressure warning switch and motor mult plugs (Fig. 24).

Disconnect the bundy pipe (1 Fig. 25), remove and discard the 'O' rings.
Remove the motor/pump special mounting bolt.
Displace and remove the motor/pump unit.
Slacken off and remove the accumulator.
Remove and discard the 'O' ring seal.
Remove the pressure warning switch. Clean the surrounding area.

Fit a new 'O' ring to the pressure warning switch.
Fit the pressure warning switch to the new motor/pump unit.
Fit a new 'O' ring to the accumulator. Fit and tighten the accumulator.
Fit the assembly to the vehicle, routing the harness through the mounting bracket.
Fit and tighten a new special bolt.
Fit a new 'O' ring to the low pressure hose; fit the hose and retaining clip.
Fit new 'O' rings to the bundy pipe and fit the pipe.

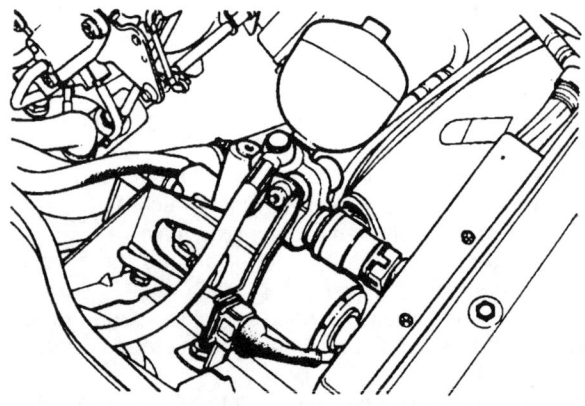

Fig. 24

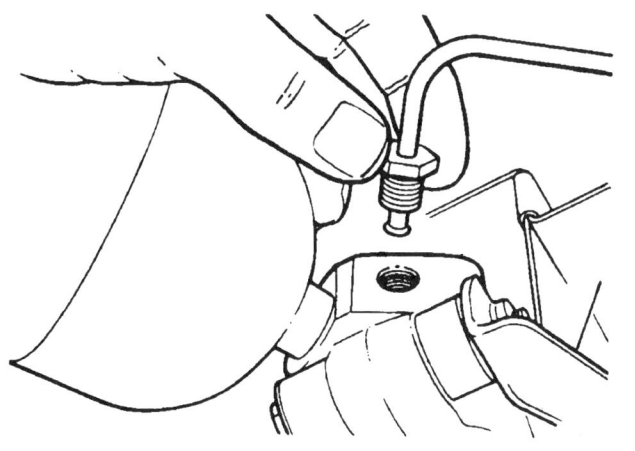

Fig. 25

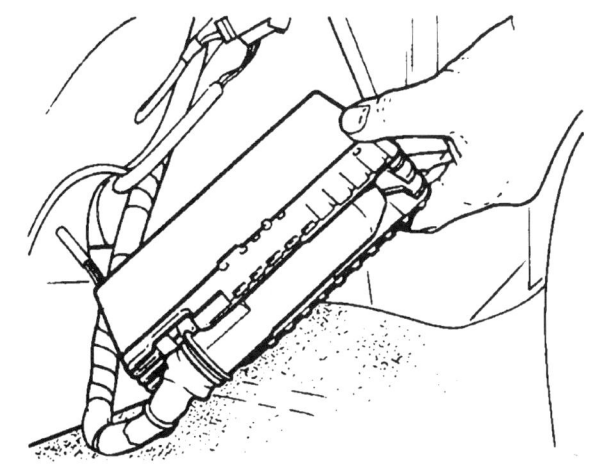

Fig. 26

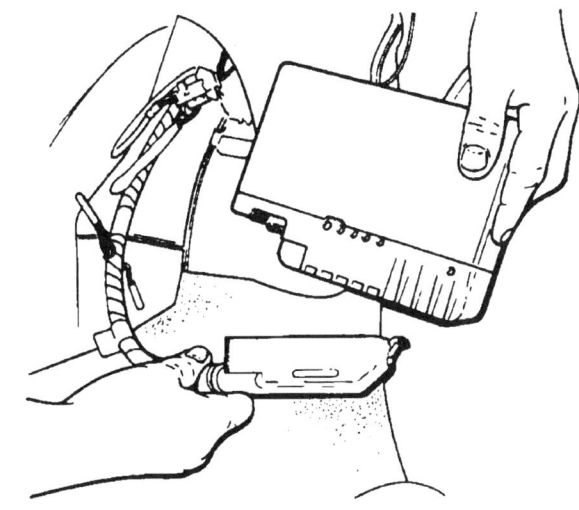

Fig. 27

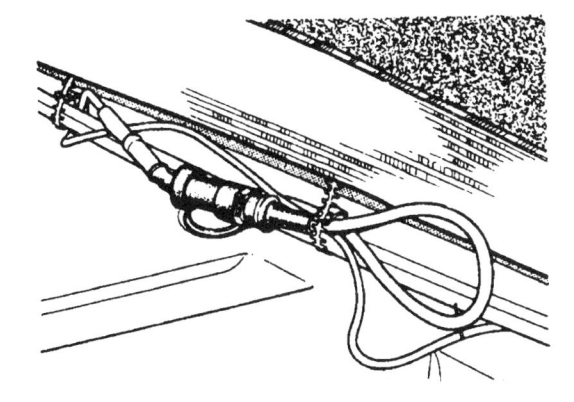

Fig. 28

Reconnect the multi plugs.
Bleed the ABS hydraulic system.
Switch on the ignition.
Wait until the accumulator is charged to the upper cut-out point, and top up the reservoir to the maximum level. Close the bonnet.

Torque Figures

Accumulator to pump	4 - 46 Nm
Bleed screw to caliper	7 - 13 Nm
High pressure pipe union	7 - 12 Nm
Pump/motor mounting bolt	7 - 9 Nm

ELECTRONIC CONTROL MODULE (ECM), RENEW

Remove the left hand side boot liner.

Release the ECM retaining strap and displace the ECM from the carrier (Fig. 26).

Disconnect the multi plug (Fig. 27) and remove the ECM.
Offer the new ECM up to position.
Connect the multi plug.
Fit the ECM to the carrier and fit the retaining strap.
Refit the boot side liner.

WHEEL SPEED SENSOR REAR, RENEW

Move the front seats fully forward. Where fitted, open the stowage compartment lid and remove the carpet.
(RH SENSOR RENEWAL ONLY). Where fitted, release the hood motor cover fasteners and remove the cover.

Displace the stowage compartment rear carpet for access to the sensor multi plug (Fig. 28).
Disconnect the multi plug.
Attach a drawstring to the sensor harness, and remove the inner grommet.
Locate a trolley jack below the outer fork of the rear wishbone.

CAUTION:

To prevent wishbone component damage,
use a suitably shaped block of wood
between the jack and wishbone.

Jack up the vehicle. Place an axle stand below the jacking spigot. Lower the car/jacking spigot on to the axle stand.
Leave the jack in position as safety measure.
Remove one road wheel nut, and mark the stud-to-wheel relationship.
Remove the remaining nuts, and remove the road wheel.

Remove the sensor securing bolt and displace the sensor from the hub (Fig. 29).
Cut and remove the harness ratchet clips.

Displace the harness outer body grommet (1 Fig. 30).
From below, carefully withdraw the harness and drawstring (2 Fig. 30).
Transfer the drawstring to the new sensor harness.
Carefully draw the new harness into position and seat the outer body grommet.

CAUTION:

Ensure that two washers (one green, one black)
are fitted to the body of the new sensor.

Route the harness to the hub, fit the sensor to the hub.
Fit and tighten the sensor securing bolt.
Secure the harness with new ratchet straps.
Lift the road wheel up to the hub, align the wheel with the marked stud and fit the wheel.

Fit and tighten the wheel nuts to 88 to 102 Nm in the sequence shown in Fig 31.

Note: It will not be possible to fully torque tighten the wheel nuts until the car has been lowered.

Take the weight of the car with the trolley jack, remove the axle stand and lower the vehicle.
Fully seat the inner body grommet.
Route the harness, remove the drawstring and connect the multi plug.
Refit the cover, carpets etc.

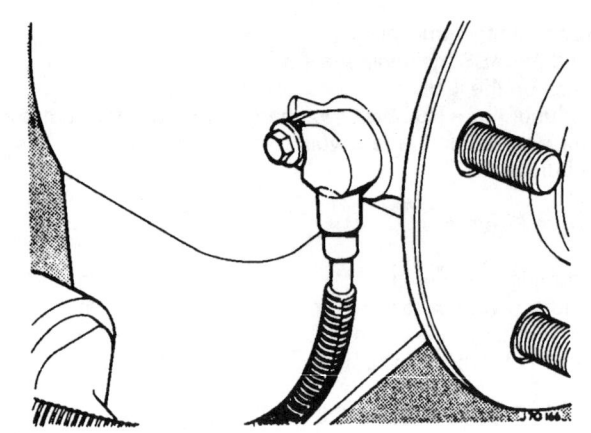

Fig. 29

WHEEL SPEED SENSOR FRONT, RENEW

Drive the vehicle on to a ramp.
Open the bonnet. Disconnect the sensor multi plug.
Displace the body grommet from the inner wing valance.
Raise the ramp.
Remove the brake disc cover plate.
Remove the retaining bracket.
Remove the upper cover securing bolt and remove the cover.

Remove the sensor securing bolt (1 Fig. 32) and displace the sensor from the hub.

Displace the harness carrier grommets (1 Fig. 33).
Displace and remove the sensor and harness.
Fit the new sensor in position, route through the inner wing and seat the harness to the carrier grommets.
Fit the sensor to the hub, and fit and tighten the securing bolt to 8 - 11 Nm.
Fit the body grommet to the inner wing (2 Fig. 33).
Refit the brake disc cover plate.
Refit the retaining bracket.
Lower the ramp.
Route the harness to the connector and connect the multi plug.
Secure the harness with new ratchet straps.
Close the bonnet.

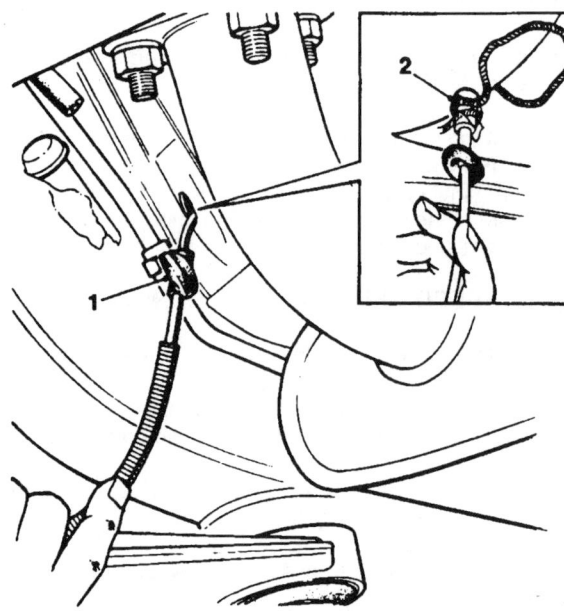

Fig. 30

ANTI-LOCKING BRAKING SYSTEM, PRESSURE TEST

Notes:
1. Read the notes relating to the ABS system bleed procedure before commencing this operation
2. Use only the Jaguar approved ABS pressure test equipment.
3. Under ideal conditions this test should be carried out at 20 degrees Celsius (room temperature).

Fig. 31

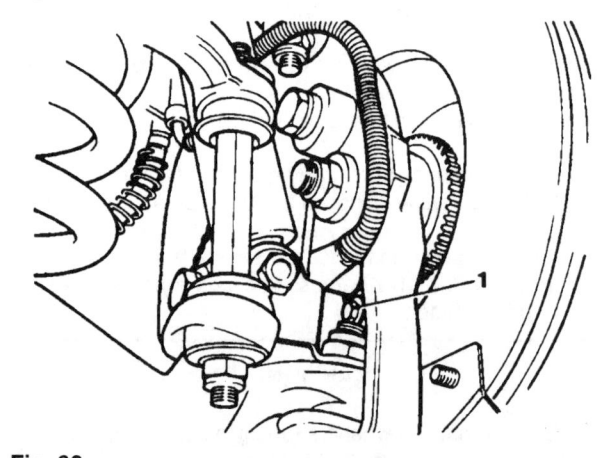

Fig. 32

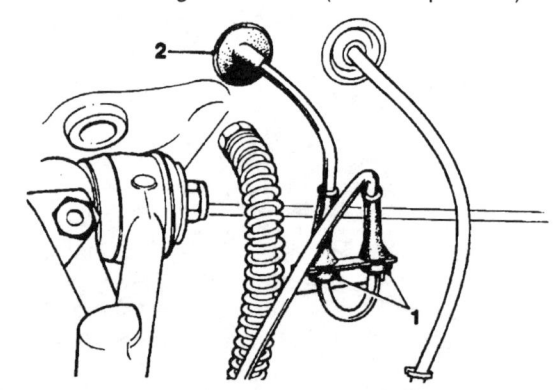

Fig. 33

Drive the car on to a ramp.
Switch the ignition to '0'.
Discharge the hydraulic accumulator by operating the brake pedal (approximately 20 times) until the pedal travel goes hard.
Open the bonnet.

Disconnect the bundy pipe from the motor/pump unit (Fig. 34).
Ensure that the '0' rings and mating faces are clean.

Fit the hydraulic test point No. 07 to the high pressure outlet (Fig. 35).

Raise the lift. Remove the bleed nipples and fit the correct test points (M10 x 1) No. 6 to both front brake calipers (Fig. 36).

Remove the bleed nipple and fit the correct test point (3/8 x 24 UNF) No. 08 to one rear brake caliper (Fig. 37).
Lower the lift.
Connect the flexible pipe (with yellow protective caps) to the 250 bar gauge.
Connect the other end to the test point on the accumulator block noting that different threads are used.

> **CAUTION:**
>
> Always connect the flexible pipe to the gauge first, then to the test point. This is because the adaptor has a one way valve fitted to it and incorrect fitment could result in loss of fluid at high pressure.

Switch the ignition on (position 2), noting the immediate reading up to approximately 80 bar.
Note the elapsed time when the brake warning and the ABS lights extinguish - should be up to a maximum of 40 seconds with an approximate 1 second differential between the two lights.
Note the reading on the pressure gauge - should be 125 bar.

FAULT DIAGNOSIS:

1. If the immediate reading shows less than 80 bar, there is an internal leakage within the accumulator. Change the accumulator.
2. If the pressure gauge does not read 125 bar when the warning lights extinguish change the pressure switch.
3. If the pressure gauge reads more than 190 bar change the pressure switch.

Note the elapsed time when the pump cuts out - up to a maximum of 60 seconds.
Note the reading on the pressure gauge - should be no more than 190 bar.

FAULT DIAGNOSIS:

If the gauge does not read 190 bar or the elapsed time is greater than 60 seconds:

1. Check the pump/motor voltage (motor multi plug pin 2). If it is above 10 volts renew the pump unit. If it is below 10 volts check for voltage drop at the pump/motor line.
2. Check for corroded contacts.
3. Check the reservoir filter: remove the hose from the pump and check for free flow through the filter hose. If free flow is evident reconnect the hose to the pump and repeat the test. If elapsed time is greater than 60 seconds renew the pump.

Ensure that the hydraulic circuit is pressurised (pump not running) and the ignition is on (position 2).
Press the brake pedal continuously.
Note the reading on the pressure gauge when the pump is activated - should be approximately 140 bar.
Stop braking.

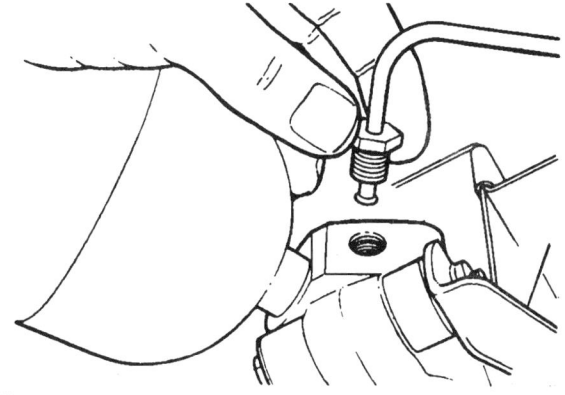

Fig. 34

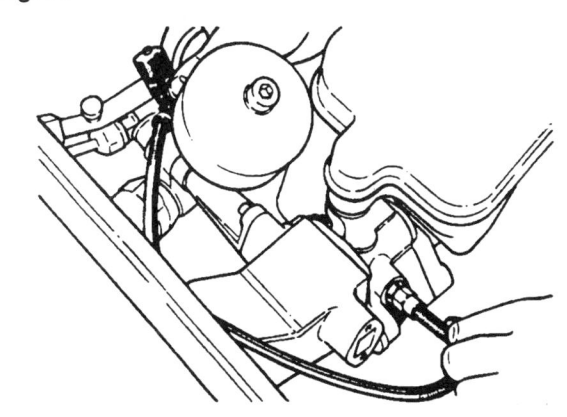

Fig. 35

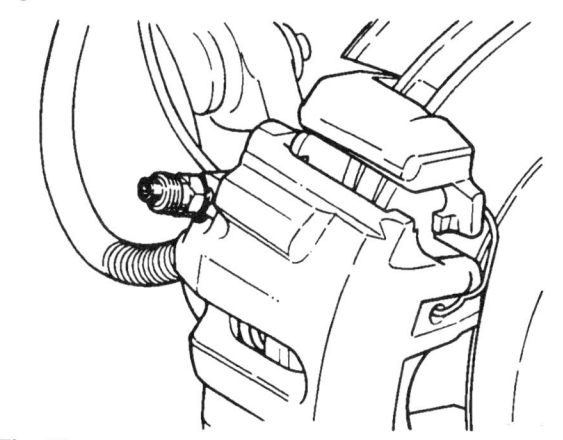

Fig. 36

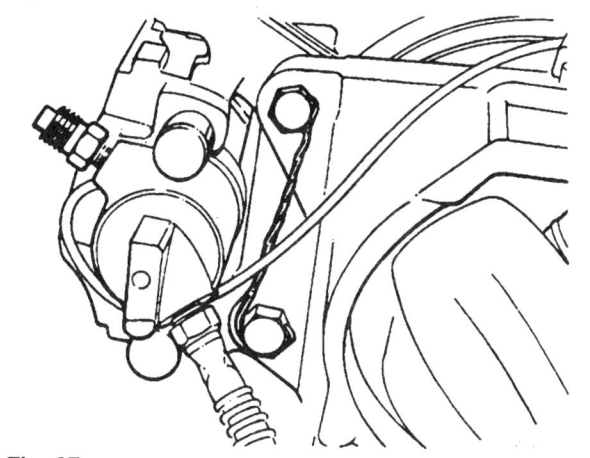

Fig. 37

The pressure should then be restored to a maximum of 190 bar.
Switch the ignition off (position 0).

FAULT DIAGNOSIS:

1. **If the gauge does not read approximately 140 bar when the pump is activated renew the pressure warning switch.**
2. **If maximum pressure is not reached (approximately 190 bar) renew the pressure warning switch.**

Remove the pump/motor fuse.
Switch the ignition on (position 2).
Press the brake pedal hard (approximately 7 times).
Note the reading on the pressure gauge when the warning light comes on - should be approximately 105 bar.
Continue pressing the pedal (approximately 20 times) until the pressure gauge reads 0 (zero) bar. Switch the ignition off (position 0).
Refit the pump/motor fuse.

FAULT DIAGNOSIS:

If the warning light does not come on at approximately 150 bar renew the pressure switch.
Switch the ignition on (position 2), and wait until maximum pressure is attained -190 bar.
Switch the ignition off (position 0), and wait 3 minutes for the pressure to stabilise.
Set the 'tell-tale' needle on the pressure gauge.
After 5 minutes has elapsed, check that the pressure leak is no more than 10 bar.

FAULT DIAGNOSIS:

If the pressure loss is more than 10 bar check for external leakage.
If no external leakage is evident renew the hydraulic actuation unit.

Discharge the system by operating the brake pedal 20 times.
Remove the gauge and flexible pipe from the motor/pump unit.
Raise the lift.

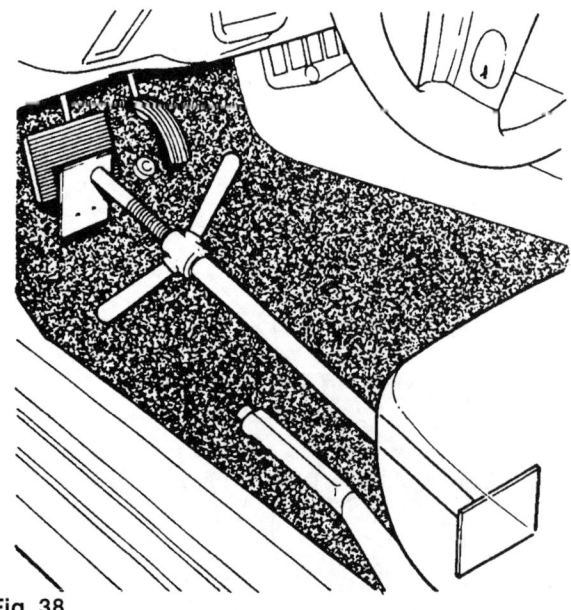

Fig. 38

Connect the second flexible pipe (with yellow protective caps) to the second pressure gauge.
Connect both flexible pipes to the front caliper test points.
Lower the lift.
Switch the ignition on (position 2) and wait until maximum pressure is attained (up to 60 seconds).
Switch the ignition off (position 0).
Fit the brake pedal operating tool (Fig. 38) from the brake pedal to the floor crossmember beneath the seat and adjust the knurled nut until both pressure gauges read 100 bar.
Wait 3 minutes for the pressure to stabilise.
Reset the 'tell-tale' needle on both pressure gauges.
After 5 minutes has elapsed, check that the pressure leak is no more than 5 bar.
Remove the brake pedal operating tool.

FAULT DIAGNOSIS:

If the pressure loss is more than 10 bar check for external leakage. If no external leakage is evident renew the hydraulic actuation unit.

Leaving one front gauge still in circuit, transfer a flexible pipe and pressure gauge to the rear adaptor.
Switch the ignition on (position 2), and wait (up to a maximum of 60 seconds) until maximum pressure is attained - 180 bar.
Switch the ignition off (position 0).
Fit the brake pedal operating tool (Fig. 38) and adjust the knurled nut until the rear pressure gauge reads 100 bar.
THE READINGS ON THE FRONT AND REAR SYSTEMS SHOULD DIFFER BY NO MORE THAN 7.5 BAR (HIGHER ON THE FRONT SYSTEM).
Wait 3 minutes for the pressure to stabilise.
Reset the 'tell-tale' needle on both pressure gauges.
After 5 minutes has elapsed, check that the pressure leak is no more than 5 bar.

FAULT DIAGNOSIS:

If the pressure loss is more than 10 bar check for external leakage.
If no external leakage is evident renew the hydraulic actuation unit.

Remove the brake pedal operating tool (Fig. 38).
Discharge the system.
Remove all test points from the circuit.
Refit the bleed nipples and high pressure hose, ensuring that new 'O' rings are fitted to the high pressure hose.
Bleed the ABS hydraulic system.
Examine the system for external leaks.

Torque Figures

Bleed screw to caliper	7 - 13 Nm
High pressure pipe union	7 - 12 Nm

BRAKES, 1993.5 MODEL YEAR ON

INTRODUCTION

For 1993.5 Model Year, the front brake assemblies and service procedures remain unchanged.
The rear brakes and handbrake, however, change to the outboard type as fitted to 93 MY saloon models.
The handbrake cable layout/linkage varies and the method of setting and adjustment differs from previous XJS models.
Service details and operations for rear brakes and some rear suspension components are as follows:

Fig. 39

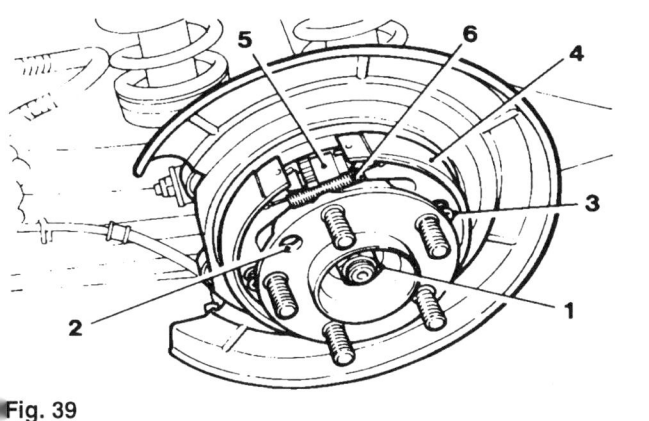

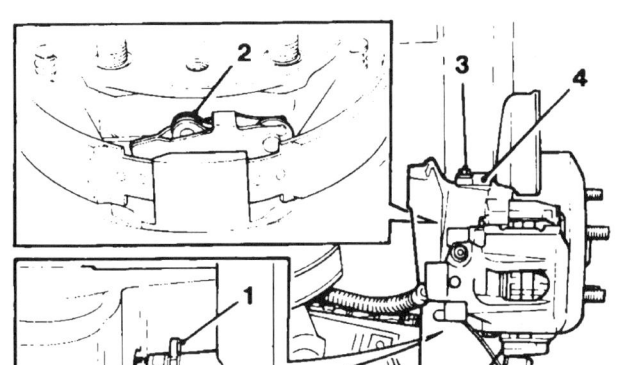

Fig. 40

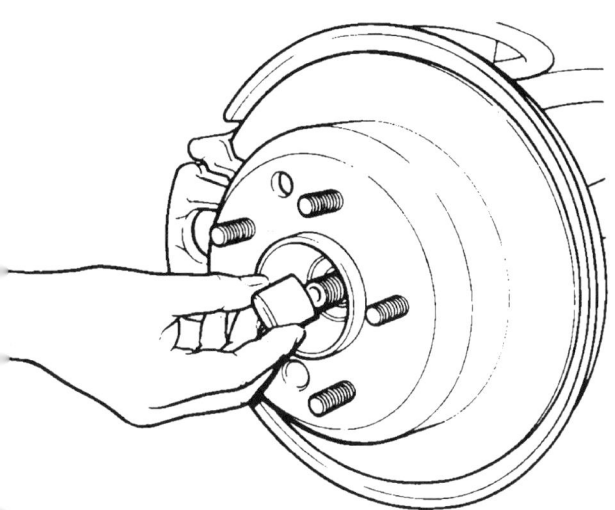

Fig. 41

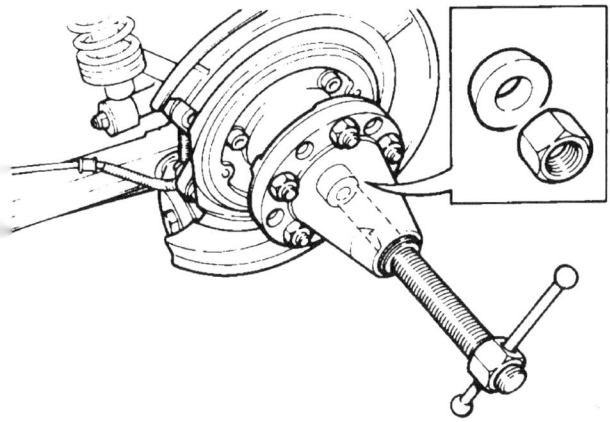

Fig. 42

REAR HUB BEARING, RENEW

Jack up the rear of the vehicle and place on stands.
Remove the roadwheel and tyre assembly.

Unscrew, but do not remove the drive shaft to hub assembly securing nut (1 Fig. 39).
Remove the rear disc.
Position the hub access hole (2 Fig. 39) to expose the front handbrake shoe retaining clip (3 Fig. 39).
Remove the retaining clip assembly.
Displace and reposition the forward shoe (4 Fig. 39) to allow removal of the upper adjuster (5 Fig. 39).
Position the hub access hole (2 Fig. 39) to expose the rear handbrake shoe retaining clip. Remove the retaining clip assembly.
Displace the front and rear shoes from the handbrake cable link.
Displace and remove the shoe/lower spring assembly from the brake backplate.
Unhook the upper and lower springs (6 Fig. 39) from the shoes and place the shoes and springs to one side.

Remove the handbrake cable securing clip (1 Fig. 40).
Reposition the handbrake link and remove the cable clevis pin (2 Fig. 40).
Cut and remove the tie strap securing the ABS harness to the hub.
Remove the ABS sensor securing bolt (3 Fig. 40).
Displace and reposition the sensor (4 Fig. 40).
Fully remove the drive shaft to hub securing nut (1 Fig. 40) and displace and remove the cone washer.

Fit a thread protector to the drive shaft (Fig. 41).
Unscrew the centre bolt of tool JD1D, hub removal tool.

Fit and tighten the tool to the hub (Fig. 42).
Tighten the tool centre bolt to withdraw the hub from the drive shaft.
Unscrew the tool securing nuts and remove the tool from the hub.
Remove the thread protector from the drive shaft.
Reposition the handbrake cable from the hub carrier.
Remove the fulcrum shaft securing nut. Remove the fulcrum shaft.
Remove the wishbone/damper tie bracket spacer washers.
Carefully displace the hub assembly and remove the shims from between the hub and wishbone. Place the hub assembly on a clean work bench.
Clean the drive shaft spline and faces.
Position hub tool JD 132-1 on a press.

Place the hub assembly to the tool/press. Ensure it is aligned correctly with the handbrake cable housing located into the tool cut out (Fig. 43).

Fit and align the tool button JD 132-2 to the hub (1 Fig. 44).
Press and remove the hub from the carrier.
Remove the hub carrier, ABS rotor and inner bearing as an assembly.
Remove the tool button JD 132-2.
Remove the ABS rotor from the hub carrier.
Remove the hub and tool assembly from the press and remove tool JD 132-1 from the hub. Place tool to one side.
Using a suitable drift, remove the outer hub seal and bearing cone.
Remove the bearing spacer and shim(s).
Reposition the hub carrier on the bench.
Using a drift remove the inner bearing cone and seal.

79

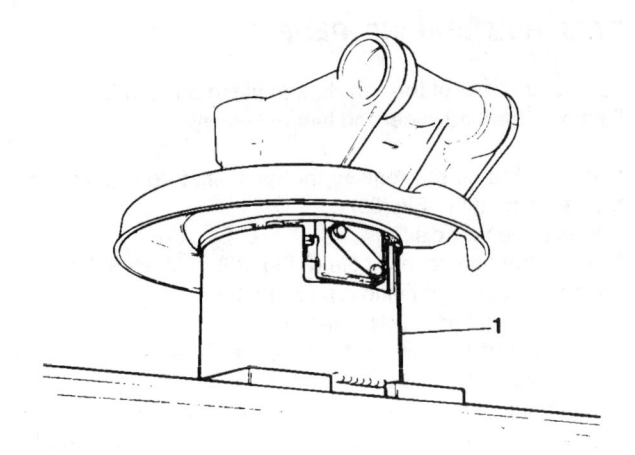

Fig. 43

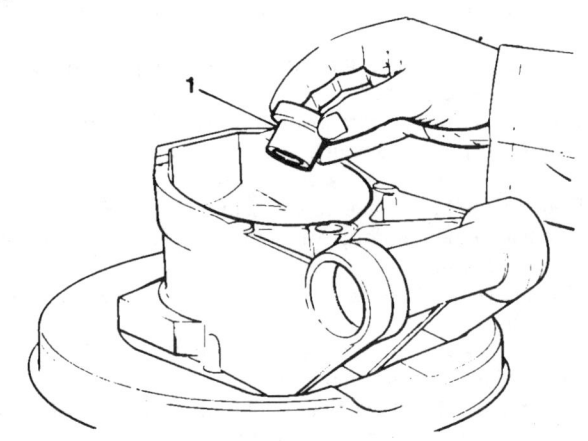

Fig. 44

Note: Always keep bearing cups and cones together in original pairs. Never mix cups and cones.

Position the hub carrier on a suitable block of wood.
Using a drift remove the outer bearing cup.
Reposition the hub carrier on the wooden block and remove the remaining, inner, bearing cup.
Thoroughly clean all components to be reused prior to reassembly.

Position the hub carrier on the press with the inner bearing side uppermost.
Position the new inner bearing cup in the hub carrier. Fit and align the bearing replacer tool JD550-4/2 to the inner bearing cup.
Position press tool 18G 134 on the bearing replacer tool.
Using the press, fully seat the new inner bearing cup to the hub carrier.
Reposition the hub carrier on the press with the outer bearing side uppermost.
Position the new outer bearing cup in the hub carrier. Fit and align the bearing replacer tool JD550-4/1 on the outer bearing cup.
Position press tool 18G 134 on the bearing replacer tool.
Using the press, fully seat the new outer bearing cup on the hub carrier.
Remove the tool assembly. Remove tool 18G 134 from the replacer tool. Place the hub carrier on the workbench.

Fit and align the new outer bearing cone to the hub carrier.
Fully seat the hub to the carrier/bearing cone assembly.
Reposition the assembly on the bench.
Fit the bearing spacer to the hub followed by the largest shim, (0.031 in.) (0.001 inch = 0.0254mm for conversion purposes).
Fit and align the new inner bearing cone to the carrier / hub assembly.

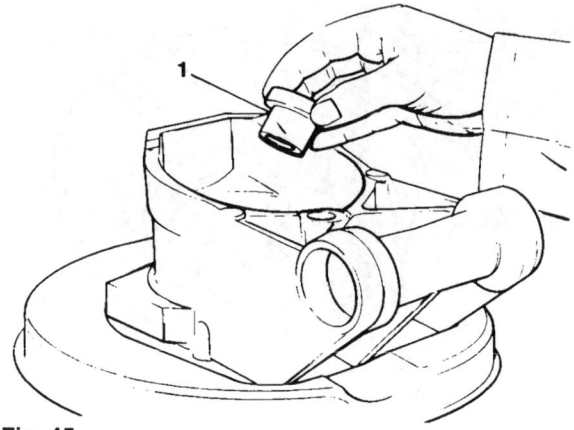

Fig. 45

Position a suitable block of wood on the press.
Fit the carrier/hub assembly on the press/wood and ensure, when using the press, the load is not put upon the wheel studs.
Fit the ABS rotor to the inner hub.
Using the press, fully seat the ABS rotor to the hub assembly.
Fit the hub securely in a vice.
Fit hub end float tool JD 15 to the hub.
Fit and tighten the hub end float dial gauge JD 13A to the hub carrier.
Using two suitable levers, insert between the hub and carrier and prise lightly back and forth to check the end float.
Make a note of the end float reading.

Note: Care must be taken not to use excess leverage as this could move the spacer and prevent a correct reading being obtained.

Remove the dial gauge JD 13A and the end float tool JD 15.
Position hub tool JD 132-1 on the press.
Align the hub assembly and fit the button JD 132-2 to the hub (Fig. 45).
Using the press, remove the hub from the carrier.
Remove the hub carrier, ABS rotor and inner bearing as an assembly.
Remove the button, tool JD 132-2.
Remove the hub and tool assembly from the press and remove tool JD132-1 from the hub. Place tool to one side.
Remove the ABS rotor and bearing cone from the hub carrier.
Remove the shim, spacer and bearing cone from the hub
Lubricate the bearings with the correct amount of grease (see DATA).
Fit and align the new outer bearing cone to the hub carrier.
Fit and fully seat a new seal to the hub carrier. Fully seat the hub to the carrier/bearing cone assembly.

Reposition the hub/carrier assembly on the bench and fit the spacer.
Place the large shim (0.031 in.) aside and calculate the shim(s) required to give a 0.003 in. pre-load.
Fit the shim(s) required to the hub.
Fit and align the new inner bearing cone to the hub.
Fully seat a new inner seal to the hub / carrier assembly.
Position a suitable block of wood on the press.
Fit the carrier/hub assembly on the press / wood and ensure, when using the press, the load is not put upon the wheel studs.
Fit the ABS rotor to the inner hub.
Using the press, fully seat the ABS rotor to the hub assembly.

Position the hub/carrier assembly on the wishbone.
Fit the shims, as removed, between the wishbone and hub assembly.

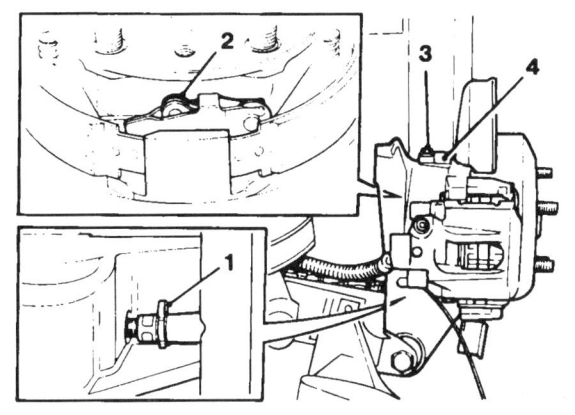

Fig. 46

Align the hub assembly to the lower wishbone.
Fit the wishbone/damper tie bracket spacer washers.
Fit and fully seat the fulcrum shaft to the wishbone / hub assembly.
Fit, but do not tighten, the fulcrum shaft securing nut.
Apply 'Loctite' to the drive shaft spline.
Pivot the hub and position the handbrake cable through the cable housing.
Position the drive shaft to the hub and pivot the hub to fully seat the shaft.
Fit the cone washer.
Fit, but do not fully tighten, the drive shaft securing nut.
Fit the handbrake link to the cable.
Fit the cable to link clevis pin.
Reposition the cable link.
Fit and seat the cable securing clip.
Position the ABS sensor to the hub / carrier assembly.
Fit and tighten the sensor securing bolt.
Fit a new tie strap to secure the sensor harness.

Lightly lubricate the adjuster with Copperslip grease.
Position the shoes on the backplate.
Ensure the forward shoe is correctly seated in the handbrake link.
Fit the lower return spring.
Align the rear shoe to the backplate.
Align the hub access hole to the rear shoe retaining clip position.
Fit and fully seat the rear shoe retaining clip.
Fit the rear shoe to handbrake link.
Fit the upper return spring. Pull both shoes outward to allow the adjuster to be fitted.
Fit and align the adjuster to the shoes.
Align the front shoe to the backplate.
Align the hub access hole to the front shoe retaining clip position.
Fit and fully seat the front shoe retaining clip.
Refit the discs.
Refit the roadwheels .
Ensure fixings/fastenings are torque tightened to the specified tolerances.
Lower the vehicle from the axle stands.

Oils / Greases / Sealants

Hub bearings Shell Retinax 'A'   11.5 cc inner bearing cone
                                 9.0 cc outer bearing cone
Brake shoe adjuster   Copperslip Grease
Drive shaft splines   Loctite 270

Service Tools

Hub remover JD 1D
Hub press tool JD 132-1
Hub press tool button JD 132-2

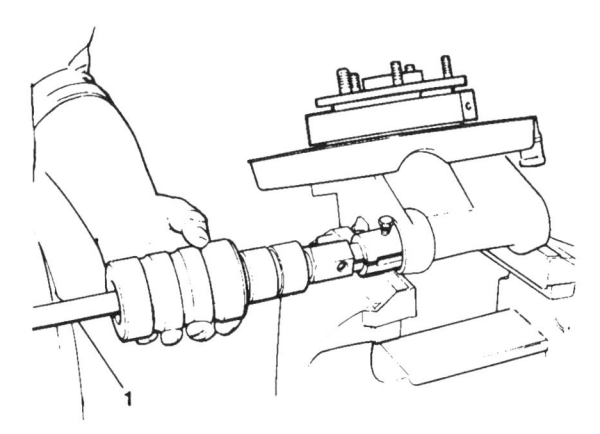

Fig. 47

Press tool handle 18G 134
Bearing replacer JD 550-4 / 1
Bearing replacer JD 550-4 / 2
Dial gauge JD 13A
End float measurement base JD 15

REAR HUB OIL SEAL, RENEW

Follow the instructions given under the preceding operation, omitting to renew the cups and cones unless they are found to be damaged.

REAR HUB CARRIER FULCRUM BEARING, RENEW

Raise the vehicle on a ramp.
Using suitable levers reposition the handbrake ratio levers inward.
Disconnect the main handbrake cable from the right hand ratio lever. Position the lever outwards toward the abutment bracket.
Remove the rear cable retaining collet and clip.
Reposition rear cable from the abutment bracket.
Lower the ramp to a suitable working position to carry out the following procedures.
Raise the rear end of the vehicle and support on stands.
Remove the roadwheel and tyre assembly.
Unscrew, but do not remove the drive shaft to hub assembly securing nut (1 Fig. 39).
Remove the rear disc.
Cut and remove the tie strap securing the ABS harness to the hub.
Remove the ABS sensor securing bolt (3 Fig. 46).
Displace and reposition the sensor (4 Fig. 46).
Remove the fulcrum shaft securing nut.
Fully remove the drive shaft to hub securing nut (1 Fig. 39) and displace and remove the cone washer.
Fit a thread protector to the drive shaft (Fig. 41).
Unscrew the centre bolt of tool JD1D, hub removal tool.
Fit and tighten the tool to the hub (Fig. 42).
Tighten the tool centre bolt to withdraw the hub from the drive shaft.
Unscrew the tool securing nuts and remove the tool from the hub.
Remove the thread protector from the drive shaft.
Remove the fulcrum shaft.
Remove the wishbone/damper tie bracket spacer washers.
Carefully displace the hub assembly and remove the shims from between the hub and wishbone.

Carefully secure the hub assembly in a vice.
Using a suitable drift, remove the fulcrum bearing tube.
Remove the shim and bearing cone from the tube.
Remove the remaining bearing cone and shim from the hub carrier.

Fit extractor tool 18G 284 AAH (1 Fig. 47) to the fulcrum bearing cup.
Tighten the tool cross bolt and ensure the tool legs locate behind the bearing cup.
Fit the slide hammer 18G 284 to the tool.
Using the slide hammer, remove the bearing cup.
Unscrew, but do not remove, the tool cross bolt.
Place the bearing cup aside.
Repeat the procedure to remove the remaining bearing cup. Place the bearing cup aside. Remove the slide hammer 18G 284 from tool 18G 284 AAH.
Remove the hub/carrier assembly from the vice.
Clean the hub carrier and fulcrum bearing faces.
Clean the cross shaft and tube.
Position the hub/carrier assembly on a press.
Fit the tool handle 18G 134 to the bearing replacer tool JD 550-6.
Fit a new bearing cup to the replacer tool.
Align the tool/bearing cup to the hub/carrier assembly.
Using the press, fully seat the new bearing cup to the hub carrier.
Remove the replacer tool.
Reposition the hub/carrier assembly on the press.
Fit the remaining new bearing cup.
Remove the replacer tool.
Remove the tool handle 18G 134 from the bearing replacer tool JD 550-6.
Move the assembly from the press and secure in position in a vice.
Apply grease to the new bearings.
Fit a new bearing cone to the fulcrum tube.
Fit the minimum thickness of shim to the fulcrum tube/bearing cone.
Fit the tube/bearing cone assembly to the hub carrier.
Fit the remaining bearing cone to the tube/carrier assembly.
Fit the fulcrum shaft.
Fit and fully seat the remaining shim to the tube.
Fit a suitable piece of tube to seat onto the shim.
Fit and tighten the cross shaft securing nut to fully seat the shims onto the cross tube. Remove the shaft securing nut.
Remove the tube.
Fit a suitable flat washer to the shaft and tighten the shaft securing nut.
Using feelers, measure and note the gap between the shim and the washer.
Remove the shaft securing nut.
Remove the washer and displace the cross shaft.
Using a suitable drift, move the cross tube from the shims.
Retrieve the shims, measure, then place aside.

Note: Calculate equal size shims to be used to give a 0.003 in. (0.0762 mm) pre-load.

Fit the new shims to the cross tube.
Fit the cross shaft.
Align a suitable tube then tighten the shaft securing nut to fully seat the shims on to the cross tube.
Unscrew and remove the nut.
Remove the tube.
Carefully move the cross shaft from the tube.
Remove the hub/carrier assembly from the vice.

Position the hub/carrier assembly on the lower wishbone.
Route the handbrake cable into position.
Fit the shims, as removed, between the wishbone and hub assembly.
Align the hub assembly to the lower wishbone.
Fit the wishbone/damper tie bracket spacer washers.

Fit and fully seat the fulcrum shaft to the wishbone/hub assembly.
Fit, but do not tighten, the fulcrum shaft securing nut.
Apply 'Loctite' to the drive shaft spline.
Pivot the hub to engage on the drive shaft.
Fit the cone washer.
Fit, but do not fully tighten, the drive shaft securing nut.

Position the ABS sensor to the hub / carrier assembly.
Fit and tighten the sensor securing bolt.
Fit a new tie strap to secure the sensor harness.
Fully tighten the drive shaft and fulcrum shaft securing nuts.
Refit the disc.
Raise the ramp.
Position the rear handbrake cable through the abutment bracket.
Fit the cable securing clip.
Position the ratio lever outwards.
Position the cable through the ratio lever.
Fit the cable retaining collet.
Move both ratio levers inward and fit the main handbrake cable to the right hand ratio lever.
Lower the ramp.
Refit the roadwheel.
Ensure fixings/fastenings are torque tightened to the specified tolerances.
Lower the vehicle from the axle stands.

Oils / Greases / Sealants

Fulcrum shaft and bearings - LM Multipurpose Grease
Drive shaft splines - Loctite 270

Service Tools

Hub remover JD 1D
Bearing cup remover 18G 284 AAH
Bearing cup replacer JD 550 6
Press tool handle 18G 134
Slide hammer 18G 284

REAR DISC, RENEW

Raise the rear of the vehicle and support on stands.
Remove the roadwheel and tyre.
Carefully remove the caliper retaining spring (1 Fig. 48).
Remove the caliper to carrier securing bolt covers (2 Fig. 48).
Remove the socket head securing bolts (3 Fig. 48).
Displace the caliper from the carrier.
Remove the brake pads (4 Fig. 48) from the caliper.
Position the caliper to one side to allow access to the carrier/hub assembly.

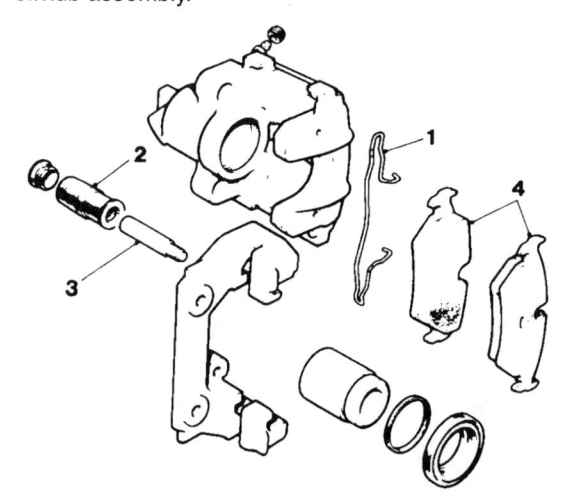

Fig. 48

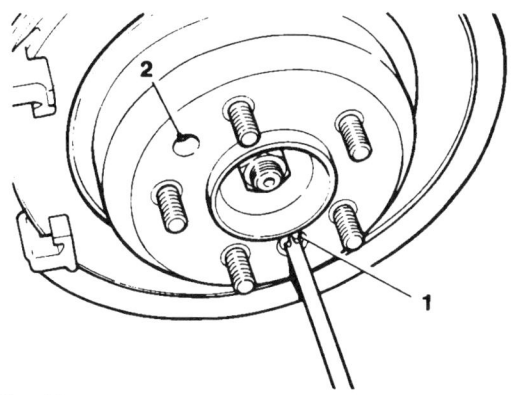

Fig. 49

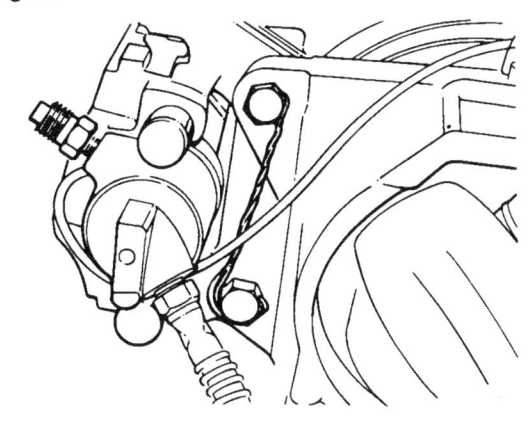

Fig. 50

Note: Do not allow the caliper to hang with the weight on the feed hose. Ensure it is placed safely and securely aside.

Cut and remove the lock wire securing the carrier to hub assembly securing bolts. Unscrew and remove the securing bolts. Remove the carrier.

Unscrew and remove the disc securing screw (1 Fig. 49). Remove the disc.
Carefully remove any brake dust / dirt from the caliper and handbrake shoe areas.

Fit a new disc.
Fit and tighten the disc securing screw (1 Fig. 49).

Operate the handbrake lever to centralize the shoes.
Align the hub access hole (2 Fig. 49) to the handbrake shoe adjuster.
Turn the adjuster anti-clockwise until the shoes contact the disc.
Turn the adjuster clockwise until the disc can be turned quite freely.
Fit the carrier to the hub assembly and tighten the carrier securing bolts.

WARNING:

BRAKE PAD/LINING DUST CAN CONTAIN ASBESTOS WHICH IF INHALED, CAN DAMAGE YOUR HEALTH. ALWAYS USE A VACUUM BRUSH TO REMOVE DRY BRAKE PAD/LINING DUST. NEVER USE AN AIR LINE.

Secure the carrier bolts with new lockwire (Fig. 50). Ensure the caliper piston is fully retracted and refit the brake pads.

Note: If the pads are worn to the minimum thickness, new pads must be fitted.

Position the caliper/pad assembly to the carrier.
Lubricate, then fit and tighten the caliper securing bolts.
Fit the caliper securing bolt covers.
Fit the caliper retaining spring.
Refit the roadwheel.
Lower the vehicle from the axle stands.
Check the brake fluid and top up as necessary.
Ensure fixings/fastenings are torque tightened to the specified tolerances.

Note: It may be necessary to run the engine to give power assistance to the brake pedal.

WARNING:

APPLICATION OF THE BRAKE PEDAL MUST BE CARRIED OUT, AS THE BRAKE WILL NOT OPERATE EFFICIENTLY UNTIL THE PADS ARE CORRECTLY POSITIONED.

Oils / Greases / Sealants

Brake fluid to a minimum DOT 4 specification.
Caliper securing bolts Molykote 111.

REAR HANDBRAKE CABLE ASSEMBLY, LH (and RH), RENEW

Raise the vehicle on a ramp.
Using suitable levers reposition the handbrake ratio levers inward.
Disconnect the main handbrake cable from the right hand ratio lever.
Position the lever outward toward the abutment bracket.
Remove the rear cable retaining collet and clip.

Reposition the rear cable from the abutment bracket.
Lower the ramp to a suitable working position to carry out the following procedures.
Raise the rear end of the vehicle and support on stands.
Remove the roadwheel and tyre assembly.

Unscrew, but do not remove the drive shaft to hub assembly securing nut (1 Fig. 51).
Remove the rear disc.
Position the hub access hole (2 Fig. 51) to expose the front handbrake shoe retaining clip (3 Fig. 51).
Remove the retaining clip assembly.
Displace and reposition the forward shoe (4 Fig. 51) to allow removal of the upper adjuster (5 Fig. 51).

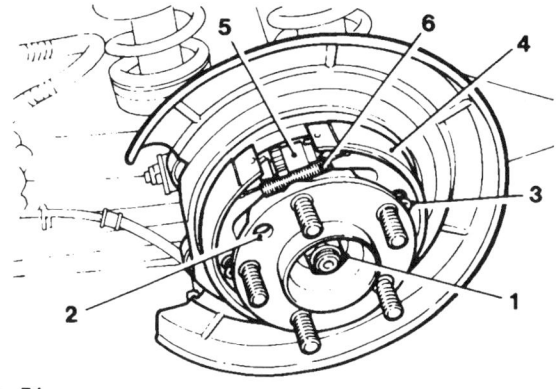

Fig. 51

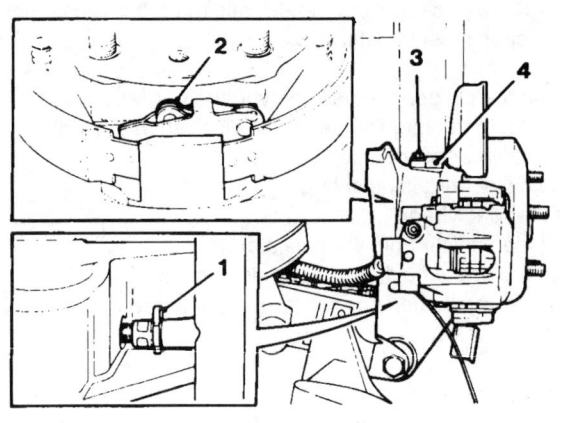

Fig. 52

Position the hub access hole (2 Fig. 51) to expose the rear handbrake shoe retaining clip. Remove the retaining clip assembly.
Displace the front and rear shoes from the handbrake cable link.
Displace and remove the shoe/lower spring assembly from the brake backplate.
Unhook the upper and lower springs (6 Fig. 51) from the shoes and place the shoes and springs to one side.

Remove the handbrake cable securing clip (1 Fig. 52).
Reposition the handbrake link and remove the cable clevis pin (2 Fig. 52).
Remove the cable from the hub assembly.

Fit the new handbrake cable to the cable housing.
Fit the handbrake link to the cable.
Fit the cable to link clevis pin.
Reposition the cable link.
Fit and seat the cable securing clip.
Lightly lubricate the adjuster.
Position the shoes on the backplate. Ensure the forward shoe is correctly seated in the handbrake link.
Fit the lower return spring.
Align the rear shoe to the backplate.
Align the hub access hole to the rear shoe retaining clip position.
Fit and fully seat the rear shoe retaining clip.
Fit the rear shoe to handbrake link.
Fit the upper return spring.
Pull both shoes outward to allow the adjuster to be fitted.
Fit and align the adjuster to the shoes.
Align the front shoe to the backplate.
Align the hub access hole to the front shoe retaining clip position.
Fit and fully seat the front shoe retaining clip.

Refit the disc.
Refit the roadwheel.
Raise the ramp.
Position the rear handbrake cable through the abutment bracket.
Fit the cable securing clip.
Position the ratio lever outwards.
Position the cable through the ratio lever.
Fit the cable retaining collet.
Move both ratio levers inward and fit the main handbrake cable to the right hand ratio lever. Lower the ramp.

Open the driver's door.
Switch on the ignition.
Power the driver's seat fully forward.
Position the seat squab fully forward.
Switch off the ignition.
Unscrew the handbrake lever cover securing screw.

Remove the cover.
Position the carpet for access to the seat belt slider rail cover.
Remove the cover.
Remove the slider rail securing bolt.
Move the slider rail and sill carpet to allow access to remove the slider rail spacer.
Unscrew the handbrake cable adjuster locknuts.
Adjust the cable until upon application of the handbrake, only three to four notches of the ratchet are felt/heard.
Tighten the locknuts.
Refit the slider rail spacer.
Reposition the sill carpet.
Refit the slider rail to position and fit and tighten the securing bolt.
Fit the securing bolt cover and reposition the floor carpet.
Fit the handbrake lever cover.
Fit and tighten the cover securing screw.
Reposition the seat squab back to the original position.
Switch on the ignition.
Power the seat back to the original position.
Switch off the ignition.
Close the door.
Ensure fixings/fastenings are torque tightened to the specified tolerances.
Lower the vehicle from the stands.

To renew the rear handbrake cable assembly, right hand, carry out the operation on the opposite rear side of the vehicle.

Oils / Greases / Sealants

Brake shoe adjuster - Copperslip Grease

HANDBRAKE, SETTING PROCEDURE

Open the driver's door.
Switch on the ignition.
Power the driver's seat fully forward.
Position the seat squab fully forward.
Switch off the ignition.
Unscrew the handbrake lever cover securing screw.
Remove the cover.
Position the carpet for access to the seat belt slider rail cover.
Remove the cover. Remove the slider rail securing bolt.
Move the slider rail and sill carpet to allow access to remove the slider rail spacer.
Unscrew the handbrake cable adjuster locknuts.
Raise the vehicle on a ramp.
Reposition the main handbrake cable rubber seal.
Move the handbrake ratio levers inward.
Disconnect the main handbrake cable from the ratio lever cable abutments and reposition the cable.
Unscrew, but do not remove the fulcrum shaft/ratio lever securing nuts.
Position the ratio levers to the abutment brackets to give a 1 mm to 3 mm (0.039 to 0.118 in.) gap.
Tighten the fulcrum shaft/ratio lever securing nuts.

Note: Ensure the ratio levers do not move whilst tightening the fulcrum shaft/ratio lever securing nuts.

Offer the handbrake cable up to the ratio lever. Fit and fully seat the cable in the abutment. Position the rubber cable seal.
Position the ratio levers inward. Fit and fully seat the cable to the opposite ratio lever abutment.
Lower the ramp to a suitable working position to carry out the following procedures.
Raise the rear end of the vehicle and support on stands.
Remove the roadwheel and tyre assembly.

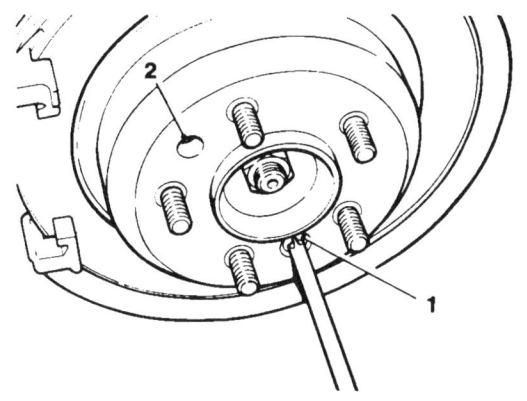

Fig. 53

Turn the hub assembly and align the disc/hub access hole (2 Fig. 53) to the handbrake shoe adjuster.
Turn the adjuster anti-clockwise until the shoes contact the disc.
Turn the adjuster clockwise until the disc can be turned quite freely.
Repeat the procedure to adjust the opposite side rear handbrake shoes.
Refit the roadwheel.
Lower the ramp.
From inside the vehicle:
Adjust the cable until upon application of the handbrake, only three to four notches of the ratchet are felt/heard. Tighten the locknuts.
Refit the slider rail spacer.
Reposition the sill carpet.
Refit the slider rail to position and fit and tighten the securing bolt.
Fit the securing bolt cover and reposition the floor carpet.
Fit the handbrake lever cover. Fit and tighten the cover securing screw.
Reposition the seat squab back to the original position.
Switch on the ignition. Power the seat back to the original position. Switch off the ignition. Close the door.
Lower the vehicle from the stands.

REAR BRAKE PADS, RENEW

Raise the rear of the vehicle and support on stands.
Remove the roadwheel and tyre assembly.

Carefully remove the caliper retaining spring (1 Fig. 54).
Remove the caliper to carrier securing bolt covers (2 Fig. 54).
Remove the socket head securing bolts (3 Fig. 54).
Displace the caliper from the carrier.
Remove the brake pads (4 Fig. 54) from the caliper.
Safely discard the worn brake pads.
Position the caliper to one side.

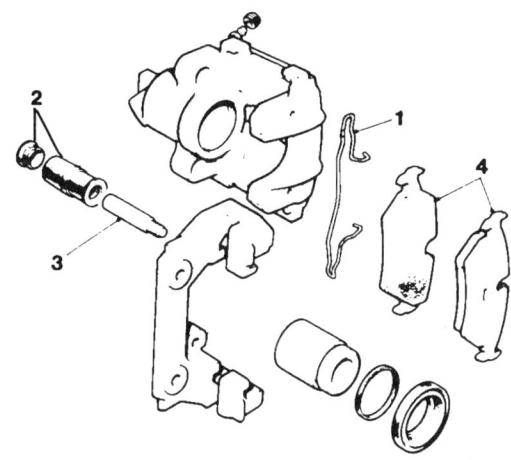

Fig. 54

Note: Do not allow the caliper to hang with the weight on the feed hose. Ensure it is placed safely and securely aside.

WARNING:

BRAKE PAD/LINING DUST CAN CONTAIN ASBESTOS WHICH IF INHALED, CAN DAMAGE YOUR HEALTH. ALWAYS USE A VACUUM BRUSH TO REMOVE DRY BRAKE PA /LINING DUST. NEVER USE AN AIR LINE.

Carefully remove any brake dust/dirt from the caliper area.

Ensure the caliper piston is fully retracted and fit new brake pads.

Note: The pad with the spring clip attached locates into the caliper piston.

Offer up the caliper/pad assembly to the carrier.
Lubricate, then fit and tighten the caliper securing bolts.
Fit the caliper securing bolt covers.
Fit the caliper retaining spring.
Repeat the procedure on the opposite rear side of the vehicle. Refit the roadwheel.
Lower the vehicle from the axle stands.
Check the brake fluid and top up as necessary.

WARNING:

IT IS ESSENTIAL THAT THE BRAKE PEDAL IS REPEATEDLY APPLIED TO ENSURE THAT THE PADS ARE FULLY IN CONTACT WITH THE DISC.

Note: It may be necessary to run the engine to give power assistance to the brake pedal.

Oils / Greases / Sealants

Brake fluid to a minimum DOT 4 specification.
Caliper securing bolts - Molykote 111.

HANDBRAKE SHOES, RENEW

Jack up the rear of the vehicle and place on stands.
Remove the roadwheel and tyre assembly.
Remove the rear disc.
Position the hub access hole (2 Fig. 55) to expose the front handbrake shoe retaining clip (3 Fig. 55). Remove the retaining clip assembly.
Displace and reposition the forward shoe (4 Fig. 55) to allow removal of the upper adjuster (5 Fig. 55).
Position the hub access hole (2 Fig. 55) to expose the rear handbrake shoe retaining clip. Remove the retaining clip assembly.
Displace the front and rear shoes from the handbrake cable link.
Displace and remove the shoe/lower spring assembly from the brake backplate.
Unhook the upper and lower springs (6 Fig. 55) from the shoes and place the shoes and springs to one side.
Carefully remove brake dust/dirt from the caliper and shoe area.

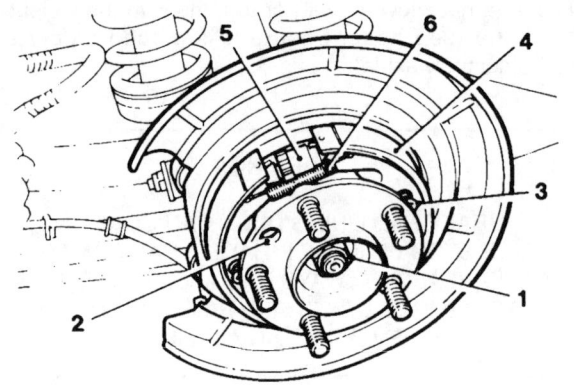

Fig. 55

> **WARNING:**
>
> BRAKE PAD/LINING DUST CAN CONTAIN ASBESTOS WHICH IF INHALED, CAN DAMAGE YOUR HEALTH. ALWAYS USE A VACUUM BRUSH TO REMOVE DRY BRAKE PAD / LINING DUST. NEVER USE AN AIR LINE.

Clean the springs, retaining pins and adjuster.
Lightly lubricate the adjuster.
Offer the shoes up to the backplate.
Ensure the forward shoe is correctly seated in the handbrake link.
Fit the lower return spring.
Align the rear shoe to the backplate. Align the hub access hole to the rear shoe retaining clip position.
Fit and fully seat the rear shoe retaining clip.
Fit the rear shoe to handbrake link.
Fit the upper return spring.
Pull both shoes outward to allow the adjuster to be fitted.
Fit and align the adjuster to the shoes.
Align the front shoe to the backplate.
Align the hub access hole to the front shoe retaining clip position.
Fit and fully seat the front shoe retaining clip.
Repeat the procedure for the opposite rear side of the vehicle.

Refit the discs.
Refit the roadwheels.
Lower the vehicle from the axle stands.

Oils / Greases / Sealants

Brake shoe adjuster - Copperslip Grease

REAR BRAKE CALIPER, RENEW

Open the bonnet and place a cloth around the brake reservoir filler cap/neck to absorb any possible spillage.
Jack up the rear of the vehicle and place on stands.
Remove the roadwheel and tyre.
Carefully remove the caliper retaining spring (1 Fig. 54).
Partially unscrew the brake pipe to feed hose union nut.
Remove the caliper to carrier securing bolt covers (2 Fig. 54).
Remove the socket head securing bolts (3 Fig. 54).
Displace the caliper from the carrier.
Remove the brake pads (4 Fig. 54) from the caliper.
Unscrew and remove the caliper from the feed hose.
Fit suitable plugs to the caliper and hose.
Cut and remove the lockwire securing the carrier to hub assembly securing bolts.

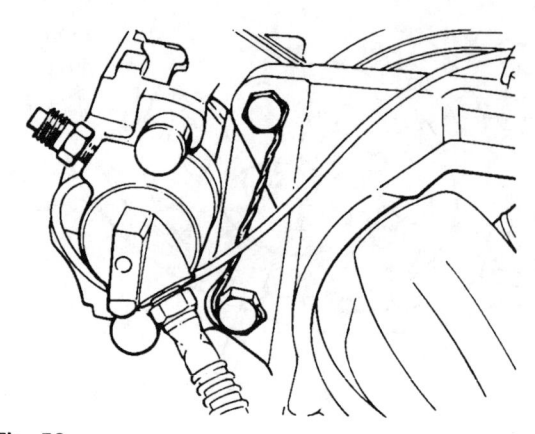

Fig. 56

> **WARNING:**
>
> BRAKE PAD/LINING DUST CAN CONTAIN ASBESTOS WHICH IF INHALED, CAN DAMAGE YOUR HEALTH. ALWAYS USE A VACUUM BRUSH TO REMOVE DRY BRAKE PAD/LINING DUST. NEVER USE AN AIR LINE.

Unscrew and remove the securing bolts.
Remove the carrier.
Carefully remove any brake dust/dirt from the carrier and disc.

Fit the carrier to the hub assembly and tighten the carrier securing bolts.
Secure the carrier bolts with new lockwire (Fig. 56).
Remove the plugs from the new caliper and the feed hose.
Fit, but do not fully tighten the caliper to the feed hose.
Ensure the caliper piston on the new caliper is fully retracted and refit the brake pads.
Note: If the pads are worn to the minimum thickness, new pads must be fitted.

Position the caliper/pad assembly to the carrier.
Lubricate, then fit and tighten the caliper securing bolts. Fit the caliper securing bolt covers.

Fully tighten the feed hose to the caliper.
Fully tighten the brake pipe to feed hose union nut.
Fit the caliper retaining spring.
Refit the roadwheel and tyre assembly.
Lower the vehicle from the axle stands.
Bleed the brakes.

Oils / Greases / Sealants

Brake fluid to a minimum DOT 4 specification.

Torque Figures (for 1993.5 MY vehicles)

(All figures in Newton Metres)

Brake disc to hub	11 - 16
Caliper to carrier	25 - 30
Caliper carrier to hub carriers	54 - 66
Fulcrum shaft nut	70 - 80
Bleed screw/nipple to caliper	7 - 13
Brake hydraulic hose to caliper	10 - 12
ABS sensor securing bolt	8,5 - 11
Handbrake lever to body	15 - 21
Handbrake cable locknuts	19 - 25
Handbrake lever cover screw	0.6 - 1.0
Wheel nut to wheel stud	88 - 102

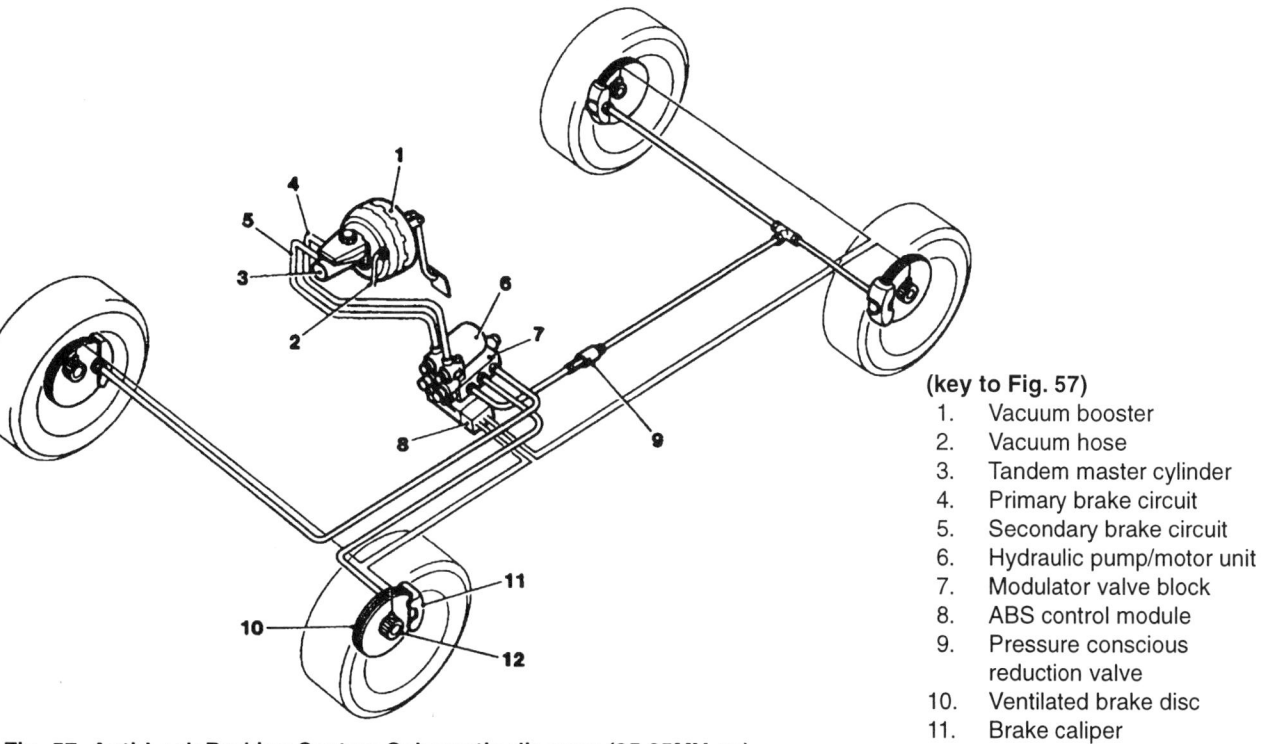

Fig. 57 Anti-Lock Braking System Schematic diagram (95.25MY on)

(key to Fig. 57)
1. Vacuum booster
2. Vacuum hose
3. Tandem master cylinder
4. Primary brake circuit
5. Secondary brake circuit
6. Hydraulic pump/motor unit
7. Modulator valve block
8. ABS control module
9. Pressure conscious reduction valve
10. Ventilated brake disc
11. Brake caliper
12. Wheel speed sensor

BRAKES, 1995.25 MODEL YEAR ON

DESCRIPTION

The anti-lock braking system (ABS) components are combined with a hydraulic booster, tandem master cylinder (TMC) and ventilated disc brakes on all four wheels, to provide a two-circuit braking system. The front brakes are fitted with four-piston calipers; the rear brakes have single-piston calipers and drums for the cable operated handbrake. The anti-lock braking system comprises the following:

- Hydraulic control module comprising electric motor driven pump, two low-pressure accumulators, modulator valve block and ABS control module (ABS CM)
- Four inductive wheel speed sensors, hub end mounted
- ABS warning indicator, mounted on the instrument panel
- Auxiliary inputs providing information to the ABS CM
- Diagnostic ISO communication BUS input/output link

The valve block houses solenoid operated valves which are activated by voltage signals from the control module. The signals are generated using wheel speed information received from the wheel speed sensors.

The valves regulate the supply of pressure individually to the front wheels and collectively to the rear wheels, as necessary, to prevent wheel locking during braking.

When the ignition is switched on, an ABS self test is initiated. During this test, the ABS warning indicator is lit for approximately 1.7 seconds and then extinguished. After this time delay the control module is ready to process signals provided from the various input sources and, using the software defined algorithm, control the electrical and hydraulic circuits. A fault is indicated if the warning indicator remains lit or comes on whilst the vehicle is being driven. Under fault conditions the system is inhibited or disabled, although conventional braking is unaffected.

The fluid level indicator lamp, mounted on the instrument panel, is lit when the brake fluid falls below the MIN mark on the brake fluid reservoir.

Fault conditions detected by the ABS CM disable the ABS until the fault is rectified. The system will be disabled when any of the following conditions occur:

- Valve failure
- Sensor failure
- Main driver failure (internal ABS CM fault)
- Redundancy error (internal ABS CM fault)
- Over-voltage/under-voltage
- Pump motor failure
- Under-voltage condition.

The input frequency from each wheel speed sensor signal is translated by the ABS CM into a comparable wheel speed. The ABS CM continually monitors the system. False wheel speed information, such as sudden speed changes or excessive speeds, is detected as a sensor malfunction. The ABS CM reacts to fault conditions in the following ways:

Inhibit - ABS is inhibited until the sensed speed returns to within an acceptable limit, whereupon ABS is restored. Conventional braking is unaffected. Depending on vehicle speed, the ABS warning indicator may come on.

Disable - ABS is disabled (switched off) and the ABS warning indicator comes on. The system will not be restored until the engine is switched off and restarted or the fault has been rectified. After the system has been disabled, the warning indicator remains on until the vehicle has reached a speed of 20 km / h (12.5 mile / h) during the first ignition cycle after fault rectification.

Vacuum booster

The vacuum booster (1 Fig. 58) is mounted on the brake pedal box and secured by three bolts. The tandem master cylinder (TMC) (2 Fig. 58) locates on two studs on the vacuum booster. Two lugs locate the fluid reservoir (3 Fig. 58) on the TMC which is secured by a split pin.

The vacuum is drawn from the inlet manifold. At the vacuum booster, the vacuum hose is connected to the vacuum chamber via an elbow connector. At the inlet manifold, the vacuum hose connector is of the push-on quick-release type.

Applied pedal force is increased by the vacuum booster which actuates the intermediate piston of the TMC.

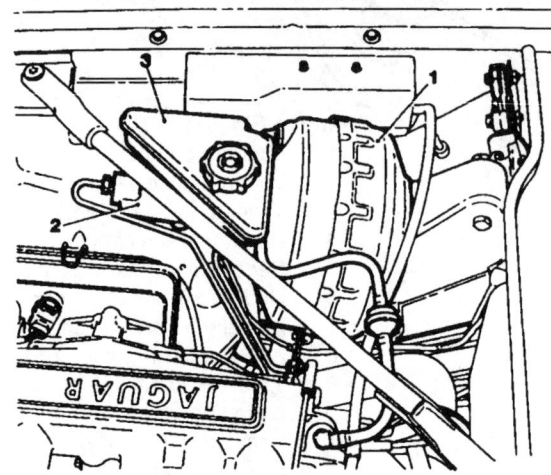

Fig. 58

The boost ratio supplied by the vacuum booster is 6.5 : 1.

Note: The vacuum booster and the TMC are supplied as a unit but are individually serviceable.

The brake fluid reservoir is fitted with a fluid level switch, which opens when fluid level is low and lights the low fluid level indicator.

Hydraulic control module

The hydraulic control module is located under the bonnet adjacent to the engine compartment bulkhead. It is secured within a steel mounting bracket at three securing points.

The hydraulic pump (1 Fig. 59) is a reciprocating two-circuit pump in which one brake circuit is assigned to each pump circuit. The pump supplies adequate pressure and volume supply to the brake circuits under anti-lock braking conditions. The pump housing incorporates a low pressure accumulator and damping chamber for each brake circuit. The pump is driven by an electric motor (2 Fig. 59) which draws 32A current at peak operation and has an internal resistance of 0.8 ohms.

The modulator valve block (3 Fig. 59) houses six solenoid valves: three normally open (NO) inlet valves and three normally closed (NC) outlet valves. This provides three outlet ports, one for each front brake and one for both rear brakes.

All electronic and power connections are made through one cable loom connector to the ABS control module (4 Fig. 59) located below the valve block.

The pump, motor and valve block are non-serviceable. If a fault occurs in any of these components, the whole hydraulic control module must be replaced. The ABS CM control module can be replaced separately.

ABS control module

The control module (CM) (4 Fig. 59), located beneath the modulator valve block (3 Fig. 59), is the system controller and processes all the information supplied from the external sensors and probes. Refer to the Control Module Connection Diagram. The signals from the four wheel speed sensors are independently processed by the CM, calculating numerical values which correspond directly to

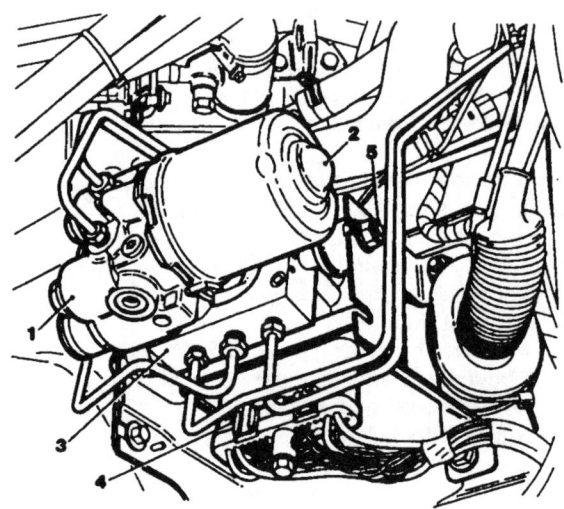

Fig. 59

(Key to Fig 60)
1. ABS control module
2. Wheel speed sensors
3. Battery voltage inputs
4. Diagnostic communication bus
5. Ignition voltage input
6. Instrument pack
7. ABS warning indicator
8. Low fluid level indicator
9. Fluid level switch
10. Brake pedal switch

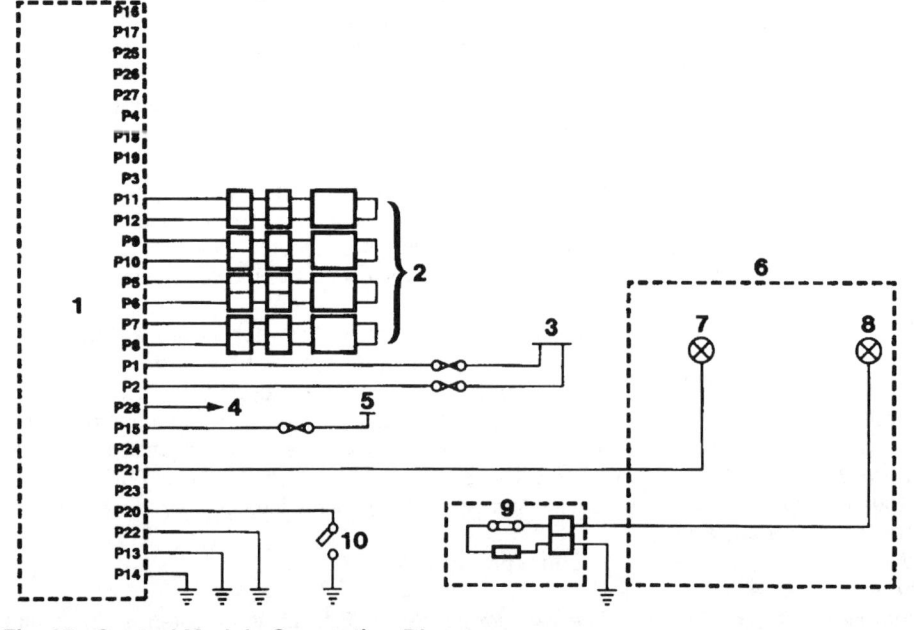

Fig. 60 Control Module Connection Diagram

the wheel speed. These values are converted into control signals for pressure modulation during ABS control.

The CM continuously monitors ABS operation, lighting the ABS warning indicator and inhibiting or disabling the system when faults are detected. In a fault condition, conventional braking is unaffected. The CM is self testing and cannot be fault diagnosed beyond 'black box' level, i.e. a faulty module. The CM houses the solenoids which operate the inlet and outlet valves of the modulator valve block. There is no electrical connection between the CM and the modulator valve block, but there is an electrical connection from the CM to the pump motor.

The CM functions include the following:

- Providing control signals for the operation of ABS solenoid valves
- Calculating wheel speed from voltage signals transmitted by the wheel speed sensors
- Monitoring of all electrical components
- On Board Diagnostics (OBD): storage of fault codes in a non-volatile memory

The fault codes generated by the CM are stored in a non-volatile memory which can be read via the OBD link. The ABS warning indicator is lit if the CM connector is loose or not fitted.

Control module connections

Control module connections, numbered 1 to 28, provide the necessary input / output signals to enable the module to control and monitor ABS operation.

Connections are as follows:

1	Battery positive feed (via fuse F22)
2	Battery positive feed (via fuse F11)
3	Not used
4	Not used
5/6	Wheel sensor, left-hand front
7/8	Wheel sensor, right-hand front
9/10	Wheel sensor, left-hand rear
11/12	Wheel sensor, right-hand rear
13	Ground
14	Ground
15	Ignition feed (via fuse F10)
16	Not used
17	Not used
18	Not used
19	Not used
20	Brake pedal switch
21	ABS warning indicator
22	ABS ground
23	Not used
24	Not used
25	Not used
26	Not used
27	Not used
28	Diagnostic ISO communication bus

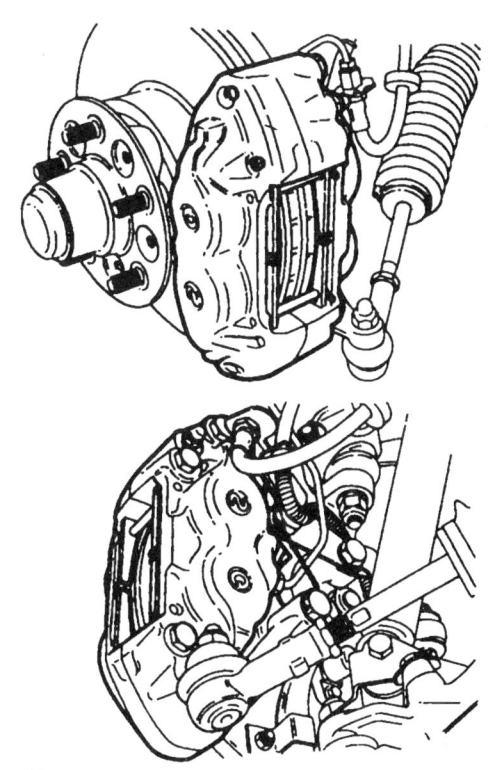

Fig. 61

Brake calipers

The front brakes are fitted with four-piston calipers acting upon 24 mm (15/16 in) thick ventilated brake discs (Fig. 61). The caliper carrier is secured by two bolts to the suspension vertical link.

The rear brakes are fitted with single-piston calipers acting upon 20 mm (25/32 in) thick ventilated brake discs (Fig. 62). The caliper carrier is secured by two bolts (wire locked) to the hub carrier.

The brake discs must be renewed when the minimum thicknesses specified below are reached:

Front brake disc - 22.9 mm (29/32 in)
Rear brake disc - 18.5 mm (47/64 in)

The rear brake caliper (1 Fig. 63) is mounted on the carrier (2 Fig. 63) by means of two guiding pins (3 Fig. 63) and a caliper retaining clip (5 Fig. 63). The guiding pins slide in bushes (4 Fig. 63) fitted to the caliper.

Note: The guiding pins are fitted with dust caps which must be fitted when reassembling the caliper.

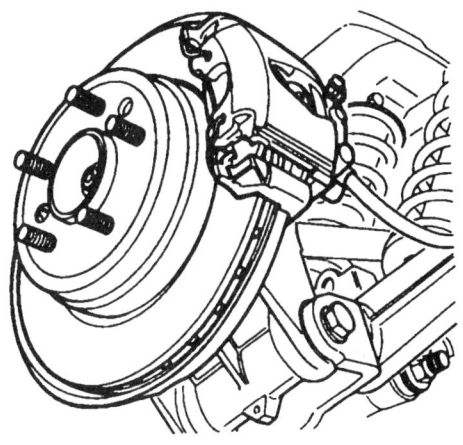

Fig. 62

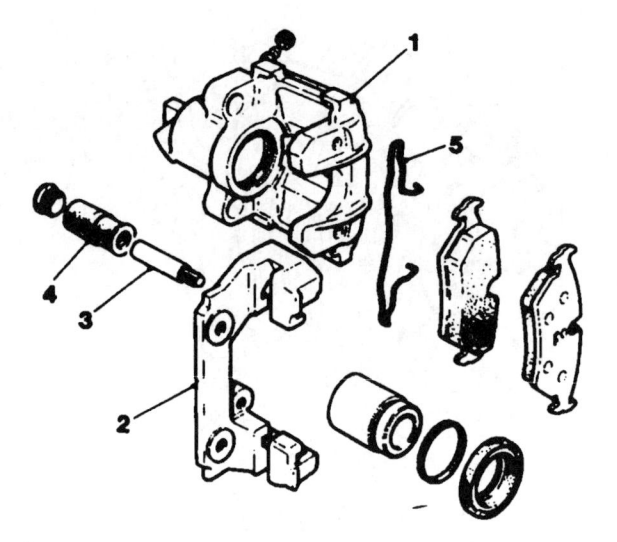

Fig. 63

Fig. 64

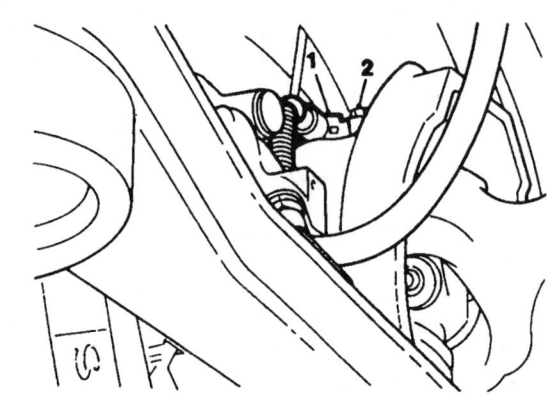

Fig. 65

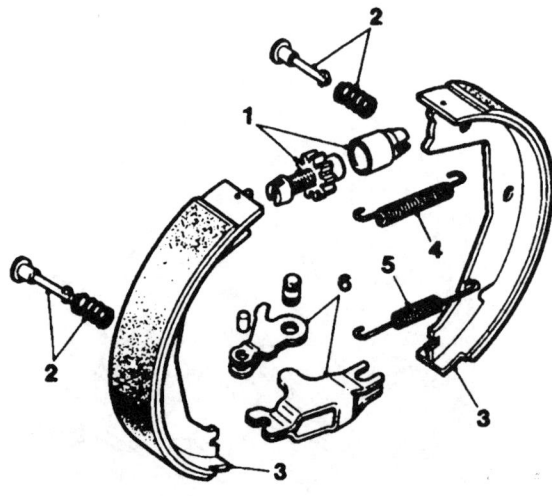

Fig. 66

Wheel speed sensors

Speed sensors are provided for each road wheel. The front sensors (1 Fig. 64) are mounted on the vertical link, while the rear sensors (1 Fig. 65) are mounted on the hub carrier. A toothed wheel, which turns with the road wheel, induces an a.c. voltage signal in the sensor. The frequency and amplitude of the a.c. voltage varies directly in relation to wheel speed, providing the control module with wheel speed information to give a comparison between the speed of each individual wheel, controlling braking as necessary.

Each sensor is monitored for open and short circuit failure, causing ABS control to be disabled on detection of a fault condition. ABS is also disabled should any sensed speed in excess of 330 km/h (205 mile/h) be detected. Similarly, ABS control is inhibited (switched off until fault condition is cleared) at speeds up to 40 km/h (25 mile/h) when frequency fluctuations are detected that are inconsistent with wheel rotation. At speeds above 40 km/h (25 mile/h), ABS control is disabled when inconsistencies are detected.

The sensor coil has a resistance value of 1100 ohms and has a voltage of 2.5V present on each connecting pin when the vehicle is stationary.

Handbrake

The handbrake comprises drum brakes on each rear wheel, cable operated by the handbrake lever which is located between the driver's seat and the inner sill.

When the handbrake lever is operated, the cable system applies equal force to both RH and LH brake shoe expander assemblies. The brake shoes expand and press against the hub assembly, locking the rear wheels.

The handbrake switch latches when the lever is operated and lights the handbrake warning indicator mounted on the instrument panel.

The drum brakes are of the duo-servo type. The expander assembly (6 Fig. 66) is mounted on the backplate mounting lug. The brake shoes locate on the expander assembly and the adjuster (1 Fig. 66). These are held in position by the upper and lower return springs (4 and 5 Fig. 66) and the hold-down springs (2 Fig. 66). The adjuster allows manual adjustment of the brake shoes.

OPERATION

The wheel speed sensors, fitted to all four road wheels, transmit wheel speed information to the control module. The module uses this information to modulate brake pressure during anti-lock braking.

Brake pedal force is increased by the vacuum booster which activates the Tandem Master Cylinder (TMC) intermediate piston. Brake fluid is supplied to the pump inlet ports on two separate circuits. The primary circuit supplies the front brakes whilst the secondary circuit supplies the rear brakes.

The rear wheels are controlled collectively on a 'select-low' principle during ABS operation. This means that if locking in either rear wheel is detected, controlled brake pressure is applied to both wheels.

A pressure conscious reduction valve (PCRV) is fitted

FAULT DIAGNOSIS

Trouble	Cause	Remedy
Long brake pedal	Brake caliper piston(s) or caliper guide pins (rear only) sticking Worn/damaged brake pads	Service or renew caliper or caliper guide pins (rear only) Renew brake pads
Vibration during braking	Worn/damaged brake pads Loose caliper mounting bolts Insufficient grease on sliding parts Foreign material or scratches on brake disc contact surface Damaged brake disc contact surface	Renew brake pads Tighten caliper mounting bolts Apply grease where necessary Clean brake disc contact surface Renew brake disc
Poor braking performance	Leak in hydraulic system Air in hydraulic system Worn/damaged brake pads Foreign material on brake pads Brake caliper piston malfunction Tandem master cylinder malfunction Vacuum booster fault Disconnected or damaged vacuum hose Low brake fluid level	Repair leak. Check all pipework connections. Refill and bleed the system Check for leaks and bleed the system Renew brake pads Examine brake pads and clean or renew as necessary Renew faulty brake caliper piston Service or renew tandem master cylinder Renew vacuum booster Renew vacuum hose Check for leaks, refill and bleed the system
Brakes pull to one side	Worn/damaged brake pads Foreign material on brake pad Failing valves in ABS valve block Abnormal wear or distortion on brake disc Incorrect tyre pressure	Renew brake pads Examine brake pads and clean or renew as necessary Renew hydraulic control module Examine front brake disc and service or renew as necessary Inflate tyre to correct pressure
Brakes do not release	No brake pedal free play Vacuum booster binding Tandem master cylinder return port faulty Faulty valve in ABS valve block	Adjust brake pedal free play Renew vacuum booster Clean return port on tandem master cylinder Renew hydraulic control module
Excessive pedal travel	Leak in hydraulic system Air in hydraulic system Worn tandem master cylinder piston seals or scored cylinder bore 'Knock back'. Excessive brake disc run-out or loose wheel bearings	Repair leak. Check all pipework connections. Refill and bleed the system Check for leaks and bleed the system Renew tandem master cylinder Check brake disc run-out and renew as necessary. Adjust wheel bearing
Brakes grab	Brake pads contaminated by grease or brake fluid Brake pads distorted, cracked or loose Loose caliper mounting bolts or guide pins (rear only)	Renew brake pads. Check pipework for leaks Renew brake pads Check caliper and repair/renew as necessary
Brakes drag	Seized or incorrectly adjusted handbrake or cable Broken or weak handbrake return springs Caliper piston(s) seized Brake pedal binding at pivot points Vacuum booster binding Tandem master cylinder faulty	Examine handbrake and repair/renew as necessary Renew handbrake return springs Examine calipers and repair/renew as necessary Examine brake pedal bushings and repair/renew as necessary Renew vacuum booster Examine tandem master cylinder and repair/renew as necessary
Hard brake pedal when pressed	Lack of vacuum at the vacuum booster Tandem master cylinder push-rod binding Frozen tandem master cylinder piston Brake caliper piston or caliper guide pins (rear only) seized	Check vacuum hose. Repair or renew as necessary Renew tandem master cylinder Renew tandem master cylinder Examine caliper and renew/repair as necessary
Excessive brake noise	Worn brake pads Bent or cracked handbrake shoes Foreign objects in brake pads or handbrake shoes Broken/loose handbrake hold-down springs or return springs Loose caliper mounting bolts	Renew brake pads Renew handbrake shoes Examine brake pads and handbrake shoes. Clean or renew as necessary Examine handbrake assembly. Repair or renew as necessary Torque-tighten caliper mounting bolt

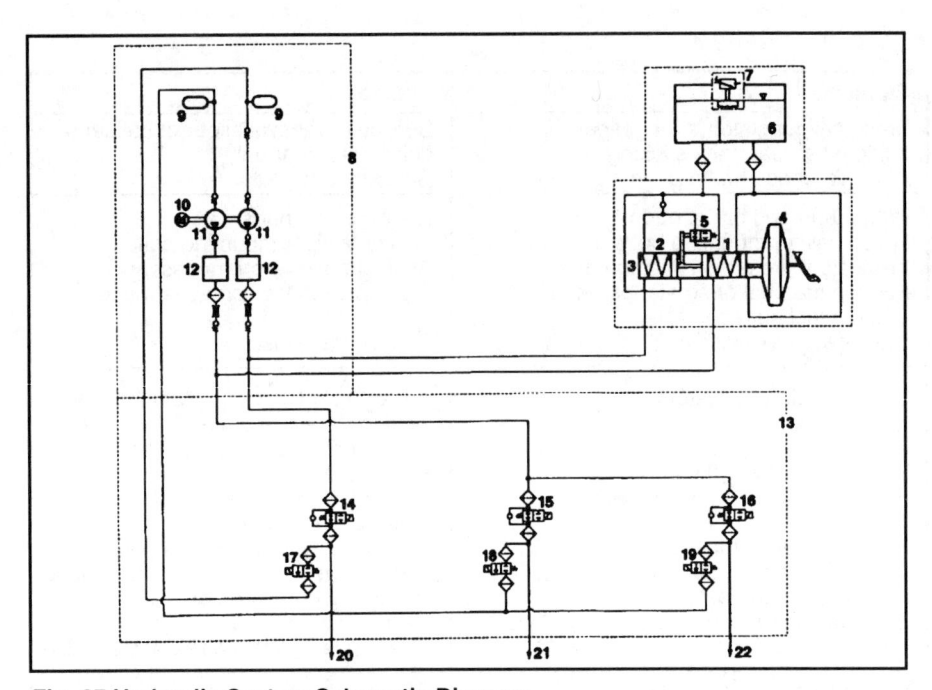

Fig. 67 Hydraulic System Schematic Diagram

(key to Fig. 67)
1. TMC 1 (primary circuit)
2. TMC 2 (secondary circuit)
3. Tandem master cylinder
4. Vacuum booster
5. Central valve
6. Fluid reservoir
7. Fluid level indicator
8. Pump motor unit
9. Low pressure accumulator
10. Electric pump motor
11. Two circuit hydraulic pump
12. Damping chamber
13. Valve block
14. Inlet valve NO
15. Inlet valve NO
16. Inlet valve NO
17. Outlet valve NC
18. Outlet valve NC
19. Outlet valve NC
20. Rear brake circuit
21. Front brake circuit (left)
22. Front brake circuit (right)

between the outlet of the valve block and the rear brake circuit. The valve is fitted to prevent over-braking at the rear wheels. Up to a threshold of 25 bar (363 lbf/in²), brake pressure to the front and rear brakes is equal. Above this threshold, the PCRV reduces pressure to the rear brakes to provide a closer balance between front and rear brakes and optimize road adhesion.

Hydraulic operation

Referring to the Hydraulic System Schematic Diagram (Fig. 67), the TMC primary circuit (item 1) applies brake pressure to the front brakes. Individual control of the front wheels is provided by solenoid valves. One pair of valves (items 15 and 18) controls the front left brake circuit (item 21) and another pair of valves (items 16 and 19) controls the front right brake circuit (item 22). The TMC secondary circuit (item 2) applies brake pressure to the rear brake circuit (item 20) via valves (items 14 and 17), on a 'select low' principle.
Should the ABS be initiated by a locking tendency of any wheel during braking, the pump unit (item 8) is started and the appropriate NO inlet valve (item 14, 15 or 16) closes in response to signals from the control module. This action prevents further increase of brake pressure by blocking the supply of brake fluid from the TMC (item 3). If excessive deceleration continues, the appropriate NC outlet valve (item 17, 18 or 19) opens, releasing brake pressure to the low pressure accumulators (item 9) until the wheel accelerates again.

From the low pressure accumulators, volume is pumped back into the TMC, forcing the brake pedal back. To optimize the friction coefficient between tyre and road surface, brake pressure is increased in small steps by closing the outlet valve and opening the inlet valve and recharging brake pressure.

During the pressure build-up phase, the volume required for replenishment is supplied by the TMC and additionally by the pump from the low pressure accumulators. Since the delivered flow is generally greater than volume flow drained from the brake circuits, the low pressure accumulators serve as intermediate accumulators to compensate for temporary volume flow peaks.

The TMC piston positions, and therefore the brake pedal, vary with the fluid displacement in the brake caliper. As controlled pressure in the brake caliper decreases and increases during ABS, the brake pedal 'cycles', informing the driver that controlled braking is in progress.

Actuation of the brake pedal, causes the central valve (item 5) in the TMC to close. This action prevents damage to the TMC piston seals.

At the end of a brake application, volume is restored to the TMC, at low pressure from the fluid reservoir (item 6).

SYSTEM FAULT INDICATIONS

Fault indication:

ABS warning indicator still illuminated after ignition switch on and ABS CM self test.

Possible causes

Fuses blown
Faulty wheel speed sensor or harness
Faulty wiring
Faulty ABS CM.

Fault diagnosis

Note: After a fault has been successfully diagnosed and corrected, the ABS warning indicator will remain lit until the fault codes have been cleared from the non-volatile memory. This can be done by using Jaguar Diagnostic Equipment or, alternatively, by driving the vehicle to a speed above 20 km/h (12.5 mile/h). If the indicator remains on after this procedure, repeat fault diagnosis.

1. Check the fuses (F11 & F22) in the battery feed lines and fuse (F10) in the ignition line. The fuses are located in the left and right-hand scuttle fuse boxes
2. Unbolt the 28-way multi-plug connector from the ABS CM
3. Measure the resistance across each wheel speed sensor. The value should be 1100R ± 50%. If not, unplug the sensor flying lead and measure again. If the value is now within range, inspect the harness

between ABS CM and sensor, otherwise renew the sensor

4. Check continuity to ground from ABS CM harness connections 13 and 14. If the value is much greater than 0.1R, renew the harness
5. Measure the voltage between ABS CM harness connection 14 and connections 1 and 2 respectively. If the value is not approximately equal to battery voltage, renew the harness
6. With the ignition switch ON, measure the voltage between ABS CM harness connections 14 and 15. If the value is not approximately equal to battery voltage, renew the harness.

Renew the ABS CM if the fault has not been located after carrying out the above procedures.

Fault indication:

ABS warning indicator illuminates at 20 km/h (12.5 mile/h).

Possible causes

Fuses blown
Faulty hydraulic control module pump/motor unit or circuitry
Faulty ABS CM.

Fault diagnosis

1. Check fuses
2. Disconnect the pump / motor unit and measure the resistance across the two pin connector. The measured value should be in the region of 0.8R. Renew the complete hydraulic control module if the measured value indicates excessive resistance or a short circuit
3. Unbolt the 28-way connector from the ABS CM and measure the voltage between harness connections 1 and 14. If the value is not approximately equal to battery voltage, renew the harness.

Renew the ABS CM if the fault has not been located after carrying out the above procedures.

Fault indication:

ABS warning indicator illuminates on 'pull-away' or during driving.

Possible causes

Faulty sensor or wiring
Faulty brake disc or wheel bearing installation giving inconsistent signals to the ABS CM.

Fault diagnosis

Check sensor installation for:

1. Security of sensor lead fixing bolt
2. Damage to sensor lead
3. Possible damage to brake disc
4. Excessive play in wheel bearing
5. Intermittent faults caused by poor harness connection or damage.

Pin-point tests

Wheel sensor

Testing between pins of the 28-way multi-plug connector, check that the resistance of each sensor coil is 1100 ohms ± 50%.

Code	Fault	Comment
5242h	Outlet valve, rear	
5250h	Inlet valve, rear	
5120h	Outlet valve, front right	
5214h	Inlet valve, front right	
5194h	Outlet valve, front left	
5198h	Inlet valve, front left	
5168h	Sensor, rear right	Sensor failure recognised by monitoring of d.c. voltage with vehicle stationary
5178h	Sensor, rear left	
5148h	Sensor front right	
5158h	Sensor, front left	
5165h	Sensor, rear right	Sensor failure recognised by monitoring of wheel speed continuity
5175h	Sensor, rear left	
5145h	Sensor, front right	
5155h	Sensor, front left	
5260h	Sensor, rear right	Sensor failure recognised by wheel speed comparison
5261h	Sensor, rear left	
5259h	Sensor, front right	
5258h	Sensor, front left	
5235h	Sensor, rear right	Sensor failure recognised by long term detection of missing sensor signal
5236h	Sensor, rear left	
5234h	Sensor, front right	
5233h	Sensor, front left	
9317h	Over-voltage	
9342h	CPU failure	
5095h	Pump motor	
5267h	Disturbance detection	

Note: See Control Module Connections for the ABS CM wheel sensor connector references.

Hydraulic pump motor

Disconnect the pump motor bi-pin connector and check that the resistance of the motor winding is approximately 0.8 ohms.

Stored fault codes

The following fault codes are stored automatically within the ABS CM and may be accessed, as an aid to fault diagnosis, using Jaguar Diagnostic Equipment.

BRAKE HYDRAULIC SYSTEM

Fluid level

WARNING:
AVOID SKIN/EYE CONTACT OR INGESTION OF BRAKE FLUID. IF SKIN OR EYES ARE ACCIDENTALLY SPLASHED WITH BRAKE FLUID, RINSE THE AFFECTED AREA IMMEDIATELY WITH PLENTY OF WATER AND SEEK MEDICAL ATTENTION IMMEDIATELY.

> **CAUTION:**
>
> Fluid must not be allowed to contact the vehicle paintwork. Remove any spilt fluid from the paintwork by rinsing away with running water. Methylated spirit (denatured alcohol) must not be used to clean the contaminated area.

Checking the fluid level

Correct brake fluid level is essential for the efficient operation of the brake system. Check that the fluid level is between the MAX and MIN marks on the fluid reservoir (Fig. 68). Top up if necessary with recommended brake fluid.

Note: The efficiency of the brakes may be impaired if fluid is used which does not meet specifications. Use ONLY brake and clutch fluid that conforms to a minimum DOT 4 specification. Do not use brake fluid that has been exposed to atmosphere for any length of time. Moisture absorbed from the atmosphere impairs the efficiency of the brake fluid.

> **WARNING:**
>
> GREAT CARE MUST BE EXERCISED WHEN SERVICING OR REPAIRING THE SYSTEM. AVOID SKIN/EYE CONTACT OR INGESTION OF BRAKE FLUID. IF SKIN OR EYES ARE ACCIDENTALLY SPLASHED WITH BRAKE FLUID, RINSE THE AFFECTED AREA IMMEDIATELY WITH PLENTY OF WATER AND SEEK MEDICAL ATTENTION. IF BRAKE FLUID IS INGESTED, SEEK MEDICAL ATTENTION IMMEDIATELY.

> **WARNING:**
>
> THROUGHOUT THE FOLLOWING MAINTENANCE/SERVICE OPERATIONS, ABSOLUTE CLEANLINESS MUST BE OBSERVED TO PREVENT FOREIGN MATTER CONTAMINATING THE BRAKE SYSTEM.

> **CAUTION:**
>
> Fluid must not be allowed to contact the vehicle paintwork. Remove any spilt fluid from the paintwork by rinsing away with running water. Methylated spirit (denatured alcohol) must not be used to clean the contaminated area.

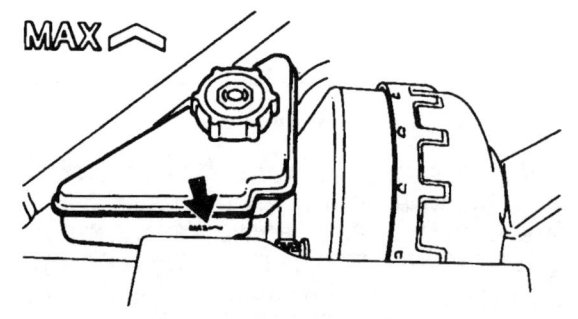

Fig. 68

> **CAUTION:**
>
> Never use methylated spirit (denatured alcohol) for component cleaning purposes. Use only a proprietary brake cleaning fluid.

BRAKE HYDRAULIC SYSTEM, BLEED

General instructions

Use a brake bleeder bottle with a clear bleeder tube. Also recommended is a filler unit with a fill pressure of 1.0 bar (14.5 lbf/in^2). If a filler unit is not used, ensure that there is sufficient brake fluid in the reservoir throughout the bleeding procedure.

Note: Always bleed the caliper furthest away from the actuation unit first. On right-hand drive vehicles, bleed in the following order: front left (FL), front right (FR), rear left (RL) and rear right (RR). On left-hand drive vehicles bleed in the following order: FR, FL, RR and RL.

System Bleeding After Brake Fluid Renewal

Ensure that the vehicle is standing level. Switch the ignition off. Check that the fluid level in the reservoir is between the MIN and MAX marks.
Connect the bleeder bottle tube to the relevant front caliper bleeder screw (Fig. 69) and open the screw (see preceding 'Note').
Bleed until new, clear, bubble free fluid is observed in the tube and then close the bleeder screw.
Repeat this procedure at each remaining caliper.
With the engine running, check brake pedal travel. If excessive, check for leaks and repeat the bleed procedure. Fill the reservoir to the MAX level.
System bleeding after tandem master cylinder renewal
Ensure that the vehicle is standing level. Switch the ignition off. Check that the fluid level in the reservoir is between the MIN and MAX marks.
Connect the bleeder bottle tube to the relevant front caliper bleeder screw (Fig. 69) and open the screw (see Note on previous page).
Actuate the brake pedal to the floor; hold for approximately two seconds and then release the pedal. Wait another two

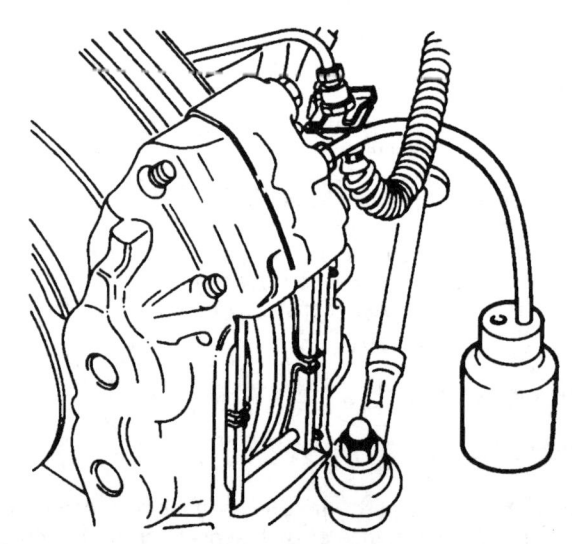

Fig. 69

seconds and actuate the brake pedal again for a further two seconds. Repeat this action 20 to 30 times until clear, bubble free brake fluid streams out.
With the brake pedal actuated, close the bleeder screw. Build up fluid pressure by pumping the pedal and then open the bleeder screw. Repeat this action three to five times.

Note: If a filler unit is not used, observe the fluid level in the reservoir and top up if necessary.

Repeat this procedure for the remaining three calipers: (Fig. 70) shows the bleeder screw of the rear left caliper.

With the engine running, check brake pedal travel. If excessive, check for leaks and repeat the bleed procedure. Fill the reservoir to the MAX level.

System bleeding after hydraulic control module renewal

Hydraulic control modules are supplied pre-filled to enable the brake system to be bled in the conventional way.
Ensure that the vehicle is standing level. Switch the ignition off. Check that the fluid level in the reservoir is between the MIN and MAX marks.
Connect the bleeder bottle tube to the relevant front caliper bleeder screw (Fig. 69) and open the screw (see Note on previous page).
Actuate the brake pedal full stroke, wait a moment and then release. Wait two to three seconds and then actuate the brake pedal full stroke again. This allows the TMC to be completely re-filled with fluid.
Repeat 20 to 30 times until the fluid in the bleeder tube is clear and bubble free.
With the brake pedal actuated, close the bleeder screw. Build up fluid pressure by pumping the pedal and then open the bleeder screw. Repeat this action three times.

Repeat this procedure for the remaining three calipers. With the engine running, check brake pedal travel. If pedal travel is excessive, check the system for leaks and repeat the bleed procedure.
Fill the reservoir to the MAX level.

Bleeding after renewal of caliper

Follow the procedure above but only at the affected caliper.

BRAKE CALIPERS, INSPECTION AND CLEANING

> **WARNING:**
>
> BRAKE LINING DUST CAN, IF INHALED, DAMAGE YOUR HEALTH. ALWAYS USE A VACUUM BRUSH TO REMOVE DRY BRAKE LINING DUST. NEVER USE AN AIR LINE.

When fitting new brake pads always take necessary precautions and remove the brake dust from around the caliper area. After renewal, pump the brake pedal several times to centralize the new brake pads.

> **CAUTION:**
>
> When cleaning brake components only use a proprietary fluid. Never use petrol. Use of petrol, paraffin or other mineral based fluids can prove dangerous.

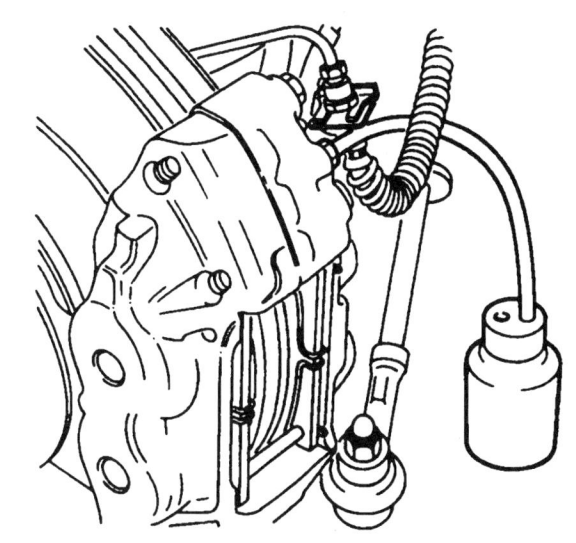

Fig. 70

Remove all brake dust from the caliper, carrier and brake disc. Thoroughly clean the pad abutment areas. Avoid damaging the piston and dust cover.
Examine all the components for signs of wear, damage and corrosion. Pay particular attention to the piston and piston bore.
Remove caliper body corrosion with a wire brush or wire wool. No attempt should be made to clean a badly corroded or scored piston bore. The caliper must be renewed.

> **CAUTION:**
>
> No attempt should be made to clean corroded bolts.

Inspect the caliper guide pins. Ensure that they are not corroded or seized and that the caliper moves freely. If they are difficult to remove or corroded in any way, they must be replaced together with new dust covers.
When reassembling always renew piston seals. Lubricate the new piston seal and fit carefully to the inner groove of the piston bore.

> **CAUTION:**
>
> Ensure that working surfaces and hands are clean. Use only brake fluid of the correct specification to lubricate the new seals when fitting.

WHEEL SPEED SENSOR, REAR, RENEW

Open the door. Move the front seats fully forward.
Remove the rear seat cushion and squab.
Displace the insulation material and release the ratchet strap securing the sensor multi-plug (1 Fig. 71) to the transmission tunnel.

Disconnect the multi-plug. Attach a drawstring to the sensor harness.
Locate a trolley jack below the outer fork of the rear wishbone.

> **CAUTION:**
>
> To prevent wishbone component damage, use a suitably shaped block of wood between the jack and wishbone.

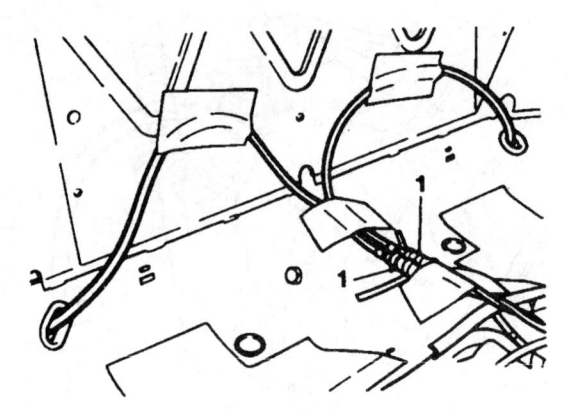

Fig. 71

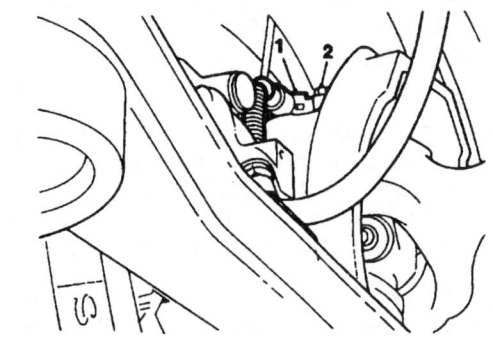

Fig. 72

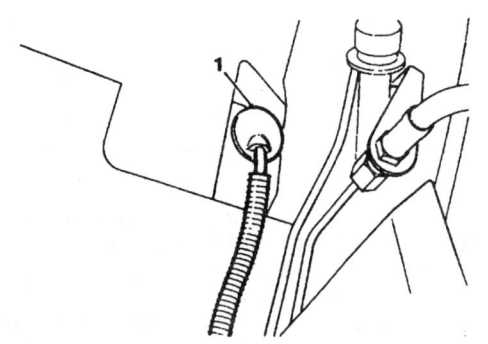

Fig. 73

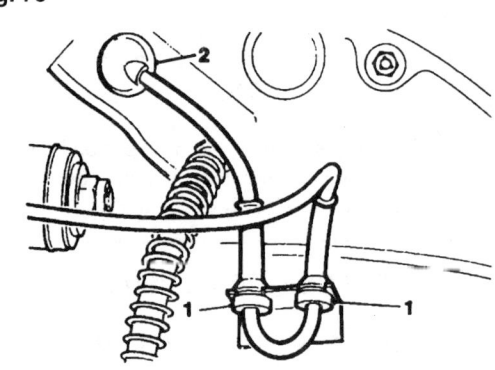

Fig. 74

Fig. 75

Jack up the vehicle. Place an axle stand below the jacking spigot. Lower the car/jacking spigot on to the axle stand. Leave the jack in position as a safety measure.
Remove the road wheel.
Remove the sensor securing bolt (2 Fig. 72) and displace the sensor (1 Fig. 72) from the hub.
Cut and remove the harness ratchet strap.
Displace the harness outer body grommet (1 Fig. 73). From below, carefully withdraw the harness and drawstring. Transfer the drawstring to the new sensor harness.

Carefully draw the new harness into position and seat the outer body grommet (1 Fig. 73).
Route the harness to the hub and fit the sensor to the hub (1 Fig. 72).
Fit and tighten the sensor securing bolt (2 Fig. 72).
Secure the harness with a new ratchet strap.
Fit the road wheel.
Route the harness, remove the drawstring and connect the multi-plug (1 Fig. 71).
Fasten the ratchet strap to secure the sensor multi-plug to the transmission tunnel.
Replace the rear seat cushion and squab.
Reposition the front seats and close the door.

Torque Figures

Wheel nut to wheel stud	88 - 102 Nm
Wheel sensor bolt	8 - 10 Nm

WHEEL SPEED SENSOR, FRONT, RENEW

Drive the vehicle on to a ramp.
Open the bonnet. Disconnect the sensor multi-plug. Cut and remove the ratchet strap if required.
Displace the body grommet (2 Fig. 74) from the inner wing valance.
Raise the ramp.
Remove the brake disc cover plate. Remove the retaining bracket.
Remove the upper cover securing bolt and remove the cover.
Remove the sensor securing bolt (2 Fig. 75) and displace the sensor (1 Fig. 75) from the hub.

Displace the harness carrier grommets from the inner-wing and caliper brackets. The inner-wing bracket is shown (1 Fig. 74). Cut and remove the harness ratchet strap. Displace and remove the sensor and harness.
Route the new sensor harness through the inner wing, seat the harness to the carrier grommets and fit a new ratchet strap.
Fit the sensor to the hub. Fit and tightening the securing bolt.
Fit the body grommet to the inner wing (2 Fig. 74).
Fit the brake disc cover plate and the retaining bracket.
Lower the ramp. Route the harness to the connector and connect the multi-plug. Secure the harness with a new ratchet strap if required.
Close the bonnet.

Torque Figures

Wheel sensor bolt	8 - 10 Nm

HYDRAULIC CONTROL MODULE, RENEW

Note: Refer to Brake Hydraulic System, Bleed before carrying out this procedure. Pay particular attention to the warnings and cautions relating to brake fluid, cleanliness and cleaning materials.

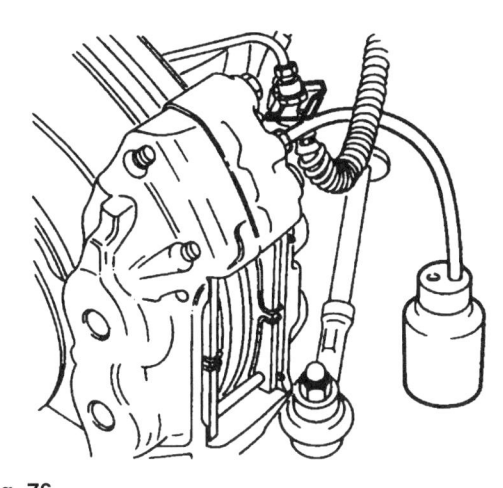

Fig. 76

Raise the vehicle.
Remove the bleed screw dust cap from the front LH caliper. Connect a bleeder tube and bottle to the bleed screw (Fig. 76) and open the bleed screw.

Fit a brake pedal hold-down tool (JDS-9013) between the brake pedal and the steering wheel. Adjust the tool to operate the brake pedal 60 mm down. This operation is necessary to prevent fluid loss from the reservoir through disconnected brake pipes.
Tighten the front LH caliper bleeder screw. Disconnect the bleed tube from the bleed screw and remove the tube and bottle. Fit the bleed screw dust cap.
Undo the securing bolt of the multi-plug connector (4 Fig. 77). The bolt will remain captive.
Disconnect the multi-plug connector and reposition safely.
Place absorbent material underneath the hydraulic control module to absorb any spillages.
Undo the tandem master cylinder (TMC) brake pipe unions at the pump unit (1 Fig. 77) and disconnect the brake pipes.
Fit plugs immediately to the brake pipes and the pump unit to prevent fluid loss.
Undo the front and rear brake pipe unions at the valve block (3 Fig. 77).
Remove the brake pipes.
Fit plugs immediately to the brake pipes and the valve block to prevent fluid loss.
Undo and remove the three securing nuts (5 Fig. 77) and remove the hydraulic control module.
Remove the absorbent material and clean the mounting bracket and surrounding area.
Fit and align a new hydraulic control module to the mounting bracket. Ensure that the mounting cup tangs fully engage the bracket slots.
Fit and tighten the securing nuts (5 Fig. 77).
Place absorbent material underneath the hydraulic control module to absorb any spillages.
Connect the front and rear brake pipes to the valve block (3 Fig. 77), removing the plugs immediately prior to connection. Tighten the pipe unions.
Connect the TMC brake pipes to the pump unit (1 Fig. 77), removing the plugs immediately prior to connection. Tighten the pipe unions.
Remove the absorbent material and clean the surrounding area.
Reposition and connect the multi-plug connector (4 Fig. 77). Tighten the securing bolt.
Ensure that all fixings are torque-tightened to specified tolerances.
Release the brake pedal hold-down tool and remove.
Bleed the system. See Brake Hydraulic System, Bleed.
Examine the hydraulic control module for leaks.

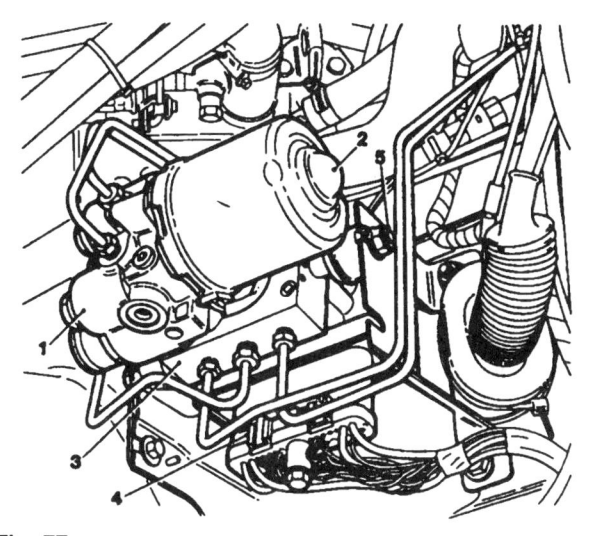

Fig. 77
Torque Figures

Hydraulic module to module bracket	18 - 26 Nm
28-way connector to control module	4 - 5.6 Nm
M12 pipe connectors	15 - 19 Nm
M10 pipe connectors	12 - 16 Nm

Oils / Greases

Jaguar Brake and Clutch Fluid or Castrol/Girling Universal to/or exceeding DOT 4 specification.

PRESSURE CONSCIOUS REDUCING VALVE, RENEW

Note: Refer to Brake Hydraulic System, Bleed before carrying out this procedure. Pay particular attention to the warnings and cautions relating to brake fluid, cleanliness and cleaning materials.

Remove the air cleaner cover and element.
Remove the bleeder screw dust cap from the front LH caliper. Connect a bleeder tube and bottle to the bleeder screw (Fig. 76) and open the bleeder screw.
Fit a brake pedal hold-down tool (JDS-9013) between the brake pedal and the steering wheel. Adjust the tool to operate the brake pedal 60 mm down. This operation is necessary to prevent fluid loss from the reservoir through disconnected brake pipes.
Tighten the front LH caliper bleeder screw. Disconnect the bleeder tube from the bleed screw and remove the tube and bottle. Fit the bleeder screw dust cap.
Place suitable absorbent material around the pressure conscious reducing valve (PCRV) area to catch any spilt fluid.
Supporting the PCRV unions, disconnect the pipe unions (1 Fig. 78) from the PCRV (2 Fig. 78). Fit plugs immediately to the brake pipes and PCRV to prevent fluid loss.

Remove the PCRV from the spring clip.
Fit the new PCRV into the spring clip. While supporting the PCRV unions, connect the pipe unions (1 Fig. 78) to the PCRV (2 Fig. 78), removing the plugs immediately prior to connection.
Tighten the pipe unions.
Remove the absorbent material and clean the surrounding area.
Fit the air cleaner element and cover.
Bleed the system. See Brake Hydraulic System, Bleed.

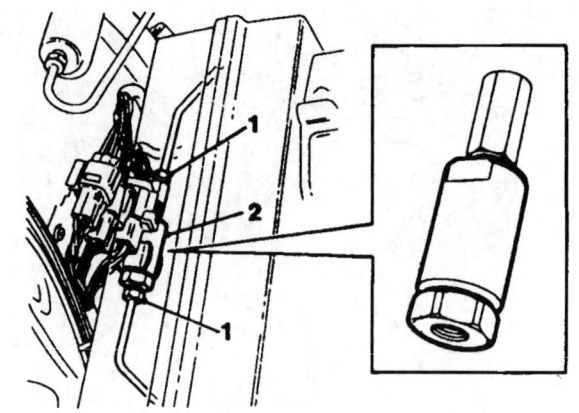

Fig. 78

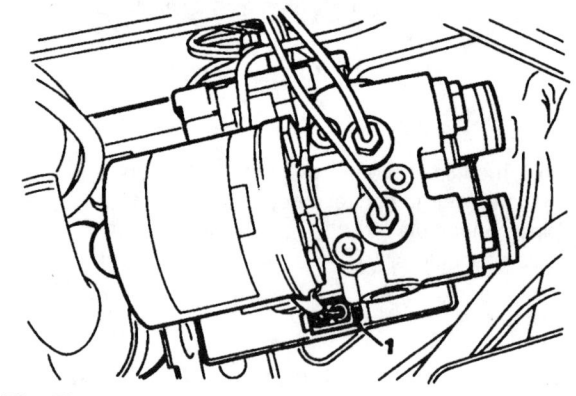

Fig. 79

Torque Figures

M12 pipe connectors	15 - 19 Nm
M10 pipe connectors	12 - 16 Nm

Oils / Greases

Jaguar Brake and Clutch Fluid or Castrol/Girling Universal to/or exceeding DOT 4 specification.

ABS Control Module, renew

Note: Refer to Brake Hydraulic System, Bleed before carrying out this procedure. Pay particular attention to the warnings and cautions relating to brake fluid, cleanliness and cleaning materials.

Remove the hydraulic control module.
Disconnect the pump electric motor to ABS CM multi-plug (1 Fig 79).
Undo and remove two securing screws and remove the ABS CM.
Clean the mating faces of the hydraulic control module and the new ABS CM.
Fit and tighten the two securing screws.
Connect the pump electric motor to ABS CM multi-plug (1 Fig 79).
Fit the hydraulic control module.

Torque Figures

Hydraulic module to module bracket	18 - 26 Nm
Control module to hydraulic module	1.8 - 2.8 Nm
28-way connector to control module	4 - 5.6 Nm
M12 pipe connectors	15 - 19 Nm
M10 pipe connectors	12 - 16 Nm

Oils / Greases

Jaguar Brake and Clutch Fluid or Castrol/Girling Universal to/or exceeding DOT 4 specification.

CHAPTER 14

BODY - 1992 MODEL YEAR ONWARDS (INCLUDING SRS SYSTEMS)

Introduction

Many improvements were made to all variants of XJS at 1992 and 1993/93.5 Model Years.

Of most note were the improvements to Body-in-White construction methods and improvements to the glasshouse at 1992, and the new bumper systems and associated changes at the later point.

Also covered in this Section are the Supplementary Restraint Systems (SRS) which have been fitted at various points from 1989 on (limited market production). Later versions benefit from airbags on both sides of the vehicle, together with 'tear-loop' type seatbelts.

Sealants

In production, the vehicle is given an under body protection. Stone chip primer is also used on the sill panels, while hot wax injection is applied to all box sections and closed members. Recommended sealants are listed in the Jaguar Body Sealing and Preservation Manual (AKM 9137).

PLENUM CHAMBER FINISHER, RENEW

Open the bonnet and remove the wiper arms.
Use a suitable tool to carefully remove the wiper pinion plastic trim covers (2 Fig. 1).

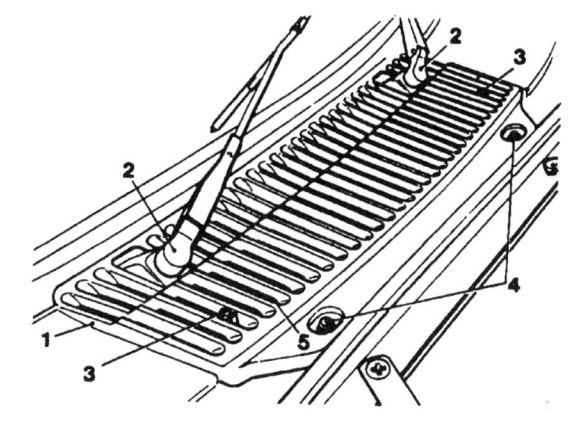

Fig. 1

Displace the washer jets (3 Fig. 1) down through the plenum chamber finisher (1 Fig. 1).
Remove the plenum chamber finisher (1 Fig. 1) by undoing the screws (4 Fig. 1) which secure the plenum chamber finisher to the body.
Remove the plenum finisher gauze from the clips (5 Fig. 1) which hold it in position.

Fit the gauze into position in the plenum chamber finisher, using the clips (5 Fig. 1) to hold it in place. Place the plenum chamber finisher in the position where it is to be mounted, and locate the washer jets (3 Fig. 1) through the finisher and into their positions.
Fix the plenum chamber finisher firmly into position using the securing screws (4 Fig. 1).
Mount the wiper pinion plastic trim covers (2 Fig. 1) firmly on to their seatings.
Fit the wiper arms and close the bonnet.

STEERING COLUMN SWITCH HOUSING, RENEW

Note: The air bag module should first be removed before proceeding to remove the steering column switch housing.

Remove the steering wheel.
Remove the steering column lower cowl.
Remove the steering column upper cowl.

Removing the steering column switch housing.
Remove the fibre optic illumination bulb cover (1 Fig. 2).
Displace both fibre optic switch leads (2 Fig. 2) from the housing.
Release both stalk switches from their mountings by squeezing their securing lugs (1 Fig. 3).

Displace them from their mountings by carefully sliding them out.
Disconnect the horn contact wire (2 Fig. 3).
Undo and remove the housing to column securing bolts (3 Fig. 3) and displace the housing (4 Fig. 3) from the steering column.
Displace the illumination bulb holder (5 Fig. 3) from the switch housing.
Remove the switch housing (4 Fig. 3) from the steering column.

Place the switch housing (4 Fig. 3) by the position where it is to be mounted and fully seat the switch illumination bulb holder (5 Fig. 3) into the switch housing.

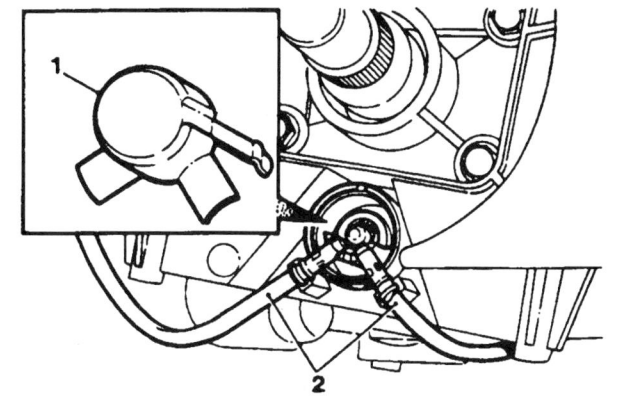

Fig. 2

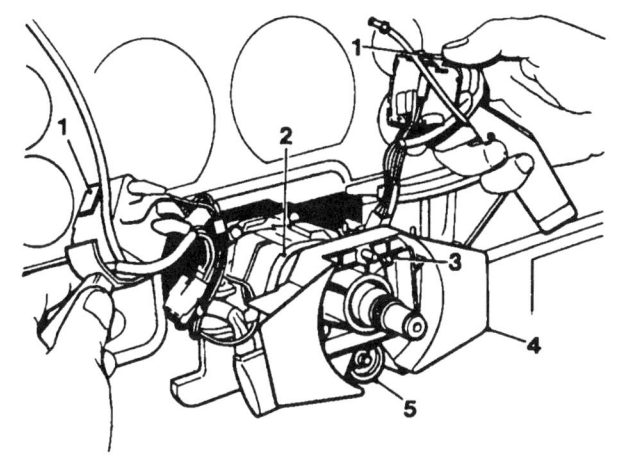

Fig. 3

Locate the switch housing on to the steering column and when it is fully seated, fix it in position by firmly tightening the securing bolts (3 Fig. 3).
Connect the horn contact wire (2 Fig. 3).
Place each stalk switch (1 Fig. 3) in position and fully seat them into their locations in the switch housing to fix them.
Locate the fibre optic switch leads into their positions in the switch housing (2 Fig. 2).
Firmly locate the fibre optic illumination bulb cover (1 Fig. 2) into its position on the switch housing.

Fit the steering column upper cowl.
Fit the steering column lower cowl.
Fit the steering wheel.

Note: The air bag module should be fitted on to the steering wheel as the final stage of fitting the steering wheel.

DOOR GLASS, RENEW

Remove the door trim pad.
Remove the front speaker.
Remove and discard the tape securing the rear upper edge

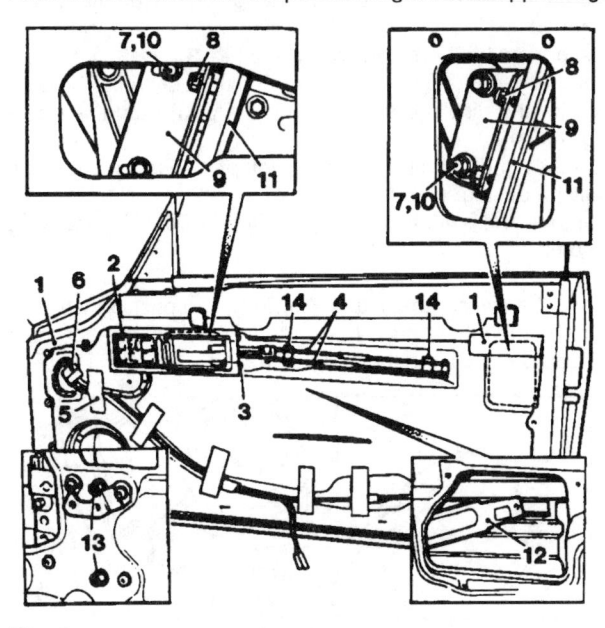

Fig. 4

of the water shedder (1 Fig. 4) to the door panel.
Remove the electric seat switch surround (2 Fig. 4).
Remove the inner door release handle (3 Fig. 4) securing screws and displace the inner door handle operating rods (4 Fig. 4) from the guide retaining clips (14 Fig. 4).
Reposition the inner door release handle to allow access to the seat switch.
Displace and remove the clips (14 Fig. 4) which retain the inner door handle operating rods (4 Fig. 4). Remove and discard the seat switch harness (5 Fig. 4) securing tape.
Undo and remove the remaining seat switch securing screw.
Disconnect the seat switch harness multi plug and remove the seat switch assembly (6 Fig. 4).
Remove and discard the tape securing the rear upper edge of the water shedder (1 Fig. 4) to the door, and reposition the shedder to give access to the door glass securing screws (7 Fig. 4).
Switch the ignition on, motor the glass down to obtain access to the glass securing bolts, and switch the ignition off.
Mark the position where the glass securing bolts (8 Fig. 4) are fixed to the glass carrier brackets (9 Fig. 4).
Take the weight of the door glass and release this from the carrier by undoing the securing bolts (8 Fig. 4).

Note: In performing this operation, care must be taken to avoid dislodging the securing nuts (10 Fig. 4) mounted on the glass.

Carefully remove the door glass.
After lubricating the runner (11 Fig. 4), fit the door glass to the door and engage the runner to the regulator arm (12 Fig. 4) when the door glass is aligned.
Align the glass carrier brackets (9 Fig. 4) with the glass and fit the glass to carrier securing bolts (8 Fig. 4) without fully tightening them.
Line up the door glass with the position marked when the glass was removed. When correctly positioned, tighten two of the glass to carrier securing bolts, one in the front and the other in the rear carrier. Switch the ignition on, motor the door glass fully up and switch the ignition off.

Note: The ignition needs to be turned on whenever the glass is to be raised or lowered and off when adjustments are being made to the glass.

Carefully close the door and note how the door glass is positioned against the aperture frame and seals.
Open the door and lower the glass down to give access to the securing bolts.
Loosen the glass to carrier bracket securing bolts (8 Fig. 4) to allow the position of the door glass to be adjusted.
Make any necessary adjustment and hold the glass in position by tightening two glass to carrier securing bolts.
Fully raise the door glass and carefully close the door to check the position of the door glass to aperture frame and seals, repeating the procedure if further adjustment is still required.
Lower the door glass and adjust in turn the upper and lower regulator stop adjusters (13 Fig. 4), slackening the stop adjuster lock nuts to enable the adjustment to be made, and tightening them after making the adjustment.
Close the door and check the glass height, finally tightening all the glass to carrier securing bolts when the glass is satisfactorily positioned.
Reposition the water shedder (1 Fig. 4) to the door and secure the front and rear edges with suitable tape.
Fit and align the electric seat switch assembly (2 Fig. 4), holding this in position with the front securing screw which should not be fully tightened. Fit and align the operating rod retaining clips (14 Fig. 4) and fully seat the door handle operating rods (4 Fig. 4) into these.
Position the inner handle (3 Fig. 4) to its location and firmly tighten the screws which secure it in position.
Tighten the seat switch securing screw.
Connect the seat switch harness (5 Fig. 4) multi plug and reposition the harness, securing it with suitable tape.

Fit the front speaker.
Fit the door trim pad.

DOOR GLASS, ADJUST

Open the door and remove the door trim pad.

If the door glass merely requires an adjustment in an upwards or downwards direction, carry out a stage 1 adjust.

Stage 1 adjust:

Partially lower the glass.
Displace the water shedder.
Slacken the regulator stop locknut (1 Fig. 5). Adjust the regulator stop stud (2 Fig. 5).

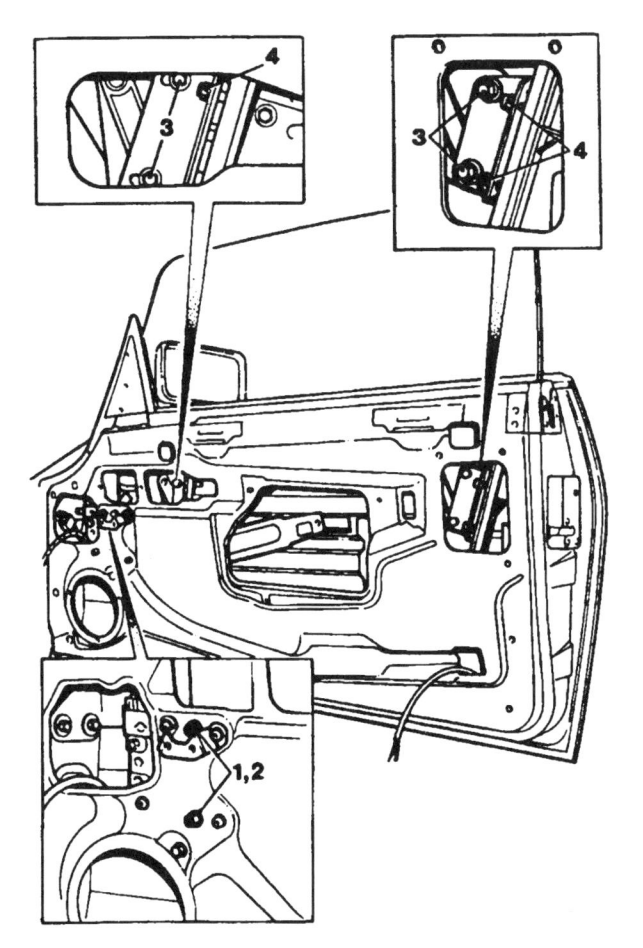

Fig. 5

Tighten the locknut (1 Fig. 5).
Raise the window glass and close the door.
Check the alignment of the glass and, if necessary, repeat the procedure until satisfied that the glass is correctly aligned.
Replace the door trim pad and close the door.
If the glass needs to be adjusted in its carrier and the runners need to be aligned with the glass, carry out a stage 2 adjust.

Stage 2 adjust:

Open the door and remove the door trim pad.
Partially lower the glass and displace the water shedder to gain access to the screws securing the glass (3 Fig. 5).
Slacken these screws and move the glass in the desired direction.
Tighten the glass securing screws (3 Fig. 5).
Fully raise the window glass and close the door.
Check the alignment of the glass and if necessary, repeat the procedure to readjust the glass.
Partially lower the glass to gain access to the glass runner securing bolts (4 Fig. 5).
Slacken these and adjust the glass.
Tighten the bolts securing the glass runner.
Complete the operation with a stage 1 adjust, as described above.

Fit the door trim pad and close the door.

DOOR CHEATER FRAME, RENEW

Switch the ignition on to lower the glass and then switch the ignition off.
Remove the door trim pad.
Remove the front speaker.

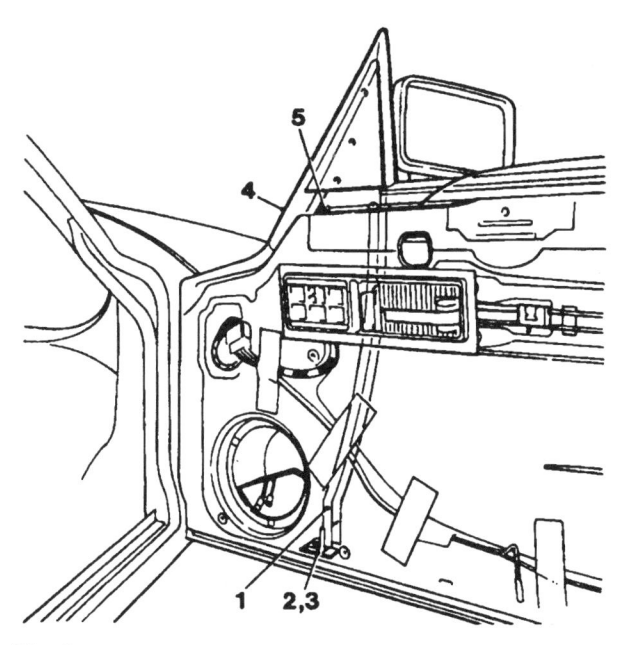

Fig. 6

Undo and remove the screws securing the cheater frame channel (1 Fig. 6) to the lower bracket (2 Fig. 6), using the speaker aperture to gain access to these screws.

Undo the bolt fixing the bracket to the door (3 Fig. 6) but do not remove this.
Remove the cheater frame inner and outer plates.
Remove the fir-tree retaining clip (4 Fig. 6) which attaches the seal to the door.
Pull the seal away from the door to release the adhesion.
Undo and remove the screws securing the cheater frame to the upper door (5 Fig. 6).
Remove the cheater frame and seal assembly from the door and pull the seal away from the frame.

Position the seal against the door cheater frame and fit these together.
Place the cheater frame and seal assembly into position in the door and hold in place with the upper and lower door securing screws (5 Fig. 6), but do not tighten these until the position of the cheater frame assembly has been adjusted.
Once the cheater frame assembly has been satisfactorily positioned, the screws securing the cheater frame to the upper and lower door (5 Fig. 6), and the bolt securing the lower bracket to the door (3 Fig. 6) should all be tightened.
Apply a suitable adhesive to the seal and fully seat this seal to the door panel.
Fit the fir-tree clip which retains the seal to the door, and fully seat this.
Fit the cheater frame outer and inner plates.
Fit and tighten the screws securing the cheater frame channel (1 Fig. 6) to the lower bracket (2 Fig. 6).

Fit the front speaker.
Fit the door trim pad.
Switch on the ignition, raise the glass, and then switch the ignition off.

DOOR TRIM PAD, RENEW

Open the door.
Using a suitable implement, ease the arm rest escutchion plate from position to gain access to the armrest upper securing screw (1 Fig. 7).

Undo and remove the armrest upper securing screw.
Undo and remove the door pocket lower securing screws (2 Fig. 7).

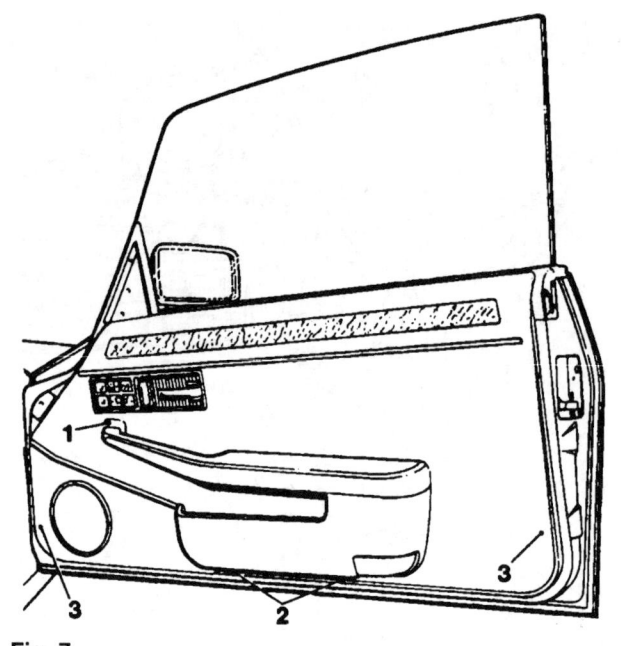

Fig. 7

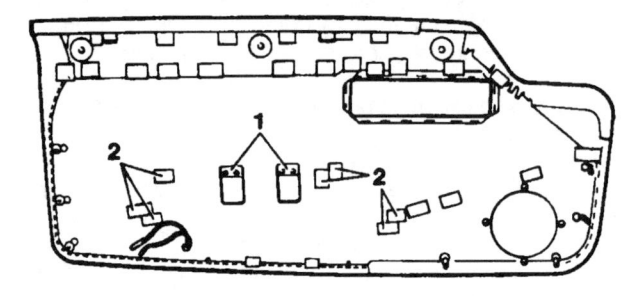

Fig. 8

Displace the trim pad to door securing clips (3 Fig. 7).
Lift the trim pad to disengage from the locating tangs and displace the trim pad from the door.
Note and disconnect the door guard lamp wires. Remove the trim pad.

Place the trim pad face downwards on a bench.
Undo and remove the veneer panel securing screws.
Displace and remove the veneer panel.
Undo and remove the screws securing the retaining brackets and remove the retaining brackets (1 Fig. 8).

Remove the remaining door pocket securing screw.
Release the door pocket retaining tangs (2 Fig. 8). Turn the trim pad over and displace and remove the door pocket from the trim pad.
Displace and remove the radio speaker grille by first removing the grille spire clips.
Displace and remove the inner waist draught seal from the trim pad.
Displace and remove the trim pad retaining clips. Put the dismantled trim pad aside.

Place the trim pad on a bench.
Fit the trim pad to door securing clips (3 Fig. 7). Fit and fully seat the inner waist draught seal.
Fit and align the speaker grill and secure this in position with spire clips.
Position the door pocket assembly.
Secure the pocket retaining tangs and fit and fully seat the new lower securing clips.
Secure the door trim pad clip retaining tangs.
Position the door trim pad retaining brackets. Fit and tighten the bracket securing screws.

102

Fit and align the veneer panel and tighten the screws which secure this.

Place the door trim pad in position against the door.
Connect the door guard lamp wires.
Locate the door trim pad securing clips against the retaining tangs and seat the trim pad fully home.
Fit and tighten the door pocket lower securing screws (2 Fig. 7).
Fit and tighten the armrest upper securing screw (1 Fig. 7).
Fully reseat the armrest escutcheon plate (1 Fig. 7).

DOOR TRIM PAD CHROME FINISHERS, RENEW

Remove the door trim pad.
Place the trim pad face downwards on a bench.
Remove and discard the tape from securing clip tangs of whichever of the two finishers is to be replaced.
Displace the centre finisher retaining tangs (1 Fig. 9) or lower finisher retaining tangs (2 Fig. 9) as appropriate.

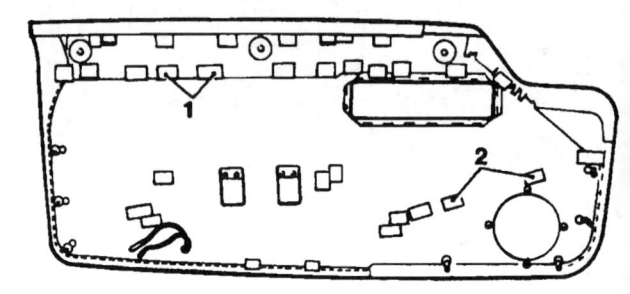

Fig. 9

Separate the finisher assembly from the trim pad and remove the finisher retaining clips.
Place the finisher to one side.

Take the centre or lower chrome trim pad finisher as appropriate and fit retaining clips to the finisher.
Fit the finisher to the door trim pad and hold it securely in position with retaining tangs.
Cover the exposed tangs with suitable adhesive tape.

Fit the door trim pad.

DOOR LOCK MECHANISM, RENEW

Remove the door trim pad.

Disconnect the multiplug connecting the harness to the door lock motor leads (1 Fig. 10) and remove and discard the tape which holds these leads to the door.
Reposition the water shedder (2 Fig. 10) to gain access to the door lock operating rods (3 Fig. 10).
Displace the door lock operating rods from the guide clips (4 Fig. 10).
Disconnect the rod which operates the outer lock mechanism from this mechanism and reposition the rod.
Disconnect the clip (5 Fig. 10) which connects the rod which operates the outer release handle to the handle and reposition this rod.
Disconnect the clip (6 Fig. 10) which connects the rod which operates the door lock to the door lock barrel and reposition this rod.
Disconnect the clip (7 Fig. 10) which connects the rod which operates the door release to the door release mechanism and reposition this rod.
Undo the screws (8 Fig. 10) securing the outer lock to the door and remove the outer lock assembly.

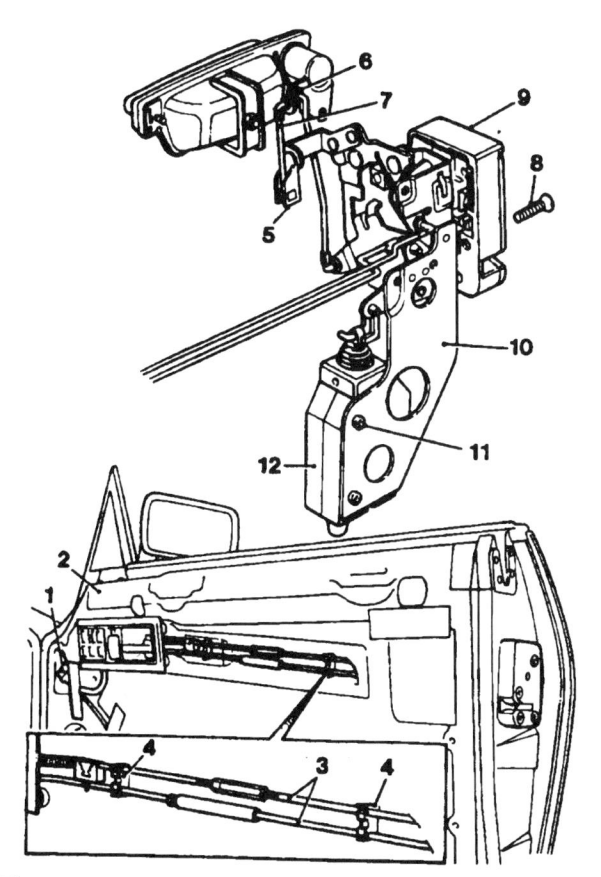

Fig. 10

Remove the motor/lock mechanism assembly (10 Fig. 10).
Undo the screws (11 Fig. 10) securing the motor to the door lock mechanism, remove the motor assembly (12 Fig. 10), and place the mechanism to one side.

Fit the motor (12 Fig. 10) to the door lock mechanism, positioning it so that the screw holes line up, and firmly fix them together with the securing screws (11 Fig. 10).
Lubricate the linkage between the motor and the door lock mechanism.
Position the motor assembly (10 Fig. 10) to the door, ensuring that no link rods or harnesses become trapped.
Fit the outer lock assembly (9 Fig. 10) to the door, fixing it firmly in position with the securing screws (8 Fig. 10).
Reposition the door lock operating rods (3 Fig. 10) and secure them with the retaining clips (5, 6 & 7 Fig. 10).
Route the leads (1 Fig. 10) to the door lock motor harness through the door panel.
Adjust the rod connected to the release lever and secure it with a retaining clip (7 Fig. 10).
Reposition the rod connected to the lock barrel and secure it with a retaining clip (6 Fig. 10).
Reposition the harness through the weather strip (2 Fig. 10) and position the weather strip against the door, securing the top edge with suitable adhesive tape.
Fit the operating rod guide clips (4 Fig. 10) to the door, fully seating them into their locations
Fit the door lock operating rods (3 Fig. 10) into the guide clips, fully seating them into position.
Connect the door lock motor harness multiple (1 Fig. 10) and secure the leads to the door lock with suitable adhesive tape.
Fit the door trim pad.

FRONT SEAT RUNNER, RENEW

Switch on the ignition and motor the seat fully forwards.
Undo the bolts securing the back of the seat runner to the floor panel (1 Fig. 11).

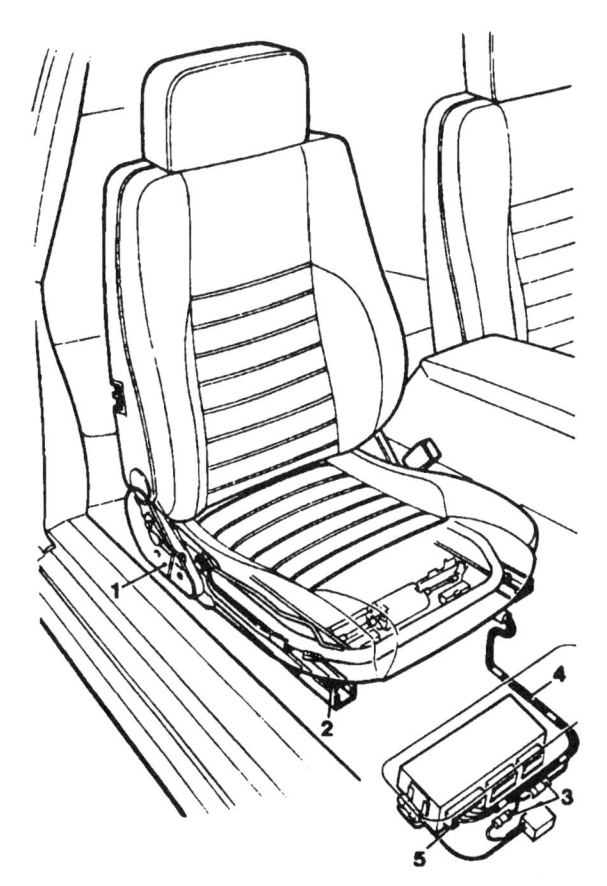

Fig. 11

Motor the seat fully backwards and switch the ignition off.
Undo the bolts securing the front of the seat runner to the floor panel (2 Fig. 11).
Reposition the footwell carpet to give access to the ECU cover panel.

Undo the screws which secure the ECU cover panel and reposition this panel to gain access to the seat harness multiplugs (3 Fig. 11).
Undo the screws which secure the seat harness plastic clamp (1 Fig. 12) and remove this.

Cut and remove the seat harness ratchet straps (4 Fig. 11).
Disconnect the seat harness multiplugs (3 Fig. 11) including the harness to ECU multiplug (5 Fig. 11). Reposition the seat to gain access to the seat belt anchorage bolt (2 Fig. 12).
Remove the seat belt anchorage bolt trim cap and undo the bolt (2 Fig. 12).
Reposition the seat belt.
Remove the seat and runner assembly and place it on a bench.

Undo the screws securing the motor drive shaft retaining plate assemblies (3 Fig. 12).
Remove the motor drive shaft retaining plate assemblies.
Cut and remove the ratchet straps (4 Fig. 12) holding the seat belt buckle harness and disconnect the seat belt buckle harness multiplug (5 Fig. 12).
Reposition the runners to give access to the bolts which secure the front of the seat frame (6 Fig. 12) and undo these bolts.
Reposition the runners to give access to the bolts which secure the rear of the seat frame (7 Fig. 12) and undo these bolts.
Reposition the runner assembly to gain access to the harness
Displace the harness multiplugs (8 Fig. 12) from the retaining clips and disconnect the multiplugs.
Remove the runner assembly.

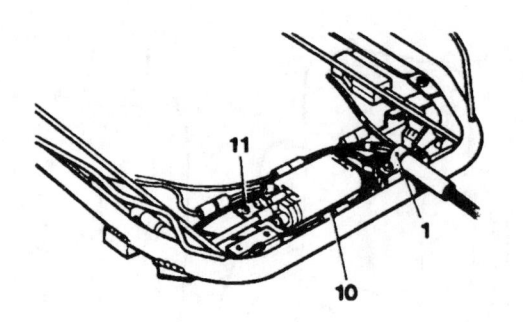

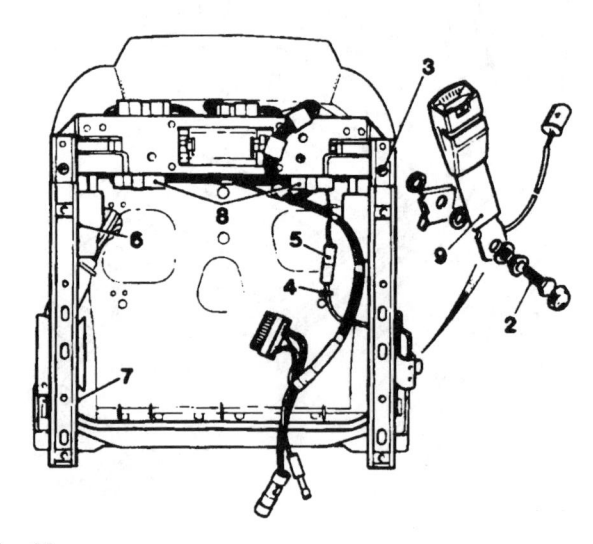

Fig. 12

Remove the trim cap from the bolt which secures the seat belt buckle and undo the bolt (2 Fig. 12).
Remove the seat belt buckle.
Remove the belt buckle bracket (9 Fig. 12).
Undo the screws which secure the seat harness plastic clamp plate (1 Fig. 12) and remove this plate. Displace the motor harness multiplug (10 Fig. 12) from its retaining clip.
Undo, without fully removing, the screws (11 Fig. 12) which secure the motor to the runner.
Remove the motor/drive and harness assembly.
Remove the clips retaining the harness.
Place the runner assembly to one side.
Fit the motor/drive and harness assembly to the runner, lining up the screws with the holes on the runner and taking care not to trap the harness.
Fix the motor/drive and harness assembly to the runner by tightening the securing screws (11 Fig. 12).
Position the harness to the retaining bracket and place the seat harness plastic clamp plate (1 Fig. 12) by the bracket so that the screw holes line up.
Firmly screw the clamp plate into position.
Fit and align the seat belt buckle bracket (9 Fig. 12).

Position the seat belt buckle against the bracket and fix with the securing bolt (2 Fig. 12) which should be tightened and covered with a trim cap.

Place the runner assembly to the seat and connect the harness multiplugs (8 Fig. 12).
Position the multiplugs into their retaining clips.
Align the runner assembly with the seat frame and fit and tighten the rear bolts (7 Fig. 12) which secure the runner to the seat frame.
Reposition the seat runners to allow access to the front securing bolts (6 Fig. 12) which should be fitted and tightened.

Connect the seat belt buckle multiplug (5 Fig. 12) and position this to the seat diaphragm where it should be secured with ratchet straps (4 Fig. 12).

104

Reposition the seat runners fully forward and fit and fully seat the motor drive shaft assemblies, manoeuvring the runners to engage in the driveshaft couplings.
Fit and tighten the screws which secure the motor drive shaft retaining plate (3 Fig. 12).

Place the seat and runner assembly in position in the vehicle.
Position the seat belt to its anchorage point on the seat and fix it in position with the anchorage bolt (2 Fig. 12) which should then be covered with a trim cap.
Reposition the seat to its mounting.
Connect the seat harness multiplugs (3 Fig. 11).
Position the harness to the bracket where it is to be clamped and secure it with the plastic clamp (1 Fig. 12) which should be lined up with its screw holes and firmly screwed into position.
Fit the bolts (2 Fig. 11) which secure the front of the seat runner to the floor and tighten these.
Switch on the ignition and motor the seat fully forwards to enable the bolts (1 Fig. 11) which secure the rear of the seat runner to the floor to be fitted and tightened.
Motor the seat fully forwards and backwards to enable the ECU to read the potentiometer switch.
Switch off the ignition.
Fully seat the ECU cover panel into position and firmly fix it in place with the securing screws.

DASH LINER - PASSENGER'S SIDE, RENEW

Open the door. Remove the fuse indicator panel from the liner. Remove the dash liner screws (1 Fig. 13) Remove the liner (2 Fig. 13).

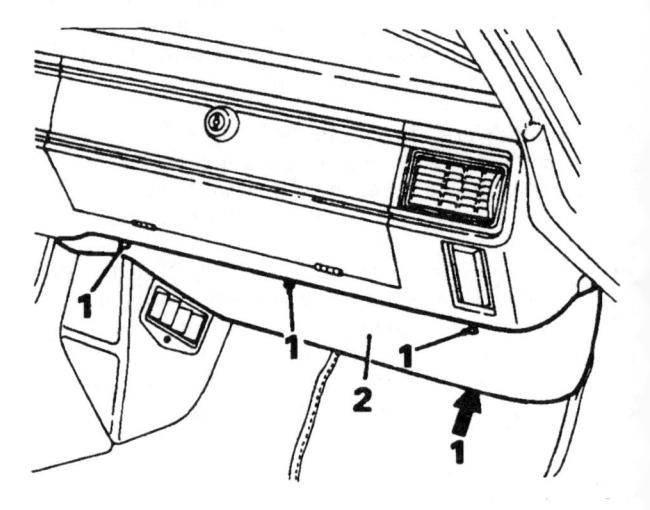

Fig. 13

Place the new liner in position.
Align the screw holes and fit and tighten the liner securing screws.
Fit the fuse indicator panel to the liner.

DASH LINER - DRIVER'S SIDE, RENEW

Open the door.
Remove the dash liner fastener (1 Fig. 14).

Pivot the liner down from the fascia underframe and remove the liner 'P' clip securing nut. Displace the 'P' clip from the underframe.
Remove the liner and place it onto a bench.
Remove the liner mounting clips securing nuts and remove the mounting clips.
Remove the 'P' clips.

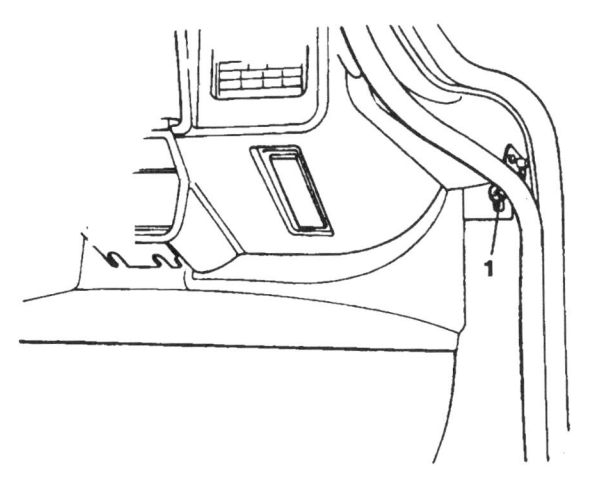

Fig. 14

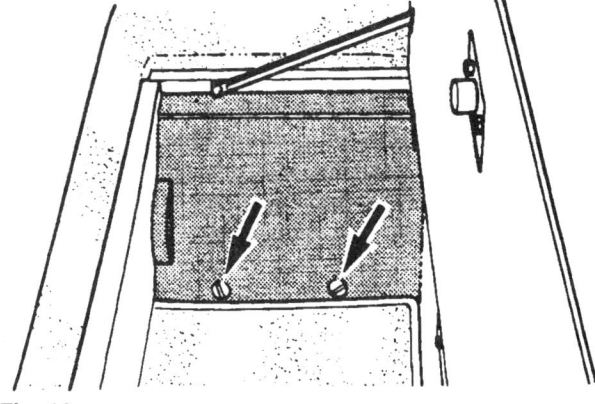

Fig. 16

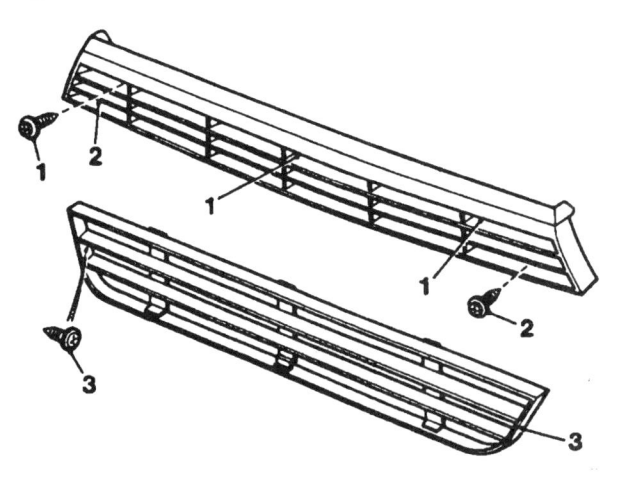

Fig. 15

Fit the 'P' clip to the new liner mounting stud and fit the mounting clips to the liner.
Fit but do not fully tighten the clips securing nuts.
Fit the liner assembly to the underframe and align the liner to the remaining trim.
Pivot the liner down from the fascia underframe and fully tighten the clips securing nuts.
Position the 'P' clip to the underframe and tighten the securing nut.
Pivot the liner to engage the underframe and secure the fastener.
Close the door.

RADIATOR GRILLE, RENEW

Undo and remove the three screws which secure the grille to the bonnet (1 Fig. 15).

Undo and remove the two screws which secure the grille to the hinge bracket (2 Fig. 15).
Remove the grille.

Fit the grille to the vehicle and firmly fix into position with the securing screws.

RADIATOR LOWER GRILLE PANEL, RENEW

Drive the car on to a ramp and raise the ramp.
Undo the two screws (3 Fig. 15) which secure the grille to the panel.

Position the grille.
Fit and tighten the screws which secure it in position.
Lower the ramp.

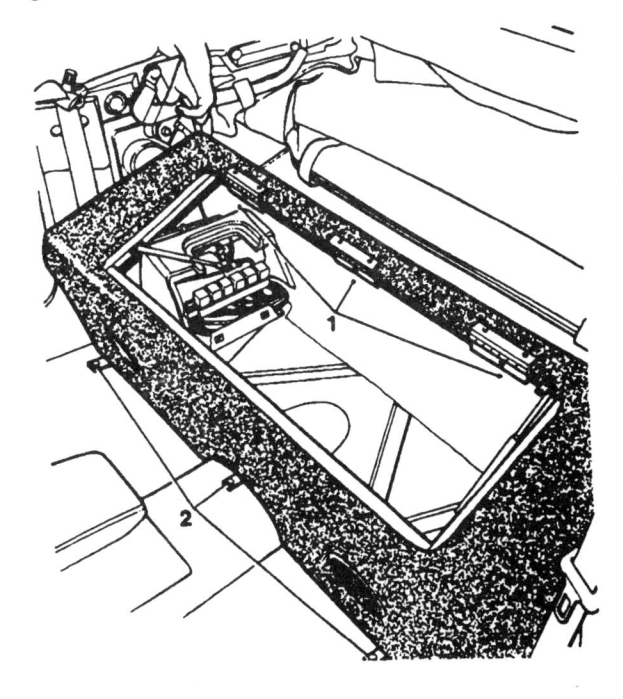

Fig. 17

REAR STOWAGE COMPARTMENT CONVERTIBLE, REMOVE AND REFIT

Switch on the ignition.
Release the hood locking catches and lower the convertible hood.
Move the seats forward.
Switch off the ignition.
Remove the rear quarter trim panels for access.
Open the stowage compartment lid and release the retaining clips (Fig. 16) on the cover of the hood motor.

Remove the cover.
Remove the carpets from the stowage compartment floor and rear panel.
Disconnect the Ty-wraps which secure the harness and hydraulic pipes to the lower edge of the internal rear vertical panel in the stowage compartment.
Carefully ease back the Velcro from the internal rear vertical panel to provide access to the three securing bolts (1 Fig. 17).
Remove the three securing bolts. Lift the footwell carpets.
Remove the screws (2 Fig. 17) which secure the stowage compartment to the floor brackets.
Disconnect the speaker wires.
Close the lid and remove the stowage compartment.

Fit the stowage compartment to the vehicle.
Fit the upper securing bolts, but do not tighten.

Fit the lower securing screws, but do not tighten. Fully tighten the securing screws and bolts.
Reposition the carpet to the footwells.
Fit the harness and hydraulic pipes to the Ty-wraps.
Connect the speaker wires.
Glue the Velcro in place, as required, over the bolt heads.
Fit the carpets to the stowage compartment. Fit the cover to the hood motor.
Close the lid of the stowage compartment. Fit the rear quarter trim panels.
Switch on the ignition.
Reposition the seats.
Raise the hood and secure the hood locking catches.
Switch off the ignition.

CONVERTIBLE HOOD, DESCRIPTION

Operation

When the OPEN (hood down) position is selected, the hood and rear quarter windows move in unison. The rear quarter windows open fully in approximately two seconds and the hood within twelve seconds.
With hood CLOSED (hood up) selected, the control module initially energizes the hydraulic pump and delays the rear quarter windows closure by approximately eight seconds.
Irrespective of how long the switch is held down, the control module will only run one cycle (twelve seconds), thus all relays will be de-energized after twelve seconds.

Safety

1. The hood cannot be raised or lowered electrically unless the handbrake is on and the gear selector in either the 'P' or 'N' position.

2. Should the hood switch be released before completion of a cycle, or if the ignition/auxiliary switch is turned off during a cycle, the system will stop immediately.

3. After interruption;

a) If the original direction is reselected, the sequence will continue to completion from where it left off (ONLY if the ignition has not been switched off).
b) If the opposite direction is selected, the timing sequence will commence at the beginning of the newly selected direction.
c) If the ignition has been switched off, the timing will commence at the beginning of its cycle.

Component Location

Control Module & Relays:
    En bloc with the pump/motor package.

Hood Switch:
    Centre console.

Hydraulic Pump Assembly:
    Convertible: Located within the stowage compartment on the right hand side concealed by a quick release trim cover.
    2 + 2: Mounted on the right hand side of the luggage compartment above the battery and concealed by a quick release trim cover.

Hydraulic Rams:
    Located underneath the rear quarter trim panels and connected to the pump by flexible nylon tubes.

HOOD HYDRAULIC CONTROL SYSTEM, RENEW

Ensure that the handbrake is on and the gearshift is in park or neutral.
Move the seats foward for access.
Release the hood handles and catches.
Turn on the ignition.
Depress the 'hood down' switch until the hood is fully lowered.
Remove the rear quarter lower trim pads and remove the rear stowage compartment lid.
Remove the rear stowage compartment (Fig. 17).
Undo and remove the bolts securing the pump mounting plate and displace the mounting plate assembly.

Undo and remove the screws securing the pump cover and remove the pump cover.
Undo and remove the cylinder (ram) upper pivot bolt and remove the spacer washers.
Undo and remove the bolts securing the cylinder pivot bracket and displace the cylinder from position (Fig. 18) - do not kink the pipes.

Repeat the operation for the opposite side cylinder.
Displace the floor carpets from the pipes. Release the pipe securing clips.
Release the hydraulic pump mounting rubbers from the plate.
Disconnect the multi-plug connector and displace and remove the complete system (Fig. 19).

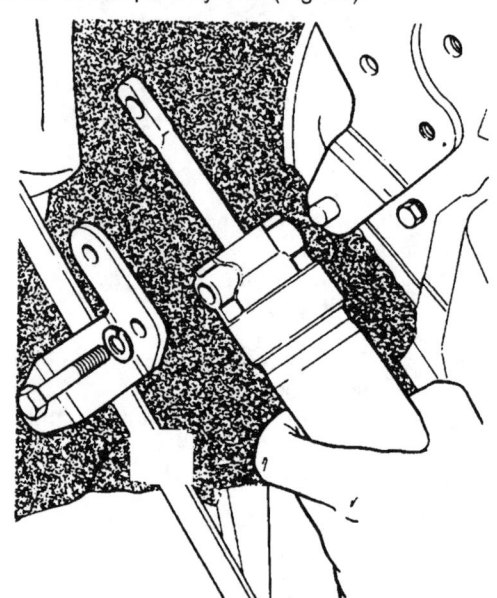

Fig. 18

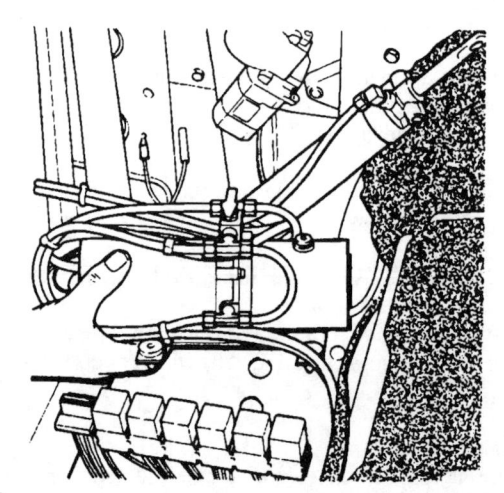

Fig. 19

Place the hydraulic system in position, route the pipes and fully seat the mounting rubbers to the plate.
Locate the cylinders to the correct position.
Reconnect the multi-plug connector.
Fit the pump cover and fit and tighten the screws which secure this cover.
Fit the cylinder pivot bracket and fit and tighten the bolts which secure this.
Position the cylinder to the hood, fit the spacers, and fit and tighten the securing bolt. Repeat the operation for the opposite side cylinder.
Position the pump mounting plate by aligning the holes and fit and tighten the bolts which secure this in position.
Secure the pipes with clips.
Apply glue to the carpet and position the carpet to the floor.
Refit the rear stowage compartment.
Refit the rear quarter trim pads.
Reposition the seats.
Depress the 'hood up' switch to close the hood.
Lock the hood with the catches and handles.
Switch off the ignition.

CONVERTIBLE HOOD HYDRAULIC SYSTEM, BLEED

Remove the Hood Hydraulic Control System from the vehicle.
Disconnect the **Left Hand** side cylinder pipes from the pump (1 Fig. 20).

Submerge the two tubes in hydraulic fluid (Univis J13) and fully extend the ram to prime the cylinder with fluid.
Invert the cylinder, fully compress the ram and reconnect the pipes.
Remove the reservoir filler plug (2 Fig. 20) and fill to highest level.
Refit the filler plug.
Select the ELECTRIC position on the manual/electric valve (3 Fig. 20).
Disconnect the Left Hand side cylinder pipes from the cylinder.
Submerge the two tubes in hydraulic fluid (Univis J13) and fully extend the Right Hand side ram.
Invert the cylinder, fully compress the ram, and reconnect the pipes to the Left Hand cylinder.
Remove the reservoir filler plug and fill to the MAX level mark, refit the filler plug.
Refit the system to the vehicle.
Power the hood several times over its full cycle ending with it fully down.
Recheck the fluid level and correct as required.

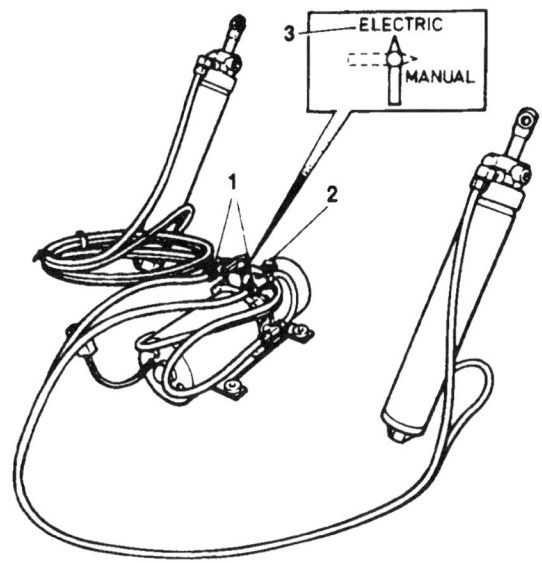

Fig. 20

CAUTION:

Always check the fluid level with the hood fully stowed, excess fluid could rupture the reservoir.

CONVERTIBLE HOOD - HEADLINING, RENEW

CAUTION:

It must be stressed that the following procedure should only be carried out by fully skilled personnel who have been trained in this type of work.

Note: This procedure describes replacement of the original FIXED headlining with a DETACHABLE type.

Removal

With the gearshift in PARK or NEUTRAL, power the hood to the closed position. As required, disconnect the vehicle battery ground lead. Protect the vehicle against dirt or damage.

Note: Selection of MANUAL on the hydraulic pump assembly will allow the hood to be moved by hand.
Remove; Rear quarter trim pads and the rear seats (2 + 2 only).
Pull the rear shelf carpet clear at the sides to reveal backlight frame pivot brackets (1 Fig. 21) and remove the fixings.

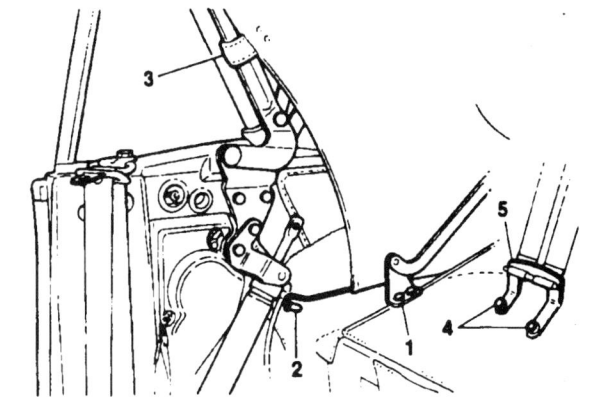

Fig. 21

Remove headlining side curtain stiffener rod from retention clip (2 Fig. 21) and disengage rod from its forward location.
Release headlining side curtain press studs at the main column (3 Fig. 21).
At the extreme rear body sides RH and LH, between the headlining and canopy, release webbing retainer fixings (4 Fig. 21).
Using a trimmers knife, cut the headlining free from the area behind the webbing (5 Fig. 21).

Position the hood to gain access to the HEADER and remove:

CAUTION:

Take care when cutting the lining free not to damage the canopy outer fabric.

Headlining retainer strip (1 Fig. 22),
Seal end plates (2 Fig. 22) and
Header location pins (3 Fig. 22).

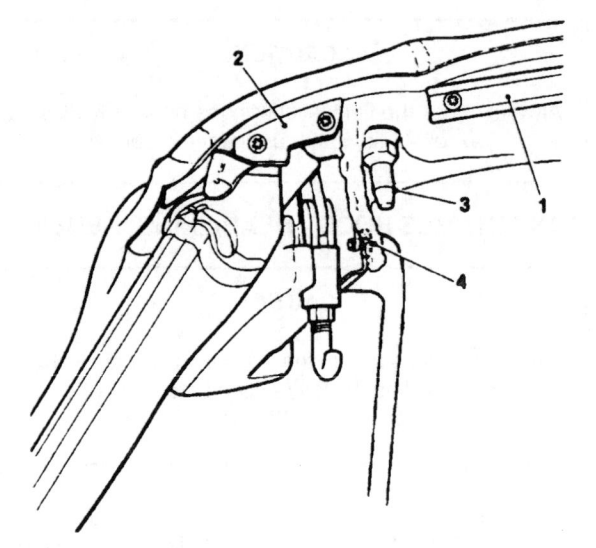

Fig. 22

Drill out rivet - headlining to link (4 Fig. 22).
Release the headlining from the HEADER and peel material back to the FIRST roof bow.
Cut the material from the FIRST bow and fold to the SECOND.
At the SECOND bow, chisel (there is no access for a drill) strap retaining rivets (1 Fig. 23), cut the lining from the bow and fold back to the MAIN bow.

At the MAIN bow remove the inner screws (2 Fig. 23) and cut lining material to leave a strip approximately 40 mm wide attached to the bow.
Remove the headlining from the vehicle.

Replacement

Apply adhesive to the MAIN bow and the 40 mm strip of the original headlining - secure the strip to the bow by folding forward and over.
Apply adhesive to the outer rear webbing (MAIN bow to rear body) and velcro strips and secure velcro to webbing (1 Fig. 24).

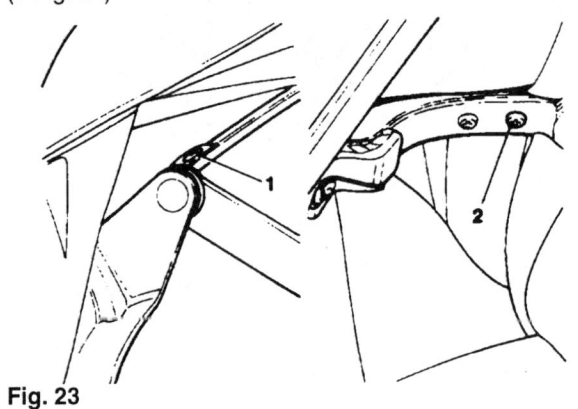

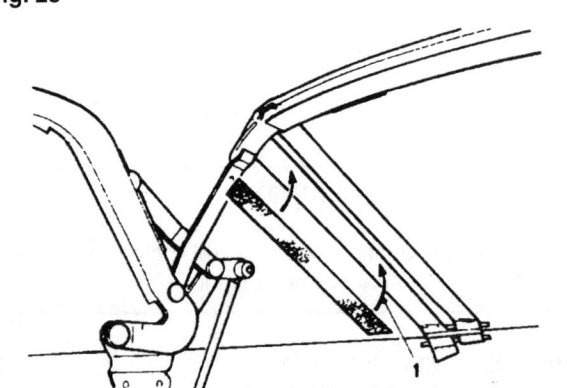

Fig. 23

Fig. 24

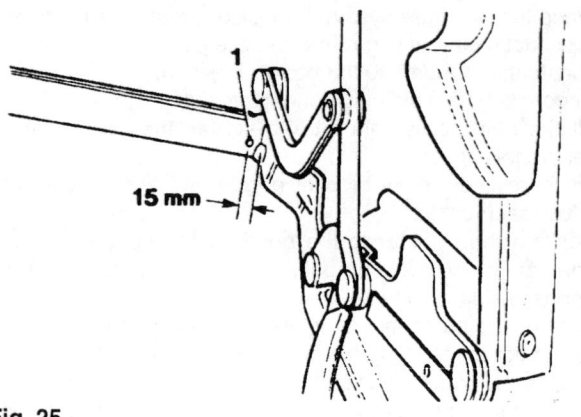

Fig. 25

Position listing rails to their respective roof bows, place centrally, mark and drill through into roof bow. Use a drill suitable for the provided self tapping screws.
The listing rails are colour coded: RED = 1st bow, YELLOW = 2nd bow, BLUE = main bow.
On the underside of the SECOND roof bow, drill one hole LH and RH, 15 mm inboard (1 Fig. 25).
These holes will provide location for fixing of the tapes which were released by chiselling from the upper surface in the removal procedure.

Feed the FIRST and SECOND roof bow listing rails through respective sleeves in the headlining taking care not to 'snag' the material.
Apply double sided adhesive tape to the MAIN bow, position the headlining 'fly' as illustrated (1 Fig. 26).

Secure headlining and listing rail to the MAIN bow (2 Fig. 26).
Thread tapes through loops on MAIN bow (1 Fig. 27), headlining (2 Fig. 27) and secure at previously drilled inboard holes (3 Fig. 27).

Please note that, in the interest of clarity the outer canopy is NOT shown.
Secure the listing rails and thus headlining at the SECOND and FIRST bow (pull the material to cover the rail ends before fixing the final screw).

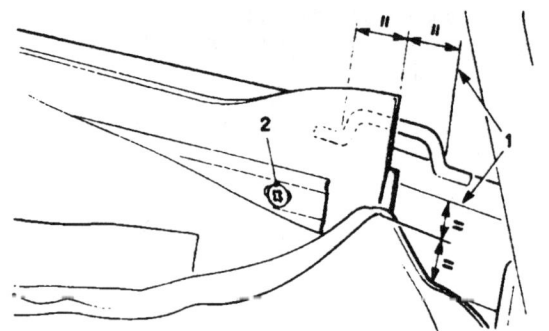

Fig. 26

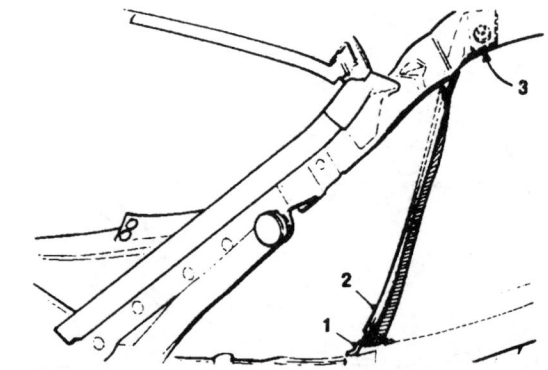

Fig. 27

Apply double sided adhesive tape at the HEADER across the retaining strip fixing holes.

Using a garnish awl or similar; pierce the headlining and fix at the side inner face adjacent to header latches.

Carefully align the headlining centre 'notch', evenly tension the material and locate to the adhesive tape.

Check tension by fully closing the hood and if acceptable, fix the headlining with the original retaining strip.

Cut off excess material above the retaining strip, if any, and replace header location pins.

At the rear, wrap the side curtains around the back of the webbing straps and secure webbing brackets to the rear body panel.

Ensuring that the rear side curtains are not creased, engage the velcro to retain in position.

Refit and tighten to the specified torque, backlight frame pivot brackets and replace the rear shelf carpet.

Engage side curtain press studs, refit stiffener rod and clips.

Verify the fit of the headlining and rectify if required.

CONVERTIBLE HOOD - CANOPY, RENEW

CAUTION:

It must be stressed that the following procedure should only be carried out by fully skilled personnel who have been trained in this type of work.

Removal

With the gearshift in PARK or NEUTRAL, power the hood to the closed position.

As required, disconnect the vehicle battery ground lead.

Protect the vehicle against dirt or damage.

Note: Selection of MANUAL on the hydraulic pump assembly will allow the hood to be moved by hand.

Remove the rear quarter trim pads.

Remove rear seats (2 + 2 only).

Pull the rear shelf carpet clear at the sides to reveal backlight frame pivot brackets and remove fixings.

Disconnect heated backlight feed wires.

Remove canopy fixing (1 Fig. 28), finisher (2 Fig. 28) and end cap (3 Fig. 28) at the 'B' / 'C' post buttress.

Note: Do not allow the end cap to fall into the body cavity.

Position the hood to gain access to the HEADER and remove the retaining strip (1 Fig. 29), seal end plates (2 Fig. 29) and header location pins (3 Fig. 29).

Release headlining adhesive at the leading edge and peel the lining back.

Note the position of, and remove drop glass seals from the carriers (upper only).

Remove seal carriers taking care to note position and shims if any.

Remove canopy rubber sealing strip (4 Fig. 29) and peel canopy from HEADER leading and side edges (5 Fig. 29).

Release canopy side tension wire from HEADER on both sides (6 Fig. 29).

Release canopy side tension wire from spring at main column.

Drill out rivet securing canopy side tie to side link on RH and LH (1 Fig. 30).

Fold the canopy back clear of the frame.

With a wax crayon or similar, mark the position of the tonneau loops on the rear panel - for future reference.

Pull the shelf carpet aside to reveal canopy tensioning cable adjusters and remove M6 nuts (1 Fig. 31), taking care to prevent rotation of the cable.

Withdraw canopy and cable from the body and fold forwards and over the MAIN roof bow.

Release webbing from backlight upper loop by pushing webbing slack through frame and pulling the plastic retainer pin clear.

Remove the canopy/backlight assembly from the vehicle.

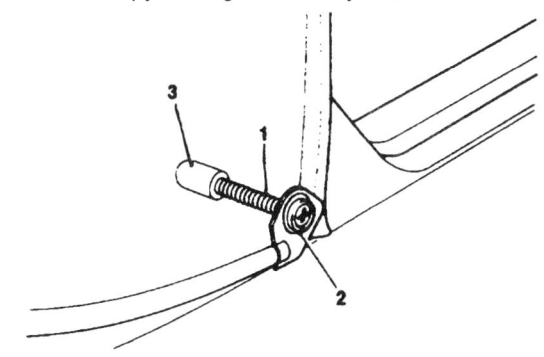

Fig. 28

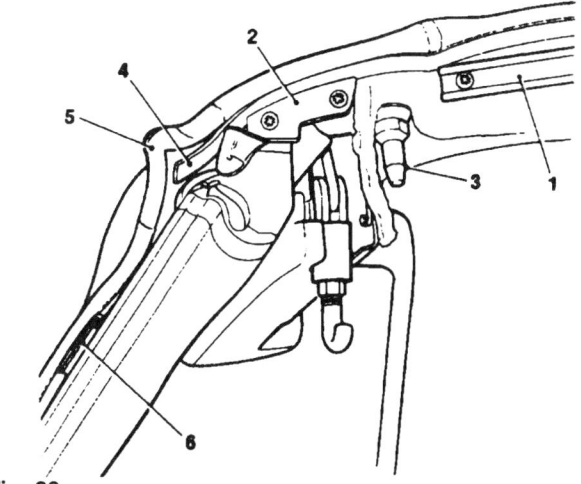

Fig. 29

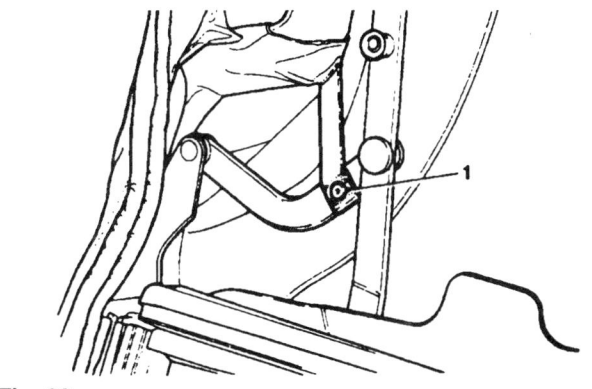

Fig. 30

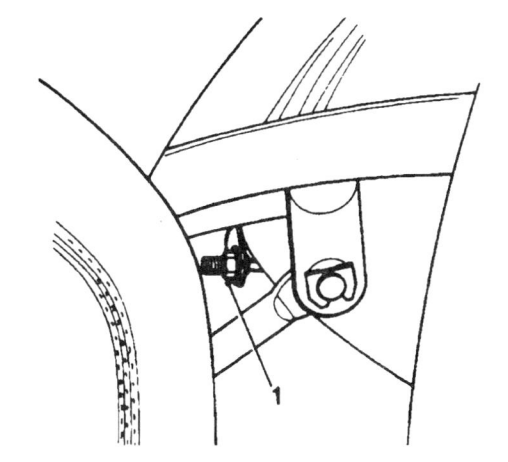

Fig. 31

109

Preparation

With the canopy on a suitable workbench, withdraw the side tension wires and cut the backlight free.

Remove interior and exterior glass seals.

At one corner of the backlight, carefully cut a hole in the bonding material through which a 'cheese wire' may be passed.

Using caution not to damage frame or glass, release the glass to frame bond.

Remove staples and original material from backlight frame. Clean bonding material from glass and frame.

Tools

It is advisable to manufacture special tools:

A (Fig. 32) used to seat the canopy material into the rear panel 'U' section and close the canopy to piping gap.
Dimension A = 135 mm; B = 32 mm; C = 16 mm; D = 65 mm; E = 15 mm

B (Fig. 33) used to lever the cable over the canopy material and thus under the 'U' section.
Dimension A = 270 mm; B = 190 mm; C = 6 mm; D = 25 mm

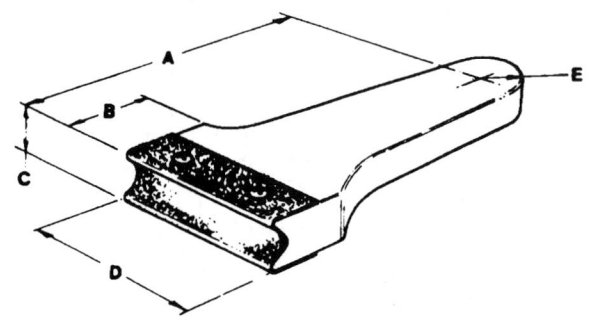

Fig. 32

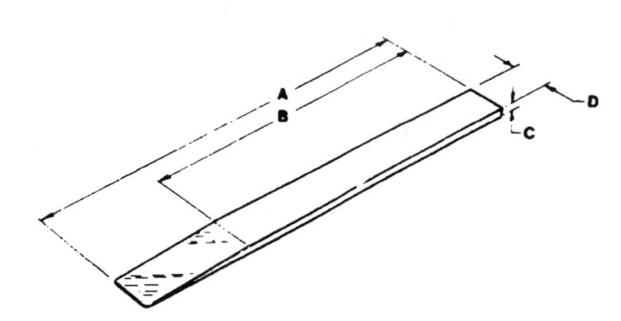

Fig. 33

Replacement

Canopy to body and frame

Lay the replacement canopy onto a clean workbench and fit the side tension wires (previously removed) large eyelet to the rear.

Position the canopy to the frame and secure side tension wires at the header and spring (main column).

Rivet side tie to link RH and LH (see Fig. 30).

Fold canopy material over MAIN roof bow to gain access to backlight webbing.

Secure backlight frame pivot brackets to the body in the most FORWARD position and attach the webbing to the frame.

Pull rear of canopy over the rear body 'U' section and fit tensioning cables through body and assemble M6 nuts by 2 or 3 threads only.

Note: The natural tendency is for the cable to stay inside the body, therefore, lever the cable over the leading edge of the 'U' section to lie outside and across the rear panel.

Position tonneau loops to previously marked locations and using special tool 'B' lever the cable over the canopy and thus into the 'U' section. It is necessary that a second person holds the cable in position on the opposite side to the lever to prevent the cable pulling out of the 'U' section (Fig. 34).

Reposition the tonneau loops if required.

Pull canopy sides evenly to the 'B'/'C' post buttress, pierce the material with a garnish awl (Fig. 35) and secure each

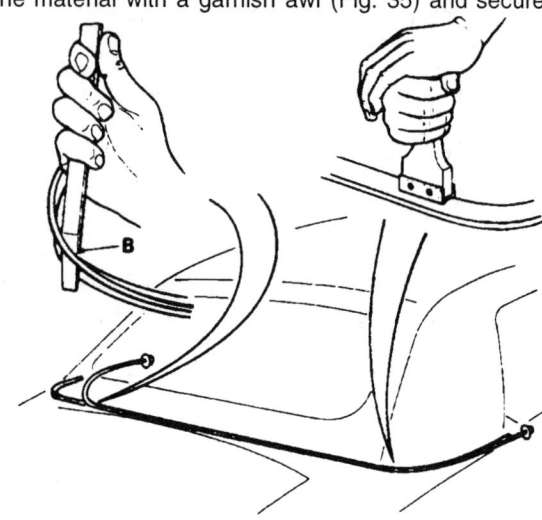

Fig. 34

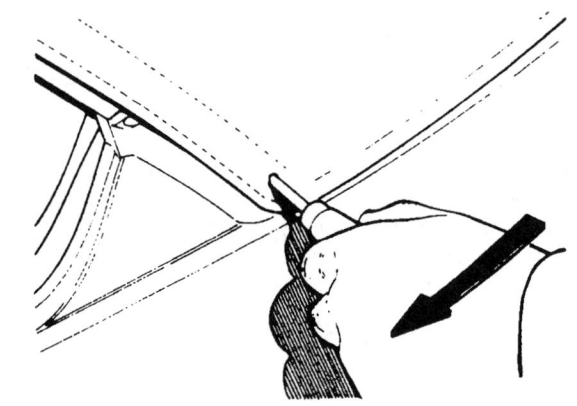

Fig. 35

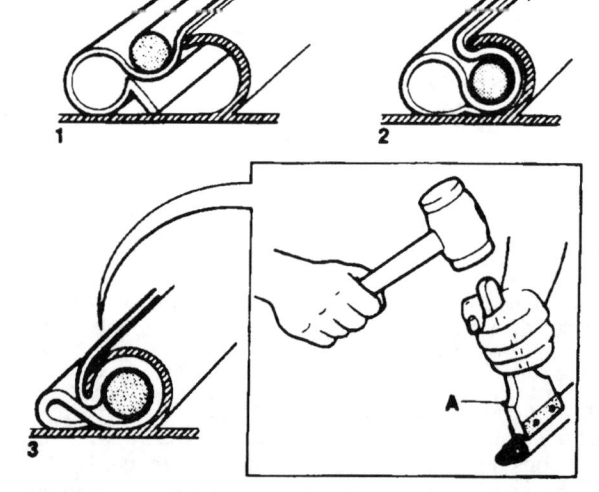

Fig. 36

side using previously removed fixings.
Using special tools 'A' and 'B' with a soft faced mallet, 'chase' the canopy material and retaining cable into the 'U' section to achieve the illustrated condition:

(1 Fig. 36) Initial fitting
(2 Fig. 36) Canopy fully seated
(3 Fig. 36) 'U' section and canopy gap fully closed.

Tighten the tensioning cable nuts evenly on each side (prevent rotation of the cable) as the previous operation is being carried out to achieve a crease-free canopy appearance.

Note: Ensure that the tonneau loops remain in the correct location during the final fitting operations.

The correct canopy fit and cable tension should result in the canopy material having even contact with the body all round.

> **CAUTION:**
>
> To avoid future canopy damage, do not omit the screw end caps.

Reposition the carpet over retaining cable adjusters.

With the hood raised for access - and to provide material tension:
Apply adhesive to the main column seal carrier flanges and canopy material; secure the canopy ensuring that the tuck under is symmetrical and the 'edge sew line' is parallel to the seal flange edge.
Apply adhesive to HEADER and canopy material; secure the canopy firstly by folding the sides under, followed by the leading edge material.

Note: Whilst lining up the material, ensure that the rain channels are straight and even side to side. Fully close the hood and check for correct material tension and appearance - rectify as required.
Fitting of the remaining items is the reversal of previous procedures, noting that special attention should be paid to seal/carrier alignment.

> **CAUTION:**
>
> Be absolutely sure that ALL aspects of the canopy fitting are acceptable before proceeding with the following operation.

Backlight Aperture

Fully latch the frame at the HEADER.
Centralize the backlight frame ensuring that an equal gap by measurement exists between the outer seams and the frame (1 Fig. 37).

Note: For 2 + 2 only, the backlight frame must be held away from the main roof bow by either inserting a distance piece between the frame and the bow or by pulling the frame away from the main roof bow toward the rear shelf. Special attention therefore must be paid to keep the frame parallel across car. These actions are required because this model has elasticated backlight frame upper attachments.

With two people working equally from both sides, staple the material to the inner vertical edges of the backlight frame (2 Fig. 37) whilst gradually relieving material tension within the frame by cutting (3 Fig. 37). **DO NOT CUT RIGHT ACROSS.**
Continue to staple for upper and lower edges, cutting to relieve tension as you go.
When the material is fixed all round, trim off excess.
Refit backlight.

All other fitting is the reversal of previous procedures.

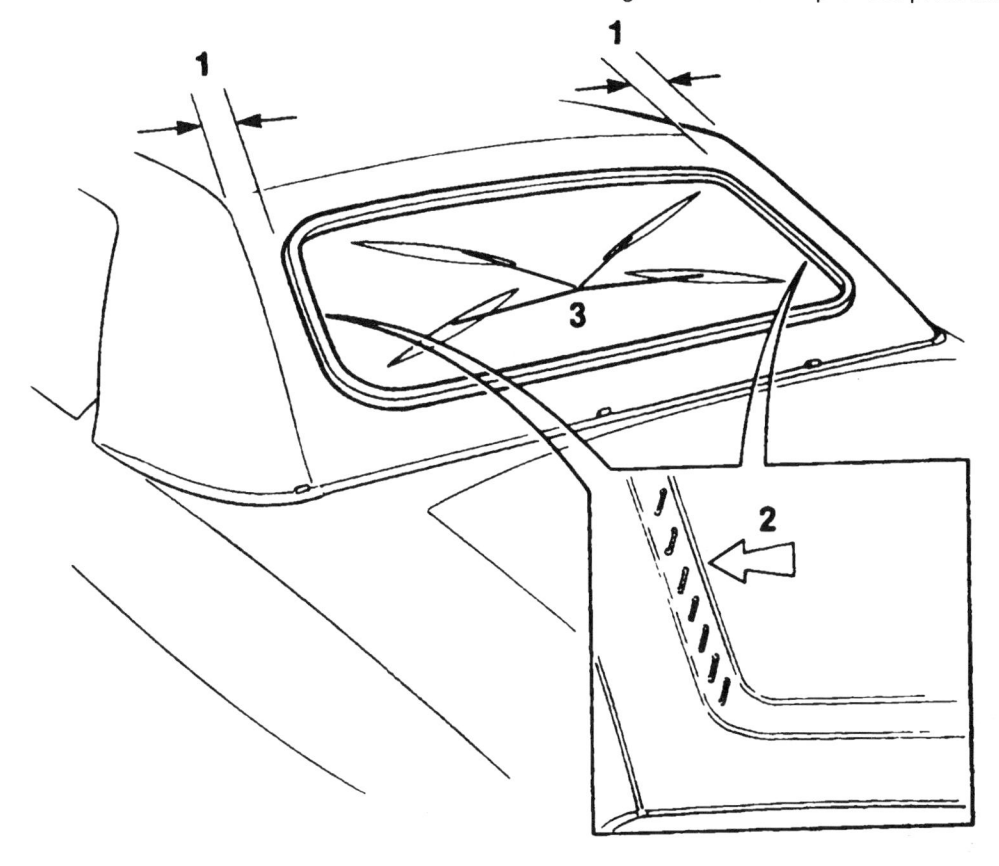

Fig. 37

CONVERTIBLE HOOD 'UP' RELAY, RENEW

Open the storage compartment lid.
Twist and release the pump cover release catches.
Displace and remove the cover.
Remove the relay (4 Fig. 38).
Fit the relay.

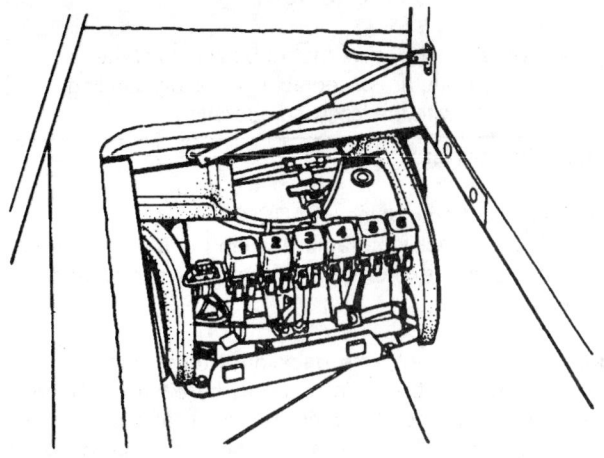

Fig. 38

Refit the cover and reposition the carpet.
Close the lid.

CONVERTIBLE HOOD 'DOWN' RELAY, RENEW

Open the storage compartment lid.
Twist and release the pump cover release catches.
Displace and remove the cover.
Remove the relay (3 Fig. 38).

Fit the relay.
Refit the cover and reposition the carpet.
Close the lid.

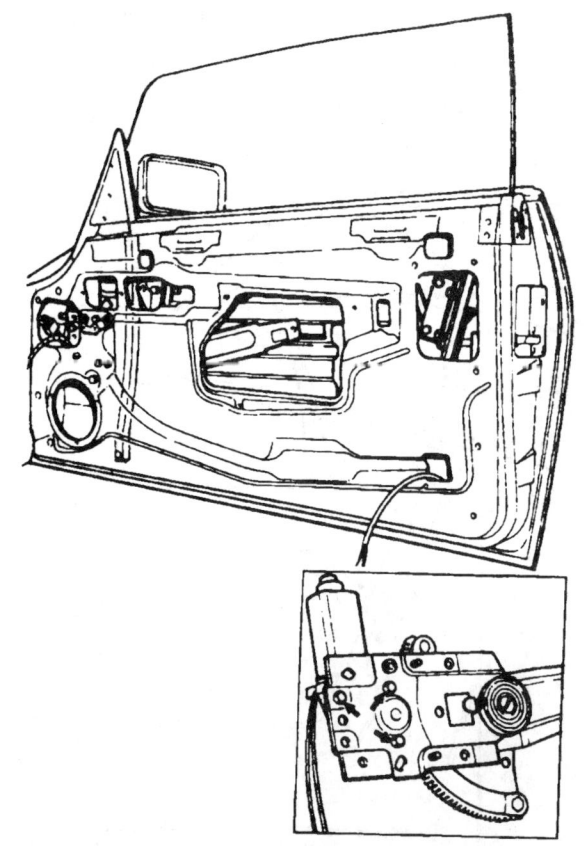

Fig. 39

DOOR - GLASS LIFT MOTOR, RENEW

As required, disconnect vehicle battery ground lead.

Remove:
Door trim pad
Door inner handle
Door glass regulator

Remove the motor (Fig. 39) from the regulator assembly

Reassembly and fitting is the reversal of this procedure noting that the glass height setting procedure must be observed.

DOOR GLASS - ADJUST

The door glass may be adjusted in two planes:

Vertically - to set or correct height and front to rear inclination.
Horizontally - to set or correct angle and pressure against seals.

These adjustments are critical to the prevention of wind noise and water ingress.
As required, disconnect vehicle battery earth lead.
Remove door trim pad, slacken and move the cheater forward and remove the water curtain aside.
To take maximum advantage of the setting procedure, slacken the regulator assembly and move it to the TOP of its mounting holes.
Two people are needed to successfully carry out the glass adjustments; one inside the vehicle to operate the regulator, adjust and secure the stops and the other outside to position and hold the glass.
With the glass partially lowered slacken the 'UP' stop stabilizer and glass mounting bracket fixings front (1 Fig. 40) and rear (2 Fig. 40).

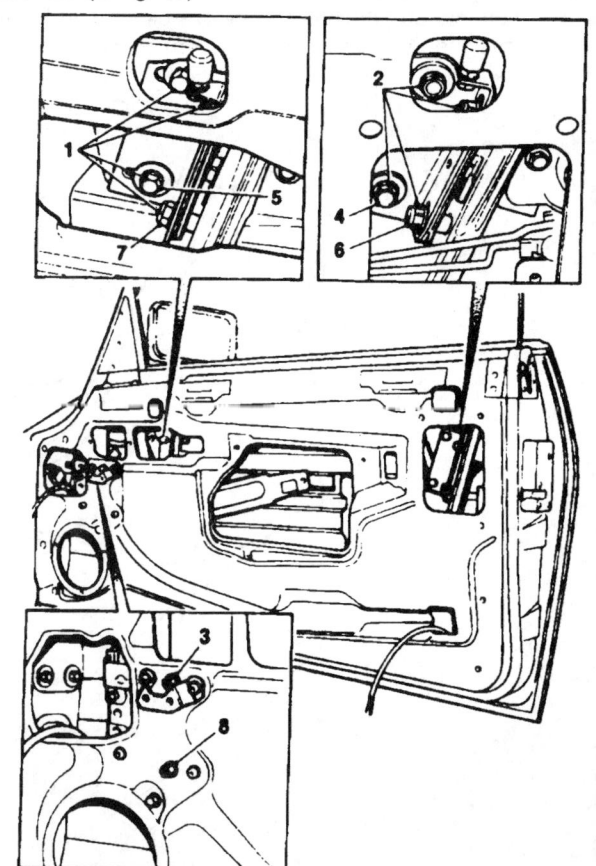

Fig. 40

112

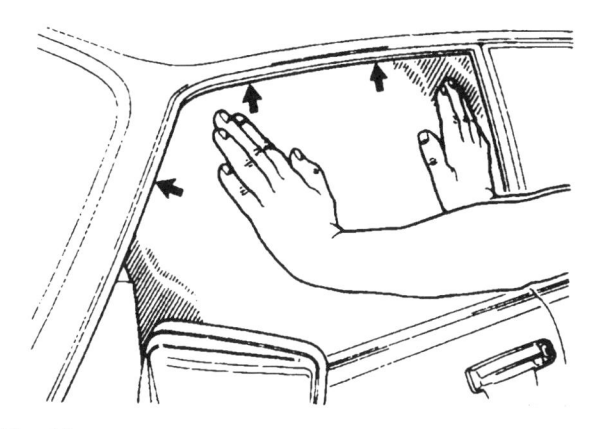

Fig. 41

Slacken the 'UP' cam quadrant stop locknut and rotate the cam anticlockwise (3 Fig. 40) to allow maximum upward glass travel.
Power the drop glass to maximum upward travel and position to contact the cantrail and 'A' post seals (Fig. 41).

Secure in order, lower rear (4 Fig. 40) and lower front (5 Fig. 40) mounting bracket fixings.
Hold the stabilizer bracket buffer against the door panel and secure in order, lower rear (6 Fig. 40) and lower front (7 Fig. 40) fixings.
Lower the drop glass to access and secure the front and rear stabilizer and glass mounting bracket upper fixings.
Verify seal contact and rectify as required.
Power the drop glass to maximum upward travel.
Reset the "UP" cam quadrant stop by rotating the cam clockwise until resistance is felt and secure the locknut.
Check the glass fully lowered position which should be with the glass upper edge at, or just below, seal lip height.
To rectify an incorrect glass to seal lower relationship, power the glass upward, release the 'DOWN' cam quadrant locknut and rotate the cam anticlockwise (8 Fig. 40).
Lower the glass to the correct level, rotate the cam quadrant clockwise until resistance is felt and secure the locknut.
Cycle the glass two or three times and verify the set position.
Power the convertible hood closed and re-check seal engagement.

Note: It is important to carry out these last two checks, the action of the hood may alter your initial settings.

Fitting is the reversal of this procedure.

REAR QUARTER GLASS - LIFT MOTOR, (CONVERTIBLE), RENEW

With the gearshift in the park or neutral position, power the hood to the fully stowed position.
As required, disconnect vehicle battery ground lead.

Remove:
    Stowage compartment
    Rear quarter trim pad
    As required, hood operating cylinder - RH ONLY (NOT necessary for 2 + 2)

Disconnect motor multiplug and release the regulator assembly fixings.
Disengage the regulator from the glass slide and remove the assembly from the vehicle (Fig. 42).

Remove the motor from the regulator assembly.
Reassembly and fitting is the reversal of this procedure noting that the glass height setting procedure must be observed.

Fig. 42

REAR QUARTER GLASS (CONVERTIBLE), ADJUST

The door glass may be adjusted in two planes:

Vertically -   to set or correct height and front to rear inclination.
Horizontally -   to set or correct angle and pressure against seals.

These adjustments are critical to the prevention of wind noise and water ingress.
As required, disconnect vehicle battery earth lead.
Remove rear quarter trim.
Two people are needed to successfully carry out the glass adjustments; one inside the vehicle to operate the regulator, adjust and secure the stops and the other outside to position and hold the glass.

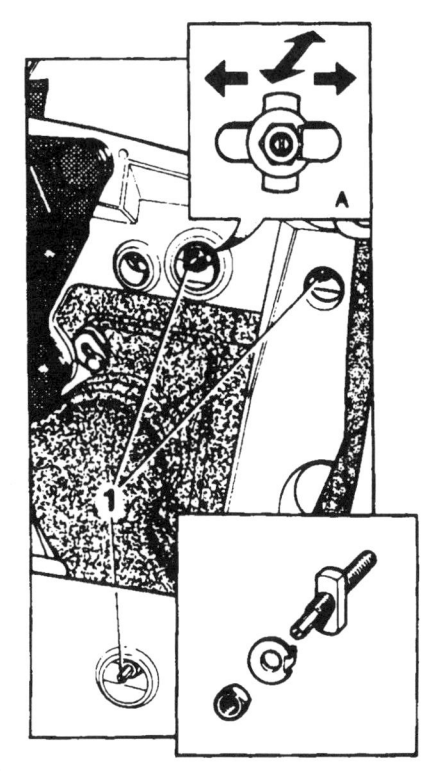

Fig. 43

Lower convertible hood and move the hydraulic selector to MANUAL (in order that that glass may be electrically raised or lowered independent of the hood).
Slacken glass adjuster locknuts (1 Fig. 43).

Power the quarter glass upwards and position using the adjuster screws relative to the correctly set door drop glass:

CRITERIA	SET AT
Height (Fig. 44)	+ 1mm
Parallel trailing edge gap (A Fig. 45)	2mm
Profile	As door

Fig. 44

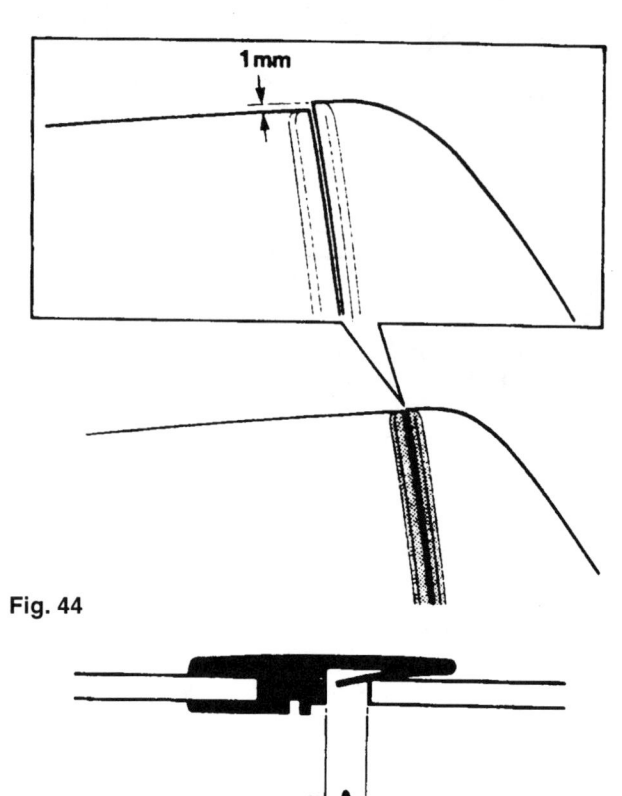

Fig. 45

Secure the adjuster locknuts.
Cycle the glass two or three times and verify the set position.
Power the convertible hood closed and recheck seal engagement and door drop glass relationship.

Note: It is important to carry out these last two checks, the action of the hood may alter your initial settings.

Fitting is the reversal of this procedure.

SUPPLEMENTARY RESTRAINT SYSTEM INTRODUCTION

General

All 1994 Model Year-on vehicles are fitted as standard with both driver and passenger side 'Supplementary Restraint System' (SRS), a system introduced at 1993.5 model year. North American and Canadian vehicles are fitted with a tear loop seat belt buckle and knee bolster on the driver side to complement those units currently fitted on the passenger side.
The passenger airbag is fitted in the area normally occupied by the glove box and when activated, exits through veneer faced deployment doors on the fascia.

WARNING:

THE VEHICLE MAY NOT COMPLY WITH LEGISLATIVE SAFETY STANDARDS IF ANY 'SRS' COMPONENT DOES NOT CONFORM TO THE ORIGINAL MANUFACTURERS SPECIFICATION FOR THAT MARKET.
THE NORTH AMERICA /CANADIAN STEERING WHEEL AND PASSENGER AIRBAG MODULES ARE NOT INTERCHANGABLE WITH THOSE OF OTHER MARKETS.

Note: To ensure operator safety during removal and handling of airbags, a mechanism is built into each module to allow it to be armed and disarmed.

The airbag modules CANNOT be removed from either the steering wheel or fascia unless that module has been disarmed.

Tear Loop Seat Belt Buckle

This unit is designed to control the rate of forward travel of the occupant towards the deployed airbag.

Knee Bolster

Both the driver and passenger side knee bolsters replace the underscuttle pads found on other market vehicles. Although similar in appearance to the non 'SRS' underscuttle pad, the knee bolster is specifically designed to work as part of the occupant restraint system and comply with USA and Canadian market legislation.

System Recognition

The following features will allow easy identification of an 'SRS' equipped vehicle.

Unique four spoke steering wheel for airbag application only.
'SRS' logo on the steering wheel centre pad.
'SRS' logo on the passenger side fascia veneer (Fig. 46).
Inclusion of an airbag symbol on the VIN plate, located at the lower left hand corner of the windscreen (Fig. 47).
Warning labels located in the vehicle interior.

Fig. 46

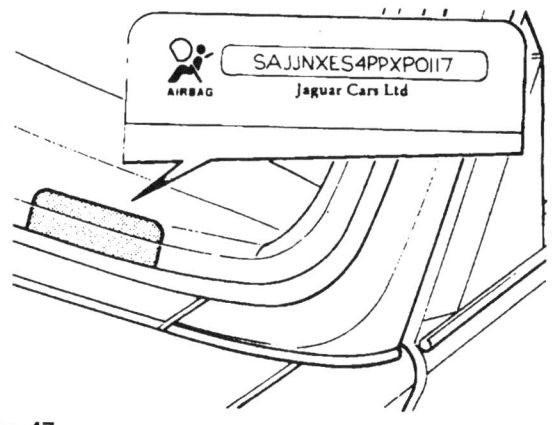

Fig. 47

WORKING PRACTICES: AIRBAG

General

Be aware of, and comply with all health and safety requirements, whether they be legislative or common sense. This applies to conditions set both for the operator and workshop. Before commencing any repair or service procedure, disconnect the vehicle battery ground connection and protect the vehicle where appropriate, from dirt or damage. Wherever possible, disarm the airbag module when working in the vicinity of it.
Use only the correct tools and equipment as described in the working text. Do not transfer an airbag module to another vehicle.

Fig. 48

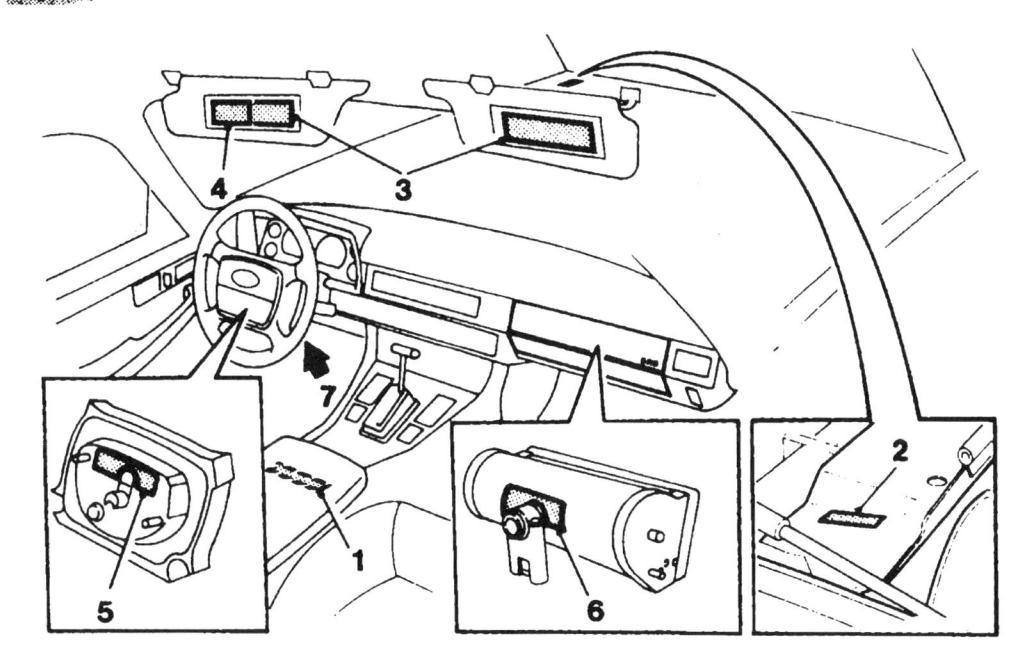

North American/Canadian specification units are NOT compatible with those of other markets and cannot be interchanged.

Handling the Airbag - Undeployed

Always wear eye and ear protection and impervious rubber gloves. **THE MODULE IS NON-SERVICEABLE; DO NOT TAMPER WITH IT.**
Do not subject the airbag module to excessive movement, sharp blows, electricity and heat.

Never carry the module against your body, hold it as shown (Fig. 48).

Note: Driver airbag shown. The passenger module should be held in a similar manner, with the deployment apperture facing either to the front or rear and at the side of the body.

The module should be stored, deployment apperture uppermost, in a secure cabinet: **NEVER** store face down, against a vertical surface, or stacked.

Handling the Airbag - Deployed

Wear eye, nose and mouth protection and impervious rubber gloves at all times. Should the materials from a deployed airbag come into contact with your eyes or skin. Wash the affected area with cool water and seek medical advice. Do not attempt to treat yourself.
Inhalation of airbag propellant residue may cause irritation to your respiratory system.
Seal the deployed module in a plastic bag in preparation for disposal.

Disposal

Contact your importer or Jaguar Service for instruction in the disposal of an undeployed module where:

> Service life has been exceeded.
> The module has been removed from a damaged vehicle.
> There is any doubt concerning the condition of the arming mechanism.

Fig. 49

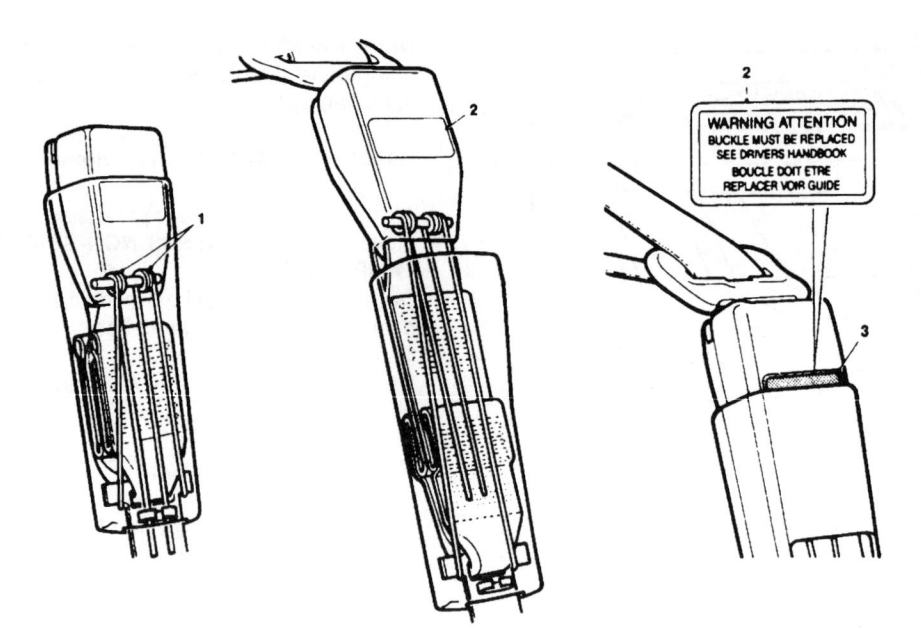

Fig. 50

If a vehicle is to be scrapped, a deployed module(s) may be disposed of with it and therefore need not be removed from the vehicle.

A deployed module which is to be disposed of separately, should be done so with regard to current local legislative requirements. If in any doubt, contact your local Environmental Agency.

Special Notes

Airbag modules for the North American/Canadian markets differ in calibration to those of other markets and therefore cannot be interchanged.

To make identification easy, an airbag module calibrated for full FMVS 208 operation (North American/Canadian) is coloured GOLD, as opposed to BLACK for all other markets. Further, with respect to the passenger side airbag, the mounting brackets are colour coded to match the module.

Should an attempt be made to ignore the colour coding, FMVS 208 modules also have unique fixing dimensions and brackets, thus further inhibiting incorrect fitting.

Warning Labels. To emphasize the need for caution and to convey maintenance information, warning labels are located at various points in the vehicle interior and in the case of label 2, under the bonnet (Fig. 49).

Label 1	Service life data, Passenger and Driver airbag (North American/Canada only).
Label 2.	Service life data, Passenger and Driver airbag (All other markets).
Label 3.	Seat belt/airbag warning, Passenger and Driver airbag only.
Label 4.	Seat belt/airbag warning, Driver airbag only.
Label 5.	Warning of misuse, Driver airbag.
Label 6.	Warning of misuse, Passenger airbag.
Label 7.	Warning, steering column removal.

TEAR LOOP SEAT BELT BUCKLE, DESCRIPTION

The mechanism within the buckle assembly is designed to release additional webbing when the stitching, which retains the webbing loops, breaks under a predetermined load.

The wires within the buckle (1 Fig. 50) have the following functions:

Protect the stitching from 'normal' loads such as heavy braking or cornering.
Control the rate of deployment.
Support the buckle assembly.

When the unit has been activated the buckle will extend from the shroud and reveal a warning label (2 Fig. 50). The extent of deployment will depend upon the severity of the load.

WARNING:

IF THE LABEL IS VISIBLE AT ALL (3 Fig. 50), THE COMPLETE ASSEMBLY MUST BE RENEWED, AS MUST ANY SEAT BELT WHICH HAS BEEN WORN IN AN ACCIDENT.

AIRBAG MODULE, DRIVER SIDE, RENEW

WARNING:

PLEASE READ THE SECTION ENTITLED 'SUPPLEMENTARY RESTRAINT SYSTEM', BEFORE PROCEEDING WITH ANY AIRBAG RELATED OPERATIONS.

Disconnect vehicle battery ground lead.
Tilt the steering wheel fully downwards.
Rotate steering wheel 90 degrees from the straight ahead to remove airbag nut cover (where fitted) and nut. Repeat for opposite side (Fig. 51).

Rotate steering wheel 180 degrees from the straight ahead to open the cover for the arming screw and third module fixing (1 Fig. 52).

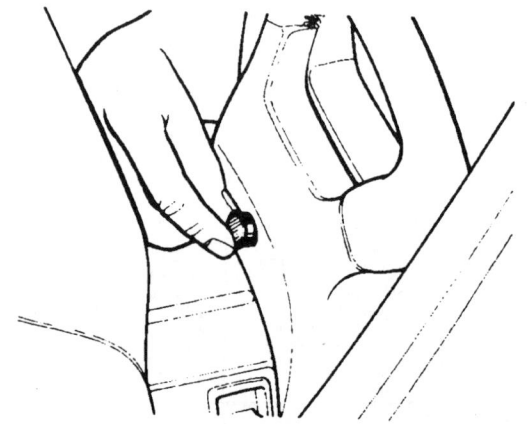

Fig. 51

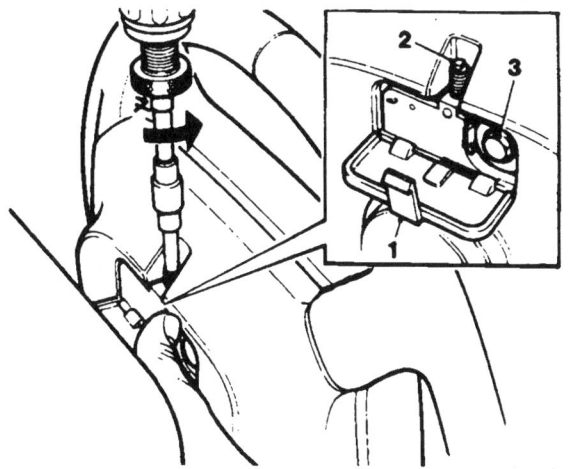

Fig. 52

Using special tool JD 159, rotate the arming screw anti-clockwise (2 Fig. 52) approximately 12 turns, or until resistance is felt.

Note: This action will also release the slide interlock for access to the third module fixing.

Do not rotate the steering wheel with the cover open.
Release the third module fixing (3 Fig. 52).
Remove the module from the vehicle and observe all safety considerations.

WARNING:

ENSURE THAT THE MODULE ARMING PIN IS IN THE DISARMED POSITION (B FIG. 53) IMMEDIATELY THAT THE ASSEMBLY IS REMOVED FROM THE VEHICLE. IF IT IS IN THE ARMED POSITION (A FIG. 53), CAREFULLY PLACE THE MODULE IN A SAFE PLACE AND CONTACT YOUR IMPORTER OR JAGUAR SERVICE.

Fitting is the reversal of this procedure noting that the arming screw and all fixings must be tightened to the specified torque.

Torque Figures

Airbag to steering wheel - Nut	9.5 - 12.5 Nm
Airbag to steering wheel - Screw	9.5 - 12.5 Nm
Arming screw	1 - 2 Nm

Fig. 53

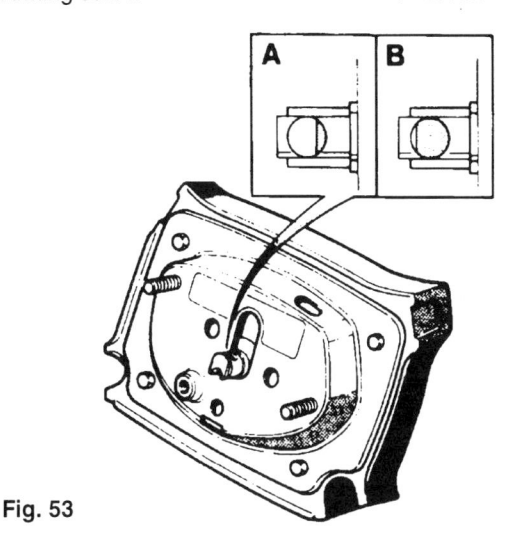

AIRBAG MODULE, PASSENGER SIDE, RENEW

WARNING:

PLEASE READ THE SECTION ENTITLED: SUPPLEMENTARY RESTRAINT SYSTEM, BEFORE PROCEEDING WITH ANY AIRBAG RELATED OPERATIONS.

Remove fascia assembly.
Slacken airbag upper M10 fixings (1 Fig. 54) and catch plate nuts M6 (2 Fig. 54).

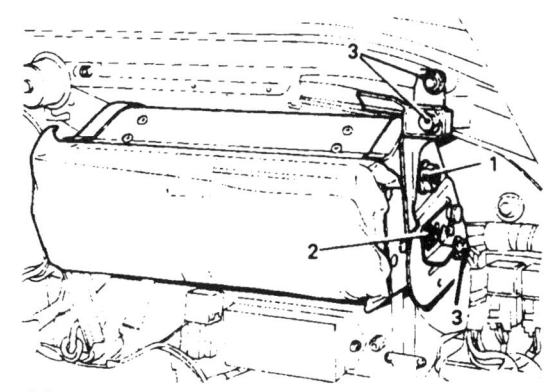

Fig. 54

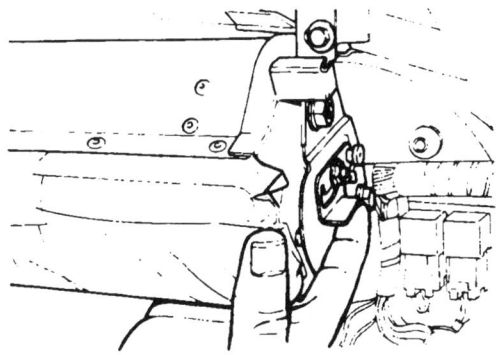

Fig. 55

Lift catch plates RH and LH (Fig. 55) and allow the airbag assembly to pivot downwards to the DISARMED position.

Remove fixings outer bracket to dash rail and crossbeam assembly (3 Fig. 54).

CAUTION:

As the catches are released the arming mechanism will apply considerable force. Do not allow the airbag assembly to 'snap' down; fully support it with both hands and ease to the disarmed position.

Remove previously slackened fixings, M10 and M6, airbag to inner bracket and remove the airbag/outer bracket assembly from the vehicle.
Check that the arming mechanism slide is fully down in the DISARMED position (1 Fig. 56).

Should the slide NOT be in the disarmed position, carefully place the airbag on a suitable work surface so that a SIDE face is towards your body and the deployment apperture is NOT facing downwards. Pull the slide downwards by finger pressure only, if this cannot be achieved, store the unit and

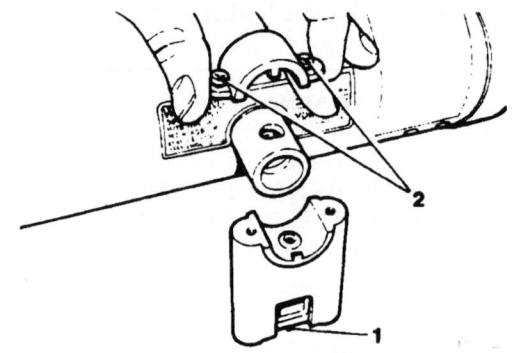

Fig. 56

contact your importer or Jaguar Service.
Only on a disarmed assembly, release split cap fixings (2 Fig. 56) and carefully remove the arming mechanism from the airbag module.

> **WARNING:**
>
> ENSURE THAT THE MODULE ARMING PIN IS IN THE DISARMED POSITION (B Fig. 57) IMMEDIATELY THAT THE ARMING MECHANISM IS DISENGAGED FROM THE MODULE. IF IT IS IN THE ARMED POSITION (A Fig. 57), CAREFULLY PLACE THE MODULE IN A SAFE PLACE AND CONTACT YOUR IMPORTER OR JAGUAR SERVICE. DO NOT TAMPER WITH THE MODULE.

Remove outer bracket and anti tamper bracket.
Reassembly and fitting is the reversal of this procedure ensuring that:
Upon assembly, the arming mechanism spigot is fully engaged onto the module.
When located in the vehicle and pivoted back to the armed position, that resistance is felt from the arming mechanism.
The anti tamper bracket fully obscures the upper outer fixing.
All fixings must be tightened to the specified torque.

Torque Figures

Airbag inner mounting bracket to body	8.5 - 10.5 Nm
Crossbeam to upper dash	23 - 31 Nm
Crossbeam to inner and outer mounting bracket.	23 - 31 Nm
Arming mechanism to module	1.7 - 2.3 Nm
Airbag outer mounting bracket to body (nut)	7.5 - 10.5 Nm
Airbag outer mounting bracket to body (bolt)	23 - 31 Nm
Airbag module to inner and outer bracket (M10)	13.5 - 18.5 Nm

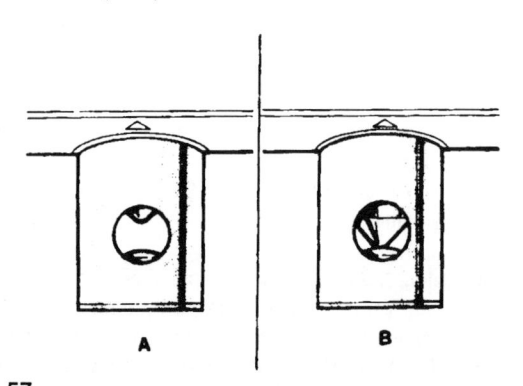

Fig. 57

UNDERFLOOR CROSS STRUT, FRONT, RENEW

> **CAUTION:**
>
> Do not lift the vehicle under the cross strut.

Remove screws, cross strut to front crossmember.
Remove jacking studs, cross strut to front floor and place the strut assembly aside.
Fitting is the reversal of this procedure noting the position of washers and spacers (Fig. 58).

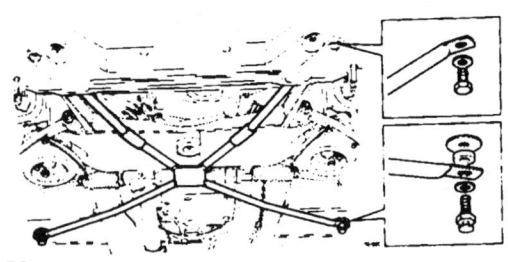

Fig. 58

Tighten fixings to the specified torque.

Torque Figures

Screw - front and rear	55-75 Nm
Jacking stud	41-55 Nm

UNDERFLOOR CROSS STRUT, REAR RENEW

> **CAUTION:**
>
> Do not lift the vehicle under the cross strut.

Remove screws, cross strut to rear longitudinal.
Remove screws, cross strut to rear floor and place the strut assembly aside.
Fitting is the reversal of this procedure, ensuring that the strut is aligned before tightening the fixings to the specified torque (Fig. 59).

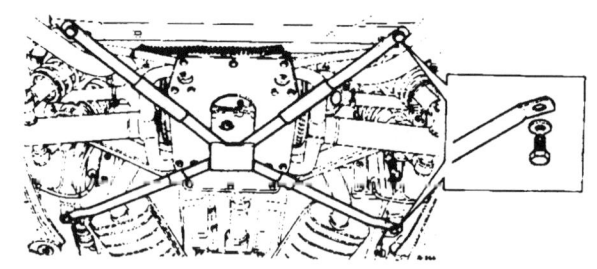

Fig. 59

Torque Figures

Screw - front and rear	55-75 Nm

WHEELARCH BAFFLE, FRONT, RENEW

Disconnect vehicle battery ground lead.
Raise front of vehicle and remove road wheel.
Release securing studs baffle to brake duct (1 Fig. 60).
Remove fixings, baffle to undertray (2 Fig. 60), inner valance (3 Fig. 60) and wing extension (4 Fig. 60).
Fitting is the reversal of this procedure.

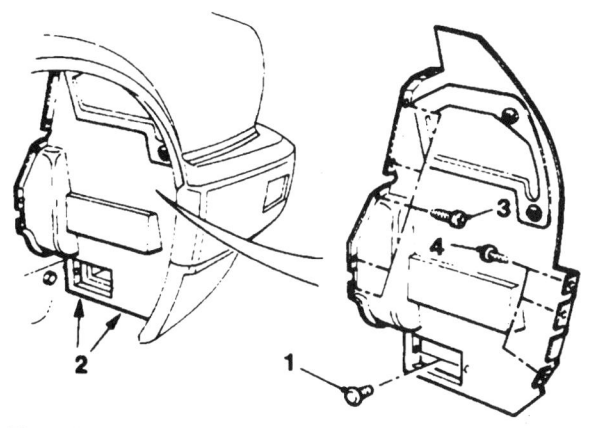

Fig. 60

CROSSMEMBER - FRONT, RENEW

Disconnect vehicle battery ground lead.
Open bonnet and protect paintwork.
Using a syringe, remove fluid from the power steering reservoir.
Remove RH and LH air cleaner element.
Remove RH and LH upper shock absorber fixings and bushings.
Attach MS 53 engine support bracket assembly to the front lifting eye and take the weight of the engine.
Remove RH and LH engine mounting upper securing nuts.
Raise the front of the vehicle and position suitable supports.
Remove front wheels noting wheel to stud relationship.
Convertible only: Remove underfloor cross strut.
All: Disconnect anti roll bar from vertical links RH and LH.
From inside engine bay, disconnect ABS harness connectors and feed cable through inner wing panel.
Release locking wire and bolts and remove front brake calipers.

Note: It is not necessary to break into the hydraulic system, but the calipers must be supported and NOT allowed to hang on the flexible hoses.

Release the fixings on the spoiler undertray at wheelarch baffle, crossmember and bumper cover and remove the undertray.

RHD only: Remove front right hand catalytic converter.
LHD only: Remove oil filter.
All: Release steering column lower shaft from steering rack pinion and lower column, and pull the shaft clear of the rack pinion.
Disconnect the power steering feed and return pipes at the pump, drain remaining fluid into a suitable container.
Fit blanking plugs to the pipes and pump.
Release earth leads from crossmember fixing (1 Fig. 61).
From below, support the crossmember assembly.
Remove crossmember to mounting rear nuts RH and LH (2 Fig. 61).
Remove crossmember forward fixings and tie down brackets (3 Fig. 61).

WARNING:

ENSURE THAT THE ENGINE IS FULLY SUPPORTED BY THE ENGINE SUPPORT BRACKET MS 53.

Lower the crossmember assembly from the vehicle ensuring that the brake calipers, pipes and steering rack are clear.
Place the crossmember assembly onto a suitable work surface.

Remove:

> Front mounting bush RH and LH.
> Road spring (using spring compressor JD 6G).
> Lower wishbone fulcrum shaft RH & LH.
> Upper wishbone fulcrum to tower bolts RH &LH (Support the hub/swivel assembly).
> Steering rack and steering arm assembly.

Note: Note position of upper wishbone fulcrum camber shims.

Remove engine mounting RH & LH.
Reassembly of the crossmember and fitting to the vehicle is the reversal of this procedure with special attention to the following:

- Renew all self locking nuts.
- Renew all bolts that were originally fitted with thread locking adhesive.
- Renew all locking wire.
- Renew all split pins (cotter pins).
- Tighten all fixings to the specified torque.
- Check steering geometry and adjust as required.
- Check operation of brakes.

BUMPER, FRONT, RENEW

Introduction

The following instructions describe the sequential procedures for the removal, strip, assembly and setting of the front bumper (all of its separate subsystems; select the appropriate procedure for the task).

CAUTION:

IT IS ESSENTIAL THAT THE ORDER OF THE SETTING PROCEDURES AND COMPONENT TIGHTENING IS ADHERED TO WHEN REPLACING COMPONENTS WHICH AFFECT BUMPER ALIGNMENT.

For all operations, disconnect the vehicle battery ground, open the bonnet, fit wing protection, raise the front of the vehicle and, as required, remove the front wheels.

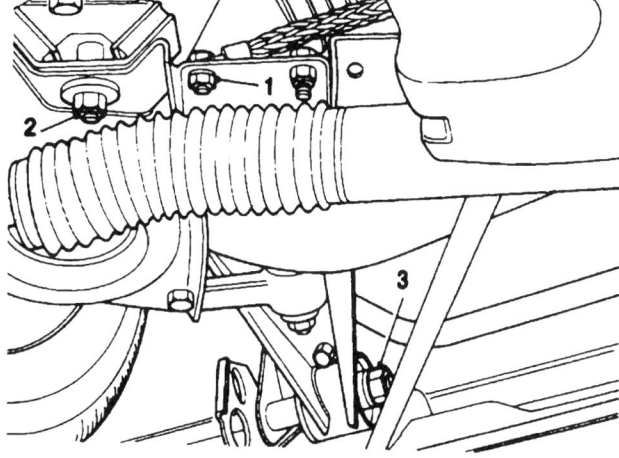

Fig. 61

Removal And Disassembly Procedures

Blade Assembly

From inside the front wheel arches RH and LH, remove the headlamp access panel. Remove quarter blade side fixings (1 Fig. 62). Release fixings, centre blade to front panel (2 Fig. 62) and pull the blade assembly forwards.

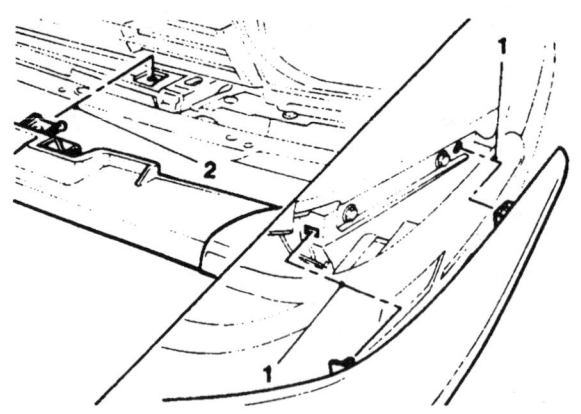

Fig. 62

Where fitted, disconnect the multiplugs and hoses to the headlamp power wash jets.
Remove the front blade assembly from the vehicle and place on a suitable workbench.

CAUTION:

Ensure that the blades are adequately supported and that paint damage cannot occur during removal. Remove the fixings on the reverse side of the blades to separate the quarters from the centre section and flip seal.

Bumper Cover

Disconnect direction indicator and foglamp multiplugs RH and LH.
Remove foglamps and brackets.
Remove fixings, bumper cover to body panel extensions (1 Fig. 63).

Release the fixings on the spoiler undertray at wheelarch baffle, crossmember and bumper cover (2 Fig. 63) and remove the undertray.
Remove fixings at the grille and crossmember to remove the lower air duct (Fig. 64).

Remove vehicle licence/registration plate and plinth.

Fig. 63

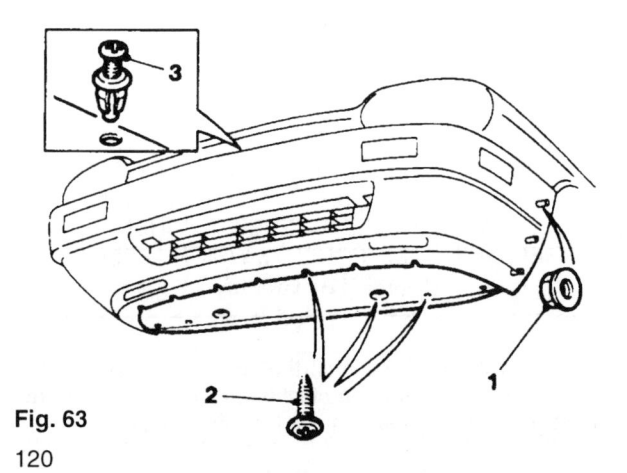

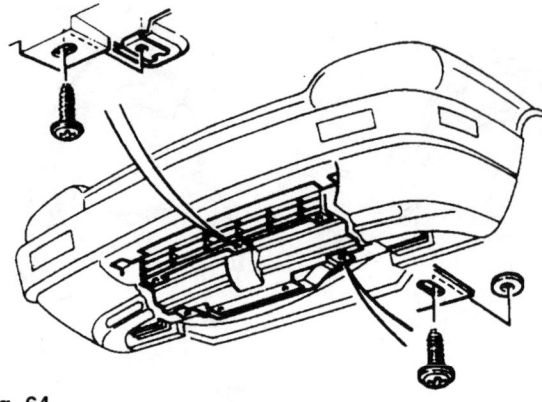

Fig. 64

Release the fixings securing bumper cover to beam (3 Fig. 63), carefully remove the bumper cover from the vehicle and place on a suitable workbench.
Remove direction indicators and spire clips. Release edge clips at the lower rear edge of the stoneguard grille and cover aperture, disengage forward lugs and pull the grille clear of the bumper cover.

Bumper Beam

Remove bolts (1 Fig. 65) securing the beam to the mounting struts noting the spacers and place the beam aside.

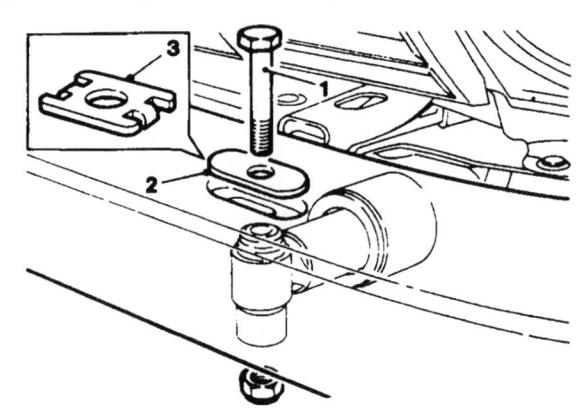

Fig. 65

Note: Specified washer fitted under the bolt head: Oval (2 Fig. 65) for 'Energy Absorbing' strut equipped vehicles, and tanged (3 Fig. 65) for 'Fixed' strut types. One washer per bumper only.

Bumper Mounting Strut

Remove the retaining nut and washer (1 Fig. 66) whilst preventing the threaded adjuster (2 Fig. 66) from rotating and withdraw the strut assembly from the chassis tube.

Note: It may be necessary to drive the strut from its rubber bush by rotating the threaded adjuster anti-clockwise. In this case, ensure that the length of thread (3 Fig. 66), is known prior to disassembly.

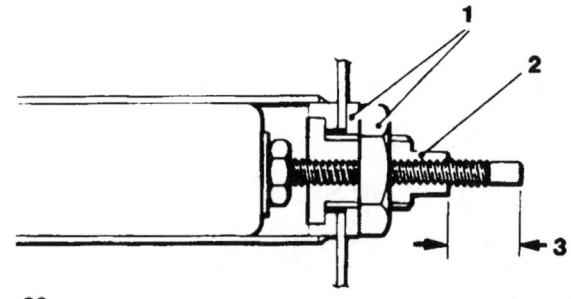

Fig. 66

With the strut assembly on a suitable workbench, remove the threaded adjuster and beam adjuster (vertical). Note the positions of the adjusters relative to the strut for initial setting of new components.

Assembly And Setting Procedures

Bumper Mounting Strut

Assembly and fitting is the reversal of the above procedure with adjusters nominally set. Insert the strut, with the beam adjuster slotted face downwards, and loosely assemble plain washer and nut (1 Fig. 66) but do not tighten.

Bumper Beam

Assemble the beam to the struts and retain with bolts and appropriate washer, do not fit spacers or nuts.

Note: Ensure correct location of bolt head and washer in beam slot.

Bumper Cover

Reassemble the cover, as required, prior to fitting and apply protection around the towing eye aperture.
Position the cover to the beam and align centre hole in the cover with the centre hole in the beam and secure with a scrivet.

Note: It is recommended that two people fit the bumper cover in order that it may be sprung over the side markers.

Working from the outside to the centre, fit the remaining scrivets.

Bumper Setting, Horizontal

Hold the strut threaded adjuster and rotate the large nut CLOCKWISE to draw the bumper assembly rearwards until the studs on the cover sides locate in the body panel extensions. Loosely assemble the cover side fixings.
The bumper cover and wheelarch should now be aligned.
Rotate the threaded adjuster ANTI-CLOCKWISE until fully seated on the inner face of the body tube, but does not begin to jack the bumper forward. Tighten the large nut to the specified torque.

Note: If sufficient rearward adjustment cannot be attained, slacken the large nut and turn the threaded adjuster CLOCKWISE 2 or 3 turns. Repeat the operations Bumper Setting Horizontal to obtain the desired condition.

Blade Assembly

Reassemble the blades as required and refit to the vehicle. Ensure that the flip seal is located against the body correctly.

Bumper Setting, Vertical

Rotate the beam adjusters, RH and LH to adjust the vertical height of the bumper assembly. The bumper cover should just contact the blade mounted buffers (Fig. 67) to give an all round 3.0mm gap.

Note: Ensure that the cover does not distort the blade vertically.

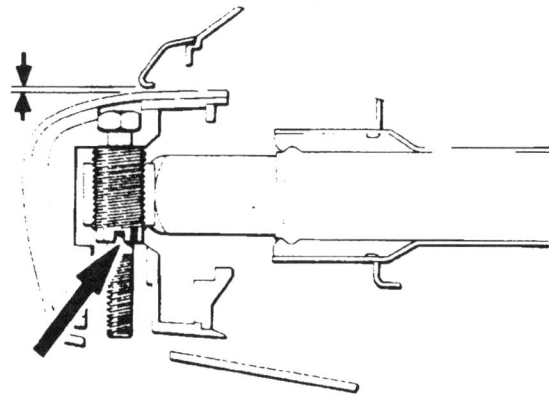

Fig. 67
Secure Fixings

Secure the cover side fixings to the specified torque.
Assemble spacers and nyloc nuts to the bumper beam bolts and tighten to the specified torque.
All other reassembly and fitting procedures are the reversal of disassembly.

BUMPER, REAR, RENEW

Introduction

The principals of fixing and adjustment are common to front and rear bumpers and many repair operations are similar.
However, the rear bumper cover, when fitted, masks the beam to strut fixings. Therefore, height adjustment may only be carried out with the cover removed.
It should also be noted that with Non Energy Absorbing struts, the beam to strut bolt head, is not captive in the beam.
For all operations open the luggage compartment, disconnect the vehicle battery ground, raise the rear of the vehicle and, as required, remove the rear wheels.
See front bumper procedures for all aspects of operations which are not detailed on this or the next page.

REMOVAL AND DISASSEMBLY PROCEDURES

Blade Assembly

Remove luggage compartment side liner RH, LH and rear protection plate to gain access to blade fixings.

Note: The blade assembly is dowelled together unlike the front which is thread secured.

Bumper Cover

See procedure for front cover noting that the lower edge is clipped to the floor pan.

Bumper Beam

See procedure for front beam noting washer types and location.

CAUTION:

Having unclipped the cover, do not attempt to gain access to the beam fixings by forcing the cover downwards.

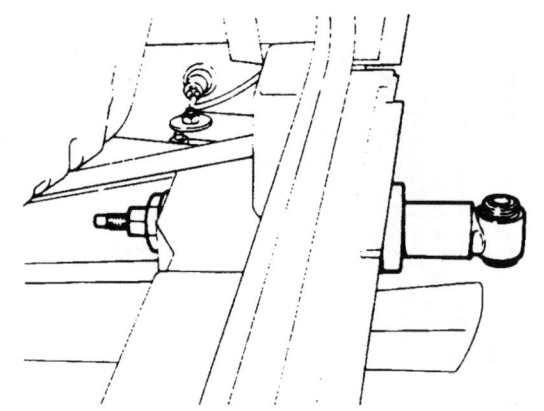

Fig. 68

Bumper Mounting Strut

See procedure for front strut, noting that access to the large nut and threaded adjuster is from inside the luggage compartment (Fig. 68).

Assembly And Setting Procedures

Bumper Mounting Strut and Initial Vertical Bumper Setting

See procedure for front strut, noting that the initial height of the bumper beam, controlled by the beam adjuster, must be set at this point.
With the strut initially located, washer and large nut engaged, set the distance from the beam adjuster top face to the rear panel flange with a simple tool, as shown, or set dimension 'A' at 59.0mm by rule measurement (Fig. 69).

Bumper Beam

As front, but the spacer and nut should be fitted and secured to the specified torque prior to fitting the cover.

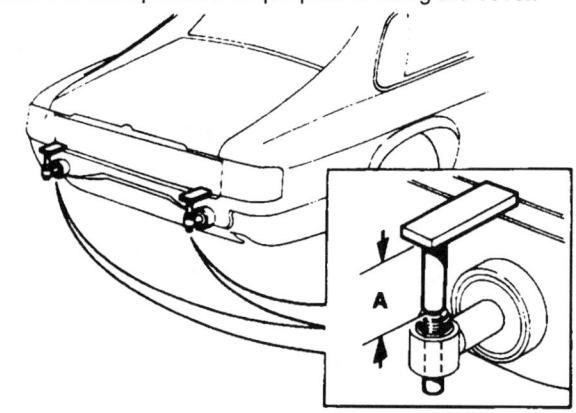

Fig. 69

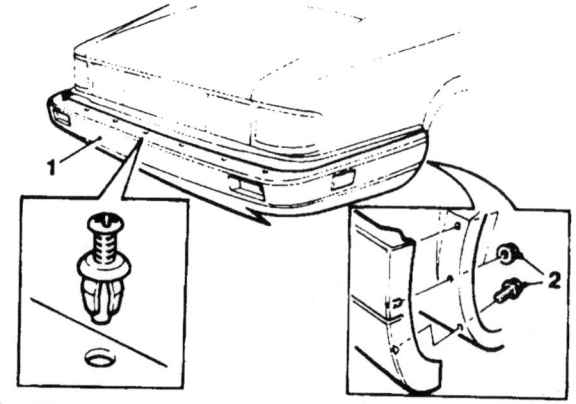

Fig. 70

Bumper Cover (1 Fig. 70)

See procedure for front cover.

Bumper Setting Horizontal

See procedure for front bumper, but secure the cover side fixings (2 Fig. 70) to the specified torque IMMEDIATELY after horizontal setting.

Blade Assembly

See procedure for front blade. To achieve the 3.0mm gap condition, apply a light downward pressure on the blade as the fixings are secured.

Note: Should the gap condition be unattainable, remove the cover and reset the beam height accordingly.

Torque figures

Bumper beam to bumper strut (All)	39-51 Nm
Bumper blade - centre to quarter	1.5-2.5 Nm
Bumper blade - centre to body	5-7 Nm
Bumper blade - quarter to body	5-7 Nm
Bumper cover to wing extension (nut)	5-7 Nm
Bumper cover to body (screw)	5-7 Nm
Bumper cover to undertray	3.5-4.5 Nm
Bumper strut to body	39-51 Nm
Fog lamp bracket to beam	5-7 Nm
Undertray to crossmember	3.5-4.5 Nm
Undertray to wheelarch baffle	1.5-2.5 Nm

Fascia board, Renew

Disconnect the vehicle battery ground lead and position the gear control lever fully rearward.

Remove:

Driver side airbag.
Steering wheel.
Upper and lower steering column cowls.
Underscuttle pad or knee bolster, driver and passenger side.
Instrument module.
Centre fascia veneer panel.
Centre fascia vent outlet.
Glove box liner (where applicable).

Through the instrument aperture, remove the upper fascia retaining nut (1 Fig. 71).

Remove RH and LH side fixing, brace to 'A' post (2 Fig. 71). Through the centre vent aperture, remove the upper fascia retaining tube nut (3 Fig. 71) (Passenger airbag only), or, through the glove box aperture, remove the upper fascia retaining nut (4 Fig. 71) (non passenger airbag vehicles).
Remove A/C control panel knobs and retaining collars.
Pull the A/C control panel clear to release centre console to fascia fixings.

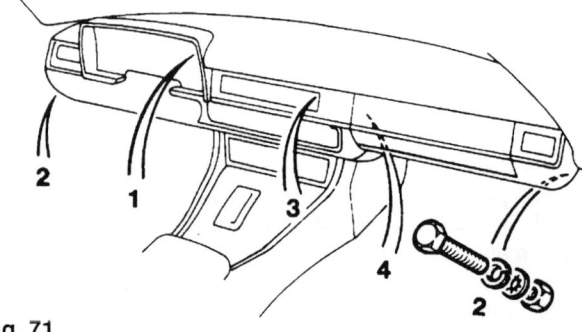

Fig. 71

> **CAUTION:**
>
> Take care not to damage fibre optic components when moving control panel.

Remove the RH and LH side 'A' post to fascia trims and fascia mounted interior lamps.
Remove the hazard warning/heated backlight switch and exterior mirror consoles (for access).
Release the switch and trip computer harness from the clips, and pull clear.
Disconnect the fog lamp, trip computer and interior lamp multiplugs.
Remove the fascia assembly from the vehicle and place on a suitably protected work bench.
Remove the trip computer and switch/veneer assembly.
Remove the RH and LH side air vent veneer and vent assembly.
Remove the dimmer control.
Remove the airbag deployment door and fascia brackets (passenger airbag only).
Remove the glove box lid (non passenger airbag).
Remove clips and retainers as required.
Reassembly and fitting is the reversal of this procedure.

REAR QUARTER GLASS, RENEW

Remove the following components:

The left and right hand rear seat cushions.
Left and right hand rear seat squabs.
Rear seat cushion centre panel.
Rear quarter trim pad veneer panel.
Rear seat squab surround trim panel.
Right hand rear quarter upper trim pad.

Displace and remove the cantrail trim and the 'B' post seat belt adjuster knob.
Undo and remove the 'B' post trim upper and lower securing screws.
Displace and remove the door waist plastic capping, the 'B' post upper trim and the seat belt adjuster trim cover.
Displace the door seal from the area adjacent to the 'B' post.
Undo and remove the seal to the 'B' post channel retention securing screws.
Displace and remove the channel, the 'B' post finisher and quarter light lower waist finisher. Unscrew and remove the quarter light securing nuts, noting that the rear window studs carry larger washers to span the aperture flange rear 'V' slots.
Carefully remove the quarter light assembly.
Remove and discard the old 'memory' foam from the quarter light aperture flange.

Fit the finishing brightwork ('banana') into the rear of the body aperture.
Fit and align new 'memory' foam to the aperture flange (B Fig. 72).

Fit and align foam strip to the new glass bottom edge and replace the large washers onto the rear studs with the nuts just started.
Slide the lower finisher on to the bottom of the encapsulation and offer the whole assembly into the body aperture.
The rear of the glass may need to be offered at an angle to engage the studs into the 'V' slots at first, then the whole is pressed into position by locating all the studs through the holes in the flange and tightening the nuts in the sequence shown (A Fig.72).
Access to the rear stud nuts is achieved via 'rat holes' through the inner bodywork (1 & 2 Fig. 73).

The remainder of the work is completed by carrying out the previous operations in reverse order.

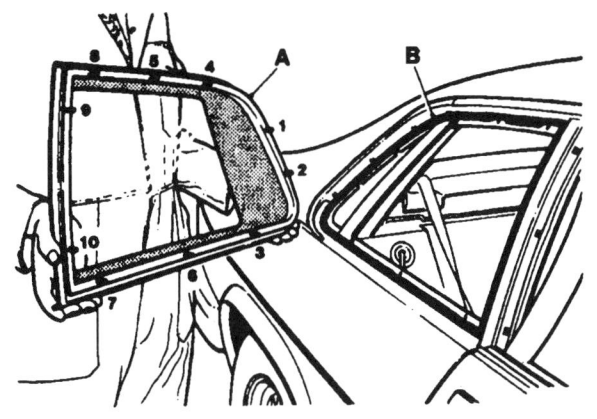

Fig. 72

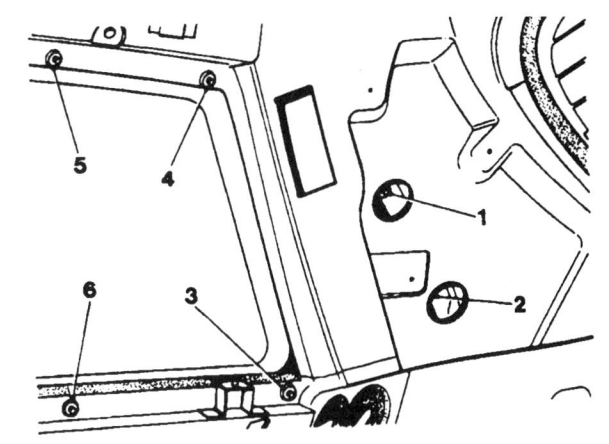

Fig. 73

Chapter 15

AIR CONDITIONING SYSTEM - 1994 MODEL YEAR ON

INTRODUCTION

This Section is primarily concerned with the 1994 Model Year introduction of HFC 134A (Hydro fluorocarbon) refrigerant. However, the opportunity has also been taken to update certain critical test and fault finding operations which are more appropriate to the new refrigerant.

Working practices and safety related procedures are relevant to all refrigerant types, but certain aspects are more critical to HFC 134A than R12; take special note of the section 'Handling refrigerant' and notes concerning moisture contamination.

GENERAL SPECIFICATIONS

REFRIGERANT		
Designation	Charge weight	Manufacturer and Type
HFC 134A	950 g +/- 50 g	ICI Klea or equivalent

COMPRESSOR		
Type & model	Configuration	Manufacturer
SD-7H15	7 Cylinder 155 cm^3 per revolution	Sanden

COMPRESSOR LUBRICATION		
Designation	System Capacity	Manufacturer and Type
Polyalkylene glycol (PAG)	120 - 150ml	Sanden

STANDARD FOR RECOVERY / RECYCLE / RECHARGE EQUIPMENT.

Feature	Requirement
Recovery rate	0.014 - 0.062 m^3 / min. (1.36 kg in 20 minutes)
Cleaning capability	15 parts per million (ppm) moisture; 4000 ppm oil; 330 ppm non condensable gases in air
Oil separator	With hermetic compressor and automatic oil return
Moisture indicator	Sight glass type, sensitive to 15 ppm minimum
Vacuum pump	2 stage 0.07 - 0.127 m^3 / min.
Filter	Replaceable with moisture indicator
Charge	Selectable charge weight and automatic delivery
Hoses	Dedicated HFC 134A port connections.
Charge pressure	Heating element to increase pressure

COMPRESSION BELT TENSION

Burroughs method	New belt 600 N. If tension is below 230 N, reset at 400 N
Clavis method	New belt 103 to 107 Hz. If tension is below 70 Hz, reset at 83 to 87 Hz

Note: Tension measured midway between compressor and air pump/idler pulleys.
For new belt, rotate engine 3 revolutions minimum and recheck tension.

CLIMATE CONTROL SYSTEM, GENERAL

The climate control system fitted to 1994-on XJS has many features which make it unique. These differences demand changes to current system maintenance and rectification working practices.

Feature:
1. Refrigerant HFC 134A (Hydro fluorocarbon), non ozone depletory.
2. PAG (polyalkylene glycol) synthetic compressor lubricating oil.
3. Dedicated and improved compressor for HFC 134A refrigerant.
4. Quick fit/release self sealing charge and discharge ports.
5. Dual pressure switch to control the compressor (incorporated into the liquid line).
6. Clamp retained '0' ring seals at the expansion valve and evaporator.
7. All aluminium evaporator matrix and pipework.
8. Aluminium receiver/dryer (without sight glass) and HFC dedicated desiccant.
9. Parallel flow extended height condenser.
10. Single muffler situated in the suction hose.
11. Improved electrical system connectors.
12. Improved system control panel.

System Recognition

The following features will facilitate easy identification of the R134A system:

1. Aluminium pipes.
2. Large diameter, quick release charge and recovery ports.
3. HFC 134A labelling on compressor. No sight glass.

> **CAUTION:**
>
> The system refrigerant HFC 134A, is NOT compatible with any other previously fitted XJS system. The PAG compressor lubrication oil is NOT compatible with previously-used mineral based oils and must be treated exactly as detailed in the following sections.

WORKING PRACTICES

General

Be aware of, and comply with all health and safety requirements, whether they be legislative or common sense. This applies to conditions set both for the operator and workshop. Before commencing any repair or service procedure, disconnect the vehicle battery ground connection and protect the vehicle where appropriate, from dirt or damage.
Work in a well ventilated, clean and tidy area.
Recovery and charge equipment must comply with, or exceed the standard detailed in General Specifications.

Handling Refrigerant

Wear eye protection at all times.
Use gloves, keep skin that may come into contact with HFC 134A covered.
Should refrigerant come into contact with your eyes or skin; wash the affected area with cool water and seek medical advice, do not attempt to treat yourself.
Avoid breathing refrigerant vapour, it may cause irritation to your respiratory system.
Never use high pressure compressed air to flush out a system. Under certain circumstances HFC 134A + compressed air + a source of combustion (welding and brazing operations in the vicinity), may result in an explosion and the release of potentially toxic compounds.
HFC 134A and CFC 12 must never come into contact with each other, they will form an inseparable mixture which can only be disposed of by incineration.
Do not vent refrigerant directly to atmosphere, always use Jaguar approved recovery equipment. Remember, HFC 134A is costly but recyclable.
Because HFC 134A is fully recyclable it may be 'cleaned' by the recovery equipment and reused following removal from a system.
Leak tests should only be carried out with an electronic analyser which is dedicated to HFC 134A. Never use a CFC 12 analyser or naked flame type.
Do not attempt to 'guess' the amount of refrigerant in a system, always recover and recharge with the correct charge weight. In this context do not depress the charge or discharge port valves to check for the presence of refrigerant.

Handling Lubricating Oil

Avoid breathing lubricant mist, it may cause irritation to your respiratory system.
Always decant fresh oil from a sealed container and do not leave oil exposed to the atmosphere for any reason other than to fill or empty a system. PAG oil is very hygroscopic (absorbs water) and will rapidly become contaminated by atmospheric moisture.
PAG oil is NOT compatible with previously used mineral based oils and must NEVER be mixed.
Do not reuse oil when it has been separated from refrigerant following a recovery cycle and dispose of the oil safely.

System Maintenance

When depressurizing a system do not vent refrigerant directly to atmosphere, always use Jaguar approved recovery equipment. Remember, HFC 134A is costly but recyclable.
Always decant compressor oil from a sealed container and do not leave oil exposed to the atmosphere for any reason other than to fill or empty a system. PAG oil is very hygroscopic and will rapidly become contaminated by atmospheric moisture.
Plug pipes and units immediately after disconnection and only unplug immediately prior to connection. Do not leave the system open to atmosphere.
It is not necessary to renew the receiver dryer whenever system has been 'opened' as previously advised. However, if a unit or part of the system is left open for more than five minutes, it may be advisable to renew the receiver dryer. This guidance is based on U.K. average humidity levels; therefore, locations with lower humidity will be less critical to moisture contamination of the unit. It must be stressed that there is not a 'safe' period for work to be carried out in: ALWAYS plug pipes and units immediately after disconnection and only remove plugs immediately prior to connection. If replacement parts are supplied without transit plugs and seals do not use the parts. Return them to your supplier.
Diagnostic equipment for pressure, mass and volume should be calibrated regularly and certified by a third party organisation.
Use extreme care when handling and securing aluminium fittings, always use a backing spanner and take special care when handling the evaporator.
Use only the correct or recommended tools for the job and apply the manufacturer's torque specifications.
Keep the working area, all components and tools clean.

SYSTEM

Air Conditioning Control Module

The electronic control module (A/CCM) is located on the right hand side of the heater unit.
A digital micro-processor within the A/CCM receives data signals from operator controlled switches. Comparison of these signals with those returned from system temperature sensors and feedback devices results in the appropriate output voltage changes needed to vary: Blower motor speed, Flap position and those Solenoids which respond to operator selected temperature demand.
The A/CCM is a non-serviceable item but may be interrogated for system test. Care must be exercised when connecting test equipment, the A/CCM may be irreparably damaged should any of the test pins be shorted or bent.

Air Conditioning Control Panel

The Control Panel (Fig. 2) contains: Fan speed/defrost rotary switch, Manual and Demist mode buttons, Temperature differential slider, Temperature rotary control, A/C On and Recirculation mode buttons. The control panel relays information to the electronic control module.

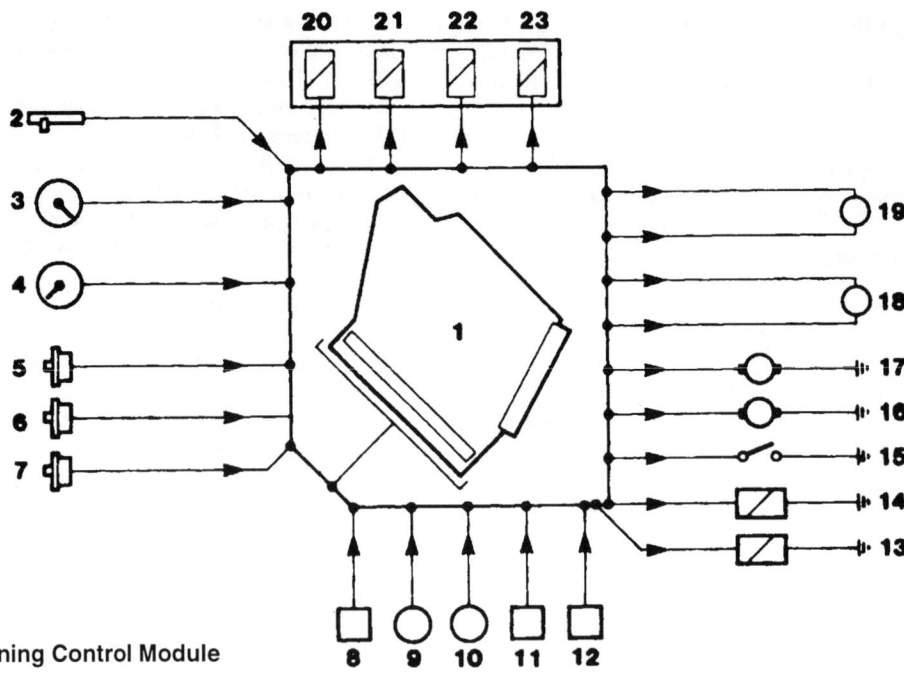

Fig. 1 Air Conditioning Control Module

(key to Fig. 1)
1. A/C control module (A/CCM)
2. Differential temperature control
3. Temperature control
4. Blower motor switch
5. Ambient temperature sensor
6. Motorized in-car aspirator
7. Evaporator temperature sensor
8. Coolant temperature switch
9. Lower flap feedback potentiometer
10. Upper flap feedback potentiometer
11. LH Blower motor feedback
12. RH blower motor feedback
13. High speed relay
14. High speed relay
15. Compressor clutch
16. Blower motor
17. Blower motor
18. Lower flap servo motor
19. Upper flap servo motor
20. Defrost vacuum solenoid
21. Auto recirculation vacuum solenoid
22. Centre vent vacuum solenoid
23. Water valve vacuum solenoid

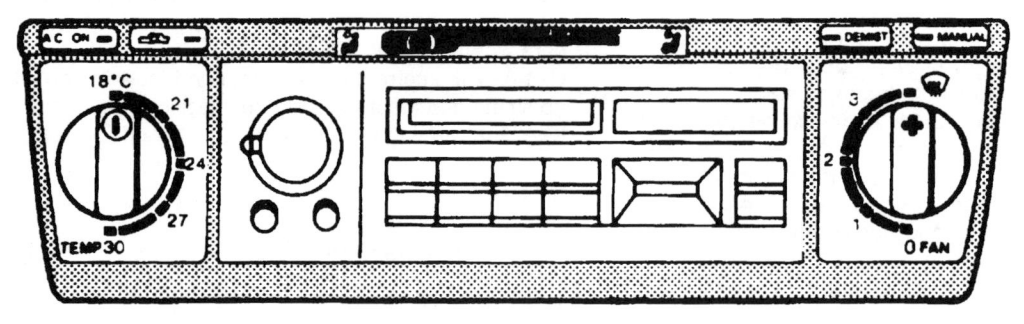

Fig. 2 Air Conditioning Control Panel

The FAN SPEED CONTROL rotary switch controls airflow from the blower motors. The switch has five positions: 0, 1, 2, 3 and DEFROST. In the 0 position the system is not operational, however, a residual signal to the control module (A/CCM) ensures that the blower flaps are closed, thus preventing outside air from entering the system.

Information regarding selection by the control switch of 1, 2 and 3 fan speeds is relayed to the A/CCM. Signals are also relayed to the A/CCM from the temperature selector feedback circuits and various sensors. Fan speed is steplessly controlled by the A/CCM, within the ranges 1, 2 and 3.

When DEFROST is selected the fans operate at maximum speed, front screen vents open fully, lower flaps close fully and maximum output is directed to the windscreen (there may be a delay of up to 30 seconds from selection to execution of this function).

The DEMIST mode button, when pressed, causes increased air flow to the front screen.

The FACE LEVEL TEMPERATURE DIFFERENTIAL sliding control is used to vary the temperature of face vent distributed air to that of footwell delivered air.

The TEMPERATURE ROTARY CONTROL is used to pre set the in car heat level in either 'automatic' or 'manual' mode. There are three temperature sensors located in the system - Exterior ambient, In-car and Evaporator.

An input voltage is supplied to the sensors from AC4-13 of the control module. The temperature sensing signal from the sensors is transmitted to the control module via AC4-4 and AC2-4 respectively. The sensors are semi-conductor devices which provide a voltage output proportional to the sensed temperature.

When pressed, the A/C ON button, causes the system to engage the Air conditioning compressor via its electromagnetic clutch.

When engaged, with indicator lamp lit, the MANUAL MODE facility provides operator selection of fan speed and

in car temperature. In car temperature will not be thermostatically corrected to a predetermined level via the system sensors.
Automatic temperature control is resumed when the button is pressed and the indicator lamp is unlit.
RECIRCULATION mode closes the blower flaps and circulates only that air which is in the vehicle. When the ignition is turned off the blower flaps revert to the fresh air position.

REFRIGERATION CYCLE:

The Compressor draws low pressure refrigerant from the evaporator and by compression, raises refrigerant temperature and pressure. High pressure, hot vaporized refrigerant enters the Condenser where it is cooled by the flow of ambient air. A change of state occurs as the refrigerant cools in the condenser and it becomes a reduced temperature high pressure liquid.
From the condenser the liquid passes into the Receiver/Drier which has three functions,

a) Storage vessel for varying system refrigerant demand.
b) Filter to remove system contaminants.
c) Moisture removal via the desiccant.

With the passage through the receiver/drier completed the still high pressure liquid refrigerant, enters the Expansion Valve where it is metered through a controlled orifice which has the effect of reducing the pressure and temperature.
The refrigerant, now in a cold atomized state, flows into the Evaporator and cools the air which is passing through the matrix.
As heat is absorbed by the refrigerant it once again changes state, into a vapour, and returns to the compressor for the cycle to be repeated (Fig. 3).

There is an automatic safety valve incorporated in the compressor which will operate should the system pressure be in excess of 41 bar. The valve will re-seat when the pressure drops below 35 bar.

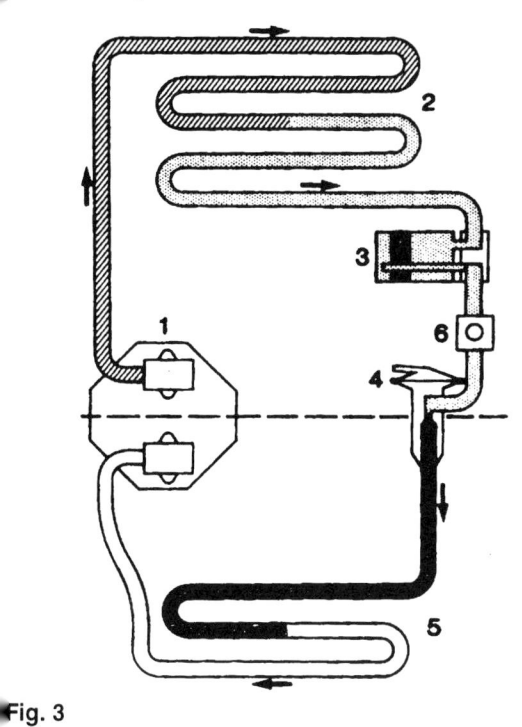

Fig. 3
(key to Fig. 3)
1. Compressor
2. Condenser
3. Receiver/Drier
4. Expansion Valve
5. Evaporator
6. Dual pressure switch

Note: The division of HIGH and LOW side is simply the system pressure differential created by the compressor discharge (pressure), suction (inlet) ports and the relative inlet and outlet ports of the expansion valve. This differential is critical to system fault diagnosis and efficiency checks.

System protection

The Dual pressure switch, located in the liquid line, cuts electrical power to the compressor clutch if the system pressure is outside of the range of 2 Bar (1st Function) to 27 Bar (2nd Function).

SYSTEM TROUBLE-SHOOTING

There are five basic symptoms associated with air conditioning fault diagnosis. A slightly different approach to problem solving will be necessary since the deletion of the sight glass. It is very important to positively identify the area of concern before starting a rectification procedure. A little time spent with your customer on problem identification, and use of the following trouble shooting guides will be beneficial.
The following conditions are not in order of priority.

NO COOLING

- Is the electrical circuit to the compressor clutch functional?
- Is the electrical circuit to the blower motor/s functional?
- Slack or broken compressor drive belt.
- Compressor partially or completely seized.
- Compressor shaft seal leak.
- Compressor valve or piston damage (may be indicated by small variation between HIGH & LOW side pressures relative to engine speed).
- Broken refrigerant pipe (causing total loss of refrigerant).
- Leak in system (causing total loss of refrigerant).
- Blocked filter in the receiver drier.
- Evaporator sensor disconnected?
- Dual pressure switch faulty?

Note: Should a leak or low refrigerant be established as the cause, follow the procedures as detailed under 'Recovery/Recycle/Recharge', and observe all refrigerant and oil handling instructions.

INSUFFICIENT COOLING

- Blower motor/s sluggish.
- Restricted blower inlet or outlet passage
- Blocked or partially restricted condenser matrix or fins.
- Blocked or partially restricted evaporator matrix.
- Blocked or partially restricted filter in the receiver drier.
- Blocked or partially restricted expansion valve.
- Partially collapsed flexible pipe.
- Expansion valve temperature sensor faulty (this sensor is integral with valve and is not serviceable).
- Excessive moisture in the system.
- Air in the system.
- Low refrigerant charge.
- Compressor clutch slipping.
- Blower flaps or distribution vents closed or partially seized.
- Water valve not closed.
- Evaporator sensor detached from evaporator.

INTERMITTENT COOLING

- Is the electrical circuit to the compressor clutch consistent?
- Is the electrical circuit to the blower motor/s consistent?
- Compressor clutch slipping.
- Faulty air distribution flap potentiometer or motor.
- Motorized in-car aspirator or evaporator temperature sensor faulty, causing temperature variations.
- Blocked or partially restricted evaporator or condenser.

NOISY SYSTEM

- Loose or damaged compressor drive belt.
- Loose or damaged compressor mountings.
- Compressor oil level low, look for evidence of leakage.
- Compressor damage caused by low oil level or internal debris.
- Blower/s motor/s noisy.
- Excessive refrigerant charge, witnessed by vibration and 'thumping' in the high pressure line (may be indicated by high HIGH & high LOW side pressures).
- Low refrigerant charge causing 'hissing' at the expansion valve (may be indicated by low HIGH side pressure).
- Excessive moisture in the system causing expansion valve noise.

Note: Electrical faults may be more rapidly traced using JDS or PDU.

INSUFFICIENT HEATING

- Water valve stuck in the closed position.
- Motorized in-car aspirator seized.
- Blend flaps stuck or seized.
- Blocked or restricted blower inlet or outlet.
- Low coolant level.
- Blower fan speed low.
- Coolant thermostat faulty or seized open.

MANIFOLD GAUGE SET

The manifold gauge set is a most important tool for fault diagnosis and system efficiency assessment. The relationship to each other of HIGH and LOW pressures and their correlation to AMBIENT and EVAPORATOR temperatures must be compared to determine system status (see Figs 5 & 6, Pressure/Temperature Graphs).

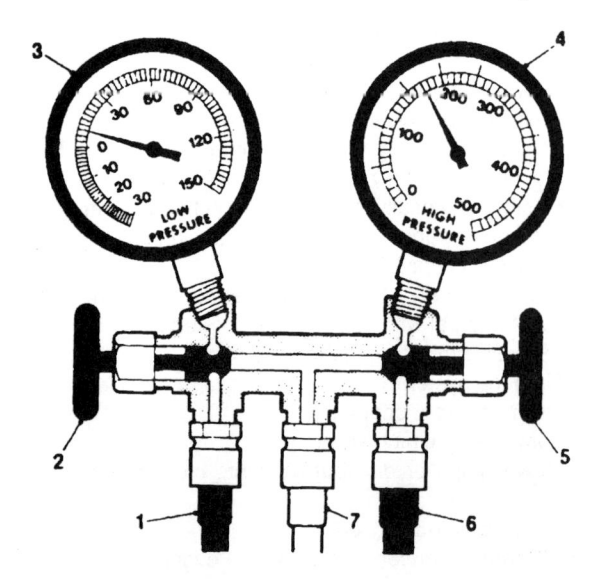

Because of the heavy reliance upon this piece of equipment for service diagnosis, ensure that the gauges are calibrated regularly and the equipment is treated with care.

The gauge set (Fig. 4) consists of a manifold fitted with:

- Low side service hose - BLUE.
- Low side hand valve - BLUE.
- Low pressure compound gauge - BLUE.
- High pressure gauge - RED.
- High side hand valve - RED.
- High side service hose - RED.
- System service hose - NEUTRAL COLOUR (commonly yellow).

Manifold

The manifold is designed to control refrigerant flow. When connected into the system, pressure is registered on both gauges at all times. During system tests both the high and low side hand valves should be closed (rotate clockwise to seat the valves). The hand valves isolate the low and the high sides from the centre (service) hose.

Low Side Pressure Gauge

This compound gauge, is designed to register positive and negative pressure and may be typically calibrated - Full Scale Deflection, 0 to 10 bar (0 to 150 lbf/in^2) pressure in a clockwise direction; 0 to 1000 mbar (0 to 30 in Hg) FSD negative pressure in a counter clockwise direction.

High Side Pressure Gauge

This pressure gauge may be typically calibrated from 0 to 30 bar (0 to 500 lbf/in^2) FSD in a clockwise direction. Depending on the manufacturer, this gauge may also be of the compound type.

SYSTEM CHECKING WITH THE MANIFOLD GAUGE SET

Evacuating the Manifold Gauge Set

Attach the centre (service) hose to a vacuum pump and start the pump. Open fully both high and low valves and allow the vacuum to remove air and moisture from the manifold set for at least five minutes.
Turn the vacuum pump off and isolate it from the centre service hose but do not open the hose to atmosphere.
Observe the manufacturer's recommendation with regard to vacuum pump oil changes.

Connecting the Manifold Gauge Set

> **CAUTION:**
>
> It is imperative that the vacuum pump is not subjected to a positive pressure of any degree. Therefore the pump must be fitted with an isolation valve at the centre (service hose) connection and this valve must be closed before the pump is switched off. This operation replaces the purge' procedure used on previous systems. Observe the manufacturer's recommendation with regard to vacuum pump oil changes.

Fig. 4 Manifold Gauge Set

CAUTION:
Only use hoses with connectors which are dedicated to HFC 134A charge ports.

Attachment of the hose quick release connectors to the high and low side system ports is straightforward, provided that the high and low valves are closed and the system is NOT operational.

Assessment of system operating efficiency and fault diagnosis may be achieved by using the facilities on your Recovery/Recharging/Recycling station, follow the manufacturers instructions implicitly and observe all safety considerations.

WARNING:
UNDER NO CIRCUMSTANCES SHOULD THE CONNECTIONS BE MADE WITH THE SYSTEM IN OPERATION OR THE VALVES OPEN. SHOULD THE VALVES BE OPEN AND A VACUUM PUMP OR REFRIGERANT CONTAINER ATTACHED, AN EXPLOSION COULD OCCUR AS A RESULT OF HIGH PRESSURE REFRIGERANT BEING FORCED BACK INTO THE VACUUM PUMP OR CONTAINER.

Stabilizing the System

Accurate test gauge data will only be attained if the system temperatures and pressures are stabilized.

Ensure that equipment and hoses cannot come into contact with engine moving parts or sources of heat.

It is recommended that a free standing air mover is placed in front of the vehicle to provide mass air flow through the condenser/cooling system see illustration below.

Start the engine, allow it to attain normal working temperature and set at fast idle (typically 1200 to 1500 rpm).

Select full air conditioning performance.

With all temperatures and pressures stable or displaying symptoms of faults, begin relevant test procedures.

Pressure/Temperature Graphs

(To obtain Bar, multiply the lbf/in^2 figure by 0.069.
To obtain kgf/cm^2 multiply the lbf/in^2 figure by 0.070)

Note: The system controls will prevent the evaporator temperature from falling below 0° C. The graph is typical of HFC 134A

SYSTEM FAULT DIAGNOSIS

Probable causes of faults may be found by comparing actual system pressures, registered on your manifold gauge set or recovery / recharge / recycle station, and the pressure to temperature relationship graphs found on the previous page. The chart below shows the interpretation that may be made by this difference. The 'Normal' condition is that which is relevant to the prevailing ambient and evaporator temperatures.

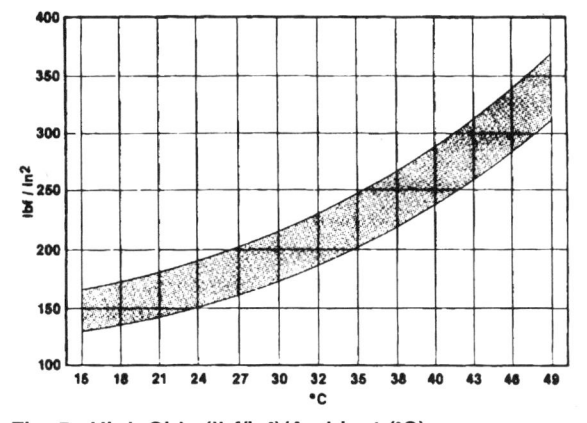

Fig. 5 High Side (lbf/in^2)/Ambient (°C)

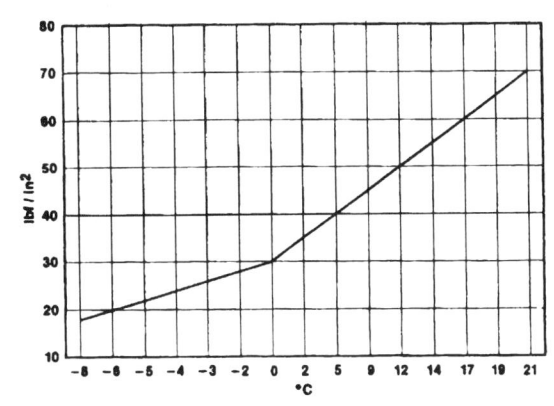

Fig. 6 Low Side (lbf/in^2)/Evaporator (°C)

Low Side Gauge	High Side Gauge	Symptom	Diagnosis
Normal	Normal	Discharge air initially cool then warms up	Moisture in system
Normal to low	Normal	As above	As above
Low	Low	Discharge air slightly cool	HFC 134A charge low
Low	Low	Discharge air warm	HFC 134A charge very low
Low	Low	Discharge air slightly cool or frost build up at expansion valve	Expansion valve stuck closed
Low	Normal to high	Discharge air slightly cool	Restriction in High side of system
High	Low	Compressor noisy	Defective reed valve
High	High	Discharge air warm and high side pipes hot	HFC 134A charge high or condenser malfunction
High	High	Discharge air warm Sweating or frost at evaporator	Expansion valve stuck open

Note: If erratic or unusual gauge movements are experienced, check the equipment against a known manifold gauge set.

GENERAL SYSTEM PROCEDURES

Leak Test

Faults associated with low refrigerant charge weight and low pressure may be caused by leakage. Leaks traced to mechanical connections may be caused by torque relaxation or joint face contamination. Evidence of oil around such areas is an indicator of leakage. When checking for non visible leaks use only a dedicated HFC 134A electronic analyser and apply the probe all round the joint / connection. Should a leak be traced to a joint, check that the fixing is secured to the correct tightening torque before any other action is taken.

Do not forget to check the compressor shaft seal and evaporator.

Note: Never use a dedicated CFC 12 or naked flame type analyser.

Charge Recovery (System depressurization)

The process of HFC 134A recovery will depend on the basic characteristics of your chosen recovery / recycle / recharge equipment, therefore, follow the manufacturers instructions carefully.

Remember that compressor oil may be drawn out of the system by this process, take note of the quantity recovered so that it may be replaced.

CAUTION:

Observe all relevant safety requirements.
Do not vent refrigerant directly to atmosphere and always use Jaguar approved recovery/recycle/recharge equipment.

Wear suitable eye and skin protection.
Do not mix HFC 134A with CFC 12.
Take note of the amount of recovered refrigerant, it will indicate the state of the system and thus the magnitude of any problem.

Evacuating the System

This process, the removal of unwanted air and moisture, is critical to the correct operation of the air conditioning system. The specific procedures will vary depending on the individual characteristics of your chosen recovery/recycle/recharge equipment and must be carried out exactly in accordance with the manufacturers instructions.

Moisture can be highly destructive and may cause internal blockages due to freezing, but more importantly, water suspended in the PAG oil will damage the compressor. Once the system has been opened for repairs, or the refrigerant charge recovered, all traces of moisture MUST be removed before recharging with new or recycled HFC 134A.

Adding Compressor Lubricating Oil

Oil may be added by three methods, two of which are direct into the system - 1) via the recovery/recycle/recharge station, 2) by proprietary oil injector.

Equipment manufacturer's instructions must be adhered to when using direct oil introduction.

The third method may be required because of rectification work to the existing compressor, or the need to fit a new compressor.

From an existing compressor, drain the oil into a measuring cylinder and record the amount. Flush the unit out with fresh PAG oil and drain thoroughly. Replenish the compressor with the same amount of PAG oil that was originally drained out and immediately plug all orifices ready for refitting to the vehicle. The transit lubricating oil must be drained and discarded from a new compressor before it may be fitted. An adjustment should be made to the system oil level by taking into account, a) the quantity found in the original compressor, and b) the quantity deposited in the recovery equipment oil separator from the charge recovery operation. Typically, 80 ml may be drained from the original compressor and 30 ml found in the oil separator; if these quantities are added together - 80 + 30 = 110 ml, then this is the amount of fresh PAG oil that must be put into the new compressor prior to fitting.

Please note that the discrepancy between this figure and the nominal capacity of 135 ml is caused by normally unrecoverable oil being trapped in components such as the receiver/drier or evaporator. The previous statements are only valid if there is NO evidence of an oil leak from the system. If oil has been lost and the fault attended to, then the compressor, whether original or replacement, should be filled with the specified quantity.

CAUTION:

Always decant fresh oil from a sealed container and do not leave oil exposed to the atmosphere. PAG oil is very hygroscopic (absorbs water) and will rapidly attract atmospheric moisture.

PAG oil must NEVER be mixed with mineral based oils.
Do not reuse oil following a recovery cycle, dispose of it safely.

Depending on the state of the air conditioning system immediately prior to charge recovery and the rate of recovery, an amount of oil will be drawn out with the refrigerant. The quantity will be approximately 30 to 40 ml; this may vary, and the figure is given only for guidance. It is most important that the oil separator vessel in the recovery equipment is clean and empty at the start of the process so that the amount drawn out may be accurately measured.

Adding Refrigerant

In order that the air conditioning system may operate efficiently it must contain a full refrigerant charge. The indications of some system defects, and the results of certain tests, will show that a low charge is the most probable cause of the fault. In such cases the charge should be recovered from the system, the weight noted, and the correct amount installed.

CAUTION:

If oil was drawn out during the recovery process, the correct amount may be added directly from your recovery / recycle /recharge station (if so equipped) prior to the 'charging process'.

Note: Never attempt to 'guess' the amount of refrigerant in a system.

Always recover and recharge with the correct charge weight, this is the only accurate method.

It must be stressed that the need to protect compressor oil from moisture is vital, observe the procedures in **HANDLING LUBRICATING OIL.**

CHAPTER 16

ELECTRICAL SYSTEM (1992 MODEL YEAR ON)

ALTERNATOR DRIVE BELT, RENEW AND ADJUST

Loosen the pivot bolts securing the air conditioning compressor.
Loosen the adjusting link securing bolt and trunnion block bolt.
Loosen the adjusting link locknut and adjust the compressor towards the engine until compressor drive belt can be removed.
Loosen the alternator pivot nut and bolt.
Loosen the adjusting link pivot bolt and the trunnion block bolt.
Loosen the adjusting link locknut and adjust the alternator towards the engine by means of the adjusting nut.
Remove the trunnion block bolt and push the alternator towards the engine until the drive belt can be removed from the pulleys.
On fitting the new belt ensure that the drive belt is adjusted to the correct tension.
A load of 1.5 kg must give a total belt deflection of 4.5 mm when applied at the mid point of the belt.

Fig. 1

Key to Fig. 1
1. To auxiliary controlled relay
2. Thermal circuit breaker
3. LH drop glass switch
4. RH drop glass switch
5. LH motor
6. RH motor

ELECTRICALLY OPERATED DOOR GLASS, DESCRIPTION

Power is supplied via auxiliary controlled load relay contacts from the main battery supply through a thermal circuit breaker. The auxiliary controlled load relay is energised when the ignition switch is closed and this energises the circuit to the window lift motor. When a control switch is operated to lower a window, current flows via contacts within the switch to the motor (circuit to earth is also via the switch). When the switch is operated to raise the window, current flows in the opposite direction through the switch and motor.

TESTING (Refer to Fig. 1)

If the drop glass fails to operate check the fuse and all connectors and ensure that all connections are clean and tight.
1. Check the thermal circuit breaker:
   Connect the white/blue and brown/blue leads together.
   Switch on the ignition.
   Operate the glass motor switch. If the glass operates as normal, the circuit breaker is faulty and should be renewed.
2. With the ignition switched on, battery voltage should be obtained at the brown/blue lead terminal of the left hand switch. Operate the switch. Battery voltage should be obtained at the red/blue lead terminal when the switch is operated in one direction, and the green/blue lead terminal when the switch is operated in the opposite direction.
   Should a zero reading be obtained at either test point, renew the switch.

Note: The same test applies to the right hand switch. The switch cable colours are red/green for one direction and green/red for the opposite direction.

If the above tests prove satisfactory, check the lift motor for continuity.

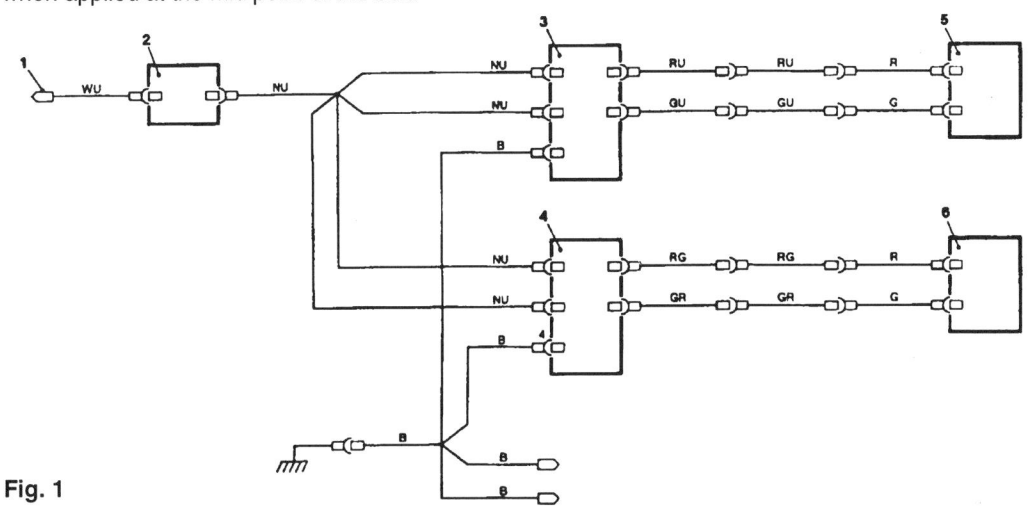

If the wiring continuity proves satisfactory, remove the lift motor for bench testing.

CIRCUIT BREAKER, RENEW

Disconnect the battery earth lead.
Remove the passenger side dash liner.
Remove nuts and shakeproof washers (1 Fig. 2) securing mounting plate to fan motor.
Ease the mounting plate from studs.

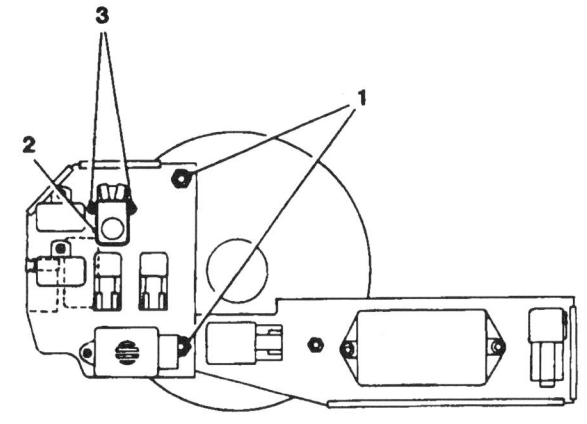

Fig. 2

Note position of cables and disconnect at lucars on relevant circuit breaker.
Remove the two screws (3 Fig. 2) securing the unit and remove the unit (2 Fig. 2).
Fitting a new circuit breaker is a reversal of the removal procedure.

GLASS LIFT SWITCHES, RENEW

Disconnect the battery earth lead.
Carefully displace the lift switchpack (Fig. 3) from the console veneer panel.

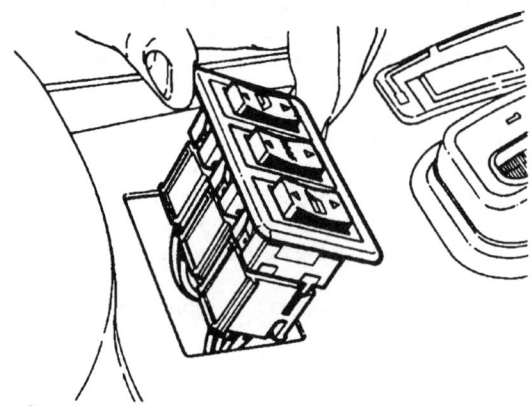

Fig. 3

Disconnect the harness multi-plug from the window lift switch.
Displace and remove the switch from the panel.
The refitting procedure is a reversal of the removal procedure.

KIEKERT CENTRAL LOCKING SYSTEM

DESCRIPTION

The system comprises an actuator in both doors and the boot lid, and is controlled by an ECM located in the passenger side 'A' post.
The doors will lock and unlock simultaneously. The boot will either open in unison with the doors or remain locked dependent on the door lock position, as follows:

Boot lock positions

1. Lock turned fully clockwise. In this position the lid locks/unlocks in unison with the door locks.
2. Lock turned fully anti-clockwise. In this position the boot lid is permanently locked and cannot be unlocked using the central locking system.
3. Lock turned to the central position. With the lock in this position the key cannot be removed.

The boot lid can be opened irrespective of the central door locking mode (locked or unlocked). This enables access to the boot if the car is centrally locked.

FRONT AND REAR PARKING LAMPS, CIRCUIT DESCRIPTION

With the master lamp switch in the parking lamp on position, current flows to the side lamp relay to energise circuit to side lamps and bulb failure units. The current flowing through the bulb failure units will cause the bulb failure warning lamp to glow for 15 to 30 seconds. If the warning lamp fails to go out then there is a bulb failure or a circuit fault in the front parking lamp, rear lamps or number plate lamps.

Fault Finding

Check the fuses and all connections, ensuring the earth connections are clean and tight.
With the master light switch in the parking lamp on position, battery voltage should be obtained at terminals 85 and 87 of the side lamp relay 2. This in turn makes the circuit to the lamp failure units 6, 11, 17 and 23.
If battery voltage is obtained at the B terminal of a lamp failure unit but a zero reading at the L terminal, renew the bulb failure unit.

HEADLAMP, ALIGNMENT

ADJUSTMENT

Headlamp beam setting should only be carried out with approved beam setting apparatus.

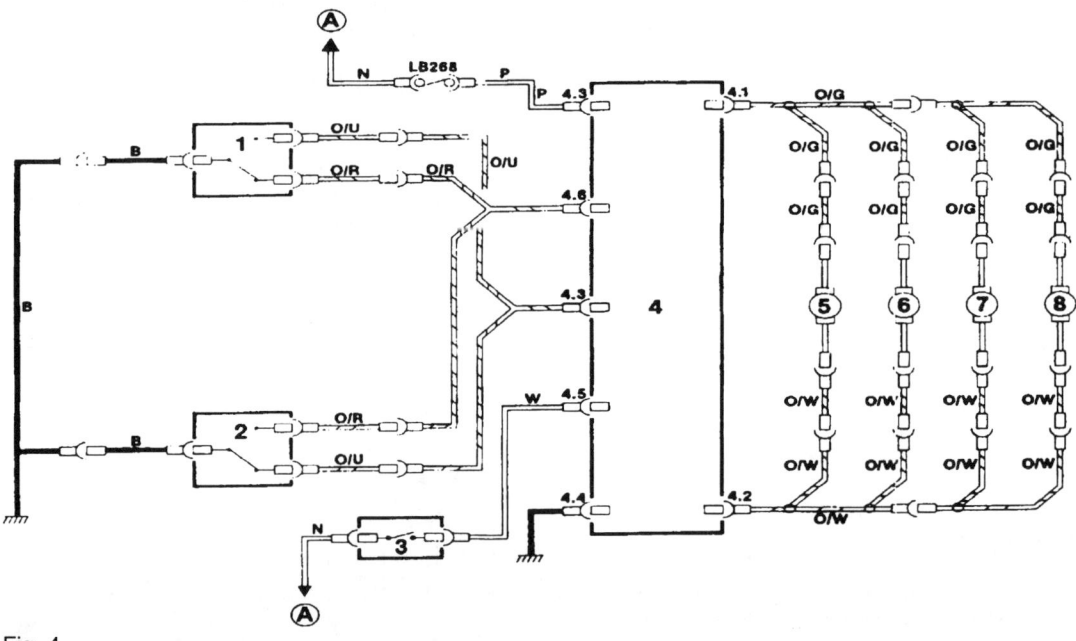

Fig. 4

Key to Fig. 4

1. RH door lock switch
2. LH door lock switch
3. Ignition switch
4. Door lock control module
5. RH door lock motor
6. LH door lock motor
7. Boot lock motor
8. Fuel filler flap lock (coupe only)

Vertical and horizontal adjustment of the headlamp beam is made with two adjusting screws set in a yellow plastic moulding sited above either front wheel arch inside the bonnet. The silver screw adjusts the horizontal alignment and the black screw adjusts the vertical alignment. For horizontal beam adjustment of the right hand headlamp turn the screw clockwise to move the beam to the right. Turn the screw anti - clockwise to move the beam to the left.

Turn in the opposite directions for adjustment of the left hand headlamp. Where a headlamp levelling motor is fitted (i.e. interposed in the drive cable), setting the beam vertically is exactly the same as for the non - headlamp beam levelling lamp. Three positions are however available for setting the levelling motor to cater for differing rear end loads.

HEADLAMP ASSEMBLY, RENEW

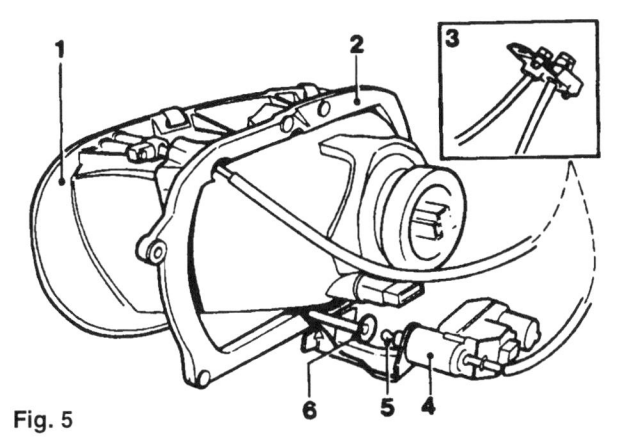

Fig. 5

Key to Fig. 5

1. Headlamp
2. Headlamp carrier
3. Manual adjuster cable
4. Motor
5. Motor drive ball
6. Headlamp carrier socket

Disconnect the battery earth lead.
Remove headlamp rim finisher.
Displace manual adjuster retaining block from inner wing.
Displace adjuster cables from block.
Remove adjuster block and place aside.
Reposition adjuster cables through inner wing.
Undo and remove headlamp carrier to body securing screws.
Displace headlamp assembly from aperture.
Disconnect headlamp multi-plug.
Remove headlamp assembly.

The refitting procedure is a reversal of the removal procedure.
Take care not to touch the bulb glass envelope (1 Fig. 6).

HEADLAMP BULB, RENEW

Disconnect the battery earth lead.
Reposition the front wheels to gain access to the plate located inside the wheel arch.
Turn fastener anti-clockwise and remove the plate.
Disconnect the multi-connector socket from the bulb and remove the rubber cover.
Disconnect the wire clip securing bulb to the headlamp unit and remove the bulb 1 Fig. 6 (not applicable to USA).
The refitting procedure is a reversal of the removal procedure. Take care not to touch the bulb glass envelope.

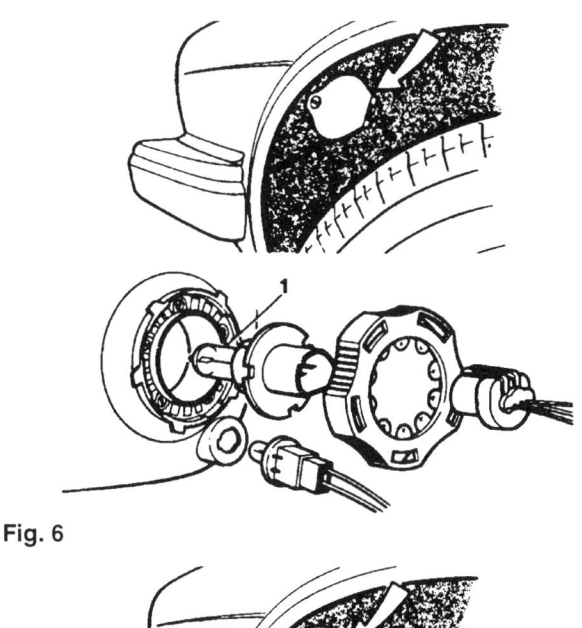

Fig. 6

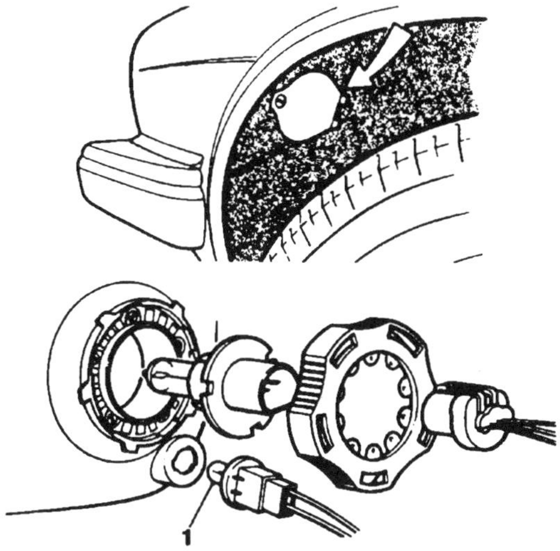

Fig. 7

FRONT PARKING LAMP BULB, RENEW

Disconnect the battery earth lead.
Reposition the front wheels to gain access to the plate located inside the wheel arch.
Turn fastener anti-clockwise and remove the plate.
Rotate the bulb holder anti-clockwise and remove from the headlamp.
Pull the capless bulb (1 Fig. 7) from the holder and replace with one of the correct type.

FOG AND REVERSE LAMP ASSEMBLY, RENEW

Disconnect the battery earth lead.

Undo and remove screws (1 Fig. 8) securing lamp assembly cover to boot lid.
Disconnect bulb wires from fog and reverse bulbs. Undo and remove nuts (2 Fig. 8) securing lamp assembly to boot lid.
Remove lamp assembly (3 Fig. 8) from boot lid.
To fit a new lamp assembly is the reversal of the removal procedure.

TAIL, STOP AND FLASHER LAMP ASSEMBLY, RENEW

Disconnect the battery earth lead.
Displace the boot rear side trim for access.
Disconnect wires from stop/tail and flasher bulbs.
Undo and remove lamp assembly to body securing nuts (1 Fig. 9).

133

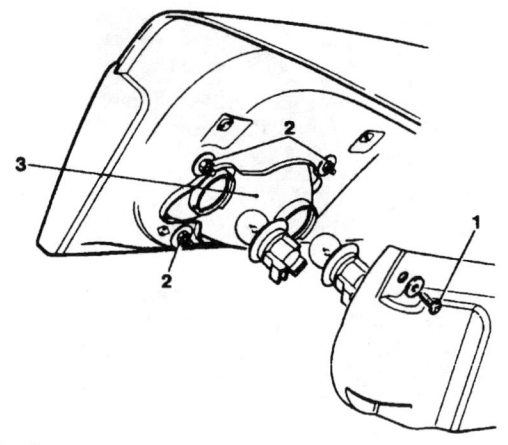

Fig. 8

Remove lamp assembly from rear wing housing.
The refitting procedure is a reversal of the removal procedure.

TAIL, STOP AND FLASHER LAMP BULB, RENEW

Displace boot rear side trim for access.
Displace bulb holder from lamp assembly (red-stop/tail; blue-flasher).
Remove bulb from holder (2 Fig. 9).

Fitting a new bulb is a reversal of the removal procedure.

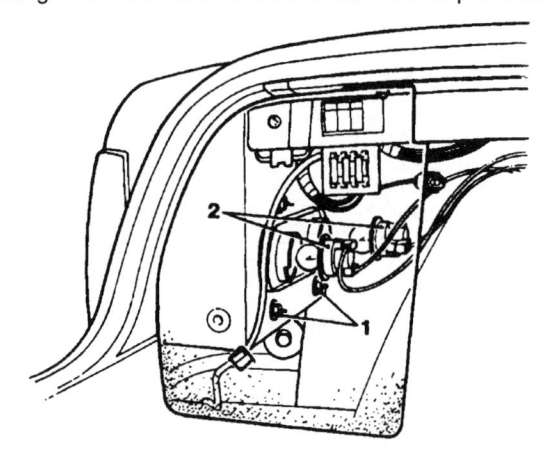

Fig. 9

FRONT FLASHER LAMP BULB, RENEW

Rotate the bulb holder anticlockwise by hand and withdraw from behind the front bumper. Remove the bulb from the holder and replace with one of the correct type. Refit the bulb holder and turn clockwise.

INSTRUMENT ILLUMINATION BULBS, RENEW

SPEEDOMETER ILLUMINATION BULB
TACHOMETER ILLUMINATION BULB
OIL AND TEMPERATURE GAUGE ILLUMINATION BULB
BATTERY AND FUEL GAUGE ILLUMINATION BULB

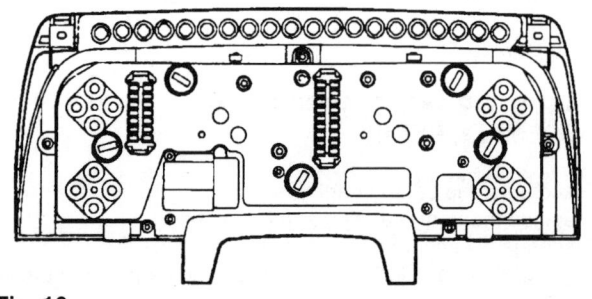

Fig. 10

Refer to Fig. 10
Remove the central finisher.
Displace and remove the side finishers.
Displace and remove the instrument panel to fascia securing screw cover plates.
Undo and remove screws securing fascia to panel.
Carefully displace the instrument panel for access.
Displace and remove the illumination bulb from the instrument panel.

Fitting a new bulb is a reversal of the removal procedure.

INSTRUMENT PACK WARNING BULBS, RENEW

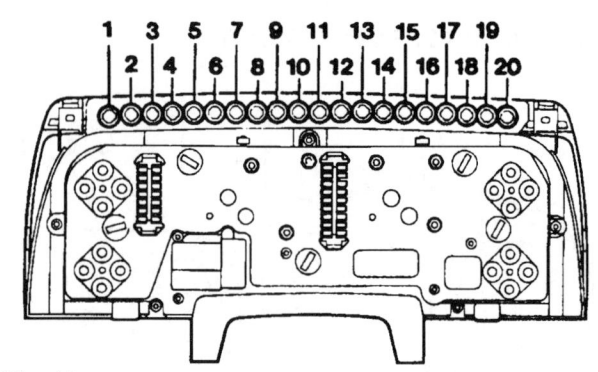

Fig. 11

Key to Fig. 11

Low washer bottle warning bulb (1)
Automatic transmission failure warning indicator bulb (3)
Flasher bulb (5 & 16)
Ignition bulb (7)
Headlamp high beam bulb (15)
Oil pressure bulb (11)
Handbrake bulb (10)
Brake warning bulb (12)
Fuel warning bulb (9)
Low coolant warning bulb (13)
Seat belt warning bulb (6)
Fog lamp warning bulb (17 & 20)
Bulb failure warning bulb (19)
Caravan warning bulb (4)
Exhaust temperature warning bulb (2)
Sport mode bulb (18)
Check engine warning bulb (8)
Anti lock brake warning bulb (14)

Disconnect the battery earth lead.
Position tilt steering column to lowest position.
Undo and remove instrument panel finishers securing screws.
Remove central finisher.
Displace and remove side finishers.
Displace and remove instrument panel to fascia securing screw cover plates.
Undo and remove panel to fascia securing screws.
Carefully displace instrument panel for access.
Displace and remove indicator bulb from instrument panel (refer to Fig. 11 for locations).

Fitting a new bulb is the reversal of the removal procedure.

VARIATIONS FOR 1993.5 MODEL YEAR - ON VEHICLES

HARNESSES

New harnesses are fitted to suit the new emissions hardware and fuel tank system.

The harnesses now have ultrasonic splices which are superior for electrical transmission. This type of splicing cannot be modified and is weakened when peeled apart. For this reason the heat-shrink sleeving should never be removed.
A new type of diagnostic connector is fitted in the boot adjacent to the 12-way fuse box.

CLIMATE CONTROL SYSTEM

An ELMOS control unit is fitted which is more reliable due to an improved design and the use of fewer components. The case is coloured brown on the XJS thereby differentiating it from the black item fitted to the saloons.
The solar sensor/alarm is modified, but functions exactly as before.
The electrical harness incorporates changes to accommodate manual recirculation, the MAX cooling function and the omission of the humidity control.

COOLING SYSTEM

An electric fan is fitted to enhance the cooling performance.
This is fitted with air flow flaps on the fan mounting assembly.

FUEL SYSTEM

The fuel tank, on vehicles fitted with the 6.0 Litre engine, has twin fuel pumps and a modified pump control module. One pump operates at engine speeds up to 2840 RPM, at which point the second pump switches in.
New in-tank electrical connecting leads are specified for interconnection with the new four-way header connector.

BATTERY

The battery is a 72 Ah item produced by Varta.

STARTER MOTOR

A Magneti Marelli 1.8 kW starter motor is fitted to give better cold start cranking.

GENERATOR

A NipponDenso 120 Amp generator is fitted to improve the balance of electrical loads and to provide a higher output at lower engine speed.
Electrical connection to the generator is by one plug and one eyelet.
The load dump module is now no longer necessary and has been omitted from the specification.

ENGINE MANAGEMENT SYSTEM

A revised Lucas Marelli system with new calibration and modifications to the hardware and harnesses has been fitted.

TRANSMISSION

The software on the 4L80E automatic transmission is modified to match the characteristics of the 6.0 litre engine.
A mode switch is fitted to all vehicles.

SUN VISOR MOUNTED ILLUMINATED VANITY MIRRORS

Sun visors, which incorporate an illuminated vanity mirror, are fitted to the driver's side and passenger's side. The sun-visors also incorporate interior courtesy lights.

SECURITY SYSTEM

A security system, which currently complies with most European and overseas legislation, is offered as a factory fit option. The remote arming/disarming control unit is a radio transmitter or infra-red controller depending on market legislation. The features offered by the new system are:

> Radio frequency arming/disarming which also controls the central door locking.
> Entry sensing on doors and boot. Ignition anti-tamper.
> Start inhibit.
> Headlamp convenience (illumination for twenty-five seconds).
> Starter disable.
> Arm and disarm audible and visual indication.
> Warn away potential thieves (reduced sound alarm).
> Remote panic alarm.
> Escalating siren response.
> Headlamp warning flash.
> Error tone.
> Security OFF (Valet) switch.

There are options available which may be initialized on the system, by the dealer, using JDS.

IN-CAR ENTERTAINMENT

An additional 3 inch co-axial, loudspeaker is added to the top of each rear quarter trim panel on the two-seater convertible. This improves the quality of the mid-range sound and provides a better sound field for the front seat occupants.

Changes to the body manufacture now provide the option to supply the Compact Disc Autochanger as a production line fitted unit or as a dealer fitted unit.

The link lead which was fitted to the rear of the radio has been omitted. A telephone muting facility has been added to the radio which operates automatically when the radio telephone is in use.
Head cleaning tapes, type Allsop-3, are now accepted by the tape player.

Changes to the Radio Data System (RDS) software are made as follows:

When the RDS data is lost and no alternative frequency is available, the radio will remain on-station and display the station frequency only; RDS will remain on in readiness for the signal to recover.
When TP is selected, the radio will check the current station for TP. If it is not a TP station the radio will search for a TP station, and while doing so, the display will show TRAFFIC, flashing on and off, during the search.

The function of the RDS and TP switches has been reversed compared to the previous radio. The RDS switch is now a momentary on/off switch and the TP switch operates after being pressed for two seconds, with a beep to confirm acceptance of on/off. The volume minimum pre-set level for a traffic announcement has been reduced because the original volume level was considered to be too loud.

MULTI-FUNCTION UNIT

The multi-function unit is located under the fascia, mounted on a bracket behind the glovebox. The unit is new and combines the functions previously performed by six separate modules, plus two new functions.

The separate modules which have been replaced by the multi-function unit are:

> Seat belt warning module
> Lights-on buffer module
> Over-speed module (Saudi Arabia only)
> Interior lamp delay module
> Heated rear window timer
> Bulb check unit

The two new functions performed by the multi-function unit are:

Ignition key-in audible warning, and
Security system interface.

A description of each function follows:

SEAT BELT WARNING

With the ignition switched on, the seat belt visual warning located in the instrument pack will be activated for approximately six seconds. With the ignition switched on and the driver's seat belt unlatched, an audible warning will sound for approximately six seconds; there is no audible warning if the seat belt is latched before the ignition is switched on. Both warning devices will reset (and cancel) when the ignition is switched off.

LIGHTS-ON WARNING

An audible warning will sound if the following three conditions exist - the light switch is in the side lamp or headlamp position, the ignition key is removed and the driver's door is opened. The alarm can be cancelled by moving the light switch to the off position or closing the driver's door or inserting the key in the ignition switch. It should be noted that when the driver's door is open and the key is in the ignition switch, then the key-in alarm will also sound.

OVER-SPEED WARNING (SAUDI ARABIA ONLY)

An audible warning will be given when the road speed is between 120 and 130 km / h.

INTERIOR LAMP DELAY

The interior lamps will operate immediately if any door is opened. When leaving the vehicle parked, the interior lamps will remain on for approximately ten seconds after the last door has been closed. On entering the vehicle, and then closing both doors, the lamps will turn off either after approximately ten seconds or when the ignition is switched on. If the door(s) are left open with the ignition key removed, the lamps will go off after approximately two minutes (to prevent draining the battery) and will reset only after the doors have been closed for approximately ten seconds.

HEATED REAR WINDOW TIMER

The heated rear window is activated, by the multi-function unit and a relay, for approximately ten minutes when the heated rear screen switch is operated. It can be switched off before the time-out period if required, by operating the switch again. The timing function will also switch off and reset if the ignition is switched off.

BULB CHECK

The following bulbs; park brake, brake fluid and catalyst overheat (Japan only) are activated for approximately 2.5 seconds when the ignition is switched on, to indicate that the bulbs are functional.

IGNITION KEY-IN WARNING

When the ignition is in the off position but the key is left in the ignition switch, the alarm will sound when the driver's door is opened. The warning can be cancelled by removing the ignition key or closing the driver's door or switching on the ignition.

SECURITY SYSTEM INTERFACE

An output from the multi-function unit to the security system is provided which reflects the state (on / off) of the interior lamp door switches.

MULTI-FUNCTION UNIT SELF DIAGNOSTICS

The Multi-Function Unit has its own diagnostic mode which is capable of identifying faults within the module, open circuit and short circuit faults in the wiring and faults in the components of the vehicle systems which supply inputs to the module and which receive signals from the module.

INPUT DIAGNOSTICS

Before starting the diagnostic mode, sit in the front of the car with the doors closed and the seat belt unlatched. Under these conditions, the audible chime on the multi-function unit should not operate in the diagnostic mode. The input diagnostic mode is activated as follows:

1. Ensure that the sidelights are switched off.
2. Press and hold in the heated rear window switch.
3. Switch on the ignition switch whilst still holding in the heated rear screen switch.
4. Release the heated rear window switch.

If the alarm chime does not sound, this indicates that there is no fault.
The following checks should now be performed:

> Latch the seat belt; the chime should sound. Unlatch the seat belt and the chime should be silenced.

> Open the driver's door; the chime should sound. Close the driver's door and the chime should be silenced.

> Open the passenger door; the chime should sound. Close the passenger door and the chime should be silenced.

Note that the vehicle road speed sensor is also being tested and could be causing a fault indication. Where this is suspected, disconnect it and look for a change in the above tests. Due to the method involved in selecting the diagnostics mode, the integrity of the following function/components is tested by default:

> key-in ignition switch,
> ignition input to the multi-function unit and
> the heated rear window switch.

OUTPUT DIAGNOSTICS

If not already in the input diagnostic mode, repeat steps 1 to 4 described above.
Switch on the sidelights to select the output diagnosis mode; all inputs will now be inhibited. As each output is selected, it should operate its circuit load and the chime will remain silent. If it fails to operate its particular circuit due to a short circuit load, open circuit etc., then the chime will sound.
Monitoring of the output functions is achieved by cycling through each, using the heated rear window switch:

All outputs inactive, but output diagnosis mode enabled. Press and release the heated rear window (HRW) switch to select the heated rear window circuit. Press and release the HRW switch to de-select the heated rear window circuit.

Press and release the HRW switch to select the interior lamp circuit. Press and release the HRW switch to de-select the interior lamp circuit.

Press and release the HRW switch to select the seat belt visual warning circuit. Press and release the HRW switch to de-select the seat belt visual warning circuit.

Press and release the HRW switch to select the bulb check circuit. This will test the circuits and warning lamps associated with low wash, park brake, brake fluid and catalyst overheat (Japan). Press and release the HRW switch to de-select the bulb check circuit.

Switch off the sidelights to de-select the output diagnostics mode. Switch off the ignition.

The multi-function unit will reset and operate normally next time the ignition is switched on.

LAMPS

INTRODUCTION

New lamps and reflectors are specified, styled to match the new design of the front and rear bumper assemblies. Access to the front and rear side marker lamps/reflectors and rear reflectors involves the removal of parts of the bumper. Removal of the bumper components is described in the Body section.

FRONT DIRECTION INDICATOR LAMPS

These are retained by two screws and are similar to those on the 1992 model year vehicles.

FRONT DIRECTION INDICATOR LAMP, BULB CHANGE

Remove the two screws which secure the lamp to the bumper cover. Remove the lamp. Release the bulb holder and remove the bulb.

Refit the bulb and the bulb holder to the lamp. Position the lamp to the bumper cover and fit the retaining screws. Check the operation of the lamp.

FRONT DIRECTION INDICATOR LAMP, RENEW

Remove the two screws which secure the lamp to the bumper cover. Remove the lamp. Disconnect the electrical connector from the bulb holder.

Connect the electrical connector to the bulb holder. Position the lamp to the bumper cover and fit the retaining screws. Check the operation of the lamp.

SIDE MARKER LAMPS / REFLECTORS (FRONT)

These lamps/reflectors mount directly to the vehicle body and not to the bumpers. However, the side marker lamp bulbs may be reached through an access panel in the front wheelarch liner and the rear side marker lamp bulbs may be reached from inside the luggage compartment.

FRONT SIDE MARKER LAMP ASSEMBLY, RENEW

Support the front of the vehicle and remove the front road wheel(s).
Referring to the Body section remove the (plated) front bumper blade assembly.
Remove the plastic 'scrivets' which secure the top of the bumper cover to the body. Remove the three body-to-bumper cover fixings from the relevant side of the car, and carefully reposition the bumper cover to give access to the marker lamp (if both lamps are to be renewed simultaneously, the bumper cover will have to be completely removed).

Reaching through the wheelarch access panel, disconnect the harness connector. Release the tangs which secure the lamp to the body, and remove the lamp.

Fit the new lamp, ensuring that the locking tangs have fully located. Reaching through the wheel arch access panel, connect the harness connector. Check that the lamp operates.

Reposition the bumper cover, and referring to the Body section, secure the cover and refit the blade, using new scrivets.

Fit the front road wheels and lower the vehicle.

REAR SIDE MARKER LAMP / REFLECTOR ASSEMBLY, RENEW

Support the rear of the vehicle and remove the rear road wheel(s).

Referring to the Body section, remove the (plated) rear quarter bumper blade assembly from the relevant side of the vehicle. Remove the three body-to-bumper cover fixings at the rear of the wheel arch.

Reposition the bumper cover to give access to the marker lamp.

From inside the boot, move the trim panel and disconnect the harness connector. Release the tangs which secure the lamp to the body, and remove the lamp from beneath the bumper cover.

Fit the new lamp to the body, ensuring that the locking tangs have fully located. From inside the boot, connect the harness connector and reposition the trim panel. Check that the lamp operates.

Referring to the Body section, secure the bumper cover and refit the rear quarter bumper blade assembly.

Fit the rear road wheel(s) and lower the vehicle.

REAR REFLECTOR LENS - RENEW

Support the rear of the vehicle and remove the rear road wheels.

Referring to the Body section, remove the bumper cover/blade assembly.

Release the spire clips which secure the reflector(s) to the bumper cover. Remove the reflector(s).

Fit the reflector(s) to the bumper cover and secure with new spire clips.

Referring to the Body section, refit the bumper cover/blade assembly.

Fit the rear road wheels and lower the vehicle.

CHAPTER 17

INSTRUMENTS - 1992 MODEL YEAR ON

DESCRIPTION

The instrument pack (Fig. 1) features the traditional layout of two large main dials with four small supplementary gauges. These are conventional analogue gauges comprising 90 degree movements for the four minor gauges and 270 degree movements for the two major gauges.

The trip meter is integrated into the tachometer and the odometer is included in the speedometer. The pulse signal required to operate the speedometer is controlled by a speed sensor situated in the differential unit. The engine speed signal received by the tachometer is derived from the ignition coil negative terminal. The voltage wave form at this point can reach as much as 400 volts when a spark is generated and it is desirable to suppress this voltage before allowing it into the wiring harness.

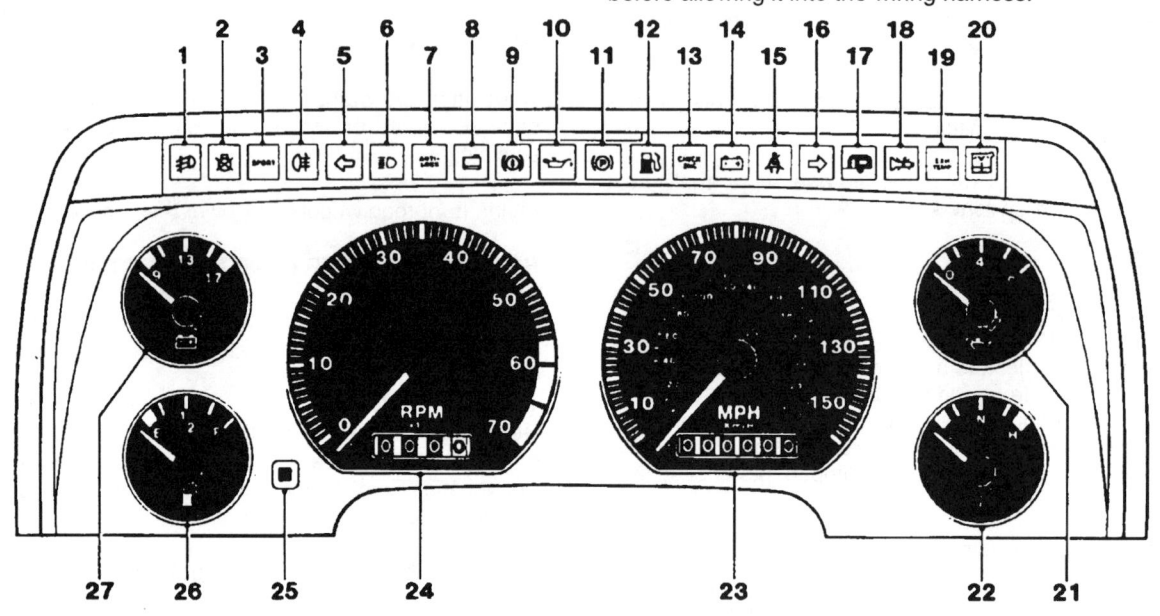

Fig. 1

Key to Fig. 1
(not all features are fitted to all models/markets)

1. Front fog lamps
2. Bulb failure
3. Sport mode
4. Rear fog guard
5. Left direction indicator
6. Headlamp main beam
7. Anti-lock braking system
8. Low coolant level
9. Low brake fluid/ABS low hydraulic pressure
10. Low oil pressure
11. Handbrake
12. Low fuel
13. Check engine
14. Ignition
15. Seat belt
16. Right direction indicator
17. Caravan DI indicator
18. Automatic transmission failure
19. Exhaust temperature
20. Low windscreen wash reservoir level
21. Oil pressure
22. Coolant temperature
23. Speedometer/odometer
24. Tachometer/trip meter
25. Trip reset
26. Fuel level
27. Battery condition

There is an array of 'secret-till-lit' warning lights which are situated in a row at the top of the instrument pack (Fig. 1).

INSTRUMENT PACK, TESTING

Fuel Gauge Tank Unit, Convertible, Calibration Limits

LEVEL	RESISTANCE (Ohms)	FUEL REMAINING (Litres)
Empty	240-250	6
Warning Light	185-215	12
Quarter Full	102-104	24.5
Half Full	68-70	43
Three Quarters Full	44-46	61.5
Full	16-18	80

Fuel Gauge Tank Unit, Coupe, Calibration Limits

LEVEL	RESISTANCE (Ohms)	FUEL REMAINING (Litres)
Empty	240-250	6
Warning Light	185-215	12
Quarter Full	102-104	26
Half Full	68-70	46
Three Quarters Full	44-46	66
Full	16-18	86

INSTRUMENT PACK CONNECTIONS (FIG. 2)

SOCKET A		SOCKET B	
PIN	CIRCUIT INPUT	PIN	CIRCUIT INPUT
01	Brake systems	01	Panel lamps (5 off), ground
02	Main beam	02	Panel lamps (5 off), positive
03	LH turn	03	Oil pressure gauge
04	Rear fog	04	Not used
05	Auto transmission sports mode	05	Supply for exhaust temperature/speed warning, oil pressure gauge, washer level
06	Bulb failure	06	Washer level
07	Front fog	07	Exhaust temperature (Japan)/Speed warning (Saudi Arabia)
08	Tachometer	08	Gear box fail
09	Fuel gauge	09	Caravan/trailer
10	Speedometer	10	RH turn
11	Check engine	11	Seat belt
12	Low fuel tell-tale	12	Ignition
13	Low fuel tell-tale	13	Supply for analogue instruments & A14 to A17, B11, B12
14	Park brake	14	Temperature gauge
15	Oil pressure tell-tale		
16	Brake systems		
17	Low coolant		
18	Anti-lock brakes		

Fig. 2

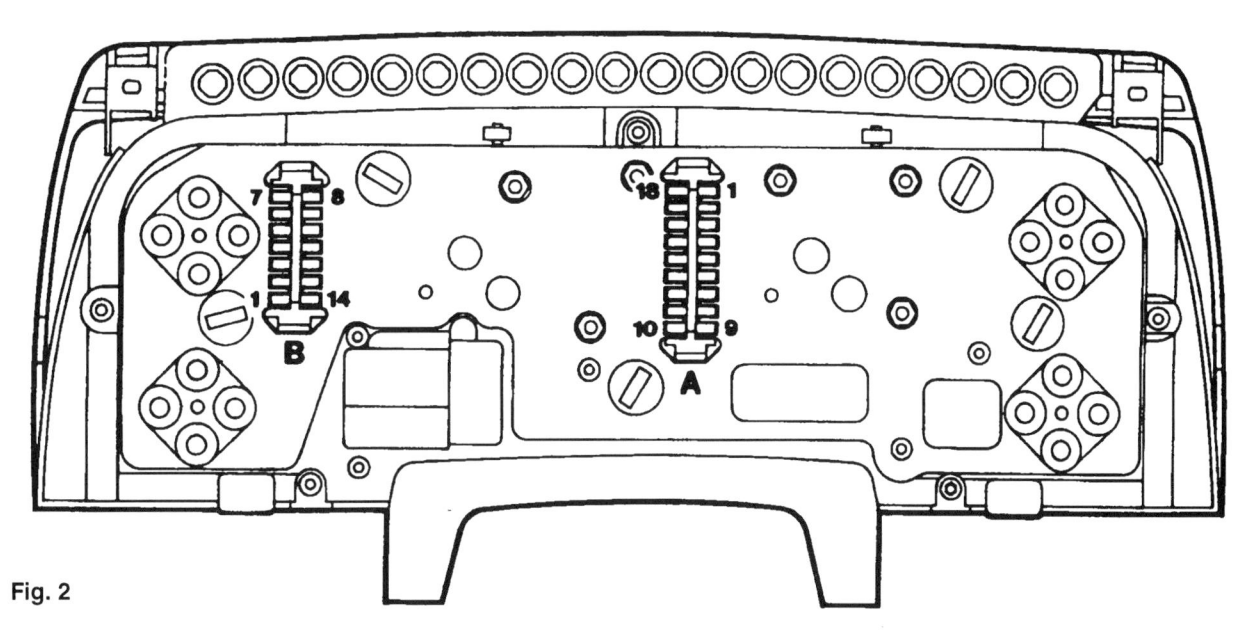

TRIP COMPUTER, DESCRIPTION

When the ignition is switched ON, the LCD back lights illuminate and the trip computer defaults to the time of day. The time button is dual function; first press gives the time of day, a subsequent press causes the elapsed time since reset to be displayed, with leading zero suppressed. After five seconds the elapsed time reverts to the time of day.

The trip computer provides information on vehicle speed, fuel usage and distance travelled, all of which are calculated by a microprocessor. It computes fuel consumption, both average and 'at the moment' usage, fuel used on a journey or period; distance travelled, average speed and time elapsed since the start of the journey or over a period. The information may be displayed in either litres and kilometres or in miles and gallons.

The unit also provides a warning of fuel failure. In the event of an engine management fault occurring, the 'check engine' warning light and the words 'check engine' are permanently displayed on the trip computer until the engine is switched off. When the fault has been signalled, the likely area of malfunction can be indicated when the vehicle is stationary. Switch off the engine, wait at least five seconds turn the ignition switch to position II (do not start the engine). The relevant failure code FF11 to FF99 is displayed.

INSTRUMENT PANEL MODULE, RENEW

Disconnect the battery earth cable.
Position the tilt steering column to the lowest position.
Remove the instrument panel finishers' securing screws and remove the central and side finishers (1 Fig. 3).

Fig. 3

Remove the instrument panel to fascia securing screw cover plates and remove the screws securing the instrument panel to the fascia (2 Fig. 3).
Ease the instrument panel forward for access and disconnect the multi-plugs.
Remove the instrument panel.

Note: To minimise the risk of damage and contamination, all repairs conducted on the Instrument Pack should be performed in a non-static dust free environment.

Position the new instrument panel adjacent to the mounting position and connect the multi-plugs.
Fully seat the instrument panel.
Fit and tighten the instrument panel to the fascia and secure the screws.
Fit the securing screw finishers.
Position and fit the side and central finishers and secure the screws.
Reposition the steering column to the original position.
Reconnect the battery earth cable.

INSTRUMENT PRINTED CIRCUIT, RENEW

Disconnect the battery earth lead. Position the tilt steering column to the lowest position. Remove the instrument panel finishers' securing screws and remove the central and side finishers (1 Fig. 3). Remove the instrument panel to fascia securing screw cover plates and remove the screws securing the instrument panel to the fascia (2 Fig. 3).

Note: To minimise the risk of damage and contamination, all repairs conducted on the Instrument Pack should be performed in a non-static dust free environment.

Place a protective cover on a workbench and position the instrument panel down onto the cover.
Remove the panel illumination bulbs (1 Fig. 4).

Remove the printed circuit securing nut rubber covers.
Remove the printed circuit to instrument panel securing nuts and remove the printed circuit from the instrument panel (2 Fig. 4).
Carefully fit the new printed circuit to the instrument panel and secure the nuts.
Refit the printed circuit securing nut rubber covers.

Note: Ensure the printed circuits are not torn or deformed.

Fit the panel illumination bulbs.

Refit the instrument panel assembly to the vehicle.
Reconnect the battery earth lead.

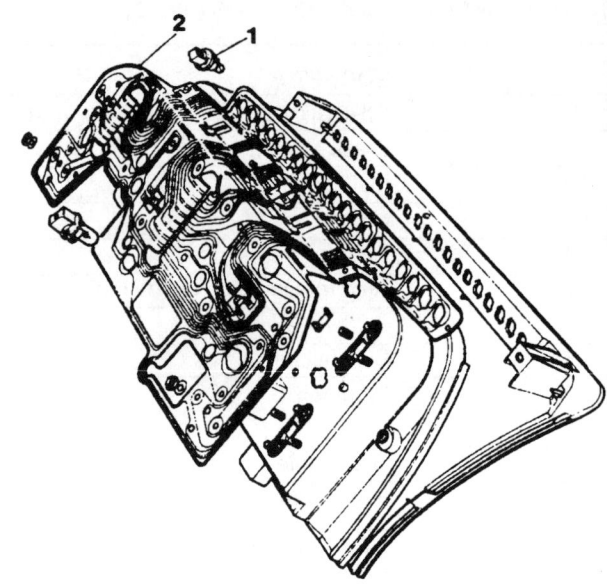

Fig. 4
SPEEDOMETER, RENEW

Disconnect the battery earth cable.
Position the tilt steering column to the lowest position.
Remove the instrument panel finishers' securing screws and remove the central and side finishers (1 Fig. 3).
Remove the instrument panel to fascia securing screw cover plates and remove the screws securing the instrument panel to the fascia (2 Fig. 3).

Note: To minimise the risk of damage and contamination, all repairs conducted on the Instrument Pack should be performed in a non-static dust free environment.

Place a protective cover on a workbench and position the instrument panel lens down onto the cover.
Remove the gauge illumination bulbs (1 Fig. 5).

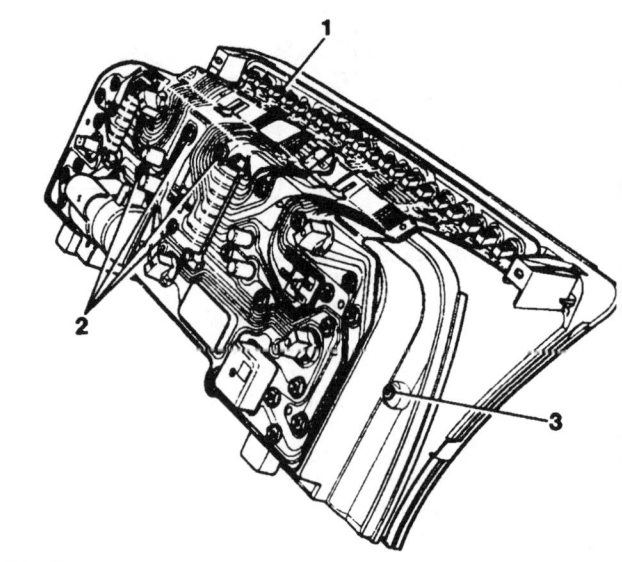

Fig. 5

Remove the printed circuit to instrument panel securing nuts (2 Fig. 5) and remove the printed circuit from the locating lugs.
Remove the screws securing the rear cover to the lens/veneer panel (3 Fig. 5).
Invert the assembly and remove complete with the printed circuit from the rear cover.
Remove the tachometer from the rear cover housing (1 Fig. 6).

Fig. 6

Fig. 7

Fig. 8

Fig. 9

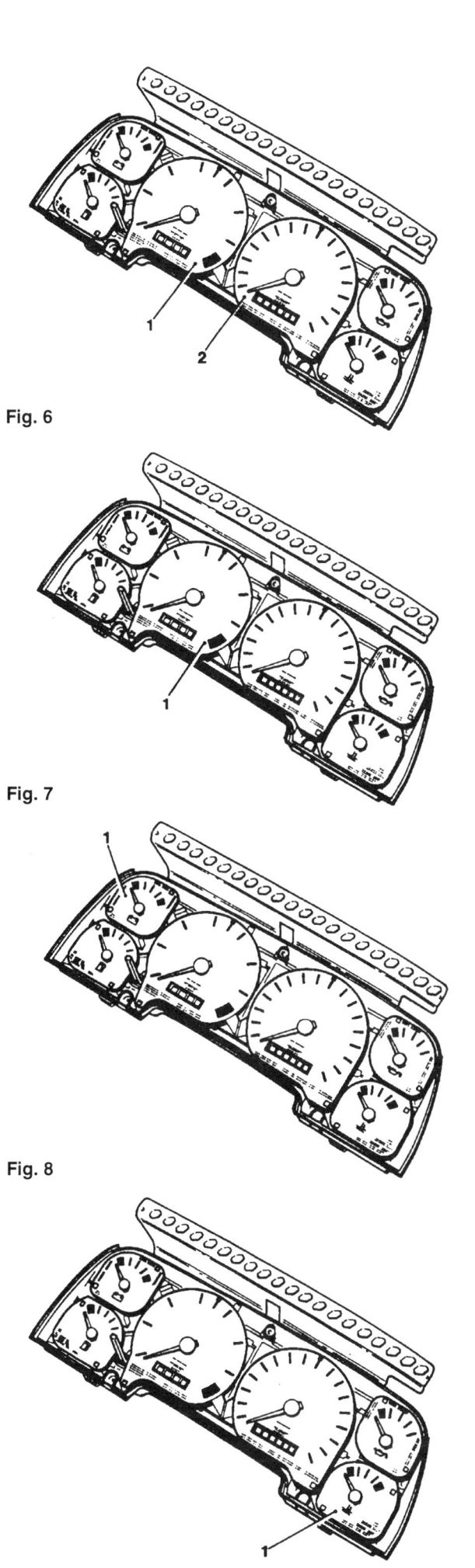

Remove the speedometer and odometer assembly from the rear cover housing (2 Fig. 6).
Disconnect the odometer to speedometer link lead and place the speedometer to one side.

Place the new speedometer to the front and connect the odometer to speedometer link lead connector plug.
Fit the speedometer and odometer assembly to the rear cover housing.
Fit the new tachometer to the rear cover housing.
Reposition the printed circuit around the rear cover and fit the lens/assembly to the rear cover housing.
Fit the rear cover to the lens/veneer panel and secure the screws.
Fit the printed circuit to the rear housing and secure the nuts.
Fit the gauge illumination bulbs.
Refit the instrument panel.
Reconnect the battery earth cable.

TACHOMETER, RENEW

The procedure is basically as for Speedometer. Refer to Fig. 7

BATTERY CONDITION INDICATOR, RENEW

The procedure is basically as for Speedometer. Refer to Fig. 8

COOLANT TEMPERATURE GAUGE, RENEW

The procedure is basically as for Speedometer. Refer to Fig. 9

FUEL GAUGE, RENEW

The procedure is basically as for Speedometer. Refer to Fig. 10

OIL PRESSURE GAUGE, RENEW

The procedure is basically as for Speedometer. Refer to Fig. 11

COOLANT TEMPERATURE TRANSMITTER, RENEW

WARNING:

DO NOT REMOVE THE CAP AT THE REMOTE HEADER TANK UNLESS THE ENGINE IS COLD.

Open the bonnet. Disconnect the battery earth lead.

Remove the radiator pressure cap to depressurize the system.
Disconnect the transmitter feed wire.
Unscrew and remove the transmitter (1 Fig. 12).
Remove and discard the seal/washer.

Fit a new seal/washer to the replacement transmitter.
Fit and tighten the transmitter.
Reconnect the transmitter feed wire.
Check/top up the coolant. Refit the pressure cap.

Note: Always top-up with the recommended strength of antifreeze, never with water only.

Reconnect the battery earth lead.
Close the bonnet.

141

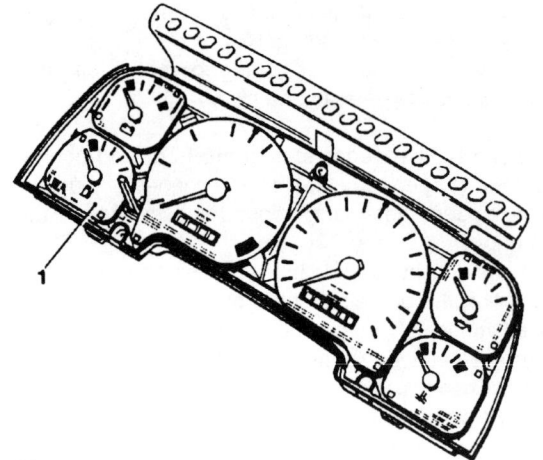

Fig. 10

Fig. 11

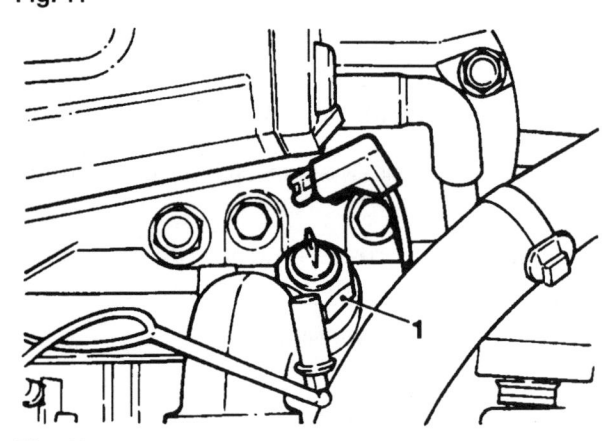

Fig. 12

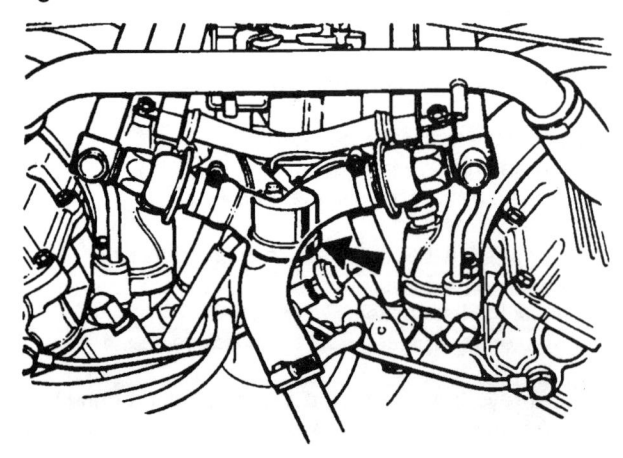

Fig. 13

Torque Figures

Transmitter to engine   49 - 54 Nm

Oils/Sealants/Lubricants

'JAGUAR UNIVERSAL' or a PHOSPHATE FREE type to B.S. 6580 antifreeze. 50% down to -36°C (-33°f); 55% down to -40°C (-40°F); 33% down to -19°C (-2.2°F).

OIL PRESSURE TRANSMITTER, RENEW

Open the bonnet.
Disconnect the battery earth lead.
Disconnect the rubber boot and the transmitter feed wire.
Unscrew and remove the transmitter (Fig 13).

Fit and tighten the new transmitter.
Reconnect the transmitter feed wire and the rubber boot.
Reconnect the battery earth lead.
Close the bonnet.

Torque Figures

Transmitter to engine   20 - 27 Nm

OIL PRESSURE WARNING LIGHT SWITCH, RENEW

Open the bonnet.
Disconnect the battery earth lead.
Disconnect the switch feed wire.
Unscrew and remove the switch (Fig 14).

Fit and tighten the new switch. Reconnect the switch feed wire.
Reconnect the battery earth lead.
Close the bonnet.

Torque Figures

Switch to engine   20 - 27 Nm

FUEL GAUGE TANK UNIT, RENEW

```
                    WARNING:

  FUEL IS HIGHLY FLAMMABLE AND GREAT CARE
  MUST BE TAKEN WHEN DRAINING THE FUEL TANK.
  NO SMOKING SIGNS MUST BE DISPLAYED NEAR
  THE WORKING AREA. DISCONNECT THE BATTERY
     LEADS BEFORE DRAINING THE TANK.
     KEEP NEARBY A CARBON DIOXIDE FIRE
   EXTINGUISHER AND DRY SAND TO SOAK
            UP ANY SPILLAGE.
    ENSURE THE AREA IS WELL VENTILATED.
     THE FUEL MUST BE DRAINED INTO AN
   AUTHORISED EXPLOSION PROOF CONTAINER.
```

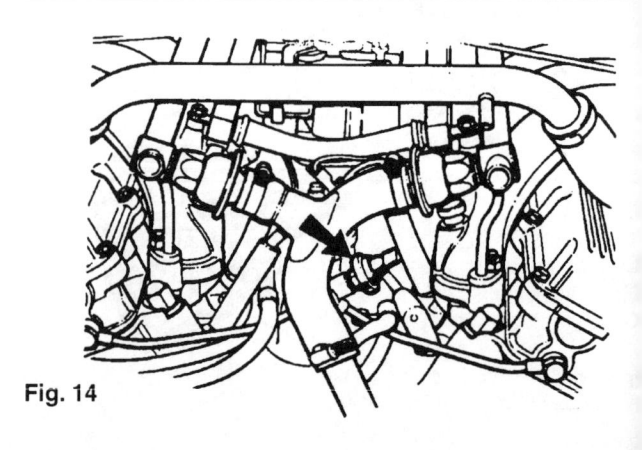

Fig. 14

Open the boot and disconnect the battery.
Drain the fuel from the tank using the approved equipment.
Remove the boot right and left hand side liners.
Remove the spare wheel trim cover and remove the spare wheel.
Remove the boot seal from the front body flange and remove the boot front liner.
Disconnect the harness to the tank unit wires.
Using the service tool 18G 1001 (Fig. 15), remove the tank unit securing ring then the tank unit assembly.

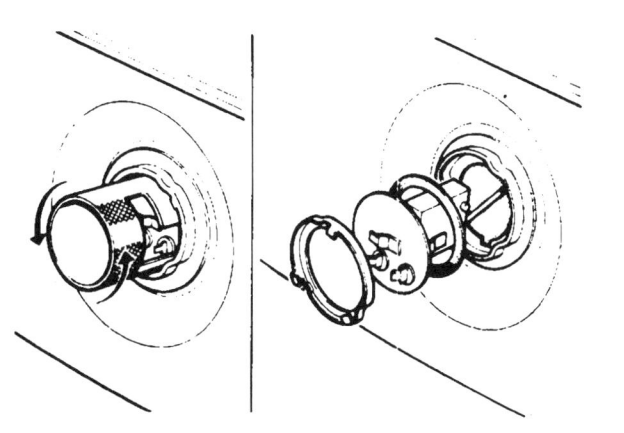

Fig. 15

Remove and discard the 'O' ring seal and clean the tank seal face.

Fit a new seal to the tank and carefully fit the tank unit, secure with the retaining ring.
Reconnect the harness to the tank unit wires.
Fit the front liner and reposition the trim over the front flange. Fully seat the boot seal to the flange. Refit the spare wheel, secure the nut and replace the wheel cover.
Fit the left and right hand front side liners.
Refill the fuel tank; check for leaks.
Close the boot.

Service Tools

18G 1001 Locking ring spanner

SPEED SENSOR AIR GAP, CHECK/ADJUST

Drive the vehicle on to a ramp. Raise the ramp.
Remove the sensor (Fig. 16) from the final drive backplate.

Fig. 16

Displace and remove the shims from the sensor.
Remove and discard the 'O' ring seal.
Clean the backplate face.

Position the car on the ramp to align one rotor tooth directly in front of the sensor mounting hole. Fit the sensor to the

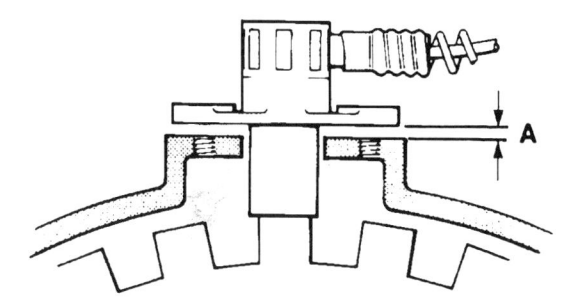

Fig. 17

backplate (without the 'O' ring), ensuring that the sensor bottoms onto the rotor tooth.
Measure and note the gap (A Fig. 17) between the sensor securing flange and the backplate.

Remove the sensor.
Fit a new 'O' ring.
Fit and align the correct size/number of shims to the sensor to give 0.010 - 0.020 in. clearance between the sensor and the rotor teeth.
Fit the sensor assembly to the backplate.
Fit and tighten the securing bolts.
Lower the ramp.

SPEED SENSOR, RENEW

Drive the vehicle on to a ramp. Open the boot.
Remove the spare wheel for access.
Remove the LH boot liner.
Disconnect the speed sensor multi plug, located under the LH rear wing (black PMHD).
Displace the harness and body grommet.
Reposition the harness from the boot into the axle area.
Raise the ramp.
Remove the sensor from the final drive backplate (Fig. 16).
Displace and remove the sensor/harness assembly.
Displace and remove the shims from the sensor. Clean the backplate face.

Remove the protection cap from the new sensor. Displace and remove the 'O' ring.
Position the car on the ramp to align one rotor tooth directly in front of the sensor mounting hole.
Fit the sensor to the backplate, ensuring that the sensor bottoms onto the rotor tooth.
Measure and note the gap (A Fig. 17) between the sensor securing flange and the backplate.
Remove the sensor. Fit a new O ring.
Fit and align the correct size/number of shims to the sensor to give 0.010 - 0.020 in. clearance between the sensor and the rotor teeth.

Fit the sensor assembly to the backplate.
Fit and tighten the securing bolts.
Reposition the harness into the boot area.
Seat the body grommet.
Lower the ramp.
Position the harness and reconnect the multi plug.
Fit the liner and spare wheel.
Close the boot.

TRIP COMPUTER, RENEW

Disconnect the battery earth lead.
Remove the combined interior light switch.
Remove the combined heated backlight and hazard warning switch.
Remove the computer to fascia securing bolts (1 Fig. 18) and carefully displace the computer from the veneer panel.

Fig. 18

Disconnect the computer from the harness multi-plug, then connect the new computer to the multi-plug.

Fit the new computer to the veneer panel and tighten the fascia securing bolts.
Refit the combined heated backlight and hazard warning switch assembly.
Refit the combined interior light switch assembly.
Reconnect the battery earth lead.

CIRCUITS - 1990 MODEL YEAR ON

WIRING COLOUR CODE

N	BROWN	Y	YELLOW
B	BLACK	O	ORANGE
W	WHITE	S	SLATE
K	PINK	L	LIGHT
G	GREEN	U	BLUE
R	RED	P	PURPLE

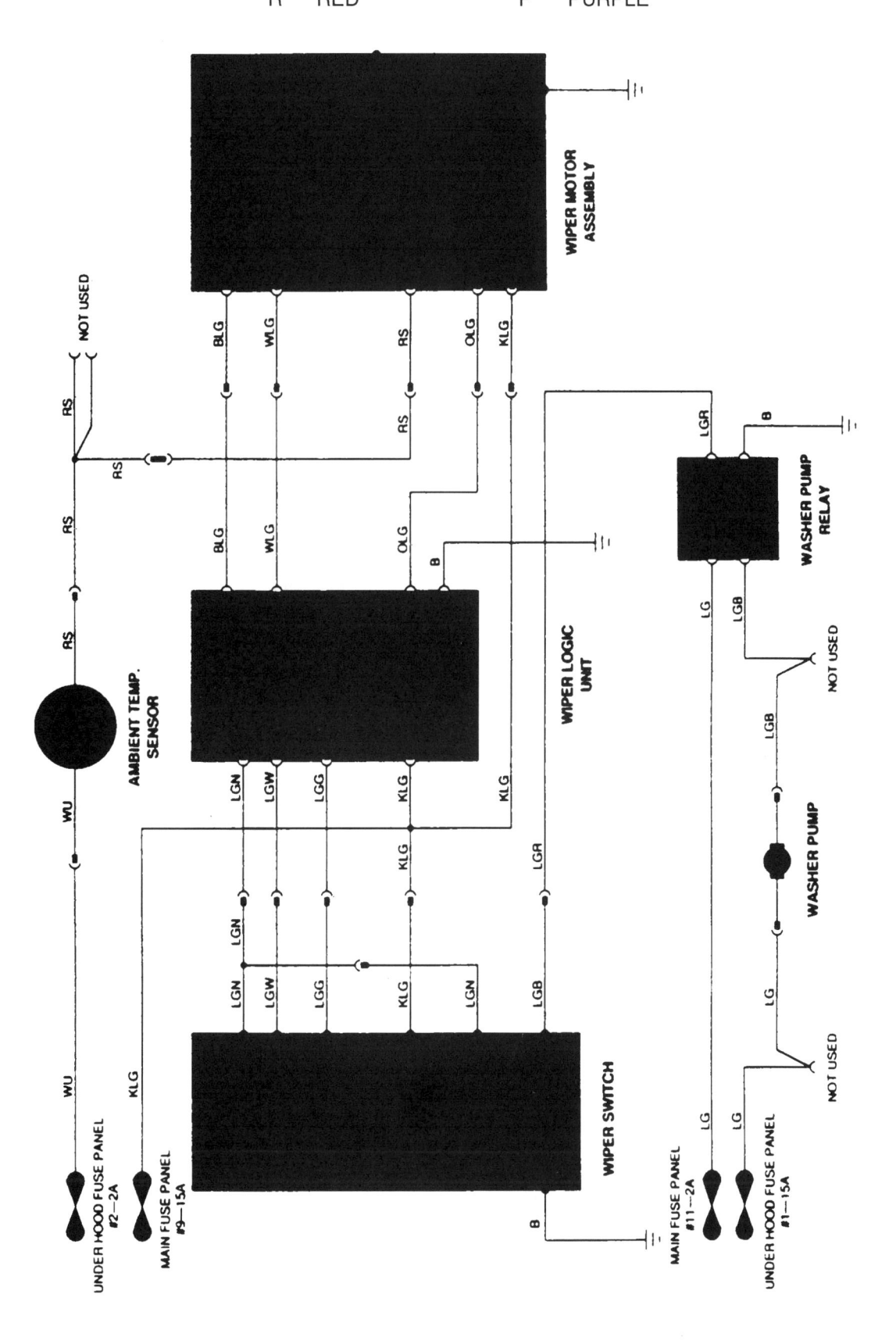

WINDSHIELD WIPER AND WASHERS

145

ANTI-LOCK BRAKING

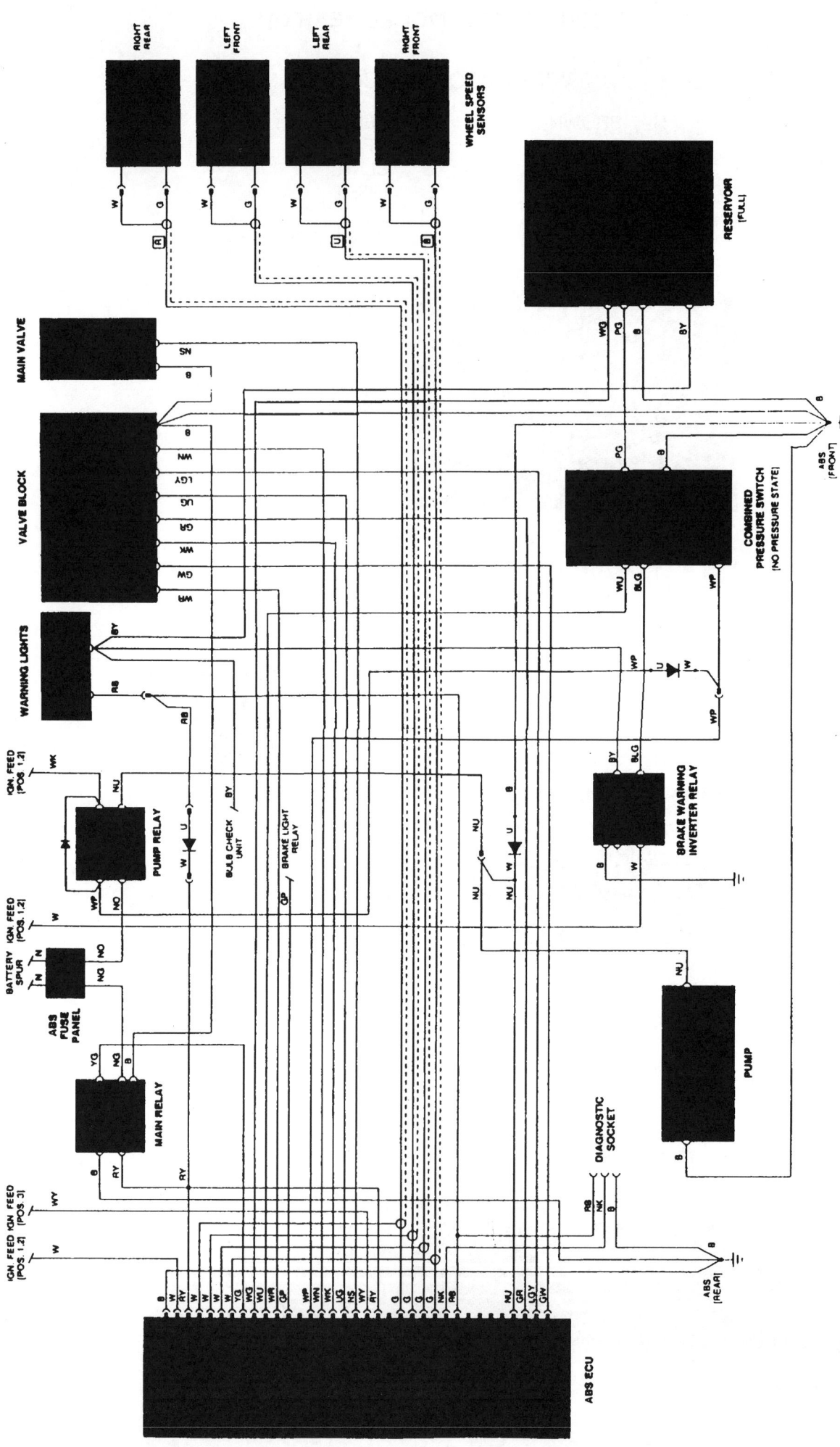

CLIMATE CONTROL SYSTEM

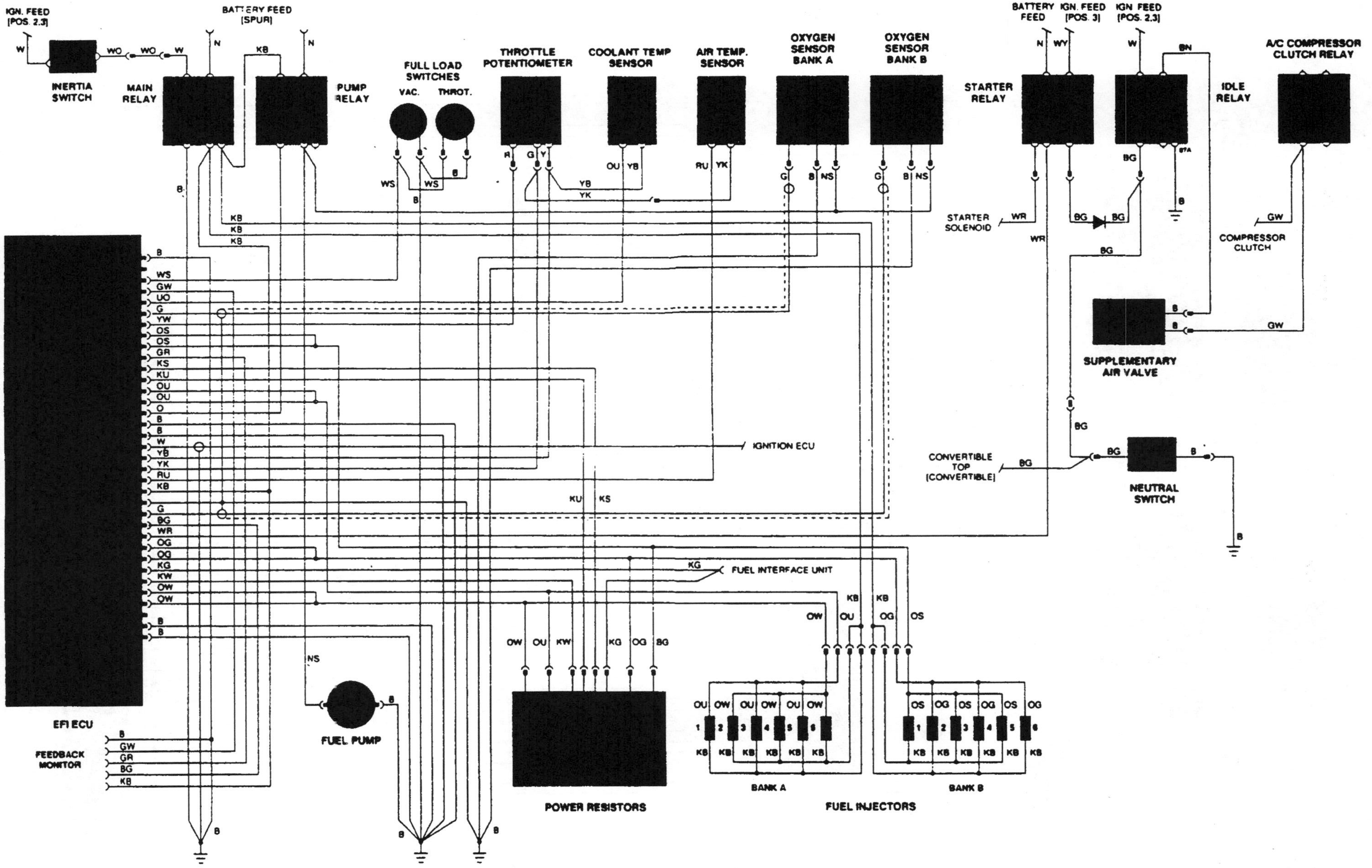

EFI AND EMMISIONS CONTROL

CIRCUITS - 1992 MODEL YEAR ON

> Please note, that due to the ever increasing sophistication of these vehicles, it has been necessary to cover Electrical Circuits from 1992 on in a separate publication. A limited selection of circuits are printed here, but full coverage will only be available by reference to the 1992 Model Year-on Service Manual (Part Number JJM 10 04 06/02) and the relevant Model Year's 'Electrical Guide'.

CIRCUIT DIAGRAMS AND HARNESSES

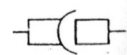

ENGINE MANAGEMENT (5.3 LITRE)

KEY TO CIRCUIT COMPONENTS

1	TDC sensor IH12
2	Flywheel sensor IH3
3	Idle switch IH16
4	Air temperature switch IH10
5	Coolant temperature sensor IH9
6	Connector to engine management ECU IH1
7	Engine management E.C.U.
8	Fuel rail temperature switch
9	LH power module IH13
10	RH power module IH14
11	Coil forward IH7
12	Coil rear IH8
13	Distributor
14	Resistor EMR1
15	V12 catalyst engine only
16	Supplementary air valve CS6
17	Anti–stall relay AS5
18	V12 non–catalyst engine only
19	Extra air valve relay XA5
20	Extra air valve relay XA4
21	15 minute timer FC1
22	Supplementary air valve XA6
23	3–way solenoid operated vac valve CL4
24	Thermal switch

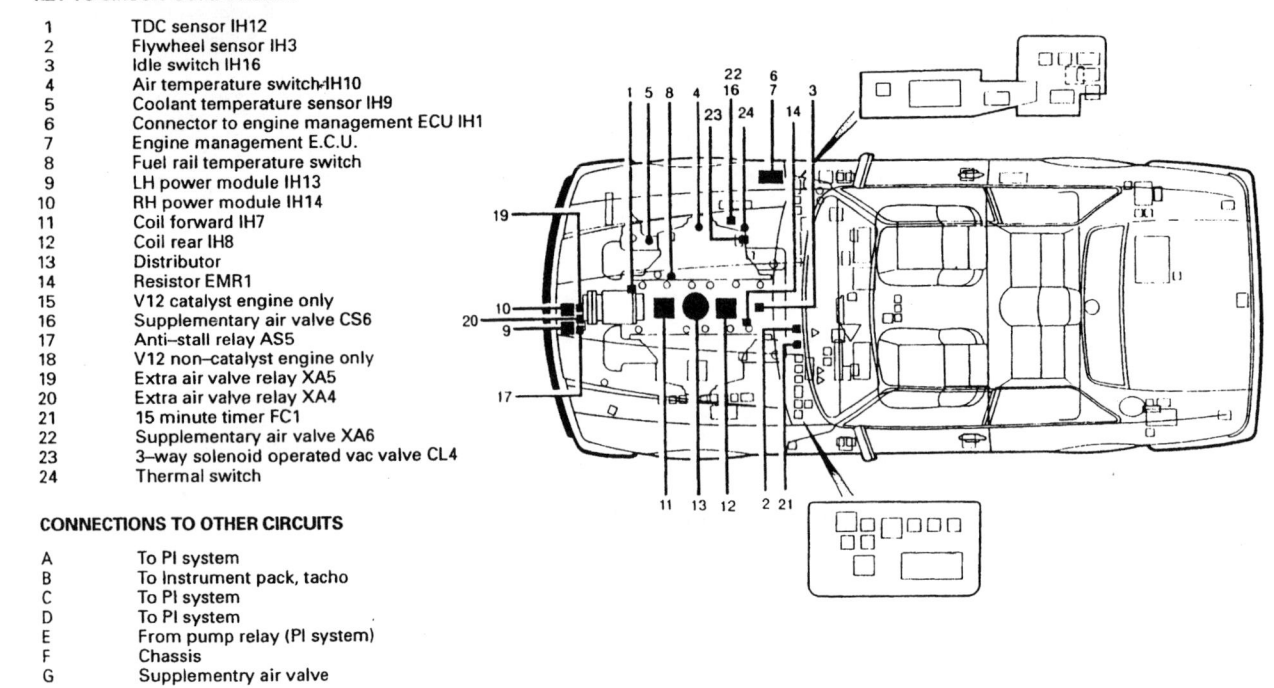

CONNECTIONS TO OTHER CIRCUITS

A	To PI system
B	To instrument pack, tacho
C	To PI system
D	To PI system
E	From pump relay (PI system)
F	Chassis
G	Supplementry air valve

November 1995 ISSUE 2

ENGINE MANAGEMENT (5.3 LITRE)

November 1995 ISSUE 2

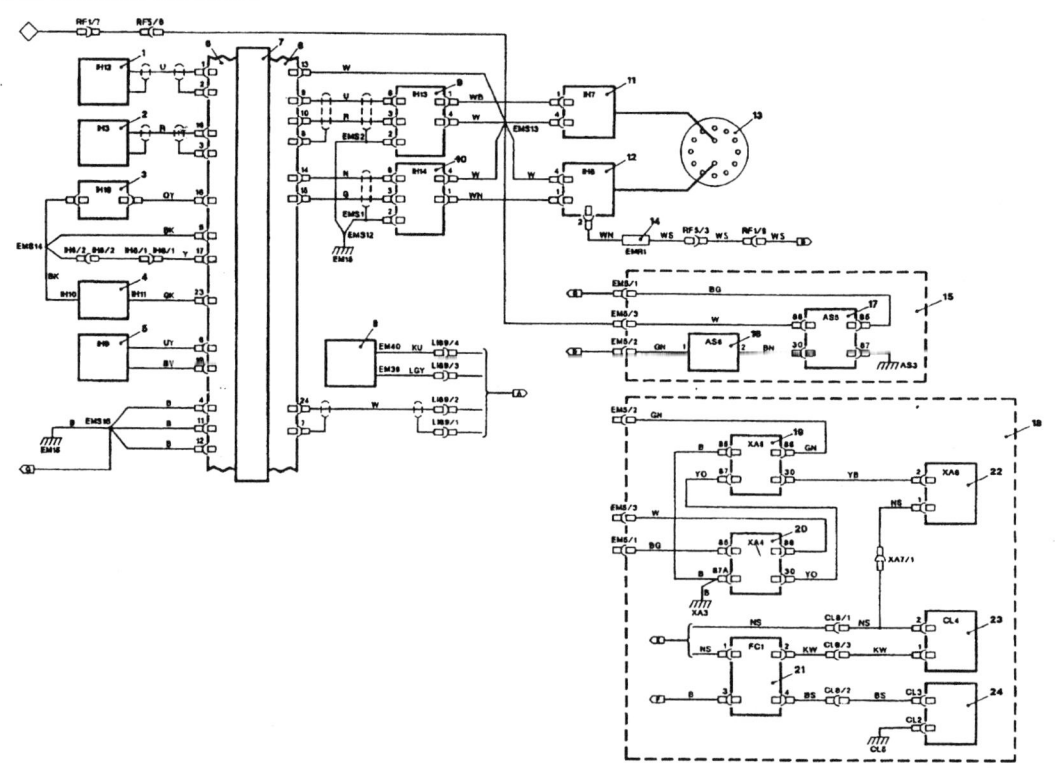

CIRCUIT DIAGRAMS AND HARNESSES

FUEL INJECTION SYSTEM (5.3 LITRE)

KEY TO CIRCUIT COMPONENTS

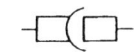

1. Inertia switch LB212
2. Diagnostic connector LI90
3. Main PI relay LI47
4. In-line fuse LI88
5. Connection on non-cat system only
6. Fuel pump relay LI46
7. A bank lambda heater LI76 (Cat system only)
8. B bank lambda heater LI63 (Cat system only)
9. Throttle pot LI136
10. Fuse 10, RH Fusebox LB268
11. Fuse 10, LH Fusebox LB263
12. Connection on non-cat system only
13. Fuel pump LI141
14. A bank lambda sensor LI75 (Cat system only)
15. B bank lambda sensor LI62 (Cat system only)
16. Air temperature sensor LI64
17. Water temperature sensor LI65
18. Connector to PI ECU
19. PI ECU LI45
20. Arrangement for Cat system only
21. 3-way solenoid vacuum valve LI137 (Cat system only)
22. Fuel rail temperature switch
23. Vacuum valve (hot start) LI145
24. Ballast resistors LI73
25. Fuel failure code reset LI91
26. Fuel control timer FC2
27. 1A injector EM19
28. 2A injector EM20
29. 3A injector EM21
30. 4A injector EM22
31. 5A injector EM23
32. 6A injector EM24
33. 1B injector EM25
34. 2B injector EM26
35. 3B injector EM27
36. 4B injector EM28
37. 5B injector EM29
38. 6B injector EM30

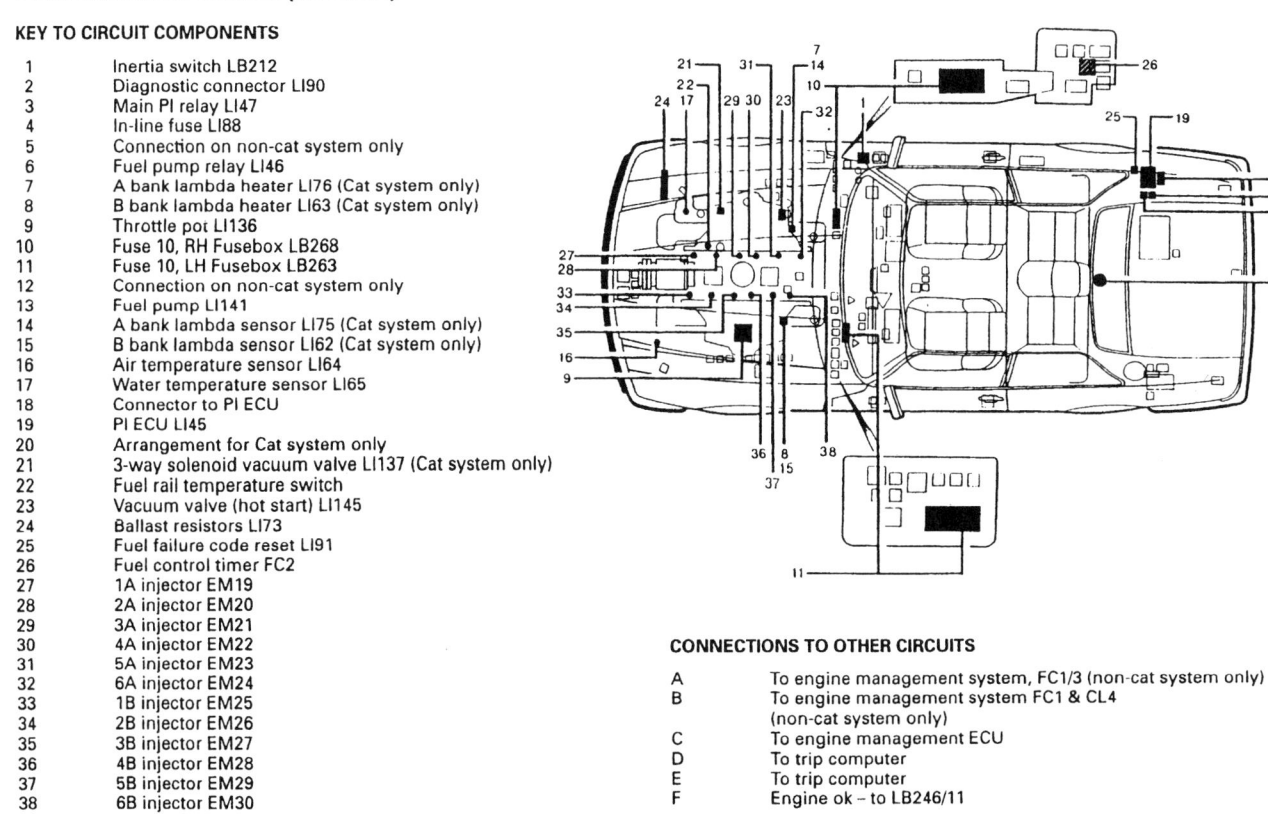

CONNECTIONS TO OTHER CIRCUITS

A — To engine management system, FC1/3 (non-cat system only)
B — To engine management system FC1 & CL4 (non-cat system only)
C — To engine management ECU
D — To trip computer
E — To trip computer
F — Engine ok – to LB246/11

FUEL INJECTION SYSTEM (5.3 LITRE)

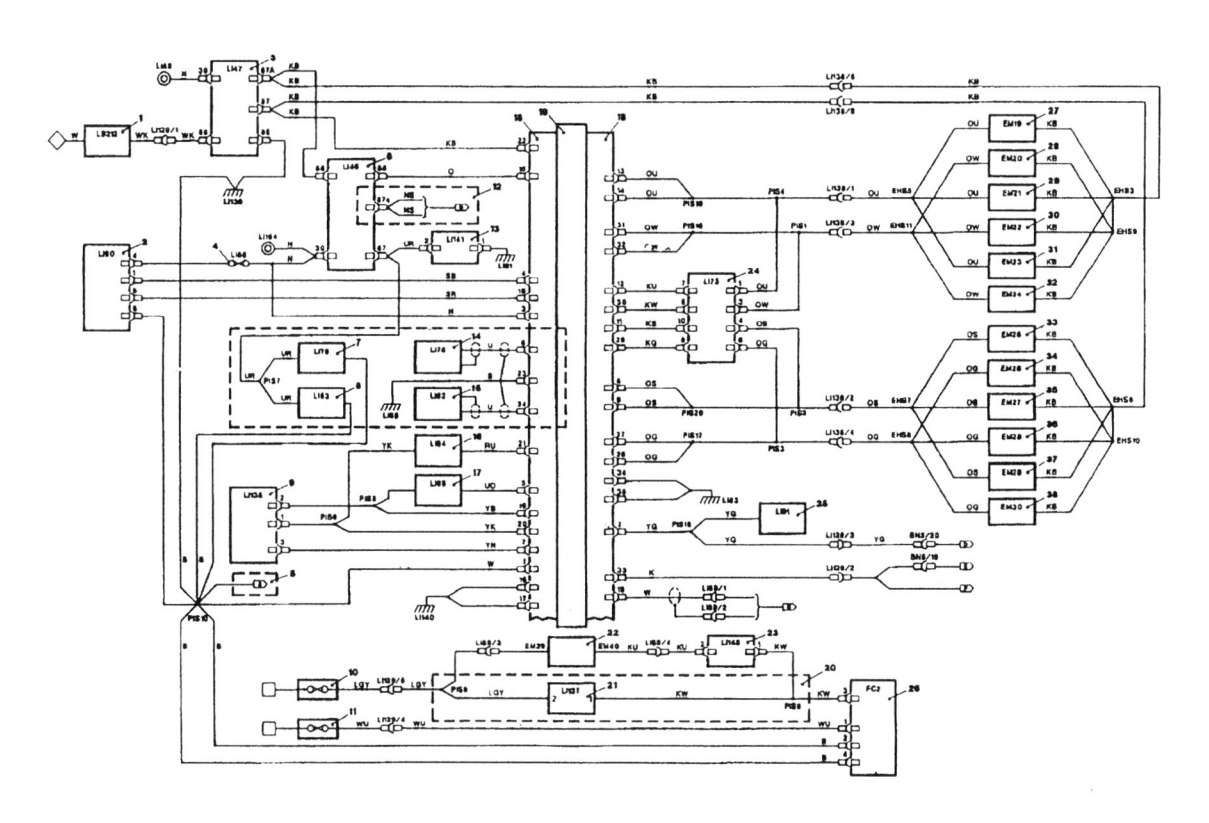

CIRCUIT DIAGRAMS AND HARNESSES

CRUISE CONTROL

KEY TO CIRCUIT COMPONENTS

1. Ignition switch LB/RB202
2. Fuse 2 (RH fusebox – LH drive/ LH fusebox – RH drive)
3. Brake pedal switches LB/RB167
4. Stop light relay RH113
5. Cruise control inhibit switch (automatic transmission only) or Link (manual transmission only)
6. Electronic control unit CC2
7. Speed interface unit RH114
8. Fuse 16 (LH fusebox – LH drive/ RH fusebox RH drive)
9. PI harness connector LI134
10. Speed warning circuit/exhaust temperature connector LB/RB211 (Saudi Arabia) / XT7 (Japan)
11. Cruise control pump LF16
12. Speed control dump valve LF13
13. Cruise control switch CC1
14. Lighting switch LL30
15. Clutch switch (left forward harness–LH drive only)
16. Clutch switch (right forward harness–manual transmission and RH drive vehicles only)

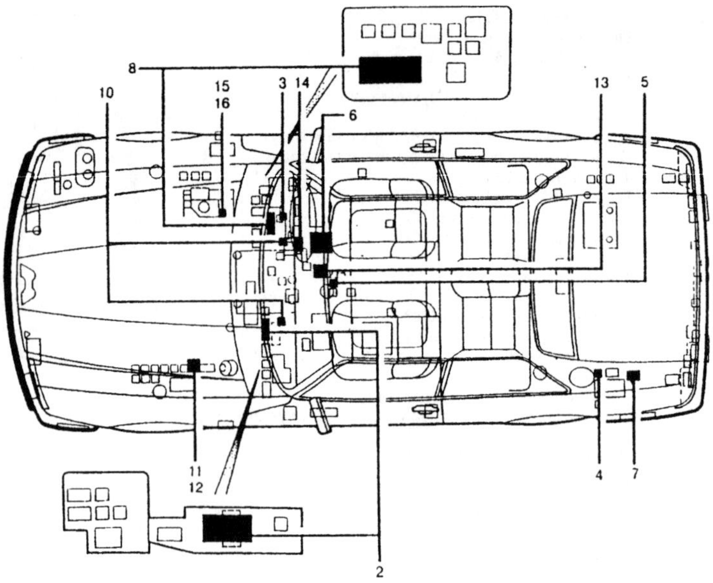

CONNECTIONS TO OTHER CIRCUITS

A. To instrument pack LB/RB246/10
B. To trip computer BN3/4
C. To reversing light switch (manual gearbox only)
D. To trip computer BN3/6

November 1995 ISSUE 2

CRUISE CONTROL

November 1995 ISSUE 2

152

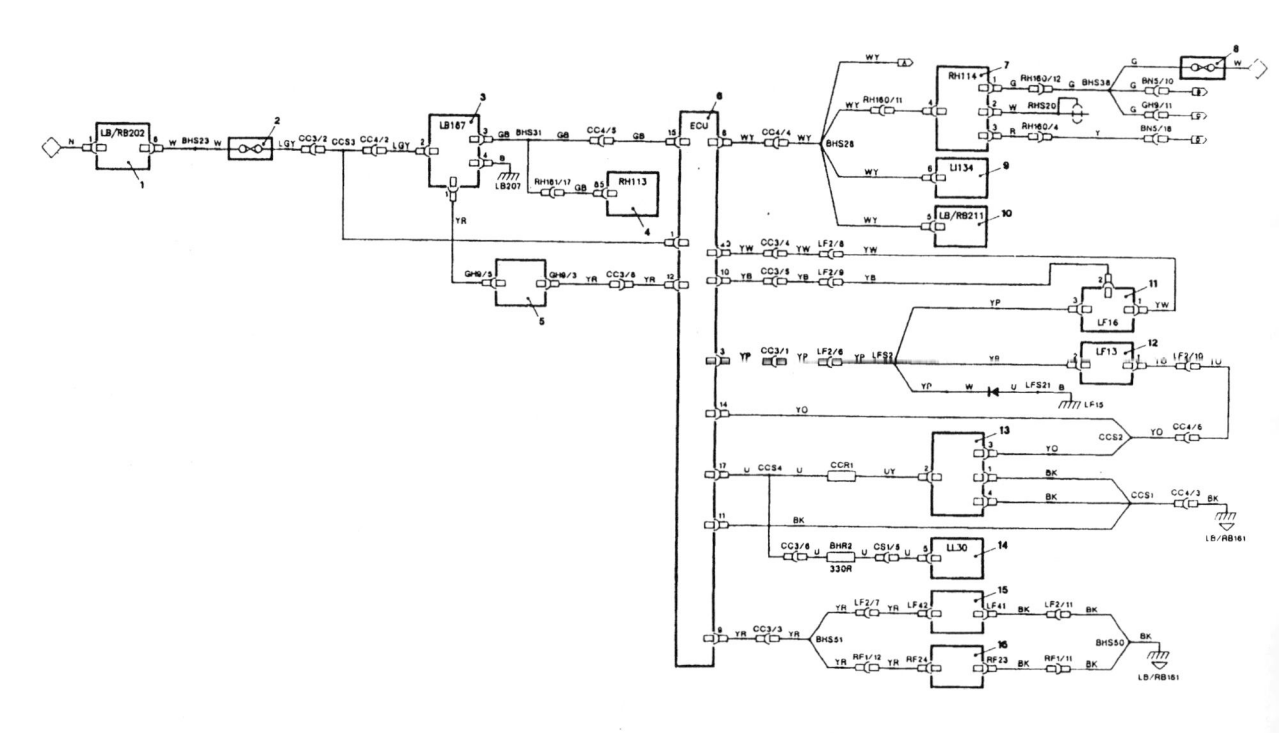

CIRCUIT DIAGRAMS AND HARNESSES

ANTI-LOCK BRAKE SYSTEM

KEY TO CIRCUIT COMPONENTS

1. Fuse 11 (RH fusebox – LH drive/LH fusebox – RH drive)
2. ABS main relay (ABS16)
3. Diagnostic connector (ABS2)
4. Electronic control unit – ECU (ABS1)
5. Main valve (ABS9)
6. Valve block (ABS8)
7. Fuse 22 (RH fusebox – LH drive/LH fusebox – RH drive)
8. ABS pump relay (LB-RB144)
9. ABS pump motor (ABS14)
10. Ignition switch (LB-RB202)
11. Battery post LB-RB192)
12. Brake warning lamp (instrument pack LB-RB246/16)
13. ABS reservoir (ABS7)
14. ABS accumulator (ABS15)
15. LH front brake pad sensor ABS6
16. RH front brake pad sensor ABS11
17. LH rear brake pad sensor ABS5
18. RH rear brake pad sensor ABS4

CONNECTIONS TO OTHER CIRCUITS

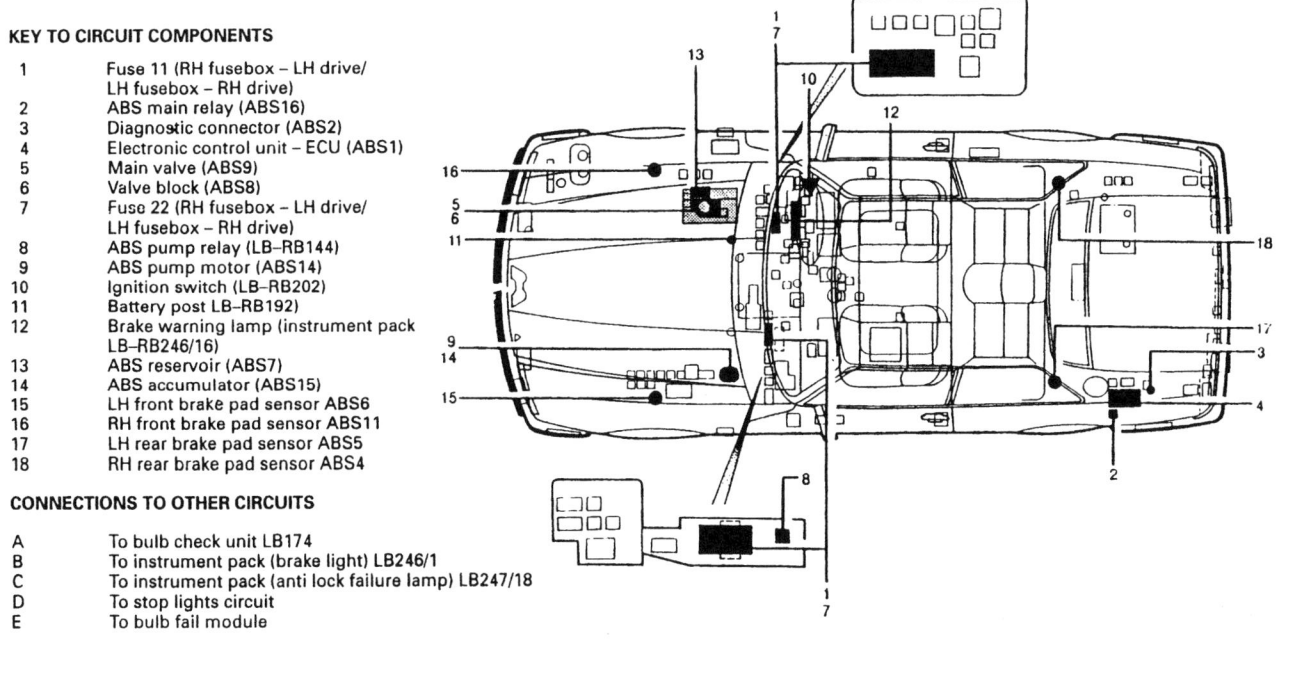

A. To bulb check unit LB174
B. To instrument pack (brake light) LB246/1
C. To instrument pack (anti lock failure lamp) LB247/18
D. To stop lights circuit
E. To bulb fail module

ANTI-LOCK BRAKE SYSTEM

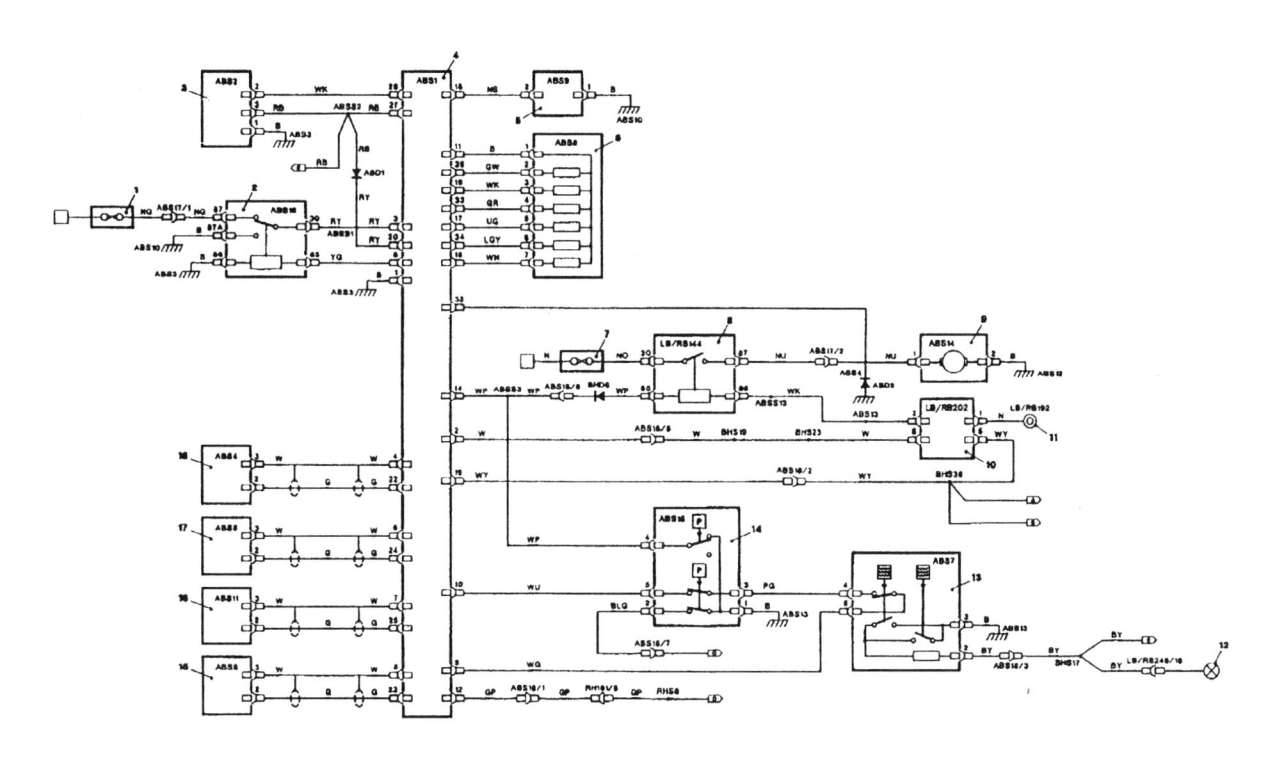

J91 370

CIRCUIT DIAGRAMS AND HARNESSES

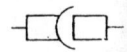

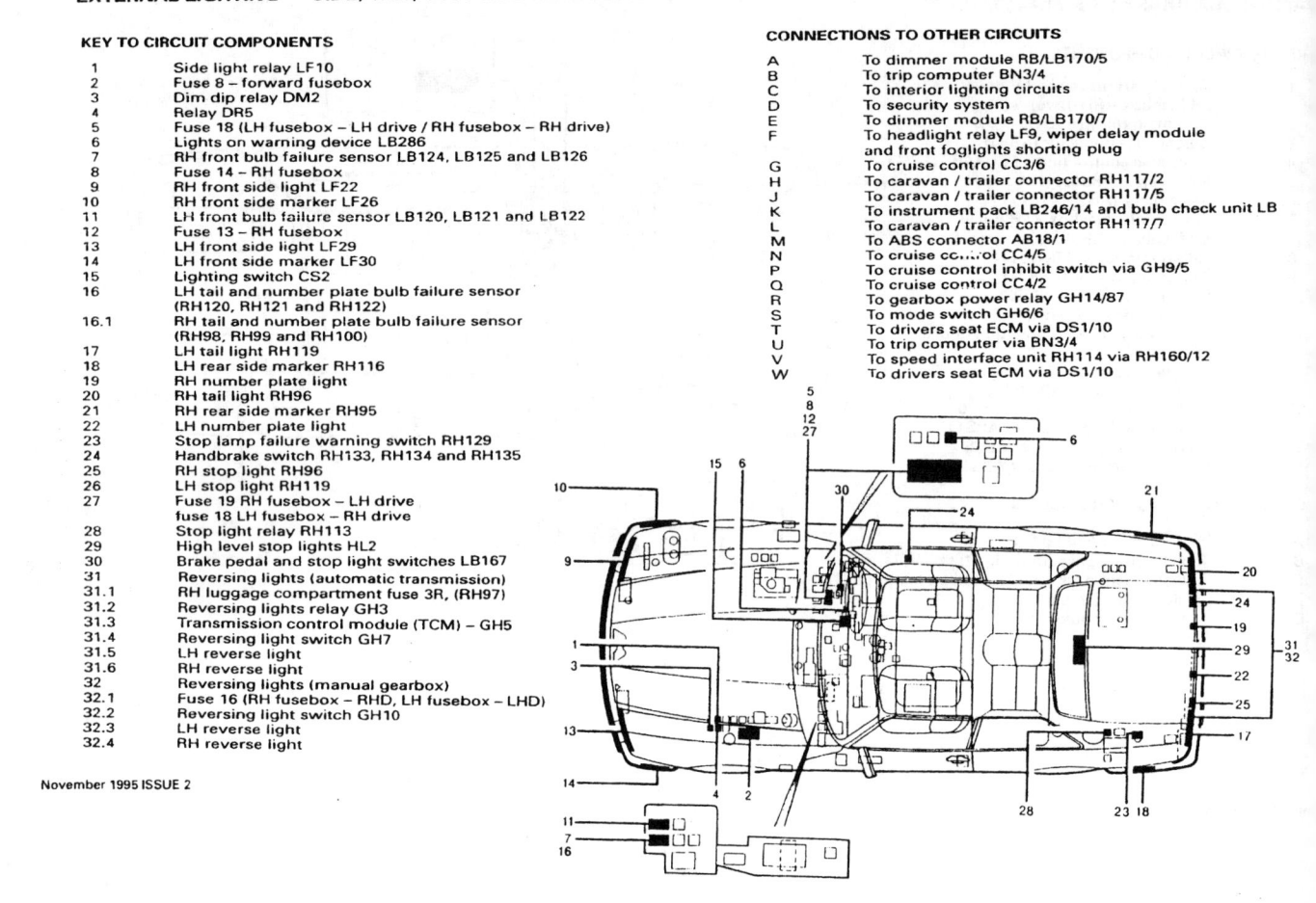

EXTERNAL LIGHTING — SIDE, TAIL, STOP AND REVERSE LAMPS

KEY TO CIRCUIT COMPONENTS

1	Side light relay LF10
2	Fuse 8 – forward fusebox
3	Dim dip relay DM2
4	Relay DR5
5	Fuse 18 (LH fusebox – LH drive / RH fusebox – RH drive)
6	Lights on warning device LB286
7	RH front bulb failure sensor LB124, LB125 and LB126
8	Fuse 14 – RH fusebox
9	RH front side light LF22
10	RH front side marker LF26
11	LH front bulb failure sensor LB120, LB121 and LB122
12	Fuse 13 – RH fusebox
13	LH front side light LF29
14	LH front side marker LF30
15	Lighting switch CS2
16	LH tail and number plate bulb failure sensor (RH120, RH121 and RH122)
16.1	RH tail and number plate bulb failure sensor (RH98, RH99 and RH100)
17	LH tail light RH119
18	LH rear side marker RH116
19	RH number plate light
20	RH tail light RH96
21	RH rear side marker RH95
22	LH number plate light
23	Stop lamp failure warning switch RH129
24	Handbrake switch RH133, RH134 and RH135
25	RH stop light RH96
26	LH stop light RH119
27	Fuse 19 RH fusebox – LH drive fuse 18 LH fusebox – RH drive
28	Stop light relay RH113
29	High level stop lights HL2
30	Brake pedal and stop light switches LB167
31	Reversing lights (automatic transmission)
31.1	RH luggage compartment fuse 3R, (RH97)
31.2	Reversing lights relay GH3
31.3	Transmission control module (TCM) – GH5
31.4	Reversing light switch GH7
31.5	LH reverse light
31.6	RH reverse light
32	Reversing lights (manual gearbox)
32.1	Fuse 16 (RH fusebox – RHD, LH fusebox – LHD)
32.2	Reversing light switch GH10
32.3	LH reverse light
32.4	RH reverse light

CONNECTIONS TO OTHER CIRCUITS

A	To dimmer module RB/LB170/5
B	To trip computer BN3/4
C	To interior lighting circuits
D	To security system
E	To dimmer module RB/LB170/7
F	To headlight relay LF9, wiper delay module and front foglights shorting plug
G	To cruise control CC3/6
H	To caravan / trailer connector RH117/2
J	To caravan / trailer connector RH117/5
K	To instrument pack LB246/14 and bulb check unit LB
L	To caravan / trailer connector RH117/7
M	To ABS connector AB18/1
N	To cruise control CC4/5
P	To cruise control inhibit switch via GH9/5
Q	To cruise control CC4/2
R	To gearbox power relay GH14/87
S	To mode switch GH6/6
T	To drivers seat ECM via DS1/10
U	To trip computer via BN3/4
V	To speed interface unit RH114 via RH160/12
W	To drivers seat ECM via DS1/10

November 1995 ISSUE 2

EXTERNAL LIGHTING — SIDE, TAIL, STOP AND REVERSE LAMPS

November 1995 ISSUE 2

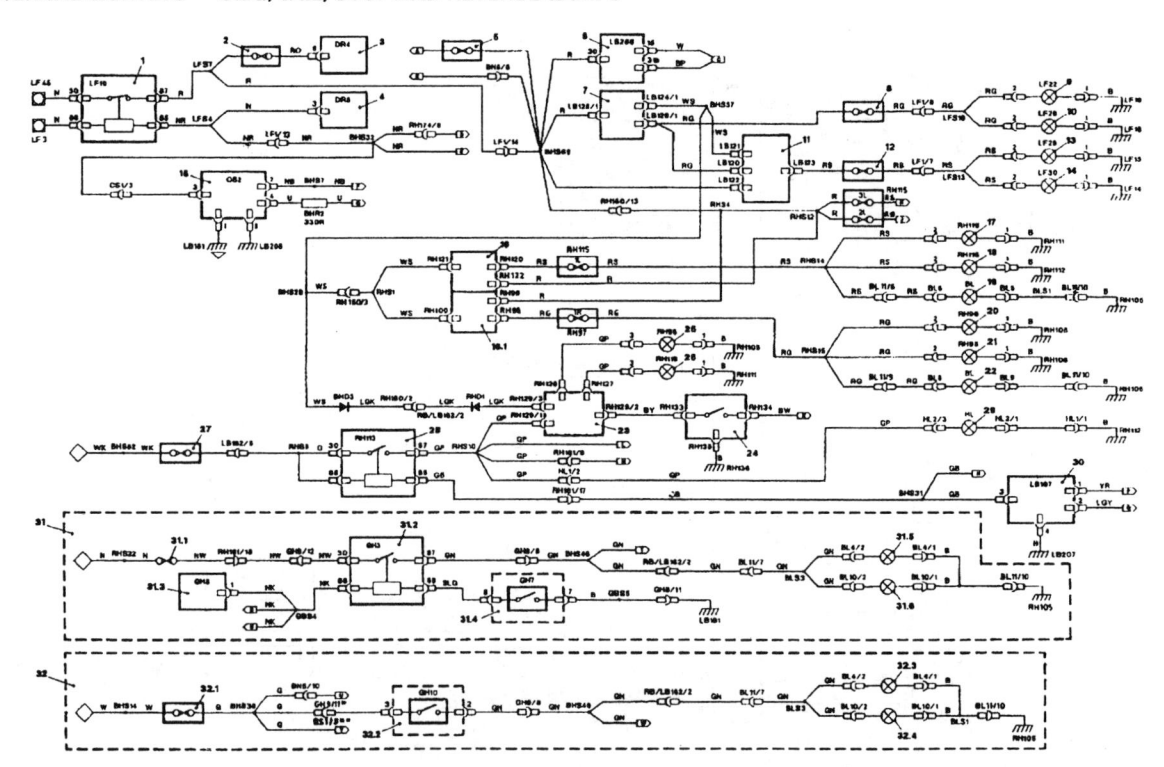

CIRCUIT DIAGRAMS AND HARNESSES

FOGLIGHTS (LHD — USA AND CANADA)

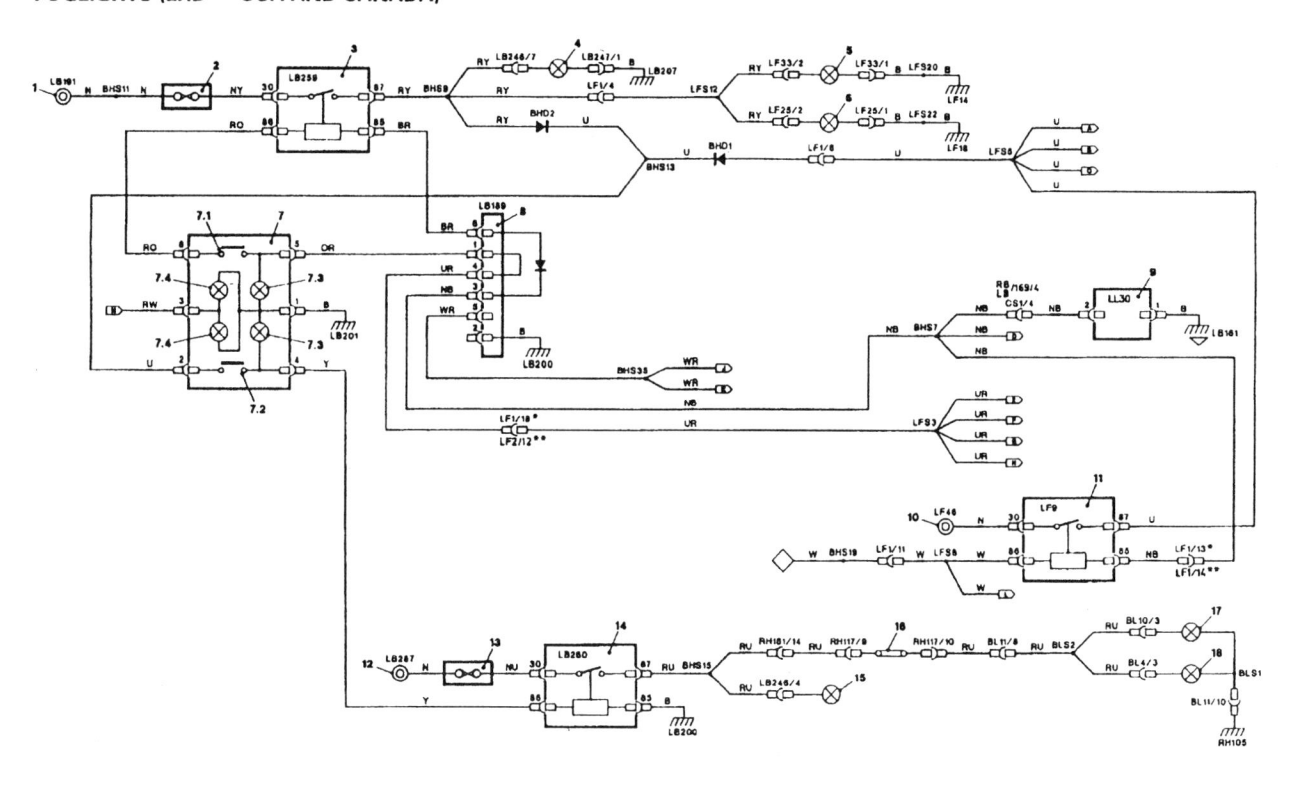

KEY TO CIRCUIT COMPONENTS

1	Battery post LB191
2	Fuse 20 (LH fusebox)
3	Front foglight relay LB259
4	Front foglights ON indicator
5	Front LH foglight LF33
6	Front RH foglight LF25
7	Front and rear foglight switches
7.1	Front foglight switch
7.2	Rear foglight switch
7.3	Foglights ON indicators
7.4	Foglights switch panel lamps
8	Shorting plug LB189 (USA / Canada configuration)
9	Lighting switch LL30
10	Battery post LF46
11	Headlights relay LF9
12	Battery post LB287
13	Fuse 17 (LH fusebox)
14	Rear foglights relay LB260
15	Rear foglights ON indicator
16	Caravan connector link (pins 9 and 10)
17	Rear RH foglight RH96
18	Rear LH foglight RH119

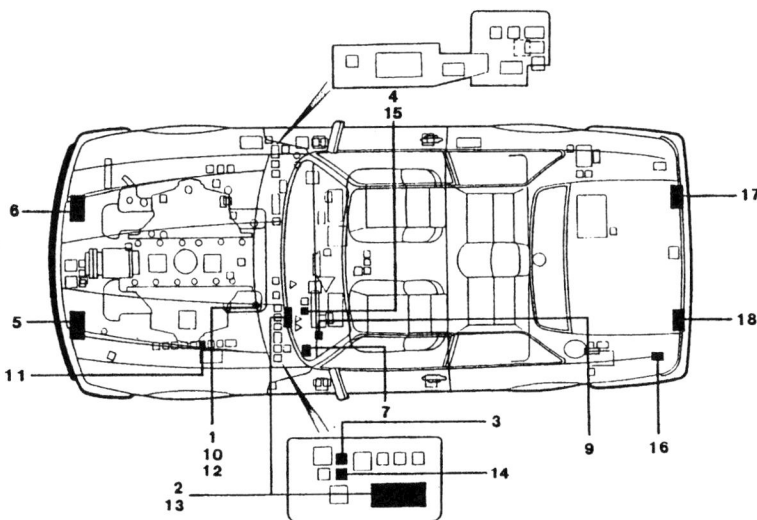

CONNECTIONS TO OTHER CIRCUITS

A	To daylight running lights connector DR5/2
B	To headlights control module LF6/56
C	To fuse 12 (forward fusebox)
D	To wiper delay module LB148/12
E	To daylight running lights connector DR4/2
F	To headlights control module LF6/56b
G	To fuse 10 (forward fusebox)
H	To fuse 9 (forward fusebox)
J	To illumination control module LB170/6
K	To radio connector LB160/11
L	To daylight running lights connector DR4/3

FOGLIGHTS (LHD — USA AND CANADA)

CIRCUIT DIAGRAMS AND HARNESSES

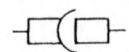

FOGLIGHTS (RHD)

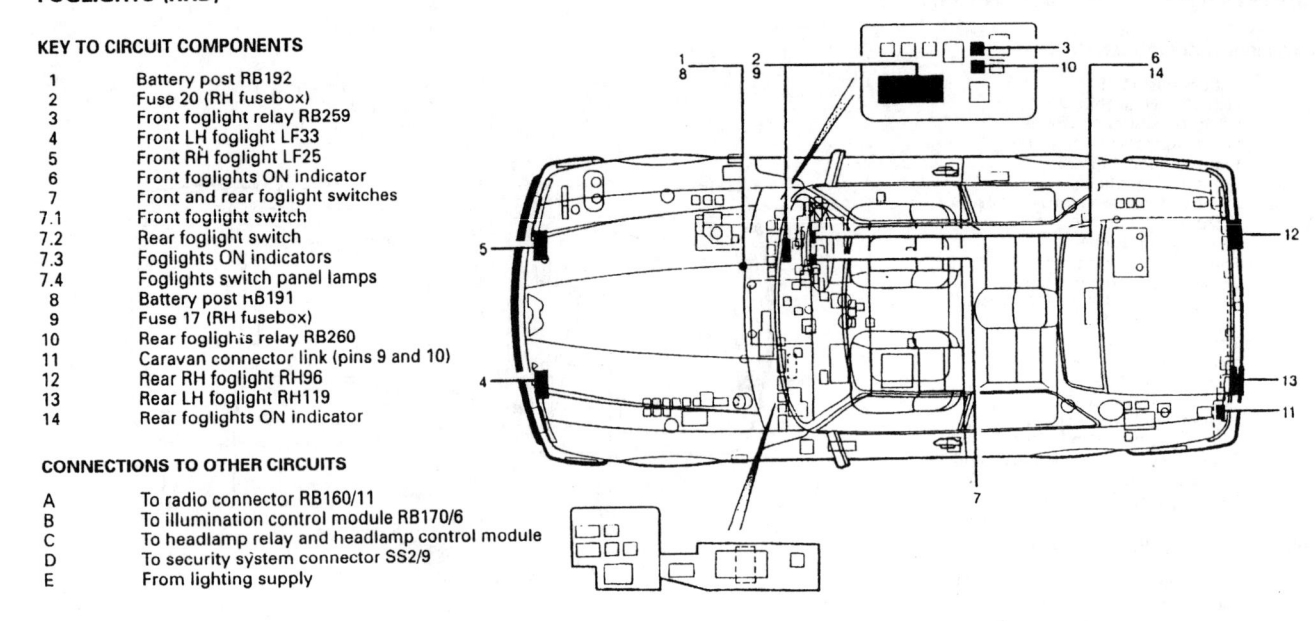

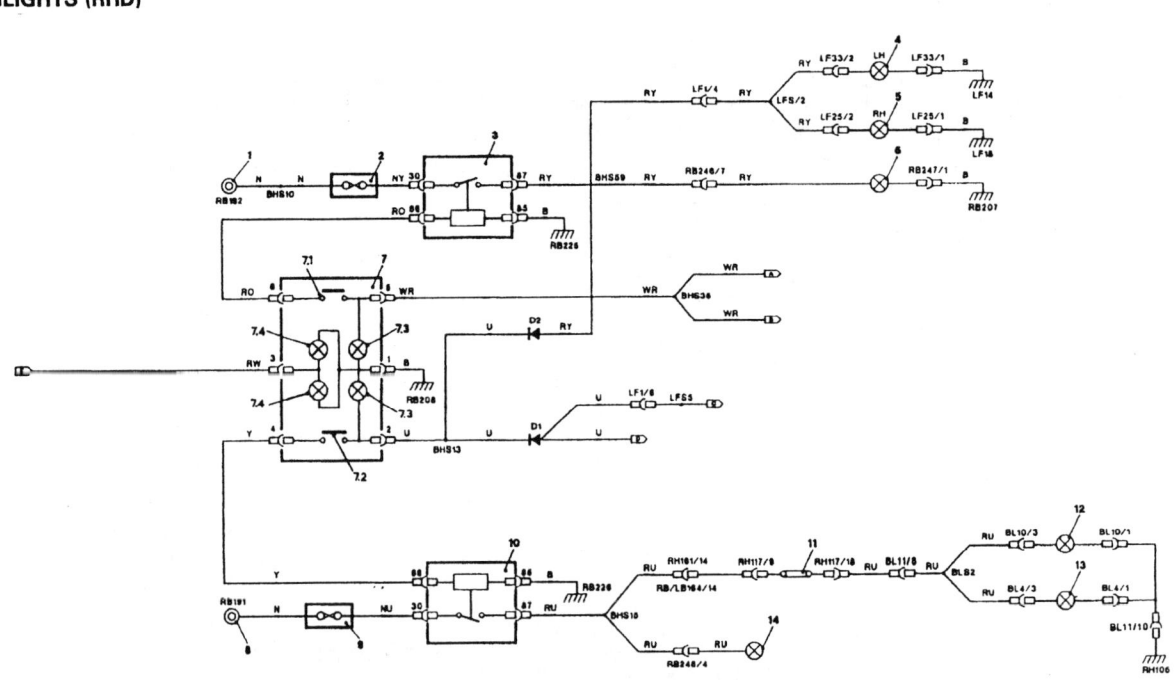

KEY TO CIRCUIT COMPONENTS

1	Battery post RB192
2	Fuse 20 (RH fusebox)
3	Front foglight relay RB259
4	Front LH foglight LF33
5	Front RH foglight LF25
6	Front foglights ON indicator
7	Front and rear foglight switches
7.1	Front foglight switch
7.2	Rear foglight switch
7.3	Foglights ON indicators
7.4	Foglights switch panel lamps
8	Battery post RB191
9	Fuse 17 (RH fusebox)
10	Rear foglights relay RB260
11	Caravan connector link (pins 9 and 10)
12	Rear RH foglight RH96
13	Rear LH foglight RH119
14	Rear foglights ON indicator

CONNECTIONS TO OTHER CIRCUITS

A	To radio connector RB160/11
B	To illumination control module RB170/6
C	To headlamp relay and headlamp control module
D	To security system connector SS2/9
E	From lighting supply

FOGLIGHTS (RHD)

CIRCUIT DIAGRAMS AND HARNESSES

FOGLIGHTS (LHD — NOT USA OR CANADA)

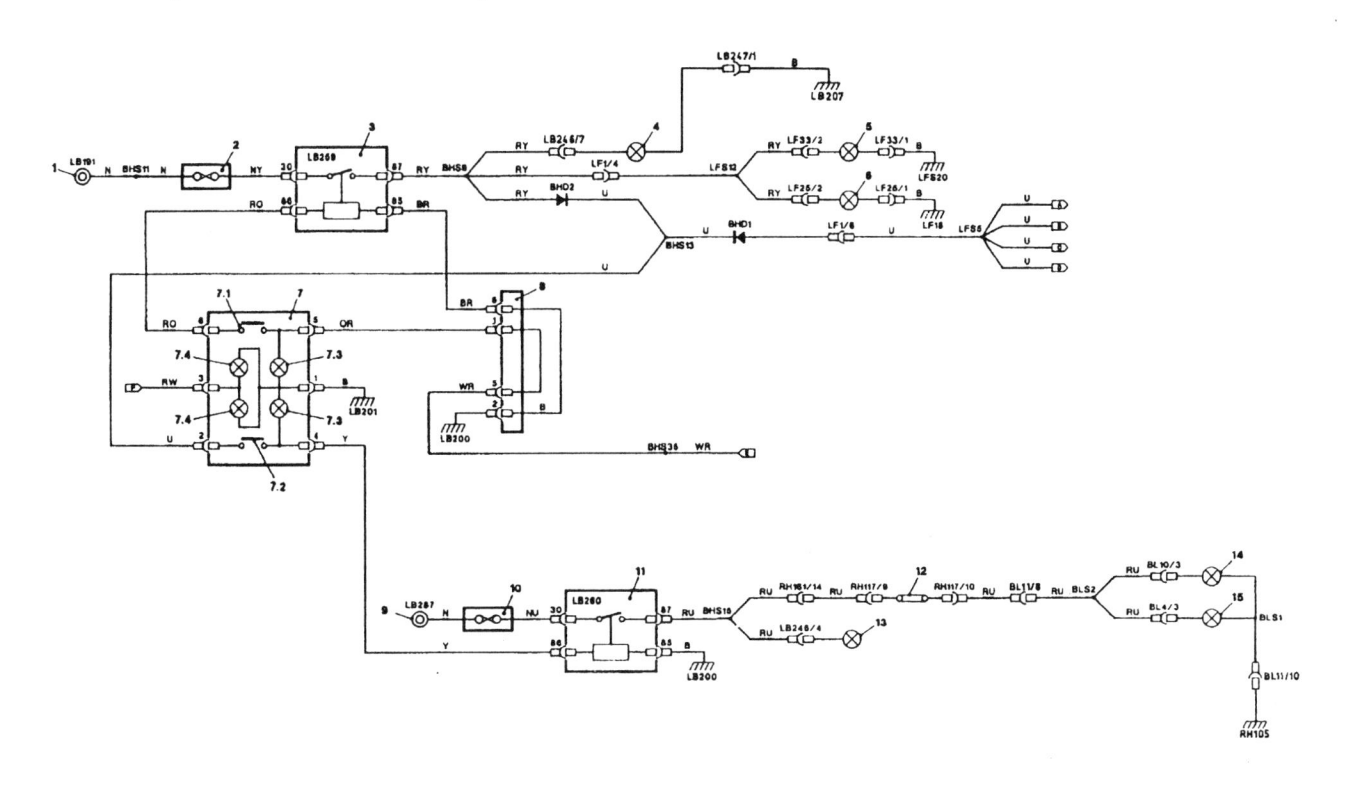

KEY TO CIRCUIT COMPONENTS

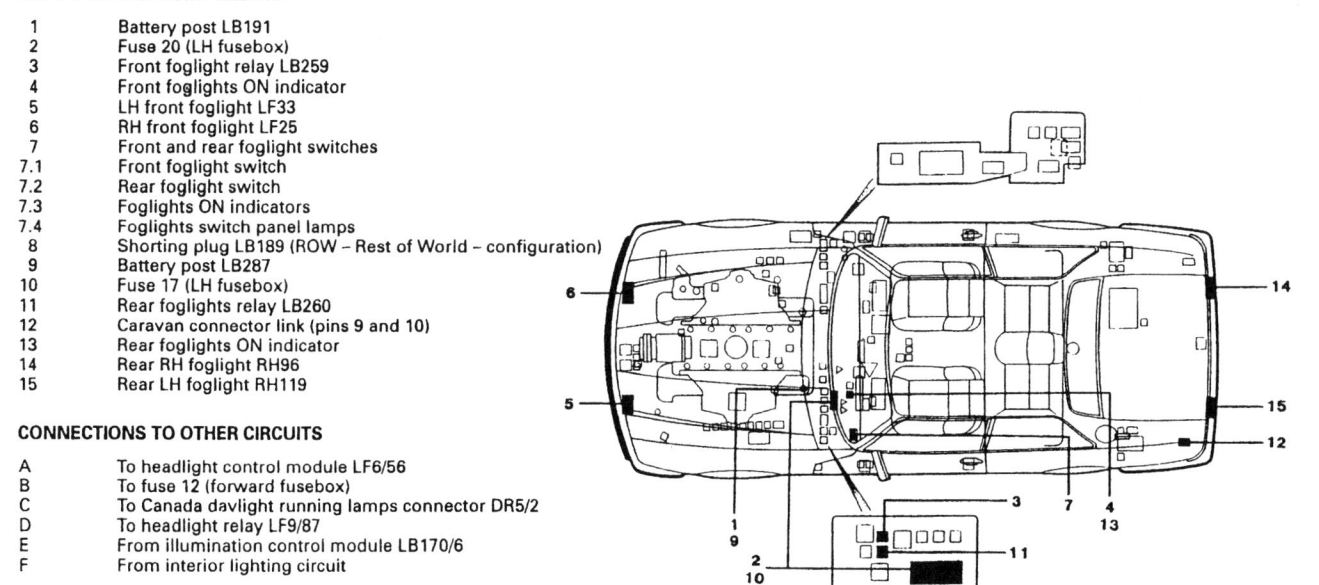

1	Battery post LB191
2	Fuse 20 (LH fusebox)
3	Front foglight relay LB259
4	Front foglights ON indicator
5	LH front foglight LF33
6	RH front foglight LF25
7	Front and rear foglight switches
7.1	Front foglight switch
7.2	Rear foglight switch
7.3	Foglights ON indicators
7.4	Foglights switch panel lamps
8	Shorting plug LB189 (ROW – Rest of World – configuration)
9	Battery post LB287
10	Fuse 17 (LH fusebox)
11	Rear foglights relay LB260
12	Caravan connector link (pins 9 and 10)
13	Rear foglights ON indicator
14	Rear RH foglight RH96
15	Rear LH foglight RH119

CONNECTIONS TO OTHER CIRCUITS

A	To headlight control module LF6/56
B	To fuse 12 (forward fusebox)
C	To Canada daylight running lamps connector DR5/2
D	To headlight relay LF9/87
E	From illumination control module LB170/6
F	From interior lighting circuit

FOGLIGHTS (LHD — NOT USA OR CANADA)

CIRCUIT DIAGRAMS AND HARNESSES

HEADLAMPS — DIM/DIP

KEY TO CIRCUIT COMPONENTS

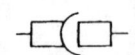

1	Fuse 8 (forward fusebox)
2	Dim/dip relay DM2
3	Dim/dip resistor DM1
4	Battery post LF46
5	Headlamp relay LF9
6	Lighting switch LL30
7	Dip switch CS2
8	Headlamp control module LF6
9	Battery post LF3
10	Fuse 9 (forward fusebox)
11	LH dipped headlamp LF29
12	Fuse 5 (forward fusebox)
13	LH main beam LF29
14	Fuse 10 (forward fusebox)
15	RH dipped headlamp LF22
16	Fuse 6 (forward fusebox)
17	RH main beam LF22

CONNECTIONS TO OTHER CIRCUITS

A	To sidelamp relay LF10
B	To bulb failure sensors, trip computer and illumination control module
C	To daylight running lights, fuse 12 – forward fusebox, and front / rear foglight switch via LF1/6
D	To illumination control module LB170/7,
E	To front foglight shorting plug LB189/4 – LHD only
F	To windscreen wiper delay module LB148/12
G	To front fog shorting plug LB189/3

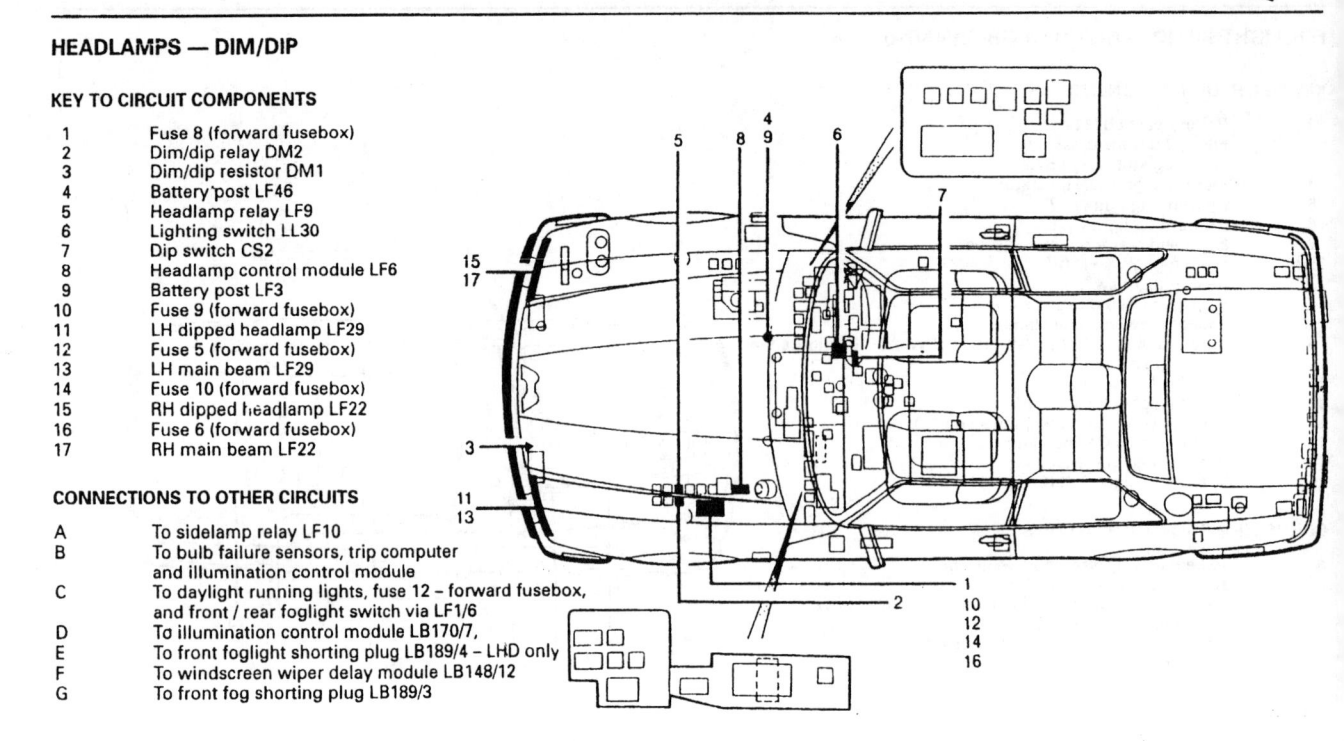

November 1995 ISSUE 2

HEADLAMPS — DIM/DIP

November 1995 ISSUE 2

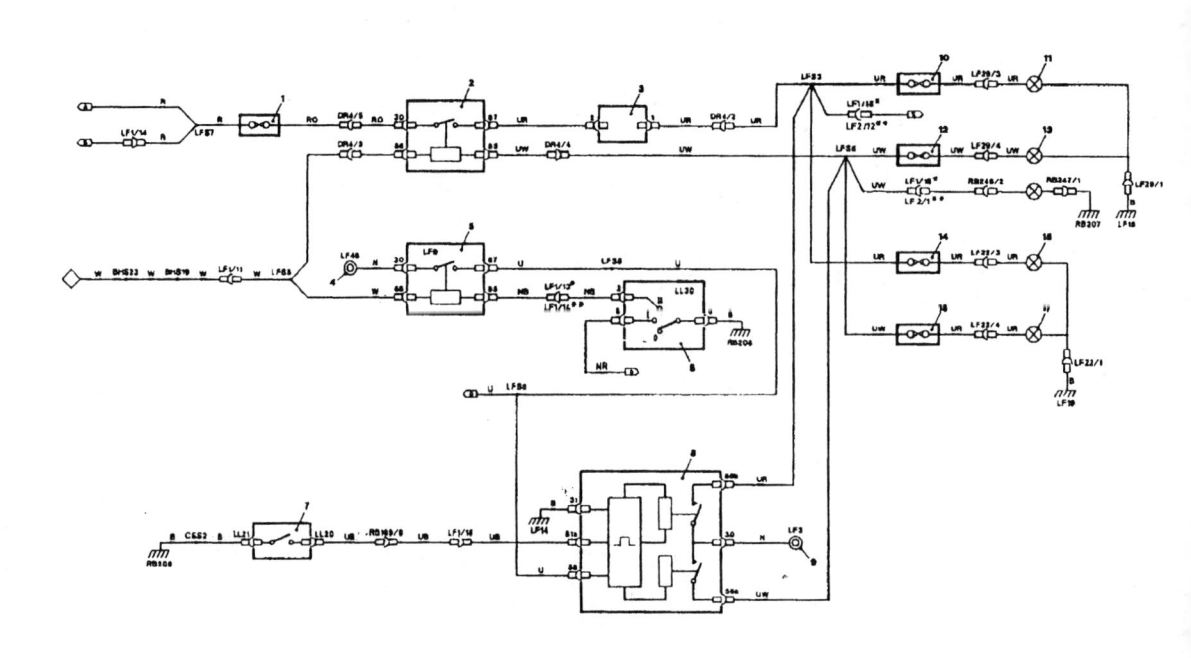

CIRCUIT DIAGRAMS AND HARNESSES

HEADLAMPS — DAYLIGHT RUNNING AND HEADLAMP LEVELLING

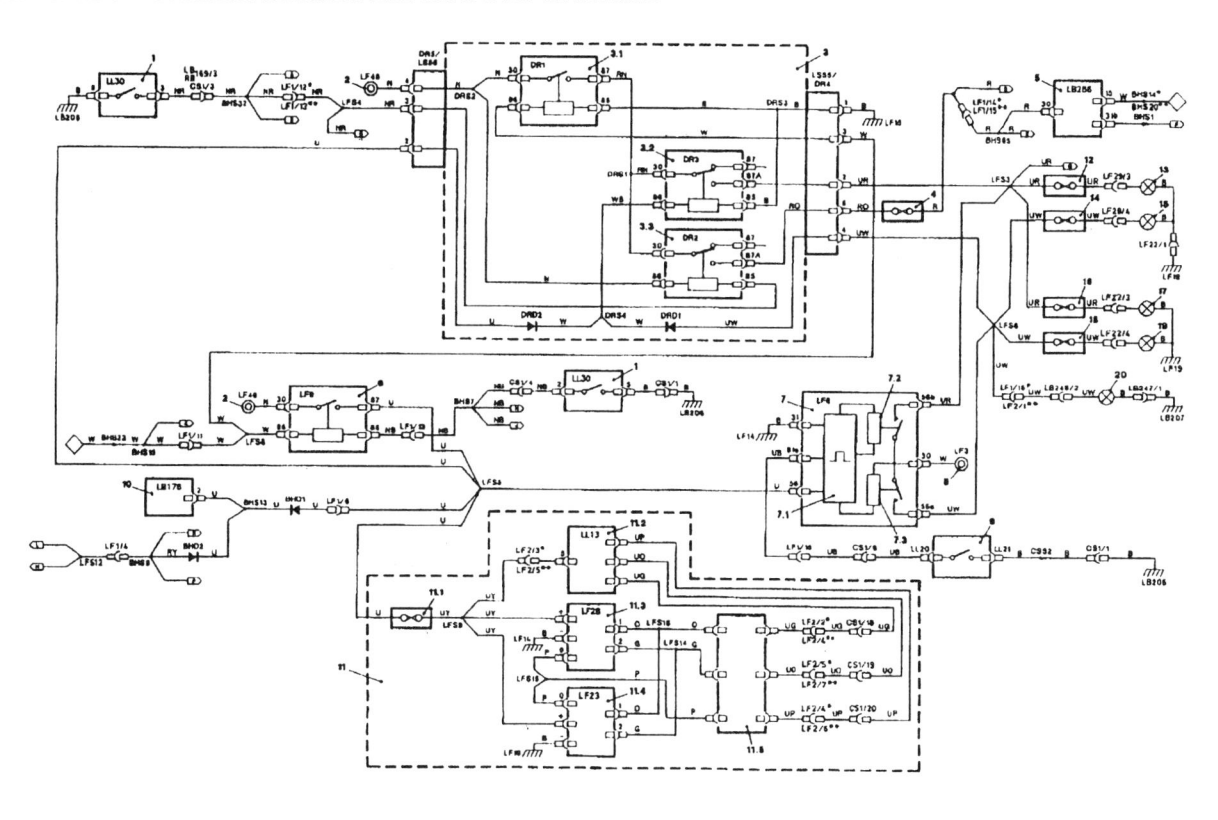

KEY TO CIRCUIT COMPONENTS

1	Lighting switch LL30
2	Battery post LF46
3	Daylight running circuit (Canada only)
3.1	Relay DR1 (ignition on)
3.2	Relay DR3 (dipped beam)
3.3	Relay DR2 (main beam)
4	Fuse 8 (forward fusebox)
5	Lights on warning LB286
6	Headlamp relay LF9
7	Headlamp control module LF6
7.1	Pulse width modulator
7.2	Dipped beam relay
7.3	Main beam relay
8	Battery post LF3
9	Headlamp change-over (dip) switch
10	Front/rear fog lamp switches LB176
11	Headlamp levelling circuit (Germany only)
11.1	Fuse 12 (forward fusebox)
11.2	Headlamp levelling switch LL13
11.3	Headlamp levelling motor LF28 (LH)
11.4	Headlamp levelling motor LF23 (RH)
11.5	Headlamp levelling diode module
12	Fuse 9 (forward fusebox)
13	LH dipped beam
14	Fuse 5 (forward fusebox)
15	LH main beam
16	Fuse 10 (forward fusebox)
17	RH dipped beam
18	Fuse 6 (forward fusebox)
19	RH main beam
20	Main beam indicator (instrument pack)

CONNECTIONS TO OTHER CIRCUITS

A	To security system RH174/5
B	To illumination control module LB170/7
C	To sidelight relay LF10/85 and DR5/3
D	To sidelight relay LF10
E	To bulb fail sensors, trip computer, internal lighting, and fuse 18 (LH fuse box – RHD, RH fusebox – LHD)
F	To internal lights and security system
G	To front fog shorting plug
H	To wiper delay module
J	To front fog shorting plug
K	To ignition switched loads
L	To LH foglamp LF33
M	To RH foglamp LF25
N	To front foglamp relay LB259/87
P	To instrument pack LB246/7

HEADLAMPS — DAYLIGHT RUNNING AND HEADLAMP LEVELLING

CIRCUIT DIAGRAMS AND HARNESSES

INTERNAL LIGHTING

KEY TO CIRCUIT COMPONENTS

1	Battery post LB/RB145
2	Fuse 4 (RH fusebox – LH drive, LH fusebox – RH drive)
3	Driver's courtesy light
4	Driver's door lamp
5	Roof light
6	Pasenger's door lamp
7	Passenger's map light
8	Interior light LH
9	Interior light RH
10	Interior lighting switch BN4
11	Interior lights delay unit LB/RB143
12	Key switch LB/RB203
13	Door switch LB/RB188
14	Driver's seat control unit DS2
15	Lights ON warning LB/RB286
16	Passenger's seat control unit
17	Door switch LB/RB140
18	Seat belt logic unit LB/RB130
19	Battery post LB/RB191
20	Fuse 11 (LH fusebox – LH drive, RH fusebox – RH drive)
21	Luggage compartment lights
22	Luggage compartment lights switch

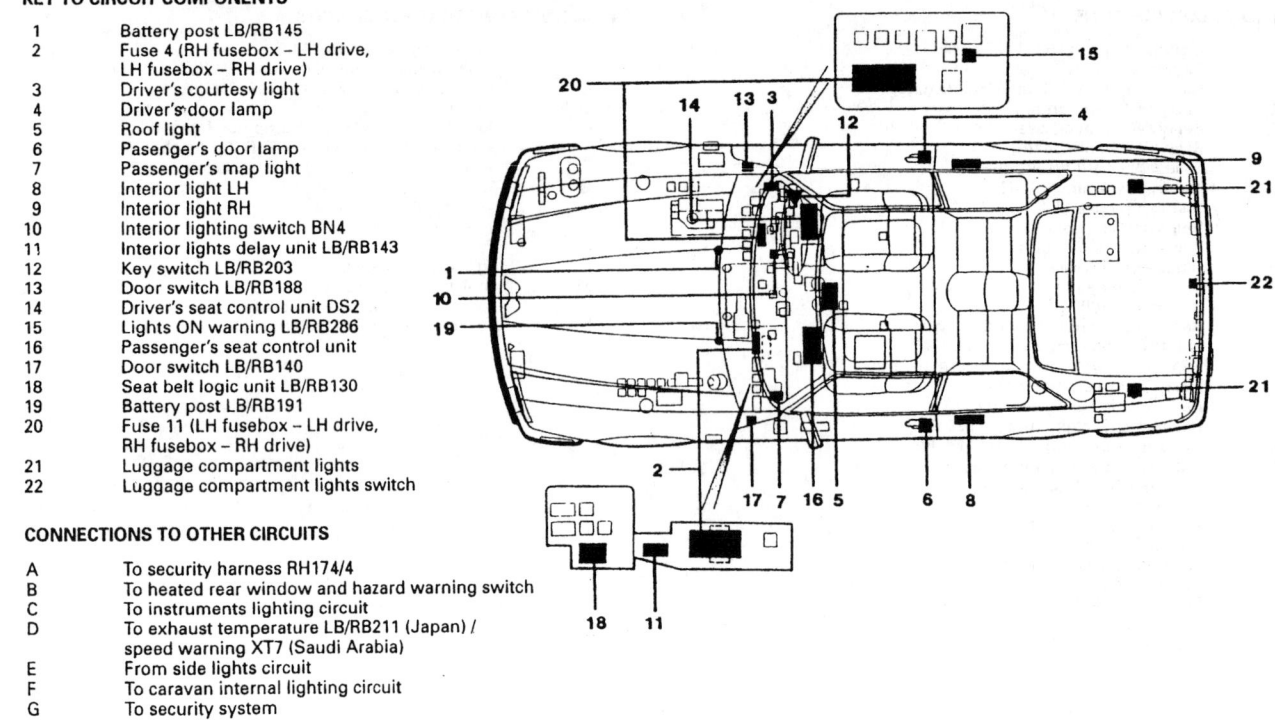

CONNECTIONS TO OTHER CIRCUITS

A	To security harness RH174/4
B	To heated rear window and hazard warning switch
C	To instruments lighting circuit
D	To exhaust temperature LB/RB211 (Japan) / speed warning XT7 (Saudi Arabia)
E	From side lights circuit
F	To caravan internal lighting circuit
G	To security system

November 1995 ISSUE 2

INTERNAL LIGHTING

November 1995 ISSUE 2

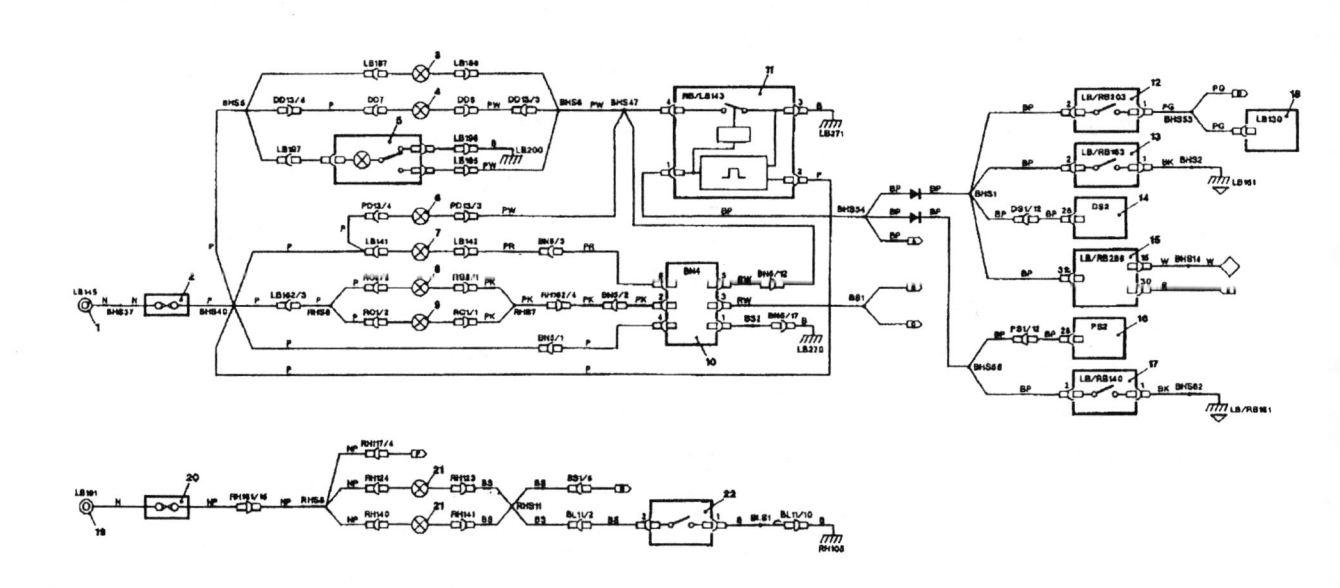

CIRCUIT DIAGRAMS AND HARNESSES

DIRECTION AND HAZARD WARNING LAMPS

KEY TO CIRCUIT COMPONENTS

1	Battery post LB / RB191
2	Fuse 19 (LH fusebox–LHD / RH fusebox RHD)
3	Flasher unit LB / RB172
4	Trailer /caravan indicator (instrument pack)
5	Fuse 5 (RH fusebox–LHD / LH fusebox RHD)
6	Hazard warning switch
7	Direction indicator switch
7.1	Turn left
7.2	Turn right
8	Rear LH direction indicator (vehicle)
9	Rear LH direction indicator (caravan)
10	Front LH direction indicator
11	Front LH direction indicator repeater
12	Turn left indicator (instrument pack)
13	Front RH direction indicator
14	Front RH direction indicator repeater
15	Rear RH direction indicator (vehicle)
16	Rear RH direction indicator (caravan)
17	Turn right indicator (instrument pack)

CONNECTIONS TO OTHER CIRCUITS

A	To seat belt logic unit
B	To internal lighting circuit

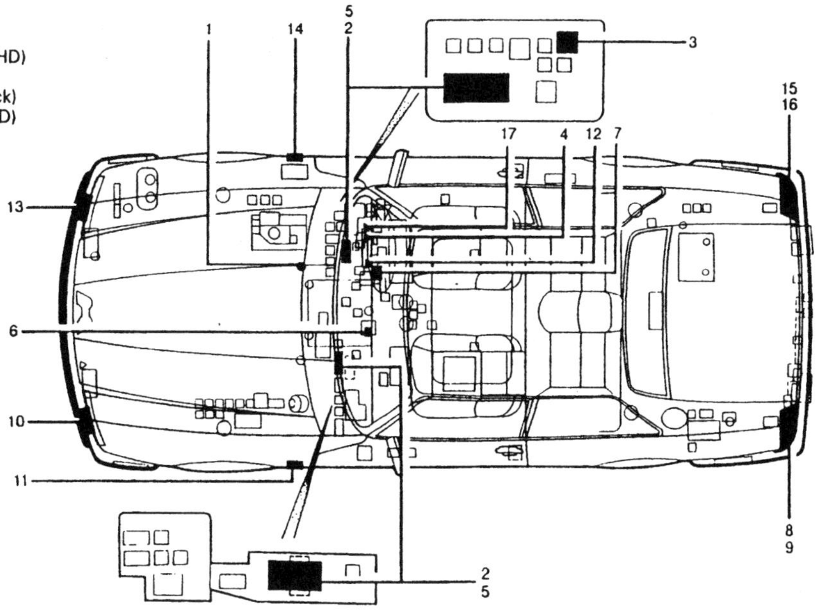

DIRECTION AND HAZARD WARNING LAMPS

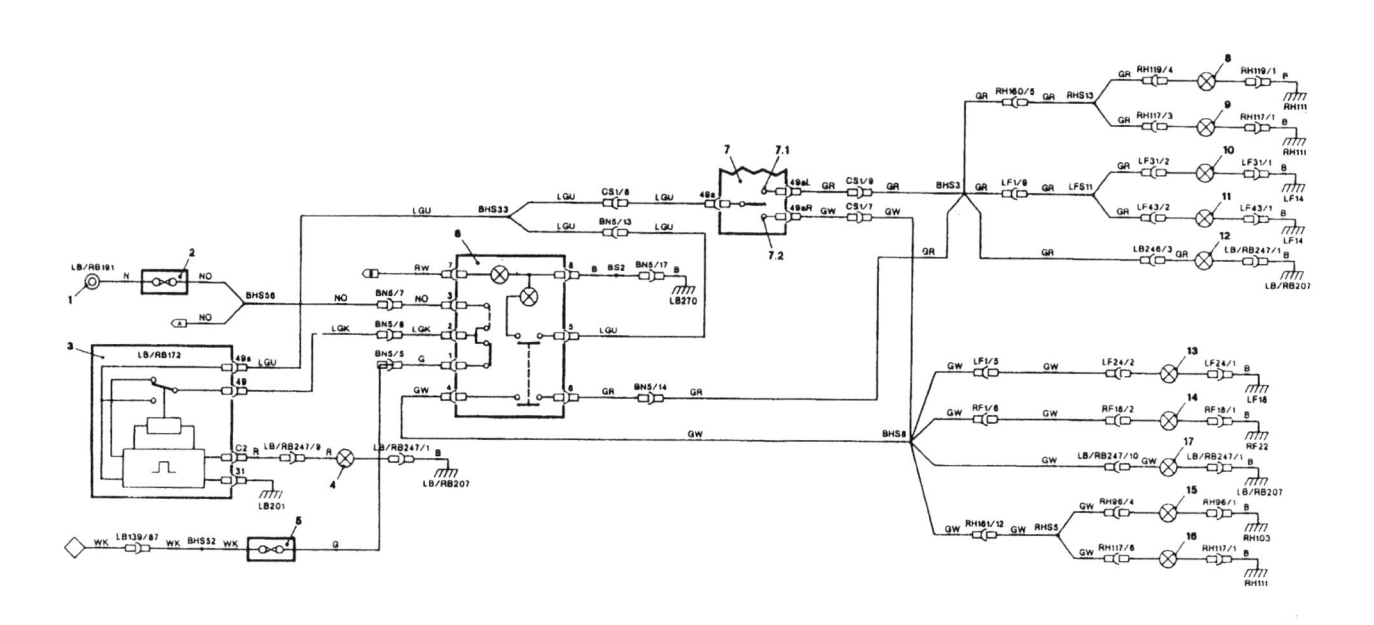

ECU and Relays Location

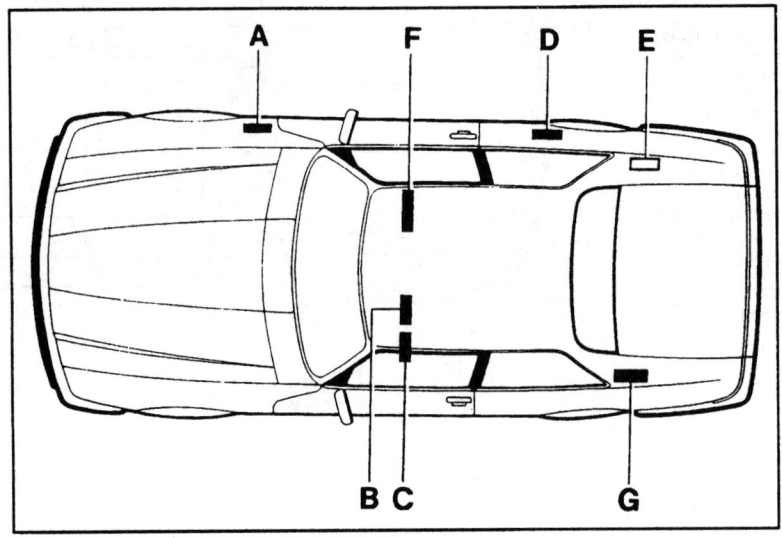

A. Engine management
 ECU (4.0L)
 Marelli ignition ECU (V12)
B. Seat drive non-memory
 ECU
C. Speed control ECU
D. Gearbox ECU
 (4.0L Automatic)
E. Fuel injection ECU
 (V12 only)
F. Seat memory ECU
 (handed – always on)
 driver's side
G. ABS ECU

ECU locations

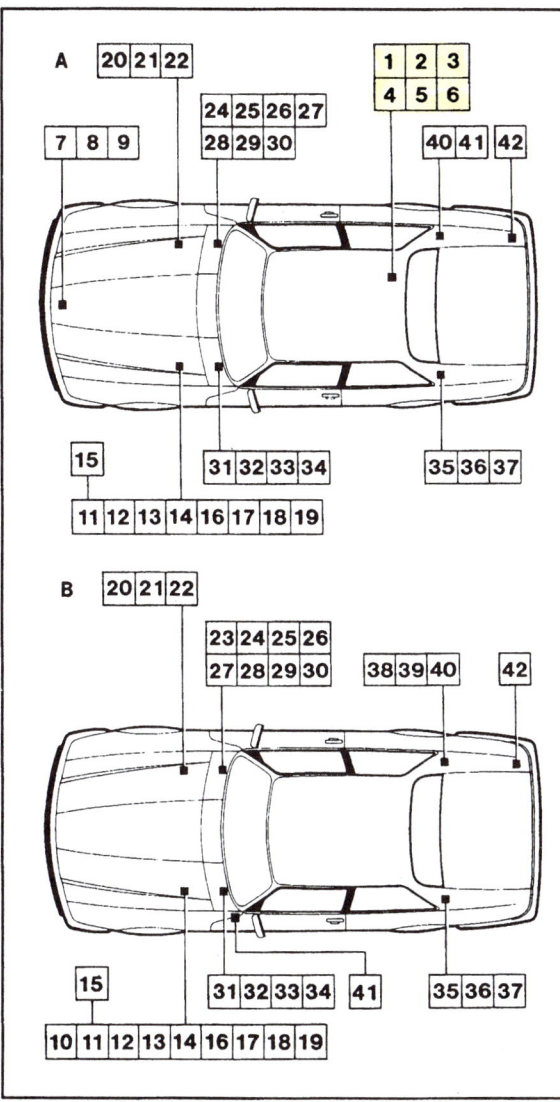

A – V12 RHD   B – 4 litre RHD

Relay Locations

Component	Case-Base Colour
1. Hood up relay	Violet-Green
2. Hood down relay	Violet-Blue
3. Quarters up RH relay	Violet-Red
4. Quarters down RH relay	Violet-Black
5. Quarters up LH relay	Violet-Yellow
6. Quarters down LH relay	Violet-Natural
7. Extra air valve 1 relay (V12 non-catalyst)	Violet-Blue
8. Extra air valve 2 relay (V12 non-catalyst)	Blue-Black
9. Anti-stall relay (V12)	Violet-Blue
10. Breather heater relay (4.0L catalyst)	Blue-Black
11. Daylight running 1 relay	Blue-Blue
12. Daylight running 2 relay	Violet-Violet
13. Daylight running 3 relay	Violet-Violet
14. Radiator fan relay	Blue-Blue
15. Dim-dip control relay	Blue-Blue
16. Sidelights relay	Blue-Blue
17. Headlight relay	Blue-Blue
18. Fan run on diode unit	Blue-Natural
19. Headlamp levelling diode module	Red-Black with red top
20. Air conditioning relay	Blue-Green
21. Horn relay	Blue-Black
22. Starter solenoid relay	Blue-Natural
23. Gearbox warning relay (4.0L)	Violet-Black
24. Ignition continuity load relay	Silver-Natural
25. Front fog lamp relay	Blue-Violet
26. Rear fog lamp relay	Blue-Brown
27. Lights-on warning unit	Black-Natural
28. Dimmer module	Red-Black with red top
29. Bulb check unit	Black-Green
30. Heated rear screen timer	Yellow-Blue
31. ABS pump relay	Yellow-Yellow
32. Auxiliary continuity load relay	Silver-Black
33. Seat heater relay	Silver-Natural
34. Interior light delay unit	Red-Red
35. ABS main relay	White-Black
36. Stop lamp relay	Blue-Yellow
37. Anti-slosh module	Black-Black
38. Reverse lights relay (4.0L)	Blue-Black
39. Gearbox power relay (4.0L)	Blue-Blue
40. Fuel pump relay	Silver-Yellow-Black
41. Main PI relay	Silver with red stripe-Red
42. Radio aerial delay relay	Green-Black

Note: Items 1 to 6 are for the convertible only. Items 11 to 13 are for Canadian market vehicles, item 15 is for UK market vehicles, item 19 is for German market vehicles. Either item 11 or item 15 occupies the one location depending on market requirement.

Fuses

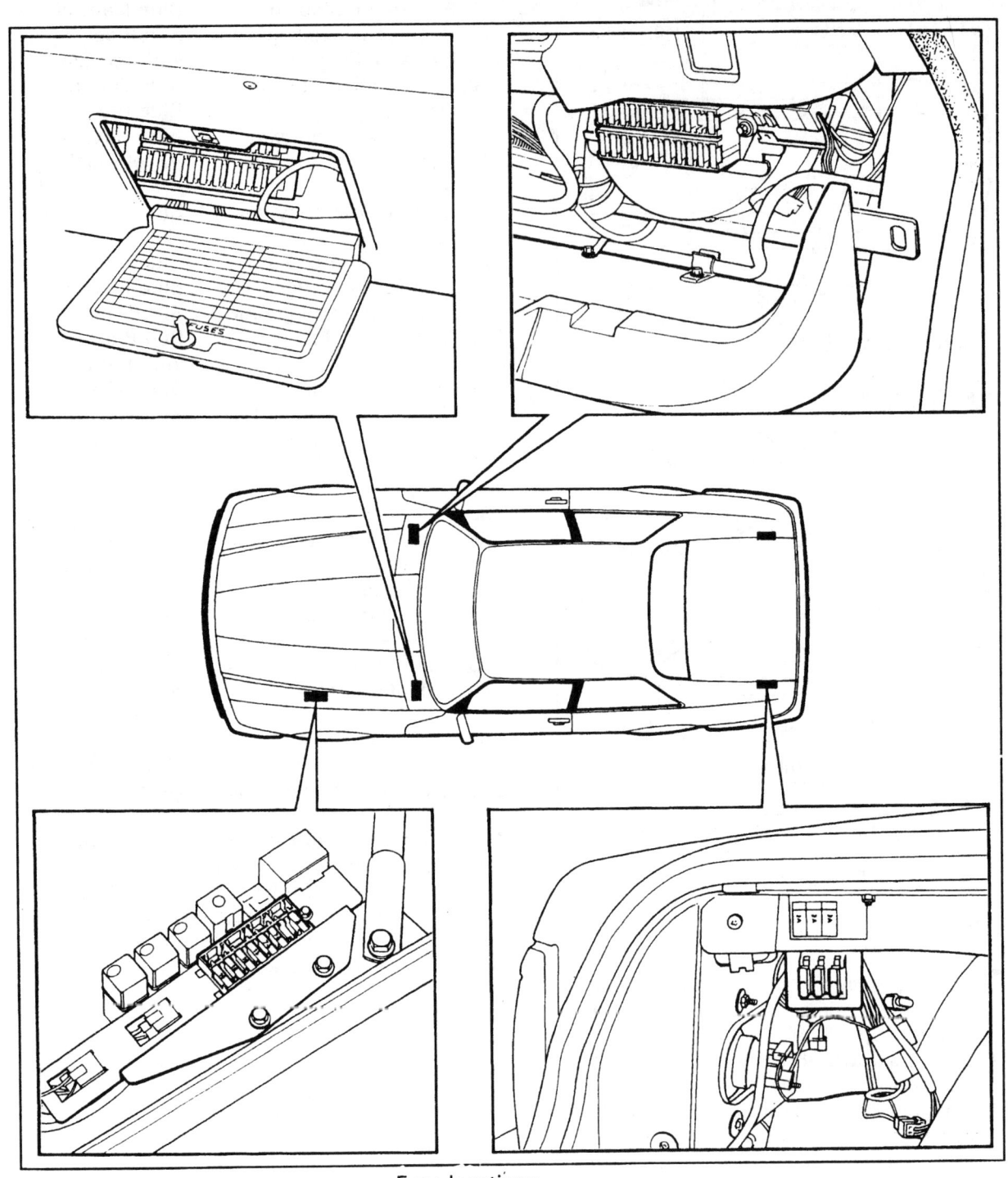

Fuse locations

Key – RHD shown, read clockwise from top left:

    Left-hand side fuse box
    Right-hand side fuse box
    Left-hand side rearward fuse box,
    (RHS rearward fuse box shown opposite)
    Left foward fuse box

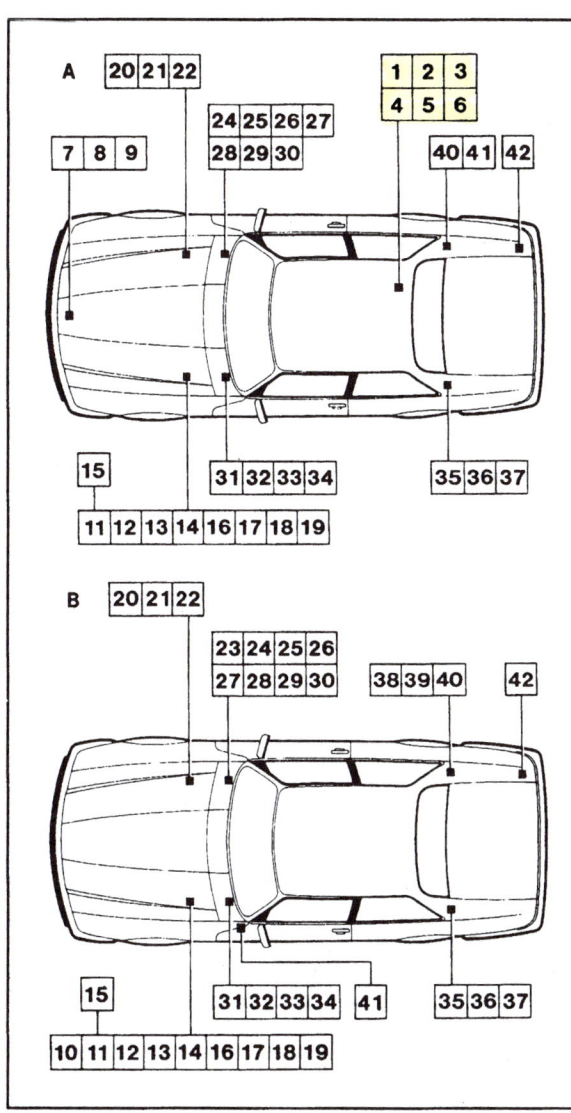

#	Component	Case-Base Colour
15.	Dim-dip control relay	Blue-Blue
16.	Sidelights relay	Blue-Blue
17.	Headlight relay	Blue-Blue
18.	Fan run on diode unit	Blue-Natural
19.	Headlamp levelling diode module	Red-Black with red top
20.	Air conditioning relay	Blue-Green
21.	Horn relay	Blue-Black
22.	Starter solenoid relay	Blue-Natural
23.	Gearbox warning relay (4.0L)	Violet-Black
24.	Ignition continuity load relay	Silver-Natural
25.	Front fog lamp relay	Blue-Violet
26.	Rear fog lamp relay	Blue-Brown
27.	Lights-on warning unit	Black-Natural
28.	Dimmer module	Red-Black with red top
29.	Bulb check unit	Black-Green
30.	Heated rear screen timer	Yellow-Blue
31.	ABS pump relay	Yellow-Yellow
32.	Auxiliary continuity load relay	Silver-Black
33.	Seat heater relay	Silver-Natural
34.	Interior light delay unit	Red-Red
35.	ABS main relay	White-Black
36.	Stop lamp relay	Blue-Yellow
37.	Anti-slosh module	Black-Black
38.	Reverse lights relay (4.0L)	Blue-Black
39.	Gearbox power relay (4.0L)	Blue-Blue
40.	Fuel pump relay	Silver-Yellow-Black
41.	Main PI relay	Silver with red stripe-Red
42.	Radio aerial delay relay	Green-Black

Note: Items 1 to 6 are for the convertible only. Items 11 to 13 are for Canadian market vehicles, item 15 is for UK market vehicles, item 19 is for German market vehicles. Either item 11 or item 15 occupies the one location depending on market requirement.

A – V12 RHD   B – 4 litre RHD

Relay Locations

Component	Case-Base Colour
1. Hood up relay	Violet-Green
2. Hood down relay	Violet-Blue
3. Quarters up RH relay	Violet-Red
4. Quarters down RH relay	Violet-Black
5. Quarters up LH relay	Violet-Yellow
6. Quarters down LH relay	Violet-Natural
7. Extra air valve 1 relay (V12 non-catalyst)	Violet-Blue
8. Extra air valve 2 relay (V12 non-catalyst)	Blue-Black
9. Anti-stall relay (V12)	Violet-Blue
10. Breather heater relay (4.0L catalyst)	Blue-Black
11. Daylight running 1 relay	Blue-Blue
12. Daylight running 2 relay	Violet-Violet
13. Daylight running 3 relay	Violet-Violet
14. Radiator fan relay	Blue-Blue

163

Fuses

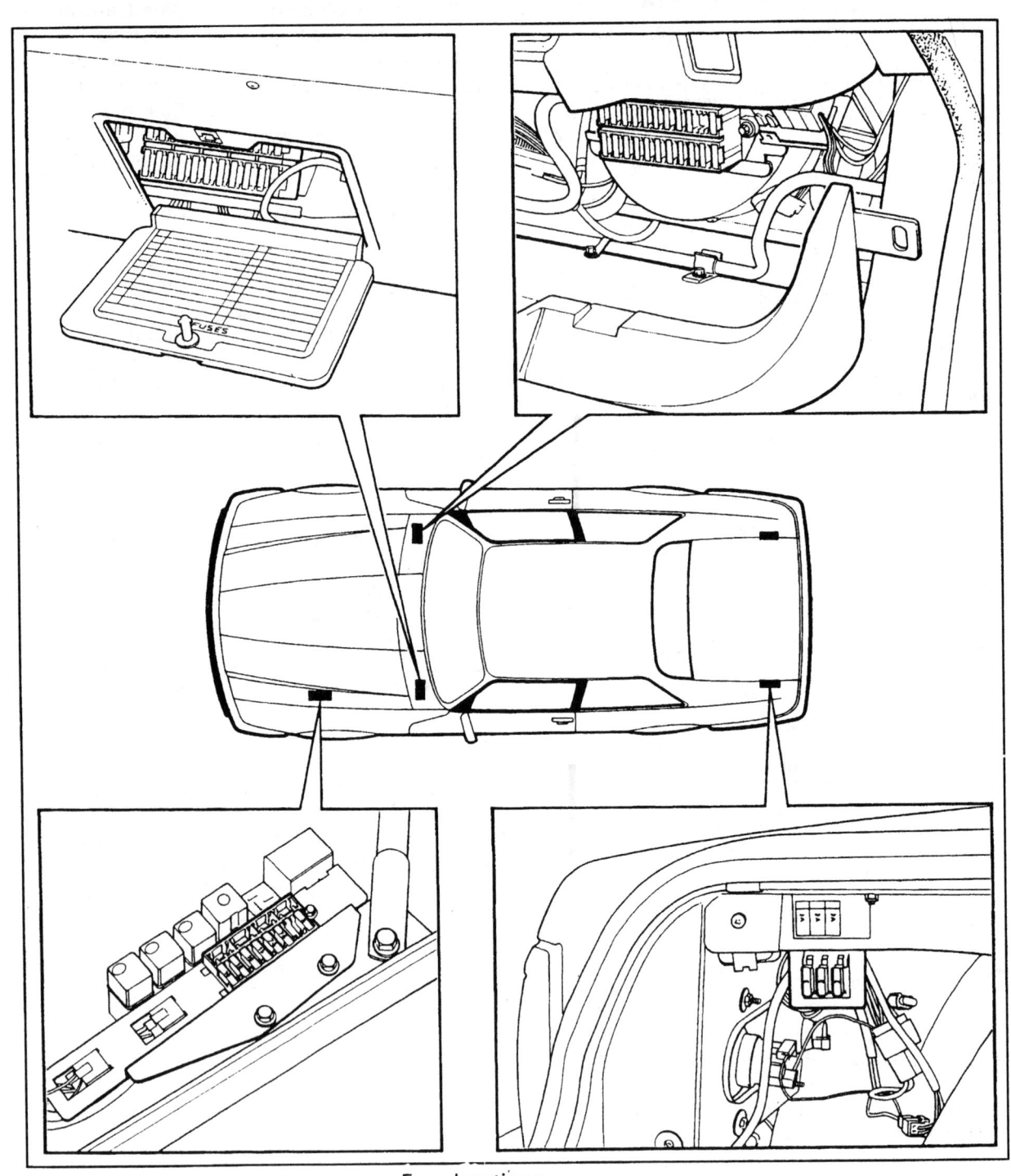

Fuse locations
Key – RHD shown, read clockwise from top left:

    Left-hand side fuse box
    Right-hand side fuse box
    Left-hand side rearward fuse box,
    (RHS rearward fuse box shown opposite)
    Left foward fuse box

Left Foward Fuse Box Chart

Fuse No	Fuse Colour Code	Value	Circuit
1 to 4			Unused
5	Red	10A	Left-hand main beam headlamp
6	Red	10A	Right-hand main beam headlamp
7	Red	10A	Radiator fan
8	Light blue	15A	Dim/dip headlamps (United Kingdom only)
9	Brown	7.5A	Left-hand dipped beam headlamp
10	Brown	7.5A	Right-hand dipped beam headlamp
11			Unused
12	Violet	3A	Headlamp levelling (German market only)

Left-Hand Side Rearward Fuse Box Chart

Fuse No	Fuse Colour Code	Value	Circuit
1	Violet	3A	Left-hand tail and left-hand number plate lamps
2	Violet	3A	Right-hand caravan/trailer tail lamp
3	Violet	3A	Left-hand caravan/trailer tail lamp

Right-Hand Side Rearward Fuse Box Chart

Fuse No	Fuse Colour Code	Value	Circuit
1	Violet	3A	Right-Hand tail & right-hand number plate lamps
2	Red	10A	Radio aerial
3	Tan	5A	Reverse lights (XJ-S 4.0 automatic)

R.H. Steering – Driver's Side Fuse Box Chart – Memory Seats

Fuse No	Fuse Colour Code	Value	Circuit
1	Light blue	15 A	Driver's seat control power – fore-aft-lumbar
2	Violet	3A	Kickdown (XJ-S V12 only)
3	Yellow	20 A	R.H. air conditioning fan
4	Pink	4 A	Radio telephone ignition feed
5	Light blue	15 A	Horns
6	Tan	5 A	Radio memory
7	Red	10 A	Radio power
8	Red	10 A	Windscreen washer pump
9	Red	10 A	Driver's seat heater
10	Black	1 A	45 second timer (XJ-S V12 only)
11	Red	10 A	Luggage compartment lamps (caravan power)

R.H. Steering – Driver's Side Fuse Box Chart – Non-Memory Seats

Fuse No	Fuse Colour Code	Value	Circuit
1	Yellow	20 A	Driver's seat control power – fore-aft-recline
2	Violet	3 A	Kickdown (XJ–S V12 only)
3	Yellow	20 A	R.H. air conditioning fan
4	Pink	4 A	Radio telephone ignition feed
5	Light blue	15 A	Horns
6	Tan	5 A	Radio memory
7	Red	10 A	Radio power
8	Red	10 A	Windscreen washer pump
9	Red	10 A	Driver's seat heater
10			
11	Red	10 A	Luggage compartment lamps (caravan power)

R.H. Steering Passenger's Side Fuse Box Chart – XJ–S V12

Fuse No	Fuse Colour Code	Value	Circuit
1	Yellow	20 A	Passenger seat control power – fore-aft-recline
2	Violet	3 A	Speed (cruise) control
3	Yellow	20 A	L.H. air conditioning fan
4	Tan	5 A	Interior lights
5	Brown	7.5 A	Direction indicators
6	Red	10 A	Central door locking
7	Red	10 A	Cigar lighter
8	Light blue	15 A	Windscreen wiper system
9	Light green	30 A	Headlamp power wash
10	Violet	3 A	Solenoid vacuum valves
11	Light green	30 A	ABS ECU

L.H. Steering – Driver's Side Fuse Box Chart – Memory Seats

Fuse No	Fuse Colour Code	Value	Circuit
1	Light blue	15 A	Driver's seat control power – fore-aft-lumbar
2	Violet	3 A	Kickdown (XJ–S V12 only)
3	Yellow	20 A	L.H. air conditioning fan
4	Pink	4 A	Radio telephone ignition feed
5	Tan	5 A	Radio telephone power feed (where fitted)
6	Tan	5 A	Radio memory
7	Red	10 A	Radio power
8	Red	10 A	Windscreen washer pump
9	Red	10 A	Driver's seat heater
10	Black	1 A	45 second timer (XJ–S V12 only)
11	Red	10 A	Luggage compartment lamps (caravan power)

L.H. Steering – Driver's Side Fuse Box Chart – Non–Memory Seats

Fuse No	Fuse Colour Code	Value	Circuit
1	Yellow	20 A	Driver's seat control power – fore-aft-recline
2	Violet	3 A	Kickdown (XJ–S V12 only)
3	Yellow	20 A	L.H. air conditioning fan
4	Pink	4 A	Radio telephone ignition feed
5	Tan	5 A	Radio telephone power feed (where fitted)
6	Tan	5 A	Radio memory
7	Red	10 A	Radio power
8	Red	10 A	Windscreen washer pump
9	Red	10 A	Driver's seat heater
10			
11	Red	10 A	Luggage compartment lamps (caravan power)

L.H. Steering Passenger's Side Fuse Box Chart – XJ–S V12

Fuse No	Fuse Colour Code	Value	Circuit
1	Yellow	20 A	Passenger's seat control power – fore-aft-recline
2	Violet	3 A	Speed (cruise) control
3	Yellow	20 A	R.H. air conditioning fan
4	Tan	5 A	Interior lights
5	Brown	7.5 A	Direction indicators
6	Red	10 A	Central door locking

Fuse No	Fuse Colour Code	Value	Circuit
7	Red	10 A	Cigar lighter
8	Light blue	15 A	Windscreen wiper system
9	Light green	30 A	Headlamp power wash
10	Violet	3 A	Solenoid vacuum valves
11	Light green	30 A	ABS ECU

Convertible only – Power Operated Hood Fuse

Fuse Colour Code	Value	Circuit	Location
Light green	30A	Power operated hood pump motor	Mounted on a bracket adjacent to the pump motor in rear stowage compartment

In-Line Fuses

Fuse Colour Code	Value	Circuit	Location
Tan	5A	Exhaust temperature warning (Japan only)	Behind the centre console adjacent to the R. H. footwell
Grey	2A	Overspeed warning (Saudi Arabia only)	Behind the centre console adjacent to the R. H. footwell
Tan	5A	P.I. engine harness diagnostic circuit	Adjacent to the diagnostic socket behind the centre console trim (R.H. side – XJS 5.3; passenger side – XJS 4.0)

© Copyright Jaguar Cars Limited and Brooklands Books Limited 1984, 1989 and 1995

This book is published by Brooklands Books Limited and based upon text and illustrations protected by copyright and first published in 1984 by Jaguar Cars Limited and may not be reproduced transmitted or copied by any means without the prior written permission of Jaguar Cars Limited and Brooklands Books Limited.

Printed in England and distributed by Brooklands Books Ltd., PO Box 146, Cobham, Surrey, KT11 1LG, England Phone: 01932 865051 Fax: 01932 868803
e mail: info@brooklands-books.com or visit our website www.brooklands-books.com

ISBN 1 85520 262X